New types of Neutrosophic Set/Logic/Probability, Neutrosophic Over-/Under-/Off-Set, Neutrosophic Refined Set, and their Extension to Plithogenic Set/Logic/Probability, with Applications

New types of Neutrosophic Set/Logic/Probability, Neutrosophic Over-/Under-/Off-Set, Neutrosophic Refined Set, and their Extension to Plithogenic Set/Logic/Probability, with Applications

Special Issue Editor

Florentin Smarandache

MDPI • Basel • Beijing • Wuhan • Barcelona • Belgrade

MDPI

Special Issue Editor
Florentin Smarandache
University of New Mexico
USA

Editorial Office
MDPI
St. Alban-Anlage 66
4052 Basel, Switzerland

This is a reprint of articles from the Special Issue published online in the open access journal *Symmetry* (ISSN 2073-8994) in 2019 (available at: https://www.mdpi.com/journal/symmetry/special_issues/Neutrosophic_Set_Logic_Probability).

For citation purposes, cite each article independently as indicated on the article page online and as indicated below:

LastName, A.A.; LastName, B.B.; LastName, C.C. Article Title. *Journal Name* **Year**, *Article Number*, Page Range.

ISBN 978-3-03921-938-4 (Pbk)
ISBN 978-3-03921-939-1 (PDF)

Contents

About the Special Issue Editor ix

Preface to "New types of Neutrosophic Set/Logic/Probability, Neutrosophic Over-/Under-/Off-Set, Neutrosophic Refined Set, and their Extension to Plithogenic Set/Logic/Probability, with Applications" xi

Yingcang Ma, Xiaohong Zhang, Florentin Smarandache, Juanjuan Zhang
The Structure of Idempotents in Neutrosophic Rings and Neutrosophic Quadruple Rings
Reprinted from: *Symmetry* **2019**, *11*, 1254, doi:10.3390/sym11101254 . . . . **1**

Xiaohong Zhang, Florentin Smarandache and Yingcang Ma
Symmetry in Hyperstructure: Neutrosophic Extended Triplet Semihypergroups and Regular Hypergroups
Reprinted from: *Symmetry* **2019**, *11*, 1217, doi:10.3390/sym11101217 . . . . **16**

Erick González Caballero, Florentin Smarandache and Maikel Leyva Vázquez
On Neutrosophic Offuninorms
Reprinted from: *Symmetry* **2019**, *11*, 1136, doi:10.3390/sym11091136 . . . . **34**

Junhui Kim, Florentin Smarandache, Jeong Gon Lee and Kul Hur
Ordinary Single Valued Neutrosophic Topological Spaces
Reprinted from: *Symmetry* **2019**, *11*, 1075, doi:10.3390/sym11091075 . . . . **60**

Jingqian Wang, Xiaohong Zhang
A New Type of Single Valued Neutrosophic Covering Rough Set Model
Reprinted from: *Symmetry* **2019**, *11*, 1074, doi:10.3390/sym11091074 . . . . **86**

Muhammad Akram, Sumera Naz and Florentin Smarandache
Generalization of Maximizing Deviation and TOPSIS Method for MADM in Simplified Neutrosophic Hesitant Fuzzy Environment
Reprinted from: *Symmetry* **2019**, *11*, 1058, doi:10.3390/sym11081058 . . . . **109**

Marcel-Ioan Bolos, Ioana-Alexandra Bradea and Camelia Delcea
Modeling the Performance Indicators of Financial Assets with Neutrosophic Fuzzy Numbers
Reprinted from: *Symmetry* **2019**, *11*, 1021, doi:10.3390/sym11081021 . . . . **135**

Qiaoyan Li, Yingcang Ma, Xiaohong Zhang and Juanjuan Zhang
Study on the Algebraic Structure of Refined Neutrosophic Numbers
Reprinted from: *Symmetry* **2019**, *11*, 954, doi:10.3390/sym11080954 . . . . **158**

Chunxin Bo, Xiaohong Zhang and Songtao Shao
Non-Dual Multi-Granulation Neutrosophic Rough Set with Applications
Reprinted from: *Symmetry* **2019**, *11*, 910, doi:10.3390/sym11070910 . . . . **171**

Mohamed Abdel-Basset, Rehab Mohamed, Abd El-Nasser H. Zaied and Florentin Smarandache
A Hybrid Plithogenic Decision-Making Approach with Quality Function Deployment for Selecting Supply Chain Sustainability Metrics
Reprinted from: *Symmetry* **2019**, *11*, 903, doi:10.3390/sym11070903 . . . . **187**

Keli Hu, Wei He, Jun Ye, Liping Zhao, Hua Peng and Jiatian Pi
Online Visual Tracking of Weighted Multiple Instance Learning via Neutrosophic Similarity-Based Objectness Estimation
Reprinted from: *Symmetry* **2019**, *11*, 832, doi:10.3390/sym11060832 . . . . **208**

Qingqing Hu, Xiaohong Zhang
Neutrosophic Triangular Norms and Their Derived Residuated Lattices
Reprinted from: *Symmetry* **2019**, *11*, 817, doi:10.3390/sym11060817 . . . . **232**

Mohammed A. Al Shumrani and Florentin Smarandache
Introduction to Non-Standard Neutrosophic Topology
Reprinted from: *Symmetry* **2019**, *11*, 706, doi:10.3390/sym11050706 . . . . **254**

Muhammad Jamil, Saleem Abdullah, Muhammad Yaqub Khan, Florentin Smarandache and Fazal Ghani
Application of the Bipolar Neutrosophic Hamacher Averaging Aggregation Operators to Group Decision Making: An Illustrative Example
Reprinted from: *Symmetry* **2019**, *11*, 698, doi:10.3390/sym11050698 . . . . **268**

Qiaoyan Li, Yingcang Ma, Xiaohong Zhang, Juanjuan Zhang
Neutrosophic Extended Triplet Group Based on Neutrosophic Quadruple Numbers
Reprinted from: *Symmetry* **2019**, *11*, 696, doi:10.3390/sym11050696 . . . . **286**

Shao Songtao, Zhang Xiaohong and Zhao Quan
Multi-Attribute Decision Making Based on Probabilistic Neutrosophic Hesitant Fuzzy Choquet Aggregation Operators
Reprinted from: *Symmetry* **2019**, *11*, 623, doi:10.3390/sym11050623 . . . . **301**

Mohamed Abdel-Basset, Mai Mohamed, Victor Chang and Florentin Smarandache
IoT and Its Impact on the Electronics Market: A Powerful Decision Support System for Helping Customers in Choosing the Best Product
Reprinted from: *Symmetry* **2019**, *11*, 611, doi:10.3390/sym11050611 . . . . **316**

Firoz Ahmad, Ahmad Yusuf Adhami and Florentin Smarandache
Neutrosophic Optimization Model and Computational Algorithm for Optimal Shale Gas Water Management under Uncertainty
Reprinted from: *Symmetry* **2019**, *11*, 544, doi:10.3390/sym11040544 . . . . **337**

Florentin Smarandache
Extended Nonstandard Neutrosophic Logic, Set, and Probability Based on Extended Nonstandard Analysis
Reprinted from: *Symmetry* **2019**, *11*, 515, doi:10.3390/sym11040515 . . . . **371**

Jihong Chen, Kai Xue, Jun Ye, Tiancun Huang, Yan Tian, Chengying Hua and Yuhua Zhu
Simplified Neutrosophic Exponential Similarity Measures for Evaluation of Smart Port Development
Reprinted from: *Symmetry* **2019**, *11*, 485, doi:10.3390/sym11040485 . . . . **396**

Mohamed Abdel-Basset, Victor Chang, Mai Mohamed and Florentin Smarandache
A Refined Approach for Forecasting Based on Neutrosophic Time Series
Reprinted from: *Symmetry* **2019**, *11*, 457, doi:10.3390/sym11040457 . . . . **408**

Ashraf Al-Quran, Nasruddin Hassan and Emad Marei
A Novel Approach to Neutrosophic Soft Rough Set under Uncertainty
Reprinted from: *Symmetry* **2019**, *11*, 384, doi:10.3390/sym11030384 . . . . **431**

Ashraf Al-Quran, Nasruddin Hassan and Shawkat Alkhazaleh
Fuzzy Parameterized Complex Neutrosophic Soft Expert Set for Decision under Uncertainty
Reprinted from: *Symmetry* **2019**, *11*, 382, doi:10.3390/sym11030382 . . . . **447**

Shahzaib Ashraf, Saleem Abdullah, Florentin Smarandache and Noor ul Amin
Logarithmic Hybrid Aggregation Operators Based on Single Valued Neutrosophic Sets and Their Applications in Decision Support Systems
Reprinted from: *Symmetry* **2019**, *11*, 364, doi:10.3390/sym11030364 . . . . **466**

Muhammad Aslam and Mohammed Albassam
Application of Neutrosophic Logic to Evaluate Correlation between Prostate Cancer Mortality and Dietary Fat Assumption
Reprinted from: *Symmetry* **2019**, *11*, 330, doi:10.3390/sym11030330 . . . . **489**

Yingcang Ma, Xiaohong Zhang, Xiaofei Yang, Juanjuan Zhang and Hu Zhao
Generalized Neutrosophic Extended Triplet Group
Reprinted from: *Symmetry* **2019**, *11*, 327, doi:10.3390/sym11030327 . . . . **496**

Muhammad Gulistan and Nasruddin Hassan
A Generalized Approach towards Soft Expert Sets via Neutrosophic Cubic Sets with Applications in Games
Reprinted from: *Symmetry* **2019**, *11*, 289, doi:10.3390/sym11020289 . . . . **512**

Lilian Shi and Yue Yuan
Hybrid Weighted Arithmetic and Geometric Aggregation Operator of Neutrosophic Cubic Sets for MADM
Reprinted from: *Symmetry* **2019**, *11*, 278, doi:10.3390/sym11020278 . . . . **538**

Chengdong Cao, Shouzhen Zeng and Dandan Luo
A Single-Valued Neutrosophic Linguistic Combined Weighted Distance Measure and Its Application in Multiple-Attribute Group Decision-Making
Reprinted from: *Symmetry* **2019**, *11*, 275, doi:10.3390/sym11020275 . . . . **548**

Wen Jiang, Zihan Zhang and Xinyang Deng
Multi-Attribute Decision Making Method Based on Aggregated Neutrosophic Set
Reprinted from: *Symmetry* **2019**, *11*, 267, doi:10.3390/sym11020267 . . . . **559**

Majid Khan, Muhammad Gulistan, Naveed Yaqoob, Madad Khan and Florentin Smarandache
Neutrosophic Cubic Einstein Geometric Aggregation Operators with Application to Multi-Criteria Decision Making Method
Reprinted from: *Symmetry* **2019**, *11*, 247, doi:10.3390/sym11020247 . . . . **572**

Muhammad Aslam and Mansour Sattam Aldosari
Inspection Strategy under Indeterminacy Based on Neutrosophic Coefficient of Variation
Reprinted from: *Symmetry* **2019**, *11*, 193, doi:10.3390/sym11020193 . . . . **596**

Aliya Fahmi, Fazli Amin, Madad Khan and Florentin Smarandache
Group Decision Making Based on Triangular Neutrosophic Cubic Fuzzy Einstein Hybrid Weighted Averaging Operators
Reprinted from: *Symmetry* **2019**, *11*, 180, doi:10.3390/sym11020180 . . . . **604**

Jun Ye and Wenhua Cui
Neutrosophic Compound Orthogonal Neural Network and Its Applications in Neutrosophic Function Approximation
Reprinted from: *Symmetry* **2019**, *11*, 147, doi:10.3390/sym11020147 . . . . **633**

Majdoleen Abu Qamar and Nasruddin Hassan
An Approach toward a Q-Neutrosophic Soft Set and Its Application in Decision Making
Reprinted from: *Symmetry* **2019**, *11*, 139, doi:10.3390/sym11020139 . . . . **642**

Muhammad Aslam
A Variable Acceptance Sampling Plan under Neutrosophic Statistical Interval Method
Reprinted from: *Symmetry* **2019**, *11*, 114, doi:10.3390/sym11010114 . . . . **660**

Chiranjibe Jana and Madhumangal Pal
A Robust Single-Valued Neutrosophic Soft Aggregation Operators in Multi-Criteria Decision Making
Reprinted from: *Symmetry* **2019**, *11*, 110, doi:10.3390/sym11010110 . . . . **667**

About the Special Issue Editor

Florentin Smarandache is Professor of Mathematics at the University of New Mexico, United States. He was awarded his MSc in Mathematics and Computer Science from the University of Craiova, Romania, PhD in Mathematics from the State University of Kishinev, and was a Postdoctor in Applied Mathematics at the Okayama University of Sciences, Japan. He is the founder of neutrosophy (generalization of dialectics), neutrosophic set, logic, probability and statistics, and since 1995, has published hundreds of papers on neutrosophic physics, superluminal and instantaneous physics, unmatter, absolute theory of relativity, redshift and blueshift due to the medium gradient and refraction index besides the Doppler effect, paradoxism, outerart, neutrosophy as a new branch of philosophy, law of included multiple-middle, multispace and multistructure, quantum paradoxes, degree of dependence and independence between neutrosophic components, refined neutrosophic set, neutrosophic over/under/offset, plithogenic set, neutrosophic triplet and duplet structures, quadruple neutrosophic structures, and Dezert–Smarandache theory (DSmT) to many peer-reviewed international journals and many books, and has presented papers and plenary lectures at many international conferences around the world.
[http://fs.unm.edu/FlorentinSmarandache.htm]

Preface to "New types of Neutrosophic Set/Logic/Probability, Neutrosophic Over-/Under-/ Off-Set, Neutrosophic Refined Set, and Their Extension to Plithogenic Set/Logic/Probability, with Applications"

Florentin Smarandache

Department of Mathematics, University of New Mexico, 705 Gurley Ave., Gallup, NM 87301, USA; smarand@unm.edu

Received: 07 November 2019; Published: 18 November 2019

Abstract: In this paper we review all thirty-seven neutrosophic and plithogenic papers published by Symmetry journal within the special issue "New types of Neutrosophic Set/Logic/Probability, Neutrosophic Over-/Under-/Off-Set, Neutrosophic Refined Set, and their Extension to Plithogenic Set/Logic/Probability, with Applications" (2019).

Keywords: neutrosophy, neutrosophic set, neutrosophic logic, neutrosophic probability, neutrosophic statistics, neutrosophic overset, neutrosophic underset, neutrosophic offset, degree of dependence and independence between components, refined neutrosophic set, law of included multiple-middle, neutrosophic bipolar and tripolar and multipolar sets, neutrosophic algebraic structures, neutrosophic triplet algebraic structures, neutrosophic extended triplet algebraic structures, quadruple neutrosophic algebraic structures, plithogeny, plithogenic set, plithogenic logic

---

The fields of neutrosophic and plithogenic sets, logic, measure, probability, and statistics have been developed and explored extensively in the last few years because of their multiple practical applications.

The neutrosophic components of truth (T), indeterminacy (I), and falsehood (F) are symmetric in form, since T is symmetric to its opposite F with respect to I, which acts as an axis of symmetry between T–I–F.

This Special Issue invited state-of-the-art papers on new topics related to neutrosophic theories and applications, including:

—Studies of corner cases of neutrosophic sets/probabilities/statistics/logics, such as:

- neutrosophic intuitionistic sets (which are different from intuitionistic fuzzy sets), neutrosophic paraconsistent sets, neutrosophic faillibilist sets, neutrosophic paradoxist sets, neutrosophic, pseudo-paradoxist sets, neutrosophic tautological sets, neutrosophic nihilist sets, neutrosophic dialetheist sets, and neutrosophic trivialist sets;
- neutrosophic intuitionistic probability and statistics, neutrosophic paraconsistent probability and statistics, neutrosophic faillibilist probability and statistics, neutrosophic paradoxist probability and statistics, neutrosophic pseudo-paradoxist probability and statistics, neutrosophic tautological probability and statistics, neutrosophic nihilist probability and statistics, neutrosophic dialetheist probability and statistics, and neutrosophic trivialist probability and statistics;

- neutrosophic paradoxist logic (or paradoxism), neutrosophic pseudo-paradoxist logic (or neutrosophic pseudo-paradoxism), and neutrosophic tautological logic (or neutrosophic tautologism) [1];

—Refined neutrosophic set components (T, I, and F), which are refined/split into many neutrosophic subcomponents: (T_1, T_2, ...; I_1, I_2, ...; F_1, F_2, ...) [2];

—Degrees of dependence and independence between neutrosophic components: T, I, and F as independent components, leave room for incomplete information (when their superior sum < 1), paraconsistent and contradictory information (when the superior sum > 1), or complete information (sum of components = 1).

For technical and engineering proposals, the classical unit interval [0,1] is used.

For single-valued neutrosophic logic, the sum of the components is:

$0 \leq t + i + f \leq 3$ when all three components are independent;

$0 \leq t + i + f \leq 2$ when two components are dependent, while the third one is independent from them;

$0 \leq t + i + f \leq 1$ when all three components are dependent.

When three or two of the components T, I, and F are independent, one leaves room for incomplete information (sum < 1), paraconsistent and contradictory information (sum > 1), or complete information (sum = 1).

If all three components T, I, and F are dependent then, similarly, one leaves room for incomplete information (sum < 1) or complete information (sum = 1).

In general, the sum of two components x and y that vary in the unitary interval [0,1] is $0 \leq x + y \leq 2 - d^{\circ}(x, y)$, in which $d^{\circ}(x, y)$ is the degree of dependence between x and y [3];

Neutrosophic overset (when some neutrosophic component is > 1) is observed, for example, when an employee works overtime and deserves a degree of membership > 1, with respect to an employee that only works regular full-time and whose degree of membership = 1.

Neutrosophic underset (when some neutrosophic component is < 0) is observed, for example, when an employee causes more damage than benefit to his company and deserves a degree of membership < 0, with respect to an employee that produces benefit to the company and has a degree of membership > 0.

Neutrosophic offset occurs when some neutrosophic components are off the interval [0,1] (i.e., some neutrosophic component > 1 and some neutrosophic component < 0).

—Then, similarly, neutrosophic logic/measure/probability/statistics, and so on, were extended to, respectively, neutrosophic over-/under-/off-logic, measure, probability, statistics, and so on [4,5];

—Neutrosophic tripolar set and neutrosophic multipolar set and, consequently, the neutrosophic tripolar graph and neutrosophic multipolar graph [6];

—N-norm and N-conorm [7];

—Neutrosophic measure and neutrosophic probability (chance that an event occurs, indeterminate chance of occurrence, chance that the event does not occur) [8];

—Law of included multiple-middle (as middle part of refined neutrosophy): (<A>; <neutA$_1$>, <neutA$_2$>, …; <antiA>) [9];

—Neutrosophic statistics (indeterminacy is introduced into classical statistics with respect to the sample/population characteristics, or with respect to the individuals that only partially belong to a sample/population, or with respect to the neutrosophic probability distributions) [10];

—Neutrosophic precalculus and neutrosophic calculus [11];

—Refined neutrosophic numbers ($a + b_1I_1 + b_2I_2 + \ldots + b_nI_n$), in which I_1, I_2, …, I_n are sub-indeterminacies of indeterminacy I; (t,i,f)-neutrosophic graphs; thesis–antithesis–neutrothesis, and neutrosynthesis; neutrosophic axiomatic system; neutrosophic dynamic systems; symbolic neutrosophic logic; (t, i, f)-neutrosophic structures; i-neutrosophic structures; refined literal indeterminacy; quadruple neutrosophic algebraic structures; and multiplication law of sub-indeterminacies [12];

—Theory of neutrosophic evolution: degrees of evolution, indeterminacy or neutrality, and involution [13];

—Plithogeny as generalization of dialectics and neutrosophy; plithogenic set/logic/probability/statistics (as generalization of fuzzy, intuitionistic fuzzy, neutrosophic set/logic/probability/statistics) [14];

—Neutrosophic psychology (neutropsyche; refined neutrosophic memory: conscious, aconscious, unconscious; neutropsychic personality; Eros/Aoristos/Thanatos; and neutropsychic crisp personality) [15];

—Neutrosophic applications in artificial intelligence, information systems, computer science, cybernetics, theory methods, mathematical algebraic structures, applied mathematics, automation, control systems, big data, engineering, electrical, electronic, philosophy, social science, psychology, biology, engineering, operational research, management science, imaging science, photographic technology, instruments, instrumentation, physics, optics, economics, mechanics, neurosciences, radiology nuclear, interdisciplinary applications, multidisciplinary sciences, and so on [16].

The Special Issue "New types of Neutrosophic Set/Logic/Probability, Neutrosophic Over-/Under-/Off-Set, Neutrosophic Refined Set, and their Extension to Plithogenic Set/Logic/Probability, with Applications" comprises 37 papers focusing on topics such as: neutrosophic set; neutrosophic rings; neutrosophic quadruple rings; idempotents; neutrosophic extended triplet group; hypergroup; semihypergroup; neutrosophic extended triplet group; neutrosophic extended triplet semihypergroup (NET-semihypergroup); NET-hypergroup; neutrosophic offset; uninorm; neutrosophic offuninorm; neutrosophic offnorm; neutrosophic offconorm; implicator; prospector; n-person cooperative game; ordinary single valued neutrosophic (co)topology; ordinary single valued neutrosophic subspace; α-level; ordinary single valued neutrosophic neighborhood system; ordinary single valued neutrosophic base; ordinary single valued neutrosophic subbase; fuzzy numbers; neutrosophic numbers; neutrosophic symmetric scenarios; performance indicators; financial assets; neutrosophic extended triplet group; neutrosophic quadruple numbers; refined neutrosophic numbers; refined neutrosophic quadruple numbers; multigranulation neutrosophic rough set; nondual; two universes; multiattribute group decision making; nonstandard analysis; extended nonstandard analysis; monad; binad; left monad closed to the right; right monad closed to the left; pierced binad; unpierced binad; nonstandard neutrosophic mobinad set; neutrosophic topology; nonstandard neutrosophic topology; visual tracking; neutrosophic weight; objectness; weighted multiple instance learning; neutrosophic triangular norms; residuated lattices; representable neutrosophic t-norms; De Morgan neutrosophic triples; neutrosophic residual implications; infinitely $\vee$-distributive; probabilistic neutrosophic hesitant fuzzy set (PNHFS); decision-making; Choquet integral; e-marketing; Internet of Things; neutrosophic set; multicriteria decision-making techniques; intuitionistic fuzzy parameters; uncertainty modeling; neutrosophic goal programming approach; shale gas water management system, and many more.

Molodtsov originated soft set theory that provided a general mathematical framework for handling uncertainties, in which one meets the data by an affixed parameterized factor during information analysis as differentiated to fuzzy as well as neutrosophic set theory. The main objective of the first paper [17] is to lay a foundation for providing a new approach of a single-valued neutrosophic soft tool that considers many problems that contain uncertainties. In the present study, new aggregation operators of single-valued neutrosophic soft numbers have, so far, not yet been applied for ranking the alternatives in decision-making problems. To this proposed work, a single-valued neutrosophic soft-weighted arithmetic averaging (SVNSWA) operator and single-valued neutrosophic soft-weighted geometric averaging (SVNSWGA) operator have been used to compare two single-valued neutrosophic soft numbers (SVNSNs) for aggregating different single-valued neutrosophic soft input arguments in neutrosophic soft environments. Then, its related properties have been investigated. Finally, a practical example for medical diagnosis problems is provided to test the feasibility and applicability of the proposed work.

The acceptance sampling plan plays an important role in maintaining the high quality of a product. The variable control chart, using classical statistics, helps in making acceptance or rejection decisions about the submitted lot of the product. Furthermore, the sampling plan, using classical statistics, assumes that complete or determinate information are available for a lot of the products. However, in some situations, data may be ambiguous, vague, imprecise, and incomplete or indeterminate. In this case, the use of neutrosophic statistics can be applied to guide the experimenters. In the second paper [18], the authors proposed a new variable sampling plan using the neutrosophic interval statistical method. The neutrosophic operating characteristic (NOC) is derived using a neutrosophic normal distribution. An optimization solution is also presented for the proposed plan under the neutrosophic interval method. The effectiveness of the proposed plan is compared with the plan under classical statistics. Tables are presented for practical use, and a real example is given to explain the neutrosophic fuzzy variable sampling plan in the industry.

A neutrosophic set was proposed as an approach to study neutral, uncertain information. It is characterized through three memberships, T, I and F, such that these independent functions stand for the truth, indeterminate, and false-membership degrees of an object. The neutrosophic set presents a symmetric form, since truth enrolment T is symmetric to its opposite false enrolment F with respect to indeterminacy enrollment I that acts as an axis of symmetry. The neutrosophic set was further extended to a Q-neutrosophic soft set, which is a hybrid model that keeps the features of the neutrosophic soft set in dealing with uncertainty and the features of a Q-fuzzy soft set that handles two-dimensional information. In the next paper [19], the authors discuss some operations of Q-neutrosophic soft sets, such as subset, equality, complement, intersection, union, AND operation, and OR operation. The authors also define the necessity and possibility operations of a Q-neutrosophic soft set. Several properties and illustrative examples are discussed. Then, the authors define the Q-neutrosophic set aggregation operator and use it to develop an algorithm for using a Q-neutrosophic soft set in decision-making issues that have indeterminate and uncertain data, followed by an illustrative real-life example.

Neural networks are powerful universal approximation tools. They have been utilized for functions/data approximation, classification, pattern recognition, as well as their various applications. Uncertain or interval values result from the incompleteness of measurements, human observations, and estimations in the real world. Thus, a neutrosophic number (NsN) can represent both certain and uncertain information in an indeterminate setting and imply a changeable interval depending on its indeterminate ranges. In NsN settings, however, existing interval neural networks cannot deal with uncertain problems with NsNs. Therefore, the next study [20] proposes a neutrosophic compound orthogonal neural network (NCONN) for the first time, containing the NsN weight values, NsN input and output, and hidden layer neutrosophic neuron functions, to approximate neutrosophic functions/NsN data. In the proposed NCONN model, single input and single output neurons are the transmission notes of NsN data, and hidden layer neutrosophic neurons are constructed by the compound functions of both the Chebyshev neutrosophic orthogonal polynomial and the neutrosophic sigmoid function. In addition, illustrative and actual examples are provided to verify the effectiveness and learning performance of the proposed NCONN model for approximating neutrosophic nonlinear functions and NsN data. The contribution of this study is that the proposed NCONN can handle the approximation problems of neutrosophic nonlinear functions and NsN data. However, the main advantage is that the proposed NCONN implies a simple learning algorithm, higher speed learning convergence, and higher learning accuracy in indeterminate/NsN environments.

In the following paper [21], a new concept of the triangular neutrosophic cubic fuzzy numbers (TNCFNs), their scores, and accuracy functions are introduced. Based on TNCFNs, some new

Einstein aggregation operators, such as triangular neutrosophic cubic fuzzy Einstein weighted averaging (TNCFEWA), triangular neutrosophic cubic fuzzy Einstein ordered weighted averaging (TNCFEOWA), and triangular neutrosophic cubic fuzzy Einstein hybrid weighted averaging (TNCFEHWA) operators are developed. Furthermore, their application to multiple-attribute decision making with triangular neutrosophic cubic fuzzy (TNCF) information is discussed. Finally, a practical example is given to verify the developed approach and to demonstrate its practicality and effectiveness.

The existing sampling plans that use the coefficient of variation (CV) are designed under classical statistics. These available sampling plans cannot be used for sentencing if the sample or the population has indeterminate, imprecise, unknown, incomplete, or uncertain data. In the next paper [22], the authors introduce the neutrosophic coefficient of variation (NCV) first. The authors design a sampling plan based on the NCV. The neutrosophic operating characteristic (NOC) function is then given and used to determine the neutrosophic plan parameters under some constraints. The neutrosophic plan parameters such as neutrosophic sample size and neutrosophic acceptance number are determined through the neutrosophic optimization solution. The efficiency of the proposed plan under the neutrosophic statistical interval method with the sampling plan under classical statistics is compared. A real example, which has indeterminate data, is given to illustrate the proposed plan.

Neutrosophic cubic sets (NCs) are a more-generalized version of neutrosophic sets (Ns) and interval neutrosophic sets (INs). Neutrosophic cubic sets are better placed to express consistent, indeterminate, and inconsistent information, which provides a better platform to deal with incomplete, inconsistent, and vague data. Aggregation operators play a key role in daily life and in relation to science and engineering problems. In the following paper [23], the authors define the algebraic and Einstein sum, multiplication and scalar multiplication, and score and accuracy functions. Using these operations, the authors defined geometric aggregation operators and Einstein geometric aggregation operators. First, they define the algebraic and Einstein operators of addition, multiplication, and scalar multiplication, then the score and accuracy function to compare neutrosophic cubic values, and afterwards the neutrosophic cubic weighted geometric operator (NCWG), neutrosophic cubic ordered weighted geometric operator (NCOWG), neutrosophic cubic Einstein weighted geometric operator (NCEWG), and neutrosophic cubic Einstein ordered weighted geometric operator (NCEOWG) over neutrosophic cubic sets. A multicriteria decision-making method is developed as an application for these operators. This method is then applied to a daily life problem.

Multiattribute decision making refers to the decision-making problem of selecting the optimal alternative or sorting the scheme when considering multiple attributes, which is widely used in engineering design, economy, management, military, and so on. But in real applications, the attribute information of many objects is often inaccurate or uncertain, so it is very important for us to find a useful and efficient method to solve the problem. A neutrosophic set is proposed from philosophical point of view to handle inaccurate information efficiently, and a single-valued neutrosophic set (SVNS) is a special case of neutrosophic set, which is widely used in actual field applications. In the next paper [24], a new method based on aggregating a single-valued neutrosophic set is proposed to solve a multiattribute decision-making problem. Firstly, a neutrosophic decision matrix is obtained by expert assessment, then a score function of single-valued neutrosophic sets (SVNSs) is defined to obtain the positive ideal solution (PIS) and the negative ideal solution (NIS). Then, all alternatives are aggregated based on the TOPSIS method to make a decision. Finally, numerical examples are given to verify the feasibility and rationality of the method.

The aim of the next paper [25] is to present a multiple-attribute group decision-making (MAGDM) framework based on a new single-valued neutrosophic linguistic (SVNL) distance measure. By unifying the idea of the weighted average and ordered weighted average into a single-valued neutrosophic linguistic distance, the authors first developed a new SVNL weighted distance measure, namely a SVNL combined and weighted distance (SVNLCWD) measure. The focal characteristics of the devised SVNLCWD are its ability to combine both the decision-makers' attitudes toward the importance as well as the weights of the arguments. Various desirable properties and families of the developed SVNLCWD were contemplated. Moreover, a MAGDM approach based on the SVNLCWD was formulated. Lastly, a real numerical example concerning a low-carbon supplier selection problem was used to describe the superiority and feasibility of the developed approach.

Neutrosophic cubic sets (NCSs) can express complex multiattribute decision-making (MADM) problems with its interval and single-valued neutrosophic numbers simultaneously. The weighted arithmetic average (WAA) and geometric average (WGA) operators are common aggregation operators for handling MADM problems. However, the neutrosophic cubic weighted arithmetic average (NCWAA) and neutrosophic cubic geometric weighted average (NCWGA) operators may result in some unreasonable aggregated values in some cases. In order to overcome the drawbacks of the NCWAA and NCWGA, a new neutrosophic cubic hybrid weighted arithmetic and geometric aggregation (NCHWAGA) operator is developed, and its suitability and effectiveness are investigated in the next paper [26]. Then, the authors established a MADM method based on the NCHWAGA operator. Finally, a MADM problem with neutrosophic cubic information was provided to illustrate the application and effectiveness of the proposed method.

An interesting approach is proposed in [27]. Games are considered to be the most attractive and healthy event between nations and peoples. Soft expert sets are helpful for capturing uncertain and vague information. By contrast, neutrosophic sets are tri-component logic sets; thus, they can deal with uncertain, indeterminate, and incompatible information where the indeterminacy is quantified explicitly and truth membership, indeterminacy membership, and falsity membership are independent of each other. Subsequently, the authors develop a combined approach and extend this concept further to introduce the notion of neutrosophic cubic soft expert sets (NCSESs) by using the concept of neutrosophic cubic soft sets, which is a powerful tool for handling uncertain information in many problems and especially in games. Then, the authors define and analyze the properties of internal neutrosophic cubic soft expert sets (INCSESs) and external neutrosophic cubic soft expert sets (ENCSESs), *P*-order, *P*-union, *P*-intersection, *P*-AND, and *P*-OR as well as *R*-order, *R*-union, *R*-intersection, *R*-AND, and *R*-OR of NCSESs. The NCSESs satisfy the laws of commutativity, associativity, De Morgan, distributivity, idempotentency, and absorption. Some conditions are derived for *P*-union and *P*-intersection of two INCSESs to be an INCSES. It is shown that *P*-union and *P*-intersection of ENCSESs need not be an ENCSES. The *R*-union and *R*-intersection of the INCSESs (resp., ENCSESs) need not be an INCSES (resp., ENCSES). Necessary conditions for the *P*-union, *R*-union, and *R*-intersection of two ENCSESs to be an ENCSES are obtained. The authors also study the conditions for *R*-intersection and *P*-intersection of two NCSESs to be an INCSES and ENCSES. Finally, for its applications in games, the developed procedure to analyze a cricket series between Pakistan and India is used. It is shown that the proposed method is suitable to be used for decision making and is as good as, or better than, existing models.

A neutrosophic extended triplet group is a new algebra structure and is different from the classical group. In the following paper [28], the notion of a generalized neutrosophic extended triplet group is proposed, and some properties are discussed. In particular, the following conclusions are strictly proved: (1) an algebraic system is a generalized neutrosophic extended triplet group if and only if it is a quasi-complete regular semigroup; (2) an algebraic system is a weak commutative generalized neutrosophic extended triplet group if and only if it is a quasi-Clifford semigroup; (3) for each $n \in$ Z+, $n \geq 2$, (Zn,$\otimes$) is a commutative generalized neutrosophic extended triplet group; (4) for each $n \in$ Z+, $n \geq 2$, (Zn,$\otimes$) is a commutative neutrosophic extended triplet group if and only if n = p1p2···pm (i.e., n has only single factor).

The next paper [29] presents an epidemiological study on the dietary fat that causes prostate cancer in an uncertain environment. To study this relationship under an indeterminate environment, data from 30 countries are selected on the prostate cancer death rate and dietary fat level in food. Neutrosophic correlation and regression lines are fitted on the data. The authors note from the neutrosophic analysis that the prostate cancer death rate increases as the dietary fat level in people increases. The neutrosophic regression coefficient also confirms this claim. From this study, the authors conclude that neutrosophic regression is a more effective model under uncertainty than the regression model under classical statistics. They also found a statistical correlation between dietary fat and prostate cancer risk.

Recently, neutrosophic sets are found to be more general and useful to express incomplete, indeterminate, and inconsistent information. The purpose of the next paper [30] is to introduce new aggregation operators based on logarithmic operations and to develop a multicriteria decision-making approach to study the interaction between the input argument under a single valued neutrosophic (SVN) environment. The main advantage of the proposed operator is that it can deal with positive interaction situations, negative interactions, or non-interaction among the criteria during the decision-making process. In this paper, the authors also defined some logarithmic operational rules on SVN sets, then proposed the single valued neutrosophic hybrid aggregation operators as a tool for multicriteria decision making (MCDM) under a neutrosophic environment and discussed some properties. Finally, detailed decision-making steps for single valued neutrosophic MCDM problems were developed, and a practical case was given to check the created approach and to illustrate its validity and superiority. Besides this, a systematic comparison with other existent methods is conducted to reveal the advantages of the proposed method. Results indicate that the proposed method is suitable and effective for the decision process to evaluate their best alternative.

In the definition of the complex neutrosophic soft expert set (CNSES), the parameters are a classical set, and the parameters have the same degree of importance, which is considered as 1. This poses a limitation in modeling some problems. The subsequent paper [31] introduces the concept of a fuzzy parameterized complex neutrosophic soft expert set (FP-CNSES) to handle this issue by assigning a degree of importance to each of the problem parameters. The authors develop FP-CNSES by establishing the concept of a weighted fuzzy parameterized complex neutrosophic soft expert set (WFP-CNSES) based on the idea that each expert has a relative weight. These new mathematical frameworks reduce the chance of unfairness in the decision-making process. Some essential operations with their properties and relevant laws related to the notion of FP-CNSES are defined and verified. The notation of mapping on fuzzy parameterized complex neutrosophic soft expert classes is defined, and some properties of fuzzy parameterized complex neutrosophic soft expert images and inverse images were investigated. FP-CNSES is used to put forth an algorithm on decision making by converting it from a complex state to a real state, and the detailed decision steps are subsequently provided. Then, the authors provide comparisons of FP-CNSES to the current methods to show the superiority of the proposed method.

To handle indeterminate and incomplete data, neutrosophic logic/set/probability have been established. The neutrosophic truth, falsehood, and indeterminacy components exhibit symmetry as the truth, and the falsehood looks the same and behaves in a symmetrical way with respect to the indeterminacy component, which serves as a line of symmetry. A soft set is a generic mathematical tool for dealing with uncertainty. A rough set is a new mathematical tool for dealing with vague, imprecise, inconsistent, and uncertain knowledge in information systems. The next paper [32] introduces a new rough set model based on neutrosophic soft sets to exploit, simultaneously, the advantages of rough sets and neutrosophic soft sets in order to handle all types of uncertainty in data. The idea of a neutrosophic right neighborhood is utilized to define the concepts of neutrosophic soft/rough (NSR) lower and upper approximations. Properties of the suggested approximations are proposed and subsequently proven. Some of the NSR set concepts such as NSR definability, NSR relations, and NSR membership functions are suggested and illustrated with examples. Further, the authors demonstrate the feasibility of the new rough set model with decision-making problems involving neutrosophic soft sets. Finally, a discussion on the features and limitations of the proposed model is provided.

The following research [33] introduces a neutrosophic forecasting approach based on neutrosophic time series (NTS). Historical data can be transformed into neutrosophic time series data to determine their truth, indeterminacy, and falsity functions. The basis for the neutrosophication process is the score and accuracy functions of historical data. In addition, neutrosophic logical relationship groups (NLRGs) are determined, and a deneutrosophication method for NTS is presented. The objective of this research is to suggest an idea of first- and high-order NTSs. By comparing this approach with other approaches, the authors conclude that the suggested approach of forecasting gets better results compared to the other existing approaches of fuzzy, intuitionistic fuzzy, and neutrosophic time series.

Smart ports represent the current trend of port development. Intelligent operations reduce the daily production cost of ports, facilitate efficient production, strengthen the risk mitigation ability, and comply with the requirements for long-term development. However, a systematic and scientific smart port evaluation method is missing to nail down the evaluation indicators of a smart port and enable accurate evaluation of a port's degree of intelligence. The next paper [34] analyzes the concept of the smart port, establishes a set of smart port evaluation indicator systems, and applies a single-valued neutrosophic exponential similarity measure in port evaluation to enable a quantitative evaluation of port integrity. This evaluation method is capable of making decisions in the event of incomplete, uncertain, and inconsistent information during general evaluation, opening up a new method for smart port evaluation and acting as a helpful tool for ports to carry out improvements during actual application.

The nonstandard analysis is extended in [35], by adding the left monad closed to the right and right monad closed to the left. Besides the pierced binad (introduced in 1998), Smarandache adds now the unpierced binad—all these in order to close the newly extended nonstandard space under nonstandard addition, nonstandard subtraction, nonstandard multiplication, nonstandard division, and nonstandard power operations. Then, the author extends the nonstandard neutrosophic logic, nonstandard neutrosophic set, and nonstandard probability on this extended nonstandard analysis space, and proves that it is a nonstandard neutrosophic lattice of the first type (endowed with a nonstandard neutrosophic partial order) as well as a nonstandard neutrosophic lattice of the second type (as algebraic structure, endowed with two binary neutrosophic laws: inf_N and sup_N). Many theorems, new terms introduced, better notations for monads and binads, and examples of nonstandard neutrosophic operations are given.

Shale gas energy is the most prominent and dominating source of power across the globe. The processes for the extraction of shale gas from shale rocks are very complex. In the next study [36], a multiobjective optimization framework is presented for an overall water management system that includes the allocation of freshwater for hydraulic fracturing and optimal management of the resulting wastewater with different techniques. The generated wastewater from the shale fracking process contains highly toxic chemicals. The optimal control of a massive amount of contaminated water is quite a challenging task. Therefore, an on-site treatment plant, underground disposal facility, and treatment plant with expansion capacity were designed to overcome environmental issues. A multiobjective trade-off between socio-economic and environmental concerns was established under a set of conflicting constraints. A solution method —the neutrosophic goal programming approach—is suggested, inspired by independent, neutral/indeterminacy thoughts of the decision-maker(s). A theoretical computational study is presented to show the validity and applicability of the proposed multiobjective shale gas water management optimization model and solution procedure. The obtained results and conclusions, along with significant contributions, are discussed in the context of shale gas supply chain planning policies over different time horizons.

Many companies have observed the significant benefits they can get via using the Internet. Since then, large companies have been able to develop business transactions with customers at anytime, anywhere, and in relation to anything, so that a more comprehensive concept than the Internet is needed. This concept is the Internet of Things (IoT). IoT will influence decision-making styles in various phases of selling, buying, and marketing processes. Therefore, every individual and company should know precisely what IoT is and how and why they should incorporate it in their operations. In [37], a smart system based on IoT was used to help companies and marketers make a powerful marketing strategy via utilizing obtained data from IoT devices. Not only this, but the proposed system can also solve the problems that companies and customers face in online shopping. Since there are different types of the same product, and also different criteria for purchasing that can be different between individuals, customers will need a decision support system to recommend them the best selection. This motivated the authors to propose a neutrosophic technique to deal with unclear and conflicting information that exists usually in the purchasing process. Therefore, the smart system and neutrosophic technique are considered as a comprehensive system, which links customers, companies, and marketers to achieve satisfaction for each of them.

Taking third-party logistics providers (3PLs) as an example, according to the characteristics of correlation between attributes in multiattribute decision making, two Choquet aggregation operators adopting probabilistic neutrosophic hesitation fuzzy elements (PNHFEs) are proposed to cope with the situations of correlation among criteria in [38]. This measure not only provides support for the correlation phenomenon between internal attributes, but it also fully concerns the incidental uncertainty of the external space. The goal is to make it easier for decision makers to cope with this uncertainty; thus, the authors establish the notion of a probabilistic neutrosophic hesitant fuzzy Choquet averaging (geometric) (PNHFCOA, PNHFCOG) operator. Based on this foundation, a method for aggregating decision makers' information is proposed, and then the optimal decision scheme is obtained. Finally, an example of selecting optimal 3PL is given to demonstrate the objectivity of the above-mentioned standpoint.

In the next paper [39], the authors explore the algebra structure based on neutrosophic quadruple numbers. Moreover, two kinds of degradation algebra systems of neutrosophic quadruple numbers are introduced. In particular, the following results are strictly proved: (1) the set of neutrosophic quadruple numbers with a multiplication operation is a neutrosophic extended triplet group; (2) the neutral element of each neutrosophic quadruple number is unique, and there are only sixteen different neutral elements in all of the neutrosophic quadruple numbers; (3) the set which has same neutral element is closed with respect to the multiplication operator; and (4) the union of the set which has same neutral element is a partition of four-dimensional space.

The following study [40] aims to introduce the notion of bipolar neutrosophic Hamacher aggregation operators and to also provide its application in real life. Then, the neutrosophic set (NS) can elaborate the incomplete, inconsistent, and indeterminate information, Hamacher aggregation operators, and extended Einstein aggregation operators to the arithmetic and geometric aggregation operators. First, the authors give the fundamental definition and operations of the neutrosophic set and the bipolar neutrosophic set. The main focus is on the Hamacher aggregation operators of bipolar neutrosophic sets, namely, bipolar neutrosophic Hamacher weighted averaging (BNHWA), bipolar neutrosophic Hamacher ordered weighted averaging (BNHOWA), and bipolar neutrosophic Hamacher hybrid averaging (BNHHA) along with their desirable properties. The prime gain of utilizing the suggested methods is that these operators progressively provide a total perspective on the issues necessary to the decision makers. These tools provide generalized, increasingly exact, and precise outcomes when compared to the current methods. Finally, as an application, the authors propose new methods for the multicriteria group decision-making issues by using various kinds of bipolar neutrosophic operators with a numerical model. This demonstrates the usefulness and practicality of this proposed approach in real life.

In the next paper [41], the authors introduce nonstandard neutrosophic topology in the extended nonstandard analysis space, called nonstandard real monad space, which is closed under neutrosophic nonstandard infimum and supremum conditions. Many classical topological concepts are extended to the nonstandard neutrosophic topology, several theorems and properties about them are proven, and many examples are presented.

Neutrosophic triangular norms (t-norms) and their residuated lattices are not only the main research object of neutrosophic set theory, but they are also the core content of neutrosophic logic. Neutrosophic implications are important operators of neutrosophic logic. Neutrosophic residual implications based on neutrosophic t-norms can be applied to the fields of neutrosophic inference and neutrosophic control. In [42], neutrosophic t-norms, neutrosophic residual implications, and the residuated lattices derived from neutrosophic t-norms are deeply investigated by Qingqing Hu and Xiaohong Zhang. First of all, the lattice and its corresponding system are proven to be a complete lattice and follow De Morgan algebra, respectively. Secondly, the notions of neutrosophic t-norms are introduced on the complete lattice discussed earlier. The basic concepts and typical examples of representable and nonrepresentable neutrosophic t-norms are obtained. Naturally, De Morgan neutrosophic triples are defined for the duality of neutrosophic t-norms and neutrosophic t-conorms with respect to neutrosophic negators. Thirdly, neutrosophic residual implications generated from neutrosophic t-norms and their basic properties are investigated. Furthermore, residual neutrosophic t-norms are proven to be infinitely ∨-distributive, and then some important properties possessed by neutrosophic residual implications are given. Finally, a method for producing neutrosophic t-norms from neutrosophic implications is presented, and the residuated lattices are constructed on the basis of neutrosophic t-norms and neutrosophic residual implications.

An online neutrosophic similarity-based object tracking with a weighted multiple instance learning algorithm (NeutWMIL) is proposed in [43]. Each training sample is extracted surrounding the object location, and the distribution of these samples is symmetric. To provide a more robust weight for each sample in the positive bag, the asymmetry of the importance of the samples is considered. The neutrosophic similarity-based object estimation with object properties (super straddling) is applied. The neutrosophic theory is a new branch of philosophy for dealing with incomplete, indeterminate, and inconsistent information. By considering the surrounding information of the object, a single valued neutrosophic set (SVNS)-based segmentation parameter selection method is proposed to produce a well-built set of superpixels that can better explain the object area at each frame. Then, the intersection and shape-distance criteria are proposed for weighting each superpixel in the SVNS domain, mainly via three membership functions, T (truth), I (indeterminacy), and F (falsity), for each criterion. After filtering out the superpixels with low responses, the newly defined neutrosophic weights are utilized for weighting each sample. Furthermore, the objectness estimation information is also applied for estimating and alleviating the problem of tracking drift. Experimental results on challenging benchmark video sequences reveal the superior performance of the algorithm when confronting appearance changes and background clutters.

Supply chain sustainability has become one of the most attractive decision management topics. There are many articles that have focused on this field, presenting many different points of view. The following research [44] is centered on the evaluation of supply chain sustainability based on two critical dimensions. The first is the importance of evaluation metrics based on economic, environmental, and social aspects, and the second is the degree of difficulty of information gathering. This paper aims to increase the accuracy of the evaluation. The proposed method is a combination of quality function deployment (QFD) with plithogenic aggregation operations. The aggregation operation is applied to aggregate, firstly, the decision maker's opinions of requirements that are needed to evaluate the supply chain sustainability; secondly, the evaluation metrics based on the requirements; and lastly, the evaluation of information gathering difficulty. To validate the proposed model, this study presented a real-world case study of Thailand's sugar industry. The results showed the most preferred and the lowest preferred metrics in order to evaluate the sustainability of the supply chain strategy.

Multiattribute decision making (MADM) is a part of management decision making and an important branch of the modern decision theory and method. MADM focuses on the decision problem of discrete and finite decision schemes. Uncertain MADM is an extension and development of classical multiattribute decision-making theory. When the attribute value of MADM is shown by neutrosophic numbers, that is, the attribute value is complex data and needs three values to express, it is called an MADM problem in which the attribute values are neutrosophic numbers. However, in practical MADM problems, to minimize errors in individual decision making, one needs to consider the ideas of many people and synthesize their opinions. Therefore, it is of great significance to study the method of attribute information aggregation. In the subsequent paper [45], the authors propose a new theory, a nondual multigranulation neutrosophic rough set (MS), to aggregate multiple attributes and solve a multiattribute group decision-making (MGDM) problem where the attribute values are neutrosophic numbers. First, the authors defined two kinds of nondual MS models, intersection-type MS and union-type MS. Additionally, their properties are studied. Then, the relationships between MS, nondual MS, neutrosophic rough set (NRS) based on a neutrosophic intersection (union) relationship, and NRS based on a neutrosophic transitive closure relation of union relationship are outlined, and a figure is given to show them directly. Finally, the definition of nondual MS on two universes is given and used to solve a MGDM problem with a neutrosophic number as the attribute value.

The following paper [46] aims to explore the algebraic structure of refined neutrosophic numbers. Firstly, the algebraic structure of neutrosophic quadruple numbers on a general field is studied. Secondly, the addition operator $\oplus$ and multiplication operator $\otimes$on refined neutrosophic numbers are proposed, and the algebraic structure is discussed. The authors reveal that the set of neutrosophic refined numbers with an additive operation is an abelian group, and the set of neutrosophic refined numbers with a multiplication operation is a neutrosophic extended triplet group. Moreover, algorithms for solving the neutral element and opposite elements of each refined neutrosophic number are given.

The next research [47] sets the basis for modeling the performance indicators of financial assets using triangular neutrosophic fuzzy numbers. This type of number allows for the modeling of financial asset performance indicators by taking into consideration all possible scenarios of their achievement. The key performance indicators (KPIs) modeled with the help of triangular fuzzy neutrosophic numbers are the return on financial assets, the financial assets risk, and the covariance between financial assets. Thus far, the return on financial assets has been studied using statistical indicators, like the arithmetic and geometric mean, or using the financial risk indicators with the help of the squared deviations from the mean and covariance. These indicators are well known as the basis of portfolio theory. This paper opens up the perspective of modeling these three mentioned statistical indicators using triangular neutrosophic fuzzy numbers because of the major advantages they have. The first advantage of the neutrosophic approach is that it includes three possible symmetric scenarios for KPI achievement, namely the scenario of certainty, the scenario of nonrealization, and the scenario of indecision, in which it cannot be appreciated whether the performance indicators are or are not achieved. The second big advantage is its data series clustering, representing the financial performance indicators by which these scenarios can be delimitated by means of neutrosophic fuzzy numbers in very good, good, or weak performance indicators. This clustering is realized by means of the linguistic criteria and measuring the belonging degree to a class of indicators using fuzzy membership functions. The third major advantage is the selection of risk mitigation analysis scenarios and the formation of optimal financial asset portfolios.

With the development of the social economy and the enlarged volume of information, the application of multiple-attribute decision making (MADM) has become increasingly complex, uncertain, and obscure. As a further generalization of the hesitant fuzzy set (HFS), the simplified neutrosophic hesitant fuzzy set (SNHFS) is an efficient tool to process vague information and contains the ideas of a single-valued neutrosophic hesitant fuzzy set (SVNHFS) and an interval neutrosophic hesitant fuzzy set (INHFS). In the following paper [48], the authors propose a decision-making approach based on the maximizing deviation method and TOPSIS (technique for order preference by similarity to ideal solution) to solve the MADM problems, in which the attribute weight information is incomplete, and the decision information is expressed in simplified neutrosophic hesitant fuzzy elements. Firstly, the authors inaugurate an optimization model on the basis of maximizing the deviation method, which is useful to determine the attribute weights. Secondly, using the idea of TOPSIS, the authors determine the relative closeness coefficient of each alternative and, based on that, they rank the considered alternatives to select the optimal one(s). Finally, the authors use a numerical example to show the detailed implementation procedure and effectiveness of the method in solving MADM problems under a simplified neutrosophic hesitant fuzzy environment.

Recently, various types of single valued neutrosophic (SVN) rough set models have been presented based on the same inclusion relation. However, there is another SVN inclusion relation in SVN sets. In the next paper [49], the authors propose a new type of SVN that covers a rough set model based on the new inclusion relation. Furthermore, graph and matrix representations of the new SVN covering approximation operators are presented. Firstly, the notion of SVN β2-covering approximation space is proposed, which is decided by the new inclusion relation. Then, a type of SVN covering rough set model under the SVN β2-covering approximation space is presented. Moreover, there is a corresponding SVN relation rough set model based on an SVN relation induced by the SVN β2-covering space, and two conditions under which the SVN β2-covering space can induce a symmetric SVN relation are presented. Thirdly, the graph and matrix representations of the new SVN covering rough set model are investigated. Finally, the authors propose a novel method for decision making (DM) problems in a paper on defect diagnosis under the new SVN covering rough set model.

In the subsequent paper [50], the authors define an ordinary single valued neutrosophic topology and obtain some of its basic properties. In addition, the authors introduce the concept of an ordinary single valued neutrosophic subspace. Next, they define the ordinary single-valued neutrosophic neighborhood system and show that an ordinary single-valued neutrosophic neighborhood system has the same properties in a classical neighborhood system. Finally, the authors introduce the concepts of an ordinary single-valued neutrosophic base and an ordinary single-valued neutrosophic sub-base, and obtain two characterizations of an ordinary single-valued neutrosophic base and one characterization of an ordinary single-valued neutrosophic sub-base.

Uninorms comprise an important kind of operator in fuzzy theory. They are obtained from the generalization of the t-norm and t-conorm axiomatic. Uninorms are theoretically remarkable, and furthermore, they have a wide range of applications. For that reason, when fuzzy sets have been generalized to others (e.g., intuitionistic fuzzy sets, interval-valued fuzzy sets, interval-valued intuitionistic fuzzy sets, or neutrosophic sets), then uninorm generalizations have emerged in those novel frameworks. Neutrosophic sets contain the notion of indeterminacy, which is caused by unknown, contradictory, and paradoxical information; thus, it includes, aside from the membership and nonmembership functions, an indeterminate membership function. Also, the relationship among them does not satisfy any restriction. Along this line of generalizations, the following paper [51] aims to extend uninorms to the framework of neutrosophic offsets, which are called neutrosophic offuninorms. Offsets are neutrosophic sets such that their domains exceed the scope of the interval [0,1]. In the paper, the definition, properties, and application areas of this new concept are provided. It is necessary to emphasize that the neutrosophic offuninorms are feasible for application in several fields, as the authors illustrate.

The symmetry of hyperoperation is expressed by hypergroups; more extensive hyperalgebraic structures than hypergroups are studied in the next paper [52]. The new concepts of neutrosophic extended triplet semihypergroup (NET semihypergroup) and neutrosophic extended triplet hypergroup (NET-hypergroup) are firstly introduced, some basic properties are obtained, and the relationships among NET-semihypergroups, regular semihypergroups, NET-hypergroups, and regular hypergroups are systematically are investigated. Moreover, pure NET-semihypergroup and pure NET-hypergroup are investigated, and a structure theorem of commutative pure NET-semihypergroup is established. Finally, a new notion of weak commutative NET-semihypergroup is proposed, some important examples are obtained by software MATLAB, and the following important result is proved: every pure and weak commutative NET-semihypergroup is a disjointed union of some regular hypergroups that are its subhypergroups.

The last paper [53] aims to reveal the structure of idempotents in neutrosophic rings and neutrosophic quadruple rings. First, all idempotents in neutrosophic rings $\langle R \cup I \rangle$ are given when R is C, R, Q, Z, or Zn. Secondly, the neutrosophic quadruple ring $\langle R \cup T \cup I \cup F \rangle$ is introduced, and all idempotents in neutrosophic quadruple rings $\langle C \cup T \cup I \cup F \rangle$, $\langle R \cup T \cup I \cup F \rangle$, $\langle Q \cup T \cup I \cup F \rangle$, $\langle Z \cup T \cup I \cup F \rangle$, and $\langle Zn \cup T \cup I \cup F \rangle$ are also given. Furthermore, the algorithms for solving the idempotents in $\langle Zn \cup I \rangle$ and $\langle Zn \cup T \cup I \cup F \rangle$ for each non-negative integer n are provided. Lastly, as a general result, if all idempotents in any ring R are known, then the structure of idempotents in neutrosophic ring $\langle R \cup I \rangle$ and neutrosophic quadruple ring $\langle R \cup T \cup I \cup F \rangle$ can be determined.

The individual articles of this book can be downloaded from here:

https://www.mdpi.com/journal/symmetry/special_issues/Neutrosophic_Set_Logic_Probability#published.

Our authors' geographical distribution (published papers) is:

China (32)
Pakistan (14)
Saudi Arabia (5)
Egypt (4)
India (4)
Korea (3)
Romania (3)
Jordan (2)
Malaysia (2)
Cuba (1)
Ecuador (1)
UK (1)
USA (1)

We found the edition and selections of papers for this book very inspiring and rewarding. We also thank the editorial staff and reviewers for their efforts and help during the process.

Funding: The author received no funding.

Conflicts of Interest: The author declares no conflicts of interest.

References

1. Smarandache, F. Definitions Derived from Neutrosophics. Available online: http://fs.unm.edu/DefinitionsDerivedFromNeutrosophics.pdf (accessed on 10 November 2019).
2. Smarandache, F. N-Valued Refined Neutrosophic Logic and Its Applications to Physics. Available online: https://arxiv.org/ftp/arxiv/papers/1407/1407.1041.pdf (accessed on 10 November 2019).
3. Smarandache, F. Degree of Dependence and Independence of the (Sub)Components of Fuzzy Set and Neutrosophic Set. *Neutrosophic Sets Syst.* **2016**, *11*, 95–97. Available online: http://fs.unm.edu/NSS/DegreeOfDependenceAndIndependence.pdf (accessed on 10 November 2019).
4. Smarandache, F. *Neutrosophic Overset, Neutrosophic Underset, and Neutrosophic Offset. Similarly for Neutrosophic Over-/Under-/OffLogic, Probability, and Statistics*; Pons Editions: Brussels, Belgium, 2016; 168p.

5. Smarandache, F. Operators on Single-Valued Neutrosophic Oversets, Neutrosophic Undersets, and Neutrosophic Offsets. *J. Math. Inform.* **2016**, *5*, 63–67. Available online: http://fs.unm.edu/SVNeutrosophicOverset-JMI.pdf (accessed on 10 November 2019).
6. Smarandache, F. *A Unifying Field in Logics: Neutrosophic Logic. Neutrosophy, Neutrosophic Set, Neutrosophic Probability and Statistics*, 6th ed.; InfoLearnQuest:, Ann Arbor, MI, USA, 2007; p. 93. Available online: http://fs.unm.edu/eBook-Neutrosophics6.pdf (accessed on 10 November 2019).
7. Smarandache, F. N-norm and N-conorm in Neutrosophic Logic and Set, and the Neutrosophic Topologies. *Crit. Rev.* **2009**, *3*, 73–83. Available online: http://fs.unm.edu/N-normN-conorm.pdf (accessed on 10 November 2019).
8. Smarandache, F. *Introduction to Neutrosophic Measure, Neutrosophic Integral, and Neutrosophic Probability*; Sitech Education: West Perth, Australia, 2013; 140p. Available online: http://fs.unm.edu/NeutrosophicMeasureIntegralProbability.pdf (accessed on 10 November 2019).
9. Smarandache, F. *Law of Included Multiple-Middle & Principle of Dynamic Neutrosophic Opposition*; EuropaNova & Education Publisher: Brussels, Belgium; Columbus, OH, USA, 2014; 135p. Available online: http://fs.unm.edu/LawIncludedMultiple-Middle.pdf (accessed on 10 November 2019).
10. Smarandache, F. *Introduction to Neutrosophic Statistics*; Sitech & Education Publishing: West Perth, Australia, 2014; 140p. Available online: http://fs.unm.edu/NeutrosophicStatistics.pdf (accessed on 10 November 2019).
11. Smarandache, F. *Neutrosophic Precalculus and Neutrosophic Calculus*; EuropaNova: Brussels, Belgium, 2015; 154p. Available online: http://fs.unm.edu/NeutrosophicPrecalculusCalculus.pdf (accessed on 10 November 2019).
12. Smarandache, F. *Symbolic Neutrosophic Theory*; EuropaNova: Bruxelles, Belgium, 2015; 194p. Available online: http://fs.unm.edu/SymbolicNeutrosophicTheory.pdf (accessed on 10 November 2019).
13. Smarandache, F. Introducing a Theory of Neutrosophic Evolution: Degrees of Evolution, Indeterminacy, and Involution. *Prog. Phys.* **2017**, *13*, 130–135. Available online: http://fs.unm.edu/neutrosophic-evolution-PP-49-13.pdf (accessed on 10 November 2019).
14. Smarandache, F. *Plithogeny, Plithogenic Set, Logic, Probability, and Statistics*; Pons: Bruxelles, Belgium, 2017; 141p. Available online: http://fs.unm.edu/Plithogeny.pdf (accessed on 10 November 2019).
15. Smarandache, F. *Neutropsychic Personality. A Mathematical Approach to Psychology*, 3rd ed.; Pons: Bruxelles, Belgium, 2018; 131p. Available online: http://fs.unm.edu/NeutropsychicPersonality-ed3.pdf (accessed on 10 November 2019).
16. Peng, X.; Dai, J. A bibliometric analysis of neutrosophic set: Two decades review from 1998 to 2017. *Artif. Intell. Rev.* **2018**, 1–57. Available online: http://fs.unm.edu/BibliometricNeutrosophy.pdf (accessed on 10 November 2019).
17. Jana, C.; Pal, M. A Robust Single-Valued Neutrosophic Soft Aggregation Operators in Multi-Criteria Decision Making. *Symmetry* **2019**, *11*, 110, doi:10.3390/sym11010110.
18. Aslam, M. A Variable Acceptance Sampling Plan under Neutrosophic Statistical Interval Method. *Symmetry* **2019**, *11*, 114, doi:10.3390/sym11010114.
19. Abu Qamar, M.; Hassan, N. An Approach toward a Q-Neutrosophic Soft Set and Its Application in Decision Making. *Symmetry* **2019**, *11*, 139, doi:10.3390/sym11020139.
20. Ye, J.; Cui, W. Neutrosophic Compound Orthogonal Neural Network and Its Applications in Neutrosophic Function Approximation. *Symmetry* **2019**, *11*, 147, doi:10.3390/sym11020147.
21. Fahmi, A.; Amin, F.; Khan, M.; Smarandache, F. Group Decision Making Based on Triangular Neutrosophic Cubic Fuzzy Einstein Hybrid Weighted Averaging Operators. *Symmetry* **2019**, *11*, 180, doi:10.3390/sym11020180.
22. Aslam, M.; Aldosari, M.S. Inspection Strategy under Indeterminacy Based on Neutrosophic Coefficient of Variation. *Symmetry* **2019**, *11*, 193, doi:10.3390/sym11020193.
23. Khan, M.; Gulistan, M.; Yaqoob, N.; Khan, M.; Smarandache, F. Neutrosophic Cubic Einstein Geometric Aggregation Operators with Application to Multi-Criteria Decision Making Method. *Symmetry* **2019**, *11*, 247, doi:10.3390/sym11020247.

24. Jiang, W.; Zhang, Z.; Deng, X. Multi-Attribute Decision Making Method Based on Aggregated Neutrosophic Set. *Symmetry* **2019**, *11*, 267, doi:10.3390/sym11020267.
25. Cao, C.; Zeng, S.; Luo, D. A Single-Valued Neutrosophic Linguistic Combined Weighted Distance Measure and Its Application in Multiple-Attribute Group Decision-Making. *Symmetry* **2019**, *11*, 275, doi:10.3390/sym11020275.
26. Shi, L.; Yuan, Y. Hybrid Weighted Arithmetic and Geometric Aggregation Operator of Neutrosophic Cubic Sets for MADM. *Symmetry* **2019**, *11*, 278, doi:10.3390/sym11020278.
27. Gulistan, M.; Hassan, N. A Generalized Approach towards Soft Expert Sets via Neutrosophic Cubic Sets with Applications in Games. *Symmetry* **2019**, *11*, 289, doi:10.3390/sym11020289.
28. Ma, Y.; Zhang, X.; Yang, X.; Zhou, X. Generalized Neutrosophic Extended Triplet Group. *Symmetry* **2019**, *11*, 327, doi:10.3390/sym11030327.
29. Aslam, M.; Albassam, M. Application of Neutrosophic Logic to Evaluate Correlation between Prostate Cancer Mortality and Dietary Fat Assumption. *Symmetry* **2019**, *11*, 330, doi:10.3390/sym11030330.
30. Ashraf, S.; Abdullah, S.; Smarandache, F. Logarithmic Hybrid Aggregation Operators Based on Single Valued Neutrosophic Sets and Their Applications in Decision Support Systems. *Symmetry* **2019**, *11*, 364, doi:10.3390/sym11030364.
31. Al-Quran, A.; Hassan, N.; Alkhazaleh, S. Fuzzy Parameterized Complex Neutrosophic Soft Expert Set for Decision under Uncertainty. *Symmetry* **2019**, *11*, 382, doi:10.3390/sym11030382.
32. Al-Quran, A.; Hassan, N.; Marei, E. A Novel Approach to Neutrosophic Soft Rough Set under Uncertainty. *Symmetry* **2019**, *11*, 384, doi:10.3390/sym11030384.
33. Abdel-Basset, M.; Chang, V.; Mohamed, M.; Smarandche, F. A Refined Approach for Forecasting Based on Neutrosophic Time Series. *Symmetry* **2019**, *11*, 457, doi:10.3390/sym11040457.
34. Chen, J.; Xue, K.; Ye, J.; Huang, T.; Tian, Y.; Hua, C.; Zhu, Y. Simplified Neutrosophic Exponential Similarity Measures for Evaluation of Smart Port Development. *Symmetry* **2019**, *11*, 485, doi:10.3390/sym11040485.
35. Smarandache, F. Extended Nonstandard Neutrosophic Logic, Set, and Probability Based on Extended Nonstandard Analysis. *Symmetry* **2019**, *11*, 515, doi:10.3390/sym11040515.
36. Ahmad, F.; Adhami, A.Y.; Smarandache, F. Neutrosophic Optimization Model and Computational Algorithm for Optimal Shale Gas Water Management under Uncertainty. *Symmetry* **2019**, *11*, 544, doi:10.3390/sym11040544.
37. Abdel-Basset, M.; Mohamed, M.; Chang, V.; Smarandache, F. IoT and Its Impact on the Electronics Market: A Powerful Decision Support System for Helping Customers in Choosing the Best Product. *Symmetry* **2019**, *11*, 611, doi:10.3390/sym11050611.
38. Shao, S.; Zhang, X.; Zhao, Q. Multi-Attribute Decision Making Based on Probabilistic Neutrosophic Hesitant Fuzzy Choquet Aggregation Operators. *Symmetry* **2019**, *11*, 623, doi:10.3390/sym11050623.
39. Li, Q.; Ma, Y.; Zhang, X.; Zhang, J. Neutrosophic Extended Triplet Group Based on Neutrosophic Quadruple Numbers. *Symmetry* **2019**, *11*, 696, doi:10.3390/sym11050696.
40. Jamil, M.; Abdullah, S.; Yaqub Khan, M.; Smarandache, F.; Ghani, F. Application of the Bipolar Neutrosophic Hamacher Averaging Aggregation Operators to Group Decision Making: An Illustrative Example. *Symmetry* **2019**, *11*, 698, doi:10.3390/sym11050698.
41. Al Shumrani, M.A.; Smarandache, F. Introduction to Non-Standard Neutrosophic Topology. *Symmetry* **2019**, *11*, 706, doi:10.3390/sym11050706.
42. Hu, Q.; Zhang, X. Neutrosophic Triangular Norms and Their Derived Residuated Lattices. *Symmetry* **2019**, *11*, 817, doi:10.3390/sym11060817.
43. Hu, K.; He, W.; Ye, J.; Zhao, L.; Peng, H.; Pi, J. Online Visual Tracking of Weighted Multiple Instance Learning via Neutrosophic Similarity-Based Objectness Estimation. *Symmetry* **2019**, *11*, 832, doi:10.3390/sym11060832.
44. Abdel-Basset, M.; Mohamed, R.; Zaied, A.E.; Smarandache, F. Supply Chain Sustainability Metrics. *Symmetry* **2019**, *11*, 903, doi:10.3390/sym11070903.
45. Bo, C.; Zhang, X.; Shao, S. Non-Dual Multi-Granulation Neutrosophic Rough Set with Applications. *Symmetry* **2019**, *11*, 910, doi:10.3390/sym11070910.
46. Li, Q.; Ma, Y.; Zhang, X.; Zhang, J. Study on the Algebraic Structure of Refined Neutrosophic Numbers. *Symmetry* **2019**, *11*, 954, doi:10.3390/sym11080954.

47. Bolos, M.I.; Bradea, I.A.; Delcea, C. Modeling the Performance Indicators of Financial Assets with Neutrosophic Fuzzy Numbers. *Symmetry* **2019**, *11*, 1021, doi:10.3390/sym11081021.
48. Akram, M.; Naz, S.; Smarandache, F. Generalization of Maximizing Deviation and TOPSIS Method for MADM in Simplified Neutrosophic Hesitant Fuzzy Environment. *Symmetry* **2019**, *11*, 1058, doi:10.3390/sym11081058.
49. Wang, J.; Zhang, X. A New Type of Single Valued Neutrosophic Covering Rough Set Model. *Symmetry* **2019**, *11*, 1074, doi:10.3390/sym11091074.
50. Kim, J.; Smarandache, F.; Lee, J.G.; Hur, K. Ordinary Single Valued Neutrosophic Topological Spaces. *Symmetry* **2019**, *11*, 1075, doi:10.3390/sym11091075.
51. Caballero, E.G.; Smarandache, F.; Leyva Vázquez, M. On Neutrosophic Offuninorms. *Symmetry* **2019**, *11*, 1136, doi:10.3390/sym11091136.
52. Zhang, X.; Smarandache, F.; Ma, Y. Symmetry in Hyperstructure: Neutrosophic Extended Triplet Semihypergroups and Regular Hypergroups. *Symmetry* **2019**, *11*, 1217, doi:10.3390/sym11101217.
53. Ma, Y.; Zhang, X.; Smarandache, F.; Zhang, J. The Structure of Idempotents in Neutrosophic Rings and Neutrosophic Quadruple Rings. *Symmetry* **2019**, *11*, 1254, doi:10.3390/sym11101254.

Florentin Smarandache

Special Issue Editor

symmetry

MDPI

Article

The Structure of Idempotents in Neutrosophic Rings and Neutrosophic Quadruple Rings

Yingcang Ma [1,*], Xiaohong Zhang [2], Florentin Smarandache [3] and Juanjuan Zhang [1]

1 School of Science, Xi'an Polytechnic University, Xi'an 710048, China; 20080712@xpu.edu.cn
2 School of Arts and Sciences, Shaanxi University of Science & Technology, Xi'an 710021, China; zhangxiaohong@sust.edu.cn
3 Department of Mathematics, University of New Mexico, Gallup, NM 87301, USA; smarand@unm.edu
* Correspondence: mayingcang@xpu.edu.cn

Received: 12 September 2019; Accepted: 5 October 2019; Published: 8 October 2019

Abstract: This paper aims to reveal the structure of idempotents in neutrosophic rings and neutrosophic quadruple rings. First, all idempotents in neutrosophic rings $\langle R \cup I \rangle$ are given when R is $\mathbb{C}, \mathbb{R}, \mathbb{Q}, \mathbb{Z}$ or $\mathbb{Z}_n$. Secondly, the neutrosophic quadruple ring $\langle R \cup T \cup I \cup F \rangle$ is introduced and all idempotents in neutrosophic quadruple rings $\langle \mathbb{C} \cup T \cup I \cup F \rangle$, $\langle \mathbb{R} \cup T \cup I \cup F \rangle$, $\langle \mathbb{Q} \cup T \cup I \cup F \rangle$, $\langle \mathbb{Z} \cup T \cup I \cup F \rangle$ and $\langle \mathbb{Z}_n \cup T \cup I \cup F \rangle$ are also given. Furthermore, the algorithms for solving the idempotents in $\langle \mathbb{Z}_n \cup I \rangle$ and $\langle \mathbb{Z}_n \cup T \cup I \cup F \rangle$ for each nonnegative integer n are provided. Lastly, as a general result, if all idempotents in any ring R are known, then the structure of idempotents in neutrosophic ring $\langle R \cup I \rangle$ and neutrosophic quadruple ring $\langle R \cup T \cup I \cup F \rangle$ can be determined.

Keywords: neutrosophic rings; neutrosophic quadruple rings; idempotents; neutrosophic extended triplet group; neutrosophic set

1. Introduction

The notions of neutrosophic set and neutrosophic logic were proposed by Smarandache [1]. In neutrosophic logic, every proposition is considered by the truth degree T, the indeterminacy degree I, and the falsity degree F, where T, I and F are subsets of the nonstandard unit interval $]0^-, 1^+[= 0^- \cup [0,1] \cup 1^+$.

Using the idea of neutrosophic set, some related algebraic structures have been studied in recent years. Among these algebraic structures, by extending classical groups, the neutrosophic triplet group (NTG) and the neutrosophic extended triplet group (NETG) have been introduced in refs. [2–4]. As an example, paper [5] shows that $(\mathbb{Z}_{p_1 p_2 \cdots p_t}, \cdot)$ is not only a semigroup, but also a NETG, where $\cdot$ the classical mod multiplication and $p_1, p_2, \cdots, p_t$ are distinct primes. After the notions were put forward, NTG and NETG have been carried out in-depth research. For example, the inclusion relations of neutrosophic sets [6], neutrosophic triplet coset [7], neutrosophic duplet semi-groups [8], AG-neutrosophic extended triplet loops [9,10], the neutrosophic set theory to pseudo-BCI algebras [11], neutrosophic triplet ring and a neutrosophic triplet field [12,13], neutrosophic triplet normed space [14], neutrosophic soft sets [15], neutrosophic vector spaces [16], and so on.

In contrast to the neutrosophic triplet ring, the neutrosophic ring $\langle R \cup I \rangle$, which is a ring generated by the ring R and the indeterminate element I ($I^2 = I$), was proposed by Vasantha and Smarandache in [17]. The concept of neutrosophic ring was further developed and studied in [18–20].

As a special kind of element in an algebraic system, the idempotent element plays a major role in describing the structure and properties of the algebra. For example, Boolean rings refer to rings in which all elements are idempotent, clean rings [21] refer to rings in which each element is clean (an element in a ring is clean, if it can be written as the sum of an idempotent element and an invertible element), and Albel ring is a ring if each element in the ring is central. From these we can see that some rings can be characterized by idempotents. Thus, it is also quite meaningful to find all idempotents in a ring. In this paper, the idempotents in neutrosophic rings and neutrosophic quadruple rings will be studied in depth, and all idempotents in them can be obtained if the idempotents in R are known. In addition, the relationship between idempotents and neutral elements will be given. The elements of each NETG can be partitioned by neutrals [10]. Therefore, as an application, if $R = \mathbb{F}$, where $\mathbb{F}$ is any field, we can divide the elements of $\langle R \cup I \rangle$ (or $\langle R \cup T \cup I \cup F \rangle$) by idempotents. As another application, in paper [22], the authors explore the idempotents and semi-idempotents in neutrosophic ring $\langle \mathbb{Z}_n \cup I \rangle$ and some open problems and conjectures are given. In this paper, we will answer partial open problems and conjectures in paper [22] and some further studies are discussed.

The outline of this paper is organized as follows. Section 2 gives the basic concepts. In Section 3, the idempotents in neutrosophic ring $\langle R \cup I \rangle$ will be explored. For neutrosophic rings $\langle \mathbb{Z}_n \cup I \rangle$, $\langle \mathbb{C} \cup I \rangle, \langle \mathbb{R} \cup I \rangle, \langle \mathbb{Q} \cup I \rangle$ and $\langle \mathbb{Z} \cup I \rangle$, all idempotents will be given. Moreover, the open problem and conjectures proposed in paper [22] about idempotents in neutrosophic ring $\langle \mathbb{Z}_n \cup I \rangle$ will be solved. In Section 4, the neutrosophic quadruple ring $\langle R \cup T \cup I \cup F \rangle$ is introduced and all idempotents in neutrosophic quadruple rings $\langle \mathbb{C} \cup T \cup I \cup F \rangle, \langle \mathbb{R} \cup T \cup I \cup F \rangle, \langle \mathbb{Q} \cup T \cup I \cup F \rangle, \langle \mathbb{Z} \cup T \cup I \cup F \rangle$ and $\langle \mathbb{Z}_n \cup T \cup I \cup F \rangle$ will be given. Finally, the summary and future work is presented in Section 5.

2. Basic Concepts

In this section, the related basic definitions and properties of neutrosophic ring $\langle R \cup I \rangle$ and NETG are provided, the details can be seen in [3,4,17,18].

Definition 1. *([17,18]) Let $(R, +, \cdot)$ be any ring. The set*

$$\langle R \cup I \rangle = \{a + bI : a, b \in R\}$$

is called a neutrosophic ring generated by R and I. Let $a_1 + b_1 I, a_2 + b_2 I \in \langle R \cup I \rangle$, The operators $\oplus$ and $\otimes$ on $\langle R \cup I \rangle$ are defined as follows:

$$(a_1 + b_1 I) \oplus (a_2 + b_2 I) = (a_1 + a_2) + (b_1 + b_2)I,$$

$$(a_1 + b_1 I) \otimes (a_2 + b_2 I) = (a_1 \cdot a_2) + (a_1 \cdot b_2 + b_1 \cdot a_2 + b_1 \cdot b_2)I.$$

Remark 1. *It is easy to verify that $(\langle R \cup I \rangle, \oplus, \otimes)$ is a ring, so $\langle R \cup I \rangle$ is named by a neutrosophic ring is reasonable.*

Remark 2. *It should be noted that the operators $+, \cdot$ are defined on ring R and $\oplus, \otimes$ are defined on neutrosophic ring $\langle R \cup I \rangle$. For simplicity of notation, we also use $+, \cdot$ to replace $\oplus, \otimes$ on ring $\langle R \cup I \rangle$. That is $a + b$ also means $a \oplus b$ if $a, b \in \langle R \cup I \rangle$. $a \cdot b$ also means $a \otimes b$ if $a, b \in \langle R \cup I \rangle$. For short $a \cdot b$ denoted by ab and $a \cdot a$ denoted by a^2.*

Example 1. *$\langle \mathbb{Z} \cup I \rangle, \langle \mathbb{Q} \cup I \rangle, \langle \mathbb{R} \cup I \rangle$ and $\langle \mathbb{C} \cup I \rangle$ are neutrosophic rings of integer, rational, real and complex numbers, respectively. $\langle \mathbb{Z}_n \cup I \rangle$ is neutrosophic ring of modulo integers. Of course, $\mathbb{Z}, \mathbb{Q}, \mathbb{R}, \mathbb{C}$ and $\mathbb{Z}_n$ are neutrosophic rings when $b = 0$.*

Definition 2. *([17,18]) Let $\langle R \cup I \rangle$ be a neutrosophic ring. $\langle R \cup I \rangle$ is said to be commutative if*

$$ab = ba, \forall a, b \in \langle R \cup I \rangle.$$

In addition, if there exists $1 \in \langle R \cup I \rangle$ such that $1 \cdot a = a \cdot 1 = a$ for all $a \in \langle R \cup I \rangle$ then we call $\langle R \cup I \rangle$ a commutative neutrosophic ring with unity.

Definition 3. *([17,18]) An element a in a neutrosophic ring $\langle R \cup I \rangle$ is called an idempotent element if $a^2 = a$.*

Definition 4. *([3,4]) Let N be a non-empty set together with a binary operation $*$. Then, N is called a neutrosophic extended triplet set if for any $a \in N$, there exists a neutral of "a" (denote by $neut(a)$), and an opposite of "a"(denote by $anti(a)$), such that $neut(a) \in N$, $anti(a) \in N$ and:*

$$a * neut(a) = neut(a) * a = a, \quad a * anti(a) = anti(a) * a = neut(a).$$

The triplet $(a, neut(a), anti(a))$ is called a neutrosophic extended triplet.

Definition 5. *([3,4]) Let $(N,*)$ be a neutrosophic extended triplet set. Then, N is called a neutrosophic extended triplet group (NETG), if the following conditions are satisfied:*
(1) $(N,*)$ is well-defined, i.e., for any $a, b \in N$, one has $a * b \in N$.
(2) $(N,*)$ is associative, i.e., $(a*b)*c = a*(b*c)$ for all $a, b, c \in N$.
A NETG N is called a commutative NETG if for all $a, b \in N, a*b = b*a$.

Proposition 1. *([4]) $(N,*)$ be a NETG. We have:*
(1) $neut(a)$ is unique for any $a \in N$.
(2) $neut(a) * neut(a) = neut(a)$ for any $a \in N$.
(3) $neut(neut(a)) = neut(a)$ for any $a \in N$.

Proposition 2. *([10]) Let $(N,*)$ is a NETG, denote the set of all different neutral element in N by $E(N)$. For any $e \in E(N)$, denote $N(e) = \{x | neut(x) = e, x \in N\}$. Then:*
(1) $N(e)$ is a classical group, and the unit element is e.
(2) For any $e_1, e_2 \in E(N), e_1 \neq e_2 \Rightarrow N(e_1) \cap N(e_2) = \emptyset$.
(3) $N = \bigcup_{e \in E(N)} N(e)$. i.e., $\bigcup_{e \in E(N)} N(e)$ is a partition of N.

3. The Idempotents in Neutrosophic Rings

In this section, we will explore the idempotents in neutrosophic rings $\langle R \cup I \rangle$. If R is $\mathbb{Z}, \mathbb{Q}, \mathbb{R}, \mathbb{C}$ or $\mathbb{Z}_n$, all idempotents in neutrosophic rings $\langle \mathbb{Z}_n \cup I \rangle, \langle \mathbb{C} \cup I \rangle, \langle \mathbb{R} \cup I \rangle, \langle \mathbb{Q} \cup I \rangle$ or $\langle \mathbb{Z} \cup I \rangle$ will be given. Moreover, we can also obtain all idempotents in neutrosophic ring $\langle R \cup I \rangle$ if all idempotents in any ring R are known. As an application, the open problem and conjectures about the idempotents of neutrosophic ring $\langle \mathbb{Z}_n \cup I \rangle$ in paper [22] will be solved. Moreover, an example is given to show how to use the idempotents to get a partition for a neutrosophic ring. The following proposition reveal the relation of a neutral element and an idempotent element.

Proposition 3. *Let G be a non-empty set, $*$ is a binary operation on G. For each $a \in G$, a is idempotent iff it is a neutral element.*

Proof. Necessity: If a is idempotent, i.e., $a * a = a$, from Definition 4, which shows that a has neutral element a and opposite element a, so a is a neutral element.

Sufficiency: If a is a neutral element, from Proposition 1(2), we have $a * a = a$, thus a is idempotent. □

Theorem 1. *The set of all idempotents in neutrosophic ring $\langle \mathbb{C} \cup I \rangle, \langle \mathbb{R} \cup I \rangle, \langle \mathbb{Q} \cup I \rangle$ or $\langle \mathbb{Z} \cup I \rangle$ is $\{0, 1, I, 1-I\}$.*

Proof. We just give the proof for $\langle \mathbb{R} \cup I \rangle$, and the same result can be obtained for $\langle \mathbb{C} \cup I \rangle, \langle \mathbb{Q} \cup I \rangle$ or $\langle \mathbb{Z} \cup I \rangle$.

Let $a + bI \in \langle \mathbb{R} \cup I \rangle$. If $a + bI$ is idempotent, so $(a+bI)^2 = a + bI$, which means

$$\begin{cases} a^2 = a \\ 2ab + b^2 = b \end{cases} \quad (1)$$

From $a^2 = a$, we can get $a = 0$ or $a = 1$. When $a = 0$, from $2ab + b^2 = b$, we can get $b = 0$ or $b = 1$. That is 0 and I are idempotents. When $a = 1$, from $2ab + b^2 = b$, we can get $b = 0$ or $b = -1$. That is 1 and $1 - I$ are idempotents. Thus, the set of all idempotents of neutrosophic ring $\langle \mathbb{R} \cup I \rangle$ is $\{0, 1, I, 1 - I\}$. □

The above theorem reveals that the set of all idempotents in neutrosophic ring $\langle R \cup I \rangle$ is $\{0, 1, I, 1 - I\}$ when R is $\mathbb{C}, \mathbb{R}, \mathbb{Q}$ or $\mathbb{Z}$. For any ring R, we have the following results.

Proposition 4. *If a is idempotent in any ring R, then aI is also idempotent in neutrosophic ring $\langle R \cup I \rangle$.*

Proof. If $a \in R$ is idempotent, i.e., $a^2 = a$, so $(aI)^2 = (0 + aI)(0 + aI) = a^2 I = aI$, thus, aI is also idempotent in neutrosophic ring $\langle R \cup I \rangle$. □

Proposition 5. *In neutrosophic ring $\langle R \cup I \rangle$, then $a - aI$ is idempotent iff a is idempotent.*

Proof. Necessity: If $a - aI$ is idempotent, i.e., $(a - aI)^2 = a - aI$, so $(a - aI)^2 = (a - aI)(a - aI) = a^2 - 2aI + a^2 I = a^2 + (a^2 - 2a)I = a - aI$, which means $a^2 = a$ and $a^2 - 2a = -a$. Thus, we have $a^2 = a$, so a is idempotent.

Sufficiency: If a is idempotent, so $(a - aI)^2 = a^2 + (a^2 - 2a)I = a - aI$, thus $a - aI$ is idempotent. □

Theorem 2. *In neutrosophic ring $\langle R \cup I \rangle$, let $a + bI \in \langle R \cup I \rangle$, then $a + bI$ is idempotent iff a is idempotent in R and $b = c - a$, where c is any idempotent element in R.*

Proof. Necessity: If $a + bI$ is idempotent, i.e., $(a + bI)^2 = a + bI$, so $(a + bI)^2 = a^2 + (2ab + b^2) = a + bI$, which means $a^2 = a$ and $2ab + b^2 = b$. From $a^2 = a$, we can get a is idempotent. From $2ab + b^2 = b$ and $a^2 = a$, we can get $(b + a)^2 = b^2 + 2ab + a^2 = b + a$, so $b + a$ is also idempotent in R, denoted by c, so $b = c - a$.

Sufficiency: If a and c are any idempotents in R, let $b = c - a$, so $(a + bI)^2 = (a + (c - a)I)^2 = a^2 + (2a(c - a) + (c - a)^2)I = a^2 + (2ac - 2a^2 + c^2 - 2ac + a^2) = a + (c - a)I = a + bI$, thus $a + bI$ is idempotent. □

Theorem 3. *If the number of different idempotents in ring R is t, then the number of different idempotents in the neutrosophic ring $\langle R \cup I \rangle$ is t^2.*

Proof. If the number of idempotents in R is t and let $a + bI \in \langle R \cup I \rangle$ is idempotent, so from Theorem 2, we can infer that a is idempotent in R, i.e., a has t different selections. When a is fixed, set $b = c - a$, where c is any idempotent in R and c also has t different selections, which means b has t different selections. Thus, $a + bI$ has $t \cdot t = t^2$ different selections, i.e., the number of all idempotents in $\langle R \cup I \rangle$ is t^2. □

From the above analysis, for any ring R, all idempotents in $\langle R \cup I \rangle$ can be determined if all idempotents in R are known. In the following, we will explore all idempotents in neutrosophic ring $\langle \mathbb{Z}_n \cup I \rangle$, i.e., when $R = \mathbb{Z}_n$.

Theorem 4. *([5]) In the algebra system $(\mathbb{Z}_n, \cdot)$ (see Appendix A), $\cdot$ is the classical mod multiplication, for each $a \in \mathbb{Z}_n$, a has neut(a) and anti(a) iff* $\gcd(\gcd(a, n), n/\gcd(a, n)) = 1$.

Theorem 5. *([5]) For an algebra system $(\mathbb{Z}_n, \cdot)$ and $n = p_1^{k_1} p_2^{k_2} \cdots p_t^{k_t}$, where each $p_i (i = 1, 2, \cdots, t)$ is a prime, then the number of different neutral elements in $\mathbb{Z}_n$ is 2^t.*

Remark 3. *From Proposition 3 and Theorem 5, we can infer that the number of all idempotents in* $\mathbb{Z}_{p_1^{k_1} p_2^{k_2} \cdots p_t^{k_t}}$ *is also* 2^t.

Example 2. *For* $(\mathbb{Z}_{36}, \cdot)$, $n = 36 = 2^2 3^2$. *From Theorem 5, the number of different neutral elements in* $\mathbb{Z}_{36}$ *is* $2^2 = 4$. *They are:*

(1) $[0]$ *has the neutral element* $[0]$.
(2) $[1], [5], [7], [11], [13], [17], [19], [23], [25], [29], [31]$ *and* $[35]$ *have the same neutral element* $[1]$.
(3) $[9]$ *and* $[27]$ *have the same neutral element* $[9]$ *being* $\gcd(9, 36) = \gcd(27, 36) = 9$.
(4) $[4]$ *and* $[8]$ *have the same neutral element being* $\gcd(4, 36) = \gcd(8, 36) = 4$. *In fact,* $[4], [8], [16], [20], [28]$ *and* $[32]$ *have the same neutral element, which is* $[28]$.

From Remark 3, the number of idempotents in $\mathbb{Z}_{36}$ *is also* 4, *which are* $[0], [1], [9]$ *and* $[28]$.

From Theorems 2 and 3 and Remark 3, it follows easily that:

Corollary 1. *In neutrosophic ring* $\langle \mathbb{Z}_n \cup I \rangle$, *let* $a + bI \in \langle \mathbb{Z}_n \cup I \rangle$, *then* $a + bI$ *is idempotent iff* $a^2 = a$ *and* $b = c - a$, *where* c *is any idempotent element in* $\mathbb{Z}_n$.

Corollary 2. *For an algebra system* $(\mathbb{Z}_n, \cdot)$ *and* $n = p_1^{k_1} p_2^{k_2} \cdots p_t^{k_t}$, *where each* $p_1, p_2, \cdots$, *and* p_k *are distinct primes. Then the number of different idempotents in* $\langle \mathbb{Z}_n \cup I \rangle$ *is* 2^{2t}.

The solving process for $\langle \mathbb{Z}_n \cup I \rangle$ is given by Algorithm 1. Just only input n, then we can get all idempotents in $\langle \mathbb{Z}_n \cup I \rangle$. The MATLAB code is provided in the Appendix B.

Example 3. *Solve all idempotents in* $\langle \mathbb{Z}_{600} \cup I \rangle$.

Since $n = 600 = 2^3 \cdot 3 \cdot 5^2$, *from Theorem 5, we can get the different neutral elements in* $\mathbb{Z}_{600}$ *are* $neut(1), neut(2^3), neut(3), neut(5^2), neut(2^3 \cdot 3), neut(2^3 \cdot 5^2), neut(3 \cdot 5^2)$ *and* $neut(0)$, *i.e., the different idempotents in* $\mathbb{Z}_{600}$ *are 1, 376, 201, 25, 576, 400, 225, 0. From Corollary 2, the number of different idempotents in neutrosophic ring* $\langle \mathbb{Z}_{600} \cup I \rangle$ *is* $2^{2 \cdot 3} = 64$.

From Algorithm 1, the set of all 64 idempotents in $\langle \mathbb{Z}_{600} \cup I \rangle$ *is:* $\{0, I, 25I, 201I, 225I, 376I, 400I, 576I, 1 + 599I, 1, 1 + 24I, 1 + 200I, 1 + 224I, 1 + 375I, 1 + 399I, 1 + 575I, 25 + 575I, 25 + 576I, 25, 25 + 176I, 25 + 200I, 25 + 351I, 25 + 375I, 25 + 551I, 201 + 399I, 201 + 400I, 201 + 424I, 201, 201 + 24I, 201 + 175I, 201 + 199I, 201 + 375I, 225 + 375I, 225 + 376I, 225 + 400I, 225 + 576I, 225, 225 + 151I, 225 + 175I, 225 + 351I, 376 + 224I, 376 + 225I, 376 + 249I, 376 + 425I, 376 + 449I, 376, 376 + 24I, 376 + 200I, 400 + 200I, 400 + 201I, 400 + 225I, 400 + 401I, 400 + 425I, 400 + 576I, 400, 400 + 176I, 576 + 24I, 576 + 25I, 576 + 49I, 576 + 225I, 576 + 249I, 576 + 400I, 576 + 424I, 576\}$.

Algorithm 1: Solving the different idempotents in $\langle \mathbb{Z}_n \cup I \rangle$

```
Input: n
Factorization of integer n, we can get n = p_1^{k_1} p_2^{k_2} ··· p_t^{k_t}.
Computing the neutral element of 1, p_1^{k_1}, p_2^{k_2}, ···, p_t^{k_t}, p_1^{k_1} p_2^{k_2}, ··· p_1^{k_1} p_t^{k_t}, ···, p_2^{k_2} p_3^{k_3} ··· p_t^{k_t}
  and p_1^{k_1} p_2^{k_2} ··· p_t^{k_t}. So, we can get all idempotents in Z_n, denoted by a_1, a_2, ···, a_{2^t}.
Let ID=[];
for i = 1 : 2^t
    a = a_i
    for j = 1 : 2^t
        b = mod(a_j − a, n);
        ID = [ID; [a, b]];
    end
end
Output: ID: all the idempotents in ⟨Z_n ∪ I⟩
```

In paper [22], the authors studied the idempotents and semi-idempotents in $\langle \mathbb{Z}_n \cup I \rangle$ and proposed some open problems and conjectures. We list partial open problems and conjectures about idempotents in $\langle \mathbb{Z}_n \cup I \rangle$ as follows and answer them.

Problem 1. *([22]) Let $S = \langle \mathbb{Z}_{pq}, +, \cdot \rangle$, where p and q are two distinct primes, be the neutrosophic ring. Can S have non-trivial idempotents other than the ones mentioned in (b) of the Theorem 6?*

Conjecture 1. *([22]) Let $S = \langle \mathbb{Z}_n, +, \cdot \rangle$ be the neutrosophic ring $n = pqr$, where p, q and r are three distinct primes.*

1. *$\mathbb{Z}_n = \mathbb{Z}_{pqr}$ has only six non-trivial idempotents associated with it.*
2. *If m_1, m_2, m_3, m_4, m_5 and m_6 are the idempotents, then, associated with each real idempotent m_i, we have seven non-trivial neutrosophic idempotents associated with it, i.e., $\{m_i + n_j I, j = 1, 2, \cdots, 7\}$, such that $m_i + n_j \equiv t$, where t_j takes the seven distinct values from the set $\{0, 1, m_k, k \neq i; k = 1, 2, 3, \cdots, 6\}.i = 1, 2, \cdots, 6$.*

Conjecture 2. *([22]) Given $\langle \mathbb{Z}_n \cup I \rangle$, where $n = p_1 p_2 \cdot p_t; t > 2$ and $p_i s$ are all distinct primes, find:*

1. *the number of idempotents in $\mathbb{Z}_n$;*
2. *the number of idempotents in $\langle \mathbb{Z}_n \cup I \rangle \backslash \mathbb{Z}_n$;*

Conjecture 3. *([22]) Prove if $\langle \mathbb{Z}_n \cup I \rangle$ and $\langle \mathbb{Z}_m \cup I \rangle$ are two neutrosophic rings where $n > m$ and $n = p^t q$ ($t > 2$, and p and q two distinct primes) and $m = p_1 p_2 \cdots p_s$ where $p_i s$ are distinct primes. $1 \leq i \leq s$, then*

1. *prove $\mathbb{Z}_n$ has a greater number of idempotents than $\mathbb{Z}_m$; and*
2. *prove $\langle \mathbb{Z}_n \cup I \rangle$ has a greater number of idempotents than $\langle \mathbb{Z}_n \cup I \rangle$.*

Theorem 6. *([22]) Let $S = \langle \mathbb{Z}_{pq}, +, \cdot \rangle$ where p and q are two distinct primes:*

(a) There are two idempotents in $\mathbb{Z}_{pq}$ say r and s.
(b) $\{r, s, rI, sI, I, r + tI, s + tI | t \in \{\mathbb{Z}_{pq} \backslash 0\}\}$ such that $r + t = s, 1$ or 0 and $s + t = 0, 1$ or r is the partial collection of idempotents of S.

For Problem 1, from Remark 3, there are four idempotents in $\mathbb{Z}_{pq}$, which are $\{1, neut(p), neut(q), neut(pq) = 0\}$. Let $r = neut(p), s = neut(q)$, so there are two non-trivial idempotents r, s in $\mathbb{Z}_{pq}$. From Corollary 1 and 2, the number of all idempotents in $\langle \mathbb{Z}_{pq} \cup I \rangle$ is $2^4 = 16$, they are $\{0 + (0 - 0)I = 0, 0 + (1 - 0)I = I, 0 + (r - 0)I = rI, 0 + (s - 0)I = sI, 1 + (0 - 1)I = 1 + (n - 1)I, 1 + (1 - 1)I = 1, 1 + (r - 1)I, 1 + (s - 1)I, r + (0 - r)I = r + (n - r)I, r + (1 - r)I = r +$

$(n+1-r)I, r+(r-r)I = r, r+(s-r)I, s+(0-s)I = s+(n-r)s, s+(1-s)I = s+(n+1-s)I, s+(r-s)I, s+(s-s)I = s\}$. So there are 14 non-trivial idempotents in $\langle \mathbb{Z}_{pq} \cup I \rangle$, but there are only include 11 non-trivial idempotents in (b) of the Theorem 6, missing $\{1+(n-1)I, 1+(r-1)I, 1+(s-1)I\}$.

For Conjecture 1, from Corollary 1 and 2, there are eight idempotents in $\mathbb{Z}_{pqr}$, which are $\{1 = m_0, neut(p) = m_1, neut(q) = m_2, neut(r) = m_3, neut(pq) = m_4, neut(pr) = m_5, neut(qr) = m_6, neut(pqr) = 0 = m_7\}$. There are six non-trivial idempotents in $\mathbb{Z}_{pqr}$. In $\langle \mathbb{Z}_n \cup I \rangle$, all idempotents are $\{m_i + (m_j - m_i)I | i, j = 0, 1, 2, \cdots, 7\}$.

For Conjecture 2, from Remark 3, the number of idempotents in $\mathbb{Z}_{p_1 p_2 \cdots p_t}$ is 2^t, and the number of idempotents in $\langle \mathbb{Z}_{p_1 p_2 \cdots p_t} \cup I \rangle \backslash \mathbb{Z}_{p_1 p_2 \cdots p_t}$ is $2^{2t} - 2^t$.

For Conjecture 3, from Remark 3, the number of idempotents in $\mathbb{Z}_n$ is 2^2, and the number of idempotents in $\mathbb{Z}_m$ is 2^s, where $n = p^t q, m = p_1 p_2 \cdot p_s$. So, if $s > 2$, $\mathbb{Z}_m$ is characterized by a larger number of idempotents than $\mathbb{Z}_n$. In similarly way, the number of idempotents in $\langle \mathbb{Z}_n \cup I \rangle$ is 2^4, and the number of idempotents in $\langle \mathbb{Z}_m \cup I \rangle$ is 2^{2s}. So, if $s > 2$, we can infer that $\langle \mathbb{Z}_m \cup I \rangle$ is characterized by a larger number of idempotents than $\langle \mathbb{Z}_n \cup I \rangle$.

As another application, we will use the idempotents to divide the elements of the neutrosophic rings $\langle R \cup I \rangle$ when $R = \mathbb{F}$.

For each NETG $(N, *)$, $a \in N$, from Proposition 1, the neutral element of a is uniquely determined. From Proposition 2, $\bigcup_{e \in E(N)} N(e)$ is a partition of N. Since the idempotents and neutral elements are same, we can use the idempotents to get a partition of N. Let us illustrate these with the following example.

Example 4. *Let $R = \mathbb{Z}_3$, which is a field. Since $n = 3$, from Theorem 5, we can get the different neutral elements in $\mathbb{Z}_3$ are $neut(1)$ and $neut(0)$, i.e., the different idempotents in $\mathbb{Z}_3$ are $1, 0$. From Corollary 2, the number of different idempotents in neutrosophic ring $\langle \mathbb{Z}_3 \cup I \rangle$ is $2^{2 \cdot 1} = 4$.*

From Algorithm 1, the set of all 4 idempotents in $\langle \mathbb{Z}_3 \cup I \rangle$ is: $\{0, 1, I, 1+2I\}$. We have $E(0) = \{0\}, E(1) = \{1, 2, 1+I, 2+2I\}$, $E(I) = \{I, 2I\}, E(1+2I) = \{1+2I, 2+I\}$. So $\langle \mathbb{Z}_3 \cup I \rangle = E(0) \cup E(1) \cup E(I) \cup E(1+2I)$.

4. The Idempotents in Neutrosophic Quadruple Rings

In the above section, we explored the idempotents in $\langle R \cup I \rangle$. In neutrosophic logic, each proposition is approximated to represent respectively the truth (T), the falsehood (F), and the indeterminacy (I). In this section, according the idea of neutrosophic ring $\langle R \cup I \rangle$, the neutrosophic quadruple ring $\langle R \cup T \cup I \cup F \rangle$ is proposed and the idempotents are given in this section.

Definition 6. *Let $(R, +, \cdot)$ be any ring. The set*

$$\langle R \cup T \cup I \cup F \rangle = \{a_1 + a_2 T + a_3 I + a_4 F : a_1, a_2, a_3, a_4 \in R\} \tag{2}$$

is called a neutrosophic quadruple ring generated by R and T, I, F. Consider the order $T \prec I \prec F$. Let $a = a_1 + a_2 T + a_3 I + a_4 F, b = b_1 + b_2 T + b_3 I + b_4 F \in \langle R \cup T \cup I \cup F \rangle$, the operators $\oplus, \otimes$ on $\langle R \cup T \cup I \cup F \rangle$ are defined as follows:

$$\begin{aligned} a \oplus b &= (a_1 + a_2 T + a_3 I + a_4 F) \oplus (b_1 + b_2 T + b_3 I + b_4 F) \\ &= a_1 + b_1 + (a_2 + b_2)T + (a_3 + b_3)I + (a_4 + b_4)F. \end{aligned} \tag{3}$$

$$\begin{aligned} a * b = \ & (a_1 + a_2 T + a_3 I + a_4 F) * (b_1, b_2 T, b_3 I, b_4 F) \\ & = a_1 b_1 + (a_1 b_2 + a_2 b_1 + a_2 b_2)T + (a_1 b_3 + a_2 b_3 + a_3 b_1 + a_3 b_2 + a_3 b_3)I \\ & + (a_1 b_4 + a_2 b_4 + a_3 b_4 + a_4 b_1 + a_4 b_2 + a_4 b_3 + a_4 b_4)F. \end{aligned} \tag{4}$$

Remark 4. *It is easy to verify that* $(\langle R \cup T \cup I \cup F\rangle, \oplus, *)$ *is a ring, moreover, it also has the same algebra structure with neutrosophic quadruple numbers (see [23–25]), so the we call* $\langle R \cup T \cup I \cup F\rangle$ *is a neutrosophic quadruple ring is reasonable.*

Remark 5. *Similarly with Remark 2, for simplicity of notation, we use* $+, \cdot$ *to replace* $\oplus, *$ *on neutrosophic quadruple ring* $\langle R \cup T \cup I \cup F\rangle$*. That is* $a+b$ *also means* $a \oplus b$ *if* $a, b \in \langle R \cup T \cup I \cup F\rangle$*. and* $a \cdot b$ *also means* $a * b$ *if* $a, b \in \langle R \cup T \cup I \cup F\rangle$*. For short* $a \cdot b$ *denoted by* ab *and* $a \cdot a$ *denoted by* a^2*.*

Example 5. $\langle \mathbb{Z} \cup T \cup I \cup F\rangle, \langle \mathbb{Q} \cup T \cup I \cup F\rangle, \langle \mathbb{R} \cup T \cup I \cup F\rangle$ *and* $\langle \mathbb{C} \cup T \cup I \cup F\rangle$ *are neutrosophic quadruple rings of integer, rational, real and complex numbers, respectively.* $\langle \mathbb{Z}_n \cup T \cup I \cup F\rangle$ *is neutrosophic quadruple ring of modulo integers. Of course,* $\mathbb{Z}, \mathbb{Q}, \mathbb{R}, \mathbb{C}$ *and* $\mathbb{Z}_n$ *are neutrosophic quadruple rings when coefficients of* T, I *and* F *equal zero.*

Definition 7. *Let* $\langle R \cup T \cup I \cup F\rangle$ *be a neutrosophic quadruple ring.* $\langle R \cup T \cup I \cup F\rangle$ *is commutative if*

$$ab = ba, \forall a, b \in \langle R \cup T \cup I \cup F\rangle.$$

In addition, if there exists $1 \in \langle R \cup T \cup I \cup F\rangle$*, such that* $1 \cdot a = a \cdot 1 = a$ *for all* $a \in \langle R \cup T \cup I \cup F\rangle$*, then* $\langle R \cup T \cup I \cup F\rangle$ *is called a commutative neutrosophic quadruple ring with unity.*

Definition 8. *An element* a *in a neutrosophic quadruple ring* $\langle R \cup T \cup I \cup F\rangle$ *is called an idempotent element if* $a^2 = a$*.*

Theorem 7. *The set of all idempotents of neutrosophic quadruple rings* $\langle \mathbb{C} \cup T \cup I \cup F\rangle, \langle \mathbb{R} \cup T \cup I \cup F\rangle, \langle \mathbb{Q} \cup T \cup I \cup F\rangle$ *and* $\langle \mathbb{Z} \cup T \cup I \cup F\rangle$ *is*

$$\{(1,0,0,0), (0,0,0,F), (0,0,I,-F), (0,0,I,0), (0,T,-I,0), (0,T,-I,F), (0,T,0,-F), (0,T,0,0),$$

$$(1,-T,0,0), (1,-T,0,F), (1,-T,I,-F), (1,-T,I,0), (1,0,-I,0), (1,0,-I,F), (1,0,0,-F), (1,0,0,0)\}.$$

Proof. We only give the proof for $\langle \mathbb{R} \cup T \cup I \cup F\rangle$, and the same result can be obtained for $\langle \mathbb{C} \cup T \cup I \cup F\rangle, \langle \mathbb{Q} \cup T \cup I \cup F\rangle$ or $\langle \mathbb{Z} \cup T \cup I \cup F\rangle$.

Let $a = a_1 + a_2T + a_3I + a_4F$, if a is idempotent in $\langle \mathbb{R} \cup T \cup I \cup F\rangle$, so $a^2 = a$, i.e., $(a_1 + a_2T + a_3I + a_4F)^2 = (a_1 + a_2T + a_3I + a_4F)$, which means

$$\begin{cases} a_1^2 = a_1, \\ 2a_1a_2 + a_2^2 = a_2, \\ 2(a_1 + a_2)a_3 + a_3^2 = a_3, \\ 2(a_1 + a_2 + a_3)a_4 + a_4^2 = a_4. \end{cases}$$

Since $a_1 \in \mathbb{R}$, so from $a_1^2 = a_1$, we can get $a_1 = 0$ or $a_1 = 1$.

Case A: if $a_1 = 0$, then from $2a_1a_2 + a_2^2 = a_2$, we can infer $a_2^2 = a_2$, so $a_2 = 0$ or $a_2 = 1$.

Case A1: if $a_1 = 0$ and $a_2 = 0$, so from $2(a_1 + a_2)a_3 + a_3^2 = a_3$, we can infer $a_3^2 = a_3$, so $a_3 = 0$ or $a_3 = 1$.

Case A11: if $a_1 = 0$, $a_2 = 0$ and $a_3 = 0$, so from $2(a_1 + a_2 + a_3)a_4 + a_4^2 = a_4$, we can infer $a_4^2 = a_4$, so $a_4 = 0$ or $a_4 = 1$.

Case A111: if $a_1 = a_2 = a_3 = a_4 = 0$, i.e., $(0,0,0,0)$ is idempotent in $\langle \mathbb{R} \cup T \cup I \cup F\rangle$.

Case A112: if $a_1 = a_2 = a_3 = 0$ and $a_4 = 1$, i.e., $(0,0,0,F)$ is idempotent in $\langle \mathbb{R} \cup T \cup I \cup F\rangle$.

Case A12: if $a_1 = a_2 = 0$ and $a_3 = 1$, so from $2(a_1 + a_2 + a_3)a_4 + a_4^2 = a_4$, we can infer $2a_4 + a_4^2 = a_4$, so $a_4 = 0$ or $a_4 = -1$.

Case A121: if $a_1 = a_2 = 0, a_3 = 1$ and $a_4 = 0$, i.e., $(0, 0, I, 0)$ is idempotent in $\langle \mathbb{R} \cup T \cup I \cup F \rangle$.

Case A122: if $a_1 = a_2 = 0, a_3 = 1$ and $a_4 = -1$, i.e., $(0, 0, I, -F)$ is idempotent in $\langle \mathbb{R} \cup T \cup I \cup F \rangle$.

Case A2: if $a_1 = 0$ and $a_2 = 1$, so from $2(a_1 + a_2)a_3 + a_3^2 = a_3$, we can infer $2a_3 + a_3^2 = a_3$, so $a_3 = 0$ or $a_3 = -1$.

Case A21: if $a_1 = 0$, $a_2 = 1$, and $a_3 = 0$, so from $2(a_1 + a_2 + a_3)a_4 + a_4^2 = a_4$, we can infer $2a_4 + a_4^2 = a_4$, so $a_4 = 0$ or $a_4 = -1$.

Case A121: if $a_1 = 0, a_2 = 1, a_3 = 0$ and $a_4 = 0$, i.e., $(0, T, 0, 0)$ is idempotent in $\langle \mathbb{R} \cup T \cup I \cup F \rangle$.

Case A112: if $a_1 = 0, a_2 = 1, a_3 = 0$ and $a_4 = -1$, i.e., $(0, T, 0, -F)$ is idempotent in $\langle \mathbb{R} \cup T \cup I \cup F \rangle$.

Case A22: if $a_1 = 0, a_2 = 1$ and $a_3 = -1$, so from $2(a_1 + a_2 + a_3)a_4 + a_4^2 = a_4$, we can infer $a_4^2 = a_4$, so $a_4 = 0$ or $a_4 = 1$.

Case A121: if $a_1 = 0, a_2 = 1, a_3 = -1$ and $a_4 = 0$, i.e., $(0, T, -I, 0)$ is idempotent in $\langle \mathbb{R} \cup T \cup I \cup F \rangle$.

Case A112: if $a_1 = 0, a_2 = 1, a_3 = -1$ and $a_4 = 1$, i.e., $(0, T, -I, F)$ is idempotent in $\langle \mathbb{R} \cup T \cup I \cup F \rangle$.

Case B: if $a_1 = 1$, then from $2a_1a_2 + a_2^2 = a_2$, we can infer $2a_2 + a_2^2 = a_2$, so $a_2 = 0$ or $a_2 = -1$.

Case B1: if $a_1 = 1$ and $a_2 = 0$, so from $2(a_1 + a_2)a_3 + a_3^2 = a_3$, we can infer $2a_3 + a_3^2 = a_3$, so $a_3 = 0$ or $a_3 = -1$.

Case B11: if $a_1 = 1$, $a_2 = 0$ and $a_3 = 0$, so from $2(a_1 + a_2 + a_3)a_4 + a_4^2 = a_4$, we can infer $2a_4 + a_4^2 = a_4$, so $a_4 = 0$ or $a_4 = -1$.

Case B111: if $a_1 = 1, a_2 = 0, a_3 = 0$ and $a_4 = 0$, i.e., $(1, 0, 0, 0)$ is idempotent in $\langle \mathbb{R} \cup T \cup I \cup F \rangle$.

Case B112: if $a_1 = 1, a_2 = 0, a_3 = 0$ and $a_4 = -1$, i.e., $(1, 0, 0, -F)$ is idempotent in $\langle \mathbb{R} \cup T \cup I \cup F \rangle$.

Case B12: if $a_1 = 1, a_2 = 0$ and $a_3 = -1$, so from $2(a_1 + a_2 + a_3)a_4 + a_4^2 = a_4$, we can infer $a_4^2 = a_4$, so $a_4 = 0$ or $a_4 = 1$.

Case B121: if $a_1 = 1, a_2 = 0, a_3 = -1$ and $a_4 = 0$, i.e., $(1, 0, -I, 0)$ is idempotent in $\langle \mathbb{R} \cup T \cup I \cup F \rangle$.

Case B122: if $a_1 = 1, a_2 = 0, a_3 = -1$ and $a_4 = 1$, i.e., $(1, 0, -I, F)$ is idempotent in $\langle \mathbb{R} \cup T \cup I \cup F \rangle$.

Case B2: if $a_1 = 1$ and $a_2 = -1$, so from $2(a_1 + a_2)a_3 + a_3^2 = a_3$, we can infer $a_3^2 = a_3$, so $a_3 = 0$ or $a_3 = 1$.

Case B21: if $a_1 = 1, a_2 = -1$, and $a_3 = 0$, so from $2(a_1 + a_2 + a_3)a_4 + a_4^2 = a_4$, we can infer $a_4^2 = a_4$, so $a_4 = 0$ or $a_4 = 1$.

Case B121: if $a_1 = 1, a_2 = -1, a_3 = 0$ and $a_4 = 0$, i.e., $(1, -T, 0, 0)$ is idempotent in $\langle \mathbb{R} \cup T \cup I \cup F \rangle$.

Case B112: if $a_1 = 1, a_2 = -1, a_3 = 0$ and $a_4 = 1$, i.e., $(1, -T, 0, F)$ is idempotent in $\langle \mathbb{R} \cup T \cup I \cup F \rangle$.

Case B22: if $a_1 = 1$, $a_2 = -1$ and $a_3 = 1$, so from $2(a_1 + a_2 + a_3)a_4 + a_4^2 = a_4$, we can infer $2a_4 + a_4^2 = a_4$, so $a_4 = 0$ or $a_4 = -1$.

Case B121: if $a_1 = 1, a_2 = -1, a_3 = 1$ and $a_4 = 0$, i.e., $(1, -T, I, 0)$ is idempotent in $\langle \mathbb{R} \cup T \cup I \cup F \rangle$.

Case B112: if $a_1 = 1$, $a_2 = -1$, $a_3 = 1$ and $a_4 = -1$, i.e., $(1, -T, I, -F)$ is idempotent in $\langle \mathbb{R} \cup T \cup I \cup F \rangle$.

From the above analysis, we can get the set of all idempotents in neutrosophic quadruple ring $\langle \mathbb{R} \cup T \cup I \cup F \rangle$ are $\{(1,0,0,0), (0,0,0,F), (0,0,I,-F), (0,0,I,0), (0,T,-I,0), (0,T,-I,F), (0,T,0,-F), (0,T,0,0), (1,-T,0,0), (1,-T,0,F), (1,-T,I,-F), (1,-T,I,0), (1,0,-I,0), (1,0,-I,F), (1,0,0,-F), (1,0,0,0)\}$. □

The above theorem reveals that the idempotents in neutrosophic quadruple ring $\langle R \cup T \cup I \cup F \rangle$ is fixed when R is $\mathbb{C}, \mathbb{R}, \mathbb{Q}$ or $\mathbb{Z}$. For any ring R, we have the following results.

Theorem 8. *For neutrosophic quadruple ring $\langle R \cup T \cup I \cup F \rangle$, $a = a_1 + a_2T + a_3I + a_4F$ is idempotent in neutrosophic quadruple ring $\langle R \cup T \cup I \cup F \rangle$ iff a_1 is idempotent in R, $a_2 = c - a_1$, $a_3 = d - (a_1 + a_2)$ and $a_4 = e - (a_1 + a_2 + a_3)$, where c, d and e are any idempotents in R.*

Proof. Necessity: If $a = a_1 + a_2T + a_3I + a_4F$ is idempotent, i.e., $(a_1 + a_2T + a_3I + a_4F)^2 = a_1 + a_2T + a_3I + a_4F$, which means

$$\begin{cases} a_1^2 = a_1, \\ 2a_1a_2 + a_2^2 = a_2, \\ 2(a_1 + a_2)a_3 + a_3^2 = a_3, \\ 2(a_1 + a_2 + a_3)a_4 + a_4^2 = a_4. \end{cases}$$

Since $a_1 \in R$, from $a_1^2 = a_1$, we can get a_1 is idempotent in R.

From $2a_1a_2 + a_2^2 = a_2$ and $a_1^2 = a_1$, we can get $(a_1 + a_2)^2 = a_1^2 + 2a_1a_2 + a_2^2 = a_1 + a_2$, so $a_1 + a_2$ is also idempotent in R, denoted by c, so $a_2 = c - a_1$.

From $2(a_1 + a_2)a_3 + a_3^2 = a_3$, and $(a_1 + a_2)^2 = a_1 + a_2$, we can get $(a_1 + a_2 + a_3)^2 = (a_1 + a_2)^2 + 2(a_1 + a_2)a_3 + a_3^2 = a_1 + a_2 + a_3$, so $a_1 + a_2 + a_3$ is also idempotent in R, denoted by d, so $a_3 = d - a_1 - a_2$.

From $2(a_1 + a_2 + a_3)a_4 + a_4^2 = a_4$, and $(a_1 + a_2 + a_3)^2 = a_1 + a_2 + a_3$, we can get $(a_1 + a_2 + a_3 + a_4)^2 = (a_1 + a_2 + a_3)^2 + 2(a_1 + a_2 + a_3)a_3 + a_4^2 = a_1 + a_2 + a_3 + a_4$, so $a_1 + a_2 + a_3 + a_4$ is also idempotent in R, denoted by e, so $a_4 = e - a_1 - a_2 - a_3$.

Sufficiency: If a_1, c, d and e are arbitrary idempotents in R, let $a_2 = c - a_1$, $a_3 = d - (a_1 + a_2)$ and $a_4 = e - (a_1 + a_2 + a_3)$. so $(a_1 + a_2T + a_3I + a_4F)^2 = (a_1 + (c - a_1)T + (d - a_1 - a_2)I + (e - a_1 - a_2 - a_3)F)^2 = a_1^2 + (2(c - a_1)a_1 + (c - a_1)^2)T + (2c(d - a_1 - a_2) + (d - a_1 - a_2)^2)I + (2d(e - a_1 - a_2 - a_3) + (e - a_1 - a_2 - a_3)^2)F = a_1 + (c - a_1)T + (d - a_1 - a_2)I + (e - a_1 - a_2 - a_3)F$. Thus, $a = a_1 + a_2T + a_3I + a_4F$ is idempotent. □

Theorem 9. *If the number of different idempotents in R is t, then the number of different idempotents in neutrosophic quadruple ring* $\langle R \cup T \cup I \cup F \rangle$ *is* t^4.

Proof. If the number of different idempotents in R is t, let $a_1 + a_2T + a_3I + a_4F \in \langle \mathbb{Z}_n \cup T \cup I \cup F \rangle$ is idempotent, so a_1 is idempotent in R, i.e., a_1 has t different selections. When a_1 is selected, $a_2 = c - a_1$, where c is idempotent, which also has t different selections. When a_1, a_2 are selected, $a_3 = d - a_1 - a_2$, where d is idempotent, which also has t different selections. When a_1, a_2, a_3 is selected, $a_4 = e - a_1 - a_2 - a_3$, where e is idempotent, which also has t different selections. Thus, the number of all selections is $t \cdot t \cdot t \cdot t = t^4$, i.e., the number of different idempotents in $\langle R \cup T \cup I \cup F \rangle$ is t^4. □

From Theorems 8 and 9 and Remark 3, it follows easily that:

Corollary 3. *In neutrosophic quadruple ring* $\langle \mathbb{Z}_n \cup T \cup I \cup F \rangle$, $a = a_1 + a_2T + a_3I + a_4F$ *is idempotent in neutrosophic quadruple ring* $\langle \mathbb{Z}_n \cup T \cup I \cup F \rangle$ *iff* a_1 *is idempotent in* $\mathbb{Z}_n$, $a_2 = c - a_1$, $a_3 = d - (a_1 + a_2)$ *and* $a_4 = e - (a_1 + a_2 + a_3)$, *where* c, d *and* e *are any idempotents in* $\mathbb{Z}_n$.

Corollary 4. *The number of different idempotents in neutrosophic quadruple ring* $\langle \mathbb{Z}_n \cup T \cup I \cup F \rangle$ *is* 2^{4t}.

The solving process for neutrosophic quadruple ring $\langle \mathbb{Z}_n \cup T \cup I \cup F \rangle$ is given by Algorithm 2. Just only input n, we can get all idempotents in $\langle \mathbb{Z}_n \cup T \cup I \cup F \rangle$. The MATLAB code is provided in the Appendix C.

Algorithm 2: Solving the different idempotents in $\langle \mathbb{Z}_n \cup T \cup I \cup F \rangle$

Input: n

1: Factorization of integer n, we can get $n = p_1^{k_1} p_2^{k_2} \cdots p_t^{k_t}$.

2: Computing the neutral element of $1, p_1^{k_1}, p_2^{k_2}, \cdots, p_t^{k_t}, p_1^{k_1} p_2^{k_2}, \cdots p_1^{k_1} p_t^{k_t}, \cdots, p_2^{k_2} p_3^{k_3} \cdots p_t^{k_t}$ and $p_1^{k_1} p_2^{k_2} \cdots p_t^{k_t}$. So, we can get all idempotents in $\mathbb{Z}_n$, denoted by $c_1, c_2, \cdots, c_{2^t}$.

3: Let ID=[];

4: for $i = 1 : 2^t$

5: $\quad a_1 = c_i$

6: $\quad$ for $j = 1 : 2^t$

7: $\quad\quad a_2 = \mathrm{mod}(c_j - a_1, n)$;

8: $\quad\quad$ for $m = 1 : 2^t$

9: $\quad\quad\quad a_3 = \mathrm{mod}(c_m - a_1 - a_2, n)$;

10: $\quad\quad\quad$ for $q = 1 : 2^t$

11: $\quad\quad\quad\quad a_4 = \mathrm{mod}(c_q - a_1 - a_2 - a_3, n)$;

12: $\quad\quad\quad\quad ID = [ID;\ [a_1, a_2, a_3, a_4]]$;

13: $\quad\quad\quad$ end

14: $\quad\quad$ end

15: $\quad$ end

16: end

Output: ID: all the idempotents in $\langle \mathbb{Z}_n \cup T \cup I \cup F \rangle$

Example 6. *Solve all idempotents in* $\langle \mathbb{Z}_{12} \cup T \cup I \cup F \rangle$.

Since $n = 12 = 2^2 \cdot 3$, *from Theorems 4 and 5, we can get the different neutral elements in* $\mathbb{Z}_{12}$ *are* $neut(1), neut(2^2), neut(3), neut(2^3 \cdot 3)$ *and* $neut(0)$, *i.e., the different idempotents in* $\mathbb{Z}_{12}$ *are* $1, 4, 9, 0$. *From Corollary 4, the number of different idempotents in neutrosophic quadruple ring* $\langle \mathbb{Z}_{12} \cup T \cup I \cup F \rangle$ *is* $2^{4 \cdot 2} = 256$.

From Algorithm 2, the set of all 256 idempotents in $\langle \mathbb{Z}_{12} \cup T \cup I \cup F \rangle$ *is:* $\{0, 1F, 4F, 9F, I+11F, I, I+3F, I+8F, 4I+8F, 4I+9F, 4I, 4I+5F, 9I+3F, 9I+4F, 9I+7F, 9I, T+11I, T+11I+F, T+11I+4F, T+11I+9F, T+11F, T, T+3F, T+8F, T+3I+8F, T+3I+9F, T+3I, T+3I+5F, T+8I+3F, T+8I+4F, T+8I+7F, T+8, 4T+8I, 4T+8I+F, 4T+8I+4F, 4T+8I+9F, 4T+9I+11F, 4T+9I, 4T+9I+3F, 4T+9I+8F, 4T+8F, 4T+9F, 4T, 4T+5F, 4T+5I+3F, 4T+5I+4F, 4T+5I+7F, 4T+5I, 9T+3I, 9T+3I+F, 9T+3I+4F, 9T+3I+9F, 9T+4I+11F, 9T+4I, 9T+4I+3F, 9T+4I+8F, 9T+7I+8F, 9T+7I+9F, 9T+7I, 9T+7I+5F, 9T+3F, 9T+4F, 9T+7F, 9T, 1+11T, 1+11T+F, 1+11T+4F, 1+11T+9F, 1+11T+I+11F, 1+11T+I, 1+11T+I+3F, 1+11T+I+8F, 1+11T+4I+8F, 1+11T+4I+9F, 1+11T+4I, 1+11T+4I+5F, 1+11T+9I+3F, 1+11T+9I+4F, 1+11T+9I+7F, 1+11T+9I, 1+11I, 1+11I+F, 1+11I+4F, 1+11I+9F, 1+11F, 1, 1+3F, 1+8F, 1+3I+8F, 1+3I+9F, 1+3I, 1+3I+5F, 1+8I+3F, 1+8I+4F, 1+8I+7F, 1+8I, 1+3T+8I, 1+3T+8I+F, 1+3T+8I+4F, 1+3T+8I+9F, 1+3T+9I+11F, 1+3T+9I, 1+3T+9I+3F, 1+3T+9I+8F, 1+3T+8F, 1+3T+9F, 1+3T, 1+3T+5F, 1+3T+5I+3F, 1+3T+5I+4F, 1+3T+5I+7F, 1+3T+5I, 1+8T+3I, 1+8T+3I+F, 1+8T+3I+4F, 1+8T+3I+9F, 1+8T+4I+11F, 1+8T+4I, 1+8T+4I+3F, 1+8T+4I+8F, 1+8T+7I+8F, 1+8T+7I+9F, 1+8T+7I, 1+8T+7I+5F, 1+8T+3F, 1+8T+4F, 1+8T+7F, 1+8T, 4+8T, 4+8T+F, 4+8T+4F, 4+8T+9F, 4+8T+I+11F, 4+8T+I, 4+8T+I+3F, 4+8T+I+8F, 4+8T+4I+8F, 4+8T+4I+9F, 4+8T+4I, 4+8T+4I+5F, 4+8T+9I+3F, 4+8T+9I+4F, 4+8T+9I+7F, 4+8T+9I, 4+9T+11I, 4+9T+11I+F, 4+9T+11I+4F, 4+9T+11I+9F, 4+9T+11F, 4+9T, 4+9T+3F, 4+9T+8F, 4+9T+3I+8F, 4+9T+3I+9F, 4+9T+3I, 4+9T+3I+5F, 4+9T+8I+3F, 4+9T+8I+4F, 4+9T+8I+7F, 4+9T+8I, 4+8I, 4+8I+F, 4+8I+4F, 4+8I+9F, 4+9I+11F, 4+9I, 4+9I+3F, 4+9I+8F, 4+8F, 4+9F, 4, 4+5F, 4+5I+3F, 4+5I+4F, 4+5I+7F, 4+5I, 4+5T+3I, 4+5T+3I+F, 4+5T+3I+4F, 4+5T+3I+9F, 4+5T+4I+11F, 4+5T+4I, 4+5T+4I+3F, 4+5T+4I+8F, 4+5T+7I+8F, 4+5T+7I+9F, 4+5T+7I, 4+5T+7I+5F, 4+5T+3F, 4+$

$5T+4F, 4+5T+7F, 4+5T, 9+3T, 9+3T+F, 9+3T+4F, 9+3T+9F, 9+3T+I+11F, 9+3T+I, 9+3T+I+3F, 9+3T+I+8F, 9+3T+4I+8F, 9+3T+4I+9F, 9+3T+4I, 9+3T+4I+5F, 9+3T+9I+3F, 9+3T+9I+4F, 9+3T+9I+7F, 9+3T+9I, 9+4T+11I, 9+4T+11I+F, 9+4T+11I+4F, 9+4T+11I+9F, 9+4T+11F, 9+4T, 9+4T+3F, 9+4T+8F, 9+4T+3I+8F, 9+4T+3I+9F, 9+4T+3I, 9+4T+3I+5F, 9+4T+8I+3F, 9+4T+8I+4F, 9+4T+8I+7F, 9+4T+8I, 9+7T+8I, 9+7T+8I+F, 9+7T+8I+4F, 9+7T+8I+9F, 9+7T+9I+11F, 9+7T+9I, 9+7T+9I+3F, 9+7T+9I+8F, 9+7T+8F, 9+7T+9F, 9+7T, 9+7T+5F, 9+7T+5I+3F, 9+7T+5I+4F, 9+7T+5I+7F, 9+7T+5I, 9+3I, 9+3I+F, 9+3I+4F, 9+3I+9F, 9+4I+11F, 9+4I, 9+4I+3F, 9+4I+8F, 9+7I+8F, 9+7I+9F, 9+7I, 9+7I+5F, 9+3F, 9+4F, 9+7F, 9.\}$

Similarly, we will use the idempotents to divide the elements of the neutrosophic rings $\langle R \cup T \cup I \cup F \rangle$ when $R = \mathbb{F}$. Let us illustrate these with the following example.

Example 7. *Let $R = \mathbb{Z}_3$, which is a field. From Example 4, the different idempotents in $\mathbb{Z}_3$ are* $1, 0$. *From Corollary 4, the number of different idempotents in neutrosophic quadruple ring $\langle \mathbb{Z}_3 \cup T \cup I \cup F \rangle$ is* $2^{4\cdot} = 16$.

From Algorithm 2, the set of all 16 *idempotents in $\langle \mathbb{Z}_3 \cup I \rangle$ is:* $E = \{0, F, I+2F, I, T+2I, T+2I+F, T+2F, T, 1+2T, 1+2T+F, 1+2T+I+2F, 1+2T+I, 1+2I, 1+2I+F, 1+2F, 1\}$. *We have* $E(0) = \{0\}$, $E(F) = \{F, 2F\}$, $E(I+2F) = \{I+2F, 2I+F\}$, $E(I) = \{I, I+F, 2I, 2I+2F\}$, $E(T+2I) = \{T+2I, 2T+I\}$, $E(T+2I+F) = \{T+2I+F, T+2I+2F, 2T+I+F, 2T+I+2F\}$, $E(T+2F) = \{T+2F, T+I+F, 2T+F, 2T+2I+2F\}$, $E(T) = \{T+F, T, T+I, T+I+2F, 2T, 2T+2F, 2T+2I, 2T+2I+F\}$, $E(1+2T) = \{1+2T, 2+T\}$, $E(1+2T+F) = \{1+2T+F, 1+2T+2F, 2+T+F, 2+T+2F\}$, $E(1+2T+I+2F) = \{1+2T+I+2F, 1+2T+2I+F, 2+T+I+2F, 2+T+2I+F\}$, $E(1+2T+I) = \{1+2T+I, 1+2T+I+F, 1+2T+2I, 1+2T+2I+2F, 2+T+I, 2+T+I+F, 2+T+2I, 2+T+2I+2F\}$, $E(1+2I) = \{1+2I, 1+T+I, 2+I, 2+2T+2I\}$, $E(1+2I+F) = \{1+2I+F, 1+2I+2F, 1+T+I+F, 1+T+I+2F, 2+I+F, 2+I+2F, 2+2T+2I+F, 2+2T+2I+2F\}$, $E(1+2F) = \{1+2F, 1+I+F, 1+T+F, 1+T+2I+2F, 2+F, 2+2I+2F, 2+2T+2F, 2+2T+I+F\}$, $E(1) = \{1, 1+F, 1+I, 1+I+2F, 1+T, 1+T+2F, 1+T+2I, 1+T+2I+F, 2, 2+2F, 2+2I, 2+2I+F, 2+2T, 2+2T+F, 2+2T+I, 2+2T+I+2F\}$. *So* $\langle \mathbb{Z}_3 \cup T \cup I \cup F \rangle = \bigcup_{e \in E} E(e)$.

5. Conclusions

In this paper, we study the idempotents in neutrosophic ring $\langle R \cup I \rangle$ and neutrosophic quadruple ring $\langle R \cup T \cup I \cup F \rangle$. We not only solve the open problem and conjectures in paper [22] about idempotents in neutrosophic ring $\langle \mathbb{Z}_n \cup I \rangle$, but also give algorithms to obtain all idempotents in $\langle \mathbb{Z}_n \cup I \rangle$ and $\langle \mathbb{Z}_n \cup T \cup I \cup F \rangle$ for each n. Furthermore, if $R = \mathbb{F}$, then the neutrosophic rings (neutrosophic quadruple rings) can be viewed as a partition divided by the idempotents. As a general result, if all idempotents in ring R are known, then all idempotents in $\langle R \cup I \rangle$ and $\langle R \cup T \cup I \cup F \rangle$ can be obtained too. Moreover, if the number of all idempotents in ring R is t, then the numbers of all idempotents in $\langle R \cup I \rangle$ and $\langle R \cup T \cup I \cup F \rangle$ are t^2 and t^4 respectively. In the following, on the one hand, we will explore semi-idempotents in neutrosophic rings, on the other hand, we will study the algebra properties of neutrosophic rings and neutrosophic quadruple rings.

Author Contributions: All authors have contributed equally to this paper.

Funding: This research was funded by National Natural Science Foundation of China (Grant No. 61976130), Discipline Construction Funding of Xi'an Polytechnic University, Instructional Science and Technology Plan Projects of China National Textile and Apparel Council (No. 2016073) and Scientific Research Program Funded by Shaanxi Provincial Education Department (Program No. 18JS042).

Acknowledgments: The authors would like to thank the reviewers for their many insightful comments and suggestions.

Conflicts of Interest: The authors declare no conflict of interest.

Appendix A. The MATLAB code for solving the idempotents in $(\mathbb{Z}_n, \cdot)$

```
function neut = solve_neut(n)

% n: nonnegative integer
% neut: all idempotents in Z_n

B=[];
digits(32);
for i=1:n
    for j=1:n
        A1(i,j)=mod((i-1)*(j-1),n);
    end
end
a1=factor(n);
a2=unique(a1);
for i=1:length(a2)
    b=length(find(a1==a2(i)));
    B(i)=a2(i)^b;
end
D=[1];
for i=1:length(a2)
    C=combnk(B,i);
    A=prod(C,2);
    D=[D;A];
end
D=mod(D,n);
for i=1:length(D)
    if D(i)==1
        neut(i)=1;
    elseif D(i)==0
        neut(i)=0;
    else
        for j=1:n
            if mod(D(i)*j,n)==D(i)
                for k=1:n
                    if mod(D(i)*k,n)==j
                        neut(i)=j;
                        break
                    end
                end
            end
        end
    end
end
neut=sort(neut);
```

Appendix B. The MATLAB code for solving the idempotents in $\langle \mathbb{Z}_n \cup I \rangle$

```
function ID = Idempotents_ZR(n)
% n: nonnegative integer
```

```
% ID: all idempotents in in neutrosophic  ring <Z_n \cup I>

neut = solve_neut(n);

neutall=[];
for i=1:length(neut)
    for j=1:length(neut)
        c1=mod(neut(j)-neut(i),n);
        neutall=[neutall;  [neut(i), c1]];
    end
end

ID=sortrows(neutall',1)';
```

Appendix C. The MATLAB code for solving the idempotents in $\langle \mathbb{Z}_n \cup T \cup I \cup F \rangle$

```
function ID = Idempotents_ZRTIF(n)
% n: nonnegative integer
% ID: all idempotents in in neutrosophic quadruple ring <Z_n\cup T\cup I\cup F>


neut = solve_neut(n);
neutall=[];
for i=1:length(neut)
    a1=neut(i);
    for j=1:length(neut)
        a2=mod(neut(j)-a1,n);
        for m=1:length(neut)
            a3=mod(neut(m)-a1-a2,n);
            for q=1:length(neut)
                a4=mod(neut(q)-a1-a2-a3,n);
                neutall=[neutall;  [a1 a2 a3 a4]];
            end
        end
    end
end

ID=sortrows(neutall',1)';
```

References

1. Smarandache, F. *Neutrosophy: Neutrosophic Probability, Set, and Logic: Analytic Synthesis and Synthetic Analysis*; American Research Press: Santa Fe, NM, USA, 1998.
2. Smarandache, F.; Ali, M. Neutrosophic triplet group. *Neural Comput. Appl.* **2018**, *29*, 595–601. [CrossRef]
3. Smarandache, F. *Neutrosophic Perspectives: Triplets, Duplets, Multisets, Hybrid Operators, Modal Logic, Hedge Algebras. And Applications*; Infinite Study: Conshohocken, PA, USA, 2017.
4. Zhang, X.; Hu, Q.; Smarandache, F.; An, X. On neutrosophic triplet groups: basic properties, NT-subgroups, and some notes. *Symmetry* **2018**, *10*, 289. [CrossRef]
5. Ma, Y.; Zhang, X.; Yang, X.; Zhou, X. Generalized neutrosophic extended triplet group. *Symmetry* **2019**, *11*, 327. [CrossRef]

6. Zhang, X.H.; Bo, C.X.; Smarandache, F.; Dai, J.H. New inclusion relation of neutrosophic sets with applications and related lattice structure. *Int. J. Mach. Learn. Cybern.* **2018**, *9*, 1753–1763. [CrossRef]
7. Bal, M.; Shalla, M.M.; Olgun, N. Neutrosophic triplet cosets and quotient groups. *Symmetry* **2017**, *10*, 126. [CrossRef]
8. Zhang, X.H.; Smarandache, F.; Liang, X.L. Neutrosophic duplet semi-group and cancellable neutrosophic triplet groups. *Symmetry* **2017**, *9*, 275. [CrossRef]
9. Wu, X.Y.; Zhang, X.H. The decomposition theorems of AG-neutrosophic extended triplet loops and strong AG-(l, l)-loops. *Mathematics* **2019**, *7*, 268. [CrossRef]
10. Zhang, X.H.; Wu, X.Y.; Mao, X.Y.; Smarandache, F.; Park, C. On neutrosophic extended triplet groups (loops) and Abel-Grassmann's groupoids (AG-groupoids). *J. Intell. Fuzzy Syst.* **2019**, in press.
11. Zhang, X.H.; Mao, X.Y.; Wu, Y.T.; Zhai, X.H. Neutrosophic filters in pseudo-BCI algebras. *Int. J. Uncertainty Quant.* **2018**, *8*, 511–526. [CrossRef]
12. Smarandache, F. Hybrid neutrosophic triplet ring in physical structures. *Bull. Am. Phys. Soc.* **2017**, *62*, 17.
13. Ali, M.; Smarandache, F.; Khan, M. Study on the development of neutrosophictriplet ring and neutrosophictriplet field. *Mathematics* **2018**, *6*, 46. [CrossRef]
14. Sahin, M.; Abdullah, K. Neutrosophic triplet normed space. *Open Phys.* **2017**, *15*, 697–704. [CrossRef]
15. Zhang, X.H.; Bo, C.X.; Smarandache, F.; Park, C. New operations of totally dependent-neutrosophic sets and totally dependent-neutrosophic soft sets. *Symmetry* **2018**, *10*, 187. [CrossRef]
16. Agboola, A.; Akinleye, S. Neutrosophic vector spaces. *Neutrosophic Sets Syst.* **2014**, *4*, 9–18.
17. Vasantha, W.B.; Smaradache, F. *Neutrosophic Rings*; Hexis: Phoenix, AZ, USA, 2006.
18. Agboola, A.A.D.; Akinola, A.D.; Oyebola, O.Y. Neutrosophic rings I. *Int. J. Math. Comb.* **2011**, *4*, 115.
19. Broumi, S.; Smarandache, F.; Maji, P.K. Intuitionistic neutrosphic soft set over rings. *Math. Stat.* **2014**, *2*, 120–126.
20. Ali, M.; Shabir, M.; Smarandache, F.; Vladareanu, L. Neutrosophic LA-semigroup rings. *Neutrosophic Sets Syst.* **2015**, *7*, 81–88.
21. Nicholson, W.K. Lifting idempotents and exchange rings. *Trans. Am. Math. Soc.* **1977**, *229*, 269–278. [CrossRef]
22. Vasantha, W.B.; Kandasamy, I.; Smarandache, F. Semi-idempotents in neutrosophic rings. *Mathematics* **2019**, *7*, 507.
23. Smarandache F. Neutrosophic quadruple numbers, refined neutrosophic quadruple numbers, absorbance law, and the multiplication of neutrosophic quadruple numbers. *Neutrosophic Sets Syst.* **2015**, *10*, 96–98.
24. Akinleye, S.A.; Smarandache, F.; Agboola, A.A.A. On neutrosophic quadruple algebraic structures. *Neutrosophic Sets Syst.* **2016**, *12*, 122–126.
25. Li, Q.; Ma, Y.; Zhang, X.; Zhang, J. Neutrosophic extended triplet group based on neutrosophic quadruple numbers. *Symmetry* **2019**, *11*, 187. [CrossRef]

symmetry

MDPI

Article

Symmetry in Hyperstructure: Neutrosophic Extended Triplet Semihypergroups and Regular Hypergroups

Xiaohong Zhang [1,*], Florentin Smarandache [2] and Yingcang Ma [3]

1 Department of Mathematics, Shaanxi University of Science & Technology, Xi'an 710021, China
2 Department of Mathematics, University of New Mexico, 705 Gurley Avenue, Gallup, NM 87301, USA; smarand@unm.edu
3 School of Science, Xi'an Polytechnic University, Xi'an 710048, China; mayingcang@xpu.edu.cn
* Correspondence: zhangxiaohong@sust.edu.cn

Received: 28 August 2019; Accepted: 25 September 2019; Published: 1 October 2019

Abstract: The symmetry of hyperoperation is expressed by hypergroup, more extensive hyperalgebraic structures than hypergroups are studied in this paper. The new concepts of neutrosophic extended triplet semihypergroup (NET- semihypergroup) and neutrosophic extended triplet hypergroup (NET-hypergroup) are firstly introduced, some basic properties are obtained, and the relationships among NET- semihypergroups, regular semihypergroups, NET-hypergroups and regular hypergroups are systematically are investigated. Moreover, pure NET-semihypergroup and pure NET-hypergroup are investigated, and a strucuture theorem of commutative pure NET-semihypergroup is established. Finally, a new notion of weak commutative NET-semihypergroup is proposed, some important examples are obtained by software MATLAB, and the following important result is proved: every pure and weak commutative NET-semihypergroup is a disjoint union of some regular hypergroups which are its subhypergroups.

Keywords: hypergroup; semihypergroup; neutrosophic extended triplet group; neutrosophic extended triplet semihypergroup (NET-semihypergroup); NET-hypergroup

---

1. Introduction and Preliminaries

As a generalization of traditional algebraic structures, hyper algebraic structures (or hypercompositional structures) have been extensively studied and applied [1–7]. Especially, hypergroups and semihypergroups are basic hyper structures which are extensions of groups and semigroups [8]. In fact, hypergroups characterize the symmetry of hyperoperations.

On the other hand, as an extension of fuzzy set and intuitionistic fuzzy set, the concept of neutrosophic set firstly proposed by F. Smarandache in [9], has been applied to many fields [10–12]. Moreover, as an application of the ideal of neutrsphic sets, a new notion of neutrosophic triplet group (NTG) was proposed by F. Smarandache and Ali in [13], while the new notion of neutrosophic extended group (NETG) was proposed by Smarandache in [14]. Furthermore, the basic properties and structural characteristics of neutrosophic extended groups (NETGs) are studied in [15,16]; the closed connection between between NETG and regular semigroup investigated, and the new notion of neutrosophic extended triplet Abel-Grassmann's Groupoid is proposed in [17]; the decomposition theorem of NETG is poved in [18]; the generalized neutrosophic extended groups are presented in [19]; the relationship and difference between NETGs and generalized groups are systematically studied in [20]. From these research results, we know that NETG is a typical algebraic system with important research value.

In this paper, we combine the two directions mentioned above to study the hyperalgebraic structures related to neutrosophic extended triplet groups (NETGs), which can be regarded as a further development of the research ideas in [21].

At first, we recall some concepts and results on hypergroups, semigroups and NETGs.

Let H be a non-empty set and $P^*(H)$ the set of all non-empty subsets of H. A map $\circ$: $H \times H \rightarrow P^*(H)$ is called (binary) hyperoperation (or hypercomposition), and $(H, \circ)$ is called a hypergroupoid. If A, $B \in P^*(H)$, $x \in H$, then

$$A \circ B = \bigcup_{a \in A, b \in B} (a \circ b),\ A \circ x = A \circ \{x\},\ x \circ B = \{x\} \circ B.$$

Definition 1. *([1–4]) Let $(H, \circ)$ be a hypergroupoid. If $(\forall x,y,z \in H)$ $(x \circ y) \circ z = x \circ (y \circ z)$, then $(H, \circ)$ is called a semihypergroup. That is,*

$$\bigcup_{u \in x \circ y} (u \circ z) = \bigcup_{v \in y \circ z} (x \circ v).$$

For a semihypergroup $(H, \circ)$, if $(\forall x,y \in H)$ $x \circ y = y \circ x$, then we call that H is commutative.

Note that, if $(H, \circ)$ is a semihypergroup, then $(A \circ B) \circ C = A \circ (B \circ C)$, $\forall A, B, C \in P^*(H)$.

Definition 2. *([1–4]) Assume that $(H, \circ)$ is a semihypergroup. (1) If $a \in H$ satisfies $(\forall x \in H)$ $|a \circ x| = |x \circ a| = 1$, then a is called to be scalar. (2) If $e \in H$ satisfies $(\forall x \in H)$ $x \circ e = e \circ x = \{x\}$, then e is called scalar identity. (3) If $e \in H$ satisfies $(\forall x \in H)$ $x \in (e \circ x) \cap (x \circ e)$, then e is called identity. (4) Let a, $b \in H$. If there exists an identity $e \in H$ satisfies $e \in (a \circ b) \cap (b \circ a)$, then b is called an inverse of a. (5) If $0 \in H$ satisfies $(\forall x \in H)$ $x \circ 0 = 0 \circ x = \{0\}$, then 0 is called zero element.*

Definition 3. *([1–4]) Let $(H, \circ)$ be a semihypergroup. (1) If $(\forall x \in H)$ $a \circ H = H \circ a = H$ (reproductive axiom), then $(H, \circ)$ is called a hypergroup. (2) If $(H, \circ)$ is a hypergroup and $(H, \circ)$ has at least one identity and each element has at least one inverse, then $(H, \circ)$ is called to be regular.*

Definition 4. *([1–4]) Let $(H, \circ)$ be a semihypergroup. If $x \in H$ satisfies $x \in x \circ H \circ x$, i.e., there exists an element $y \in H$, $x \in x \circ y \circ x$, then x is said to be regular. If $(\forall x \in H)$ x is regular, then $(H, \circ)$ is called to be regular.*

Note that, Every regular semigroup is a regular semihypergroup, and every hypergroup is a regular semihypergroup.

Definition 5. *([14]) Let N be a non-empty set, and * a binary operation on N. If $(\forall a \in N)$ there exist $neut(a) \in N$, $anti(a) \in N$ satisfy*

$$neut(a)*a = a*neut(a) = a, \text{ and}$$
$$anti(a)*a = a*anti(a) = neut(a).$$

Then N is called a neutrosophic extended triplet set (NETS). Moreover, for $a \in N$, (a, neut(a), anti(a)) is called a neutrosophic extend triplet, neut(a) is called an extend neutral of "a", and anti(a) is called an opposite of "a".

For a neutrosophic extended triplet set N, $a \in N$, the set of $neut(a)$ is denoted by $\{neut(a)\}$, and the set of $anti(a)$ is denoted by $\{anti(a)\}$.

Definition 6. *([13,14]) Let (N, *) be a NETS. If (N,*) is a semigroup, then (N, *) is called to be a neutrosophic extended triplet group (NETG).*

About some basic properties of neutrosophic extended triplet groups, plesse see [15,17,20].

2. Neutrosophic Extended Triplet Semihypergroups (NET-Semihypergroups) and Neutrosophic Extended Triplet Hypergroups (NET-Hypergroups)

In this section, we propose the new concepts of neutrosophic extended triplet semihypergroup (NET-semihypergroup) and neutrosophic extended triplet hypergroup (NET-hypergroup), and give some typical examples to illustrate their wide representativeness.

Definition 7. *Let (H,*) be a semihypergroup (i.e., * be a binary hyperoperation on nonempty set H such that $(x*y)*z = x*(y*z)$, for all x, y, z∈H). (H,*) is called a neutrosophic extended triplet semihypergroup (shortened form, NET-semihypergroup), if for every x∈H, there exist neut(x) and anti(x) such that*

$$x \in (neut(x)*x) \cap (x*neut(x)), \text{ and}$$

$$neut(x) \in (anti(x)*x) \cap (x*anti(x)).$$

Here, we call that (x, neut(x), anti(x)) to be a hyper-neutrosophic-triplet.

Example 1. *Denote H = {a, b, c}, define hyperoperations * on H as shown in Table 1. We can verify that (H, *) is semihypergroup by software MATLAB (see Figure 1).*

Table 1. The hyperoperation * on *H*.

*	a	b	c
a	a	{a, b}	{a, b, c}
b	a	{a, b}	{a, b, c}
c	a	{a, b}	c

Moreover,

$$a \in (a*a) \cap (a*a);$$
$$b \in (b*b) \cap (b*b);$$
$$c \in (c*c) \cap (c*c).$$

*This means that (H, *) is neutrosophic extended triplet semihypergroup (NET-semihypergroup) and (a, a, a), (b, b, b), (c, c, c) are hyper-neutrosophic-triplets.*

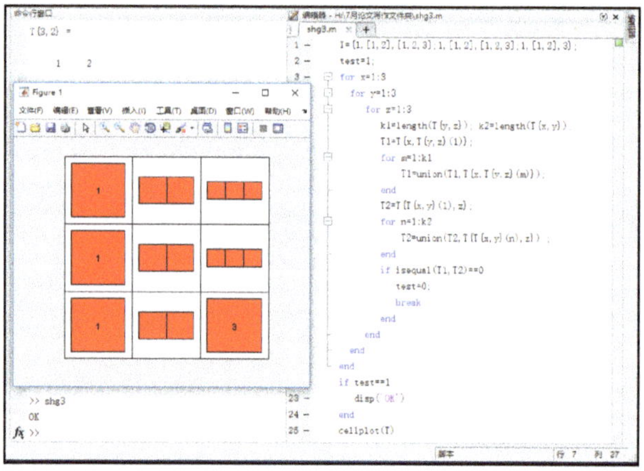

Figure 1. A program by Matlab to verify hyperoperation.

Example 2. *Denote H = {a, b, c, d}, define hyperoperations * on H as shown in Table 2. We can verify that (H, *) is semihypergroup by software MATLAB (see Figure 2).*

Table 2. The hyperoperation * on *H*.

*	a	b	c	d
a	{a, b}	{a, b}	{c, d}	{c, d}
b	{a, b}	{a, b}	{c, d}	{c, d}
c	{c, d}	{c, d}	a	b
d	{c, d}	{c, d}	b	a

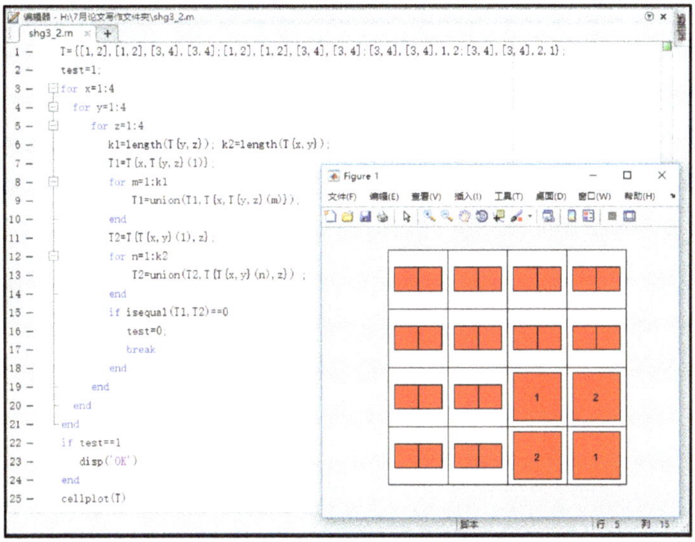

Figure 2. Verify hyperoperation by Matlab.

Moreover,

$$A\in(a*a)\cap(a*a);\ a\in(b*a)\cap(a*b),\ b\in(b*a)\cap(a*b).$$

$$b\in(b*b)\cap(b*b).$$

$$c\in(a*c)\cap(c*a),\ a\in(c*c)\cap(c*c);\ c\in(b*c)\cap(c*b),\ b\in(d*c)\cap(c*d).$$

$$d\in(a*d)\cap(d*a),\ a\in(d*d)\cap(d*d);\ d\in(b*d)\cap(d*b),\ b\in(c*d)\cap(d*c).$$

*This means that (H, *) is neutrosophic extended triplet semihypergroup (NET-semihypergroup) and (a, a, a), (a, b, b), (b, b, b), (c, a, c), (c, b, d), (d, a, d), (d, b, c) are hyper-neutrosophic-triplets.*

Remark 1. From Example 2 we know that *neut*(*x*) may be not unique for an element *x* in a neutrosophic extended triplet semihypergroup (NET-semihypergroup). In fact, in Example 2, we have

$$\{neut(a)\} = \{a, b\},\ neut(b) = b,\ \{neut(c)\} = \{a, b\},\ \{neut(d)\} = \{a, b\}.$$

Example 3. *Let H be the set of all nonnegative integers, and define a hyperoperation * on H as following:*

$$x*y = \{z\in H \mid z\geq max\{x,y\}\}.$$

For examples,

$$3*5 = \{5, 6, 7, 8, \ldots\};\ 9*9 = \{9, 10, 11, 12, \ldots\};\ 2019*0 = \{2019, 2020, 2021, 2022, \ldots\}.$$

*Then (H, *) is a commutative semihypergroup. Moreove, for any x∈H, we have*

$$x\in (x*x)\cap(x*x);\ x\in (x*x)\cap(x*x).$$

*This means that (H, *) is a neutrosophic extended triplet semihypergroup (NET-semihypergroup). In fact, we have*

$$neut(0)=0;\ \{neut(1)\}=\{0,1\};\ \{neut(2)\}=\{0, 1, 2\};\ \{neut(3)\}=\{0, 1, 2, 3\}\ \ldots$$

Example 4. *Let R be the set of all real numbers, and Z the set of integers. We use the modulo of real numbers (that we denote by mod$_R$) in the following way:*

$$\forall\ a, b \in R, \text{ then } a = b\ (mod_R\ 6), \text{ if and only if } a - b = 6n, \text{ where } n \text{ is an integer.}$$

For examples, 14.73 = 2.73 (mod$_R$ 6), since 14.73 − 2.73 = 12 = 6 × 2; but 18≠15 (mod$_R$ 6), since 18 - 15 = 3 ≠ 6n with n integer. Now, we define a hyperoperation # on R as following:

$$a\#b = \{x \in R \mid x = 4ab\ (mod_R\ 6)\}.$$

Then (R, #) is a commutative semihypergroup, since a#b = b#a = 4ab (mod$_R$ 6), and associative because:

$$(a\#b)\#c = (4ab)\#c = 4(4ab)c = 16abc\ (mod_R\ 6), \text{ and}$$
$$a\#(b\#c) = a\#(4bc) = 4a(4bc) = 16abc\ (mod_R\ 6).$$

Moreove, for any a∈R, we have

(1) *when a=0, (a, 6m, r) are hyper-triplets for any integer number m and real number r;*
(2) *when a≠ 0, $\left(a, \frac{1}{4} + \frac{3m}{2a}, \frac{1}{16a} + \frac{3m}{8a} + \frac{3n}{2a}\right)$ are hyper-neutrosophic-triplets for any integer numbers m, n.*

This means that (R, #) is a neutrosophic extended triplet semihypergroup (NET-semihypergroup), and infinitely many neut(a) and infinitely many anti(a) for any element a in R.

Remark 2. The following example shows that a sub-semihypergroup of a NET-semihypergroup may be not a NET-semihypergroup.

Example 5. *Denote H = {a, b, c, d, e}, define hyperoperations * on H as shown in Table 3. We can verify that (H, *) is semihypergroup by software MATLAB (see Figure 3).*

Table 3. The hyperoperation * on *H*.

*	*a*	*b*	*c*	*d*	*e*
a	*a*	*a*	*a*	*d*	{*a, b, c, d, e*}
b	*a*	{*a, b*}	{*a, c*}	*d*	{*a, b, c, d, e*}
c	*a*	*a*	*a*	*d*	{*a, b, c, d, e*}
d	*d*	*d*	*d*	*d*	{*a, b, c, d, e*}
e	{*a, b, c, d, e*}	{*a, b, c, d, e*}	{*a, b, c, d, e*}	{*a, b, c, d, e*}	{*a, b, c, d, e*}

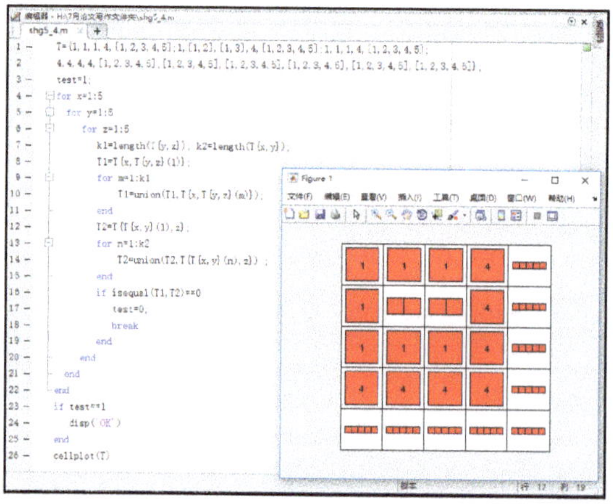

Figure 3. Verify the hyperoperation by Matlab.

*Moreover, (a, a, a), (a, e, e), (b, b, b), (b, e, e), (c, e, e), (d, d, d), (d, e, e), (e, e, e), (e, a, e), (e, b, e), (e, c, e), (e, d, e) are hyper-neutrosophic-triplets.This means that (H, *) is a NET-semihypergroup. For S={a, b, c }⊆H, (S, *) is sub-semihypergroup of (H, *). But, (S, *) is not a NET-semihypergroup.*

Remark 3. For the traditional algebraic structures, we have the conclusion that any group must be a neutrosophic extended triplet group (NETG). For hyper algebraic structures, we know from Example 1 that a NET-semihypergroup is not necessarily a hypergroup (since $a*H \neq H$ in Example 1). Moreover, the following example shows that a hypergroup may be not a NET-semihypergroup. Therefore, hypergroup and NET-semihypergroup are two non-inclusion hyperalgebraic systems.

Example 6. *Denote H = {1, 2, 3}, define hyperoperations * on H as shown in Table 4. We can verify that (H, *) is semihypergroup by software MATLAB.*

Table 4. The hyperoperation * on *H*.

*	1	2	3
1	2	2	{1, 3}
2	{1, 2, 3}	{2, 3}	{1, 2, 3}
3	2	{1, 2, 3}	{1, 3}

Moreover,

$$1*H = H*1 = H,\ 2*H = H*2 = H,\ 3*H = H*3 = H.$$

*This means that (H, *) is a hypergroup. But, for 1∈H, we cannot find x,y∈H such that* $1 \in (x*1) \cap (1*x)$, *and* $x \in (y*1) \cap (1*y)$. *That is, (H, *) is not a NET- semihypergroup.*

Definition 8. *Let (H,*) be a semihypergroup. (H,*) is called a neutrosophic extended triplet hypergroup (shortened form, NET-hypergroup), if (H,*) is both a NET-semihypergroup and a hypergroup.*

Obviously, the NET-semihypergroups in Example 2 and Example 5 are all NET-hypergroups. And, the following propostion is true (the proof is omitted).

Proposition 1. *Every regular hypergroup is a NET-hypergroup.*

The NET-hypergroup in Example 2 is not a regular hypergroup, it shows that the inverse of Proposition 1 is not true.

Proposition 2. *Let (H,*) be a NET-semihypergroup (or a NET-hypergroup). Then (H,*) is a regular semihypergroup.*

Proof. Assume that (*H*,*) is a NET-semihypergroup. For any $x \in H$, by Definition 7 we get that there exist *neut*(*x*) and *anti*(*x*) such that

$$x \in (neut(x)*x) \cap (x*neut(x)), \text{ and } neut(x) \in (anti(x)*x) \cap (x*anti(x)).$$

Then,

$$x \in neut(x)*x \subseteq (x*anti(x))*x.$$

That is, $x \in x*anti(x)*x$. From this, by Definition 4, we know that (*H*,*) is a regular semihypergroup. If (*H*,*) is a NET-hypergroup, by Definition 8, it follows that (*H*,*) is a NET-semihypergroup. Then, by the proof above, (*H*,*) is a regular semihypergroup. □

The following example shows that the inverse of Proposition 2 is not true. Moreover, it also shows that a regular semihypergroup may be not a hypergroup.

Example 7. *Denote H = {a, b, c}, define hyperoperations * on H as shown in Table 5. We can verify that (H, *) is semihypergroup.*

Table 5. The hyperoperation * on *H*.

*	*a*	*b*	*c*
a	*a*	*a*	*a*
b	{*a*, *b*, *c*}	{*a*, *b*, *c*}	{*a*, *b*, *c*}
c	{*a*, *b*, *c*}	{*a*, *b*}	{*a*, *b*}

*Moreover, a ∈ a*a*a; b ∈ b*b*b; c ∈ c*a*c.This means that (H, *) is a regular semihypergroup. But it is not a NET-semihypergroup, since there is not any x ∈ H such that c ∈ x*c and c ∈ c*x.Obviously, (H, *) is not a hypergroup.*

Therefore, the relationships among semihypergroup, NET-semihypergroup, NET-hypergroup, (regular) hypergroup and regular semihypergroup can be expressed by Figure 4.

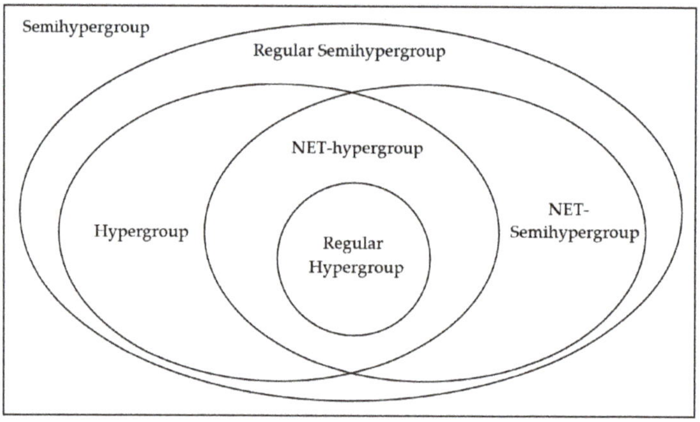

Figure 4. The relationships among some kinds of semihypergroups.

For basic properties of NET-semihypergroups and NET-hypergroups, we can get following results.

Theorem 1. *Let (H,*) be a semihypergroup. Then*

(1) *if (H,*) is commutative NET-semihypergroup, then for any x∈H and hyper-neutrosophic-triplet (x, neut(x), anti(x)), there exists p∈neut(x)*neut(x) and q∈anti(x)*neut(x) such that (x, p, q) is also a hyper-neutrosophic-triplet.*

(2) *if (H,*) is commutative NET-semihypergroup, then for any x∈H and neut(x)∈{neut(x)}, there exists p∈neut(x)*neut(x) such that p∈{neut(x)}.*

(3) *if (H,*) is NET-semihypergroup and x∈H is scalar, then |{neut(x)}|=1, that is, the neutral element of x is unique; Moreover, if x is scalar, then neut(x)*neut(x)=neut(x).*

(4) *if (H,*) is commutative hypergroup, then (H,*) is NET-hypergroup.*

Proof. (1) Assume that *x*∈*H* and (*x*, *neut*(*x*), *anti*(*x*)) is a hyper-neutrosophic-triplet. By Definition 7:

$$x \in (neut(x)*x) \cap (x*neut(x)), \text{ and } neut(x) \in (anti(x)*x) \cap (x*anti(x)).$$

Since $(H, *)$ is commutative, then:

$$x \in neut(x)*x \subseteq neut(x)*(neut(x)*x) = (neut(x)*neut(x))*x = x*(neut(x)*neut(x)).$$

This means that there exists $p \in neut(x)*neut(x)$ such that $x \in p*x = x*p$. Moreover:

$$p \in neut(x)*neut(x) \subseteq (x*anti(x))*neut(x) = x*(anti(x)*neut(x)) = (anti(x)*neut(x))*x.$$

It follows that there exists $q \in anti(x)*neut(x)$ such that $p \in q*x = x*q$. By Definition 7 we know that (x, p, q) is also a hyper-neutrosophic-triplet.

(2) It follows from (1).

(3) Suppose that $x \in H$ and x is scalar. Using Definition 2, $|x*a|=|a*x|=1$ for any $a \in H$. From this, for a hyper-neutrosophic-triplet $(x, neut(x), anti(x))$, applying Definition 7, we have:

$$x = neut(x)*x = x*neut(x), \text{ and } neut(x) = anti(x)*x = x*anti(x).$$

Assume $p_1, p_2 \in \{neut(x)\}$, then there exists $q_1, q_2 \in H$ such that:

$$x = p_1*x = x*p_1, p_1 = q_1*x = x*q_1; x = p_2*x = x*p_2, p_2 = q_2*x = x*q_2.$$

Then:

$$p_1 = q_1*x = q_1*(x*p_2) = (q_1*x)*p_2 = p_1*p_2;$$
$$p_2 = x*q_2 = (x*p_1)*q_2 = (x*(q_1*x))*q_2 = (x*q_1)*(x*q_2) = p_1*p_2.$$

It follows that $p_1 = p_2$ and $p_1 = p_1*p_1$. That is, $|\{neut(x)\}|=1$ and $neut(x)*neut(x) = neut(x)$.

(4) Let $(H, *)$ be a commutative hypergroup. By Definition 3, for any $x \in H$, $x*H = H*x = H$. Then, for any $x \in H$, there exists $h \in H$ such that $x = h*x = x*h$. Moreover, for $h \in H$, there exists $u \in H$ such that $h = u*x = x*u$. Thus, (x, h, u) is a hyper-neutrosophic-triplet, and it means that $(H, *)$ is a NET-semihypergroup by Definition 7. On the other hand, since $(H, *)$ is a hypergroup, so $(H, *)$ is a NET-hypergroup by Definition 8.

3. Pure NET-semihypergroups and Regular hypergroups

In this section, we discuss some properties of NET-semihypergroups. We'll propose the new notion of pure NET-semihypergroup, investigate the structure of pure NET-semihypergroups.

Definition 9. *Let $(H,*)$ be a NET-semihypergroup. $(H,*)$ is called a pure NET-semihypergroup, if for every $x \in H$, there exist neut(x) and anti(x) such that*

$$x = (neut(x)*x) \cap (x*neut(x)), \text{ and } neut(x) = (anti(x)*x) \cap (x*anti(x)).$$

Obviously, the following proposition is true and the proof is omitted.

Proposition 3. *(1) Every neutrosophic extended triplet group (NETG) is pure NET-semihypergroup. (2) If $(H,*)$ is a pure NET-semihypergroup and the hyper operation $*$ is commutative, then for every $x \in H$, there exists y, $z \in H$ such that*

$$x = y*x = x*y, \text{ and } y = z*x = x*z.$$

Example 8. *Denote $H = \{a, b, c\}$, define hyperoperations $*$ on H as shown in Table 6. We can verify that $(H, *)$ is semihypergroup.*

Table 6. The hyperoperation * on *H*.

*	a	b	c
a	a	{a, b, c}	{a, b, c}
b	{a, b, c}	c	b
c	{a, b, c}	b	c

Moreover,

$$a=(a^*a)\cap(a^*a);\ b=(c^*b)\cap(b^*c),\ c=(b^*b)\cap(b^*b);\ c=(c^*c)\cap(c^*c).$$

*This means that (H, *) is a pure NET-semihypergroup.*

Example 9. *Denote H = {a, b, c, d, e}, define hyperoperations * on H as shown in Table 7. We can verify that (H, *) is semihypergroup.*

Table 7. The hyperoperation * on *H*.

*	a	b	c	d	e
a	a	{a, b, c}	{a, b, c}	d	a
b	{a, b, c}	b	c	d	b
c	{a, b, c}	c	b	d	c
d	d	d	d	d	d
e	a	b	c	d	e

Moreover:

$$a=(a^*a)\cap(a^*a);\ b=(b^*b)\cap(b^*b);\ c=(b^*c)\cap(c^*b),\ b=(c^*c)\cap(c^*c);\ d=(d^*d)\cap(d^*d);\ e=(e^*e)\cap(e^*e).$$

*This means that (H, *) is a pure NET-semihypergroup.*

Remark 4. From Example 8 and Example 9, we have:

$$a=(a^*a)\cap(a^*a);$$
$$a\in(b^*a)\cap(a^*b),\ b\in(b^*a)\cap(a^*b);\ a\in(c^*a)\cap(a^*c),\ c\in(c^*a)\cap(a^*c).$$

This means that $\{neut(a)\} = \{a, b, c\}$. But, $b\in\{neut(a)\}$ and $c\in\{neut(a)\}$ are different to $a\in\{neut(a)\}$, since one is "∈" and the other is "=". In order to clearly express the difference between the two kinds of neutral elements, we introduce a new concept: pure neutral element.

Definition 10. *Let (H,*) be a NET-semihypergroup and x∈H. An element y∈H is called a pure neutral element of the element x, if there exist z∈H such that:*

$$x = y^*x = x^*y, \text{ and } y = z^*x = x^*z.$$

Here, we denote y by pneut(x).

Proposition 4. *Let (H,*) be a NET-semihypergroup and x∈H. If there exists a pure neutral element of x, then the pure neutral element of x, that is, pneut(x), is unique.*

Proof. Assume that there exists two pure neutral elements y_1, y_2 for $x\in H$. Then there exists $z_1, z_2 \in H$ such that:

$$x = y_1{}^*x = x^*y_1, \text{ and } y_1 = z_1{}^*x = x^*z_1;$$
$$x = y_2{}^*x = x^*y_2, \text{ and } y_2 = z_2{}^*x = x^*z_2.$$

Therefore,

$$y_1 = z_1*x = z_1*(x*y_2) = (z_1*x)*y_2 = y_1*y_2;$$
$$y_2 = x*z_2 = (x*y_1)*z_2 = (x*(z_1*x))*z_2 = (x*z_1)*(x*z_2) = y_1*y_2.$$

Hence, $y_1 = y_2$. That is, $pneut(x)$ is unique. □

By the proof of Proposition 4, we know that $y_1 = y_2 = y_1*y_2$, it follows that $y_1 = y_1*y_1$. Therefore, we have the following corollary.

Corollary 1. *Let (H,*) be a NET-semihypergroup and x∈H. If there exists a pure neutral element of x, then the pure neutral element of x is idempotent, that is,* $pneut(x)*pneut(x)= pneut(x)$.

Remark 5. From Proposition 4, we know that the pure neutral element of an elemetn x is unique when there exists one pure neutral element of x. Particularly, for commutative pure NET- semihypergroups, applying Proposition 3 (2), we get following Proposition 5 (the proof is omitted).

Proposition 5. *Let (H,*) be a commutative pure NET-semihypergroup. Then for any x∈H, pneut(x) is unique.*

Proposition 6. *Let (H,*) be a commutative pure NET-semihypergroup. Then for any x, y∈H,* $pneut(x*y)= pneut(x)*pneut(y)$ *when* $|x*y|=1$. *Moreover, if* $pneut(x) = z_1*x = x*z_1$ *and* $pneut(y) = z_2*y = y*z_2$, $z_1, z_2 \in H$, *then:*

$$pneut(x*y) = (z_1*z_2)*(x*y) = (x*y)*(z_1*z_2).$$

Proof. Assume that $x, y \in H$ and $|x*y|=1$. Since $(H, *)$ be a commutative pure NET-semihypergroup, then:

$$\begin{aligned}
&(x*y)*(pneut(x)*pneut(y)) = (x*y)*(pneut(y)*pneut(x))\\
&= x*(y*pneut(y))*pneut(x)\\
&= x*y*pneut(x)\\
&= (x*pneut(x))*y\\
&= x*y;\\
&(pneut(x)*pneut(y))*(x*y) = (pneut(y)*pneut(x))*(x*y)\\
&= pneut(y)*(pneut(x)*x)*y\\
&= pneut(y)*x*y\\
&= x*(pneut(y)*y)\\
&= x*y.
\end{aligned}$$

On the other hand, assume that $(x, pneut(x), anti(x))$ and $(y, pneut(y), anti(y))$ are hyper-neutrosophic-triplets, then:

$$\begin{aligned}
&(x*y)*(anti(x)*anti(y)) = (x*y)*(anti(y)*anti(x))\\
&= x*(y*anti(y))*anti(x)\\
&= x*pneut(y)*anti(x)\\
&= (x*anti(x))*pneut(y)\\
&= pneut(x)*pneut(y);\\
&(anti(x)*anti(y))*(x*y) = (anti(x)*anti(y))*(y*x)\\
&= anti(x)*(anti(y)*y)*x\\
&= anti(x)*pneut(y)*x\\
&= (anti(x)*x)*pneut(y)\\
&= pneut(x)*pneut(y).
\end{aligned}$$

Applying Proposition 5 we get that $pneut(x*y)= pneut(x)*pneut(y)$.

Morocover, assume $pneut(x) = z_1{}^*x = x^*z_1$, $pneut(y) = z_2{}^*y = y^*z_2$. Then, by commutativity of the hyper operation *:

$$\begin{aligned}(z_1{}^*z_2)^*(x^*y) &= (z_1{}^*x)^*(z_2{}^*y)\\ &= pneut(x)^*pneut(y)\\ &= pneut(x^*y);\\ (x^*y)^*(z_1{}^*z_2) &= (x^*z_1)^*(y^*z_2)\\ &= pneut(x)^*pneut(y)\\ &= pneut(x^*y).\end{aligned}$$

Therefore, the proof is completed. □

Theorem 2. *Let (H,*) be a commutative pure NET-semihypergroup and H satisfies:*

$$\forall x, y \in H, pneut(x)=pneut(y) \Rightarrow |x^*y|=1. \tag{C1}$$

Define a binary relation ≈ on H as following:

$$\forall x, y \in H, x \approx y \text{ if and only if } pneut(x)=pneut(y).$$

Then:

(1) *The binary relation is a equivalent relation on H;*
(2) *For any x∈H, [x]≈ is a sub-NET-semihypergroup of H, where [x]≈ is the equivalent class of x based on equivalent relation ≈;*
(3) *For any x∈H, [x]≈ is a regular hypergroupe.*

Proof. (1) It is obviously.

(2) Assume $a, b \in [x]_\approx$, then $pneut(a)= pneut(b)=pneut(x)$. Applying Proposition 6 and Corollary 1, we have

$$\begin{aligned}pneut(a^*b) &= pneut(a)^*pneut(b)\\ &= pneut(x)^*pneut(x)\\ &= pneut(x).\end{aligned}$$

This means that [x]≈ is closed on the hyper operation *.

Moreover, by Corollary 1, we have $pneut(x)^*pneut(x) = pneut(x)$. From this and using Proposition 5, we get that $pneut(pneut(x)) = pneut(x)$. It follows that $pneut(a) \in [x]\approx$ for any $a \in [x]\approx$. Moreover, assume that $a \in [x]\approx$, by the definition of commutative pure NET-semihypergroup, there exists $r \in H$ such that:

$$pneut(a) = r^*a = a^*r.$$

It follows that:

$$pneut(a) = (r^*pneut(a))^*a = a^*(r^*pneut(a)). \tag{C2}$$

Applying Proposition 6 and Corollary 1:

$$\begin{aligned}&pneut(r^*pneut(a))\\ &= pneut(r)^*\, pneut(pneut(a))\\ &= pneut(r)^*\, pneut(a)\\ &= pneut(r^*a)\\ &= pneut(pneut(a)\\ &= pneut(a).\end{aligned}$$

That is, *pneut*(*r***pneut*(*a*)) = *pneut*(*a*) = *pneut*(*x*). This means that *r***pneut*(*a*) ∈[*x*]≈. Therefore, by (C2), there exists *anti*(*a*) (see Definition 7), it is in [*x*]≈. This means that [*x*]≈ is a sub-NET-semihypergroup of *H*.

(3) For any *x*∈*H*, from (2) we know that [*x*]≈ is a sub-NET-semihypergroup of *H*. By the definition of ≈, for any *a*∈[*x*]≈, *pneut*(*a*) = *pneut*(*x*). Then, *a**[*x*]≈**a* =[*x*]≈, and *pneut*(*x*) is a (local) identity in [*x*]≈. By Definition 3, we get that [*x*]≈ is a regular hypergroup. □

From Theorem 2 we know that for a commutative pure NET-semihypergroup (it satisfies the condition in Theorem 2), it is a union of some regular hypergroups. The following picture (Figure 5) shows this special structure.

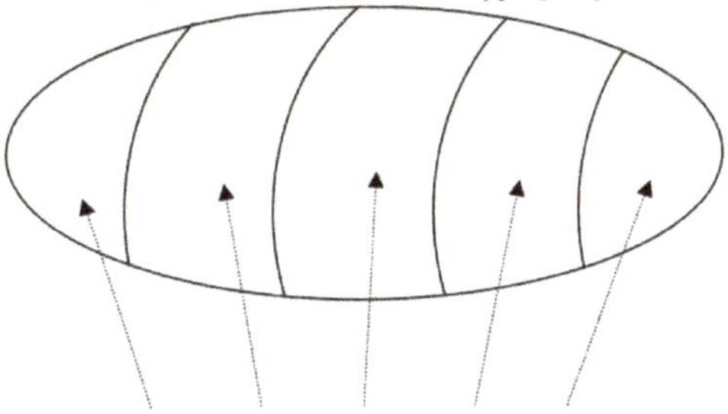

Figure 5. The structure of a commutative pure NET-semihypergroups.

Example 10. *Denote H = {a, b, c, d, e}, define hyperoperations * on H as shown in Table 8. We can verify that (H, *) is commutative pure NET-semihypergroup.*

Table 8. The hyperoperation * on *H*.

*	*a*	*b*	*c*	*d*	*e*
a	*a*	{*a*, *b*, *c*}	{*a*, *b*, *c*}	*d*	{*a*, *d*, *e*}
b	{*a*, *b*, *c*}	*b*	*c*	*d*	{*b*, *c*, *d*, *e*}
c	{*a*, *b*, *c*}	*c*	*b*	*d*	{*b*, *c*, *d*, *e*}
d	*d*	*d*	*d*	*d*	*d*
e	{*a*, *d*, *e*}	{*b*, *c*, *d*, *e*}	{*b*, *c*, *d*, *e*}	*d*	*e*

Moreover:

$$H_1=\{a\}=[a]_{\approx};$$
$$H_2=\{b, c\}=[b]_{\approx} =[c]_{\approx};$$
$$H_3=\{d\}=[d]_{\approx};$$
$$H_4=\{e\}=[e]_{\approx};$$

and $H= H_1 \cup H_2 \cup H_3 \cup H_4$*, where,* H_i *(i=1, 2, 3, 4) are regular hypergroups.*

Remark 6. The above example shows that a commutative pure NET-semihypergroup may be not a hypergroup (since *d***H*≠*H* in Example 10).

4. Weak Commutative NET-Semihypergroups and Their Structures

In this section, we discuss generalized commutativity in NET-semihypergroups. We propose a new notion of weak commutative NET-semihypergroup, and prove the structure theorem of weak commutative pure NET-semihypergroup (WCP-NET-semihypergroup), which can be regarded as a generalization of Cliffod Theorem in semigroup theory.

Definition 11. *Let (H,*) be a NET-semihypergroup. (H,*) is called a weak commutative NET- semihypergroup, if for every x∈H, every hyper-neutrosophic-triplet (x, neut(x), anti(x)), the following conditions are satisfied:*

(H,) is called a weak commutative pure NET-semihypergroup (shortly, WCP-NET-semihypergroup), if it both weak commutative and pure.*

Obviously, the following proposition is true and the proof is omitted.

Proposition 7. *Every commutative NET-semihypergroup is weak commutative.*

The following examples show that there exists some weak commutative NET- semihypergroups which are not commutative.

Example 11. *Denote H = {1, 2, 3, 4, 5, 6, 7, 8}, define hyperoperations * on H as shown in Table 9. We can verify that (H, *) is NET-semihypergroup.*

Table 9. The hyperoperation * on *H*.

*	1	2	3	4	5	6	7	8
1	1	{1, 2}	1	1	1	1	1	1
2	{1, 2}	2	1	1	1	1	1	1
3	1	1	3	4	5	6	7	8
4	1	1	4	3	8	7	6	5
5	1	1	5	7	3	8	4	6
6	1	1	6	8	7	3	5	4
7	1	1	7	5	6	4	8	3
8	1	1	8	6	4	5	3	7

*Moreover, (1, 1, 1), (2, 2, 2), (3, 3, 3), (4, 3, 4), (5, 3, 5), (6, 3, 6), (7, 3, 8) and (8, 3, 7) are hyper-neutrosophic-triplets, and (∀x∈H) 1*x = x*1, 2*x = x*2 and 3*x = x*3, 7*8 = 8*7. This means that (H, *) is a weak commutative NET-semihypergroup. Since 4*5 ≠ 5*4, (H, *) is not commutative.*

Remark 7. The above example shows that there exists WCP-NET-semihypergroup (by Definition 9, we know that the NET-semihypergroup in Example 11 is pure).

Example 12. *Denote H = {1, 2, 3, 4, 5, 6, 7, 8, 9}, define hyperoperations * on H as shown in Table 10. We can verify that (H, *) is NET-semihypergroup.*

Table 10. The hyperoperation * on *H*.

*	1	2	3	4	5	6	7	8	9
1	2	{1, 3}	3	1	1	1	1	1	1
2	{1, 3}	2	{1, 3}	{1, 2, 3}	{1, 2, 3}	{1, 2, 3}	{1, 2, 3}	{1, 2, 3}	{1, 2, 3}
3	3	{1, 3}	1	{1, 3}	{1, 3}	{1, 3}	{1, 3}	{1, 3}	{1, 3}
4	1	{1, 2, 3}	{1, 3}	4	5	6	7	8	9
5	1	{1, 2, 3}	{1, 3}	5	4	9	8	7	6
6	1	{1, 2, 3}	{1, 3}	6	8	4	9	5	7
7	1	{1, 2, 3}	{1, 3}	7	9	8	4	6	5
8	1	{1, 2, 3}	{1, 3}	8	6	7	5	9	4
9	1	{1, 2, 3}	{1, 3}	9	7	5	6	4	8

Moreover, (1, 2, 1), (2, 2, 2), (3, 1, 3), (4, 4, 4), (5, 4, 5), (6, 4, 6), (7, 4, 7), (8, 4, 9) and (9, 4, 8) are hyper-neutrosophic-triplets, and $(\forall x \in H)$ $2*x = x*2$, $1*x = x*1$ *and* $4*x = x*4$, $8*9 = 9*8$. *This means that (H, *) is a weak commutative NET-semihypergroup. Since* $5*6 \neq 6*5$, *(H, *) is not commutative.*

Proposition 8. *Let (H,*) be a weak commutative pure NET-semihypergroup (WCP-NET-semihypergroup). Then for any* $x \in H$, *there exists a pure neutral element of x, and pneut(x) is unique,* $pneut(x)*pneut(x)= pneut(x)$.

Proof. For any $x \in H$. Since $(H, *)$ is pure, by Definition 9, there exists hyper-neutrosophic-triplet $(x, neut(x), anti(x))$ such that

$$x = (neut(x)*x) \cap (x*neut(x)), \text{ and } neut(x) = (anti(x)*x) \cap (x*anti(x)).$$

Moreover, since $(H, *)$ is weak commutative, by Definition 11, $neut(x)*x = x*neut(x)$, and $anti(x)*x = x*anti(x)$. Thus

$$x = neut(x)*x = x*neut(x), \text{ and } neut(x) = anti(x)*x = x*anti(x).$$

Therefore, by Definition 10, $neut(x)$ is a pure neutral element of x. Applying Proposition 4 we know that pure neutral element of x is unique. Moreover, using Corollary 1, $pneut(x)*pneut(x)= pneut(x)$. □

Proposition 9. *Let (H,*) be a weak commutative pure NET-semihypergroup (WCP-NET-semihypergroup). Then for any* $x, y \in H$, $pneut(x*y)= pneut(x)*pneut(y)$ *when* $|x*y|=1$. *Moreover, if* $pneut(x) = z_1*x = x*z_1$ *and* $pneut(y) = z_2*y = y*z_2$, $z_1, z_2 \in H$, *then*

$$pneut(x*y) = (z_2*z_1)*(x*y) = (x*y)*(z_2*z_1).$$

Proof. Since $(H, *)$ be a WCP-NET-semihypergroup, then for any $x, y \in H$ and $|x*y|=1$, $pneut(x)*y = y*pneut(x)$ by Definition 11. Then

$$x*y)*(pneut(x)*pneut(y)) = (x*y)*(pneut(y)*pneut(x)) = x*y*pneut(x) = (x*pneut(x))*y = x*y;$$
$$(pneut(x)*pneut(y))*(x*y) = (pneut(y)*pneut(x))*(x*y) = pneut(y)*x*y = x*(pneut(y)*y) = x*y.$$

On the other hand, let $(x, pneut(x), anti(x))$ and $(y, pneut(y), anti(y))$ are hyper-neutrosophic-triplets, then

$$\begin{aligned} & xx*y)*(anti(y)*anti(x)) \\ & = x*(y*anti(y))*anti(x) \\ & = x*pneut(y)*anti(x) \\ & = pneut(y)*x*anti(x) \\ & = pneut(y)*pneut(x) \\ & = pneut(x)*pneut(y); \end{aligned}$$
$$(anti(y)*anti(x))*(x*y) = anti(y)*(anti(x)*x)*y = anti(y)*pneut(x)*y = pneut(x)*anti(y)*y = pneut(x)*pneut(y).$$

Thus, $pneut(x)*pneut(y)$ is a pure neutral element of $x*y$ by Definition 7 and Definition 10. Applying Proposition 8 we get that $pneut(x*y)= pneut(x)*pneut(y)$.

Moroeover, assume $pneut(x) = z_1*x = x*z_1$, $pneut(y) = z_2*y = y*z_2$. Then, by weak commutativity (Definition 11) we have

$$(z_2*z_1)*(x*y) = z_2*(z_1*x)*y = z_2*pneut(x)*y = pneut(x)*(z_2*y) = pneut(x)*pneut(y) = pneut(x*y);$$
$$(x*y)*(z_2*z_1) = x*(y*z_2)*z_1 = x*pneut(y)*z_1 = (x*z_1)*pneut(y) = pneut(x)*pneut(y) = pneut(x*y).$$

Therefore, the proof is completed. □

Theorem 3. *Let (H,*) be a WCP-NET-semihypergroup and H satisfies*

$$(\forall x, y\in H, pneut(x)=pneut(y)\Rightarrow |x^*y|=1. \quad \text{(C1)}$$

Define a binary relation $\approx$ *on H as following:*

$$\forall x, y\in H, x\approx y \text{ if and only if } pneut(x)=pneut(y).$$

Then

(1) *The binary relation* $\approx$ *is a equivalent relation on H;*
(2) *For any* $x\in H$, $[x]_\approx$ *is a sub-NET-semihypergroup of H, where* $[x]_\approx$ *is the equivalent class of x based on equivalent relation* $\approx$;
(3) *For any* $x\in H$, $[x]_\approx$ *is a regular hypergroupe.*

Proof. (1) From the definition of $\approx$, by Proposition 8 and Proposition 9, we know that the binary relation $\approx$ is a equivalent relation.

(2) Suppose $a, b\in[x]_\approx$. By the definition of $\approx$, $pneut(a) = pneut(b) = pneut(x)$. Using Proposition 8 and Proposition 9, we have

$$pneut(a^*b) = pneut(a)^*pneut(b) = pneut(x)^*pneut(x) = pneut(x).$$

It follows that $[x]_\approx$ is closed on the hyper operation *.

And, applying Proposition 8, we have $pneut(x)^*pneut(x) = pneut(x)$. From this and using Proposition 8, we get that $pneut(pneut(x)) = pneut(x)$. It follows that $pneut(a)\in[x]_\approx$ for any $a\in[x]_\approx$. Moreover, assume that $a\in[x]_\approx$, by the definition of WCP-NET-semihypergroup, there exists $r\in H$ such that $pneut(a) = r^*a = a^*r$. Thus (by Proposition 9)

$$\begin{aligned} pneut(a) &= (r^*pneut(a))^*a = a^*(r^*pneut(a)) \\ &\Rightarrow r^*pneut(a) \in\{anti(a)\}. \\ &pneut(r^*pneut(a)) \\ &= pneut(r)^* pneut(pneut(a)) \\ &= pneut(r)^* pneut(a) \\ &= pneut(r^*a) \\ &= pneut(pneut(a)) \\ &= pneut(a). \end{aligned}$$

That is, $pneut(r^*pneut(a)) = pneut(a) = pneut(x)$. This means that $r^*pneut(a) \in[x]_\approx$. Combining this and $r^*pneut(a)\in\{anti(a)\}$, we know that there exists $anti(a)$ which is in $[x]_\approx$. This means that $[x]_\approx$ is a sub-NET- semihypergroup of H.

(3) Assume $x\in H$, from (2) we know that $[x]_\approx$ is a sub-NET-semihypergroup of H. By the definition of $\approx$, for any $a\in[x]_\approx$, $pneut(a) = pneut(x)$. From the proof of (2), there exists $anti(a)\in\{anti(a)\}$ and $anti(a)\in[x]_\approx$. Then, $[x]_\approx \subseteq a^*[x]_\approx{}^*a$. Obviously, $a^*[x]_\approx{}^*a \subseteq [x]_\approx$. Thus, $a^*[x]_\approx{}^*a=[x]_\approx$.

On the other hand, $pneut(x)$ is a (local) identity in $[x]_\approx$. Therefore, by Definition 3, we get that $[x]_\approx$ is a regular hypergroup. □

Example 13. *Denote H = {1, 2, 3, 4, 5, 6, 7, 8, 9, 10, 11}, define hyperoperations * on H as shown in Table 11. We can verify that (H, *) is WCP-NET-semihypergroup, and not commutative.*

Table 11. The hyperoperation * on *H*.

*	1	2	3	4	5	6	7	8	9	10	11
1	1	{1,2,3}	{1,2,3}	1	1	1	1	1	1	1	1
2	{1,2,3}	3	2	3	2	{1,2,3}	{1,2,3}	{1,2,3}	{1,2,3}	{1,2,3}	{1,2,3}
3	{1,2,3}	2	3	2	3	{1,2,3}	{1,2,3}	{1,2,3}	{1,2,3}	{1,2,3}	{1,2,3}
4	1	3	2	5	4	{6,7,8, 9,10,11}	{6,7,8, 9,10,11}	{6,7,8, 9,10,11}	{6,7,8, 9,10,11}	{6,7,8, 9,10,11}	{6,7,8, 9,10,11}
5	1	2	3	4	5	{6,7,8, 9,10,11}	{6,7,8, 9,10,11}	{6,7,8, 9,10,11}	{6,7,8, 9,10,11}	{6,7,8, 9,10,11}	{6,7,8, 9,10,11}
6	1	{1,2,3}	{1,2,3}	{6,7,8, 9,10,11}	{6,7,8, 9,10,11}	6	7	8	9	10	11
7	1	{1,2,3}	{1,2,3}	{6,7,8, 9,10,11}	{6,7,8, 9,10,11}	7	6	11	10	9	8
8	1	{1,2,3}	{1,2,3}	{6,7,8, 9,10,11}	{6,7,8, 9,10,11}	8	10	6	11	7	9
9	1	{1,2,3}	{1,2,3}	{6,7,8, 9,10,11}	{6,7,8, 9,10,11}	9	11	10	6	8	7
10	1	{1,2,3}	{1,2,3}	{6,7,8, 9,10,11}	{6,7,8, 9,10,11}	10	8	9	7	11	6
11	1	{1,2,3}	{1,2,3}	{6,7,8, 9,10,11}	{6,7,8, 9,10,11}	11	9	7	8	6	10

Moreover,

$$H_1 = \{1\} = [1]_{\approx};$$
$$H2 = \{2, 3\} = [2]_{\approx} = [3]_{\approx};$$
$$H3 = \{4, 5\} = [4]_{\approx} = [5]_{\approx};$$
$$H4 = \{6, 7, 8, 9, 10, 11\} = [6]_{\approx} = [7]_{\approx} = [8]_{\approx} = [9]_{\approx} = [10]_{\approx} = [11]_{\approx};$$

and $H = H_1 \cup H_2 \cup H_3 \cup H_4$, *where,* H_i *(i=1, 2, 3, 4) are regular hypergroups.*

5. Conclusions

In this paper, we propose some new notions of neutrosophic extended triplet semihypergroup (NET-semihypergroup), neutrosophic extended triplet hypergroup (NET-hypergroup), pure NET-semihypergroup and weak commutative NET-semihypergroup, investigate some basic properties and the relationships among them (see Figure 6), study their close connections with regular hypergroups and regular semihypergroups. Particularly, we prove two structure theorems of commutative pure NET-semihypergroup (CP-NET-semihypergroup) and weak commutative pure NET-semihypergroup (WCP-NET-semihypergroup) under the condition (C1) (see Theorem 2 and Theorem 3). From these results, we know that NET-semihypergroup is a hyperalgebraic structure independent of hypergroup, and NET-semihypergroup is also a generalization of group concept in hyperstructures. The research results in this paper show that NET-semihypergroups and NET- hypergroups have important theoretical research value, which greatly enriches the traditional theory of hyperalgebraic structures.

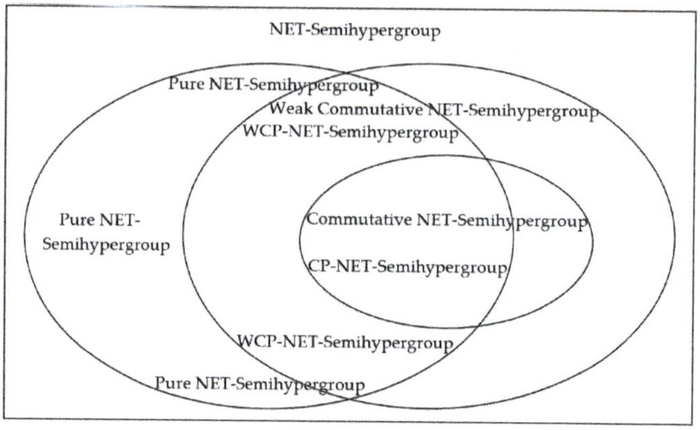

Figure 6. The relationships among some kinds of NET-semihypergroups.

In the future, we will investigate the combinations of NET-semihypergroups and related algebraic systems ([22–24]).

Author Contributions: F.S. and X.Z. proposed the idea of the research, X.Z. and Y.M. wrote the paper.

Funding: This research was funded by National Natural Science Foundation of China (Grant Nos. 61573240 and 61976130).

Conflicts of Interest: The authors declare no conflict of interest.

References

1. Corsini, P.; Leoreanu, V. *Applications of Hyperstructure Theory*; Springer: Berlin/Heidelberg, Germany, 2003.
2. Wall, H.S. Hypergroups. *Am. J. Math.* **1937**, *59*, 77–98. [CrossRef]
3. Freni, D. A new characterization of the derived hypergroup via strongly regular equivalences. *Commun. Algebra* **2002**, *30*, 3977–3989. [CrossRef]
4. Davvaz, B. *Semihypergroup Theory*; Elsevier: Amsterdam, The Netherlands, 2016.
5. Jafarabadi, H.M.; Sarmin, N.H.; Molaei, M.R. Simple semihypergroups. *Aust. J. Basic Appl. Sci.* **2011**, *5*, 51–55.
6. Salvo, M.D.; Freni, D.; Faro, G.L. Fully simple semihypergroups. *J. Algebra* **2014**, *399*, 358–377. [CrossRef]
7. Jafarabadi, H.M.; Sarmin, N.H.; Molaei, M.R. Completely simple and regular semihypergroups. *Bull. Malays. Math. Sci. Soc.* **2012**, *35*, 335–343.
8. Howie, J.M. *Fundamentals of Semigroup Theory*; Oxford University Press: Oxford, UK, 1995.
9. Smarandache, F. Neutrosophic set—A generialization of the intuituionistics fuzzy sets. *Int. J. Pure Appl. Math.* **2005**, *3*, 287–297.
10. Zhang, X.H.; Bo, C.X.; Smarandache, F.; Dai, J.H. New inclusion relation of neutrosophic sets with applications and related lattice structure. *Int. J. Mach. Learn. Cybern.* **2018**, *9*, 1753–1783. [CrossRef]
11. Zhang, X.H.; Bo, C.X.; Smarandache, F.; Park, C. New operations of totally dependent-neutrosophic sets and totally dependent-neutrosophic soft sets. *Symmetry* **2018**, *10*, 187. [CrossRef]
12. Zhang, X.H.; Mao, X.Y.; Wu, Y.T.; Zhai, X.H. Neutrosophic filters in pseudo-BCI algebras. *Int. J. Uncertain. Quantif.* **2018**, *8*, 511–526. [CrossRef]
13. Smarandache, F.; Ali, M. Neutrosophic triplet group. *Neural Comput. Appl.* **2018**, *29*, 595–601. [CrossRef]
14. Smarandache, F. *Neutrosophic Perspectives: Triplets, Duplets, Multisets, Hybrid Operators, Modal Logic, Hedge Algebras. And Applications*; Pons Publishing House: Brussels, Belgium, 2017.
15. Zhang, X.H.; Hu, Q.Q.; Smarandache, F.; An, X.G. On neutrosophic triplet groups: Basic properties, NT-subgroups, and some notes. *Symmetry* **2018**, *10*, 289. [CrossRef]
16. Jaíyéolá, T.G.; Smarandache, F. Some sesults on neutrosophic triplet group and their applications. *Symmetry* **2018**, *10*, 202. [CrossRef]

17. Zhang, X.H.; Wu, X.Y.; Mao, X.Y.; Smarandache, F.; Park, C. On neutrosophic extended triplet groups (Loops) and Abel-Grassmann's Groupoids (AG-Groupoids). *J. Intell. Fuzzy Syst.* **2019**. [CrossRef]
18. Wu, X.Y.; Zhang, X.H. The decomposition theorems of AG-neutrosophic extended triplet loops and strong AG-(*l*, *l*)-loops. *Mathematics* **2019**, *7*, 268. [CrossRef]
19. Ma, Y.C.; Zhang, X.H.; Yang, X.F.; Zhou, X. Generalized neutrosophic extended triplet group. *Symmetry* **2019**, *11*, 327. [CrossRef]
20. Zhang, X.H.; Wang, X.J.; Smarandache, F.; Jaíyéolá, T.G.; Lian, T.Y. Singular neutrosophic extended triplet groups and generalized groups. *Cogn. Syst. Res.* **2019**, *57*, 32–40. [CrossRef]
21. Gulistan, M.; Nawaz, S.; Hassan, N. Neutrosophic triplet non-associative semihypergroups with application. *Symmetry* **2018**, *10*, 613. [CrossRef]
22. Zhang, X.H.; Borzooei, R.A.; Jun, Y.B. Q-filters of quantum B-algebras and basic implication algebras. *Symmetry* **2018**, *10*, 573. [CrossRef]
23. Omidi, S.; Davvaz, B.; Zhan, J.M. An investigation on ordered algebraic hyperstructures. *Acta Math. Sin. Engl. Ser.* **2017**, *33*, 1107–1124. [CrossRef]
24. Zhang, X.H.; Mao, X.Y.; Smarandache, F.; Park, C. On homomorphism theorem for perfect neutrosophic extended triplet groups. *Information* **2018**, *9*, 237. [CrossRef]

symmetry

MDPI

Article

On Neutrosophic Offuninorms

Erick González Caballero [1,*], Florentin Smarandache [2] and Maikel Leyva Vázquez [3]

1 Center of Studies in Mathematic for Technical Sciences, Technological University of Havana, Havana, Havana 19390, Cuba
2 Math & Science Department, University of New Mexico, Gallup, NM 87301, USA; fsmarandache@gmail.com
3 Facultad de Ciencias Matemáticas y Físicas, University of Guayaquil, Guayas, Guayaquil 090514, Ecuador; mleyvaz@gmail.com
* Correspondence: erickgc@yandex.com

Received: 29 July 2019; Accepted: 22 August 2019; Published: 6 September 2019

Abstract: Uninorms comprise an important kind of operator in fuzzy theory. They are obtained from the generalization of the t-norm and t-conorm axiomatic. Uninorms are theoretically remarkable, and furthermore, they have a wide range of applications. For that reason, when fuzzy sets have been generalized to others—e.g., intuitionistic fuzzy sets, interval-valued fuzzy sets, interval-valued intuitionistic fuzzy sets, or neutrosophic sets—then uninorm generalizations have emerged in those novel frameworks. Neutrosophic sets contain the notion of indeterminacy—which is caused by unknown, contradictory, and paradoxical information—and thus, it includes, aside from the membership and non-membership functions, an indeterminate-membership function. Also, the relationship among them does not satisfy any restriction. Along this line of generalizations, this paper aims to extend uninorms to the framework of neutrosophic offsets, which are called neutrosophic offuninorms. Offsets are neutrosophic sets such that their domains exceed the scope of the interval [0,1]. In the present paper, the definition, properties, and application areas of this new concept are provided. It is necessary to emphasize that the neutrosophic offuninorms are feasible for application in several fields, as we illustrate in this paper.

Keywords: neutrosophic offset; uninorm; neutrosophic offuninorm; neutrosophic offnorm; neutrosophic offconorm; implicator; prospector; n-person cooperative game

---

1. Introduction

Uninorms extend the t-norm and t-conorm axiomatic in fuzzy theory. They retain the axioms of commutativity, associativity, and monotony. Alternatively, they generalize the boundary condition, where the neutral element is any number lying in [0,1]. Thus, t-norm and t-conorm are special cases of uninorms, t-norms have 1 as their neutral element and the neutral element of t-conorms is 0, see [1–3].

Uninorms are theoretically important, and moreover they have also been used as operators in several areas of application; for example, in image processing, to aggregate group decision criteria, among others, see [4–8]. An exhaustive search on uninorm applications made by the authors of this paper yielded more than six hundred scientific articles that have been written in the last five years devoted to this subject.

Rudas et al. in [9] report that uninorms have been applied in diverse applications ranging, e.g., from defining Gross Domestic Product index in economics, to fusing sequences of DNA and RNA or combining information on taxonomies or dendograms in biology, and in the fusion of data provided by sensors of robotics in data mining, and in knowledge-based and intelligent systems. Particularly, they offer many examples in Decision Making, Utility Theory, Fuzzy Inference Systems, Multisensor Data Fusion, network aggregation in sensor networks, image approximation, hardware implementation of parametric operations, in Fuzzy Systems, and as software tools for aggregation problems.

Depaire et al. in [10] proposed a new approach to apply uninorms in Importance Performance Analysis, which is a useful technique to evaluate elements in marketing programs. They proved that their approach was superior when compared with regression and that it matched well with the customer satisfaction theory.

A very recent paper written by Modley et al. in [11] applied uninorms in the market basket analysis. Also, Appel et al. proposed a method based on cross-ratio uninorms as a mechanism to aggregate in the Sentiment Analysis; see [12].

Kamiset al. in [13] implement a geo-uninorm operator in a consensus model. They utilized them to derive a consistently based preference relation from a given reciprocal preference relation. Whereas, Wu et al. in [14] and Ureña et al. in [15] applied uninorms in trust propagation and aggregation methods for group decision making in a social network.

Bordignon and Gomide in [16], introduce a learning approach to train uninorm-based hybrid neural networks using extreme learning concepts. According to them, uninorms bring flexibility and generality to fuzzy neuron models. Wang in [17] and Yang in [18] applied uninorms as a basis to define logics.

Other areas of application can be consulted in González-Hidalgo et al. [19] where uninorms were utilized in edge detection of image processing, in fuzzy morphological associative memories (see [20]), and was also applied in time series prediction.

It is well-known that the minimum is the biggest t-norm and the maximum is the lowest t-conorm, thus they are not compensatory operators; whereas uninorms compensate when the truth values are situated on both sides of the neutral element. The compensation property could be the key factor in the wide range of uninorm applicability, mainly in decision making. Zimmermann experimentally proved in [21], many years before the introduction of uninorms, that often human beings do not make decisions interpreting AND like a t-norm and OR like a t-conorm, but that compensatory operators are more adequate to model human aggregations to signify AND and OR in some situations. The use of means as aggregators to define membership functions can be seen in [22]. However, when the aggregated values are situated on one side with respect to the neutral element, then uninorms operate either like a t-norm or a t-conorm.

Uninorms have been extended to other theories more general than fuzzy logic, due to their applicability. Let us mention intuitionistic fuzzy sets, interval-valued fuzzy sets, and interval-valued intuitionistic fuzzy sets; where the generalizations consist of the inclusion of an independent non-membership function or an interval-valued membership function, or both [23]. They have also been generalized as multi-polar aggregators in [24].

Following this trend, the authors of this paper defined the neutrosophic uninorms, such that the uninorms were extended to the neutrosophy framework [25]. Neutrosophy is the philosophical discipline that studies theories, entities, objects, phenomena, among others, related to neutrality [26]. In particular, neutrosophic sets contain three independent functions, namely, a membership function, a non-membership function, and additionally, an indeterminate-membership function. The last one represents what is unknown, contradictory, and paradoxical. Furthermore, these elements can be intervals.

In addition, the relationship among these three functions has no restriction, contrary to the intuitionist fuzzy sets, which must fulfill the constraint that the sum of the membership truth value with the non-membership truth value of an element to the set does not exceed the unit.

Neutrosophy theory has been used in a wide spectrum of applications such as in image processing, decision making, clustering, among others [27–30]. Therefore, it is not difficult to appreciate the applicability of neutrosophic uninorms.

More recently, other concepts have been defined within the neutrosophy framework, which further generalizes the traditional membership functions, including the axiomatic in probability theory.

They are the undersets, oversets, and offsets, where the basic idea is that negative truth values or truth values greater than 1 are permitted in the calculus [31].

A recurring example in literature is that concerning employment, where the truth value of a worker's effectiveness is measured in working hours. Those workers who have met all of their working hours established for the week will be an effectiveness truth value of 1, those workers who have only partially met their working hours have a truth value between 0 and 1, and other workers who have not attended work throughout the week have the truth value of 0. In addition, those who have performed voluntary overtime after meeting their established hours have a truth value greater than 1, and finally, the workers who have not attended work throughout the week and, moreover, have caused losses to the company, must have a negative truth value.

Other examples take into consideration the relationship between two variables or more, where a negative value represents that they are inversely related, whereas a direct relationship is represented by positive values [31].

The aim of this paper is to extend for the first time the theory of uninorms to the offsets framework—we call them neutrosophic offuninorms—in such a way that they are a generalization of both n-offnorms and n-offconorms equivalently, as fuzzy uninorms generalize both t-norms and t-conorms.

In this paper, definitions and also properties of neutrosophic offuninorms will be given. Additionally, we will emphasize the relationship between these new operators and the aggregation functions used in the well-known medical expert system MYCIN [32], as well as define logical implicators in offset fields and solve voting cooperative games.

In particular, the association of the proposed theory with the aggregation functions used in MYCIN supports the hypothesis that neutrosophic offuninorms are more than an interesting theoretical approach. Historically, within the fuzzy logic framework, some authors have accepted the idea of extending the uninorms domain to [a, b], in order to include the aggregation functions used in MYCIN, [33,34]. This proposal is an important precedent for this investigation because uninorms were there adapted to offsets in the fuzzy theory context. The relationship between uninorms and the PROSPECTOR operator, as well as their application, can be consulted in [35], where they were used in e-arning.

Authors in [33,34] also emphasize that this generalization has important practical advantages because it allows us to naturally apply uninorms in fields like Artificial Neural Networks and Cognitive Maps. These elements certainly suggest that the proposed theory can be applied in fields like Artificial Neural Networks based on neutrosophic sets and in neutrosophic cognitive maps, [36,37].

Let us observe that when uninorms have been extended to other domains they have preserved the property of compensation. Further, we shall prove that offuninorms are not the exception; consequently, the applicability of offuninorms is practically guaranteed. In the discussion section, we insist on this aspect and the advantages that offuninorms have over other generalizations.

This paper is divided as follows. It begins with a preliminary section where concepts such as neutrosophic sets, neutrosophic offsets, neutrosophic uninorms, among other useful aspects, are discussed in order to develop the content of this article. The section on neutrosophic offuninorms is devoted to exposing definitions and properties of these novel operators. Next, the applications section is where the three possible areas of application of this theory are explained. We then finish with the sections of discussion and conclusions.

2. Preliminaries

This section contains the main definitions necessary to develop the theory proposed in this paper. We begin with Definitions 1 and 2, which introduce the neutrosophic sets. These sets are characterized by an independent indeterminacy-membership function that models the unknown, contradictions, inconsistencies in information and so on. Additionally, we have the classic membership and non-membership functions, which are not necessarily dependent on each other.

Definition 1. *Let X be a space of points (objects), with a generic element in X denoted by x. A Neutrosophic Set A in X is characterized by a truth-membership function $T_A(x)$, an indeterminacy-membership function $I_A(x)$, and a falsity-membership function $F_A(x)$. $T_A(x)$, $I_A(x)$, and $F_A(x)$ are real standard or nonstandard subsets of $]^-0, 1^+[$. There is no restriction on the sum of $T_A(x)$, $I_A(x)$, and $F_A(x)$, thus, $^-0 \leq \inf T_A(x) + \inf I_A(x) + \inf F_A(x) \leq \sup T_A(x) + \sup I_A(x) + \sup F_A(x) \leq 3^+$ (see [26]).*

The neutrosophic sets are useful in their nonstandard form only in philosophy, in order to make a distinction between absolute truth (truth in all possible worlds—according to Leibniz) and relative truth (truth in at least one world), but not in technical applications, thus the *Single-Valued Neutrosophic Sets* are defined, see Definition 2.

Definition 2. *Let X be a space of points (objects), with a generic element in X denoted by x. A Single-Valued Neutrosophic Set A in X is characterized by a truth-membership function $T_A(x)$, an indeterminacy-membership function $I_A(x)$, and a falsity-membership function $F_A(x)$. $T_A(x)$, $I_A(x)$, and $F_A(x)$ are elements of [0,1]. There is no restriction on the sum of $T_A(x)$, $I_A(x)$, and $F_A(x)$, thus, $0 \leq T_A(x) + I_A(x) + F_A(x) \leq 3$ (see [38]).*

The domain of the single-valued neutrosophic sets does not surpass the limits of the interval [0,1]. This is a classical condition imposed in previous theories such as probability and fuzzy sets. Despite the past, Smarandache in 2007 proposed the membership >1 and <0 and illustrated this proposal; see [39] (pp. 92–93) and the example given in the introduction of this paper. In the following, the *Single-Valued Neutrosophic Oversets*, *Single-Valued Neutrosophic Undersets*, and *Single-Valued Neutrosophic Offsets* are formally defined.

Definition 3. *Let X be a universe of discourse and the neutrosophic set $A_1 \subset X$. Let T(x), I(x), F(x) be the functions that describe the degree of membership, indeterminate-membership, and non-membership respectively, of a generic element $x \in X$, with respect to the neutrosophic set A_1:*

T, I, F: $X \to [0, \Omega]$, where $\Omega > 1$ is called overlimit, T(x), I(x), $F(x) \in [0, \Omega]$. A Single-Valued Neutrosophic Overset A_1 is defined as $A_1 = \{(x, T(x), I(x), F(x)), x \in X\}$, such that there exists at least one element in A_1 that has at least one neutrosophic component that is bigger than 1, and no element has neutrosophic components that are smaller than 0 (see [31]).

Definition 4. *Let X be a universe of discourse and the neutrosophic set $A_2 \subset X$. Let T(x), I(x), F(x) be the functions that describe the degree of membership, indeterminate-membership, and non-membership, respectively, of a generic element $x \in X$, with respect to the neutrosophic set A_2:*

T, I, F: $X \to [\Psi, 1]$, where $\Psi < 0$ is called underlimit, T(x), I(x), $F(x) \in [\Psi, 1]$. A Single-Valued Neutrosophic Underset A_2 is defined as $A_2 = \{(x, T(x), I(x), F(x)), x \in X\}$, such that there exists at least one element in A_2 that has at least one neutrosophic component that is smaller than 0, and no element has neutrosophic components that are bigger than 1 (see [31]).

Definition 5. *Let X be a universe of discourse and the neutrosophic set $A_3 \subset X$. Let T(x), I(x), F(x) be the functions that describe the degree of membership, indeterminate-membership, and non-membership respectively, of a generic element $x \in X$, with respect to the neutrosophic set A_3:*

T, I, F: $X \to [\Psi, \Omega]$, where $\Psi < 0 < 1 < \Omega$, Ψ is called underlimit, while Ω is called overlimit, T(x), I(x), $F(x) \in [\Psi, \Omega]$. A Single-Valued Neutrosophic Offset A_3 is defined as $A_3 = \{(x, T(x), I(x), F(x)), x \in X\}$, such that there exists at least one element in A_3 that has at least one neutrosophic component that is bigger than 1, and at least another neutrosophic component that is smaller than 0 (see [31]).

Let us note that the oversets, undersets, and offsets cover the three possible cases to characterize. Now, the logical operations over these kinds of sets have to be redefined, in view that the classical ones cannot always be straightforwardly extended to these domains. This is the case of complement given by Smarandache in [31], whereas the union and intersection definitions do not change with respect to those of single-valued neutrosophic sets. This is summarized below:

Let X be a universe of discourse, $A = \{(x, \langle T_A(x), I_A(x), F_A(x)\rangle), x \in X\}$ and $B = \{(x, \langle T_B(x), I_B(x), F_B(x)\rangle), x \in X\}$ be two single-valued neutrosophic oversets/undersets/offsets.

$T_A, I_A, F_A, T_B, I_B, F_B$: $X \to [\Psi, \Omega]$, where $\Psi \le 0 < 1 \le \Omega$, Ψ is the underlimit, whilst Ω is the overlimit, $T_A(x)$, $I_A(x)$, $F_A(x)$,$T_B(x)$, $I_B(x)$, $F_B(x) \in [\Psi, \Omega]$. Let us remark that the three cases are here comprised, viz., overset when $\Psi = 0$ and $\Omega > 1$, underset when $\Psi < 0$ and $\Omega = 1$, and offset when $\Psi < 0$ and $\Omega > 1$.

Then, the main operators are defined as follows:

$A \cup B = \{(x, \langle \max(T_A(x), T_B(x)),\ \min(I_A(x), I_B(x)),\ \min(F_A(x), F_B(x))\rangle), x \in X\}$ is the union.

$A \cap B = \{(x, \langle \min(T_A(x), T_B(x)),\ \max(I_A(x), I_B(x)),\ \max(F_A(x), F_B(x))\rangle), x \in X\}$ is the intersection,

$C(A) = \{(x, \langle F_A(x),\ \Psi + \Omega - I_A(x), T_A(x)\rangle), x \in X\}$ is the neutrosophic complement of the neutrosophic set.

Let us remark that when $\Psi = 0$ and $\Omega = 1$, the precedent operators convert in the classical ones. With regard to logical operators, e.g., n-norms and n-conorms, their redefinitions in the offsets framework are not so evident. Below, definitions of *offnegation*, *neutrosophic component n-offnorm*, and *neutrosophic component n-offconorm* are provided.

One *offnegation* can be defined as in Equation (1).

$$\overset{\neg}{O}\langle T, I, F\rangle = \langle F, \Psi_I + \Omega_I - I, T\rangle \tag{1}$$

Definition 6. *Let c be a neutrosophic component (T_O, I_O or F_O). c: $M_O \to [\Psi, \Omega]$, where $\Psi \le 0$ and $\Omega \ge 1$. The neutrosophic component n-offnorm $N_O^n : [\Psi, \Omega]^2 \to [\Psi, \Omega]$ satisfies the following conditions for any elements x, y, and z $\in M_O$:*

i. $N_O^n(c(x), \Psi) = \Psi$, $N_O^n(c(x), \Omega) = c(x)$ *(Overbounding Conditions),*
ii. $N_O^n(c(x), c(y)) = N_O^n(c(y), c(x))$ *(Commutativity),*
iii. *If* $c(x) \le c(y)$ *then* $N_O^n(c(x), c(z)) \le N_O^n(c(y), c(z))$ *(Monotonicity),*
iv. $N_O^n\big(N_O^n(c(x), c(y)), c(z)\big) = N_O^n\big(c(x), N_O^n(c(y), c(z))\big)$ *(Associativity).*

To simplify the notation, sometimes we use $\langle T_1, I_1, F_1\rangle \underset{O}{\wedge} \langle T_2, I_2, F_2\rangle = \langle T_1 \underset{O}{\wedge} T_2, I_1 \underset{O}{\vee} I_2, F_1 \underset{O}{\vee} F_2\rangle$ instead of $N_O^n(\cdot,\cdot)$.

Let us remark that the definition of the neutrosophic component n-offnorm is valid for every one of the components, thus, we have to apply it three times. Also, Definition 6 contains the definition of n-norm when $\Psi = 0$ and $\Omega = 1$.

Proposition 1. *Let $N_O^n(\cdot,\cdot)$ be a neutrosophic component n-offnorm, then, for any elements x, y $\in M_O$ we have $N_O^n(c(x), c(y)) \le \min(c(x), c(y))$.*

Proof. Because of the monotonicity of the neutrosophic component n-offnorm and one of the overbounding conditions, we have $N_O^n(c(x), c(y)) \le N_O^n(c(x), \Omega) = c(x)$, hence $N_O^n(c(x), c(y)) \le c(x)$ and similarly $N_O^n(c(x), c(y)) \le c(y)$ can be proved, therefore, $N_O^n(c(x), c(y)) \le \min(c(x), c(y))$. □

See that Proposition 1 maintains this property of the n-norms. Likewise to the definition of the neutrosophic component n-offnorm, in Definition 7 it is described the *neutrosophic component n-offconorm.*

Definition 7. *Let c be a neutrosophic component (T_O, I_O or F_O). c: $M_O \to [\Psi, \Omega]$, where $\Psi \leq 0$ and $\Omega \geq 1$. The neutrosophic component n-offconorm $N_O^{co}: [\Psi, \Omega]^2 \to [\Psi, \Omega]$ satisfies the following conditions for any elements x, y, and z $\in M_O$:*

i. $N_O^{co}(c(x), \Omega) = \Omega$, $N_O^{co}(c(x), \Psi) = c(x)$ *(Overbounding Conditions),*
ii. $N_O^{co}(c(x), c(y)) = N_O^{co}(c(y), c(x))$ *(Commutativity),*
iii. *If* $c(x) \leq c(y)$ *then* $N_O^{co}(c(x), c(z)) \leq N_O^{co}(c(y), c(z))$ *(Monotonicity),*
iv. $N_O^{co}\left(N_O^{co}(c(x), c(y)), c(z)\right) = N_O^{co}\left(c(x), N_O^{co}(c(y), c(z))\right)$ *(Associativity).*

To simplify the notation sometimes we use $\langle T_1, I_1, F_1\rangle \underset{O}{\vee} \langle T_2, I_2, F_2\rangle = \langle T_1 \underset{O}{\vee} T_2, I_1 \underset{O}{\wedge} I_2, F_1 \underset{O}{\wedge} F_2\rangle$ instead of $N_O^{co}(\cdot,\cdot)$.

Proposition 2. *Let $N_O^{co}(\cdot,\cdot)$ be a neutrosophic component n-offconorm, then, for any elements x, y$\in M_O$ we have $N_O^{co}(c(x), c(y)) \geq max(c(x), c(y))$.*

Proof. The proof is equivalent to the proof of Proposition 1. □

In this paper, we use the notion of lattice, based on the poset denoted by $\leq_O$, where $\langle T_1, I_1, F_1\rangle \leq_O \langle T_2, I_2, F_2\rangle$ if and only if $T_2 \geq T_1$, $I_2 \leq I_1$ and $F_2 \leq F_1$, where the infimum and the supremum of the set are $\langle \Psi, \Omega, \Omega\rangle$ and $\langle \Omega, \Psi, \Psi\rangle$, respectively.

One property that is preserved of n-norms is that the minimum is the biggest neutrosophic component n-offnorm for T_O, as it is demonstrated in Proposition 1. Proposition 2 proved that the maximum is the smallest neutrosophic component n-offconorm for I_O and F_O when we consider $\leq_O$.

Evidently, the minimum is a neutrosophic component n-offnorm and the maximum is a neutrosophic component n-offconorm; see Example 1.

Example 1. *An example of a pair offAND/offOR is,* $c(x) \underset{ZO}{\wedge} c(y) = min(c(x), c(y))$ *and* $c(x) \underset{ZO}{\vee} c(y) = max(c(x), c(y))$*, respectively.*

Example 2. *A pair of offAND/offOR is,* $c(x) \underset{LO}{\wedge} c(y) = max(\Psi, c(x) + c(y) - \Omega)$ *and* $c(x) \underset{LO}{\vee} c(y) = min(\Omega, c(x) + c(y))$*, respectively.*

Example 2 extends the Łukasiewicz t-norm and t-conorm to the neutrosophic offsets. Let us remark that the simple product t-norm and its dual t-conorm cannot be extended to this new domain.

Finally, we recall the definition of neutrosophic uninorms that appeared in [25], see Definition 8.

Definition 8. *A neutrosophic uninorm $\boldsymbol{U_N}$ is a commutative, increasing, and associative mapping,* $U_N: (]^-0, 1^+[\times]^-0, 1^+[\times]^-0, 1^+[)^2 \to]^-0, 1^+[\times]^-0, 1^+[\times]^-0, 1^+[$, *such that* $U_N\left(x\langle T_x, I_x, F_x\rangle, y\langle T_y, I_y, F_y\rangle\right) = \langle U_N T(x,y), U_N I(x,y), U_N F(x,y)\rangle$, *where $\boldsymbol{U_N}T$ means the degree of membership, $\boldsymbol{U_N}I$ the degree of indeterminacy, and $\boldsymbol{U_N}F$ the degree of non-membership of both x and y. Additionally, there exists a neutral element* $e \in]^-0, 1^+[\times]^-0, 1^+[\times]^-0, 1^+[$, *where* $\forall x \in]^-0, 1^+[\times]^-0, 1^+[\times]^-0, 1^+[$, $U_N(e, x) = x$.

Let us observe that this definition can be restricted to single-valued neutrosophic sets. Neutrosophic uninorms generalize n-norms, n-conorms, uninorms in *L**-fuzzy set theory, and fuzzy uninorms.

3. On Neutrosophic Offuninorms

This section contains the core of the present paper. It is devoted to exposing the definitions and properties of the neutrosophic offuninorms.

Definition 9. *Let c be a neutrosophic component (T_O, I_O or F_O). c: $M_O \rightarrow [\Psi, \Omega]$, where $\Psi \leq 0$ and $\Omega \geq 1$. The neutrosophic component n-offuninorm $N_O^u : [\Psi, \Omega]^2 \rightarrow [\Psi, \Omega]$ satisfies the following conditions for any elements x, y, and z $\in M_O$:*

i. *There exists $c(e) \in M_O$, such that $N_O^u(c(x), c(e)) = c(x)$ (Identity),*
ii. *$N_O^u(c(x), c(y)) = N_O^u(c(y), c(x))$ (Commutativity),*
iii. *If $c(x) \leq c(y)$ then $N_O^u(c(x), c(z)) \leq N_O^u(c(y), c(z))$ (Monotonicity),*
iv. *$N_O^u\left(N_O^u(c(x), c(y)), c(z)\right) = N_O^u\left(c(x), N_O^u(c(y), c(z))\right)$ (Associativity).*

The definition of a neutrosophic uninorm is an especial case of neutrosophic offuninorm when $\Psi = 0$ and $\Omega = 1$ (see Definition 8) and, additionally, we are dealing with single-valued neutrosophic sets.

It is easy to prove that the neutral element *e* is unique.

Let *c* be a neutrosophic component (T_O, I_O or F_O). *c*: $M_O \rightarrow [\Psi, \Omega]$, where $\Psi \leq 0$ and $\Omega \geq 1$. Let us define four useful functions, $\varphi_1 : [\Psi, c(e)] \rightarrow [\Psi, \Omega]$, $\varphi_1^{-1} : [\Psi, \Omega] \rightarrow [\Psi, c(e)]$, $\varphi_2 : [c(e), \Omega] \rightarrow [\Psi, \Omega]$, and $\varphi_2^{-1} : [\Psi, \Omega] \rightarrow [c(e), \Omega]$, defined in Equations (2)–(5), respectively.

$$\varphi_1(c(x)) = \left(\frac{\Omega - \Psi}{c(e) - \Psi}\right)(c(x) - \Psi) + \Psi \tag{2}$$

$$\varphi_1^{-1}(c(x)) = \left(\frac{c(e) - \Psi}{\Omega - \Psi}\right)(c(x) - \Psi) + \Psi \tag{3}$$

$$\varphi_2(c(x)) = \left(\frac{\Omega - \Psi}{\Omega - c(e)}\right)(c(x) - c(e)) + \Psi \tag{4}$$

$$\varphi_2^{-1}(c(x)) = \left(\frac{\Omega - c(e)}{\Omega - \Psi}\right)(c(x) - \Psi) + c(e) \tag{5}$$

where, the superscript −1 means it is an inverse mapping. If the condition $c(e) \in (\Psi, \Omega)$ is fulfilled, then the degenerate cases $\Omega = \Psi$, $c(e) = \Psi$ and $c(e) = \Omega$ are excluded. Therefore, $\varphi_1(c(x))$ and $\varphi_2(c(x))$ are well-defined non-constant linear functions. Thus, they are bijective and have inverse mappings defined in Equations (3) and (5), respectively, in the sense that for $c(x) \in [\Psi, \Omega]$, then $\varphi_1\left(\varphi_1^{-1}(c(x))\right) = c(x)$ and $\varphi_2\left(\varphi_2^{-1}(c(x))\right) = c(x)$. Whereas, for $c(x) \in [\Psi, c(e)]$, we have $\varphi_1^{-1}(\varphi_1(c(x))) = c(x)$ and for $c(x) \in [c(e), \Omega]$, $\varphi_2^{-1}(\varphi_2(c(x))) = c(x)$. These properties can be easily verified. Also, it is trivial that they are non-decreasing mappings.

Additionally, let $U_C, U_D : [\Psi, \Omega]^2 \rightarrow [\Psi, \Omega]$ be two operators defined by Equations (6) and (7), respectively,

$$U_C(c(x), c(y)) = \begin{cases} \varphi_1^{-1}\left(\varphi_1(c(x)) \overset{\wedge}{_O} \varphi_1(c(y))\right), & \text{if } c(x), c(y) \in [\Psi, c(e)] \\ \varphi_2^{-1}\left(\varphi_2(c(x)) \overset{\vee}{_O} \varphi_2(c(y))\right), & \text{if } c(x), c(y) \in [c(e), \Omega] \\ \min(c(x), c(y)), & \text{otherwise} \end{cases} \tag{6}$$

$$U_D(c(x),c(y)) = \begin{cases} \varphi_1^{-1}\left(\varphi_1(c(x)) \hat{O} \varphi_1(c(y))\right), \text{ if } c(x),c(y) \in [\Psi,\ c(e)] \\ \varphi_2^{-1}\left(\varphi_2(c(x)) \overset{\vee}{O} \varphi_2(c(y))\right), \text{ if } c(x),c(y) \in [c(e),\ \Omega] \\ \max(c(x),c(y)), \text{ otherwise} \end{cases} \quad (7)$$

where, $\hat{O}$ denotes a neutrosophic component n-offnorm and $\overset{\vee}{O}$ denotes a neutrosophic component n-offconorm.

Lemma 1. *Let c be a neutrosophic component (T_O, I_O or F_O). c: $M_O \rightarrow [\Psi, \Omega]$, where $\Psi \leq 0$ and $\Omega \geq 1$. Given $\hat{O}$ a neutrosophic component n-offnorm and $\overset{\vee}{O}$ a neutrosophic component n-offconorm, let us consider $U_C(c(x),c(y))$ and $U_D(c(x),c(y))$ the operators defined in Equations (6) and (7) for $c(e) \in (\Psi,\ \Omega)$. They are commutative, non-decreasing, and c(e) is the neutral element.*

Proof.

i. Commutativity is evidently satisfied due to the commutativity of $\hat{O}$, $\overset{\vee}{O}$, *min*, and *max*.

ii. $\varphi_1(\cdot)$, $\varphi_1^{-1}(\cdot)$, $\varphi_2(\cdot)$, $\varphi_2^{-1}(\cdot)$, $\hat{O}$, $\overset{\vee}{O}$, *min* and *max* are non-decreasing mappings, thus both $U_C(\cdot,\ \cdot)$ and $U_D(\cdot,\ \cdot)$ satisfy monotonicity.

iii. To prove *c(e)* is the neutral element, we have two cases, which are the following:

- If $c(x) \in [\Psi,\ c(e)]$, then, $U_C(c(e),c(x)) = U_D(c(e),c(x)) = \varphi_1^{-1}\left(\varphi_1(c(e)) \hat{O} \varphi_1(c(x))\right) = \varphi_1^{-1}\left(\Omega \hat{O} \varphi_1(c(x))\right) = \varphi_1^{-1}(\varphi_1(c(x))) = c(x)$.
- If $c(x) \in [c(e),\ \Omega]$, then $U_C(c(e),c(x)) = U_D(c(e),c(x)) = \varphi_2^{-1}\left(\varphi_2(c(e)) \overset{\vee}{O} \varphi_2(c(x))\right)\varphi_2^{-1}\left(\Psi \overset{\vee}{O} \varphi_2(c(x))\right) = \varphi_2^{-1}(\varphi_2(c(x))) = c(x)$.

Therefore, identity is satisfied. □

Lemma 2. *Let c be a neutrosophic component (T_O, I_O, or F_O). c: $M_O \rightarrow [\Psi, \Omega]$, where $\Psi \leq 0$ and $\Omega \geq 1$. Given $\hat{O}$ a neutrosophic component n-offnorm and $\overset{\vee}{O}$ a neutrosophic component n-offconorm, let us consider $U_C(c(x),c(y))$ and $U_D(c(x),c(y))$ the operators defined in Equations (6) and (7) for $c(e) \in (\Psi,\ \Omega)$. They are associative.*

Proof. Four cases are possible:

i. Let $c(x), c(y), c(z) \in [\Psi,\ c(e)]$, then $U_C(U_C(c(x),c(y)),c(z)) = \varphi_1^{-1}\left(\varphi_1\left(\varphi_1^{-1}\left(\varphi_1(c(x)) \hat{O} \varphi_1(c(y))\right)\right) \hat{O} \varphi_1(c(z))\right) = \varphi_1^{-1}\left(\left[\varphi_1(c(x)) \hat{O} \varphi_1(c(y))\right] \hat{O} \varphi_1(c(z))\right) = \varphi_1^{-1}\left(\varphi_1(c(x)) \hat{O} \left[\varphi_1(c(y)) \hat{O} \varphi_1(c(z))\right]\right) = \varphi_1^{-1}\left(\varphi_1(c(x)) \hat{O} \left[\varphi_1\left(\varphi_1^{-1}\left[\varphi_1(c(y)) \hat{O} \varphi_1(c(z))\right]\right)\right]\right) = \varphi_1^{-1}\left(\varphi_1(c(x)) \hat{O} \varphi_1(U_C(c(y),c(z)))\right) = U_C(c(x),U_C(c(y),c(z)))$.

ii. Let $c(x), c(y), c(z) \in [c(e),\ \Omega]$, $U_C(U_C(c(x),c(y)),c(z)) = \varphi_2^{-1}\left(\varphi_2\left(\varphi_2^{-1}\left(\varphi_2(c(x)) \overset{\vee}{O} \varphi_2(c(y))\right)\right) \overset{\vee}{O} \varphi_2(c(z))\right) = \varphi_2^{-1}\left(\left[\varphi_2(c(x)) \overset{\vee}{O} \varphi_2(c(y))\right] \overset{\vee}{O} \varphi_2(c(z))\right) =$

$\varphi_2^{-1}\Big(\varphi_2(c(x)) \overset{\vee}{O} \Big[\varphi_2(c(y)) \overset{\vee}{O} \varphi_2(c(z))\Big]\Big) = \varphi_2^{-1}\Big(\varphi_2(c(x)) \overset{\vee}{O} \Big[\varphi_2\Big(\varphi_2^{-1}\Big[\varphi_2(c(y)) \overset{\vee}{O} \varphi_2(c(z))\Big]\Big)\Big]\Big) =$
$\varphi_2^{-1}\Big(\varphi_2(c(x)) \overset{\vee}{O} \varphi_2(U_C(c(y), c(z)))\Big) = U_C(c(x), U_C(c(y), c(z)))$. These proofs are also valid for U_D.

iii. Let $c(x), c(y) \in [\Psi,\ c(e)]$ and $c(z) \in [c(e),\ \Omega]$, $U_C(U_C(c(x), c(y)), c(z)) = \min(U_C(c(x), c(y)), c(z)) = U_C(c(x), c(y))$. Also, we have $U_C(c(x), U_C(c(y), c(z))) = U_C(c(x), \min(c(y), c(z))) = U_C(c(x), c(y))$, then, it is associative.

iv. Let $c(x), c(y) \in [c(e),\ \Omega]$ and $c(z) \in [\Psi,\ c(e)]$, then $U_C(U_C(c(x), c(y)), c(z)) = \min(U_C(c(x), c(y)), c(z)) = c(z)$. In addition, $U_C(c(x), (U_C(c(y), c(z)))) = U_C(c(x), \min(c(y), c(z))) = U_C(c(x), c(z)) = \min(c(x), c(z)) = c(z)$.

Thus, U_C satisfies the associativity.
Similarly, associativity of U_D can be proved.

Let us remark that we applied the properties, $c(x) \hat{O} c(y) \le \min(c(x),\ c(y))$ and $c(x) \overset{\vee}{O} c(y) \ge \max(c(x),\ c(y))$, as well as $U_C(c(x), c(y)) \le c(e)$ if $c(x),\ c(y) \in [\Psi,\ c(e)]$ and $U_C(c(x), c(y)) \ge c(e)$ if $c(x),\ c(y) \in [c(e),\ \Omega]$. □

Proposition 3. *Let c be a neutrosophic component (T_O, I_O, or F_O). c: $M_O \to [\Psi,\ \Omega]$, where $\Psi \le 0$ and $\Omega \ge 1$. Given $\hat{O}$ a neutrosophic component n-offnorm and $\overset{\vee}{O}$ a neutrosophic component n-offconorm, let us consider $U_C(c(x), c(y))$ and $U_D(c(x), c(y))$ the operators defined in Equations 6 and 7 for $c(e) \in (\Psi,\ \Omega)$. Then, $U_C(c(x), c(y))$ and $U_D(c(x), c(y))$ are neutrosophic component n-offuninorms and they satisfy the conditions $U_C(\Psi,\ \Omega) = \Psi$ and $U_D(\Psi,\ \Omega) = \Omega$,* i.e., U_C is a conjunctive neutrosophic component n-offuninorm, and U_D is a disjunctive neutrosophic component n-offuninorm.

Proof. Since Lemma 1, they are commutative, non-decreasing operators, and *c(e)* is the neutral element. Since Lemma 2, they are associative operators. Moreover, it is easy to verify that $U_C(\Psi,\ \Omega) = \Psi$ and $U_D(\Psi,\ \Omega) = \Omega$. □

Example 3. *Two neutrosophic component n-offuninorms can be defined as:*

$$U_{ZC}(c(x), c(y)) = \begin{cases} \varphi_1^{-1}\Big(\varphi_1(c(x)) \overset{\wedge}{ZO} \varphi_1(c(y))\Big), \text{ if } c(x), c(y) \in [\Psi,\ c(e)] \\ \varphi_2^{-1}\Big(\varphi_2(c(x)) \overset{\vee}{ZO} \varphi_2(c(y))\Big), \text{ if } c(x), c(y) \in [c(e),\ \Omega] \\ \min(c(x), c(y)), \text{ otherwise} \end{cases}$$

$$U_{ZD}(c(x), c(y)) = \begin{cases} \varphi_1^{-1}\Big(\varphi_1(c(x)) \overset{\wedge}{ZO} \varphi_1(c(y))\Big), \text{ if } c(x), c(y) \in [\Psi,\ c(e)] \\ \varphi_2^{-1}\Big(\varphi_2(c(x)) \overset{\vee}{ZO} \varphi_2(c(y))\Big), \text{ if } c(x), c(y) \in [c(e),\ \Omega] \\ \max(c(x), c(y)), \text{ otherwise} \end{cases}$$

where $\overset{\wedge}{ZO}$ and $\overset{\vee}{ZO}$ were defined in the Example 1; $c(e) \in (\Psi, \Omega)$.

Then two examples of n-offuninorms are: $U_1(\langle T_1, I_1, F_1\rangle, \langle T_2, I_2, F_2\rangle) = \langle U_{ZC}(T_1, T_2), U_{ZD}(I_1, I_2), U_{ZD}(F_1, F_2)\rangle$ and $U_2(\langle T_1, I_1, F_1\rangle, \langle T_2, I_2, F_2\rangle) = \langle U_{ZD}(T_1, T_2), U_{ZC}(I_1, I_2), U_{ZC}(F_1, F_2)\rangle$.
They satisfy $U_1(\langle\Psi,\ \Omega,\ \Omega\rangle, \langle\Omega, \Psi, \Psi\rangle) = \langle\Psi,\ \Omega, \Omega\rangle$ and $U_2(\langle\Psi,\ \Omega,\ \Omega\rangle, \langle\Omega, \Psi, \Psi\rangle) = \langle\Omega,\ \Psi, \Psi\rangle$.

Example 4. *Two neutrosophic component n-offuninorms can be defined as*

$$U_{LC}(c(x),c(y)) = \begin{cases} \varphi_1^{-1}\left(\varphi_1(c(x)) \underset{LO}{\wedge} \varphi_1(c(y))\right), \text{ if } c(x),c(y) \in [\Psi,\ c(e)] \\ \varphi_2^{-1}\left(\varphi_2(c(x)) \underset{LO}{\vee} \varphi_2(c(y))\right), \text{ if } c(x),c(y) \in [c(e),\ \Omega] \\ \min(c(x),c(y)), \text{ otherwise} \end{cases}$$

$$U_{LD}(c(x),c(y)) = \begin{cases} \varphi_1^{-1}\left(\varphi_1(c(x)) \underset{LO}{\wedge} \varphi_1(c(y))\right), \text{ if } c(x),c(y) \in [\Psi,\ c(e)] \\ \varphi_2^{-1}\left(\varphi_2(c(x)) \underset{LO}{\vee} \varphi_2(c(y))\right), \text{ if } c(x),c(y) \in [c(e),\ \Omega] \\ \max(c(x),c(y)), \text{ otherwise} \end{cases}$$

where $\underset{LO}{\wedge}$ *and* $\underset{LO}{\vee}$ *were defined in the Example 2;* $c(e)\in(\Psi,\Omega)$.

Now, two examples of n-offuninorms are: $U_3(\langle T_1,I_1,F_1\rangle,\langle T_2,I_2,F_2\rangle) = \langle U_{LC}(T_1,T_2),U_{LD}(I_1,I_2),U_{LD}(F_1,F_2)\rangle$ and $U_4(\langle T_1,I_1,F_1\rangle,\langle T_2,I_2,F_2\rangle) = \langle U_{LD}(T_1,T_2),U_{LC}(I_1,I_2),U_{LC}(F_1,F_2)\rangle$.

They satisfy, $U_3(\langle\Psi,\ \Omega,\ \Omega\rangle,\langle\Omega,\Psi,\Psi\rangle) = \langle\Psi,\ \Omega,\Omega\rangle$ and $U_4(\langle\Psi,\Omega,\Omega\rangle,\langle\Omega,\Psi,\Psi\rangle) = \langle\Omega,\Psi,\Psi\rangle$.

Remark 1. *The neutrosophic components n-offuninorms defined by Equations (6) and (7) are idempotent, i.e.,* $N_O^n(c(x),c(x)) = c(x)$, *if and only if they are defined from idempotent neutrosophic component n-offnorms and n-offconorms. Moreover, they are Archimedean, i.e., they satisfy both,* $N_O^u(c(x),c(x)) <_O c(x)$ *when* $\Psi < c(x) < c(e)$ *and* $c(x) <_O N_O^u(c(x),c(x))$ *when* $c(e) < c(x) < \Omega$, *if and only if the neutrosophic component n-offnorm and n-offconorm are Archimedean. Let us observe that* $<_O$ *is the order < defined in the real line when* $c(x)$ *is* $T_O(x)$ *and it is > when* $c(x)$ *is* $I_O(x)$ *or* $F_O(x)$.

Proposition 4. *Let c be a neutrosophic component* (T_O, I_O *or* F_O). $c\colon M_O\to[\Psi,\Omega]$, *where* $\Psi<0$ *and* $\Omega>1$, *and let a neutrosophic component n-offuninorm* $N_O^u : [\Psi,\ \Omega]^2 \to [\Psi,\ \Omega]$. *Then, for every* $x, y \in M_O$, *a neutrosophic component n-offnorm and a neutrosophic component n-offconorm are defined by Equations (8) and (9).*

$$c(x) \underset{UO}{\wedge} c(y) = \varphi_1\left(N_O^u\left(\varphi_1^{-1}(c\,(x)),\varphi_1^{-1}(c(y))\right)\right) \tag{8}$$

$$c(x) \underset{UO}{\vee} c(y) = \varphi_2\left(N_O^u\left(\varphi_2^{-1}(c(x)),\varphi_2^{-1}(c(y))\right)\right) \tag{9}$$

Proof. Evidently, both operators are commutative, since N_O^u is. Also, it is non-decreasing since N_O^u and the functions in Equations (2)–(5) are. They are associative because of the associativity of N_O^u.

It is easy to verify that the overbounding conditions $\Omega \underset{UO}{\wedge} c(y) = c(y)$ and $\Psi \underset{UO}{\vee} c(y) = c(y)$ are also satisfied.

Additionally, we have $\Psi \underset{UO}{\wedge} c(y) = \varphi_1\left(N_O^u\left(\varphi_1^{-1}(\Psi),\varphi_1^{-1}(c(y))\right)\right) = \varphi_1\left(N_O^u\left(\Psi,\varphi_1^{-1}(c(y))\right)\right) \leq \varphi_1\left(N_O^u(\Psi,c(e))\right) = \varphi_1(\Psi) = \Psi$, then, $\Psi \underset{UO}{\wedge} c(y) = \Psi$; also, $\Omega \underset{UO}{\vee} c(y) = \varphi_2\left(N_O^u\left(\varphi_2^{-1}(\Omega),\varphi_2^{-1}(c(y))\right)\right) = \varphi_2\left(N_O^u\left(\Omega,\varphi_2^{-1}(c(y))\right)\right) \geq \varphi_2\left(N_O^u\left(\Omega,\varphi_2^{-1}(\Psi)\right)\right) = \varphi_2\left(N_O^u(\Omega,c(e))\right) = \varphi_2(\Omega) = \Omega$, then, $\Omega \underset{UO}{\vee} c(y) = \Omega$. □

Proposition 5. *Let (T_O, I_O, or F_O), c_O: $M_O \to [\Psi, \Omega]$ and (T, I, or F), c_N: $MN \to [0,1]$ be a neutrosophic component n-offset and a neutrosophic component, respectively. There exists a bijective mapping such that every neutrosophic component n-offuninormis transformed into a neutrosophic component uninorm and vice versa.*

Proof. Let us define the function $\varphi_3 : [\Psi, \Omega] \to [0, 1]$ and its inverse $\varphi_3^{-1} : [0, 1] \to [\Psi, \Omega]$, expressed in Equations (10) and (11), respectively.

$$\varphi_3(c(x)) = \frac{c(x) - \Psi}{\Omega - \Psi} \tag{10}$$

$$\varphi_3^{-1}(c(x)) = (\Omega - \Psi)c(x) + \Psi \tag{11}$$

Evidently, they are increasing bijective mappings.

If $\hat{U}_N(\cdot,\cdot)$ is a neutrosophic uninorm, then we can define the neutrosophic component n-offuninorm $\hat{N}_O^u(\cdot,\cdot)$ as follows:

$$\hat{N}_O^u(c_O(x), c_O(y)) = \varphi_3^{-1}\big(\hat{U}_N(\varphi_3(c_O(x)), \varphi_3(c_O(y)))\big)$$

Conversely, if we have $\hat{N}_O^u(\cdot,\cdot)$, we can define $\hat{U}_N(\cdot,\cdot)$ as follows:

$$\hat{U}_N(c_N(x), c_N(y)) = \varphi_3\big(\hat{N}_O^u\big(\varphi_3^{-1}(c_N(x)), \varphi_3^{-1}(c_N(y))\big)\big)$$

Then, it is easy to prove that $\hat{N}_O^u(c_O(x), c_O(y))$ is a neutrosophic component n-offuninorm and $\hat{U}_N(c_N(x), c_N(y))$ is a neutrosophic component uninorm. Moreover, the relationship between the components of their neutral elements $c_O(e_O)$ and $c_N(e_N)$ is $c_N(e_N) = \varphi_3(c_O(e_O))$ and thus $c_O(e_O) = \varphi_3^{-1}(c_N(e_N))$. □

Let us remark that we maintain the definition of inverse mapping that we explained in Equations (3) and (5).

In agreement with Proposition 5, many predefined neutrosophic uninorms can be used to define n-offuninorms. In turn, fuzzy uninorms can be used to define neutrosophic uninorms, thus, it is simply necessary to find examples in the field of fuzzy uninorms; see further Section 4.1. First, let us make reference to some properties of n-offuninorms.

Proposition 6. *Let c be a neutrosophic component (T_O, I_O or F_O). c: $M_O \to [\Psi, \Omega]$, where $\Psi \leq 0$ and $\Omega \geq 1$. Given the neutrosophic component n-offuninorm $N_O^u : [\Psi, \Omega]^2 \to [\Psi, \Omega]$ and the offuninorm $U_O : [\Psi, \Omega]^3 \times [\Psi, \Omega]^3 \to [\Psi, \Omega]^3$ defined from $N_O^u(\cdot, \cdot)$, $U_O(\langle T_O(x), I_O(x), F_O(x)\rangle, \langle T_O(y), I_O(y), F_O(y)\rangle) = \langle N_O^u(T_O(x), T_O(y)), N_O^u(I_O(x), I_O(y)), N_O^u(F_O(x), F_O(y))\rangle$, satisfies the following properties for any $x = \langle T_O(x), I_O(x), F_O(x)\rangle$, denoting $\Psi_O = \langle \Psi, \Omega, \Omega\rangle$ and $\Omega_O = \langle \Omega, \Psi, \Psi\rangle$:*

1. $U_O(\Psi_O, \Psi_O) = \Psi_O$ *and* $U_O(\Omega_O, \Omega_O) = \Omega_O$.
2. *If $c(e) \neq \Psi, \Omega$, then,* $U_O(\Psi_O, \Omega_O) = U_O(U_O(\Psi_O, \Omega_O), x)$
3. *If $c(e) \neq \Psi, \Omega$, then either $U_O(\Psi_O, \Omega_O) = \Psi_O$ or $U_O(\Psi_O, \Omega_O) = \Omega_O$ or $U_O(\Psi_O, \Omega_O)$ is $\leq_O$-incomparable respect to $e = \langle T_O(e), I_O(e), F_O(e)\rangle$.*
4. *If there exists $y = \langle T_O(y), I_O(y), F_O(y)\rangle$, such that either $x \leq_O e \leq_O y$ or $y \leq_O e \leq_O x$, then, $min(x, y) \leq_O U_O(x, y) \leq_O max(x, y)$.*

Proof.

1. Since $N_O^u(\Psi, c(e)) = \Psi$ and $N_O^u(\Omega, c(e)) = \Omega$ and considering that $N_O^u(\Psi, \cdot)$ and $N_O^u(\Omega, \cdot)$ are non-decreasing, the result is trivial. Then, $U_O(\Psi_O, \Psi_O) = \Psi_O$ and $U_O(\Omega_O, \Omega_O) = \Omega_O$.
2. First suppose $c(x) \leq c(e)$, then $N_O^u(\Psi, c(x)) \leq N_O^u(\Psi, c(e)) = \Psi$, therefore $N_O^u(\Psi, c(x)) = \Psi$, thus $N_O^u(\Psi, \Omega) = N_O^u\big(N_O^u(\Psi, c(x)), \Omega\big) = N_O^u\big(\Omega, N_O^u(\Psi, c(x))\big) = N_O^u\big(N_O^u(\Omega, \Psi), c(x)\big)$

$= N_O^u(N_O^u(\Psi, \Omega), c(x))$. See that we applied the commutativity and associativity of $N_O^u(\cdot, \cdot)$. Now, suppose $c(e) \leq c(x)$, then $N_O^u(c(x), \Omega) \geq N_O^u(c(x), \Omega) = \Omega$, therefore, $N_O^u(c(x), \Omega) = \Omega$, and $N_O^u(\Psi, \Omega) = N_O^u(\Psi, N_O^u(c(x), \Omega)) = N_O^u(N_O^u(\Omega, \Psi), c(x))$. Suppose x and $e = T_O(e), I_O(e), F_O(e)$ are $\leq_O$-incomparable, i.e., $x \nleq_O e$ and $e \nleq_O x$. Then,

$$x \underset{O}{\wedge} e = \min(T_O(x), T_O(e)), \max(I_O(x), I_O(e)), \max(F_O(x), F_O(e)) \leq_O x$$
$$\leq_O \max(T_O(x), T_O(e)), \min(I_O(x), I_O(e)), \min(F_O(x), F_O(e)) = x \underset{O}{\vee} e$$

Then, according to the previous results we have $U_O(\Psi_O, \Omega_O) = U_O\left(U_O(\Psi_O, \Omega_O), x \underset{O}{\wedge} e\right) = U_O\left(U_O(\Psi_O, \Omega_O), x \underset{O}{\vee} e\right)$, thus, for the increasing condition of $U_O(\cdot, \cdot)$ it is satisfied $U_O(\Psi_O, \Omega_O) = U_O(U_O(\Psi_O, \Omega_O), x)$. Then, we proved $U_O(\Psi_O, \Omega_O) = U_O(U_O(\Psi_O, \Omega_O), x)$.

3. Suppose $U_O(\Psi_O, \Omega_O)$ is $\leq_O$-comparable respect to e, then, if $U_O(\Psi_O, \Omega_O) \leq_O e$ since the previous proof $U_O(\Psi_O, \Omega_O) = U_O(U_O(\Psi_O, \Omega_O), \Psi_O) = \Psi_O$. If $e \leq_O U_O(\Psi_O, \Omega_O)$ then $U_O(\Psi_O, \Omega_O) = U_O(U_O(\Psi_O, \Omega_O), \Omega_O) = \Omega_O$.
4. Let us assume without loss of generality that $x \leq_O e \leq_O y$, then, $x = U_O(x, e) \leq_O U_O(x, y) \leq_O U_O(e, y) = y$. □

When c_1: $M_O \to [\Psi_1, \Omega_1]$ and c_2: $M_O \to [\Psi_2, \Omega_2]$ are two neutrosophic components, such that $\Psi_1 \neq \Psi_2$ or $\Omega_1 \neq \Omega_2$, satisfying that at least one of Ψ_1 and Ψ_2 is smaller than 0, or at least one of Ω_1 and Ω_2 is bigger than 1, then, a neutrosophic component n-offuninorm aggregates both of them, according to the interpretation we have to obtain.

For example, if c_1: MO→[−1,1] and c2: MO→[0,1], and the first one means the relationship between two variables like the linear regression coefficient and the second one represents a classical probability, if we need to obtain the aggregation in [−1,1] in the framework of variable relationships, then after transforming $c_2 : M_O \to [0, 1]$ to $\hat{c}_2 : M_O \to [-1, 1]$, we aggregate c_1 and $\hat{c}_2$ using $N_O^u : [-1, 1]^2 \to [-1, 1]$, only in the case that it makes sense to rescale c_2, otherwise, because [0,1] ⊂[−1,1], we can apply $N_O^u : [-1, 1]^2 \to [-1, 1]$ over c_1 and c_2.

However, if we need to obtain a classical probabilistic interpretation, then we aggregate c_2: $M_O \to [0,1]$ and $\hat{c}_1 : M_O \to [0, 1]$, where $\hat{c}_1$ is a transformation obtained from c_1: $M_O \to [-1, 1]$.

Example 5. *Let us revisit Example 3 with* $U_1 : [-0.7, 1.2]^3 \times [-0.7, 1.2]^3 \to [-0.7, 1.2]^3$ *and neutral element* $e = \langle -0.5, 0, 0 \rangle$, *defined as* $U_1(\langle T_1, I_1, F_1 \rangle, \langle T_2, I_2, F_2 \rangle) = \langle U_{ZC}(T_1, T_2), U_{ZD}(I_1, I_2), U_{ZD}(F_1, F_2) \rangle$. Then, we have:

$$U_{LC}(T_O(x), T_O(y)) = \begin{cases} \max(T_O(x), T_O(y)), & \text{if } T_O(x), T_O(y) \in [-0.5, 1.2] \\ \min(T_O(x), T_O(y)), & \text{otherwise} \end{cases}$$

$$U_{LD}(I_O(x), I_O(y)) = \begin{cases} \min(I_O(x), I_O(y)), & \text{if } I_O(x), I_O(y) \in [-0.7, 0] \\ \max(I_O(x), I_O(y)), & \text{otherwise} \end{cases}$$

$$U_{LD}(F_O(x), F_O(y)) = \begin{cases} \min(F_O(x), F_O(y)), & \text{if } F_O(x), F_O(y) \in [-0.7, 0] \\ \max(F_O(x), F_O(y)), & \text{otherwise} \end{cases}$$

Let us aggregate the elements of $A = \{(x_1, \langle 1.2, 0.4, -0.1 \rangle), (x_2, \langle 0.2, 0.3, -0.7 \rangle)\}$ by using $U_1(\cdot, \cdot)$, then, $U_1((x_1, \langle 1.2, 0.4, -0.1 \rangle), (x_2, \langle 0.2, 0.3, -0.7 \rangle)) = \langle U_{LC}(1.2, 0.2), U_{LD}(0.4, 0.3), U_{LD}(-0.1, -0.7) \rangle = \langle 1.2, 0.4, -0.7 \rangle$.

4. Applications

In the following, we illustrate the applicability of the present investigation aided by three areas of application.

4.1. N-Offuninorms and MYCIN

Let us start with the parameterized Silvert uninorms, see [40]:

$$u_{N\lambda}(c_N(x), c_N(y)) = \begin{cases} \frac{\lambda c_N(x)c_N(y)}{\lambda c_N(x)c_N(y)+(1-c_N(x))(1-c_N(y))}, & \text{if } (c_N(x),\ c_N(y)) \in [0,\ 1]^2 \backslash \{(0,\ 1), (1,\ 0)\} \\ 0, & \text{otherwise} \end{cases}$$

where $\lambda > 0$ and $c_N(e_\lambda) = \frac{1}{\lambda+1}$. To convert this family to the equivalent one defined into $[-1, 1]$ we have to apply the Equations in Proposition 5. Then, it is obtained $u_{O\lambda}(c_O(x), c_O(y)) = \begin{cases} \frac{(\lambda-1)(1+c_O(x)c_O(y))+(\lambda+1)(c_O(x)+c_O(y))}{(\lambda+1)(1+c_O(x)c_O(y))+(\lambda-1)(c_O(x)+c_O(y))}, & \text{if } (c_O(x),\ c_O(y)) \in [-1,\ 1]^2 \backslash \{(-1,\ 1), (1,\ -1)\} \\ 0, & \text{otherwise} \end{cases}$ where $c_O(e_\lambda) = \frac{1-\lambda}{1+\lambda}$.

Let us note that $\lim_{\lambda\to 0^+} c_O(e_\lambda) = 1$ and $\lim_{\lambda\to +\infty} c_O(e_\lambda) = -1$. Therefore, the closer λ approximates to 0, the closer $u_{O\lambda}(\cdot,\cdot)$ performs like a neutrosophic component n-offnorm; whereas, the greater λ, the closer $u_{O\lambda}(\cdot,\cdot)$ performs like a neutrosophic component n-offconorm.

An additional consequence of these assertions is that inequalities $0<\lambda_1<\lambda_2$ imply $u_{O\lambda_1}(c_O(x), c_O(y)) < u_{O\lambda_2}(c_O(x), c_O(y))$.

Applying Equations (2)–(5) to the conditions of the present example, the following transformations are obtained:

$\varphi_{1\lambda}(c_O(x)) = (1+\lambda)c_O(x) + \lambda$, $\varphi_{1\lambda}^{-1}(c_O(x)) = \frac{c_O(x)-\lambda}{1+\lambda}$, $\varphi_{2\lambda}(c_O(x)) = \frac{(1+\lambda)c_O(x)-1}{\lambda}$ and $\varphi_{2\lambda}^{-1}(c_O(x)) = \frac{\lambda c_O(x)+1}{1+\lambda}$.

Then, a neutrosophic component n-offnorm and a neutrosophic component n-offconorm are defined from Equations (8) and (9), as follows:

$c(x) \underset{\lambda O}{\wedge} c(y) = \varphi_{1\lambda}\left(u_{O\lambda}\left(\varphi_{1\lambda}^{-1}(c(x)), \varphi_{1\lambda}^{-1}(c(y))\right)\right)$ and $c(x) \underset{\lambda O}{\vee} c(y) = \varphi_{2\lambda}\left(u_{O\lambda}\left(\varphi_{2\lambda}^{-1}(c(x)), \varphi_{2\lambda}^{-1}(c(y))\right)\right)$, respectively.

Other properties of $u_{O\lambda}(\cdot,\cdot)$ are the following:

1. $u_{O\lambda}(c_O(x), -c_O(x)) = \begin{cases} \frac{\lambda-1}{1+\lambda}, & \text{if } c_O(x) \in (-1,\ 1) \\ -1, & \text{otherwise} \end{cases}$
2. $u_{O\lambda}(\cdot,\cdot)$ is Archimedean. To prove it, given $c_O(x) < c_O(e_\lambda)$, then $u_{O\lambda}(c_O(x), c_O(x)) \leq u_{O\lambda}(c_O(x), c_O(e_\lambda)) = c_O(x)$ and if $c_O(x) > c_O(e_\lambda)$, $u_{O\lambda}(c_O(x), c_O(x)) \geq u_{O\lambda}(c_O(x), c_O(e_\lambda)) = c_O(x)$.

To prove those inequalities are strict, let us suppose the equation $u_{O\lambda}(c_O(x), c_O(x)) = \frac{(\lambda-1)\left(1+c_O^2(x)\right)+2(\lambda+1)c_O(x)}{(\lambda+1)\left(1+c_O^2(x)\right)+2(\lambda-1)c_O(x)} = c_O(x)$ holds, or equivalently $(\lambda-1)\left(1+c_O^2(x)\right) + 2(\lambda+1)c_O(x) = c_O(x)\left[(\lambda+1)\left(1+c_O^2(x)\right)+2(\lambda-1)c_O(x)\right]$, thus, $(\lambda-1)\left(1-c_O^2(x)\right) + (\lambda+1)c_O(x)\left(1-c_O^2(x)\right) = 0$ and finally, $\left(1-c_O^2(x)\right)(\lambda-1+(\lambda+1)c_O(x)) = 0$, hence the solutions are $c_O(x) = \pm 1$ and $c_O(x) = c_O(e_\lambda)$. Then, we conclude it is Archimedean.

A remarkable case is λ = 1, which converts into Equation (12).

$$u_{O1}(c_O(x), c_O(y)) = \begin{cases} \frac{c_O(x)+c_O(y)}{1+c_O(x)c_O(y)}, & \text{if } (c_O(x),\ c_O(y)) \in [-1,\ 1]^2 \backslash \{(-1,\ 1), (1,\ -1)\} \\ -1, & \text{otherwise} \end{cases} \quad (12)$$

$u_{O1}(\cdot,\cdot)$ is the function called PROSPECTOR which aggregates hypothesis values or Certainty Factors (CF) related to MYCIN, the well-known medical Expert System; nevertheless, the function used in MYCIN is undefined for the arguments (−1, 1) and (1, −1), see [32–34]. Summarizing, we can say that PROSPECTOR is a neutrosophic component n-offuninorm, such that $c_O(e_1) = 0$, which is an effective and widely used aggregation operator.

$u_{O1}(\cdot,\cdot)$ means the combination of the CFs of two independent experts about the hypothesis H. CF = –1.0 means expert has 100% evidence against H and CF = 1.0 means he or she has 100% evidence to support H. The smaller the CF, the greater the evidence against H; the larger the CF, the greater the evidence supporting H; whereas evidence with degree close to 0 means a borderline degree of evidence. Here, $u_{O1}(c_O(x), -c_O(x)) = 0$, where $u_{O1}(-1, 1) = u_{O1}(1, -1) = -1$ for meaning that the 100% contradiction is assessed as 100% against H. The original $u_{O1}(\cdot,\cdot)$ in [32] accepts they are undefined.

Another function is the *Modified Combining Function* C(x,y), see [34], defined as

$$C(x,y) = \begin{cases} x + y(1-x), \text{ if } \min(x,y) \geq 0 \\ \frac{x+y}{1-\min(|x|,|y|)}, \text{ if } \min(x,y) < 0 < \max(x,y) \\ x + y(1+x), \text{ if } \max(x,y) \leq 0 \end{cases}$$

The components n-offnorm and n-offconorm obtained from the PROSPECTOR are the following: $c_O(x) \underset{1O}{\wedge} c_O(y) = \frac{4(c_O(x)+c_O(y)-2)}{4+(c_O(x)-1)(c_O(y)-1)} + 1$ and $c_O(x) \underset{1O}{\vee} c_O(y) = \frac{4(c_O(x)+c_O(y)+2)}{4+(c_O(x)+1)(c_O(y)+1)} - 1$, respectively, see Figures 1 and 2.

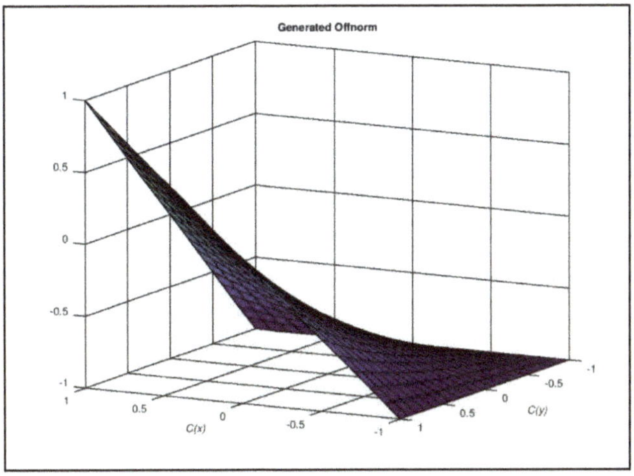

Figure 1. Depiction of the neutrosophic component n-offnorm generated by $u_{O1}(\cdot,\cdot)$.

Hitherto we mostly calculated on neutrosophic components, nevertheless n-offuninorms have to be defined for the three components altogether. For example, given $x, y \in [-1, 1]^3$, $U_{N\lambda}(x,y) = \langle u_{O\lambda_1}(T_O(x), T_O(y)), u_{O\lambda_2}(I_O(x), I_O(y)), u_{O\lambda_3}(F_O(x), F_O(y))\rangle$ is an n-offuninorm, which evidently it is not conjunctive, neither is it disjunctive, see that $U_{N\lambda}(\langle -1, 1, 1\rangle, \langle 1, -1, -1\rangle) = \langle -1, -1, -1\rangle$.

Conjunctive and disjunctive neutrosophic component n-offuninorms were illustrated in Example 3; see also Example 5. Example 6 is a hypothetical example to explain the use of this theory in a real-life situation.

Example 6. *Three physicians, denoted by A, B, and C, have to emit a criterion about a patient's disease which suffers from somewhat confusing symptoms. They agree that the Certainty Factor is the better way to express* *their opinions. They use single-valued neutrosophic offsets, instead of a simple CF to increase the accuracy of the criteria.*

After a discussion, they are convinced that it is most likely that the patient has either a thyroid disease or an infectious one. The treatment for each disease is different each other. Therefore, they have two hypotheses; one is H_T which means the patient has thyroid disease and H_I that patient has an infectious disease.

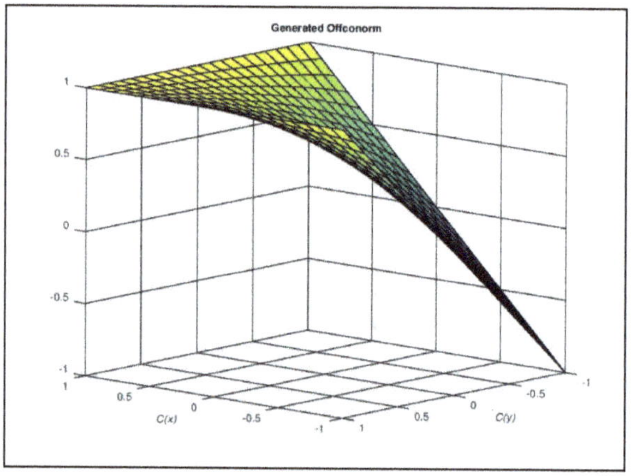

Figure 2. Depiction of the neutrosophic component n-offconorm generated by $u_{O1}(\cdot,\cdot)$.

Physician A thinks that the probability they are dealing with a thyroid disease is $A_T = <-0.6, 0.4, 0.6>$ and that it is an infectious disease is $A_I = <0.8, -0.5, -0.8>$, thus, A is 60% against H_T and 40% undecided about it; however, A is 80% in favor of H_I and 50% sure about it.

Similarly, we have that B's criteria are, $B_T = <-0.1, -0.2,0.1>$ and $B_I = <0.1, 0.8, -0.1>$, whereas C's criteria are $C_T = <0.7, 0.1, -0.2>$ and $C_I = <-0.6, -0.3, 0.7>$.

To decide what is the strongest hypothesis, H_T or H_I, they select the well-known PROSPECTOR function used in MYCIN (see Equation (12)) for each component.

Thus, for H_T we have an aggregated value equal to <0.073684, 0.31064, 0.53043> and for H_I it is <0.46667, 0.23529, −0.32>, therefore, evidently, the infectious disease is the strongest hypothesis, because $\langle 0.073684, 0.31064, 0.53043\rangle <_O \langle 0.46667, 0.23529, -0.32\rangle$.

Despite we proved in Proposition 5 that neutrosophic uninorms are mathematically equivalent to offuninorms, it is worthwhile to remark that the reason for using an interval different of [0, 1] is that it could be useful to model real-life problems. The present example is a good one to explain that reason. The advantages arise from the accuracy and compactness of an expert's information. In this example, from an expert's viewpoint, it is easier to express opinions in the scale [−1, 1] with the aforementioned meaning than in the scale [0, 1], which is less clear. Information compactness is given because of only a single offset is semantically equivalent to at least two neutrosophic sets.

Additionally, because of the significance of functions like $u_{O1}(\cdot,\cdot)$ and $C(x,y)$, which were used as aggregation functions in that well-known expert system, some authors have extended the domain of fuzzy uninorms to any interval [a, b], not necessarily restricted to a = 0 and b = 1; see [33,34].

This fact supports the usefulness of the present work, where for the first time the precedent ideas on extending the truth values beyond the scope of [0, 1] naturally associate with the offset concept maintaining the original definitions of the aggregation functions used in MYCIN.

Another powerful reason is the applicability of $u_{O1}(\cdot,\cdot)$ and $C(x,y)$, and hence of the fuzzy uninorms defined in [a, b], as threshold functions of artificial neurons in Artificial Neural Networks, as well as to Fuzzy Cognitive Maps, which are used in fields like decision making, forecasting, and strategic planning [33].

Such applications of uninorms in the fuzzy domain can be explored in the framework of neutrosophy theory, e.g., in Artificial Neural Networks based on neutrosophic sets, in Neutrosophic Cognitive Maps, among others [36,37].

4.2. *N-Offuninorms and Implicators*

Fuzzy uninorms are used to define implicators (see [41], pp. 151–160). This application was extended to neutrosophic uninorms ([25]). To extend the implication operator in the offuninorm framework, first, we need to consider the notion of offimplication, which has been defined symbolically.

The *Symbolic Neutrosophic Offlogic Operators* or briefly the *Symbolic Neutrosophic Offoperators* extend the Symbolic Neutrosophic Logic Operators, where every one of T, I, F has an under and an over version (see [31], pp. 132–139).

T_O = Over Truth,
T_U = Under Truth;
I_O = Over Indeterminacy,
I_U = Under Indeterminacy;
F_O = Over Falsehood,
F_U = Under Falsehood.

Let $S_N = \{T_O, T, T_U, I_O, I, I_U, F_O, F, F_U\}$ be the set of neutrosophic symbols, an order is defined in S_N as follows: if '<' denotes "more important than", we have the following order, $T_U < I_U < F_U < F < I < T < F_O < I_O < T_O$, where $-\infty < T_U < I_U < F_U < 0$, $0 \leq F < I < T \leq 1$ and $1 < F_O < I_O < T_O < +\infty$; see Figure 3. Let us note that the proposed order is not the unique one, it depends on the decision maker's objective.

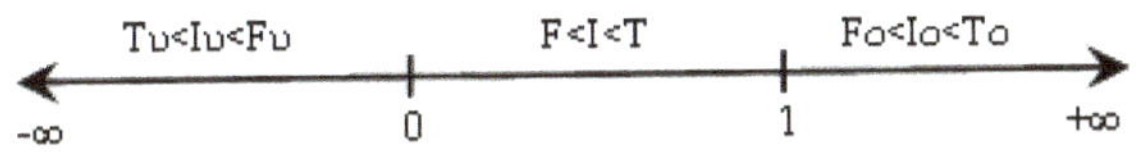

Figure 3. Ordered symbolic neutrosophic components in the neutrosophic offlogic.

Let us observe that I is the center of the elements according to <. For every $\alpha \in S_N$, the *symbolic neutrosophic offcomplement* is denoted by $C_O(\alpha)$ and it is defined as the symmetric element respect to the median centered in I, e.g., $C_{SO}(F_O) = F_U$ and $C_{SO}(F) = T$, hence, given $\alpha \in S_N$ its *symbolic neutrosophic offnegation* is $\underset{SO}{\neg}\alpha = C_{SO}(\alpha)$.

Additionally, for any $\alpha, \beta \in S_N$ the *symbolic neutrosophic offconjunction* is defined as $\alpha \underset{SO}{\wedge} \beta = \min(\alpha, \beta)$, the *symbolic neutrosophic offdisjunction* is defined as $\alpha \underset{SO}{\vee} \beta = \max(\alpha, \beta)$, whereas the *symbolic neutrosophic offimplication* is defined in Equation (13).

$$\alpha \underset{SO}{\rightarrow} \beta = \left(\underset{SO}{\neg}\alpha\right) \underset{SO}{\vee} \beta \tag{13}$$

In this paper, we redefine some of the symbolic neutrosophic offoperators to the continuous quantitative domain. Given $\overline{\alpha} \in [\Psi, \Omega]$, where $\Psi < 0$ or $\Omega > 1$, the *neutrosophic offnegation* is defined by Equation (14).

$$\underset{O}{\neg}\overline{\alpha} = \begin{cases} \min\{\Omega, 1 - \overline{\alpha}\}, \text{ if } \overline{\alpha} \leq 0.5 \\ \max\{\Psi, 1 - \overline{\alpha}\}, \text{ if } \overline{\alpha} > 0.5 \end{cases} \tag{14}$$

The *neutrosophic offnegation* satisfies the following properties:

1. It is a non-increasing operator, which extends the classical negation operator in fuzzy logic theory. It is strictly decreasing when $\Omega + \Psi = 1$.
2. It extends the notion of symbolic neutrosophic offnegation because satisfies the following properties:

   2.1 It is centered in 0.5, i.e., $\underset{O}{\neg}\ 0.5 = 0.5$, therefore I = 0.5.

   2.2 If $\overline{\alpha} \in [0, 1]$, then $\underset{O}{\neg}\ \overline{\alpha} \in [0, 1]$, $\underset{O}{\neg}\ 0 = 1$ and $\underset{O}{\neg}\ 1 = 0$, which is the usual negation operator in fuzzy logic.

   2.3 If $\overline{\alpha} < 0$, then $\underset{O}{\neg}\ \overline{\alpha} \geq 1$. $\underset{O}{\neg}\ \overline{\alpha} = 1$ only when $\Omega = 1$.

   2.4 If $\overline{\alpha} > 1$, then $\underset{O}{\neg}\ \overline{\alpha} \leq 0$. $\underset{O}{\neg}\ \overline{\alpha} = 0$ only when $\Psi = 0$.

   2.5 When $\Omega + \Psi = 1$, we have $\underset{O}{\neg}\ \Psi = \Omega$ and $\underset{O}{\neg}\ \Omega = \Psi$.

3. If $\Omega + \Psi = 1$, then $\underset{O}{\neg}\ \underset{O}{\neg}\ \overline{\alpha} = \overline{\alpha}$, for every $\overline{\alpha} \in [\Psi, \Omega]$.

The precedent properties are easy to demonstrate.

Hence, the definition of *offimplication* $\underset{O}{\rightarrow} : [\Psi, \Omega]^3 \times [\Psi, \Omega]^3 \rightarrow [\Psi, \Omega]^3$ is defined in Equation (15), for every $\overline{\alpha}, \overline{\beta} \in [\Psi, \Omega]^3$.

$$\overline{\alpha} \underset{O}{\rightarrow} \overline{\beta} = \langle N_O^{co}\left(\underset{O}{\neg}\ T_O(\overline{\alpha}), T_O(\overline{\beta})\right), N_O^{n_1}\left(\underset{O}{\neg}\ I_O(\overline{\alpha}), I_O(\overline{\beta})\right), N_O^{n_2}\left(\underset{O}{\neg}\ F_O(\overline{\alpha}), F_O(\overline{\beta})\right)\rangle \tag{15}$$

where, $N_O^{n_i}(\cdot, \cdot)$ i = 1, 2 are neutrosophic components n-offnorms, $N_O^{co}(\cdot, \cdot)$ is a neutrosophic component n-offconorm, and $\underset{O}{\neg}$ is the offnegation defined in Equation (14).

Equation (15) is generalized by using offuninorms, see Equation (16).

$$\overline{\alpha} \underset{UO}{\rightarrow} \overline{\beta} = \langle N_O^{u_1}\left(\underset{O}{\neg}\ T_O(\overline{\alpha}), T_O(\overline{\beta})\right), N_O^{u_2}\left(\underset{O}{\neg}\ I_O(\overline{\alpha}), I_O(\overline{\beta})\right), N_O^{u_3}\left(\underset{O}{\neg}\ F_O(\overline{\alpha}), F_O(\overline{\beta})\right)\rangle \tag{16}$$

where $N_O^{u_i}(\cdot, \cdot)$ for i = 1, 2, and 3 are neutrosophic components n-offuninorms.

Example 7. *One illustrative example of Equation (16) is obtained revisiting Section 4.1, by defining the following neutrosophic component n-offnorm:*

$$u_O(c_O(x), c_O(y)) = \begin{cases} \frac{3(c_O(x)+1)(c_O(y)+1)}{(c_O(x)+1)(c_O(y)+1)+(2-c_O(x))(2-c_O(y))} - 1, & \text{if } (c_O(x), c_O(y)) \in [-1, 2]^2 \backslash \{(-1, 2), (2, -1)\} \\ -1, & \text{otherwise} \end{cases}$$

This is the transformation of Silvert uninorms to the domain $[-1, 2]^2$ applying the functions in Equations (10) and (11), and the transformation in Proposition 5. Also, let us take $U_{ZD}(c(x), c(y))$ of Example 3. See that [−1, 2] is symmetric respect to 0.5, and the neutral element is 0.5.

Then, we study the offuninorm defined in the following equation: $U_O(\overline{\alpha}, \overline{\beta}) = \langle U_{ZD}(T_O(\overline{\alpha}), T_O(\overline{\beta})), u_O(I_O(\overline{\alpha}), I_O(\overline{\beta})), u_O(F_O(\overline{\alpha}), F_O(\overline{\beta}))\rangle$ for $\overline{\alpha} = \langle T_O(\overline{\alpha}), I_O(\overline{\alpha}), F_O(\overline{\alpha})\rangle$ and $\overline{\beta} = \langle T_O(\overline{\beta}), I_O(\overline{\beta}), F_O(\overline{\beta})\rangle$ in $[-1, 2]^3$.

Thus, we define the offimplication generated by $U_O(\cdot, \cdot)$ according to Equation (16) as follows:

$$\overline{\alpha} \underset{U_O}{\rightarrow} \overline{\beta} = \langle U_{ZD}\left(\underset{O}{\neg} T_O(\overline{\alpha}),\ T_O(\overline{\beta})\right),\ u_O\left(\underset{O}{\neg} I_O(\overline{\alpha}),\ I_O(\overline{\beta})\right), u_O\left(\underset{O}{\neg} F_O(\overline{\alpha}),\ F_O(\overline{\beta})\right)\rangle.$$

where in this case we have $U_{ZD}\left(T_O(\overline{\alpha}),\ T_O(\overline{\beta})\right) = \begin{cases} \min\left(T_O(\overline{\alpha}),\ T_O(\overline{\beta})\right), \text{ if } T_O(\overline{\alpha}),\ T_O(\overline{\beta}) \in \left[-1,\ \frac{1}{2}\right] \\ \max\left(T_O(\overline{\alpha}),\ T_O(\overline{\beta})\right), \text{ otherwise} \end{cases}$,

see Figure 4, and $u_O(\cdot,\cdot)$ models the neutrosophic n-components I_O and F_O, see Figure 5.

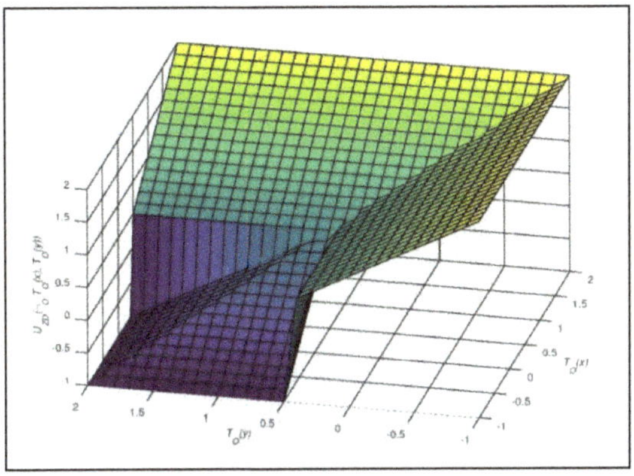

Figure 4. Depiction of the neutrosophic n-offimplication generated by U_{ZD} for T_O.

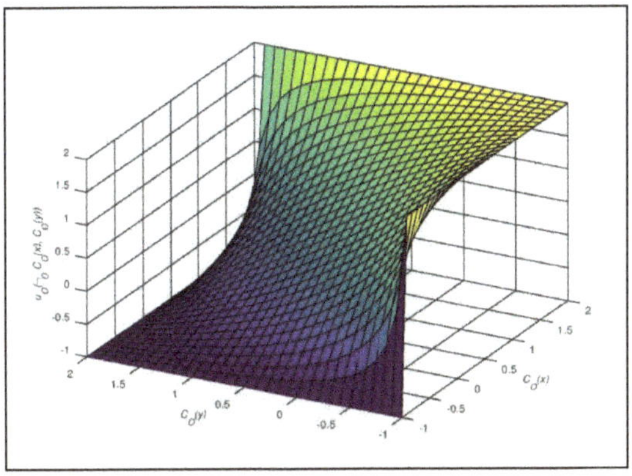

Figure 5. Depiction of the neutrosophic n-offimplication generated by u_O for both, I_O and F_O.

This offimplicator satisfies the overbounding conditions $\langle -1,\ 2,\ 2\rangle \underset{U_O}{\rightarrow} \langle -1,\ 2,\ 2\rangle = \langle -1,\ 2,\ 2\rangle \underset{U_O}{\rightarrow} \langle 2,\ -1,\ -1\rangle = \langle 2,\ -1,\ -1\rangle \underset{U_O}{\rightarrow} \langle 2,\ -1,\ -1\rangle = \langle 2,\ -1,\ -1\rangle$, whereas, $\langle 2,\ -1,\ -1\rangle \underset{U_O}{\rightarrow} \langle -1,\ 2,\ 2\rangle = \langle -1,\ 2,\ 2\rangle$.

Also, $\langle 0,\ 1,\ 1\rangle \underset{U_O}{\rightarrow} \langle 0,\ 1,\ 1\rangle = \langle 1,\ 0,\ 0\rangle \underset{U_O}{\rightarrow} \langle 1,\ 0,\ 0\rangle = \langle 1,\ 0.5,\ 0.5\rangle$, $\langle 0,\ 1,\ 1\rangle \underset{U_O}{\rightarrow} \langle 1,\ 0,\ 0\rangle = \langle 1,\ -0.4,\ -0.4\rangle$ and $\langle 1,\ 0,\ 0\rangle \underset{U_O}{\rightarrow} \langle 0,\ 1,\ 1\rangle = \langle 0,\ 1.4,\ 1.4\rangle$. Additionally, $\langle 0.5,\ 0.5,\ 0.5\rangle \underset{U_O}{\rightarrow} \langle 0.5,\ 0.5,\ 0.5\rangle = \langle 0.5,\ 0.5,\ 0.5\rangle$ because 0.5 is the neutral element of every neutrosophic component n-offuninorm and $\underset{O}{\neg}\ 0.5 = 0.5$.

It is easy to check that substituting $u_O(\cdot,\cdot)$ by $U_{ZC}(\cdot,\cdot)$ in $\underset{U_O}{\rightarrow}$, we obtain the more classical equations $\langle 0, 1, 1\rangle \underset{U_O}{\rightarrow} \langle 0, 1, 1\rangle = \langle 1, 0, 0\rangle \underset{U_O}{\rightarrow} \langle 1, 0, 0\rangle = \langle 0, 1, 1\rangle \underset{U_O}{\rightarrow} \langle 1, 0, 0\rangle = \langle 1, 0, 0\rangle$ and $\langle 1, 0, 0\rangle \underset{U_O}{\rightarrow} \langle 0, 1, 1\rangle = \langle 0, 1, 1\rangle$.

4.3. N-Offuninorms and Voting Games

The applicability of uninorms to solve group decision problems is evident. However, the use of them as part of a game theory solution is not so obvious. This subsection is devoted to solving voting games based on n-offuninorms.

A *cooperative game with transferable utility* consists of a pair (N,v), where N = {1, 2, ... ,n} is a non-empty set of *players*,n ∈ ℕ and v: $2^N \rightarrow \mathbb{R}$, i.e., $v(\cdot)$ is a function of the power set of N such that each *coalition* or $S \subseteq$ N is associated with a real number. v is called *characteristic function* and $v(S)$ represents the conjoint payoff of players in S. Additionally, $v(\emptyset) = 0$ (see [42], p. 2).

A *simple game* models voting situations. It is a cooperative game such that for every coalition S, either $v(S) = 0$ or $v(S) = 1$, and $v(N) = 1$ (see [42], p. 7).

One solution is the Shapley–Shubik index, which is the Shapley value to simple games (see [42], pp. 6–7). The equation of Shapley value is the following:

$$\phi_i(v) = \sum_{S \subseteq N \setminus \{i\}} \frac{|S|!(|N| - |S| - 1)!}{|N|!}[v(S \cup \{i\}) - v(S)] \quad (17)$$

where $|S|$ is the cardinality of coalition S, $|N|$ is the cardinality of the set of players or *grand coalition* and $\phi_i(v)$ is the value assigned to player i in the game.

This is the unique solution which satisfies the following axioms:

- $\sum_{i \in N} \phi_i(v) = v(N)$ (Efficiency),
- If $i, j \in$N are interchangeable in v, then $\phi_i(v) = \phi_j(v)$ (Symmetry),
- If i is such that for every coalition S the equation $v(S \cup \{i\}) = v(S)$ holds, then $\phi_i(v) = 0$ (Dummy),
- Given v and w two games over N, then $\phi_i(v + w) = \phi_i(v) + \phi_i(w)$ (Additivity).

This value is the sum of the terms $[v(S \cup \{i\}) - v(S)]$, which mean the marginal contribution of player i to the coalitions S, multiplied by $\frac{|S|!(|N|-|S|-1)!}{|N|!}$ which is the probability that $|S| - 1$ players precede player i in the game and $|N| - |S|$ players follow him or her. Thus, the Shapley value of i is the expected marginal contribution of i to the game (see [42], p. 7). The result of the Shapley–Shubik index is interpreted as a measure of each player's power.

In the present paper we basically study voting games with some additional features. We call them *voting n-offgames*. A *voting n-offgame* consists in a pair (N, v), where N = {1, 2, ... , n} is the set of players; the characteristic function v: $2^N \rightarrow \{1, \ldots, 2^n\} \times \{1, \ldots, 2^n\} \times \{1, \ldots, 2^n\}$ is such that for any coalition S we have $v(S) = (k, l, 2^n\text{-}k+1)$ and $v(\emptyset) = (2^n, 2^n, 1)$.

The n-offgame is interpreted in the following way:

1. Experts forecast that voters will rank coalition S in the kth position of their preference, also they cannot decide if S will be ranked in the lth position. The first place or k = 1 corresponds to the preferred coalition of all and so on. Additionally, the n-offgame must satisfy the following rules:
2. Given any two coalitions S_1 and S_2, $S_1 \neq S_2$, we have the first component that both $v(S_1)$ and $v(S_2)$ are different. Thus, every coalition is associated with a unique number in the order of preference.
3. $v(S) = (k, k, 2^n - k+1)$ means experts have no doubt that coalition S will be voted in the kth position.

Let us observe that it is not a simple game. This game can be interpreted as a multicriteria decision-making problem, where its solution is a measure of every player's power in the game according to the forecasted experts' ranking of the coalitions. Each coalition can represent a bloc of political parties.

Shapley value can be the solution to voting n-offgames, in the form given in Equation (18):

$$\phi_i(v) = -\sum_{S \subseteq N\setminus\{i\}} \frac{|S|!(|N|-|S|-1)!}{|N|!}[v(S\cup\{i\}) - v(S)] \quad (18)$$

Let us note that the minus sign in the expression was taken for convenience because the rank we applied is decreasing respect to the coalition´s significance. Additionally, $v(S\cup\{i\}) - v(S)$ is the difference between two 3-tuple values, thus the operation $(k_1, l_1, 2^n - k_1+1)-(k_2, l_2, 2^n-k_2+1):= (k_1- k_2, l_1- l_2, k_2- k_1)$ is defined. Equation (18) means the expected number of places won or lost in voter preference, as predicted by experts.

Apparently, Shapley value cannot be the solution to this problem because $v(\emptyset) \neq 0$ and $v(\cdot)$ is not a game. However, if we take in that $v(S) = (k,l,2^n - k+1)$ in fact represents three games, namely, $v_1(S)=k$, $v_2(S)= l$, and $v_3(S) =2^n- k+1$, one per component and additionally taking into account they are linear transformations of three games with characteristic functions w_1, w_2,, and w_3; where $w_1(S)= 2^n- v_1(S)$, $w_2(S)= 2^n- v_2(S)$, and $w_3(S) = 1 - v_3(S)$, then, the marginal contributions of the three pairs, $w_1(\cdot)$ and $v_1(\cdot)$, $w_2(\cdot)$ and $v_2(\cdot)$, $w_3(\cdot)$ and $v_3(\cdot)$, are the same except for the sign. Thus, these three pairs have the same Shapley value except for the sign and therefore this property is extended to $v(\cdot)$ and $w(\cdot)$.

Shapley value is a rational solution to the game, nevertheless, it can differ from actual human behavior, as Zhang et al. suggested in [43] to model restrictions in game decisions according to the human behavior based on fuzzy uninorms. Therefore, we propose n-offuninorms to explore other behaviors in human decision making by recursively applying an n-offuninorm to every pair of values $\frac{|S|!(|N|-|S|-1)!}{|N|!}[v(S) - v(S\cup\{i\})]$ in the set of $S \subseteq N\setminus\{i\}$.

Here we explore n-offuninorms defined on [−L, L], $L = 2^n-1$ and with the PROSPECTOR parameterized function with $\lambda > 0$ and neutral element $e = L\left(\frac{1-\lambda}{1+\lambda}\right)$, see Equation (19).

$$U_{O\lambda}(c(x), c(y)) = \varphi_3^{-1}\left(\frac{\lambda\varphi_3(c(x))\varphi_3(c(y))}{\lambda\varphi_3(c(x))\varphi_3(c(y)) + (1-\varphi_3(c(x)))(1-\varphi_3(c(y)))}\right) \quad (19)$$

where $\varphi_3(\cdot)$ and $\varphi_3^{-1}(\cdot)$ are those defined in Equations (10) and (11), respectively, and now they are $\varphi_3(c(x)) = \frac{c(x)+L}{2L}$ and $\varphi_3^{-1}(c(x)) = 2Lc(x) - L$.

Thus the Algorithm for solving voting n-offgames can be described as follows:

Algorithm 1. Algorithm for solving voting n-offgames

1. Given (N, v) a voting n-offgame. Fix $\lambda > 0$.
2. Fix player $i = 1$.
3. Let S_j be the set of coalitions not containing i, and $j = 1, 2, \ldots, 2^{n-1}$. Let us take $a_{i1} = v(S_1)$ and $a_{i2} = v(S_2)$ and calculate $a_{prev} = U_{O\lambda}\left(\frac{|S_1|!(n-|S_1|-1)!}{n!}[v(S_1) - v(S_1\cup\{i\})], \frac{|S_2|!(n-|S_2|-1)!}{n!}[v(S_2) - v(S_2\cup\{i\})]\right)$, fix $j = 3$ and go to step 4.
4. If $j<2^{n-1}$, calculate $a_{curr} = U_{O\lambda}\left(a_{prev}, \frac{|S_j|!(n-|S_j|-1)!}{n!}\left[v(S_j) - v(S_j\cup\{i\})\right]\right)$. $a_{prev} = a_{curr}$ and $j = j + 1$. Repeat this step. Else, if $j = 2^{n-1}$, $\pi_i(v) = a_{curr}$. Go to Step 5.
5. If $i< n$, then $i = i+ 1$ and go to Step 3. Else Finish.

Let us point out that in the precedent algorithm the associativity of n-offuninorms was used. Moreover, the algebraic sum in Shapley value and the n-offuninorms yield to somewhat similar results. Thus, for $U_{o\lambda}(\cdot,\cdot)$ with $\lambda = 1$, we have that $x, y < 0$ imply both $U_{o\lambda}(x,y)<\min(x,y)$ and $x+y< \min(x,y)$, whereas when $x, y > 0$, we have $U_{o\lambda}(x,y)>\max(x,y)$ and $x+y> \max(x,y)$. For x,y satisfying $x\cdot y<0$, then both $U_{o\lambda}(x,y)$ and $x+y$ are compensatory operators, and finally 0 is the neutral element of them. For $\lambda \neq 1$ and hence $e \neq 0$, we obtain other behavioral effects. Let us also recall that $U_{o\lambda}(\cdot,\cdot)$ is a neutrosophic uninorm transformation, which is described as symmetric summation by Silvert in [40].

Example 8. *Let us consider the 3-person voting n-offgame (N, v), where N = {1, 2, 3} and experts predict that coalitions will be ranked according to the positions shown in Table 1.*

Table 1. Position assigned to the coalitions of the 3-person voting n-offgame.

Coalition	Ranking
∅	(8,8,1)
{1}	(3,2,6)
{2}	(4,3,5)
{3}	(7,6,2)
{1,2}	(2,3,7)
{1,3}	(5,6,4)
{2,3}	(6,5,3)
{1,2,3}	(1,1,8)

According to Table 1, the grand coalition N has (1,1,8) as ranking value, i.e., experts think this coalition will undoubtedly be ranked in the first place or k = 1. $v(\emptyset)$ = (8,8,1) because it is axiomatically predetermined, which means that to not negotiate at all is the worst option, whereas $v(\{2,3\})$ = (6,5,3) means this coalition shall be ranked in the sixth place and maybe in the fifth one, but never in the third place.

Thus, to calculate each player´s power according to our approach we have to apply the precedent algorithm. We fixed λ=1 in $U_{O\lambda}$ therefore $c(e)$= 0, which is defined in [−7, 7].

Table 2 contains the detailed calculus of the Shapley value in Equation (18) and the proposed algorithm to resolve the precedent voting n-offgame.

Table 2. Shapley value and n-offuninorm based solutions to the 3-person voting n-offgame. The final values are written in bold font.

Player*i*	SSuch That $i \notin S$	$v(S)-v(S\cup\{i\})$	$v(S)-v(S\cup\{i\})$ Multiplied by the Probability	Partial Summations of the Shapley Value	Partial Aggregation with U_{o1}
1	∅	(5,6,−5)	(5/3, 2,−5/3)	(5/3,2,−5/3)	(5/3,2,−5/3)
	{2}	(2,0,−2)	(1/3,0,−1/3)	(2,2,−2)	(1.9776,2.0000,−1.9776)
	{3}	(2,0,−2)	(1/3,0,−1/3)	(7/3,2,−7/3)	(2.2802,2.0000,−2.2802)
	{2,3}	(5,4,−5)	(5/3,4/3,−5/3)	**(4,10/3,−4)**	**(3.6628,3.1613,−3.6628)**
2	∅	(4,5,−4)	(4/3,5/3,−4/3)	(4/3,5/3,−4/3)	(4/3,5/3,−4/3)
	{1}	(1,−1,−1)	(1/6,-1/6,−1/6)	(3/2,3/2,−3/2)	(1.4932,1.5086,−1.4932)
	{3}	(1,1,−1)	(1/6,1/6,−1/6)	(5/3,5/3,−5/3)	(1.6515,1.6667,−1.6515)
	{1,3}	(4,5,−4)	(4/3,5/3,−4/3)	**(3,10/3,−3)**	**(2.8565,3.1545,−2.8565)**
3	∅	(1,2,−1)	(1/3,2/3,−1/3)	(1/3,2/3,−1/3)	(1/3,2/3,−1/3)
	{1}	(-2,−4,2)	(-1/3,−2/3,1/3)	(0,0,0)	(0,0,0)
	{2}	(-2,−2,2)	(-1/3,−1/3,1/3)	(-1/3,−1/3,1/3)	(−1/3,−1/3,1/3)
	{1,2}	(1,2,−1)	(1/3,2/3,−1/3)	**(0,1/3,0)**	**(0,0.33485,0)**

According to the results summarized in Table 1, we have that the expected value of places gains by player 1 is 4 with the Shapley value solution and 3.6628 with U_{o1}, whereas the results for player 2 are 3 and 2.8565, respectively, and for player 3 are 0 and 0. Therefore, player 1 is the most powerful of them, followed by player 2 and 3 in this order. Thus, the proposed approach and Shapley value are similar.

Table 3 contains the voting n-offgame solutions comparing U_{o1} with $c(e)$ = 0, $U_{o99/101}$ with $c(e)$ = 7/100 and $U_{o101/99}$ with $c(e)$ = −7/100.

Table 3. Solutions of the 3-person voting n-offgame applying $U_{o\lambda}$ with λ = 1, 99/101 and 101/99, respectively.

Player	Solution with U_{o1}	Solution with $U_{o99/101}$	Solution with $U_{o101/99}$
1	(3.6628,3.1613,−3.6628)	(3.5079,2.9919,−3.8129)	(3.8129,3.3262,−3.5079)
2	(2.8565,3.1545,−2.8565)	(2.6793,2.9849,−3.0293)	(3.0293,3.3196,−2.6793)
3	(0,0.33485,0)	(−0.20994,0.12509,−0.20994)	(0.20994,0.54402,0.20994)

The solutions in Table 3 prove that the greater λ, the greater the solution values. Thus, when λ is increased, its associated solution models more optimistic behavior with respect to the first component, which is compensated with more pessimistic behavior with respect to the third component.

The advantages of the proposed approach are more evident when it is compared with a classical one restricted to {0, 1}. Here we used a semantic represented with natural numbers and we calculated directly on them. In contrast, for applying classical definitions in {0, 1}, we would need to define eight Boolean functions, one per element. What is more, some operations such as marginal contributions, which is an algebraic difference, cannot be directly applied in the logic sense.

In case we would need to extend the approaches to the continuous gradation, then a continuous ranking can be modeled with the identity line $I_d(x) = x$, but in the classical approach, eight memberships functions would have to be considered, where the simplest ones are triangular (see Figure 6). From Figure 6 we can infer that there exists a transformation between both models; however, the proposed model is the simplest one.

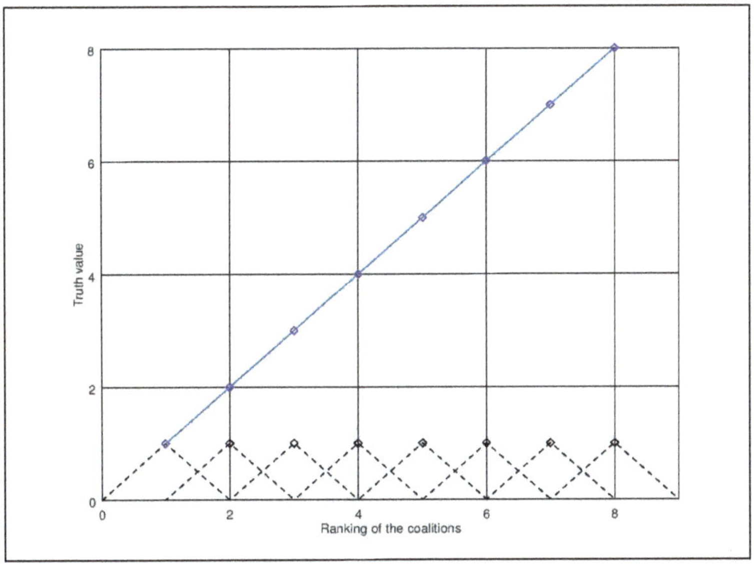

Figure 6. Depiction of two kinds of 3-person game modeling. Classical [0, 1] is represented in dashed lines and triangular membership functions, whereas the solid line represents the solution based on offsets. The points represent the Boolean restrictions.

5. Discussion

Neutrosophic oversets, undersets, and offsets are concepts of a novel and non-conventional theory of uncertainty. Historically, the convention of restricting logic to the interval [0, 1] has dominated fuzzy logic and its generalizations. Possibly this is a legacy of probability and mathematical logic, where, semantically speaking, 0 and 1 have been considered the two extreme opposite sides. Therefore, oversets, undersets, and offsets can be understood as controversial subjects. Nevertheless, Smarandache in [31] illustrates with some examples that such sets, of which their domains surpass the scope of [0, 1], could be useful to represent knowledge in a valid semantic.

This is a recent theory that needs more developing and the scientific community's acknowledgment of its usefulness. One of our aims with this paper is to demonstrate that this theory can be useful. To achieve this end, we introduced the uninorm theory in the neutrosophic offset framework. This union is manifold advantageous, the most evident one being that we have provided a new aggregator operator to these sets. As we mentioned in the introduction, there exists a wide variety of fuzzy uninorm applications, namely, Decision Making [9,14,15], DNA and RNA fusion [9], logic [17], Artificial Neural Networks [16], among others. Uninorm is more flexible than t-norm and t-conorm because it includes the compensatory property in some cases, which is more realistic for modeling human decision making, as was experimentally proved by Zimmermann in [21].

Also, uninorms have enriched other theories when they were generalized to other frameworks. In L^*-fuzzy set theory [23], uninorms also aggregate independent non-membership functions to achieve more precision. Moreover, neutrosophic uninorms aggregate the indeterminate-membership functions [25].

Additionally, some authors have associated uninorms with non-conventional theories. In [33,34] we can find some attempts to extend uninorm domains to an interval [a, b]. The reason is that the PROSPECTOR function related to the MYCIN Expert System is one very important milestone in Artificial Intelligence history. The point is that the PROSPECTOR function is basically a uninorm except it is defined in the interval [−1, 1], thus, we can consider intervals greater than [0, 1]. They have argued that there exist two reasons to maintain the interval [−1, 1]—the first one is the importance of the PROSPECTOR function, the second one is the facility to interchange information among users and decision makers in form of degrees to accept or reject hypotheses.

The second non-conventional approach is the bipolar or Multi-Polar uninorms defined in [24]. The world is (and some people are) is evidently multi-polar; in case of bipolarity they are modeled in [−1, 1]. Especially in [24], we have a multi-polar space consisting of an ordered pair of (k, x), where k∈{1, 2, … , n} represents a category or class and x∈(0, 1], with the convention **0** = (k, 0) for every category. This is a more complex representation that takes a unique interval [−n, n] where, for x∈[−n, n], the function round(x) represents the category and its fractional part represents the degree of membership to that category. This is a real extension of bipolarity in [−1, 1] to multi-polarity. In [31] (pp. 127, 130) Tripolar offsets and Multi-polar offsets are defined. We illustrated in Example 8 that considering the semantic values belong to {−n,−n+1, … , 0, 1, … , n} could be advantageous.

The definition of uninorm-based implicators is not new in literature, they can be seen in [41] (pp. 151–160) for fuzzy uninorms, in [17] it is extended for type 2 fuzzy sets, in [24] for L^*-fuzzy set theory, and in [25] for neutrosophic uninorms. In the present paper, uninorm-based offimplicators are defined, however, we only counted on symbolic offimplication operators (see [31], p. 139). To extend this definition to a continuous framework, we had to extend the symbolic offnegation to a continuous one.

Finally, we preferred to illustrate a voting game solution instead of a group decision method because the relationship of offuninorms with the latter subject is predictable. However, to find any game theory associated with uninorms is uncommon in literature. One remarkable example can be seen in [43], where a behavioral approach has been made to certain kind of games, where uninorms model the humans' restrictions to make the division of gains among the players.

In the present paper, another approach is proposed where an indeterminacy component is taken into account. Also, we proved that modeling with a natural number semantic is simpler than to utilize the classical [0, 1] interval, because of the fact that n membership functions can be substituted by a linear identity function. We basically defined the voting game solution since the Shapley–Shubik index components (see [42], pp. 6–7), where we only changed the algebraic sum by offuninorms. The classical approaches such as the Shapley–Shubik index are interested in a rational and fair solution; nevertheless, many times that does not occur in real negotiations and then behavioral solutions are needed.

6. Conclusions

This paper was devoted to defining for the first time the theory of neutrosophic offuninorms, which is a generalization of both the neutrosophic offnorms and neutrosophic offconorms, where the neutral element lays in the interval [Ψ, Ω]. The properties of these novel operators were proved. Moreover, we defined neutrosophic offuninorms from neutrosophic offnorms and neutrosophic offconorms and vice versa, we also proved their properties. Additionally, we proved the relationship between neutrosophic offuninorms and neutrosophic uninorms.

One of the purposes of this paper is to show the convenience of applying offsets, and to prove that they are not only simple theoretical concepts; furthermore, they are also necessary to define new concepts. This need is demonstrated in this paper by associating offsets with the PROSPECTOR aggregation function, where it is recommendable to extend its domain to the interval [−1, 1]. Some authors in fuzzy logic have suggested the advantages to calculate in the domains [a, b] instead of the classical [0, 1]. Therefore, the use of the idea of the offset in uninorms has some precedence in fuzzy logic.

Additionally, we recommend offsets because they permit more accuracy and compactness. We showed that it is possible to define offimplication operators based on offuninorms. A future direction of this research is to solve problems by using artificial neural networks based on neutrosophic offuninorms, such that neutrosophic offuninorms are utilized as the threshold functions in the neurons or in neutrosophic cognitive maps. For the first time, solutions to cooperative games are defined in the neutrosophic framework—this is an area that it is worthy of development.

Author Contributions: Conceptualization, F.S. and E.G.-C.; methodology, E.G.-C.; software, M.L.-V. and E.G.-C.; validation, E.G.-C., F.S. and M.L.-V.; formal analysis, E.G.-C.; investigation, E.G.-C. and F.S.; resources, F.S. and M.L.-V.; data curation, M.L.-V.; writing—original draft preparation, E.G.-C.; writing—review and editing, M.L.-V. and F.S.; visualization, E.G.-C.; supervision, F.S.; project administration, E.G.-C.; funding acquisition, F.S.

Funding: This research received no external funding.

Conflicts of Interest: The authors declare no conflict of interest.

References

1. Yager, R.R.; Rybalov, A. Uninorm aggregation operators. *Fuzzy Sets Syst.* **1996**, *80*, 111–120. [CrossRef]
2. Fodor, R.; Yage, R.; Rybalov, A. Structure of uninorms. *Internat. J. Uncertain. Fuzziness Knowl.-Based Syst.* **1997**, *5*, 411–427. [CrossRef]
3. Calvo, T.; Kolesárová, A.; Komorníková, M.; Mesiar, R. Aggregation Operators: Basic Concepts, Issues and Properties. In *Aggregation Operators: New Trends and Applications*; Calvo, T., Mayor, G., Mesiar, R., Eds.; Physica-Verlag: Heidelberg, Germany, 2002; Volume 97, pp. 29–32.
4. Yager, R.R. Uninorms in fuzzy system modeling. *Fuzzy Sets Syst.* **2001**, *122*, 167–175. [CrossRef]
5. Yager, R.R.; Kreinovich, V. Universal approximation theorem for uninorm-based fuzzy systems modeling. *Fuzzy Sets Syst.* **2003**, *140*, 331–339. [CrossRef]
6. Jenei, S. A New Class of Uninorm Aggregation Operations for Fuzzy Theory. In Proceedings of the International Conference on Artificial Intelligence and Soft Computing, Zakopane, Poland, 16–20 June 2019; Springer: Cham, Switzerland, 2019; pp. 296–303.
7. Fodor, J.; Rudas, I.J.; Bede, B. Uninorms and absorbing norms with applications to Image Processing. In Proceedings of the 4th Serbian-Hungarian Joint Symposium on Intelligent Systems, SISY, Subotica, Serbia, 29–30 September 2006; pp. 59–72.
8. Xie, A. On Extended Representable Uninorms and Their Extended Fuzzy Implications (Coimplications). *Symmetry* **2017**, *8*, 160. [CrossRef]
9. Rudas, I.J.; Pap, E.; Fodor, J. Information aggregation in intelligent systems: An application oriented approach. *Knowl.-Based Syst.* **2013**, *38*, 3–13. [CrossRef]
10. Depaire, B.; Vanhoof, K.; Wets, G. The Application of Uninorms in Importance-Performance Analysis. In Proceedings of the 7th WSEAS International Conference on Fuzzy Systems, Cavtat, Croatia, 12–14 June 2016; pp. 1–7.

11. Moodley, R.; Chiclana, F.; Caraffini, F.; Carter, J. Application of uninorms to market basket analysis. *Int. J. Intell. Syst.* **2018**, *34*, 39–49. [CrossRef]
12. Appel, O.; Chiclana, F.; Carter, J.; Fujita, H. Cross-ratio uninorms as an effective aggregation mechanism in Sentiment Analysis. *Knowl.-Based Syst.* **2017**, *124*, 16–22. [CrossRef]
13. Kamis, N.H.; Chiclana, F.; Levesley, J. Geo-uninorm consistency control module for preference similarity network hierarchical clustering based consensus model. *Knowl.-Based Syst.* **2018**, *162*, 103–114. [CrossRef]
14. Wu, J.; Xiong, R.; Chiclana, F. Uninorm trust propagation and aggregation methods for group decision making in social network with four tuple information. *Knowl.-Based Syst.* **2016**, *96*, 29–39. [CrossRef]
15. Ureña, R.; Kou, G.; Dong, Y.; Chiclana, F.; Herrera-Viedma, E. A review on trust propagation and opinion dynamics in social networks and group decision making frameworks. *Inf. Sci.* **2019**, *478*, 461–475. [CrossRef]
16. Bordignon, F.; Gomide, F. Uninorm based evolving neural networks and approximation capabilities. *Neurocomputing* **2014**, *127*, 13–20. [CrossRef]
17. Wang, S. A Proof of the Standard Completeness for the Involutive Uninorm Logic. *Symmetry* **2019**, *11*, 445. [CrossRef]
18. Yang, E. Fixpointed Idempotent Uninorm (Based) Logics. *Mathematics* **2019**, *7*, 107. [CrossRef]
19. González-Hidalgo, M.; Massanet, S.; Mir, A.; Ruiz-Aguilera, D. A new edge detector based on uninorms. In Proceedings of the 15th International Conference on Information Processing and Management of Uncertainty in Knowledge-Based Systems, Montpellier, France, 15–19 July 2014; Springer International Publishing: Cham, Switzerland, 2014; pp. 184–193.
20. Sussner, P.; Schuster, T. Interval-valued fuzzy morphological associative memories: Some theoretical aspects and applications. *Inf. Sci.* **2018**, *438*, 127–144. [CrossRef]
21. Zimmermann, H.J.; Zysno, P. Latent connectives in human decision making. *Fuzzy Sets Syst.* **1980**, *4*, 37–51. [CrossRef]
22. Marasini, D.; Quatto, P.; Ripamonti, E. Fuzzy analysis of students' ratings. *Eval. Rev.* **2016**, *40*, 122–141. [CrossRef] [PubMed]
23. Deschrijver, G.; Kerre, E.E. Uninorms in L^*-fuzzy set theory. *Fuzzy Sets Syst.* **2004**, *148*, 243–262. [CrossRef]
24. Mesiarová-Zemánková, A. Multi-polar t-conorms and uninorms. *Inf. Sci.* **2015**, *301*, 227–240. [CrossRef]
25. González, E.; Leyva, M.; Smarandache, F. On Neutrosophic Uninorms. *Neutrosophic Sets Syst.* Under review.
26. Smarandache, F. *A Unifying Field in Logics: Neutrosophic Logic. Neutrosophy Set, Neutrosophic Probability and Statistics*, 4th ed.; American Research Press: Rehoboth, DE, USA, 2005; pp. 125–128.
27. Guo, Y.; Cheng, H.D. New neutrosophic approach to image segmentation. *Pattern Recognit.* **2009**, *42*, 587–595. [CrossRef]
28. Guo, Y.; Sengur, A. NCM: Neutrosophic c-means clustering algorithm. *Pattern Recognit.* **2015**, *48*, 2710–2724. [CrossRef]
29. Abdel-Basset, M.; Mohamed, M.; Smarandache, F. An extension of neutrosophic AHP–SWOT analysis for strategic planning and decision-making. *Symmetry* **2018**, *10*, 116. [CrossRef]
30. Abdel-Basset, M.; Mohamed, M.; Smarandache, F. A hybrid neutrosophic group ANP-TOPSIS framework for supplier selection problems. *Symmetry* **2018**, *10*, 226. [CrossRef]
31. Smarandache, F. *Neutrosophic Overset, Neutrosophic Underset, and Neutrosophic Offset:Similarly for Neutrosophic Over-/Under-/Off-Logic, Probability, and Statistics*; Pons Editions: Brussels, Belgium, 2016; pp. 23–31.
32. Shortliffe, E.H. *Computer-Based Medical Consultations: MYCIN*; Elsevier: New York, NY, USA, 1976; pp. 98–108.
33. Tsadiras, A.K.; Margaritis, K.G. The MYCIN certainty factor handling function as uninorm operator and its use as a threshold function in artificial neurons. *Fuzzy Sets Syst.* **1998**, *93*, 263–274. [CrossRef]
34. De Baets, B.; Fodor, J. Van Melle's combining function in MYCIN is a representable uninorm: An alternative proof. *Fuzzy Sets Syst.* **1999**, *104*, 133–136. [CrossRef]
35. Karacı, A. Fuzzy Logic based Assessment Model Proposal for Online Problem-based Learning. *IJCA* **2015**, *117*, 5–8. [CrossRef]
36. Kraipeerapun, P.; Fung, C.C. Binary classification using ensemble neural networks and interval neutrosophic sets. *Neurocomputing* **2009**, *72*, 2845–2856. [CrossRef]
37. Pérez-Teruel, K.; Leyva-Vázquez, M. Neutrosophic Logic for Mental Model Elicitation and Analysis. *Neutrosophic Sets Syst.* **2014**, *2*, 31–33.
38. Wang, H.; Smarandache, F.; Zhang, Y.Q.; Sunderraman, R. Single valued neutrosophic sets. *Multispace Multistruct.* **2010**, *4*, 410–413.

39. Smarandache, F. *A Unifying Field in Logics: Neutrosophic Logic. Neutrosophy, Neutrosophic Set, Neutrosophic Probability and Statistics*, 6th ed.; InfoLearnQuest: Ann Arbor, MI, USA, 2007; pp. 92–93.
40. Silvert, W. Symmetric Summation: A Class of Operations on Fuzzy Sets. *IEEE Trans. Syst. Man Cybernet* **1979**, *9*, 657–659.
41. Baczynski, M.; Jayaram, B. *Fuzzy Implications*; Springer: Berlin, Germany, 2008; Volume 231, pp. 151–160.
42. Roth, A.E. Introduction to the Shapley value. In *The Shapley Value: Essays in Honor of Lloyd S. Shapley*; Roth, A.E., Ed.; Cambridge University Press: New York, NY, USA, 1988; pp. 2, 6–7.
43. Zhang, Y.; Luo, X.; Leung, H.F. Games Played under Fuzzy Constraints. *Int. J. Intell. Syst.* **2016**, *31*, 347–378. [CrossRef]

symmetry

MDPI

Article

Ordinary Single Valued Neutrosophic Topological Spaces

Junhui Kim [1,*], Florentin Smarandache [2], Jeong Gon Lee [3] and Kul Hur [3]

[1] Department of Mathematics Education, Wonkwang University, 460, Iksan-daero, Iksan-Si 54538, Jeonbuk, Korea
[2] Department of Mathematics and Science, University of New Mexico, 705 Gurley Ave., Gallup, NM 87301, USA
[3] Department of Applied Mathematics, Wonkwang University, 460, Iksan-daero, Iksan-Si 54538, Jeonbuk, Korea
* Correspondence: junhikim@wku.ac.kr

Received: 31 July 2019 ; Accepted: 21 August 2019; Published: 27 August 2019

Abstract: We define an ordinary single valued neutrosophic topology and obtain some of its basic properties. In addition, we introduce the concept of an ordinary single valued neutrosophic subspace. Next, we define the ordinary single valued neutrosophic neighborhood system and we show that an ordinary single valued neutrosophic neighborhood system has the same properties in a classical neighborhood system. Finally, we introduce the concepts of an ordinary single valued neutrosophic base and an ordinary single valued neutrosophic subbase, and obtain two characterizations of an ordinary single valued neutrosophic base and one characterization of an ordinary single valued neutrosophic subbase.

Keywords: ordinary single valued neutrosophic (co)topology; ordinary single valued neutrosophic subspace; α-level; ordinary single valued neutrosophic neighborhood system; ordinary single valued neutrosophic base; ordinary single valued neutrosophic subbase

1. Introduction

In 1965, Zadeh [1] introduced the concept of fuzzy sets as the generalization of an ordinary set. In 1986, Chang [2] was the first to introduce the notion of a fuzzy topology by using fuzzy sets. After that, many researchers [3–13] have investigated several properties in fuzzy topological spaces.

However, in their definitions of fuzzy topology, fuzziness in the notion of openness of a fuzzy set was absent. In 1992, Samanta et al. [14,15] introduced the concept of gradation of openness (closedness) of fuzzy sets in X in two different ways, and gave definitions of a smooth topology and a smooth co-topology on X satisfying some axioms of gradation of openness and some axioms of gradation of closedness of fuzzy sets in X, respectively. After then, Ramadan [16] defined level sets of a smooth topology and smooth continuity, and studied some of their properties. Demirci [17] defined a smooth neighborhood system and a smooth Q-neighborhood system, and investigated their properties. Chattopadhyay and Samanta [18] introduced a fuzzy closure operator in smooth topological spaces. In addition, they defined smooth compactness in the sense of Lowen [8,9], and obtained its properties. Peters [19] gave the concept of initial smooth fuzzy structures and found its properties. He [20] also introduced a smooth topology in the sense of Lowen [8] and proved that the collection of smooth topologies forms a complete lattice. Al Tahan et al. [21] defined a topology such that the hyperoperation is pseudocontinuous, and showed that there is no relation in general between pseudotopological and strongly pseudotopological hypergroupoids. In addition, Onassanya and Hošková-Mayerová [22] investigated some topological properties of α-level subsets' topology of a fuzzy subset. Moreover, Çoker and Demirci [23], and Samanta and Mondal [24,25] defined intuitionistic gradation of openness (in short IGO) of fuzzy sets in Šostak's sense [26] by using intuitionistic fuzzy sets introduced by Atanassov [27]. They mainly dealt with intuitionistic gradation of openness of fuzzy sets in the sense of Chang. However, in 2010, Lim et al. [28] investigated intuitionistic smooth topological spaces in Lowen's sense. Recently, Kim et al. [29] studied continuities and neighborhood systems in intuitionistic smooth topological spaces. In addition, Choi et al. [30] studied an interval-valued smooth topology by gradation of openness of interval-valued fuzzy sets introduced by Gorzalczany [31] and Zadeh [32], respectively. In particular, Ying [33] introduced the concept of the topology (called a fuzzifying topology) considering the degree of openness of an ordinary subset of a set. In 2012, Lim et al. [34] studied general properties in ordinary smooth topological spaces. In addition, they [35–37] investigated closures, interiors and compactness in ordinary smooth topological spaces.

In 1998, Smarandache [38] defined the concept of a neutrusophic set as the generalization of an intuitionistic fuzzy set. Salama et al. [39] introduced the concept of a neutrosophic crisp set and neutrosophic crisp relation (see [40] for a neutrosophic crisp set theory). After that, Hur et al. [41,42] introduced categories $\mathbf{NSet}(H)$ and **NCSet** consisting of neutrosophic sets and neutrosophic crisp sets, respectively, and investigated them in a topological universe view-point. Smarandache [43] defined the notion of neutrosophic topology on the non-standard interval and Lupiáñez proved that Smarandache's definitions of neutrsophic topology are not suitable as extensions of the intuitionistic fuzzy topology (see Proposition 3 in [44,45]). In addition, Salama and Alblowi [46] defined a neutrosophic topology and obtained some of its properties. Salama et al. [47] defined a neutrosophic crisp topology and studied some of its properties. Wang et al. [48] introduced the notion of a single valued neutrosophic set. Recently, Kim et al. [49] studied a single valued neutrosophic relation, a single valued neutrosophic equivalence relation and a single valued neutrosophic partition.

In this paper, we define an ordinary single valued neutrosophic topology and obtain some of its basic properties. In addition, we introduce the concept of an ordinary single valued neutrosophic subspace. Next, we define the ordinary single valued neutrosophic neighborhood system and we show that an ordinary single valued neutrosophic neighborhood system has the same properties in a classical neighborhood system. Finally, we introduce the concepts of an ordinary single valued neutrosophic base and an ordinary single valued neutrosophic subbase, and obtain two characterizations of an ordinary single valued neutrosophic base and one characterization of an ordinary single valued neutrosophic subbase.

2. Preliminaries

In this section, we introduce the concepts of single valued neutrosophic set, the complement of a single valued neutrosophic set, the inclusion between two single valued neutrosophic sets, the union and the intersection of them.

Definition 1 ([43]). *Let X be a non-empty set. Then, A is called a neutrosophic set (in sort, NS) in X, if A has the form $A = (T_A, I_A, F_A)$, where*

$$T_A : X \to]^-0, 1^+[, \quad I_A : X \to]^-0, 1^+[, \quad F_A : X \to]^-0, 1^+[.$$

Since there is no restriction on the sum of $T_A(x)$, $I_A(x)$ and $F_A(x)$, for each $x \in X$,

$$^-0 \leq T_A(x) + I_A(x) + F_A(x) \leq 3^+.$$

Moreover, for each $x \in X$, $T_A(x)$ (resp., $I_A(x)$ and $F_A(x)$) represent the degree of membership (resp., indeterminacy and non-membership) of x to A.

From Example 2.1.1 in [17], we can see that every IFS (intutionistic fuzzy set) A in a non-empty set X is an NS in X having the form

$$A = (T_A, 1 - (T_A + F_A), F_A),$$

where $(1 - (T_A + F_A))(x) = 1 - (T_A(x) + F_A(x))$.

Definition 2 ([43]). *Let A and B be two NSs in X. Then, we say that A is contained in B, denoted by $A \subset B$, if, for each $x \in X$, $inf\ T_A(x) \leq inf\ T_B(x)$, $sup\ T_A(x) \leq sup\ T_B(x)$, $inf\ I_A(x) \geq inf\ I_B(x)$, $sup\ I_A(x) \geq sup\ I_B(x)$, $inf\ F_A(x) \geq inf\ F_B(x)$ and $sup\ F_A(x) \geq sup\ F_B(x)$.*

Definition 3 ([48]). *Let X be a space of points (objects) with a generic element in X denoted by x. Then, A is called a single valued neutrosophic set (in short, SVNS) in X, if A has the form $A = (T_A, I_A, F_A)$, where $T_A,\ I_A,\ F_A : X \to [0,1]$.*

In this case, $T_A,\ I_A,\ F_A$ are called truth-membership function, indeterminacy-membership function, falsity-membership function, respectively, and we will denote the set of all SVNSs in X as $SVNS(X)$.

Furthermore, we will denote the empty SVNS (resp. the whole SVNS] in X as 0_N (resp. 1_N) and define by $0_N(x) = (0,1,1)$ (resp. $1_N = (1,0,0)$), for each $x \in X$.

Definition 4 ([48]). *Let $A \in SVNS(X)$. Then, the complement of A, denoted by A^c, is an SVNS in X defined as follows: for each $x \in X$,*

$$T_{A^c}(x) = F_A(x),\ I_{A^c}(x) = 1 - I_A(x) \text{ and } F_{A^c}(x) = T_A(x).$$

Definition 5 ([50]). *Let $A,\ B \in SVNS(X)$. Then,*

(i) A is said to be contained in B, denoted by $A \subset B$, if, for each $x \in X$,

$$T_A(x) \leq T_B(x),\ I_A(x) \geq I_B(x) \text{ and } F_A(x) \geq F_B(x),$$

(ii) A is said to be equal to B, denoted by $A = B$, if $A \subset B$ and $B \subset A$.

Definition 6 ([51]). *Let $A,\ B \in SVNS(X)$. Then,*

(i) the intersection of A and B, denoted by $A \cap B$, is a SVNS in X defined as:

$$A \cap B = (T_A \wedge T_B, I_A \vee I_B, F_A \vee F_B),$$

where $(T_A \wedge T_B)(x) = T_A(x) \wedge T_B(x)$, $(F_A \vee F_B) = F_A(x) \vee F_B(x)$, for each $x \in X$,

(ii) the union of A and B, denoted by $A \cup B$, is an SVNS in X defined as:

$$A \cup B = (T_A \vee T_B, I_A \wedge I_B, F_A \wedge F_B).$$

Remark 1. *Definitions 5 and 6 are different from the corresponding definitions in [48].*

Result 1 ([51], Proposition 2.1). *Let $A,\ B \in SVNS(X)$. Then,*

(1) *$A \subset A \cup B$ and $B \subset A \cup B$,*

(2) *$A \cap B \subset A$ and $A \cap B \subset B$,*

(3) *$(A^c)^c = A$,*

(4) *$(A \cup B)^c = A^c \cap B^c$, $(A \cap B)^c = A^c \cup B^c$.*

The following are immediate results of Definitions 5 and 6.

Proposition 1. *Let $A,\ B,\ C \in SVNS(X)$. Then,*

(1) (Commutativity) *$A \cup B = B \cup A$, $A \cap B = B \cap A$,*

(2) (Associativity) $A \cup (B \cup C) = (A \cup B) \cup C$, $A \cap (B \cap C) = (A \cap B) \cap C$,
(3) (Distributivity) $A \cup (B \cap C) = (A \cup B) \cap (A \cup C)$, $A \cap (B \cup C) = (A \cap B) \cup (A \cap C)$,
(4) (Idempotency) $A \cup A = A$, $A \cap A = A$,
(5) (Absorption) $A \cup (A \cap B) = A$, $A \cap (A \cup B) = A$,
(6) (DeMorgan's laws) $(A \cup B)^c = A^c \cap B^c$, $(A \cap B)^c = A^c \cup B^c$,
(7) $A \cap 0_N = 0_N$, $A \cup 1_N = 1_N$,
(8) $A \cup 0_N = A$, $A \cap 1_N = A$.

Definition 7 (see [46]). *Let $\{A_\alpha\}_{\alpha \in \Gamma} \subset SVNS(X)$. Then,*

(i) the union of $\{A_\alpha\}_{\alpha \in \Gamma}$, denoted by $\bigcup_{\alpha \in \Gamma} A_\alpha$, is a single valued neutrosophic set in X defined as follows: for each $x \in X$,

$$(\bigcup_{\alpha \in \Gamma} A_\alpha)(x) = (\bigvee_{\alpha \in \Gamma} T_{A_\alpha}(x), \bigwedge_{\alpha \in \Gamma} I_{A_\alpha}(x), \bigwedge_{\alpha \in \Gamma} F_{A_\alpha}(x)),$$

(ii) the intersection of $\{A_\alpha\}_{\alpha \in \Gamma}$, denoted by $\bigcap_{\alpha \in \Gamma} A_\alpha$, is a single valued neutrosophic set in X defined as follows: for each $x \in X$,

$$(\bigcap_{\alpha \in \Gamma} A_\alpha)(x) = (\bigwedge_{\alpha \in \Gamma} T_{A_\alpha}(x), \bigvee_{\alpha \in \Gamma} I_{A_\alpha}(x), \bigvee_{\alpha \in \Gamma} F_{A_\alpha}(x)).$$

The following are immediate results of the above definition.

Proposition 2. *Let $A \in SVNS(X)$ and let $\{A_\alpha\}_{\alpha \in \Gamma} \subset SVNS(X)$. Then,*
(1) (Generalized Distributivity)

$$A \cup (\bigcap_{\alpha \in \Gamma} A_\alpha) = \bigcap_{\alpha \in \Gamma} (A \cup A_\alpha), \quad A \cap (\bigcup_{\alpha \in \Gamma} A_\alpha) = \bigcup_{\alpha \in \Gamma} (A \cap A_\alpha),$$

(2) (Generalized DeMorgan's laws)

$$(\bigcup_{\alpha \in \Gamma} A_\alpha)^c = \bigcap_{\alpha \in \Gamma} A_\alpha^c, \quad (\bigcap_{\alpha \in \Gamma} A_\alpha)^c = \bigcup_{\alpha \in \Gamma} A_\alpha^c.$$

3. Ordinary Single Valued Neutrosophic Topology

In this section, we define an ordinary single valued neutrosophic topological space and obtain some of its properties. Throughout this paper, we denote the set of all subsets (resp. fuzzy subsets) of a set X as 2^X (resp. I^X).

For $T_\alpha, I_\alpha, F_\alpha \in I$, $\alpha = (T_\alpha, I_\alpha, F_\alpha) \in I \times I \times I$ is called a single valued neutrosophic value. For two single valued neutrosophic values α and β,

(i) $\alpha \leq \beta$ iff $T_\alpha \leq T_\beta$, $I_\alpha \geq I_\beta$ and $F_\alpha \geq F_\beta$,
(ii) $\alpha < \beta$ iff $T_\alpha < T_\beta$, $I_\alpha > I_\beta$ and $F_\alpha > F_\beta$.

In particular, the form $\alpha^* = (\alpha, 1-\alpha, 1-\alpha)$ is called a single valued neutrosophic constant.

We denote the set of all single valued neutrosophic values (resp. constant) as **SVNV** (resp. **SVNC**) (see [49]).

Definition 8. *Let X be a nonempty set. Then, a mapping $\tau = (T_\tau, I_\tau, F_\tau) : 2^X \to I \times I \times I$ is called an ordinary single valued neutrosophic topology (in short, osvnt) on X if it satisfies the following axioms: for any $A, B \in 2^X$ and each $\{A_\alpha\}_{\alpha \in \Gamma} \subset 2^X$,*

(OSVNT1) $\tau(\phi) = \tau(X) = (1, 0, 0)$,
(OSVNT2) $T_\tau(A \cap B) \geq T_\tau(A) \wedge T_\tau(B)$, $I_\tau(A \cap B) \leq I_\tau(A) \vee I_\tau(B)$, $F_\tau(A \cap B) \leq F_\tau(A) \vee F_\tau(B)$,
(OSVNT3) $T_\tau(\bigcup_{\alpha \in \Gamma} A_\alpha) \geq \bigwedge_{\alpha \in \Gamma} T_\tau(A_\alpha)$, $I_\tau(\bigcup_{\alpha \in \Gamma} A_\alpha) \leq \bigvee_{\alpha \in \Gamma} I_\tau(A_\alpha)$, $F_\tau(\bigcup_{\alpha \in \Gamma} A_\alpha) \leq \bigvee_{\alpha \in \Gamma} F_\tau(A_\alpha)$.

The pair (X, τ) is called an ordinary single valued neutrosophic topological space (in short, osvnts). We denote the set of all ordinary single valued neutrosophic topologies on X as $OSVNT(X)$.

Let $2 = \{0, 1\}$ and let $\tau : 2^X \to 2 \times 2 \times 2$ satisfy the axioms in Definition 8. Since we can consider as $(1, 0, 0) = 1$ and $(0, 1, 1) = 0$, $\tau \in T(X)$, where $T(X)$ denotes the set of all classical topologies on X. Thus, we can see that $T(X) \subset OSVNT(X)$.

Example 1. *(1) Let $X = \{a, b, c\}$. Then, $2^X = \{\phi, X, \{a\}, \{b\}, \{c\}, \{a, b\}, \{a, c\}, \{b, c\}\}$. We define the mapping $\tau : 2^X \to I \times I \times I$ as follows:*

$$\tau(\phi) = \tau(X) = (1, 0, 0),$$
$$\tau(\{a\}) = (0.7, 0.3, 0.4), \tau(\{b\}) = (0.6, 0.2, 0.3), \tau(\{c\}) = (0.8, 0.1, 0.2),$$
$$\tau(\{a, b\}) = (0.6, 0.3, 0.4), \tau(\{b, c\}) = (0.7, 0.1, 0.2), \tau(\{a, c\}) = (0.8, 0.2, 0.3).$$

Then, we can easily see that $\tau \in OSVNT(X)$.

(2) Let X be a nonempty set. We define the mapping $\tau_\phi : 2^X \to I \times I \times I$ as follows: for each $A \in 2^X$,

$$\tau_\phi(A) = \begin{cases} (1, 0, 0) & \text{if either } A = \phi \text{ or } A = X, \\ (0, 1, 1) & \text{otherwise.} \end{cases}$$

Then, clearly, $\tau_\phi \in OSVT(X)$.

In this case, τ_ϕ (resp. (X, τ_ϕ)) is called the ordinary single valued neutrosophic indiscrete topology on X (resp. the ordinary single valued neutrosophic indiscrete space].

(3) Let X be a nonempty set. We define the mapping $\tau_X : 2^X \to I \times I \times I$ as follows: for each $A \in 2^X$,

$$\tau_X(A) = (1, 0, 0).$$

Then, clearly, $\tau_X \in OSVNT(X)$.

In this case, τ_X (resp. (X, τ_X)) is called the ordinary single valued neutrosophic discrete topology on X (resp. the ordinary single valued neutrosophic discrete space].

*(4) Let X be a set and let $\alpha = (T_\alpha, I_\alpha, F_\alpha) \in$ **SVNV** be fixed, where $T_\alpha \in I_1$ and I_α, $F_\alpha \in I_0$. We define the mapping $\tau : 2^X \to I \times I \times I$ as follows: for each $A \in 2^X$,*

$$\tau(A) = \begin{cases} (1, 0, 0) & \text{if either } A = \phi \text{ or } A^c \text{ is finite,} \\ \alpha & \text{otherwise.} \end{cases}$$

Then, we can easily see that $\tau \in OSVNT(X)$.

In this case, τ is called the α-ordinary single valued neutrosophic finite complement topology on X and will be denoted by $OSVNCof(X)$. $OSVNCof(X)$ is of interest only when X is an infinite set because if X is finite, then $OSVNCof(X) = \tau_\phi$.

*(5) Let X be an infinite set and let $\alpha = (T_\alpha, I_\alpha, F_\alpha) \in$ **SVNV** be fixed, where $T_\alpha \in I_1$ and I_α, $F_\alpha \in I_0$. We define the mapping $\tau : 2^X \to I \times I \times I$ as follows: for each $A \in 2^X$,*

$$\tau(A) = \begin{cases} (1, 0, 0) & \text{if either } A = \phi \text{ or } A^c \text{ is countable,} \\ \alpha & \text{otherwise.} \end{cases}$$

Then, clearly, $\tau \in OSVNT(X)$.

In this case, τ is called the α-ordinary single valued neutrosophic countable complement topology on X and is denoted by $OSVNCoc(X)$.

*(6) Let T be the topology generated by $\mathcal{S} = \{(a,b] : a, b \in \mathbb{R}, a < b\}$ as a subbase, let T_0 be the family of all open sets of $\mathbb{R}$ with respect to the usual topology on $\mathbb{R}$ and let $\alpha = (T_\alpha, I_\alpha, F_\alpha) \in$ **SVNV** be fixed, where $T_\alpha \in I_1$ and I_α, $F_\alpha \in I_0$. We define the mapping $\tau : 2^{\mathbb{R}} \to I \times I \times I$ as follows: for each $A \in I^{\mathbb{R}}$,*

$$\tau(A) = \begin{cases} (1,0,0) & \text{if } A \in T_0, \\ \alpha & \text{if } A \in T \setminus T_0, \\ (0,1,1) & \text{otherwise.} \end{cases}$$

Then, we can easily see that $\tau \in OSVNT(X)$.

(7) Let $T \in T(X)$. We define the mapping $\tau_T : 2^X \to I \times I \times I$ as follows : for each $A \in 2^X$,

$$\tau_T(A) = \begin{cases} (1,0,0) & \text{if } A \in T, \\ (0,1,1) & \text{otherwise.} \end{cases}$$

Then, it is easily seen that $\tau_T \in OSVNT(X)$. Moreover, we can see that if T is the classical indiscrete topology, then $\tau_T = \tau_\phi$ and if T is the classical discrete topology, then $\tau_T = \tau_X$.

Remark 2. *(1) If $I = 2$, then we can think that Definition 8 also coincides with the known definition of classical topology.*

(2) Let (X, τ) be an osvnsts. We define two mappings $[\,]\tau$, $<\ >\tau : 2^X \to I \times I \times I$, respectively, as follows : for each $A \in 2^X$,

$$([\,]\tau)(A) = (T_\tau(A), I_\tau(A), 1 - T_\tau(A)),\ (<\ >\tau)(A) = (1 - F_\tau(A), I_\tau(A), F_\tau(A)).$$

Then, we can easily see that $[\,]\tau$, $<\ >\tau \in OSVNT(X)$.

Definition 9. *Let X be a nonempty set. Then, a mapping $\mathcal{C} = (\mu_{\mathcal{C}}, \nu_{\mathcal{C}}) : 2^X \to I \times I \times I$ is called an ordinary single valued neutrosophic cotopology (in short, osvnct) on X if it satisfies the following conditions: for any $A, B \in 2^X$ and each $\{A_\alpha\}_{\alpha\in\Gamma} \subset 2^X$,*

(OSVNCT1) $\mathcal{C}(\phi) = \mathcal{C}(X) = (1,0,0)$,

(OSVNCT2) $T_{\mathcal{C}}(A \cup B) \geq T_{\mathcal{C}}(A) \wedge T_{\mathcal{C}}(B)$, $I_{\mathcal{C}}(A \cup B) \leq I_{\mathcal{C}}(A) \vee I_{\mathcal{C}}(B)$, $F_{\mathcal{C}}(A \cup B) \leq F_{\mathcal{C}}(A) \vee F_{\mathcal{C}}(B)$,

(OSVNCT3) $T_{\mathcal{C}}(\bigcap_{\alpha\in\Gamma} A_\alpha) \geq \bigwedge_{\alpha\in\Gamma} T_{\mathcal{C}}(A_\alpha)$, $I_{\mathcal{C}}(\bigcap_{\alpha\in\Gamma} A_\alpha) \leq \bigvee_{\alpha\in\Gamma} I_{\mathcal{C}}(A_\alpha)$, $F_{\mathcal{C}}(\bigcap_{\alpha\in\Gamma} A_\alpha) \leq \bigvee_{\alpha\in\Gamma} F_{\mathcal{C}}(A_\alpha)$.

The pair $(X, \mathcal{C})$ is called an ordinary single valued neutrosophic cotopological space (in short, osvncts).

The following is an immediate result of Definitions 8 and 9.

Proposition 3. *We define two mappings $f : OSVNT(X) \to OSVNCT(X)$ and $g : OSVNCT(X) \to OSVNT(X)$ respectively as follows:*

$$[f(\tau)](A) = \tau(A^c) \quad \text{for any } \tau \in OSVNT(X) \text{ and any } A \in 2^X$$

and

$$[g(\mathcal{C})](A) = \mathcal{C}(A^c) \quad \text{for any } \mathcal{C} \in OSVNCT(X) \text{ and any } A \in 2^X.$$

Then, f and g are well-defined. Moreover, $g \circ f = 1_{OSVNT(X)}$ and $f \circ g = 1_{OSVNCT(X)}$.

Remark 3. *(1) For each $\tau \in OSVNT(X)$ and each $\mathcal{C} \in OSVNCT(X)$, let $f(\tau) = \mathcal{C}_\tau$ and $g(\mathcal{C}) = \tau_{\mathcal{C}}$. Then, from Proposition 3, we can see that $\tau_{\mathcal{C}_\tau} = \tau$ and $\mathcal{C}_{\tau_{\mathcal{C}}} = \mathcal{C}$.*

(2) Let $(X, \mathcal{C})$ be an osvncts. We define two mappings $[\,]\mathcal{C}$, $<\ >\mathcal{C} : 2^X \to I \times I \times I$, respectively, as follows: for each $A \in 2^X$,

$$([\,]\mathcal{C})(A) = (T_{\mathcal{C}}(A), I_{\mathcal{C}}(A), 1 - T_{\mathcal{C}}(A)),\ (<\ >\mathcal{C})(A) = (1 - F_{\mathcal{C}}(A), I_{\mathcal{C}}(A), F_{\mathcal{C}}(A)).$$

Then, we can easily see that $[\,]\mathcal{C}$, $<\ >\mathcal{C} \in OSVNCT(X)$.

Definition 10. *Let $\tau_1, \tau_2 \in OSVNT(X)$ and let $\mathcal{C}_1, \mathcal{C}_2 \in OSVNCT(X)$.*

(i) We say that τ_1 is finer than τ_2 or τ_2 is coarser than τ_1, denoted by $\tau_2 \preceq \tau_1$, if $\tau_2(A) \leq \tau_1(A)$, i.e., for each $A \in 2^X$,

$$T_{\tau_2}(A) \leq T_{\tau_1}(A),\ I_{\tau_2}(A) \geq I_{\tau_1}(A),\ F_{\tau_2}(A) \geq F_{\tau_1}(A).$$

(ii) We say that $\mathcal{C}_1$ is finer than $\mathcal{C}_2$ or $\mathcal{C}_2$ is coarser than $\mathcal{C}_1$, denoted by $\mathcal{C}_2 \preceq \mathcal{C}_1$, if $\mathcal{C}_2(A) \leq \mathcal{C}_1(A)$, i.e., for each $A \in 2^X$,

$$T_{\mathcal{C}_2}(A) \leq T_{\mathcal{C}_1}(A),\ I_{\mathcal{C}_2}(A) \geq I_{\mathcal{C}_1}(A),\ F_{\mathcal{C}_2}(A) \geq F_{\mathcal{C}_1}(A).$$

We can easily see that τ_1 is finer than τ_2 if and only if $\mathcal{C}_{\tau_1}$ is finer than $\mathcal{C}_{\tau_2}$, and $(OSVNT(X), \preceq)$ and $(OSVNCT(X), \preceq)$ are posets, respectively.

From Example 1 (2) and (3), it is obvious that τ_ϕ is the coarsest ordinary single valued neutrosophic topology on X and τ_X is the finest ordinary single valued neutrosophic topology on X.

Proposition 4. *If $\{\tau_\alpha\}_{\alpha\in\Gamma} \subset OSVNT(X)$, then $\bigcap_{\alpha\in\Gamma} \tau_\alpha \in OSVNT(X)$, where $[\bigcap_{\alpha\in\Gamma} \tau_\alpha](A) = (\bigwedge_{\alpha\in\Gamma} T_{\tau_\alpha}(A), \bigvee_{\alpha\in\Gamma} I_{\tau_\alpha}(A), \bigvee_{\alpha\in\Gamma} F_{\tau_\alpha}(A)),\ \forall\, A \in 2^X$.*

Proof. Let $\tau = \bigcap_{\alpha\in\Gamma} \tau_\alpha$ and let $\alpha \in \Gamma$. Since $\tau_\alpha \in OSVNT(X)$, $\tau_\alpha(X) = \tau_\alpha(\phi) = (1, 0, 0)$, i.e.,

$$T_{\tau_\alpha}(X) = T_{\tau_\alpha}(\phi) = 1, \quad I_{\tau_\alpha}(X) = I_{\tau_\alpha}(\phi) = 0, \quad F_{\tau_\alpha}(X) = F_{\tau_\alpha}(\phi) = 0.$$

Then, $T_\tau(X) = \bigwedge_{\alpha\in\Gamma} T_{\tau_\alpha}(X) = 1$, $I_\tau(X) = \bigvee_{\alpha\in\Gamma} I_{\tau_\alpha}(X) = 0 = F_\tau(X)$. Similarly, we have $T_\tau(\phi) = 1$, $I_\tau(\phi) = 0 = F_\tau(\phi)$. Thus, the condition (OSVNT1) holds.

Let $A, B \in 2^X$. Then,

$$\begin{aligned} T_\tau(A \cap B) &= \textstyle\bigwedge_{\alpha\in\Gamma} T_{\tau_\alpha}(A \cap B) && \text{[By the definition of } \tau] \\ &\geq \textstyle\bigwedge_{\alpha\in\Gamma}(T_{\tau_\alpha}(A) \wedge T_{\tau_\alpha}(B)) && [\text{Since } \tau_\alpha \in OSVNT(X)] \\ &= (\textstyle\bigwedge_{\alpha\in\Gamma} T_{\tau_\alpha}(A)) \wedge (\textstyle\bigwedge_{\alpha\in\Gamma} T_{\tau_\alpha}(B)) \\ &= T_\tau(A) \wedge T_\tau(B) && \text{[By the definition of } \tau] \end{aligned}$$

and

$$\begin{aligned} I_\tau(A \cap B) &= \textstyle\bigvee_{\alpha\in\Gamma} I_{\tau_\alpha}(A \cap B) && \text{[By the definition of } \tau] \\ &\leq \textstyle\bigvee_{\alpha\in\Gamma}(I_{\tau_\alpha}(A) \vee I_{\tau_\alpha}(B)) && [\text{Since } \tau_\alpha \in OSVNT(X)] \\ &= (\textstyle\bigvee_{\alpha\in\Gamma} I_{\tau_\alpha}(A)) \vee (\textstyle\bigvee_{\alpha\in\Gamma} I_{\tau_\alpha}(B)) \\ &= I_\tau(A) \vee I_\tau(B). && \text{[By the definition of } \tau] \end{aligned}$$

Similarly, we have $F_\tau(A \cap B) \leq F_\tau(A) \vee F_\tau(B)$. Thus, the condition (OSVNT2) holds:

Now, let $\{A_j\}_{j\in J} \subset 2^X$. Then,

$$\begin{aligned} T_\tau(\textstyle\bigcup_{j\in J} A_j) &= \textstyle\bigwedge_{\alpha\in\Gamma} T_{\tau_\alpha}(\textstyle\bigcup_{j\in J} A_j) && \text{[By the definition of } \tau] \\ &\geq \textstyle\bigwedge_{\alpha\in\Gamma}(\textstyle\bigwedge_{j\in J} T_{\tau_\alpha}(A_j)) && [\text{Since } \tau_\alpha \in OSVNT(X)] \\ &= \textstyle\bigwedge_{j\in J}(\textstyle\bigwedge_{\alpha\in\Gamma} T_{\tau_\alpha}(A_j)) \\ &= \textstyle\bigwedge_{j\in J}[\textstyle\bigcap_{\alpha\in\Gamma} T_{\tau_\alpha}](A_j) && \text{[By the definition of } \tau] \\ &= \textstyle\bigvee_{j\in J} T_\tau(A_j) \end{aligned}$$

and

$$\begin{aligned} I_\tau(\bigcup_{j\in J} A_j) &= \bigvee_{\alpha\in\Gamma} I_{\tau_\alpha}(\bigcup_{j\in J} A_j) && \text{[By the definition of } \tau] \\ &\leq \bigvee_{\alpha\in\Gamma}(\bigvee_{j\in J} I_{\tau_\alpha}(A_j)) && [\text{Since } \tau_\alpha \in OSVNT(X)] \\ &= \bigvee_{j\in J}(\bigvee_{\alpha\in\Gamma} I_{\tau_\alpha}(A_j)) \\ &= \bigvee_{j\in J}[\bigcup_{\alpha\in\Gamma} I_{\tau_\alpha}](A_j) && \text{[By the definition of } \tau] \\ &= \bigvee_{j\in J} I_\tau(A_j). \end{aligned}$$

Similarly, we have $F_\tau(\bigcup_{j\in J} A_j) \leq \bigvee_{j\in J} F_\tau(A_j)$. Thus, the condition (OSVNT3) holds. This completes the proof. □

From Definition 10 and Proposition 4, we have the following.

Proposition 5. *$(OSVNT(X), \preceq)$ is a meet complete lattice with the least element τ_ϕ and the greatest element τ_X.*

Definition 11. *Let (X, τ) be an osvnts and let $\alpha \in$ **SVNV**. We define two sets $[\tau]_\alpha$ and $[\tau]^*_\alpha$ as follows, respectively:*
(i) $[\tau]_\alpha = \{A \in 2^X : T_\tau(A) \geq T_\alpha,\ I_\tau(A) \leq I_\alpha,\ I_\tau(A) \leq F_\alpha\}$,
(ii) $[\tau]^*_\alpha = \{A \in 2^X : T_\tau(A) > T_\alpha,\ I_\tau(A) < I_\alpha,\ F_\tau(A) < F_\alpha\}$.

In this case, $[\tau]_\alpha$ (resp. $[\tau]^*_\alpha$) is called the α-level (resp. strong α-level] of τ. If $\alpha = (0,1,1)$, then $[\tau]_{(0,1,1)} = 2^X$, i.e., $[\tau]_{(0,1,1)}$ is the classical discrete topology on X and if $\alpha = (1,0,0)$, then $[\tau]^*_{(1,0,0)} = \phi$. Moreover, we can easily see that for any $\alpha \in$ **SVNV**, $[\tau]^*_\alpha \subset [\tau]_\alpha$.

Lemma 1. *Let $\tau \in OSVNT(X)$ and let $\alpha, \beta \in$ **SVNV**. Then,*
(1) $[\tau]_\alpha \in T(X)$,
(2) *if $\alpha \leq \beta$, then $[\tau]_\beta \subset [\tau]_\alpha$,*
(3) $[\tau]_\alpha = \bigcap_{\beta<\alpha} [\tau]_\beta$, *where $\alpha \in I_0 \times I_1 \times I_1$,*
$(1)'$ $[\tau]^*_\alpha \in T(X)$, *where $\alpha \in I_1 \times I_0 \times I_0$,*
$(2)'$ *if $\alpha \leq \beta$, then $[\tau]^*_\beta \subset [\tau]^*_\alpha$,*
$(3)'$ $[\tau]^*_\alpha = \bigcup_{\beta>(\alpha} [\tau]^*_\beta$, *where $\alpha \in I_1 \times I_0 \times I_0$.*

Proof. The proofs of (1), $(1)'$, (2) and $(2)'$ are obvious from Definitions 8 and 11.

(3) From (2), $\{[\tau]_\alpha\}_{\alpha\in I_0\times I_1\times I_1}$ is a descending family of classical topologies on X. Then, clearly, $[\tau]_\alpha \subset \bigcap_{\beta<\alpha}[\tau]_\beta$, for each $\alpha \in I_0 \times I_1 \times I_1$.

Suppose $A \notin [\tau]_\alpha$. Then, $T_\tau(A) < T_\alpha$ or $I_\tau(A) > I_\alpha$ or $F_\tau(A) > F_\alpha$. Thus,

$$\text{there exists } T_\beta \in I_0 \text{ such that } T_\tau(A) < T_\beta < T_\alpha$$

or

$$\text{there exists } I_\beta \in I_1 \text{ such that } I_\tau(A) > I_\beta > I_\alpha$$

or

$$\text{there exists } F_\beta \in I_1 \text{ such that } F_\tau(A) > F_\beta > F_\alpha.$$

Thus, $A \notin [\tau]_\beta$, for some $\beta \in$ **SVNV** such that $\beta < \alpha$, i.e., $A \notin \bigcap_{\beta<\alpha} [\tau]_\beta$. Hence, $\bigcap_{\beta<\alpha} [\tau]_\beta \subset [\tau]_\alpha$.

Therefore, $[\tau]_\alpha = \bigcap_{\beta<\alpha} [\tau]_\beta$.

$(3)'$ The proof is similar to (3). □

Remark 4. *From (1) and (2) in Lemma 1, we can see that, for each* $\tau \in OSVNT(X)$, $\{[\tau]_\alpha\}_{\alpha \in \mathbf{SVNV}}$ *is a family of descending classical topologies called the* α*-level classical topologies on* X *with respect to* τ.

The following is an immediate result of Lemma 1.

Corollary 1. *Let* (X, τ) *be an osvnts. Then,* $[\tau]_{\alpha^*} = \bigcap_{\beta<\alpha} [\tau]_{\beta^*}$ *for each* $\alpha^* \in \mathbf{SVNC}$*, where* $\alpha \in I_0$.

Lemma 2. (1) *Let* $\{\tau_\alpha\}_{\alpha \in \mathbf{SVNV}}$ *be a descending family of classical topologies on* X *such that* $\tau_{(0,1,1)}$ *is the classical discrete topology on* X. *We define the mapping* $\tau : 2^X \to I \times I \times I$ *as follows: for each* $A \in 2^X$,

$$\tau(A) = \Big(\bigvee_{A \in \tau_\alpha} T_\alpha, \bigwedge_{A \in \tau_\alpha} I_\alpha, \bigwedge_{A \in \tau_\alpha} F_\alpha \Big).$$

Then, $\tau \in OSVNT(X)$.
(2) *If* $\tau_\alpha = \bigcap_{\beta<\alpha} \tau_\alpha$*, for each* $\alpha \in \mathbf{SVNV}$ $(\alpha \in I_0 \times I_1 \times I_1)$*, then* $[\tau]_\alpha = \tau_\alpha$.
(3) *If* $\tau_\alpha = \bigcup_{\beta>\alpha} \tau_\beta$*, for each* $\alpha \in \mathbf{SVNV}$ $(\alpha \in I_1 \times I_0 \times I_0)$*, then* $[\tau]^*_\alpha = \tau_\alpha$.

Proof. The proof is similar to Lemma 3.9 in [28]. □

The following is an immediate result of Lemma 2.

Corollary 2. *Let* $\{\tau_{\alpha^*}\}_{\alpha \in I_0}$ *be a descending family of classical topologies on* X *such that* $\tau_{(0,1,1)}$ *is the classical discrete topology on* X. *We define the mapping* $\tau : 2^X \to I \times I \times I$ *as follows: for each* $A \in 2^X$,

$$\tau(A) = \Big(\bigvee_{A \in \tau_{\alpha^*}} \alpha, \bigwedge_{A \in \tau_{\alpha^*}} (1-\alpha), \bigwedge_{A \in \tau_{\alpha^*}} (1-\alpha) \Big).$$

Then, $\tau \in OSVNT(X)$ *and* $[\tau]_{\alpha^*} = \bigcap_{\beta<\alpha} \tau_{\beta^*} = \tau_{\alpha^*}$ $\forall \alpha \in I_0$.

From Lemmas 1 and 2, we have the following result.

Proposition 6. *Let* $\tau \in OSVNT(X)$ *and let* $[\tau]_\alpha$ *be the* α*-level classical topology on* X *with respect to* τ. *We define the mapping* $\eta : 2^X \to I \times I \times I$ *as follows: for each* $A \in 2^X$,

$$\eta(A) = \Big(\bigvee_{A \in [\tau]_\alpha} T_\alpha, \bigwedge_{A \in [\tau]_\alpha} I_\alpha, \bigwedge_{A \in [\tau]_\alpha} F_\alpha \Big).$$

Then, $\eta = \tau$.

The fact that an ordinary single valued neutrosophic topological space fully determined by its decomposition in classical topologies is restated in the following theorem.

Theorem 1. *Let* $\tau_1, \tau_2 \in OSVNT(X)$. *Then,* $\tau_1 = \tau_2$ *if and only if* $[\tau_1]_\alpha = [\tau_2]_\alpha$ *for each* $\alpha \in \mathbf{SVNV}$*, or alternatively, if and only if* $[\tau_1]^*_\alpha = [\tau_2]^*_\alpha$ *for each* $\alpha \in \mathbf{SVNV}$.

Remark 5. *In a similar way, we can construct an ordinary single valued neutrosophic cotopology* $\mathcal{C}$ *on a set* X*, by using the* α*-levels,*

$$[\mathcal{C}]_\alpha = \{A \in I^X : T_c(A) \geq T_\alpha,\ I_c(A) \leq I_\alpha,\ F_c(A) \leq F_\alpha\}$$

and

$$[\mathcal{C}]^*_\alpha = \{A \in I^X : T_c(A) > T_\alpha,\ I_c(A) < I_\alpha,\ F_c(A) < F_\alpha\},$$

for each $\alpha \in \mathbf{SVNV}$.

Definition 12. *Let $T \in T(X)$ and let $\tau \in OSVNT(X)$. Then, τ is said to be compatible with T if $T = S(\tau)$, where $S(\tau) = \{A \in 2^X : T_\tau(A) > 0, I_\tau(A) < 1, F_\tau(A) < 1\}$.*

Example 2. *(1) Let τ_ϕ be the ordinary single valued neutrosophic indiscrete topology on a nonempty set X and let T_0 be the classical indiscrete topology on X. Then, clearly,*

$$S(\tau_\phi) = \{A \in 2^X : T_{\tau_\phi}(A) > 0, I_{\tau_\phi}(A) < 1, F_{\tau_\phi}(A) < 1\} = \{\phi, X\} = T_0.$$

Thus, τ_ϕ is compatible with T_0.

(2) Let τ_X be the ordinary single valued neutrosophic discrete topology on a nonempty set X and let T_1 be the classical discrete topology on X. Then, clearly,

$$S(\tau_X) = \{A \in 2^X : T_{\tau_X}(A) > 0, I_{\tau_X}(A) < 1, F_{\tau_X}(A) < 1\} = 2^X = T_1.$$

Thus, τ_X is compatible with T_1.

*(3) Let X be a nonempty set and let $\alpha \in$ **SVNV** be fixed, where $\alpha \in I_0 \times I_1 \times I_1$. We define the mapping $\tau : 2^X \to I \times I \times I$ as follows: for each $A \in 2^X$,*

$$\tau(A) = \begin{cases} (1,0,0) & \text{if either } A = \phi \text{ or } A = X, \\ \alpha & \text{otherwise.} \end{cases}$$

Then, clearly, $\tau \in OSVNT(X)$ and τ is compatible with T_1.

Furthermore, every classical topology can be considered as an ordinary single valued neutrosophic topology in the sense of the following result.

Proposition 7. *Let (X, τ) be a classical topological space and and let $\alpha \in$ **SVNV** be fixed, where $\alpha \in I_0 \times I_1 \times I_1$. Then, there exists $\tau^\alpha \in OSVNT(X)$ such that τ^α is compatible with T. Moreover, $[\tau^\alpha]_\alpha = \tau$.*

In this case, τ^α is called the α-th ordinary single valued neutrosophic topology on X and (X, τ^α) is called the α-th ordinary single valued neutrosophic topological space.

Proof. Let $\alpha \in$ **SVNV** be fixed, where $\alpha \in I_0 \times I_1 \times I_1$ and we define the mapping $\tau^\alpha : 2^X \to I \times I \times I$ as follows: for each $A \in 2^X$,

$$\tau^\alpha(A) = \begin{cases} (1,0,0) & \text{if either } A = \phi \text{ or } A = X, \\ \alpha & \text{if } A \in \tau \setminus \{\phi, X\}, \\ (0,1,1) & \text{otherwise.} \end{cases}$$

Then, we can easily see that $\tau^\alpha \in OSVNT(X)$ and $[\tau^\alpha]_\alpha = \tau$. Moreover, by the definition of τ^α,

$$S(\tau^\alpha) = \{A \in 2^X : T_{\tau^\alpha}(A) > 0, I_{\tau^\alpha}(A) < 1, F_{\tau^\alpha}(A) < 1\} = \tau.$$

Thus, τ^α is compatible with τ. □

Proposition 8. *Let (X, T) be a classical topological space, let $C(T)$ be the set of all osvnts on X compatible with T, let $\widetilde{T} = T \setminus \{\phi, X\}$ and let $(I \times I \times I)^{\widetilde{T}}_{(0,1,1)}$ be the set of all mappings $f : \widetilde{T} \to I \times I \times I$ satisfying the following conditions: for any $A, B \in \widetilde{T}$ and each $(A_j)_{j \in J} \subset \widetilde{T}$,*

(1) $f(A) \neq (0,1,1)$,

(2) $T_f(A \cap B) \geq T_f(A) \wedge T_f(B)$, $I_f(A \cap B) \leq I_f(A) \vee T_f(B)$, $F_f(A \cap B) \leq F_f(A) \vee F_f(B)$,

(3) $T_f(\bigcup_{j \in J} A_j) \geq \bigwedge_{j \in J} T_f(A_j)$, $I_f(\bigcup_{j \in J} A_j) \leq \bigvee_{j \in J} I_f(A_j)$, $F_f(\bigcup_{j \in J} A_j) \leq \bigvee_{j \in J} F_f(A_j)$.

Then, there is a one-to-one correspondence between $C(T)$ *and* $(I \times I \times I)^{\widetilde{T}}_{(0,1,1)}$.

Proof. We define the mapping $F : (I \times I \times I)^{\widetilde{T}}_{(0,1,1)} \to C(T)$ as follows: for each $f \in (I \times I \times I)^{\widetilde{T}}_{(0,1,1)}$,

$$F(f) = \tau_f,$$

where $\tau_f : 2^X \to I \times I \times I$ is the mapping defined by: for each $A \in 2^X$,

$$\tau_f(A) = \begin{cases} (1,0,0) & \text{if either } A = \phi \text{ or } A = X, \\ f(A) & \text{if } A \in \widetilde{T}, \\ (0,1,1) & \text{otherwise.} \end{cases}$$

Then, we easily see that $\tau_f \in C(T)$.

Now, we define the mapping $G : C(T) \to (I \times I \times I)^{\widetilde{T}}_{(0,1,1)}$ as follows: for each $\tau \in C(T)$,

$$G(\tau) = f_\tau,$$

where $f_\tau : \widetilde{T} \to I \times I \times I$ is the mapping defined by: for each $A \in \widetilde{T}$,

$$f_\tau(A) = \tau(A).$$

Then, clearly, $f_\tau \in (I \times I \times I)^{\widetilde{T}}_{(0,1,1)}$. Furthermore, we can see that $F \circ G = id_{C(T)}$ and $G \circ F = id_{(I \times I \times I)^{\widetilde{T}}_{(0,1,1)}}$. Thus, $C(T)$ is equipotent to $I \times I \times I)^{\widetilde{T}}_{(0,1,1)}$. This completes the proof. □

Proposition 9. *Let* (X, τ) *be an osvnts and let* $Y \subset X$. *We define the mapping* $\tau_Y : 2^Y \to I \times I \times I$ *as follows: for each* $A \in 2^Y$,

$$\tau_Y(A) = \Big(\bigvee_{B \in 2^X,\ A = B \cap Y} T_\tau(B), \bigwedge_{B \in 2^X,\ A = B \cap Y} I_\tau(B), \bigwedge_{B \in 2^X,\ A = B \cap Y} F_\tau(B)\Big).$$

Then, $\tau_Y \in OSVNT(Y)$ *and for each* $A \in 2^Y$,

$$T_{\tau_Y}(A) \geq T_\tau(A),\ I_{\tau_Y}(A) \leq I_\tau(A),\ F_{\tau_Y}(A) \leq F_\tau(A).$$

In this case, (Y, τ_Y) is called an ordinary single valued neutrosophic subspace of (X, τ) and τ_Y is called the induced ordinary single valued neutrosophic topology on A by τ.

Proof. It is obvious that the condition (OSVNT1) holds, i.e., $\tau_Y(\phi) = \tau_Y(Y) = (1,0,0)$.

Let $A, B \in 2^Y$. Then, by proof of Proposition 5.1 in [34], $T_{\tau_Y}(A \cap B) \geq T_{\tau_Y}(A) \wedge T_{\tau_Y}(B)$.

Let us show that $I_{\tau_Y}(A \cap B) \leq I_{\tau_Y}(A) \vee I_{\tau_Y}(B)$. Then,

$$\begin{aligned} I_{\tau_Y}(A) \vee I_{\tau_Y}(B) &= (\textstyle\bigwedge_{C_1 \in 2^X,\ A = Y \cap C_1} I_\tau(C_1)) \vee (\bigwedge_{C_2 \in 2^X,\ B = Y \cap C_2} I_\tau(C_2)) \\ &= \textstyle\bigwedge_{C_1,\ C_1 \in 2^X,\ A \cap B = Y \cap (C_1 \cap C_2)} [I_\tau(C_1) \vee I_\tau(C_2)] \\ &\geq \textstyle\bigwedge_{C_1,\ C_1 \in 2^X,\ A \cap B = Y \cap (C_1 \cap C_2)} I_\tau(C_1 \cap C_2) \\ &= I_{\tau_Y}(A \cap B). \end{aligned}$$

Similarly, we have $F_{\tau_Y}(A \cap B) \leq F_{\tau_Y}(A) \vee F_{\tau_Y}(B)$. Thus, the condition (OSVNT2) holds.

Now, let $\{A_\alpha\}_{\alpha \in \Gamma} \subset 2^Y$. Then, by the proof of Proposition 5.1 in [34], $T_{\tau_Y}(\bigcup_{\alpha \in \Gamma} A_\alpha) \geq \bigwedge_{\alpha \in \Gamma} T_{\tau_Y}(A_\alpha)$. On the other hand,

$$\begin{aligned} I_{\tau_Y}(\bigcup_{\alpha\in\Gamma} A_\alpha) &= \bigwedge_{B_\alpha\in 2^X,\ (\bigcup_{\alpha\in\Gamma} B_\alpha)\cap Y=\bigcup_{\alpha\in\Gamma} A_\alpha} I_\tau(\bigcup_{\alpha\in\Gamma} B_\alpha) \\ &\leq \bigwedge_{B_\alpha\in 2^X,\ (\bigcup_{\alpha\in\Gamma} B_\alpha)\cap Y=\bigcup_{\alpha\in\Gamma} A_\alpha} [\bigwedge_{\alpha\in\Gamma} I_\tau(B_\alpha)] \\ &= \bigwedge_{\alpha\in\Gamma} [\bigwedge_{B_\alpha\in 2^X,\ (\bigcup_{\alpha\in\Gamma} B_\alpha)\cap Y=\bigcup_{\alpha\in\Gamma} A_\alpha} I_\tau(B_\alpha)] \\ &= \bigwedge_{\alpha\in\Gamma} I_{\tau_Y}(A_\alpha). \end{aligned}$$

Similarly, we have $F_{\tau_Y}(\bigcup_{\alpha\in\Gamma} A_\alpha) \leq \bigwedge_{\alpha\in\Gamma} F_{\tau_Y}(A_\alpha)$. Thus, the condition (OSVNT3) holds. Thus, $\tau_Y \in OSVNT(Y)$.

Furthermore, we can easily see that for each $A \in 2^Y$,

$$T_{\tau_Y}(A) \geq T_\tau(A), \quad I_{\tau_Y}(A) \leq I_\tau(A), \quad F_{\tau_Y}(A) \leq F_\tau(A).$$

This completes the proof. □

The following is an immediate result of Proposition 9.

Corollary 3. *Let (Y, τ_Y) be an ordinary single valued neutrosaophic subspace of (X, τ) and let $A \in 2^Y$.*

(1) $\mathcal{C}_Y(A) = (\bigvee_{B\in 2^X, A=B\cap Y} T_\mathcal{C}(B), \bigwedge_{B\in 2^X, A=B\cap Y} I_\mathcal{C}(B), \bigwedge_{B\in 2^X, A=B\cap Y} F_\mathcal{C}(B))$, where $\mathcal{C}_Y(A) = \tau_Y(Y - A)$.

(2) If $Z \subset Y \subset X$, then $\tau_Z = (\tau_Y)_Z$.

4. Ordinary Single Valued Neutrosophic Neighborhood Structures of a Point

In this section, we define an ordinary single valued neutrosophic neighborhood system of a point, and prove that it has the same properties in a classical neighborhood system.

Definition 13. *Let (X, τ) be an osvnts and let $x \in X$. Then, a mapping $\mathcal{N}_x : 2^X \to I \times I \times I$ is called the ordinary single valued neutrosophic neighborhood system of x if, for each $A \in 2^X$,*

$$A \in \mathcal{N}_x := \exists B(B \in \tau) \wedge (x \in B \subset A)),$$

i.e.,

$$[A \in \mathcal{N}_x] = \mathcal{N}_x(A) = (\bigvee_{x\in B\subset A} T_\tau(B), \bigwedge_{x\in B\subset A} I_\tau(B), \bigwedge_{x\in B\subset A} F_\tau(B)).$$

Lemma 3. *Let (X, τ) be an osvnts and let $A \in 2^X$. Then,*

$$\bigwedge_{x\in A} \bigvee_{x\in B\subset A} T_\tau(B) = T_\tau(A),$$

$$\bigvee_{x\in A} \bigwedge_{x\in B\subset A} I_\tau(B) = I_\tau(A)$$

and

$$\bigvee_{x\in A} \bigwedge_{x\in B\subset A} F_\tau(B) = F_\tau(A).$$

Proof. By Theorem 3.1 in [33], it is obvious that $\bigwedge_{x\in A} \bigvee_{x\in B\subset A} T_\tau(B) = T_\tau(A)$.

On the other hand, it is clear that $\bigvee_{x\in A} \bigwedge_{x\in B\subset A} I_\tau(B) \geq I_\tau(A)$. Now, let $\mathcal{B}_x = \{B \in 2^X : x \in B \subset A\}$ and let $f \in \Pi_{x\in A}\mathcal{B}_x$. Then, clearly, $\bigcup_{x\in A} f(x) = A$. Thus,

$$\bigvee_{x\in A} I_\tau(f(x)) \leq I_\tau(\bigcup_{x\in A} f(x)) = I_\tau(A).$$

Thus,

$$\bigvee_{x\in A} \bigwedge_{x\in B\subset A} I_\tau(B) = \bigwedge_{f\in\Pi_{x\in A}} \bigvee_{x\in A} I_\tau(f(x)) \leq I_\tau(A).$$

Hence, $\bigvee_{x \in A} \bigwedge_{x \in B \subset A} I_\tau(B) = I_\tau(A)$. Similarly, we have

$$\bigvee_{x \in A} \bigwedge_{x \in B \subset A} F_\tau(B) = F_\tau(A).$$

□

Theorem 2. *Let (X, τ) be an osvnts, let $A \in 2^X$ and let $x \in X$. Then,*

$$\models (A \in \tau) \leftrightarrow \forall x(x \in A \rightarrow \exists B(B \in \mathcal{N}_x) \wedge (B \subset A)),$$

i.e.,

$$[A \in \tau] = [\forall x(x \in A \rightarrow \exists B(B \in \mathcal{N}_x) \wedge (B \subset A))],$$

i.e.,

$$[A \in \tau] = (\bigwedge_{x \in A} \bigvee_{B \subset A} T_{\mathcal{N}_x}(B), \bigvee_{x \in A} \bigwedge_{B \subset A} I_{\mathcal{N}_x}(B), \bigvee_{x \in A} \bigwedge_{B \subset A} F_{\mathcal{N}_x}(B)).$$

Proof. From Theorem 3.1 in [33], it is clear that $T_\tau(A) = \bigwedge_{x \in A} \bigvee_{B \subset A} T_{\mathcal{N}_x}(B)$.
On the other hand,

$$\begin{aligned} I_\tau(A) &= \textstyle\bigvee_{x \in A} \bigwedge_{x \in C \subset A} I_\tau(C) && \text{[By Lemma 3]} \\ &= \textstyle\bigvee_{x \in A} \bigwedge_{B \subset A} \bigwedge_{x \in C \subset B} I_\tau(C) \\ &= \textstyle\bigvee_{x \in A} \bigwedge_{B \subset A} I_{\mathcal{N}_x}(B). && \text{[By Definition 13]} \end{aligned}$$

Similarly, we have $F_\tau(A) = \bigvee_{x \in A} \bigwedge_{B \subset A} F_{\mathcal{N}_x}(B)$. This completes the proof. □

Definition 14. *Let $\mathcal{A}$ be a single valued neutrosophic set in a set 2^X. Then, $\mathcal{A}$ is said to be normal if there is $A_0 \in 2^X$ such that $\mathcal{A}(A_0) = (1, 0, 0)$.*
We will denote the set of all normal single valued neutrosophic sets in 2^X as $(I \times I \times I)_N^{2^X}$.

From the following result, we can see that an ordinary single valued neutrosophic neighborhood system has the same properties in a classical neighborhood system.

Theorem 3. *Let (X, τ) be an osvnts and let $\mathcal{N} : X \rightarrow (I \times I \times I)_N^{2^X}$ be the mapping given by $\mathcal{N}(x) = \mathcal{N}_x$, for each $x \in X$. Then, $\mathcal{N}$ has the following properties:*

(1) *for any $x \in X$ and $A \in 2^X$, $\models A \in \mathcal{N}_x \rightarrow x \in A$,*
(2) *for any $x \in X$ and $A, B \in 2^X$, $\models (A \in \mathcal{N}_x) \wedge (B \in \mathcal{N}_x) \rightarrow A \cap B \in \mathcal{N}_x$,*
(3) *for any $x \in X$ and $A, B \in 2^X$, $\models (A \subset B) \rightarrow (A \in \mathcal{N}_x \rightarrow B \in \mathcal{N}_x)$,*
(4) *for any $x \in X$, $\models (A \in \mathcal{N}_x) \rightarrow \exists C((C \in \mathcal{N}_x) \wedge (C \subset A) \wedge \forall y(y \in C \rightarrow C \in \mathcal{N}_y))$.*

Conversely, if a mapping $\mathcal{N} : X \rightarrow (I \times I \times I)_N^{2^X}$ satisfies the above properties (2) *and* (3)*, then there is an ordinary single valued neutrosophic topology $\tau : 2^X \rightarrow I \times I \times I$ on X defined as follows: for each $A \in 2^X$,*

$$A \in \tau := \forall x(x \in A \rightarrow A \in \mathcal{N}_x),$$

i.e.,

$$[A \in \tau] = \tau(A) = (\bigwedge_{x \in A} T_{\mathcal{N}_x}(A), \bigvee_{x \in A} I_{\mathcal{N}_x}(A), \bigvee_{x \in A} F_{\mathcal{N}_x}(A)).$$

In particular, if $\mathcal{N}$ also satisfies the above properties (1) and (4), then, for each $x \in X$, $\mathcal{N}_x$ is an ordinary single valued neutrosophic neighborhood system of x with respect to τ.

Proof. (1) Since $A \in 2^X$, we can consider A as a special single valued neutrosophic set in x represented by $A = (\chi_A, \chi_{A^c}, \chi_{A^c})$. Then,

$$[x \in A] = A(x) = (\chi_A(x), \chi_{A^c}(x), \chi_{A^c}(x)) = (1,0,0).$$

On the other hand,

$$[A \in \mathcal{N}_x] = (\bigvee_{x \in C \subset A} T_\tau(C), \bigwedge_{x \in C \subset A} I_\tau(C), \bigwedge_{x \in C \subset A} F_\tau(C)) \leq (1,0,0).$$

Thus, $[A \in \mathcal{N}_x] \leq [x \in A]$.
(2) By the definition of $\mathcal{N}_x$,

$$[A \cap B \in \mathcal{N}_x] = (\bigvee_{x \in C \subset A \cap B} T_\tau(C), \bigwedge_{x \in C \subset A \cap B} I_\tau(C)), \bigwedge_{x \in C \subset A \cap B} F_\tau(C)).$$

From the proof of Theorem 3.2 (2) in [33], it is obvious that

$$T_{\mathcal{N}_x}(A \cap B) \geq T_{\mathcal{N}_x}(A) \wedge T_{\mathcal{N}_x}(B).$$

Thus, it is sufficient to show that $I_{\mathcal{N}_x}(A \cap B) \leq I_{\mathcal{N}_x}(A) \vee I_{\mathcal{N}_x}(B)$:

$$\begin{aligned} I_{\mathcal{N}_x}(A \cap B) &= \bigwedge_{x \in C \subset A \cap B} I_\tau(C) = \bigwedge_{x \in C_1 \subset A,\ x \in C_2 \subset B} I_\tau(C_1 \cap C_2) \\ &\leq \bigwedge_{x \in C_1 \subset A,\ x \in C_2 \subset B} (I_\tau(C_1) \vee I_\tau(C_2)) \\ &= \bigwedge_{x \in C_1 \subset A} I_\tau(C_1) \vee \bigwedge_{x \in C_2 \subset B} I_\tau(C_2) \\ &= I_{\mathcal{N}_x}(A) \vee I_{\mathcal{N}_x}(B). \end{aligned}$$

Similarly, we have $F_{\mathcal{N}_x}(A \cap B) \leq F_{\mathcal{N}_x}(A) \vee F_{\mathcal{N}_x}(B)$. On the other hand,

$$[(A \in \mathcal{N}_x) \wedge (B \in \mathcal{N}_x)] = (T_{\mathcal{N}_x}(A) \wedge T_{\mathcal{N}_x}(B), I_{\mathcal{N}_x}(A) \vee I_{\mathcal{N}_x}(B), F_{\mathcal{N}_x}(A) \vee F_{\mathcal{N}_x}(B)).$$

Thus, $[A \cap B \in \mathcal{N}_x] \geq [(A \in \mathcal{N}_x) \wedge (B \in \mathcal{N}_x)]$.
(3) From the definition of $\mathcal{N}_x$, we can easily show that $[A \in \mathcal{N}_x] \leq [B \in \mathcal{N}_x]$.
(4) It is clear that

$$\begin{aligned} &[\exists C((C \in \mathcal{N}_x) \wedge (C \subset A) \wedge \forall y(y \in C \to C \in \mathcal{N}_y))] \\ &= (\bigvee_{C \subset A}[T_{\mathcal{N}_x}(C) \wedge \bigwedge_{y \in C} T_{\mathcal{N}_y}(C)], \bigwedge_{C \subset A}[I_{\mathcal{N}_x}(C) \vee \bigvee_{y \in C} I_{\mathcal{N}_y}(C)], \\ &\qquad \bigwedge_{C \subset A}[F_{\mathcal{N}_x}(C) \vee \bigvee_{y \in C} F_{\mathcal{N}_y}(C)]). \end{aligned}$$

Then, by the proof of Theorem 3.2 (4) in [33], it is obvious that

$$\bigvee_{C \subset A} [T_{\mathcal{N}_x}(C) \wedge \bigwedge_{y \in C} T_{\mathcal{N}_y}(C)] \geq T_{\mathcal{N}_x}(A).$$

From Lemma 3, $\bigvee_{y \in C} I_{\mathcal{N}_y}(C) = \bigvee_{y \in C} \bigwedge_{y \in D \subset C} I_\tau(D) = I_\tau(C)$. Thus,

$$\begin{aligned} \bigwedge_{C \subset A}[I_{\mathcal{N}_x}(C) \vee \bigvee_{y \in C} I_{\mathcal{N}_y}(C)] &= \bigwedge_{C \subset A}[I_{\mathcal{N}_x}(C) \vee I_\tau(C)] = \bigwedge_{C \subset A} I_\tau(C) \\ &\leq \bigwedge_{x \in C \subset A} I_\tau(C) = I_{\mathcal{N}_x}(A). \end{aligned}$$

Similarly, we have $\bigwedge_{C \subset A}[F_{\mathcal{N}_x}(C) \vee \bigvee_{y \in C} F_{\mathcal{N}_y}(C)] \leq \bigwedge_{x \in C \subset A} F_\tau(C) = F_{\mathcal{N}_x}(A)$. Thus,

$$[\exists C((C \in \mathcal{N}_x) \wedge (C \subset A) \wedge \forall y(y \in C \to C \in \mathcal{N}_y))] \geq [A \in \mathcal{N}_x].$$

Conversely, suppose $\mathcal{N}$ satisfies the above properties (2) and (3) and let $\tau : 2^X \to I \times I \times I$ be the mapping defined as follows: for each $A \in 2^X$,

$$\tau(A) = (\bigwedge_{x \in A} T_{\mathcal{N}_x}(A), \bigvee_{x \in A} I_{\mathcal{N}_x}(A), \bigvee_{x \in A} F_{\mathcal{N}_x}(A)).$$

Then, clearly, $\tau(\phi) = (1,0,0)$. Since $\mathcal{N}_x$ is single valued neutrosophic normal, there is $A_0 \in 2^X$ such that $\mathcal{N}_x(A_0) = (1,0,0)$. Thus, $\mathcal{N}_x(X) = (1,0,0)$. Thus,

$$\tau(X) = (\bigwedge_{x\in X} T_{\mathcal{N}_x}(X), \bigvee_{x\in X} I_{\mathcal{N}_x}(X), \bigvee_{x\in X} F_{\mathcal{N}_x}(X)) = (1,0,0).$$

Hence, τ satisfies the axiom (OSVNT1).

From the proof of Theorem 3.2 in [33], it is clear that $T_\tau(A \cap B) \geq T_\tau(A) \wedge T_\tau(B)$.

On the other hand,

$$\begin{aligned} I_\tau(A \cap B) &= \bigvee_{x\in A\cap B} I_{\mathcal{N}_x}(A \cap B) \leq \bigvee_{x\in A\cap B}(I_{\mathcal{N}_x}(A) \vee I_{\mathcal{N}_x}(B)) \\ &= \bigvee_{x\in A\cap B} I_{\mathcal{N}_x}(A) \vee \bigvee_{x\in A\cap B} I_{\mathcal{N}_x}(B) \\ &\leq \bigvee_{x\in A} I_{\mathcal{N}_x}(A) \vee \bigvee_{x\in B} I_{\mathcal{N}_x}(B) \\ &= I_\tau(A) \vee I_\tau(B). \end{aligned}$$

Similarly, we have $F_\tau(A \cap B) \leq F_\tau(A) \vee F_\tau(B)$. Then, τ satisfies the axiom (OSVNT2). Moreover, we can easily see that τ satisfies the axiom (OSVNT3). Thus, $\tau \in OSVNT(X)$.

Now, suppose $\mathcal{N}$ satisfies additionally the above properties (1) and (4). Then, from the proof of Theorem 3.2 in [33], we have $T_{\mathcal{N}_x}(A) = \bigvee_{x\in B\subset A} T_\tau(B)$ for each $x \in X$ and each $A \in 2^X$.

Let $x \in X$ and let $A \in 2^X$. Then, by property (4),

$$I_{\mathcal{N}_x}(A) \geq \bigwedge_{C\subset A} [I_{\mathcal{N}_x}(C) \vee \bigvee_{y\in C} I_{\mathcal{N}_y}(C)].$$

From the property (1), $I_{\mathcal{N}_x}(C) = 1$ for any $x \notin C$. Thus,

$$\begin{aligned} I_{\mathcal{N}_x}(A) &\geq \bigwedge_{x\in C\subset A}[I_{\mathcal{N}_x}(C) \vee \bigvee_{y\in C} I_{\mathcal{N}_y}(C)] \\ &\geq \bigwedge_{x\in C\subset A} \bigvee_{y\in C} I_{\mathcal{N}_y}(C) \\ &= \bigwedge_{x\in B\subset A} I_\tau(B). \end{aligned}$$

Now, suppose $x \in C \subset A$. Then, clearly, $\bigvee_{y\in C} I_{\mathcal{N}_y}(C) \geq I_{\mathcal{N}_x}(C) \geq I_{\mathcal{N}_x}(A)$.

Thus,

$$\bigwedge_{x\in B\subset A} I_\tau(B) = \bigwedge_{x\in C\subset A} \bigvee_{y\in C} I_{\mathcal{N}_y}(C) \geq I_{\mathcal{N}_x}(A).$$

Thus, $I_{\mathcal{N}_x}(A) = \bigwedge_{x\in B\subset A} I_\tau(B)$. Similarly, we have $F_{\mathcal{N}_x}(A) = \bigwedge_{x\in B\subset A} F_\tau(B)$. This completes the proof. □

5. Ordinary Single Valued Neutrosophic Bases and Subbases

In this section, we define an ordinary single valued neutrosophic base and subbase for an ordinary single valued neutrosophic topological space, and investigated general properties. Moreover, we obtain two characterizations of an ordinary single valued neutrosophic base and one characterization of an ordinary single valued neutrosophic subbase.

Definition 15. *Let (X, τ) be an osvnts and let $\mathcal{B} : 2^X \to I \times I \times I$ be a mapping such that $\mathcal{B} \leq \tau$, i.e., $T_\mathcal{B} \leq T_\tau$, $I_\mathcal{B} \geq I_\tau$, $F_\mathcal{B} \geq F_\tau$. Then, $\mathcal{B}$ is called an ordinary single valued neutrosophic base for τ if, for each $A \in 2^X$,*

$$T_\tau(A) = \bigvee_{\{B_\alpha\}_{\alpha\in\Gamma}\subset 2^X,\ A=\bigcup_{\alpha\in\Gamma} B_\alpha} \bigwedge_{\alpha\in\Gamma} T_\mathcal{B}(B_\alpha),$$

$$I_\tau(A) = \bigwedge_{\{B_\alpha\}_{\alpha\in\Gamma}\subset 2^X,\ A=\bigcup_{\alpha\in\Gamma} B_\alpha} \bigvee_{\alpha\in\Gamma} I_\mathcal{B}(B_\alpha),$$

$$F_\tau(A) = \bigwedge_{\{B_\alpha\}_{\alpha\in\Gamma}\subset 2^X,\ A=\bigcup_{\alpha\in\Gamma} B_\alpha} \bigvee_{\alpha\in\Gamma} F_\mathcal{B}(B_\alpha).$$

Example 3. *(1) Let X be a set and let $\mathcal{B} : 2^X \to I \times I \times I$ be the mapping defined by:*

$$\mathcal{B}(\{x\}) = (1,0,0) \ \forall x \in X.$$

Then, $\mathcal{B}$ is an ordinary single valued neutrosophic base for τ_X.

*(2) Let $X = \{a,b,c\}$, let $\alpha \in$ **SVNV** be fixed, where $\alpha \in I_1 \times I_0 \times I_0$ and let $\mathcal{B} : 2^X \to I \times I \times I$ be the mapping as follows: for each $A \in 2^X$,*

$$\mathcal{B}(A) = \begin{cases} (1,0,0) & \text{if either } A = \{a,b\} \text{ or } \{b,c\} \text{ or } X, \\ \alpha & \text{otherwise.} \end{cases}$$

Then, $\mathcal{B}$ is not an ordinary single valued neutrosophic base for an osvnt on X.

Suppose that $\mathcal{B}$ is an ordinary single valued neutrosophic base for an osvnt τ on X. Then, clearly, $\mathcal{B} \leq \tau$. Moreover, $\tau(\{a,b\}) = \tau(\{b,c\}) = (1,0,0)$. Thus,

$$T_\tau(\{b\}) = T_\tau(\{a,b\} \cap \tau(\{b,c\}) \geq T_\tau(\{a,b\} \wedge T_\tau(\{b,c\} = 1$$

and

$$I_\tau(\{b\}) = I_\tau(\{a,b\} \cap \tau(\{b,c\}) \leq I_\tau(\{a,b\} \wedge I_\tau(\{b,c\} = 0.$$

Similarly, we have $F_\tau(\{b\}) = 0$. Thus, $\tau(\{b\}) = (1,0,0)$. On the other hand, by the definition of $\mathcal{B}$,

$$T_\tau(\{b\}) = \bigvee_{\{A_\alpha\}_{\alpha\in\Gamma}\subset 2^X,\ \{b\}=\bigcup_{\alpha\in\Gamma} A_\alpha} \bigwedge_{\alpha\in\Gamma} T_\mathcal{B}(A_\alpha) = T_\alpha$$

and

$$I_\tau(\{b\}) = \bigwedge_{\{A_\alpha\}_{\alpha\in\Gamma}\subset 2^X,\ \{b\}=\bigcup_{\alpha\in\Gamma} A_\alpha} \bigvee_{\alpha\in\Gamma} I_\mathcal{B}(A_\alpha) = I_\alpha.$$

Similarly, we have $F_\tau(\{b\}) = F_\alpha$. This is a contradiction. Hence, $\mathcal{B}$ is not an ordinary single valued neutrosophic base for an osvnt on X

Theorem 4. *Let (X,τ) be an osvnts and let $\mathcal{B} : 2^X \to I \times I \times I$ be a mapping such that $\mathcal{B} \leq \tau$. Then, $\mathcal{B}$ is an ordinary single valued neutrosophic base for τ if and only if for each $x \in X$ and each $A \in 2^X$,*

$$T_{\mathcal{N}_x}(A) \leq \bigvee_{x\in B\subset A} T_\mathcal{B}(B),$$

$$I_{\mathcal{N}_x}(A) \geq \bigwedge_{x\in B\subset A} I_\mathcal{B}(B),$$

$$F_{\mathcal{N}_x}(A) \geq \bigwedge_{x\in B\subset A} F_\mathcal{B}(B).$$

Proof. ($\Rightarrow$): Suppose $\mathcal{B}$ is an ordinary single valued neutrosophic base for τ. Let $x \in X$ and let $A \in 2^X$. Then, by Theorem 4.4 in [34], it is obvious that $T_{\mathcal{N}_x}(A) \leq \bigvee_{x\in B\subset A} T_\mathcal{B}(B)$. On the other hand,

$$\begin{aligned} I_{\mathcal{N}_x}(A) &= \textstyle\bigwedge_{x\in B\subset A} I_\tau(B) && \text{[By Definition 13]} \\ &= \textstyle\bigwedge_{x\in B\subset A} \bigwedge_{\{B_\alpha\}_{\alpha\in\Gamma}\subset 2^X,\ B=\bigcup_{\alpha\in\Gamma} B_\alpha} \bigvee_{\alpha\in\Gamma} I_\mathcal{B}(B_\alpha). && \text{[By Definition 15]} \end{aligned}$$

If $x \in B \subset A$ and $B = \bigcup_{\alpha\in\Gamma} B_\alpha$, then there is $\alpha_0 \in \Gamma$ such that $x \in B_{\alpha_0}$. Thus,

$$\bigvee_{\alpha\in\Gamma} I_\mathcal{B}(B_\alpha) \geq I_\mathcal{B}(B_{\alpha_0}) \geq \bigwedge_{x\in B\subset A} I_\mathcal{B}(B).$$

Thus, $I_{\mathcal{N}_x}(A) \geq \bigwedge_{x\in B\subset A} I_\mathcal{B}(B)$. Similarly, we have $F_{\mathcal{N}_x}(A) \geq \bigwedge_{x\in B\subset A} F_\mathcal{B}(B)$. Hence, the necessary condition holds.

($\Leftarrow$): Suppose the necessary condition holds. Then, by Theorem 4.4 in [34], it is clear that

$$T_\tau(A) = \bigvee_{\{B_\alpha\}_{\alpha\in\Gamma}\subset 2^X,\ A=\bigcup_{\alpha\in\Gamma} B_\alpha} \bigwedge_{\alpha\in\Gamma} T_{\mathcal{B}}(B_\alpha).$$

Let $A \in 2^X$. Suppose $A = \bigcup_{\alpha\in\Gamma} B_\alpha$ and $\{B_\alpha\} \subset 2^X$. Then,

$$\begin{aligned} I_\tau(A) &\leq \textstyle\bigvee_{\alpha\in\Gamma} I_\tau(B_\alpha) && \text{[By the axiom (OSVNT3)]} \\ &\leq \textstyle\bigvee_{\alpha\in\Gamma} I_{\mathcal{B}}(B_\alpha). && \text{[Since } \mathcal{B} \leq \tau\text{]} \end{aligned}$$

Thus,

$$I_\tau(A) \leq \bigwedge_{\{B_\alpha\}_{\alpha\in\Gamma}\subset 2^X,\ A=\bigcup_{\alpha\in\Gamma} B_\alpha} \bigvee_{\alpha\in\Gamma} I_{\mathcal{B}}(B_\alpha). \tag{1}$$

On the other hand,

$$\begin{aligned} I_\tau(A) &= \textstyle\bigvee_{x\in A} \bigwedge_{x\in B\subset A} I_\tau(B) && \text{[By Lemma 3]} \\ &= \textstyle\bigvee_{x\in A} I_{\mathcal{N}_x}(A) && \text{[By Definition 13]} \\ &= \textstyle\bigvee_{x\in A} \bigwedge_{x\in B\subset A} I_{\mathcal{B}}(B) && \text{[By the hypothesis]} \\ &= \textstyle\bigwedge_{f\in\Pi_{x\in A}\mathcal{B}_x} \bigvee_{x\in A} I_{\mathcal{B}}(f(x)), \end{aligned}$$

where $\mathcal{B}_x = \{B \in 2^X : x \in B \subset A\}$. Furthermore, $A = \bigcup_{x\in A} f(x)$ for each $f \in \Pi_{x\in A}\mathcal{B}_x$. Thus,

$$\bigwedge_{f\in\Pi_{x\in A}\mathcal{B}_x} \bigvee_{x\in A} I_{\mathcal{B}}(f(x)) = \bigwedge_{\{B_\alpha\}_{\alpha\in\Gamma}\subset 2^X,\ A=\bigcup_{\alpha\in\Gamma} B_\alpha} \bigvee_{\alpha\in\Gamma} I_{\mathcal{B}}(B_\alpha).$$

Hence,

$$I_\tau(A) \geq \bigwedge_{\{B_\alpha\}_{\alpha\in\Gamma}\subset 2^X,\ A=\bigcup_{\alpha\in\Gamma} B_\alpha} \bigvee_{\alpha\in\Gamma} I_{\mathcal{B}}(B_\alpha). \tag{2}$$

By (1) and (2), $I_\tau(A) = \bigwedge_{\{B_\alpha\}_{\alpha\in\Gamma}\subset 2^X,\ A=\bigcup_{\alpha\in\Gamma} B_\alpha} \bigvee_{\alpha\in\Gamma} I_{\mathcal{B}}(B_\alpha)$. Similarly, we have $F_\tau(A) = \bigwedge_{\{B_\alpha\}_{\alpha\in\Gamma}\subset 2^X,\ A=\bigcup_{\alpha\in\Gamma} B_\alpha} \bigvee_{\alpha\in\Gamma} F_{\mathcal{B}}(B_\alpha)$. Therefore, $\mathcal{B}$ is an ordinary single valued neutrosophic base for τ. □

Theorem 5. *Let $\mathcal{B} : 2^X \to I \times I \times I$ be a mapping. Then, $\mathcal{B}$ is an ordinary single valued neutrosophic base for some oist τ on X if and only if it has the following conditions:*

(1) $\bigvee_{\{B_\alpha\}_{\alpha\in\Gamma}\subset 2^X,\ X=\bigcup_{\alpha\in\Gamma} B_\alpha} \bigwedge_{\alpha\in\Gamma} T_{\mathcal{B}}(B_\alpha) = 1$,
$\bigwedge_{\{B_\alpha\}_{\alpha\in\Gamma}\subset 2^X,\ X=\bigcup_{\alpha\in\Gamma} B_\alpha} \bigvee_{\alpha\in\Gamma} I_{\mathcal{B}}(B_\alpha) = 0$,
$\bigwedge_{\{B_\alpha\}_{\alpha\in\Gamma}\subset 2^X,\ X=\bigcup_{\alpha\in\Gamma} B_\alpha} \bigvee_{\alpha\in\Gamma} F_{\mathcal{B}}(B_\alpha) = 0$,

(2) *for any $A_1,\ A_2 \in 2^X$ and each $x \in A_1 \cap A_2$,*

$$T_{\mathcal{B}}(A_1) \wedge T_{\mathcal{B}}(A_2) \leq \bigvee_{x\in A\subset A_1\cap A_2} T_{\mathcal{B}}(A),$$

$$I_{\mathcal{B}}(A_1) \vee I_{\mathcal{B}}(A_2) \geq \bigwedge_{x\in A\subset A_1\cap A_2} I_{\mathcal{B}}(A),$$

$$F_{\mathcal{B}}(A_1) \vee F_{\mathcal{B}}(A_2) \geq \bigwedge_{x\in A\subset A_1\cap A_2} F_{\mathcal{B}}(A).$$

In fact, $\tau : 2^X \to I \times I \times I$ is the mapping defined as follows: for each $A \in 2^X$,

$$T_\tau(A) = \begin{cases} 1 & \text{if } A = \phi \\ \bigvee_{\{B_\alpha\}_{\alpha\in\Gamma}\subset 2^X,\ A=\bigcup_{\alpha\in\Gamma} B_\alpha} \bigwedge_{\alpha\in\Gamma} T_{\mathcal{B}}(B_\alpha) & \text{otherwise,} \end{cases}$$

$$I_\tau(A) = \begin{cases} 0 & \text{if } A = \phi \\ \bigwedge_{\{B_\alpha\}_{\alpha\in\Gamma}\subset 2^X,\ A=\bigcup_{\alpha\in\Gamma} B_\alpha} \bigvee_{\alpha\in\Gamma} I_{\mathcal{B}}(B_\alpha) & \text{otherwise,} \end{cases}$$

$$F_\tau(A) = \begin{cases} 0 & if\ A = \phi \\ \bigwedge_{\{B_\alpha\}_{\alpha\in\Gamma}\subset 2^X,\ A=\bigcup_{\alpha\in\Gamma} B_\alpha} \bigvee_{\alpha\in\Gamma} F_\mathcal{B}(B_\alpha) & otherwise. \end{cases}$$

In this case, τ is called an ordinary single valued neutrosophic topology on X induced by $\mathcal{B}$.

Proof. ($\Rightarrow$): Suppose $\mathcal{B}$ is an ordinary single valued neutrosophic base for some *osvnt* τ on X. Then, by Definition 15 and the axiom (OSVNT1),

$$\bigvee_{\{B_\alpha\}_{\alpha\in\Gamma}\subset 2^X,\ X=\bigcup_{\alpha\in\Gamma} B_\alpha} \bigwedge_{\alpha\in\Gamma} T_\mathcal{B}(B_\alpha) = T_\tau(X) = 1,$$

$$\bigwedge_{\{B_\alpha\}_{\alpha\in\Gamma}\subset 2^X,\ X=\bigcup_{\alpha\in\Gamma} B_\alpha} \bigvee_{\alpha\in\Gamma} I_\mathcal{B}(B_\alpha)) = I_\tau(X) = 0,$$

$$\bigwedge_{\{B_\alpha\}_{\alpha\in\Gamma}\subset 2^X,\ X=\bigcup_{\alpha\in\Gamma} B_\alpha} \bigvee_{\alpha\in\Gamma} F_\mathcal{B}(B_\alpha)) = F_\tau(X) = 0.$$

Thus, condition (1) holds.

Let $A_1,\ A_2 \in 2^X$ and let $x \in A_1 \cap A_2$. Then, by the proof of Theorem 4.2 in [33], it is obvious that $T_\mathcal{B}(A_1) \wedge T_\mathcal{B}(A_2) \leq \bigvee_{x\in A\subset A_1\cap A_2} T_\mathcal{B}(A)$. On the other hand,

$$I_\mathcal{B}(A_1) \vee I_\mathcal{B}(A_2) \geq I_\tau(A_1) \vee I_\tau(A_2) \geq I_\tau(A_1 \cap A_2) \geq I_{\mathcal{N}_x}(A_1 \cap A_2) \geq \bigwedge_{x\in A\subset A_1\cap A_2} I_\mathcal{B}(A).$$

Thus,

$$I_\mathcal{B}(A_1) \vee I_\mathcal{B}(A_2) \geq \bigwedge_{x\in A\subset A_1\cap A_2} I_\mathcal{B}(A).$$

Similarly, we have

$$F_\mathcal{B}(A_1) \vee F_\mathcal{B}(A_2) \geq \bigwedge_{x\in A\subset A_1\cap A_2} F_\mathcal{B}(A).$$

Thus, condition (2) holds.

($\Leftarrow$): Suppose the necessary conditions (1) and (2) are satisfied. Then, by the proof of Theorem 4.2 in [33], we can see that the following hold:

$T_\tau(X) = T_\tau(\phi) = 1$,
$T_\tau(A \cap B) \geq T_\tau(A) \wedge T_\tau(B)$ for any $A,\ B \in 2^X$

and

$T_\tau(\bigcup_{\alpha\in\Gamma} A_\alpha) \geq \bigwedge_{\alpha\in\Gamma} T_\tau(A_\alpha)$ for each $\{A_\alpha\}_{\alpha\in\Gamma} \subset 2^X$.

From the definition of τ, it is obvious that $I_\tau(X) = I_\tau(\phi) = 0$. Similarly, we have $F_\tau(X) = F_\tau(\phi) = 0$. Thus, τ satisfies the axiom (OSVNT1).

Let $\{A_\alpha\}_{\alpha\in\Gamma} \subset 2^X$ and let $\mathcal{B}_\alpha = \{\{B_{\delta_\alpha} : \delta_\alpha \in \Gamma_\alpha\} : \bigcup_{\delta_\alpha\in\Gamma_\alpha} B_{\delta_\alpha} = A_\alpha\}$. Let $f \in \Pi_{\alpha\in\Gamma}\mathcal{B}_\alpha$. Then, clearly, $\bigcup_{\alpha\in\Gamma} \bigcup_{B_{\delta_\alpha}\in f(\alpha)} B_{\delta_\alpha} = \bigcup_{\alpha\in\Gamma} A_\alpha$. Thus,

$$\begin{aligned} I_\tau(\textstyle\bigcup_{\alpha\in\Gamma} A_\alpha) &= \textstyle\bigwedge_{\bigcup_{\delta\in\Gamma} B_\delta=\bigcup_{\alpha\in\Gamma} A_\alpha} \bigvee_{\delta\in\Gamma} I_\mathcal{B}(B_\delta) \\ &\leq \textstyle\bigwedge_{f\in\Pi_{\alpha\in\Gamma}\mathcal{B}_\alpha} \bigvee_{\alpha\in\Gamma} \bigvee_{B_{\delta_\alpha}\in f(\alpha)} I_\mathcal{B}(B_{\delta_\alpha}) \\ &= \textstyle\bigvee_{\alpha\in\Gamma} \bigwedge_{\{B_{\delta_\alpha}:\delta_\alpha\in\Gamma_\alpha\}\in\mathcal{B}_\alpha} \bigvee_{\delta_\alpha\in\Gamma_\alpha} I_\mathcal{B}(B_{\delta_\alpha}) \\ &= \textstyle\bigvee_{\alpha\in\Gamma} I_\tau(A_\alpha). \end{aligned}$$

Similarly, we have $F_\tau(\bigcup_{\alpha\in\Gamma} A_\alpha) \leq \bigvee_{\alpha\in\Gamma} F_\tau(A_\alpha)$. Thus, τ satisfies the axiom (OSVNT3).

Now, let $A,\ B \in 2^X$ and suppose $I_\tau(A) < I_\alpha$ and $I_\tau(B) < I_\alpha$ for $\alpha \in$ **SVNV**. Then, there are $\{A_{\alpha_1} : \alpha_1 \in \Gamma_1\}$ and $\{B_{\alpha_2} : \alpha_2 \in \Gamma_2\}$ such that $\bigcup_{\alpha_1\in\Gamma_1} A_{\alpha_1} = A$, $\bigcup_{\alpha_2\in\Gamma_2} B_{\alpha_2} = B$ and $I_\mathcal{B}(A_{\alpha_1}) < I_\alpha$ for each $\alpha_1 \in \Gamma_1$, $I_{\mathcal{B}}(B_{\alpha_2}) < I_\alpha$ for each $\alpha_2 \in \Gamma_2$. Let $x \in A \cap B$. Then, there are $\alpha_{1x} \in \Gamma_1$ and $\alpha_{2x} \in \Gamma_2$ such that $x \in A_{\alpha_{1x}} \cap B_{\alpha_{2x}}$. Thus, from the assumption,

$$I_\alpha > I_{\mathcal{B}}(A_{\alpha_{1x}}) \vee I_{\mathcal{B}}(B_{\alpha_{2x}}) \geq \bigwedge_{x \in C \subset A_{\alpha_{1x}} \cap B_{\alpha_{2x}}} I_{\mathcal{B}}(C).$$

Moreover, there is C_x such that $x \in C_x \subset A_{\alpha_{1x}} \cap B_{\alpha_{2x}} \subset A \cap B$ and $I_{\mathcal{B}}(C_x) < I_\alpha$. Since $\bigcup_{x \in A \cap B} C_x = A \cap B$, we obtain

$$I_\alpha \geq \bigvee_{x \in A \cap B} I_{\mathcal{B}}(C_x) \geq \bigwedge_{\bigcup_{\alpha \in \Gamma} B_\alpha = A \cap B} \bigvee_{\alpha \in \Gamma} I_{\mathcal{B}}(B_\alpha) = I_\tau(A \cap B).$$

Now, let $I_\beta = I_\tau(A) \vee I_\tau(B)$ and let n be any natural number, where $I_\beta \in I$. Then, $I_\tau(A) < I_\beta + 1/n$ and $I_\tau(B) < I_\beta + 1/n$. Thus, $I_\tau(A \cap B) \leq I_\beta + 1/n$. Thus, $I_\tau(A \cap B) \leq I_\beta = I_\tau(A) \vee I_\tau(B)$. Similarly, we have $F_\tau(A \cap B) \leq F_\tau(A) \vee F_\tau(B)$. Hence, τ satisfies the axiom (OSVNT2). This completes the proof. □

Example 4. *(1) Let $X = \{a, b, c\}$ and let $\alpha \in$ **SVNV** be fixed, where $\alpha \in I_1 \times I_0 \times I_0$. We define the mapping $\mathcal{B} : 2^X \to I \times I \times I$ as follows: for each $A \in 2^X$,*

$$T_{\mathcal{B}}(A) = \begin{cases} 1 & \text{if } A = \{b\} \text{ or } \{a,b\} \text{ or } \{b,c\} \\ T_\alpha & \text{otherwise,} \end{cases}$$

$$I_{\mathcal{B}}(A) = \begin{cases} 0 & \text{if } A = \{b\} \text{ or } \{a,b\} \text{ or } \{b,c\} \\ I_\alpha & \textit{otherwise,} \end{cases}$$

$$F_{\mathcal{B}}(A) = \begin{cases} 0 & \text{if } A = \{b\} \text{ or } \{a,b\} \text{ or } \{b,c\} \\ F_\alpha & \textit{otherwise.} \end{cases}$$

Then, we can easily see that $\mathcal{B}$ satisfies conditions (1) and (2) in Theorem 5. Thus, $\mathcal{B}$ is an ordinary single valued neutrosophic base for an osvnt τ on X. In fact, $\tau : 2^X \to I \times I \times I$ is defined as follows: for each $A \in 2^X$,

$$T_\tau(A) = \begin{cases} 1 & \text{if } A \in \{\phi, \{b\}, \{a,b\}, \{b,c\}, X\} \\ T_\alpha & \text{otherwise,} \end{cases}$$

$$I_\tau(A) = \begin{cases} 0 & \text{if } A \in \{\phi, \{b\}, \{a,b\}, \{b,c\}, X\} \\ I_\alpha & \text{otherwise,} \end{cases}$$

$$F_\tau(A) = \begin{cases} 0 & \text{if } A \in \{\phi, \{b\}, \{a,b\}, \{b,c\}, X\} \\ F_\alpha & \text{otherwise.} \end{cases}$$

*(2) Let $\alpha \in$ **SVNV** be fixed, where $\alpha \in I_1 \times I_0 \times I_0$. We define the mapping $\mathcal{B} : 2^{\mathbb{R}} \to I \times I \times I$ as follows: for each $A \in 2^{\mathbb{R}}$,*

$$T_{\mathcal{B}}(A) = \begin{cases} 1 & \text{if } A = (a,b) \text{ for } a,\ b \in \mathbb{R} \text{ with } a \leq b \\ T_\alpha & \text{otherwise,} \end{cases}$$

$$I_{\mathcal{B}}(A) = \begin{cases} 0 & \text{if } A = (a,b) \text{ for } a,\ b \in \mathbb{R} \text{ with } a \leq b \\ I_\alpha & \text{otherwise,} \end{cases}$$

$$F_{\mathcal{B}}(A) = \begin{cases} 0 & \text{if } A = (a,b) \text{ for } a,\ b \in \mathbb{R} \text{ with } a \leq b \\ F_\alpha & \text{otherwise.} \end{cases}$$

Then, it can be easily seen that $\mathcal{B}$ satisfies the conditions (1) and (2) in Theorem 5. Thus, $\mathcal{B}$ is an ordinary single valued neutrosophic base for an osvnt τ_α on $\mathbb{R}$.

In this case, τ_α is called the α-ordinary single valued neutrosophic usual topology on $\mathbb{R}$.

*(3) Let $\alpha \in$ **SVNV** be fixed, where $\alpha \in I_1 \times I_0 \times I_0$. We define the mapping $\mathcal{B} : 2^{\mathbb{R}} \to I \times I \times I$ as follows: for each $A \in 2^{\mathbb{R}}$,*

$$T_{\mathcal{B}}(A) = \begin{cases} 1 & \text{if } A = [a,b) \text{ for } a,\ b \in \mathbb{R} \text{ with } a \leq b \\ T_\alpha & \text{otherwise,} \end{cases}$$

$$I_{\mathcal{B}}(A) = \begin{cases} 0 & \text{if } A = [a,b) \text{ for } a,\ b \in \mathbb{R} \text{ with } a \leq b \\ I_\alpha & \text{otherwise,} \end{cases}$$

$$F_{\mathcal{B}}(A) = \begin{cases} 0 & \text{if } A = [a,b) \text{ for } a,\ b \in \mathbb{R} \text{ with } a \leq b \\ F_\alpha & \text{otherwise.} \end{cases}$$

Then, we can easily see that $\mathcal{B}$ satisfies the conditions (1) and (2) in Theorem 5. Thus, $\mathcal{B}$ is an ordinary single valued neutrosophic base for an osvnt τ_l on $\mathbb{R}$.

In this case, τ_l is called the α-ordinary single valued neutrosophic lower-limit topology on $\mathbb{R}$.

Definition 16. *Let $\tau_1,\ \tau_2 \in OSVNT(X)$, and let $\mathcal{B}_1$ and $\mathcal{B}_1$ be ordinary single valued neutrosophic bases for τ_1 and τ_2, respectively. Then, $\mathcal{B}_1$ and $\mathcal{B}_1$ are said to be equivalent if $\tau_1 = \tau_2$.*

Theorem 6. *Let $\tau_1,\ \tau_2 \in OSVNT(X)$, and let $\mathcal{B}_1$ and $\mathcal{B}_1$ be ordinary single valued neutrosophic bases for τ_1 and τ_2 respectively. Then, τ_1 is coarser than τ_2, i.e.,*

$$T_{\tau_1} \leq T_{\tau_2},\ I_{\tau_1} \geq I_{\tau_2},\ F_{\tau_1} \geq F_{\tau_2}$$

if and only if for each $A \in 2^X$ and each $x \in A$,

$$T_{\mathcal{B}_1}(A) \leq \bigvee_{x \in B \subset A} T_{\mathcal{B}_2}(B), \quad I_{\mathcal{B}_1}(A) \geq \bigwedge_{x \in B \subset A} I_{\mathcal{B}_2}(B), \quad F_{\mathcal{B}_1}(A) \geq \bigwedge_{x \in B \subset A} F_{\mathcal{B}_2}(B).$$

Proof. ($\Rightarrow$): Suppose τ_1 is coarser than τ_2. For each $x \in X$, let $x \in A \in 2^X$. Then, by Theorem 4.8 in [34], $T_{\mathcal{B}_1}(A) \leq \bigvee_{x \in B \subset A} T_{\mathcal{B}_2}(B)$. On the other hand,

$$\begin{aligned} I_{\mathcal{B}_1}(A) &\geq I_{\tau_1}(A) && [\text{since } \mathcal{B}_1 \text{ is an ordinary single valued neutrosophic base for } \tau_1] \\ &\geq I_{\tau_2}(A) && [\text{By the hypothesis}] \\ &= \textstyle\bigwedge_{\{A_\alpha\}_{\alpha \in \Gamma} \subset 2^X,\ A = \bigcup_{\alpha \in \Gamma} A_\alpha} \bigvee_{\alpha \in \Gamma} I_{\mathcal{B}_2}(A_\alpha). && [\text{Since } \mathcal{B}_2 \text{ is an ordinary single valued neutrosophic base for } \tau_2] \end{aligned}$$

Since $x \in A$ and $A = \bigcup_{\alpha \in \Gamma} A_\alpha$, there is $\alpha_0 \in \Gamma$ such that $x \in A_{\alpha_0}$. Thus,

$$\bigwedge_{\{A_\alpha\}_{\alpha \in \Gamma} \subset 2^X,\ A = \bigcup_{\alpha \in \Gamma} A_\alpha} \bigvee_{\alpha \in \Gamma} I_{\mathcal{B}_2}(A_\alpha) \geq I_{\mathcal{B}_2}(A_{\alpha_0}) \geq \bigwedge_{x \in B \subset A} I_{\mathcal{B}_2}(B).$$

Thus, $I_{\mathcal{B}_1}(A) \geq \bigwedge_{x \in B \subset A} I_{\mathcal{B}_2}(B)$. Similarly, we have $F_{\mathcal{B}_1}(A) \geq \bigwedge_{x \in B \subset A} F_{\mathcal{B}_2}(B)$.

($\Leftarrow$): Suppose the necessary condition holds. Then, by Theorem 4.8 in [34], $T_{\tau_1} \leq T_{\tau_2}$. Let $A \in 2^X$. Then,

$$\begin{aligned} I_{\tau_1}(A) &= \textstyle\bigvee_{x \in A} \bigwedge_{x \in B \subset A} I_{\mathcal{B}_1}(B) && [\text{By Lemma 3}] \\ &\geq \textstyle\bigvee_{x \in A} \bigwedge_{x \in B \subset A} \bigwedge_{x \in C \subset B} I_{\mathcal{B}_2}(C) && [\text{By the hypothesis}] \\ &= \textstyle\bigwedge_{x \in C \subset A} \bigvee_{x \in A} I_{\mathcal{B}_2}(C) \\ &= \textstyle\bigwedge_{\{C_x\}_{x \in A} \subset 2^X,\ A = \bigcup_{x \in A} C_x} \bigvee_{x \in A} I_{\mathcal{B}_2}(C_x) \\ &= I_{\tau_2}(A). \end{aligned}$$

Thus, $I_{\tau_1} \geq I_{\tau_2}$. Similarly, we have $F_{\tau_1} \geq F_{\tau_2}$. Thus, τ_1 is coarser than τ_2. This completes the proof. □

The following is an immediate result of Definition 16 and Theorem 6.

Corollary 4. *Let $\mathcal{B}_1$ and $\mathcal{B}_2$ be ordinary single valued neutrosophic bases for two ordinary single valued neutrosophic topologies on a set X, respectively. Then,*

$\mathcal{B}_1$ and $\mathcal{B}_2$ are equivalent if and only if the following two conditions hold:

(1) for each $B_1 \in 2^X$ and each $x \in B_1$,

$$T_{\mathcal{B}_1}(B_1) \leq \bigvee_{x \in B_2 \subset B_1} T_{\mathcal{B}_2}(B_2),$$

$$I_{\mathcal{B}_1}(B_1) \geq \bigwedge_{x \in B_2 \subset B_1} I_{\mathcal{B}_2}(B_2),$$

$$F_{\mathcal{B}_1}(B_1) \geq \bigwedge_{x \in B_2 \subset B_1} F_{\mathcal{B}_2}(B_2),$$

(2) for each $B_2 \in 2^X$ and each $x \in B_2$,

$$T_{\mathcal{B}_2}(B_2) \leq \bigvee_{x \in B_1 \subset B_2} T_{\mathcal{B}_1}(B_1),$$

$$I_{\mathcal{B}_2}(B_2) \geq \bigwedge_{x \in B_1 \subset B_2} I_{\mathcal{B}_1}(B_1),$$

$$F_{\mathcal{B}_2}(B_2) \geq \bigwedge_{x \in B_1 \subset B_2} F_{\mathcal{B}_1}(B_1).$$

It is obvious that every ordinary single valued neutrosophic topology itself forms an ordinary single valued neutrosophic base. Then, the following provides a sufficient condition for one to see if a mapping $\mathcal{B} : 2^X \to I \times I \times I$ such that $T_{\mathcal{B}} \leq T_\tau$, $I_{\mathcal{B}} \geq I_\tau$ and $F_{\mathcal{B}} \geq F_\tau$ is an ordinary single valued neutrosophic base for $\tau \in OSVNT(X)$.

Proposition 10. *Let (X, τ) be an osvnts and let $\mathcal{B} : 2^X \to I \times I \times I$ be a mapping such that $T_{\mathcal{B}} \leq T_\tau$, $I_{\mathcal{B}} \geq I_\tau$ and $F_{\mathcal{B}} \geq F_\tau$. For each $A \in 2^X$ and each $x \in A$, suppose $T_\tau(A) \leq \bigvee_{x \in B \subset A} T_{\mathcal{B}}(B)$, $I_\tau(A) \geq \bigwedge_{x \in B \subset A} I_{\mathcal{B}}(B)$ and $F_\tau(A) \geq \bigwedge_{x \in B \subset A} F_{\mathcal{B}}(B)$. Then, $\mathcal{B}$ is an ordinary single valued neutrosophic base for τ.*

Proof. From the proof of Proposition 4.10 in [34], it is clear that the first part of the condition (1) of Theorem 5 holds, i.e., $\bigvee_{\{B_\alpha\}_{\alpha \in \Gamma} \subset 2^X,\ X = \bigcup_{\alpha \in \Gamma} B_\alpha} \bigwedge_{\alpha \in \Gamma} T_{\mathcal{B}}(B_\alpha) = 1$. On the other hand,

$$\begin{aligned}
&\bigwedge_{\{B_\alpha\}_{\alpha \in \Gamma} \subset 2^X,\ X = \bigcup_{\alpha \in \Gamma} B_\alpha} \bigvee_{\alpha \in \Gamma} I_{\mathcal{B}}(B_\alpha) \\
&\geq \bigwedge_{\{B_\alpha\}_{\alpha \in \Gamma} \subset 2^X,\ X = \bigcup_{\alpha \in \Gamma} B_\alpha} \bigvee_{\alpha \in \Gamma} I_\tau(B_\alpha) && [\text{since } I_{\mathcal{B}} \geq I_\tau] \\
&\geq \bigwedge_{\{B_\alpha\}_{\alpha \in \Gamma} \subset 2^X,\ X = \bigcup_{\alpha \in \Gamma} B_\alpha} I_\tau(\textstyle\bigcup_{\alpha \in \Gamma} B_\alpha) && [\text{by the axiom (OSVNT3)}] \\
&= I_\tau(X) \\
&= \bigvee_{x \in X} \bigwedge_{x \in B \subset X} I_\tau(B) && [\text{By Lemma 3}] \\
&\geq \bigvee_{x \in X} \bigwedge_{x \in B \subset X} \bigwedge_{x \in C \subset B} I_{\mathcal{B}}(C) && [\text{By the hypothesis}] \\
&= \bigwedge_{x \in C \subset X} \bigvee_{x \in X} I_{\mathcal{B}}(C) \\
&= \bigwedge_{\{B_\alpha\}_{\alpha \in \Gamma} \subset 2^X,\ X = \bigcup_{\alpha \in \Gamma} B_\alpha} \bigvee_{\alpha \in \Gamma} I_{\mathcal{B}}(B_\alpha).
\end{aligned}$$

Since $\tau \in OSVNT(X)$, $I_\tau(X) = 0$. Thus, $\bigwedge_{\{B_\alpha\}_{\alpha \in \Gamma} \subset 2^X,\ X = \bigcup_{\alpha \in \Gamma} B_\alpha} \bigvee_{\alpha \in \Gamma} I_{\mathcal{B}}(B_\alpha) = 0$. Similarly, we have $\bigwedge_{\{B_\alpha\}_{\alpha \in \Gamma} \subset 2^X,\ X = \bigcup_{\alpha \in \Gamma} B_\alpha} \bigvee_{\alpha \in \Gamma} F_{\mathcal{B}}(B_\alpha) = 0$. Thus, condition (1) of Theorem 5 holds.

Now, let A_1, $A_2 \in 2^X$ and let $x \in A_1 \cap A_2$. Then, by the proof of Proposition 4.10 in [34], it is obvious that $T_{\mathcal{B}}(A_1) \wedge T_{\mathcal{B}}(A_2) \leq \bigvee_{x \in A \subset A_1 \cap A_2} T_{\mathcal{B}}(A)$. On the other hand,

$$I_{\mathcal{B}}(A_1) \vee I_{\mathcal{B}}(A_2) \geq I_{\tau}(A_1) \vee I_{\tau}(A_2) \quad \text{[Since } I_{\mathcal{B}} \geq I_{\tau}\text{]}$$
$$\geq I_{\tau}(A_1 \cap A_2) \quad \text{[by the axiom (OSVNT2)]}$$
$$\geq \bigwedge_{x \in A \subset A_1 \cap A_2} I_{\mathcal{B}}(A). \quad \text{[by the hypothesis]}$$

Similarly, we have $F_{\mathcal{B}}(A_1) \vee F_{\mathcal{B}}(A_2) \geq \bigwedge_{x \in A \subset A_1 \cap A_2} F_{\mathcal{B}}(A)$. Thus, condition (2) of Theorem 5 holds. Thus, by Theorem 5, $\mathcal{B}$ is an ordinary single valued neutrosophic base for τ. This completes the proof. □

Definition 17. *Let (X, τ) be an osvnts and let $\simeq : 2^X \to I \times I \times I$ be a mapping. Then, φ is called an ordinary single valued neutrosophic subbase for τ, if $\varphi^{\sqcap}$ is an ordinary single valued neutrosophic base for τ, where $\varphi^{\sqcap} : 2^X \to I \times I \times I$ is the mapping defined as follows: for each $A \in 2^X$,*

$$T_{\varphi^{\sqcap}}(A) = \bigvee_{\{B_\alpha\} \sqsubset 2^X,\ A = \bigcap_{\alpha \in \Gamma} B_\alpha} \bigwedge_{\alpha \in \Gamma} T_{\simeq}(B_\alpha),$$

$$I_{\varphi^{\sqcap}}(A) = \bigwedge_{\{B_\alpha\} \sqsubset 2^X,\ A = \bigcap_{\alpha \in \Gamma} B_\alpha} \bigvee_{\alpha \in \Gamma} I_{\simeq}(B_\alpha),$$

$$F_{\varphi^{\sqcap}}(A) = \bigwedge_{\{B_\alpha\} \sqsubset 2^X,\ A = \bigcap_{\alpha \in \Gamma} B_\alpha} \bigvee_{\alpha \in \Gamma} F_{\simeq}(B_\alpha),$$

where $\sqsubset$ stands for "a finite subset of".

Example 5. *Let $\alpha \in$ **SVNV** be fixed, where $\alpha \in I_1 \times I_0 \times I_0$. We define the mapping $\simeq : 2^{\mathbb{R}} \to I \times I \times I$ as follows: for each $A \in 2^{\mathbb{R}}$,*

$$T_{\simeq}(A) = \begin{cases} 1 & \text{if } A = (a, \infty) \text{ or } (-\infty, b) \text{ or } (a, b) \\ T_\alpha & \text{otherwise,} \end{cases}$$

$$I_{\simeq}(A) = \begin{cases} 0 & \text{if } A = (a, \infty) \text{ or } (-\infty, b) \text{ or } (a, b) \\ I_\alpha & \text{otherwise,} \end{cases}$$

$$F_{\simeq}(A) = \begin{cases} 0 & \text{if } A = (a, \infty) \text{ or } (-\infty, b) \text{ or } (a, b) \\ F_\alpha & \text{otherwise,} \end{cases}$$

where $a,\ b \in \mathbb{R}$ such that $a < b$. Then, we can easily see that $\simeq$ is an ordinary single valued neutrosophic subbase for the α-ordinary single valued neutrosophic usual topology $\mathcal{U}_\alpha$ on $\mathbb{R}$.

Theorem 7. *Let $\simeq : 2^X \to I \times I \times I$ be a mapping. Then, $\simeq$ is an ordinary single valued neutrosophic subbase for some osvnt if and only if*

$$\bigvee_{\{B_\alpha\}_{\alpha \in \Gamma} \subset 2^X,\ X = \bigcup_{\alpha \in \Gamma} B_\alpha} \bigwedge_{\alpha \in \Gamma} T_{\simeq}(B_\alpha) = 1,$$

$$\bigwedge_{\{B_\alpha\}_{\alpha \in \Gamma} \subset 2^X,\ X = \bigcup_{\alpha \in \Gamma} B_\alpha} \bigvee_{\alpha \in \Gamma} I_{\simeq}(B_\alpha) = 0,$$

$$\bigwedge_{\{B_\alpha\}_{\alpha \in \Gamma} \subset 2^X,\ X = \bigcup_{\alpha \in \Gamma} B_\alpha} \bigvee_{\alpha \in \Gamma} F_{\simeq}(B_\alpha) = 0.$$

Proof. ($\Rightarrow$): Suppose $\simeq$ is an ordinary single valued neutrosophic subbase for some *osvnt*. Then, by Definition 17, it is clear that the necessary condition holds.

($\Leftarrow$): Suppose the necessary condition holds. We only show that $\varphi^{\sqcap}$ satisfies the condition (2) in Theorem 5. Let $A,\ B \in 2^X$ and $x \in A \cap B$. Then, by the proof of Theorem 4.3 in [33], it is obvious that $T_{\varphi^{\sqcap}}(A) \wedge T_{\varphi^{\sqcap}}(B) \leq \bigvee_{x \in C \subset A \cap B} T_{\varphi^{\sqcap}}(C)$. On the other hand,

$$
\begin{aligned}
&I_{\varphi^{\sqcap}}(A) \vee I_{\varphi^{\sqcap}}(B) \\
&= (\bigwedge_{\sqcap_{\alpha_1 \in \Gamma_1} B_{\alpha_1} = A} \bigvee_{\alpha_1 \in \Gamma_1} I_{\simeq}(B_{\alpha_1})) \vee (\bigwedge_{\sqcap_{\alpha_2 \in \Gamma_2} B_{\alpha_2} = B} \bigvee_{\alpha_2 \in \Gamma_2} I_{\simeq}(B_{\alpha_2})) \\
&= \bigwedge_{\sqcap_{\alpha_1 \in \Gamma_1} B_{\alpha_1} = A} \bigwedge_{\sqcap_{\alpha_2 \in \Gamma_2} B_{\alpha_2} = B} (\bigvee_{\alpha_1 \in \Gamma_1} I_{\simeq}(B_{\alpha_1}) \vee \bigvee_{\alpha_2 \in \Gamma_2} I_{\simeq}(B_{\alpha_2})) \\
&\geq \bigwedge_{\sqcap_{\alpha \in \Gamma} B_{\alpha} = A \cap B} \bigvee_{\alpha \in \Gamma} I_{\simeq}(B_{\alpha}) \\
&= I_{\varphi^{\sqcap}}(A \cap B).
\end{aligned}
$$

Since $x \in A \cap B$, $I_{\varphi^{\sqcap}}(A) \vee I_{\varphi^{\sqcap}}(B) \geq I_{\varphi^{\sqcap}}(A \cap B) \geq \bigwedge_{x \in C \subset A \cap B} I_{\varphi^{\sqcap}}(C)$. Similarly, we have $F_{\varphi^{\sqcap}}(A) \vee F_{\varphi^{\sqcap}}(B) \geq F_{\varphi^{\sqcap}}(A \cap B) \geq \bigwedge_{x \in C \subset A \cap B} F_{\varphi^{\sqcap}}(C)$. Thus, $\varphi^{\sqcap}$ satisfies the condition (2) in Theorem 5. This completes the proof. □

Example 6. *Let $X = \{a, b, c, d, e\}$ and let $\alpha \in$ **SVNV** be fixed, where $\alpha \in I_1 \times I_0 \times I_0$. We define the mapping $\simeq : 2^X \to I \times I \times I$ as follows: for each $A \in 2^X$,*

$$T_{\simeq}(A) = \begin{cases} 1 & \text{if } A \in \{\{a\}, \{a,b,c\}, \{b,c,d\}, \{c,e\}\} \\ T_{\alpha} & \text{otherwise,} \end{cases}$$

$$I_{\simeq}(A) = \begin{cases} 0 & \text{if } A \in \{\{a\}, \{a,b,c\}, \{b,c,d\}, \{c,e\}\} \\ I_{\alpha} & \text{otherwise,} \end{cases}$$

$$F_{\simeq}(A) = \begin{cases} 0 & \text{if } A \in \{\{a\}, \{a,b,c\}, \{b,c,d\}, \{c,e\}\} \\ F_{\alpha} & \text{otherwise.} \end{cases}$$

Then, $X = \{a\} \cup \{b,c,d\} \cup \{c,e\}$,

$$
\begin{aligned}
&T_{\varphi^{\sqcap}}(\{a\}) = T_{\varphi^{\sqcap}}(\{b,c,d\}) = T_{\varphi^{\sqcap}}(\{c,e\}) = 1, \\
&I_{\varphi^{\sqcap}}(\{a\}) = I_{\varphi^{\sqcap}}(\{b,c,d\}) = I_{\varphi^{\sqcap}}(\{c,e\}) = 0. \\
&F_{\varphi^{\sqcap}}(\{a\}) = F_{\varphi^{\sqcap}}(\{b,c,d\}) = F_{\varphi^{\sqcap}}(\{c,e\}) = 0.
\end{aligned}
$$

Thus,

$$\bigvee_{\{B_\alpha\}_{\alpha \in \Gamma} \subset 2^X,\ X = \bigcup_{\alpha \in \Gamma} B_\alpha} \bigwedge_{\alpha \in \Gamma} T_{\simeq}(B_\alpha) = 1,$$

$$\bigwedge_{\{B_\alpha\}_{\alpha \in \Gamma} \subset 2^X,\ X = \bigcup_{\alpha \in \Gamma} B_\alpha} \bigvee_{\alpha \in \Gamma} I_{\simeq}(B_\alpha) = 0,$$

$$\bigwedge_{\{B_\alpha\}_{\alpha \in \Gamma} \subset 2^X,\ X = \bigcup_{\alpha \in \Gamma} B_\alpha} \bigvee_{\alpha \in \Gamma} F_{\simeq}(B_\alpha) = 0.$$

Thus, by Theorem 7, $\simeq$ is an ordinary single valued neutrosophic subbase for some osvnt.

The following is an immediate result of Corollary 4 and Theorem 7.

Proposition 11. *$\simeq_1$, $\simeq_2 : 2^X \to I \times I \times I$ be two mappings such that*

$$\bigvee_{\{B_\alpha\}_{\alpha \in \Gamma} \subset 2^X,\ X = \bigcup_{\alpha \in \Gamma} B_\alpha} \bigwedge_{\alpha \in \Gamma} T_{\simeq_1}(B_\alpha) = 1,$$

$$\bigwedge_{\{B_\alpha\}_{\alpha \in \Gamma} \subset 2^X,\ X = \bigcup_{\alpha \in \Gamma} B_\alpha} \bigvee_{\alpha \in \Gamma} I_{\simeq_1}(B_\alpha) = 0,$$

$$\bigwedge_{\{B_\alpha\}_{\alpha \in \Gamma} \subset 2^X,\ X = \bigcup_{\alpha \in \Gamma} B_\alpha} \bigvee_{\alpha \in \Gamma} F_{\simeq_1}(B_\alpha) = 0$$

and

$$\bigvee_{\{B_\alpha\}_{\alpha \in \Gamma} \subset 2^X,\ X = \bigcup_{\alpha \in \Gamma} B_\alpha} \bigwedge_{\alpha \in \Gamma} T_{\simeq_2}(B_\alpha) = 1,$$

$$\bigwedge_{\{B_\alpha\}_{\alpha\in\Gamma}\subset 2^X,\ X=\bigcup_{\alpha\in\Gamma} B_\alpha}\ \bigvee_{\alpha\in\Gamma} I_{\simeq_2}(B_\alpha)=0,$$

$$\bigwedge_{\{B_\alpha\}_{\alpha\in\Gamma}\subset 2^X,\ X=\bigcup_{\alpha\in\Gamma} B_\alpha}\ \bigvee_{\alpha\in\Gamma} F_{\simeq_2}(B_\alpha)=0.$$

Suppose the two conditions hold:
(1) *for each* $S_1 \in 2^X$ *and each* $x \in S_1$,

$$T_{\simeq_1}(S_1) \leq \bigvee_{x\in S_2\subset S_1} T_{\simeq_2}(S_2),\ I_{\simeq_1}(S_1) \geq \bigwedge_{x\in S_2\subset S_1} I_{\simeq_2}(S_2),\ F_{\simeq_1}(S_1) \geq \bigwedge_{x\in S_2\subset S_1} F_{\simeq_2}(S_2),$$

(2) *for each* $S_2 \in 2^X$ *and each* $x \in S_2$,

$$T_{\simeq_2}(S_2) \leq \bigvee_{x\in S_1\subset S_2} T_{\simeq_1}(S_1),\ I_{\simeq_2}(S_2) \geq \bigwedge_{x\in S_1\subset S_2} I_{\simeq_1}(S_1),\ f_{\simeq_2}(S_2) \geq \bigwedge_{x\in S_1\subset S_2} f_{\simeq_1}(S_1).$$

Then, $\simeq_1$ *and* $\simeq_2$ *are ordinary single valued neutrosophic subbases for the same ordinary single valued neutrosophic topology on* X.

6. Conclusions

In this paper, we defined an ordinary single valued neutrosophic topology and level set of an *osvnst* to study some topological characteristics of neutrosophic sets and obtained some their basic properties. In addition, we defined an ordinary single valued neutrosophic subspace. Next, the concepts of an ordinary single valued neutrosophic neighborhood system and an ordinary single valued neutrosophic base (or subbase) were introduced and studied. Their results are summarized as follows:

First, an ordinary single valued neutrosophic neighborhood system has the same properties in a classical neighborhood system (see Theorem 3).

Second, we found two characterizations of an ordinary single valued neutrosophic base (see Theorems 4 and 5).

Third, we obtained one characterization of an ordinary single valued neutrosophic subbase (see Theorem 7).

Finally, we expect that this paper can be a guidance for the research of separation axioms, compactness, connectedness, etc. in ordinary single valued neutrosophic topological spaces. In addition, one can deal with single valued neutrosophic topology from the viewpoint of lattices.

Author Contributions: All authors have contributed equally to this paper in all aspects. This paper was organized by the idea of Hur Kul. Junhui Kim and Jeong Gon Lee analyzed the related papers with this research, and they also wrote the paper. Florentin Smarandache checked the overall contents and mathematical accuracy.

Funding: This research received no external funding.

Acknowledgments: This paper was supported by Wonkwang University in 2017 (Junhui Kim).

Conflicts of Interest: The authors declare no conflict of interest.

References

1. Zadeh, L.A. Fuzzy sets. *Inf. Control* **1965**, *8*, 338–353. [CrossRef]
2. Chang, C.L. Fuzzy topological spaces. *J. Math. Anal. Appl.* **1968**, *24*, 182–190. [CrossRef]
3. El-Gayyar, M.K.; Kerre, E.E.; Ramadan, A.A. On smooth topological space II: Separation axioms. *Fuzzy Sets Syst.* **2001**, *119*, 495–504. [CrossRef]
4. Ghanim, M.H.; Kerre, E.E.; Mashhour, A.S. Separation axioms, subspaces and sums in fuzzy topology. *J. Math. Anal. Appl.* **1984**, *102*, 189–202. [CrossRef]
5. Kandil, A.; El Etriby, A.M. On separation axioms in fuzzy topological space. *Tamkang J. Math.* **1987**, *18*, 49–59.

6. Kandil, A.; Elshafee, M.E. Regularity axioms in fuzzy topological space and FR_i-proximities. *Fuzzy Sets Syst.* **1988**, *27*, 217–231. [CrossRef]
7. Kerre, E.E. Characterizations of normality in fuzzy topological space. *Simon Steven* **1979**, *53*, 239–248.
8. Lowen, R. Fuzzy topological spaces and fuzzy compactness. *J. Math. Anal. Appl.* **1976**, *56*, 621–633. [CrossRef]
9. Lowen, R. A comparison of different compactness notions in fuzzy topological spaces. *J. Math. Anal.* **1978**, *64*, 446–454. [CrossRef]
10. Lowen, R. Initial and final fuzzy topologies and the fuzzy Tychonoff Theorem. *J. Math. Anal.* **1977**, *58*, 11–21. [CrossRef]
11. Pu, P.M.; Liu, Y.M. Fuzzy topology I. Neighborhood structure of a fuzzy point. *J. Math. Anal. Appl.* **1982**, *76*, 571–599.
12. Pu, P.M.; Liu, Y.M. Fuzzy topology II. Products and quotient spaces. *J. Math. Anal. Appl.* **1980**, *77*, 20–37.
13. Yalvac, T.H. Fuzzy sets and functions on fuzzy spaces. *J. Math. Anal.* **1987**, *126*, 409–423. [CrossRef]
14. Chattopadhyay, K.C.; Hazra, R.N.; Samanta, S.K. Gradation of openness: Fuzzy topology. *Fuzzy Sets Syst.* **1992**, *49*, 237–242. [CrossRef]
15. Hazra, R.N.; Samanta, S.K.; Chattopadhyay, K.C. Fuzzy topology redefined. *Fuzzy Sets Syst.* **1992**, *45*, 79–82. [CrossRef]
16. Ramaden, A.A. Smooth topological spaces. *Fuzzy Sets Syst.* **1992**, *48*, 371–375. [CrossRef]
17. Demirci, M. Neighborhood structures of smooth topological spaces. *Fuzzy Sets Syst.* **1997**, *92*, 123–128. [CrossRef]
18. Chattopadhyay, K.C.; Samanta, S.K. Fuzzy topology: Fuzzy closure operator, fuzzy compactness and fuzzy connectedness. *Fuzzy Sets Syst.* **1993**, *54*, 207–212. [CrossRef]
19. Peeters, W. Subspaces of smooth fuzzy topologies and initial smooth fuzzy structures. *Fuzzy Sets Syst.* **1999**, *104*, 423–433. [CrossRef]
20. Peeters, W. The complete lattice $(S(X), \preceq)$ of smooth fuzzy topologies. *Fuzzy Sets Syst.* **2002**, *125*, 145–152. [CrossRef]
21. Al Tahan, M.; Hošková-Mayerová, Š.; Davvaz, B. An overview of topological hypergroupoids. *J. Intell. Fuzzy Syst.* **2018**, *34*, 1907–1916. [CrossRef]
22. Onasanya, B.O.; Hošková-Mayerová, Š. Some topological and algebraic properties of α-level subsets' topology of a fuzzy subset. *An. Univ. Ovidius Constanta* **2018**, *26*, 213–227. [CrossRef]
23. Çoker, D.; Demirci, M. An introduction to intuitionistic fuzzy topological spaces in Šostak's sense. *Busefal* **1996**, *67*, 67–76.
24. Samanta, S.K.; Mondal, T.K. Intuitionistic gradation of openness: Intuitionistic fuzzy topology. *Busefal* **1997**, *73*, 8–17.
25. Samanta, S.K.; Mondal, T.K. On intuitionistic gradation of openness. *Fuzzy Sets Syst.* **2002**, *131*, 323–336.
26. Šostak, A. On a fuzzy topological structure. In *Proceedings of the 13th Winter School on Abstract Analysis, Section of Topology. Circolo Matematico di Palermo, Palermo*; Rendiconti del Circolo Matematico di Palermo; Circolo Matematico di Palermo: Palermo, Italy, 1985; pp. 89–103.
27. Atanassov, K. Intuitionistic fuzzy sets. *Fuzzy Sets Syst.* **1986**, *20*, 87–96. [CrossRef]
28. Lim, P.K.; Kim, S.R.; Hur, K. Intuitionisic smooth topological spaces. *J. Korean Inst. Intell. Syst.* **2010**, *20*, 875–883. [CrossRef]
29. Kim, S.R.; Lim, P.K.; Kim, J.; Hur, K. Continuities and neighborhood structures in intuitionistic fuzzy smooth topological spaces. *Ann. Fuzzy Math. Inform.* **2018**, *16*, 33–54. [CrossRef]
30. Choi, J.Y.; Kim, S.R.; Hur, K. Interval-valued smooth topological spaces. *Honam Math. J.* **2010**, *32*, 711–738. [CrossRef]
31. Gorzalczany, M.B. A method of inference in approximate reasoning based on interval-valued fuzzy sets. *Fuzzy Sets Syst.* **1987**, *21*, 1–17. [CrossRef]
32. Zadeh, L.A. The concept of a linguistic variable and its application to approximate reasoning I. *Inform. Sci.* **1975**, *8*, 199–249. [CrossRef]
33. Ying, M.S. A new approach for fuzzy topology(I). *Fuzzy Sets Syst.* **1991**, *39*, 303–321. [CrossRef]
34. Lim, P.K.; Ryou, B.G.; Hur, K. Ordinary smooth topological spaces. *Int. J. Fuzzy Log. Intell. Syst.* **2012**, *12*, 66–76. [CrossRef]
35. Lee, J.G.; Lim, P.K.; Hur, K. Some topological structures in ordinary smooth topological spaces. *J. Korean Inst. Intell. Syst.* **2012**, *22*, 799–805. [CrossRef]

36. Lee, J.G.; Lim, P.K.; Hur, K. Closures and interiors redefined, and some types of compactness in ordinary smooth topological spaces. *J. Korean Inst. Intell. Syst.* **2013**, *23*, 80–86. [CrossRef]
37. Lee, J.G.; Hur, K.; Lim, P.K. Closure, interior and compactness in ordinary smooth topological spaces. *Int. J. Fuzzy Log. Intell. Syst.* **2014**, *14*, 231–239. [CrossRef]
38. Smarandache, F. *Neutrosophy, Neutrisophic Property, Sets, and Logic*; American Research Press: Rehoboth, DE, USA, 1998.
39. Salama, A.A.; Broumi, S.; Smarandache, F. Some types of neutrosophic crisp sets and neutrosophic crisp relations. *I.J. Inf. Eng. Electron. Bus.* **2014**. Available online: http://www.mecs-press.org/ (accessed on 10 February 2019).
40. Salama, A.A.; Smarandache, F. *Neutrosophic Crisp Set Theory*; The Educational Publisher Columbus: Columbus, OH, USA, 2015.
41. Hur, K.; Lim, P.K.; Lee, J.G.; Kim, J. The category of neutrosophic crisp sets. *Ann. Fuzzy Math. Inform.* **2017**, *14*, 43–54. [CrossRef]
42. Hur, K.; Lim, P.K.; Lee, J.G.; Kim, J. The category of neutrosophic sets. *Neutrosophic Sets Syst.* **2016**, *14*, 12–20.
43. Smarandache, F. *A Unifying Field in Logics: Neutrosophic Logic. Neutrosophy, Neutrosophic Set, Neutrosophic Probability and Statistics*, 6th ed.; InfoLearnQuest: Ann Arbor, MI, USA, 2007. Available online: http://fs.gallup.unm.edu/eBook-neutrosophics6.pdf (accessed on 10 February 2019).
44. Lupiáñez, F.G. On neutrosophic topology. *Kybernetes* **2008**, *37*, 797–800. [CrossRef]
45. Lupiáñez, F.G. On neutrosophic sets and topology. *Procedia Comput. Sci.* **2017**, *120*, 975–982. [CrossRef]
46. Salama, A.A.; Alblowi, S.A. Neutrosophic set and neutrosophic topological spaces. *IOSR J. Math.* **2012**, *3*, 31–35. [CrossRef]
47. Salama, A.A.; Smarandache, F.; Kroumov, V. Neutrosophic crisp sets and neutrosophic crisp topological spaces. *Neutrosophic Sets Syst.* **2014**, *2*, 25–30.
48. Wang, H.; Smarandache, F.; Zhang, Y.Q.; Sunderraman, R. Single valued neutrosophic sets. *Multispace Multistruct.* **2010**, *4*, 410–413.
49. Kim, J.; Lim, P.K.; Lee, J.G.; Hur, K. Single valued neutrosophic relations. *Ann. Fuzzy Math. Inform.* **2018**, *16*, 201–221. [CrossRef]
50. Ye, J. A multicriteria decision-making method using aggregation operators for simplified neutrosophic sets. *J. Intell. Fuzzy Syst.* **2014**, *26*, 2450–2466.
51. Yang, H.L.; Guo, Z.L.; Liao, X. On single valued neutrosophic relations. *J. Intell. Fuzzy Syst.* **2016**, *30*, 1045–1056. [CrossRef]

symmetry

MDPI

Article

A New Type of Single Valued Neutrosophic Covering Rough Set Model

Jingqian Wang [1] and Xiaohong Zhang [2,*]

[1] School of Electrical and Control Engineering, Shaanxi University of Science and Technology, Xi'an 710021, China

[2] School of Arts and Sciences, Shaanxi University of Science and Technology, Xi'an 710021, China

* Correspondence: zhangxiaohong@sust.edu.cn or zxhonghz@263.net

Received: 8 July 2019; Accepted: 21 August 2019; Published: 24 August 2019

Abstract: Recently, various types of single valued neutrosophic (SVN) rough set models were presented based on the same inclusion relation. However, there is another SVN inclusion relation in SVN sets. In this paper, we propose a new type of SVN covering rough set model based on the new inclusion relation. Furthermore, the graph and matrix representations of the new SVN covering approximation operators are presented. Firstly, the notion of SVN β^2-covering approximation space is proposed, which is decided by the new inclusion relation. Then, a type of SVN covering rough set model under the SVN β^2-covering approximation space is presented. Moreover, there is a corresponding SVN relation rough set model based on a SVN relation induced by the SVN β^2-covering, and two conditions under which the SVN β^2-covering can induce a symmetric SVN relation are presented. Thirdly, the graph and matrix representations of the new SVN covering rough set model are investigated. Finally, we propose a novel method for decision making (DM) problems in paper defect diagnosis under the new SVN covering rough set model.

Keywords: single valued neutrosophic set; covering; symmetric relation; graph representation; matrix representation; paper defect diagnosis

1. Introduction

Rough set theory, as a tool to deal with various types of data in data mining, was proposed by Pawlak [1,2] in 1982. Then rough set theory has been extended to generalize rough sets based on other notions such as binary relations [3], neighborhood systems [4], and coverings [5].

Covering-based rough sets [6–9] were proposed to deal with the type of covering data. In application, they have been applied to knowledge reduction [10,11], decision rule synthesis [12,13], and other fields [14,15]. In theory, covering-based rough set theory has been connected with matroid theory [16–18], lattice theory [19,20], and fuzzy set theory [21–23]. Zadeh's fuzzy set theory [24] addresses the problem of how to understand and manipulate imperfect knowledge. It has been used in various applications [25–28]. Recent investigations have attracted more attention on combining covering-based rough set and fuzzy set theories. There are many fuzzy covering rough set models proposed by researchers, such as Ma [29] and Yang et al. [30].

Smarandache [31] and Wang et al. [32] presented single valued neutrosophic (SVN) sets, which can be regarded as an extension of intuitionistic fuzzy sets [33]. Both neutrosophic sets and rough sets can deal with partial and uncertainty information [34]. Therefore, it is necessary to combine them. Recently, Mondal and Pramanik [35] presented the concept of rough neutrosophic set. Yang et al. [36] presented a SVN rough set model based on SVN relations. Wang and Zhang [37] presented two types of SVN covering rough set models. All these SVN rough set models are presented based on an inclusion relation which is named type-1 inclusion relation and denoted by $\subseteq_1$. The definition of $\subseteq_1$ is shown as follows; for any $A, B \in SVN(U)$,

$$A \subseteq_1 B \text{ iff } T_A(x) \leq T_B(x),\ I_B(x) \leq I_A(x) \text{ and } F_B(x) \leq F_A(x) \text{ for all } x \in U.$$

Under the type-1 inclusion relation, for two SVN numbers $\alpha = \langle a, b, c \rangle$ and $\beta = \langle d, e, f \rangle$, we have $\alpha \leq_1 \beta \Leftrightarrow a \leq d, b \geq e$ and $c \geq f$. The definition of SVN β-covering approximation space is presented as follows (see the work by the authors of [37]).

Let U be a universe and $SVN(U)$ be the SVN power set of U. For a SVN number $\beta = \langle a, b, c \rangle$, we call $\widehat{\mathbf{C}} = \{C_1, C_2, \cdots, C_m\}$, with $C_i \in SVN(U)(i = 1, 2, ..., m)$, a SVN β-covering of U, if for all $x \in U$, $C_i \in \widehat{\mathbf{C}}$ exists such that $C_i(x) \geq_1 \beta$. We also call $(U, \widehat{\mathbf{C}})$ a SVN β-covering approximation space.

However, there exists another inclusion relation in the work by the authors of [38], which is called type-2 inclusion relation and denoted by $\subseteq_2$. The definition of $\subseteq_2$ is introduced as follows; for any $A, B \in SVN(U)$,

$$A \subseteq_2 B \text{ iff } T_A(x) \leq T_B(x),\ I_A(x) \leq I_B(x) \text{ and } F_B(x) \leq F_A(x) \text{ for all } x \in U.$$

Under the type-2 inclusion relation, for two SVN numbers $\alpha = \langle a, b, c \rangle$ and $\beta = \langle d, e, f \rangle$, we have $\alpha \leq_2 \beta \Leftrightarrow a \leq d, b \leq e$ and $c \geq f$.

In the definition of SVN β-covering approximation space, if $C_i(x) \geq_1 \beta$ is replaced by $C_i(x) \geq_2 \beta$, there will be a new SVN covering approximation space (we call it a SVN β^2-covering approximation space in this paper). In Example 1 in this paper, we find the following statements.

(1) Let $\beta = \langle 0.5, 0.1, 0.8 \rangle$. Then $\widehat{\mathbf{C}}$ is a SVN β^2-covering of U, but it is not a SVN β-covering of U.
(2) Let $\beta = \langle 0.5, 0.3, 0.8 \rangle$. Then $\widehat{\mathbf{C}}$ is a SVN β-covering of U, but it is not a SVN β^2-covering of U.

That is to say, the SVN β^2-covering approximation space is a new SVN covering approximation space, which is different from the SVN β-covering approximation space. Since different inclusion relations ($\subseteq_1$ and $\subseteq_2$) have different union and intersection operations, the SVN β^2-covering approximation space has different union and intersection operations from the SVN β-covering approximation space. Hence, notions and corresponding SVN covering rough set models of SVN β-covering approximation space do not apply to SVN β^2-covering approximation space, which is the justification for studying this topic. Therefore, the investigation of the SVN β^2-covering approximation space and its corresponding SVN covering rough set model is very important. It not only can manage some issues that the SVN β-covering approximation space can not deal with, but also constructs a new type of SVN covering rough set model. This is our motivation of this research.

In this paper, we present some new concepts in SVN β^2-covering approximation space, as well as their properties. Then the type-2 SVN covering rough set model under the SVN β^2-covering approximation space is proposed. On the one hand, the graph and matrix representations of the type-2 SVN covering rough set model are investigated respectively. Moreover, some relationships between the type-2 SVN covering rough set model and other SVN rough set models are presented. One the other hand, we present a method to DM problems in paper defect diagnosis, which is an important topic in paper making industries, under the type-2 SVN covering rough set model. Many researchers have studied decision making (DM) problems by rough set models [39–42]. Hence, the proposed DM method is compared with other methods which are presented by Liu [43], Ye [44], Yang et al. [36], and Wang et al. [37] respectively.

The rest of this paper is organized as follows. Section 2 reviews some fundamental definitions about covering-based rough sets and SVN sets. In Section 3, some concepts and properties in SVN β^2-covering approximation space are studied. The relationship between the SVN β-covering approximation space and the SVN β^2-covering approximation space is presented. In Section 4, we present the type-2 SVN covering rough set model. Some relationships between the type-2 SVN covering rough set model and other SVN rough set models are presented. Moreover, a SVN relation can be induced by the SVN β^2-covering, so a corresponding SVN relation rough set model and two conditions under which the SVN β^2-covering can induce a symmetric SVN relation are presented. In Section 5, some new graphs and graph operations are presented. Based on this, the graph representation of the type-2 SVN covering approximation operators is shown. In Section 6, some new matrices and matrix operations are also presented, and the matrix representation of the type-2 SVN covering approximation operators is presented. In Section 7, a novel method to paper defect diagnosis is presented under the type-2 SVN covering rough set model. Moreover, the proposed method is compared with other methods. This paper is concluded and further work is indicated in Section 8.

2. Basic Definitions

Suppose U is a nonempty and finite set called universe.

Definition 1. *(Covering [45,46]) Let U be a universe and $\mathbf{C}$ be a family of subsets of U. If none of subsets in $\mathbf{C}$ is empty and $\cup\mathbf{C} = U$, then $\mathbf{C}$ is called a covering of U.*

The pair $(U, \mathbf{C})$ is called a covering approximation space.

Definition 2. *(SVN set [32]) Let U be a nonempty fixed set. A SVN set A in U is defined as an object of the following form.*

$$A = \{\langle x, T_A(x), I_A(x), F_A(x)\rangle : x \in U\},$$

where $T_A : U \to [0,1]$ is called the degree of truth-membership of the element $x \in U$ to A, $I_A : U \to [0,1]$ is called the degree of indeterminacy-membership of the element $x \in U$ to A, $F_A(x) : U \to [0,1]$ is called the degree of falsity-membership. They satisfy $0 \le T_A(x) + I_A(x) + F_A(x) \le 3$ for all $x \in U$. The family of all SVN sets in U is denoted by $SVN(U)$. For convenience, a SVN number is represented by $\alpha = \langle a, b, c\rangle$, where $a, b, c \in [0,1]$ and $a + b + c \le 3$.

For the inclusion relation of neutrosophic sets, there are two different definitions in the literature. An original definition is proposed by Smarandache [31,47], we call it type-1 inclusion relation in this paper, denoted by $\subseteq_1$. For set theory, union and intersection operations are corresponding to inclusion relation. Hence, there are corresponding union and intersection operations defined as follows; for any $A, B \in SVN(U)$,

(1) $A \subseteq_1 B$ iff $T_A(x) \le T_B(x)$, $I_B(x) \le I_A(x)$ and $F_B(x) \le F_A(x)$ for all $x \in U$;
(2) $A \cap_1 B = \{\langle x, T_A(x) \wedge T_B(x), I_A(x) \vee I_B(x), F_A(x) \vee F_B(x)\rangle : x \in U\}$;
(3) $A \cup_1 B = \{\langle x, T_A(x) \vee T_B(x), I_A(x) \wedge I_B(x), F_A(x) \wedge F_B(x)\rangle : x \in U\}$.

Specially, for two SVN numbers, $\alpha = \langle a, b, c\rangle$ and $\beta = \langle d, e, f\rangle$, $\alpha \le_1 \beta \Leftrightarrow a \le d$, $b \ge e$ and $c \ge f$.

Under the type-1 inclusion relation, Wang and Zhang [37] presented the definition of SVN β-covering approximation space.

Definition 3. *[37] Let U be a universe and $SVN(U)$ be the SVN power set of U. For a SVN number $\beta = \langle a, b, c\rangle$, we call $\widehat{\mathbf{C}} = \{C_1, C_2, \cdots, C_m\}$, with $C_i \in SVN(U) (i = 1, 2, ..., m)$, a SVN β-covering of U, if for all $x \in U$, $C_i \in \widehat{\mathbf{C}}$ exists such that $C_i(x) \ge_1 \beta$. We also call $(U, \widehat{\mathbf{C}})$ a SVN β-covering approximation space.*

Definition 4. *[37] Let $\widehat{\mathbf{C}}$ be a SVN β-covering of U and $\widehat{\mathbf{C}} = \{C_1, C_2, \ldots, C_m\}$. For any $x \in U$, the SVN β-neighborhood $\widetilde{\mathbb{N}}_x^\beta$ of x induced by $\widehat{\mathbf{C}}$ can be defined as*

$$\widetilde{\mathbb{N}}_x^\beta = \cap_1\{C_i \in \widehat{\mathbf{C}} : C_i(x) \ge_1 \beta\}. \tag{1}$$

Another one is used in some papers [32,38], we call it type-2 inclusion relation in this paper, denote it by $\subseteq_2$. Hence, the type-2 inclusion relation, corresponding union and intersection operations are shown as follows; for any $A, B \in SVN(U)$,

(1) $A \subseteq_2 B$ iff $T_A(x) \le T_B(x)$, $I_A(x) \le I_B(x)$ and $F_B(x) \le F_A(x)$ for all $x \in U$;
(2) $A \cap_2 B = \{\langle x, T_A(x) \wedge T_B(x), I_A(x) \wedge I_B(x), F_A(x) \vee F_B(x)\rangle : x \in U\}$;

(3) $A \cup_2 B = \{\langle x, T_A(x) \vee T_B(x), I_A(x) \vee I_B(x), F_A(x) \wedge F_B(x)\rangle : x \in U\}$.

Specially, for two SVN numbers $\alpha = \langle a, b, c\rangle$ and $\beta = \langle d, e, f\rangle$, $\alpha \leq_2 \beta \Leftrightarrow a \leq d, b \leq e$ and $c \geq f$.

For the above inclusion relations of neutrosophic sets, the following operations use the same definition in this paper [32,36].

(4) $A = B$ iff $A \subseteq_1 B$ and $B \subseteq_1 A$, iff $A \subseteq_2 B$ and $B \subseteq_2 A$;
(5) $A' = \{\langle x, F_A(x), 1 - I_A(x), T_A(x)\rangle : x \in U\}$;
(6) $A \oplus B = \{\langle x, T_A(x) + T_B(x) - T_A(x) \cdot T_B(x), I_A(x) \cdot I_B(x), F_A(x) \cdot F_B(x)\rangle : x \in U\}$.

3. SVN β^2-Covering Approximation Space

In this section, the definition of SVN β^2-covering approximation space is presented. There are two basic concepts—SVN β^2-covering and SVN β^2-neighborhood—in this new approximation space.

Definition 5. *Let U be a universe and $SVN(U)$ be the SVN power set of U. For a SVN number $\beta = \langle a, b, c\rangle$, we call $\widehat{C} = \{C_1, C_2, \cdots, C_m\}$, with $C_i \in SVN(U)(i = 1, 2, ..., m)$, a SVN β^2-covering of U, if for all $x \in U$, $C_i \in \widehat{C}$ exists such that $C_i(x) \geq_2 \beta$. We also call $(U, \widehat{C})$ a SVN β^2-covering approximation space.*

In Definition 5, if $C_i(x) \geq_2 \beta$ is replaced by $C_i(x) \geq_1 \beta$, then $\widehat{C} = \{C_1, C_2, \cdots, C_m\}$ is called a SVN β-covering of U in [37]. By the definitions of $\geq_1$ and $\geq_2$, we know if $\widehat{C}$ is a SVN β^2-covering of U, then $\widehat{C}$ is not necessarily a SVN β-covering. To show the difference between SVN β-covering and SVN β^2-covering, we use the work presented by the authors of [37] in the following example.

Example 1. *Let $U = \{x_1, x_2, x_3, x_4, x_5\}$, $\widehat{C} = \{C_1, C_2, C_3, C_4\}$ and $\beta = \langle 0.5, 0.1, 0.8\rangle$. We can see that $\widehat{C}$ is a SVN β^2-covering of U in Table 1, but it is not a SVN β-covering of U.*

Table 1. The tabular representation of $\widehat{C}$ in [37].

U	C_1	C_2	C_3	C_4
x_1	$\langle 0.7, 0.2, 0.5\rangle$	$\langle 0.6, 0.2, 0.4\rangle$	$\langle 0.4, 0.1, 0.5\rangle$	$\langle 0.1, 0.5, 0.6\rangle$
x_2	$\langle 0.5, 0.3, 0.2\rangle$	$\langle 0.5, 0.2, 0.8\rangle$	$\langle 0.4, 0.5, 0.4\rangle$	$\langle 0.6, 0.1, 0.7\rangle$
x_3	$\langle 0.4, 0.5, 0.2\rangle$	$\langle 0.2, 0.3, 0.6\rangle$	$\langle 0.5, 0.2, 0.4\rangle$	$\langle 0.6, 0.3, 0.4\rangle$
x_4	$\langle 0.6, 0.1, 0.7\rangle$	$\langle 0.4, 0.5, 0.7\rangle$	$\langle 0.3, 0.6, 0.5\rangle$	$\langle 0.5, 0.3, 0.2\rangle$
x_5	$\langle 0.3, 0.2, 0.6\rangle$	$\langle 0.7, 0.3, 0.5\rangle$	$\langle 0.6, 0.3, 0.5\rangle$	$\langle 0.8, 0.1, 0.2\rangle$

Conversely, if $\widehat{C}$ is a SVN β-covering of U, then $\widehat{C}$ is not necessarily a SVN β^2-covering. In Example 1, suppose $\beta = \langle 0.5, 0.3, 0.8\rangle$. Then $\widehat{C}$ is a SVN β-covering of U, but it is not a SVN β^2-covering of U.

By the definition of SVN β-neighborhood, the notion of SVN β^2-neighborhood is presented in the following definition.

Definition 6. *Let $\widehat{C}$ be a SVN β^2-covering of U and $\widehat{C} = \{C_1, C_2, \ldots, C_m\}$. For any $x \in U$, the SVN β^2-neighborhood $\widetilde{\mathbb{N}}_x^{\beta^2}$ of x induced by $\widehat{C}$ can be defined as*

$$\widetilde{\mathbb{N}}_x^{\beta^2} = \cap_2\{C_i \in \widehat{C} : C_i(x) \geq_2 \beta\}. \tag{2}$$

Note that $C_i(x)$ is a SVN number $\langle T_{C_i}(x), I_{C_i}(x), F_{C_i}(x)\rangle$. Hence, $C_i(x) \geq_2 \beta$ means $T_{C_i}(x) \geq a$, $I_{C_i}(x) \geq b$ and $F_{C_i}(x) \leq c$, where SVN number $\beta = \langle a, b, c\rangle$.

Remark 1. *Let $\widehat{C}$ be a SVN β^2-covering of U, $\beta = \langle a, b, c\rangle$ and $\widehat{C} = \{C_1, C_2, \ldots, C_m\}$. For any $x \in U$,*

$$\widetilde{\mathbb{N}}_x^{\beta^2} = \cap_2\{C_i \in \widehat{C} : T_{C_i}(x) \geq a,\ I_{C_i}(x) \geq b,\ F_{C_i}(x) \leq c\}. \tag{3}$$

Example 2. *Let $(U,\widehat{C})$ be a SVN β^2-covering approximation space in Example 1 and $\beta = \langle 0.5, 0.1, 0.8\rangle$. Then*

$$\widetilde{N}^{\beta^2}_{x_1} = C_1 \cap_2 C_2, \widetilde{N}^{\beta^2}_{x_2} = C_1 \cap_2 C_2 \cap_2 C_4, \widetilde{N}^{\beta^2}_{x_3} = C_3 \cap_2 C_4, \widetilde{N}^{\beta^2}_{x_4} = C_1 \cap_2 C_4, \widetilde{N}^{\beta^2}_{x_5} = C_2 \cap_2 C_3 \cap_2 C_4.$$

Hence, all SVN β^2-neighborhoods are shown in Table 2:

Table 2. The tabular representation of $\widetilde{N}^{\beta^2}_{x_k}$ $(k = 1,2,3,4,5)$.

$\widetilde{N}^{\beta^2}_{x_k}$	x_1	x_2	x_3	x_4	x_5
$\widetilde{N}^{\beta^2}_{x_1}$	$\langle 0.6,0.2,0.5\rangle$	$\langle 0.5,0.2,0.8\rangle$	$\langle 0.2,0.3,0.6\rangle$	$\langle 0.4,0.1,0.7\rangle$	$\langle 0.3,0.2,0.6\rangle$
$\widetilde{N}^{\beta^2}_{x_2}$	$\langle 0.1,0.2,0.6\rangle$	$\langle 0.5,0.1,0.8\rangle$	$\langle 0.2,0.3,0.6\rangle$	$\langle 0.4,0.1,0.7\rangle$	$\langle 0.3,0.1,0.6\rangle$
$\widetilde{N}^{\beta^2}_{x_3}$	$\langle 0.1,0.1,0.6\rangle$	$\langle 0.4,0.1,0.7\rangle$	$\langle 0.5,0.2,0.4\rangle$	$\langle 0.3,0.3,0.5\rangle$	$\langle 0.6,0.1,0.5\rangle$
$\widetilde{N}^{\beta^2}_{x_4}$	$\langle 0.1,0.2,0.6\rangle$	$\langle 0.5,0.1,0.7\rangle$	$\langle 0.4,0.3,0.4\rangle$	$\langle 0.5,0.1,0.7\rangle$	$\langle 0.3,0.1,0.6\rangle$
$\widetilde{N}^{\beta^2}_{x_5}$	$\langle 0.1,0.1,0.6\rangle$	$\langle 0.4,0.1,0.8\rangle$	$\langle 0.2,0.2,0.6\rangle$	$\langle 0.3,0.3,0.7\rangle$	$\langle 0.6,0.1,0.5\rangle$

According to Definitions 3–6, we know that "Let $\widehat{C}$ be a SVN β^2-covering of U. If $\widehat{C}$ is also a SVN β-covering of U, then $\widetilde{N}^{\beta^2}_{x}$ and $\widetilde{N}^{\beta}_{x}$ have no inclusion relations ($\subseteq_1$ and $\subseteq_2$) for all $x \in U$. To explain this statement, the following example is presented.

Example 3. *Let $U = \{x_1, x_2, x_3, x_4, x_5\}$, $\widehat{C} = \{C_1, C_2, C_3, C_4\}$ and $\beta = \langle 0.5, 0.2, 0.8\rangle$, where $\widehat{C}$ is shown in Table 1 of Example 1. By Definitions 3 and 5, we know that $\widehat{C}$ is a SVN β^2-covering and also a SVN β-covering of U. Then*

$$\widetilde{N}^{\beta^2}_{x_1} = C_1 \cap_2 C_2, \widetilde{N}^{\beta^2}_{x_2} = C_1 \cap_2 C_2, \widetilde{N}^{\beta^2}_{x_3} = C_3 \cap_2 C_4, \widetilde{N}^{\beta^2}_{x_4} = C_4, \widetilde{N}^{\beta^2}_{x_5} = C_2 \cap_2 C_3.$$
$$\widetilde{N}^{\beta}_{x_1} = C_1 \cap_1 C_2, \widetilde{N}^{\beta}_{x_2} = C_2 \cap_1 C_4, \widetilde{N}^{\beta}_{x_3} = C_3, \widetilde{N}^{\beta}_{x_4} = C_1, \widetilde{N}^{\beta}_{x_5} = C_4.$$

Hence, all SVN β^2-neighborhoods and SVN β^2-neighborhoods are shown in Tables 3 and 4 respectively: By Tables 3 and 4, we see that for all $x_k \in U$ $(k = 1,2,3,4,5)$

- $\widetilde{N}^{\beta^2}_{x_k} \subseteq_1 \widetilde{N}^{\beta}_{x_k}$ *is not established, since* $\widetilde{N}^{\beta^2}_{x_2} \not\subseteq_1 \widetilde{N}^{\beta}_{x_2}$.
- $\widetilde{N}^{\beta}_{x_k} \subseteq_1 \widetilde{N}^{\beta^2}_{x_k}$ *is not established, since* $\widetilde{N}^{\beta}_{x_3} \not\subseteq_1 \widetilde{N}^{\beta^2}_{x_3}$.
- $\widetilde{N}^{\beta^2}_{x_k} \subseteq_2 \widetilde{N}^{\beta}_{x_k}$ *is not established, since* $\widetilde{N}^{\beta^2}_{x_4} \not\subseteq_2 \widetilde{N}^{\beta}_{x_4}$.
- $\widetilde{N}^{\beta}_{x_k} \subseteq_2 \widetilde{N}^{\beta^2}_{x_k}$ *is not established, since* $\widetilde{N}^{\beta}_{x_5} \not\subseteq_2 \widetilde{N}^{\beta^2}_{x_5}$.

Hence, $\widetilde{N}^{\beta^2}_{x_k}$ and $\widetilde{N}^{\beta}_{x_k}$ have no inclusion relations ($\subseteq_1$ and $\subseteq_2$) for all $x_k \in U$.

In a SVN β^2-covering approximation space $(U, \widehat{C})$, we present the following properties of the SVN β^2-neighborhood.

Table 3. The tabular representation of $\widetilde{N}^{\beta^2}_{x_k}$ $(k = 1,2,3,4,5)$.

$\widetilde{N}^{\beta^2}_{x_k}$	x_1	x_2	x_3	x_4	x_5
$\widetilde{N}^{\beta^2}_{x_1}$	$\langle 0.6,0.2,0.5\rangle$	$\langle 0.5,0.2,0.8\rangle$	$\langle 0.2,0.3,0.6\rangle$	$\langle 0.4,0.1,0.7\rangle$	$\langle 0.3,0.2,0.6\rangle$
$\widetilde{N}^{\beta^2}_{x_2}$	$\langle 0.6,0.2,0.5\rangle$	$\langle 0.5,0.2,0.8\rangle$	$\langle 0.2,0.3,0.6\rangle$	$\langle 0.4,0.1,0.7\rangle$	$\langle 0.3,0.2,0.6\rangle$
$\widetilde{N}^{\beta^2}_{x_3}$	$\langle 0.1,0.1,0.6\rangle$	$\langle 0.4,0.1,0.7\rangle$	$\langle 0.5,0.2,0.4\rangle$	$\langle 0.3,0.3,0.5\rangle$	$\langle 0.6,0.1,0.5\rangle$
$\widetilde{N}^{\beta^2}_{x_4}$	$\langle 0.1,0.5,0.6\rangle$	$\langle 0.6,0.1,0.7\rangle$	$\langle 0.6,0.3,0.4\rangle$	$\langle 0.5,0.3,0.2\rangle$	$\langle 0.8,0.1,0.2\rangle$
$\widetilde{N}^{\beta^2}_{x_5}$	$\langle 0.4,0.1,0.5\rangle$	$\langle 0.4,0.2,0.8\rangle$	$\langle 0.2,0.2,0.6\rangle$	$\langle 0.3,0.5,0.7\rangle$	$\langle 0.6,0.3,0.5\rangle$

Table 4. The tabular representation of $\widetilde{\mathbb{N}}^{\beta}_{x_k}$ $(k = 1, 2, 3, 4, 5)$.

$\widetilde{\mathbb{N}}^{\beta}_{x_k}$	x_1	x_2	x_3	x_4	x_5
$\widetilde{\mathbb{N}}^{\beta}_{x_1}$	$\langle 0.6, 0.2, 0.5\rangle$	$\langle 0.5, 0.3, 0.8\rangle$	$\langle 0.2, 0.5, 0.6\rangle$	$\langle 0.4, 0.5, 0.7\rangle$	$\langle 0.3, 0.3, 0.6\rangle$
$\widetilde{\mathbb{N}}^{\beta}_{x_2}$	$\langle 0.1, 0.5, 0.6\rangle$	$\langle 0.5, 0.2, 0.8\rangle$	$\langle 0.2, 0.3, 0.6\rangle$	$\langle 0.4, 0.5, 0.7\rangle$	$\langle 0.7, 0.3, 0.5\rangle$
$\widetilde{\mathbb{N}}^{\beta}_{x_3}$	$\langle 0.4, 0.1, 0.5\rangle$	$\langle 0.4, 0.5, 0.4\rangle$	$\langle 0.5, 0.2, 0.4\rangle$	$\langle 0.3, 0.6, 0.5\rangle$	$\langle 0.6, 0.3, 0.5\rangle$
$\widetilde{\mathbb{N}}^{\beta}_{x_4}$	$\langle 0.7, 0.2, 0.5\rangle$	$\langle 0.5, 0.3, 0.2\rangle$	$\langle 0.4, 0.5, 0.2\rangle$	$\langle 0.6, 0.1, 0.7\rangle$	$\langle 0.3, 0.2, 0.6\rangle$
$\widetilde{\mathbb{N}}^{\beta}_{x_5}$	$\langle 0.1, 0.5, 0.6\rangle$	$\langle 0.6, 0.1, 0.7\rangle$	$\langle 0.6, 0.3, 0.4\rangle$	$\langle 0.5, 0.3, 0.2\rangle$	$\langle 0.8, 0.1, 0.2\rangle$

Proposition 1. *Let $\widehat{\mathbf{C}}$ be a SVN β^2-covering of U and $\widehat{\mathbf{C}} = \{C_1, C_2, \ldots, C_m\}$. Then $\widetilde{\mathbb{N}}^{\beta^2}_x(x) \geq_2 \beta$ for each $x \in U$.*

Proof. For any $x \in U$, $\widetilde{\mathbb{N}}^{\beta^2}_x(x) = (\underset{C_i(x)\geq_2\beta}{\cap_2} C_i)(x) \geq_2 \beta$. □

Proposition 2. *Let $\widehat{\mathbf{C}}$ be a SVN β^2-covering of U and $\widehat{\mathbf{C}} = \{C_1, C_2, \ldots, C_m\}$. For all $x, y, z \in U$, if $\widetilde{\mathbb{N}}^{\beta^2}_x(y) \geq_2 \beta$, $\widetilde{\mathbb{N}}^{\beta^2}_y(z) \geq_2 \beta$, then $\widetilde{\mathbb{N}}^{\beta^2}_x(z) \geq_2 \beta$.*

Proof. Let $I = \{1, 2, \cdots, m\}$. Since $\widetilde{\mathbb{N}}^{\beta^2}_x(y) \geq_2 \beta$, for any $i \in I$, if $C_i(x) \geq_2 \beta$, then $C_i(y) \geq_2 \beta$. Since $\widetilde{\mathbb{N}}^{\beta^2}_y(z) \geq_2 \beta$, for any $i \in I$, $C_i(z) \geq_2 \beta$ when $C_i(y) \geq_2 \beta$. Then, for any $i \in I$, $C_i(x) \geq_2 \beta$ implies $C_i(z) \geq_2 \beta$. Therefore, $\widetilde{\mathbb{N}}^{\beta^2}_x(z) \geq_2 \beta$. □

Proposition 3. *Let $\widehat{\mathbf{C}}$ be a SVN β^2-covering of U and $\widehat{\mathbf{C}} = \{C_1, C_2, \ldots, C_m\}$. For two SVN numbers β_1, β_2, if $\beta_1 \leq_2 \beta_2 \leq_2 \beta$, then $\widetilde{\mathbb{N}}^{\beta_1^2}_x \subseteq_2 \widetilde{\mathbb{N}}^{\beta_2^2}_x$ for all $x \in U$.*

Proof. For all $x \in U$, since $\beta_1 \leq_2 \beta_2 \leq_2 \beta$, $\{C_i \in \widehat{\mathbf{C}} : C_i(x) \geq_2 \beta_1\} \supseteq \{C_i \in \widehat{\mathbf{C}} : C_i(x) \geq_2 \beta_2\}$. Hence, $\widetilde{\mathbb{N}}^{\beta_1^2}_x = \cap_2\{C_i \in \widehat{\mathbf{C}} : C_i(x) \geq_2 \beta_1\} \subseteq_2 \cap_2\{C_i \in \widehat{\mathbf{C}} : C_i(x) \geq_2 \beta_2\} = \widetilde{\mathbb{N}}^{\beta_2^2}_x$ for all $x \in U$. □

Proposition 4. *Let $\widehat{\mathbf{C}}$ be a SVN β^2-covering of U. For any $x, y \in U$, $\widetilde{\mathbb{N}}^{\beta^2}_x(y) \geq_2 \beta$ if and only if $\widetilde{\mathbb{N}}^{\beta^2}_y \subseteq_2 \widetilde{\mathbb{N}}^{\beta^2}_x$.*

Proof. Suppose the SVN number $\beta = \langle a, b, c\rangle$.

($\Rightarrow$): Since $\widetilde{\mathbb{N}}^{\beta^2}_x(y) \geq_2 \beta$,

$$T_{\widetilde{\mathbb{N}}^{\beta^2}_x}(y) = T_{\underset{\substack{T_{C_i}(x)\geq a\\ I_{C_i}(x)\geq b\\ F_{C_i}(x)\leq c}}{\cap_2} C_i}(y) = \bigwedge_{\substack{T_{C_i}(x)\geq a\\ I_{C_i}(x)\geq b\\ F_{C_i}(x)\leq c}} T_{C_i}(y) \geq a,\ I_{\widetilde{\mathbb{N}}^{\beta^2}_x}(y) = I_{\underset{\substack{T_{C_i}(x)\geq a\\ I_{C_i}(x)\geq b\\ F_{C_i}(x)\leq b}}{\cap_2} C_i}(y) = \bigwedge_{\substack{T_{C_i}(x)\geq a\\ I_{C_i}(x)\geq b\\ F_{C_i}(x)\leq c}} I_{C_i}(y) \geq b,$$

and

$$F_{\widetilde{\mathbb{N}}^{\beta^2}_x}(y) = F_{\underset{\substack{T_{C_i}(x)\geq a\\ I_{C_i}(x)\geq b\\ F_{C_i}(x)\leq c}}{\cap_2} C_i}(y) = \bigvee_{\substack{T_{C_i}(x)\geq a\\ I_{C_i}(x)\geq b\\ F_{C_i}(x)\leq c}} F_{C_i}(y) \leq c.$$

Then,

$\{C_i \in \widehat{\mathbf{C}} : T_{C_i}(x) \geq a, I_{C_i}(x) \geq b, F_{C_i}(x) \leq c\} \subseteq \{C_i \in \widehat{\mathbf{C}} : T_{C_i}(y) \geq a, I_{C_i}(y) \geq b, F_{C_i}(y) \leq c\}$.

Therefore, for each $z \in U$,

$$T_{\widetilde{\mathbb{N}}_x^{\beta^2}}(z) = \bigwedge_{\substack{T_{C_i}(x)\geq a\\ I_{C_i}(x)\geq b\\ F_{C_i}(x)\leq c}} T_{C_i}(z) \geq \bigwedge_{\substack{T_{C_i}(y)\geq a\\ I_{C_i}(y)\geq b\\ F_{C_i}(y)\leq c}} T_{C_i}(z) = T_{\widetilde{\mathbb{N}}_y^{\beta^2}}(z),$$

$$I_{\widetilde{\mathbb{N}}_x^{\beta^2}}(z) = \bigwedge_{\substack{T_{C_i}(x)\geq a\\ I_{C_i}(x)\geq b\\ F_{C_i}(x)\leq c}} I_{C_i}(z) \geq \bigwedge_{\substack{T_{C_i}(y)\geq a\\ I_{C_i}(y)\geq b\\ F_{C_i}(y)\leq c}} I_{C_i}(z) = I_{\widetilde{\mathbb{N}}_y^{\beta^2}}(z),$$

$$F_{\widetilde{\mathbb{N}}_x^{\beta^2}}(z) = \bigvee_{\substack{T_{C_i}(x)\geq a\\ I_{C_i}(x)\geq b\\ F_{C_i}(x)\leq c}} F_{C_i}(z) \leq \bigvee_{\substack{T_{C_i}(y)\geq a\\ I_{C_i}(y)\geq b\\ F_{C_i}(y)\leq c}} F_{C_i}(z) = F_{\widetilde{\mathbb{N}}_y^{\beta^2}}(z).$$

Hence, $\widetilde{\mathbb{N}}_y^{\beta^2} \subseteq_2 \widetilde{\mathbb{N}}_x^{\beta^2}$.

($\Leftarrow$): For any $x, y \in U$, since $\widetilde{\mathbb{N}}_y^{\beta^2} \subseteq_2 \widetilde{\mathbb{N}}_x^{\beta^2}$,

$$T_{\widetilde{\mathbb{N}}_x^{\beta^2}}(y) \geq T_{\widetilde{\mathbb{N}}_y^{\beta^2}}(y) \geq a,\ I_{\widetilde{\mathbb{N}}_x^{\beta^2}}(y) \geq I_{\widetilde{\mathbb{N}}_y^{\beta^2}}(y) \geq b \text{ and } F_{\widetilde{\mathbb{N}}_x^{\beta^2}}(y) \leq F_{\widetilde{\mathbb{N}}_y^{\beta^2}}(y) \leq c.$$

Therefore, $\widetilde{\mathbb{N}}_x^{\beta^2}(y) \geq_2 \beta$. □

4. A Type of SVN Covering Rough Set Model Based on a New Inclusion Relation

In this section, we propose a type of SVN covering rough set model on the basis of the SVN β^2-neighborhoods, which is decided by a type-2 inclusion relation. Then, we investigate some properties of the new lower and upper SVN covering approximation operators. Finally, some relationships between this model and some other rough set models are presented.

4.1. Characteristics of the New Type of SVN Covering Rough Set Model Based on the New Inclusion Relation

Definition 7. *Let $(U, \widehat{\mathbf{C}})$ be a SVN β^2-covering approximation space. For each $A \in SVN(U)$, where $A = \{\langle x, T_A(x), I_A(x), F_A(x)\rangle : x \in U\}$, we define the type-2 SVN covering upper approximation $\widetilde{\mathbb{C}}^2(A)$ and lower approximation $\underset{\sim}{\mathbb{C}}^2(A)$ of A as*

$$\begin{aligned} \widetilde{\mathbb{C}}^2(A) &= \{\langle x, \vee_{y\in U}[T_{\widetilde{\mathbb{N}}_x^{\beta^2}}(y) \wedge T_A(y)], \vee_{y\in U}[I_{\widetilde{\mathbb{N}}_x^{\beta^2}}(y) \wedge I_A(y)], \wedge_{y\in U}[F_{\widetilde{\mathbb{N}}_x^{\beta^2}}(y) \vee F_A(y)]\rangle : x \in U\},\\ \underset{\sim}{\mathbb{C}}^2(A) &= \{\langle x, \wedge_{y\in U}[F_{\widetilde{\mathbb{N}}_x^{\beta^2}}(y) \vee T_A(y)], \wedge_{y\in U}[(1 - I_{\widetilde{\mathbb{N}}_x^{\beta^2}}(y)) \vee I_A(y)], \vee_{y\in U}[T_{\widetilde{\mathbb{N}}_x^{\beta^2}}(y) \wedge F_A(y)]\rangle : x \in U\}. \end{aligned} \tag{4}$$

If $\widetilde{\mathbb{C}}^2(A) \neq \underset{\sim}{\mathbb{C}}^2(A)$, then A is called the type-2 SVN covering rough set.

Example 4. *Let $(U, \widehat{\mathbf{C}})$ be a SVN β^2-covering approximation space in Example 1, $\beta = \langle 0.5, 0.1, 0.8\rangle$ and $A = \frac{(0.6,0.3,0.5)}{x_1} + \frac{(0.4,0.5,0.1)}{x_2} + \frac{(0.3,0.2,0.6)}{x_3} + \frac{(0.5,0.3,0.4)}{x_4} + \frac{(0.7,0.2,0.3)}{x_5}$. Then all SVN β^2-neighborhoods $\widetilde{\mathbb{N}}_{x_k}^{\beta^2}$ $(k = 1,2,3,4,5)$ are shown in Table 2 of Example 2. By Definition 7, we have*

$$\widetilde{\mathbb{C}}^2(A) = \{\langle x_1, 0.6, 0.2, 0.5\rangle, \langle x_2, 0.4, 0.2, 0.6\rangle, \langle x_3, 0.6, 0.3, 0.5\rangle, \langle x_4, 0.5, 0.2, 0.6\rangle, \langle x_5, 0.6, 0.3, 0.5\rangle\},$$

$$\underset{\sim}{\mathbb{C}}^2(A) = \{\langle x_1, 0.6, 0.8, 0.5\rangle, \langle x_2, 0.6, 0.8, 0.4\rangle, \langle x_3, 0.4, 0.7, 0.5\rangle, \langle x_4, 0.4, 0.7, 0.4\rangle, \langle x_5, 0.6, 0.8, 0.3\rangle\}.$$

Let the SVN universe set be $U = \{\langle x, 1, 1, 0\rangle : x \in U\}$ and the SVN empty set be $\emptyset = \{\langle x, 0, 0, 1\rangle : x \in U\}$, which are decided by the type-2 inclusion relation $\subseteq_2$. Some basic properties of the type-2 SVN covering upper and lower approximation operators are presented in the following proposition.

Proposition 5. *Let $\widehat{\mathbb{C}}$ be a SVN β^2-covering of U. Then the type-2 SVN covering upper and lower approximation operators in Definition 7 satisfy the following properties for all $A, B \in SVN(U)$.*

(1) $\underset{\sim}{\mathbb{C}}^2(U) = U, \widetilde{\mathbb{C}}^2(\emptyset) = \emptyset$;

(2) $\widetilde{\mathbb{C}}^2(A') = (\underset{\sim}{\mathbb{C}}^2(A))', \underset{\sim}{\mathbb{C}}^2(A') = (\widetilde{\mathbb{C}}^2(A))'$;

(3) *If* $A \subseteq_2 B$, *then* $\underset{\sim}{\mathbb{C}}^2(A) \subseteq_2 \underset{\sim}{\mathbb{C}}^2(B), \widetilde{\mathbb{C}}^2(A) \subseteq_2 \widetilde{\mathbb{C}}^2(B)$;

(4) $\underset{\sim}{\mathbb{C}}^2(A \cap_2 B) = \underset{\sim}{\mathbb{C}}^2(A) \cap_2 \underset{\sim}{\mathbb{C}}^2(B), \widetilde{\mathbb{C}}^2(A \cup_2 B) = \widetilde{\mathbb{C}}^2(A) \cup_2 \widetilde{\mathbb{C}}^2(B)$;

(5) $\underset{\sim}{\mathbb{C}}^2(A \cup_2 B) \supseteq_2 \underset{\sim}{\mathbb{C}}^2(A) \cup_2 \underset{\sim}{\mathbb{C}}^2(B), \widetilde{\mathbb{C}}^2(A \cap_2 B) \subseteq_2 \widetilde{\mathbb{C}}^2(A) \cap_2 \widetilde{\mathbb{C}}^2(B)$.

Proof.(1) Since the SVN universe set is $U = \{\langle x, 1, 1, 0\rangle : x \in U\}$ and the SVN empty set is $\emptyset = \{\langle x, 0, 0, 1\rangle : x \in U\}$,

$$\begin{aligned}\underset{\sim}{\mathbb{C}}^2(U) &= \{\langle x, \wedge_{y\in U}[F_{\widetilde{\mathbb{N}}_x^{\beta 2}}(y) \vee T_U(y)], \wedge_{y\in U}[(1 - I_{\widetilde{\mathbb{N}}_x^{\beta 2}}(y)) \vee I_U(y)], \vee_{y\in U}[T_{\widetilde{\mathbb{N}}_x^{\beta 2}}(y) \wedge F_U(y)]\rangle : x \in U\} \\ &= \{\langle x, 1, 1, 0\rangle : x \in U\} \\ &= U,\end{aligned}$$

and

$$\begin{aligned}\widetilde{\mathbb{C}}^2(\emptyset) &= \{\langle x, \vee_{y\in U}[T_{\widetilde{\mathbb{N}}_x^{\beta 2}}(y) \wedge T_\emptyset(y)], \vee_{y\in U}[I_{\widetilde{\mathbb{N}}_x^{\beta 2}}(y) \wedge I_\emptyset(y)], \wedge_{y\in U}[F_{\widetilde{\mathbb{N}}_x^{\beta 2}}(y) \vee F_\emptyset(y)]\rangle : x \in U\} \\ &= \{\langle x, 0, 0, 1\rangle : x \in U\} \\ &= \emptyset;\end{aligned}$$

(2)

$$\begin{aligned}\widetilde{\mathbb{C}}^2(A') &= \{\langle x, \vee_{y\in U}[T_{\widetilde{\mathbb{N}}_x^{\beta 2}}(y) \wedge T_{A'}(y)], \vee_{y\in U}[I_{\widetilde{\mathbb{N}}_x^{\beta 2}}(y) \wedge I_{A'}(y)], \wedge_{y\in U}[F_{\widetilde{\mathbb{N}}_x^{\beta 2}}(y) \vee F_{A'}(y)]\rangle : x \in U\} \\ &= \{\langle x, \vee_{y\in U}[T_{\widetilde{\mathbb{N}}_x^{\beta 2}}(y) \wedge F_A(y)], \vee_{y\in U}[I_{\widetilde{\mathbb{N}}_x^{\beta 2}}(y) \wedge (1 - I_A(y))], \wedge_{y\in U}[F_{\widetilde{\mathbb{N}}_x^{\beta 2}}(y) \vee T_A(y)]\rangle : x \in U\} \\ &= (\underset{\sim}{\mathbb{C}}^2(A))'.\end{aligned}$$

If we replace A by A' in this proof, we can also prove $\underset{\sim}{\mathbb{C}}^2(A') = (\widetilde{\mathbb{C}}^2(A))'$.

(3) Since $A \subseteq_2 B$, $T_A(x) \leq T_B(x)$, $I_A(x) \leq I_B(x)$ and $F_B(x) \leq F_A(x)$ for all $x \in U$. Therefore

$$\begin{aligned}T_{\underset{\sim}{\mathbb{C}}^2(A)}(x) &= \wedge_{y\in U}[F_{\widetilde{\mathbb{N}}_x^{\beta 2}}(y) \vee T_A(y)] \leq \wedge_{y\in U}[F_{\widetilde{\mathbb{N}}_x^{\beta 2}}(y) \vee T_B(y)] = T_{\underset{\sim}{\mathbb{C}}^2(B)}(x), \\ I_{\underset{\sim}{\mathbb{C}}^2(A)}(x) &= \wedge_{y\in U}[(1 - I_{\widetilde{\mathbb{N}}_x^{\beta 2}}(y)) \vee I_A(y)] \leq \wedge_{y\in U}[(1 - I_{\widetilde{\mathbb{N}}_x^{\beta 2}}(y)) \vee I_B(y)] = I_{\underset{\sim}{\mathbb{C}}^2(B)}(x), \\ F_{\underset{\sim}{\mathbb{C}}^2(A)}(x) &= \vee_{y\in U}[T_{\widetilde{\mathbb{N}}_x^{\beta 2}}(y) \wedge F_A(y)] \geq \vee_{y\in U}[T_{\widetilde{\mathbb{N}}_x^{\beta 2}}(y) \wedge F_B(y)] = F_{\underset{\sim}{\mathbb{C}}^2(B)}(x).\end{aligned}$$

Hence, $\underset{\sim}{\mathbb{C}}^2(A) \subseteq_2 \underset{\sim}{\mathbb{C}}^2(B)$. In the same way, there is $\widetilde{\mathbb{C}}^2(A) \subseteq_2 \widetilde{\mathbb{C}}^2(B)$;

(4) Since

$$\begin{aligned}&\underset{\sim}{\mathbb{C}}^2(A \cap_2 B) \\ =\ &\{\langle x, \wedge_{y\in U}[F_{\widetilde{\mathbb{N}}_x^{\beta 2}}(y) \vee T_{A\cap_2 B}(y)], \wedge_{y\in U}[(1 - I_{\widetilde{\mathbb{N}}_x^{\beta 2}}(y)) \vee I_{A\cap_2 B}(y)], \vee_{y\in U}[T_{\widetilde{\mathbb{N}}_x^{\beta 2}}(y) \wedge F_{A\cap_2 B}(y)]\rangle : x \in U\} \\ =\ &\{\langle x, \wedge_{y\in U}[F_{\widetilde{\mathbb{N}}_x^{\beta 2}}(y) \vee (T_A(y) \wedge T_B(y))], \wedge_{y\in U}[(1 - I_{\widetilde{\mathbb{N}}_x^{\beta 2}}(y)) \vee (I_A(y) \wedge I_B(y))], \vee_{y\in U}[T_{\widetilde{\mathbb{N}}_x^{\beta 2}}(y) \wedge (F_A(y) \\ &\vee F_B(y))]\rangle : x \in U\} \\ =\ &\{\langle x, \wedge_{y\in U}[(F_{\widetilde{\mathbb{N}}_x^{\beta 2}}(y) \vee T_A(y)) \wedge (F_{\widetilde{\mathbb{N}}_x^{\beta 2}}(y) \vee T_B(y))], \wedge_{y\in U}[((1 - I_{\widetilde{\mathbb{N}}_x^{\beta 2}}(y)) \vee I_A(y)) \wedge (1 - I_{\widetilde{\mathbb{N}}_x^{\beta 2}}(y)) \vee \\ &I_B(y))], \vee_{y\in U}[(T_{\widetilde{\mathbb{N}}_x^{\beta 2}}(y) \wedge F_A(y)) \vee (T_{\widetilde{\mathbb{N}}_x^{\beta 2}}(y) \wedge F_B(y))]\rangle : x \in U\} \\ =\ &\underset{\sim}{\mathbb{C}}^2(A) \cap_2 \underset{\sim}{\mathbb{C}}^2(B).\end{aligned}$$

Similarly, we can obtain $\widetilde{\mathbb{C}}^2(A \cup_2 B) = \widetilde{\mathbb{C}}^2(A) \cup_2 \widetilde{\mathbb{C}}^2(B)$;

(5) Since $A \subseteq_2 (A \cup_2 B)$, $B \subseteq_2 (A \cup_2 B)$, $(A \cap_2 B) \subseteq_2 A$ and $(A \cap_2 B) \subseteq_2 B$,

$$\underset{\sim}{\mathbb{C}}^2(A) \subseteq_2 \underset{\sim}{\mathbb{C}}^2(A \cup_2 B), \underset{\sim}{\mathbb{C}}^2(B) \subseteq_2 \underset{\sim}{\mathbb{C}}^2(A \cup_2 B), \widetilde{\mathbb{C}}^2(A \cap_2 B) \subseteq_2 \widetilde{\mathbb{C}}^2(A) \text{ and } \widetilde{\mathbb{C}}^2(A \cap_2 B) \subseteq_2 \widetilde{\mathbb{C}}^2(B).$$

Hence, $\underset{\sim}{\mathbb{C}}^2(A \cup_2 B) \supseteq_2 \underset{\sim}{\mathbb{C}}^2(A) \cup_2 \underset{\sim}{\mathbb{C}}^2(B)$, $\widetilde{\mathbb{C}}^2(A \cap_2 B) \subseteq_2 \widetilde{\mathbb{C}}^2(A) \cap_2 \widetilde{\mathbb{C}}^2(B)$.

□

4.2. Relationships between the New Model and Some Other Rough Set Models

In this subsection, we investigate some relationships between the type-2 SVN covering rough set model and other two SVN rough set models respectively. Among these two SVN rough set models, one is a SVN covering rough set model and the other is a SVN relation rough set model.

Wang and Zhang [37] presented the type-1 SVN covering rough set model under a SVN β-covering approximation space, which is related to the type-1 inclusion relation. We consider whether the type-1 SVN covering approximate operators and the type-2 SVN covering approximate operators presented in Section 4.1 have inclusion relations.

Definition 8. *[37] Let $(U, \widehat{\mathbf{C}})$ be a SVN β-covering approximation space. For each $A \in SVN(U)$, where $A = \{\langle x, T_A(x), I_A(x), F_A(x)\rangle : x \in U\}$, we define the type-1 SVN covering upper approximation $\widetilde{\mathbb{C}}(A)$ and lower approximation $\underset{\sim}{\mathbb{C}}(A)$ of A as*

$$\begin{aligned}
\widetilde{\mathbb{C}}(A) &= \{\langle x, \vee_{y\in U}[T_{\widetilde{\mathbb{N}}_x^\beta}(y) \wedge T_A(y)], \vee_{y\in U}[I_{\widetilde{\mathbb{N}}_x^\beta}(y) \wedge I_A(y)], \wedge_{y\in U}[F_{\widetilde{\mathbb{N}}_x^\beta}(y) \vee F_A(y)]\rangle : x \in U\}, \\
\underset{\sim}{\mathbb{C}}(A) &= \{\langle x, \wedge_{y\in U}[F_{\widetilde{\mathbb{N}}_x^\beta}(y) \vee T_A(y)], \wedge_{y\in U}[(1 - I_{\widetilde{\mathbb{N}}_x^\beta}(y)) \vee I_A(y)], \vee_{y\in U}[T_{\widetilde{\mathbb{N}}_x^\beta}(y) \wedge F_A(y)]\rangle : x \in U\}.
\end{aligned} \tag{5}$$

If $\widetilde{\mathbb{C}}(A) \neq \underset{\sim}{\mathbb{C}}(A)$, then A is called the type-1 SVN covering rough set.

Let $\widehat{\mathbf{C}}$ be a SVN β^2-covering of U and also be a SVN β-covering of U. By Definitions 7 and 8, we know that the type-1 SVN covering approximate operators ($\widetilde{\mathbb{C}}$ and $\underset{\sim}{\mathbb{C}}$) and the type-2 SVN covering approximate operators ($\widetilde{\mathbb{C}}^2$ and $\underset{\sim}{\mathbb{C}}^2$) are related to all SVN β-neighborhoods ($\widetilde{\mathbb{N}}_x^\beta$, for any $x \in U$) and SVN β^2-neighborhoods ($\widetilde{\mathbb{N}}_x^{\beta^2}$, for any $x \in U$), respectively. By Example 3, we know that $\widetilde{\mathbb{N}}_x^{\beta^2}$ and $\widetilde{\mathbb{N}}_x^\beta$ have no inclusion relations ($\subseteq_1$ and $\subseteq_2$) for all $x \in U$. Hence, the type-1 SVN covering approximate operators and the type-2 SVN covering approximate operators also have no inclusion relations ($\subseteq_1$ and $\subseteq_2$).

In the work by the authors of [36], a SVN relation R on U is defined as $R = \{\langle (x,y), T_R(x,y), I_R(x,y), F_R(x,y)\rangle : (x,y) \in U \times U\}$, where $T_R : U \times U \to [0,1]$, $I_R : U \times U \to [0,1]$ and $F_R : U \times U \to [0,1]$. If for any $x, y \in U$, $T_R(x,y) = T_R(y,x)$, $I_R(x,y) = I_R(y,x)$ and $F_R(x,y) = F_R(y,x)$, then R is called a symmetric SVN relation.

For a SVN β^2-covering $\widehat{\mathbf{C}}$ of U, one can use the SVN β^2-covering $\widehat{\mathbf{C}}$ induce a SVN relation $R_{\widehat{\mathbf{C}}}$ on U as

$$R_{\widehat{\mathbf{C}}} = \{\langle (x,y), T_{R_{\widehat{\mathbf{C}}}}(x,y), I_{R_{\widehat{\mathbf{C}}}}(x,y), F_{R_{\widehat{\mathbf{C}}}}(x,y)\rangle : (x,y) \in U \times U\},$$

where

$$T_{R_{\widehat{\mathbf{C}}}}(x,y) = T_{\widetilde{N}_x^{\beta 2}}(y), \; I_{R_{\widehat{\mathbf{C}}}}(x,y) = I_{\widetilde{N}_x^{\beta 2}}(y), \; F_{R_{\widehat{\mathbf{C}}}}(x,y) = F_{\widetilde{N}_x^{\beta 2}}(y) \text{ for any } x, y \in U.$$

The following two propositions present two conditions under which $R_{\widehat{\mathbf{C}}}$ is a symmetric SVN relation.

Proposition 6. *Let $\widehat{\mathbf{C}}$ be a SVN β^2-covering of U, and $R_{\widehat{\mathbf{C}}}$ be the induced SVN relation on U by $\widehat{\mathbf{C}}$. If $\widetilde{N}_x^{\beta^2}(y) = \widetilde{N}_y^{\beta^2}(x)$ for any $x, y \in U$, then $R_{\widehat{\mathbf{C}}}$ is a symmetric SVN relation.*

Proof. Since $\widetilde{N}_x^{\beta^2}(y) = \widetilde{N}_y^{\beta^2}(x)$ for any $x, y \in U$, $T_{\widetilde{N}_x^{\beta 2}}(y) = T_{\widetilde{N}_y^{\beta 2}}(x)$, $I_{\widetilde{N}_x^{\beta 2}}(y) = I_{\widetilde{N}_y^{\beta 2}}(x)$ and $F_{\widetilde{N}_x^{\beta 2}}(y) = F_{\widetilde{N}_y^{\beta 2}}(x)$. Hence, $T_{R_{\widehat{C}}}(x, y) = T_{R_{\widehat{C}}}(y, x)$, $I_{R_{\widehat{C}}}(x, y) = I_{R_{\widehat{C}}}(y, x)$ and $F_{R_{\widehat{C}}}(x, y) = F_{R_{\widehat{C}}}(y, x)$, i.e., $R_{\widehat{C}}$ is a symmetric SVN relation. □

Proposition 7. *Let $\widehat{\mathbf{C}}$ be a SVN β^2-covering of U, $R_{\widehat{C}}$ be the induced SVN relation on U by $\widehat{\mathbf{C}}$, and $C \in \widehat{\mathbf{C}}$. If $|\widehat{\mathbf{C}}| = 1$ and $C(x) = C(y)$ for any $x, y \in U$, then $R_{\widehat{C}}$ is a symmetric SVN relation, where $|\widehat{\mathbf{C}}|$ denotes the cardinality of $\widehat{\mathbf{C}}$.*

Proof. Since $|\widehat{\mathbf{C}}| = 1$, C is the only one element of $\widehat{\mathbf{C}}$. Since $C(x) = C(y)$ for any $x, y \in U$, $\widetilde{N}_x^{\beta^2}(y) = \widetilde{N}_y^{\beta^2}(x)$. Hence, $T_{R_{\widehat{C}}}(x, y) = T_{R_{\widehat{C}}}(y, x)$, $I_{R_{\widehat{C}}}(x, y) = I_{R_{\widehat{C}}}(y, x)$ and $F_{R_{\widehat{C}}}(x, y) = F_{R_{\widehat{C}}}(y, x)$, i.e., $R_{\widehat{C}}$ is a symmetric SVN relation. □

Then, the type-2 SVN covering rough set model defined in Section 4.1 can be viewed as a SVN relation rough set model.

Definition 9. *Let $\widehat{\mathbf{C}}$ be a SVN β^2-covering of U, and $R_{\widehat{C}}$ be the induced SVN relation on U by $\widehat{\mathbf{C}}$. For any $A \in SVN(U)$, the upper approximation $\widetilde{R}_{\widehat{C}}(A)$ and lower approximation $\underset{\sim}{R}_{\widehat{C}}(A)$ of A are defined as*

$$\widetilde{R}_{\widehat{C}}(A) = \{\langle x, \vee_{y \in U}[T_{R_{\widehat{C}}}(x, y) \wedge T_A(y)], \vee_{y \in U}[I_{R_{\widehat{C}}}(x, y) \wedge I_A(y)], \wedge_{y \in U}[F_{R_{\widehat{C}}}(x, y) \vee F_A(y)]\rangle : x \in U\},$$
$$\underset{\sim}{R}_{\widehat{C}}(A) = \{\langle x, \wedge_{y \in U}[F_{R_{\widehat{C}}}(x, y) \vee T_A(y)], \wedge_{y \in U}[(1 - I_{R_{\widehat{C}}}(x, y)) \vee I_A(y)], \vee_{y \in U}[T_{R_{\widehat{C}}}(x, y) \wedge F_A(y)]\rangle : x \in U\}.$$

Remark 2. *Let $\widehat{\mathbf{C}}$ be a SVN β^2-covering of U, and $R_{\widehat{C}}$ be the induced SVN relation on U by $\widehat{\mathbf{C}}$. Then*

$$\widetilde{R}_{\widehat{C}}(A) = \widetilde{\mathbb{C}}^2(A),$$
$$\underset{\sim}{R}_{\widehat{C}}(A) = \underset{\sim}{\mathbb{C}}^2(A).$$

5. Graph Representation of the Type-2 SVN Covering Rough Set Model

In this section, the graph representation of the type-2 SVN covering rough set model is presented. Firstly, some new graphs and graph operations are presented. Then, we show the graph representation of the type-2 SVN covering approximation operators defined in Definition 7. The order of elements in U is given.

A graph is a pair $G = (V, E)$ consisting of a nonempty set V of vertices and a set E of edges such that $E \subseteq U \times U$. We shall often write $V(G)$ for V and $E(G)$ for E, particularly when several graphs are being considered. Two vertices are adjacent if there is an edge with them as ends. A graph $G = (V, E)$ is called bipartite if the vertex set V can be divided into two disjoint sets V_1 and V_2, such that every edge connects a vertex in V_1 to one in V_2. One often writes$G = (V_1 \cup V_2, E)$ to denote a bipartite graph whose partition has the partite sets V_1 and V_2. A complete bipartite graph is a simple bipartite graph such that two vertices are adjacent if and only if they are in different partite sets. A weighted graph is a graph with numerical labels on the edges.

Firstly, the graph representation of the SVN β^2-covering $\widehat{\mathbf{C}}$ is defined in the following definition.

Definition 10. *Let $\widehat{\mathbf{C}}$ be a SVN β^2-covering of U with $U = \{x_1, x_2, \cdots, x_n\}$ and $\widehat{\mathbf{C}} = \{C_1, C_2, \cdots, C_m\}$. For any $A \in SVN(U)$, we define a completely weighted bipartite graph $G(A) = (U \cup V, E)$, named completely weighted bipartite graph associated with A, where $V = \{T_A, I_A, F_A\}$, the weight $w(T_A, x_k) = T_A(x_k)$, $w(I_A, x_k) = I_A(x_k)$ and $w(F_A, x_k) = F_A(x_k)$ $(k = 1, 2, \cdots, n)$. For the SVN β-covering $\widehat{\mathbf{C}}$, there are m completely weighted bipartite graphs $G(C_i)$ $(i = 1, 2, \cdots, m)$, and all $G(C_i)$ are called the graph representation of the SVN β^2-covering $\widehat{\mathbf{C}}$.*

Example 5. *Let $(U, \widehat{\mathbb{C}})$ be a SVN β^2-covering approximation space in Example 1 and $\beta = \langle 0.5, 0.1, 0.8 \rangle$. Then $G(C_i)$ $(i = 1,2,3,4)$ are the graph representation of the SVN β^2-covering $\widehat{\mathbb{C}}$. All $G(C_i)$ $(i = 1,2,3,4)$ are shown in Figures 1 and 2.*

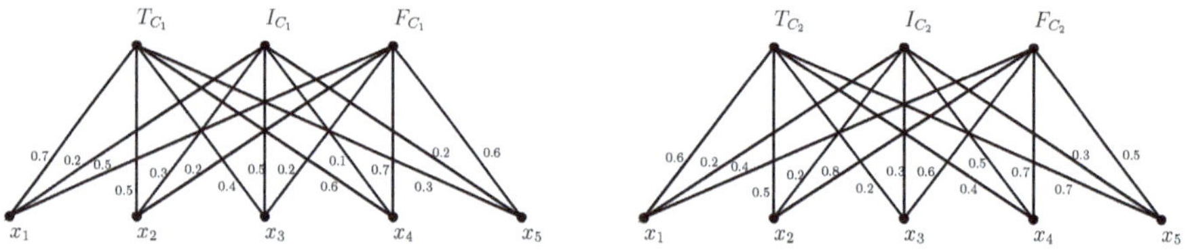

(a) $G(C_1)$ **(b)** $G(C_2)$

Figure 1. $G(C_1)$ and $G(C_2)$.

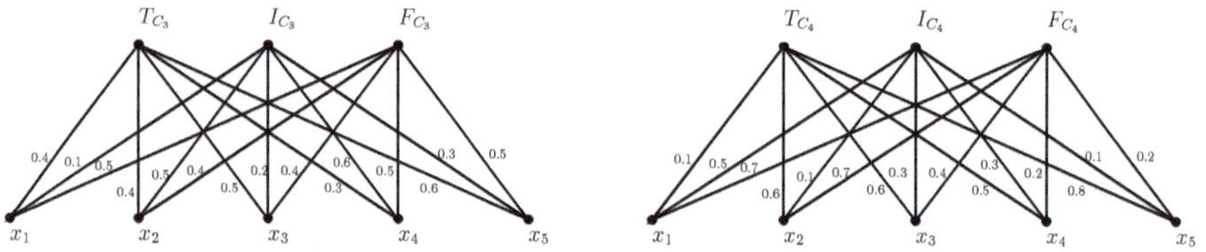

(a) $G(C_3)$ **(b)** $G(C_4)$

Figure 2. $G(C_3)$ and $G(C_4)$.

An intersection operation about $G(A)$ and $G(B)$ is presented in the following definition, for any $A, B \in SVN(U)$.

Definition 11. *Let $\widehat{\mathbb{C}}$ be a SVN β^2-covering of U with $U = \{x_1, x_2, \cdots, x_n\}$ and $\widehat{\mathbb{C}} = \{C_1, C_2, \cdots, C_m\}$. For any $A, B, D \in SVN(U)$, we define a completely weighted bipartite graph $G(D) = G(A) \cap_2 G(B)$ associated with D, where $G(D) = (U \cup \{T_D, I_D, F_D\}, E)$ and*

$$w(T_D, x_k) = w(T_A, x_k) \wedge w(T_B, x_k),\ w(I_D, x_k) = w(I_A, x_k) \wedge w(I_A, x_k) \text{ and}$$
$$w(F_D, x_k) = w(F_A, x_k) \vee w(F_B, x_k)\ (k = 1, 2, \cdots, n).$$

Based on Definition 11 and the definition of $A \cap_2 B$, the relationship between $G(A \cap_2 B)$ and $G(A) \cap_2 G(B)$ can be obtained for any $A, B \in SVN(U)$.

Lemma 1. *Let $\widehat{\mathbb{C}}$ be a SVN β^2-covering of U. Then $G(A \cap_2 B) = G(A) \cap_2 G(B)$ for any $A, B \in SVN(U)$.*

Proof. According to the definition of $A \cap_2 B$ and Definition 11, it is immediate. □

By Definition 6, $\widetilde{\mathbb{N}}_{x_k}^{\beta^2} \in SVN(U)$ for any $x_k \in U$. Hence, any $G(\widetilde{\mathbb{N}}_{x_k}^{\beta^2})$ is a completely weighted bipartite graph, which can be represented in the following proposition.

Proposition 8. *Let $\widehat{\mathbb{C}}$ be a SVN β^2-covering of U with $U = \{x_1, x_2, \cdots, x_n\}$, $\widehat{\mathbb{C}} = \{C_1, C_2, \cdots, C_m\}$ and $\beta = \langle a, b, c \rangle$. Then*

$$G(\widetilde{\mathbb{N}}_{x_k}^{\beta^2}) = \cap_2 \{G(C_i) : w(T_{C_i}, x_k) \geq a, w(I_{C_i}, x_k) \geq b, w(F_{C_i}, x_k) \leq c\}.$$

Proof. By Definitions 6 and 11, and Lemma 1, it is immediate. □

Example 6. *Let* $(U, \widehat{\mathbb{C}})$ *be a SVN* β^2*-covering approximation space in Example 1 and* $\beta = \langle 0.5, 0.1, 0.8 \rangle$*. By Proposition 8, we have*

$$G(\widetilde{\mathbb{N}}^{\beta^2}_{x_1}) = G(C_1) \cap_2 G(C_2), G(\widetilde{\mathbb{N}}^{\beta^2}_{x_2}) = G(C_1) \cap_2 G(C_2) \cap_2 G(C_4), G(\widetilde{\mathbb{N}}^{\beta^2}_{x_3}) = G(C_3) \cap_2 G(C_4),$$
$$G(\widetilde{\mathbb{N}}^{\beta^2}_{x_4}) = G(C_1) \cap_2 G(C_4), G(\widetilde{\mathbb{N}}^{\beta^2}_{x_5}) = G(C_2) \cap_2 G(C_3) \cap_2 G(C_4).$$

Then all $G(\widetilde{\mathbb{N}}^{\beta^2}_{x_k})$ *are shown in Figures 3, 4, and 5a.*

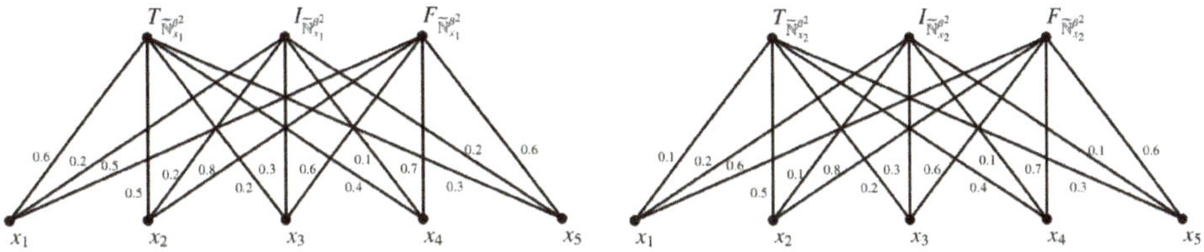

(**a**) $G(\widetilde{\mathbb{N}}^{\beta^2}_{x_1})$ (**b**) $G(\widetilde{\mathbb{N}}^{\beta^2}_{x_2})$

Figure 3. $G(\widetilde{\mathbb{N}}^{\beta^2}_{x_1})$ and $G(\widetilde{\mathbb{N}}^{\beta^2}_{x_2})$.

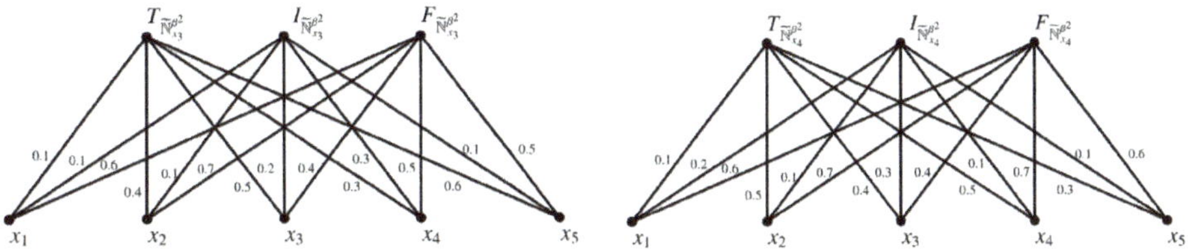

(**a**) $G(\widetilde{\mathbb{N}}^{\beta^2}_{x_3})$ (**b**) $G(\widetilde{\mathbb{N}}^{\beta^2}_{x_4})$

Figure 4. $G(\widetilde{\mathbb{N}}^{\beta^2}_{x_3})$ and $G(\widetilde{\mathbb{N}}^{\beta^2}_{x_4})$.

Finally, the type-2 SVN covering upper approximation $\widetilde{\mathbb{C}}^2(A)$ and lower approximation $\underset{\sim}{\mathbb{C}}^2(A)$ of A are represented by graphs.

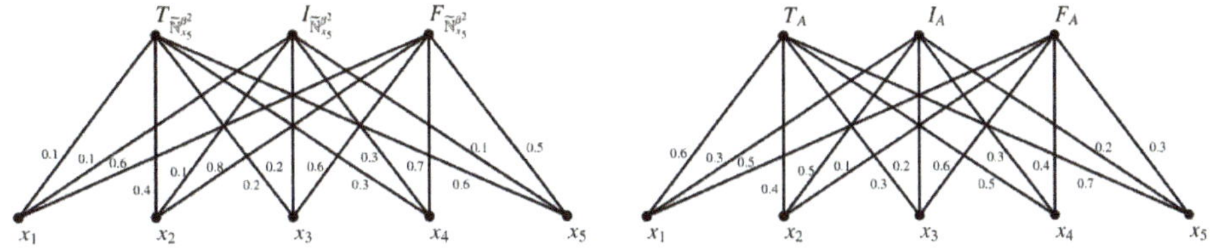

(**a**) $G(\widetilde{\mathbb{N}}^{\beta^2}_{x_5})$ (**b**) $G(A)$

Figure 5. $G(\widetilde{\mathbb{N}}^{\beta^2}_{x_5})$ and $G(A)$.

Theorem 1. *Let* $\widehat{\mathbb{C}}$ *be a SVN* β^2*-covering of* U *with* $U = \{x_1, x_2, \cdots, x_n\}$*. For each* $A \in SVN(U)$*,* $G(\widetilde{\mathbb{C}}^2(A))$ *and* $G(\underset{\sim}{\mathbb{C}}^2(A))$ *are completely weighted bipartite graphs, where* $G(\widetilde{\mathbb{C}}^2(A)) = (U \cup$

$\{T_{\widetilde{\mathbb{C}}^2(A)}, I_{\widetilde{\mathbb{C}}^2(A)}, F_{\widetilde{\mathbb{C}}^2(A)}\}, E_1)$, $G(\underset{\sim}{\mathbb{C}}^2(A)) = (U \cup \{T_{\underset{\sim}{\mathbb{C}}^2(A)}, I_{\underset{\sim}{\mathbb{C}}^2(A)}, F_{\underset{\sim}{\mathbb{C}}^2(A)}\}, E_2)$ *and the weight of any edge is listed as follows.*

$$\begin{aligned}
w(T_{\widetilde{\mathbb{C}}^2(A)}, x_k) &= \bigvee_{j=1}^{n} [w(T_{\widetilde{\mathbb{N}}_{x_k}^{\beta 2}}, x_j) \wedge w(T_A, x_j)] \ (1 \le k \le n),\\
w(I_{\widetilde{\mathbb{C}}^2(A)}, x_k) &= \bigvee_{j=1}^{n} [w(I_{\widetilde{\mathbb{N}}_{x_k}^{\beta 2}}, x_j) \wedge w(I_A, x_j)] \ (1 \le k \le n),\\
w(F_{\widetilde{\mathbb{C}}^2(A)}, x_k) &= \bigwedge_{j=1}^{n} [w(F_{\widetilde{\mathbb{N}}_{x_k}^{\beta 2}}, x_j) \vee w(F_A, x_j)] \ (1 \le k \le n),\\
w(T_{\underset{\sim}{\mathbb{C}}^2(A)}, x_k) &= \bigwedge_{j=1}^{n} [w(F_{\widetilde{\mathbb{N}}_{x_k}^{\beta 2}}, x_j) \vee w(T_A, x_j)] \ (1 \le k \le n),\\
w(I_{\underset{\sim}{\mathbb{C}}^2(A)}, x_k) &= \bigwedge_{j=1}^{n} [(1 - w(I_{\widetilde{\mathbb{N}}_{x_k}^{\beta 2}}, x_j)) \vee w(I_A, x_j)] \ (1 \le k \le n),\\
w(F_{\underset{\sim}{\mathbb{C}}^2(A)}, x_k) &= \bigvee_{j=1}^{n} [w(T_{\widetilde{\mathbb{N}}_{x_k}^{\beta 2}}, x_j) \wedge w(F_A, x_j)] \ (1 \le k \le n).
\end{aligned} \tag{6}$$

Proof. According to Definition 7, we know $\widetilde{\mathbb{C}}^2(A) \in SVN(U)$ and $\underset{\sim}{\mathbb{C}}^2(A) \in SVN(U)$ for any $A \in SVN(U)$. Hence, $G(\widetilde{\mathbb{C}}^2(A))$ and $G(\underset{\sim}{\mathbb{C}}^2(A))$ are completely weighted bipartite graphs by Definition 10. According to Definitions 7 and 10, $G(\widetilde{\mathbb{C}}^2(A)) = (U \cup \{T_{\widetilde{\mathbb{C}}^2(A)}, I_{\widetilde{\mathbb{C}}^2(A)}, F_{\widetilde{\mathbb{C}}^2(A)}\}, E_1)$, $G(\underset{\sim}{\mathbb{C}}^2(A)) = (U \cup \{T_{\underset{\sim}{\mathbb{C}}^2(A)}, I_{\underset{\sim}{\mathbb{C}}^2(A)}, F_{\underset{\sim}{\mathbb{C}}^2(A)}\}, E_2)$ and the weight of any edge is shown as follows.

$$\begin{aligned}
w(T_{\widetilde{\mathbb{C}}^2(A)}, x_k) &= T_{\widetilde{\mathbb{C}}^2(A)}(x_k) = \bigvee_{j=1}^{n} [T_{\widetilde{\mathbb{N}}_{x_k}^{\beta 2}}(x_j) \wedge T_A(x_j)] = \bigvee_{j=1}^{n} [w(T_{\widetilde{\mathbb{N}}_{x_k}^{\beta 2}}, x_j) \wedge w(T_A, x_j)] \ (1 \le k \le n),\\
w(I_{\widetilde{\mathbb{C}}^2(A)}, x_k) &= I_{\widetilde{\mathbb{C}}^2(A)}(x_k) = \bigvee_{j=1}^{n} [I_{\widetilde{\mathbb{N}}_{x_k}^{\beta 2}}(x_j) \wedge I_A(x_j)] = \bigvee_{j=1}^{n} [w(I_{\widetilde{\mathbb{N}}_{x_k}^{\beta 2}}, x_j) \wedge w(I_A, x_j)] \ (1 \le k \le n),\\
w(F_{\widetilde{\mathbb{C}}^2(A)}, x_k) &= F_{\widetilde{\mathbb{C}}^2(A)}(x_k) = \bigwedge_{j=1}^{n} [F_{\widetilde{\mathbb{N}}_{x_k}^{\beta 2}}(x_j) \vee F_A(x_j)] = \bigwedge_{j=1}^{n} [w(F_{\widetilde{\mathbb{N}}_{x_k}^{\beta 2}}, x_j) \vee w(F_A, x_j)] \ (1 \le k \le n),\\
w(T_{\underset{\sim}{\mathbb{C}}^2(A)}, x_k) &= T_{\underset{\sim}{\mathbb{C}}^2(A)}(x_k) = \bigwedge_{j=1}^{n} [F_{\widetilde{\mathbb{N}}_{x_k}^{\beta 2}}(x_j) \vee T_A(x_j)] = \bigwedge_{j=1}^{n} [w(F_{\widetilde{\mathbb{N}}_{x_k}^{\beta 2}}, x_j) \vee w(T_A, x_j)] \ (1 \le k \le n),\\
w(I_{\underset{\sim}{\mathbb{C}}^2(A)}, x_k) &= I_{\underset{\sim}{\mathbb{C}}^2(A)}(x_k) = \bigwedge_{j=1}^{n} [(1 - I_{\widetilde{\mathbb{N}}_{x_k}^{\beta 2}}(x_j)) \vee I_A(x_j)] = \bigwedge_{j=1}^{n} [(1 - w(I_{\widetilde{\mathbb{N}}_{x_k}^{\beta 2}}, x_j)) \vee w(I_A, x_j)] \ (1 \le k \le n),\\
w(F_{\underset{\sim}{\mathbb{C}}^2(A)}, x_k) &= F_{\underset{\sim}{\mathbb{C}}^2(A)}(x_k) = \bigvee_{j=1}^{n} [T_{\widetilde{\mathbb{N}}_{x_k}^{\beta 2}}(x_j) \wedge F_A(x_j)] = \bigvee_{j=1}^{n} [w(T_{\widetilde{\mathbb{N}}_{x_k}^{\beta 2}}, x_j) \wedge w(F_A, x_j)] \ (1 \le k \le n).
\end{aligned}$$

□

Example 7. *Let* $(U, \widehat{\mathbf{C}})$ *be a SVN* β^2*-covering approximation space in Example 1 and* $\beta = \langle 0.5, 0.1, 0.8 \rangle$, $A = \frac{(0.6,0.3,0.5)}{x_1} + \frac{(0.4,0.5,0.1)}{x_2} + \frac{(0.3,0.2,0.6)}{x_3} + \frac{(0.5,0.3,0.4)}{x_4} + \frac{(0.7,0.2,0.3)}{x_5}$. $G(A)$ *is shown in Figure 5b. Based on Theorem 1 and all* $G(\widetilde{\mathbb{N}}_{x_k}^{\beta^2})$ $(k = 1, 2, \cdots, 5)$ *in Example 6 and* $G(\widetilde{\mathbb{C}}^2(A))$ *and* $G(\underset{\sim}{\mathbb{C}}^2(A))$ *are obtained in Figure 6.*

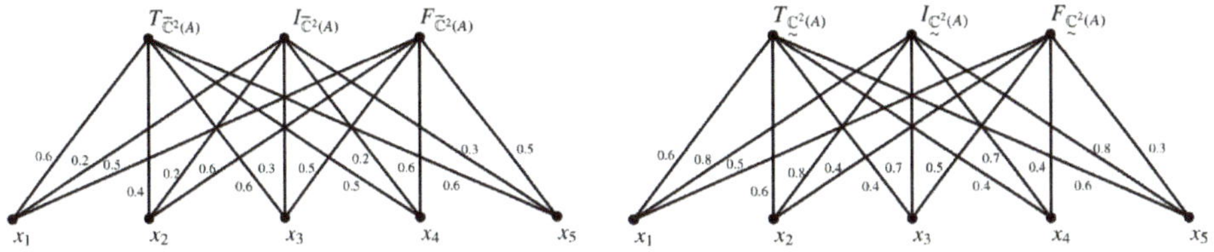

(a) $G(\widetilde{\mathbb{C}}^2(A))$ **(b)** $G(\underset{\sim}{\mathbb{C}}^2(A))$

Figure 6. $G(\widetilde{\mathbb{C}}^2(A))$ and $G(\underset{\sim}{\mathbb{C}}^2(A))$.

6. Matrix Representation of the Type-2 SVN Covering Rough Set Model

In this section, the matrix representation of the type-2 SVN covering rough set model is investigated. Firstly, some new matrices and matrix operations are presented. Then, we show the matrix representation of the type-2 SVN approximation operators defined in Definition 7. The order of elements in U is given.

Two new matrices about a SVN β^2-covering are presented in the following definition.

Definition 12. *Let $\widehat{\mathbb{C}}$ be a SVN β^2-covering of U with $U = \{x_1, x_2, \cdots, x_n\}$ and $\widehat{\mathbb{C}} = \{C_1, C_2, \cdots, C_m\}$. Then $M_{\widehat{\mathbb{C}}} = (C_j(x_i))_{n\times m}$ is named a matrix representation of $\widehat{\mathbb{C}}$, and $M_{\widehat{\mathbb{C}}}^{\beta^2} = (s_{ij})_{n\times m}$ is called a β^2-matrix representation of $\widehat{\mathbb{C}}$, where*

$$s_{ij} = \begin{cases} 1, & C_j(x_i) \geq_2 \beta; \\ 0, & \text{otherwise.} \end{cases}$$

Example 8. *Let $(U, \widehat{\mathbb{C}})$ be a SVN β^2-covering approximation space in Example 1 and $\beta = \langle 0.5, 0.1, 0.8\rangle$. Then*

$$M_{\widehat{\mathbb{C}}} = \begin{pmatrix} \langle 0.7,0.2,0.5\rangle & \langle 0.6,0.2,0.4\rangle & \langle 0.4,0.1,0.5\rangle & \langle 0.1,0.5,0.6\rangle \\ \langle 0.5,0.3,0.2\rangle & \langle 0.5,0.2,0.8\rangle & \langle 0.4,0.5,0.4\rangle & \langle 0.6,0.1,0.7\rangle \\ \langle 0.4,0.5,0.2\rangle & \langle 0.2,0.3,0.6\rangle & \langle 0.5,0.2,0.4\rangle & \langle 0.6,0.3,0.4\rangle \\ \langle 0.6,0.1,0.7\rangle & \langle 0.4,0.5,0.7\rangle & \langle 0.3,0.6,0.5\rangle & \langle 0.5,0.3,0.2\rangle \\ \langle 0.3,0.2,0.6\rangle & \langle 0.7,0.3,0.5\rangle & \langle 0.6,0.3,0.5\rangle & \langle 0.8,0.1,0.2\rangle \end{pmatrix}, \quad M_{\widehat{\mathbb{C}}}^{\beta^2} = \begin{pmatrix} 1 & 1 & 0 & 0 \\ 1 & 1 & 0 & 1 \\ 0 & 0 & 1 & 1 \\ 1 & 0 & 0 & 1 \\ 0 & 1 & 1 & 1 \end{pmatrix}.$$

In order to calculate all $\widetilde{\mathbb{N}}_x^{\beta^2}$ (for any $x \in U$) by matrices, the following operation is presented.

Definition 13. *Let $A = (a_{ik})_{n\times m}$ and $B = (\langle b_{kj}^+, b_{kj}, b_{kj}^-\rangle)_{1\leq k\leq m, 1\leq j\leq l}$ be two matrices. We define $D = A \star B = (\langle d_{ij}^+, d_{ij}, d_{ij}^-\rangle)_{1\leq i\leq n, 1\leq j\leq l}$, where*

$$\langle d_{ij}^+, d_{ij}, d_{ij}^-\rangle = \langle \wedge_{k=1}^m [(1-a_{ik}) \vee b_{kj}^+], \wedge_{k=1}^m [(1-a_{ik}) \vee b_{kj}], 1 - \wedge_{k=1}^m [(1-a_{ik}) \vee (1-b_{kj}^-)]\rangle. \quad (7)$$

Based on Definitions 12 and 13, all $\widetilde{\mathbb{N}}_x^{\beta^2}$ (for any $x \in U$) can be obtained by matrix operations.

Proposition 9. *Let $\widehat{\mathbb{C}}$ be a SVN β^2-covering of U with $U = \{x_1, x_2, \cdots, x_n\}$ and $\widehat{\mathbb{C}} = \{C_1, C_2, \cdots, C_m\}$. Then*

$$M_{\widehat{\mathbb{C}}}^{\beta^2} \star M_{\widehat{\mathbb{C}}}^T = (\widetilde{\mathbb{N}}_{x_i}^{\beta^2}(x_j))_{1\leq i\leq n, 1\leq j\leq n}, \quad (8)$$

where $M_{\widehat{\mathbb{C}}}^T$ is the transpose of $M_{\widehat{\mathbb{C}}}$

Proof. Suppose $M_{\widehat{\mathbb{C}}}^T = (C_k(x_j))_{m\times n}$, $M_{\widehat{\mathbb{C}}}^{\beta^2} = (s_{ik})_{n\times m}$ and $M_{\widehat{\mathbb{C}}}^{\beta^2} \star M_{\widehat{\mathbb{C}}}^T = (\langle d_{ij}^+, d_{ij}, d_{ij}^-\rangle)_{1\leq i\leq n, 1\leq j\leq n}$. Since $\widehat{\mathbb{C}}$ is a SVN β^2-covering of U, for each i $(1 \leq i \leq n)$, there exists k $(1 \leq k \leq m)$ such that $s_{ik} = 1$.

Then

$$
\begin{aligned}
& \langle d_{ij}^{+}, d_{ij}, d_{ij}^{-}\rangle \\
= & \langle \wedge_{k=1}^{m}[(1-s_{ik}) \vee T_{C_k}(x_j)], \wedge_{k=1}^{m}[(1-s_{ik}) \vee I_{C_k}(x_j)], 1-\wedge_{k=1}^{m}[(1-s_{ik}) \vee (1-F_{C_k}(x_j))]\rangle \\
= & \langle \wedge_{s_{ik}=1}[(1-s_{ik}) \vee T_{C_k}(x_j)], \wedge_{s_{ik}=1}[(1-s_{ik}) \vee I_{C_k}(x_j)], 1-\wedge_{s_{ik}=1}[(1-s_{ik}) \vee (1-F_{C_k}(x_j))]\rangle \\
= & \langle \wedge_{s_{ik}=1} T_{C_k}(x_j), \wedge_{s_{ik}=1} I_{C_k}(x_j), 1-\wedge_{s_{ik}=1}(1-F_{C_k}(x_j))\rangle \\
= & \langle \wedge_{C_k(x_i)\geq_2\beta} T_{C_k}(x_j), \wedge_{C_k(x_i)\geq_2\beta} I_{C_k}(x_j), 1-\wedge_{C_k(x_i)\geq_2\beta}(1-F_{C_k}(x_j))\rangle \\
= & \langle \wedge_{C_k(x_i)\geq_2\beta} T_{C_k}(x_j), \wedge_{C_k(x_i)\geq_2\beta} I_{C_k}(x_j), \vee_{C_k(x_i)\geq_2\beta}(F_{C_k}(x_j))\rangle \\
= & (\cap_{2C_k(x_i)\geq_2\beta} C_k)(x_j) \\
= & \widetilde{\mathbb{N}}_{x_i}^{\beta^2}(x_j), 1 \leq i,j \leq n.
\end{aligned}
$$

Hence, $M_{\widehat{\mathbf{C}}}^{\beta^2} \star M_{\widehat{\mathbf{C}}}^{T} = (\widetilde{\mathbb{N}}_{x_i}^{\beta^2}(x_j))_{1\leq i\leq n, 1\leq j\leq n}$. □

Example 9. *Let $(U, \widehat{\mathbf{C}})$ be a SVN β^2-covering approximation space in Example 1 and $\beta = \langle 0.5, 0.1, 0.8\rangle$. According to $M_{\widehat{\mathbf{C}}}^{\beta^2}$ and $M_{\widehat{\mathbf{C}}}^{T}$ in Example 8, we have*

$$
\begin{aligned}
& M_{\widehat{\mathbf{C}}}^{\beta^2} \star M_{\widehat{\mathbf{C}}}^{T} \\
= & \begin{pmatrix} 1 & 1 & 0 & 0 \\ 1 & 1 & 0 & 1 \\ 0 & 0 & 1 & 1 \\ 1 & 0 & 0 & 1 \\ 0 & 1 & 1 & 1 \end{pmatrix} \star \begin{pmatrix} \langle 0.7,0.2,0.5\rangle & \langle 0.6,0.2,0.4\rangle & \langle 0.4,0.1,0.5\rangle & \langle 0.1,0.5,0.6\rangle \\ \langle 0.5,0.3,0.2\rangle & \langle 0.5,0.2,0.8\rangle & \langle 0.4,0.5,0.4\rangle & \langle 0.6,0.1,0.7\rangle \\ \langle 0.4,0.5,0.2\rangle & \langle 0.2,0.3,0.6\rangle & \langle 0.5,0.2,0.4\rangle & \langle 0.6,0.3,0.4\rangle \\ \langle 0.6,0.1,0.7\rangle & \langle 0.4,0.5,0.7\rangle & \langle 0.3,0.6,0.5\rangle & \langle 0.5,0.3,0.2\rangle \\ \langle 0.3,0.2,0.6\rangle & \langle 0.7,0.3,0.5\rangle & \langle 0.6,0.3,0.5\rangle & \langle 0.8,0.1,0.2\rangle \end{pmatrix}^{T} \\
= & \begin{pmatrix} \langle 0.6,0.2,0.5\rangle & \langle 0.5,0.2,0.8\rangle & \langle 0.2,0.3,0.6\rangle & \langle 0.4,0.1,0.7\rangle & \langle 0.3,0.2,0.6\rangle \\ \langle 0.1,0.2,0.6\rangle & \langle 0.5,0.1,0.8\rangle & \langle 0.2,0.3,0.6\rangle & \langle 0.4,0.1,0.7\rangle & \langle 0.3,0.1,0.6\rangle \\ \langle 0.1,0.1,0.6\rangle & \langle 0.4,0.1,0.7\rangle & \langle 0.5,0.2,0.4\rangle & \langle 0.3,0.3,0.5\rangle & \langle 0.6,0.1,0.5\rangle \\ \langle 0.1,0.2,0.6\rangle & \langle 0.5,0.1,0.7\rangle & \langle 0.4,0.3,0.4\rangle & \langle 0.5,0.1,0.7\rangle & \langle 0.3,0.1,0.6\rangle \\ \langle 0.1,0.1,0.6\rangle & \langle 0.4,0.1,0.8\rangle & \langle 0.2,0.2,0.6\rangle & \langle 0.3,0.3,0.7\rangle & \langle 0.6,0.1,0.5\rangle \end{pmatrix} \\
= & (N_{x_i}^{\beta^2}(x_j))_{1\leq i\leq 5, 1\leq j\leq 5}.
\end{aligned}
$$

There are two operations in the work by the authors of [37], which can be used to calculate $\widetilde{\mathbb{C}}^2(A)$ and $\underset{\sim}{\mathbb{C}}^2(A)$ (for any $A \in SVN(U)$) by matrices.

Definition 14. *[37] Let $A = (\langle c_{ij}^{+}, c_{ij}, c_{ij}^{-}\rangle)_{m\times n}$ and $B = (\langle d_j^{+}, d_j, d_j^{-}\rangle)_{n\times 1}$ be two matrices. We define $C = A \circ B = (\langle e_i^{+}, e_i, e_i^{-}\rangle)_{m\times 1}$ and $D = A \diamond B = (\langle f_i^{+}, f_i, f_i^{-}\rangle)_{m\times 1}$, where*

$$
\begin{aligned}
\langle e_i^{+}, e_i, e_i^{-}\rangle &= \langle \vee_{j=1}^{n}(c_{ij}^{+} \wedge d_j^{+}), \vee_{j=1}^{n}(c_{ij} \wedge d_j), \wedge_{j=1}^{n}(c_{ij}^{-} \vee d_j^{-})\rangle, \\
\langle f_i^{+}, f_i, f_i^{-}\rangle &= \langle \wedge_{j=1}^{n}(c_{ij}^{-} \vee d_j^{+}), \wedge_{j=1}^{n}[(1-c_{ij}) \vee d_j], \vee_{j=1}^{n}(c_{ij}^{+} \wedge d_j^{-})\rangle.
\end{aligned} \tag{9}
$$

According to Proposition 9 and Definition 14, the set representations of $\widetilde{\mathbb{C}}^2(A)$ and $\underset{\sim}{\mathbb{C}}^2(A)$ (for any $A \in SVN(U)$) can be converted to matrix representations. $A = (a_i)_{n\times 1}$ with $a_i = \langle T_A(x_i), I_A(x_i), F_A(x_i)\rangle$ is the vector representation of A. $\widetilde{\mathbb{C}}^2(A)$ and $\underset{\sim}{\mathbb{C}}^2(A)$ are also vector representations.

Theorem 2. *Let $\widehat{C}$ be a SVN β^2-covering of U with $U = \{x_1, x_2, \cdots, x_n\}$ and $\widehat{C} = \{C_1, C_2, \cdots, C_m\}$. Then for any $A \in SVN(U)$,*

$$\widetilde{\mathbb{C}}^2(A) = (M_{\widehat{C}}^{\beta^2} \star M_{\widehat{C}}^T) \circ A, \\ \underset{\sim}{\mathbb{C}}^2(A) = (M_{\widehat{C}}^{\beta^2} \star M_{\widehat{C}}^T) \diamond A. \tag{10}$$

Proof. According to Proposition 9, Definitions 7 and 14, for any x_i $(i = 1, 2, \cdots, n)$,

$$\begin{aligned}((M_{\widehat{C}}^{\beta^2} \star M_{\widehat{C}}^T) \circ A)(x_i) &= \langle \vee_{j=1}^n (T_{\widetilde{\mathbb{N}}_{x_i}^{\beta^2}}(x_j) \wedge T_A(x_j)), \vee_{j=1}^n (I_{\widetilde{\mathbb{N}}_{x_i}^{\beta^2}}(x_j) \wedge I_A(x_j)), \wedge_{j=1}^n (F_{\widetilde{\mathbb{N}}_{x_i}^{\beta^2}}(x_j) \vee F_A(x_j))\rangle \\ &= (\widetilde{\mathbb{C}}^2(A))(x_i),\end{aligned}$$

and

$$\begin{aligned}((M_{\widehat{C}}^{\beta^2} \star M_{\widehat{C}}^T) \diamond A)(x_i) &= \langle \wedge_{j=1}^n (F_{\widetilde{\mathbb{N}}_{x_i}^{\beta^2}}(x_j) \vee T_A(x_j)), \wedge_{j=1}^n [(1 - I_{\widetilde{\mathbb{N}}_{x_i}^{\beta^2}}(x_j)) \vee I_A(x_j)], \vee_{j=1}^n (T_{\widetilde{\mathbb{N}}_{x_i}^{\beta^2}}(x_j) \wedge F_A(x_j))\rangle \\ &= (\underset{\sim}{\mathbb{C}}^2(A))(x_i).\end{aligned}$$

Hence, $\widetilde{\mathbb{C}}^2(A) = (M_{\widehat{C}}^{\beta^2} \star M_{\widehat{C}}^T) \circ A$, $\underset{\sim}{\mathbb{C}}^2(A) = (M_{\widehat{C}}^{\beta^2} \star M_{\widehat{C}}^T) \diamond A$. □

Example 10. *Let $(U, \widehat{C})$ be a SVN β^2-covering approximation space in Example 1, $\beta = \langle 0.5, 0.1, 0.8\rangle$ and $A = \frac{(0.6,0.3,0.5)}{x_1} + \frac{(0.4,0.5,0.1)}{x_2} + \frac{(0.3,0.2,0.6)}{x_3} + \frac{(0.5,0.3,0.4)}{x_4} + \frac{(0.7,0.2,0.3)}{x_5}$. Then*

$$\begin{aligned}&\widetilde{\mathbb{C}}^2(A) \\ =\ &(M_{\widehat{C}}^{\beta^2} \star M_{\widehat{C}}^T) \circ A \\ =\ &\begin{pmatrix} \langle 0.6,0.2,0.5\rangle & \langle 0.5,0.2,0.8\rangle & \langle 0.2,0.3,0.6\rangle & \langle 0.4,0.1,0.7\rangle & \langle 0.3,0.2,0.6\rangle \\ \langle 0.1,0.2,0.6\rangle & \langle 0.5,0.1,0.8\rangle & \langle 0.2,0.3,0.6\rangle & \langle 0.4,0.1,0.7\rangle & \langle 0.3,0.1,0.6\rangle \\ \langle 0.1,0.1,0.6\rangle & \langle 0.4,0.1,0.7\rangle & \langle 0.5,0.2,0.4\rangle & \langle 0.3,0.3,0.5\rangle & \langle 0.6,0.1,0.5\rangle \\ \langle 0.1,0.2,0.6\rangle & \langle 0.5,0.1,0.7\rangle & \langle 0.4,0.3,0.4\rangle & \langle 0.5,0.1,0.7\rangle & \langle 0.3,0.1,0.6\rangle \\ \langle 0.1,0.1,0.6\rangle & \langle 0.4,0.1,0.8\rangle & \langle 0.2,0.2,0.6\rangle & \langle 0.3,0.3,0.7\rangle & \langle 0.6,0.1,0.5\rangle \end{pmatrix} \circ \begin{pmatrix} \langle 0.6,0.3,0.5\rangle \\ \langle 0.4,0.5,0.1\rangle \\ \langle 0.3,0.2,0.6\rangle \\ \langle 0.5,0.3,0.4\rangle \\ \langle 0.7,0.2,0.3\rangle \end{pmatrix} \\ =\ &\begin{pmatrix} \langle 0.6,0.2,0.5\rangle \\ \langle 0.4,0.2,0.6\rangle \\ \langle 0.6,0.3,0.5\rangle \\ \langle 0.5,0.2,0.6\rangle \\ \langle 0.6,0.3,0.5\rangle \end{pmatrix},\end{aligned}$$

$$\begin{aligned}&\underset{\sim}{\mathbb{C}}^2(A) \\ =\ &(M_{\widehat{C}}^{\beta^2} \star M_{\widehat{C}}^T) \diamond A \\ =\ &\begin{pmatrix} \langle 0.6,0.2,0.5\rangle & \langle 0.5,0.2,0.8\rangle & \langle 0.2,0.3,0.6\rangle & \langle 0.4,0.1,0.7\rangle & \langle 0.3,0.2,0.6\rangle \\ \langle 0.1,0.2,0.6\rangle & \langle 0.5,0.1,0.8\rangle & \langle 0.2,0.3,0.6\rangle & \langle 0.4,0.1,0.7\rangle & \langle 0.3,0.1,0.6\rangle \\ \langle 0.1,0.1,0.6\rangle & \langle 0.4,0.1,0.7\rangle & \langle 0.5,0.2,0.4\rangle & \langle 0.3,0.3,0.5\rangle & \langle 0.6,0.1,0.5\rangle \\ \langle 0.1,0.2,0.6\rangle & \langle 0.5,0.1,0.7\rangle & \langle 0.4,0.3,0.4\rangle & \langle 0.5,0.1,0.7\rangle & \langle 0.3,0.1,0.6\rangle \\ \langle 0.1,0.1,0.6\rangle & \langle 0.4,0.1,0.8\rangle & \langle 0.2,0.2,0.6\rangle & \langle 0.3,0.3,0.7\rangle & \langle 0.6,0.1,0.5\rangle \end{pmatrix} \diamond \begin{pmatrix} \langle 0.6,0.3,0.5\rangle \\ \langle 0.4,0.5,0.1\rangle \\ \langle 0.3,0.2,0.6\rangle \\ \langle 0.5,0.3,0.4\rangle \\ \langle 0.7,0.2,0.3\rangle \end{pmatrix} \\ =\ &\begin{pmatrix} \langle 0.6,0.8,0.5\rangle \\ \langle 0.6,0.8,0.4\rangle \\ \langle 0.4,0.7,0.5\rangle \\ \langle 0.4,0.7,0.6\rangle \\ \langle 0.6,0.8,0.3\rangle \end{pmatrix}.\end{aligned}$$

7. An Application to DM Problems in Paper Defect Diagnosis

Under the type-2 SVN covering rough set model, we present a novel approach to DM problems in paper defect diagnosis in this section.

7.1. The Problem of DM in Paper Defect Diagnosis

Let $U = \{x_k : k = 1, 2, \cdots, n\}$ be the set of papers and $V = \{y_i | i = 1, 2, \cdots, m\}$ be the m main symptoms (for example, spot, steak, and so on) for a paper defect B. Assume that an inspector R evaluate every paper x_k $(k = 1, 2, \cdots, n)$.

Assume that the inspector R believes each paper $x_k \in U$ $(k = 1, 2, \cdots, n)$ has a symptom value C_i $(i = 1, 2, \cdots, m)$ denoted by $C_i(x_k) = \langle T_{C_i}(x_k), I_{C_i}(x_k, F_{C_i}(x_k) \rangle$, where $T_{C_i}(x_k) \in [0, 1]$ is the degree that inspector R confirms paper x_k has symptom y_i, $I_{C_i}(x_k) \in [0, 1]$ is the degree that inspector R is not sure paper x_k has symptom y_i, $F_{C_i}(x_k) \in [0, 1]$ is the degree that inspector R confirms paper x_k does not have symptom y_i, and $T_{C_i}(x_k) + I_{C_i}(x_k) + F_{C_i}(x_k) \leq 3$.

Let $\beta = \langle a, b, c \rangle$ be the critical value. If any paper $x_k \in U$, there is at least one symptom $y_i \in V$ such that the symptom value C_i for the paper x_k is not less than β (i.e., $C_i(x_k) \geq_2 \beta$), respectively, then $\widehat{\mathbf{C}} = \{C_1, C_2, \cdots, C_m\}$ is a SVN β^2-covering of U for some SVN number β.

If d is a possible degree, e is an indeterminacy degree and f is an impossible degree of the paper defect B of every paper $x_k \in U$ that is diagnosed by the inspector R, denoted by $A(x_k) = \langle d, e, f \rangle$, then the decision maker (the inspector R) for the DM problem needs to know how to evaluate whether the papers $x_k \in U$ have the paper defect B or not.

7.2. The DM Algorithm

In this subsection, we give an approach for the problem of DM with the above characterizations using the type-2 SVN covering rough set model. According to the characterizations of the DM problem in Section 7.1, we construct the SVN decision information system and present the Algorithm 1 of DM under the framework of the type-2 SVN covering rough set model.

Algorithm 1 The DM algorithm under the type-2 SVN covering rough set model

Input: SVN decision information system $(U, \widehat{\mathbf{C}}, \beta, A)$.
Output: The score ordering for all alternatives.

- **Step 1**: Compute the SVN β^2-neighborhood $\widetilde{\mathbb{N}}_x^{\beta^2}$ of x induced by $\widehat{\mathbf{C}}$, for all $x \in U$ according to Definition 6;
- **Step 2**: Compute the SVN covering upper approximation $\widetilde{\mathbb{C}}^2(A)$ and lower approximation $\underset{\sim}{\mathbb{C}}^2(A)$ of A, according to Definition 7;
- **Step 3**: Compute $\widetilde{R}_A^2 = \widetilde{\mathbb{C}}^2(A) \oplus \underset{\sim}{\mathbb{C}}^2(A)$;
- **Step 4**: Compute

$$s(x) = \frac{T_{\widetilde{R}_A^2}(x)}{\sqrt{(T_{\widetilde{R}_A^2}(x))^2 + (I_{\widetilde{R}_A^2}(x))^2 + (F_{\widetilde{R}_A^2}(x))^2}};$$

- **Step 5**: Rank all the alternatives $s(x)$ by using the principle of numerical size and select the paper that is more likely to be sick with the paper defect B.

According to the above process, we can get the DM according to the ranking. In Step 4, $S(x)$ is the cosine similarity measure between $\widetilde{R}_A(x)$ and the ideal solution $(1, 0, 0)$, which is proposed by Ye [44].

7.3. An Applied Example

Example 11. *Assume that* $U = \{x_1, x_2, x_3, x_4, x_5\}$ *is a set of papers. According to the paper defects' symptoms, we write* $V = \{y_1, y_2, y_3, y_4\}$ *to be four main symptoms (spot, steak, crater, and fracture) for a paper defect B. Assume that the inspector R evaluates every paper* x_k *(*$k = 1, 2, \cdots, 5$*) as shown in Table 1.*

Let $\beta = \langle 0.5, 0.1, 0.8\rangle$ *be the critical value. Then,* $\widehat{\mathbb{C}} = \{C_1, C_2, C_3, C_4\}$ *is a SVN* β^2*-coverings of U. Assume that the inspector R diagnosed the value* $A = \frac{(0.6,0.3,0.5)}{x_1} + \frac{(0.4,0.5,0.1)}{x_2} + \frac{(0.3,0.2,0.6)}{x_3} + \frac{(0.5,0.3,0.4)}{x_4} + \frac{(0.7,0.2,0.3)}{x_5}$ *of the paper defect B of every paper.*

Step 1: $\widetilde{\mathbb{N}}^{\beta^2}_{x_k}$ *(*$k = 1, 2, 3, 4, 5$*) are shown in Table 2.*

Step 2:

$$\widetilde{\mathbb{C}}^2(A) = \{\langle x_1, 0.6, 0.2, 0.5\rangle, \langle x_2, 0.4, 0.2, 0.6\rangle, \langle x_3, 0.6, 0.3, 0.5\rangle, \langle x_4, 0.5, 0.2, 0.6\rangle, \langle x_5, 0.6, 0.3, 0.5\rangle\},$$
$$\underset{\sim}{\mathbb{C}}^2(A) = \{\langle x_1, 0.6, 0.8, 0.5\rangle, \langle x_2, 0.6, 0.8, 0.4\rangle, \langle x_3, 0.4, 0.7, 0.5\rangle, \langle x_4, 0.4, 0.7, 0.4\rangle, \langle x_5, 0.6, 0.8, 0.3\rangle\}.$$

Step 3:

$$\begin{aligned} & \widetilde{R}^2_A \\ = \ & \widetilde{\mathbb{C}}^2(A) \oplus \underset{\sim}{\mathbb{C}}^2(A) \\ = \ & \{\langle x_1, 0.84, 0.16, 0.25\rangle, \langle x_2, 0.76, 0.16, 0.24\rangle, \langle x_3, 0.76, 0.21, 0.25\rangle, \langle x_4, 0.70, 0.14, 0.24\rangle, \langle x_5, 0.84, 0.24, 0.15\rangle\}. \end{aligned}$$

Step 4: *We can obtain* $s(x_k)$ *(*$k = 1, 2, \cdots, 5$*) in Table 5.*

Step 5: *According to the principle of numerical size, we have*

$$x_3 < x_4 < x_2 < x_1 < x_5.$$

Therefore, the inspector R diagnoses the paper x_5 *as more likely to be sick with the paper defect B.*

Table 5. $s(x_k)$ $(k = 1, 2, \cdots, 5)$.

U	x_1	x_2	x_3	x_4	x_5
$s(x_k)$	0.943	0.935	0.919	0.929	0.948

7.4. A Comparison Analysis

To validate the feasibility of the proposed DM method, a comparative study is conducted with other methods. These methods which were introduced in Liu [43], Ye [44], Yang et al. [36], and Wang et al. [37] are compared with the proposed approach using SVN information system.

Because Table 1 is the same as in the work by the authors of [37] and the counting processes of the methods presented by Liu [43], Ye [44], Yang et al. [36], and Wang et al. [37], are shown in the work by the authors of [37], so we do not show these counting processes in this paper. For Example 11, the results of them are calculated as follows.

- In Liu's method, we suppose the weight vector of the criteria is $\mathbf{w} = (0.35, 0.25, 0.3, 0.1)$ and $\gamma = 1$. Hence, we get

$$s(n_1) = 0.735, s(n_2) = 0.706, s(n_3) = 0.660, s(n_4) = 0.596, s(n_5) = 0.734.$$

According to the cosine similarity degrees $s(n_k)$ $(k = 1, 2, \cdots, 5)$, we obtain

$$x_4 < x_3 < x_2 < x_5 < x_1.$$

- In Ye's method, we suppose the weight vector of the criteria is $\mathbf{w} = (0.35, 0.25, 0.3, 0.1)$. Then

$$W_1(x_1, A^*) = 0.677, W_2(x_2, A^*) = 0.608, W_3(x_3, A^*) = 0.580, W_4(x_4, A^*) = 0.511,$$
$$W_5(x_5, A^*) = 0.666.$$

According to all $s(n_{x_k}, n^*)$ $(k = 1, 2, \cdots, 5)$, we obtain

$$x_4 < x_3 < x_2 < x_5 < x_1.$$

- In Yang's method, we suppose paper defect $B \in SVN(V)$ and $B = \frac{(0.3,0.6,0.5)}{y_1} + \frac{(0.7,0.2,0.1)}{y_2} + \frac{(0.6,0.4,0.3)}{y_3} + \frac{(0.8,0.4,0.5)}{y_4}$. Let $n^* = \langle 1, 0, 0 \rangle$. We get

$$\overline{\widetilde{R}}(B) = \{\langle x_1, 0.6, 0.2, 0.4 \rangle, \langle x_2, 0.6, 0.2, 0.4 \rangle, \langle x_3, 0.6, 0.3, 0.4 \rangle, \langle x_4, 0.5, 0.4, 0.5 \rangle, \langle x_5, 0.8, 0.3, 0.5 \rangle\},$$
$$\underline{\widetilde{R}}(B) = \{\langle x_1, 0.5, 0.6, 0.5 \rangle, \langle x_2, 0.3, 0.6, 0.5 \rangle, \langle x_3, 0.3, 0.5, 0.5 \rangle, \langle x_4, 0.6, 0.6, 0.5 \rangle, \langle x_5, 0.6, 0.6, 0.5 \rangle\}.$$

Then,

$$s(n_{x_1}, n^*) = 0.960, s(n_{x_2}, n^*) = 0.951, s(n_{x_3}, n^*) = 0.945, s(n_{x_4}, n^*) = 0.918, s(n_{x_5}, n^*) = 0.948.$$

According to all $s(n_{x_k}, n^*)$ $(k = 1, 2, \cdots, 5)$, we obtain

$$x_4 < x_3 < x_5 < x_2 < x_1.$$

- In Wang's method, we do not use $\beta = \langle 0.5, 0.1, 0.8 \rangle$ in Example 11, and the reason is explained later. We suppose $\beta' = \langle 0.5, 0.3, 0.8 \rangle$ in Wang's method. Then

$$\widetilde{\mathbb{C}}(A) = \{\langle x_1, 0.6, 0.3, 0.5 \rangle, \langle x_2, 0.4, 0.3, 0.6 \rangle, \langle x_3, 0.6, 0.5, 0.5 \rangle, \langle x_4, 0.5, 0.3, 0.6 \rangle, \langle x_5, 0.6, 0.5, 0.5 \rangle\},$$
$$\underset{\sim}{\mathbb{C}}(A) = \{\langle x_1, 0.6, 0.5, 0.5 \rangle, \langle x_2, 0.6, 0.5, 0.4 \rangle, \langle x_3, 0.4, 0.4, 0.5 \rangle, \langle x_4, 0.4, 0.5, 0.4 \rangle, \langle x_5, 0.6, 0.4, 0.3 \rangle\}.$$

Hence,

$$s(x_1) = 0.945, s(x_2) = 0.937, s(x_3) = 0.922, s(x_4) = 0.909, s(x_5) = 0.958.$$

According to all $s(x_k)$ $(k = 1, 2, \cdots, 5)$, we obtain

$$x_4 < x_3 < x_2 < x_1 < x_5.$$

All results are shown in Table 6 and Figure 7.

Liu [43] and Ye [44] presented the methods by SVN theory. The method developed by Liu [43] is based on the Hammer SVN number aggregation (HSVNNWA) operator, the ranking order willed be changed by different $\mathbf{w}$ and γ. The parameter γ can be regarded as an attitude of the decision maker's preferences. For Example 11, we set the weight vector of the criteria is $\mathbf{w} = (0.35, 0.25, 0.3, 0.1)$ and $\gamma = 1$, then we obtain $x_4 < x_3 < x_2 < x_5 < x_1$. The method developed by Ye [44] is based on the weighted correlation coefficient $W_k(x_k, A^*)$ or the weighted cosine similarity measure $M_k(x_k, A^*)$, where A^* is the ideal alternative. We can get two ranking orders of x_k $(k = 1, 2, 3, 4, 5)$ by the values of $W_k(x_k, A^*)$ and $M_k(x_k, A^*)$, respectively. Then, we find that these two kinds of ranking orders are the same. Hence, we only show $W_k(x_k, A^*)$ in this paper. In Table 6 and Figure 7, there are the same ranking results of their methods.

Table 6. The results utilizing the different methods of Example 11.

Methods	The Final Ranking	The Paper Is Most Sick With the Paper Defect B
Liu [43]	$x_4 < x_3 < x_2 < x_5 < x_1$	x_1
Ye [44]	$x_4 < x_3 < x_2 < x_5 < x_1$	x_1
Yang et al. [36]	$x_4 < x_3 < x_5 < x_2 < x_1$	x_1
Wang et al. [37]	$x_4 < x_3 < x_2 < x_1 < x_5$	x_5
This paper	$x_3 < x_4 < x_2 < x_1 < x_5$	x_5

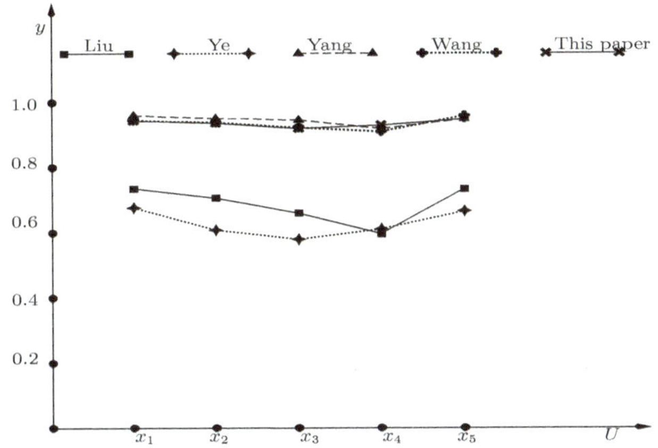

Figure 7. The chat of different values of patient in utilizing different methods in Example 11.

Yang et al. [36] and Wang et al. [37] used different SVN rough set models to make a decision. The method presented by Yang et al. [36] is based on a SVN relation rough set model on two-universes. That is to say, the DM problems with SVN information can be dealt with by Yang's method when it induces a SVN relation on two-universes. In Example 11, we obtain a SVN relation on two universes from Table 1. The method presented by Wang et al. [37] based on the type-1 SVN covering rough set model. That is to say, the DM problems with SVN information can be dealt with by Wang's method when it can induce a SVN β-covering. In Example 11, we suppose $\beta = \langle 0.5, 0.1, 0.8 \rangle$. However, $\widehat{\mathbf{C}}$ is not a SVN β-covering of U when $\beta = \langle 0.5, 0.1, 0.8 \rangle$. Hence, the method presented by Wang et al. can not be used in Example 11 when $\beta = \langle 0.5, 0.1, 0.8 \rangle$. Let's re-assume $\beta' = \langle 0.5, 0.3, 0.8 \rangle$. Then $\widehat{\mathbf{C}}$ is a SVN β'-covering of U. Hence, the method presented by Wang et al. can be used in Example 11 when $\beta' = \langle 0.5, 0.3, 0.8 \rangle$.

In this paper, we present the type-2 SVN covering rough set model based on SVN β^2-coverings. Under the type-2 SVN covering rough set model, a novel method for DM problems with SVN information is presented. The contributions of our proposed method are summarized as follows.

(1) The DM problems with SVN information can be dealt with by our proposed method when it can induce a SVN β^2-covering. The method presented by Wang et al. [37] can not be used in Example 11 when $\beta = \langle 0.5, 0.1, 0.8 \rangle$. But our proposed method can deal with Example 11 when $\beta = \langle 0.5, 0.1, 0.8 \rangle$. Hence, our proposed method complements Wang's.

(2) It is a new viewpoint to use SVN sets and rough sets in paper defect diagnosis.

Using different methods, the obtained results may be different. To achieve the most accurate results, further diagnosis is necessary for combination with other hybrid methods.

8. Conclusions

This paper investigates a new type of SVN covering rough set model, which can be seen as a new bridge linking SVN sets and covering-based rough sets. Comparing the existing literatures [36,37,48,49], the main contributions of this paper are concluded as follows.

(1) By introducing some definitions and properties in SVN β^2-covering approximation spaces, we present the type-2 SVN covering rough set model based on the type-2 inclusion relation. The existing literatures [36,37,48,49] used the type-1 inclusion relation to study the combination of SVN sets and rough sets. Hence, this paper presents a new and interesting viewpoint to study the combination of SVN sets and rough sets.

(2) It would be tedious and complicated to use set representation to calculate the new SVN covering approximation operators. Therefore, the graph and matrix representations of these new SVN covering approximation operators make it possible to calculate them. We are the first to study the equivalent representation of the SVN rough set model by graph theory. By these graph and matrix representations, calculations will become algorithmic and can be easily implemented by computers.

(3) Paper defect diagnosis is important in paper making industries. We propose a method to paper defect diagnosis under the type-2 SVN covering rough set model. The proposed DM method is compared with other methods which are presented by Liu [43], Ye [44], Yang et al. [36], and Wang et al. [37], respectively.

Further study will be deserved by the following research topics. On the one hand, the type-2 inclusion relation or graph theory can be considered into other SVN rough set models [34,36,48,49] in future research. On the other hand, neutrosophic sets and related algebraic structures [50–55] will be connected with the research content of this paper in further research.

Author Contributions: All authors have contributed equally to this paper. The idea of this whole thesis was put forward by X.Z., he also completed the preparatory work of the paper. J.W. analyzed the existing work of the rough sets and SVN sets and wrote the paper.

Funding: This work is supported by the National Natural Science Foundation of China under Grant Nos. 61976130 and 61573240, and the Natural Science Foundation of Education Department of Shaanxi Province, China, under Grant No. 19JK0506.

Conflicts of Interest: The authors declare no conflicts of interest.

References

1. Pawlak, Z. Rough sets. *Int. J. Comput. Inf. Sci.* **1982**, *11*, 341–356. [CrossRef]
2. Pawlak, Z. *Rough Sets: Theoretical Aspects of Reasoning About Data*; Kluwer Academic Publishers: Boston, MA, USA, 1991.
3. Kondo, M. On the structure of generalized rough sets. *Inf. Sci.* **2006**, *176*, 589–600. [CrossRef]
4. Wang, C.; Chen, D.; Wu, C.; Hu, Q. Data compression with homomorphism in covering information systems. *Int. J. Approx. Reason.* **2011**, *52*, 519–525. [CrossRef]
5. Wang, S.; Zhu, W.; Zhu, Q.; Min, F. Characteristic matrix of covering and its application to Boolean matrix decomposition. *Inf. Sci.* **2014**, *263*, 186–197. [CrossRef]
6. Bartol, W.; Miro, J.; Pioro, K.; Rossello, F. On the coverings by tolerance classes. *Inf. Sci.* **2004**, *166*, 193–211. [CrossRef]
7. Bianucci, D.; Cattaneo, G.; Ciucci, D. Entropies and co-entropies of coverings with application to incomplete information systems. *Fundam. Inf.* **2007**, *75*, 77–105.
8. Zhu, W. Relationship among basic concepts in covering-based rough sets. *Inf. Sci.* **2009**, *179*, 2478–2486. [CrossRef]
9. Zhu, W. Topological approaches to covering rough sets. *Inf. Sci.* **2007**, *177*, 1499–1508. [CrossRef]
10. Wang, J.; Zhang, X. Matrix approaches for some issues about minimal and maximal descriptions in covering-based rough sets. *Int. J. Approx. Reason.* **2019**, *104*, 126–143. [CrossRef]

11. Yao, Y.; Zhao, Y. Attribute reduction in decision-theoretic rough set models. *Inf. Sci.* **2008**, *178*, 3356–3373. [CrossRef]
12. Li, F.; Yin, Y. Approaches to knowledge reduction of covering decision systems based on information theory. *Inf. Sci.* **2009**, *179*, 1694–1704. [CrossRef]
13. Wu, W. Attribute reduction based on evidence theory in incomplete decision systems. *Inf. Sci.* **2008**, *178*, 1355–1371. [CrossRef]
14. Lang, G.; Li, Q.; Cai, M.; Fujita, H.; Zhang, H. Related families-based methods for updating reducts under dynamic object sets. *Knowl. Inf. Syst.* **2019**, *60*, 1081–1104. [CrossRef]
15. Wang, J.; Zhu, W. Applications of bipartite graphs and their adjacency matrices to covering-based rough sets. *Fundam. Inf.* **2017**, *156*, 237–254. [CrossRef]
16. Li, X.; Yi, H.; Liu, S. Rough sets and matroids from a lattice-theoretic viewpoint. *Inf. Sci.* **2016**, *342*, 37–52. [CrossRef]
17. Wang, J.; Zhang, X. Four operators of rough sets generalized to matroids and a matroidal method for attribute reduction. *Symmetry* **2018**, *10*, 418. [CrossRef]
18. Wang, J.; Zhu, W.; Wang, F.; Liu, G. Conditions for coverings to induce matroids. *Int. J. Mach. Learn. Cybern.* **2014**, *5*, 947–954. [CrossRef]
19. Chen, J.; Li, J.; Lin, Y.; Lin, G.; Ma, Z. Relations of reduction between covering generalized rough sets and concept lattices. *Inf. Sci.* **2015**, *304*, 16–27. [CrossRef]
20. Zhang, X.; Dai, J.; Yu, Y. On the union and intersection operations of rough sets based on various approximation spaces. *Inf. Sci.* **2015**, *292*, 214–229. [CrossRef]
21. D'eer, L.; Cornelis, C. A comprehensive study of fuzzy covering-based rough set models: Definitions, properties and interrelationships. *Fuzzy Sets Syst.* **2018**, *336*, 1–26. [CrossRef]
22. Wang, J.; Zhang, X. Two types of intuitionistic fuzzy covering rough sets and an application to multiple criteria group decision making. *Symmetry* **2018**, *10*, 462. [CrossRef]
23. Zhang, X.; Miao, D.; Liu, C.; Le, M. Constructive methods of rough approximation operators and multigranulation rough sets. *Knowl.-Based Syst.* **2016**, *91*, 114–125. [CrossRef]
24. Zadeh, L.A. Fuzzy sets. *Inf. Control* **1965**, *8*, 338–353. [CrossRef]
25. Jankowski, J.; Kazienko, P.; Watróbski, J.; Lewandowska, A.; Ziemba, P.; Zioło, M. Fuzzy multi-objective modeling of effectiveness and user experience in online advertising. *Expert Syst. Appl.* **2016**, *65*, 315–331. [CrossRef]
26. Boloş, M.I.; Bradea, I.A.; Delcea, C. A fuzzy logic algorithm for optimizing the investment decisions within companies. *Symmetry* **2019**, *11*, 186. [CrossRef]
27. Pozna, C.; Minculete, N.; Precup, R.E.; Kóczy, L.T.; Ballagi, Á. Signatures: Definitions, operators and applications to fuzzy modeling. *Fuzzy Sets Syst.* **2012**, *201*, 86–104. [CrossRef]
28. Vrkalovic, S.; Lunca, E.C.; Borlea, I.D. Model-free sliding mode and fuzzy controllers for reverse osmosis desalination plants. *Int. J. Artif. Intell.* **2018**, *16*, 208–222.
29. Ma, L. Two fuzzy covering rough set models and their generalizations over fuzzy lattices. *Fuzzy Sets Syst.* **2016**, *294*, 1–17. [CrossRef]
30. Yang, B.; Hu, B. On some types of fuzzy covering-based rough sets. *Fuzzy Sets Syst.* **2017**, *312*, 36–65. [CrossRef]
31. Smarandache, F. *Neutrosophy, Neutrosophic Probability, Set, and Logic*; American Research Press: Rehoboth, DE, USA, 1998.
32. Wang, H.; Smarandache, F.; Zhang, Y.; Sunderraman, R. Single valued neutrosophic sets. *Multispace Multistruct.* **2010**, *4*, 410–413.
33. Atanassov, K. Intuitionistic fuzzy sets. *Fuzzy Sets Syst.* **1986**, *20*, 87–96. [CrossRef]
34. Al-Quran, A.; Hassan, N.; Marei, E. A novel approach to neutrosophic soft rough set under uncertainty. *Symmetry* **2019**, *11*, 384. [CrossRef]
35. Mondal, K.; Pramanik, S. Rough neutrosophic multi-attribute decision-making based on grey relational analysis. *Neutrosophic Sets Syst.* **2015**, *7*, 8–17.
36. Yang, H.; Zhang, C.; Guo, Z.; Liu, Y.; Liao, X. A hybrid model of single valued neutrosophic sets and rough sets: Single valued neutrosophic rough set model. *Soft Comput.* **2017**, *21*, 6253–6267. [CrossRef]
37. Wang, J.; Zhang, X. Two types of single valued neutrosophic covering rough sets and an application to decision making. *Symmetry* **2018**, *10*, 710. [CrossRef]

38. Borzooei, R.A.; Farahani, H.; Moniri, M. Neutrosophic deductive filters on BL-algebras. *J. Intell. Fuzzy Syst.* **2014**, *26*, 2993–3004.
39. Akram, M.; Ali, G.; Alshehri, N.O. A new multi-attribute decision-making method based on m-polar fuzzy soft rough sets. *Symmetry* **2017**, *9*, 271. [CrossRef]
40. Zhan, J.; Ali, M.I.; Mehmood, N. On a novel uncertain soft set model: Z-soft fuzzy rough set model and corresponding decision making methods. *Appl. Soft Comput.* **2017**, *56*, 446–457. [CrossRef]
41. Zhan, J.; Alcantud, J.C.R. A novel type of soft rough covering and its application to multicriteria group decision making. *Artif. Intell. Rev.* **2018**, *4*, 1–30. [CrossRef]
42. Zhang, Z. An approach to decision making based on intuitionistic fuzzy rough sets over two universes. *J. Oper. Res. Soc.* **2013**, *64*, 1079–1089.
43. Liu, P. The aggregation operators based on archimedean t-conorm and t-norm for single-valued neutrosophic numbers and their application to decision making. *Int. J. Fuzzy Syst.* **2016**, *18*, 849–863. [CrossRef]
44. Ye, J. Multicriteria decision-making method using the correlation coefficient under single-valued neutrosophic environment. *Int. J. Gen. Syst.* **2013**, *42*, 386–394. [CrossRef]
45. Pomykala, J.A. Approximation operations in approximation space. *Bull. Pol. Acad. Sci.* **1987**, *35*, 653–662.
46. Bonikowski, Z.; Bryniarski, E.; Wybraniec-Skardowska, U. Extensions and intentions in the rough set theory. *Inf. Sci.* **1998**, *107*, 149–167. [CrossRef]
47. Smarandache, F. Neutrosophic set—A generalization of the intuitionistic fuzzy set. *J. Def. Resour. Manag.* **2010**, *1*, 38–42.
48. Bao, Y.; Yang, H. On single valued neutrosophic refined rough set model and its application. *J. Intell. Fuzzy Syst.* **2018**, *33*, 1235–1248. [CrossRef]
49. Guo, Z.; Liu, Y.; Yang, H. A novel rough set model in generalized single valued neutrosophic approximation spaces and its application. *Symmetry* **2017**, *9*, 119. [CrossRef]
50. Wu, X.; Zhang, X. The decomposition theorems of AG-neutrosophic extended triplet loops and strong AG-(l, l)-loops. *Mathematics* **2019**, *7*, 268. [CrossRef]
51. Zhang, X.; Borzooei, R.A.; Jun, Y. Q-filters of quantum B-algebras and basic implication algebras. *Symmetry* **2018**, *10*, 573. [CrossRef]
52. Zhang, X.; Mao, X.; Wu, Y.; Zhai, X. Neutrosophic filters in pseudo-BCI algebras. *Int. J. Uncertain. Quantif.* **2018**, *8*, 511–526. [CrossRef]
53. Zhang, X.; Hu, Q.; Smarandache, F.; An, X. On neutrosophic triplet groups: basic properties, NT-subgroups, and some notes. *Symmetry* **2018**, *10*, 289. [CrossRef]
54. Ma, Y.; Zhang, X.; Yang, X.; Zhou, X. Generalized neutrosophic extended triplet group. *Symmetry* **2019**, *11*, 327. [CrossRef]
55. Zhang, X.; Wang, X.; Smarandache, F.; Jaíyéolá, T.G.; Lian, T. Singular neutrosophic extended triplet groups and generalized groups. *Cognit. Syst. Res.* **2019**, *57*, 32–40. [CrossRef]

symmetry

MDPI

Article

Generalization of Maximizing Deviation and TOPSIS Method for MADM in Simplified Neutrosophic Hesitant Fuzzy Environment

Muhammad Akram [1,*], Sumera Naz [2] and Florentin Smarandache [3]

[1] Department of Mathematics, University of the Punjab, New Campus, Lahore 54590, Pakistan
[2] Department of Mathematics, Government College Women University Faisalabad, Punjab 38000, Pakistan
[3] Science Department 705 Gurley Ave., University of New Mexico Mathematics, Gallup, NM 87301, USA
* Correspondence: m.akram@pucit.edu.pk

Received: 25 June 2019; Accepted: 1 August 2019; Published: 17 August 2019

Abstract: With the development of the social economy and enlarged volume of information, the application of multiple-attribute decision-making (MADM) has become increasingly complex, uncertain, and obscure. As a further generalization of hesitant fuzzy set (HFS), simplified neutrosophic hesitant fuzzy set (SNHFS) is an efficient tool to process the vague information and contains the ideas of a single-valued neutrosophic hesitant fuzzy set (SVNHFS) and an interval neutrosophic hesitant fuzzy set (INHFS). In this paper, we propose a decision-making approach based on the maximizing deviation method and TOPSIS (Technique for Order Preference by Similarity to Ideal Solution) to solve the MADM problems, in which the attribute weight information is incomplete, and the decision information is expressed in simplified neutrosophic hesitant fuzzy elements. Firstly, we inaugurate an optimization model on the basis of maximizing deviation method, which is useful to determine the attribute weights. Secondly, using the idea of the TOPSIS, we determine the relative closeness coefficient of each alternative and based on which we rank the considered alternatives to select the optimal one(s). Finally, we use a numerical example to show the detailed implementation procedure and effectiveness of our method in solving MADM problems under simplified neutrosophic hesitant fuzzy environment.

Keywords: simplified neutrosophic hesitant fuzzy set; multi-attribute decision-making; maximizing deviation; TOPSIS

---

1. Introduction

The concept of neutrosophy was originally introduced by Smarandache [1] from a philosophical viewpoint. Gradually, it has been discovered that without a specific description, it is not easy to apply neutrosophic sets in real applications because a truth-membership, an indeterminacy-membership, and a falsity-membership degree, in non-standard unit interval $]0^-, 1^+[$, are independently assigned to each element in the set. After analyzing this difficulty, Smarandache [2] and Wang [3] initiated the notion of a single-valued neutrosophic set (SVNS) and made the first ever neutrosophic publication. Ye [4] developed the concept of simplified neutrosophic set (SNS). SNS, a subclass of a neutrosophic set, contains the ideas of a SVNS and an interval neutrosophic set (INS), which are very useful in real science and engineering applications with incomplete, indeterminate, and inconsistent information existing commonly in real situations. Torra and Narukawa [5] put forward the concept of HFS as another extension of fuzzy set [6]. HFS is an effective tool to represent vague information in the process of MADM, as it permits the element membership degree to a set characterized by a few possible values in $[0, 1]$ and can be accurately described in terms of the judgment of the experts.

Ye [7] introduced SVNHFS as an extension of SVNS in the spirit of HFS and developed the single-valued neutrosophic hesitant fuzzy weighted averaging and weighted geometric operator. The SVNHFS represents some uncertain, incomplete, and inconsistent situations where each element has certain different values characterized by truth-membership hesitant, indeterminacy-membership hesitant, and falsity-membership hesitant function. For instance, when the opinion of three experts is required for a certain statement, they may state that the possibility that the statement is true is $\{0.3, 0.5, 0.8\}$, and the statement is false is $\{0.1, 0.4\}$, and the degree that they are not sure is $\{0.2, 0.7, 0.8\}$. For single-valued neutrosophic hesitant fuzzy notation, it can be expressed as $\{\{0.3, 0.5, 0.8\}, \{0.1, 0.4\}, \{0.2, 0.7, 0.8\}\}$. Liu and Luo [8] discussed the certainty function, score function, and accuracy function of SVNHFS and proposed the single-valued neutrosophic hesitant fuzzy ordered weighted averaging operator and hybrid weighted averaging operator. Sahin and Liu [9] proposed the correlation coefficient with single-valued neutrosophic hesitant fuzzy information and successfully applied it to decision-making problems. Li and Zhang [10] introduced Choquet aggregation operators with single-valued neutrosophic hesitant fuzzy information for MADM. Juan-Juan et al. [11] developed a decision-making technique using geometric weighted Choquet integral Heronian mean operator for SVNHFSs. Wang and Li [12] developed the generalized prioritized weighted average operator, the generalized prioritized weighted geometric operator with SVNHFS, and further developed an approach on the basis of the proposed operators to solve MADM problems. Recently, Akram et al. [13–16] and Naz et al. [17–19] put forward certain novel decision-making techniques in the frame work of extended fuzzy set theory. Furthermore, Liu and Shi [20] proposed the concept of INHFS by combining INS with HFS and developed the generalized weighted operator, generalized ordered weighted operator, and generalized hybrid weighted operator with the proposed interval neutrosophic hesitant fuzzy information. Ye [21] and Kakati et al. [22] proposed the correlation coefficients and Choquet integrals, respectively, with INHFS. Mahmood et al. [23] discussed the vector similarity measures with SNHFS. In practical terms, the SNHFS measures the truth-membership, the indeterminacy-membership and the falsity-membership degree by SVNHFSs and INHFSs. The classical sets, fuzzy sets, intuitionistic fuzzy sets, SVNSs, INSs, SNSs, and HFSs are the particular situations of SNHFSs. In modeling vague and uncertain information, SNHFS is more flexible and practice.

In the theory of decision analysis, MADM is one of the most important branches and several beneficial models and approaches have been developed related to decision analysis. However, due to limited time, lack of data or knowledge, and the limited expertise of the expert about the problem, MADM process under simplified neutrosophic hesitant fuzzy circumstances, encounters the situations where the information about attribute weights is completely unknown or incompletely known. The existing approaches are not suitable to handle these situations. Furthermore, among some useful MADM methodologies, the maximizing deviation method and the TOPSIS provide a ranking approach, which is measured by the farthest distance from the negative-ideal solution (NIS) and the shortest distance from the positive-ideal solution (PIS). For all these, in this paper, we propose an innovative approach of maximizing deviation and TOPSIS to objectively determine the attribute weights and rank the alternatives with completely unknown or partly known attribute weights. We propose the new distance measure and discuss the application of SNHFSs to MADM. In the framework of TOPSIS, we construct a novel generalized method under the simplified neutrosophic hesitant fuzzy environment. As compared to the existing work, the SNHFSs availably depict more general decision-making situations.

The paper is structured as follows: Section 2 establishes a simplified neutrosophic hesitant fuzzy MADM based on maximizing deviations and TOPSIS. In Section 3, a numerical example is given to demonstrate the effectiveness of our model and method and finally we draw conclusions in Section 4.

SVNHFS as a more flexible general formal framework extends the concept of fuzzy set [6], intuitionistic fuzzy set [24], SVNS [3] and HFS [25]. Ye [7] proposed the following definition of SVNHFS.

Definition 1. [7] *Let Z be a fixed set, a SVNHFS* $\mathfrak{n}$ *on Z is defined as*

$$\mathfrak{n} = \{\langle z, \mathfrak{t}(z), \mathfrak{i}(z), \mathfrak{f}(z)\rangle | z \in Z\}$$

where $\mathfrak{t}(z), \mathfrak{i}(z), \mathfrak{f}(z)$ *are the sets of a few values in* $[0,1]$*, representing the possible truth-membership hesitant degree, indeterminacy-membership hesitant degree and falsity-membership hesitant degree of the element* z *to* $\mathfrak{n}$*, respectively;* $\mathfrak{t}(z) = \{\gamma_1, \gamma_2, \ldots, \gamma_l\}, \gamma_1, \gamma_2, \ldots, \gamma_l$ *are the elements of* $\mathfrak{t}(z)$*;* $\mathfrak{i}(z) = \{\delta_1, \delta_2, \ldots, \delta_p\}, \delta_1, \delta_2, \ldots, \delta_p$ *are the elements of* $\mathfrak{i}(z)$*;* $\mathfrak{f}(z) = \{\eta_1, \eta_2, \ldots, \eta_q\}, \eta_1, \eta_2, \ldots, \eta_q$ *are the elements of* $\mathfrak{f}(z)$*, for every* $z \in Z$*; and* l, p, q *denote, respectively, the numbers of the hesitant fuzzy elements in* $\mathfrak{t}, \mathfrak{i}, \mathfrak{f}$.

For simplicity, the expression $\mathfrak{n}(z) = \{\mathfrak{t}(z), \mathfrak{i}(z), \mathfrak{f}(z)\}$ is called a single-valued neutrosophic hesitant fuzzy element (SVNHFE), which we represent by simplified symbol $\mathfrak{n} = \{\mathfrak{t}, \mathfrak{i}, \mathfrak{f}\}$.

Definition 2. [7] *Let* $\mathfrak{n}$, $\mathfrak{n}_1$ *and* $\mathfrak{n}_2$ *be three SVNHFEs. Then their operations are defined as follows:*

1. $\mathfrak{n}_1 \oplus \mathfrak{n}_2 = \bigcup_{\gamma_1 \in \mathfrak{t}_1, \delta_1 \in \mathfrak{i}_1, \eta_1 \in \mathfrak{f}_1, \gamma_2 \in \mathfrak{t}_2, \delta_2 \in \mathfrak{i}_2, \eta_2 \in \mathfrak{f}_2} \{\{\gamma_1 + \gamma_2 - \gamma_1\gamma_2\}, \{\delta_1\delta_2\}, \{\eta_1\eta_2\}\};$
2. $\mathfrak{n}_1 \otimes \mathfrak{n}_2 = \bigcup_{\gamma_1 \in \mathfrak{t}_1, \delta_1 \in \mathfrak{i}_1, \eta_1 \in \mathfrak{f}_1, \gamma_2 \in \mathfrak{t}_2, \delta_2 \in \mathfrak{i}_2, \eta_2 \in \mathfrak{f}_2} \{\{\gamma_1\gamma_2\}, \{\delta_1 + \delta_2 - \delta_1\delta_2\}, \{\eta_1 + \eta_2 - \eta_1\eta_2\}\};$
3. $\varsigma\mathfrak{n} = \bigcup_{\gamma \in \mathfrak{t}, \delta \in \mathfrak{i}, \eta \in \mathfrak{f}} \{\{1 - (1-\gamma)^{\varsigma}\}, \{\delta^{\varsigma}\}, \{\eta^{\varsigma}\}\};\ \varsigma > 0$
4. $\mathfrak{n}^{\varsigma} = \bigcup_{\gamma \in \mathfrak{t}, \delta \in \mathfrak{i}, \eta \in \mathfrak{f}} \{\{\gamma^{\varsigma}\}, \{1 - (1-\delta)^{\varsigma}\}, \{1 - (1-\eta)^{\varsigma}\}\}\ \varsigma > 0.$

2. TOPSIS and Maximizing Deviation Method for Simplified Neutrosophic Hesitant Fuzzy Multi-Attribute Decision-Making

In this section, we propose the normalization technique and the distance measures of SNHFSs and based on this we develop further a new decision-making approach based on maximum deviation and TOPSIS under simplified neutrosophic hesitant fuzzy circumstances to explore the application of SNHFSs to MADM.

2.1. TOPSIS and Maximizing Deviation Method for Single-Valued Neutrosophic Hesitant Fuzzy Multi-Attribute Decision-Making

In this subsection, we only use SVNHFSs in SNHFSs and develop a new decision-making approach, by combining the idea of SVNHFSs with maximizing deviation, to solve a MADM problem in single-valued neutrosophic hesitant fuzzy environment.

2.1.1. Description of the MADM Problem

Consider a MADM problem containing a discrete set of m alternatives $\{A_1, A_2, \ldots, A_m\}$ and a set of all attributes $P = \{P_1, P_2, \ldots, P_n\}$. The evaluation information of the ith alternative with respect to the jth attribute is a SVNHFE $\mathfrak{n}_{ij} = \langle \mathfrak{t}_{ij}, \mathfrak{i}_{ij}, \mathfrak{f}_{ij}\rangle$, where $\mathfrak{t}_{ij}$, $\mathfrak{i}_{ij}$ and $\mathfrak{f}_{ij}$ indicate the preference degree, uncertain degree, and falsity degree, respectively, of the decision maker facing the ith alternative that satisfied the jth attribute. Then the single-valued neutrosophic hesitant fuzzy decision matrix (SVNHFDM) $\mathcal{N}$, can be constructed as follows:

$$\mathcal{N} = \begin{bmatrix} \mathfrak{n}_{11} & \mathfrak{n}_{12} & \cdots & \mathfrak{n}_{1n} \\ \mathfrak{n}_{21} & \mathfrak{n}_{22} & \cdots & \mathfrak{n}_{2n} \\ \vdots & \vdots & \ddots & \vdots \\ \mathfrak{n}_{m1} & \mathfrak{n}_{m2} & \cdots & \mathfrak{n}_{mn} \end{bmatrix}$$

Assume that each attribute has different importance, the weight vector of all attributes is defined as $w = (w_1, w_2, \ldots, w_n)^t$, where $0 \leq w_j \leq 1$ and $\sum_{j=1}^{n} w_j = 1$ with w_j representing the importance degree of the attribute P_j. Due to the complexity of the practical decision-making problems, the attribute weights information is frequently incomplete. For ease, let $\Im$ be the set of the known information about attribute weights, which we can construct by the following forms, for $i \neq j$:

(i) $w_i \geq w_j$ (weak ranking);
(ii) $w_i - w_j \geq \alpha_i, \alpha_i > 0$ (strict ranking);
(iii) $w_i - w_j \geq w_k - w_l$, for $j \neq k \neq l$ (ranking of differences);
(iv) $w_i \geq \alpha_i w_j$, $0 \leq \alpha_i \leq 1$ (ranking with multiples);
(v) $\alpha_i \leq w_i \leq \alpha_i + \xi_i, 0 \leq \alpha_i \leq \alpha_i + \xi_i \leq 1$ (interval form).

In the comparison of SVNHFEs, the number of their corresponding element may be unequal. To handle this situation, we normalize the SVNHFEs as follows:

Suppose that $\mathfrak{n} = \{\mathfrak{t}, \mathfrak{i}, \mathfrak{f}\}$ is a SVNHFE, then $\bar{\gamma} = \varpi\gamma^+ + (1-\varpi)\gamma^-$, $\bar{\delta} = \varpi\delta^+ + (1-\varpi)\delta^-$ and $\bar{\eta} = \varpi\eta^+ + (1-\varpi)\eta^-$ are the added truth-membership, the indeterminacy-membership and the falsity-membership degree, respectively, where γ^- and γ^+ are the minimum and the maximum elements of $\mathfrak{t}$, respectively, δ^- and δ^+ are the minimum and the maximum elements of $\mathfrak{i}$, respectively, η^- and η^+ are the minimum and the maximum elements of $\mathfrak{f}$, respectively, and $\varpi \in [0,1]$ is a parameter assigned by the expert according to his risk preference.

For the normalization of SVNHFE, different values of ϖ produce different results for the added truth-membership, the indeterminacy-membership and the falsity-membership degree. Usually, there are three cases of the preference of the expert:

- If $\varpi = 0$, the pessimist expert may add the minimum truth-membership degree γ^-, the minimum indeterminacy-membership degree δ^- and the minimum falsity-membership degree η^-.
- If $\varpi = 0.5$, the neutral expert may add the truth-membership degree $\frac{\gamma^-+\gamma^+}{2}$, the indeterminacy-membership degree $\frac{\delta^-+\delta^+}{2}$ and the falsity-membership degree $\frac{\eta^-+\eta^+}{2}$.
- If $\varpi = 1$, the optimistic expert may add the maximum truth-membership degree γ^-, the maximum indeterminacy-membership degree δ^- and the maximum falsity-membership degree η^-.

For instance, if we have two SVNHFEs $\mathfrak{n}_1 = \{\mathfrak{t}_1, \mathfrak{i}_1, \mathfrak{f}_1\} = \{\{0.3, 0.5\}, \{0.4, 0.6, 0.8\}, \{0.5, 0.7\}\}$, $\mathfrak{n}_2 = \{\mathfrak{t}_2, \mathfrak{i}_2, \mathfrak{f}_2\} = \{\{0.1, 0.4, 0.5\}, \{0.6, 0.7\}, \{0.2, 0.6, 0.9\}\}$. Here $\#\mathfrak{t}_1 = 2$, $\#\mathfrak{i}_1 = 3$, $\#\mathfrak{f}_1 = 2$, $\#\mathfrak{t}_2 = 3$, $\#\mathfrak{i}_2 = 2$ and $\#\mathfrak{f}_2 = 3$. Clearly, $\#\mathfrak{t}_1 \neq \#\mathfrak{t}_2$, $\#\mathfrak{i}_1 \neq \#\mathfrak{i}_2$, and $\#\mathfrak{f}_1 \neq \#\mathfrak{f}_2$. The truth-membership and the falsity-membership degree of $\mathfrak{n}_1$, while the indeterminacy-membership degree of $\mathfrak{n}_2$ need to be pre-treated.

If $\varpi = 0$, then we may add the minimum truth-membership degree or the indeterminacy-membership degree or the falsity-membership degree for the target object. For the SVNHFE $\mathfrak{n}_1$, the truth-membership and falsity-membership degree of $\mathfrak{n}_1$ can be attained as $\{0.3, 0.3, 0.5\}$ and $\{0.5, 0.5, 0.7\}$, i.e., $\mathfrak{n}_1$ can be normalized as $\mathfrak{n}_1 = \{\{0.3, 0.3, 0.5\}, \{0.4, 0.6, 0.8\}, \{0.5, 0.5, 0.7\}\}$. For the SVNHFE $\mathfrak{n}_2$, the indeterminacy-membership degree of $\mathfrak{n}_2$ can be obtained as {0.6,0.6,0.7}, i.e., $\mathfrak{n}_2$ is normalized as $\mathfrak{n}_2 = \{\{0.1, 0.4, 0.5\}, \{0.6, 0.6, 0.7\}, \{0.2, 0.6, 0.9\}\}$.

If $\varpi = 0.5$, then we may add the average truth-membership degree or the indeterminacy-membership degree or the falsity-membership degree for the target object. For the SVNHFE $\mathfrak{n}_1$, the truth-membership and falsity-membership degree of $\mathfrak{n}_1$ can be attained as $\{0.3, 0.4, 0.5\}$ and $\{0.5, 0.6, 0.7\}$, i.e., $\mathfrak{n}_1$ can be normalized as $\mathfrak{n}_1 = \{\{0.3, 0.4, 0.5\}, \{0.4, 0.6, 0.8\}, \{0.5, 0.6, 0.7\}\}$. For the SVNHFE $\mathfrak{n}_2$, the indeterminacy-membership degree of $\mathfrak{n}_2$ can be obtained as {0.6,0.65,0.7}, i.e., $\mathfrak{n}_2$ is normalized as $\mathfrak{n}_2 = \{\{0.1, 0.4, 0.5\}, \{0.6, 0.65, 0.7\}, \{0.2, 0.6, 0.9\}\}$.

If $\varpi = 1$, then we may add the maximum truth-membership degree or the indeterminacy-membership degree or the falsity-membership degree for the normalization. For the SVNHFE $\mathfrak{n}_1$, the truth-membership and falsity-membership degree of $\mathfrak{n}_1$

can be attained as $\{0.3, 0.5, 0.5\}$ and $\{0.5, 0.7, 0.7\}$, i.e., $\mathfrak{n}_1$ is normalized as $\mathfrak{n}_1 = \{\{0.3, 0.5, 0.5\}, \{0.4, 0.6, 0.8\}, \{0.5, 0.7, 0.7\}\}$. For the SVNHFE $\mathfrak{n}_2$, the indeterminacy-membership degree of $\mathfrak{n}_2$ can be attained as $\{0.6, 0.7, 0.7\}$, i.e., $\mathfrak{n}_2$ is normalized as $\mathfrak{n}_2 = \{\{0.1, 0.4, 0.5\}, \{0.6, 0.7, 0.7\}, \{0.2, 0.6, 0.9\}\}$.

The algorithm for the normalization of SVNHFEs is given in Algorithm 1.

Algorithm 1 The algorithm for the normalization of SVNHFEs.

INPUT: Two SVNHFEs $\mathfrak{n}_1 = (\mathfrak{t}_1, \mathfrak{i}_1, \mathfrak{f}_1)$, $\mathfrak{n}_2 = (\mathfrak{t}_2, \mathfrak{i}_2, \mathfrak{f}_2)$ and the value of ϖ.
OUTPUT: The normalization of $\mathfrak{n}_1 = (\mathfrak{t}_1, \mathfrak{i}_1, \mathfrak{f}_1)$ and $\mathfrak{n}_2 = (\mathfrak{t}_2, \mathfrak{i}_2, \mathfrak{f}_2)$.

Count the number of elements of $\mathfrak{n}_1$ and $\mathfrak{n}_2$, i.e., $\#\mathfrak{t}_1, \#\mathfrak{i}_1, \#\mathfrak{f}_1, \#\mathfrak{t}_2, \#\mathfrak{i}_2, \#\mathfrak{f}_2$;
Determine the minimum and the maximum of the elements of $\mathfrak{n}_1$ and $\mathfrak{n}_2$;
$\mathfrak{t} = \arg\min_{i=1,2} \#\mathfrak{t}_i$, $\mathfrak{i} = \arg\min_{i=1,2} \#\mathfrak{i}_i$, $\mathfrak{f} = \arg\min_{i=1,2} \#\mathfrak{f}_i$;
if $\#\mathfrak{t}_1 = \#\mathfrak{t}_2$ **then break;**
else if $\mathfrak{t} = \#\mathfrak{t}_1$ **then**
    $n = \#\mathfrak{t}_2 - \#\mathfrak{t}_1$;
    Determine the value of $\bar{\gamma}$ for $\mathfrak{t}_1$;
    for i=1:1:n **do**
        $\mathfrak{t}_1 = \mathfrak{t}_1 \cup \bar{\gamma}$;
    end for
else
    $n = \#\mathfrak{t}_1 - \#\mathfrak{t}_2$;
    Determine the value of $\bar{\gamma}$ for $\mathfrak{t}_2$;
    for i=1:1:n **do**
        $\mathfrak{t}_2 = \mathfrak{t}_2 \cup \bar{\gamma}$;
    end for
end if
if $\#\mathfrak{i}_1 = \#\mathfrak{i}_2$ **then break;**
else if $\mathfrak{i} = \#\mathfrak{i}_1$ **then**
    $n = \#\mathfrak{i}_2 - \#\mathfrak{i}_1$;
    Determine the value of $\bar{\delta}$ for $\mathfrak{i}_1$;
    for i=1:1:n **do**
        $\mathfrak{i}_1 = \mathfrak{i}_1 \cup \bar{\delta}$;
    end for
else
    $n = \#\mathfrak{i}_1 - \#\mathfrak{i}_2$;
    Determine the value of $\bar{\delta}$ for $\mathfrak{i}_2$;
    for i=1:1:n **do**
        $\mathfrak{i}_2 = \mathfrak{i}_2 \cup \bar{\delta}$;
    end for
end if
if $\#\mathfrak{f}_1 = \#\mathfrak{f}_2$ **then break;**
else if $\mathfrak{f} = \#\mathfrak{f}_1$ **then**
    $n = \#\mathfrak{f}_2 - \#\mathfrak{f}_1$;
    Determine the value of $\bar{\eta}$ for $\mathfrak{f}_1$;
    for i=1:1:n **do**
        $\mathfrak{f}_1 = \mathfrak{f}_1 \cup \bar{\eta}$;
    end for
else
    $n = \#\mathfrak{f}_1 - \#\mathfrak{f}_2$;
    Determine the value of $\bar{\eta}$ for $\mathfrak{f}_2$;
    for i=1:1:n **do**
        $\mathfrak{f}_2 = \mathfrak{f}_2 \cup \bar{\eta}$;
    end for
end if

2.1.2. The Distance Measures for SVNHFSs

Definition 3. *Let $\mathfrak{n}_1 = \{\mathfrak{t}_1, \mathfrak{i}_1, \mathfrak{f}_1\}$ and $\mathfrak{n}_2 = \{\mathfrak{t}_2, \mathfrak{i}_2, \mathfrak{f}_2\}$ be two normalized SVNHFEs, then the single-valued neutrosophic hesitant fuzzy Hamming distance between $\mathfrak{n}_1$ and $\mathfrak{n}_2$ can be defined as follows:*

$$d_1(\mathfrak{n}_1, \mathfrak{n}_2) = \frac{1}{3}\left(\frac{1}{\#\mathfrak{t}}\sum_{\varsigma=1}^{\#\mathfrak{t}}\left|\gamma_1^{\sigma(\varsigma)} - \gamma_2^{\sigma(\varsigma)}\right| + \frac{1}{\#\mathfrak{i}}\sum_{\varsigma=1}^{\#\mathfrak{i}}\left|\delta_1^{\sigma(\varsigma)} - \delta_2^{\sigma(\varsigma)}\right| + \frac{1}{\#\mathfrak{f}}\sum_{\varsigma=1}^{\#\mathfrak{f}}\left|\eta_1^{\sigma(\varsigma)} - \eta_2^{\sigma(\varsigma)}\right|\right), \tag{1}$$

where $\#\mathfrak{t} = \#\mathfrak{t}_1 = \#\mathfrak{t}_2$, $\#\mathfrak{i} = \#\mathfrak{i}_1 = \#\mathfrak{i}_2$ and $\#\mathfrak{f} = \#\mathfrak{f}_1 = \#\mathfrak{f}_2$. $\gamma_i^{\sigma(\varsigma)}$, $\delta_i^{\sigma(\varsigma)}$ and $\eta_i^{\sigma(\varsigma)}$ are the ςth largest values in γ_i, δ_i and η_i, respectively $(i = 1, 2)$.

In addition, the single-valued neutrosophic hesitant fuzzy Euclidean distance is defined as:

$$d_2(\mathfrak{n}_1, \mathfrak{n}_2) = \sqrt{\frac{1}{3}\left(\frac{1}{\#\mathfrak{t}}\sum_{\varsigma=1}^{\#\mathfrak{t}}\left(\gamma_1^{\sigma(\varsigma)} - \gamma_2^{\sigma(\varsigma)}\right)^2 + \frac{1}{\#\mathfrak{i}}\sum_{\varsigma=1}^{\#\mathfrak{i}}\left(\delta_1^{\sigma(\varsigma)} - \delta_2^{\sigma(\varsigma)}\right)^2 + \frac{1}{\#\mathfrak{f}}\sum_{\varsigma=1}^{\#\mathfrak{f}}\left(\eta_1^{\sigma(\varsigma)} - \eta_2^{\sigma(\varsigma)}\right)^2\right)}. \tag{2}$$

By using the geometric distance model of [26], the above distances can be generalized as follows:

$$d(\mathfrak{n}_1, \mathfrak{n}_2) = \left(\frac{1}{3}\left(\frac{1}{\#\mathfrak{t}}\sum_{\varsigma=1}^{\#\mathfrak{t}}\left|\gamma_1^{\sigma(\varsigma)} - \gamma_2^{\sigma(\varsigma)}\right|^\alpha + \frac{1}{\#\mathfrak{i}}\sum_{\varsigma=1}^{\#\mathfrak{i}}\left|\delta_1^{\sigma(\varsigma)} - \delta_2^{\sigma(\varsigma)}\right|^\alpha + \frac{1}{\#\mathfrak{f}}\sum_{\varsigma=1}^{\#\mathfrak{f}}\left|\eta_1^{\sigma(\varsigma)} - \eta_2^{\sigma(\varsigma)}\right|^\alpha\right)\right)^{\frac{1}{\alpha}}, \tag{3}$$

where α is constant and $\alpha > 0$. Based on the value of α, the relationship among $d(\mathfrak{n}_1, \mathfrak{n}_2), d_1(\mathfrak{n}_1, \mathfrak{n}_2)$ and $d_2(\mathfrak{n}_1, \mathfrak{n}_2)$ can be deduced as:

- *If $\alpha = 1$, then the distance $d(\mathfrak{n}_1, \mathfrak{n}_2) = d_1(\mathfrak{n}_1, \mathfrak{n}_2)$.*
- *If $\alpha = 2$, then the distance $d(\mathfrak{n}_1, \mathfrak{n}_2) = d_2(\mathfrak{n}_1, \mathfrak{n}_2)$.*

Therefore, the distance $d(\mathfrak{n}_1, \mathfrak{n}_2)$ is a generalization of the single-valued neutrosophic hesitant fuzzy Hamming distance $d_1(\mathfrak{n}_1, \mathfrak{n}_2)$ and the single-valued neutrosophic hesitant fuzzy Euclidean distance $d_2(\mathfrak{n}_1, \mathfrak{n}_2)$.

Theorem 1. *Let $\mathfrak{n}_1 = \{\mathfrak{t}_1, \mathfrak{i}_1, \mathfrak{f}_1\}$ and $\mathfrak{n}_2 = \{\{1\}, \{0\}, \{0\}\}$ be two SVNHFEs, then the generalized distance $d(\mathfrak{n}_1, \mathfrak{n}_2')$ can be calculated as:*

$$d(\mathfrak{n}_1, \mathfrak{n}_2') = \left(\frac{1}{3}\left(\frac{1}{\#\mathfrak{t}_1}\sum_{\gamma\in\mathfrak{t}_1}(1-\gamma)^\alpha + \frac{1}{\#\mathfrak{i}_1}\sum_{\delta\in\mathfrak{i}_1}\delta^\alpha + \frac{1}{\#\mathfrak{f}_1}\sum_{\eta\in\mathfrak{f}_1}\eta^\alpha\right)\right)^{\frac{1}{\alpha}}$$

where $\mathfrak{n}_2'$ is the normalization outcome of $\mathfrak{n}_2$ by the comparison of $\mathfrak{n}_1$ and $\mathfrak{n}_2$.

Proof. Using (3), the generalized distance $d(\mathfrak{n}_1, \mathfrak{n}_2')$ can be calculated as:

$$\begin{aligned}
d(\mathfrak{n}_1, \mathfrak{n}_2') &= \left(\frac{1}{3}\left(\frac{1}{\#\mathfrak{t}}\sum_{\varsigma=1}^{\#\mathfrak{t}}\left|\gamma_1^{\sigma(\varsigma)} - \gamma_2^{\sigma(\varsigma)}\right|^{\alpha} + \frac{1}{\#\mathfrak{i}}\sum_{\varsigma=1}^{\#\mathfrak{i}}\left|\delta_1^{\sigma(\varsigma)} - \delta_2^{\sigma(\varsigma)}\right|^{\alpha} + \frac{1}{\#\mathfrak{f}}\sum_{\varsigma=1}^{\#\mathfrak{f}}\left|\eta_1^{\sigma(\varsigma)} - \eta_2^{\sigma(\varsigma)}\right|^{\alpha}\right)\right)^{\frac{1}{\alpha}} \\
&= \left(\frac{1}{3}\left(\frac{1}{\#\mathfrak{t}}\sum_{\varsigma=1}^{\#\mathfrak{t}}\left|\gamma_1^{\sigma(\varsigma)} - 1\right|^{\alpha} + \frac{1}{\#\mathfrak{i}}\sum_{\varsigma=1}^{\#\mathfrak{i}}\left|\delta_1^{\sigma(\varsigma)} - 0\right|^{\alpha} + \frac{1}{\#\mathfrak{f}}\sum_{\varsigma=1}^{\#\mathfrak{f}}\left|\eta_1^{\sigma(\varsigma)} - 0\right|^{\alpha}\right)\right)^{\frac{1}{\alpha}} \\
&= \left(\frac{1}{3}\left(\frac{1}{\#\mathfrak{t}}\sum_{\varsigma=1}^{\#\mathfrak{t}}\left(1 - \gamma_1^{\sigma(\varsigma)}\right)^{\alpha} + \frac{1}{\#\mathfrak{i}}\sum_{\varsigma=1}^{\#\mathfrak{i}}\left(\delta_1^{\sigma(\varsigma)}\right)^{\alpha} + \frac{1}{\#\mathfrak{f}}\sum_{\varsigma=1}^{\#\mathfrak{f}}\left(\eta_1^{\sigma(\varsigma)}\right)^{\alpha}\right)\right)^{\frac{1}{\alpha}} \\
&= \left(\frac{1}{3}\left(\frac{1}{\#\mathfrak{t}_1}\sum_{\varsigma=1}^{\#\mathfrak{t}_1}\left(1 - \gamma_1^{\sigma(\varsigma)}\right)^{\alpha} + \frac{1}{\#\mathfrak{i}_1}\sum_{\varsigma=1}^{\#\mathfrak{i}_1}\left(\delta_1^{\sigma(\varsigma)}\right)^{\alpha} + \frac{1}{\#\mathfrak{f}_1}\sum_{\varsigma=1}^{\#\mathfrak{f}_1}\left(\eta_1^{\sigma(\varsigma)}\right)^{\alpha}\right)\right)^{\frac{1}{\alpha}} \\
&= \left(\frac{1}{3}\left(\frac{1}{\#\mathfrak{t}_1}\sum_{\gamma\in\mathfrak{t}_1}(1-\gamma)^{\alpha} + \frac{1}{\#\mathfrak{i}_1}\sum_{\delta\in\mathfrak{i}_1}\delta^{\alpha} + \frac{1}{\#\mathfrak{f}_1}\sum_{\eta\in\mathfrak{f}_1}\eta^{\alpha}\right)\right)^{\frac{1}{\alpha}}.
\end{aligned}$$

□

Theorem 2. *Let $\mathfrak{n}_1 = \{\mathfrak{t}_1, \mathfrak{i}_1, \mathfrak{f}_1\}$ and $\mathfrak{n}_2 = \{\{0\}, \{1\}, \{1\}\}$ be two SVNHFEs, then the generalized distance $d(\mathfrak{n}_1, \mathfrak{n}_2')$ can be calculated as:*

$$d(\mathfrak{n}_1, \mathfrak{n}_2') = \left(\frac{1}{3}\left(\frac{1}{\#\mathfrak{t}_1}\sum_{\gamma\in\mathfrak{t}_1}\gamma^{\alpha} + \frac{1}{\#\mathfrak{i}_1}\sum_{\delta\in\mathfrak{i}_1}(1-\delta)^{\alpha} + \frac{1}{\#\mathfrak{f}_1}\sum_{\eta\in\mathfrak{f}_1}(1-\eta)^{\alpha}\right)\right)^{\frac{1}{\alpha}}.$$

where $\mathfrak{n}_2'$ is the normalization outcome of $\mathfrak{n}_2$ by the comparison of $\mathfrak{n}_1$ and $\mathfrak{n}_2$.

Proof. Using (3), the generalized distance $d(\mathfrak{n}_1, \mathfrak{n}_2')$ can be calculated as:

$$\begin{aligned}
d(\mathfrak{n}_1, \mathfrak{n}_2') &= \left(\frac{1}{3}\left(\frac{1}{\#\mathfrak{t}}\sum_{\varsigma=1}^{\#\mathfrak{t}}\left|\gamma_1^{\sigma(\varsigma)} - \gamma_2^{\sigma(\varsigma)}\right|^{\alpha} + \frac{1}{\#\mathfrak{i}}\sum_{\varsigma=1}^{\#\mathfrak{i}}\left|\delta_1^{\sigma(\varsigma)} - \delta_2^{\sigma(\varsigma)}\right|^{\alpha} + \frac{1}{\#\mathfrak{f}}\sum_{\varsigma=1}^{\#\mathfrak{f}}\left|\eta_1^{\sigma(\varsigma)} - \eta_2^{\sigma(\varsigma)}\right|^{\alpha}\right)\right)^{\frac{1}{\alpha}} \\
&= \left(\frac{1}{3}\left(\frac{1}{\#\mathfrak{t}}\sum_{\varsigma=1}^{\#\mathfrak{t}}\left|\gamma_1^{\sigma(\varsigma)} - 0\right|^{\alpha} + \frac{1}{\#\mathfrak{i}}\sum_{\varsigma=1}^{\#\mathfrak{i}}\left|\delta_1^{\sigma(\varsigma)} - 1\right|^{\alpha} + \frac{1}{\#\mathfrak{f}}\sum_{\varsigma=1}^{\#\mathfrak{f}}\left|\eta_1^{\sigma(\varsigma)} - 1\right|^{\alpha}\right)\right)^{\frac{1}{\alpha}} \\
&= \left(\frac{1}{3}\left(\frac{1}{\#\mathfrak{t}}\sum_{\varsigma=1}^{\#\mathfrak{t}}\left(\gamma_1^{\sigma(\varsigma)}\right)^{\alpha} + \frac{1}{\#\mathfrak{i}}\sum_{\varsigma=1}^{\#\mathfrak{i}}\left(1 - \delta_1^{\sigma(\varsigma)}\right)^{\alpha} + \frac{1}{\#\mathfrak{f}}\sum_{\varsigma=1}^{\#\mathfrak{f}}\left(1 - \eta_1^{\sigma(\varsigma)}\right)^{\alpha}\right)\right)^{\frac{1}{\alpha}} \\
&= \left(\frac{1}{3}\left(\frac{1}{\#\mathfrak{t}_1}\sum_{\varsigma=1}^{\#\mathfrak{t}_1}\left(\gamma_1^{\sigma(\varsigma)}\right)^{\alpha} + \frac{1}{\#\mathfrak{i}_1}\sum_{\varsigma=1}^{\#\mathfrak{i}_1}\left(1 - \delta_1^{\sigma(\varsigma)}\right)^{\alpha} + \frac{1}{\#\mathfrak{f}_1}\sum_{\varsigma=1}^{\#\mathfrak{f}_1}\left(1 - \eta_1^{\sigma(\varsigma)}\right)^{\alpha}\right)\right)^{\frac{1}{\alpha}} \\
&= \left(\frac{1}{3}\left(\frac{1}{\#\mathfrak{t}_1}\sum_{\gamma\in\mathfrak{t}_1}\gamma^{\alpha} + \frac{1}{\#\mathfrak{i}_1}\sum_{\delta\in\mathfrak{i}_1}(1-\delta)^{\alpha} + \frac{1}{\#\mathfrak{f}_1}\sum_{\eta\in\mathfrak{f}_1}(1-\eta)^{\alpha}\right)\right)^{\frac{1}{\alpha}}.
\end{aligned}$$

□

2.1.3. Computation of Optimal Weights Using Maximizing Deviation Method

Case I: Completely unknown attribute weight information

Construct an optimization model on the basis of the approach of maximizing deviation to determine the attributes optimal relative weights with SVNHFS. For the attribute $P_j \in Z$, the deviation of the alternative A_i to all the other alternatives can be represented as:

$$D_{ij}(w) = \sum_{k=1}^{m} d(\mathfrak{n}_{ij}, \mathfrak{n}_{kj})w_j,\ i = 1,2,\ldots,m,\ j = 1,2,\ldots,n$$

where $d(\mathfrak{n}_{ij}, \mathfrak{n}_{kj}) = \left(\frac{1}{3}\left(\frac{1}{\#\mathfrak{t}}\sum_{\varsigma=1}^{\#\mathfrak{t}}\left|\gamma_{ij}^{\sigma(\varsigma)} - \gamma_{kj}^{\sigma(\varsigma)}\right|^{\alpha} + \frac{1}{\#\mathfrak{i}}\sum_{\varsigma=1}^{\#\mathfrak{i}}\left|\delta_{ij}^{\sigma(\varsigma)} - \delta_{kj}^{\sigma(\varsigma)}\right|^{\alpha} + \frac{1}{\#\mathfrak{f}}\sum_{\varsigma=1}^{\#\mathfrak{f}}\left|\eta_{ij}^{\sigma(\varsigma)} - \eta_{kj}^{\sigma(\varsigma)}\right|^{\alpha}\right)\right)^{\frac{1}{\alpha}}$.

Let

$$D_j(w) = \sum_{i=1}^{m} D_{ij}(w) = \sum_{i=1}^{m}\sum_{k=1}^{m} w_j\left(\frac{1}{3}\left(\frac{1}{\#\mathfrak{t}}\sum_{\varsigma=1}^{\#\mathfrak{t}}\left|\gamma_{ij}^{\sigma(\varsigma)} - \gamma_{kj}^{\sigma(\varsigma)}\right|^{\alpha} + \frac{1}{\#\mathfrak{i}}\sum_{\varsigma=1}^{\#\mathfrak{i}}\left|\delta_{ij}^{\sigma(\varsigma)} - \delta_{kj}^{\sigma(\varsigma)}\right|^{\alpha} + \frac{1}{\#\mathfrak{f}}\sum_{\varsigma=1}^{\#\mathfrak{f}}\left|\eta_{ij}^{\sigma(\varsigma)} - \eta_{kj}^{\sigma(\varsigma)}\right|^{\alpha}\right)\right)^{\frac{1}{\alpha}},$$

$j = 1,2,\ldots,n$. Then $D_j(w)$ indicates the deviation value of all alternatives to other alternatives for the attribute $P_j \in Z$.

On the basis of the above analysis, to select the weight vector w which maximizes all deviation values for all the attributes, a non-linear programming model is constructed as follows:

$$(M-1)\begin{cases} \max D(w) = \sum_{j=1}^{n}\sum_{i=1}^{m}\sum_{k=1}^{m} w_j\left(\frac{1}{3}\left(\frac{1}{\#\mathfrak{t}}\sum_{\varsigma=1}^{\#\mathfrak{t}}\left|\gamma_{ij}^{\sigma(\varsigma)} - \gamma_{kj}^{\sigma(\varsigma)}\right|^{\alpha} + \frac{1}{\#\mathfrak{i}}\sum_{\varsigma=1}^{\#\mathfrak{i}}\left|\delta_{ij}^{\sigma(\varsigma)} - \delta_{kj}^{\sigma(\varsigma)}\right|^{\alpha} + \frac{1}{\#\mathfrak{f}}\sum_{\varsigma=1}^{\#\mathfrak{f}}\left|\eta_{ij}^{\sigma(\varsigma)} - \eta_{kj}^{\sigma(\varsigma)}\right|^{\alpha}\right)\right)^{\frac{1}{\alpha}} \\ s.t.\ w_j \geq 0, j = 1,2,\ldots,n,\ \sum_{j=1}^{n} w_j^2 = 1 \end{cases}$$

To solve the above model, we construct the Lagrange function:

$$L(w,\xi) = \sum_{j=1}^{n}\sum_{i=1}^{m}\sum_{k=1}^{m}\left(\frac{1}{3}\left(\frac{1}{\#\mathfrak{t}}\sum_{\varsigma=1}^{\#\mathfrak{t}}\left|\gamma_{ij}^{\sigma(\varsigma)} - \gamma_{kj}^{\sigma(\varsigma)}\right|^{\alpha} + \frac{1}{\#\mathfrak{i}}\sum_{\varsigma=1}^{\#\mathfrak{i}}\left|\delta_{ij}^{\sigma(\varsigma)} - \delta_{kj}^{\sigma(\varsigma)}\right|^{\alpha} + \frac{1}{\#\mathfrak{f}}\sum_{\varsigma=1}^{\#\mathfrak{f}}\left|\eta_{ij}^{\sigma(\varsigma)} - \eta_{kj}^{\sigma(\varsigma)}\right|^{\alpha}\right)\right)^{\frac{1}{\alpha}} w_j + \frac{\xi}{2}\left(\sum_{j=1}^{n} w_j^2 - 1\right)$$

where ξ is a real number, representing the Lagrange multiplier variable. Then we compute the partial derivatives of L and let:

$$\begin{aligned} \frac{\partial L}{\partial w_j} &= \sum_{i=1}^{m}\sum_{k=1}^{m}\left(\frac{1}{3}\left(\frac{1}{\#\mathfrak{t}}\sum_{\varsigma=1}^{\#\mathfrak{t}}\left|\gamma_{ij}^{\sigma(\varsigma)} - \gamma_{kj}^{\sigma(\varsigma)}\right|^{\alpha} + \frac{1}{\#\mathfrak{i}}\sum_{\varsigma=1}^{\#\mathfrak{i}}\left|\delta_{ij}^{\sigma(\varsigma)} - \delta_{kj}^{\sigma(\varsigma)}\right|^{\alpha} + \frac{1}{\#\mathfrak{f}}\sum_{\varsigma=1}^{\#\mathfrak{f}}\left|\eta_{ij}^{\sigma(\varsigma)} - \eta_{kj}^{\sigma(\varsigma)}\right|^{\alpha}\right)\right)^{\frac{1}{\alpha}} + \xi w_j = 0 \\ \frac{\partial L}{\partial \xi} &= \frac{1}{2}\left(\sum_{j=1}^{n} w_j^2 - 1\right) = 0 \end{aligned}$$

By solving above equations, an exact and simple formula for determining the attribute weights can be obtained as follows:

$$w_j^* = \frac{\sum_{i=1}^{m}\sum_{k=1}^{m}\left(\frac{1}{3}\left(\frac{1}{\#\mathfrak{t}}\sum_{\varsigma=1}^{\#\mathfrak{t}}\left|\gamma_{ij}^{\sigma(\varsigma)} - \gamma_{kj}^{\sigma(\varsigma)}\right|^{\alpha} + \frac{1}{\#\mathfrak{i}}\sum_{\varsigma=1}^{\#\mathfrak{i}}\left|\delta_{ij}^{\sigma(\varsigma)} - \delta_{kj}^{\sigma(\varsigma)}\right|^{\alpha} + \frac{1}{\#\mathfrak{f}}\sum_{\varsigma=1}^{\#\mathfrak{f}}\left|\eta_{ij}^{\sigma(\varsigma)} - \eta_{kj}^{\sigma(\varsigma)}\right|^{\alpha}\right)\right)^{\frac{1}{\alpha}}}{\sqrt{\sum_{j=1}^{n}\left[\sum_{i=1}^{m}\sum_{k=1}^{m}\left(\frac{1}{3}\left(\frac{1}{\#\mathfrak{t}}\sum_{\varsigma=1}^{\#\mathfrak{t}}\left|\gamma_{ij}^{\sigma(\varsigma)} - \gamma_{kj}^{\sigma(\varsigma)}\right|^{\alpha} + \frac{1}{\#\mathfrak{i}}\sum_{\varsigma=1}^{\#\mathfrak{i}}\left|\delta_{ij}^{\sigma(\varsigma)} - \delta_{kj}^{\sigma(\varsigma)}\right|^{\alpha} + \frac{1}{\#\mathfrak{f}}\sum_{\varsigma=1}^{\#\mathfrak{f}}\left|\eta_{ij}^{\sigma(\varsigma)} - \eta_{kj}^{\sigma(\varsigma)}\right|^{\alpha}\right)\right)^{\frac{1}{\alpha}}\right]^2}}$$

Because the weights of the attributes should satisfy the normalization condition, so we obtain the normalized attribute weights:

$$w_j = \frac{\sum_{i=1}^{m}\sum_{k=1}^{m}\left(\frac{1}{3}\left(\frac{1}{\#\mathfrak{t}}\sum_{\varsigma=1}^{\#\mathfrak{t}}\left|\gamma_{ij}^{\sigma(\varsigma)}-\gamma_{kj}^{\sigma(\varsigma)}\right|^{\alpha}+\frac{1}{\#\mathfrak{i}}\sum_{\varsigma=1}^{\#\mathfrak{i}}\left|\delta_{ij}^{\sigma(\varsigma)}-\delta_{kj}^{\sigma(\varsigma)}\right|^{\alpha}+\frac{1}{\#\mathfrak{f}}\sum_{\varsigma=1}^{\#\mathfrak{f}}\left|\eta_{ij}^{\sigma(\varsigma)}-\eta_{kj}^{\sigma(\varsigma)}\right|^{\alpha}\right)\right)^{\frac{1}{\alpha}}}{\sum_{j=1}^{n}\sum_{i=1}^{m}\sum_{k=1}^{m}\left(\frac{1}{3}\left(\frac{1}{\#\mathfrak{t}}\sum_{\varsigma=1}^{\#\mathfrak{t}}\left|\gamma_{ij}^{\sigma(\varsigma)}-\gamma_{kj}^{\sigma(\varsigma)}\right|^{\alpha}+\frac{1}{\#\mathfrak{i}}\sum_{\varsigma=1}^{\#\mathfrak{i}}\left|\delta_{ij}^{\sigma(\varsigma)}-\delta_{kj}^{\sigma(\varsigma)}\right|^{\alpha}+\frac{1}{\#\mathfrak{f}}\sum_{\varsigma=1}^{\#\mathfrak{f}}\left|\eta_{ij}^{\sigma(\varsigma)}-\eta_{kj}^{\sigma(\varsigma)}\right|^{\alpha}\right)\right)^{\frac{1}{\alpha}}} \quad (4)$$

Case II: Partly known attribute weight information

However, there are some situations that the information about the weight vector is partially known instead of completely known. For such situations, on the basis of the set of the known weight information, $\Im$, the constrained optimization model can be designed as:

$$(M-2)\begin{cases}\max D(w)=\sum_{j=1}^{n}\sum_{i=1}^{m}\sum_{k=1}^{m}w_j\left(\frac{1}{3}\left(\frac{1}{\#\mathfrak{t}}\sum_{\varsigma=1}^{\#\mathfrak{t}}\left|\gamma_{ij}^{\sigma(\varsigma)}-\gamma_{kj}^{\sigma(\varsigma)}\right|^{\alpha}+\frac{1}{\#\mathfrak{i}}\sum_{\varsigma=1}^{\#\mathfrak{i}}\left|\delta_{ij}^{\sigma(\varsigma)}-\delta_{kj}^{\sigma(\varsigma)}\right|^{\alpha}+\frac{1}{\#\mathfrak{f}}\sum_{\varsigma=1}^{\#\mathfrak{f}}\left|\eta_{ij}^{\sigma(\varsigma)}-\eta_{kj}^{\sigma(\varsigma)}\right|^{\alpha}\right)\right)^{\frac{1}{\alpha}}\\ s.t.\ w\in\Im,\ w_j\geq 0,\ j=1,2,\ldots,n,\ \sum_{j=1}^{n}w_j=1\end{cases}$$

where $\Im$ is also a set of constraint conditions that the weight value w_j should satisfy according to the requirements in real situations. The model $(M-2)$ is a linear programming model. By solving this model, we obtain the optimal solution $w=(w_1,w_2,\ldots,w_n)^t$, which can be used as the attributes weight vector.

2.1.4. TOPSIS Method

Recently, several MADM techniques are established such as TOPSIS [27], TODIM [28], VIKOR [29], MULTIMOORA [30] and minimum deviation method [31]. TOPSIS method is attractive as limited subjective input is required from experts. It is quite well known that TOPSIS is a useful and easy approach helping an expert choose the optimal alternative according to both the minimal distance from the positive-ideal solution and the maximal distance from the negative-ideal solution. Therefore, after attaining the weight of attributes by using the maximizing deviation method, in this section, we develop a MADM approach based on TOPSIS model under single-valued neutrosophic hesitant fuzzy circumstances. The PIS A^+, and the NIS A^- can be computed as:

$$A^+ = \{\mathfrak{n}_1^+,\mathfrak{n}_2^+,\ldots,\mathfrak{n}_n^+\} \quad (5)$$
$$= \{\{\{1\},\{0\},\{0\}\},\{\{1\},\{0\},\{0\}\},\ldots,\{\{1\},\{0\},\{0\}\}\}. \quad (6)$$

$$A^- = \{\mathfrak{n}_1^-,\mathfrak{n}_2^-,\ldots,\mathfrak{n}_n^-\} \quad (7)$$
$$= \{\{\{0\},\{1\},\{1\}\},\{\{0\},\{1\},\{1\}\},\ldots,\{\{0\},\{1\},\{1\}\}\}. \quad (8)$$

Based on Equation (3), Theorems 1 and 2, the separation measures d_i^+ and d_i^- of each alternative from the single-valued neutrosophic hesitant fuzzy PIS A^+ and the NIS A^-, respectively, are determined as:

$$d_i^+ = \sum_{j=1}^{n} d(\mathfrak{n}_{ij}^{'},\mathfrak{n}_j^+)w_j = \sum_{j=1}^{n} d(\mathfrak{n}_{ij}^{'},\{\{1\},\{0\},\{0\}\})w_j \quad (9)$$

$$= \sum_{j=1}^{n} w_j\left(\frac{1}{3}\left(\frac{1}{\#\mathfrak{t}_{ij}^{'}}\sum_{\gamma\in\mathfrak{t}_{ij}^{'}}(1-\gamma)^{\alpha}+\frac{1}{\#\mathfrak{i}_{ij}^{'}}\sum_{\delta\in\mathfrak{i}_{ij}^{'}}\delta^{\alpha}+\frac{1}{\#\mathfrak{f}_{ij}^{'}}\sum_{\eta\in\mathfrak{f}_{ij}^{'}}\eta^{\alpha}\right)\right)^{\frac{1}{\alpha}}, \quad (10)$$

$$d_i^- = \sum_{j=1}^{n} d(\mathfrak{n}_{ij}', \mathfrak{n}_j^-) w_j = \sum_{j=1}^{n} d(\mathfrak{n}_{ij}', \{\{0\}, \{1\}, \{1\}\}) w_j \tag{11}$$

$$= \sum_{j=1}^{n} w_j \left(\frac{1}{3} \left(\frac{1}{\#\mathfrak{t}_{ij}'} \sum_{\gamma \in \mathfrak{t}_{ij}'} \gamma^\alpha + \frac{1}{\#\mathfrak{i}_{ij}'} \sum_{\delta \in \mathfrak{i}_{ij}'} (1-\delta)^\alpha + \frac{1}{\#\mathfrak{f}_{ij}'} \sum_{\eta \in \mathfrak{f}_{ij}'} (1-\eta)^\alpha \right) \right)^{\frac{1}{\alpha}}, \tag{12}$$

where $i = 1, 2, \ldots, m$.

The relative closeness coefficient of an alternative A_i with respect to the single-valued neutrosophic hesitant fuzzy PIS A^+ can be defined as follows:

$$RC(A_i) = \frac{d_i^-}{d_i^+ + d_i^-} \tag{13}$$

where $0 \le RC(A_i) \le 1$, $i = 1, 2, \ldots, m$. The ranking orders of all alternatives can be determined according to the closeness coefficient $CR(A_i)$ and select the best one(s) from a set of appropriate alternatives.

The scheme of the proposed MADM technique is given in Figure 1. The detailed algorithm is constructed as follows:

Step 1. Construct the decision matrix $\mathcal{N} = [\mathfrak{n}_{ij}]_{m \times n}$ for the MADM problem, where the entries $\mathfrak{n}_{ij} (i = 1, 2, \ldots, m; j = 1, 2, \ldots, n)$ are SVNHFEs, given by the decision makers, for the alternative A_i according to the attribute P_j.

Step 2. On the basis of Equation (4) determine the attribute weights $w = (w_1, w_2, \ldots, w_m)^t$, if the attribute weights information is completely unknown, and turn to Step 4. Otherwise go to Step 3.

Step 3. Use model (M-2) to determine the attribute weights $w = (w_1, w_2, \ldots, w_m)^t$, if the information about the attribute weights is partially known.

Step 4. Based on Equations (6) and (8), we determine the corresponding single-valued neutrosophic hesitant fuzzy PIS A^+ and the single-valued neutrosophic hesitant fuzzy NIS A^-, respectively.

Step 5. Based on Equations (10) and (12), we compute the separation measures d_i^+ and d_i^- of each alternative A_i from the single-valued neutrosophic hesitant fuzzy PIS A^+ and the single-valued neutrosophic hesitant fuzzy NIS A^-, respectively.

Step 6. Based on Equation (13), we determine the relative closeness coefficient $RC(A_i)$ $(i = 1, 2, \ldots, m)$ of each alternative A_i to the single-valued neutrosophic hesitant fuzzy PIS A^+.

Step 7. Rank the alternatives A_i $(i = 1, 2, \ldots, m)$ based on the relative closeness coefficients $RC(A_i)$ $(i = 1, 2, \ldots, m)$ and select the optimal one(s).

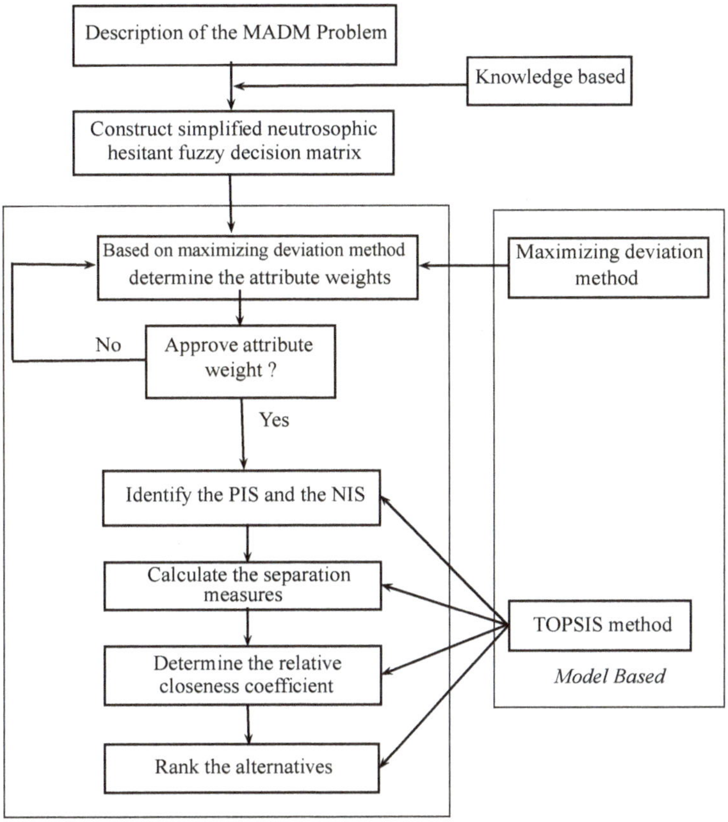

Figure 1. The scheme of the developed approach for MADM.

2.2. *TOPSIS and Maximizing Deviation Method for Interval Neutrosophic Hesitant Fuzzy Multi-Attribute Decision-Making*

In this subsection, we only use INHFSs in SNHFSs and put forward a novel decision-making approach, by combining the idea of INHFSs with maximizing deviation, to solve a MADM problem in interval neutrosophic hesitant fuzzy environment.

Definition 4 ([20]). *Let Z be a fixed set, an INHFS* $\tilde{\mathfrak{n}}$ *on Z is defined as:*

$$\tilde{\mathfrak{n}} = \{\langle z, \tilde{\mathfrak{t}}(z), \tilde{\mathfrak{i}}(z), \tilde{\mathfrak{f}}(z)\rangle | z \in Z\}$$

where $\tilde{\mathfrak{t}}(z), \tilde{\mathfrak{i}}(z), \tilde{\mathfrak{f}}(z)$ *are sets of some interval-values in* $[0,1]$, *indicating the possible truth-membership hesitant degree, indeterminacy-membership hesitant degree and falsity-membership hesitant degree of the element z to* $\tilde{\mathfrak{n}}$, *respectively;* $\tilde{\mathfrak{t}}(z) = \{\tilde{\gamma}_1, \tilde{\gamma}_2, \ldots, \tilde{\gamma}_l\}$, $\tilde{\gamma}_1, \tilde{\gamma}_2, \ldots, \tilde{\gamma}_l$ *are the elements of* $\tilde{\mathfrak{t}}(z)$; $\tilde{\mathfrak{i}}(z) = \{\tilde{\delta}_1, \tilde{\delta}_2, \ldots, \tilde{\delta}_p\}$, $\tilde{\delta}_1, \tilde{\delta}_2, \ldots, \tilde{\delta}_p$ *are the elements of* $\tilde{\mathfrak{i}}(z)$; $\tilde{\mathfrak{f}}(z) = \{\tilde{\eta}_1, \tilde{\eta}_2, \ldots, \tilde{\eta}_q\}$, $\tilde{\eta}_1, \tilde{\eta}_2, \ldots, \tilde{\eta}_q$ *are the elements of* $\tilde{\mathfrak{f}}(z)$, *for every* $z \in Z$; *and* l, p, q *denote, respectively, the numbers of the interval-valued hesitant fuzzy elements in* $\tilde{\mathfrak{t}}, \tilde{\mathfrak{i}}, \tilde{\mathfrak{f}}$.

For convenience, the expression $\tilde{\mathfrak{n}}(z) = \{\tilde{\mathfrak{t}}(z), \tilde{\mathfrak{i}}(z), \tilde{\mathfrak{f}}(z)\}$ is called an interval neutrosophic hesitant fuzzy element (INHFE), which we represent by simplified symbol $\tilde{\mathfrak{n}} = \{\tilde{\mathfrak{t}}, \tilde{\mathfrak{i}}, \tilde{\mathfrak{f}}\}$.

Similar to Section 2.1, we consider a MADM problem, where $A = \{A_1, A_2, \ldots, A_m\}$ is a discrete set of m alternatives and $P = \{P_1, P_2, \ldots, P_n\}$ is a set of n attributes. The evaluation information of the ith alternative with respect to the jth attribute is an INHFE $\tilde{\mathfrak{n}}_{ij} = \langle \tilde{\mathfrak{t}}_{ij}, \tilde{\mathfrak{i}}_{ij}, \tilde{\mathfrak{f}}_{ij} \rangle$, where $\tilde{\mathfrak{t}}_{ij}$, $\tilde{\mathfrak{i}}_{ij}$ and $\tilde{\mathfrak{f}}_{ij}$ indicate the interval-valued preference degree, interval-valued uncertain degree, and interval-valued falsity degree, respectively, of the expert facing the ith alternative that satisfied the jth attribute. Then the interval neutrosophic hesitant fuzzy decision matrix (INHFDM) $\tilde{\mathcal{N}}$, can be constructed as follows:

$$\tilde{\mathcal{N}} = \begin{bmatrix} \tilde{\mathfrak{n}}_{11} & \tilde{\mathfrak{n}}_{12} & \ldots & \tilde{\mathfrak{n}}_{1n} \\ \tilde{\mathfrak{n}}_{21} & \tilde{\mathfrak{n}}_{22} & \ldots & \tilde{\mathfrak{n}}_{2n} \\ \vdots & \vdots & \ddots & \vdots \\ \tilde{\mathfrak{n}}_{m1} & \tilde{\mathfrak{n}}_{m2} & \ldots & \tilde{\mathfrak{n}}_{mn} \end{bmatrix}$$

In the comparison of INHFEs, the number of their corresponding element may be unequal. To handle this situation, we normalize the INHFEs as follows:

Suppose that $\tilde{\mathfrak{n}} = \{\tilde{\mathfrak{t}}, \tilde{\mathfrak{i}}, \tilde{\mathfrak{f}}\}$ is an INHFE, then $\bar{\tilde{\gamma}} = \varpi\tilde{\gamma}^+ + (1-\varpi)\tilde{\gamma}^-$, $\bar{\tilde{\delta}} = \varpi\tilde{\delta}^+ + (1-\varpi)\tilde{\delta}^-$ and $\bar{\tilde{\eta}} = \varpi\tilde{\eta}^+ + (1-\varpi)\tilde{\eta}^-$ are the added truth-membership, the indeterminacy-membership and the falsity-membership degree, respectively, where $\tilde{\gamma}^-$, $\tilde{\gamma}^+$, $\tilde{\delta}^-$, $\tilde{\delta}^+$ and $\tilde{\eta}^-$, $\tilde{\eta}^+$ are the minimum and the maximum elements of $\tilde{\mathfrak{t}}$, $\tilde{\mathfrak{i}}$ and $\tilde{\mathfrak{f}}$, respectively, and $\varpi \in [0,1]$ is a parameter assigned by the expert according to his risk preference.

For the normalization of INHFE, different values of ϖ produce different results for the added truth-membership, the indeterminacy-membership and the falsity-membership degree. Usually, there are three cases of the preference of the expert:

- If $\varpi = 0$, the pessimist expert may add the minimum truth-membership degree $\tilde{\gamma}^-$, the minimum indeterminacy-membership degree $\tilde{\delta}^-$ and the minimum falsity-membership degree $\tilde{\eta}^-$.
- If $\varpi = 0.5$, the neutral expert may add the truth-membership degree $\frac{\tilde{\gamma}^- + \tilde{\gamma}^+}{2}$, the indeterminacy-membership degree $\frac{\tilde{\delta}^- + \tilde{\delta}^+}{2}$ and the falsity-membership degree $\frac{\tilde{\eta}^- + \tilde{\eta}^+}{2}$.
- If $\varpi = 1$, the optimistic expert may add the maximum truth-membership degree $\tilde{\gamma}^+$, the maximum indeterminacy-membership degree $\tilde{\delta}^+$ and the maximum falsity-membership degree $\tilde{\eta}^+$.

The algorithm for the normalization of INHFEs is given in Algorithm 2.

Algorithm 2 The algorithm for the normalization of INHFEs.

INPUT: Two INHFEs $\tilde{n}_1 = (\tilde{t}_1, \tilde{i}_1, \tilde{f}_1)$ and $\tilde{n}_2 = (\tilde{t}_2, \tilde{i}_2, \tilde{f}_2)$ and the value of $\tilde{\omega}$.

OUTPUT: The normalization of $\tilde{n}_1 = (\tilde{t}_1, \tilde{i}_1, \tilde{f}_1)$ and $\tilde{n}_2 = (\tilde{t}_2, \tilde{i}_2, \tilde{f}_2)$.

1: Count the number of elements of $\tilde{n}_1$ and $\tilde{n}_2$, i.e., $\#\tilde{t}_1, \#\tilde{i}_1, \#\tilde{f}_1, \#\tilde{t}_2, \#\tilde{i}_2, \#\tilde{f}_2$;
2: Determine the minimum and the maximum of the elements of $\tilde{n}_1$ and $\tilde{n}_2$;
3: $\tilde{t} = \arg\min_{i=1,2} \#\tilde{t}_i$, $\tilde{i} = \arg\min_{i=1,2} \#\tilde{i}_i$, $\tilde{f} = \arg\min_{i=1,2} \#\tilde{f}_i$
4: **if** $\#\tilde{t}_1 = \#\tilde{t}_2$ **then break**;
5: **else if** $\tilde{t} = \#\tilde{t}_1$ **then**
6: $\quad n = \#\tilde{t}_2 - \#\tilde{t}_1$;
7: $\quad$ Determine the value of $\tilde{\gamma}$ for $\tilde{t}_1$;
8: $\quad$ **for** i=1:1:n **do**
9: $\qquad \tilde{t}_1 = \tilde{t}_1 \cup \tilde{\gamma}$;
10: $\quad$ **end for**
11: **else**
12: $\quad n = \#\tilde{t}_1 - \#\tilde{t}_2$;
13: $\quad$ Determine the value of $\tilde{\gamma}$ for $\tilde{t}_2$;
14: $\quad$ **for** i=1:1:n **do**
15: $\qquad \tilde{t}_2 = \tilde{t}_2 \cup \tilde{\gamma}$;
16: $\quad$ **end for**
17: **end if**
18: **if** $\#\tilde{i}_1 = \#\tilde{i}_2$ **then break**;
19: **else if** $\tilde{i} = \#\tilde{i}_1$ **then**
20: $\quad n = \#\tilde{i}_2 - \#\tilde{i}_1$;
21: $\quad$ Determine the value of $\tilde{\delta}$ for $\tilde{i}_1$;
22: $\quad$ **for** i=1:1:n **do**
23: $\qquad \tilde{i}_1 = \tilde{i}_1 \cup \tilde{\delta}$;
24: $\quad$ **end for**
25: **else**
26: $\quad n = \#\tilde{i}_1 - \#\tilde{i}_2$;
27: $\quad$ Determine the value of $\tilde{\delta}$ for $\tilde{i}_2$;
28: $\quad$ **for** i=1:1:n **do**
29: $\qquad \tilde{i}_2 = \tilde{i}_2 \cup \tilde{\delta}$;
30: $\quad$ **end for**
31: **end if**
32: **if** $\#\tilde{f}_1 = \#\tilde{f}_2$ **then break**;
33: **else if** $\tilde{f} = \#\tilde{f}_1$ **then**
34: $\quad n = \#\tilde{f}_2 - \#\tilde{f}_1$;
35: $\quad$ Determine the value of $\tilde{\eta}$ for $\tilde{f}_1$;
36: $\quad$ **for** i=1:1:n **do**
37: $\qquad \tilde{f}_1 = \tilde{f}_1 \cup \tilde{\eta}$;
38: $\quad$ **end for**
39: **else**
40: $\quad n = \#\tilde{f}_1 - \#\tilde{f}_2$;
41: $\quad$ Determine the value of $\tilde{\eta}$ for $\tilde{f}_2$;
42: $\quad$ **for** i=1:1:n **do**
43: $\qquad \tilde{f}_2 = \tilde{f}_2 \cup \tilde{\eta}$;
44: $\quad$ **end for**
45: **end if**

2.2.1. The Distance Measures for INHFSs

Definition 5. *Let $\tilde{n}_1 = \{\tilde{t}_1, \tilde{i}_1, \tilde{f}_1\}$ and $\tilde{n}_2 = \{\tilde{t}_2, \tilde{i}_2, \tilde{f}_2\}$ be two normalized INHFEs, then we define the interval neutrosophic hesitant fuzzy Hamming distance between $\tilde{n}_1$ and $\tilde{n}_2$ as follows:*

$$
\begin{aligned}
\tilde{d}_1(\tilde{n}_1, \tilde{n}_2) &= \frac{1}{6}\left(\frac{1}{\#\tilde{t}} \sum_{\varsigma=1}^{\#\tilde{t}} \left(\left| \tilde{\gamma}_1^{\sigma(\varsigma)^L} - \tilde{\gamma}_2^{\sigma(\varsigma)^L} \right| + \left| \tilde{\gamma}_1^{\sigma(\varsigma)^U} - \tilde{\gamma}_2^{\sigma(\varsigma)^U} \right| \right) + \frac{1}{\#\tilde{i}} \sum_{\varsigma=1}^{\#\tilde{i}} \left(\left| \tilde{\delta}_1^{\sigma(\varsigma)^L} - \tilde{\delta}_2^{\sigma(\varsigma)^L} \right| \right. \right. \\
&\quad \left. + \left| \tilde{\delta}_1^{\sigma(\varsigma)^U} - \tilde{\delta}_2^{\sigma(\varsigma)^U} \right| \right) + \frac{1}{\#\tilde{f}} \sum_{\varsigma=1}^{\#\tilde{f}} \left(\left| \tilde{\eta}_1^{\sigma(\varsigma)^L} - \tilde{\eta}_2^{\sigma(\varsigma)^L} \right| + \left| \tilde{\eta}_1^{\sigma(\varsigma)^U} - \tilde{\eta}_2^{\sigma(\varsigma)^U} \right| \right) \bigg),
\end{aligned}
$$

where $\#\tilde{\mathfrak{t}} = \#\tilde{\mathfrak{t}}_1 = \#\tilde{\mathfrak{t}}_2$, $\#\tilde{\mathfrak{i}} = \#\tilde{\mathfrak{i}}_1 = \#\tilde{\mathfrak{i}}_2$ *and* $\#\tilde{\mathfrak{f}} = \#\tilde{\mathfrak{f}}_1 = \#\tilde{\mathfrak{f}}_2$. $\tilde{\gamma}_i^{\sigma(\varsigma)}$, $\tilde{\delta}_i^{\sigma(\varsigma)}$ *and* $\eta_i^{\sigma(\varsigma)}$ *are the* ς*th largest values in* $\tilde{\gamma}_i$, $\tilde{\delta}_i$ *and* $\tilde{\eta}_i$, *respectively* $(i = 1, 2)$.

In addition, the interval neutrosophic hesitant fuzzy Euclidean distance is defined as:

$$\tilde{d}_2(\tilde{n}_1,\tilde{n}_2) = \left(\frac{1}{6}\left(\frac{1}{\#\tilde{\mathfrak{t}}}\sum_{\varsigma=1}^{\#\tilde{\mathfrak{t}}}\left(\left(\tilde{\gamma}_1^{\sigma(\varsigma)^L}-\tilde{\gamma}_2^{\sigma(\varsigma)^L}\right)^2+\left(\tilde{\gamma}_1^{\sigma(\varsigma)^U}-\tilde{\gamma}_2^{\sigma(\varsigma)^U}\right)^2\right)+\frac{1}{\#\tilde{\mathfrak{i}}}\sum_{\varsigma=1}^{\#\tilde{\mathfrak{i}}}\left(\left(\tilde{\delta}_1^{\sigma(\varsigma)^L}-\tilde{\delta}_2^{\sigma(\varsigma)^L}\right)^2\right.\right.\right.$$

$$\left.\left.\left.+\left(\tilde{\delta}_1^{\sigma(\varsigma)^U}-\tilde{\delta}_2^{\sigma(\varsigma)^U}\right)^2\right)+\frac{1}{\#\tilde{\mathfrak{f}}}\sum_{\varsigma=1}^{\#\tilde{\mathfrak{f}}}\left(\left(\tilde{\eta}_1^{\sigma(\varsigma)^L}-\tilde{\eta}_2^{\sigma(\varsigma)^L}\right)^2+\left(\tilde{\eta}_1^{\sigma(\varsigma)^U}-\tilde{\eta}_2^{\sigma(\varsigma)^U}\right)^2\right)\right)\right)^{\frac{1}{2}}.$$

By using the geometric distance model of [26], the above distances can be generalized as follows:

$$\tilde{d}(\tilde{n}_1,\tilde{n}_2) = \left(\frac{1}{6}\left(\frac{1}{\#\tilde{\mathfrak{t}}}\sum_{\varsigma=1}^{\#\tilde{\mathfrak{t}}}\left(\left(\tilde{\gamma}_1^{\sigma(\varsigma)^L}-\tilde{\gamma}_2^{\sigma(\varsigma)^L}\right)^\alpha+\left(\tilde{\gamma}_1^{\sigma(\varsigma)^U}-\tilde{\gamma}_2^{\sigma(\varsigma)^U}\right)^\alpha\right)+\frac{1}{\#\tilde{\mathfrak{i}}}\sum_{\varsigma=1}^{\#\tilde{\mathfrak{i}}}\left(\left(\tilde{\delta}_1^{\sigma(\varsigma)^L}-\tilde{\delta}_2^{\sigma(\varsigma)^L}\right)^\alpha\right.\right.\right.$$

$$\left.\left.\left.+\left(\tilde{\delta}_1^{\sigma(\varsigma)^U}-\tilde{\delta}_2^{\sigma(\varsigma)^U}\right)^\alpha\right)+\frac{1}{\#\tilde{\mathfrak{f}}}\sum_{\varsigma=1}^{\#\tilde{\mathfrak{f}}}\left(\left(\tilde{\eta}_1^{\sigma(\varsigma)^L}-\tilde{\eta}_2^{\sigma(\varsigma)^L}\right)^\alpha+\left(\tilde{\eta}_1^{\sigma(\varsigma)^U}-\tilde{\eta}_2^{\sigma(\varsigma)^U}\right)^\alpha\right)\right)\right)^{\frac{1}{\alpha}},$$

where α *is constant and* $\alpha > 0$. *Based on the value of* α, *the relationship among* $\tilde{d}(\tilde{n}_1,\tilde{n}_2), \tilde{d}_1(\tilde{n}_1,\tilde{n}_2)$ *and* $\tilde{d}_2(\tilde{n}_1,\tilde{n}_2)$ *can be deduced as:*

- *If* $\alpha = 1$, *then the distance* $\tilde{d}(\tilde{n}_1,\tilde{n}_2) = \tilde{d}_1(\tilde{n}_1,\tilde{n}_2)$.
- *If* $\alpha = 2$, *then the distance* $\tilde{d}(\tilde{n}_1,\tilde{n}_2) = \tilde{d}_2(\tilde{n}_1,\tilde{n}_2)$.

Therefore, the distance $\tilde{d}(\tilde{n}_1,\tilde{n}_2)$ *is a generalization of the interval neutrosophic hesitant fuzzy Hamming distance* $\tilde{d}_1(\tilde{n}_1,\tilde{n}_2)$ *and the interval neutrosophic hesitant fuzzy Euclidean distance* $\tilde{d}_2(\tilde{n}_1,\tilde{n}_2)$.

Theorem 3. *Let* $\tilde{n}_1 = \{\tilde{\mathfrak{t}}_1, \tilde{\mathfrak{i}}_1, \tilde{\mathfrak{f}}_1\}$ *and* $\tilde{n}_2 = \{\{[1,1]\},\{[0,0]\},\{[0,0]\}\}$ *be two INHFEs, then the generalized distance* $\tilde{d}(\tilde{n}_1,\tilde{n}_2')$ *can be calculated as:*

$$\tilde{d}(\tilde{n}_1,\tilde{n}_2') = \left(\frac{1}{6}\left(\frac{1}{\#\tilde{\mathfrak{t}}_1}\sum_{\tilde{\gamma}\in\tilde{\mathfrak{t}}_1}\left(\left(1-\tilde{\gamma}^L\right)^\alpha+\left(1-\tilde{\gamma}^U\right)^\alpha\right)+\frac{1}{\#\tilde{\mathfrak{i}}_1}\sum_{\tilde{\delta}\in\tilde{\mathfrak{i}}_1}((\tilde{\delta}^L)^\alpha+(\tilde{\delta}^U)^\alpha)+\frac{1}{\#\tilde{\mathfrak{f}}_1}\sum_{\tilde{\eta}\in\tilde{\mathfrak{f}}_1}((\tilde{\eta}^L)^\alpha+(\tilde{\eta}^U)^\alpha)\right)\right)^{\frac{1}{\alpha}}.$$

where $\tilde{n}_2'$ *is the normalization outcome of* $\tilde{n}_2$ *by the comparison of* $\tilde{n}_1$ *and* $\tilde{n}_2$.

Theorem 4. *Let* $\tilde{n}_1 = \{\tilde{\mathfrak{t}}_1, \tilde{\mathfrak{i}}_1, \tilde{\mathfrak{f}}_1\}$ *and* $\tilde{n}_2 = \{\{[0,0]\},\{[1,1]\},\{[1,1]\}\}$ *be two INHFEs, then the generalized distance* $\tilde{d}(\tilde{n}_1,\tilde{n}_2')$ *can be calculated as:*

$$\tilde{d}(\tilde{n}_1,\tilde{n}_2') = \left(\frac{1}{6}\left(\frac{1}{\#\tilde{\mathfrak{t}}_1}\sum_{\tilde{\gamma}\in\tilde{\mathfrak{t}}_1}((\tilde{\gamma}^L)^\alpha+(\tilde{\gamma}^U)^\alpha)+\frac{1}{\#\tilde{\mathfrak{i}}_1}\sum_{\tilde{\delta}\in\tilde{\mathfrak{i}}_1}\left(\left(1-\tilde{\delta}^L\right)^\alpha+\left(1-\tilde{\delta}^U\right)^\alpha\right)+\frac{1}{\#\tilde{\mathfrak{f}}_1}\sum_{\tilde{\eta}\in\tilde{\mathfrak{f}}_1}\left(\left(1-\tilde{\eta}^L\right)^\alpha+\left(1-\tilde{\eta}^U\right)^\alpha\right)\right)\right)^{\frac{1}{\alpha}}.$$

where $\tilde{n}_2'$ *is the normalization outcome of* $\tilde{n}_2$ *by the comparison of* $\tilde{n}_1$ *and* $\tilde{n}_2$.

2.2.2. Computation of Optimal Weights Using Maximizing Deviation Method

Case I: Completely unknown information on attribute weights

Using the maximizing deviation method, we construct an optimization model to determine the attributes optimal relative weights in interval neutrosophic hesitant fuzzy setting. For the attribute $P_j \in Z$, the deviation of the alternative A_i to all the other alternatives can be represented as:

$$\tilde{D}_{ij}(w) = \sum_{k=1}^{m} \tilde{d}(\tilde{n}_{ij}, \tilde{n}_{kj})w_j,\ i = 1,2,\ldots,m,\ j = 1,2,\ldots,n$$

where

$$\tilde{d}(\tilde{\mathfrak{n}}_{ij},\tilde{\mathfrak{n}}_{kj}) = \left(\frac{1}{6}\left(\frac{1}{\#\tilde{\mathfrak{t}}}\sum_{\varsigma=1}^{\#\tilde{\mathfrak{t}}}\left(\left|\tilde{\gamma}_{ij}^{\tilde{\sigma}(\varsigma)^L}-\tilde{\gamma}_{kj}^{\tilde{\sigma}(\varsigma)^L}\right|^{\alpha}+\left|\tilde{\gamma}_{ij}^{\tilde{\sigma}(\varsigma)^U}-\tilde{\gamma}_{kj}^{\tilde{\sigma}(\varsigma)^U}\right|^{\alpha}\right)+\frac{1}{\#\tilde{\mathfrak{i}}}\sum_{\varsigma=1}^{\#\tilde{\mathfrak{i}}}\left(\left|\tilde{\delta}_{ij}^{\tilde{\sigma}(\varsigma)^L}-\tilde{\delta}_{kj}^{\tilde{\sigma}(\varsigma)^L}\right|^{\alpha}+\left|\tilde{\delta}_{ij}^{\tilde{\sigma}(\varsigma)^U}-\tilde{\delta}_{kj}^{\tilde{\sigma}(\varsigma)^U}\right|^{\alpha}\right)+\frac{1}{\#\tilde{\mathfrak{f}}}\sum_{\varsigma=1}^{\#\tilde{\mathfrak{f}}}\left(\left|\tilde{\eta}_{ij}^{\tilde{\sigma}(\varsigma)^L}-\tilde{\eta}_{kj}^{\tilde{\sigma}(\varsigma)^L}\right|^{\alpha}+\left|\tilde{\eta}_{ij}^{\tilde{\sigma}(\varsigma)^U}-\tilde{\eta}_{kj}^{\tilde{\sigma}(\varsigma)^U}\right|^{\alpha}\right)\right)\right)^{\frac{1}{\alpha}}.$$

Let

$$\tilde{D}_j(w)=\sum_{i=1}^{m}\tilde{D}_{ij}(w) = \sum_{i=1}^{m}\sum_{k=1}^{m}w_j\left(\frac{1}{6}\left(\frac{1}{\#\tilde{\mathfrak{t}}}\sum_{\varsigma=1}^{\#\tilde{\mathfrak{t}}}\left(\left|\tilde{\gamma}_{ij}^{\tilde{\sigma}(\varsigma)^L}-\tilde{\gamma}_{kj}^{\tilde{\sigma}(\varsigma)^L}\right|^{\alpha}+\left|\tilde{\gamma}_{ij}^{\tilde{\sigma}(\varsigma)^U}-\tilde{\gamma}_{kj}^{\tilde{\sigma}(\varsigma)^U}\right|^{\alpha}\right)+\frac{1}{\#\tilde{\mathfrak{i}}}\sum_{\varsigma=1}^{\#\tilde{\mathfrak{i}}}\left(\left|\tilde{\delta}_{ij}^{\tilde{\sigma}(\varsigma)^L}-\tilde{\delta}_{kj}^{\tilde{\sigma}(\varsigma)^L}\right|^{\alpha}+\left|\tilde{\delta}_{ij}^{\tilde{\sigma}(\varsigma)^U}-\tilde{\delta}_{kj}^{\tilde{\sigma}(\varsigma)^U}\right|^{\alpha}\right)+\frac{1}{\#\tilde{\mathfrak{f}}}\sum_{\varsigma=1}^{\#\tilde{\mathfrak{f}}}\left(\left|\tilde{\eta}_{ij}^{\tilde{\sigma}(\varsigma)^L}-\tilde{\eta}_{kj}^{\tilde{\sigma}(\varsigma)^L}\right|^{\alpha}+\left|\tilde{\eta}_{ij}^{\tilde{\sigma}(\varsigma)^U}-\tilde{\eta}_{kj}^{\tilde{\sigma}(\varsigma)^U}\right|^{\alpha}\right)\right)\right)^{\frac{1}{\alpha}},$$

$j=1,2,\ldots,n$. Then $D_j(w)$ represents the deviation value of all alternatives to other alternatives for the attribute $P_j \in Z$.

On the basis of the analysis above, to select the weight vector w which maximizes all deviation values for all the attributes, a non-linear programming model is constructed as follows:

$$(M-3)\begin{cases}\max \tilde{D}(w)=\sum_{j=1}^{n}\sum_{i=1}^{m}\sum_{k=1}^{m}w_j\left(\frac{1}{6}\left(\frac{1}{\#\tilde{\mathfrak{t}}}\sum_{\varsigma=1}^{\#\tilde{\mathfrak{t}}}\left(\left|\tilde{\gamma}_{ij}^{\tilde{\sigma}(\varsigma)^L}-\tilde{\gamma}_{kj}^{\tilde{\sigma}(\varsigma)^L}\right|^{\alpha}+\left|\tilde{\gamma}_{ij}^{\tilde{\sigma}(\varsigma)^U}-\tilde{\gamma}_{kj}^{\tilde{\sigma}(\varsigma)^U}\right|^{\alpha}\right)+\frac{1}{\#\tilde{\mathfrak{i}}}\sum_{\varsigma=1}^{\#\tilde{\mathfrak{i}}}\left(\left|\tilde{\delta}_{ij}^{\tilde{\sigma}(\varsigma)^L}-\tilde{\delta}_{kj}^{\tilde{\sigma}(\varsigma)^L}\right|^{\alpha}+\left|\tilde{\delta}_{ij}^{\tilde{\sigma}(\varsigma)^U}-\tilde{\delta}_{kj}^{\tilde{\sigma}(\varsigma)^U}\right|^{\alpha}\right)+\frac{1}{\#\tilde{\mathfrak{f}}}\sum_{\varsigma=1}^{\#\tilde{\mathfrak{f}}}\left(\left|\tilde{\eta}_{ij}^{\tilde{\sigma}(\varsigma)^L}-\tilde{\eta}_{kj}^{\tilde{\sigma}(\varsigma)^L}\right|^{\alpha}+\left|\tilde{\eta}_{ij}^{\tilde{\sigma}(\varsigma)^U}-\tilde{\eta}_{kj}^{\tilde{\sigma}(\varsigma)^U}\right|^{\alpha}\right)\right)\right)^{\frac{1}{\alpha}}\\ s.t.\ w_j\geq 0, j=1,2,\ldots,n,\ \sum_{j=1}^{n}w_j^2=1\end{cases}$$

To solve the above model, we construct the Lagrange function:

$$L(w,\xi) = \sum_{j=1}^{n}\sum_{i=1}^{m}\sum_{k=1}^{m}\left(\frac{1}{6}\left(\frac{1}{\#\tilde{\mathfrak{t}}}\sum_{\varsigma=1}^{\#\tilde{\mathfrak{t}}}\left(\left|\tilde{\gamma}_{ij}^{\tilde{\sigma}(\varsigma)^L}-\tilde{\gamma}_{kj}^{\tilde{\sigma}(\varsigma)^L}\right|^{\alpha}+\left|\tilde{\gamma}_{ij}^{\tilde{\sigma}(\varsigma)^U}-\tilde{\gamma}_{kj}^{\tilde{\sigma}(\varsigma)^U}\right|^{\alpha}\right)+\frac{1}{\#\tilde{\mathfrak{i}}}\sum_{\varsigma=1}^{\#\tilde{\mathfrak{i}}}\left(\left|\tilde{\delta}_{ij}^{\tilde{\sigma}(\varsigma)^L}-\tilde{\delta}_{kj}^{\tilde{\sigma}(\varsigma)^L}\right|^{\alpha}+\left|\tilde{\delta}_{ij}^{\tilde{\sigma}(\varsigma)^U}-\tilde{\delta}_{kj}^{\tilde{\sigma}(\varsigma)^U}\right|^{\alpha}\right)+\frac{1}{\#\tilde{\mathfrak{f}}}\sum_{\varsigma=1}^{\#\tilde{\mathfrak{f}}}\left(\left|\tilde{\eta}_{ij}^{\tilde{\sigma}(\varsigma)^L}-\tilde{\eta}_{kj}^{\tilde{\sigma}(\varsigma)^L}\right|^{\alpha}+\left|\tilde{\eta}_{ij}^{\tilde{\sigma}(\varsigma)^U}-\tilde{\eta}_{kj}^{\tilde{\sigma}(\varsigma)^U}\right|^{\alpha}\right)\right)\right)^{\frac{1}{\alpha}}w_j+\frac{\xi}{2}\left(\sum_{j=1}^{n}w_j^2-1\right)$$

where ξ is a real number, representing the Lagrange multiplier variable. Then we compute the partial derivatives of L and let:

$$\frac{\partial L}{\partial w_j} = \sum_{i=1}^{m}\sum_{k=1}^{m}\left(\frac{1}{6}\left(\frac{1}{\#\tilde{\mathfrak{t}}}\sum_{\varsigma=1}^{\#\tilde{\mathfrak{t}}}\left(\left|\tilde{\gamma}_{ij}^{\tilde{\sigma}(\varsigma)^L}-\tilde{\gamma}_{kj}^{\tilde{\sigma}(\varsigma)^L}\right|^{\alpha}+\left|\tilde{\gamma}_{ij}^{\tilde{\sigma}(\varsigma)^U}-\tilde{\gamma}_{kj}^{\tilde{\sigma}(\varsigma)^U}\right|^{\alpha}\right)+\frac{1}{\#\tilde{\mathfrak{i}}}\sum_{\varsigma=1}^{\#\tilde{\mathfrak{i}}}\left(\left|\tilde{\delta}_{ij}^{\tilde{\sigma}(\varsigma)^L}-\tilde{\delta}_{kj}^{\tilde{\sigma}(\varsigma)^L}\right|^{\alpha}+\left|\tilde{\delta}_{ij}^{\tilde{\sigma}(\varsigma)^U}-\tilde{\delta}_{kj}^{\tilde{\sigma}(\varsigma)^U}\right|^{\alpha}\right)+\frac{1}{\#\tilde{\mathfrak{f}}}\sum_{\varsigma=1}^{\#\tilde{\mathfrak{f}}}\left(\left|\tilde{\eta}_{ij}^{\tilde{\sigma}(\varsigma)^L}-\tilde{\eta}_{kj}^{\tilde{\sigma}(\varsigma)^L}\right|^{\alpha}+\left|\tilde{\eta}_{ij}^{\tilde{\sigma}(\varsigma)^U}-\tilde{\eta}_{kj}^{\tilde{\sigma}(\varsigma)^U}\right|^{\alpha}\right)\right)\right)^{\frac{1}{\alpha}}+\xi w_j=0$$

$$\frac{\partial L}{\partial \xi} = \frac{1}{2}\left(\sum_{j=1}^{n}w_j^2-1\right)=0$$

By solving the above equations, to determining the attribute weights, an exact and simple formula can be obtained as follows:

$$w_j^{*}=\frac{\sum_{i=1}^{m}\sum_{k=1}^{m}\left(\frac{1}{6}\left(\frac{1}{\#\tilde{\mathfrak{t}}}\sum_{\varsigma=1}^{\#\tilde{\mathfrak{t}}}\left(\left|\tilde{\gamma}_{ij}^{\tilde{\sigma}(\varsigma)^L}-\tilde{\gamma}_{kj}^{\tilde{\sigma}(\varsigma)^L}\right|^{\alpha}+\left|\tilde{\gamma}_{ij}^{\tilde{\sigma}(\varsigma)^U}-\tilde{\gamma}_{kj}^{\tilde{\sigma}(\varsigma)^U}\right|^{\alpha}\right)+\frac{1}{\#\tilde{\mathfrak{f}}}\sum_{\varsigma=1}^{\#\tilde{\mathfrak{f}}}\left(\left|\tilde{\delta}_{ij}^{\tilde{\sigma}(\varsigma)^L}-\tilde{\delta}_{kj}^{\tilde{\sigma}(\varsigma)^L}\right|^{\alpha}+\left|\tilde{\delta}_{ij}^{\tilde{\sigma}(\varsigma)^U}-\tilde{\delta}_{kj}^{\tilde{\sigma}(\varsigma)^U}\right|^{\alpha}\right)+\frac{1}{\#\tilde{\mathfrak{f}}}\sum_{\varsigma=1}^{\#\tilde{\mathfrak{f}}}\left(\left|\tilde{\eta}_{ij}^{\tilde{\sigma}(\varsigma)^L}-\tilde{\eta}_{kj}^{\tilde{\sigma}(\varsigma)^L}\right|^{\alpha}+\left|\tilde{\eta}_{ij}^{\tilde{\sigma}(\varsigma)^U}-\tilde{\eta}_{kj}^{\tilde{\sigma}(\varsigma)^U}\right|^{\alpha}\right)\right)\right)^{\frac{1}{\alpha}}}{\sqrt{\sum_{j=1}^{n}\left[\sum_{i=1}^{m}\sum_{k=1}^{m}\left(\frac{1}{6}\left(\frac{1}{\#\tilde{\mathfrak{t}}}\sum_{\varsigma=1}^{\#\tilde{\mathfrak{t}}}\left(\left|\tilde{\gamma}_{ij}^{\tilde{\sigma}(\varsigma)^L}-\tilde{\gamma}_{kj}^{\tilde{\sigma}(\varsigma)^L}\right|^{\alpha}+\left|\tilde{\gamma}_{ij}^{\tilde{\sigma}(\varsigma)^U}-\tilde{\gamma}_{kj}^{\tilde{\sigma}(\varsigma)^U}\right|^{\alpha}\right)+\frac{1}{\#\tilde{\mathfrak{f}}}\sum_{\varsigma=1}^{\#\tilde{\mathfrak{f}}}\left(\left|\tilde{\delta}_{ij}^{\tilde{\sigma}(\varsigma)^L}-\tilde{\delta}_{kj}^{\tilde{\sigma}(\varsigma)^L}\right|^{\alpha}+\left|\tilde{\delta}_{ij}^{\tilde{\sigma}(\varsigma)^U}-\tilde{\delta}_{kj}^{\tilde{\sigma}(\varsigma)^U}\right|^{\alpha}\right)+\frac{1}{\#\tilde{\mathfrak{f}}}\sum_{\varsigma=1}^{\#\tilde{\mathfrak{f}}}\left(\left|\tilde{\eta}_{ij}^{\tilde{\sigma}(\varsigma)^L}-\tilde{\eta}_{kj}^{\tilde{\sigma}(\varsigma)^L}\right|^{\alpha}+\left|\tilde{\eta}_{ij}^{\tilde{\sigma}(\varsigma)^U}-\tilde{\eta}_{kj}^{\tilde{\sigma}(\varsigma)^U}\right|^{\alpha}\right)\right)\right)^{\frac{1}{\alpha}}\right]^2}}$$

As the weights of the attributes should satisfy the normalization condition, so we obtain the normalized attribute weights:

$$w_j=\frac{\sum_{i=1}^{m}\sum_{k=1}^{m}\left(\frac{1}{6}\left(\frac{1}{\#\tilde{\mathfrak{t}}}\sum_{\varsigma=1}^{\#\tilde{\mathfrak{t}}}\left(\left|\tilde{\gamma}_{ij}^{\tilde{\sigma}(\varsigma)^L}-\tilde{\gamma}_{kj}^{\tilde{\sigma}(\varsigma)^L}\right|^{\alpha}+\left|\tilde{\gamma}_{ij}^{\tilde{\sigma}(\varsigma)^U}-\tilde{\gamma}_{kj}^{\tilde{\sigma}(\varsigma)^U}\right|^{\alpha}\right)+\frac{1}{\#\tilde{\mathfrak{f}}}\sum_{\varsigma=1}^{\#\tilde{\mathfrak{f}}}\left(\left|\tilde{\delta}_{ij}^{\tilde{\sigma}(\varsigma)^L}-\tilde{\delta}_{kj}^{\tilde{\sigma}(\varsigma)^L}\right|^{\alpha}+\left|\tilde{\delta}_{ij}^{\tilde{\sigma}(\varsigma)^U}-\tilde{\delta}_{kj}^{\tilde{\sigma}(\varsigma)^U}\right|^{\alpha}\right)+\frac{1}{\#\tilde{\mathfrak{f}}}\sum_{\varsigma=1}^{\#\tilde{\mathfrak{f}}}\left(\left|\tilde{\eta}_{ij}^{\tilde{\sigma}(\varsigma)^L}-\tilde{\eta}_{kj}^{\tilde{\sigma}(\varsigma)^L}\right|^{\alpha}+\left|\tilde{\eta}_{ij}^{\tilde{\sigma}(\varsigma)^U}-\tilde{\eta}_{kj}^{\tilde{\sigma}(\varsigma)^U}\right|^{\alpha}\right)\right)\right)^{\frac{1}{\alpha}}}{\sum_{j=1}^{n}\sum_{i=1}^{m}\sum_{k=1}^{m}\left(\frac{1}{6}\left(\frac{1}{\#\tilde{\mathfrak{t}}}\sum_{\varsigma=1}^{\#\tilde{\mathfrak{t}}}\left(\left|\tilde{\gamma}_{ij}^{\tilde{\sigma}(\varsigma)^L}-\tilde{\gamma}_{kj}^{\tilde{\sigma}(\varsigma)^L}\right|^{\alpha}+\left|\tilde{\gamma}_{ij}^{\tilde{\sigma}(\varsigma)^U}-\tilde{\gamma}_{kj}^{\tilde{\sigma}(\varsigma)^U}\right|^{\alpha}\right)+\frac{1}{\#\tilde{\mathfrak{f}}}\sum_{\varsigma=1}^{\#\tilde{\mathfrak{f}}}\left(\left|\tilde{\delta}_{ij}^{\tilde{\sigma}(\varsigma)^L}-\tilde{\delta}_{kj}^{\tilde{\sigma}(\varsigma)^L}\right|^{\alpha}+\left|\tilde{\delta}_{ij}^{\tilde{\sigma}(\varsigma)^U}-\tilde{\delta}_{kj}^{\tilde{\sigma}(\varsigma)^U}\right|^{\alpha}\right)+\frac{1}{\#\tilde{\mathfrak{f}}}\sum_{\varsigma=1}^{\#\tilde{\mathfrak{f}}}\left(\left|\tilde{\eta}_{ij}^{\tilde{\sigma}(\varsigma)^L}-\tilde{\eta}_{kj}^{\tilde{\sigma}(\varsigma)^L}\right|^{\alpha}+\left|\tilde{\eta}_{ij}^{\tilde{\sigma}(\varsigma)^U}-\tilde{\eta}_{kj}^{\tilde{\sigma}(\varsigma)^U}\right|^{\alpha}\right)\right)\right)^{\frac{1}{\alpha}}} \quad (14)$$

Case II: Partly known information on attribute weights

However, there are some situations that the information about the weight vector is partially known. For such situations, using the set of the known weight information, $\Im$, the constrained optimization model can be designed as:

$$(M-4)\begin{cases}\max \tilde{D}(w)=\sum_{j=1}^{n}\sum_{i=1}^{m}\sum_{k=1}^{m}w_j\left(\frac{1}{6}\left(\frac{1}{\#\tilde{\mathfrak{t}}}\sum_{\varsigma=1}^{\#\tilde{\mathfrak{t}}}\left(\left|\tilde{\gamma}_{ij}^{\tilde{\sigma}(\varsigma)^L}-\tilde{\gamma}_{kj}^{\tilde{\sigma}(\varsigma)^L}\right|^{\alpha}+\left|\tilde{\gamma}_{ij}^{\tilde{\sigma}(\varsigma)^U}-\tilde{\gamma}_{kj}^{\tilde{\sigma}(\varsigma)^U}\right|^{\alpha}\right)+\frac{1}{\#\tilde{i}}\sum_{\varsigma=1}^{\#\tilde{i}}\left(\left|\tilde{\delta}_{ij}^{\tilde{\sigma}(\varsigma)^L}-\tilde{\delta}_{kj}^{\tilde{\sigma}(\varsigma)^L}\right|^{\alpha}+\left|\tilde{\delta}_{ij}^{\tilde{\sigma}(\varsigma)^U}-\tilde{\delta}_{kj}^{\tilde{\sigma}(\varsigma)^U}\right|^{\alpha}\right)+\frac{1}{\#\tilde{\mathfrak{f}}}\sum_{\varsigma=1}^{\#\tilde{\mathfrak{f}}}\left(\left|\tilde{\eta}_{ij}^{\tilde{\sigma}(\varsigma)^L}-\tilde{\eta}_{kj}^{\tilde{\sigma}(\varsigma)^L}\right|^{\alpha}+\left|\tilde{\eta}_{ij}^{\tilde{\sigma}(\varsigma)^U}-\tilde{\eta}_{kj}^{\tilde{\sigma}(\varsigma)^U}\right|^{\alpha}\right)\right)\right)^{\frac{1}{\alpha}}\\ s.t.\ w\in\Im,\ w_j\geq 0,\ j=1,2,\ldots,n,\ \sum_{j=1}^{n}w_j=1\end{cases}$$

where $\Im$ is also a set of constraint conditions that the weight value w_j should satisfy according to the requirements in real situations. By solving the linear programming model $(M-4)$, we obtain the optimal solution $w=(w_1,w_2,\ldots,w_n)^t$, which can be used as the weight vector of attributes.

In interval neutrosophic hesitant fuzzy environment, the PIS $\tilde{A}^+$, and the NIS $\tilde{A}^-$ can be defined as follows:

$$\begin{aligned}\tilde{A}^+ &= \{\tilde{\mathfrak{n}}_1^+,\tilde{\mathfrak{n}}_2^+,\ldots,\tilde{\mathfrak{n}}_n^+\}\\ &= \{\{\{[1,1]\},\{[0,0]\},\{[0,0]\}\},\{\{[1,1]\},\{[0,0]\},\{[0,0]\}\},\ldots,\{\{[1,1]\},\{[0,0]\},\{[0,0]\}\}\}.\end{aligned}$$

$$\begin{aligned}\tilde{A}^- &= \{\tilde{\mathfrak{n}}_1^-,\tilde{\mathfrak{n}}_2^-,\ldots,\tilde{\mathfrak{n}}_n^-\}\\ &= \{\{\{[0,0]\},\{[1,1]\},\{[1,1]\}\},\{\{[0,0]\},\{[1,1]\},\{[1,1]\}\},\ldots,\{\{[0,0]\},\{[1,1]\},\{[1,1]\}\}\}.\end{aligned}$$

On the basis of Equation (14), Theorems 3 and 4, the separation measures $\tilde{d}_i^+$ and $\tilde{d}_i^-$ of each alternative from the interval neutrosophic hesitant fuzzy PIS $\tilde{A}^+$ and the interval neutrosophic hesitant fuzzy NIS $\tilde{A}^-$, respectively, are determined as:

$$\tilde{d}_i^+ = \sum_{j=1}^{n} \tilde{d}(\tilde{n}'_{ij}, \tilde{n}_j^+) w_j = \sum_{j=1}^{n} \tilde{d}(\tilde{n}'_{ij}, \{\{[1,1]\},\{[0,0]\},\{[0,0]\}\}) w_j \tag{15}$$

$$= \sum_{j=1}^{n} w_j \left(\frac{1}{6} \left(\frac{1}{\#\tilde{\Upsilon}'_{ij}} \sum_{\tilde{\gamma}\in\tilde{\Upsilon}'_{ij}} \left(\left(1-\tilde{\gamma}^L\right)^\alpha + \left(1-\tilde{\gamma}^U\right)^\alpha \right) + \frac{1}{\#\tilde{\mathrm{I}}'_{ij}} \sum_{\tilde{\delta}\in\tilde{\mathrm{I}}'_{ij}} \left((\tilde{\delta}^L)^\alpha + (\tilde{\delta}^U)^\alpha \right) + \frac{1}{\#\tilde{\mathrm{F}}'_{ij}} \sum_{\tilde{\eta}\in\tilde{\mathrm{F}}'_{ij}} \left((\tilde{\eta}^L)^\alpha + (\tilde{\eta}^U)^\alpha \right) \right) \right)^{\frac{1}{\alpha}} \tag{16}$$

$$\tilde{d}_i^- = \sum_{j=1}^{n} \tilde{d}(\tilde{n}'_{ij}, \tilde{n}_j^-) w_j = \sum_{j=1}^{n} \tilde{d}(\tilde{n}'_{ij}, \{\{[0,0]\},\{[1,1]\},\{[1,1]\}\}) w_j \tag{17}$$

$$= \sum_{j=1}^{n} w_j \left(\frac{1}{6} \left(\frac{1}{\#\tilde{\Upsilon}'_{ij}} \sum_{\tilde{\gamma}\in\tilde{\Upsilon}'_{ij}} \left((\tilde{\gamma}^L)^\alpha + (\tilde{\gamma}^U)^\alpha \right) + \frac{1}{\#\tilde{\mathrm{I}}'_{ij}} \sum_{\tilde{\delta}\in\tilde{\mathrm{I}}'_{ij}} \left(\left(1-\tilde{\delta}^L\right)^\alpha + \left(1-\tilde{\delta}^U\right)^\alpha \right) + \frac{1}{\#\tilde{\mathrm{F}}'_{ij}} \sum_{\tilde{\eta}\in\tilde{\mathrm{F}}'_{ij}} \left(\left(1-\tilde{\eta}^L\right)^\alpha + \left(1-\tilde{\eta}^U\right)^\alpha \right) \right) \right)^{\frac{1}{\alpha}}, \tag{18}$$

where $i = 1, 2, \ldots, m$. The relative closeness coefficient of an alternative $\tilde{A}_i$ with respect to the PIS $\tilde{A}^+$ is defined as:

$$RC(\tilde{A}_i) = \frac{\tilde{d}_i^-}{\tilde{d}_i^+ + \tilde{d}_i^-} \tag{19}$$

where $0 \leq RC(\tilde{A}_i) \leq 1$, $i = 1, 2, \ldots, m$. The ranking orders of all alternatives can be determined according to the closeness coefficient $CR(\tilde{A}_i)$ and select the optimal one(s) from a set of appropriate alternatives.

3. An Illustrative Example

To examine the validity and feasibility of developed decision-making approach in this section, we give a smartphone accessories supplier selection problem in realistic scenario as follows: In the smartphone fields, the Chinese market is the immense one in the world and the competition of smartphone field is so fierce that several companies could not avoid the destiny of bankrupt. In the Chinese market, a firm, who does not want to be defeated must choose the excellent accessories suppliers to fit its supply requirements and technology strategies. A new smartphone design firm called "Hua Xin" incorporated company, who wants to choose a few accessories suppliers for guaranteeing the productive throughput. For simplicity, we assume only one kind of accessory known as Central Processing Unit (CPU), which is used as an essential part in smartphones. The firm determines five CPU suppliers (alternatives) $A_i (i = 1, 2, \ldots, 5)$ through the analysis of their planned level of effort and the market investigation. The evaluation criteria are (1) P_1 : cost; (2) P_2 : technical ability; (3) P_3 : product performance; (4) P_4 : service performance. Because the uncertainty of the information, the evaluation information given by the three experts is expressed as SVNHFEs. The SVNHFDM is given in Table 1. The hierarchical structure of constructed decision-making problem is depicted in Figure 2.

Table 1. Single-valued neutrosophic hesitant fuzzy decision matrix.

	P_1	P_2
A_1	{{0.2},{0.3,0.5},{0.1,0.2,0.3}}	{{0.6,0.7},{0.1,0.3},{0.2,0.4}}
A_2	{{0.1},{0.3},{0.5,0.6}}	{{0.4},{0.3,0.5},{0.5,0.6}}
A_3	{{0.6,0.7},{0.2,0.3},{0.1,0.2}}	{{0.1,0.2},{0.3},{0.6,0.7}}
A_4	{{0.2,0.3},{0.1,0.2},{0.5,0.6}}	{{0.3,0.4},{0.2,0.3},{0.5,0.6,0.7}}
A_5	{{0.7},{0.4,0.5},{0.2,0.4,0.5}}	{{0.6},{0.1,0.7},{0.3,0.5}}
	P_3	P_4
A_1	{{0.2,0.3},{0.4},{0.7,0.8}}	{{0.4},{0.1,0.3},{0.5,0.7,0.9}}
A_2	{{0.1,0.3},{0.4},{0.5,0.6,0.8}}	{{0.6,0.8},{0.2},{0.3,0.5}}
A_3	{{0.2,0.3},{0.1,0.2},{0.6,0.7}}	{{0.2,0.3},{0.4},{0.2,0.5,0.6}}
A_4	{{0.2,0.4},{0.3},{0.1,0.2}}	{{0.6},{0.2},{0.3,0.5}}
A_5	{{0.3},{0.5},{0.1,0.4}}	{{0.5},{0.1,0.2},{0.3,0.4}}

Take $\varpi = 0.5$, $\alpha = 2$, and we normalize the SVNHFDM by using Algorithm 1. The normalized SVNHFDM is given in Table 2.

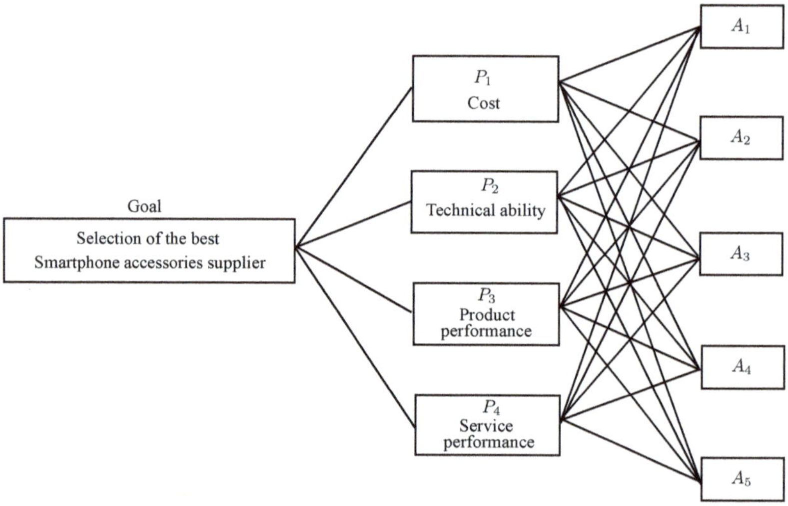

Figure 2. The smartphone accessories supplier selection hierarchical structure.

Table 2. Normalized single-valued neutrosophic hesitant fuzzy decision matrix.

	P_1	P_2
A_1	{{0.2,0.2},{0.3,0.5},{0.1,0.2,0.3}}	{{0.6,0.7},{0.1,0.3},{0.2,0.3,0.4}}
A_2	{{0.1,0.1},{0.3,0.3},{0.5,0.55,0.6}}	{{0.4,0.4},{0.3,0.5},{0.5,0.55,0.6}}
A_3	{{0.6,0.7},{0.2,0.3},{0.1,0.15,0.2}}	{{0.1,0.2},{0.3,0.3},{0.6,0.65,0.7}}
A_4	{{0.2,0.3},{0.1,0.2},{0.5,0.55,0.6}}	{{0.3,0.4},{0.2,0.3},{0.5,0.6,0.7}}
A_5	{{0.7,0.7},{0.4,0.5},{0.2,0.4,0.5}}	{{0.6,0.6},{0.1,0.7},{0.3,0.4,0.5}}
	P_3	P_4
A_1	{{0.2,0.3},{0.4,0.4},{0.7,0.75,0.8}}	{{0.4,0.4},{0.1,0.3},{0.5,0.7,0.9}}
A_2	{{0.1,0.3},{0.4,0.4},{0.5,0.6,0.8}}	{{0.6,0.8},{0.2,0.2},{0.3,0.4,0.5}}
A_3	{{0.2,0.3},{0.1,0.2},{0.6,0.65,0.7}}	{{0.2,0.3},{0.4,0.4},{0.2,0.5,0.6}}
A_4	{{0.2,0.4},{0.3,0.3},{0.1,0.15,0.2}}	{{0.6,0.6},{0.2,0.2},{0.3,0.4,0.5}}
A_5	{{0.3,0.3},{0.5,0.5},{0.1,0.25,0.4}}	{{0.5,0.5},{0.1,0.2},{0.3,0.35,0.4}}

Now to obtain the optimal accessory supplier, we use the developed method, which contains the following two cases:
Case 1: The information of the attribute weights is completely unknown, then the MADM approach related to accessory supplier selection includes the following steps:

Step 1: On the basis of Equation (4), we get the optimal weight vector:

$$w = (0.2994, 0.2367, 0.2521, 0.2118)^T$$

Step 2: Based on the decision matrix of Table 2, we get the normalization of the reference points A^+ and A^- as follows:

$$\begin{aligned} A^+ &= \{n_1^+, n_2^+, n_3^+, n_4^+\} \\ &= \{\{\{1,1\},\{0,0\},\{0,0,0\}\},\{\{1,1\},\{0,0\},\{0,0,0\}\},\{\{1,1\},\{0,0\},\{0,0,0\}\},\{\{1,1\},\{0,0\},\{0,0,0\}\}\}, \end{aligned}$$

$$\begin{aligned} A^- &= \{n_1^-, n_2^-, n_3^-, n_4^-\} \\ &= \{\{\{0,0\},\{1,1\},\{1,1,1\}\},\{\{0,0\},\{1,1\},\{1,1,1\}\},\{\{0,0\},\{1,1\},\{1,1,1\}\},\{\{0,0\},\{1,1\},\{1,1,1\}\}\}. \end{aligned}$$

Step 3: On the basis of Equations (10) and (12), we determine the geometric distances $d_i^+ = d(A_i, A^+)$ and $d_i^- = d(A_i, A^-)$ for the alternative $A_i (i = 1, 2, \ldots, 5)$ as shown in Table 3.

Step 4: Use Equation (13) to determine the relative closeness of each alternative A_i with respect to the single-valued neutrosophic hesitant fuzzy PIS A^+:

$$RC(A_1) = 0.5251, RC(A_2) = 0.4896, RC(A_3) = 0.5394, RC(A_4) = 0.5600, RC(A_5) = 0.5927.$$

Step 5: On the basis of the relative closeness coefficients $RC(A_i)$, rank the alternatives $A_i (i = 1, 2, \ldots, 5)$: $A_5 \succ A_4 \succ A_3 \succ A_1 \succ A_2$. Thus, the optimal alternative (CPU supplier) is A_5.

Table 3. The geometric distances for alternatives.

Geometric Distance	A_1	A_2	A_3	A_4	A_5
$d_i^+ = d(A_i, A^+)$	0.5142	0.5434	0.4974	0.4781	0.4279
$d_i^- = d(A_i, A^-)$	0.5685	0.5212	0.5824	0.6086	0.6226

Case 2: The information of the attribute weights is partly known, and the known weight information is as follows:

$$\Im = \{0.15 \le w_1 \le 0.2, 0.16 \le w_2 \le 0.18, 0.3 \le w_3 \le 0.35, 0.3 \le w_4 \le 0.45, \sum_{j=1}^{4} w_j = 1\}$$

Step 1: Use the model (M-2) to establish the single-objective programming model as follows:

$$(M-2) \begin{cases} \max D(w) = 5.6368w_1 + 4.4554w_2 + 4.7465w_3 + 3.9864w_4 \\ s.t.\ \ w \in \Im, w_j \ge 0, j = 1,2,3,4, \sum_{j=1}^{4} w_j = 1 \end{cases}$$

By solving this model, we obtain the attributes weight vector:

$$w = (0.2000, 0.1600, 0.3400, 0.3000)^T$$

Step 2: According to the decision matrix of Table 2, the normalization of the reference points A^+ and A^- can be obtained as follows:

$$\begin{aligned} A^+ &= \{n_1^+, n_2^+, n_3^+, n_4^+\} \\ &= \{\{\{1,1\},\{0,0\},\{0,0,0\}\},\{\{1,1\},\{0,0\},\{0,0,0\}\},\{\{1,1\},\{0,0\},\{0,0,0\}\},\{\{1,1\},\{0,0\},\{0,0,0\}\}\}, \end{aligned}$$

$$\begin{aligned} A^- &= \{n_1^-, n_2^-, n_3^-, n_4^-\} \\ &= \{\{\{0,0\},\{1,1\},\{1,1,1\}\},\{\{0,0\},\{1,1\},\{1,1,1\}\},\{\{0,0\},\{1,1\},\{1,1,1\}\},\{\{0,0\},\{1,1\},\{1,1,1\}\}\}. \end{aligned}$$

Step 3: Based on Equations (10) and (12), we determine the geometric distances $d(A_i, A^+)$ and $d(A_i, A^-)$ for the alternative $A_i (i = 1, 2, \ldots, 5)$ as shown in Table 4.

Step 4: Use Equation (13) to determine the relative closeness of each alternative A_i with respect to the single-valued neutrosophic hesitant fuzzy PIS A^+:

$$RC(A_1) = 0.4972, RC(A_2) = 0.5052, RC(A_3) = 0.5199, RC(A_4) = 0.5808, RC(A_5) = 0.5883.$$

Step 5: Based on the relative closeness coefficients $RC(A_i)$, rank the alternatives $A_i (i = 1, 2, \ldots, 5)$: $A_5 \succ A_4 \succ A_3 \succ A_2 \succ A_1$. Thus, the optimal alternative (CPU supplier) is A_5.

Taking $\varpi = 0.5$, we normalize the single-valued neutrosophic hesitant fuzzy decision matrix and compute the closeness coefficient of the alternatives with the different values of α. The comparison results are given in Figure 3.

Table 4. The geometric distances for alternatives.

Geometric Distance	A_1	A_2	A_3	A_4	A_5
$d(A_i, A^+)$	0.5446	0.5244	0.5220	0.4534	0.4341
$d(A_i, A^-)$	0.5385	0.5355	0.5652	0.6281	0.6202

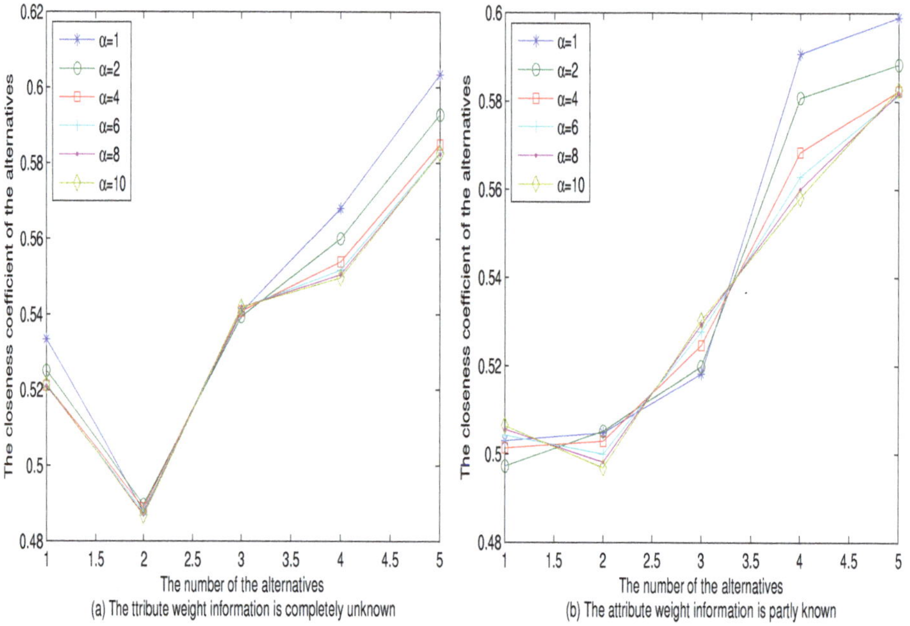

Figure 3. Comparison of the closeness coefficient of the alternative.

The analysis process under interval neutrosophic hesitant fuzzy circumstances:

In the above smartphone accessories supplier selection problem, if the information provided by the experts is indicated in INHFEs, as in Table 5. Then, to choose the optimal CPU supplier, we proceed to use the developed approach.

Take $\varpi = 0.5$, $\alpha = 2$, and we normalize the INHFDM by using Algorithm 2. The normalized INHFDM is given in Table 6.

Case 1: The information of the attribute weights is completely unknown , then the MADM method of accessory supplier selection consists of the following steps:

Step 1: On the basis of Equation (14), we get the optimal weight vector:

$$w = \{0.2963, 0.2562, 0.2388, 0.2087\}$$

Step 2: According to the decision matrix of Table 6, the normalization of the reference points $\tilde{A}^+$ and $\tilde{A}^-$ can be obtained as follows:

$$\begin{aligned}\tilde{A}^+ &= \{\tilde{n}_1^+, \tilde{n}_2^+, \tilde{n}_3^+, \tilde{n}_4^+\} \\ &= \{\{\{[1,1],[1,1]\},\{[0,0],[0,0]\},\{[0,0],[0,0],[0,0]\}\},\{\{[1,1],[1,1]\},\{[0,0],[0,0]\},\{[0,0],[0,0],[0,0]\}\}, \\ &\quad \{\{[1,1],[1,1]\},\{[0,0],[0,0]\},\{[0,0],[0,0],[0,0]\}\},\{\{[1,1],[1,1]\},\{[0,0],[0,0]\},\{[0,0],[0,0],[0,0]\}\}\},\end{aligned}$$

$$\begin{aligned}\tilde{A}^- &= \{\tilde{n}_1^-, \tilde{n}_2^-, \tilde{n}_3^-, \tilde{n}_4^-\} \\ &= \{\{\{[0,0],[0,0]\},\{[1,1],[1,1]\},\{[1,1],[1,1],[1,1]\}\},\{\{[0,0],[0,0]\},\{[1,1],[1,1]\},\{[1,1],[1,1],[1,1]\}\}, \\ &\quad \{\{[0,0],[0,0]\},\{[1,1],[1,1]\},\{[1,1],[1,1],[1,1]\}\},\{\{[0,0],[0,0]\},\{[1,1],[1,1]\},\{[1,1],[1,1],[1,1]\}\}\}.\end{aligned}$$

Step 3: Based on Equations (15) and (17), we determine the geometric distances $\tilde{d}(A_i, A^-)$ and $\tilde{d}(A_i, A^+)$ for the alternative $A_i(i = 1, 2, \ldots, 5)$ as shown in Table 7.

Step 4: Use Equation (19) to determine the relative closeness of each alternative $\tilde{A}_i$ with respect to the interval neutrosophic hesitant fuzzy PIS $\tilde{A}^+$:

$$RC(\tilde{A}_1) = 0.5169, RC(\tilde{A}_2) = 0.4592, RC(\tilde{A}_3) = 0.4969, RC(\tilde{A}_4) = 0.5368, RC(\tilde{A}_5) = 0.5643.$$

Step 5: Based on the relative closeness coefficients $RC(\tilde{A}_i)$, rank the alternatives $A_i(i = 1, 2, \ldots, 5)$: $A_5 \succ A_4 \succ A_1 \succ A_3 \succ A_2$. Thus, the optimal alternative (CPU supplier) is A_5.

Case 2: The information of the attribute weights is partly known, and the known weight information is given as follows:

$$\Im = \{0.15 \le w_1 \le 0.2, 0.16 \le w_2 \le 0.18, 0.3 \le w_3 \le 0.35, 0.3 \le w_4 \le 0.45, \sum_{j=1}^{4} w_j = 1\}$$

Step 1: Use the model (M-4) to establish the single-objective programming model as follows: $(M-4)\begin{cases} \max D(w) = 4.5556w_1 + 4.2000w_2 + 3.3222w_3 + 3.3111w_4 \\ s.t.\ w \in \Im, w_j \ge 0, j = 1,2,3,4,\ \sum_{j=1}^{4} w_j = 1 \end{cases}$

By solving this model, we obtain the weight vector of attributes:

$$w = \{0.2000, 0.1800, 0.3200, 0.3000\}$$

Step 2: According to the decision matrix of Table 6, we can obtain the normalization of the reference points $\tilde{A}^+$ and $\tilde{A}^-$ as follows:

$$\begin{aligned}\tilde{A}^+ &= \{\tilde{n}_1^+, \tilde{n}_2^+, \tilde{n}_3^+, \tilde{n}_4^+\} \\ &= \{\{\{[1,1],[1,1]\},\{[0,0],[0,0]\},\{[0,0],[0,0],[0,0]\}\},\{\{[1,1],[1,1]\},\{[0,0],[0,0]\},\{[0,0],[0,0],[0,0]\}\}, \\ &\quad \{\{[1,1],[1,1]\},\{[0,0],[0,0]\},\{[0,0],[0,0],[0,0]\}\},\{\{[1,1],[1,1]\},\{[0,0],[0,0]\},\{[0,0],[0,0],[0,0]\}\}\},\end{aligned}$$

$$\begin{aligned}\tilde{A}^- &= \{\tilde{n}_1^-, \tilde{n}_2^-, \tilde{n}_3^-, \tilde{n}_4^-\} \\ &= \{\{\{[0,0],[0,0]\},\{[1,1],[1,1]\},\{[1,1],[1,1],[1,1]\}\},\{\{[0,0],[0,0]\},\{[1,1],[1,1]\},\{[1,1],[1,1],[1,1]\}\}, \\ &\quad \{\{[0,0],[0,0]\},\{[1,1],[1,1]\},\{[1,1],[1,1],[1,1]\}\},\{\{[0,0],[0,0]\},\{[1,1],[1,1]\},\{[1,1],[1,1],[1,1]\}\}\}.\end{aligned}$$

Step 3: Use Equations (15) and (17) to determine the geometric distances $\tilde{d}(A_i, A^+)$ and $\tilde{d}(A_i, A^-)$ for the alternative $A_i(i = 1, 2, \ldots, 5)$ as shown in Table 8.

Step 4: Use Equation (19) to determine the relative closeness of each alternative $\tilde{A}_i$ with respect to the interval neutrosophic hesitant fuzzy PIS $\tilde{A}^+$:

$$RC(\tilde{A}_1) = 0.4955, RC(\tilde{A}_2) = 0.4729, RC(\tilde{A}_3) = 0.4803, RC(\tilde{A}_4) = 0.5536, RC(\tilde{A}_5) = 0.5607.$$

Step 5: According to the relative closeness coefficients $RC(\tilde{A}_i)$, rank the alternatives $A_i(i = 1, 2, \ldots, 5)$: $A_5 \succ A_4 \succ A_1 \succ A_3 \succ A_2$. Thus, the optimal alternative (CPU supplier) is A_5.

Taking $\varpi = 0.5$, we normalize the interval neutrosophic hesitant fuzzy decision matrix and compute the closeness coefficient of the alternatives with the different values of α. The comparison results are given in Figure 4.

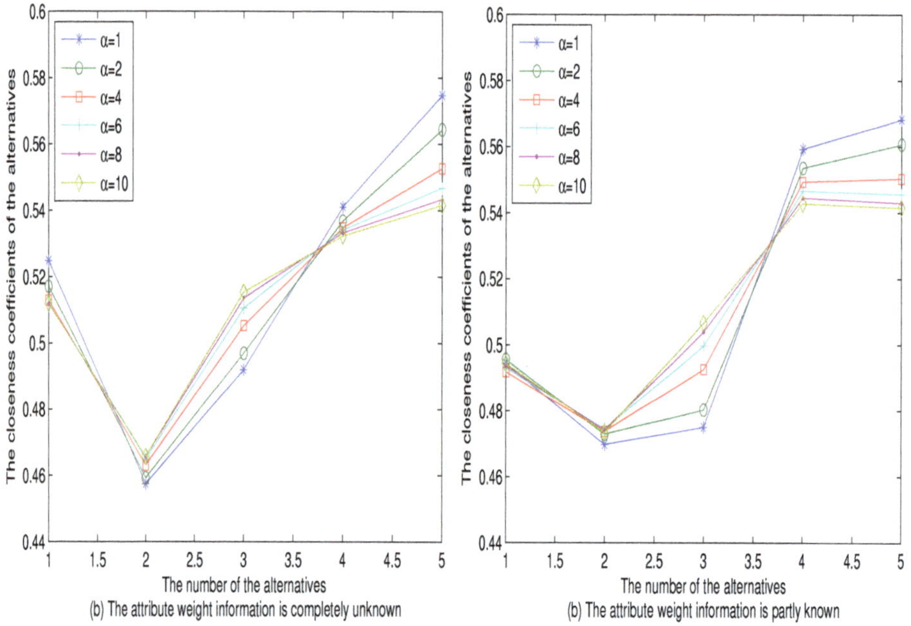

Figure 4. Comparison of the closeness coefficient of the alternative.

Table 5. Interval neutrosophic hesitant fuzzy decision matrix.

	P_1	P_2
A_1	{{[0.2,0.3]},{[0.3,0.4],[0.5,0.7]},{[0.1,0.3],[0.2,0.5],[0.3,0.6]}}	{{[0.6,0.8],[0.7,0.9]},{[0.1,0.2],[0.3,0.5]},{[0.2,0.3],[0.4,0.5]}}
A_2	{{[0.1,0.3]},{[0.3,0.5]},{[0.5,0.7],[0.6,0.8]}}	{{[0.4,0.6]},{[0.3,0.4],[0.5,0.6]},{[0.5,0.7],[0.6,0.8]}}
A_3	{{[0.6,0.7],[0.7,0.8]},{[0.2,0.4],[0.3,0.5]},{[0.1,0.3],[0.2,0.4]}}	{{[0.1,0.3],[0.2,0.4]},{[0.3,0.6]},{[0.6,0.8],[0.7,0.9]}}
A_4	{{[0.2,0.5],[0.3,0.4]},{[0.1,0.3],[0.2,0.3]},{[0.5,0.6],[0.6,0.7]}}	{{[0.3,0.5],[0.4,0.6]},{[0.2,0.3],[0.3,0.4]},{[0.5,0.7],[0.6,0.8],[0.7,0.9]}}
A_5	{{[0.7,0.8]},{[0.4,0.6],[0.5,0.7]},{[0.2,0.3],[0.4,0.6],[0.5,0.7]}}	{{[0.6,0.8]},{[0.1,0.3],[0.7,0.8]},{[0.3,0.4],[0.5,0.6}}
	P_3	P_4
A_1	{{[0.2,0.4],[0.3,0.5]},{[0.4,0.5]},{[0.7,0.8],[0.8,0.9]}}	{{[0.4,0.6]},{[0.1,0.2],[0.3,0.4]},{[0.5,0.6],[0.7,0.8],[0.8,0.9]}}
A_2	{{[0.1,0.3],[0.3,0.5]},{[0.4,0.6]},{[0.5,0.6],[0.6,0.7],[0.8,0.9]}}	{{[0.6,0.7],[0.8,0.9]},{[0.2,0.5]},{[0.3,0.5],[0.5,0.7]}}
A_3	{{[0.2,0.3],[0.3,0.4]},{[0.1,0.3],[0.2,0.4]},{[0.6,0.8],[0.7,0.9]}}	{{[0.2,0.4],[0.3,0.5]},{[0.4,0.6]},{[0.2,0.3],[0.5,0.7],[0.6,0.8]}}
A_4	{{[0.2,0.3],[0.4,0.5]},{[0.3,0.6]},{[0.1,0.4],[0.2,0.5]}}	{{[0.6,0.8]},{[0.2,0.3]},{[0.3,0.4],[0.5,0.6]}}
A_5	{{[0.3,0.5]},{[0.5,0.6]},{[0.1,0.3],[0.4,0.5]}}	{{[0.5,0.7]},{[0.1,0.3],[0.2,0.5]},{[0.3,0.5],[0.4,0.8]}}

Table 6. Normalized interval neutrosophic hesitant fuzzy decision matrix.

	P_1	P_2
A_1	{{[0.2,0.3],[0.2,0.3]},{[0.3,0.4],[0.5,0.7]},{[0.1,0.3],[0.2,0.5],[0.3,0.6]}}	{{[0.6,0.8],[0.7,0.9]},{[0.1,0.2],[0.3,0.5]},{[0.2,0.3],[0.3,0.4],[0.4,0.5]}}
A_2	{{[0.1,0.3],[0.1,0.3]},{[0.3,0.5],[0.3,0.5]},{[0.5,0.7],[0.55,0.75],[0.6,0.8]}}	{{[0.4,0.6],[0.4,0.6]},{[0.3,0.4],[0.5,0.6]},{[0.5,0.7],[0.55,0.75],[0.6,0.8]}}
A_3	{{[0.6,0.7],[0.7,0.8]},{[0.2,0.4],[0.3,0.5]},{[0.1,0.3],[0.15,0.35],[0.2,0.4]}}	{{[0.1,0.3],[0.2,0.4]},{[0.3,0.6],[0.3,0.6]},{[0.6,0.8],[0.65,0.85],[0.7,0.9]}}
A_4	{{[0.2,0.5],[0.3,0.4]},{[0.1,0.3],[0.2,0.3]},{[0.5,0.6],[0.55,0.65],[0.6,0.7]}}	{{[0.3,0.5],[0.4,0.6]},{[0.2,0.3],[0.3,0.4]},{[0.5,0.7],[0.6,0.8],[0.7,0.9]}}
A_5	{{[0.7,0.8],[0.7,0.8]},{[0.4,0.6],[0.5,0.7]},{[0.2,0.3],[0.4,0.6],[0.5,0.7]}}	{{[0.6,0.8],[0.6,0.8]},{[0.1,0.3],[0.7,0.8]},{[0.3,0.4],[0.4,0.5],[0.5,0.6}}
	P_3	P_4
A_1	{{[0.2,0.4],[0.3,0.5]},{[0.4,0.5],[0.4,0.5]},{[0.7,0.8],[0.75,0.85],[0.8,0.9]}}	{{[0.4,0.6],[0.4,0.6]},{[0.1,0.2],[0.3,0.4]},{[0.5,0.6],[0.7,0.8],[0.8,0.9]}}
A_2	{{[0.1,0.3],[0.3,0.5]},{[0.4,0.6],[0.4,0.6]},{[0.5,0.6],[0.6,0.7],[0.8,0.9]}}	{{[0.6,0.7],[0.8,0.9]},{[0.2,0.5],[0.2,0.5]},{[0.3,0.5],[0.4,0.6],[0.5,0.7]}}
A_3	{{[0.2,0.3],[0.3,0.4]},{[0.1,0.3],[0.2,0.4]},{[0.6,0.8],[0.65,0.85],[0.7,0.9]}}	{{[0.2,0.4],[0.3,0.5]},{[0.4,0.6],[0.4,0.6]},{[0.2,0.3],[0.5,0.7],[0.6,0.8]}}
A_4	{{[0.2,0.3],[0.4,0.5]},{[0.3,0.6],[0.3,0.6]},{[0.1,0.4],[0.15,0.45],[0.2,0.5]}}	{{[0.6,0.8],[0.6,0.8]},{[0.2,0.3],[0.2,0.3]},{[0.3,0.4],[0.4,0.5],[0.5,0.6]}}
A_5	{{[0.3,0.5],[0.3,0.5]},{[0.5,0.6],[0.5,0.6]},{[0.1,0.3],[0.25,0.4],[0.4,0.5]}}	{{[0.5,0.7],[0.5,0.7]},{[0.1,0.3],[0.2,0.5]},{[0.3,0.5],[0.35,0.65],[0.4,0.8]}}

Table 7. The geometric distances for alternatives.

Geometric Distance	A_1	A_2	A_3	A_4	A_5
$\tilde{d}(A_i, A^+)$	0.5169	0.5711	0.5361	0.4952	0.4625
$\tilde{d}(A_i, A^-)$	0.5531	0.4849	0.5295	0.5740	0.5991

Table 8. The geometric distances for alternatives.

Geometric Distance	A_1	A_2	A_3	A_4	A_5
$\tilde{d}(A_i, A^+)$	0.5406	0.5562	0.5569	0.4752	0.4653
$\tilde{d}(A_i, A^-)$	0.5310	0.4990	0.5147	0.5894	0.5938

Comparative Analysis

Zhao et al. [31] generalized the minimum deviation method to accommodate hesitant fuzzy values for solving the decision-making problems. We have used this approach on the above illustrative example and compared the decision results with proposed approach of this paper for SNHFSs. In the approach of Zhao et al., assume that the subjective preference values to all the alternatives $A_j (j = 1, 2, 3, 4, 5)$ assigned by the experts are: $s_1 = \{\{0.3, 0.4\}, \{0.2, 0.5\}, \{0.1, 0.3, 0.7\}\}, s_2 = \{\{0.2, 0.7\}, \{0.1, 0.9\}, \{0.3, 0.6\}\}$, $s_3 = \{\{0.8\}, \{0.5, 0.8\}, \{0.4, 0.7, 0.9\}\}, s_4 = \{\{0.1, 0.4\}, \{0.6\}, \{0.5, 0.7, 0.8\}\}$ and $s_5 = \{\{0.3\}, \{0.4, 0.6\}$, $\{0.2, 0.4\}\}$. Also $\tilde{s}_1 = \{\{[0.3, 0.5], [0.4, 0.6]\}, \{[0.2, 0.3], [0.5, 0.7]\}, \{[0.1, 0.2], [0.3, 0.4], [0.7, 0.9]\}\}$, $\tilde{s}_2 = \{\{[0.2, 0.3], [0.7, 0.9]\}, \{[0.1, 0.4], [0.7, 0.9]\}, \{[0.3, 0.4], [0.6, 0.8]\}\}, \tilde{s}_3 = \{\{[0.8, 0.9]\}, \{[0.5, 0.6], [0.8, 0.9]\}$, $\{[0.4, 0.6], [0.7, 0.9], [0.6, 0.7]\}\}, \tilde{s}_4 = \{\{[0.1, 0.4], [0.4, 0.5]\}, \{[0.6, 0.7]\}, \{[0.5, 0.7], [0.7, 0.8], [0.8, 0.9]\}\}$ and $\tilde{s}_5 = \{\{[0.3, 0.5]\}, \{[0.4, 0.5], [0.6, 0.8]\}, \{[0.2, 0.3], [0.4, 0.7]\}\}$.

The results corresponding to these approaches are summarized in Table 9.

Table 9. Comparative analysis.

Methods	Score of Alternatives					Ranking of Alternatives
Zhao et al. [31] for SVNHFS	0.4431	0.4025	0.4941	0.5073	0.5691	$A_5 \succ A_4 \succ A_3 \succ A_1 \succ A_2$
Our proposed method for SVNHFS	0.5251	0.4896	0.5394	0.5600	0.5927	$A_5 \succ A_4 \succ A_3 \succ A_1 \succ A_2$
Zhao et al. [31] for INHFS	0.4559	0.4206	0.4255	0.5334	0.5791	$A_5 \succ A_4 \succ A_1 \succ A_3 \succ A_2$
Our proposed method for INHFS	0.5169	0.4592	0.4969	0.5368	0.5643	$A_5 \succ A_4 \succ A_1 \succ A_3 \succ A_2$

From this comparative study, the results obtained by the approach [31] coincide with the proposed one which validates the proposed approach. The main reason is that in approach [31], the subjective preferences are taken into account to serve as decision information and will have a positive effect on the final decision results. Hence, the proposed approach can be suitably used to solve the MADM problems. The advantages of our proposed method are as follows: (1) The developed approach has good flexibility and extension. (2) The SNHFSs of developed approach availably depicts increasingly general decision-making situations. (3) With the aid of the maximizing deviation and TOPSIS, the developed approach uses the satisfaction level of the alternative to the ideal solutions to make the decision.

4. Conclusions

SNHFS is a suitable tool for dealing with the obscurity of an expert's judgments over alternatives according to attributes. SNHFSs are useful for representing the hesitant assessments of the experts, and remains the edge of SNSs and HFSs, which accommodates an increasingly complex MADM situation. SNHFS (by combining SNS and HFS) as an extended format represents some general hesitant scenarios. In this paper, firstly we have developed the normalization method and the distance measures of SNHFSs and further, to obtain the attribute optimal relative weights, we have proposed a decision-making approach called the maximizing deviation method with SNHFSs including SVNHFSs and INHFSs. Secondly, we have developed a new approach based on TOPSIS to solve MADM problems under SNHFS environment (SVNHFS and INHFS). Finally, we have illustrated the applicability and effectiveness of the developed method with a smartphone accessories supplier selection problem. In future work, we will extend the proposed approach of SNHFSs to other areas, such as pattern recognition, medical diagnosis, clustering analysis, and image processing.

Author Contributions: M.A. and S.N. developed the theory and performed the computations. F.S. verified the analytical methods.

Conflicts of Interest: The authors declare that they have no conflict of interest regarding the publication of the research article.

References

1. Smarandache, F. *A Unifying Field in Logics. Neutrosophy: Neutrosophic Probability, Set and Logic*; American Research Press: Rehoboth, DE, USA, 1999.
2. Smarandache, F. *Neutrosophy. Neutrosophic Probability, Set, and Logic, ProQuest Information & Learning*; American Research Press: Ann Arbor, MI, USA, 1998; Volume 105, pp. 118–123.
3. Wang, H.; Smarandache, F.; Zhang, Y.Q.; Sunderraman, R. Single-valued neutrosophic sets. *Multispace Multistruct.* **2010**, *4*, 410–413.
4. Ye, J. A multi-criteria decision making method using aggregation operators for simplified neutrosophic sets. *J. Intell. Fuzzy Syst.* **2014**, *26*, 2459–2466.
5. Torra, V.; Narukawa, Y. On hesitant fuzzy sets and decision. In Proceedings of the 18th IEEE International Conference on Fuzzy Systems, Jeju Island, Korea, 20–24 August 2009; pp. 1378–1382.
6. Zadeh, L.A. Fuzzy sets. *Inf. Control* **1965**, *8*, 338–353. [CrossRef]
7. Ye, J. Multiple-attribute decision making method under a single valued neutrosophic hesitant fuzzy environment. *J. Intell. Syst.* **2015**, *24*, 23–36. [CrossRef]
8. Liu, C.F.; Luo, Y.S. New aggregation operators of single-valued neutrosophic hesitant fuzzy set and their application in multi-attribute decision making. *Pattern Anal. Appl.* **2019**, *22*, 417–427 . [CrossRef]
9. Sahin, R.; Liu, P. Correlation coefficient of single-valued neutrosophic hesitant fuzzy sets and its applications in decision making. *Neural Comput. Appl.* **2017**, *28*, 1387–1395. [CrossRef]
10. Li, X.; Zhang, X. Single-valued neutrosophic hesitant fuzzy Choquet aggregation operators for multi-attribute decision making. *Symmetry* **2018**, *10*, 50. [CrossRef]
11. Juan-juan, P.; Jian-qiang, W.; Jun-hua, H. Multi-criteria decision making approach based on single-valued neutrosophic hesitant fuzzy geometric weighted choquet integral heronian mean operator. *J. Intell. Fuzzy Syst.* **2018**, 1–14. [CrossRef]
12. Wang, R.; Li, Y. Generalized single-valued neutrosophic hesitant fuzzy prioritized aggregation operators and their applications to multiple criteria decision making. *Information* **2018**, *9*, 10. [CrossRef]
13. Akram, M.; Adeel, A.; Alcantud, J.C.R. Group decision making methods based on hesitant *N*-soft sets. *Expert Syst. Appl.* **2019**, *115*, 95–105. [CrossRef]
14. Akram, M.; Adeel, A. TOPSIS approach for MAGDM based on interval-valued hesitant fuzzy *N*-soft environment. *Int. J. Fuzzy Syst.* **2019**, *21*, 993–1009. [CrossRef]
15. Akram, M.; Adeel, A.; Alcantud, J.C.R. Hesitant Fuzzy N-Soft Sets: A New Model with Applications in Decision-Making. *J. Intell. Fuzzy Syst.* **2019**., 2019. [CrossRef]
16. Akram, M.; Naz, S. A Novel Decision-Making Approach under Complex Pythagorean Fuzzy Environment. *Math. Comput. Appl.* **2019**, *24*, 73. [CrossRef]
17. Naz, S.; Ashraf, S.; Akram, M. A novel approach to decision making with Pythagorean fuzzy information. *Mathematics* **2018**, *6*, 95. [CrossRef]
18. Naz, S.; Ashraf, S.; Karaaslan, F. Energy of a bipolar fuzzy graph and its application in decision making. *Italian J. Pure Appl. Math.* **2018**, *40*, 339–352.
19. Naz, S; Akram, M. Novel decision making approach based on hesitant fuzzy sets and graph theory. *Comput. Appl. Math.* **2018**. [CrossRef]
20. Liu, P.; Shi, L. The generalized hybrid weighted average operator based on interval neutrosophic hesitant set and its application to multiple attribute decision making. *Neural Comput. Appl.* **2015**, *26*, 457–471. [CrossRef]

21. Ye, J. Correlation coefficients of interval neutrosophic hesitant fuzzy sets and its application in a multiple attribute decision making method. *Informatica* **2016**, *27*, 179–202. [CrossRef]
22. Kakati, P.; Borkotokey, S.; Mesiar, R.; Rahman, S. Interval neutrosophic hesitant fuzzy Choquet integral in multi-criteria decision making. *J. Intell. Fuzzy Syst.* **2018**, 1–19. [CrossRef]
23. Mahmood, T.; Ye, J.; Khan, Q. Vector similarity measures for simplified neutrosophic hesitant fuzzy set and their applications. *J. Inequal. Spec. Funct.* **2016**, *7*, 176–194.
24. Atanassov, K.T. Intuitionistic fuzzy sets. *Fuzzy Sets Syst.* **1986**, *20*, 87–96. [CrossRef]
25. Torra, V. Hesitant fuzzy sets. *Int. J. Intell. Syst.* **2010**, *25*, 529–539. [CrossRef]
26. Xu, Z. Some similarity measures of intuitionistic fuzzy sets and their applications to multiple attribute decision making. *Fuzzy Optim. Decis. Mak.* **2007**, *6*, 109–121. [CrossRef]
27. Xu, Z.; Zhang, X. Hesitant fuzzy multi-attribute decision making based on TOPSIS with incomplete weight information. *Knowl.-Based Syst.* **2013**, *52*, 53–64. [CrossRef]
28. Wei, C.; Ren, Z.; Rodriguez, R.M. A hesitant fuzzy linguistic TODIM method based on a score function. *Int. J. Comput. Intell. Syst.* **2015**, *8*, 701–712. [CrossRef]
29. Liao, H.; Xu, Z.; Zeng, X.J. Hesitant fuzzy linguistic VIKOR method and its application in qualitative multiple criteria decision making. *IEEE Trans. Fuzzy Syst.* **2015**, *23*, 1343–1355. [CrossRef]
30. Gou, X.J.; Liao, H.C.; Xu, Z.S.; Herrera, F. Double hierarchy hesitant fuzzy linguistic term set and MULTIMOORA method: A case of study to evaluate the implementation status of haze controlling measures. *Inf. Fusion* **2017**, *38*, 22–34. [CrossRef]
31. Zhao, H.; Xu, Z.; Wang, H.; Liu, S. Hesitant fuzzy multi-attribute decision making based on the minimum deviation method. *Soft Comput.* **2017**, *21*, 3439–3459. [CrossRef]

symmetry

MDPI

Article

Modeling the Performance Indicators of Financial Assets with Neutrosophic Fuzzy Numbers

Marcel-Ioan Bolos [1], Ioana-Alexandra Bradea [2] and Camelia Delcea [2,*]

[1] Department of Finance and Banks, University of Oradea, 410087 Oradea, Romania
[2] Department of Informatics and Cybernetics, Bucharest University of Economic Studies, 010552 Bucharest, Romania
* Correspondence: camelia.delcea@softscape.ro; Tel.: +40-769-652-813

Received: 18 July 2019; Accepted: 5 August 2019; Published: 7 August 2019

Abstract: This research sets the basis for modeling the performance indicators of financial assets using triangular neutrosophic fuzzy numbers. This type of number allows for the modeling of financial assets performance indicators by taking into consideration all the possible scenarios of their achievement. The key performance indicators (KPIs) modeled with the help of triangular fuzzy neutrosophic numbers are the return on financial assets, the financial assets risk, and the covariance between financial assets. Thus far, the return on financial assets has been studied using statistical indicators, like the arithmetic and geometric mean, or using the financial risk indicators with the help of the squared deviations from the mean and covariance. These indicators are well known as the basis of portfolio theory. This paper opens the perspective of modeling these three mentioned statistical indicators using triangular neutrosophic fuzzy numbers due to the major advantages they have. The first advantage of the neutrosophic approach is that it includes three possible symmetric scenarios of the KPIs achievement, namely the scenario of certainty, the scenario of non-realization, and the scenario of indecision, in which it cannot be appreciated whether the performance indicators are or are not achieved. The second big advantage is its data series clustering, representing the financial performance indicators by which these scenarios can be delimitated by means of neutrosophic fuzzy numbers in very good, good or weak performance indicators. This clustering is realized by means of the linguistic criteria and measuring the belonging degree to a class of indicators using fuzzy membership functions. The third major advantage is the selection of risk mitigation analysis scenarios and the formation of financial assets' optimal portfolios.

Keywords: fuzzy numbers; neutrosophic numbers; neutrosophic symmetric scenarios; performance indicators; financial assets

1. Introduction

Financial markets specialists have shown a particular interest for financial assets lately, both due to returns they generate for investors and also because they can predict the future evolution of financial performance [1]. Researchers in financial markets, starting with Markowittz [2], developed the modern theory of financial asset portfolios, focusing on the financial assets return (R_a) and on the financial assets risk (σ_a^2). An extremely important issue in the modern theory of financial assets portfolios is the risk diversification; in the famous words of Markowitz: "Don't put all your eggs in one basket." These studies, regarding the risk of diversification, have led to the foundation of Markowitz's efficient frontier, demonstrating that the financial assets portfolio risk is much lower than the individual risk of each financial assets' category [2].

In order to evaluate financial asset returns, the literature has enabled two categories of models. The first model is known as market return model, and it is based on the financial asset market price at two moments of time (t_1) and (t_0). The market-based return calculation is as follows:

$$R_a = \frac{P_1 - P_0 + D_1}{P_0}$$

where P_1 represents the market price at time (t_1), P_0 represents the market price at time (t_0), and D_1 represents the dividend at time (t_1).

The second model for evaluating returns on financial assets is based on the market portfolio (M), as well as on the return on the risk-free financial asset (R_f) as a result of applying the Capital Assets Price Model (CAPM). The financial assets return is, in this case:

$$R_a = R_f + \beta(R_M - R_f),$$

where R_M is the portfolio return, and β is the volatility coefficient of the financial asset, which is determined by a relation of the form:

$$\beta = \frac{\sigma_{aM}}{\sigma_M^2},$$

where σ_{aM} is the covariance between the asset and the portfolio and σ_M^2 represents the portfolio risk. If the financial asset return is considered to be fairly measured using the CAPM model, then two situations are encountered in regard to the investors' decision to trade the financial asset on the capital market. In the first situation, if $R_a < R_a(CAPM)$, then R_a has to decrease by increasing P_0—as such, investors will buy financial assets. In the second situation, if $R_a < R_a(CAPM)$, then R_a should increase by subtracting P_0—as such, the investors will sell the financial assets.

In order to assess financial asset risk, the following statistical indicators are used: The squared deviations from the mean, given by the formula:

$$\sigma_a^2 = \frac{1}{N-1}\sum_{i=1}^{N}(R_{ai} - \overline{R})^2$$

and the variance:

$$\sigma_a = \sqrt{\frac{1}{N-1}\sum_{i=1}^{N}(R_{ai} - \overline{R})^2}$$

to express the deviation of the financial asset over a period of time from the average return value. The higher the value of the deviation is, the higher the risk assumed by the investors will be. Otherwise, the lower the deviation value is, the lower the risk.

Regardless of the model used to evaluate financial asset return, there is a certain degree of probability that characterizes the achievement of financial asset return. Similarly, also the risk assumed by investors can be manifested with different intensities (this risk may take maximum or minimum values, and there is also an area where the risk intensity is uncertain). This degree of uncertainty for the obtaining of the financial asset return, noted as $(D_u(R_a))$, can be grouped into three main categories:

- The first category: A high degree of obtaining financial asset returns, denoted by $\mu(D_u(\sigma a, Ra))$, which is the value of the financial asset return that can be achieved with a high probability, approximated by a professional judgment of around 50%. For each financial asset, this degree has specific values.
- The second category: A low degree of obtaining financial asset returns, denoted by $\vartheta(D_u(\sigma a,\ Ra))$. Here, there is no prospect of achieving the financial asset return. The causes that lead to these situations are various: The risk assumed by investors is appropriate to the value of the realizable asset's return, the expected return should record high values above the market level, or the

dynamics of the exchange rate market is not known. The degree of probability for this situation approximated by professional judgement is 30%, and it is specific to each financial asset.

- The third category: The degree for obtaining the financial asset return is uncertain, denoted by $\lambda(D_u(\sigma a, R_a))$, corresponding to the situation in which the realization or non-realization of the return is uncertain or not appreciated. This area of uncertainty is approximated at 20%, and it is also specific to each category of financial assets.

The introduction of these criteria for assessing financial asset return allows for the analysis of these performance indicators in line with the real needs of investors. They can thus select portfolios of financial assets for which the likelihood of achieving profitability is known. Additionally, by introducing these specific notions, the basis for the use of neutrosophic fuzzy numbers is created in the field of modeling financial decision-making to form financial assets portfolios.

The aim of this paper is to properly model the indicators from portfolio theory using triangular neutrosophic fuzzy numbers (considering the major advantages they provide) while solving the problems that arise in the classical approach—these being several limitations which might appear in the case of the financial assets' performance indicators use.

2. State of the Art

Neutrosophic fuzzy numbers represent a quite new research area that has captured the attention of researchers worldwide since 2013. Querying the WoS (Web of Science) database for the keywords "neutrosophic fuzzy numbers," the search results indicated 184 ISI Web of Science articles. Analyzing by publication year, it can be observed that, since the appearance of this research topic, interest has increased exponentially each and every year, starting from two ISI articles published in 2013 to seven ISI articles in both 2014 and 2015, 18 in 2016, 41 in 2017, and 77 in 2018. Researchers from all over the world have started to use this niche of fuzzy intelligence, but the majority of these studies have come from China, India, USA, Turkey and Pakistan.

Most of these articles are included in the following categories: Computer science and artificial intelligence (47%), multidisciplinary sciences (20%), information systems (10%), automation control systems (7%), management (2%), economics (1%), and business (0.5%). The first articles published in 2013 targeted color image segmentation with applicability in image processing, pattern recognition and computer vision [3] and the topic of neutrosophic fuzzy classifications [4]. The advantage of using the neutrosophic set was shown by Ali and Smarandache [5] in their paper, where they studied the complexity of this topic and revealed that neutrosophic fuzzy numbers can handle imprecise, indeterminate, inconsistent, and incomplete information.

The most cited articles that have used neutrosophic fuzzy numbers as a research methodology to target the decision-making problem. Pramanik et al. [6] used an extension of the interval neutrosophic set, namely the interval bipolar neutrosophic set, in order to develop a multi-attribute decision-making strategy. Ye [7] introduced the concept of simplified neutrosophic sets in order to solve a multicriteria decision-making problem. Zhang et al. [8] also proposed an interval neutrosophic set to address a multicriteria decision-making problem. The paper published by Liu and Teng [9] presented a new method based on a single-valued neutrosophic normalized weighted Bonferroni mean that demonstrated its effectiveness for solving decision-making problems. Peng et al. [10] developed a new outranking approach for multi-criteria decision-making problems. This method was developed in a simplified neutrosophic environment where the truth-membership degree, indeterminacy-membership degree, and falsity-membership degree were subsets in [0,1].

None of the studies indexed in WoS have targeted portfolio theory or the finance domain. Even if there have been a lot of studies that have focused on solving multi-criteria decision-making problems, none of them have addressed this portfolio theory or the finance domain. There are only five articles that have approached the economic area of research. The first of them is the study proposed by Bausys et al. [11], in which the complex proportional assessment method (COPRAS) was used in the context of single value neutrosophic sets in order to select the location of a liquefied natural gas terminal. In the same year, Bausys and Zavadskas [12] published another article in which was created an extension of the Vlse Kriterijumska Optimizacija Kompromisno Resenje (VIKOR) method for the solution of multicriteria decision-making problems. The applicability and efficiency of neutrosophic fuzzy numbers were studied by Nabadan et al. [13]. The authors compared neutrosophic fuzzy numbers to other fuzzy methods used in the decision-making field, and their results showed higher values of efficiency for this new approach. The fourth article in economic domain targeted the tourism area [14] by creating a decision support model for satisfactory restaurants using social information. The model used neutrosophic fuzzy numbers to sign online ratings, and it used the Bonferroni mean to consider interdependence among criteria. Tian et al. [15] used a life cycle assessment technique to develop an innovative multi-criteria group decision-making approach that incorporates aggregation operators and the Technique for Order of Preference by Similarity to Ideal Solution (TOPSIS)-based QUALIFLEX method. This study managed to offer a solution for green product design selection problems using neutrosophic linguistic information. As for the performance indicators used in financial analysis, a series of recent papers have addressed optimal portfolio selection, corporate entrepreneurship, indicators selection or portfolio optimization [16–22].

Thus, our study has an innovative approach, as it introduces in the literature the modeling of the financial performance indicators with the help of neutrosophic fuzzy numbers with the portfolio theory. The financial performance indicators modeled using neutrosophic fuzzy numbers refer to: The financial return of assets (neutrosophic return), financial risk (neutrosophic risk), and financial covariance (neutrosophic covariance).

This research paper solves the identified problem that consist of the limitations of the classical approach for the financial assets' performance indicators. Performance indicators of financial assets are modeled using classical statistical indicators, namely arithmetic mean, geometric mean, squared deviation from the mean, and variance. These indicators show a series of limitations, including the following:

- Not taking into account possible scenarios for achieving performance indicators of financial assets; as such, investor cannot assess the chance of achieving them.
- Not allowing data series stratification to delimit the values of the financial performance indicators—the very good, good and low values—so that the investor can select the scenarios best suited to his/her investment profile.
- Not allowing the selection of performance indicators scenarios that characterize financial assets for analyzing financial risk mitigation or building optimal portfolios of financial assets.

These limitations in financial assets' performance indicators make them subject to some degree of rigidity in substantiating decisions and also affect their capacity to respond properly to information needed by capital market investors.

In order to solve these shortcomings, the proposed solution introduces the modeling of financial assets performance indicators using triangular neutrosophic fuzzy numbers due to the numerous advantages they present, namely:

- Allow for the consideration of all possible achievement scenarios for the financial assets' performance indicators like the scenario of certainty, the scenario of non-realization, and the scenario of indecision—these scenarios derive from the neutrosophic components such as truth, indeterminacy and falsehood, which are symmetric in form, as the truth is opposite to false, with respect to the indeterminacy [23].
- Allow for financial performance indicators' data series stratification or clustering. The delimitation of these data series was done using linguistic criteria with assigned values such as: very good, good and weak.
- Allow for the selection of analysis scenarios to mitigate financial risk or to form optimal portfolios of financial assets, both of which could lead to a better substantiation of financial asset decisions.

The methodology used in this paper is aimed at combining portfolio theory with the fuzzy intelligence and neutrosophic numbers in order to enable the decision-making process for investors. The proposed model allows for the modeling of financial assets performance indicators by taking into consideration all possible scenarios of their achievement. Additionally, the model clusters the data series, representing the financial performance indicators, by delimitating these scenarios by means of neutrosophic fuzzy numbers in very good, good or weak performance indicators. This clustering is realized with the help of the linguistic criteria which belong to degrees of a class of indicators using fuzzy membership functions. This methodology offers the possibility to form financial assets' optimal portfolios that are characterized by low risk and high return.

The effective neutrosophic fuzzy modeling is presented in the following paragraphs of the paper.

3. Establishment of Neutrosophic Numbers for the Financial Assets Risk and Return

Financial asset return and risk are manifested by varying intensities, depending on the particularities of the assets that form a portfolio. These intensities may take high, low, or even uncertain values. Neutrosophic fuzzy numbers are formed and defined separately for each of the two indicators mentioned above [24,25].

Definition 1. *Let the financial asset return on the financial market be* (R_a)*, and let* $F\,[0,1]$ *be the rules set for all fuzzy triangular numbers. The fuzzy number* $(\widetilde{Ra})$ *is considered the triangular neutrosophic fuzzy number of the financial assets return:*

$$\widetilde{Ra} = \left\{ \langle \widetilde{r_a}, \mu_{\widetilde{Ra}}, \vartheta_{\widetilde{Ra}},\ \lambda_{\widetilde{Ra}} \rangle / r_a \in R_a \right\},$$

where $\mu_{\widetilde{Ra}} : R_a \to [0,1]$; $\vartheta_{\widetilde{Ra}} : R_a \to [0,1]$ *and* $\lambda_{\widetilde{Ra}} : R_a \to [0,1]$, *for which the membership functions are defined according to the relations depicted in Figure 1.*

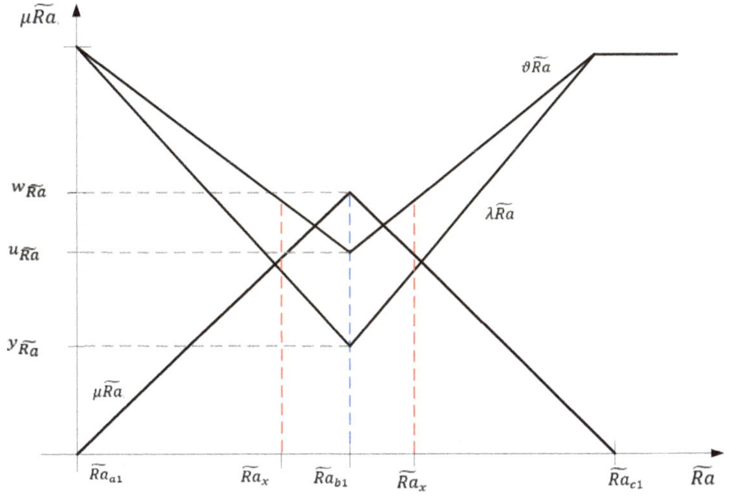

Figure 1. The neutrosophic fuzzy triangular number of financial assets return.

The truth membership, which is the membership function for the financial assets return with the highest degree of realization $(\mu\widetilde{Ra}_{(x)})$, is the following:

$$\mu\widetilde{Ra}_{(x)} = \begin{cases} \frac{w_{\widetilde{Ra}}(\widetilde{Ra}_x - \widetilde{Ra}_{a1})}{\widetilde{Ra}_{b1} - \widetilde{Ra}_{a1}} \; for \; \widetilde{Ra}_{a1} \leq \widetilde{Ra}_x \leq \widetilde{Ra}_{b1} \\ w_{\widetilde{Ra}} \; for \; \widetilde{Ra}_x = \widetilde{Ra}_{b1} \\ \frac{w_{\widetilde{Ra}}(\widetilde{Ra}_{c1} - \widetilde{Ra}_x)}{\widetilde{Ra}_{c1} - \widetilde{Ra}_{b1}} \; for \; \widetilde{Ra}_{b1} \leq \widetilde{Ra}_x \leq \widetilde{Ra}_{c1} \\ 0 \; for \; others \; values \; out \; of \; range \left[\widetilde{Ra}_{c1}; \widetilde{Ra}_{a1}\right] \end{cases} \tag{1}$$

The indeterminacy membership, which is the membership function for the financial assets return with the medium degree of realization $(\vartheta\widetilde{Ra}_{(x)})$, is the following:

$$\vartheta\widetilde{Ra}_{(x)} = \begin{cases} \frac{u_{\widetilde{Ra}}(\widetilde{Ra}_x - \widetilde{Ra}_{a1}) + \widetilde{Ra}_{b1} - \widetilde{Ra}_x}{\widetilde{Ra}_{b1} - \widetilde{Ra}_{a1}} \; for \; \widetilde{Ra}_{a1} \leq \widetilde{Ra}_x \leq \widetilde{Ra}_{b1} \\ u_{Ra} \; for \; Ra_x = Ra_{b1} \\ \frac{u_{\widetilde{Ra}}(\widetilde{Ra}_{c1} - \widetilde{Ra}_x) + \widetilde{Ra}_x - \widetilde{Ra}_{b1}}{\widetilde{Ra}_{c1} - \widetilde{Ra}_{b1}} \; for \; \widetilde{Ra}_{b1} \leq \widetilde{Ra}_x \leq \widetilde{Ra}_{c1} \\ 0 \; for \; others \; values \; out \; of \; range \left[\widetilde{Ra}_{c1}; \widetilde{Ra}_{a1}\right] \end{cases} \tag{2}$$

The falsity membership, which is the membership function for the financial assets return with low degree of realization $(\lambda\widetilde{Ra}_{(x)})$, is the following:

$$\lambda\widetilde{Ra}_{(x)} = \begin{cases} \frac{y_{\widetilde{Ra}}(\widetilde{Ra}_x - \widetilde{Ra}_{a1}) + \widetilde{Ra}_{b1} - \widetilde{Ra}_x}{\widetilde{Ra}_{b1} - \widetilde{Ra}_{a1}} \; for \; \widetilde{Ra}_{a1} \leq \widetilde{Ra}_x \leq \widetilde{Ra}_{b1} \\ \lambda_{\widetilde{R_a}} \; for \; \widetilde{Ra}_x = \widetilde{Ra}_{b1} \\ \frac{y_{\widetilde{Ra}}(\widetilde{Ra}_{c1} - \widetilde{Ra}_x) + \widetilde{Ra}_x - \widetilde{Ra}_{b1}}{\widetilde{Ra}_{c1} - \widetilde{Ra}_{b1}} \; for \; \widetilde{Ra}_{b1} \leq \widetilde{Ra}_x \leq \widetilde{Ra}_{c1} \\ 0 \; for \; others \; values \; out \; of \; range \left[\widetilde{Ra}_{c1}; \widetilde{Ra}_{a1}\right] \end{cases} \tag{3}$$

Definition 2. *Let the financial asset risk on the financial market be* (σ_a), *and let* $F\,[0,1]$ *be the rules set for all triangular fuzzy numbers. The fuzzy number* $(\widetilde{\sigma_a})$ *is considered the triangular neutrosophic fuzzy number of the financial assets risk:*

$$\widetilde{\sigma_a} = \left\{ \langle \widetilde{\sigma_a}, \mu_{\widetilde{\sigma_a}}, \vartheta_{\widetilde{\sigma_a}}, \lambda_{\widetilde{\sigma_a}} \rangle / \widetilde{\sigma_a} \in \sigma_A \right\},$$

where $\mu_{\widetilde{\sigma_a}} : \sigma_A \to [0,1]$; $\vartheta_{\widetilde{\sigma_a}} : \sigma_A \to [0,1]$ *and* $\lambda_{\widetilde{\sigma_a}} : \sigma_A \to [0,1]$, *for which the membership functions are defined according to the relations depicted in Figure 2.*

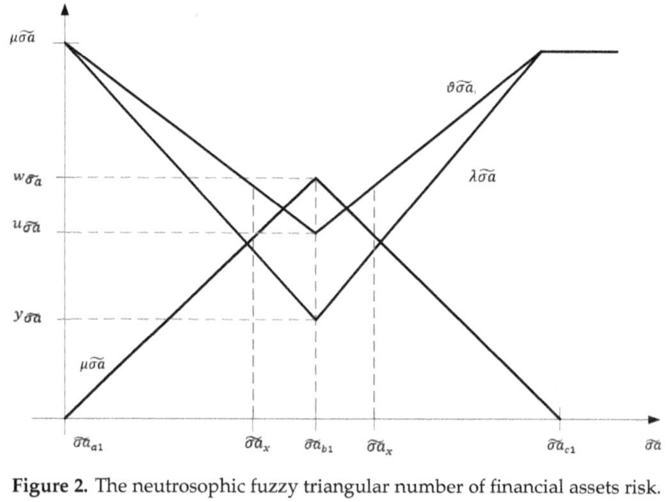

Figure 2. The neutrosophic fuzzy triangular number of financial assets risk.

The truth membership, which is the membership function for the financial assets risk with the highest degree of realization $(\mu\widetilde{\sigma a}_{(x)})$, is the following:

$$\mu\widetilde{\sigma a}_{(x)} = \begin{cases} \frac{w_{\widetilde{\sigma a}}(\widetilde{\sigma a}_x - \widetilde{\sigma a}_{a1})}{\widetilde{\sigma a}_{b1} - \widetilde{\sigma a}_{a1}} \; for \; \widetilde{\sigma a}_{a1} \le \widetilde{\sigma a}_x \le \widetilde{\sigma a}_{b1} \\ w_{\widetilde{\sigma a}} \; for \; \widetilde{\sigma a}_x = \widetilde{\sigma a}_{b1} \\ \frac{w_{\widetilde{\sigma a}}(\widetilde{\sigma a}_{c1} - \widetilde{\sigma a}_x)}{\widetilde{\sigma a}_{c1} - \widetilde{\sigma a}_{b1}} \; for \; \widetilde{\sigma a}_{b1} \le \widetilde{\sigma a}_x \le \widetilde{\sigma a}_{c1} \\ 0 \; for \; others \; values \; out \; of \; range \; [\widetilde{\sigma a}_{c1}; \widetilde{\sigma a}_{a1}] \end{cases} \tag{4}$$

The indeterminacy membership, which is the membership function for the financial assets risk with the medium degree of realization $(\vartheta\widetilde{\sigma a}_{(x)})$, is the following:

$$\vartheta\widetilde{\sigma a}_{(x)} = \begin{cases} \frac{u_{\widetilde{\sigma a}}(\widetilde{\sigma a}_x - \widetilde{\sigma a}_{a1}) + \widetilde{\sigma a}_{b1} - \widetilde{\sigma a}_x}{\widetilde{\sigma a}_{b1} - \widetilde{\sigma a}_{a1}} \; for \; \widetilde{\sigma a}_{a1} \le \widetilde{\sigma a}_x \le \widetilde{\sigma a}_{b1} \\ u_{\widetilde{\sigma a}} \; for \; \widetilde{\sigma a}_x = \widetilde{\sigma a}_{b1} \\ \frac{u_{\widetilde{\sigma a}}(\widetilde{\sigma a}_{c1} - \widetilde{\sigma a}_x) + \widetilde{\sigma a}_x - \widetilde{\sigma a}_{b1}}{\widetilde{\sigma a}_{c1} - \widetilde{\sigma a}_{b1}} \; for \; \widetilde{\sigma a}_{b1} \le \widetilde{\sigma a}_x \le \widetilde{\sigma a}_{c1} \\ 0 \; for \; others \; values \; out \; of \; range \; [\widetilde{\sigma a}_{c1}; \widetilde{\sigma a}_{a1}] \end{cases} \tag{5}$$

The falsity membership, which is the membership function for the financial assets risk with low degree of realization $\lambda\widetilde{\sigma a}_{(x)}$, is the following:

$$\lambda\widetilde{\sigma a}_{(x)} = \begin{cases} \frac{y_{\widetilde{\sigma a}}(\widetilde{\sigma a}_x - \widetilde{\sigma a}_{a1}) + \widetilde{\sigma a}_{b1} - \widetilde{\sigma a}_x}{\widetilde{\sigma a}_{b1} - \widetilde{\sigma a}_{a1}} \; for \; \widetilde{\sigma a}_{a1} \le \widetilde{\sigma a}_x \le \widetilde{\sigma a}_{b1} \\ \lambda_{\widetilde{\sigma a}} \; for \; \widetilde{\sigma a}_x = \widetilde{\sigma a}_{b1} \\ \frac{y_{\widetilde{\sigma a}}(\widetilde{\sigma a}_{c1} - \widetilde{\sigma a}_x) + \widetilde{\sigma a}_x - \widetilde{\sigma a}_{b1}}{\widetilde{\sigma a}_{c1} - \widetilde{\sigma a}_{b1}} \; for \; \widetilde{\sigma a}_{b1} \le \widetilde{\sigma a}_x \le \widetilde{\sigma a}_{c1} \\ 0 \; for \; [\widetilde{\sigma a}_{c1}; \widetilde{\sigma a}_{a1}] \end{cases} \tag{6}$$

The two neutrosophic numbers that characterize the return and the risk of a financial asset are as follows:

$$\widetilde{Ra}_i = \langle(\widetilde{Ra}_{ai}, \widetilde{Ra}_{bi}, \widetilde{Ra}_{ci}); w\widetilde{Ra}, u\widetilde{Ra}, y\widetilde{Ra}\rangle, \text{ for } i = \overline{1, n} \tag{7}$$

$$\widetilde{\sigma a}_i = \langle(\widetilde{\sigma a}_{ai}, \widetilde{\sigma a}_{bi}, \widetilde{\sigma a}_{ci}); w\widetilde{\sigma a}, u\widetilde{\sigma a}, y\widetilde{\sigma a}\rangle, \text{ for } i = \overline{1, n} \tag{8}$$

Definition 3. *Two specific neutrosophic numbers are considered for the two financial assets that define the financial asset return:*

$$\widetilde{Ra}_1 = \langle(\widetilde{Ra}_{a1}, \widetilde{Ra}_{b1}, \widetilde{Ra}_{c1}); w\widetilde{Ra}_1, u\widetilde{Ra}_1, y\widetilde{Ra}_1\rangle$$

for the first financial asset (A_1)*;*

$$\widetilde{Ra}_2 = \langle(\widetilde{Ra}_{a2}, \widetilde{Ra}_{b2}, \widetilde{Ra}_{c2}); w\widetilde{Ra}_2, u\widetilde{Ra}_2, y\widetilde{Ra}_2\rangle$$

for the second financial asset (A_2) *and the parameter* $\gamma \neq 0$, $\gamma \in R$.

The following arithmetical operations are valid:

1. Addition:

$$\widetilde{Ra}_1 + \widetilde{Ra}_2 = \langle(\widetilde{Ra}_{a1} + \widetilde{Ra}_{a2}, \widetilde{Ra}_{b1} + \widetilde{Ra}_{b2}, \widetilde{Ra}_{c1} + \widetilde{Ra}_{c2}) w\widetilde{Ra}_1 \wedge w\widetilde{Ra}_2, u\widetilde{Ra}_1 \vee u\widetilde{Ra}_2, y\widetilde{Ra}_1 \vee y\widetilde{Ra}_2\rangle \tag{9}$$

2. Subtraction:

$$\widetilde{Ra}_1 - \widetilde{Ra}_2 = \langle(\widetilde{Ra}_{a1} - \widetilde{Ra}_{a2}, \widetilde{Ra}_{b1} - \widetilde{Ra}_{b2}, \widetilde{Ra}_{c1} - \widetilde{Ra}_{c2}) w\widetilde{Ra}_1 \wedge w\widetilde{Ra}_2, u\widetilde{Ra}_1 \vee u\widetilde{Ra}_2, y\widetilde{Ra}_1 \vee y\widetilde{Ra}_2\rangle \tag{10}$$

3. Multiplication:

- for $\widetilde{Ra}_{c1} > 0$ and $\widetilde{Ra}_{c2} > 0$;

$$\begin{aligned}&\widetilde{Ra}_1 \times \widetilde{Ra}_2\\&= \langle(\widetilde{Ra}_{a1} X \widetilde{Ra}_{a2}, \widetilde{Ra}_{b1} X \widetilde{Ra}_{b2}, \widetilde{Ra}_{c1} X \widetilde{Ra}_{c2}) w\widetilde{Ra}_1 \wedge w\widetilde{Ra}_2, u\widetilde{Ra}_1 \vee u\widetilde{Ra}_2, y\widetilde{Ra}_1 \vee y\widetilde{Ra}_2\rangle\end{aligned} \tag{11}$$

- for $\widetilde{Ra}_{c1} < 0$ and $\widetilde{Ra}_{c2} > 0$;

$$\begin{aligned}&\widetilde{Ra}_1 \times \widetilde{Ra}_2\\&= \langle(\widetilde{Ra}_{a1} X \widetilde{Ra}_{c2}, \widetilde{Ra}_{b1} X \widetilde{Ra}_{b2}, \widetilde{Ra}_{c1} X \widetilde{Ra}_{a2}); w\widetilde{Ra}_1 \wedge w\widetilde{Ra}_2, u\widetilde{Ra}_1 \vee u\widetilde{Ra}_2, y\widetilde{Ra}_1 \vee y\widetilde{Ra}_2\rangle\end{aligned} \tag{12}$$

- for $\widetilde{Ra}_{c1} < 0$ and $\widetilde{Ra}_{c2} < 0$;

$$\widetilde{Ra}_1 \times \widetilde{Ra}_2 = \langle(\widetilde{Ra}_{c1} \times \widetilde{Ra}_{c2}, \widetilde{Ra}_{b1} \times \widetilde{Ra}_{b2}, \widetilde{Ra}_{a1} \times \widetilde{Ra}_{a2}); w\widetilde{Ra}_1 \wedge w\widetilde{Ra}_2, u\widetilde{Ra}_1 \vee u\widetilde{Ra}_2, y\widetilde{Ra}_1 \vee y\widetilde{Ra}_2\rangle \tag{13}$$

4. Division:

- for $\widetilde{Ra}_{c1} > 0$ and $\widetilde{Ra}_{c2} > 0$;

$$\widetilde{Ra}_1 / \widetilde{Ra}_2 = \langle(\widetilde{Ra}_{a1} / \widetilde{Ra}_{c2}, \widetilde{Ra}_{b1} / \widetilde{Ra}_{b2}, \widetilde{Ra}_{c1} / \widetilde{Ra}_{a2}); w\widetilde{Ra}_1 \wedge w\widetilde{Ra}_2, u\widetilde{Ra}_1 \vee u\widetilde{Ra}_2, y\widetilde{Ra}_1 \vee y\widetilde{Ra}_2\rangle \tag{14}$$

- for $\widetilde{Ra}_{c1} < 0$ and $\widetilde{Ra}_{c2} > 0$;

$$\widetilde{Ra}_1 / \widetilde{Ra}_2 = \langle(\widetilde{Ra}_{c1} / \widetilde{Ra}_{c2}, \widetilde{Ra}_{b1} / \widetilde{Ra}_{b2}, \widetilde{Ra}_{a1} / \widetilde{Ra}_{a2}); w\widetilde{Ra}_1 \wedge w\widetilde{Ra}_2, u\widetilde{Ra}_1 \vee u\widetilde{Ra}_2, y\widetilde{Ra}_1 \vee y\widetilde{Ra}_2\rangle \tag{15}$$

- for $\widetilde{Ra}_{c1} < 0$ and $\widetilde{Ra}_{c2} < 0$;

$$\widetilde{Ra}_1 / \widetilde{Ra}_2 = \langle(\widetilde{Ra}_{c1} / \widetilde{Ra}_{a2}, \widetilde{Ra}_{b1} / \widetilde{Ra}_{b2}, \widetilde{Ra}_{a1} / \widetilde{Ra}_{c2}); w\widetilde{Ra}_1 \wedge w\widetilde{Ra}_2, u\widetilde{Ra}_1 \vee u\widetilde{Ra}_2, y\widetilde{Ra}_1 \vee y\widetilde{Ra}_2\rangle \tag{16}$$

5. Scalar multiplication:

- for $\gamma > 0$;

$$\gamma \times \widetilde{Ra}_1 = \langle(\gamma \times \widetilde{Ra}_{a1}, \gamma \times \widetilde{Ra}_{b1}, \gamma \times \widetilde{Ra}_{c1}); w\widetilde{Ra_1}, u\widetilde{Ra}_1, y\widetilde{Ra}_1\rangle \tag{17}$$

- for $\gamma < 0$;

$$\gamma \times \widetilde{Ra}_1 = \langle(\gamma \times \widetilde{Ra}_{c1}, \gamma \times \widetilde{Ra}_{b1}, \gamma \times \widetilde{Ra}_{a1}); w\widetilde{Ra_1}, u\widetilde{Ra}_1, y\widetilde{Ra}_1\rangle \tag{18}$$

6. The inverse of a neutrosophic number:

$$\widetilde{Ra}_1^{-1} = \langle(1/\widetilde{Ra}_{c1}, 1/\widetilde{Ra}_{b1}, 1/\widetilde{Ra}_{a1}); w\widetilde{Ra_1}, u\widetilde{Ra}_1, y\widetilde{Ra}_1\rangle \tag{19}$$

It is important to notice that the arithmetic operations with the neutrosophic numbers defined for the financial asset return $\widetilde{Ra}_i = \langle(\widetilde{Ra}_{ai}, \widetilde{Ra}_{bi}, \widetilde{Ra}_{ci}); w\widetilde{Ra}, u\widetilde{Ra}, y\widetilde{Ra}\rangle$ are the same as the specific arithmetic operations for the financial asset risk: $\widetilde{\sigma a}_i = \langle(\widetilde{\sigma a}_{ai}, \widetilde{\sigma a}_{bi}, \widetilde{\sigma a}_{ci}); w\widetilde{\sigma a}, u\widetilde{\sigma a}, y\widetilde{\sigma a}\rangle$ for any $i = \overline{1, n}$.

Definition 4. *Let the neutrosophic number defined for the financial asset return be of the following form:*

$$\widetilde{Ra}_i = \langle(\widetilde{Ra}_{ai}, \widetilde{Ra}_{bi}, \widetilde{Ra}_{ci}); w\widetilde{Ra}, u\widetilde{Ra}, y\widetilde{Ra}\rangle$$

for any $i = \overline{1, n}$. *To assure the comparability of the neutrosophic fuzzy numbers, a score function is introduced [4,5]:*

$$S_{a1} = \frac{1}{8}\left[\widetilde{Ra}_{a1} + \widetilde{Ra}_{b1} + \widetilde{Ra}_{c1}\right] + (2 + w\widetilde{Ra_1} - u\widetilde{Ra_1} - y\widetilde{Ra_1}) \quad (20)$$

This is an accuracy function of the form:

$$A_{a1} = \frac{1}{8}\left[\widetilde{Ra}_{a1} + \widetilde{Ra}_{b1} + \widetilde{Ra}_{c1}\right] - (2 + w\widetilde{Ra_1} - u\widetilde{Ra_1} - y\widetilde{Ra_1}) \quad (21)$$

Definition 5. *There are two neutrosophic numbers that define the financial asset return:*

$$\widetilde{Ra_1} = \langle(\widetilde{Ra}_{a1}, \widetilde{Ra}_{b1}, \widetilde{Ra}_{c1}); w\widetilde{Ra_1}, u\widetilde{Ra_1}, y\widetilde{Ra_1}\rangle$$

for the first asset (A_1) *and*

$$\widetilde{Ra_2} = \langle(\widetilde{Ra}_{a2}, \widetilde{Ra}_{b2}, \widetilde{Ra}_{c2}); w\widetilde{Ra_2}, u\widetilde{Ra_2}, y\widetilde{Ra_2}\rangle$$

for the second asset (A_2). *As such:*

- *If* $S(\widetilde{Ra_1}) < S(\widetilde{Ra_2})$, *then* $\widetilde{Ra_1}$ *, and it is noted that* $\widetilde{Ra_1} < \widetilde{Ra_2}$;
- *If* $S(\widetilde{Ra_1}) > S(\widetilde{Ra_2})$, *then* $\widetilde{Ra_1}$ *is higher than* $\widetilde{Ra_2}$, *and it is noted that* $\widetilde{Ra_1} > \widetilde{Ra_2}$;
- *If* $S(\widetilde{Ra_1}) = S(\widetilde{Ra_2})$ *are distinguished two cases:*
    - *If* $A(\widetilde{Ra_1}) < A(\widetilde{Ra_2})$, *then* $\widetilde{Ra_1}$ *is smaller than* $\widetilde{Ra_2}$, *and it is noted that* $\widetilde{Ra_1} < \widetilde{Ra_2}$;
    - *If* $A(\widetilde{Ra_1}) = A(\widetilde{Ra_2})$, *then the neutrosophic numbers are equal, and it is noted that* $\widetilde{Ra_1} = \widetilde{Ra_2}$;

4. Modeling the Financial Assets Return Using the Neutrosophic Fuzzy Numbers

The financial assets return, as mentioned above, is the most relevant performance indicator, because it provides information on the earnings that investors can obtain over a limited period of time as a result of asset ownership. In literature, the model for determining the financial assets return is based on: The capital gain $\frac{P_1 - P_0}{P_0}$, formed by the stock price differences at time (t_1) and (t_0), as well as on the return on invested capital, represented by the ratio between the dividend at time (t_1) and the price at time (t_0). $\frac{D_1}{P_0}$, illustrates a remuneration form for the invested capital. The market model that evaluates the financial asset return thus becomes $Ra = \frac{P_1 - P_0 + D_1}{P_0}$, where the gains are the exchange rate differences and the dividend. For each financial asset (A_i), there are different return values over a time horizon [0, t] at different time moments of the form:

$$Ra : \begin{pmatrix} t_0 & t_1 & t_2 & t_3 & \ldots & t_{k-1} & t_k & \ldots & t_n \\ Ra_0 & Ra_1 & Ra_2 & Ra_3 & \ldots & Ra_{k-1} & Ra_k & \ldots & Ra_n \end{pmatrix} t = \overline{1, n} \quad (22)$$

The formed data series is modeled using neutrosophic fuzzy numbers due to the many advantages they have: The possibility of stratification for the financial asset; clustering the return values according to linguistic criteria such as high, medium or small financial asset return; the possibility of selection the return category desired by the investor in order to maximize his profit; or analyzing the financial asset return by means of probability grades. The neutrosophic fuzzy numbers built for the financial asset return, on the above data series, are presented in Figure 3.

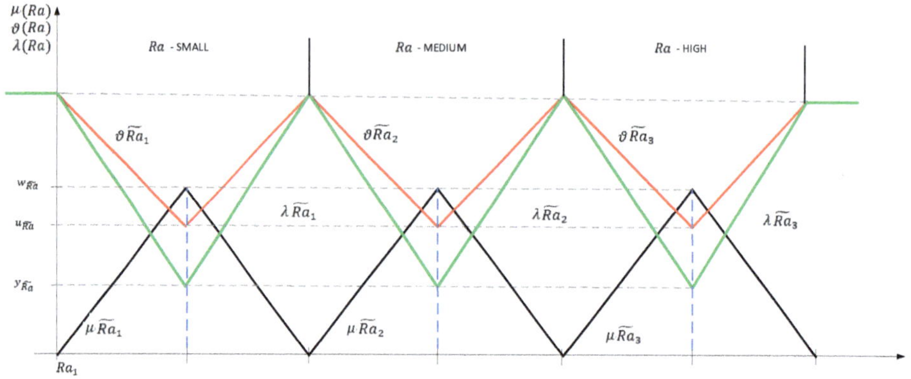

Figure 3. Modeling financial assets return with neutrosophic numbers.

The financial assets return is thus characterized by three neutrosophic numbers:

$$\begin{array}{l} \widetilde{Ra_1} = \langle(\widetilde{Ra_{a1}}, \widetilde{Ra_{b1}}, \widetilde{Ra_{c1}}); w\widetilde{Ra}, u\widetilde{Ra}, y\widetilde{Ra}\rangle \text{ for } \widetilde{Ra} \in \left[\widetilde{Ra_{a1}}, \widetilde{Ra_{c1}}\right] \\ \widetilde{Ra_2} = \langle(\widetilde{Ra_{a2}}, \widetilde{Ra_{b2}}, \widetilde{Ra_{c2}}); w\widetilde{Ra}, u\widetilde{Ra}, y\widetilde{Ra}\rangle \text{ for } \widetilde{Ra} \in \left[\widetilde{Ra_{a2}}, \widetilde{Ra_{c2}}\right] \\ \widetilde{Ra_3} = \langle(\widetilde{Ra_{a3}}, \widetilde{Ra_{b3}}, \widetilde{Ra_{c3}}); w\widetilde{Ra}, u\widetilde{Ra}, y\widetilde{Ra}\rangle \text{ for } \widetilde{Ra} \in \left[\widetilde{Ra_{a3}}, \widetilde{Ra_{c3}}\right] \end{array} \quad (23)$$

Definition 6. *The following conditions lead to a function called a weighting function f: [0,1] → R:*

(a) *It is monotone increasing ∀x, y ∈ R and x ≤ y results that f(x) ≤ f(y).*
(b) *Checks the normality condition:* $\int_0^1 f(\alpha)d\alpha = 1$

The weighting function is used in the calculation of the neutrosophic fuzzy numbers main indicators—the arithmetic mean and the squared deviation from the mean and the covariance. The most frequently used weighting function is $f(\alpha) = 2\alpha$, which meets the conditions imposed above, namely:

(a) It is monotone increasing ∀ $\alpha_1, \alpha_2 \in R$ with $\alpha_1 \leq \alpha_2$ results that $f(\alpha_1) \leq f(\alpha_2)$. From this condition, it follows that $2\alpha_1 \leq 2\alpha_2$, and, as such, the $\alpha_1 \leq \alpha_2$ is a condition fulfilled.
(b) Checks the normality $\int_0^1 f(\alpha)d\alpha = \int_0^1 2\alpha d\alpha = 2\frac{\alpha^2}{2}\Big|_0^1 = 1$.

The required conditions are fulfilled, and, thus, the function $f(\alpha) = 2\alpha$ is a weighting function for calculating the specific indicators of the triangular neutrosophic fuzzy numbers. The weighting function $f(\alpha) = 2\alpha$ is part of a weighting class of the form: $f(\alpha) = (n+1)\alpha^n$ with $n \in N$. This class of functions is used to determine the main statistical indicators of fuzzy numbers.

Definition 7. *This is considered the triangular neutrosophic fuzzy number that defines the financial asset return of the form:*

$$\widetilde{Ra_i} = \langle(\widetilde{Ra_{ai}}, \widetilde{Ra_{bi}}, \widetilde{Ra_{ci}}); w\widetilde{Ra}, u\widetilde{Ra}, y\widetilde{Ra}\rangle, \text{ for every } i = \overline{1,n}.$$

It is considered that the set $\left[\widetilde{R_a}\right]^\alpha = \left[\widetilde{Ra_1}(\alpha); \widetilde{Ra_2}(\alpha)\right]$ *for every* $\alpha \in R$ *is the level set of the triangular neutrosophic fuzzy numbers, where:*

$$\widetilde{Ra_1}(\alpha) = \langle(\widetilde{Ra_{b1}} - \widetilde{Ra_{a1}})\alpha + \widetilde{Ra_{a1}}; w\widetilde{Ra}, u\widetilde{Ra}, y\widetilde{Ra}\rangle$$

and $\widetilde{Ra_2}(\alpha) = \langle\widetilde{Ra_{c1}} - (\widetilde{Ra_{c1}} - \widetilde{Ra_{b1}})\alpha; w\widetilde{Ra}, u\widetilde{Ra}, y\widetilde{Ra}\rangle$. *See Figure 4.*

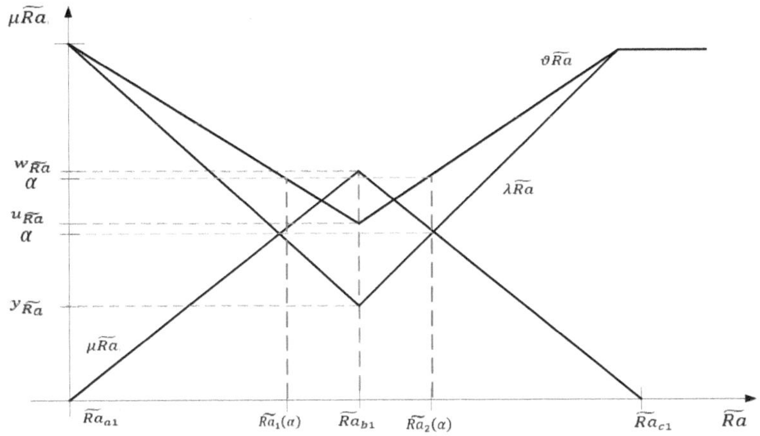

Figure 4. Levels sets for the triangular neutrosophic fuzzy number—financial asset return.

Definition 8. *The medium value of a neutrosophic fuzzy number,* $\widetilde{Ra_i} = \langle(\widetilde{Ra_{ai}}, \widetilde{Ra_{bi}}, \widetilde{Ra_{ci}}); w\widetilde{Ra}, u\widetilde{Ra}, y\widetilde{Ra}\rangle$, *is given by:*

$$E_f(\widetilde{Ra_i}) = \frac{1}{2}\int_0^1 \langle(\widetilde{Ra_1}(\alpha) + \widetilde{Ra_2}(\alpha)); w\widetilde{Ra}, u\widetilde{Ra}, y\widetilde{Ra}\rangle f(\alpha)d\alpha \tag{24}$$

When the weighting function is $f(\alpha) = 2\alpha$, *the medium value of the neutrosophic fuzzy number will be:* $E_f(\widetilde{Ra_i}) = \int_0^1 \langle(\widetilde{Ra_1}(\alpha) + \widetilde{Ra_2}(\alpha)); w\widetilde{Ra}, u\widetilde{Ra}, y\widetilde{Ra}\rangle\alpha d\alpha$.

Proposition 1. *The medium value of a triangular neutrosophic fuzzy number of the form,* $\widetilde{Ra_i} = \langle(\widetilde{Ra_{ai}}, \widetilde{Ra_{bi}}, \widetilde{Ra_{ci}}); w\widetilde{Ra}, u\widetilde{Ra}, y\widetilde{Ra}\rangle$, *is given by the relationship:*

$$E_f(\widetilde{Ra_i}) = \langle(\frac{1}{6}(\widetilde{Ra_{a1}} + \widetilde{Ra_{c1}}) + \frac{2}{3}\widetilde{Ra_{b1}}); w\widetilde{Ra}, u\widetilde{Ra}, y\widetilde{Ra}\rangle \tag{25}$$

Demonstration 1. *According to Definition 8, the medium value of the neutrosophic fuzzy number is calculated with the relation:*

$$E_f(\widetilde{Ra_i}) = \frac{1}{2}\int_0^1 \langle(\widetilde{Ra_1}(\alpha) + \widetilde{Ra_2}(\alpha)); w\widetilde{Ra}, u\widetilde{Ra}, y\widetilde{Ra}\rangle f(\alpha)d\alpha$$

Which is computed as:

$$\begin{aligned}
&\langle(\widetilde{Ra_1}(\alpha) + \widetilde{Ra_2}(\alpha)); w\widetilde{Ra}, u\widetilde{Ra}, y\widetilde{Ra}\rangle \\
&= \langle(\widetilde{Ra_{b1}} - \widetilde{Ra_{a1}})\alpha + \widetilde{Ra_{a1}} + \widetilde{Ra_{c1}} - (\widetilde{Ra_{c1}} - \widetilde{Ra_{b1}})\alpha; w\widetilde{Ra}, u\widetilde{Ra}, y\widetilde{Ra}\rangle \\
&= \langle(\widetilde{Ra_{b1}}\alpha - \widetilde{Ra_{a1}}\alpha + \widetilde{Ra_{a1}} + \widetilde{Ra_{c1}} - \widetilde{Ra_{c1}}\alpha + \widetilde{Ra_{b1}}\alpha); w\widetilde{Ra}, u\widetilde{Ra}, y\widetilde{Ra}\rangle
\end{aligned}$$

$$\langle(\widetilde{Ra_1}(\alpha) + \widetilde{Ra_2}(\alpha)); w\widetilde{Ra}, u\widetilde{Ra}, y\widetilde{Ra}\rangle = \langle(\widetilde{Ra_{a1}}(1-\alpha) + 2\widetilde{Ra_{b1}}\alpha + \widetilde{Ra_{c1}}(1-\alpha)); w\widetilde{Ra}, u\widetilde{Ra}, y\widetilde{Ra}\rangle$$

The formula above can be rewritten as follows:

$$\langle(\widetilde{Ra_1}(\alpha) + \widetilde{Ra_2}(\alpha)); w\widetilde{Ra}, u\widetilde{Ra}, y\widetilde{Ra}\rangle = \langle((1-\alpha)(\widetilde{Ra_{a1}} + \widetilde{Ra_{c1}}) + 2\widetilde{Ra_{b1}}\alpha); w\widetilde{Ra}, u\widetilde{Ra}, y\widetilde{Ra}\rangle$$

The medium value of a triangular neutrosophic fuzzy number becomes:

$$\begin{aligned}
E_f(\widetilde{Ra_i}) &= \tfrac{1}{2}\int_0^1 \langle(\widetilde{Ra_1}(\alpha) + \widetilde{Ra_2}(\alpha)); w\widetilde{Ra}, u\widetilde{Ra}, y\widetilde{Ra}\rangle f(\alpha)d\alpha \\
&= \tfrac{1}{2}\int_0^1 \left[\langle((1-\alpha)(\widetilde{Ra_{a1}} + \widetilde{Ra_{c1}}) + 2\widetilde{Ra_{b1}}\alpha); w\widetilde{Ra}, u\widetilde{Ra}, y\widetilde{Ra}\rangle\right] 2\alpha d\alpha
\end{aligned}$$

$$E_f(\widetilde{Ra_i}) = \langle(\widetilde{Ra}_{a1} + \widetilde{Ra}_{c1}); w\widetilde{Ra}, u\widetilde{Ra}, y\widetilde{Ra}\rangle \int_0^1 (\alpha - \alpha^2)d\alpha + 2\langle\widetilde{Ra}_{b1}; w\widetilde{Ra}, u\widetilde{Ra}, y\widetilde{Ra}\rangle \int_0^1 \alpha^2 d\alpha$$

$$= \langle(\widetilde{Ra}_{a1} + \widetilde{Ra}_{c1}); w\widetilde{Ra}, u\widetilde{Ra}, y\widetilde{Ra}\rangle \frac{\alpha^2}{2}\Big|_0^1 - \langle(\widetilde{Ra}_{a1} + \widetilde{Ra}_{c1}); w\widetilde{Ra}, u\widetilde{Ra}, y\widetilde{Ra}\rangle \frac{\alpha^3}{3}\Big|_0^1 + 2\langle\widetilde{Ra}_{b1}; w\widetilde{Ra}, u\widetilde{Ra}, y\widetilde{Ra}\rangle \frac{\alpha^3}{3}\Big|_0^1;$$

$$E_f(\widetilde{Ra_i}) = \frac{1}{6}\langle(\widetilde{Ra}_{a1} + \widetilde{Ra}_{c1}); w\widetilde{Ra}, u\widetilde{Ra}, y\widetilde{Ra}\rangle + \frac{2}{3}\langle\widetilde{Ra}_{b1}; w\widetilde{Ra}, u\widetilde{Ra}, y\widetilde{Ra}\rangle;$$

$$E_f(\widetilde{Ra_i}) = \langle(\frac{1}{6}(\widetilde{Ra}_{a1} + \widetilde{Ra}_{c1}) + \frac{2}{3}\widetilde{Ra}_{b1}); w\widetilde{Ra}, u\widetilde{Ra}, y\widetilde{Ra}\rangle$$

Example 1. *There are three financial assets (A_1, A_2, A_3), to which three specific triangular neutrosophic fuzzy numbers are attached for the financial assets return:*

$$\widetilde{Ra}_1 = \langle(0.2\ 0.3\ 0.5); 0.5,\ 0.2,\ 0.3\rangle \text{ for } \widetilde{Ra} \in [0,2;0,5];$$

$$\widetilde{Ra}_2 = \langle(0.1\ 0.2\ 0.3); 0.6,\ 0.3,\ 0.2\rangle \text{ for } \widetilde{Ra} \in [0,1;0,3];$$

$$\widetilde{Ra}_2 = \langle(0.3\ 0.4\ 0.6); 0.4,\ 0.3,\ 0.3\rangle \text{ for } \widetilde{Ra} \in [0,3;0,6];$$

In order to determine the average return on financial assets resulting from the model using the neutrosophic fuzzy numbers, was applied the result obtained from Proposition 1, which stipulates that:

$$E_f(\widetilde{Ra_i}) = \langle(\frac{1}{6}(\widetilde{Ra}_{a1} + \widetilde{Ra}_{c1}) + \frac{2}{3}\widetilde{Ra}_{b1}); w\widetilde{Ra}, u\widetilde{Ra}, y\widetilde{Ra}\rangle$$

Thus, it is obtained:

$$E_f(\widetilde{Ra}_1) = \langle(\frac{1}{6}(0.2 + 0.5) + \frac{2}{3} \times 0.3); 0.5,\ 0.2,\ 0.3\rangle$$

$$E_f(\widetilde{Ra}_1) = \langle(\frac{1}{6}0.7 + \frac{2}{3}0.3); 0.5,\ 0.2,\ 0.3\rangle$$

$$E_f(\widetilde{Ra}_1) = \langle 0.316; 0.5,\ 0.2,\ 0.3\rangle;$$

$$E_f(\widetilde{Ra}_2) = \langle(\frac{1}{6}(0.1 + 0.3) + \frac{2}{3} \times 0.2); 0.6,\ 0.3,\ 0.2\rangle$$

$$E_f(\widetilde{Ra}_2) = \langle(\frac{1}{6}0.4 + \frac{2}{3}0.2); 0.6,\ 0.3,\ 0.2\rangle$$

$$E_f(\widetilde{Ra}_2) = \langle 0.199; 0.5,\ 0.2,\ 0.3\rangle;$$

$$E_f(\widetilde{Ra}_3) = \langle(\frac{1}{6}(0.3 + 0.6) + \frac{2}{3} \times 0.4); 0.4,\ 0.3,\ 0.3\rangle$$

$$E_f(\widetilde{Ra}_3) = \langle(\frac{1}{6}0.9 + \frac{2}{3}0.4); 0.4,\ 0.3,\ 0.3\rangle$$

$$E_f(\widetilde{Ra}_3) = \langle 0.416; 0.4,\ 0.3,\ 0.3\rangle;$$

In conclusion, it can be argued that the financial assets return modeled by neutrosophic fuzzy numbers ensures the determination of the average value of the asset's return at the same time as establishing the degrees of truth, falsity, and indeterminacy in the process of obtaining the profitability expected by the investors.

5. Modelling the Financial Assets Risk Using the Neutrosophic Fuzzy Numbers

As mentioned above, the risk of financial assets is represented by the squared deviation from the mean (σ_a^2) which means the return on financial assets deviation from the average recorded value. Thus, the financial risk will be higher when the deviation of the financial return from the average is higher. Contrary to this, the risk exposure will be low.

Definition 9. *The financial asset risk modeled by the neutrosophic fuzzy number* $\tilde{\sigma a}_i = \langle(\tilde{\sigma a}_{ai}, \tilde{\sigma a}_{bi}, \tilde{\sigma a}_{ci}); w\tilde{\sigma a}, u\tilde{\sigma a}, y\tilde{\sigma a}\rangle$ *is given by the neutrosophic fuzzy number variance, determined by the formula:*

$$\widetilde{\sigma_f a_i} = \frac{1}{2}\int_0^1\left[\langle(\widetilde{Ra}_1(\alpha) - E_f(\widetilde{Ra}_i))^2; w\widetilde{Ra}, u\widetilde{Ra}, y\widetilde{Ra}\rangle + \langle(\widetilde{Ra}_2(\alpha) - E_f(\widetilde{Ra}_i))^2; w\widetilde{Ra}, u\widetilde{Ra}, y\widetilde{Ra}\rangle\right]f(\alpha)d\alpha \tag{26}$$

If the weighting function is $f(\alpha) = 2\alpha$, *then the neutrosophic fuzzy number variance is of the form:*

$$\widetilde{\sigma_f a_i} = \int_0^1\left[\langle(\widetilde{Ra}_1(\alpha) - E_f(\widetilde{Ra}_i))^2; w\widetilde{Ra}, u\widetilde{Ra}, y\widetilde{Ra}\rangle + \langle(\widetilde{Ra}_2(\alpha) - E_f(\widetilde{Ra}_i))^2; w\widetilde{Ra}, u\widetilde{Ra}, y\widetilde{Ra}\rangle\right]\alpha d\alpha \tag{27}$$

Proposition 2. *The financial asset variance modeled by the neutrosophic fuzzy number can be determined by the formula:*

$$\widetilde{\sigma_f a_i} = \frac{1}{2}\int_0^1 \langle\left(\widetilde{Ra}_1^2(\alpha) + \widetilde{Ra}_2^2(\alpha)\right); w\widetilde{Ra}, u\widetilde{Ra}, y\widetilde{Ra}\rangle f(\alpha)d\alpha - \frac{1}{2}\langle E_f^2(\widetilde{Ra}_i); w\widetilde{Ra}, u\widetilde{Ra}, y\widetilde{Ra}\rangle \int_0^1 f(\alpha)d\alpha \tag{28}$$

Demonstration 2. *It is known that:*

$$\widetilde{\sigma_f a_i} = \frac{1}{2}\int_0^1\left[\langle(\widetilde{Ra}_1(\alpha) - E_f(\widetilde{Ra}_i))^2; w\widetilde{Ra}, u\widetilde{Ra}, y\widetilde{Ra}\rangle + \frac{1}{2}\langle(\widetilde{Ra}_2(\alpha) - E_f(\widetilde{Ra}_i))^2; w\widetilde{Ra}, u\widetilde{Ra}, y\widetilde{Ra}\rangle\right]f(\alpha)d\alpha$$

From the relation illustrated above is obtained that:

$$\begin{aligned}\widetilde{\sigma_f a_i} = \frac{1}{2}\int_0^1\Big[&\langle(\widetilde{Ra}_1^2(\alpha) - 2\widetilde{Ra}_1(\alpha)E_f(\widetilde{Ra}_i) + E_f^2(\widetilde{Ra}_i)); w\widetilde{Ra}, u\widetilde{Ra}, y\widetilde{Ra}\rangle + \\ &\langle\left(\widetilde{Ra}_2^2(\alpha) - 2\widetilde{Ra}_2(\alpha)E_f(\widetilde{Ra}_i) + E_f^2(\widetilde{Ra}_i)\right); w\widetilde{Ra}, u\widetilde{Ra}, y\widetilde{Ra}\rangle\Big]f(\alpha)d\alpha\end{aligned}$$

$$\begin{aligned}\widetilde{\sigma_f a_i} &= \frac{1}{2}\int_0^1 \langle\left(\widetilde{Ra}_1^2(\alpha) + \widetilde{Ra}_2^2(\alpha)\right); w\widetilde{Ra}, u\widetilde{Ra}, y\widetilde{Ra}\rangle f(\alpha)d\alpha \\ &-2E_f(\widetilde{Ra}_i)\int_0^1 \langle\left(\frac{\widetilde{Ra}_1(\alpha)+\widetilde{Ra}_2(\alpha)}{2}\right); w\widetilde{Ra}, u\widetilde{Ra}, y\widetilde{Ra}\rangle f(\alpha)d\alpha \\ &+2\frac{1}{2}\langle E_f^2(\widetilde{Ra}_i); w\widetilde{Ra}, u\widetilde{Ra}, y\widetilde{Ra}\rangle \int_0^1 f(\alpha)d\alpha \\ &= \frac{1}{2}\int_0^1 \langle\left(\widetilde{Ra}_1^2(\alpha) + \widetilde{Ra}_2^2(\alpha)\right); w\widetilde{Ra}, u\widetilde{Ra}, y\widetilde{Ra}\rangle f(\alpha)d\alpha \\ &-2\langle\left(E_f(\widetilde{Ra}_i)E_f(\widetilde{Ra}_i)\right); w\widetilde{Ra}, u\widetilde{Ra}, y\widetilde{Ra}\rangle \int_0^1 f(\alpha)d\alpha \\ &+2\frac{1}{2}\langle\left(E_f^2(\widetilde{Ra}_i)\right); w\widetilde{Ra}, u\widetilde{Ra}, y\widetilde{Ra}\rangle \int_0^1 f(\alpha)d\alpha\end{aligned}$$

$$\begin{aligned}\widetilde{\sigma_f a_i} = \frac{1}{2}\int_0^1 &\langle\left(\widetilde{Ra}_1^2(\alpha) + \widetilde{Ra}_2^2(\alpha)\right); w\widetilde{Ra}, u\widetilde{Ra}, y\widetilde{Ra}\rangle f(\alpha)d\alpha - \\ 2\langle\left(E_f^2(\widetilde{Ra}_i)\right); w\widetilde{Ra}, u\widetilde{Ra}, y\widetilde{Ra}\rangle &\int_0^1 f(\alpha)d\alpha + \langle\left(E_f^2(\widetilde{Ra}_i)\right); w\widetilde{Ra}, u\widetilde{Ra}, y\widetilde{Ra}\rangle \int_0^1 f(\alpha)d\alpha;\end{aligned}$$

$$\begin{aligned}\widetilde{\sigma_f a_i} &= \frac{1}{2}\int_0^1 \langle\left(\widetilde{Ra}_1^2(\alpha) + \widetilde{Ra}_2^2(\alpha)\right); w\widetilde{Ra}, u\widetilde{Ra}, y\widetilde{Ra}\rangle f(\alpha)d\alpha \\ &-\frac{1}{2}\langle\left(E_f^2(\widetilde{Ra}_i)\right); w\widetilde{Ra}, u\widetilde{Ra}, y\widetilde{Ra}\rangle \int_0^1 f(\alpha)d\alpha\end{aligned}$$

Proposition 3. *The neutrosophic fuzzy number variance of the form* $\widetilde{\sigma a}_i = \langle(\widetilde{\sigma a}_{ai}, \widetilde{\sigma a}_{bi}, \widetilde{\sigma a}_{ci}); w\widetilde{\sigma a}, u\widetilde{\sigma a}, y\widetilde{\sigma a}\rangle$ *is given by:*

$$\begin{aligned}\sigma_f \widetilde{a}_i = &\langle \tfrac{1}{4}\Big[(\widetilde{Ra}_{b1} - \widetilde{Ra}_{a1})^2 + (\widetilde{Ra}_{c1} - \widetilde{Ra}_{b1})^2\Big]; w\widetilde{Ra}, u\widetilde{Ra}, y\widetilde{Ra}\rangle \\ &+ \tfrac{2}{3}\langle\Big[\widetilde{Ra}_{a1}(\widetilde{Ra}_{b1} - \widetilde{Ra}_{a1}) - \widetilde{Ra}_{c1}(\widetilde{Ra}_{c1} - \widetilde{Ra}_{b1})\Big]; w\widetilde{Ra}, u\widetilde{Ra}, y\widetilde{Ra}\rangle \\ &+ \langle \tfrac{1}{2}(\widetilde{Ra}_{a1}^2 + \widetilde{Ra}_{c1}^2); w\widetilde{Ra}, u\widetilde{Ra}, y\widetilde{Ra}\rangle - \langle \tfrac{1}{2} E_f^2(\widetilde{Ra}_i); w\widetilde{Ra}, u\widetilde{Ra}, y\widetilde{Ra}\rangle\end{aligned} \tag{29}$$

Demonstration 3. *From Proposition Number 2, it is known the computing relationship for the neutrosophic fuzzy number variance leads to:*

$$\sigma_f \widetilde{a}_i = \frac{1}{2}\int_0^1 \langle\Big(\widetilde{Ra}_1^2(\alpha) + \widetilde{Ra}_2^2(\alpha)\Big); w\widetilde{Ra}, u\widetilde{Ra}, y\widetilde{Ra}\rangle f(\alpha)d\alpha - \frac{1}{2}\langle E_f^2(\widetilde{Ra}_i); w\widetilde{Ra}, u\widetilde{Ra}, y\widetilde{Ra}\rangle \int_0^1 f(\alpha)d\alpha$$

From Definition 7, it is also known the fact that the triangular fuzzy number level sets are of the form:

$$\widetilde{Ra}_1(\alpha) = \langle(\widetilde{Ra}_{b1} - \widetilde{Ra}_{a1})\alpha + \widetilde{Ra}_{a1}; w\widetilde{Ra}, u\widetilde{Ra}, y\widetilde{Ra}\rangle$$
$$\widetilde{Ra}_2(\alpha) = \langle\widetilde{Ra}_{c1} - (\widetilde{Ra}_{c1} - \widetilde{Ra}_{b1})\alpha; w\widetilde{Ra}, u\widetilde{Ra}, y\widetilde{Ra}\rangle$$

After computing, the following expression is obtained:

$$\begin{aligned}\langle\Big(\widetilde{Ra}_1^2(\alpha) + \widetilde{Ra}_2^2(\alpha)\Big); w\widetilde{Ra}, u\widetilde{Ra}, y\widetilde{Ra}\rangle = \\ \Big[(\widetilde{Ra}_{b1} - \widetilde{Ra}_{a1})\alpha + \widetilde{Ra}_{a1}; w\widetilde{Ra}, u\widetilde{Ra}, y\widetilde{Ra}\Big]^2 + \\ \Big[\widetilde{Ra}_{c1} - (\widetilde{Ra}_{c1} - \widetilde{Ra}_{b1})\alpha; w\widetilde{Ra}, u\widetilde{Ra}, y\widetilde{Ra}\Big]^2\end{aligned}$$

$$\begin{aligned}&\langle\Big(\widetilde{Ra}_1^2(\alpha) + \widetilde{Ra}_2^2(\alpha)\Big); w\widetilde{Ra}, u\widetilde{Ra}, y\widetilde{Ra}\rangle = \\ &= \langle(\widetilde{Ra}_{b1} - \widetilde{Ra}_{a1})^2\alpha^2 + \widetilde{Ra}_{a1}^2 + 2\widetilde{Ra}_{a1}(\widetilde{Ra}_{b1} - \widetilde{Ra}_{a1})\alpha;\ w\widetilde{Ra}, u\widetilde{Ra}, y\widetilde{Ra}\rangle \\ &+ \langle(\widetilde{Ra}_{c1} - \widetilde{Ra}_{b1})^2\alpha^2 + \widetilde{Ra}_{c1}^2 - 2\widetilde{Ra}_{c1}(\widetilde{Ra}_{c1} - \widetilde{Ra}_{b1})\alpha;\ w\widetilde{Ra}, u\widetilde{Ra}, y\widetilde{Ra}\rangle\end{aligned}$$

$$\begin{aligned}&\langle\Big(\widetilde{Ra}_1^2(\alpha) + \widetilde{Ra}_2^2(\alpha)\Big); w\widetilde{Ra}, u\widetilde{Ra}, y\widetilde{Ra} = \\ &= \langle\alpha^2\Big[(\widetilde{Ra}_{b1} - \widetilde{Ra}_{a1})^2 + (\widetilde{Ra}_{c1} - \widetilde{Ra}_{b1})^2\Big] \\ &+ \alpha\Big[2\widetilde{Ra}_{a1}(\widetilde{Ra}_{b1} - \widetilde{Ra}_{a1}) - 2\widetilde{Ra}_{c1}(\widetilde{Ra}_{c1} - \widetilde{Ra}_{b1})\Big] \\ &+ \widetilde{Ra}_{a1}^2 + \widetilde{Ra}_{c1}^2; w\widetilde{Ra}, u\widetilde{Ra}, y\widetilde{Ra}\rangle\end{aligned}$$

The above expression can be written as:

$$\begin{aligned}&\langle\Big(\widetilde{Ra}_1^2(\alpha) + \widetilde{Ra}_2^2(\alpha)\Big); w\widetilde{Ra}, u\widetilde{Ra}, y\widetilde{Ra}\rangle \\ &= \langle\alpha^2\Big[(\widetilde{Ra}_{b1} - \widetilde{Ra}_{a1})^2 + (\widetilde{Ra}_{c1} - \widetilde{Ra}_{b1})^2\Big] \\ &+ 2\alpha\Big[\widetilde{Ra}_{a1}(\widetilde{Ra}_{b1} - \widetilde{Ra}_{a1}) - \widetilde{Ra}_{c1}(\widetilde{Ra}_{c1} - \widetilde{Ra}_{b1})\Big] \\ &+ \widetilde{Ra}_{a1}^2 + \widetilde{Ra}_{c1}^2; w\widetilde{Ra}, u\widetilde{Ra}, y\widetilde{Ra}\rangle\end{aligned}$$

By replacing the values for the level set in the variance expression, we get:

$$\begin{aligned}\widetilde{\sigma_f a_i} &= \tfrac{1}{2}\int_0^1 \langle\left(\widetilde{Ra}_1^2(\alpha)+\widetilde{Ra}_2^2(\alpha)\right); w\widetilde{Ra}, u\widetilde{Ra}, y\widetilde{Ra}\rangle f(\alpha)d\alpha \\ &-\tfrac{1}{2}\langle E_f^2(\widetilde{Ra}_i); w\widetilde{Ra}, u\widetilde{Ra}, y\widetilde{Ra}\rangle \int_0^1 f(\alpha)d\alpha \\ &= \tfrac{1}{2}\int_0^1 \langle\alpha^2\left[\left(\widetilde{Ra}_{b1}-\widetilde{Ra}_{a1}\right)^2+\left(\widetilde{Ra}_{c1}-\widetilde{Ra}_{b1}\right)^2\right]; w\widetilde{Ra}, u\widetilde{Ra}, y\widetilde{Ra}\rangle 2\alpha d\alpha \\ &+\tfrac{1}{2}\int_0^1 \langle 2\alpha\left[\widetilde{Ra}_{a1}\left(\widetilde{Ra}_{b1}-\widetilde{Ra}_{a1}\right)-\widetilde{Ra}_{c1}\left(\widetilde{Ra}_{c1}-\widetilde{Ra}_{b1}\right)\right]; w\widetilde{Ra}, u\widetilde{Ra}, y\widetilde{Ra}\rangle 2\alpha d\alpha \\ &+\tfrac{1}{2}\int_0^1 \langle \widetilde{Ra}_{a1}^2+\widetilde{Ra}_{c1}^2; w\widetilde{Ra}, u\widetilde{Ra}, y\widetilde{Ra}\rangle 2\alpha d\alpha \\ &-\tfrac{1}{2}\langle E_f^2(\widetilde{Ra}_i); w\widetilde{Ra}, u\widetilde{Ra}, y\widetilde{Ra}\rangle \int_0^1 2\alpha d\alpha\end{aligned}$$

$$\begin{aligned}\widetilde{\sigma_f a_i} &= \langle\left(\widetilde{Ra}_{b1}-\widetilde{Ra}_{a1}\right)^2+\left(\widetilde{Ra}_{c1}-\widetilde{Ra}_{b1}\right)^2; w\widetilde{Ra}, u\widetilde{Ra}, y\widetilde{Ra}\rangle \tfrac{\alpha^4}{4}\Big|_0^1 \\ &+2\langle\left[\widetilde{Ra}_{a1}\left(\widetilde{Ra}_{b1}-\widetilde{Ra}_{a1}\right)-\widetilde{Ra}_{c1}\left(\widetilde{Ra}_{c1}-\widetilde{Ra}_{b1}\right)\right]; w\widetilde{Ra}, u\widetilde{Ra}, y\widetilde{Ra}\rangle \tfrac{\alpha^3}{3}\Big|_0^1 \\ &+\langle\left(\widetilde{Ra}_{a1}^2+\widetilde{Ra}_{c1}^2\right); w\widetilde{Ra}, u\widetilde{Ra}, y\widetilde{Ra}\rangle \tfrac{\alpha^2}{2}\Big|_0^1 - \langle E_f^2(\widetilde{Ra}_i); w\widetilde{Ra}, u\widetilde{Ra}, y\widetilde{Ra}\rangle \tfrac{\alpha^2}{2}\Big|_0^1\end{aligned}$$

After making the calculations, we get:

$$\begin{aligned}\widetilde{\sigma_f a_i} &= \langle\tfrac{1}{4}\left[\left(\widetilde{Ra}_{b1}-\widetilde{Ra}_{a1}\right)^2+\left(\widetilde{Ra}_{c1}-\widetilde{Ra}_{b1}\right)^2\right]; w\widetilde{Ra}, u\widetilde{Ra}, y\widetilde{Ra}\rangle \\ &+\langle\tfrac{2}{3}\left[\widetilde{Ra}_{a1}\left(\widetilde{Ra}_{b1}-\widetilde{Ra}_{a1}\right)-\widetilde{Ra}_{c1}\left(\widetilde{Ra}_{c1}-\widetilde{Ra}_{b1}\right)\right]; w\widetilde{Ra}, u\widetilde{Ra}, y\widetilde{Ra}\rangle \\ &+\langle\tfrac{1}{2}\left(\widetilde{Ra}_{a1}^2+\widetilde{Ra}_{c1}^2\right); w\widetilde{Ra}, u\widetilde{Ra}, y\widetilde{Ra}\rangle - \langle\tfrac{1}{2}E_f^2(\widetilde{Ra}_i); w\widetilde{Ra}, u\widetilde{Ra}, y\widetilde{Ra}\rangle\end{aligned}$$

Example 2. *There are three financial assets (A_1, A_2, A_3) to which are attached three triangular neutrosophic fuzzy numbers for the financial assets return:*

$$\widetilde{Ra}_1 = \langle(0.2\ 0.3\ 0.5); 0.5,\ 0.2,\ 0.3\rangle \text{ for } \widetilde{Ra} \in [0.2; 0.5]$$

$$\widetilde{Ra}_2 = \langle(0.1\ 0.2\ 0.3); 0.6,\ 0.3,\ 0.2\rangle \text{ for } \widetilde{Ra} \in [0.1; 0.3]$$

$$\widetilde{Ra}_2 = \langle(0.3\ 0.4\ 0.6); 0.4,\ 0.3,\ 0.3\rangle \text{ for } \widetilde{Ra} \in [0.3; 0.6]$$

These triangular neutrosophic fuzzy numbers are known as the neutrosophic numbers. Using the form computed in the previous example, we get:

$$E_f(\widetilde{Ra}_1) = \langle 0.316; 0.5,\ 0.2,\ 0.3\rangle$$

$$E_f(\widetilde{Ra}_2) = \langle 0.199; 0.5,\ 0.2,\ 0.3\rangle$$

$$E_f(\widetilde{Ra}_3) = \langle 0.416; 0.4,\ 0.3,\ 0.3\rangle$$

In order to determine the financial assets variance modeled by neutrosophic numbers, the following relation is used, according to Proposition 3:

$$\begin{aligned}\widetilde{\sigma_f a_i} &= \langle\tfrac{1}{4}\left[\left(\widetilde{Ra}_{b1}-\widetilde{Ra}_{a1}\right)^2+\left(\widetilde{Ra}_{c1}-\widetilde{Ra}_{b1}\right)^2\right]; w\widetilde{Ra}, u\widetilde{Ra}, y\widetilde{Ra}\rangle \\ &+\langle\tfrac{2}{3}\left[\widetilde{Ra}_{a1}\left(\widetilde{Ra}_{b1}-\widetilde{Ra}_{a1}\right)-\widetilde{Ra}_{c1}\left(\widetilde{Ra}_{c1}-\widetilde{Ra}_{b1}\right)\right]; w\widetilde{Ra}, u\widetilde{Ra}, y\widetilde{Ra}\rangle \\ &+\langle\tfrac{1}{2}\left(\widetilde{Ra}_{a1}^2+\widetilde{Ra}_{c1}^2\right); w\widetilde{Ra}, u\widetilde{Ra}, y\widetilde{Ra}\rangle - \langle\tfrac{1}{2}E_f^2(\widetilde{Ra}_i); w\widetilde{Ra}, u\widetilde{Ra}, y\widetilde{Ra}\rangle\end{aligned}$$

By replacing the data in the above expression, this is obtained:

$$\begin{aligned}\widetilde{\sigma_f}a_1 = &\langle \tfrac{1}{4}\left[(0.3-0.2)^2+(0.5-0.3)^2\right]; 0.5,\ 0.2,\ 0.3\rangle \\ &+\langle \tfrac{2}{3}(0.2(0.3-0.2)-0.5(0.5-0.2)); 0.5,\ 0.2,\ 0.3\rangle \\ &+\langle \tfrac{1}{2}(0.2^2+0.5^2); 0.5,\ 0.2,\ 0.3\rangle - \langle \tfrac{1}{2}(0.316)^2; 0.5,\ 0.2,\ 0.3\rangle\end{aligned}$$

$$\begin{aligned}\widetilde{\sigma_f}a_1 = &\langle \tfrac{1}{4}(0.01+0.04); 0.5,\ 0.2,\ 0.3\rangle + \langle \tfrac{2}{3}(0.02-0.15); 0.5,\ 0.2,\ 0.3\rangle \\ &+\langle \tfrac{1}{2}0.29; 0.5,\ 0.2,\ 0.3\rangle - \langle \tfrac{1}{2}0,099; 0.5,\ 0.2,\ 0.3\rangle;\end{aligned}$$

$$\widetilde{\sigma_f}a_1 = \langle (0.0125-0.086+0,145-0.049; 0.5,\ 0.2,\ 0.3)\rangle;$$

$$\widetilde{\sigma_f}a_1 = \langle 0.0225; 0.5,\ 0.2,\ 0.3\rangle$$

$$\begin{aligned}\widetilde{\sigma_f a_2} = &\langle \tfrac{1}{4}\left[(0.2-0.1)^2+(0.3-0.2)^2\right]; 0.6,\ 0.3,\ 0.2\rangle \\ &+\langle \tfrac{2}{3}(0.1(0.2-0.1)-0.3(0.3-0.1)); 0.6,\ 0.3,\ 0.2\rangle \\ &+\langle \tfrac{1}{2}(0.1^2+0.3^2); 0.6,\ 0.3,\ 0.2\rangle - \langle \tfrac{1}{2}(0,199)^2; 0.6,\ 0.3,\ 0.2\rangle\end{aligned}$$

$$\begin{aligned}\widetilde{\sigma_f a_2} = &\langle \tfrac{1}{4}(0.01+0.01); 0.6,\ 0.3,\ 0.2\rangle + \langle \tfrac{2}{3}(0.01-0.03); 0.6,\ 0.3,\ 0.2\rangle + \\ &\langle \tfrac{1}{2}0.10; 0.6,\ 0.3,\ 0.2\rangle - \langle \tfrac{1}{2}0,039; 0.6,\ 0.3,\ 0.2\rangle;\end{aligned}$$

$$\widetilde{\sigma_f a_2} = \langle (0.005-0.013+0,05-0.0195; 0.6,\ 0.3,\ 0.2)\rangle;$$

$$\widetilde{\sigma_f a_2} = \langle 0.0180; 0.6,\ 0.3,\ 0.2\rangle$$

$$\begin{aligned}\widetilde{\sigma_f}a_3 = &\langle \tfrac{1}{4}\left[(0.4-0.3)^2+(0.6-0.4)^2\right]; 0.4,\ 0.3,\ 0.3\rangle \\ &+\langle \tfrac{2}{3}(0.3(0.4-0.3)-0.6(0.6-0.4)); 0.4,\ 0.3,\ 0.3\rangle \\ &+\langle \tfrac{1}{2}(0.3^2+0.6^2); 0.4,\ 0.3,\ 0.3\rangle - \langle \tfrac{1}{2}(0,416)^2; 0.4,\ 0.3,\ 0.3\rangle;\end{aligned}$$

$$\begin{aligned}\widetilde{\sigma_f}a_3 = &\langle \tfrac{1}{4}(0.01+0.04); 0.4,\ 0.3,\ 0.3\rangle \\ &+\langle \tfrac{2}{3}(0.03-0,12); 0.4,\ 0.3,\ 0.3\rangle \\ &+\langle \tfrac{1}{2}0.45; 0.4,\ 0.3,\ 0.3\rangle - \langle \tfrac{1}{2}0.173; 0.4,\ 0.3,\ 0.3\rangle;\end{aligned}$$

$$\widetilde{\sigma_f}a_3 = \langle (0.0125-0.059+0,225-0.086; 0.4,\ 0.3,\ 0.3)\rangle$$

$$\widetilde{\sigma_f}a_3 = \langle 0.0925; 0.4,\ 0.3,\ 0.3\rangle$$

In conclusion we can state that the triangular neutrosophic fuzzy number variance depends on the values of $\widetilde{Ra_{a1}}; \widetilde{Ra_{b1}}; \widetilde{Ra_{c1}}$, all of which are part of its level sets and are also on the average value of the return of the financial asset $E_f^2(\widetilde{Ra_i})$.

6. Determination of Covariance Using the Triangular Neutrosophic Fuzzy Numbers

The covariance between two triangular neutrosophic fuzzy numbers that model the return of two financial assets defines the intensity of the links between two fuzzy numbers and the way they mutually influence their profitability. There may be three possible situations:

- If the return of two financial assets increases, $\widetilde{Ra_1} > 0$ and $\widetilde{Ra_2} > 0$, both financial assets show a positive growth trend. As such, we can say that the return on financial assets is positively correlated.
- If the return of two financial assets registers different growth trends, $\widetilde{Ra_1} > 0$ and $\widetilde{Ra_2} < 0$, or $\widetilde{Ra_1} < 0$ and $\widetilde{Ra_2} > 0$, we can say that the return on financial assets isn't correlated.
- If the return of two financial assets shows negative growth trends, $\widetilde{Ra_1} < 0$ and $\widetilde{Ra_2} < 0$, both returns on financial assets decrease. As such, we can say they are negatively correlated.

Definition 10. *There are two triangular neutrosophic fuzzy numbers that define the return of two financial assets:*

$$\widetilde{Ra_1} = \langle (\widetilde{Ra_{a1}}, \widetilde{Ra_{b1}}, \widetilde{Ra_{c1}}); w\widetilde{Ra_1}, u\widetilde{Ra_1}, y\widetilde{Ra_1}\rangle,$$

and

$$\widetilde{Ra_2} = \langle(\widetilde{Ra_{a2}}, \widetilde{Ra_{b2}}, \widetilde{Ra_{c2}}); w\widetilde{Ra_1}, u\widetilde{Ra_1}, y\widetilde{Ra_1}\rangle.$$

In addition, their level sets, of the form:

$$\left[\widetilde{R_a}\right]^{\alpha} = \left[\widetilde{Ra_1}(\alpha); \widetilde{Ra_2}(\alpha)\right]$$

for every $\alpha \in [0, 1]$ where:

- For the neutrosophic level set, $\left[\widetilde{R_{a1}}\right]^{\alpha}$ will have:

$$\widetilde{Ra_{11}}(\alpha) = \langle(\widetilde{Ra_{b11}} - \widetilde{Ra_{a11}})\alpha + \widetilde{Ra_{a11}}; w\widetilde{Ra_1}, u\widetilde{Ra_1}, y\widetilde{Ra_1}\rangle \tag{30}$$

$$\widetilde{Ra_{12}}(\alpha) = \langle\widetilde{Ra_{c11}} - (\widetilde{Ra_{c11}} - \widetilde{Ra_{b11}})\alpha; w\widetilde{Ra_1}, u\widetilde{Ra_1}, y\widetilde{Ra_1}\rangle \tag{31}$$

- For the neutrosophic level set, $\left[\widetilde{R_{a2}}\right]^{\alpha}$ will have:

$$\widetilde{Ra_{21}}(\alpha) = \langle(\widetilde{Ra_{b21}} - \widetilde{Ra_{a21}})\alpha + \widetilde{Ra_{a21}}; w\widetilde{Ra_2}, u\widetilde{Ra_2}, y\widetilde{Ra_2}\rangle \tag{32}$$

$$\widetilde{Ra_{22}}(\alpha) = \langle\widetilde{Ra_{c21}} - (\widetilde{Ra_{c21}} - \widetilde{Ra_{b21}})\alpha; w\widetilde{Ra_2}, u\widetilde{Ra_2}, y\widetilde{Ra_2}\rangle \tag{33}$$

The covariance between two triangular neutrosophic fuzzy numbers $\widetilde{Ra_1}$ and $\widetilde{Ra_2}$ can be given by the following relationship:

$$\begin{aligned} cov(\widetilde{Ra_1}, \widetilde{Ra_2}) &= \tfrac{1}{2}\int_0^1 \big[\langle(\widetilde{Ra_{11}}(\alpha) - E_f(\widetilde{Ra_1}))(\widetilde{Ra_{21}}(\alpha) \\ &-E_f(\widetilde{Ra_2})); w\widetilde{Ra_1} \wedge w\widetilde{Ra_2}, u\widetilde{Ra_1} \vee u\widetilde{Ra_2}, y\widetilde{Ra_1} \vee y\widetilde{Ra_2}\rangle \\ &+\langle(\widetilde{Ra_{12}}(\alpha) - E_f(\widetilde{Ra_1}))(\widetilde{Ra_{22}}(\alpha) \\ &-E_f(\widetilde{Ra_2})); w\widetilde{Ra_1} \wedge w\widetilde{Ra_2}, u\widetilde{Ra_1} \vee u\widetilde{Ra_2}, y\widetilde{Ra_1} \vee y\widetilde{Ra_2}\rangle\big] f(\alpha)d\alpha \end{aligned} \tag{34}$$

If the weighting function is $f(\alpha) = 2\alpha$, then the covariance between the two neutrosophic fuzzy numbers will be written as follows:

$$\begin{aligned} cov(\widetilde{Ra_1}, \widetilde{Ra_2}) &= \int_0^1 \big[\langle(\widetilde{Ra_{11}}(\alpha) - E_f(\widetilde{Ra_1}))(\widetilde{Ra_{21}}(\alpha) \\ &-E_f(\widetilde{Ra_2})); w\widetilde{Ra_1} \wedge w\widetilde{Ra_2}, u\widetilde{Ra_1} \vee u\widetilde{Ra_2}, y\widetilde{Ra_1} \vee y\widetilde{Ra_2}\rangle \\ &+\langle(\widetilde{Ra_{12}}(\alpha) - E_f(\widetilde{Ra_1}))(\widetilde{Ra_{22}}(\alpha) \\ &-E_f(\widetilde{Ra_2})); w\widetilde{Ra_1} \wedge w\widetilde{Ra_2}, u\widetilde{Ra_1} \vee u\widetilde{Ra_2}, y\widetilde{Ra_1} \vee y\widetilde{Ra_2}\rangle\big] \alpha d\alpha \end{aligned} \tag{35}$$

Proposition 4. *The covariance between two triangular neutrosophic fuzzy numbers can be determined by the following relationship:*

$$\begin{aligned} &cov(\widetilde{Ra_1}, \widetilde{Ra_2}) \\ &= \tfrac{1}{2}\int_0^1 \langle(\widetilde{Ra_{11}}(\alpha)\widetilde{Ra_{21}}(\alpha) \\ &+\widetilde{Ra_{12}}(\alpha)\widetilde{Ra_{22}}(\alpha)); w\widetilde{Ra_1} \wedge w\widetilde{Ra_2}, u\widetilde{Ra_1} \vee u\widetilde{Ra_2}, y\widetilde{Ra_1} \vee y\widetilde{Ra_2}\rangle f(\alpha)d\alpha \\ &+ \int_0^1 \langle(E_f(\widetilde{Ra_1})E_f(\widetilde{Ra_2}))w\widetilde{Ra_1} \wedge w\widetilde{Ra_2}, u\widetilde{Ra_1} \vee u\widetilde{Ra_2}, y\widetilde{Ra_1} \vee y\widetilde{Ra_2};\rangle f(\alpha)d\alpha \end{aligned} \tag{36}$$

Demonstration 4.

$$\begin{aligned} cov(\widetilde{Ra_1}, \widetilde{Ra_2}) &= \tfrac{1}{2}\int_0^1 \widetilde{Ra_{11}}(\alpha)\widetilde{Ra_{21}}(\alpha) - (\widetilde{Ra_{11}}(\alpha) + \widetilde{Ra_{12}}(\alpha))E_f(\widetilde{Ra_2}) - (\widetilde{Ra_{21}} \\ &+\widetilde{Ra_{22}}(\alpha))E_f(\widetilde{Ra_1}) + \widetilde{Ra_{12}}(\alpha)\widetilde{Ra_{22}}(\alpha) \\ &+2E_f(\widetilde{Ra_1})E_f(\widetilde{Ra_2}); w\widetilde{Ra_1} \wedge w\widetilde{Ra_2}, u\widetilde{Ra_1} \vee u\widetilde{Ra_2}, y\widetilde{Ra_1} \vee y\widetilde{Ra_2} f(\alpha)d\alpha \end{aligned}$$

$$\langle\big(\widetilde{Ra_{11}}(\alpha)+\widetilde{Ra_{12}}(\alpha)\big)E_f(\widetilde{Ra_2});w\widetilde{Ra_1},u\widetilde{Ra_1},y\widetilde{Ra_1}\rangle=\langle(\widetilde{Ra}_{b11}-\widetilde{Ra}_{a11})\alpha+\widetilde{Ra}_{a11};w\widetilde{Ra_1},u\widetilde{Ra_1},y\widetilde{Ra_1}\rangle+$$

$$+\ \langle\widetilde{Ra}_{c11}-(\widetilde{Ra}_{c11}-\widetilde{Ra}_{b11})\alpha;w\widetilde{Ra_1},u\widetilde{Ra_1},y\widetilde{Ra_1}\rangle=\langle\big((1-\alpha)\widetilde{Ra}_{a11}+(1-\alpha)\widetilde{Ra}_{c11}\big);w\widetilde{Ra_1},u\widetilde{Ra_1},y\widetilde{Ra_1}\rangle$$

$$\langle\big(\widetilde{Ra_{21}}+\widetilde{Ra_{22}}(\alpha)\big)E_f(\widetilde{Ra_1});w\widetilde{Ra_2},u\widetilde{Ra_2},y\widetilde{Ra_2}\rangle=\langle(\widetilde{Ra}_{b21}-\widetilde{Ra}_{a21})\alpha+\widetilde{Ra}_{a21};w\widetilde{Ra_2},u\widetilde{Ra_2},y\widetilde{Ra_2}\rangle;$$

$$+\langle\widetilde{Ra}_{c21}-(\widetilde{Ra}_{c21}-\widetilde{Ra}_{b21})\alpha;w\widetilde{Ra_2},u\widetilde{Ra_2},y\widetilde{Ra_2}\rangle=\langle\big((1-\alpha)\widetilde{Ra}_{a21}+(1-\alpha)\widetilde{Ra}_{c21}\big);w\widetilde{Ra_2},u\widetilde{Ra_2},y\widetilde{Ra_2}\rangle$$

$$\begin{aligned}&\langle\big(\widetilde{Ra_{11}}(\alpha)+\widetilde{Ra_{12}}(\alpha)\big)E_f(\widetilde{Ra_2});w\widetilde{Ra_1},u\widetilde{Ra_1},y\widetilde{Ra_1}\\&\quad-(\widetilde{Ra_{21}}+\widetilde{Ra_{22}}(\alpha))E_f(\widetilde{Ra_1});w\widetilde{Ra_2},u\widetilde{Ra_2},y\widetilde{Ra_2}\rangle\\&\quad=\big[((1-\alpha)\widetilde{Ra}_{a11}+(1-\alpha)\widetilde{Ra}_{c11});w\widetilde{Ra_1}\wedge w\widetilde{Ra_2},u\widetilde{Ra_1}\vee u\widetilde{Ra_2},y\widetilde{Ra_1}\vee y\widetilde{Ra_2}\big]E_f(\widetilde{Ra_2})\\&\quad-\big[((1-\alpha)\widetilde{Ra}_{a21}+(1-\alpha)\widetilde{Ra}_{c21});w\widetilde{Ra_1}\wedge w\widetilde{Ra_2},u\widetilde{Ra_1}\vee u\widetilde{Ra_2},y\widetilde{Ra_1}\vee y\widetilde{Ra_2}\big]E_f(\widetilde{Ra_1})\\&\quad=\langle 0;w\widetilde{Ra_1}\wedge w\widetilde{Ra_2},u\widetilde{Ra_1}\vee u\widetilde{Ra_2},y\widetilde{Ra_1}\vee y\widetilde{Ra_2}\rangle\end{aligned}$$

The expression for covariance between two triangular neutrosophic fuzzy numbers can be written under these conditions as:

$$\begin{aligned}cov(\widetilde{Ra_1},\widetilde{Ra_2})=&\tfrac{1}{2}\int_0^1\langle(\widetilde{Ra_{11}}(\alpha)\widetilde{Ra_{21}}(\alpha)\\&+\widetilde{Ra_{12}}(\alpha)\widetilde{Ra_{22}}(\alpha));w\widetilde{Ra_1}\wedge w\widetilde{Ra_2},u\widetilde{Ra_1}\vee u\widetilde{Ra_2},y\widetilde{Ra_1}\vee y\widetilde{Ra_2}\rangle f(\alpha)d\alpha\\&+\int_0^1\langle(E_f(\widetilde{Ra_1})E_f(\widetilde{Ra_2}))w\widetilde{Ra_1}\wedge w\widetilde{Ra_2},u\widetilde{Ra_1}\vee u\widetilde{Ra_2},y\widetilde{Ra_1}\vee y\widetilde{Ra_2};\rangle f(\alpha)d\alpha\end{aligned}$$

Proposition 5. *The covariance between two triangular neutrosophic fuzzy numbers can be determined as:*

$$\begin{aligned}cov(\widetilde{Ra_1},\widetilde{Ra_2})\ =&\ \langle(\tfrac{1}{4}\big[(\widetilde{Ra}_{b11}-\widetilde{Ra}_{a11})(\widetilde{Ra}_{b21}-\widetilde{Ra}_{a21})\\&+(\widetilde{Ra}_{c11}-\widetilde{Ra}_{b11})(\widetilde{Ra}_{c21}-\widetilde{Ra}_{b21})\big]\\&+\tfrac{1}{3}\big\{\big[\widetilde{Ra}_{a21}(\widetilde{Ra}_{b11}-\widetilde{Ra}_{a11})+\widetilde{Ra}_{a11}(\widetilde{Ra}_{b21}-\widetilde{Ra}_{a21})\big]\\&-\big[\widetilde{Ra}_{c11}(\widetilde{Ra}_{c21}-\widetilde{Ra}_{b21})+\widetilde{Ra}_{c21}(\widetilde{Ra}_{c11}-\widetilde{Ra}_{b11})\big]\big\}\\&+\tfrac{1}{2}(\widetilde{Ra}_{a11}\widetilde{Ra}_{a21}+\widetilde{Ra}_{c11}\widetilde{Ra}_{c21})\\&+\tfrac{1}{2}E_f(\widetilde{Ra_1})E_f(\widetilde{Ra_2}));w\widetilde{Ra_1}\wedge w\widetilde{Ra_2},u\widetilde{Ra_1}\vee u\widetilde{Ra_2},y\widetilde{Ra_1}\vee y\widetilde{Ra_2}\rangle\end{aligned}$$

Demonstration 5. *It is known from Proposition. 4 that the covariance can be determined using the relationship:*

$$\begin{aligned}cov(\widetilde{Ra_1},\widetilde{Ra_2})\ =&\ \tfrac{1}{2}\int_0^1\langle(\widetilde{Ra_{11}}(\alpha)\widetilde{Ra_{21}}(\alpha)\\&+\widetilde{Ra_{12}}(\alpha)\widetilde{Ra_{22}}(\alpha));w\widetilde{Ra_1}\wedge w\widetilde{Ra_2},u\widetilde{Ra_1}\vee u\widetilde{Ra_2},y\widetilde{Ra_1}\vee y\widetilde{Ra_2}\rangle f(\alpha)d\alpha\\&+\int_0^1\langle(E_f(\widetilde{Ra_1})E_f(\widetilde{Ra_2}))w\widetilde{Ra_1}\wedge w\widetilde{Ra_2},u\widetilde{Ra_1}\vee u\widetilde{Ra_2},y\widetilde{Ra_1}\vee y\widetilde{Ra_2};\rangle f(\alpha)d\alpha\end{aligned}$$

At the same time, we know the expressions for the level sets as follows:

- For the neutrosophic number's level set, $\left[\widetilde{R_{a1}}\right]^\alpha$ will have:

$$\widetilde{Ra_{11}}(\alpha)=\langle(\widetilde{Ra}_{b11}-\widetilde{Ra}_{a11})\alpha+\widetilde{Ra}_{a11};w\widetilde{Ra_1},u\widetilde{Ra_1},y\widetilde{Ra_1}\rangle$$

$$\widetilde{Ra_{12}}(\alpha)=\langle\widetilde{Ra}_{c11}-(\widetilde{Ra}_{c11}-\widetilde{Ra}_{b11})\alpha;w\widetilde{Ra_1},u\widetilde{Ra_1},y\widetilde{Ra_1}\rangle$$

- For the neutrosophic number's level set, $\left[\widetilde{R_{a2}}\right]^\alpha$ will have:

$$\widetilde{Ra_{21}}(\alpha)=\langle(\widetilde{Ra}_{b21}-\widetilde{Ra}_{a21})\alpha+\widetilde{Ra}_{a21};w\widetilde{Ra_2},u\widetilde{Ra_2},y\widetilde{Ra_2}\rangle$$

$$\widetilde{Ra_{22}}(\alpha)=\langle\widetilde{Ra}_{c21}-(\widetilde{Ra}_{c21}-\widetilde{Ra}_{b21})\alpha;w\widetilde{Ra_2},u\widetilde{Ra_2},y\widetilde{Ra_2}\rangle$$

From Proposition 1, we know that the average values of the neutrosophic triangular fuzzy numbers by which the financial assets return is modeled:

$$E_f(\widetilde{Ra_1}) = \langle\left(\frac{1}{6}(\widetilde{Ra}_{a11} + \widetilde{Ra}_{c11}) + \frac{2}{3}\widetilde{Ra}_{b11}\right); w\widetilde{Ra}_1, u\widetilde{Ra}_1, y\widetilde{Ra}_1\rangle$$

$$E_f(\widetilde{Ra_2}) = \langle\left(\frac{1}{6}(\widetilde{Ra}_{a21} + \widetilde{Ra}_{c21}) + \frac{2}{3}\widetilde{Ra}_{b21}\right); w\widetilde{Ra}_2, u\widetilde{Ra}_2, y\widetilde{Ra}_2\rangle$$

By replacing in the covariance expression, we obtain:

$$\begin{aligned} cov(\widetilde{Ra}_1, \widetilde{Ra}_2) &= \tfrac{1}{2}\int_0^1 \langle(\widetilde{Ra_{11}}(\alpha)\widetilde{Ra_{21}}(\alpha) \\ &+\widetilde{Ra_{12}}(\alpha)\widetilde{Ra_{22}}(\alpha)\big); w\widetilde{Ra}_1 \wedge w\widetilde{Ra}_2, u\widetilde{Ra}_1 \vee u\widetilde{Ra}_2, y\widetilde{Ra}_1 \vee y\widetilde{Ra}_2\rangle f(\alpha)d\alpha \\ &+ \int_0^1 \langle\big(E_f(\widetilde{Ra}_1)E_f(\widetilde{Ra}_2)\big)w\widetilde{Ra}_1 \wedge w\widetilde{Ra}_2, u\widetilde{Ra}_1 \vee u\widetilde{Ra}_2, y\widetilde{Ra}_1 \vee y\widetilde{Ra}_2; \rangle f(\alpha)d\alpha \end{aligned}$$

$$\begin{aligned} cov(\widetilde{Ra}_1, \widetilde{Ra}_2) &= \tfrac{1}{2}\int_0^1 \langle(\widetilde{Ra}_{b11} - \widetilde{Ra}_{a11})\alpha + \widetilde{Ra}_{a11}; w\widetilde{Ra}_1, u\widetilde{Ra}_1, y\widetilde{Ra}_1\rangle\langle(\widetilde{Ra}_{b21} - \widetilde{Ra}_{a21})\alpha \\ &+\widetilde{Ra}_{a21}; w\widetilde{Ra}_2, u\widetilde{Ra}_2, y\widetilde{Ra}_2\rangle + \langle\widetilde{Ra}_{c11} - (\widetilde{Ra}_{c11} - \widetilde{Ra}_{b11})\alpha; w\widetilde{Ra}_1, u\widetilde{Ra}_1, y\widetilde{Ra}_1\rangle\langle\widetilde{Ra}_{c21} \\ &-(\widetilde{Ra}_{c21} - \widetilde{Ra}_{b21})\alpha; w\widetilde{Ra}_2, u\widetilde{Ra}_2, y\widetilde{Ra}_2\rangle f(\alpha)d\alpha \\ &+ \int_0^1 \langle\big(E_f(\widetilde{Ra}_1)E_f(\widetilde{Ra}_2)\big)w\widetilde{Ra}_1 \wedge w\widetilde{Ra}_2, u\widetilde{Ra}_1 \vee u\widetilde{Ra}_2, y\widetilde{Ra}_1 \vee y\widetilde{Ra}_2; \rangle f(\alpha)d\alpha \end{aligned}$$

$$\begin{aligned} cov(\widetilde{Ra}_1, \widetilde{Ra}_2) &= \tfrac{1}{2}\int_0^1 \langle\big((\widetilde{Ra}_{b11} - \widetilde{Ra}_{a11})(\widetilde{Ra}_{b21} - \widetilde{Ra}_{a21})\alpha^2 \\ &+\big[\widetilde{Ra}_{a21}(\widetilde{Ra}_{b11} - \widetilde{Ra}_{a11}) + \widetilde{Ra}_{a11}(\widetilde{Ra}_{b21} - \widetilde{Ra}_{a21})\big]\alpha + \widetilde{Ra}_{a11}\widetilde{Ra}_{a21} \\ &+\alpha^2(\widetilde{Ra}_{c11} - \widetilde{Ra}_{b11})(\widetilde{Ra}_{c21} - \alpha\big[\widetilde{Ra}_{c11}(\widetilde{Ra}_{c21} - \widetilde{Ra}_{b21}) + \widetilde{Ra}_{c21}(\widetilde{Ra}_{c11} - \widetilde{Ra}_{b11})\big] \\ &+\widetilde{Ra}_{c11}\widetilde{Ra}_{c21}; w\widetilde{Ra}_1 \wedge w\widetilde{Ra}_2, u\widetilde{Ra}_1 \vee u\widetilde{Ra}_2, y\widetilde{Ra}_1 \vee y\widetilde{Ra}_2\rangle f(\alpha)d\alpha \\ &+ \int_0^1 \langle\big(E_f(\widetilde{Ra}_1)E_f(\widetilde{Ra}_2)\big); w\widetilde{Ra}_1 \wedge w\widetilde{Ra}_2, u\widetilde{Ra}_1 \vee u\widetilde{Ra}_2, y\widetilde{Ra}_1 \vee y\widetilde{Ra}_2; \rangle f(\alpha)d\alpha \end{aligned}$$

$$\begin{aligned} cov(\widetilde{Ra}_1, \widetilde{Ra}_2) &= \langle(\widetilde{Ra}_{b11} - \widetilde{Ra}_{a11})(\widetilde{Ra}_{b21} - \widetilde{Ra}_{a21})\tfrac{\alpha^4}{4}\Big|_0^1 \\ &+\big[\widetilde{Ra}_{a21}(\widetilde{Ra}_{b11} - \widetilde{Ra}_{a11}) \\ &++\widetilde{Ra}_{a11}(\widetilde{Ra}_{b21} - \widetilde{Ra}_{a21})\big]\tfrac{\alpha^3}{3}\Big|_0^1 + \widetilde{Ra}_{a11}\widetilde{Ra}_{a21}\tfrac{\alpha^3}{2}\Big|_0^1 \\ &+(\widetilde{Ra}_{c11} - \widetilde{Ra}_{b11})(\widetilde{Ra}_{c21} - \widetilde{Ra}_{b21})\tfrac{\alpha^4}{4}\Big|_0^1 \\ &-\big[\widetilde{Ra}_{c11}(\widetilde{Ra}_{c21} - \widetilde{Ra}_{b21}) + \widetilde{Ra}_{c21}(\widetilde{Ra}_{c11} - \widetilde{Ra}_{b11})\big]\tfrac{\alpha^3}{3}\Big|_0^1 \\ &+\widetilde{Ra}_{c11}\widetilde{Ra}_{c21}\tfrac{\alpha^3}{2}\Big|_0^1; w\widetilde{Ra}_1 \wedge w\widetilde{Ra}_2, u\widetilde{Ra}_1 \vee u\widetilde{Ra}_2, y\widetilde{Ra}_1 \vee y\widetilde{Ra}_2\rangle \\ &+E_f(\widetilde{Ra}_1)E_f(\widetilde{Ra}_2)\tfrac{\alpha^3}{2}\Big|_0^1; w\widetilde{Ra}_1 \wedge w\widetilde{Ra}_2, u\widetilde{Ra}_1 \vee u\widetilde{Ra}_2, y\widetilde{Ra}_1 \vee y\widetilde{Ra}_2\rangle \end{aligned}$$

The above equation can be written as:

$$\begin{aligned} cov(\widetilde{Ra}_1, \widetilde{Ra}_2) &= \langle\tfrac{1}{4}\big[(\widetilde{Ra}_{b11} - \widetilde{Ra}_{a11})(\widetilde{Ra}_{b21} - \widetilde{Ra}_{a21}) \\ &+(\widetilde{Ra}_{c11} - \widetilde{Ra}_{b11})(\widetilde{Ra}_{c21} - \widetilde{Ra}_{b21})\big] \\ &+\tfrac{1}{3}\big\{\big[\widetilde{Ra}_{a21}(\widetilde{Ra}_{b11} - \widetilde{Ra}_{a11}) + \widetilde{Ra}_{a11}(\widetilde{Ra}_{b21} - \widetilde{Ra}_{a21})\big] \\ &-\big[\widetilde{Ra}_{c11}(\widetilde{Ra}_{c21} - \widetilde{Ra}_{b21}) + \widetilde{Ra}_{c21}(\widetilde{Ra}_{c11} - \widetilde{Ra}_{b11})\big]\big\} \\ &+\tfrac{1}{2}(\widetilde{Ra}_{a11}\widetilde{Ra}_{a21} + \widetilde{Ra}_{c11}\widetilde{Ra}_{c21}) \\ &+\tfrac{1}{2}E_f(\widetilde{Ra}_1)E_f(\widetilde{Ra}_2)\big); w\widetilde{Ra}_1 \wedge w\widetilde{Ra}_2, u\widetilde{Ra}_1 \vee u\widetilde{Ra}_2, y\widetilde{Ra}_1 \vee y\widetilde{Ra}_2\rangle \end{aligned}$$

In conclusion, for the triangular neutrosophic fuzzy numbers, the expressions for the following statistical indicators were obtained:

(a) The financial asset return:

$$E_f(\widetilde{Ra}_i) = \langle\left(\frac{1}{6}(\widetilde{Ra}_{a1} + \widetilde{Ra}_{c1}) + \frac{2}{3}\widetilde{Ra}_{b1}\right); w\widetilde{Ra}, u\widetilde{Ra}, y\widetilde{Ra}\rangle$$

(b) The variance, or the financial asset's risk:

$$\widetilde{\sigma_f a_i} = \langle\Big(\tfrac{1}{4}\Big[(\widetilde{Ra}_{b1} - \widetilde{Ra}_{a1})^2 + (\widetilde{Ra}_{c1} - \widetilde{Ra}_{b1})^2\Big] \\ +\tfrac{2}{3}\Big[\widetilde{Ra}_{a1}(\widetilde{Ra}_{b1} - \widetilde{Ra}_{a1}) - \widetilde{Ra}_{c1}(\widetilde{Ra}_{c1} - \widetilde{Ra}_{b1})\Big] + \tfrac{1}{2}(\widetilde{Ra}_{a1}^2 + \widetilde{Ra}_{c1}^2) - \tfrac{1}{2}E_f^2(\widetilde{Ra}_i)\Big); w\widetilde{Ra}, u\widetilde{Ra}, y\widetilde{Ra}\rangle$$

(c) The covariance between two financial assets modeled with triangular neutrosophic fuzzy numbers:

$$\begin{aligned} cov(\widetilde{Ra}_1, \widetilde{Ra}_2) &= \langle\Big(\tfrac{1}{4}\Big[(\widetilde{Ra}_{b11} - \widetilde{Ra}_{a11})(\widetilde{Ra}_{b21} - \widetilde{Ra}_{a21}) \\ &+(\widetilde{Ra}_{c11} - \widetilde{Ra}_{b11})(\widetilde{Ra}_{c21} - \widetilde{Ra}_{b21})\Big] \\ &+\tfrac{1}{3}\Big\{\Big[\widetilde{Ra}_{a21}(\widetilde{Ra}_{b11} - \widetilde{Ra}_{a11}) + \widetilde{Ra}_{a11}(\widetilde{Ra}_{b21} - \widetilde{Ra}_{a21})\Big] \\ &-\Big[\widetilde{Ra}_{c11}(\widetilde{Ra}_{c21} - \widetilde{Ra}_{b21}) + \widetilde{Ra}_{c21}(\widetilde{Ra}_{c11} - \widetilde{Ra}_{b11})\Big]\Big\} \\ &+\tfrac{1}{2}(\widetilde{Ra}_{a11}\widetilde{Ra}_{a21} + \widetilde{Ra}_{c11}\widetilde{Ra}_{c21}) \\ &+\tfrac{1}{2}E_f(\widetilde{Ra}_1)E_f(\widetilde{Ra}_2)\Big); w\widetilde{Ra}_1 \wedge w\widetilde{Ra}_2, u\widetilde{Ra}_1 \vee u\widetilde{Ra}_2, y\widetilde{Ra}_1 \vee y\widetilde{Ra}_2\rangle \end{aligned}$$

Example 3. *There are two financial assets, (A_1, A_2), to which two triangular neutrosophic fuzzy numbers are attached:*

$$\widetilde{Ra}_1 = \langle(0.2\ 0.3\ 0.5); 0.5,\ 0.2,\ 0.3\rangle \text{ for } \widetilde{Ra} \in [0,2;0,5]$$

$$\widetilde{Ra}_2 = \langle(0.1\ 0.2\ 0.3); 0.6,\ 0.3,\ 0.2\rangle \text{ for } \widetilde{Ra} \in [0,1;0,3]$$

In order to determine the covariance between two financial assets to determine the $cov(\widetilde{Ra}_1, \widetilde{Ra}_2)$ and the variance–covariance matrix, the following formula can be used:

$$\begin{aligned} cov(\widetilde{Ra}_1, \widetilde{Ra}_2) &= \langle\Big(\tfrac{1}{4}\Big[(\widetilde{Ra}_{b11} - \widetilde{Ra}_{a11})(\widetilde{Ra}_{b21} - \widetilde{Ra}_{a21}) \\ &+(\widetilde{Ra}_{c11} - \widetilde{Ra}_{b11})(\widetilde{Ra}_{c21} - \widetilde{Ra}_{b21})\Big] \\ &+\tfrac{1}{3}\Big\{\Big[\widetilde{Ra}_{a21}(\widetilde{Ra}_{b11} - \widetilde{Ra}_{a11}) + \widetilde{Ra}_{a11}(\widetilde{Ra}_{b21} - \widetilde{Ra}_{a21})\Big] \\ &-\Big[\widetilde{Ra}_{c11}(\widetilde{Ra}_{c21} - \widetilde{Ra}_{b21}) + \widetilde{Ra}_{c21}(\widetilde{Ra}_{c11} - \widetilde{Ra}_{b11})\Big]\Big\} \\ &+\tfrac{1}{2}(\widetilde{Ra}_{a11}\widetilde{Ra}_{a21} + \widetilde{Ra}_{c11}\widetilde{Ra}_{c21}) \\ &+\tfrac{1}{2}E_f(\widetilde{Ra}_1)E_f(\widetilde{Ra}_2)\Big); w\widetilde{Ra}_1 \wedge w\widetilde{Ra}_2, u\widetilde{Ra}_1 \vee u\widetilde{Ra}_2, y\widetilde{Ra}_1 \vee y\widetilde{Ra}_2\rangle \end{aligned}$$

By replacing in the formula, we can obtain:

$$\begin{aligned} cov(\widetilde{Ra}_1, \widetilde{Ra}_2) &= \langle\tfrac{1}{4}[(0.3 - 0.2)(0.2 - 0.1) + (0.5 - 0.3)(0.3 - 0.2)] \\ &+\tfrac{1}{3}[0.1(0.3 - 0.2) + 0.2(0.2 - 0.1)] \\ &-[0.5(0.3 - 0.2) + 0.3(0.5 - 0.3)] \\ &+\tfrac{1}{2}(0.2 * 0.1 + 0.5 * 0.3) + \tfrac{1}{2}0,316 * 0,199; 0.5 \wedge 0.6,\ 0.2 \vee 0.3, 0.3 \vee 0.2\rangle \end{aligned}$$

$$\begin{aligned} cov(\widetilde{Ra}_1, \widetilde{Ra}_2) &= \langle\tfrac{1}{4}[0.01 + 0.02] + \tfrac{1}{3}[0.01 + 0.02] - [0.05 + 0.06] + \tfrac{1}{2}(0.02 + 0.15) \\ &+\tfrac{1}{2}0.062; 0.6,\ 0.2,\ 0.2\rangle \end{aligned}$$

$$cov(\widetilde{Ra}_1, \widetilde{Ra}_2) = \langle 0.0075 - 0.053 + 0.085 + 0.031; 0.6,\ 0.2,\ 0.2\rangle$$

$$cov(\widetilde{Ra}_1, \widetilde{Ra}_2) = \langle 0.0705; 0.6,\ 0.2,\ 0.2\rangle$$

The variance–covariance matrix is then:

$$\Omega = \begin{pmatrix} \langle 0.0225; 0.5,\ 0.2,\ 0.3\rangle & \langle 0.0705; 0.6,\ 0.2,\ 0.2\rangle \\ \langle 0.0705; 0.6,\ 0.2,\ 0.2\rangle & \langle 0.0180; 0.6,\ 0.3,\ 0.2\rangle \end{pmatrix}$$

Modeling the performance indicators leads to the following results:

- The financial asset return:

$$E_f(\widetilde{Ra_i}) = \langle\Big(\frac{1}{6}(\widetilde{Ra_{a1}} + \widetilde{Ra_{c1}}) + \frac{2}{3}\widetilde{Ra_{b1}}\Big); w\widetilde{Ra}, u\widetilde{Ra}, y\widetilde{Ra}\rangle$$

- The variance of the financial asset risk:

$$\begin{aligned}\widetilde{\sigma_f a_i} = \langle\Big(&\tfrac{1}{4}\Big[(\widetilde{Ra_{b1}} - \widetilde{Ra_{a1}})^2 + (\widetilde{Ra_{c1}} - \widetilde{Ra_{b1}})^2\Big]\\ &+\tfrac{2}{3}\Big[\widetilde{Ra_{a1}}(\widetilde{Ra_{b1}} - \widetilde{Ra_{a1}}) - \widetilde{Ra_{c1}}(\widetilde{Ra_{c1}} - \widetilde{Ra_{b1}})\Big] + \tfrac{1}{2}(\widetilde{Ra_{a1}}^2 + \widetilde{Ra_{c1}}^2) - \tfrac{1}{2}E_f^2(\widetilde{Ra_i})\Big); w\widetilde{Ra}, u\widetilde{Ra}, y\widetilde{Ra}\rangle\end{aligned}$$

- The covariance between two financial assets modelled with the triangular neutrosophic fuzzy numbers:

$$\begin{aligned}cov(\widetilde{Ra_1}, \widetilde{Ra_2}) &= \langle\Big(\tfrac{1}{4}\Big[(\widetilde{Ra_{b11}} - \widetilde{Ra_{a11}})(\widetilde{Ra_{b21}} - \widetilde{Ra_{a21}})\\ &+(\widetilde{Ra_{c11}} - \widetilde{Ra_{b11}})(\widetilde{Ra_{c21}} - \widetilde{Ra_{b21}})\Big]\\ &+\tfrac{1}{3}\Big\{\Big[\widetilde{Ra_{a21}}(\widetilde{Ra_{b11}} - \widetilde{Ra_{a11}}) + \widetilde{Ra_{a11}}(\widetilde{Ra_{b21}} - \widetilde{Ra_{a21}})\Big]\\ &-\Big[\widetilde{Ra_{c11}}(\widetilde{Ra_{c21}} - \widetilde{Ra_{b21}}) + \widetilde{Ra_{c21}}(\widetilde{Ra_{c11}} - \widetilde{Ra_{b11}})\Big]\Big\}\\ &+\tfrac{1}{2}(\widetilde{Ra_{a11}}\widetilde{Ra_{a21}} + \widetilde{Ra_{c11}}\widetilde{Ra_{c21}})\\ &+\tfrac{1}{2}E_f(\widetilde{Ra_1})E_f(\widetilde{Ra_2})\Big); w\widetilde{Ra_1} \wedge w\widetilde{Ra_2}, u\widetilde{Ra_1} \vee u\widetilde{Ra_2}, y\widetilde{Ra_1} \vee y\widetilde{Ra_2}\rangle\end{aligned}$$

Conclusion: The covariance of two financial assets determined according to the above formula shows that there is a weak link between the two financial assets modeled by triangular neutrosophic fuzzy numbers $cov(\widetilde{Ra_1}, \widetilde{Ra_2}) = \langle 0.0705; 0.6,\ 0.2,\ 0.2\rangle$. The covariance is positive, resulting in the fact that the financial returns of the two assets increase and register a favorable trend.

7. Conclusions

The performance indicators of the financial assets are represented by the return on financial assets, the financial assets risk, and the covariance between them, the latter of which, as mentioned above, indicates the intensity of the links between the return on financial assets. Modeling these three performance indicators of the financial assets has been achieved with the help of triangular neutrosophic fuzzy numbers, which presents a number of advantages:

- The neutrosophic approach of these three financial assets performance indicators must take all the possible scenarios for their achievement into account—these are the scenario of certainty, the scenario of non-realization, and the scenario of indecision (in which it cannot be appreciated whether the performance indicators are or are not achieved). All three scenarios have attached performance, non-execution, or uncertainty ratios according to the investor's professional judgment.
- The possibility of stratification or the clustering of the financial asset return values according to linguistic criteria such as very good, good or weak performance indicators. This method of stratification can be also applied in the calculation/determination of financial risk. Stratification/clustering takes place with the help of triangular neutrosophic fuzzy numbers.
- The possibility of selecting the desired return/risk group in order to maximize the investor earnings, analyzing the profitability of the financial asset by means of probability grades or other purposes desired by investors, etc.

The results obtained by modeling with triangular neutrosophic fuzzy numbers: The financial assets return, the financial risk, and the covariance between two financial assets were tested on three practical examples in order to confirm their applicability. For future research, two aspects for modelling with the help of neutrosophic fuzzy intelligence are being considered: The mitigation of portfolio risk and optimal portfolios.

Author Contributions: Conceptualization, M.-I.B., I.-A.B. and C.D.; data curation, C.D.; formal analysis, M.-I.B. and I.-A.B.; investigation, M.-I.B., I.-A.B. and C.D.; methodology, M.-I.B. and I.-A.B.; supervision, M.-I.B.; Validation, I.-A.B.; visualization, C.D.; writing—original draft, M.-I.B. and I.-A.B.; writing—review & editing, C.D.

Funding: This research received no external funding.

Conflicts of Interest: The authors declare no conflict of interest.

References

1. Boloș, M.-I.; Bradea, I.-A.; Delcea, C. A Fuzzy Logic Algorithm for Optimizing the Investment Decisions within Companies. *Symmetry* **2019**, *11*, 186. [CrossRef]
2. Markowitz, H. Portfolio Selection. *J. Financ.* **1952**, *7*, 77.
3. Guo, Y.; Sengur, A. A Novel Color Image Segmentation Approach Based on Neutrosophic Set and Modified Fuzzy c-Means. *Circuits Syst. Signal Process.* **2013**, *32*, 1699–1723. [CrossRef]
4. Ansari, A.Q.; Biswas, R.; Aggarwal, S. Neutrosophic classifier: An extension of fuzzy classifer. *Appl. Soft Comput.* **2013**, *13*, 563–573. [CrossRef]
5. Ali, M.; Smarandache, F. Complex neutrosophic set. *Neural Comput. Appl.* **2017**, *28*, 1817–1834. [CrossRef]
6. Pramanik, S.; Dey, P.; Smarandache, F. Correlation Coefficient Measures of Interval Bipolar Neutrosophic Sets for Solving Multi-Attribute Decision Making Problems. *Neutrosophic Sets Syst.* **2018**, *19*, 70–79.
7. Ye, J. A multicriteria decision-making method using aggregation operators for simplified neutrosophic sets. *J. Intell. Fuzzy Syst.* **2014**, *26*, 2459–2466.
8. Zhang, H.; Wang, J.; Chen, X. Interval Neutrosophic Sets and Their Application in Multicriteria Decision Making Problems. *Sci. World J.* **2014**, *2014*, 1–15. [CrossRef]
9. Liu, P.; Teng, F. Multiple attribute decision making method based on normal neutrosophic generalized weighted power averaging operator. *Int. J. Mach. Learn. Cybern.* **2018**, *9*, 281–293. [CrossRef]
10. Peng, J.; Wang, J.; Zhang, H.; Chen, X. An outranking approach for multi-criteria decision-making problems with simplified neutrosophic sets. *Appl. Soft Comput.* **2014**, *25*, 336–346. [CrossRef]
11. Bausys, R.; Zavadskas, E.K.; Kaklauskas, A. Application of Netrosophic Set to Multi-criteria Decision Making by COPRAS. *Econ. Comput. Econ. Cybern. Stud. Res.* **2015**, *49*, 91–105.
12. Bausys, R.; Zavadskas, E.K. Multi-criteria Decision-Making Approach by Vikor under Interval Neutrosophic Set Environment. *Econ. Comput. Econ. Cybern. Stud. Res.* **2015**, *49*, 33–48.
13. Nădăban, S.; Dzitac, S.; Dzitac, I. Fuzzy TOPSIS: A General View. *Procedia Comput. Sci.* **2016**, *91*, 823–831. [CrossRef]
14. Zhang, H.; Ji, P.; Wang, J.; Chen, X. A novel decision support model for satisfactory restaurants utilizing social information: A case study of TripAdvisor.com. *Tour. Manag.* **2017**, *59*, 281–297. [CrossRef]
15. Tian, Z.; Wang, J.; Wang, J.; Zhang, H. Simplified Neutrosophic Linguistic Multi-criteria Group Decision-Making Approach to Green Product Development. *Group Decis. Negot.* **2017**, *26*, 597–627. [CrossRef]
16. Palmowski, Z.; Stettner, Ł.; Sulima, A. Optimal Portfolio Selection in an Itô–Markov Additive Market. *Risks* **2019**, *7*, 34. [CrossRef]
17. Jain, P.; Jain, S. Can Machine Learning-Based Portfolios Outperform Traditional Risk-Based Portfolios? The Need to Account for Covariance Misspecification. *Risks* **2019**, *7*, 74. [CrossRef]
18. Dreżewski, R.; Doroz, K. An Agent-Based Co-Evolutionary Multi-Objective Algorithm for Portfolio Optimization. *Symmetry* **2017**, *9*, 168. [CrossRef]
19. Agapie, A.; Vizitiu, C.; Cristache, S.; Năstase, M.; Crăciun, L.; Molănescu, A. Analysis of Corporate Entrepreneurship in Public R&D Institutions. *Sustainability* **2018**, *10*, 2297.
20. Lin, A.J.; Chang, H.-Y. Business Sustainability Performance Evaluation for Taiwanese Banks—A Hybrid Multiple-Criteria Decision-Making Approach. *Sustainability* **2019**, *11*, 2236. [CrossRef]
21. Lassala, C.; Apetrei, A.; Sapena, J. Sustainability Matter and Financial Performance of Companies. *Sustainability* **2017**, *9*, 1498. [CrossRef]

22. Boloș, M.I.; Bradea, I.A.; Delcea, C. The Development of a Fuzzy Logic System in a Stochastic Environment with Normal Distribution Variables for Cash Flow Deficit Detection in Corporate Loan Policy. *Symmetry* **2019**, *11*, 548. [CrossRef]
23. Abdel-Basset, M.; Chang, V.; Mohamed, M.; Smarandche, F. A Refined Approach for Forecasting Based on Neutrosophic Time Series. *Symmetry* **2019**, *11*, 457. [CrossRef]
24. Subas, Y. Neutrosophic Numbers and Their Application to Multi Attribute Decision Making Problems. Master's Thesis, Kilis 7 Aralık University, Graduate School of Natural and Applied Science, Ankara, Turkey, 2005.
25. Sahin, M.; Kargin, A.; Smarandache, F. Generalized Single Valued Triangular Neutrosophic Numbers and Aggregation Operators for Application to Multi-attribute Group Decision Making. *New Trends Neutrosophic Theory Appl.* **2018**, *2*, 51–84.

symmetry

MDPI

Article

Study on the Algebraic Structure of Refined Neutrosophic Numbers

Qiaoyan Li [1], Yingcang Ma [1,*], Xiaohong Zhang [2] and Juanjuan Zhang [1]

[1] School of Science, Xi'an Polytechnic University, Xi'an 710048, China
[2] School of Arts and Sciences, Shaanxi University of Science & Technology, Xi'an 710021, China
* Correspondence: mayingcang@126.com

Received: 28 May 2019; Accepted: 22 July 2019; Published: 27 July 2019

Abstract: This paper aims to explore the algebra structure of refined neutrosophic numbers. Firstly, the algebra structure of neutrosophic quadruple numbers on a general field is studied. Secondly, The addition operator $\oplus$ and multiplication operator $\otimes$ on refined neutrosophic numbers are proposed and the algebra structure is discussed. We reveal that the set of neutrosophic refined numbers with an additive operation is an abelian group and the set of neutrosophic refined numbers with a multiplication operation is a neutrosophic extended triplet group. Moreover, algorithms for solving the neutral element and opposite elements of each refined neutrosophic number are given.

Keywords: neutrosophic extended triplet group; neutrosophic quadruple numbers; refined neutrosophic numbers; refined neutrosophic quadruple numbers; neutrosophic set

1. Introduction

The notion of neutrosophic set was proposed by F. Smarandache [1], which is an extension of fuzzy set and in order to solve real-world problems. A neutrosophic set has three membership functions, and each membership degree is a real standard or non-standard subset of the nonstandard unit interval $]0^-, 1^+[= 0^- \cup [0,1] \cup 1^+$.

In recent years, the idea of neutrosophic set has been applicable in related algebraic structures. Among these algebraic structures, Smarandache and Ali [2] proposed the algebraic system neutrosophic triplet group (NTG), which is an extension of the classical group but the neutral element is different from the classical algebraic unit element. To regard the unit element as a special neutral element, the neutrosophic extended triplet group (NETG) has been proposed [3,4] and the classical group is regarded as a special case of a NETG. Moreover, some research papers have carried out in-depth research based on NTG (NETG). For example, the inclusion relations of neutrosophic sets [5], neutrosophic triplet coset [6], neutrosophic duplet semi-groups [7], generalized neutrosophic extended triplet group [8], AG-neutrosophic extended triplet loops [9,10], the neutrosophic set theory to pseudo-BCI algebras [11], neutrosophic triplet ring and a neutrosophic triplet field [12,13], neutrosophic triplet normed space [14], neutrosophic soft sets [15], neutrosophic vector spaces [16] and so on have been studied.

As an example of NETG, Ma [8] revealed that for each $n \in Z^+, n \geq 2, (Z_n, \otimes)$ is a commutative NETG if and only if the factorization of n is a product of single factors. As another example, Ma [17] showed that the set of neutrosophic quadruple numbers with a multiplication operation is a NETG. The concept of neutrosophic numbers of the form $a + bI$, where I is the indeterminacy with $I^n = I$, and, a and are real or complex numbers. If I into many types of indeterminacies $I_1, I_2, \cdots, I_q$, in [18], Smarandache extended the neutrosophic numbers $a + bI$ into refined neutrosophic numbers of the form $a + b_1 I_1 + b_2 I_2 + \cdots + b_n I_n$, where $a, b_1, b_2, \cdots, b_n$ are real or complex numbers and considered the refined neutrosophic set based on these refined neutrosophic numbers. The notion of neutrosophic quadruple number, which has form: $NQ = a + bT + cI + dF$ where a, b, c, d are real (or complex) numbers; and T is the truth/membership/probability; I is the indeterminacy; and F is the false/membership/improbability are called Neutrosophic Quadruple (Real, respectively, Complex) Numbers. "a" is called the known part of NQ, while $bT + cI + dF$ is called the unknown part of NQ. Similar to refined neutrosophic numbers, if T can be split into many types of truths, $T_1, T_2, \cdots, T_p$, I into many types of indeterminacies, $I_1, I_2, \cdots, I_r$, and F into many types of falsities, $F_1, F_2, \cdots, F_r$, we can get the refined neutrosophic quadruple numbers. We know that the set of neutrosophic quadruple numbers with a multiplication operation is a NETG. In this paper, we explore the algebra structure of refined neutrosophic numbers (refined neutrosophic quadruple numbers) and give new examples of NETG. In fact, the solving method of the neutral element and opposite elements for each refined neutrosophic number is different from the solving method for each neutrosophic quadruple number.

The paper is organized as follows. Section 2 gives the basic concepts. In Section 3, we show that the set of neutrosophic quadruple numbers on the general field with a multiplication operation also consists of a NETG. In Section 4, the algebra structure of refined neutrosophic numbers and refined neutrosophic quadruple numbers are studied. Finally, the summary and future work is presented in Section 5.

2. Basic Concepts

In this section, we provide the related basic definitions and properties of NETG, neutrosophic quadruple numbers, and refined neutrosophic numbers (for details, see [3,4,18–20]).

Definition 1 ([3,4]). *Let N be a non-empty set together with a binary operation $*$. Then, N is called a neutrosophic extended triplet set if, for any $a \in N$, there exists a neutral of "a" (denote by $neut(a)$), and an opposite of "a"(denote by $anti(a)$), such that $neut(a) \in N$, $anti(a) \in N$ and:*

$$a * neut(a) = neut(a) * a = a, \quad a * anti(a) = anti(a) * a = neut(a).$$

The triplet $(a, neut(a), anti(a))$ is called a neutrosophic extended triplet.

Definition 2 ([3,4]). *Let $(N, *)$ be a neutrosophic extended triplet set. Then, N is called a neutrosophic extended triplet group (NETG), if the following conditions are satisfied:*

(1) $(N, *)$ is well-defined, i.e., for any $a, b \in N$, one has $a * b \in N$.
(2) $(N, *)$ is associative, i.e., $(a * b) * c = a * (b * c)$ for all $a, b, c \in N$.

A NETG N is called a commutative NETG if for all $a, b \in N, a * b = b * a$.

Proposition 1 ([4]). *$(N, *)$ be a NETG. We have:*

(1) $neut(a)$ is unique for any $a \in N$.
(2) $neut(a) * neut(a) = neut(a)$ for any $a \in N$.
(3) $neut(neut(a)) = neut(a)$ for any $a \in N$.

Definition 3 ([18,19]). *A neutrosophic number is a number of the form (a, bI), where I is the indeterminacy with $I^2 = I$, and a and b are real or complex numbers. A refined neutrosophic number is a number of the form $(a_0, a_1I_1, a_2I_2, \cdots, a_nI_n)$, where $I_1, I_2, \cdots, I_n$ are different types of indeterminacies, and $a_0, a_1, a_2, \cdots, a_n$ are real or complex numbers. The set NN defined by*

$$NN = \{(a, bI) | a, b \in \mathbb{R} \text{ or } \mathbb{C}\}. \tag{1}$$

is called a neutrosophic set of neutrosophic numbers. The set RNN defined by

$$RNN = \{(a_0, aI_1, a_2I_2, \cdots, a_nI_n) | a_0, a_1, a_2, \cdots, a_n \in \mathbb{R} \text{ or } \mathbb{C}\}. \tag{2}$$

is called a neutrosophic set of refined neutrosophic numbers.

Definition 4 ([18,20]). *A neutrosophic quadruple number is a number of the form (a, bT, cI, dF), where T, I and F have their usual neutrosophic logic meanings, i.e., truth, indeterminacy and false, respectively, and $a, b, c, d \in \mathbb{R}$ or $\mathbb{C}$. The set NQ defined by*

$$NQ = \{(a, bT, cI, dF) | a, b, c, d \in \mathbb{R} \text{ or } \mathbb{C}\}. \tag{3}$$

is called a neutrosophic set of quadruple numbers. For a neutrosophic quadruple number (a, bT, cI, dF), a is called the known part and (bT, cI, dF) is called the unknown part. The set RNQ defined by

$$\begin{aligned} RNQ = \; & \{(a, b_1T_1, b_2T_2, \cdots, b_pT_p, c_1I_1, c_2I_2, \cdots, c_qI_q, d_1F_1, d_2F_2, \cdots, d_rF_r) | \\ & a, b_1, b_2, \cdots, b_p, c_1, c_2, \cdots, c_q, d_1, d_2, \cdots, d_r \in \mathbb{R} \text{ or } \mathbb{C}\}. \end{aligned} \tag{4}$$

is called a neutrosophic set of refined neutrosophic quadruple numbers.

Definition 5 ([18,20]). *Let N be a set, endowed with a total order $a \prec b$, named "a prevailed by b", "a less stronger than b" or "a less preferred than b". We consider $a \preceq b$ as "a prevailed by or equal to b", "a less stronger than or equal to b", or "a less preferred than or equal to b".*

For any elements $a, b \in N$, with $a \preceq b$, one has the absorbance law:

$$a \cdot b = b \cdot a = \text{absorb}(a, b) = \max(a, b) = b. \tag{5}$$

which means that the bigger element absorbs the smaller element. Clearly,

$$a \cdot a = a^2 = \text{absorb}(a, a) = \max(a, a) = a. \tag{6}$$

and

$$a_1 \cdot a_2 \cdots a_n = \max(a_1, a_2, \cdots, a_n). \tag{7}$$

Analogously, we say that "$a \succ b$" and we read: "a prevails to b", "a is stronger than b" or "a is preferred to b". In addition, $a \succeq b$, and we read: "a prevails or is equal to b", "a is stronger than or equal to b", or "a is preferred or equal to b".

Definition 6 ([18,20]). *Consider the set $\{T, I, F\}$. Suppose in an optimistic way we consider the prevalence order $T \succ I \succ F$. Then, we have: $TI = IT = \max(T, I) = T$, $TF = FT = \max(T, F) = T$, $IF = FI = \max(I, F) = I$, $TT = T^2 = T$, $II = I^2 = I$, $FF = F^2 = F$.*

Analogously, suppose in a pessimistic way we consider the prevalence order $T \prec I \prec F$. Then, we have: $TI = IT = \max(T, I) = I$, $TF = FT = \max(T, F) = F$, $IF = FI = \max(I, F) = F$, $TT = T^2 = T$, $II = I^2 = I$, $FF = F^2 = F$.

Definition 7 ([18,20]). *Let $a = (a_1, a_2T, a_3I, a_4F), b = (b_1, b_2T, b_3I, b_4F) \in NQ$. Suppose, in an pessimistic way, the neutrosophic expert considers the prevalence order $T \prec I \prec F$. Then, the multiplication operation is defined as follows:*

$$\begin{aligned} a * b = \; & (a_1, a_2T, a_3I, a_4F) * (b_1, b_2T, b_3I, b_4F) \\ & = (a_1b_1, (a_1b_2 + a_2b_1 + a_2b_2)T, (a_1b_3 + a_2b_3 + a_3b_1 + a_3b_2 + a_3b_3)I, \\ & \quad (a_1b_4 + a_2b_4 + a_3b_4 + a_4b_1 + a_4b_2 + a_4b_3 + a_4b_4)F). \end{aligned} \tag{8}$$

Suppose in an optimistic way the neutrosophic expert considers the prevalence order $T \succ I \succ F$. Then,

$$\begin{aligned} a \star b = \; & (a_1, a_2T, a_3I, a_4F) \star (b_1, b_2T, b_3I, b_4F) \\ & = (a_1b_1, (a_1b_2 + a_2b_1 + a_2b_2 + a_3b_2 + a_4b_2 + a_2b_3 + a_2b_4)T, \\ & \quad (a_1b_3 + a_3b_1 + a_3b_3 + a_3b_4 + a_4b_3)I, (a_1b_4 + a_4b_1 + a_4b_4)F). \end{aligned} \tag{9}$$

Proposition 2 ([18,20]). *Let* $NQ = \{(a, bT, cI, dF) : a, b, c, d \in \mathbb{R} \text{ or } \mathbb{C}\}$. *We have:*

(1) $(NQ, *)$ *is a commutative monoid.*
(2) $(NQ, \star)$ *is a commutative monoid.*

Theorem 1. *[17] For the algebra system* $(NQ, *)$*(or* $(NQ, \star)$*), for every element* $a \in NQ$*, there exists the neutral element* $neut(a)$ *and opposite element* $anti(a)$*, which means that the algebra system* $(NQ, *)$ *(or* $(NQ, \star)$*) is a NETG.*

3. The Algebra Structure of Neutrosophic Quadruple Numbers on General Field

From the above section, we can see that the neutrosophic quadruple numbers are defined on number field $\mathbb{R}$ or $\mathbb{C}$. In this section, the notions of the neutrosophic quadruple numbers on a general field are introduced and the algebra structure of the neutrosophic quadruple numbers on general field is explored.

Let $(\mathbb{F}, +, \cdot)$ be a field, and 0 and 1 are the unit elements for operator $+$ and $\cdot$, respectively. For every $a \in \mathbb{F}$, $-a$ is the inverse element of a for operator $+$, and a^{-1} is the inverse element of a for operator $\cdot$. In the following, field $(\mathbb{F}, +, \cdot)$ is denoted by $\mathbb{F}$ for short and $a \cdot b$ is denoted by ab.

Definition 8. *Let* $\mathbb{F}$ *be a field; a neutrosophic quadruple number is a number of the form* (a, bT, cI, dF)*, where* T, I, F *have their usual neutrosophic logic meanings, i.e., truth, indeterminacy and false, respectively, and* $a, b, c, d \in \mathbb{F}$*. The set* NQF *defined by*

$$NQF = \{(a, bT, cI, dF) | a, b, c, d \in \mathbb{F}\}. \tag{10}$$

is called a neutrosophic set of quadruple numbers on field $\mathbb{F}$*.*

Definition 9. *Let* $a = (a_1, a_2T, a_3I, a_4F), b = (b_1, b_2T, b_3I, b_4F) \in NQF$*, then the addition operator is defined as follows:*

$$a \oplus b = (a_1 + b_1, (a_2 + b_2)T, (a_3 + b_3)I, (a_4 + b_4)F). \tag{11}$$

Definition 10. *Let* $a = (a_1, a_2T, a_3I, a_4F), b = (b_1, b_2T, b_3I, b_4F) \in NQF$*. Suppose, in an pessimistic way, the neutrosophic expert considers the prevalence order* $T \prec I \prec F$*. Then, the multiplication operation is defined as follows:*

$$\begin{aligned} a * b = \; & (a_1, a_2T, a_3I, a_4F) * (b_1, b_2T, b_3I, b_4F) \\ & = (a_1b_1, (a_1b_2 + a_2b_1 + a_2b_2)T, (a_1b_3 + a_2b_3 + a_3b_1 + a_3b_2 + a_3b_3)I, \\ & \quad (a_1b_4 + a_2b_4 + a_3b_4 + a_4b_1 + a_4b_2 + a_4b_3 + a_4b_4)F). \end{aligned} \tag{12}$$

Suppose in an optimistic way the neutrosophic expert considers the prevalence order $T \succ I \succ F$*. Then,*

$$\begin{aligned} a \star b = \; & (a_1, a_2T, a_3I, a_4F) \star (b_1, b_2T, b_3I, b_4F) \\ & = (a_1b_1, (a_1b_2 + a_2b_1 + a_2b_2 + a_3b_2 + a_4b_2 + a_2b_3 + a_2b_4)T, \\ & \quad (a_1b_3 + a_3b_1 + a_3b_3 + a_3b_4 + a_4b_3)I, (a_1b_4 + a_4b_1 + a_4b_4)F). \end{aligned} \tag{13}$$

Theorem 2. $(NQF, \oplus)$ *is an abelian group.*

Proof. It is obvious. □

Theorem 3. *For the algebra system* $(NQF, *)$ *(or* $(NQF, \star)$*), for every element* $a \in NQF$*, there exists the neutral element* $neut(a)$ *and opposite element* $anti(a)$*, thus the algebra system* $(NQF, *)$ *(or* $(NQF, \star)$*) is a NETG.*

The proof's method is similar to the proof when $\mathbb{F} = \mathbb{R}$ in [17]. The detailed proof is omitted. For algebra system $(NQF, *)$, Table 1 gives all the subsets which have the same neutral element, and the corresponding neutral element and opposite elements. In the following, from two examples, we show that how to solve the the neutral element and opposite elements of each element for algebra system $(NQF, *)$ on different fields.

Example 1. *Let* $\mathbb{F} = Z_5 = \{[0],[1],[2],[3],[4]\}$, *then* $(Z_5, +, \cdot)$ *is a field, where* $+$ *and* $\cdot$ *are the classical mod addition and multiplication, respectively. For algebra system* $(NQF, *)$, *if* $a = (a_1, a_2T, a_3I, a_4F) = ([2],[4]T,[3]I,[1]F)$, *i.e.,* $a_1 \neq [0], a_1 + a_2 \neq [0], a_1 + a_2 + a_3 \neq [0], a_1 + a_2 + a_3 + a_4 = [0]$, *then, from Table 1, we can get* $neut(a) = ([1],[0],[0],[4]F)$. *Let* $anti(a) = (c_1, c_2T, c_3I, c_4F)$, *so* $c_1 = a_1^{-1} = [3], c_2 = [3], c_3 = [3], c_4 \in Z_5$, *thus* $anti(a) = ([3],[3]T,[3]I,c_4F)$, *where* $c_4 \in Z_5$. *Thus, we can easily get the neutral element and opposite elements of each neutrosophic quadruple number on general field. For more examples, see the following:*

1. *Let* $b = ([1],[2]T,[1]I,[3]F)$, *then* $neut(b) = ([1],[0],[0],[0])$ *and* $anti(b) = ([1],[1]T,[2]I,[4]F)$.
2. *Let* $c = ([0],[0],[1]I,[4]F)$, *then* $neut(c) = ([0],[0],[1],[4])$ *and* $anti(c) = (c_1, c_2T, c_3I, c_4F)$, *where* $c_1 \oplus c_2 \oplus c_3 = [1], c_4 \in Z_5$.
3. *Let* $d = ([0],[1]T,[1]I,[1]F)$, *then* $neut(d) = ([0],[1]T,[0],[0])$ *and* $anti(d) = (c_1, c_2T, [2]I, [4]F)$, *where* $c_1 \oplus c_2 = [1]$.

Table 1. The corresponding neutral element and opposite elements for $(NQF, *)$.

The Subset of NQF	Neutral Element	Opposite Elements (c_1, c_2T, c_3I, c_4F)
$\{(0,0,0,0)\}$	$(0,0,0,0)$	$c_i \in \mathbb{F}$
$\{(0,0,0,a_4F) \mid a_4 \neq 0\}$	$(0,0,0,F)$	$c_1+c_2+c_3+c_4 = a_4^{-1}$
$\{(0,0,a_3I,-a_3F) \mid a_3 \neq 0\}$	$(0,0,I,-F)$	$c_1+c_2+c_3 = a_3^{-1}, c_4 \in \mathbb{F}$
$\{(0,0,a_3I,a_4F) \mid a_3 \neq 0, a_3+a_4 \neq 0\}$	$(0,0,I,0)$	$c_1+c_2+c_3 = a_3^{-1}, c_4 = -(a_4a_3^{-1}(a_3+a_4)^{-1})$
$\{(0,a_2T,-a_2I,0) \mid a_2 \neq 0\}$	$(0,T,-I,0)\}$	$c_1+c_2 = a_2^{-1}, c_3, c_4 \in \mathbb{F}$
$\{(0,a_2T,-a_2I,a_4F) \mid a_2 \neq 0, a_4 \neq 0\}$	$(0,T,-I,F)$	$c_1+c_2 = a_2^{-1}, c_3+c_4 = a_4^{-1}+(-a_2^{-1})$
$\{(0,a_2T,a_3I,a_4F) \mid a_2 \neq 0, a_2+a_3 \neq 0, a_2+a_3+a_4 = 0\}$	$(0,T,0,-F)$	$c_1+c_2 = a_2^{-1}, c_3 = -(a_3a_2^{-1}(a_2+a_3)^{-1})$, $c_4 \in \mathbb{F}$
$\{(0,a_2T,a_3I,a_4F) \mid a_2 \neq 0, a_2+a_3 \neq 0, a_2+a_3+a_4 \neq 0\}$	$(0,T,0,0)$	$c_1+c_2 = a_2^{-1}, c_3 = -(a_3a_2^{-1}(a_2+a_3)^{-1})$, $c_4 = -(a_4(a_2+a_3)^{-1}(a_2+a_3+a_4)^{-1})$
$\{(a_1,-a_1T,0,0) \mid a_1 \neq 0\}$	$(1,-T,0,0)\}$	$c_1 = a_1^{-1}, c_2, c_3, c_4 \in \mathbb{F}$
$\{(a_1,-a_1T,0,a_4F) \mid a_1 \neq 0, a_4 \neq 0\}$	$(1,-T,0,F)$	$c_1 = a_1^{-1}, c_2+c_3+c_4 = a_4^{-1}+(-a_1^{-1})$
$\{(a_1,-a_1T,a_3I,-a_3F) \mid a_1 \neq 0, a_3 \neq 0\}$	$(1,-T,I,-F)$	$c_1 = a_1^{-1}, c_2+c_3 = a_3^{-1}-a_1^{-1}, c_4 \in \mathbb{F}$
$\{(a_1,-a_1T,a_3I,a_4F) \mid a_1 \neq 0, a_3 \neq 0, a_3+a_4 \neq 0\}$	$(1,-T,I,0)$	$c_1 = a_1^{-1}, c_2+c_3 = a_3^{-1}+(-a_1^{-1})$, $c_4 = -(a_4a_3^{-1}(a_3+a_4)^{-1})$
$\{(a_1,a_2T,a_3I,0) \mid a_1 \neq 0, a_1+a_2 \neq 0, a_1+a_2+a_3 = 0\}$	$(1,0,-I,0)$	$c_1 = a_1^{-1}, c_2 = -(a_2a_1^{-1}(a_1+a_2)^{-1})$, $c_3, c_4 \in \mathbb{F}$
$\{(a_1,a_2T,a_3I,a_4F) \mid a_1 \neq 0, a_1+a_2 \neq 0, a_1+a_2+a_3 = 0, a_4 \neq 0\}$	$(1,0,-I,F)$	$c_1 = a_1^{-1}, c_2 = -(a_2a_1^{-1}(a_1+a_2)^{-1})$, $c_3+c_4 = a_4^{-1}+(-(a_1+a_2)^{-1})$
$\{(a_1,a_2T,a_3I,a_4F) \mid a_1 \neq 0, a_1+a_2 \neq 0, a_1+a_2+a_3 \neq 0, a_1+a_2+a_3+a_4 = 0\}$	$(1,0,0,-F)$	$c_1 = a_1^{-1}, c_2 = -(a_2a_1^{-1}(a_1+a_2)^{-1})$, $c_3 = -(a_3(a_1+a_2)^{-1}(a_1+a_2+a_3)^{-1})$, $c_4 \in \mathbb{F}$
$\{(a_1,a_2T,a_3I,a_4F) \mid a_1 \neq 0, a_1+a_2 \neq 0, a_1+a_2+a_3 \neq 0, a_1+a_2+a_3+a_4 \neq 0\}$	$(1,0,0,0)$	$c_1 = a_1^{-1}, c_2 = -(a_2a_1^{-1}(a_1+a_2)^{-1})$, $c_3 = -(a_3(a_1+a_2)^{-1}(a_1+a_2+a_3)^{-1})$, $c_4 = -(a_4(a_1+a_2+a_3)^{-1}(a_1+a_2+a_3+a_4)^{-1})$

Example 2. *Let $\mathbb{F}_4 = \{0, 1, x, y\}$, the operators $+$ and $\cdot$ on $\mathbb{F}_4$ is defined by Table 2.*

Table 2. The operators $+$ and $\cdot$ on $\mathbb{F}_4$.

$+$	0	1	x	y	$\cdot$	0	1	x	y
0	0	1	x	y	0	0	0	0	0
1	1	0	y	x	1	0	1	x	y
x	x	y	0	1	x	0	x	y	1
y	y	x	1	0	y	0	y	1	x

Then, $(\mathbb{F}_4, +, \cdot)$ is a field. Set $NQF = \{(a, bT, cI, dF) | a, b, c, d \in \mathbb{F}_4\}$. We have:

1. *Let $a = (0, 0, xI, xF)$, then $neut(a) = (0, 0, I, F)$ and $anti(a) = (c_1, c_2T, c_3I, c_4F)$, where $c_1 + c_2 + c_3 = y, c_4 \in \mathbb{F}_4$.*
2. *Let $b = (0, xT, xI, yF)$, then $neut(b) = (0, T, I, F)$ and $anti(b) = (c_1, c_2T, c_3I, c_4F)$, where $c_1 + c_2 = y, c_3 + c_4 = 1$.*
3. *Let $c = (x, xT, 0, 0)$, then $neut(c) = (1, T, 0, 0)$ and $anti(c) = (y, c_2T, c_3I, c_4F)$, where $c_2, c_3, c_4 \in \mathbb{F}_4$.*

In the same way, for algebra system $(NQF, \star)$, Table 3 gives all the subsets which have the same neutral element, and the corresponding neutral element and opposite elements.

Table 3. The corresponding neutral element and opposite elements for $(NQF, \star)$.

The Subset of NQF	Neutral Element	Opposite Elements (c_1, c_2T, c_3I, c_4F)
$\{(0,0,0,0)\}$	$(0,0,0,0)$	$c_i \in \mathbb{F}$
$\{(0, a_2T, 0, 0) \| a_2 \neq 0\}$	$(0, T, 0, 0)$	$c_1 + c_2 + c_3 + c_4 = a_2^{-1}$
$\{(0, -a_3T, a_3I, 0) \| a_3 \neq 0\}$	$(0, -T, I, 0)$	$c_1 + c_3 + c_4 = a_3^{-1}, c_2 \in \mathbb{F}$
$\{(0, a_2T, a_3I, a_4F) \| a_3 \neq 0, a_2 + a_3 \neq 0\}$	$(0, 0, I, 0)$	$c_1 + c_3 + c_4 = a_3^{-1}$, $c_2 = -(a_2a_3^{-1}(a_2 + a_3)^{-1})$
$\{(0, 0, -a_4I, a_4F) \| a_4 \neq 0\}$	$(0, 0, -I, F)\}$	$c_1 + c_4 = a_4^{-1}, c_2, c_3 \in \mathbb{F}$
$\{(0, a_2T, -a_4I, a_4F) \| a_2 \neq 0, a_4 \neq 0\}$	$(0, T, -I, F)$	$c_1 + c_4 = a_4^{-1}, c_2 + c_3 = a_2^{-1} + (-a_4^{-1})$
$\{(0, a_2T, a_3I, a_4F) \| a_4 \neq 0, a_3 + a_4 \neq 0, a_2 + a_3 + a_4 = 0\}$	$(0, -T, 0, F)$	$c_1 + c_4 = a_4^{-1}, c_3 = -(a_3a_4^{-1}(a_3 + a_4)^{-1})$, $c_2 \in \mathbb{F}$
$\{(0, a_2T, a_3I, a_4F) \| a_4 \neq 0, a_3 + a_4 \neq 0, a_2 + a_3 + a_4 \neq 0\}$	$(0, 0, 0, F)$	$c_1 + c_4 = a_4^{-1}, c_3 = -(a_3a_4^{-1}(a_3 + a_4)^{-1})$, $c_2 = -(a_2(a_3 + a_4)^{-1}(a_2 + a_3 + a_4)^{-1})$
$\{(a_1, 0, 0, -a_1F) \| a_1 \neq 0\}$	$(1, 0, 0, -F)\}$	$c_1 = a_1^{-1}, c_2, c_3, c_4 \in \mathbb{F}$
$\{(a_1, a_2T, 0, -a_1F) \| a_1 \neq 0, a_2 \neq 0\}$	$(1, T, 0, -F)$	$c_1 = a_1^{-1}, c_2 + c_3 + c_4 = a_2^{-1} + (-a_1^{-1})$
$\{(a_1, -a_3T, a_3I, -a_1F) \| a_1 \neq 0, a_3 \neq 0\}$	$(1, -T, I, -F)$	$c_1 = a_1^{-1}, c_3 + c_4 = a_3^{-1} + (-a_1^{-1}), c_4 \in \mathbb{F}$
$\{(a_1, a_2T, a_3I, -a_1F) \| a_1 \neq 0, a_3 \neq 0, a_2 + a_3 \neq 0\}$	$(1, 0, I, -F)$	$c_1 = a_1^{-1}, c_3 + c_4 = a_3^{-1} + (-a_1^{-1})$, $c_2 = -(a_2a_3^{-1}(a_2 + a_3)^{-1})$
$\{(a_1, 0, a_3I, a_4F) \| a_1 \neq 0, a_1 + a_4 \neq 0, a_1 + a_3 + a_4 = 0\}$	$(1, 0, -I, 0)$	$c_1 = a_1^{-1}, c_4 = -(a_4a_1^{-1}(a_1 + a_4)^{-1})$, $c_2, c_3 \in \mathbb{F}$
$\{(a_1, a_2T, a_3I, a_4F) \| a_1 \neq 0, a_1 + a_4 \neq 0, a_1 + a_3 + a_4 = 0, a_2 \neq 0\}$	$(1, T, -I, 0)$	$c_1 = a_1^{-1}, c_4 = -(a_4a_1^{-1}(a_1 + a_4)^{-1})$, $c_2 + c_3 = a_2^{-1} + (-(a_1 + a_4)^{-1})$
$\{(a_1, a_2T, a_3I, a_4F) \| a_1 \neq 0, a_1 + a_4 \neq 0, a_1 + a_3 + a_4 \neq 0, a_1 + a_2 + a_3 + a_4 = 0\}$	$(1, -T, 0, 0)$	$c_1 = a_1^{-1}, c_4 = -(a_4a_1^{-1}(a_1 + a_4)^{-1})$, $c_3 = -(a_3(a_1 + a_4)^{-1}(a_1 + a_3 + a_4)^{-1})$, $c_2 \in \mathbb{F}$
$\{(a_1, a_2T, a_3I, a_4F) \| a_1 \neq 0, a_1 + a_4 \neq 0, a_1 + a_3 + a_4 \neq 0, a_1 + a_2 + a_3 + a_4 \neq 0\}$	$(1, 0, 0, 0)$	$c_1 = a_1^{-1}, c_4 = -(a_4a_1^{-1}(a_1 + a_4)^{-1})$, $c_3 = -(a_3(a_1 + a_4)^{-1}(a_1 + a_3 + a_4)^{-1})$, $c_2 = -(a_2(a_1 + a_3 + a_4)^{-1}(a_1 + a_2 + a_3 + a_4)^{-1})$

4. The Algebra Structure of Refined Neutrosophic Numbers on General Field

In the above section, we reveal that the algebra structure of the neutrosophic quadruple numbers on general field. In this section, we explore the the algebra structure of the refined neutrosophic numbers (refined neutrosophic quadruple numbers) on general field.

Definition 11. *Let $\mathbb{F}$ be a field; a refined n-ary neutrosophic number is a number of the form $(a_0, a_1I_1, a_2I_2, \cdots, a_nI_n)$, where $I_1, I_2, \cdots,$ and I_n are different types of indeterminacies, and $a_0, a_1, a_2, \cdots, a_n \in \mathbb{F}$. The set RNF_n defined by*

$$RNF_n = \{(a_0, a_1I_1, a_2I_2, \cdots, a_nI_n) | a_0, a_1, a_2, \cdots, a_n \in \mathbb{F}\}. \tag{14}$$

is called a refined neutrosophic set on field $\mathbb{F}$.

Definition 12. *Let $x = (a_0, a_1I_1, a_2I_2, \cdots, a_nI_n), y = (b_0, b_1I_1, b_2I_2, \cdots, b_nI_n) \in RNF_n$, then the addition operator on RNF_n is defined as follows:*

$$x \oplus y = (a_0 + b_0, (a_1 + b_1)I_1, (a_2 + b_2)I_2, \cdots, (a_n + b_n)I_n). \tag{15}$$

Definition 13. *Let $x = (a_0, a_1I_1, a_2I_2, \cdots, a_nI_n), y = (b_0, b_1I_1, b_2I_2, \cdots, b_nI_n) \in RNF_n$, the neutrosophic expert considers the prevalence order $I_1 \prec I_2 \prec \cdots \prec I_n$. Then, the multiplication operation is defined as follows:*

$$\begin{aligned} x * y = & (a_0, a_1I_1, a_2I_2, \cdots, a_nI_n) * (b_0, b_1I_1, b_2I_2, \cdots, b_nI_n) \\ & = (a_0b_0, (a_0b_1 + a_1b_1 + a_1b_2)I_1, (a_0b_2 + a_1b_2 + a_2b_0 + a_2b_1 + a_2b_2)I_2, \\ & \cdots, (a_0b_n + a_1b_n + a_2b_n + \cdots + a_{n-1}b_n + a_nb_0 + a_nb_1 + \cdots + a_nb_n)I_n). \end{aligned} \tag{16}$$

The neutrosophic expert considers the prevalence order $I_1 \succ I_2 \succ \cdots \succ I_n$. Then,

$$\begin{aligned} x \star y = & (a_0, a_1I_1, a_2I_2, \cdots, a_nI_n) \star (b_0, b_1I_1, b_2I_2, \cdots, b_nI_n) \\ & = (a_0b_0, (a_0b_1 + a_1b_1 + \cdots + a_nb_1 + a_1b_0 + a_1b_2 + a_1b_3 + \cdots + a_1b_n)I_1, \cdots, \\ & (a_0b_{n-1} + a_{n-1}b_0 + a_{n-1}b_{n-1} + a_{n-1}b_n + a_nb_{n-1})I_{n-1}, (a_0b_n + a_nb_0 + a_nb_n)I_n). \end{aligned} \tag{17}$$

Theorem 4. *$(RNF_n, \oplus)$ is an abelian group.*

Proof. The proof is obvious. □

Theorem 5. *For the algebra system $(RNF_n, *)$ (or $(RNF_n, \star)$), for every element $a \in RNF_n$, there exists the neutral element $neut(a)$ and opposite element $anti(a)$, thus the algebra system $(RNF_n, *)$ (or $(RNF_n, \star)$) is a NETG.*

Proof. We use applied mathematical induction for n and only discuss the algebra system $(RNF_n, *)$. The algebra system $(RNF_n, \star)$ has a similar proof.

If $n = 2$, for refined 2-ary neutrosophic set, which is same as neutrosophic binary numbers set in [17], from Theorem 7 in [17], we can see that for every element $a \in RNF_2$, there exists the neutral element $neut(a)$ and opposite element $anti(a)$, thus the algebra system $(RNF_2, *)$ is a NETG.

Assume that the refined n-ary neutrosophic set RNF_n is a NETG. That is, for every element $a \in RNF_n$, there exists the neutral element $neut(a)$ and opposite element $anti(a)$. In the following, we prove that for the refined $(n+1)$-ary neutrosophic set, which is a NETG.

For each $a = (a_0, a_1I_1, a_2I_2, \cdots, a_{n+1}I_{n+1}) \in RNF_{n+1}$, let $a' = (a_0, a_1I_1, a_2I_2, \cdots, a_nI_n)$, being $a' \in RNF_n$, then, from the above assumption condition, $neut(a')$ and $anti(a')$ exist and let $neut(a') = (b_0, b_1I_1, b_2I_2, \cdots, b_nI_n)$, $anti(a') = (c_0, c_1I_1, c_2I_2, \cdots, c_nI_n)$. We prove for $a = (a_0, a_1I_1, a_2I_2, \cdots, a_{n+1}I_{n+1})$, $neut(a)$ and $anti(a)$ exist. We discuss from the different cases of a_{n+1}.

Case A: If $a_{n+1} = 0$, Being $(a_0, a_1I_1, a_2I_2, \cdots, a_nI_n, 0) * (b_0, b_1I_1, b_2I_2, \cdots, b_nI_n, b_{n+1}I_{n+1}) = (a_0, a_1I_1, a_2I_2, \cdots, a_nI_n, 0)$, that is $(a_0 + a_1 + \cdots + a_n)b_{n+1} = 0$, thus we discuss from $a_0 + a_1 + \cdots + a_n = 0$ or $a_0 + a_1 + \cdots + a_n \neq 0$.

Case A1: If $a_{n+1} = 0, a_0 + a_1 + \cdots + a_n = 0$, so $(a_0, a_1I_1, a_2I_2, \cdots, a_nI_n, 0) * (c_0, c_1I_1, c_2I_2, \cdots, c_nI_n, c_{n+1}I_{n+1}) = (b_0, b_1I_1, b_2I_2, \cdots, b_nI_n, b_{n+1}I_{n+1})$, that is $b_{n+1} = 0$ and c_{n+1} can be chosen arbitrarily in $\mathbb{F}$.

Case A2: If $a_{n+1} = 0, a_0 + a_1 + \cdots + a_n \neq 0$, so from $(a_0 + a_1 + \cdots + a_n)b_{n+1} = 0$, we have $b_{n+1} = 0$, from $(a_0, a_1I_1, a_2I_2, \cdots, a_nI_n, 0) * (c_0, c_1I_1, c_2I_2, \cdots, c_nI_n, c_{n+1}I_{n+1}) = (b_0, b_1I_1, b_2I_2, \cdots, b_nI_n, 0)$, thus $c_{n+1} = 0$.

Case B: If $a_{n+1} \neq 0$, being $(a_0, a_1I_1, a_2I_2, \cdots, a_nI_n, a_{n+1}I_{n+1}) * (b_0, b_1I_1, b_2I_2, \cdots, b_nI_n, b_{n+1}I_{n+1}) = (a_0, a_1I_1, a_2I_2, \cdots, a_nI_n, a_{n+1}I_{n+1})$, that is $(a_0 + a_1 + \cdots + a_{n+1})b_{n+1} + a_{n+1}(b_0 + b_1 + \cdots + b_n) = a_{n+1}$, that is $(a_0 + a_1 + \cdots + a_{n+1})b_{n+1} = a_{n+1}(1 - b_0 - b_1 - \cdots - b_n)$, we discuss from $a_0 + a_1 + \cdots + a_{n+1} = 0$ or $a_0 + a_1 + \cdots + a_{n+1} \neq 0$.

Case B1: If $a_{n+1} \neq 0, a_0 + a_1 + \cdots + a_{n+1} = 0$, we have $b_0 + b_1 + \cdots + b_n = 1$, so from $(a_0, a_1I_1, a_2I_2, \cdots, a_nI_n, a_{n+1}I_{n+1}) * (c_0, c_1I_1, c_2I_2, \cdots, c_nI_n, c_{n+1}I_{n+1}) = (b_0, b_1I_1, b_2I_2, \cdots, b_nI_n, b_{n+1}I_{n+1})$, that is $b_{n+1} = (c_0 + c_1 + \cdots + c_n)a_{n+1} = b_{n+1}$ and c_{n+1} can be chosen arbitrarily in $\mathbb{F}$.

Case B2: If $a_{n+1} \neq 0, a_0 + a_1 + \cdots + a_{n+1} \neq 0$, we have $b_{n+1} = a_{n+1}(1 - b_0 - b_1 - \cdots - b_n)(a_0 + a_1 + a_2 \cdots a_n)^{-1}$, so from $(a_0, a_1I_1, a_2I_2, \cdots, a_nI_n, a_{n+1}I_{n+1}) * (c_0, c_1I_1, c_2I_2, \cdots, c_nI_n, c_{n+1}I_{n+1}) = (b_0, b_1I_1, b_2I_2, \cdots, b_nI_n, b_{n+1}I_{n+1})$, that is and $c_{n+1} = (b_{n+1} - a_{n+1}(c_0 + c_1 + \cdots + c_n))(a_0 + a_1 + \cdots + a_{n+1})^{-1}$.

From the above analysis, we can see that, for each $a \in RNF_n$, which has the neutral element $neut(a)$ and opposite element $anti(a)$, from the mathematical induction method, we can obtain that the algebra system $(RNF, *)$ is a NETG. □

For algebra system $(RNF_{n+1}, *)$, if $a = (a_0, a_1I_1, a_2I_2, \cdots, a_{n+1}I_{n+1})$, let $a' = (a_0, a_1I_1, a_2I_2, \cdots, a_nI_n) \in RNF_n$, if we have $neut(a') = (b_0, b_1I_1, b_2I_2, \cdots, b_nI_n)$, $anti(a') = (c_0, c_1I_1, c_2I_2, \cdots, c_nI_n)$. Then, the corresponding neutral element and opposite elements of a are given in Table 4 according to the different cases of a_{n+1}.

Table 4. The corresponding neutral element and opposite elements for $(RNF_{n+1}, *)$

The Subset $\{(a_0, a_1I_1, \cdots, a_{n+1}I_{n+1})\}$	**Neutral Element** $(b_0, b_1I_1, \cdots, b_nI_n, b_{n+1}I_{n+1})$	**Opposite Elements** $(c_0, c_1I_1, \cdots, c_nI_n, c_{n+1}I_{n+1})$
$a_{n+1} = 0$, $a_0 + a_1 + \cdots + a_n = 0$	$b_{n+1} = 0$	$c_{n+1} \in \mathbb{F}$
$a_{n+1} = 0$, $a_0 + a_1 + \cdots + a_n \neq 0$	$b_{n+1} = 0$	$c_{n+1} = 0$
$a_{n+1} \neq 0$, $a_0 + a_1 + \cdots + a_n + a_{n+1} = 0$	$b_{n+1} = (c_0 + c_1 + \cdots + c_n)a_{n+1}$	$c_{n+1} \in \mathbb{F}$
$a_{n+1} \neq 0$, $a_0 + a_1 + \cdots + a_n + a_{n+1} \neq 0$	$b_{n+1} = a_{n+1}(1 - b_0 - b_1 - \cdots - b_n)(a_0 + a_1 + a_2 + \cdots + a_n)^{-1})$	$c_{n+1} = (b_{n+1} - a_{n+1}(c_0 + c_1 + \cdots + c_n))(a_0 + a_1 + \cdots + a_{n+1})^{-1}$

For algebra system $(RNF_n, *)$, according to the results in Tables 3 and 4 in [17], we can easily obtain the neutral element and opposite elements when $n = 1, 2$ on general fields. In Table 1, we can get the neutral element and opposite elements when $n = 3$ on general field. Thus, from Theorem 5, we can get the neutral element and opposite elements of each element in RNF_n step-by-step. The solving method is given by Algorithm 1 and the following example is used to explain the algorithm.

Algorithm 1 Solving the neutral element and opposite elements of each element in $(RNF_n, *)$.

Input: $a = (a_0, a_1I_1, a_2I_2, \cdots, a_nI_n), n \geq 3$
1: $i = 3$;
2: For $i = 3 : n$
3: $A_i = (a_0, a_1I_1, a_2I_2, \cdots, a_iI_i)$;
4: If $i == 3$
5: Obtain $neut(A_i), anti(A_i)$ by Table 1;
6: else
7: Obtain $neut(A_i), anti(A_i)$ by Table 4 combining the values of $neut(A_{i-1})$ and $anti(A_{i-1})$;
8: end
9: Save $neut(A_i), anti(A_i)$;
10: end
Output: $neut(a) = neut(A_n), anti(a) = anti(A_n)$

Example 3. *For algebra system $(RNF_2, *)$, and set $\mathbb{F} = \mathbb{R}$. If $a = (a_0, a_1I_1) = (0, -I_1)$, from Table 3 in [17], we can get $neut(a) = (0, I_1)$ and $anti(a) = (c_0, c_1I_1)$, where $c_0 + c_1 = -1$.*

In the following, we use two methods to solve the the neutral element and opposite elements of $b = (a_0, a_1I_1, a - 2I_2) = (0, -I_1, I_2) \in RNF_3$ and we get the same results.

1. *Algorithm 1: From Table 2, being $a_3 \neq 0$ and $a_0 + a_1 + a_2 = 0$, thus $b_2 = (c_0 + c_1) \cdot 1 = -1$, that is $neut(b) = (0, I_1, -I_2)$ and $anti(a) = (c_0, c_1I_1, c_2I_2)$, where $c_0 + c_1 = -1$ and c_2 can be chosen arbitrarily in $\mathbb{R}$.*
2. *Rsults from Table 4 in [17]: Being $a_2 \neq 0$ and $a_0 + a_1 + a_2 = 0$, thus $neut(b) = (0, I_1, -I_2)$ and $anti(a) = (c_0, c_1I_1, c_2I_2)$, where $c_0 + c_1 = -1$ and c_2 can be chosen arbitrarily in $\mathbb{R}$.*

Example 4. *For algebra system $(RNF_3, *)$, and set $\mathbb{F} = \mathbb{R}$. If $a = (a_0, a_1I_1, a_2I_2) = (1, -I_1, I_2) \in RNF_3$, from Table 4 in [17], we can get $neut(a) = (1, -I_1, I_2)$ and $anti(a) = (1, c_1I_1, c_2I_2)$, where $c_1 + c_2 = 0$.*

In the same way, we use two methods to solve the the neutral element and opposite elements of $b = (a_0, a_1I_1, a_2I_2, a_3I_3) = (1, -I_1, I_2, I_3) \in RNF_4$.

1. *Algorithm 1: From Table 2, being $a_3 \neq 0$ and $a_0 + a_1 + a_2 + a_3 \neq 0$, thus $b_3 = 0$, that is $neut(b) = (0, -I_1, I_2, 0)$ and $c_3 = (0 - 1 \cdot 1 \cdot \frac{1}{2}) = -\frac{1}{2}$, thus $anti(a) = (1, c_1I_1, c_2I_2, -\frac{1}{2}I_3)$, where $c_1 + c_2 = 0$.*
2. *Results from Table 1 in [17]: Being $a_0 \neq 0, a_2 \neq 0$ and $a_2 + a_3 \neq 0$, thus $neut(b) = (1, -I_1, I_2, 0)$ and $c_0 = 1, c_1 + c_2 = 0, c_3 = -\frac{1}{1 \cdot (1+1)} = -\frac{1}{2}$, thus $anti(a) = (1, c_1I_1, c_2I_2, -\frac{1}{2}I_3)$, where $c_1 + c_2 = 0$.*

For algebra system $(RNF_{n+1}, \star)$, set $a = (a_0, a_1I_1, a_2I_2, \cdots, a_{n+1}I_{n+1})$, being the order $I_1 \succ I_2 \succ \cdots \succ I_n$, thus we should obtain the neutral element and opposite elements of $a' = (a_0, a_2I_2, \cdots, a_{n+1}I_{n+1})$. Knowing that $neut(a') = (b_0, b_2I_2, b_3I_3, \cdots, b_nI_n)$, $anti(a') = (c_0, c_2I_2, c_3I_3, \cdots, c_nI_n)$, then the corresponding neutral element and opposite elements of a are given in Table 5 according to the different cases of a_1.

Table 5. The corresponding neutral element and opposite elements for $(RNF, \star)$.

The Subset $\{(a_0, a_1I_1, \cdots, a_{n+1}I_{n+1})\}$	**Neutral Element** $(b_0, b_1I_1, \cdots, b_nI_n, b_{n+1}I_{n+1})$	**Opposite Elements** $(c_0, c_1I_1, \cdots, c_nI_n, c_{n+1}I_{n+1})$
$a_1 = 0$, $a_0 + a_2 + \cdots + a_{n+1} = 0$	$b_1 = 0$	$c_1 \in \mathbb{F}$
$a_1 = 0$, $a_0 + a_2 + \cdots + a_{n+1} \neq 0$	$b_1 = 0$	$c_1 = 0$
$a_1 \neq 0$, $a_0 + a_1 + \cdots + a_n + a_{n+1} = 0$	$b_1 = (c_0 + c_2 + \cdots + c_{n+1})a_1$	$c_1 \in \mathbb{F}$
$a_1 \neq 0$, $a_0 + a_1 + \cdots + a_n + a_{n+1} \neq 0$	$b_1 = a_1(1 - (b_0 + b_2 + \cdots + b_{n+1}))(a_0 + a_1 + \cdots + a_{n+1})^{-1}$	$c_1 = (b_1 - a_1(c_0 + c_2 + \cdots + c_{n+1}))(a_0 + a_1 + \cdots + a_{n+1})^{-1}$

Similarly, we also can get the neutral element and opposite elements of each element in $(RNF_n, \star)$ step-by-step. The solving method is given by Algorithm 2 and the following example is used to explain the algorithm.

Algorithm 2 Solving the neutral element and opposite elements of each element in $(RNF_n, \star)$.

Input: $a = (a_0, a_1I_1, a_2I_2, \cdots, a_nI_n), n \geq 3$
$i = n - 2$;
While $i \geq 1$
  $A_i = (a_0, a_iI_i, \cdots, a_{n-1}I_{n-1}, a_nI_n)$;
    If $i == n - 2$
      Obtain $neut(A_i), anti(A_i)$ by Table 3;
    else
      Obtain $neut(A_i), anti(A_i)$ by Table 5 combining the values of $neut(A_{i+1})$ and $anti(A_{i+1})$;
    end
  Save $neut(A_i), anti(A_i)$;
  $i = i - 1$;
end
Output: $neut(a) = neut(A_1), anti(a) = anti(A_1)$

Example 5. *For algebra system $(RNF_6, \star)$, and set $\mathbb{F} = \mathbb{R}$, $a = (0, 0, -2I_2, -I_3, I_4, 0)$, solve the neutral element and opposite elements of a.*

According Algorithm 2 for algebra system $(RNF_6, \star)$: Firstly, we solve the neutral element and opposite elements of $a' = (0, -I_3, I_4, 0)$ from Table 3, and then solve the neutral element and opposite elements of $a'' = (0, -2I_2, -I_3, I_4, 0)$ from Table 5, lastly, we solve the neutral element and opposite elements of a from Table 5.

1. *From Table 3: $neut(a') = (0, -I_3, I_4, 0)$ and $anti(a') = (c_0, c_3I_3, c_4I_4, c_5I_5)$, where $c_0 + c_4 + c_5 = 1$, $c_3 \in \mathbb{R}$.*
2. *From Table 5 and combining the results of the above step: Being $-2 \neq 0$ and $0 + (-2) + (-1) + 1 + 0 \neq 0$, thus $neut(a'') = (0, I_2, -I_3, I_4, 0)$ and $anti(a'') = (c_0, c_2I_2, c_3I_3, c_4I_4, c_5I_5)$, where $c_0 + c_4 + c_5 = 1$, $c_2 + c_3 = -\frac{3}{2}$.*
3. *From Table 5 and combining the results of the above step: Being $0 = 0$ and $0 + 0 + (-2) + (-1) + 1 + 0 \neq 0$, thus $neut(a) = (0, 0, I_2, -I_3, I_4, 0)$ and $anti(a) = (c_0, 0, c_2I_2, c_3I_3, c_4I_4, c_5I_5)$, where $c_0 + c_4 + c_5 = 1, c_2 + c_3 = -\frac{3}{2}$.*

Similarly, we explore the the algebra structure of the refined neutrosophic quadruple numbers on general field in the following.

Definition 14. *Let $\mathbb{F}$ be a field; a refined neutrosophic quadruple number is a number of the form $(a, b_1T_1, b_2T_2, \cdots, b_pT_p, c_1I_1, c_2I_2, \cdots, c_qI_q, d_1F_1, d_2F_2, \cdots, d_rF_r)$, where $a, b_1, b_2, \cdots, b_p, c_1, c_2, \cdots, c_q, d_1, d_2, \cdots, d_r \in \mathbb{F}$. The set $RNQF_{pqr}$ defined by*

$$\begin{aligned} RNQF_{pqr} = \ & \{(a, b_1T_1, b_2T_2, \cdots, b_pT_p, c_1I_1, c_2I_2, \cdots, c_qI_q, d_1F_1, d_2F_2, \cdots, d_rF_r)| \\ & a, b_1, b_2, \cdots, b_p, c_1, c_2, \cdots, c_q, d_1, d_2, \cdots, d_r \in \mathbb{F}\}. \end{aligned} \tag{18}$$

is called a refined neutrosophic quadruple set on field $\mathbb{F}$.

Definition 15. *Let $x = (a, b_1T_1, b_2T_2, \cdots, b_pT_p, c_1I_1, c_2I_2, \cdots, c_qI_q, d_1F_1, d_2F_2, \cdots, d_rF_r)$, $y = (e, f_1T_1, f_2T_2, \cdots, f_pT_p, g_1I_1, g_2I_2, \cdots, g_qI_q, h_1F_1, h_2F_2, \cdots, h_rF_r) \in RNQF_{pqr}$, then the addition operator is defined as follows:*

$$\begin{aligned} x \oplus y = \ & (a+e, (b_1+f_1)T_1, (b_2+f_2)T_2, \cdots, (b_p+f_p)T_p, (c_1+g_1)I_1, (c_2+g_2)I_2, \cdots, \\ & (c_q+g_q)I_q, (d_1+h_1)F_1, (d_2+h_2)F_2, \cdots, (d_r+h_r)F_r). \end{aligned} \tag{19}$$

Definition 16. *Let $x = (a, b_1T_1, b_2T_2, \cdots, b_pT_p, c_1I_1, c_2I_2, \cdots, c_qI_q, d_1F_1, d_2F_2, \cdots, d_rF_r)$, $y = (e, f_1T_1, f_2T_2, \cdots, f_pT_p, g_1I_1, g_2I_2, \cdots, g_qI_q, h_1F_1, h_2F_2, \cdots, h_rF_r) \in RNQF_{pqr}$; the neutrosophic expert considers the prevalence order $T_1 \prec T_2 \prec \cdots \prec T_p \prec I_1 \prec I_2 \prec \cdots \prec I_q \prec F_1 \prec F_2 \prec \cdots \prec F_r$. Then, the multiplication operation is defined as follows:*

$$\begin{aligned} x * y \ & = (a, b_1T_1, b_2T_2, \cdots, b_pT_p, c_1I_1, c_2I_2, \cdots, c_qI_q, d_1F_1, d_2F_2, \cdots, d_rF_r) \\ & \quad *(e, f_1T_1, f_2T_2, \cdots, f_pT_p, g_1I_1, g_2I_2, \cdots, g_qI_q, h_1F_1, h_2F_2, \cdots, h_rF_r) \\ & = (ae, (af_1+b_1e+b_1f_1)T_1, (af_2+b_1f_2+b_2e+b_2f_1+b_2f_2)T_2, \\ & \quad \cdots, (ah_r+b_1h_r+b_2h_r+\cdots+d_{r-1}h_r+d_re+d_rf_1+\cdots+d_rh_r)F_r). \end{aligned} \tag{20}$$

The neutrosophic expert considers the prevalence order $T_1 \succ T_2 \succ \cdots \succ T_p \succ I_1 \succ I_2 \succ \cdots \succ I_q \succ F_1 \succ F_2 \succ \cdots \succ F_r$. Then,

$$\begin{aligned} x \star y \ & = (a, b_1T_1, b_2T_2, \cdots, b_pT_p, c_1I_1, c_2I_2, \cdots, c_qI_q, d_1F_1, d_2F_2, \cdots, d_rF_r) \\ & \quad \star(e, f_1T_1, f_2T_2, \cdots, f_pT_p, g_1I_1, g_2I_2, \cdots, g_qI_q, h_1F_1, h_2F_2, \cdots, h_rF_r) \\ & = (ae, (af_1+b_1f_1+\cdots+d_rf_1+b_1e+b_1f_2+b_1f_3+\cdots+b_1h_r)T_1, \\ & \quad \cdots, (ah_{r-1}+d_{r-1}e+d_{r-1}h_{r-1}+d_{r-1}h_r+d_rh_{r-1})F_{r-1}, (ah_r+d_re+d_rh_r)F_r). \end{aligned} \tag{21}$$

Similarly, we also have the following results.

Theorem 6. *$(RNQF_{pqr}, \oplus)$ is an abelian group.*

Theorem 7. *For the algebra system $(RNQF_{pqr}, *)$ (or $(RNQF_{pqr}, \star)$), for every element $a \in RNQF_{pqr}$, there exists the neutral element $neut(a)$ and opposite element $anti(a)$, thus the algebra system $(RNQF_{pqr}, *)$ (or $(RNQF_{pqr}, \star)$) is a NETG.*

Example 6. *For algebra system $(RNQF_{213}, *)$, and set $\mathbb{F} = \mathbb{R}$, $a = (1, 0, 2T_2, -3I_1, 2F_1, 0, -2F_3)$, solve the neutral element and opposite elements of a.*

According Algorithm 1 for algebra system $(RNF_{213}, *)$, firstly, we solve the neutral element and opposite elements of $a' = (1, 0, 2T_2, -3I_1)$ from Table 1. We then solve the neutral element and opposite elements of $a'' = (1, 0, 2T_2, -3I_1, 2F_1)$ from Table 4. Next we solve the neutral element and opposite elements of $a''' = (1, 0, 2T_2, -3I_1, 2F_1, 0)$ from Table 4. Finally, we solve the neutral element and opposite elements of a from Table 4.

1. *From Table 1, $neut(a') = (1, 0, 0, -I_1)$ and $anti(a') = (1, 0, -\frac{2}{3}T_2, c_4I_1)$, where $c_4 \in \mathbb{R}$.*
2. *From Table 4 and combining the results of the above step: Being $2 \neq 0$ and $1 + 0 + 2 + (-3) + 2 \neq 0$, thus $neut(a'') = (1, 0, 0, -I_1, F_1)$ and $anti(a'') = (1, 0, -\frac{2}{3}T_2, c_4I_1, c_5F_1)$, where $c_4 + c_5 = \frac{1}{6}$.*

3. *From Table 4 and combining the results of the above step: Being* $0 = 0$ *and* $1 + 0 + 2 + (-3) + 2 + 0 \neq 0$, *thus* $neut(a''') = (1, 0, 0, -I_2, F_1, 0)$ *and* $anti(a''') = (1, 0, -\frac{2}{3}T_2, c_4 I_1, c_5 F_1, 0)$, *where* $c_4 + c_5 = \frac{1}{6}$.
4. *From Table 4 and combining the results of the above step: Being* $-2 \neq 0$ *and* $1 + 0 + 2 + (-3) + 2 + 0 + (-2) = 0$, *thus* $neut(a) = (1, 0, 0, -I_1, F_1, 0, -F_3)$ *and* $anti(a) = (1, 0, -\frac{2}{3}T_2, c_4 I_1, c_5 F_1, 0, c_7 F_3)$, *where* $c_4 + c_5 = \frac{1}{6}, c_7 \in \mathbb{R}$.

Example 7. *For algebra system* $(RNQF_{213}, *)$, *and set* $\mathbb{F} = Z_5 = \{[0], [1], [2], [3], [4]\}$, $a = ([1], [0], [2]T_2, [2]I_1, [2]F_1, 0, [3]F_3)$, *solve the neutral element and opposite elements of a.*

Similar to Example 6, according Algorithm 1 for algebra system $(RNF_{213}, *)$, *firstly, we solve the neutral element and opposite elements of* $a' = ([1], [0], [2]T_2, [2]I_1)$ *from Table 1. We then solve the neutral element and opposite elements of* $a'' = ([1], [0], [2]T_2, [2]I_1, [2]F_1)$ *from Table 4. Next, we solve the neutral element and opposite elements of* $a''' = ([1], [0], [2]T_2, [2]I_1, [2]F_1, [0])$ *from Table 4. Finally, we solve the neutral element and opposite elements of a from Table 4.*

1. *From Table 1,* $neut(a') = ([1], [0], [0], [4]I_1)$ *and* $anti(a') = ([1], [0], [1]T_2, c_4 I_1)$, *where* $c_4 \in \mathbb{F}$.
2. *From Table 4 and combining the results of the above step: Being* $[2] \neq [0]$ *and* $[1] + [0] + [2] + [2] + [2] \neq [0]$, *thus* $neut(a'') = ([1], [0], [0], [4]I_1, [1]F_1)$ *and* $anti(a'') = ([1], [0], [1]T_2, c_4 I_1, c_5 F_1)$, *where* $c_4 + c_5 = [1]$.
3. *From Table 4 and combining the results of the above step: Being* $[0] = [0]$ *and* $[1] + [0] + [2] + [2] + [2] + [0] \neq 0$, *thus* $neut(a''') = ([1], [0], [0], [4]I_1, [1]F_1, [0])$ *and* $anti(a''') = ([1], [0], [1]T_2, c_4 I_1, c_5 F_1, [0])$, *where* $c_4 + c_5 = [1]$.
4. *From Table 4 and combining the results of the above step: Being* $[3] \neq [0]$ *and* $[1] + [0] + [2] + [2] + [2] + [0] + [3] = 0$, *thus* $neut(a) = ([1], [0], [0], [4]I_1, [1]F_1, [0], [4]F_3)$ *and* $anti(a) = ([1], [0], [1]T_2, c_4 I_1, c_5 F_1, [0], c_7 F_3)$, *where* $c_4 + c_5 = [1], c_7 \in \mathbb{F}$.

5. Conclusions

In this paper, we study the algebra structure of $(NQF(RNF_n, RNQF_n), \oplus)$, $(NQF(RNF_n, RNQF_n), *)$ and $(NQF(RNF_n, RNQF_n), \star)$, and we prove that $(NQF(RNF_n, RNQF_n), *)$ (or $(NQF(RNF_n, RNQF_n), \star)$) is a neutrosophic extended triplet group, and provide new examples of neutrosophic extended triplet group and the neutral element and opposite elements of each refined n-ary neutrosophic number (refined neutrosophic quadruple number) can be obtained by given algorithms. In the following, we can explore the algebra structure of $(NQF(RNF_n, RNQF_n), \oplus, *)$ or $(NQF(RNF_n, RNQF_n), \oplus, \star)$. We can also explore the relation of neutrosophic quadruple numbers and other algebra systems in papers [21–23]. Moreover, on the one hand, we will discuss the neutrosophic quadruple numbers based on some particular ring which can form a neutrosophic extended triplet group, while, on the other hand, we will introduce a new operation $\circ$ in order to guarantee $(NQF(RNF_n, RNQF_n), *, \circ)$ is a neutrosophic triplet ring.

Author Contributions: All authors contributed equally to this paper.

Funding: This research was funded by National Natural Science Foundation of China (Grant No. 11501435), Discipline Construction Funding of Xi'an Polytechnic University, Instructional Science and Technology Plan Projects of China National Textile and Apparel Council (No. 2016073) and Scientific Research Program Funded by Shaanxi Provincial Education Department (Program No. 18JS042).

Acknowledgments: The authors would like to thank the reviewers for their many insightful comments and suggestions.

Conflicts of Interest: The authors declare no conflicts of interest.

References

1. Smarandache, F. *A Unifying Field in Logics. Neutrosophy: Neutrosophic Probability, Set and Logic*; American Research Press: Rehoboth, TX, USA, 1998.
2. Smarandache, F.; Ali, M. Neutrosophic triplet group. *Neural Comput. Appl.* **2018**, *29*, 595–601. [CrossRef]

3. Smarandache, F. *Neutrosophic Perspectives: Triplets, Duplets, Multisets, Hybrid Operators, Modal Logic, Hedge Algebras. And Applications*; Pons Publishing House: Brussels, Belgium, 2017.
4. Zhang, X.; Hu, Q.; Smarandache, F.; An, X. On Neutrosophic Triplet Groups: Basic Properties, NT-Subgroups, and Some Notes. *Symmetry* **2018**, *10*, 289. [CrossRef]
5. Zhang, X.H.; Bo, C.X.; Smarandache, F.; Dai, J.H. New inclusion relation of neutrosophic sets with applications and related lattice structure. *Int. J. Mach. Learn. Cybern.* **2018**, *9*, 1753–1763. [CrossRef]
6. Bal, M.; Shalla, M.M.; Olgun, N. Neutrosophic Triplet Cosets and Quotient Groups. *Symmetry* **2017**, *10*, 126. [CrossRef]
7. Zhang, X.H.; Smarandache, F.; Liang, X.L. Neutrosophic Duplet Semi-Group and Cancellable Neutrosophic Triplet Groups. *Symmetry* **2017**, *9*, 275. [CrossRef]
8. Ma, Y.; Zhang, X.; Yang, X.; Zhou, X. Generalized Neutrosophic Extended Triplet Group. *Symmetry* **2019**, *11*, 327. [CrossRef]
9. Wu, X.Y.; Zhang, X.H. The decomposition theorems of AG-neutrosophic extended triplet loops and strong AG-(l, l)-loops. *Mathematics* **2019**, *7*, 268, doi:10.3390/math7030268. [CrossRef]
10. Zhang, X.H.; Wu, X.Y.; Mao, X.Y.; Smarandache, F.; Park, C. On Neutrosophic Extended Triplet Groups (Loops) and Abel-Grassmann's Groupoids (AG-Groupoids). *J. Intell. Fuzzy Syst.* **2019**, in press.
11. Zhang, X.H.; Mao, X.Y.; Wu, Y.T.; Zhai, X.H. Neutrosophic filters in pseudo-BCI algebras. *Int. J. Uncertain. Quant.* **2018**, *8*, 511–526. [CrossRef]
12. Smarandache, F. Hybrid Neutrosophic Triplet Ring in Physical Structures. *Bull. Am. Phys. Soc.* **2017**, *62*, 17.
13. Ali, M.; Smarandache, F.; Khan, M. Study on the development of neutrosophictriplet ring and neutrosophictriplet field. *Mathematics* **2018**, *6*, 46. [CrossRef]
14. Sahin, M.; Abdullah, K. Neutrosophic triplet normed space. *Open Phys.* **2017**, *15*, 697–704. [CrossRef]
15. Zhang, X.H.; Bo, C.X.; Smarandache, F.; Park, C. New operations of totally dependent-neutrosophic sets and totally dependent-neutrosophic soft sets. *Symmetry* **2018**, *10*, 187. [CrossRef]
16. Agboola, A.; Akinleye, S. Neutrosophic Vector Spaces. *Neutrosophic Sets Syst.* **2014**, *4*, 9–18.
17. Li, Q.; Ma, Y.; Zhang, X.; Zhang, J. Neutrosophic Extended Triplet Group Based on Neutrosophic Quadruple Numbers. *Symmetry* **2019**, *11*, 187. [CrossRef]
18. Smarandache, F. Neutrosophic Quadruple Numbers, Refined Neutrosophic Quadruple Numbers, Absorbance Law, and the Multiplication of Neutrosophic Quadruple Numbers. *Neutrosophic Sets Syst.* **2015**, *10*, 96–98.
19. Smarandache, F. (T, I, F)-Neutrosophic Structures. *Neutrosophic Sets Syst.* **2015**, *8*, 3–10. [CrossRef]
20. Akinleye1, S.A.; Smarandache, F.; Agboola, A.A.A. On Neutrosophic Quadruple Algebraic Structures. *Neutrosophic Sets Syst.* **2016**, *12*, 122–126.
21. Zhang, X.H. Fuzzy anti-grouped filters and fuzzy normal filters in pseudo-BCI algebras. *J. Intell. Fuzzy Syst.* **2017**, *33*, 1767–1774. [CrossRef]
22. Zhang, X.H.; Borzooei, R.A.; Jun, Y.B. Q-filters of quantum B-algebras and basic implication algebras. *Symmetry* **2018**, *10*, 573, doi.10.3390/sym10110573. [CrossRef]
23. Zhang, X.H.; Park, C.; Wu, S.P. Soft set theoretical approach to pseudo-BCI algebras. *J. Intell. Fuzzy Syst.* **2018**, *34*, 559–568. [CrossRef]

symmetry

MDPI

Article

Non-Dual Multi-Granulation Neutrosophic Rough Set with Applications

Chunxin Bo [1,2], Xiaohong Zhang [1,*] and Songtao Shao [1]

1 School of Arts and Sciences, Shaanxi University of Science and Technology, Xi'an 710021, China
2 School of Mathematics Science, Liaocheng University, Liaocheng 252000, China
* Correspondence: zhangxiaohong@sust.edu.cn

Received: 17 May 2019; Accepted: 25 June 2019; Published: 12 July 2019

Abstract: Multi-attribute decision-making (MADM) is a part of management decision-making and an important branch of the modern decision theory and method. MADM focuses on the decision problem of discrete and finite decision schemes. Uncertain MADM is an extension and development of classical multi-attribute decision making theory. When the attribute value of MADM is shown by neutrosophic number, that is, the attribute value is complex data and needs three values to express, it is called the MADM problem in which the attribute values are neutrosophic numbers. However, in practical MADM problems, to minimize errors in individual decision making, we need to consider the ideas of many people and synthesize their opinions. Therefore, it is of great significance to study the method of attribute information aggregation. In this paper, we proposed a new theory—non-dual multi-granulation neutrosophic rough set (MS)—to aggregate multiple attribute information and solve a multi-attribute group decision-making (MGDM) problem where the attribute values are neutrosophic numbers. First, we defined two kinds of non-dual MS models, intersection-type MS and union-type MS. Additionally, their properties are studied. Then the relationships between MS, non-dual MS, neutrosophic rough set (NRS) based on neutrosophic intersection (union) relationship, and NRS based on neutrosophic transitive closure relation of union relationship are outlined, and a figure is given to show them directly. Finally, the definition of non-dual MS on two universes is given and we use it to solve a MGDM problem with a neutrosophic number as the attribute value.

Keywords: multi-granulation neutrosophic rough set; non-dual; two universes; multi-attribute group decision making

1. Introduction

Fuzzy sets and rough sets are widely used to solve uncertain problems [1–4]. However, all these theories have their own deficiency, such as in a voting, you may support, not support, be neutral, or abstain from voting, so Smarandache present the definition of the neutrosophic set (NS) [5]. NS is an extensional model of the fuzzy set and intuitionistic fuzzy set. But the original definition of NS is not convenient to solve real-world problems, thus Wang et al. proposed a single-valued neutrosophic set (SVNS) [6]. After that, SVNS was extended and used in many fields. Peng et al. [7] defined simplified NS and obtained some properties. Peng et al. [8] proposed the definition of probability multi-valued NS and studied its properties. Deli et al. [9] defined bipolar NS and studied its properties. Zhang et al. [10] analyzed new inclusion relationships of SVNS and discussed its lattice structure. As an extension of fuzzy sets and rough sets, many scholars combined them and got some results [11–13]. Yang et al. [14] combined SVNS and rough set, then produced a single-valued neutrosophic rough set and discussed its properties. Now NSs and NRSs have been used widely in decision-making problems [15–19].

From the perspective of particle computing, the above rough set theories are essentially defined in a single particle space, and the lower and upper approximations (ULA) of the target concept is shown by the information particles in the particle space induced by a single binary relationship. However, Qian et al. think that, in decision analysis problems, the relationship between the multiple decision makers may be independent of each other, so multiple binary relations are needed to approximate the target. Therefore, they put forward the concept of a multi-granularity rough set (MRS) model [20], and define the optimistic MRS model and pessimistic MRS model, respectively. The biggest difference between MRS and classical rough sets is that MRS can use the knowledge in a multi-granular space to approximate the target. Additionally, because it analyzes the problem from multiple angles and levels, it can obtain a more reasonable and satisfactory solution for the problem, so it has a better application prospect in many practical decision-making problems. Yao et al. [21] studied the rough set models under the multi-granulation approximation space. Now the MRS model has been used widely and has produced some interesting results [22–28].

The ULA operator of most MRS models is dual, there are few articles studying the non-dual MRS model or hybrid MRS model [29,30]. Zhang et al. [31] put forward non-dual MRS (union-type MRS and intersection-type MRS) models and outline the relationships between MRSs. In this paper, we put forward non-dual MS (intersection-type MS and union-type MS) models and study their properties. Then we show the relationships between MS, non-dual MS, NRS-based neutrosophic union relation, and NRS-based neutrosophic intersection relation. Finally, we propose non-dual MS models on two universes and use it to solve MGDM problems with neutrosophic numbers as the attribute values.

The structure of this article is as follows. In Section 2, some basic notions and operations of NRS and MS are introduced. In Section 3, the concepts of non-dual MS are put forward and their qualities are investigated. In Section 4, the relationships between MS, non-dual MS, neutrosophic rough set (NRS) based on neutrosophic intersection (union) relationship, and NRS based on neutrosophic transitive closure relation of union relationship are discussed. In Section 5, non-dual MS models on two universes are proposed and an application to solve the MGDM problem where the attribute values are neutrosophic numbers is outlined. Finally, Section 6 provides our conclusions and outlook.

2. Preliminary

In this section, we look at several basic concepts of NRS and MS.

Definition 1 ([6]). *A SVNS A is denoted by*

$$A = \{(x, T_A(x), I_A(x), F_A(x)) | x \in X\} \tag{1}$$

where $T_A(x)$ represents truth-membership function, $I_A(x)$ represents indeterminacy-membership function, $F_A(x)$ represenst falsity-membership function and $T_A(x)$, $I_A(x)$, $F_A(x) \in [0, 1]$. Additionally, they satisfy the condition $0 \leq T_A(x) + I_A(x) + F_A(x) \leq 3$.

In this paper, "SVNS" is abbreviated to "NS" and we use the symbol NS(U) to denote the set of all NSs in U.

Definition 2 ([6]). *For any two NSs A and B, the inclusion relation, union, intersection, and complement operations are defined:*

(1) $A \subseteq B$ *iff* $\forall x \in U$, $T_A(x) \leq T_B(x)$, $I_A(x) \geq I_B(x)$, $F_A(x) \geq F_B(x)$;
(2) $A \cup B = \{(x, T_A(x) \vee T_B(x), I_A(x) \wedge I_B(x), F_A(x) \wedge F_B(x)) \mid x \in U\}$;
(3) $A \cap B = \{(x, T_A(x) \wedge T_B(x), I_A(x) \vee I_B(x), F_A(x) \vee F_B(x)) \mid x \in U\}$;
(4) $A^c = \{(x, F_A(x), 1\text{-}I_A(x), T_A(x)) \mid x \in U\}$.

Definition 3 [14]). *Suppose (U, R) is a neutrosophic approximation space (NAS). ∀ A ∈ NS(U), the LUA of A, denoted by* $\underline{R}(A)$ *and* $\overline{R}(A)$*, are defined as: ∀ x ∈ U,*

$$\underline{R}(A) = \underset{y\in U}{\cap}(R^c(x,y) \cup A(y)), \overline{R}(A) = \underset{y\in U}{\cup}(R(x,y) \cap A(y))$$

The pair $\left(\underline{R}(A), \overline{R}(A)\right)$ *is called the SVNRS of A.*

Definition 4 ([28]). *Suppose (U, R_i) is a multi-granulation neutrosophic approximation space (MAS). A ∈ NS(U), the optimistic ULA of A, represented by* $\underline{MS^o}(A)$ *and* $\overline{MS}^o(A)$*, are defined:*

$$\underline{MS^o}(A)(x) = \overset{m}{\underset{i=1}{\cup}}\left(\underset{y\in U}{\cap}(R_i{}^c(x,y) \cup A(y))\right) \tag{2}$$

$$\overline{MS}^o(A)(x) = \overset{m}{\underset{i=1}{\cap}}\left(\underset{y\in U}{\cup}(R_i(x,y) \cap A(y))\right) \tag{3}$$

Then the pair $\left(\underline{MS^o}(A), \overline{MS}^o(A)\right)$ *is called an optimistic* MS *when* $\underline{MS^o}(A) \neq \overline{MS}^o(A)$.

Definition 5 ([28]). *Suppose (U, R_i) is a MAS. ∀A ∈ NS(U), the pessimistic ULA of A, represented by* $\underline{MS^p}(A)$ *and* $\overline{MS}^p(A)$*, are defined:*

$$\underline{MS^p}(A)(x) = \overset{m}{\underset{i=1}{\cap}}\left(\underset{y\in U}{\cap}(R_i{}^c(x,y) \cup A(y))\right) \tag{4}$$

$$\overline{MS}^p(A)(x) = \overset{m}{\underset{i=1}{\cup}}\left(\underset{y\in U}{\cup}(R_i(x,y) \cap A(y))\right) \tag{5}$$

Then the pair $\left(\underline{MS^p}(A), \overline{MS}^p(A)\right)$ *is called a pessimistic* MS *when* $\underline{MS^o}(A) \neq \overline{MS}^o(A)$.

Proposition 1 ([28]). *Suppose (U, R_i) be a MAS. ∀ A, B ∈ NS(U), then*

(1) $\underline{MS^o}(A) = \sim \overline{MS}^o(\sim A), \underline{MS^p}(A) = \sim \overline{MS}^p(\sim A)$.
(2) $\overline{MS}^o(A) = \sim \underline{MS^o}(\sim A), \overline{MS}^p(A) = \sim \underline{MS^p}(\sim A)$.
(3) $\underline{MS^o}(A \cap B) = \underline{MS^o}(A) \cap \underline{MS^o}(B), \underline{MS^p}(A \cap B) = \underline{MS^p}(A) \cap \underline{MS^p}(B)$.
(4) $\overline{MS}^o(A \cup B) = \overline{MS}^o(A) \cup \overline{MS}^o(B), \overline{MS}^p(A \cup B) = \overline{MS}^p(A) \cup \overline{MS}^p(B)$.
(5) $A \subseteq B \Rightarrow \underline{MS^o}(A) \subseteq \underline{MS^o}(B), \underline{MS^p}(A) \subseteq \underline{MS^p}(B)$.
(6) $A \subseteq B \Rightarrow \overline{MS}^o(A) \subseteq \overline{MS}^o(B), \overline{MS}^p(A) \subseteq \overline{MS}^p(B)$.
(7) $\underline{MS^o}(A) \cup \underline{MS^o}(B) \subseteq \underline{MS^o}(A \cup B), \underline{MS^p}(A) \cup \underline{MS^p}(B) \subseteq \underline{MS^p}(A \cup B)$.
(8) $\overline{MS}^o(A \cap B) \subseteq \overline{MS}^o(A) \cap \overline{MS}^o(B), \overline{MS}^p(A \cap B) \subseteq \overline{MS}^p(A) \cap \overline{MS}^p(B)$.

Definition 6 ([14]). *If A and B are two neutrosophic numbers in U, the operation of A and B is defined as follows:*

$$(1)\ \lambda A = \left(1 - (1 - T_A)^\lambda, (I_A)^\lambda, (F_A)^\lambda\right) \tag{6}$$

$$(2)\ A \oplus B = (T_A + T_B - T_A \cdot T_B, I_A \cdot I_B, F_A \cdot F_B) \tag{7}$$

Definition 7 ([10]). *Let (t, i, f) be a neutrosophic number, the type-3 score function and type-3 accuracy function are defined:*

$$(1)\ s : D^* \to [0,1], s(t,i,f) = \frac{t + (1-f)}{2} \tag{8}$$

$$(2)\ h : D^* \to [0,1], h(t,i,f) = \frac{t}{t + (1-f)} \tag{9}$$

Definition 8 ([10]). *Let (t_1, i_1, f_1) and (t_2, i_2, f_2) be two neutrosophic numbers. Then*

(1) *If $s(t_1, i_1, f_1) < s(t_2, i_2, f_2)$, then $(t_1, i_1, f_1) \prec (t_2, i_2, f_2)$.*
(2) *If $s(t_1, i_1, f_1) = s(t_2, i_2, f_2), h(t_1, i_1, f_1) < h(t_2, i_2, f_2)$, then $(t_1, i_1, f_1) \prec (t_2, i_2, f_2)$.*
(3) *If $s(t_1, i_1, f_1) = s(t_2, i_2, f_2)$, $h(t_1, i_1, f_1) = h(t_2, i_2, f_2)$, $i_1 < i_2$, then $(t_1, i_1, f_1) \prec (t_2, i_2, f_2)$; if $i_1 = i_2$, then $(t_1, i_1, f_1) = (t_2, i_2, f_2)$.*

3. Non-Dual Multi-Granulation Neutrosophic Rough Set

In this section, we introduce non-dual MS (intersection-type MS and union-type MS) models and study their properties.

Definition 9. *Let tuple ordered set (U, R_i) $(1 \le i \le m)$ be a MAS. For any $A \in NS(U)$, the intersection-type ULA $\underline{MS}^{(\cap)}(A)$ and $\overline{MS}^{(\cap)}(A)$ in (U, R_i) are defined:*

$$\underline{MS}^{(\cap)}(A)(x) = \bigcap_{i=1}^{m}\left(\bigcap_{y\in U}(R_i^c(x,y) \cup A(y))\right) \tag{10}$$

$$\overline{MS}^{(\cap)}(A)(x) = \bigcap_{i=1}^{m}\left(\bigcup_{y\in U}(R_i(x,y) \cap A(y))\right) \tag{11}$$

Obviously, $\underline{MS}^{(\cap)}(A)$ and $\overline{MS}^{(\cap)}(A)$ are two NSs of U. Furthermore, A is called a definable NS on (U, R_i) when $\underline{MS}^{(\cap)}(A) = \overline{MS}^{(\cap)}(A)$. Otherwise, the pair $\left(\underline{MS}^{(\cap)}(A), \overline{MS}^{(\cap)}(A)\right)$ is called intersection-type MS.

Definition 10. *Let tuple ordered set (U, R_i) $(1 \le i \le m)$ be a MAS. For any $A \in NS(U)$, the union-type ULA $\underline{MS}^{(\cup)}(A)$ and $\overline{MS}^{(\cup)}(A)$ in (U, R_i) are defined:*

$$\underline{MS}^{(\cup)}(A)(x) = \bigcup_{i=1}^{m}\left(\bigcap_{y\in U}(R_i^c(x,y) \cup A(y))\right), \tag{12}$$

$$\overline{MS}^{(\cup)}(A)(x) = \bigcup_{i=1}^{m}\left(\bigcup_{y\in U}(R_i(x,y) \cap A(y))\right). \tag{13}$$

Obviously, $\underline{MS}^{(\cup)}(A)$ and $\overline{MS}^{(\cup)}(A)$ are two NSs of U. Furthermore, A is called a definable NS on (U, R_i) when $\underline{MS}^{(\cup)}(A) = \overline{MS}^{(\cup)}(A)$. Otherwise, the pair $\left(\underline{MS}^{(\cup)}(A), \overline{MS}^{(\cup)}(A)\right)$ is called union-type MS.

Proposition 2. *Let (U, R_i) be a MAS, R_i $(1 \le i \le m)$ be the neutrosophic relations on U. For any $A, B \in NS(U)$, we have*

(1) $\underline{MS}^{(\cap)}(A \cap B) = \underline{MS}^{(\cap)}(A) \cap \underline{MS}^{(\cap)}(B), \overline{MS}^{(\cap)}(A \cup B) = \overline{MS}^{(\cap)}(A) \cup \overline{MS}^{(\cap)}(B)$;
(2) $\underline{MS}^{(\cup)}(A \cap B) = \underline{MS}^{(\cup)}(A) \cap \underline{MS}^{(\cup)}(B), \overline{MS}^{(\cup)}(A \cup B) = \overline{MS}^{(\cup)}(A) \cup \overline{MS}^{(\cup)}(B)$;
(3) $A \subseteq B \Rightarrow \underline{MS}^{(\cap)}(A) \subseteq \underline{MS}^{(\cap)}(B)$, $A \subseteq B \Rightarrow \overline{MS}^{(\cap)}(A) \subseteq \overline{MS}^{(\cap)}(B)$;
(4) $A \subseteq B \Rightarrow \underline{MS}^{(\cup)}(A) \subseteq \underline{MS}^{(\cup)}(B)$, $A \subseteq B \Rightarrow \overline{MS}^{(\cup)}(A) \subseteq \overline{MS}^{(\cup)}(B)$;
(5) $\underline{MS}^{(\cap)}(A \cup B) \supseteq \underline{MS}^{(\cap)}(A) \cup \underline{MS}^{(\cap)}(B), \overline{MS}^{(\cap)}(A \cap B) \subseteq \overline{MS}^{(\cap)}(A) \cap \overline{MS}^{(\cap)}(B)$;
(6) $\underline{MS}^{(\cup)}(A \cup B) \supseteq \underline{MS}^{(\cup)}(A) \cup \underline{MS}^{(\cup)}(B), \overline{MS}^{(\cup)}(A \cap B) \subseteq \overline{MS}^{(\cup)}(A) \cap \overline{MS}^{(\cup)}(B)$.

Proof. (1) By Definition 9, we have

$$\begin{aligned}\underline{MS}^{(\cap)}(A\cap B) &= \bigcap_{i=1}^{m}\left(\bigcap_{y\in U}(R_i^c(x,y)\cup(A\cap B)(y))\right)\\ &= \bigcap_{i=1}^{m}\left(\bigcap_{y\in U}((R_i^c(x,y)\cup A(y))\cap(R_i^c(x,y)\cup B(y)))\right)\\ &= \left(\bigcap_{i=1}^{m}\left(\bigcap_{y\in U}(R_i^c(x,y)\cup A(y))\right)\right)\cap\left(\bigcap_{i=1}^{m}\left(\bigcap_{y\in U}(R_i^c(x,y)\cup B(y))\right)\right)\\ &= \underline{MS}^{(\cap)}(A)\cap\underline{MS}^{(\cap)}(B).\end{aligned}$$

Similarly, by Definition 9, we can get

$$\overline{MS}^{(\cap)}(A\cup B)=\overline{MS}^{(\cap)}(A)\cup\overline{MS}^{(\cap)}(B).$$

(2) By Definition 10, we have

$$\begin{aligned}\underline{MS}^{(\cup)}(A\cap B) &= \bigcup_{i=1}^{m}\left(\bigcap_{y\in U}(R_i^c(x,y)\cup(A\cap B)(y))\right)\\ &= \bigcup_{i=1}^{m}\left(\bigcap_{y\in U}((R_i^c(x,y)\cup A(y))\cap(R_i^c(x,y)\cup B(y)))\right)\\ &= \left(\bigcup_{i=1}^{m}\left(\bigcap_{y\in U}(R_i^c(x,y)\cup A(y))\right)\right)\cap\left(\bigcup_{i=1}^{m}\left(\bigcap_{y\in U}(R_i^c(x,y)\cup B(y))\right)\right)\\ &= \underline{MS}^{(\cup)}(A)\cap\underline{MS}^{(\cup)}(B).\end{aligned}$$

Similarly, by Definition 10, we can get

$$\overline{MS}^{(\cup)}(A\cup B)=\overline{MS}^{(\cup)}(A)\cup\overline{MS}^{(\cup)}(B).$$

(3) Suppose $A\subseteq B$, then $T_A(x)\le T_B(x)$, $I_A(x)\ge I_B(x)$, $F_A(x)\ge F_B(x)$,

$$T_{\underline{MS}^{(\cap)}(A)}(x)=\bigwedge_{i=1}^{m}\bigwedge_{y\in U}\left(F_{R_i}(x,y)\vee T_A(y)\right)\le\bigwedge_{i=1}^{m}\bigwedge_{y\in U}\left(F_{R_i}(x,y)\vee T_B(y)\right)=T_{\underline{MS}^{(\cap)}(B)}(x).$$

$$I_{\underline{MS}^{(\cap)}(A)}(x)=\bigvee_{i=1}^{m}\bigvee_{y\in U}\left[\left(1-I_{R_i}(x,y)\right)\wedge I_A(y)\right]\ge\bigvee_{i=1}^{m}\bigvee_{y\in U}\left[\left(1-I_{R_i}(x,y)\right)\wedge I_B(y)\right]=I_{\underline{MS}^{(\cap)}(B)}(x).$$

$$F_{\underline{MS}^{(\cap)}(A\cup B)}(x)=\bigvee_{i=1}^{m}\bigvee_{y\in U}\left[T_{R_i}(x,y)\wedge F_A(y)\right]\ge\bigvee_{i=1}^{m}\bigvee_{y\in U}\left[T_{R_i}(x,y)\wedge F_B(y)\right]=F_{\underline{MS}^{(\cap)}(B)}(x).$$

Hence, $\underline{MS}^{(\cap)}(A)\subseteq\underline{MS}^{(\cap)}(B)$.
Similarly, we can get $\overline{MS}^{(\cap)}(A)\subseteq\overline{MS}^{(\cap)}(B)$.
(4) The proof is similar with (3).
(5) According to Definition 9, we have

$$\begin{aligned}T_{\underline{MS}^{(\cap)}(A\cup B)}(x) &= \bigwedge_{i=1}^{m}\bigwedge_{y\in U}\left[F_{R_i}(x,y)\vee(T_A(y)\vee T_B(y))\right]=\bigwedge_{i=1}^{m}\bigwedge_{y\in U}\left[\left(F_{R_i}(x,y)\vee T_A(y)\right)\vee\left(F_{R_i}(x,y)\vee T_B(y)\right)\right]\\ &\ge\left[\bigwedge_{i=1}^{m}\bigwedge_{y\in U}\left(F_{R_i}(x,y)\vee T_A(y)\right)\right]\vee\left[\bigwedge_{i=1}^{m}\bigwedge_{y\in U}\left(F_{R_i}(x,y)\vee T_B(y)\right)\right]=T_{\underline{MS}^{(\cap)}(A)}(x)\vee T_{\underline{MS}^{(\cap)}(B)}(x).\end{aligned}$$

$$\begin{aligned}I_{\underline{MS}^{(\cap)}(A\cup B)}(x) &= \bigvee_{i=1}^{m}\bigvee_{y\in U}\left[\left(1-I_{R_i}(x,y)\right)\wedge(I_A(y)\wedge I_B(y))\right]\\ &= \bigvee_{i=1}^{m}\bigvee_{y\in U}\left[\left(\left(1-I_{R_i}(x,y)\right)\wedge I_A(y)\right)\wedge\left(\left(1-I_{R_i}(x,y)\right)\wedge I_B(y)\right)\right]\\ &\le\left[\bigvee_{i=1}^{m}\bigvee_{y\in U}\left(\left(1-I_{R_i}(x,y)\right)\wedge I_A(y)\right)\right]\wedge\left[\bigvee_{i=1}^{m}\bigvee_{y\in U}\left(\left(1-I_{R_i}(x,y)\right)\wedge I_B(y)\right)\right]\\ &= I_{\underline{MS}^{(\cap)}(A)}(x)\wedge I_{\underline{MS}^{(\cap)}(B)}(x).\end{aligned}$$

$$F_{\underline{MS}^{(\cap)}(A\cup B)}(x) = \overset{m}{\underset{i=1}{\vee}} \underset{y\in U}{\vee}\left[T_{R_i}(x,y)\wedge(F_A(y)\wedge F_B(y))\right] = \overset{m}{\underset{i=1}{\vee}} \underset{y\in U}{\vee}\left[\left(T_{R_i}(x,y)\wedge F_A(y)\right)\wedge\left(T_{R_i}(x,y)\wedge F_B(y)\right)\right]$$
$$\leq \left[\overset{m}{\underset{i=1}{\vee}} \underset{y\in U}{\vee}\left(T_{R_i}(x,y)\wedge F_A(y)\right)\right]\wedge\left[\overset{m}{\underset{i=1}{\vee}} \underset{y\in U}{\vee}\left(T_{R_i}(x,y)\wedge F_B(y)\right)\right] = F_{\underline{MS}^{(\cap)}(A)}(x)\wedge F_{\underline{MS}^{(\cap)}(B)}(x).$$

Hence, $\underline{MS}^{(\cap)}(A\cup B)\supseteq\underline{MS}^{(\cap)}(A)\cup\underline{MS}^{(\cap)}(B)$.

Additionally, we have

$$T_{\overline{MS}^{(\cap)}(A\cap B)}(x) = \overset{m}{\underset{i=1}{\wedge}} \underset{y\in U}{\vee}\left[T_{R_i}(x,y)\wedge(T_A(y)\wedge T_B(y))\right] = \overset{m}{\underset{i=1}{\wedge}} \underset{y\in U}{\vee}\left[\left(T_{R_i}(x,y)\wedge T_A(y)\right)\wedge\left(T_{R_i}(x,y)\wedge T_B(y)\right)\right]$$
$$\leq \left[\overset{m}{\underset{i=1}{\wedge}} \underset{y\in U}{\vee}\left(T_{R_i}(x,y)\wedge T_A(y)\right)\right]\wedge\left[\overset{m}{\underset{i=1}{\wedge}} \underset{y\in U}{\vee}\left(T_{R_i}(x,y)\wedge T_B(y)\right)\right] = T_{\overline{MS}^{(\cap)}(A)}(x)\wedge T_{\overline{MS}^{(\cap)}(B)}(x).$$

$$I_{\overline{MS}^{(\cap)}(A\cap B)}(x) = \overset{m}{\underset{i=1}{\vee}} \underset{y\in U}{\wedge}\left[I_{R_i}(x,y)\vee(I_A(y)\vee I_B(y))\right] = \overset{m}{\underset{i=1}{\vee}} \underset{y\in U}{\wedge}\left[\left(I_{R_i}(x,y)\vee I_A(y)\right)\vee\left(I_{R_i}(x,y)\vee I_A(y)\right)\right]$$
$$\leq \left[\overset{m}{\underset{i=1}{\vee}} \underset{y\in U}{\wedge}\left(I_{R_i}(x,y)\vee I_A(y)\right)\right]\wedge\left[\overset{m}{\underset{i=1}{\vee}} \underset{y\in U}{\wedge}\left(I_{R_i}(x,y)\vee I_B(y)\right)\right] = I_{\overline{MS}^{(\cap)}(A)}(x)\wedge I_{\overline{MS}^{(\cap)}(B)}(x).$$

$$F_{\overline{MS}^{(\cap)}(A\cap B)}(x) = \overset{m}{\underset{i=1}{\vee}} \underset{y\in U}{\wedge}\left[F_{R_i}(x,y)\vee(F_A(y)\vee F_B(y))\right] = \overset{m}{\underset{i=1}{\vee}} \underset{y\in U}{\wedge}\left[\left(F_{R_i}(x,y)\vee F_A(y)\right)\vee\left(F_{R_i}(x,y)\vee F_A(y)\right)\right]$$
$$\geq \left[\overset{m}{\underset{i=1}{\vee}} \underset{y\in U}{\wedge}\left(F_{R_i}(x,y)\vee F_A(y)\right)\right]\vee\left[\overset{m}{\underset{i=1}{\vee}} \underset{y\in U}{\wedge}\left(F_{R_i}(x,y)\vee F_B(y)\right)\right] = F_{\overline{MS}^{(\cap)}(A)}(x)\vee F_{\overline{MS}^{(\cap)}(B)}(x).$$

Hence, $\overline{MS}^{(\cap)}(A\cap B)\subseteq\overline{MS}^{(\cap)}(A)\cap\overline{MS}^{(\cap)}(B)$.

(6) According to Definition 10, we have

$$T_{\underline{MS}^{(\cup)}(A\cup B)}(x) = \overset{m}{\underset{i=1}{\vee}} \underset{y\in U}{\wedge}\left[F_{R_i}(x,y)\vee(T_A(y)\vee T_B(y))\right] = \overset{m}{\underset{i=1}{\vee}} \underset{y\in U}{\wedge}\left[\left(F_{R_i}(x,y)\vee T_A(y)\right)\vee\left(F_{R_i}(x,y)\vee T_B(y)\right)\right]$$
$$\geq \left[\overset{m}{\underset{i=1}{\vee}} \underset{y\in U}{\wedge}\left(F_{R_i}(x,y)\vee T_A(y)\right)\right]\vee\left[\overset{m}{\underset{i=1}{\vee}} \underset{y\in U}{\wedge}\left(F_{R_i}(x,y)\vee T_B(y)\right)\right] = T_{\underline{MS}^{(\cup)}(A)}(x)\vee T_{\underline{MS}^{(\cup)}(B)}(x).$$

$$\begin{aligned}I_{\underline{MS}^{(\cup)}(A\cup B)}(x) &= \overset{m}{\underset{i=1}{\wedge}} \underset{y\in U}{\vee}\left[\left(1-I_{R_i}(x,y)\right)\wedge(I_A(y)\wedge I_B(y))\right]\\ &= \overset{m}{\underset{i=1}{\wedge}} \underset{y\in U}{\vee}\left[\left(\left(1-I_{R_i}(x,y)\right)\wedge I_A(y)\right)\wedge\left(\left(1-I_{R_i}(x,y)\right)\wedge I_B(y)\right)\right]\\ &\leq \left[\overset{m}{\underset{i=1}{\wedge}} \underset{y\in U}{\vee}\left(\left(1-I_{R_i}(x,y)\right)\wedge I_A(y)\right)\right]\wedge\left[\overset{m}{\underset{i=1}{\wedge}} \underset{y\in U}{\vee}\left(\left(1-I_{R_i}(x,y)\right)\wedge I_B(y)\right)\right]\\ &= I_{\underline{MS}^{(\cup)}(A)}(x)\wedge I_{\underline{MS}^{(\cup)}(B)}(x).\end{aligned}$$

$$F_{\underline{MS}^{(\cup)}(A\cup B)}(x) = \overset{m}{\underset{i=1}{\wedge}} \underset{y\in U}{\vee}\left[T_{R_i}(x,y)\wedge(F_A(y)\wedge F_B(y))\right] = \overset{m}{\underset{i=1}{\wedge}} \underset{y\in U}{\vee}\left[\left(T_{R_i}(x,y)\wedge F_A(y)\right)\wedge\left(T_{R_i}(x,y)\wedge F_B(y)\right)\right]$$
$$\leq \left[\overset{m}{\underset{i=1}{\wedge}} \underset{y\in U}{\vee}\left(T_{R_i}(x,y)\wedge F_A(y)\right)\right]\wedge\left[\overset{m}{\underset{i=1}{\wedge}} \underset{y\in U}{\vee}\left(T_{R_i}(x,y)\wedge F_B(y)\right)\right] = F_{\underline{MS}^{(\cup)}(A)}(x)\wedge F_{\underline{MS}^{(\cup)}(B)}(x).$$

Hence, $\underline{MS}^{(\cup)}(A\cup B)\supseteq\underline{MS}^{(\cup)}(A)\cup\underline{MS}^{(\cup)}(B)$.

Additionally, we have

$$T_{\overline{MS}^{(\cup)}(A\cap B)}(x) = \overset{m}{\underset{i=1}{\vee}} \underset{y\in U}{\vee}\left[T_{R_i}(x,y)\wedge(T_A(y)\wedge T_B(y))\right] = \overset{m}{\underset{i=1}{\vee}} \underset{y\in U}{\vee}\left[\left(T_{R_i}(x,y)\wedge T_A(y)\right)\wedge\left(T_{R_i}(x,y)\wedge T_B(y)\right)\right]$$
$$\leq \left[\overset{m}{\underset{i=1}{\vee}} \underset{y\in U}{\vee}\left(T_{R_i}(x,y)\wedge T_A(y)\right)\right]\wedge\left[\overset{m}{\underset{i=1}{\vee}} \underset{y\in U}{\vee}\left(T_{R_i}(x,y)\wedge T_B(y)\right)\right] = T_{\overline{MS}^{(\cup)}(A)}(x)\wedge T_{\overline{MS}^{(\cup)}(B)}(x).$$

$$I_{\overline{MS}^{(\cup)}(A\cap B)}(x) = \overset{m}{\underset{i=1}{\wedge}} \underset{y\in U}{\wedge}\left[I_{R_i}(x,y)\vee(I_A(y)\vee I_B(y))\right] = \overset{m}{\underset{i=1}{\wedge}} \underset{y\in U}{\wedge}\left[\left(I_{R_i}(x,y)\vee I_A(y)\right)\vee\left(I_{R_i}(x,y)\vee I_A(y)\right)\right]$$
$$\leq \left[\overset{m}{\underset{i=1}{\wedge}} \underset{y\in U}{\wedge}\left(I_{R_i}(x,y)\vee I_A(y)\right)\right]\wedge\left[\overset{m}{\underset{i=1}{\wedge}} \underset{y\in U}{\wedge}\left(I_{R_i}(x,y)\vee I_B(y)\right)\right] = I_{\overline{MS}^{(\cup)}(A)}(x)\wedge I_{\overline{MS}^{(\cup)}(B)}(x).$$

$$F_{\overline{MS}^{(\cup)}(A\cap B)}(x) = \overset{m}{\underset{i=1}{\wedge}} \underset{y\in U}{\wedge}\left[F_{R_i}(x,y)\vee(F_A(y)\vee F_B(y))\right] = \overset{m}{\underset{i=1}{\wedge}} \underset{y\in U}{\wedge}\left[\left(F_{R_i}(x,y)\vee F_A(y)\right)\vee\left(F_{R_i}(x,y)\vee F_A(y)\right)\right]$$
$$\geq \left[\overset{m}{\underset{i=1}{\wedge}} \underset{y\in U}{\wedge}\left(F_{R_i}(x,y)\vee F_A(y)\right)\right]\vee\left[\overset{m}{\underset{i=1}{\wedge}} \underset{y\in U}{\wedge}\left(F_{R_i}(x,y)\vee F_B(y)\right)\right] = F_{\overline{MS}^{(\cup)}(A)}(x)\vee F_{\overline{MS}^{(\cup)}(B)}(x).$$

Hence, $\overline{MS}^{(\cup)}(A \cap B) \subseteq \overline{MS}^{(\cup)}(A) \cap \overline{MS}^{(\cup)}(B)$. □

4. The Relationships between Multi-Granulation Neutrosophic Rough Set Models

In this section, we discuss the relationships between MS, non-dual MS, neutrosophic rough set (NRS) based on neutrosophic intersection (union) relationship, and NRS based on neutrosophic transitive closure relation of union relationship and show it by a relational graph.

Definition 11. *Suppose U is a non-empty finite universe, and R_i ($1 \le i \le m$) is the binary NR on U. The ULA based on neutrosophic union relationship, represented by $\underline{\overset{m}{\underset{i=1}{\cup}} R_i(A)}$ and $\overline{\overset{m}{\underset{i=1}{\cup}} R_i(A)}$, are defined:*

$$\underline{\overset{m}{\underset{i=1}{\cup}} R_i(A)(x)} = \underset{y \in U}{\cap}\left(\left(\overset{m}{\underset{i=1}{\cup}} R_i(x,y)\right)^c \cup A(y)\right), \tag{14}$$

$$\overline{\overset{m}{\underset{i=1}{\cup}} R_i(A)(x)} = \underset{y \in U}{\cup}\left(\left(\overset{m}{\underset{i=1}{\cup}} R_i(x,y)\right) \cap A(y)\right). \tag{15}$$

Definition 12. *Suppose U is a non-empty finite universe, and R_i ($1 \le i \le m$) is the binary NR on U. The ULA based on neutrosophic intersection relationship, represented by $\underline{\overset{m}{\underset{i=1}{\cap}} R_i(A)}$ and $\overline{\overset{m}{\underset{i=1}{\cap}} R_i(A)}$, are defined:*

$$\underline{\overset{m}{\underset{i=1}{\cap}} R_i(A)(x)} = \underset{y \in U}{\cap}\left(\left(\overset{m}{\underset{i=1}{\cap}} R_i(x,y)\right)^c \cup A(y)\right), \tag{16}$$

$$\overline{\overset{m}{\underset{i=1}{\cap}} R_i(A)(x)} = \underset{y \in U}{\cup}\left(\left(\overset{m}{\underset{i=1}{\cap}} R_i(x,y)\right) \cap A(y)\right). \tag{17}$$

Definition 13 ([32]). *Suppose R is a neutrosophic relation in U. The minimal transitive neutrosophic relation containing R is called transitive closure of R, denoted by t(R).*

Proposition 3. *Suppose R is a neutrosophic relation in U. Then* $t(R) = \overset{\infty}{\underset{k=1}{\cup}} R^k$. *Where* $R^k = R \bullet R \bullet R \bullet \cdots$, $(R \bullet S)(x,z) = \underset{y \in Y}{\cup}(R(x,y) \cap S(y,z))$.

Definition 14. *Suppose (U, R) is neutrosophic approximation space. Suppose U is a non-empty finite universe, R_i ($1 \le i \le m$) is neutrosophic relations on U, and t(R) denotes the transitive closure of the union of neutrosophic relations R_i on U.$\forall A \in NS(U)$, the ULA of A, denoted by $\underline{t(R)}(A)$ and $\overline{t(R)}(A)$, are defined as: $\forall\, x \in U$,*

$$\underline{t(R)}(A)(x) = \underset{y \in U}{\cap}\left[t(R)^c(x,y) \cup A(y)\right], \overline{t(R)}(A)(x) = \underset{y \in U}{\cup}(t(R)(x,y) \cap A(y)).$$

Proposition 4. *Let (U, R_i) be a MAS, R_i ($1 \le i \le m$) be neutrosophic relations on U. For any $A \in NS$ (U), we have*

(1) $\underline{t(R)}(A) \subseteq \underline{MS}^p(A) = \underline{MS}^{(\cap)}(A) = \underline{\overset{m}{\underset{i=1}{\cup}} R_i(A)} \subseteq \underline{R_i}(A) \subseteq \underline{MS}^o(A) = \underline{MS}^{(\cup)}(A) \subseteq \underline{\overset{m}{\underset{i=1}{\cap}} R_i(A)} \subseteq X$;

(2) $X \subseteq \overline{\overset{m}{\underset{i=1}{\cap}} R_i(A)} \subseteq \overline{MS}^o(A) = \overline{MS}^{(\cap)}(A) \subseteq \overline{R_i}(A) \subseteq \overline{MS}^p(A) = \overline{MS}^{(\cup)}(A) = \overline{\overset{m}{\underset{i=1}{\cup}} R_i(A)} \subseteq \overline{t(R)}(A)$.

Proof. (1) According to Definition 4, Definition 5, Definition 9, and Definition 10, we can get $\overline{MS}^{o}(A)=\overline{MS}^{(\cap)}(A)$, $\overline{MS}^{p}(A)=\overline{MS}^{(\cup)}(A)$. Let $R = R_1 \cup R_2 \cup \cdots \cup R_m$, $t(R) = R \cup R_2 \cup \cdots$, then $(t(R))^c = R^c \cap (R_2)^c \cap \cdots$, so $(t(R))^c \subseteq R^c$, thus

$$\begin{aligned}
\underline{t(R)}(A)(x) &= \underset{y\in U}{\cap}\left[t(R)^c(x,y)\cup A(y)\right] \\
&\subseteq \underset{y\in U}{\cap}[R^c(x,y)\cup A(y)] = \underline{\overset{m}{\underset{i=1}{\cup}} R_i}(A)(x) \\
&= \underset{y\in U}{\cap}[(R_1{}^c\cap R_2{}^c\cap\cdots\cap R_m{}^c)(x,y)\cup A(y)] \\
&= \underset{y\in U}{\cap}[(R_1{}^c(x,y)\cup A(y))\cap(R_2{}^c(x,y)\cup A(y))\cap\cdots\cap(R_m{}^c(x,y)\cup A(y))] \\
&= \left[\underset{y\in U}{\cap}(R_1{}^c(x,y)\cup A(y))\right]\cap\left[\underset{y\in U}{\cap}(R_2{}^c(x,y)\cup A(y))\right]\cap\cdots\cap\left[\underset{y\in U}{\cap}(R_m{}^c(x,y)\cup A(y))\right] \\
&= \overset{m}{\underset{i=1}{\cap}}\left(\underset{y\in U}{\cap}(R_i{}^c(x,y)\cup A(y))\right) = \underline{MS}^{p}(A)(x) = \underline{MS}^{(\cap)}(A) \\
&\subseteq \underset{y\in U}{\cap}(R_i{}^c(x,y)\cup A(y)) = \underline{R_i}(A)(x).
\end{aligned}$$

Additionally, we have

$$\begin{aligned}
\underline{R_i}(A)(x) &= \underset{y\in U}{\cap}(R_i{}^c(x,y)\cup A(y)) \\
&\subseteq \overset{m}{\underset{i=1}{\cup}}\left(\underset{y\in U}{\cap}(R_i{}^c(x,y)\cup A(y))\right) = \underline{MS}^{o}(A)(x) = \underline{MS}^{(\cup)}(A)(x) \\
&\subseteq \underset{y\in U}{\cap}(R_1{}^c(x,y)\cup R_2{}^c(x,y)\cup\cdots\cup R_m{}^c(x,y)\cup A(y)) \\
&= \underset{y\in U}{\cap}((R_1{}^c\cup R_2{}^c\cup\cdots\cup R_m{}^c)(x,y)\cup A(y)) \\
&= \underset{y\in U}{\cap}\left(\left(\overset{m}{\underset{i=1}{\cap}} R_i(x,y)\right)^c(x,y)\cup A(y)\right) = \underline{\overset{m}{\underset{i=1}{\cap}} R_i}(A) \subseteq X.
\end{aligned}$$

Then we get the proof.

(2) According to Definition 4, Definition 5, Definition 9, and Definition 10, we can get $\underline{MS}^{p}(A) = \underline{MS}^{(\cap)}(A)$, $\underline{MS}^{o}(A) = \underline{MS}^{(\cup)}(A)$. Let $R = R_1 \cup R_2 \cup \cdots \cup R_m$, $t(R) = R \cup R_2 \cup \cdots$, then $R \subseteq t(R)$, thus

$$\begin{aligned}
X &\subseteq \overline{\overset{m}{\underset{i=1}{\cap}} R_i}(A)(x) = \underset{y\in U}{\cup}\left(\left(\overset{m}{\underset{i=1}{\cap}} R_i(x,y)\right)\cap A(y)\right) \\
&\subseteq \underset{y\in U}{\cup}(R_i(x,y)\cap A(y)) = \overline{R_i}(A)(x) \\
&\subseteq \overset{m}{\underset{i=1}{\cup}}\left(\underset{y\in U}{\cup}(R_i(x,y)\cap A(y))\right) = \overline{MS}^{p}(A)(x) = \overline{MS}^{(\cup)}(A)(x) \\
&= \underset{y\in U}{\cup}\left(\left(\overset{m}{\underset{i=1}{\cup}} R_i(x,y)\right)\cap A(y)\right) = \overline{\overset{m}{\underset{i=1}{\cup}} R_i}(A)(x).
\end{aligned}$$

Additionally, we have

$$\begin{aligned}
\overline{\overset{m}{\underset{i=1}{\cup}} R_i}(A)(x) &= \underset{y\in U}{\cup}\left(\left(\overset{m}{\underset{i=1}{\cup}} R_i(x,y)\right)\cap A(y)\right) \\
&= \underset{y\in U}{\cup}(R(x,y)\cap A(y)) \\
&\subseteq \underset{y\in U}{\cup}\left(\left(R\cup R^2\cup\cdots\right)(x,y)\cap A(y)\right) \\
&= \overline{t(R)}(A)(x).
\end{aligned}$$

Then we get the proof. □

The above results show that the four kinds of lower and upper approximations equipped with the inclusion relation $\subseteq$ can construct a lattice. This fact can be described by Figure 1, where $i \neq j$, each node denotes an approximation or a concept, and each diagonal line connects two approximations, the lower node is a subset of the upper node.

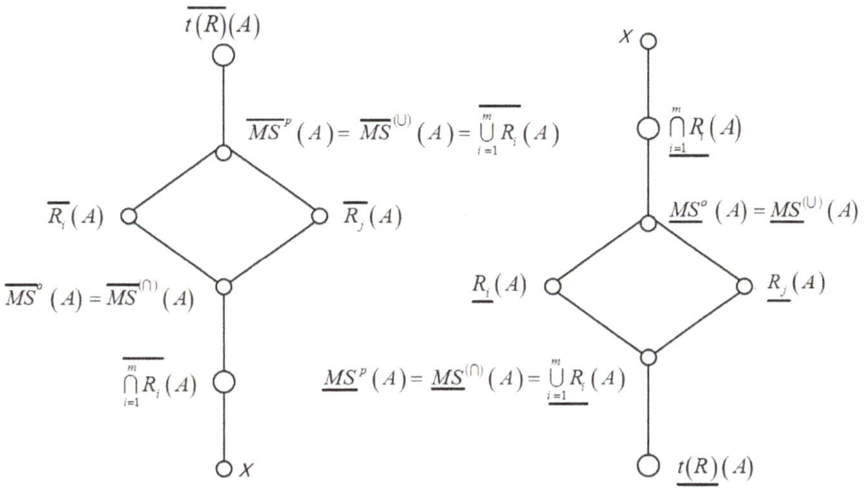

Figure 1. The relationships between neutrosophic rough lower and upper approximations.

5. The Application of Non-Dual Multi-Granulation Neutrosophic Rough Set on Two Universes in MGDM

In this section, we propose the concept of non-dual MS on two universes and we talk about the relationship between non-dual MS on two universes and non-dual MS on a single universe. Additionally, we used non-dual MS on two universes to deal with a MGDM problem where the attribute values are neutrosophic numbers.

Definition 14 ([28]). *Suppose U, V are two non-empty finite universes, and $R_i \in NS\ (U \times V)\ (1 \le i \le m)$ is binary NR. We call (U, V, R_i) the MAS on two universes.*

Definition 15. *Let tuple ordered set (U, V, R_i) $(1 \le i \le m)$ be a MAS on two universes. For any $A \in NS(U)$, the intersection-type ULA $\underline{MNRS}^{(\cap)}(A)$ and $\overline{MNRS}^{(\cap)}(A)$ in (U, V, R_i) are defined:*

$$\underline{MNRS}^{(\cap)}(A)(x) = \bigcap_{i=1}^{m}\left(\bigcap_{y \in V}(R_i^c(x,y) \cup A(y))\right), \tag{18}$$

$$\overline{MNRS}^{(\cap)}(A)(x) = \bigcap_{i=1}^{m}\left(\bigcup_{y \in V}(R_i(x,y) \cap A(y))\right). \tag{19}$$

Obviously, $\underline{MNRS}^{(\cap)}(A)$ and $\overline{MNRS}^{(\cap)}(A)$ are two NSs. Furthermore, A is called a definable NS on (U, V, R_i) when $\underline{MNRS}^{(\cap)}(A) = \overline{MNRS}^{(\cap)}(A)$. Otherwise, the pair $\left(\underline{MNRS}^{(\cap)}(A), \overline{MNRS}^{(\cap)}(A)\right)$ are called intersection-type MS *on two universes.*

Definition 16. *Let tuple ordered set (U, V, R_i) $(1 \le i \le m)$ be a MAS on two universes. For any $A \in NS(U)$, the union-type ULA $\underline{MNRS}^{(\cup)}(A)$ and $\overline{MNRS}^{(\cup)}(A)$ in (U, V, R_i) are defined:*

$$\underline{MNRS}^{(\cup)}(A)(x) = \bigcup_{i=1}^{m}\left(\bigcap_{y \in V}(R_i^c(x,y) \cup A(y))\right), \tag{20}$$

$$\overline{MNRS}^{(\cup)}(A)(x) = \bigcup_{i=1}^{m}\left(\bigcup_{y\in V}(R_i(x,y) \cap A(y))\right). \tag{21}$$

Obviously, $\underline{MNRS}^{(\cup)}(A)$ *and* $\overline{MNRS}^{(\cup)}(A)$ *are two NSs. Furthermore, A is called a definable NS on (U, V, R_i) when* $\underline{MNRS}^{(\cup)}(A) = \overline{MNRS}^{(\cup)}(A)$. *Otherwise, the pair* $\left(\underline{MNRS}^{(\cup)}(A), \overline{MNRS}^{(\cup)}(A)\right)$ *are called union-type* MS *on two universes.*

Proposition 5. *Let (U, V, R_i) be a MAS on two universes. For any A, B ∈ NS(U), we have*

(1) $\underline{MNRS}^{(\cap)}(A\cap B) = \underline{MNRS}^{(\cap)}(A) \cap \underline{MNRS}^{(\cap)}(B)$, $\overline{MNRS}^{(\cap)}(A\cup B) = \overline{MNRS}^{(\cap)}(A) \cup \overline{MNRS}^{(\cap)}(B)$;

(2) $\underline{MNRS}^{(\cup)}(A\cap B) = \underline{MNRS}^{(\cup)}(A) \cap \underline{MNRS}^{(\cup)}(B)$, $\overline{MNRS}^{(\cup)}(A\cup B) = \overline{MNRS}^{(\cup)}(A) \cup \overline{MNRS}^{(\cup)}(B)$;

(3) $A \subseteq B \Rightarrow \underline{MNRS}^{(\cap)}(A) \subseteq \underline{MNRS}^{(\cap)}(B)$, $A \subseteq B \Rightarrow \overline{MNRS}^{(\cap)}(A) \subseteq \overline{MNRS}^{(\cap)}(B)$;

(4) $A \subseteq B \Rightarrow \underline{MNRS}^{(\cup)}(A) \subseteq \underline{MNRS}^{(\cup)}(B)$, $A \subseteq B \Rightarrow \overline{MNRS}^{(\cup)}(A) \subseteq \overline{MNRS}^{(\cup)}(B)$;

(5) $\underline{MNRS}^{(\cap)}(A\cup B) \supseteq \underline{MNRS}^{(\cap)}(A) \cup \underline{MNRS}^{(\cap)}(B)$, $\overline{MNRS}^{(\cap)}(A\cap B) \subseteq \overline{MNRS}^{(\cap)}(A) \cap \overline{MNRS}^{(\cap)}(B)$;

(6) $\underline{MNRS}^{(\cup)}(A\cup B) \supseteq \underline{MNRS}^{(\cup)}(A) \cup \underline{MNRS}^{(\cup)}(B)$, $\overline{MNRS}^{(\cup)}(A\cap B) \subseteq \overline{MNRS}^{(\cup)}(A) \cap \overline{MNRS}^{(\cup)}(B)$.

Proof. The proof is similar with Proposition 2. □

Remark 1. *Note that if the two universes are the same, then the intersection-type (union-type)* MS *on two universes degenerates into the intersection-type (union-type)* MS *on a single universe in Section 3.*

Next, we will use the non-dual MSs to solve the MGDM problems where the attribute values are neutrosophic numbers. For a multiple attribute group decision making problem, let $U = \{x_1, x_2, \ldots, x_n\}$ be the decision set and $V = \{y_1, y_2, \ldots, y_m\}$ be the criteria set, R_l represent l evaluation experts. Here, $R_l \in NR(U \times V)$ is NRs from U to V, where $\forall(x_i, y_j) \in U \times V$, $R_l(x_i, y_j)$ denotes the degree of membership about criteria set y_j ($y_j \in V$) with respect to x_i ($x_i \in U$). In the following, we show the process about the non-dual MSs on two universes to solve MGDM problems with neutrosophic numbers as attribute values.

Step 1 Calculate non-dual multi-granulation neutrosophic rough ULA $\underline{MNRS}^{(\cap)}(A)$, $\overline{MNRS}^{(\cap)}(A)$, $\underline{MNRS}^{(\cup)}(A)$, and $\overline{MNRS}^{(\cup)}(A)$.

Step 2 Calculate the sum of non-dual multi-granulation neutrosophic rough ULA $MNRS^{(\cap)} = \lambda\underline{MNRS}^{(\cap)}(A) \oplus (1-\lambda)\overline{MNRS}^{(\cap)}(A)$, $MNRS^{(\cup)} = \lambda\underline{MNRS}^{(\cup)}(A) \oplus (1-\lambda)\overline{MNRS}^{(\cup)}(A)$, $\lambda \in [0, 1]$ according to Definition 6.

Step 3 Make a descending order according to Definitions 7 and 8 for the multi-granulation neutrosophic rough sets in step 2 and use the Borda number scoring method in reference [33] to make a total rank.

In practice, the parameter λ represents a decision maker's preference for risk. In general, the higher the parameter λ is, the more likely the decision maker is to be risk-prone. The smaller the parameter λ is, the less risk the decision maker prefers. Therefore, the value of the parameter λ is determined by the decision maker's preference or by an advance empirical study.

Next, we show the algorithm to calculate the ULA of a union-type multi-granulation neutrosophic rough set.

Algorithm 1 The lower approximation of a union-type multi-granulation neutrosophic rough set

Define the method to acquire a complement for a matrix A:
each neutrosophic number in matrix A do complement the operator according to the following Formula:
$a^c = (F_a, 1 - I_a, T_a)$.
Return matrix C.
Define the method for two matrixes to do union operator:
the union of B and C is the neutrosophic number of each row in C to do union operator with the corresponding neutrosophic number in B according to the Formula (22)

$$a \vee_3 b = (\max(T_a, T_b), \min(I_a, I_b), \min(F_a, F_b)). \tag{22}$$

Return matrix D.
Define the method for one matrix to do intersection operator:
the neutrosophic numbers of each row in D do intersection operator according to the Formula (23)

$$a \wedge_3 b = (\min(T_a, T_b), \max(I_a, I_b), \max(F_a, F_b)). \tag{23}$$

Return matrix E.
Define the method for one matrix to do union operator:
the neutrosophic numbers of each row in E do union operator according to the Formula (22).
Return matrix F.
For the number of iterations is h,
Transfer the method of acquire complement, assign X.
Get Y.
Transfer the method for two matrixes to do union operator, assign Y, Z.
Get M.
Transfer the method to do intersection operator, assign M.
Get N.
End for.
Combine h matrixes N.
Get P.
Transfer the method for one matrix to do union operator, assign P.
Get Q.

X, Y, M are matrixes which line numbers are m, column number is n, and every membership is a neutrosophic number. Z is a matrix which line number is 1, column number is n, and every membership is a neutrosophic number. N and Q are matrixes which line numbers are m, column number is 1, and every membership is a neutrosophic number. P is a matrix which line number is m, column number is h, and every membership is a neutrosophic number.

The lower approximation of a union-type multi-granulation neutrosophic rough set is the transpose of matrix Q.

Algorithm 2 The upper approximation of a union-type multi-granulation neutrosophic rough set

Define the method for two matrixes to do intersection operator:
the intersection of B and C is the neutrosophic number of each row in C to do intersection operator with the corresponding neutrosophic number in B according to the Formula (23).
Return matrix D.
Define the method for one matrix to do union operator:
the neutrosophic numbers of each row in D do union operator according to the Formula (22).
Return matrix E.
For the number of iterations is h,
Transfer the method for two matrixes to do intersection operator, assign Y, Z.
Get M.
Transfer the method for one matrix to do union operator, assign M.
Get N.
End for.
Combine h matrixes N.
Get P.
Transfer the method for one matrix to do intersection operator, assign P.
Get Q.

Y, M are matrixes which line numbers are m, column number is n, and every membership is a neutrosophic number. Z is a matrix which line number is 1, column number is n, and every membership is a neutrosophic number. N and Q are matrixes which line numbers are m, column number is 1, and every membership is a neutrosophic number. P is a matrix which line number is m, column number is h, and every membership is a neutrosophic number.

The upper approximation of a union-type multi-granulation neutrosophic rough set is the transpose of matrix Q.

With the same method we can get the ULA of an intersection-type multi-granulation neutrosophic rough set. Then, to decide the value of λ, we calculate the sum of ULA of the union-type MS and intersection-type MS according to Formula (6) and (7), and rank them according to Definition 7.

Next, we show an example.

Example 1. *We consider the decision making problem adapted from reference [34]. Suppose $U = \{x_1, x_2, x_3\}$ is a criterion set, where x_1 represents the ability of salesman, x_2 represents the overall condition of the stable supplier, and x_3 represents the position of high flow. Let $V = \{y_1, y_2, y_3, y_4, y_5\}$ be the decision set, where y_1 represents shop 1, y_2 represents shop 2, y_3 represents shop 3, y_4 represents shop 4, and y_5 represents shop 5.*

Assume there are three experts. They provide their evaluations shown in Tables 1–3 based on their knowledge and experience. The data of the three tables were adapted from Table 2 in reference [34]. We take the first positive membership and negative membership of the intuitionistic fuzzy set of interval values y_1–y_5 in Table 2 as the true membership and false membership of the neutrosophic set, respectively, and the second negative membership as the uncertain membership of the neutrosophic set. Let $A = \{(0.9, 0.1, 0.2), (0.7, 0.7, 0.3), (0.5, 0.8, 0.6)\}$.

Table 1. Neutrosophic relation R_1.

R_1	x_1	x_2	x_3
y_1	(0.75, 0.14, 0.09)	(0.86, 0.04, 0.01)	(0.66, 0.30, 0.29)
y_2	(0.44, 0.33, 0.29)	(0.51, 0.09, 0.04)	(0.54, 0.29, 0.27)
y_3	(0.54, 0.09, 0.08)	(0.66, 0.14, 0.06)	(0.54, 0.36, 0.34)
y_4	(0.56, 0.19, 0.14)	(0.50, 0.20, 0.12)	(0.44, 0.26, 0.23)
y_5	(0.33, 0.31, 0.30)	(0.43, 0.16, 0.02)	(0.21, 0.61, 0.60)

Table 2. Neutrosophic relation R_2.

R_2	x_1	x_2	x_3
y_1	(0.71, 0.10, 0.08)	(0.57, 0.01, 0.00)	(0.56, 0.09, 0.09)
y_2	(0.39, 0.54, 0.43)	(0.59, 0.11, 0.01)	(0.44, 0.19, 0.18)
y_3	(0.52, 0.17, 0.07)	(0.63, 0.04, 0.02)	(0.37, 0.54, 0.51)
y_4	(0.31, 0.09, 0.08)	(0.52, 0.31, 0.09)	(0.41, 0.29, 0.27)
y_5	(0.10, 0.61, 0.59)	(0.33, 0.33, 0.13)	(0.19, 0.09, 0.07)

Table 3. Neutrosophic relation R_3.

R_3	x_1	x_2	x_3
y_1	(0.89, 0.06, 0.05)	(0.86, 0.01, 0.01)	(0.77, 0.21, 0.20)
y_2	(0.61, 0.30, 0.27)	(0.76, 0.09, 0.01)	(0.56, 0.27, 0.25)
y_3	(0.64, 0.20, 0.10)	(0.63, 0.01, 0.01)	(0.59, 0.33, 0.29)
y_4	(0.68, 0.16, 0.04)	(0.59, 0.03, 0.02)	(0.57, 0.36, 0.31)
y_5	(0.39, 0.23, 0.10)	(0.34, 0.30, 0.19)	(0.29, 0.59, 0.49)

By Definitions 15 and 16, we can compute

$$\underline{MNRS}^{(\cup)}(A) = \begin{array}{l}\{(y_1, 0.50, 0.70, 0.56), (y_2, 0.50, 0.71, 0.44), (y_3, 0.51, 0.70, 0.37), (y_4, 0.50, 0.70, 0.41), \\ (y_5, 0.60, 0.70, 0.30)\}\end{array},$$

$$\overline{MNRS}^{(\cup)}(A) = \begin{array}{l}\{(y_1, 0.89, 0.10, 0.20), (y_2, 0.70, 0.30, 0.27), (y_3, 0.66, 0.10, 0.20), (y_4, 0.68, 0.10, 0.20), \\ (y_5, 0.43, 0.23, 0.20)\}\end{array},$$

$$\underline{MNRS}^{(\cap)}(A) = \begin{array}{l}\{(y_1, 0.50, 0.80, 0.60), (y_2, 0.50, 0.80, 0.56), (y_3, 0.50, 0.70, 0.59), (y_4, 0.50, 0.74, 0.57), \\ (y_5, 0.50, 0.80, 0.30)\}\end{array},$$

$$\overline{MNRS}^{(\cap)}(A) = \begin{array}{l}\{(y_1, 0.71, 0.14, 0.20), (y_2, 0.51, 0.54, 0.30), (y_3, 0.63, 0.20, 0.20), (y_4, 0.52, 0.19, 0.20), \\ (y_5, 0.33, 0.61, 0.30)\}\end{array},$$

Let $\lambda = 0.3$, then

$$MNRS^{(\cup)}(A) = \begin{array}{l}\{(y_1, 0.8267, 01793, 0.2723), (y_2, 0.6525, 0.3885, 0.2896), (y_3, 0.6183, 0.1793, 0.2258), \\ (y_4, 0.6341, 0.1793, 0.2480), (y_5, 0.4874, 0.3212, 0.2236)\}\end{array},$$

$$MNRS^{(\cap)}(A) = \begin{array}{l}\{(y_1, 0.6585, 0.2362, 0.2780), (y_2, 0.5070, 0.6076, 0.3617), (y_3, 0.5950, 0.2912, 0.2767), \\ (y_4, 0.5141, 0.2857, 0.2738), (y_5, 0.3863, 0.6617, 0.3000)\}\end{array},$$

Then, according to Definition 6, we can get

$$s^{(\cup)}(y_1) = 0.5259, s^{(\cup)}(y_2) = 0.6814, s^{(\cup)}(y_3) = 0.6963, s^{(\cup)}(y_4) = 0.6930, s^{(\cup)}(y_5) = 0.6319.$$

So, the ranking result for union-type MS is: $y_1 \prec y_5 \prec y_2 \prec y_4 \prec y_3$.

$$s^{(\cap)}(y_1) = 0.6903, s^{(\cap)}(y_2) = 0.5726, s^{(\cap)}(y_3) = 0.6592, s^{(\cap)}(y_4) = 0.6201, s^{(\cap)}(y_5) = 0.5432.$$

So, the ranking result for intersection-type MS is: $y_5 \prec y_2 \prec y_4 \prec y_3 \prec y_1$.

Using the Borda counting method, score 4, 3, 2, 1, and 0 for the first, second, third, fourth, and fifth place, respectively, then we can get

$$B(x_1) = 4, B(x_2) = 3, B(x_3) = 7, B(x_4) = 5, B(x_5) = 1.$$

So, when $\lambda = 0.3$, the best choice, shop 3, is chosen.

6. Conclusions

The multi-granulation neutrosophic rough set is a useful tool for MGDM problems. In this paper, we proposed non-dual MSs and study their operators and properties. Then we discussed the relationship between NRS, optimistic (pessimistic) MS, non-dual MS, NRS based on intersection (union) NRs, and NRS based on transitive closure relationship of union NRs, we used Figure 1 to show the relationship. Furthermore, we proposed a non-dual MS on two universes and talk about the relationship between non-dual MS on two universes and non-dual MS on a single universe, and we used non-dual MS on two universes to solve a MGDM problem where the attribute values were neutrosophic numbers.

For future orientation, we will research other types of fusions of MRSs and NSs. Additionally, we will study the applications of the concepts in this paper to totally-dependent neutrosophic sets and some algebraic systems and discuss in relation to other algorithms [35–44].

Author Contributions: X.Z. and C.B. initiated the research and wrote the paper, S.S. participated in some of the research work.

Funding: This work was supported by the National Natural Science Foundation of China (Grant No. 61573240).

Conflicts of Interest: The authors declare no conflicts of interest.

References

1. Atanassov, K.T. Type-1 Fuzzy Sets and Intuitionistic Fuzzy Sets. *Algorithms* **2017**, *10*, 106. [CrossRef]
2. Bisht, K.; Joshi, D.K.; Kumar, S. Dual Hesitant Fuzzy Set-Based Intuitionistic Fuzzy Time Series Forecasting. In *Ambient Communication and Computer Systems*; Springer: Singapore, 2018; pp. 317–329.
3. Kumar, K.; Garg, H. TOPSIS method based on the connection number of set pair analysis under interval-valued intuitionistic fuzzy set environment. *Comput. Appl. Math.* **2018**, *37*, 1319–1329. [CrossRef]
4. Maji, P. Advances in Rough Set Based Hybrid Approaches for Medical Image Analysis. In *International Joint Conference on Rough Sets*; Springer: Cham, Germany, 2017; pp. 25–33.
5. Smarandache, F. Neutrosophic set—A generialization of theintuitionistics fuzzy sets. *Int. J. Pure Appl. Math.* **2005**, *24*, 287–297.
6. Wang, H.B.; Smarandache, F.; Zhang, Y.; Sunderraman, R. Single valued neutrosophic sets. *Multispace Multistruct.* **2010**, *4*, 410–413.
7. Peng, J.; Wang, J.; Wang, J.; Zhang, H.; Chen, H. Simplified neutrosophic sets and their applications in multi-criteria group decision-making problems. *Int. J. Syst. Sci.* **2016**, *47*, 2342–2358. [CrossRef]
8. Peng, H.; Zhang, H.; Wang, J. Probability multi-valued neutrosophic sets and its application in multi-criteria group decision-making problems. *Neural Compt. Appl.* **2018**, *30*, 563–583. [CrossRef]
9. Deli, I.; Ali, M.; Smarandache, F. Bipolarneutrosophic sets and their application based on multi-criteria decision making problems. In Proceedings of the 2015 International Conference on Advanced Mechatronic Systems, Beijing, China, 22–24 August 2015; pp. 249–254.
10. Zhang, X.; Bo, C.X.; Smarandache, F.; Dai, J.H. New inclusion relation of neutrosophic sets withapplications and related lattice structure. *Int. J. Mach. Learn. Cybern.* **2018**, *9*, 1753–1763. [CrossRef]
11. Dubois, D.; Prade, H. Rough fuzzy sets and fuzzy rough sets. *Int. J. Gen. Syst.* **1990**, *17*, 191–209. [CrossRef]
12. Cornelis, C.; De Cock, M.; Kerre, E.E. Intuitionistic fuzzy rough sets: At the crossroads of imperfect knowledge. *Expert Syst.* **2003**, *20*, 260–270. [CrossRef]
13. Zhan, J.; Ali, M.I.; Mehmood, N. On a novel uncertain soft set model: Z-soft fuzzy rough set model and corresponding decision making methods. *Appl. Soft Comput.* **2017**, *56*, 446–457. [CrossRef]
14. Yang, H.L.; Zhang, C.L.; Guo, Z.L.; Liu, Y.L.; Liao, X. A hybrid model of single valued neutrosophic sets and rough sets: Single valued neutrosophic rough set model. *Soft Comput.* **2017**, *21*, 6253–6267. [CrossRef]
15. Garg, H. An improved score function for ranking neutrosophic sets and its application to decision-making process. *Int. J. Uncertain. Quantif.* **2016**, *6*, 377–385.

16. Li, Y.Y.; Zhang, H.; Wang, J.Q. Linguistic neutrosophic sets and their application in multicriteria decision-making problems. *Int. J. Uncertain. Quantif.* **2017**, *7*, 135–154. [CrossRef]
17. Abdel-Basset, M.; Mohamed, M. The role of single valued neutrosophic sets and rough sets in smart city: Imperfect and incomplete information systems. *Measurement* **2018**, *124*, 47–55. [CrossRef]
18. Yang, H.L.; Bao, Y.L.; Guo, Z.L. Generalized Interval Neutrosophic Rough Sets and its Application in Multi-Attribute Decision Making. *Filomate* **2018**, *32*, 11–33. [CrossRef]
19. Zhang, C.; Zhai, Y.; Li, D.; Mu, Y. Steam turbine fault diagnosis based on single-valued neutrosophicmultigranulation rough sets over two universes. *J. Intell. Fuzzy Syst.* **2016**, *31*, 2829–2837. [CrossRef]
20. Qian, Y.H.; Liang, J.Y.; Yao, Y.Y.; Dang, C.Y. MGRS: A multi-granulation rough set. *Inf. Sci.* **2010**, *180*, 949–970. [CrossRef]
21. Yao, Y.; She, Y. Rough set models in multigranulation spaces. *Inf. Sci.* **2016**, *327*, 40–56. [CrossRef]
22. Kumar, S.S.; Inbarani, H.H. Optimistic multi-granulation rough set based classification for medical diagnosis. *Procedia. Comput. Sci.* **2015**, *47*, 374–382. [CrossRef]
23. Majumdar, P.; Samanta, S.K. On similarity and entropy of neutrosophic sets. *J. Int. Fuzzy Syst.* **2014**, *26*, 1245–1252.
24. Kang, Y.; Wu, S.; Li, Y.; Liu, J.; Chen, B. A variable precision grey-based multi-granulation rough set model and attribute reduction. *Knowl. Based Syst.* **2018**, *148*, 131–145. [CrossRef]
25. Sun, B.Z.; Ma, W.M.; Qian, Y.H. Multigranulation fuzzy rough set over two universes and its application to decision making. *Knowl. Based Syst.* **2017**, *123*, 61–74. [CrossRef]
26. Pan, W.; She, K.; Wei, P. Multi-granulation fuzzy preference relation rough set for ordinal decision system. *Fuzzy Sets Syst.* **2017**, *312*, 87–108. [CrossRef]
27. Huang, B.; Guo, C.; Zhuang, Y.; Li, H.; Zhou, X. Intuitionistic fuzzy multi-granulation rough sets. *Inf. Sci.* **2014**, *277*, 299–320. [CrossRef]
28. Bo, C.X.; Zhang, X.; Shao, S.T.; Smarandache, F. New multi-granulation neutrosophic rough set with applications. *Symmetry* **2018**, *10*, 578. [CrossRef]
29. Xu, W.; Li, W.; Zhang, X. Generalized multigranulation rough sets and optimal granularity selection. *Granul. Comput.* **2017**, *2*, 271–288. [CrossRef]
30. Lin, H.; Wang, Q.; Lu, X.; Li, H. Hybrid multi-granulation rough sets of variable precision based on tolerance. *J. Int. Fuzzy Syst.* **2016**, *31*, 717–725. [CrossRef]
31. Zhang, X.; Miao, D.; Liu, C.; Le, M. Constructive methods of rough approximation operators and multi-granulation rough sets. *Knowl. Based Syst.* **2016**, *91*, 114–125. [CrossRef]
32. Yang, H.L.; Guo, Z.L.; Liao, X. On single valued neutrosophic relations. *J. Intell. Fuzzy Syst.* **2016**, *30*, 1045–1056. [CrossRef]
33. Zarghami, M. Soft computing of the Borda count by fuzzy linguistic quantifiers. *Appl. Soft Comput.* **2011**, *11*, 1067–1073. [CrossRef]
34. Chai, J.; Liu, J.N.K.; Xu, Z. A rule-based group decision model for warehouse evaluation under interval-valued Intuitionistic fuzzy environments. *Expert Syst. Appl.* **2013**, *40*, 1959–1970. [CrossRef]
35. Zhang, X.H.; Bo, C.X.; Smarandache, F.; Park, C. New operations of totally dependent-neutrosophic sets and totally dependent-neutrosophic soft sets. *Symmetry* **2018**, *10*, 187. [CrossRef]
36. Zhang, X.H.; Mao, X.Y.; Wu, Y.T.; Zhai, X.H. Neutrosophic filters in pseudo-BCI algebras. *Int. J. Uncertain. Quantif.* **2018**, *8*, 511–526. [CrossRef]
37. Ma, Y.C.; Zhang, X.H.; Yang, X.F.; Zhou, X. Generalized neutrosophic extended triplet group. *Symmetry* **2019**, *11*, 327. [CrossRef]
38. Wu, X.Y.; Zhang, X.H. The decomposition theorems of AG-neutrosophic extended triplet loops and strong AG-(l, l)-loops. *Mathematics* **2019**, *7*, 268. [CrossRef]
39. Zhang, X.H.; Borzooei, R.A.; Jun, Y.B. Q-filters of quantum B-algebras and basic implication algebras. *Symmetry* **2018**, *10*, 573. [CrossRef]
40. Abu Qamar, M.; Hassan, N. An Approach toward a Q-Neutrosophic Soft Set and Its Application in Decision Making. *Symmetry* **2019**, *11*, 139. [CrossRef]
41. Badamchizadeh, M.A.; Nikdel, A.; Kouzehgar, M. Comparison of genetic algorithm and particle swarm optimization for data fusion method based on Kalman filter. *Int. J. Artifi. Intell.* **2010**, *5*, 67–78.

42. Pozna, C.; Precup, R.E.; Tar, J.K.; Skrjanc, L.; Preitle, S. New results in modelling derived from Bayesian filtering. *Knowl. Based Syst.* **2010**, *23*, 182–194. [CrossRef]
43. Jankowski, J.; Kazienko, P.; Wątróbski, J.; Lewandowska, A.; Zimba, P.; Zioło, M. Fuzzy multi-objective modeling of effectiveness and user experience in online advertising. *Expert Syst. Appl.* **2016**, *65*, 315–331. [CrossRef]
44. Zhang, X.H.; Wang, X.J.; Smarandache, F.; Jaiyeola, T.G.; Lian, T.Y. Singular neutrosophic extended triplet groups and generalized groups. *Cognitive Systems Research* **2019**, *57*, 32–40. [CrossRef]

symmetry

Article

A Hybrid Plithogenic Decision-Making Approach with Quality Function Deployment for Selecting Supply Chain Sustainability Metrics

Mohamed Abdel-Basset [1,*], Rehab Mohamed [1], Abd El-Nasser H. Zaied [1] and Florentin Smarandache [2]

[1] Department of Operations Research, Faculty of Computers and Informatics, Zagazig University, Sharqiyah 44159, Egypt
[2] Math & Science Department, University of New Mexico, Gallup, NM 87301, USA
* Correspondence: analyst_mohamed@zu.edu.eg

Received: 20 June 2019; Accepted: 4 July 2019; Published: 11 July 2019

Abstract: Supply chain sustainability has become one of the most attractive decision management topics. There are many articles that have focused on this field presenting many different points of view. This research is centred on the evaluation of supply chain sustainability based on two critical dimensions. The first is the importance of evaluation metrics based on economic, environmental and social aspects, and the second is the degree of difficulty of information gathering. This paper aims to increase the accuracy of the evaluation. The proposed method is a combination of quality function deployment (QFD) with plithogenic aggregation operations. The aggregation operation is applied to aggregate: Firstly, the decision maker's opinions of requirements that are needed to evaluate the supply chain sustainability; secondly, the evaluation metrics based on the requirements; and lastly, the evaluation of information gathering difficulty. To validate the proposed model, this study presented a real world case study of Thailand's sugar industry. The results showed the most preferred and the lowest preferred metrics in order to evaluate the sustainability of the supply chain strategy.

Keywords: supply chain sustainability metrics; plithogeny; aggregation operations; neutrosophic set; quality function deployment

1. Introduction

Supply chain sustainability has been one of the most attractive and dynamic research topics in the domain of supply chain management for a long time. The influence of manufacturing activities to global warming and the consumption of natural resources assisted the researchers in considering the importance of the supply chain operation's sustainability [1]. As a result of increasing competition, globalization, technological growth and huge customer expectations, the sustainable supply chain is a significant goal to each supply chain in every field. The supply chain sustainability can be described as the capability of operating the business with the long term goal of preserving economic, environment and societal welfare [2]. A more general definition of sustainable supply chain could be the management of supply chain activities in order to improve the profitability by taking into consideration the environmental impacts and social aspects. Therefore, supply chain sustainability guarantees success and achievements of the whole supply chain management in the long term. Under the uncertainty component, supply chain sustainability became a more important goal for companies. This explains why measuring supply chain sustainability means to identify possible strategic decisions under various situations [3].

The evaluation of supply chain sustainability is an interesting topic based on metrics in economic, environmental and social scopes. Measuring sustainability of the supply chain guides firms in the direction of risk elimination and standards/guidelines following [4]. Moreover, the advantages of evaluating supply chain (SC) sustainability are reducing costs, increasing competence, supporting competitive advantages and improving operational performance [5]. The challenges of measuring supply chain sustainability are [6]: The managerial and organizational absence of the inter-organizational metrics; the variety of the organization's goals and objectives producing different measures; and the difficulty in non-traditional data gathering that reduce the SC performance.

There are several studies in supply chain sustainability assessment including supply chain sustainability risk and assessment [7], literature reviews [8], multi-objective mathematical models for sustainable supply chain management [9] and decision making models for a sustainable supply chain [10]. Evaluating supply chain sustainability is a multi-criteria decision making (MCDM) problem, therefore the evaluation metrics may be the criteria, and the alternatives may be selected based on these sets of metrics. There are some limitations of SC sustainability studies, such as the fact that the researchers do not consider the difficulty of collecting the information for the metrics that will measure the sustainability. In addition, only a few studies use the linguistic variables to evaluate the metrics, leading to less consideration on the uncertainty or lack of information [11]. Also, there is the matter of the decision maker's priorities and contradiction degree between metrics which leads to less accuracy of results. In the comical industry, Rajeev (2019) proposed a framework to describe the evolution of a sustainable supply chain [12].

In this research, most of these limitations were processed by the proposed MCDM model that assists in metrics selection and the weighting of sustainable supply chain. The proposed model is based on a combination of plithogenic aggregation operations with quality function deployment (QFD). The details of the model have been explained in Section 3.

QFD is one of the most popular techniques to improve quality in order to meet customer expectations. This tool combines all customer needs in every aspect of the product, transforming them into technical requirements so they can meet their expectations [13]. QFD records great results in many fields, such as rating engineering characteristics [14], the design of building structures [15], service level measurements [16], industry development [17], product development [18], or supplier selection problems [19].

Plithogeny refers to the creation, development and progression of new entities from composition of contradictory or non-contradictory multiple old entities [20]. It was introduced by Smarandache in 2017 as a generalization of neutrosophy. A plithogenic set (as a generalization of crisp, fuzzy, intuitionistic fuzzy, and neutrosophic sets) is a set whose elements are characterised by the attribute values. Each attribute value has its contradiction degree values $c(v_j, v_D)$ between v_j and the dominant (most important) attribute value v_D. The contradiction degree between attributes assists the model to gain more accurate results. The plithogenic set, logic, probability and statistics that were also introduced by Smarandache in 2017, are obtained from plithogeny, and they are generalizations of neutrosophic sets, logic, probability and statistics respectively.

The rest of this papers is organized as follows: In Section 2, there is a literature review of sustainable supply chain, quality function deployment, clarification of plithogenic sets, and a recapitulation of neutrosophic sets. Section 3 presents the proposed model to evaluate the sustainable supply chain. In Section 4, a real world case is studied in order to evaluate the proposed model. Section 5 discusses the results of this case. Finally, the conclusion and suggestions for future works end Section 6.

2. Literature Review

2.1. Supply Chain Sustainability

In the supply chain management field, there are many considerations that need to be taken into account to minimize the negative influence of business production to environment and social effects. These considerations pushed for the strategic developments plans for sustainability [11]. Three dimensions of sustainability are considered in the supply chain, derived from customer and stakeholder desires, which are economic, environmental and social aspects to manage raw materials, information and finance flows [21]. Other definitions of supply chain sustainability is the integration of an organization's economic, social and environmental dimensions by coordinating the business process in order to improve the organization's performance in the long term [22]. A more focused definition could be supply chain management strategies and activities concerning social and environmental aspects, correlated to the production, distribution, design and supply of products and services [23]. The evaluation of supply chain sustainability metrics are attributes and requirements used to measure the supply chain performance considering economic, social and environmental features [24]. Table 1 summarizes some of the studies on supply chain sustainability metrics and frameworks.

Table 1. Studies about supply chain sustainability metrics.

Authors	Scope	Methodology	Metrics
Akshay Jadhav, Stuart Orr, Mohsin Malik (2018) [25]	Supply chain orientation (SCO)	Literature review analysis (SEM analysis)	Co2 emission management, community engagement, supplier codes of conduct, waste elimination, energy usage efficiency, water usage efficiency, and recycled materials practices, among others.
Elkafi Hassini, ChiragSurti, CorySearcy (2012) [4]	Developing supply chain sustainability metrics	Literature review	Percent of suppliers, Percent of contracts, Percent of purchase orders, Level of stake-holder trust by category
Yazdani, Morteza, Cengiz Kahraman, Pascale Zarate, and Sezi Cevik Onar (2019) [26]	Ranking of supply chain sustainability indicators	Multi-attribute decision making (QFD and GRA)	Quality, managing environmental systems, supply chain elasticity, business social liability, transportation service situation, and financial constancy.
Qorri, Ardian, Zlatan Mujkić, and Andrzej Kraslawski (2018) [27]	Measuring supply chain sustainability performance	Literature review	Number of contributors, products, geographical encompassing, strategic goals, methods, tools, among others
Searcy, Cory, Shane M. Dixon, and W. Patrick Neumann (2016) [28]	Analysis of performance indicators in supply chain sustainability	Literature review and report analysis	Employees number, profits, supplier estimation, trainingcost, among others
Chen, Rong-Hui, Yuanhsu Lin, and Ming-Lang Tseng [29]	Sustainable development indicators in the structure minerals industry in China	Combines fuzzy set theory, the Delphi method, discrete multi-criteria method	Solid waste, Eco-efficiency, Health and safety, Energy use, Investments, Land use and rehabilitation, among others
Haghighi, S. Motevali, S. A. Torabi, and R. Ghasemi [30]	Evaluation of Sustainable Supply Chain Networks	Data envelopment analysis technique	Time delivery, Supplier rejection rate, Amount of Pollution, Customers' satisfaction, Service quality, among others

2.2. *Quality Function Deployment (QFD)*

Quality function deployment (QFD) originated in Japan in the 1960s. QFD establishes quality measurement for the improvement and design, rather than just quality control in manufacturing processes [31]. The QFD method is the link that connects the customer voice to the design requirement in order to respond these expectations effectively. As illustrated in Figure 1, the components of QFD are as follows [32]:

- *Area (1)*: The customers' requirements (what region) that consists of two indicators: The customers' requirements and the importance of each of them α_i.
- *Area (2)*: The quality characteristics or design specifications (how region), composed of two parts: The design specifications and the way of development.
- *Area (3)*: The relationship between customer requirements and design specifications (what versus how region) by score $C_{ij} = \{0, 1, 3, \ldots, 9\}$.

- *Area (4)*: This area is a combination of the value of the design specification, the acceptance level of it, and the score
$$S_j = \sum(\alpha_i * C_{ij}) \quad (1)$$
- *Area (5)*: The comparison of the product and competitors, and how much it satisfies customer needs.
- *Area (6)*: The comparison between each design specification, and how much their improvement may affect each other.

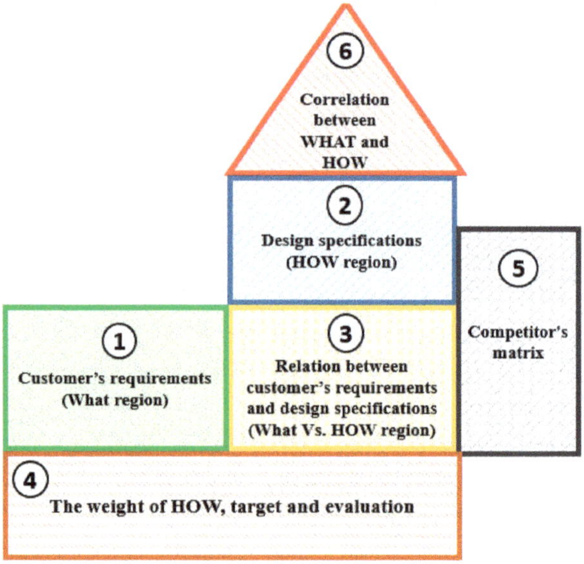

Figure 1. QFD structure.

There are several studies that combine QFD with other techniques to evaluate supply chain sustainability, such as: A hybrid QFD–ANP approach to design a sustainable maritime supply chain [33]; integration of QFD and grey relational analysis (GRA) in order to solve compound decision making complications [26]; QFD and MCDM techniques in supplier selection problems [34]. Dursun et al. (2018) considered the competition factor in the process of new product development using QFD [35]. A combination of best-worst method (BWM) and QFD was proposed in order to determine the relation between customer requirements and engineering characteristics in Mei et al. (2018) [36].

2.3. Plithogenic Set Characteristics

Plithogeny is the formation, construction, development, germination, and evolution of new entities from combinations of contradictory (dissimilar) or non-contradictory multiple old entities [37]. A plithogenic set *(P, A, V, d, c)* is a set that includes numerous elements described by a number of attributes $A = \{\alpha_1, \alpha_2, \ldots, \alpha_m\}$, $m \geq 1$, which has a values $V = \{v_1, v_2, \ldots, v_n\}$, for $n \geq 1$. There are two main features of each attribute's value, V. The first is the appurtenance degree function $d(x,v)$ of the element x, with respect to some given criteria [38]. The contradiction (dissimilarity) degree function $c(v,D)$ is the second one, which is realized between each attribute value and the most important (dominant) one. The contradiction degree function is mainly the key element of the plithogenic aggregation operations (intersection, union, complement, inclusion, and equality) that increase the accuracy of aggregation.

Let A be a non-empty set of uni-dimensional attributes $A = \{\alpha_1, \alpha_2, \ldots, \alpha_m\}$, $m \geq 1$, and let $\alpha \in A$ be an attribute with its value spectrum the set S, where S can be defined as a finite discrete set,

$S = \{s_1, s_2, \ldots, s_l\}$, $1 \leq l < \infty$, or infinitely countable set $S = \{s_1, s_2, \ldots, s_\infty\}$, or infinitely uncountable (continuum) set $S =]a,b[$, $a < b$, where $] \ldots [$ is any open, semi-open, or closed interval from the set of real numbers or from other general sets [39].

Let V be a non-empty subset of S, where V is the range of all attributes of α's values defined by the experts based on the application, $V = \{v_1, v_2, \ldots, v_n\}$ for $n \geq 1$. In the set V, there is a dominant attribute value which is determined by the experts based on preferences and the nature of the application.

Each attribute value in V has its appurtenance degree $d(x,v)$ with respect to some criteria. The degree of appurtenance may be a fuzzy, or intuitionistic fuzzy, or neutrosophic degree of appurtenance to the plithogenic set. Therefore, the appurtenance degree $d(x,v)$ of attribute value v is:

$$\forall x \in P, d: P \times V \rightarrow \mathrm{P}([0,1]^z), \tag{2}$$

Therefore, $d(x, v)$ is a subset of $[0, 1]^z$, and $\mathrm{P}([0, 1]^z)$ is the power set of $[0, 1]^z$, where z = 1, 2, 3, for fuzzy, intuitionistic fuzzy, and neutrosophic degrees of appurtenance respectively [19].

Let c: $V \times V \rightarrow [0, 1]$ be the attribute value contradiction degree function $c(v_1, v_2)$, representing the dissimilarity between two attribute values v_1 and v_2, and satisfying the following axioms:

$c(v_1, v_1) = 0$, contradiction degree between the attribute values and itself is zero.

$c(v_1, v_2) = c(v_2, v_1)$, contradiction degree function can be fuzzy C_F, intuitionistic attribute value contradiction function (C_{IF}: $V \times V \rightarrow [0, 1]^2$), or a neutrosophic attribute value contradiction function (C_N: $V \times V \rightarrow [0, 1]^3$).

2.4. Neutrosophic Set

Neutrosophy is a new branch of philosophy (generalization of dialectics and Yin Yang Chinese philosophy), introduced by Florentin Smarandache in 1980, which studies the origin, nature, and scope of neutralities, as well as their interactions with different ideational spectra. Neutrosophy is the foundation of neutrosophic logic, neutrosophic probability, neutrosophic sets, and neutrosophic statistics. Neutrosophic set definitions are clearly stated in the following:

Definition 1. *[40] Let X be a universal set of objects, consisting of non-specific elements defined as x. A neutrosophic set N ⊂ X reflects a set such that each element x from N is characterized by* $T_N(x)$*–the truth-membership function,* $I_N(x)$*–the indeterminacy-membership function, and* $F_N(x)$*–the falsity-membership function.* $T_N(x)$*,* $I_N(x)$ *and* $F_N(x)$ *are subsets of* $[0^-, 1^+]$*, so the three neutrosophic components are* $T_N(x) \in [0^-, 1^+]$*,* $I_N(x) \in [0^-, 1^+]$ *and* $F_N(x) \in [0^-, 1^+]$*.* $I_N(x)$ *is depicts uncertainty, indeterminate, unidentified, or error values. The sum of the three components is* $0^- \leq T_N(x) + I_N(x) + F_N(x) \leq 3^+$.

Definition 2. *[41] Let X be a space of points and x ∈ X. A neutrosophic set N in X is recognized by a truth-membership function* $T_N(x)$*, an indeterminacy-membership function* $I_N(x)$ *and a falsity-membership function* $F_N(x)$*, where* $T_N(x)$*,* $I_N(x)$ *and* $F_N(x)$ *are subsets of]-0, 1+[.* $T_N(x)$*:X →]-0, 1+[,* $I_N(x)$*:X →]-0, 1+[and* $F_N(x)$*:X →]-0,1+[. There is no restriction on the summation of membership functions. Therefore, 0− ≤ sup* $T_N(x)$ *+ sup* $I_N(x)$ *+ sup* $F_N(x)$ *≤ 3+.*

Definition 3. *[42] Let* $a = \langle (a1, a2, a3); \alpha, \theta, \beta \rangle$ *be a single valued triangular neutrosophic set, with truth membership* $T_a(x)$*, indeterminate membership* $I_a(x)$*, and falsity membership function* $F_a(x)$ *as follows:*

$$T_a(x) = \begin{cases} \alpha_a\left(\frac{x-a_1}{a_2-a_1}\right) if\ a_1 \leq x \leq a_2 \\ \alpha_a \quad if\ x = a_2 \\ o \quad otherwise \end{cases} \tag{3}$$

$$I_a(x) = \begin{cases} \frac{(a_2 - x + \theta_a(x - a_1))}{(a_2 - a_1)} & if\ a_1 \le x \le a_2 \\ \theta_a \quad if\ x = a_2 \\ \frac{(x - a_2 + \theta_a(a_3 - x))}{(a_3 - a_2)} & otherwise \end{cases} \tag{4}$$

$$F_a(x) = \begin{cases} \frac{(a_2 - x + \beta_a(x - a_1))}{(a_2 - a_1)} & if\ a_1 \le x \le a_2 \\ \beta_a \quad if x = a_2 \\ \frac{(x - a_2 + \beta_a(a_3 - x))}{(a_3 - a_2)} & if\ a_2 < x \le a_3 \\ 1 \quad otherwise \end{cases} \tag{5}$$

where α_a, θ_a, $\beta_a \in [0,1]$. *They represent the highest truth membership degree, the lowest indeterminacy membership degree, and the lowest falsity membership degree, respectively.*

3. Proposed Model

In this paper, the authors proposed a model to evaluate the supply chain sustainability metrics based on a combination of quality function deployment and plithogenic aggregation operations. This model combines the benefits of the QFD method to link customer needs with design requirements and plithogenic aggregation operator features. The usefulness of this model derives from the plithogenic aggregation operation, because this technique ensures more accurate results and takes into consideration the degree of uncertainty, which is defective in other studies of the same problem. The steps of the proposed model have been explained in detail in this section and it is shown in Figure 2.

- **Step 1:** First of all, decision makers (DM) identify a series of requirements to appraise the supply chain sustainability. The most popular requirements of supply chain sustainability evaluation are summarized in Table 2 or the DM can identify other requirements based on their strategy. These requirements must reflect economic, social and environmental features which is called triple bottom line (TPL).
  - The decision makers measure the importance of each requirement based on the supply chain strategy using linguistic terms.
  - The linguistic scale is defined to describe the assessment of each requirement by the DM. In this model, the scale is suggested as a triangular neutrosophic scale, as shown in Table 3.
- **Step 2:** Using plithogenic aggregation operations, the decision maker's opinions are aggregated based on the contradiction degree of each requirement. This step increases the accuracy of results.
  - Define contradiction degree c of each requirement with respect to the dominant.
  - Plithogenic neutrosophic set intersection is defined as following:

$$\begin{aligned} &((a_{i1}, a_{i2}, a_{i3}), 1 \le i \le n) \wedge_P ((b_{i1}, b_{i2}, b_{i3}), 1 \le i \le n) \\ &= \left(\left(a_{i1} \wedge_F b_{i1}, \tfrac{1}{2}(a_{i2} \wedge_F b_{i2}) + \tfrac{1}{2}(a_{i2} \vee_F b_{i2}), a_{i2} \vee_F b_{i3}\right)\right), 1 \le i \le n. \end{aligned} \tag{6}$$

  where $\wedge_F$ and $\vee_F$ are fuzzy t-norm and t-conorm respectively.
  - The neutrosophic number is transformed into a crisp number using the following equation:

$$\mathrm{S}\ (\mathrm{a}) = \frac{1}{8}(a_1 + b_1 + c_1) \times (2 + \alpha - \theta - \beta) \tag{7}$$

- **Step 3:** In order to find the best requirement considering the set of criteria, the distance of each requirement is found from the best and worst solutions.

- The best (positive) ideal solution S^+ and worst (negative) ideal solution S^- require definition.
- For example, in price requirements, the lowest value is desired (best ideal solution); on the other side, the maximum value is the worst ideal solution. In the opposite of that, in profit requirements, the maximum value is positive and the lowest value is negative.
- The formula of Euclidean distance is used to find the distance of each requirement to the ideal positive and ideal negative solutions, as shown in Equations (8) and (9) [50].

$$D_i^+ = \left[\sum_{j=1}^{m} \left(V_i - V_j^+ \right)^2 \right]^{0.5} \tag{8}$$

$$D_i^- = \left[\sum_{j=1}^{m} \left(V_i - V_j^- \right)^2 \right]^{0.5} \tag{9}$$

- The superior alternative has the smallest distance from the positive ideal solution S^+ and the worst alternative has a larger distance from the negative ideal solution S^-.

❖ **Step 4:** The performance score of each requirement is found in order to weight each of them based on Equation (10).

$$P_i = \frac{S_i^-}{S_i^+ - S_i^-} \tag{10}$$

- The performance score is normalized to find the weight of each requirement that satisfies two constraints which are $0 \le w_i \le 1$ and $\sum w_i = 1$.

❖ **Step 5:** The decision makers define a combination of metrics by considering the requirements selected previously in step 1 and the TPL. Some of the economic metrics are cost reduction, transaction costs, environmental costs, service level, or sales. The environmental metrics are environmental policies, recycling of waste, air pollution emission, solid waste, water consumption, and so on. Finally, the social diminution consists of working conditions, employee satisfaction, government relationships, employee training, and reputation, among others.

- The DMs define the relation between each metric and explain each requirement using linguistic terms as in Table 3.

❖ **Step 6:** Steps 2–4 are repeated on the evaluation metrics. As in Step 2, the plithogenic aggregation operation is used to combine all decision makers' judgments about defined metrics. Then, Equations (8) and (9) are used to establish the distance of every metric from the best and worst solutions. The importance of each metric is determined using the performance score as in Equation (10).

❖ **Step 7:** As proposed in Osiro, Lauro et al. 2018 [11], the limitations of other studies that do not consider the hardness of data gathering of each metric need to be addressed. In this step, the difficulty in regards to three dimensions are evaluated in relation to information accessibility, human resources and time needed for assessment, and other required resources [51].

- The difficulty of assessment metrics data collecting based on three dimensions explained by linguistic variables are evaluated.
- The assessment based on the contradiction degree to obtain accuracy of results are aggregated, and thenits crisp value is found.
- Their performance score of data collecting difficulty based on the distance of best and worst solutions are found.

❖ **Step 8:** In this final step, the goal is to categorize the set of supply chain sustainability metrics.

- The performance degree found in Step 4 (the importance of each metric) and Step 7 (the difficulty of data gathering) using Equation (11) are normalized as proposed in (Osiro, Lauro et al. 2018) [11].

$$v_n = \frac{1}{1 + e^{-\frac{v - \overline{v}}{\sigma_v}}} \tag{11}$$

where v_n is the normalized value, $v - \overline{v}$ is the difference between the value and the mean, and σ_v is the standard deviation.

- This is the result if supply chain sustainability evaluation metrics are categorized according to Figure 3 based on the importance of each metric and its difficulty of data gathering.

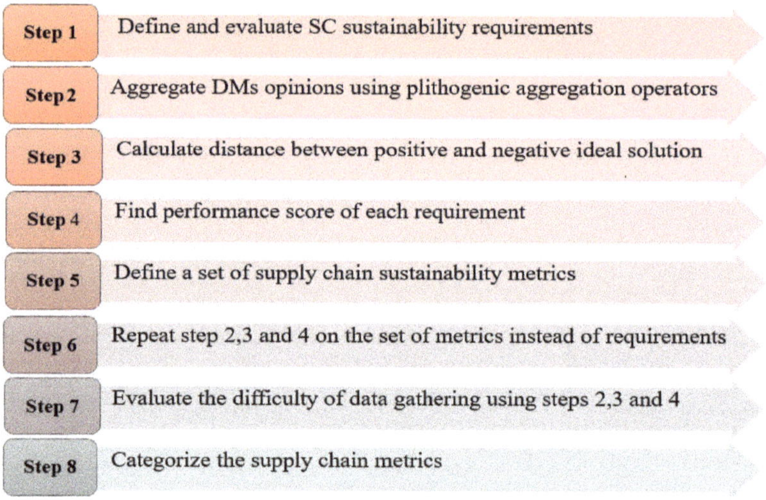
Step 1
Define and evaluate SC sustainability requirements
Step 2
Aggregate DMs opinions using plithogenic aggregation operators
Step 3
Calculate distance between positive and negative ideal solution
Step 4
Find performance score of each requirement
Step 5
Define a set of supply chain sustainability metrics
Step 6
Repeat step 2,3 and 4 on the set of metrics instead of requirements
Step 7
Evaluate the difficulty of data gathering using steps 2,3 and 4
Step 8
Categorize the supply chain metrics

Figure 2. Steps of the proposed model.

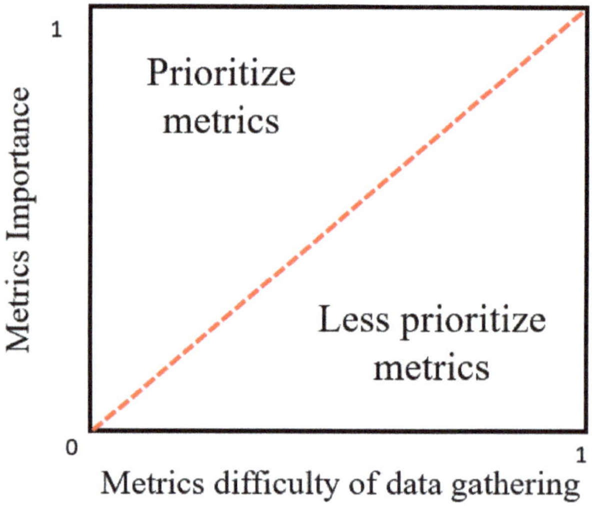
1
Metrics Importance
Prioritize metrics
Less prioritize metrics
0
1
Metrics difficulty of data gathering

Figure 3. Two dimentions model categorization.

Table 2. Popular requirements of supply chain sustainability evaluations.

	Requirement	Author
1	Cost/profit	Govindan, Kannan, Roohollah Khodaverdi, and Ahmad Jafarian [43]
2	Product quality	Osiro, Lauro, Francisco R. Lima-Junior, and Luiz Cesar R. Carpinetti [44]
3	Environmental influences	Huang, Samuel H., and Harshal Keskar [45]
4	Stability and constancy	Kannan, Devika, et al. [46]
5	Information Technology	Katsikeas, Constantine S., Nicholas G. Paparoidamis, and Eva Katsikea [47]
6	Social aspects	Mani, V., Rajat Agrawal, and Vinay Sharma [48]
7	Delivery accuracy	Chang, Betty, Chih-Wei Chang, and Chih-Hung Wu [49]

Table 3. Linguistic scale.

Linguistic Variable	Triangular Neutrosophic Scale
Nothing (N)	((0.10, 0.30,0.35), 0.1,0.2,0.15)
Very Low (VL)	((0.15,0.25,0.10), 0.6,0.2,0.3)
Low (L)	((0.40,0.35,0.50), 0.6,0.1,0.2)
Medium (M)	(0.65,0.60,0.70),0.8,0.1,0.1)
High (H)	((0.70,0.65,0.80),0.9,0.2,0.1)
Very high (VH)	((0.90,0.85,0.90),0.7,0.2,0.2)
Absolute (A)	((0.95,0.90,0.95),0.9,0.10,0.10)

4. Real World Case Study

In this paper, the proposed model has been illustrated in an application on Thailand's sugar industry in order to measure the overall sustainability of this supply chain (Figure 4). The sugar industry in Thailand is considered one of the most important economic pillars. Thailand is the fourth largest sugar producer and second largest exporter in the world. In this application, four decision makers (DMs) were assisted by their experience in solving such cases to evaluate the sustainability of Thailand's sugar industry. They are experienced in manufacturing (DM_1), quality control (DM_2), finance and purchasing (DM_3), and environmental expert (DM_4). The main goal of this case is to evaluate Thailand's sugar industry supply chain sustainability metrics based on their significance and difficulty degree of data gathering. Initially, the four experts identified a group of seven requirements for Thailand's sugar industry supply chain sustainability evaluation. They are: Profit (R_1), costs (R_2), delivery reliability (R_3), product development (R_4), environmental aspects (R_5), product quality (R_6), health and security (R_7).

- ➢ The requirements by the DMs based on linguistic variables in Table 3 are evaluated as a triangular neutrosophic value. The evaluation is shown in Table 4.
- ➢ As explained in Step 2, the plithogenic aggregation operation is used to combine all decision makers judgments about the requirements based on the contradiction degree of each one as in Table 5.
- ➢ Using Equation (6), the aggregation results are shown in Table 6 and then their crisp value is found using Equation (7).
- ➢ In **Steps 3 and 4**, the distance of positive and negative ideal solutions are found using the Euclidean distance as in Equations (8) and (9). Then, the performance degree is measured as

mentioned in Equation (10) to find the weight vector of the seven requirements as shown in Table 7.

- The decision makers define a set of supply chain sustainability metrics with respect to economic, environmental, and social dimensions (Table 8). Then, based on linguistic variables in Table 3, the DMs evaluate them as specified in Table 9.
- Table 10 shows the metrics aggregation using a plithogenic aggregation operator according to each requirement and based on the contradiction degree in Table 11. Then, their crisp values can be found.
- As in Steps 3 and 4, using the Euclidean distance as in Equations (8) and (9), the distance of positive and negative ideal solutions are found. Then, as mentioned in Equation (10), the performance degree is measured to find the weight vector of the metrics. Their results are shown in Table 12.
- As proposed in Osiro, Lauro et al. 2018 [11], the limitations of neglecting the difficulty of data gathering of every metric is addressed. In this step, the decision makers evaluate the difficulty in regards to three dimensions: Information accessibility, human resources and time needed for assessments, and other required resources, as shown in Table 13.
- Then, the decision makers evaluations were aggregated using a plithogenic aggregation equation as shown in Table 14:
- Equations (8) and (9) were used to find the distance of every metric from the positive ideal solution and negative ideal solution. Then, Equation (10) was used to calculate the performance degree, as shown in the fourth column in Table 15.
- Finally, the performance score was normalized using Equation (11) that relates to metrics importance and difficulty of data gathering. The normalization results are shown in Table 16.
- Figure 5 shows the Thailand sugar industry supply chain sustainability metrics distribution categorized in two regions which are the prioritized metrics and less prioritized metrics based on the four decision maker's evaluation.

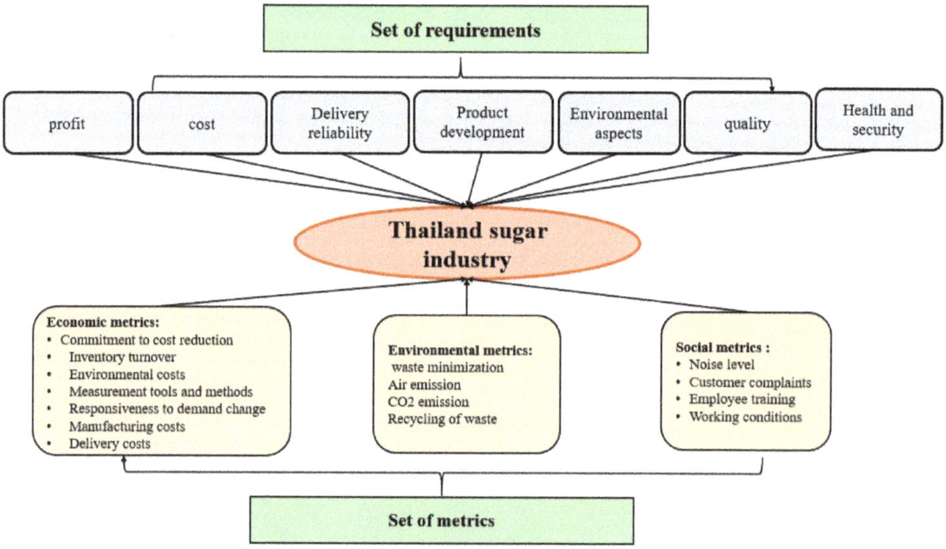

Figure 4. Thailand's sugar industry sustainability requirements and metrics.

Table 4. The evaluation of the requirements by four DMs.

Requirement	DM_1	DM_2	DM_3	DM_4
R_1	VH	H	H	M
R_2	H	VH	M	H
R_3	M	H	H	VH
R_4	H	M	M	H
R_5	H	VH	H	M
R6	H	VH	VH	M
R_7	H	M	H	M

Table 5. The requirements contradiction degree.

Requirement	R_1	R_2	R_3	R_4	R_5	R_6	R_7
Contradiction degree	0	1/7	2/7	3/7	4/7	5/7	6/7

Table 6. The aggregation results of requirements.

Requirement	$DM_1 \wedge_p DM_2 \wedge_p DM_3 \wedge_p DM_4$	Crisp Value
R_1	((0.29,0.69,1),0.45,0.18,0.36)	0.4727
R_2	((0.42,0.69,0.97),0.57,0.18,0.32)	0.1664
R_3	((0.56,0.69,0.91),0.68,0.18,0.24)	0.6130
R_4	((0.61,0.63,0.8),0.6,0.15,0.12)	0.5942
R_5	((0.79,0.69,0.75),0.86,0.18,0.1)	0.7192
R6	((0.92,0.74,0.68),0.9,0.18,0.06)	0.7781
R_7	((0.93,0.63,0.43),0.98,0.15,0.01)	0.7015

Table 7. The weights of requirements based on positive and negative distances.

Requirement	Positive Distance	Negative Distance	Performance Score	Weight	Ranking
R_1	0.2323	0.40872	0.6376	0.1551	4
R_2	0.4674	0.538	0.5351	0.1302	5
R_3	0.0914	0.0208	0.1854	0.0451	7
R_4	0.1048	0.0398	0.2752	0.0669	6
R_5	0.0146	0.0852	0.8537	0.2077	2
R6	0.0736	0.1442	0.6621	0.1611	3
R_7	0.0027	0.0677	0.9616	0.2339	1
total	-	-	4.1107	1	

Table 8. Economic, environmental and social metrics.

Dimension	Metrics
Economic	Commitment to cost reduction (I_1)
	Inventory turnover (I_2)
	Environmental costs (I_3)
	Measurement tools and methods (I_4)
	Responsiveness to demand change (I_5)
	Manufacturing cost (I_6)
	Delivery cost (I_7)
Environmental	Waste minimization(I_8)
	Air emission (I_9)
	CO_2 emission (I_{10})
	Recycling of waste (I_{11})
Social	Noise level (I_{12})
	Customer complaints (I_{13})
	Employee training (I_{14})
	Working conditions (I_{15})

Table 9. The evaluation of metrics.

Metrics	R_1	R_2	R_3	R_4	R_5	R_6	R_7
I_1	VH	A	M	M	L	M	L
I_2	L	M	H	L	L	M	L
I_3	H	VH	L	M	VH	L	H
I_4	M	M	H	VH	H	VH	H
I_5	H	H	VH	L	VL	M	VL
I_6	M	VH	L	VH	VL	VH	VL
I_7	VH	VH	VH	M	VL	H	L
I_8	L	VH	L	L	VH	L	VH
I_9	VL	VL	L	M	VH	L	VH
I_{10}	VL	VL	L	H	VH	M	VH
I_{11}	M	L	L	VL	VH	L	VH
I_{12}	L	L	L	VL	M	L	H
I_{13}	H	L	M	L	M	L	H
I_{14}	H	VH	L	VH	VL	H	H
I_{15}	H	H	M	M	L	H	H

Table 10. The aggregation results of the metrics.

	$R_1 \wedge_p R_2 \wedge_p R_3 \wedge_p R_4$	$R_5 \wedge_p R_6 \wedge_p R_7$	$R_1 \ldots R_7$	CRISP
I_1	((0.36,0.74,1),0.4,0.13,0.42)	((0.1,0.43,0.9),0.29,0.1,0.4)	((0.036,0.59,1),0.12,0.25,0.65)	0.248
I_2	((0.12,0.49,0.96),0.34,0.13,0.5)	((0.15,0.4,0.9),0.3,0.1,0.45)	((0.03,0.5,0.99),0.13,0.1,0.69)	0.2479
I_3	((0.52,0.48,0.87),0.43,0.15,0.39)	((0.6,0.63,0.9),0.5,0.2,0.34)	((0.4,0.6,0.97),0.28,0.17,0.54)	0.3866
I_4	((0.46,0.7,0.94),0.6,0.15,0.32)	((0.6,0.7,0.94),0.7,0.2,0.25)	((0.4,0.7,0.97),0.5,0.18,0.41)	0.4968
I_5	((0.43,0.63,0.92),0.6,0.18,0.3)	((0.1,0.3,0.45),0.5,0.18,0.4)	((0.43,0.5,0.8),0.43,0.18,0.46)	0.3871
I_6	((0.55,0.7,0.87),0.56,0.15,0.3)	((0.2,0.4,0.62),0.5,0.2,0.37)	((0.3,0.6,0.8),0.45,0.18,0.41)	0.3953
I_7	((0.78,0.5,0.87),0.88,0.18,0.24)	((0.3,0.4,0.7),0.7,0.14,0.24)	((0.5,0.45,0.8),0.76,0.16,0.28)	0.5191
I_8	((0.5,0.48,0.64),0.6,0.13,0.2)	((0.75,0.7,0.8),0.7,0.18,0.2)	((0.61,0.6,0.73),0.61,0.16,0.2)	0.5484
I_9	((0.37,0.73,0.32),0.68,0.15,0.21)	((0.8,0.7,0.8),0.7,0.18,0.18)	((0.6,0.73,0.53),0.7,0.17,0.18)	0.5464
I_{10}	((0.45,0.38,0.28),0.77,0.18,0.16)	((0.9,0.8,0.8),0.8,0.18,0.13)	((0.72,0.6,0.48),0.8,0.18,0.12)	0.5607
I_{11}	((0.57,0.4,0.28),0.79,0.2,0.11)	((0.9,0.7,0.85),0.8,0.18,0.1)	((0.8,0.6,0.46),0.84,0.19,0.07)	0.587
I_{12}	((0.57,0.33,0.29),0.8,0.13,0.09)	((0.8,0.6,0.5),0.9,0.15,0.05)	((0.78,0.5,0.28),0.9,0.14,0.04)	0.5168
I_{13}	((0.82,0.49,0.3),0.93,0.13,0.57)	((0.8,0.6,0.5),0.93,0.15,0.2)	((0.9,0.53,0.24),0.97,0.14,0.2)	0.5448
I_{14}	((0.17,0.7,0.46),0.95,0.13,0.02)	((0.8,0.6,0.2),0.97,0.2,0.03)	((0.8,0.6,0.17),0.99,0.2,0.007)	0.5521
I_{15}	((0.97,0.63,0.4),0.99,0.15,0.003)	((0.9,0.6,0.4),0.98,0.2,0.01)	((0.99,0.61,0.14),1,0.17,0.001)	0.6153

Table 11. Contradiction degree of metrics.

Metrics	I_1	I_2	I_3	I_4	I_5	I_6	I_7	I_8	I_9	I_{10}	I_{11}	I_{12}	I_{13}	I_{14}	I_{15}
Contradiction degree	0	$\frac{1}{15}$	$\frac{2}{15}$	$\frac{3}{15}$	$\frac{4}{15}$	$\frac{5}{15}$	$\frac{6}{15}$	$\frac{7}{15}$	$\frac{8}{15}$	$\frac{9}{15}$	$\frac{10}{15}$	$\frac{11}{15}$	$\frac{12}{15}$	$\frac{13}{15}$	$\frac{14}{15}$

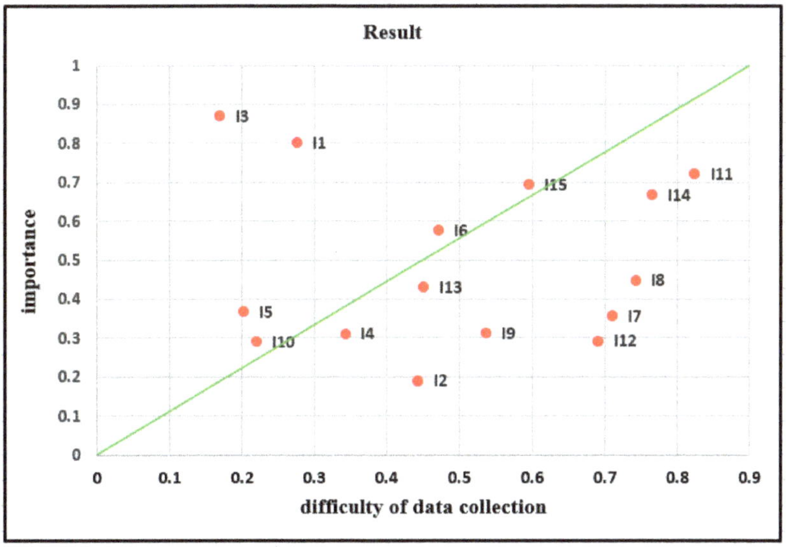

Figure 5. The categorization of Thailand's sugar industry sustainability.

Table 12. The performance score based on positive and negative distances.

Metrics	Positive Distance	Negative Distance	Performance Score
I_1	0.1114	0.5139	**0.8218**
I_2	0.4509	0.1115	0.1983
I_3	0.0272	0.3753	**0.9324**
I_4	0.2651	0.137	0.3407
I_5	0.3748	0.2467	0.3969
I_6	0.264	0.3666	0.5814
I_7	0.3878	0.2428	0.385
I_8	0.2135	0.189	0.4696
I_9	0.4151	0.2155	0.3417
I_{10}	0.4294	0.2012	0.3191
I_{11}	0.1749	0.4557	0.7226
I_{12}	0.3855	0.182	0.3207
I_{13}	0.1854	0.154	0.4537
I_{14}	0.2098	0.4208	0.6673
I_{15}	0.0835	0.2559	**0.754**

Table 13. The evaluation of information gathering difficulty.

Metrics	Information Availability	Human Resource and Time	Additional Resource Required
I_1	H	L	VH
I_2	M	H	H
I_3	H	L	L
I_4	M	H	H
I_5	L	M	L
I_6	H	H	L
I_7	H	L	VH
I_8	M	L	VH
I_9	H	M	L
I_{10}	H	M	L
I_{11}	M	L	VH
I_{12}	H	VH	L
I_{13}	VH	L	L
I_{14}	H	VH	VH
I_{15}	H	H	H

Table 14. The aggregation results of information gathering difficulty.

Metrics	A $\wedge_p$ B $\wedge_p$ C	Crisp Value
I_1	((0.25,0.68,1),0.38,0.18,0.42)	**0.4294**
I_2	((0.37,0.64,0.96),0.69,0.18,0.25)	0.5565
I_3	((0.2,0.4,0.85),0.42,0.13,0.35)	**0.3516**
I_4	((0.47,0.64,0.92),0.75,0.18,0.2)	0.6014
I_5	((0.29,0.42,0.73),0.49,0.1,0.28)	**0.3798**
I_6	((0.43,0.5,0.76),0.66,0.15,0.2)	0.488
I_7	((0.65,0.68,0.83),0.66,0.18,0.22)	0.6102
I_8	((0.7,0.67,0.77),0.68,0.15,0.19)	0.626
I_9	((0.56,0.49,0.6),0.74,0.13,0.14)	0.5094
I_{10}	((0.6,0.49,0.55),0.78,0.13,0.12)	0.506
I_{11}	((0.82,0.67,0.63),0.8,0.15,0.11)	**0.6731**
I_{12}	((0.77,0.55,0.53),0.84,0.15,0.09)	0.6013
I_{13}	((0.78,0.48,0.38),0.83,0.13,0.07)	0.5392
I_{14}	((0.97,0.8,0.72),0.85,0.2,0.04)	**0.8124**
I_{15}	((0.94,0.65,0.55),0.99,0.2,0.01	**0.7437**

Table 15. The performance score of information gathering difficulty.

Metrics	Ideal Positive	Ideal Negative	Performance Degree
I_1	0.7619	0.3594	0.1739
I_2	0.6988	0.6338	0.352
I_3	0.6988	0.3594	0.02197
I_4	0.6988	0.6338	0.2496
I_5	0.6338	0.3594	0.0743
I_6	0.6988	0.3594	0.3789
I_7	0.7619	0.3594	0.6231
I_8	0.7619	0.3594	0.6624
I_9	0.6988	0.3594	0.442
I_{10}	0.6988	0.3594	0.1003
I_{11}	0.7619	0.3594	0.7794
I_{12}	0.7619	0.3594	0.601
I_{13}	0.7619	0.3594	0.4467
I_{14}	0.7619	0.6988	0.6923
I_{15}	0.6988	0.6988	0.5

Table 16. Normalization of importance and data gathering difficulty.

Metrics	Importance	Difficulty of Data Collection
I_1	0.8031	0.2762
I_2	0.1913	0.4434
I_3	0.8711	0.1692
I_4	0.3119	0.3429
I_5	0.3694	0.2018
I_6	0.5763	0.4710
I_7	0.3569	0.7095
I_8	0.4495	0.7418
I_9	0.3129	0.5361
I_{10}	0.2911	0.2197
I_{11}	0.7216	0.8233
I_{12}	0.2926	0.6903
I_{13}	0.4316	0.4509
I_{14}	0.6682	0.7648
I_{15}	0.6958	0.595

5. Results and Discussion

Based on the decision makers evaluations, the prioritize metrics to evaluate Thailand's sugar industry sustainability are: Commitment to cost reduction (I_1), environmental costs (I_3), responsiveness to demand change (I_5), manufacturing costs (I_6), working conditions (I_{15}) and CO_2 emission (I_{10}). In this case, the decision makers were considering the economic aspects more than social or environmental dimensions. The importance of commitment to cost reduction (I_1), environmental costs (I_3) and recycling of wastes (I_{11}) gained the highest importance in value compared to other metrics, as shown in Figure 6. On the other side, recycling of the waste (I_{11}), employee training (I_{14}) and waste minimization (I_8) were the most difficult gathering information metrics. However, the CO_2 emission (I_{10}), environmental costs (I_3) and responsiveness to demand changes (I_5) were the most available information that had the lowest difficulty of data gathering degree as in Figure 7.

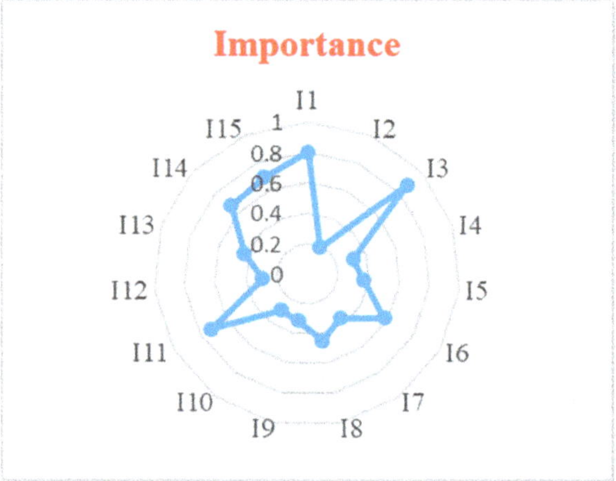

Figure 6. Metrics evaluation based on their importance.

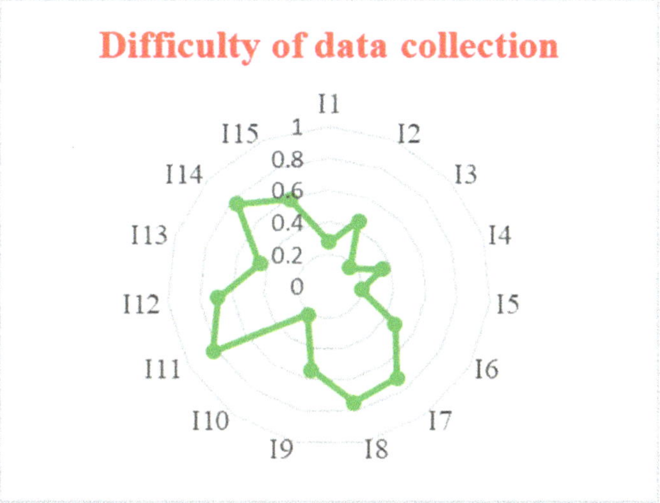

Figure 7. Metrics evaluation based on difficulty of data collection.

As the results of distribution of the Thailand sugar industry sustainability metrics were categorized based on the significance and difficulty of data gathering, this study found that the most preferred metrics were: Environmental cost (I_3), commitment to cost reduction (I_1) and realization to demand change (I_5), respectively, which were all economic metrics. It can be concluded that the economic metrics were more critical than the environmental and social metrics. These results are based on the evaluation of four decision makers, which means that it is not a general result for similar applications.

The main point in this paper is the plithogenic aggregation operation to group the decision maker's opinions in a more accurate manner. The plithogenic aggregation is taking into consideration the contradiction degree that mainly increases the accuracy of the aggregation. This study aggregated the decision maker's assessment of supply chain sustainability requirements, the evaluation of metrics importance, and measuring the difficulty of information gathering. Also, the difficulty of data gathering was not considered in many articles, but the authors found it to be a critical diminution to be measured.

Similar studies on this topic differ based on the model and the nature of the problem. As Ignatius, Joshua, et al. [52] proposed, the ANP-QFD approach which mainly considered environmental indicators in order to ensure a green building structure. On the other hand, Jamalnia, Aboozar, et al. [53] considered economic metrics such as costs, raw material and labour availability to solve a facility location problem using the QFD method. Khodakarami, Mohsen, et al. [54] and Izadikhah, et al. [55] used data envelopment analysis (DEA) for the evaluation of supply chain sustainability by taking into account mostly economic and environmental metrics. Ahmadi, et al. [56] used the best worst method (BWM) to evaluate the social sustainability of supply chain by considering social indicators rather than economic or environmental aspects.

6. Conclusions and Future Works

Due to strict government requests and the huge stress from the public, the consideration of supply chain sustainability was increased [36]. Sustainable development is one of the most significant conditions of saving resources and keeping the supply chain phases operating efficiently [57]. This explains why SSCM became one of the most important competitive strategies in organizations [58]. This study proposed an efficient combination of plithogenic aggregation operations with the quality function deployment method. The advantage of this combination is the improvement of the accuracy of the results while aggregating the assessments of the decision makers. QFD produced great results in supply chain sustainability evaluation. However, it does not consider the depth of data gathering difficulty which loosen the accuracy of the results. The supply chain sustainability evaluation is a critical topic that needs to be studied with high accuracy by considering economic, environmental and social dimensions.

This study observed the proposed combined model in a major real world case study, which was Thailand's sugar industry. Based on the nature of this supply chain strategy, its sustainability requirements was defined and measured by four decision makers. Their opinions were aggregated using plithogenic aggregation operator based on the contradiction degree to maximize the accuracy of the aggregation. In the same way, the measurement metrics was defined based on the previous requirement and included economic, social and environmental dimensions evaluated by the DMs. The results showed that the importance of commitment to cost reduction (I_1), environmental costs (I_3) and recycling of wastes (I_{11}) gained the most important metrics compared to the rest of metrics. Finally, the difficulty of data gathering of these metrics was measured and aggregated. For each dimension of evaluation, the distance of the metrics to the best and worst ideal solutions was calculated to find the importance and the degree of information gathering difficulty. The result of this point was recycling of the waste (I_{11}), employee training (I_{14}) and waste minimization (I_8) were the most difficult gathering information metrics. Moreover, the performance degree of the metrics was intended and normalized to distribute the metrics based on the two defined dimensions. It can be concluded that the most preferred metrics based in both dimensions are environmental cost (I_3), commitment to cost reduction (I_1) and realization to demand change (I_5).

Real contributions of this proposed methodology:

- The main contribution of this proposed model lies in providing accurate results of the decision makers assessments based on the contradiction degree while applying the aggregation.
- The plithogenic aggregation operation allows the DMs to consider several experts opinions in order to maximize the efficiency of the decision making.
- Also, it measures the supply chain sustainability based on two major aspects, the significance of the metrics and its level of difficulty of data gathering, which is really a critical point that affects the evaluations.
- Using a triangular neutrosophic linguistic scale to evaluate the requirement, metrics and information availability improved the level of consideration to uncertainty, because it confirms the best representation by using the three membership degrees positive, negative and boundary areas of decision making.
- The proposed methodology is efficient and has a high accuracy degree in decision making problems, Therefore, it is a great tool that may help firms in their estimation of customer needs in addition to evaluating the supply chain sustainability requirements.

In future research directions, this model may be used in assessment of other supply chain strategies to evaluate their sustainability. In addition, the plithogenic aggregation operators could be combined with other techniques to evaluate the supply chain sustainability. Finally, more evaluation dimensions could be added to the importance and difficulty of information gathering to measure supply chain sustainability.

Author Contributions: All authors have contributed equally to this paper. The individual responsibilities and contribution of all authors can be described as follows: The idea of this whole paper was put forward by M.A.-B. and R.M., A.E.-N.H.Z. completed the preparatory work of the paper. F.S. analyzed the existing work. The revision and submission of this paper was completed by M.A.-B.

Funding: "This research received no external funding".

Acknowledgments: The authors would like to thank the anonymous referees, Chief-Editor, and support Editors for their constructive suspensions and propositions that have helped to improve the quality of this research.

Conflicts of Interest: The authors declare no conflicts of interest.

References

1. Ageron, B.; Gunasekaran, A.; Spalanzani, A. Sustainable supply management: An empirical study. *Int. J. Prod. Econ.* **2012**, *140*, 168–182. [CrossRef]
2. Hassini, E.; Surti, C.; Searcy, C. A literature review and a case study of sustainable supply chains with a focus on metrics. *Int. J. Prod. Econ.* **2012**, *140*, 69–82. [CrossRef]
3. Fera, M.; Fruggiero, F.; Lambiase, A.; Macchiaroli, R.; Miranda, S. The role of uncertainty in supply chains under dynamic modeling. *Int. J. Ind. Eng. Comput.* **2017**, *8*, 119–140. [CrossRef]
4. Taticchi, P.; Tonelli, F.; Pasqualino, R. Performance measurement of sustainable supply chains: A literature review and a research agenda. *Int. J. Product. Perform. Manag.* **2013**, *62*, 782–804. [CrossRef]
5. Chithambaranathan, P.; Subramanian, N.; Gunasekaran, A.; Palaniappan, P.K. Service supply chain environmental performance evaluation using grey based hybrid MCDM approach. *Int. J. Prod. Econ.* **2015**, *166*, 163–176. [CrossRef]
6. Hervani, A.A.; Helms, M.M.; Sarkis, J. Performance measurement for green supply chain management. *Benchmarking Int. J.* **2004**, *14*, 330–353. [CrossRef]
7. Xu, M.; Cui, Y.; Hu, M.; Xu, X.; Zhang, Z.; Liang, S.; Qu, S. Supply chain sustainability risk and assessment. *J. Clean. Prod.* **2018**, *225*, 857–867. [CrossRef]
8. Martins, C.L.; Pato, M.V. Supply chain sustainability: A tertiary literature review. *J. Clean. Prod.* **2019**. [CrossRef]
9. Vafaeenezhad, T.; Tavakkoli-Moghaddam, R.; Cheikhrouhou, N. Multi-objective mathematical modeling for sustainable supply chain management in the paper industry. *Comput. Ind. Eng.* **2019**. [CrossRef]
10. Tseng, M.L.; Wu, K.J.; Hu, J.; Wang, C.H. Decision-making model for sustainable supply chain finance under uncertainties. *Int. J. Prod. Econ.* **2018**, *205*, 30–36. [CrossRef]
11. Osiro, L.; Lima-Junior, F.R.; Carpinetti, L.C. A group decision model based on quality function deployment and hesitant fuzzy for selecting supply chain sustainability metrics. *J. Clean. Prod.* **2018**, *183*, 964–978. [CrossRef]
12. Rajeev, A.; Pati, R.K.; Padhi, S.S. Sustainable supply chain management in the chemical industry: Evolution, opportunities, and challenges. *Resour. Conserv. Recycl.* **2019**, *149*, 275–291.
13. Akkawuttiwanich, P.; Yenradee, P. Fuzzy QFD approach for managing SCOR performance indicators. *Comput. Ind. Eng.* **2018**, *122*, 189–201. [CrossRef]
14. Li, S.; Tang, D.; Wang, Q. Rating engineering characteristics in open design using a probabilistic language method based on fuzzy QFD. *Comput. Ind. Eng.* **2019**. [CrossRef]
15. Eleftheriadis, S.; Duffour, P.; Mumovic, D. Participatory decision-support model in the context of building structural design embedding BIM with QFD. *Adv. Eng. Inform.* **2018**, *38*, 695–711. [CrossRef]
16. Lee, C.K.; Ru, C.T.; Yeung, C.L.; Choy, K.L.; Ip, W.H. Analyze the healthcare service requirement using fuzzy QFD. *Comput. Ind.* **2015**, *74*, 1–15. [CrossRef]
17. Eldermann, M.; Siirde, A.; Gusca, J. QFD framework for selection of industry development scenarios. *Energy Procedia* **2017**, *128*, 230–233. [CrossRef]
18. Zaim, S.; Sevkli, M.; Camgöz-Akdağ, H.; Demirel, O.F.; Yayla, A.Y.; Delen, D. Use of ANP weighted crisp and fuzzy QFD for product development. *Expert Syst. Appl.* **2014**, *41*, 4464–4474. [CrossRef]
19. Abdel-Basset, M.; Manogaran, G.; Mohamed, M.; Chilamkurti, N. Three-way decisions based on neutrosophic sets and AHP-QFD framework for supplier selection problem. *Future Gener. Comput. Syst.* **2018**, *89*, 19–30. [CrossRef]
20. Smarandache, F. *Plithogeny, Plithogenic Set, Logic, Probability, and Statistics*; Infinite Study, 2017; p. 141.
21. Ahi, P.; Searcy, C. Assessing sustainability in the supply chain: A triple bottom line approach. *Appl. Math. Model.* **2015**, *39*, 2882–2896. [CrossRef]
22. Closs, D.J.; Speier, C.; Meacham, N. Sustainability to support end-to-end value chains: The role of supply chain management. *J. Acad. Mark. Sci.* **2011**, *39*, 101–116. [CrossRef]
23. Haake, H.; Seuring, S. Sustainable procurement of minor items–exploring limits to sustainability. *Sustain. Dev.* **2009**, *17*, 284–294. [CrossRef]
24. Linton, J.D.; Klassen, R.; Jayaraman, V. Sustainable supply chains: An introduction. *J. Oper. Manag.* **2007**, *25*, 1075–1082. [CrossRef]

25. Jadhav, A.; Orr, S.; Malik, M. The role of supply chain orientation in achieving supply chain sustainability. *Int. J. Prod. Econ.* **2008**. [CrossRef]
26. Yazdani, M.; Kahraman, C.; Zarate, P.; Onar, S.C. A fuzzy multi attribute decision framework with integration of QFD and grey relational analysis. *Expert Syst. Appl.* **2019**, *115*, 474–485. [CrossRef]
27. Qorri, A.; Mujkić, Z.; Kraslawski, A. A conceptual framework for measuring sustainability performance of supply chains. *J. Clean. Prod.* **2018**, *189*, 570–584. [CrossRef]
28. Searcy, C.; Dixon, S.M.; Neumann, W.P. The use of work environment performance indicators in corporate social responsibility reporting. *J. Clean. Prod.* **2016**, *112*, 2907–2921. [CrossRef]
29. Chen, R.H.; Lin, Y.; Tseng, M.L. Multicriteria analysis of sustainable development indicators in the construction minerals industry in China. *Resour. Policy* **2015**, *46*, 123–133. [CrossRef]
30. Haghighi, S.M.; Torabi, S.A.; Ghasemi, R. An integrated approach for performance evaluation in sustainable supply chain networks (with a case study). *J. Clean. Prod.* **2016**, *137*, 579–597. [CrossRef]
31. Akao, Y.; Mazur, G.H. The leading edge in QFD: Past, present and future. *Int. J. Qual. Reliab. Manag.* **2003**, *20*, 20–35. [CrossRef]
32. Moubachir, Y.; Bouami, D. A new approach for the transition between QFD phases. *Procedia CIRP* **2015**, *26*, 82–86. [CrossRef]
33. Lam, J.S. Designing a sustainable maritime supply chain: A hybrid QFD-ANP approach. *Transp. Res. Part E: Logist. Transp. Rev.* **2015**, *78*, 70–81. [CrossRef]
34. Yazdani, M.; Chatterjee, P.; Zavadskas, E.K.; Zolfani, S.H. Integrated QFD-MCDM framework for green supplier selection. *J. Clean. Prod.* **2017**, *142*, 3728–3740. [CrossRef]
35. Dursun, M.; Arslan, Ö. An Integrated Decision Framework for Material Selection Procedure: A Case Study in a Detergent Manufacturer. *Symmetry* **2018**, *10*, 657. [CrossRef]
36. Meksavang, P.; Shi, H.; Lin, S.M.; Liu, H.C. An Extended Picture Fuzzy VIKOR Approach for Sustainable Supplier Management and Its Application in the Beef Industry. *Symmetry* **2019**, *11*, 468. [CrossRef]
37. Smarandache, F. Plithogenic Set, an Extension of Crisp, Fuzzy, Intuitionistic Fuzzy, and Neutrosophic Sets–Revisited. *Neutrosophic Sets Syst.* **2018**, *21*, 153–166.
38. Smarandache, F. Physical Plithogenic Set. *APS Meeting Abstracts.* 2018.
39. Smarandache, F. Extension of Soft Set to Hypersoft Set, and then to Plithogenic Hypersoft Set. *Neutrosophic Sets Syst.* **2018**, *24*, 168–170.
40. Liu, P.; Wang, Y. Multiple attribute decision-making method based on single-valued neutrosophic normalized weighted Bonferroni mean. *Neural Comput. Appl.* **2014**, *25*, 2001–2010. [CrossRef]
41. Abdel-Baset, M.; Hezam, I.M.; Smarandache, F. Neutrosophic goal programming. *Neutrosophic Sets Syst.* **2016**, 11.
42. Rivieccio, U. Neutrosophic logics: Prospects and problems. *Fuzzy Sets Syst.* **2008**, *159*, 1860–1868. [CrossRef]
43. Govindan, K.; Khodaverdi, R.; Jafarian, A. A fuzzy multi criteria approach for measuring sustainability performance of a supplier based on triple bottom line approach. *J. Clean. Prod.* **2003**, *47*, 345–354. [CrossRef]
44. Osiro, L.; Lima-Junior, F.R.; Carpinetti, L.C. A fuzzy logic approach to supplier evaluation for development. *Int. J. Prod. Econ.* **2014**, *153*, 95–112. [CrossRef]
45. Kannan, D.; Khodaverdi, R.; Olfat, L.; Jafarian, A.; Diabat, A. Integrated fuzzy multi criteria decision making method and multi-objective programming approach for supplier selection and order allocation in a green supply chain. *J. Clean. Prod.* **2013**, *47*, 355–367. [CrossRef]
46. Huang, S.H.; Keskar, H. Comprehensive and configurable metrics for supplier selection. *Int. J. Prod. Econ.* **2007**, *105*, 510–523. [CrossRef]
47. Katsikeas, C.S.; Paparoidamis, N.G.; Katsikea, E. Supply source selection criteria: The impact of supplier performance on distributor performance. *Ind. Mark. Manag.* **2004**, *33*, 755–764. [CrossRef]
48. Mani, V.; Agrawal, R.; Sharma, V. Supplier selection using social sustainability: AHP based approach in India. *Int. Strateg. Manag. Rev.* **2014**, *2*, 98–112. [CrossRef]
49. Chang, B.; Chang, C.W.; Wu, C.H. Fuzzy DEMATEL method for developing supplier selection criteria. *Expert Syst. Appl.* **2011**, *38*, 1850–1858. [CrossRef]
50. Liao, H.; Xu, Z.; Zeng, X.J. Distance and similarity measures for hesitant fuzzy linguistic term sets and their application in multi-criteria decision making. *Inf. Sci.* **2014**, *271*, 125–142. [CrossRef]
51. Lima-Junior, F.R.; Carpinetti, L.C. A multicriteria approach based on fuzzy QFD for choosing criteria for supplier selection. *Comput. Ind. Eng.* **2016**, *101*, 269–285. [CrossRef]

52. Ignatius, J.; Rahman, A.; Yazdani, M.; Šaparauskas, J.; Haron, S.H. An integrated fuzzy ANP–QFD approach for green building assessment. *J. Civ. Eng. Manag.* **2016**, *22*, 551–563. [CrossRef]
53. Jamalnia, A.; Mahdiraji, H.A.; Sadeghi, M.R.; Hajiagha, S.H.; Feili, A. An integrated fuzzy QFD and fuzzy goal programming approach for global facility location-allocation problem. *Int. J. Inf. Technol. Decis. Mak.* **2014**, *13*, 263–290. [CrossRef]
54. Khodakarami, M.; Shabani, A.; Saen, R.F.; Azadi, M. Developing distinctive two-stage data envelopment analysis models: An application in evaluating the sustainability of supply chain management. *Measurement* **2015**, *70*, 62–74. [CrossRef]
55. Izadikhah, M.; Saen, R.F. Evaluating sustainability of supply chains by two-stage range directional measure in the presence of negative data. *Transp. Res. Part D Transp. Environ.* **2016**, *49*, 110–126. [CrossRef]
56. Ahmadi, H.B.; Kusi-Sarpong, S.; Rezaei, J. Assessing the social sustainability of supply chains using Best Worst Method. *Resour. Conserv. Recycl.* **2017**, *126*, 99–106. [CrossRef]
57. Matić, B.; Jovanović, S.; Das, D.K.; Zavadskas, E.K.; Stević, Ž.; Sremac, S.; Marinković, M. A New Hybrid MCDM Model: Sustainable Supplier Selection in a Construction Company. *Symmetry* **2019**, *11*, 353. [CrossRef]
58. Abdel-Basset, M.; Mohamed, M.; Smarandache, F. A hybrid neutrosophic group ANP-TOPSIS framework for supplier selection problems. *Symmetry* **2018**, *10*, 226. [CrossRef]

symmetry

MDPI

Article

Online Visual Tracking of Weighted Multiple Instance Learning via Neutrosophic Similarity-Based Objectness Estimation

Keli Hu [1], Wei He [2], Jun Ye [1], Liping Zhao [1], Hua Peng [1,3] and Jiatian Pi [4,*]

1 Department of Computer Science and Engineering, Shaoxing University, Shaoxing 312000, China; ancimoon@gmail.com (K.H.); yejun@usx.edu.cn (J.Y.); zhaoliping_jian@126.com (L.Z.); penghua_47@163.com (H.P.)
2 Key Laboratory of Wireless Sensor Network & Communication, Shanghai Institute of Microsystem and Information Technology, Chinese Academy of Sciences, Shanghai 200050, China; wei.he@mail.sim.ac.cn
3 College of Information Science and Engineering, Jishou University, Jishou 416000, China
4 College of Computer and Information Science, Chongqing Normal University, Chongqing 400047, China
* Correspondence: pijiatian@cqnu.edu.cn

Received: 30 May 2019; Accepted: 20 June 2019; Published: 25 June 2019

Abstract: An online neutrosophic similarity-based objectness tracking with a weighted multiple instance learning algorithm (NeutWMIL) is proposed. Each training sample is extracted surrounding the object location, and the distribution of these samples is symmetric. To provide a more robust weight for each sample in the positive bag, the asymmetry of the importance of the samples is considered. The neutrosophic similarity-based objectness estimation with object properties (super straddling) is applied. The neutrosophic theory is a new branch of philosophy for dealing with incomplete, indeterminate, and inconsistent information. By considering the surrounding information of the object, a single valued neutrosophic set (SVNS)-based segmentation parameter selection method is proposed, to produce a well-built set of superpixels which can better explain the object area at each frame. Then, the intersection and shape-distance criteria are proposed for weighting each superpixel in the SVNS domain, mainly via three membership functions, T (truth), I (indeterminacy), and F (falsity), for each criterion. After filtering out the superpixels with low response, the newly defined neutrosophic weights are utilized for weighting each sample. Furthermore, the objectness estimation information is also applied for estimating and alleviating the problem of tracking drift. Experimental results on challenging benchmark video sequences reveal the superior performance of our algorithm when confronting appearance changes and background clutters.

Keywords: visual tracking; neutrosophic weight; objectness; weighted multiple instance learning

1. Introduction

The task for visual object tracking is to estimate the object location at each frame in a video sequence. Such a kind of visual analysis has been widely studied in computer vision due to its application in many fields, e.g., video surveillance, human–computer interaction, autonomous driving, and traffic monitoring [1,2]. While a lot of effort has been made, and numerous tracking algorithms have been proposed in past decades, it is still a very challenging task due to the factors of illumination variation, scale variation, occlusion, deformation, motion blur, fast motion, rotation, and background clutters, etc. Most of these factors can cause appearance changes, which are very challenging for the visual tracker.

To adjust these challenges adaptively, it is quite important for a robust object tracker to employ an effective appearance model. In general, the existing appearance models can be mainly divided into two categories, the generative model [3–9] and the discriminative model [10–18]. For the generative model, the tracking is cast as finding the most similar region to the learned object appearance. The tracking result at each frame can be used to update the appearance model. The MeanShift tracker [3] is one of the most influential methods due to its high efficiency and its robustness when confronting the challenges of radical changes and deformation. However, the MeanShift tracker is very sensitive when there is a similar color surrounding the target, because only the histogram feature is applied for building the appearance model. To improve the performance, some other features are introduced into the MeanShift tracking technology, e.g., cross-bin metric [5], background information [6], and depth feature [7]. The IVT tracker [4] employs incremental principal component analysis to represent and update the appearance model. Unlike generative models, both positive and negative samples are utilized by discriminative models, and the object tracking is posed as the task of binary classification to discriminate the target from its surroundings.

Generally, the robustness of generative models is inferior to discriminative models to some extent. Generative models always utilize the foreground information to model the object appearance, and the surrounding information is not considered. Thus, generative models are more likely to drift when handling appearance changes in complex environments. For discriminative models, the positive and negative samples extracted in the first frame are utilized for initialization. For a new arriving frame, the object location is estimated by the afore-trained classifier. The classifier will be updated by using the newly collected positive and negative samples. A boosting method was used for online feature selection in [10]. Only the sample located at the estimated object location is employed as the positive sample. Some negative examples are extracted around the neighborhood of the estimated location. This often causes drift and error accumulation problems when the tracked location drifts a little from the real location. In [11], only the samples extracted from the first frame were labeled for training the classifier, and the samples generated in the subsequent frames were all unlabeled. The SemiBoost tracker [11] may drift in interframe motion on account of the smooth motion assumption.

To tackle the problem of imprecise sampling, online multiple instance learning (MIL) was firstly proposed for object tracking in [12,13]. The MIL tracker uses the positive and negative bags to update the classifier at each frame. The positive bag contains several instances extracted from the close neighborhood of the object location. It has been revealed that the MIL tracker alleviates the drift problem, however, the limitation still exists. Several trackers have been developed within the framework of MIL [14–18]. A more effective and efficient algorithm was proposed, using important prior information, such as instance labels and the most correct positive instance [14], due to the fact that discriminative information about the importance of the positive examples is not considered in the MIL tracker. By the assumption that the tracking result in the current frame is the most positive sample, Zhang et al. [15] employed the Euclidean distance between each positive instance and the estimated object location as the importance of each positive sample. The weight distribution of the samples is centrally symmetric. In [16], the chaotic theory was introduced into MIL-based tracking. The fractal dimension of the dynamic model was adjusted as instance weight. There are still drawbacks for the proposed weighting algorithms. Once the tracked result drifts from the real object location, the assumption of the most positive sample will not be satisfied. Weights will be wrongly utilized to update the classifier. The objectness measure [19] is applied for judging the importance of each instance [17]. The objectness measure [20] owns the ability of judging the probability of a given window containing a whole object. Instead of using the distance measure [15], Yang et al. [17] integrated the objectness estimation into the calculation of the instance importance. Experimental results have demonstrated the power of the objectness measure once it is introduced into the MIL-based tracker. As we know, the robustness of the objectness estimation highly depends on the segmentation result [20,21]. For the application of visual tracking, the environment change occurs almost at each frame. It is essential to propose a scheme to let the segmentation result adaptively adjust different scenes. In addition, some noisy superpixels can distract the objectness-based instance weighting.

Thus, a method for filtering the noisy superpixel should also be seriously considered. However, much uncertain information must be considered when we try to tackle these two serious issues.

Neutrosophic set (NS) [22] is a new branch of philosophy to deal with the origin, nature, and scope of neutralities. It has been widely used in dealing with uncertain information [23]. Due to this, the NS theory has been successfully introduced into many applications, such as medical diagnosis [24], skeleton extraction [25], image segmentation [26–31], and object tracking [7–9,32]. SVNS (single valued neutrosophic set) [33] is a subclass of the neutrosophic set with a finite interval for practical usage. Both the cosine and tangent similarity measures were applied for medical diagnosis in [24]. For the application of image segmentation, the source image is usually transformed into the NS domain, and will be described by the T, I, and F membership set [26,31]. The cosine similarity measure was also utilized in [26,31]. Furthermore, the neutrosophic set-based MeanShift and c-means clustering methods were proposed to earn a more robust segmentation result [27,28,30].

Guided by the above idea, we propose an online visual tracking of weighted multiple instance learning via a neutrosophic similarity-based objectness estimation. First, to produce a well-built set of superpixels at each frame, we propose a method of neutrosophic set-based segmentation parameter selection. The information surrounding object location is taken into consideration. Second, the intersection and shape-distance criteria are utilized for evaluating each superpixel, and three membership functions, T, I, and F, are proposed for each criterion. Then, the neutrosophic set-based multi-criteria similarity score is utilized to facilitate the superpixel filter. The importance of each training instance is evaluated by the estimation of the filtered objectness measure. Third, the information of the objectness estimation is applied for alleviating the problem of tracking drift. Empirical results on challenging video sequences demonstrate our NeutWMIL tracker can robustly track the target. To our own knowledge, this is the first time the NS theory has been introduced into the visual object tracking of a discriminative model.

The paper is organized as follows: in Section 2, we introduce related work. Section 3 introduces our tracking system, where the basic flow is first given, and then the principle of our method and its advantages are illustrated in the following subsections. Section 4 gives the detailed experiment setup and results. Finally, Section 5 concludes the paper.

2. Related Work

Benefitting from the discriminative model, many trackers based on such kind of framework have been proposed [10–18]. Several studies revealed that the method for weighting the training samples plays quite an important role for improving the robustness of a tracker [14–18]. The asymmetry of the importance of the samples is considered, and the objectness measure [19] is applied for judging the importance of each instance [17], which is collected for training the multiple instance learning-based tracker. The results revealed that the performance of the MIL tracker was highly improved [17]. However, the parameters for segmentation were not adaptively updated by considering the surroundings of the tracked object, and all the regions were considered equally for calculating the objectness.

To overcome the above problem, we tried to utilize the NS theory to deal with the related uncertainty problems, due to the ability of the NS theory of handling uncertain information [23], as well as the requirement for enhancing the robustness of the visual trackers. Several works have been done introducing the NS theory into the tracking issue [7–9,32]. In order to fuse both the color and depth features, three membership functions, T, I, and F, were proposed to deal with the uncertainty problem for judging the robustness of each feature [7]. The single valued neutrosophic cross-entropy measure [34] was finally utilized for feature fusion. By considering the drawbacks of the traditional MeanShift tracker [3], Hu et al. [8] integrated the background information into the bin importance of the histogram feature by using the neutrosophic descriptions regarding the uncertain issue of feature enhancement. Furthermore, Hu et al. [9] proposed the element-weighted neutrosophic correlation coefficient and utilized it to improve the CAMShift tracker. It has been revealed that the proposed trackers are more robust than traditional ones when the neutrosophic theory is utilized [7–9]. Fan et al. [32] proposed a neutrosophic hough transform-based track initiation method to solve the uncertain track initiation problem, and the results demonstrated that it is superior to the traditional hough transform-based track initiation in a complex surveillance environment.

This work is quite different from the proposed NS-base trackers in that it is the first time the NS theory has been introduced into the visual object tracking of discriminative model. Though many methods for image segmentation in NS domain have been proposed, this is also the first work that has proposed to tackle the problem of segmentation parameter selection and superpixel filtering, for the purpose of enhancing the robustness of a tracker.

3. Problem Formulation

3.1. System Overview

The basic flow of the online visual tracking of weighted multiple instance learning via neutrosophic similarity-based objectness estimation is shown in Figure 1. Before the tracking process, the classifier of multiple instance learning is initialized with the same method as the weighted multiple instance learning [15]. Suppose $l_t^* \in R^2$ is the center of the object location at frame t, the object location is represented by a rectangular bounding box. To produce robust objectness [19,20] information for tracking, N segmentation results of frame t are firstly calculated by utilizing N parameter tuples for segmentation. Then we used the SVNS-based segmentation parameter selection method to select the parameter tuple, which can achieve the best segmentation result for the representation of the object area. When the next frame arrives, we calculated its segmentation result by using the selected segmentation parameter tuple. Each region in the segmentation result corresponds to a superpixel. The neutrosophic set-based superpixel filter was employed to filter out the noisy superpixel when measuring the objectness. The sliding window method was used to calculate two confidence maps. The classifier confidence map was calculated based on the afore-trained classifier. The Neut-Objectness confidence map was calculated based on the filtered objectness measure. Finally, each maximum value of the two kinds of confidence maps was utilized to decide $l_{t+1}^* \in R^2$, which is the center of the object location at frame t+1. The scale of the rectangular bounding box was fixed in this work. To update the classifier parameters, the filtered objectness measure was applied for weighting the training samples. We first cropped M positive instances within the circular region centering at l_{t+1}^* to form a positive bag $\mathbf{X}^+ = \left\{x \| \|l_{t+1}(x) - l_{t+1}^*\| < \alpha\right\}$, and the distribution is symmetric, then L negative instances were cropped from an annular region with radius $\alpha < \xi < \beta$ to form a negative bag $\mathbf{X}^- = \left\{x | \alpha < \|l_{t+1}(x) - l_{t+1}^*\| < \beta\right\}$, where $l_{t+1}(x)$ is the location of instance x at frame t+1. The instances in the positive bag were weighted by the filtered objectness measure, and the negative instances were weighted equally. All the instances were represented by W Haar-like features. Each Harr-like feature corresponded to a weak classifier. All the weighted instances were employed to train the W weak classifiers, and finally K weak classifiers were chosen for constructing a strong classifier $\mathbf{H}_k$. For each frame, we chose the updated $\mathbf{H}_k$ as the new classifier for tracking. For a new arriving frame, the above procedures were repeated.

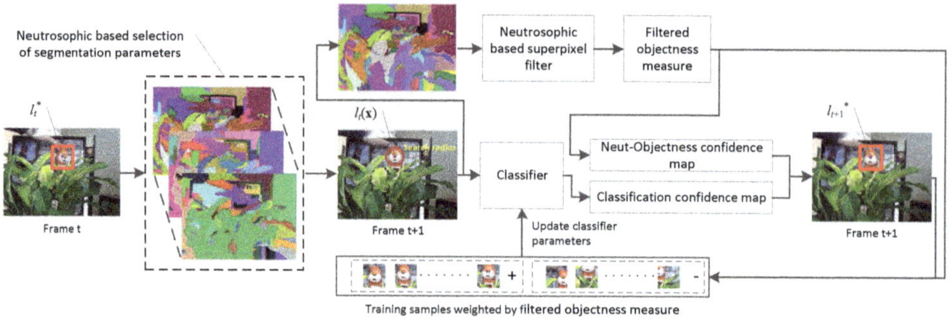

Figure 1. The flow chart of online visual tracking with NeutWMIL.

3.2. Objectness Measure

The objectness measure [20] is always utilized in the research area of object detection for quantifying the degree of an image window containing an object. The color contrast, edge density, and superpixel straddling (SS) methods were proposed in [20], and it has been proven that the SS measure is suitable for object tracking [17]. The SS-based objectness for a window O_w can be calculated by:

$$ss(O_W, \mathbf{T}_i) = 1 - \sum_{s \in \mathbf{S}(\mathbf{T}_i)} \frac{\min(|s \backslash O_W|, |s \cap O_W|)}{|O_W|}, \quad (1)$$

where $\mathbf{S}(\mathbf{T}_i)$ is the superpixel set obtained by the literature [21], with the segmentation parameter tuple $\mathbf{T}_i$. For each superpixel s, $|s \backslash O_W|$ computes its area outside O_w and $|s \cap O_W|$ calculates its area inside O_w. From Equation (1), we can get determine that a superpixel contributes less when it straddles the window O_w. The superpixels inside the window O_w contribute the most. $ss(O_W, \mathbf{T}_i)$ achieves to 1 when there is not any superpixel straddling the window O_w.

As seen in Equation (1), the segmentation parameter tuple $\mathbf{T}_i$ is also an important parameter. $\mathbf{T}_i$ is a set of parameters that can affect the segmentation result greatly. Different segmentation algorithms always relate to different parameters. For the efficient graph-based image segmentation algorithm [21] employed in this work, each tuple $\mathbf{T}_i = \{\sigma, k, m\}$ contained three parameters including σ (used to smooth the input image before segmenting it), k (value for the threshold function), and m (minimum component size enforced by post-processing). As shown in Figure 2, when we use the same segmentation algorithm [21] but employ different segmentation parameter tuples, the segmentation results are quite different. In addition, with the three different segmentation results, the corresponding superpixel sets are also quite different from each other. From Equation (1), we can find that the calculation of SS-based objectness highly depends on the shape and distribution of the superpixels in the image. To apply the objectness measure for weighting the training samples, we then proposed a neutrosophic set-based scheme to handle such a problem.

(a) original image (b) segmentation results with $\mathbf{T}_1$ (c) segmentation results with $\mathbf{T}_2$ (d) segmentation results with $\mathbf{T}_3$

Figure 2. Segmentation results with different segmentation parameter tuples. $\mathbf{T}_1 = \{0.4, 450, 150\}$, $\mathbf{T}_2 = \{0.5, 250, 400\}$, $\mathbf{T}_3 = \{0.6, 550, 250\}$.

3.3. Neutrosophic Set-Based Segmentation Parameter Selection

To apply the objectness information for enhancing the tracker, a suitable segmentation parameter tuple should first be chosen. A well-built superpixel set is one of the most important preconditions for producing reliable objectness measures for the application of visual tracking. A well-built superpixel set should have the ability to enhance the objectness response at the object location. In addition, the symmetric surrounding information should also be considered seriously. The neutrosophic theory has shown its ability to deal with uncertain situations, and the neutrosophic similarity score between SVNSs is applied in this work.

The neutrosophic theory was firstly proposed by Smarandache [22]. The original neutrosophic theory is difficult to use for tackling practical problems. SVNS [33] is a subclass of the neutrosophic set with a finite interval, and it is proposed for practical usage. Let $\mathbf{A} = \{A_1, A_2, \ldots, A_m\}$, which denotes a set of alternatives. For a multiple criteria neutrosophic situation, the alternatives A_i can be represented as:

$$A_i = \{\langle C_j, T_{Cj}(A_i), I_{Cj}(A_i), F_{Cj}(A_i)\rangle\},\ i = 1, \ldots, m; j = 1, \ldots, n, \tag{2}$$

where $\mathbf{C} = \{C_1, C_2, \ldots, C_n\}$ is a set of criteria, $T_{Cj}(A_i)$ denotes the degree to which the alternative A_i satisfies the criterion C_j, $I_{Cj}(A_i)$ indicates the indeterminacy degree to which the alternative A_i satisfies or does not satisfy the criterion C_j, $F_{Cj}(A_i)$ indicates the degree to which the alternative A_i does not satisfy the criterion C_j, $T_{Cj}(A_i) \in [0,1]$, $I_{Cj}(A_i) \in [0,1]$, $F_{Cj}(A_i) \in [0,1]$.

Suppose $A^* = \{\langle C_j, T_{Cj}(A^*), I_{Cj}(A^*), F_{Cj}(A^*)\rangle\}$ is an ideal alternative with the criteria C_j. The cosine similarity score between A_i and A^* is defined by [35]:

$$S_{\cos}(A_i, A^*) = \sum_{j=1}^{n} w_j \frac{T_{Cj}(A_i)T_{Cj}(A^*) + I_{Cj}(A_i)I_{Cj}(A^*) + F_{Cj}(A_i)F_{Cj}(A^*)}{\sqrt{T_{Cj}{}^2(A_i) + I_{Cj}{}^2(A_i) + F_{Cj}{}^2(A_i)}\sqrt{T_{Cj}{}^2(A^*) + I_{Cj}{}^2(A^*) + F_{Cj}{}^2(A^*)}}. \tag{3}$$

The corresponding tangent similarity score is defined as [24]:

$$S_{\tan}(A_i, A^*) = \sum_{j=1}^{n} w_j \tan\left[\frac{\pi}{12}\left(\begin{array}{c}|T_{Cj}(A_i) - T_{Cj}(A^*)| + \\ |I_{Cj}(A_i) - I_{Cj}(A^*)| + |F_{Cj}(A_i) - F_{Cj}(A^*)|\end{array}\right)\right], \tag{4}$$

where w_j is the weight for each criterion, and $w_j \in [0,1]$, $\sum_j w_j = 1$. Both the cosine and tangent measures have been successfully employed for medical diagnoses [24] and some visual analysis missions [7,26]. In this work, the neutrosophic similarity score was utilized for fusing information, and these two measures were tested separately in the experimental section.

As we want to use the objectness measure to weight the training samples, the weights of the samples close to the object location should be enhanced. Considering such a problem, the object location objectness enhancing criterion was proposed. For the segmentation parameter tuple $\mathbf{T}_k$, the corresponding membership functions $T_O(\mathbf{T}_k)$ (truth), $I_O(\mathbf{T}_k)$ (indeterminacy), and $F_O(\mathbf{T}_k)$ (falsity) were defined as:

$$T_O(\mathbf{T}_i) = ss(O_{l*}, \mathbf{T}_i); \tag{5}$$

$$I_O(\mathbf{T}_i) = 1 - \min\left(\frac{1}{C}\frac{\sum_{j=1}^{C}\left|ss(O_j(r), \mathbf{T}_i) - ss(O_{l*}, \mathbf{T}_i)\right|}{ss(O_{l*}, \mathbf{T}_i)}, 1\right); \tag{6}$$

$$F_O(\mathbf{T}_i) = 1 - T_O(\mathbf{T}_i), \tag{7}$$

where O_{l^*} is the rectangular window corresponding to the object location and $\mathbf{T}_i$ is the the i-th segmentation parameter tuple. As shown in Figure 3, $O_j(r)$ is a square window centered at the j-th pixel of the C uniform sampled pixels on the boundary of the square window with the edge length of

$2r+1$(pixel) centered at l^*, the distribution of $O_j(r)$ is symmetric, l^* is the center of the object location, and $O_j(r)$ has the same size as the objects.

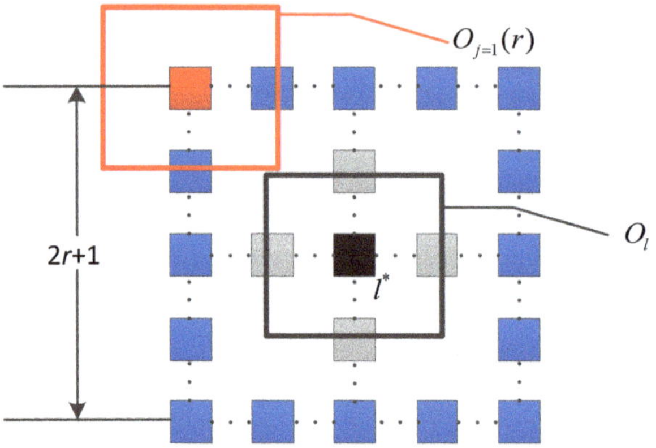

Figure 3. The illustration of $O_j(r)$ mentioned in Equation (6).

Let $A^* = \{\langle C_O, 1, 0, 0\rangle\}$ denote the ideal alternative with the object location objectness enhancing criterion. Substituting Equations (5)–(7) into Equation (3), we obtain the cosine similarity score between the choice of $\mathbf{T}_i$ and the ideal choice:

$$ltup_{\cos}(\mathbf{T}_i) = \frac{T_O(A_i)}{\sqrt{T_O{}^2(\mathbf{T}_i) + I_O{}^2(\mathbf{T}_i) + F_O{}^2(\mathbf{T}_i)}}. \quad (8)$$

When substituting Equation (5) into Equation (4), we obtain the corresponding tangent similarity score:

$$ltup_{\tan}(\mathbf{T}_i) = \tan\left[\frac{\pi}{12}\left(\left|I_O(\mathbf{T}_i)\right| + 2\left|F_O(\mathbf{T}_i)\right|\right)\right]. \quad (9)$$

Suppose we have N segmentation parameter tuples waiting for selection, the tuple with the maximum similarity score is chosen as the right tuple $\mathbf{T}_{sel}$. For the cosine and tangent measures, the selections may be different from each other.

3.4. Neutrosophic Set-Based Superpixel Filter

As shown in Equation (1) and Figure 2, all the superpixels were taken into the consideration for the estimation of the objectness. However, such a simple method may bring in some noise for the objectness estimation. For instance, the superpixel far from the object or the superpixel that is too large size may disturb the objectness of the tracked object.

By considering the uncertain information, the intersection and shape–distance criteria were considered in this work. For the intersection criteria, the corresponding true, indeterminate, and faulty membership functions are defined as follows:

$$T_{int}(i) = \max\left(\frac{|s_i \cap O_l|}{|s_i|}, \frac{|s_i \cap O_l|}{|O_l|}\right); \quad (10)$$

$$I_{int}(i) = \frac{|s_i \backslash O_l|}{|s_i|}; \quad (11)$$

$$F_{int}(i) = 1 - T_{int}(i), \quad (12)$$

where O_l is the rectangular window, which corresponds to the object location calculated by the afore-trained classifier before the modification, suppose $\mathbf{S}(\mathbf{T}_{sel})$ is the superpixel set obtained by the literature [21] with the selected segmentation tuple $\mathbf{T}_{sel}$, and s_i is the i-th superpixel included in $\mathbf{S}(\mathbf{T}_{sel})$.

A superpixel is more likely to satisfy the intersection criteria when it is located in the estimated object location. The corresponding uncertain probability increases with the area outside the object location.

To enhance the robustness of the neutrosophic set-based method for judging a superpixel, the shape and the distance information were considered. Using the shape–distance criteria, we can further give the definitions:

$$T_{sd}(i) = \min\left(f\left(\frac{w_{si}}{w_{ol}}\right), f\left(\frac{h_{si}}{h_{ol}}\right)\right); \tag{13}$$

$$I_{sd}(i) = 1 - e^{-|\mathbf{x}_{si}-l|/D}; \tag{14}$$

$$F_{sd}(i) = 1 - T_{sd}(i), \tag{15}$$

where w_{si} and h_{si} are the width and height, respectively, of the tight rectangular bounding box of the super pixel s_i, w_{ol} and h_{ol} are the width and height corresponding to the object window O_l, $\mathbf{x}_{si}$ is the centroid location of s_i, l is the center of the O_l, and D is half the length of the O_l diagonal. The function $f(x)$ in $T_{sd}(i)$ is defined as:

$$f(x) = \begin{cases} \frac{1}{2}erfc(5x-7) & x \geq 1 \\ \frac{1}{2}erfc(-2.2x+0.2) & 0 \leq x < 1 \end{cases}. \tag{16}$$

The domain of $f(x)$ is positive real numbers, and then $f(x)$ is manually designed for the purpose of decreasing the value of $T_{sd}(i)$ when the width or the height of s_i is larger or smaller than the w_{ol} or h_{ol}. As seen in Figure 4, when $x > 1$, the response of $f(x)$ decreases slowly in the intervals of [1,1.2] and [1.6,1.8], but decreases sharply during the interval of [1.2,1.6]. The reason for such a design is that we wanted to keep the information of those superpixels with a relative similar size to the object, and try to discard the superpixels with a much larger width or height than the object. As shown in Figure 4, the response has decreased at the value of less than 0.1 when x equals 1.6. However, we choose a different solution when $x < 1$. The response of $f(x)$ decreases much slower than in the interval of $x > 1$, because a small superpixel may be one of the real parts of the object. We tried to keep the information of such superpixels.

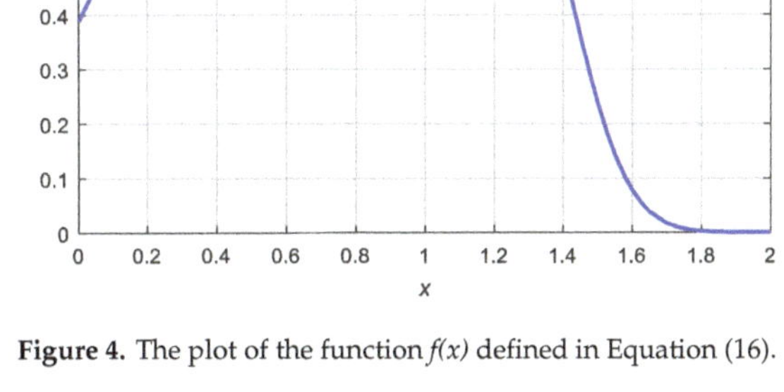

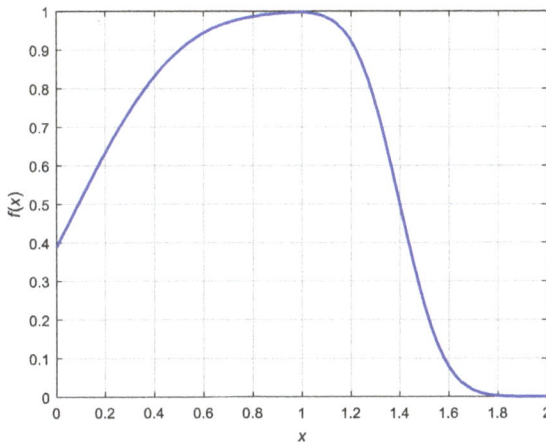

Figure 4. The plot of the function $f(x)$ defined in Equation (16).

As seen in Equation (13) and Equation (14), the shape factor is mainly considered when calculating the truth response, and the distance factor is primarily considered as the indeterminate information.

Let $A^* = \{\langle C_{int}, 1, 0, 0\rangle, \langle C_{sd}, 1, 0, 0\rangle\}$ denote the ideal alternative with the intersection and shape–distance criteria. Substituting Equations (10)–(16) into Equation (3), we obtain the cosine similarity score between the choice of s_i and the ideal superpixel:

$$ls_{\cos}(i) = w_{cint}\frac{T_{int}(i)}{\sqrt{T_{int}{}^2(i) + I_{int}{}^2(i) + F_{int}{}^2(i)}} + w_{csd}\frac{T_{sd}(i)}{\sqrt{T_{sd}{}^2(i) + I_{sd}{}^2(i) + F_{sd}{}^2(i)}}, \tag{17}$$

where $w_{cint}, w_{csd} \in [0,1]$, $w_{cint} + w_{csd} = 1$. When substituting Equation (5) into Equation (4), we obtain the corresponding tangent similarity score:

$$ls_{\tan}(i) = w_{tint}\tan\left[\frac{\pi}{12}\left(\left|I_{int}(i)\right| + 2\left|F_{int}(i)\right|\right)\right] + w_{tsd}\tan\left[\frac{\pi}{12}\left(\left|I_{sd}(i)\right| + 2\left|F_{sd}(i)\right|\right)\right], \tag{18}$$

where $w_{tint}, w_{tsd} \in [0,1]$, $w_{tint} + w_{tsd} = 1$.

By employing the similarity score, we finally give the definition of the neutrosophic set-based superpixel filter:

$$\mathbf{H}_i = \begin{cases} 1 & ls(i) > \gamma \\ 0 & else \end{cases}, \tag{19}$$

where $\mathbf{H}_i$ is the filter response for the superpixel s_i, γ is a threshold parameter, and $ls(i)$ is the cosine or tangent similarity score calculated by Equation (17) or Equation (18).

The results of the superpixel filter are applied for estimating the neutrosophic set-based objectness, then the definition of the filtered objectness measure is given by:

$$nss(O_W, \mathbf{T}_{sel}) = 1 - \sum_{s_i \in \mathbf{S}(\mathbf{T}_{sel})} \frac{\min(|s_i \backslash O_W|, |s_i \cap O_W|)}{|O_W|}\mathbf{H}_i, \tag{20}$$

where $\mathbf{S}(\mathbf{T}_{sel})$ is the superpixel set obtained by using the selected segmentation parameter tuple $\mathbf{T}_{sel}$ and O_W is a rectangular window with the same size as the object.

3.5. Object Localization

For the task of visual tracking, the precision of the object location given by the tracker is very important for the following tracking procedure. An inaccurate location may disturb the updating of the classifier, and the tracking result may drift from the real object location in the following frames, even leading to failure. To improve the robustness of the object localization, both the Neut-Objectness confidence map and the classification confidence map were employed in this work.

The classification confidence map is calculated by using the afore-trained classifier $\mathbf{H}_k$. All the windows whose center is located within the circle area for searching are employed as candidates, suppose sr denotes the searching radius. The scale of each window is the same as the tracked object. The $\mathbf{H}_k$ response of each candidate window finally forms the classification map.

The neutrosophic-based objectness measure is utilized to modify the location l, which fully depends on $\mathbf{H}_k$. We calculated the Neut-Objectness confidence map by using the similar manner of the calculation of the classification confidence map, but the response of the filtered objectness measure is employed instead of the $\mathbf{H}_k$ response. Let l_{nss} denote the center location of the candidate window with a maximum value within the Neut-Objectness confidence map, and the corresponding response is $nss(O_w(l_{nss}), \mathbf{T}_{sel})$. The fused object location is calculated by

$$l^* = \begin{cases} \lambda l + (1-\lambda) l_{nss} & if \;\; nss(O_w(l_{nss}), \mathbf{T}_{sel}) > \tau_1, \;\; and \;\; \mathbf{H}_k(l_{nss}) > \tau_2 \\ l & else \end{cases}, \tag{21}$$

where l denotes the location, corresponding to the center of the candidate window with the maximum value within the classification confidence map, $\mathbf{H}_k(l_{nss})$ is the response of $\mathbf{H}_k$ for the window centered at l_{nss}, λ is the ratio parameter, and τ_1 and τ_2 are threshold parameters for $\lambda,\tau_1,\tau_2\in[0,1]$.

As seen in Equation (21), we modify the location calculated by the afore-trained classifier only when the maximum of the Neut-Objectness confidence map and the $\mathbf{H}_k$ response at the corresponding location achieves a relative high value. Such a method can effectively remove the interference, which may be caused by an objectness estimation that is not stable enough.

3.6. Weighted Multiple Instance Learning

In this work, we considered the importance of the instances in the positive bag $\mathbf{X}^+$ during the learning process. The weight of the j-th positive instance in $\mathbf{X}^+$ is obtained by using the filtered objectness measure:

$$w_j = \begin{cases} nss\left(O_{Wj}, \mathbf{T}_{sel}\right) & if \quad nss(O_w(l_{nss}), \mathbf{T}_{sel}) > \tau_1 \\ 1 & else \end{cases}, \tag{22}$$

where O_{Wj} is the window corresponding to the j-th instance in $\mathbf{X}^+$. Then, the positive bag probability is computed by [15,17]:

$$p\left(y = 1 \middle| \mathbf{X}^+\right) = \sum_{j=1}^{M} w_j p\left(y_1 = 1 \middle| x_{1j}\right), \tag{23}$$

where $p\left(y_1 = 1 \middle| x_{1j}\right)$ is the posterior probability for the positive sample x_{1j}, and x_{1j} denotes the j-th instance in the positive bag.

Comparing the weight methods employed in WMIL [15] and ONMIL [17], we used the neutrosophic set-based objectness measure to calculate the weight of each positive instance. In the WMIL tracker, the weight is computed mainly based on the Euclidean distance between the center of the instance and the estimated object location. When the tracking result drifts from the real object location, those real positive instances will be assigned as a relative low weight because of the long distance, which is contrary to the fact. For the ONMIL tracker, the traditional objectness estimation with superpixel straddling [20] is directly employed as the weight of each instance. As we have discussed above, the traditional objectness estimation is highly relevant to the segmentation results. When the scene changes during the tracking process, a weak objectness measure may be obtained if an inappropriate set of superpixels is used. In addition, for the task of visual tracking, some superpixels may also disturb the objectness response of the tracked object area. In our method, we first employed an online selection method of segmentation parameter tuples to produce a well-built result of a superpixel set. Secondly, we proposed a neutrosophic set-based superpixel filter to enhance the objectness estimation for the tracking application. Finally, when calculating the weight for each instance, the filtered objectness measurements were utilized for weighting when the corresponding response results were robust enough.

The posterior probability of labeling x_{ij} to be positive is defined as [13]:

$$p\left(y = 1 \middle| x_{ij}\right) = \sigma\left(\mathbf{H}_k\left(x_{ij}\right)\right), \tag{24}$$

where $i\in\{0,1\}$, as has been mentioned above, x_{1j} denotes the j-th instance in the positive bag, x_{0j} denotes the j-th instance in the negative bag, and σ is the sigmoid function, $\sigma(x) = 1/1 + e^{-x}$.

The strong classifier $\mathbf{H}_k$ in Equation (24) is defined as:

$$\mathbf{H}_k\left(x_{ij}\right) = \ln\left(\frac{p\left(\mathbf{f}(x_{ij})\middle|y = 1\right)p(y = 1)}{p\left(\mathbf{f}(x_{ij})\middle|y = 0\right)p(y = 0)}\right) = \sum_{k=1}^{K} h_k\left(x_{ij}\right), \tag{25}$$

where $f(x_{ij})$ is a set of Haar-like features corresponding to the weak classifier $h_k(x_{ij})$, $f(x_{ij}) = (f_1(x), \ldots, f_K(x))^{\mathrm{T}}$. We assume the features in $f(x_{ij})$ are independent and assume uniform prior p(y = 0) = p(y = 1) as MIL tracker [13]. Then, the $h_k(x_{ij})$ is described as [13]:

$$h_k\left(x_{ij}\right) = \ln\left(\frac{p\left(f_k(x_{ij})\middle|y=1\right)}{p\left(f_k(x_{ij})\middle|y=0\right)}\right). \tag{26}$$

The conditional distribution in $h_k(\cdot)$ is also defined as a Gaussian function as the MIL tracker, that is:

$$\begin{aligned} p\left(f_k(x_{ij})\middle|y=1\right) &\sim N(\mu_1, \sigma_1) \\ p\left(f_k(x_{ij})\middle|y=0\right) &\sim N(\mu_0, \sigma_0) \end{aligned} \tag{27}$$

Like the WMIL tracker [15], the parameters μ_1 and σ_1 are updated as follows:

$$\begin{aligned} &\mu_1 \leftarrow \eta\mu_1 + (1-\eta)\overline{\mu} \\ &\sigma_1 \leftarrow \sqrt{\eta(\sigma_1)^2 + (1-\eta)\frac{1}{M}\sum_{j=0|y_i=1}^{M-1}\left(f_k(x_{ij}) - \overline{\mu}\right)^2 + \eta(1-\eta)(\mu_1 - \overline{\mu})^2} \end{aligned} \tag{28}$$

where η is the learning rate, M is the number of positive samples, and $\overline{\mu} = \frac{1}{M}\sum_{j=0|y_i=1}^{M-1} f_k(x_{ij})$ is the average of the k-th feature extracted from the M positive samples. The parameters μ_0 and σ_0 are updated by employing the same rules.

Similar to the WMIL tracker, the bag log-likelihood function is defined as [15]:

$$\updownarrow(\mathbf{H}) = \sum_{i=0}^{1}\left(y_i \log\left(\sum_{j=0}^{M-1} w_j p\left(y=1\middle|x_{1j}\right)\right) + (1-y_i)\log\left(\sum_{j=M}^{M+L-1}\left(1 - p\left(y=1\middle|x_{0j}\right)\right)\right)\right), \tag{29}$$

where L is the number of negative samples, w_j is the weight of the j-th positive instance defined in Equation (22), y_i is the label of the training bag, y_i equals to 1 when the bag is positive, and y_i is set as 0 when the bag is negative.

As the method utilized in MIL [13], WMIL [15], and ONMIL trackers [17], our tracker maintains W weak classifiers in the pool $\Phi = \{h_1, h_2, \ldots, h_W\}$. At each frame, the K weak classifier with strong classification ability is selected to form $\mathbf{H}_k$. In the WMIL tracker, a more efficient criterion was proposed. Similar rules are employed here. The scheme for selecting K weak classifier is given below [15].

$$h_k = \underset{h\in\Phi}{\operatorname{argmax}}\langle h, \nabla\updownarrow(\mathbf{H})\rangle\Big|_{\mathbf{H}=\mathbf{H}_{k-1}}, \tag{30}$$

where [15]:

$$\nabla\updownarrow(\mathbf{H})(x_{ij}) = y_i\frac{w_j\sigma(\mathbf{H}(x_{ij}))(1-\sigma(\mathbf{H}(x_{ij})))}{\sum_{m=0}^{M-1} w_m\sigma(\mathbf{H}(x_{im}))} - (1-y_i)\frac{\sigma(\mathbf{H}(x_{ij}))(1-\sigma(\mathbf{H}(x_{ij})))}{\sum_{m=M}^{M+L-1}(1-\sigma(\mathbf{H}(x_{im})))}, \tag{31}$$

where w_j and w_m are the weights of the corresponding samples, and they are calculated by Equation (22).

Finally, the main steps of the proposed NeutWMIL tracker are shown in Algorithm 1.

Algorithm 1 Online neutrosophic similarity-based objectness tracking with weighted multiple instance learning algorithm (NeutWMIL)

Initialization:
(1) Initialize the region of the tracked object in the first frame.
(2) Initialize the MIL-based classifier $\mathbf{H}_k$ by employing the training bags surrounding the object location.
Online tracking:
(1) Select suitable segmentation parameter tuple $\mathbf{T}_{sel}$ by utilizing the method of neutrosophic set-based segmentation parameter selection.
(2) Read a new frame of the video sequence.
(3) Calculate the superpixel set for the current frame with the selected tuple $\mathbf{T}_{sel}$.
(4) Compute the estimation of the filtered objectness surrounding the location obtained by the afore-trained classifier $\mathbf{H}_k$, and get the final object location l^* in this frame by Equation (21).
(5) Compute the neutrosophic set-based weight by Equation (22).
(6) Crop M positive instances within the circular region centering at to form a positive bag, then crop L negative instances from an annular region to form a negative bag.
(7) Update the parameters of W weak classifiers by Equation (28).
(8) Select K weak classifiers by Equation (30) and form the current classifier $\mathbf{H}_k$.
(9) Go to step (1) until the end.

4. Experiments

In this section, we compared the proposed NeutWMIL tracker with six trackers on 20 challenging video sequences. The six trackers were the NeutanWMIL, ONWMIL, WMIL [15], MIL [13], OAB [10], and SemiB[11] trackers. Specifically, the NeutWMIL and the NeutanWMIL trackers were two kinds of the proposed tracking algorithm. The only difference is that the NeutWMIL tracker employed the cosine similarity measure, and the NeutanWMIL tracker used the tangent measure when the neutrosophic similarity estimation was needed during the tracking process. For the ONWMIL tracker, we implemented it using a similar instance weighting method to that proposed in [17]. The objectness estimation was directly applied for weighting each instance for the ONWMIL tracker, and the segmentation parameters were kept constant. The instance-weighting scheme was the only difference when compared with the NeutWMIL and the NeutanWMIL trackers. The source codes of the WMIL, MIL, OAB, and SemiB trackers are all publicly available, and the default parameter settings were utilized.

4.1. Parameter Setting

For the NuetWMIL, NeutanWMIL, and ONWMIL trackers, all the parameters related to the WMIL tracker were set as they are mentioned in the publicly available source codes. For instance, we set $\alpha = 4$, the search radius sr was set to 25, and β ws set to $1.5sr$, the number of the cropped negative instances L was set to 50, the number of the weak classifiers $W = 150$, and the strong classifier $\mathbf{H}_k$ maintained $K = 15$ weak classifiers. Three segmentation parameter tuples were chosen—$\mathbf{T}_1 = \{0.4, 450, 150\}$, $\mathbf{T}_2 = \{0.5, 500, 200\}$, and $\mathbf{T}_3 = \{0.6, 550, 250\}$. When we performed the tuple selection algorithm, the parameter r defined in Equation (6) was set to 8, and C was set to 4, which means the four corners of the square window with an edge length of 17 pixels were considered for evaluating the indeterminate estimation. For the neutrosophic set-based superpixel filter, when calculating the similarity score of each superpixel by Equations (17)–(18), we set $w_{cint} = w_{csd} = w_{tint} = w_{tsd} = 0.5$, which meant each criterion was treated equally. The threshold parameter γ in Equation (19) decided how many superpixels could pass the filter. To fully use the superpixel information and to filter the noisy superpixel in the meantime, a near median value 0.4 was set to γ. For the object location step, the location estimated by the trained classifier was more robust than the filtered objectness-based result statistically. However, when a relatively high response of the filtered objectness was received, as well as the response of the classifier, the objectness results were usually robust enough, and a more accurate result could be achieved by fusing these two kinds of results. By considering this, for the parameters in Equation (21), we set $\tau_1 = 0.3$ and $\tau_2 = 0.3$, and then the fusing ratio λ is set to 0.3. Finally, all parameters were kept constant for all the experiments.

4.2. Evaluation Criteria

We used two kinds of evaluation criteria, one was center location error, and the other was the success ratio based on the overlap metric. For the center location error metric, the Euclidean distance between the estimated object location and the manually labeled ground truth was considered. The overlap score is defined as:

$$p_i = \frac{|ROI'_i \cap ROI_i|}{|ROI'_i \cup ROI_i|}, \tag{32}$$

where ROI'_i is the object area estimated by the tracker in the i-th frame, ROI_i corresponds to the ground truth. Giving a threshold u, we can say the result is correct if $p_i > u$. Suppose FNs is the number of the frames, then the success ratio is calculated by:

$$R = \sum_{i=1}^{FNs} q_i / FNs, \tag{33}$$

where:

$$q_i = \begin{cases} 1 & if \quad p_i > u \\ 0 & \quad\quad else \end{cases}. \tag{34}$$

4.3. Quantitative Analysis

Details of each tested video sequence are given in Table 1. 14,163 frames are evaluated here. The challenge degree grew with the increase in the NC value. The average center location errors for the tested trackers are shown in Table 2. We can see that the NeutWMIL tracker ran the best in 13 sequences, and it performed the second best in four sequences. The NeutanWMIL tracker performed the best in four sequences and the second best in six sequences. The results of the corresponding average overlap ratio are given in Table 3. We can see the NeutWMIL tracker had the best performance in 12 sequences and the second best in six sequences. The NeutanWMIL tracker performed the best in five sequences and the second best in eight sequences. Figure 5 shows the plot of average success of the evaluated trackers through all the sequence. We can see that the NeutWMIL tracker performed the best, and the NeutanWMIL tracker had the second best performance. We can also find that there is a big gap between the NeutWMIL tracker and the WMIL tracker, as well as the MIL tracker. Based on the last line of both Tables 2 and 3, a conclusion can also be drawn that the proposed NeutWMIL tracker performs the best, and the proposed NeutanWMIL tracker performs the second best.

Table 1. An overview of the 20 tested video sequences. (Total number of evaluated frames is 14,163)

Sequence	IV	SV	OCC	DEF	MB	FM	IPR	OPR	OV	BC	LR	FNs	NC
Freeman1		Y					Y	Y				326	3
Mountain-Bike							Y	Y		Y		228	3
Vase		Y				Y	Y					271	3
Sylvester	Y						Y	Y				1345	3
Rubik		Y	Y				Y	Y				1997	4
Gym		Y		Y			Y	Y				767	4
Football			Y				Y	Y		Y		362	4
Boy		Y			Y	Y	Y	Y				602	5
Couple		Y		Y		Y		Y		Y		140	5
BlurBody		Y		Y	Y	Y	Y					334	5
Basketball	Y	Y		Y				Y		Y		725	5
Doll	Y	Y	Y				Y	Y				3872	5
FleetFace		Y		Y	Y	Y	Y	Y				707	6
Coke	Y		Y			Y	Y	Y		Y		291	6
David	Y	Y	Y	Y	Y		Y	Y				471	7
ClifBar		Y	Y		Y	Y	Y		Y	Y		472	7
Tiger1	Y		Y	Y	Y	Y	Y	Y				354	7
Biker		Y	Y		Y	Y		Y	Y		Y	142	7
Tiger2	Y		Y	Y	Y	Y	Y	Y	Y			365	8
Soccer	Y	Y	Y		Y	Y	Y	Y		Y		392	8

Note: IV: Illumination Variation, SV: Scale Variation, OCC: Occlusion, DEF: Deformation, MB: Motion Blur, FM: Fast Motion, IPR: In-Plane Rotation, OPR: Out-of-Plane Rotation, OV: Out-of-View, BC: Background Clutters, and LR: Low Resolution, FNs: Frames, NC: Number of types of challenge.

Table 2. The average center location errors (in pixels) for the compared trackers (bold red fonts indicate the best performance, while the italic blue fonts indicate the second best ones).

Sequence	Neut-WMIL	Neutan-WMIL	ON-WMIL	WMIL	MIL	OAB	SemiB
Freeman1	**14.30**	16.70	17.80	*15.64*	17.06	66.12	54.69
MountainBike	**7.89**	15.02	29.80	120.05	*8.07*	12.96	44.39
Vase	21.97	21.53	22.62	*21.08*	**15.04**	34.58	32.05
Sylvester	**7.75**	*8.79*	8.96	18.37	17.49	12.18	22.75
Rubik	**14.49**	41.30	80.97	84.44	*22.56*	33.74	53.82
Gym	**11.47**	29.60	62.04	123.95	20.71	*15.24*	23.60
Football	12.86	*12.54*	16.79	14.38	**11.66**	171.91	96.91
Boy	7.54	*7.38*	19.63	7.88	108.24	**3.43**	56.03
Couple	*9.70*	**7.88**	35.68	35.92	34.80	33.86	102.71
BlurBody	**33.91**	36.44	*35.07*	85.45	81.99	59.44	108.71
Basketball	*10.76*	**10.22**	17.69	25.65	107.05	145.94	158.50
Doll	**28.18**	82.91	*47.29*	74.80	70.37	127.33	52.73
FleetFace	**29.66**	50.11	69.94	109.24	50.68	*44.27*	69.49
Coke	*23.54*	32.47	43.24	46.60	113.62	**17.64**	50.93
David	**18.97**	38.47	46.67	20.22	24.34	71.55	55.41
ClifBar	17.85	**9.65**	*10.50*	20.08	23.47	32.87	74.98
Tiger1	*14.52*	**13.94**	25.10	73.96	84.24	42.01	60.78
Biker	**10.30**	*11.51*	36.35	20.15	27.54	92.87	93.26
Tiger2	**17.28**	*19.19*	50.53	40.29	*21.93*	58.31	68.10
Soccer	**28.35**	88.16	57.80	101.44	51.88	99.71	92.07
average	**17.06**	*27.69*	36.72	52.98	45.64	58.80	68.59

Table 3. The average overlap ratio for the compared trackers (bold red fonts indicate the best performance, while the italic blue fonts indicate the second best ones).

Sequence	Neut-WMIL	Neutan-WMIL	ON-WMIL	WMIL	MIL	OAB	SemiB
Freeman1	**0.299**	*0.281*	0.243	*0.281*	0.260	0.201	0.170
MountainBike	**0.705**	0.607	0.554	0.380	*0.701*	*0.621*	0.258
Vase	**0.314**	*0.312*	0.308	0.307	0.312	0.275	0.236
Sylvester	**0.696**	*0.675*	0.657	0.547	0.514	0.612	0.478
Rubik	**0.553**	0.457	0.191	0.236	*0.490*	0.385	0.288
Gym	**0.474**	0.366	0.172	0.074	0.399	*0.438*	0.269
Football	0.574	*0.583*	0.412	0.508	**0.604**	0.237	0.149
Boy	0.662	*0.670*	0.375	0.591	0.296	**0.780**	0.272
Couple	*0.594*	**0.617**	0.462	0.456	0.486	0.279	0.078
BlurBody	**0.525**	*0.496*	0.506	0.298	0.281	0.375	0.193
Basketball	*0.655*	**0.667**	0.502	0.529	0.213	0.037	0.048
Doll	**0.356**	0.164	*0.275*	0.141	0.231	0.051	0.270
FleetFace	*0.568*	0.552	0.459	0.305	**0.579**	0.560	0.415
Coke	*0.521*	0.448	0.277	0.234	0.047	**0.551**	0.176
David	**0.407**	0.309	0.234	*0.395*	0.346	0.212	0.238
ClifBar	0.426	**0.508**	*0.467*	0.403	0.373	0.264	0.224
Tiger1	*0.657*	**0.671**	0.517	0.128	0.168	0.526	0.303
Biker	*0.437*	**0.445**	0.246	0.246	0.259	0.241	0.244
Tiger2	**0.610**	*0.571*	0.309	0.292	0.505	0.250	0.202
Soccer	**0.321**	0.101	0.204	0.145	*0.221*	0.103	0.101
average	**0.52**	*0.47*	0.37	0.32	0.36	0.35	0.23

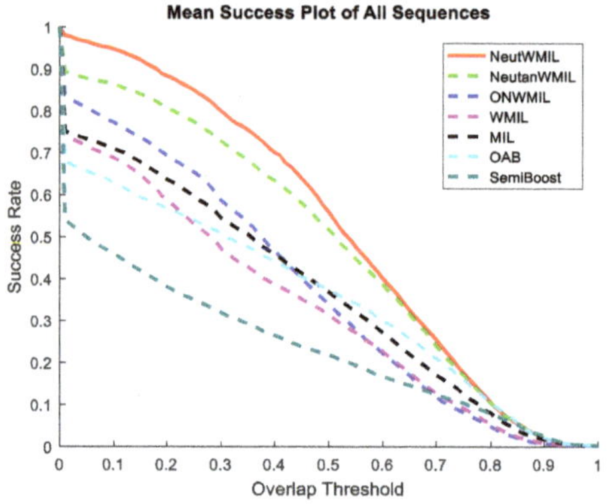

Figure 5. The plot of average success for all the tested sequences.

4.4. Qualitative Analysis

We chose thirteen representative sequences among the tested sequences for the Qualitative analysis. As seen in Tables 1–3, the names of the selected 13 sequences are shown in bold font in each table. Based on the challenge degree of these sequences, they were separated into three groups for analysis. The success and center position error plots for each sequence can be seen in Figures 6–8. Some sampled tracking results are shown in Figures 9–11.

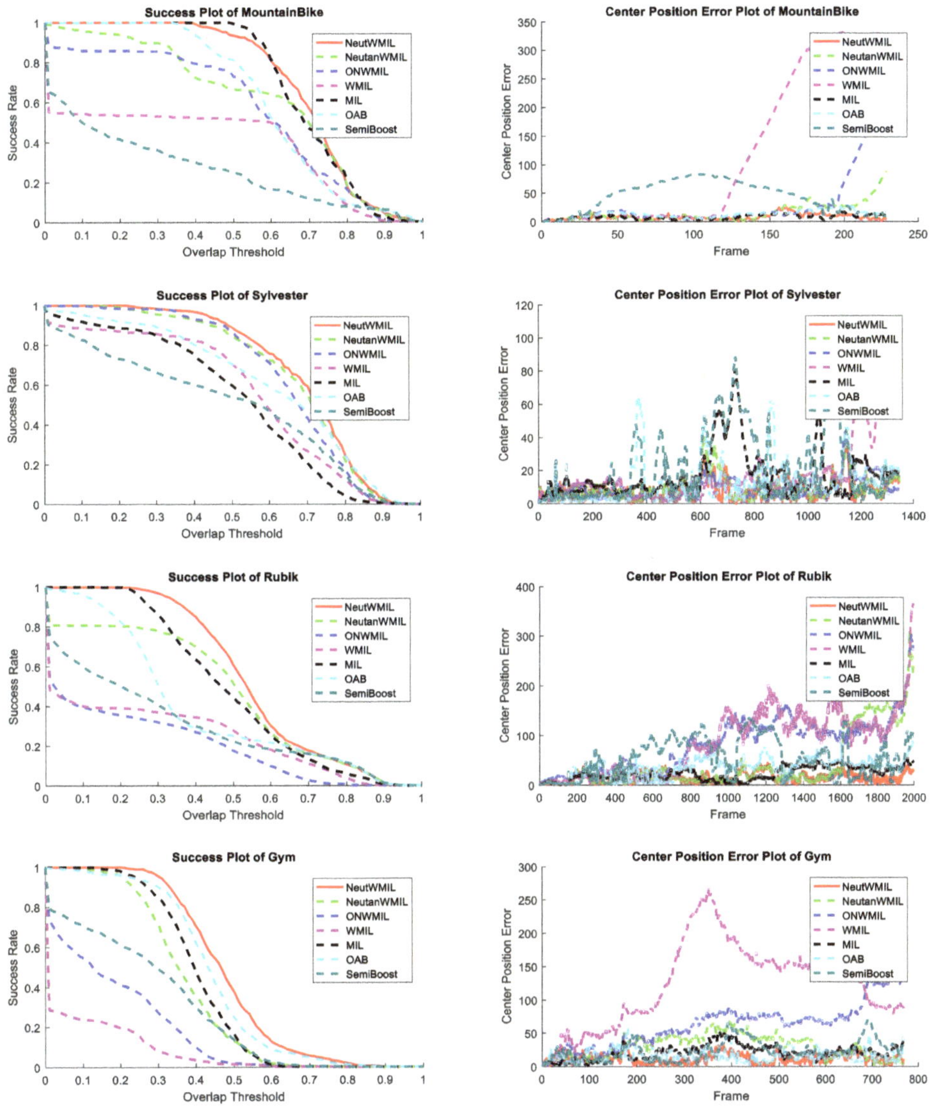

Figure 6. Success and center position error plots of tested sequences *MountainBike*, *Sylvester*, *Rubik*, and *Gym*.

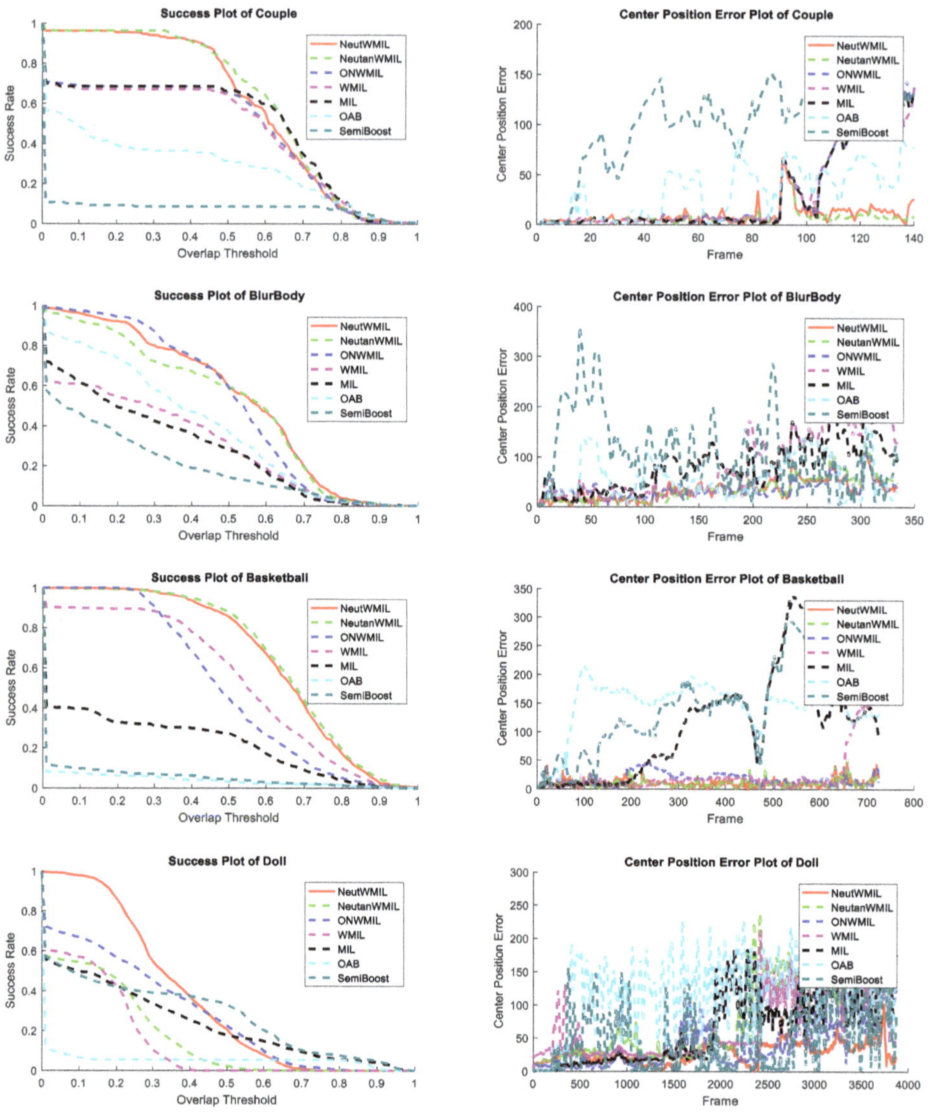

Figure 7. Success and center position error plots of tested sequences *Couple*, *BlurBody*, *Basketball*, and *Doll*.

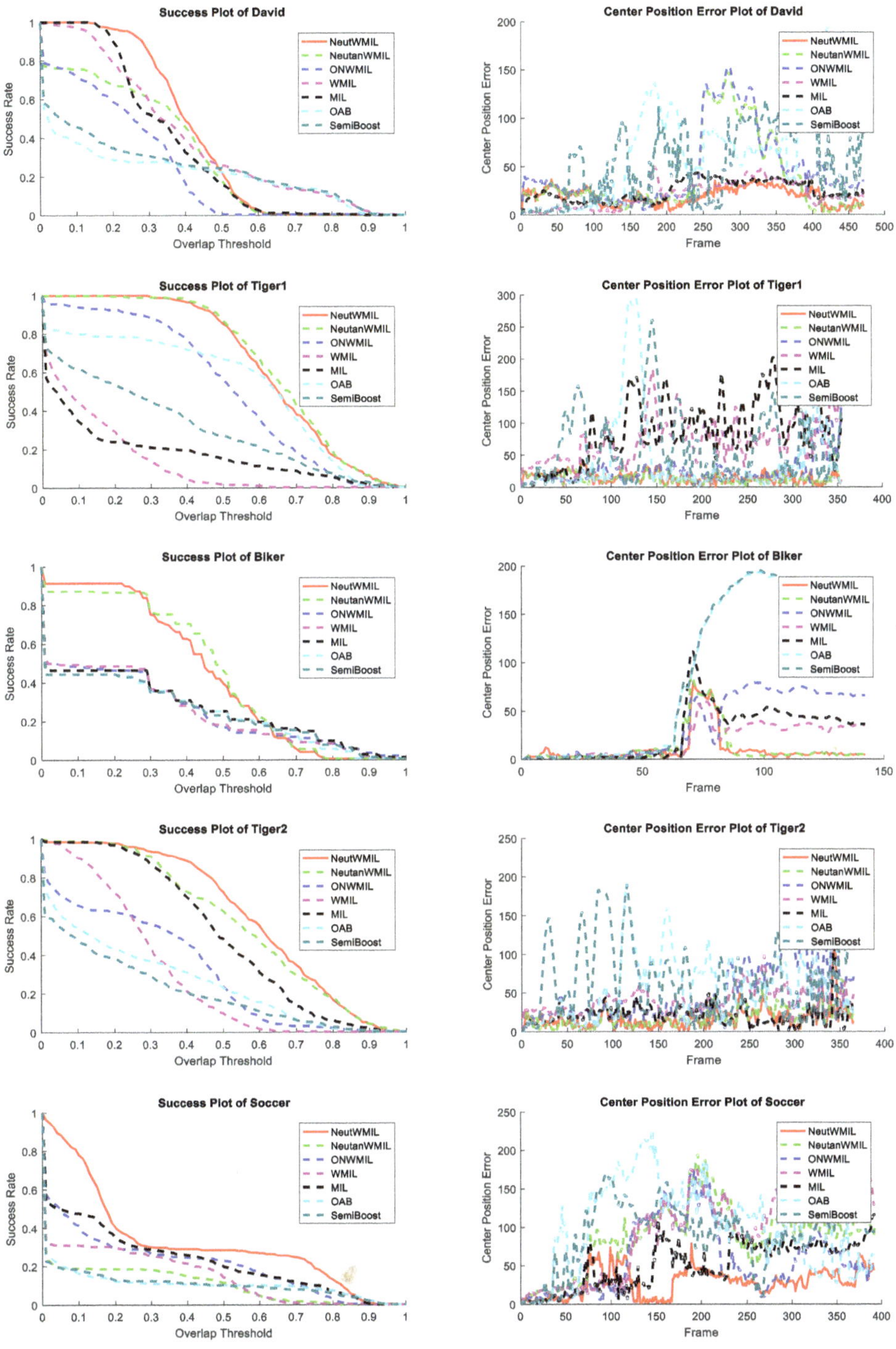

Figure 8. Success and center position error plots of tested sequences *David*, *Tiger1*, *Biker*, *Tiger2*, and *Soccer*.

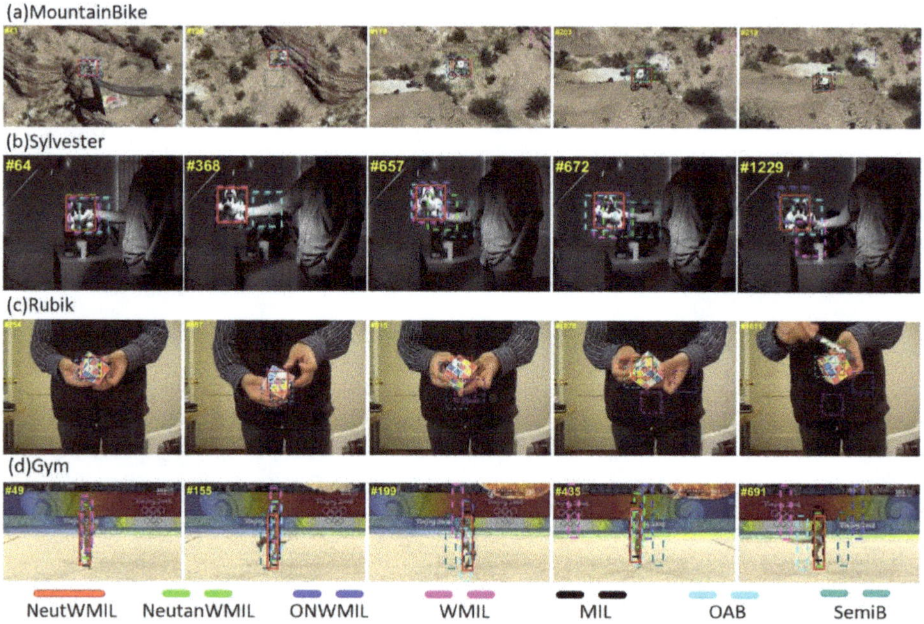

Figure 9. Sampled tracking results for tested sequences *MountainBike* (**a**), *Sylvester* (**b**), *Rubik* (**c**), and *Gym* (**d**).

Figure 10. Sampled tracking results for tested sequences *Couple* (**a**), *BlurBody* (**b**), *Basketball* (**c**), and *Doll* (**d**).

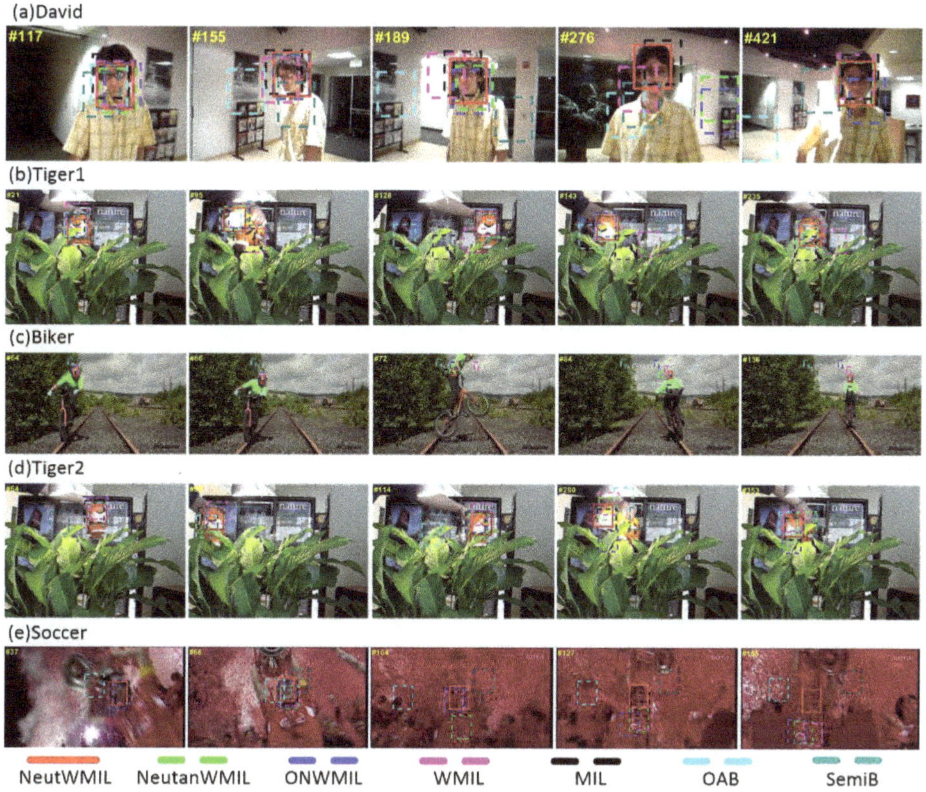

Figure 11. Sampled tracking results for tested sequences *David* (**a**), *Tiger1* (**b**), *Biker* (**c**), *Tiger2* (**d**), and *Soccer* (**e**).

4.4.1. MountainBike, Sylvester, Rubik, and Gym

As shown in Table 1, the first two sequences were with three types of challenges. The main challenges were in-plane and out-plane rotation, and there was a background clutter challenge in the *MountainBike* sequence, and illumination variation challenge in the *Sylvester* sequence. As seen in Figure 9a, the SemiB tracker yielded a severe drift problem at frames #43, #126, and #179. The WMIL tracker failed due to the background clutter, as shown by frames #126 and #179. When the biker passed by the area with more challenging background clutter, the NeutanWMIL and ONWMIL trackers produced wrong object locations, as seen in frame #219. The OAB and SemiB trackers drifted away on account of the in-plane and out-plane challenge, as shown by frames #64 and #368 in Figure 9b. Due to the rotation and illumination variation challenges, the trackers drifted from the object except the NeutWMIL tracker, as shown by frames #657 and #672. The center position errors of all the trackers at each frame are shown in Figure 6. We can see the NeutWMIL tracker outperformed all the others, the results in Tables 2 and 3 also support this conclusion. For the last two sequences, besides the scale variation, in-plane and out-plane rotation challenges, the occlusion was another challenging issue for the *Rubik* sequence. For the *Gym* sequence, there was also a deformation challenge. All the trackers drifted away due to the in-plane and out-plane rotation, as shown by frames #254, #507, #815, and #1078 in Figure 9c, except the NeutWMIL and NeutanWMIL trackers. As shown in Figure 6, the plot of center position error shows that NeutanWMIL tracker performed well until fame #1600, and the drift was mainly caused by the occlusion, as shown by frame #1611. Considering the deformation challenge, all the trackers except the NeutWMIL tracker were affected in some degree in the *Gym* sequence. Drifts can be seen in Figure 9d, the plot in Figure 6 also reveals the drift problem.

4.4.2. Couple, BlurBody, Basketball, and Doll

This group of sequences was more challenging than the previous one. The fast motion of the camera and background clutter were the most challenging problems in the *Couple* sequence. As shown in Figure 7, Table 2, and Table 3, the NeutanWMIL tracker performed the best in the *Couple* sequence. The OAB and SemiB trackers could not adaptively adjust to these challenges, and they drifted from the 'couple', as shown in Figure 10a. As seen in Figure 7, we can see that all the trackers drifted away severely near frame #90, because a very fast motion occurred to the camera. However, as shown in Figure 10a, the 'couple' was re-tracked by the NeutWMIL and NeutanWMIL trackers. For the *BlurBody* sequence, besides the fast motion, the most challenging problems were motion blur and scale variation. As shown in Figure 7, the SemiB tracker drifted away at the very beginning. The related screenshots can be seen by frames #12 and #16 in Figure 10b. As shown by frames #124 and #203, the WMIL, MIL, and OAB trackers drifted for the scale variation and motion blur challenges. From the plots shown in Figure 7, we can find that the NeutWMIL and NeutanWMIL trackers performed the best, the ONWMIL tracker also performed well to some extent. There was background clutter, out-plane rotation, deformation, and scale variation challenges in the *Basketball* sequence. The OAB and SemiB trackers failed quickly when a player appeared nearby the target, as shown by frame #39 in Figure 10c. The MIL tracker also drifted from the target player after he passed by several other players, as shown by frames #258 and #468. The WMIL tracker ran well until the player passed by several other players with similar color, as shown by frame #664. As shown in Figure 7, Table 2, and Table 3, the NeutanWMIL tracker ran the best for both the *BlurBody* and *Basketball* sequences, but the gap between the NeutanWMIL tracker and the NeutWMIL tracker is very small. Challenges like illumination variation, scale variation, occlusion, and rotation were included in the *Doll* sequence. The WMIL, OAB, and SemiB trackers drifted away quickly, as shown by frames #185, #225, and #742. The NeutanWMIL and ONWMIL trackers could not adaptively adjust to these challenges and drifted from the doll, as shown by frames #1414 and #2324. The proposed NeutWMIL tracker can yield more stable and more accurate results than the other six trackers.

4.4.3. David, Tiger1, Biker, Tiger2, and Soccer

This group had the most challenging sequences. Seven types of challenge were included in the *David* sequence. Details can be seen in Table 1. The OAB tracker was distracted by the wall because of the similar color, the SemiB tracker drifted from the target mainly due to the challenge of illumination variation, as shown by frames #117 and #155 in Figure 11a. The NeutanWMIL and ONWMIL trackers were also distracted by the wall because of the illumination change and the scale variation, as well as the similar texture between the painting and the target. As seen in Figure 7, Figure 11a, Table 2, and Table 3, the NeutWMIL tracker produced a more robust estimation of object location. For the sequences of *Tiger1* and *Tiger2*, seven types of challenge were included in the *Tiger1* sequence, and *Tiger2* consisted of eight types of challenge. The SemiB, OAB, WMIL, MIL, and ONWMIL trackers could not adjust to these challenges well, as shown by frames #21, #95, #128, #143, and #235 in Figure 11b, as well as frames #64, #99, #280, and #353 shown in Figure 11d. The performance gap between the NeutanWMIL tracker and the NeutWMIL tracker was very small in the sequences of *Tiger1* and *Tiger2*, as shown in Figure 7, Table 2, and Table 3. The *Biker* sequence mainly contained challenges like fast motion, scale variation, out-plane rotation, out of view, and low resolution. The OAB and SemiB trackers drifted from the target mainly due to the challenges of scale variation and head rotation, as shown by frames #64 and #66 in Figure 11c. All the trackers failed nearby frame #70, because of a very fast move from the target, as shown by frame #72 in Figure 11c. However, the NeutanWMIL and NeutWMIL trackers snapped to the target again when the target was located within the searching area, as shown by frames #84 and #136. The *Soccer* sequence had the most challenging problems among the tested sequences. As shown by frame #37 in Figure 11e, the OAB and SemiB trackers drifted away because of the challenges of motion blur and rotation. As seen in Figure 8, we can find the estimation results of the NeutWMIL and NeutanWMIL trackers were not stable between the frame #70 and #120. This is mainly due to the severe occlusion during such a period. The ONWMIL, WMIL, and MIL trackers drifted away after such a long-term occlusion. When the target appeared, the NeutWMIL snapped to the target again, and performed the best in the following frames, as shown by frames #127 and #155.

4.5. Discussion

With the above analysis, we can find that the proposed NeutWMIL tracker had the best performance. This was mainly due to three contributions of this work. First, by employing the proposed object location objectness enhancing criterion, a more appropriate tuple of segmentation parameter was selected, and then a more suitable superpixel set for measuring the objectness was produced, which helped the tracker to adjust for different scenes during the tracking process. Secondly, the intersection and shape–distance criteria were proposed for constructing the superpixel filter, which helped the tracker filter out those superpixels which may disturb the tracker, and more reliable sample weights were produced for updating the classifier. Thirdly, the object location was finally decided by fusing the information of the Neut-Objectness confidence map and the classification confidence map, which could also enhance the robustness of the tracker.

As shown in Figure 5, Table 2, and Table 3, we can see that there is a big gap between the proposed NeutWMIL and the WMIL tracker. The main reason for this is the NeutWMIL tracker utilized the neutrosophic set-based objectness weighting algorithm. Weights that are more robust can be produced when there is a small drift from the real object location. The comparison result also shows that the NeutWMIL tracker performed better than the NeutanWMIL tracker. As we have mentioned, the only difference between these two trackers is the usage of different neutrosophic similarity measures. The cosine similarity measure is applied for the NeutWMIL tracker, and the tangent similarity measure is employed by the NeutanWMIL tracker. Regarding Equations (8) and (9) and Equations (17) and (18), it can be found that the truth element is enhanced for the cosine measure, and the I and F elements are only applied for decreasing the similarity response. Such a scheme seems more suitable for the judging application in this work. However, all the T, I, and F elements are treated equally for the tangent measure. As shown in Figure 5, Table 2, and Table 3, a relatively big gap exists between the NeutWMIL tracker and the ONWMIL tracker. This result reveals that the proposed weighting method contributed much more to the tracker, rather than using the objectness estimation directly.

5. Conclusions

In this paper, we presented a novel algorithm of neutrosophic similarity-based objectness tracking via weighted multiple instance learning. This is the first time the NS theory has been introduced into the visual object tracking of a discriminative model. To produce more reliable sample weights, object location objectness enhancing criterion was first proposed for the selection of segmentation parameter tuples. Then, the intersection and the shape–distance criteria were proposed for filtering out the unreliable superpixels. Three membership functions, T, I, and F, for each criterion were given by considering uncertain issues. Finally, we proposed a method of filtered objectness measure and also utilized such a neutrosophic set-based measure to modify the location calculated by the afore-trained classifier. Experimental results on challenging video sequences demonstrated the superiority of the proposed tracker to state-of-the-art discriminative trackers in accuracy and stability. Moreover, our parameter selection technique can be extended to any other segmentation algorithm if its performance can be affected by some parameters. Our sample weighting scheme can be also extended to other discriminative trackers to easily help them update the classifier using robust sample weights.

Author Contributions: Conceptualization, investigation and writing, K.H.; review and editing, W.H., J.Y., L.Z., H.P., and J.P.

Funding: This work was supported in part by the National Natural Science Foundation of China under Grant 61603258, Grant 61703280, and Grant 61662025, and in part by the Natural Science Foundation of Zhejiang Province under Grant LY19F020015.

Conflicts of Interest: The authors declare no conflict of interest.

References

1. Yilmaz, A.; Javed, O.; Shah, M. Object tracking: A survey. *ACM Comput. Surv.* **2006**, *38*, 13. [CrossRef]
2. Yang, H.; Shao, L.; Zheng, F.; Wang, L.; Song, Z. Recent advances and trends in visual tracking: A review. *Neurocomputing* **2011**, *74*, 3823–3831. [CrossRef]
3. Comaniciu, D.; Ramesh, V.; Meer, P. Kernel-based object tracking. *IEEE Trans. Pattern Anal. Mach. Intell.* **2003**, *25*, 564–577. [CrossRef]
4. Ross, D.; Lim, J.; Lin, R.-S.; Yang, M.-H. Incremental learning for robust visual tracking. *Int. J. Comput. Vis.* **2008**, *77*, 125–141. [CrossRef]
5. Leichter, I. Mean shift trackers with cross-bin metrics. *IEEE Trans. Pattern Anal. Mach. Intell.* **2011**, *34*, 695–706. [CrossRef] [PubMed]
6. Vojir, T.; Noskova, J.; Matas, J. Robust scale-adaptive mean-shift for tracking. *Pattern Recogn. Lett.* **2014**, *49*, 250–258. [CrossRef]
7. Hu, K.; Ye, J.; Fan, E.; Shen, S.; Huang, L.; Pi, J. A novel object tracking algorithm by fusing color and depth information based on single valued neutrosophic cross-entropy. *J. Intell. Fuzzy Syst.* **2017**, *32*, 1775–1786. [CrossRef]
8. Hu, K.; Fan, E.; Ye, J.; Fan, C.; Shen, S.; Gu, Y. Neutrosophic similarity score based weighted histogram for robust mean-shift tracking. *Information* **2017**, *8*, 122. [CrossRef]
9. Hu, K.; Fan, E.; Ye, J.; Pi, J.; Zhao, L.; Shen, S. Element-weighted neutrosophic correlation coefficient and its application in improving camshift tracker in rgbd video. *Information* **2018**, *9*, 126. [CrossRef]
10. Grabner, H.; Grabner, M.; Bischof, H. *Real-Time Tracking Via On-Line Boosting*; Chantler, M., Fisher, B., Trucco, M., Eds.; BMVA Press: Graz, Austria, 2006; pp. 6.1–6.10.
11. Grabner, H.; Leistner, C.; Bischof, H. *Semi-Supervised On-Line Boosting for Robust Tracking*; European Conference on Computer Vision (ECCV); Forsyth, D., Torr, P., Zisserman, A., Eds.; Springer Berlin Heidelberg: Marseille, France, 2008; pp. 234–247.
12. Babenko, B.; Ming-Hsuan, Y.; Belongie, S. Visual tracking with online multiple instance learning. In Proceedings of the IEEE Conference on Computer Vision Pattern Recognition (CVPR) 2009, Miami, FL, USA, 20–25 June 2009; pp. 983–990.
13. Babenko, B.; Ming-Hsuan, Y.; Belongie, S. Robust object tracking with online multiple instance learning. *IEEE Trans. Pattern Anal. Mach. Intell.* **2011**, *33*, 1619–1632. [CrossRef]
14. Kaihua, Z.; Lei, Z.; Ming-Hsuan, Y. Real-time object tracking via online discriminative feature selection. *IEEE Trans. Image Process.* **2013**, *22*, 4664–4677.
15. Zhang, K.; Song, H. Real-time visual tracking via online weighted multiple instance learning. *Pattern Recogn.* **2013**, *46*, 397–411. [CrossRef]
16. Abdechiri, M.; Faez, K.; Amindavar, H. Visual object tracking with online weighted chaotic multiple instance learning. *Neurocomputing* **2017**, *247*, 16–30. [CrossRef]
17. Yang, H.; Qu, S.; Zhu, F.; Zheng, Z. Robust objectness tracking with weighted multiple instance learning algorithm. *Neurocomputing* **2018**, *288*, 43–53. [CrossRef]
18. Hu, K.; Zhang, X.; Gu, Y.; Wang, Y. Fusing target information from multiple views for robust visual tracking. *IET Comput. Vis.* **2014**, *8*, 86–97. [CrossRef]
19. Alexe, B.; Deselaers, T.; Ferrari, V. What is an object? In Proceedings of the 2010 IEEE Computer Society Conference on Computer Vision and Pattern Recognition, San Francisco, CA, USA, 13–18 June 2010; pp. 73–80.
20. Alexe, B.; Deselaers, T.; Ferrari, V. Measuring the objectness of image windows. *IEEE Trans. Pattern Anal. Mach. Intell.* **2012**, *34*, 2189–2202. [CrossRef] [PubMed]
21. Felzenszwalb, P.F.; Huttenlocher, D.P. Efficient graph-based image segmentation. *Int. J. Comput. Vision* **2004**, *59*, 167–181. [CrossRef]
22. Smarandache, F. *Neutrosophy: Neutrosophic Probability, Set and Logic*; American Research Press: Rehoboth, MA, USA, 1998; p. 105.

23. Peng, X.; Dai, J. A bibliometric analysis of neutrosophic set: Two decades review from 1998 to 2017. *Artif. Intell. Rev.* **2018**, *52*, 1–57. [CrossRef]
24. Ye, J.; Fu, J. Multi-period medical diagnosis method using a single valued neutrosophic similarity measure based on tangent function. *Comput. Methods Programs Biomed.* **2016**, *123*, 142–149. [CrossRef]
25. Guo, Y.; Sengur, A. A novel 3d skeleton algorithm based on neutrosophic cost function. *Appl. Soft Comput. J.* **2015**, *36*, 210–217. [CrossRef]
26. Guo, Y.; Şengür, A.; Ye, J. A novel image thresholding algorithm based on neutrosophic similarity score. *Meas. J. Int. Meas. Confed.* **2014**, *58*, 175–186. [CrossRef]
27. Guo, Y.; Şengür, A.; Akbulut, Y.; Shipley, A. An effective color image segmentation approach using neutrosophic adaptive mean shift clustering. *Meas. J. Int. Meas. Confed.* **2018**, *119*, 28–40. [CrossRef]
28. Guo, Y.; Xia, R.; Şengür, A.; Polat, K. A novel image segmentation approach based on neutrosophic c-means clustering and indeterminacy filtering. *Neural Comput. Appl.* **2017**, *28*, 3009–3019. [CrossRef]
29. Rashno, A.; Koozekanani, D.D.; Drayna, P.M.; Nazari, B.; Sadri, S.; Rabbani, H.; Parhi, K.K. Fully automated segmentation of fluid/cyst regions in optical coherence tomography images with diabetic macular edema using neutrosophic sets and graph algorithms. *IEEE Trans. Biomed. Eng.* **2018**, *65*, 989–1001. [CrossRef] [PubMed]
30. Ashour, A.S.; Guo, Y.; Kucukkulahli, E.; Erdogmus, P.; Polat, K. A hybrid dermoscopy images segmentation approach based on neutrosophic clustering and histogram estimation. *Appl. Soft Comput.* **2018**, *69*, 426–434. [CrossRef]
31. Guo, Y.; Şengür, A. A novel image segmentation algorithm based on neutrosophic similarity clustering. *Appl. Soft Comput. J.* **2014**, *25*, 391–398. [CrossRef]
32. Fan, E.; Xie, W.; Pei, J.; Hu, K.; Li, X. Neutrosophic hough transform-based track initiation method for multiple target tracking. *IEEE Access* **2018**, *6*, 16068–16080. [CrossRef]
33. Wang, H.; Smarandache, F.; Zhang, Y.; Sunderraman, R. Single valued neutrosophic sets. *Multispace Multistructure* **2010**, *4*, 410–413.
34. Ye, J. Single valued neutrosophic cross-entropy for multicriteria decision making problems. *Appl. Math. Model.* **2014**, *38*, 1170–1175. [CrossRef]
35. Ye, J. Vector similarity measures of simplified neutrosophic sets and their application in multicriteria decision making. *Int. J. Fuzzy Syst.* **2014**, *16*, 204–211.

symmetry

MDPI

Article

Neutrosophic Triangular Norms and Their Derived Residuated Lattices

Qingqing Hu and Xiaohong Zhang *

School of Arts and Sciences, Shaanxi University of Science & Technology, Xi'an 710021, China; huqingqing122@163.com

* Correspondence: zhangxiaohong@sust.edu.cn or zxhonghz@263.net

Received: 21 May 2019; Accepted: 14 June 2019; Published: 20 June 2019

Abstract: Neutrosophic triangular norms (t-norms) and their residuated lattices are not only the main research object of neutrosophic set theory, but also the core content of neutrosophic logic. Neutrosophic implications are important operators of neutrosophic logic. Neutrosophic residual implications based on neutrosophic t-norms can be applied to the fields of neutrosophic inference and neutrosophic control. In this paper, neutrosophic t-norms, neutrosophic residual implications, and the residuated lattices derived from neutrosophic t-norms are investigated deeply. First of all, the lattice and its corresponding system are proved to be a complete lattice and a De Morgan algebra, respectively. Second, the notions of neutrosophic t-norms are introduced on the complete lattice discussed earlier. The basic concepts and typical examples of representable and non-representable neutrosophic t-norms are obtained. Naturally, De Morgan neutrosophic triples are defined for the duality of neutrosophic t-norms and neutrosophic t-conorms with respect to neutrosophic negators. Third, neutrosophic residual implications generated from neutrosophic t-norms and their basic properties are investigated. Furthermore, residual neutrosophic t-norms are proved to be infinitely $\vee$-distributive, and then some important properties possessed by neutrosophic residual implications are given. Finally, a method for producing neutrosophic t-norms from neutrosophic implications is presented, and the residuated lattices are constructed on the basis of neutrosophic t-norms and neutrosophic residual implications.

Keywords: neutrosophic sets; neutrosophic triangular norms; residuated lattices; representable neutrosophic t-norms; De Morgan neutrosophic triples; neutrosophic residual implications; infinitely $\vee$-distributive

1. Introduction

Neutrosophic sets were firstly proposed by Smarandache [1] from a philosophical point of view in 1998, which is a generalization of fuzzy sets and intuitionistic fuzzy sets. However, it is difficult to apply neutrosophic sets to solve practical problems since the values of their three functions with respect to truth, indeterminacy and falsity lie in $]0-,1+[$. The definition of single-valued neutrosophic sets were introduced by Wang [2], whose values belong to [0,1]. With the development of neutrosophic set theory, single-valued neutrosophic sets and their applications have been investigated by more scholars. Single-valued neutrosophic sets were successfully applied to various decision making problems [3–8]. In addition, Zhang et al. studied the neutrosophic logic algebras and discussed neutrosophic filters and neutrosophic triplet groups, which are the important foundation of the development of neutrosophic logic theory [9–12]. To facilitate research, "single-valued neutrosophic sets" are abbreviated as "neutrosophic sets" in this paper. For neutrosophic sets, the truth-membership, indeterminacy-membership and falsity-membership are not restricted to each other, which is different from intuitionistic fuzzy sets. Picture fuzzy sets proposed by Cuong [13] in 2013 is a direct generalization of intuitionistic fuzzy sets, because their positive membership, neutral membership and negative membership are not independent completely. It is worth noting that picture fuzzy sets can be regarded as special neutrosophic sets [14,15], and also can be called standard neutrosophic sets [1,2,16–19].

Fuzzy logic plays a vital role in fuzzy set theory. T-norms, t-conorms, negators and implications are very important fuzzy logic operators. T-norms were originally defined by Menger [20], and then Schweizer and Sklar [21,22] redefined the t-norms which have been used to today. From the perspective of fuzzy logic, t-norms are the extension of intersection operation of fuzzy sets [23]. The algebraic properties of t-norms, for example, continuity, archimedean, strict, nilpotent and so on, are discussed in some papers [24–28]. Hu et al. studied t-norm extension operations [29]. Wang et al. discussed the lattice structure of algebra of fuzzy values [30]. The t-norms which satisfy the residual principle are an important class of t-norms, because they can produce fuzzy implications and constitute the residuated lattices [31–34]. Type-2 t-norms (t-conorms) and their residual operators on type-2 fuzzy sets were investigated by Li [25], which promote the development of fuzzy reference system. Intuitionistic fuzzy t-norms on intuitionistic fuzzy sets (L^*-fuzzy sets) were proposed by Deschrijver et al., they discussed t-representable intuitionistic fuzzy t-norms and their residual operators [35,36]. Picture fuzzy sets are particular L-fuzzy sets [37]. Picture fuzzy t-norms on picture fuzzy sets were introduced in [17,38,39], some basic picture fuzzy logic connectives and their properties for picture fuzzy sets are investigated in [40,41]. Some classes of representable picture fuzzy t-norms and representable picture fuzzy t-conorms on picture fuzzy sets and De Morgan picture operator triples in picture fuzzy logic are discussed [42]. Furthermore, a picture inference system is proposed by Son [43]. The residual operations, residual implications of uninorms were discussed by Baets [44]. Wang proposed the notions of residual implications (co-implications) of pseudo t-norms, left and right uninorms and studied some properties of infinitely $\vee$-distributive ($\wedge$-distributive) pseudo t-norms, left and right uninorms [45–47]. Then Liu introduced semi-uninorms and their residual implications [48].

Neutrosophic t-norms, neutrosophic t-conorms, neutrosophic negators and neutrosophic implications are important neutrosophic logic operators for neutrosophic sets. It is a very meaningful topic to discuss neutrosophic t-norms and their residual implications on neutrosophic sets. In the last few years, although Alkhazaleh discusses some neutrosophic t-norms and t-conorms in [49], Liu proposes aggregation operators based on Archimedean t-norms and t-conorms for neutrosophic numbers in [5], Smarandache discussed neutrosophic norms (n-norms), n-valued refined neutrosophic logic and its applications in physics [50,51], there are a few papers about basic neutrosophic logic connectives and their properties and neutrosophic logic inference systems and their applications in the field of control. Therefore, it is necessary to study neutrosophic logic operators and their properties, especially the application of neutrosophic residual implications in neutrosophic inference and neutrosophic control.

To achieve these goals, the definitions of neutrosophic t-norms should be given firstly. We can study neutrosophic logic and neutrosophic inference systems further only if neutrosophic t-norms and their residual implications are studied thoroughly. Thus, it is the main task of this paper to study neutrosophic t-norms and their residual implications. Section 2 presents some basic notions. In Section 3, the lattice structure of neutrosophic sets is analyzed and constructed systematically based on the first type inclusion relation on neutrosophic sets. In particular, we combine some basic algebraic operations: Union, intersection and complement and their related properties to prove that the system $(D^*; \vee_1, \wedge_1, ^c, 0_{D^*}, 1_{D^*})$ is a De Morgan algebra. In Section 4, we introduce neutrosophic t-norms (t-conorms), representable neutrosophic t-norms (t-conorms) and De Morgan neutrosophic triples. In addition, we present some important theorems and typical examples. In Section 5, the definitions of neutrosophic residual implications (co-implications) are obtained and their basic properties are discussed deeply. Moreover, residual neutrosophic t-norms (t-conorms) are proved to be infinitely $\vee$-distributive ($\wedge$-distributive), and then some important results related to residual neutrosophic t-norms and neutrosophic residual implications are given. Section 6 shows a method for obtaining neutrosophic t-norms from neutrosophic implications, and then proves that the system $(D^*; \vee_1, \wedge_1, \otimes, \rightarrow, 0_{D^*}, 1_{D^*})$ is a residuated lattice. In Section 7, we conclude the paper.

2. Preliminaries

Some basic concepts in fuzzy set theory will be reviewed in this section.

Definition 1. *([52]) Let U be a nonempty set. An intuitionistic fuzzy set M in U is characterized by a membership function $\mu_M(u)$ and a non-membership function $\nu_M(u)$. Then, an intuitionistic fuzzy set M can be denoted by*

$$M = \{(u, \mu_M(u), \nu_M(u)) \mid u \in U\},$$

where $\mu_M(u)$: $U \rightarrow [0,1]$ and $\nu_M(u)$: $U \rightarrow [0,1]$ with the condition $0 \leq \mu_M(u) + \nu_M(u) \leq 1$, for all $u \in U$. Here $\mu_M(u), \nu_M(u) \in [0,1]$ denote the membership and the non-membership functions of the intuitionistic fuzzy set M, respectively.

Definition 2. *([1]) Let U be a nonempty set. A neutrosophic set M in U is characterized by three functions: Truth-membership function $T_M(u)$, indeterminacy-membership function $I_M(u)$, and falsity-membership function $F_M(u)$. Here, $T_M(u)$: $U \rightarrow]^-0, 1^+[$, $I_M(u)$: $U \rightarrow]^-0, 1^+[$, and $F_M(u)$: $U \rightarrow]^-0, 1^+[$ with the condition $^-0 \leq \sup T_M(u) + \sup I_M(u) + \sup F_M(u) \leq 3^+$, for all $u \in U$.*

However, there are a lot of limitations in solving practical problems with neutrosophic sets, because their notions are given from a philosophical perspective. Thus, the concepts of single-valued neutrosophic sets are given by Wang et al. as follows.

Definition 3. *([2]) Let U be a nonempty set. A single-valued neutrosophic set M in U is characterized by three functions: Truth-membership function $T_M(u)$, indeterminacy-membership function $I_M(u)$, and falsity-membership function $F_M(u)$. Then, a single-valued neutrosophic set M can be denoted by*

$$M = \{\langle u, T_M(u), I_M(u), F_M(u)\rangle | u \in U\},$$

where $T_M(u), I_M(u), F_M(u) \in [0,1]$ with the condition $0 \leq T_M(u) + I_M(u) + F_M(u) \leq 3$, for all $u \in U$.

So far, scholars have described the inclusion relation of neutrosophic sets from three different angles. The first definition is proposed by Smarandache [1,18,53] and denoted as $\subseteq_1$; the second one is mentioned in [2,14,54] and denoted by $\subseteq_2$; the third one is presented in [14,38,39,54] and denoted by $\subseteq_3$. Furthermore, based on the correlation between union, intersection operations and inclusion relation, we can obtain three different types of union, intersection operations and their properties. In this paper, we consider the first type inclusion relation.

Definition 4. *([1,18,53]) Let U be a nonempty set. Suppose that $M = \{\langle u, T_M(u), I_M(u), F_M(u)\rangle | u \in U\}$ and $N = \{\langle u, T_N(u), I_N(u), F_N(u)\rangle | u \in U\}$ are two neutrosophic sets in U. The first type inclusion relation $\subseteq_1$ and its basic algebraic operations are defined as follows:*

(1) $M \subseteq_1 N$ *if and only if* $T_M(u) \leq T_M(u)$, $I_M(u) \geq I_N(u)$, $F_N(u) \geq F_N(u)$, *for all* $u \in U$;
(2) $M \cup_1 N = \{\langle u, \max(T_M(u), T_N(u)), \min(I_M(u), I_N(u)), \min(F_M(u), F_N(u))\rangle | u \in U\}$;
(3) $M \cap_1 N = \{\langle u, \min(T_M(u), T_N(u)), \max(I_M(u), I_N(u)), \max(F_M(u), F_N(u))\rangle | u \in U\}$;
(4) $M^c = \{\langle u, I_M(u), 1 - F_M(u), T_M(u)\rangle | u \in U\}$.

Definition 5. *([34]) Let $(L; \leq_L)$ be a complete lattice. A t-norm on $(L; \leq_L)$ is a commutative, associative, increasing mapping $\mathscr{T} : L^2 \rightarrow L$, which satisfies $\mathscr{T}(1_L, u) = u$, for all $u \in L$.*
A t-conorm on $(L; \leq_L)$ is a commutative, associative, increasing mapping $\mathscr{S} : L^2 \rightarrow L$, which satisfies $\mathscr{S}(0_L, u) = u$, for all $u \in L$.

Example 1. *([27]) Some basic t-norms and their residual implications on* $([0,1];\leq)$ *(Table 1) are defined as follows, for all* $u,v\in[0,1]$,

Table 1. Some basic t-norms and their residual implications on $([0,1];\leq)$.

t-Norms	Residual Implications
$T_M(u,v)=\min(u,v)$	$I_{GD}(u,v)=\begin{cases}1 & \text{if } u\leq v,\\ v & \text{otherwise}\end{cases}$
$T_P(u,v)=u\cdot v$	$I_{GG}(u,v)=\begin{cases}1 & \text{if } u\leq v,\\ \frac{v}{u} & \text{otherwise}\end{cases}$
$T_{LK}(u,v)=\max(u+v-1,0)$	$I_{LK}(u,v)=\min(1,1-u+v)$
$T_D(u,v)=\begin{cases}0 & \text{if } (u,v)\in[0,1[^2,\\ \min(u,v) & \text{otherwise}\end{cases}$	$I_{WB}(u,v)=\begin{cases}1 & \text{if } (u,v)\in[0,1[^2,\\ v & \text{otherwise}\end{cases}$
$T_{nM}(u,v)=\begin{cases}0 & \text{if } u+v\leq 1,\\ \min(u,v) & \text{otherwise}\end{cases}$	$I_{FD}(u,v)=\begin{cases}1 & \text{if } u\leq v,\\ \max(1-u,v) & \text{otherwise}\end{cases}$

Example 2. *([27]) Some basic t-conorms and their residual co-implications on* $([0,1];\leq)$ *(Table 2) are defined by, for all* $u,v\in[0,1]$,

Table 2. Some basic t-conorms and their residual co-implications on $([0,1];\leq)$.

t-Conorms	Residual Co-Implications
$S_M(u,v)=\max(u,v)$	$J_{GD}(u,v)=\begin{cases}0 & \text{if } u\geq v,\\ v & \text{otherwise}\end{cases}$
$S_P(u,v)=u+v-u\cdot v$	$J_{GG}(u,v)=\begin{cases}0 & \text{if } u\geq v,\\ \frac{v-u}{1-u} & \text{otherwise}\end{cases}$
$S_{LK}(u,v)=\min(u+v,1)$	$J_{LK}(u,v)=\max(0,v-u)$
$S_D(u,v)=\begin{cases}1 & \text{if } (u,v)\in]0,1]^2,\\ \max(u,v) & \text{otherwise}\end{cases}$	$J_{WB}(u,v)=\begin{cases}0 & \text{if } (u,v)\in]0,1]^2,\\ v & \text{otherwise}\end{cases}$
$S_{nM}(u,v)=\begin{cases}1 & \text{if } u+v\geq 1,\\ \max(u,v) & \text{otherwise}\end{cases}$	$J_{FD}(u,v)=\begin{cases}0 & \text{if } u\geq v,\\ \min(1-u,v) & \text{otherwise}\end{cases}$

3. The Lattice Structure of $(D^*;\leq_1)$

Now we consider the set D^* defined by,

$$D^*=\{u=(u_1,u_2,u_3)|u_1,u_2,u_3\in[0,1]\}.$$

As defined above, if $u\in D^*$, then u has three components: The first component u_1, the second component u_2 and the third component u_3.

The order relation $\leq_1$ on D^* can also be defined by, for all $u,v\in D^*$,

$$u\leq_1 v \text{ if and only if } u_1\leq v_1, u_2\geq v_2, u_3\geq v_3.$$

Proposition 1. $(D^*;\leq_1)$ *is a partially ordered set.*

Proof.

(1) Reflexivity: $u\leq_1 u$, for all $u\in D^*$.

(2) Anti-symmetry: If $u\leq_1 v$ and $v\leq_1 u$, then it is obvious that $u=v$, for all $u,v\in D^*$.

(3) Transitivity: If $u\leq_1 v$ and $v\leq_1 w$, then $u_1\leq w_1, u_2\geq w_2, u_3\geq w_3$, that is, $u\leq_1 w$, for all $u,v,w\in D^*$.

□

Proposition 2. *The operations $\wedge_1$ and $\vee_1$ are defined by, for all $u, v \in D^*$,*

$$u \wedge_1 v = \begin{cases} u & \text{if } u \leq_1 v, \\ v & \text{if } v \leq_1 u, \\ (\min(u_1, v_1), \max(u_2, v_2), \max(u_3, v_3)) & \text{otherwise.} \end{cases}$$

$$u \vee_1 v = \begin{cases} u & \text{if } v \leq_1 u, \\ v & \text{if } u \leq_1 v, \\ (\max(u_1, v_1), \min(u_2, v_2), \min(u_3, v_3)) & \text{otherwise.} \end{cases}$$

Then $u \wedge_1 v$ is called the greatest lower bound of u, v, denoted by $\inf(u, v)$; $u \vee_1 v$ is called the least upper bound of u, v, denoted by $\sup(u, v)$. That is, $(D^*; \leq_1)$ is a lattice.

Proof. According to the definitions above, if either $u \leq_1 v$ or $v \leq_1 u$, $u \wedge_1 v$ is the greatest lower bound of u and v. Thus, $u \wedge_1 v = \inf(u, v)$. Similarly, $u \vee_1 v = \sup(u, v)$ can be obtained.

Now, we assume that neither $u \leq_1 v$ nor $v \leq_1 u$. According to the definitions above, $u \wedge_1 v = (\min(u_1, v_1), \max(u_2, v_2), \max(u_3, v_3))$, $u \vee_1 v = (\max(u_1, v_1), \min(u_2, v_2), \min(u_3, v_3))$.

(i) To prove $u \wedge_1 v = \inf(u, v)$, we denote

$$\kappa = (\kappa_1, \kappa_2, \kappa_3) = (\min(u_1, v_1), \max(u_2, v_2), \max(u_3, v_3)).$$

Since $u_1 \geq \min(u_1, v_1) = \kappa_1$, $u_2 \leq \max(u_2, v_2) = \kappa_2$, $u_3 \leq \max(u_3, v_3) = \kappa_3$, $\kappa \leq_1 u$. Similarly, we have $\kappa \leq_1 v$. Thus, κ is the lower bound of u and v. Furthermore, κ is the greatest lower bound of u and v. In fact, assume $a = (a_1, a_2, a_3) \in D^*$ with the condition $a \leq_1 u$ and $a \leq_1 v$. Then, $a_1 \leq u_1, a_2 \geq u_2, a_3 \geq u_3$ and $a_1 \leq v_1, a_2 \geq v_2, a_3 \geq v_3$. Therefore, $a_1 \leq \min(u_1, v_1) = \kappa_1$, $a_2 \geq \max(u_2, v_2) = \kappa_2$, $a_3 \geq \max(u_3, v_3) = \kappa_3$. Hence, $a \leq_1 \kappa$. To sum up, $\kappa = (\min(u_1, v_1), \max(u_2, v_2), \max(u_3, v_3))$ is the greatest lower bound of u and v.

(ii) To prove $u \vee_1 v = \sup(u, v)$, we denote

$$\omega = (\omega_1, \omega_2, \omega_3) = (\max(u_1, v_1), \min(u_2, v_2), \min(u_3, v_3)).$$

Since $u_1 \leq \max(u_1, v_1) = \omega_1$, $u_2 \geq \min(u_2, v_2) = \omega_2$, $u_3 \geq \min(u_3, v_3) = \omega_3$, $u \leq_1 \omega$. Similarly, we have $v \leq_1 \omega$. Thus, ω is the upper bound of u and v. Furthermore, ω is the least upper bound of u and v. In fact, assume $b = (b_1, b_2, b_3) \in D^*$ with the condition $b \geq_1 u$ and $b \geq_1 v$. Then, $b_1 \geq u_1, b_2 \leq u_2, b_3 \leq u_3$ and $b_1 \geq v_1, b_2 \leq v_2, b_3 \leq v_3$. Therefore, $b_1 \geq \max(u_1, v_1) = \omega_1$, $b_2 \leq \min(u_2, v_2) = \omega_2$, $b_3 \leq \min(u_3, v_3) = \omega_3$. Hence, $b \geq_1 \omega$. To sum up, $\omega = (\max(u_1, v_1), \min(u_2, v_2), \min(u_3, v_3))$ is the least upper bound of u and v.

(i) and (ii) show that $u \wedge_1 v = \inf(u, v)$, $u \vee_1 v = \sup(u, v)$, for all $u, v \in D^*$. Then $(D^*; \leq_1)$ is a lattice. □

The first, second and third projection mapping pr_1, pr_2 and pr_3 on D^* are defined as follows, $pr_1(u) = u_1$, $pr_2(u) = u_2$ and $pr_3(u) = u_3$, for all $u \in D^*$.

Proposition 3. *$(D^*; \leq_1)$ is a complete lattice.*

Proof. Let B be a nonempty subset of D^*, we have

$$\inf B = (\inf pr_1 B, \inf pr_2 B, \inf pr_3 B),$$

where $\inf pr_1 B = \inf\{u_1 \mid u_1 \in [0,1], \exists u = (u_1, u_2, u_3) \in B\}$, $\inf pr_2 B = \sup\{u_2 \mid u_2 \in [0,1], \exists u = (u_1, u_2, u_3) \in B\}$, $\inf pr_3 B = \sup\{u_3 \mid u_3 \in [0,1], \exists u = (u_1, u_2, u_3) \in B\}$.

And

$$\sup B = (\sup pr_1 B, \sup pr_2 B, \sup pr_3 B),$$

where $\sup pr_1 B = \sup\{u_1 \mid u_1 \in [0,1], \exists u = (u_1, u_2, u_3) \in B\}$, $\sup pr_2 B = \inf\{u_2 \mid u_2 \in [0,1], \exists u = (u_1, u_2, u_3) \in B\}$, $\sup pr_3 B = \inf\{u_3 \mid u_3 \in [0,1], \exists u = (u_1, u_2, u_3) \in B\}$. □

The maximum and minimum of D^* are denoted by $1_{D^*} = (1,0,0)$ and $0_{D^*} = (0,1,1)$, respectively.

Note that, if u and v are incomparable with respect to $\leq_1$, for all $u, v \in D^*$, then the relationship between u and v can be denoted as $u \parallel_{\leq_1} v$.

Obviously, each neutrosophic set $M = \{\langle u, T_M(u), I_M(u), F_M(u)\rangle \mid u \in U\}$ corresponds to a D^*-fuzzy set. That is, there exits a mapping

$$M: U \rightarrow D^*; u \longmapsto (T_M(u), I_M(u), F_M(u)).$$

Based on the relationship between neutrosophic sets and D^*-fuzzy sets, the triple formed by the three membership degrees of neutrosophic sets is an element of D^*. Therefore, we can obtain more compact formulas for neutrosophic sets, analyze and extend some operators defined in the fuzzy case for neutrosophic sets by using the lattice $(D^*; \leq_1)$.

For example, the intersection of two neutrosophic sets M and N in a universe U is defined as

$$M \cap_1 N = \{\langle u, \min(T_M(u), T_N(u)), \max(I_M(u), I_N(u)), \max(F_M(u), F_N(u))\rangle | u \in U\}.$$

Using the lattice $(D^*; \leq_1)$, we can get, for all $u \in U$,

$$(M \cap_1 N)(u) = (\min(T_M(u), T_N(u)), \max(I_M(u), I_N(u)), \max(F_M(u), F_N(u))) = M(u) \wedge_1 N(u).$$

Definition 6. *The complement of u is defined by, for all $u \in D^*$,*

$$u^c = (u_3, 1 - u_2, u_1).$$

Proposition 4. *Let $u, v, w \in D^*$. Then*

(1) $u \wedge_1 u = u$, $u \vee_1 u = u$;
(2) $u \wedge_1 v = v \wedge_1 u$, $u \vee_1 v = v \vee_1 u$;
(3) $(u \wedge_1 v) \wedge_1 w = u \wedge_1 (v \wedge_1 w)$, $(u \vee_1 v) \vee_1 w = u \vee_1 (v \vee_1 w)$;
(4) $u \wedge_1 (v \vee_1 u) = u$, $u \vee_1 (v \wedge_1 u) = u$;
(5) $u \leq_1 v$ *if and only if* $u \vee_1 v = v$, $u \wedge_1 v = u$;
(6) $(u^c)^c = u$.

Proposition 5. *Let $u, v \in D^*$. Then*

(1) $(u \wedge_1 v)^c = u^c \vee_1 v^c$;
(2) $(u \vee_1 v)^c = u^c \wedge_1 v^c$.

Proof. (1) Suppose $u, v \in D^*$. If $u \leq_1 v$, that is, $u_1 \leq v_1, u_2 \geq v_2, u_3 \geq v_3$. By Definition 6, we have $u^c \geq_1 v^c$. Thus, $(u \wedge_1 v)^c = u^c \vee_1 v^c$. Similarly, if $u \geq_1 v$, then $u^c \leq_1 v^c$ and $(u \wedge_1 v)^c = u^c \vee_1 v^c$. If $u \parallel_{\leq_1} v$, then $u \wedge_1 v = (\min(u_1, v_1), \max(u_2, v_2), \max(u_3, v_3))$. Thus, $(u \wedge_1 v)^c = (\max(u_3, v_3), 1 - \max(u_2, v_2), \min(u_1, v_1))$. Since $u^c \vee_1 v^c = (u_3, 1 - u_2, u_1) \vee_1 (v_3, 1 - v_2, v_1) = (\max(u_3, v_3), \min(1 - u_2, 1 - v_2), \min(u_1, v_1)) = (\max(u_3, v_3), 1 - \max(u_2, v_2), \min(u_1, v_1))$. Hence, $(u \wedge_1 v)^c = u^c \vee_1 v^c$.
(2) Similarly, we can get $(u \vee_1 v)^c = u^c \wedge_1 v^c$. □

Proposition 6. *The system $(D^*; \wedge_1, \vee_1, ^c, 0_{D^*}, 1_{D^*})$ is a De Morgan algebra.*

Proof. By Propositions 1–5 and the definition of the generalized De Morgan algebra [14,55], we can get that $(D^*; \wedge_1, \vee_1, ^c, 0_{D^*}, 1_{D^*})$ is a generalized De Morgan algebra. Furthermore, we can prove that $(D^*; \wedge_1, \vee_1, ^c, 0_{D^*}, 1_{D^*})$ is a distributive lattice, that is, for all $u, v, w \in D^*$ such that $u \wedge_1 (v \vee_1 w) = (u \wedge_1 v) \vee_1 (u \wedge_1 w)$.

(1) For all $u, v, w \in D^*$, if any two of them are comparable, then there are six situations as follows:

Case 1: If $u \leq_1 v \leq_1 w$, then $u \wedge_1 (v \vee_1 w) = u \wedge_1 w = u$, $(u \wedge_1 v) \vee_1 (u \wedge_1 w) = u \vee_1 u = u$.
Case 2: If $u \leq_1 w \leq_1 v$, then $u \wedge_1 (v \vee_1 w) = u \wedge_1 v = u$, $(u \wedge_1 v) \vee_1 (u \wedge_1 w) = u \vee_1 u = u$.
Case 3: If $w \leq_1 u \leq_1 v$, then $u \wedge_1 (v \vee_1 w) = u \wedge_1 v = u$, $(u \wedge_1 v) \vee_1 (u \wedge_1 w) = u \vee_1 w = u$.
Case 4: If $v \leq_1 u \leq_1 w$, then $u \wedge_1 (v \vee_1 w) = u \wedge_1 w = u$, $(u \wedge_1 v) \vee_1 (u \wedge_1 w) = v \vee_1 u = u$.
Case 5: If $w \leq_1 v \leq_1 u$, then $u \wedge_1 (v \vee_1 w) = u \wedge_1 v = v$, $(u \wedge_1 v) \vee_1 (u \wedge_1 w) = v \vee_1 w = v$.
Case 6: If $v \leq_1 w \leq_1 u$, then $u \wedge_1 (v \vee_1 w) = u \wedge_1 w = w$, $(u \wedge_1 v) \vee_1 (u \wedge_1 w) = v \vee_1 w = w$.

Thus, $u \wedge_1 (v \vee_1 w) = (u \wedge_1 v) \vee_1 (u \wedge_1 w)$.

(2) For all $u, v, w \in D^*$, if at least two of them are not comparable, then $u \wedge_1 (v \vee_1 w) = u \wedge_1 (\max(v_1, w_1), \min(v_2, w_2), \min(v_3, w_3)) = (\min(u_1, \max(v_1, w_1)), \max(u_2, \min(v_2, w_2)), \max(u_3, \min(v_3, w_3))) = (\max(\min(u_1, v_1), \min(u_1, w_1)), \min(\max(u_2, v_2), \max(u_2, w_2)), \min(\max(u_3, v_3), \max(u_3, w_3))) = (\min(u_1, v_1), \max(u_2, v_2), \max(u_3, v_3)) \vee (\min(u_1, w_1), \max(u_2, w_2), \max(u_3, w_3)) = (u \wedge_1 v) \vee_1 (u \wedge_1 w)$.

Therefore, $(D^*; \wedge_1, \vee_1, ^c, 0_{D^*}, 1_{D^*})$ is a De Morgan algebra. □

Considering the second type inclusion relation on neutrosophic sets which is dual of the first type inclusion relation, we get that $(D^*; \wedge_2, \vee_2, ^c, (0,0,1), (1,1,0))$ is also a De Morgan algebra.

From this, Proposition 2.2 (see [14]) can be easily proved by using Proposition 6. That is, Proposition 2.2 (see [14]) is a corollary of Proposition 6.

In short, in combination with the conclusions given in [14], we find that neutrosophic net is different from intuitionistic fuzzy set.

4. Neutrosophic t-Norms and De Morgan Neutrosophic Triples

Section 3 proposes that $(D^*; \leq_1)$ is a complete lattice, Section 4 will introduce the notions of neutrosophic t-norms (t-conorms) on $(D^*; \leq_1)$.

Definition 7. *A neutrosophic t-norm is a function $\mathscr{T}$: $(D^*)^2 \rightarrow D^*$ that satisfies the following conditions, for all $u, v, w \in D^*$:*

(NT1) $\mathscr{T}(u, v) = \mathscr{T}(v, u)$;
(NT2) $\mathscr{T}(u, \mathscr{T}(v, w)) = \mathscr{T}(v, \mathscr{T}(u, w))$;
(NT3) $\mathscr{T}(u, v) \leq_1 \mathscr{T}(u', v')$, *where* $u \leq_1 u'$, $v \leq_1 v'$;
(NT4) $\mathscr{T}(u, 1_{D^*}) = u$.

Definition 8. *A neutrosophic t-conorm is a function $\mathscr{S}$: $(D^*)^2 \rightarrow D^*$ that satisfies the following conditions, for all $u, v, w \in D^*$:*

(NS1) $\mathscr{S}(u, v) = \mathscr{S}(v, u)$;
(NS2) $\mathscr{S}(u, \mathscr{S}(v, w)) = \mathscr{S}(v, \mathscr{S}(u, w))$;
(NS3) $\mathscr{S}(u, v) \leq_1 \mathscr{S}(u', v')$, *where* $u \leq_1 u'$, $v \leq_1 v'$;
(NS4) $\mathscr{S}(u, 0_{D^*}) = u$.

Some basic neutrosophic t-norms (t-conorms) on $(D^*; \leq_1)$ are presented as follows:

Example 3. *Some neutrosophic t-norms are defined by, for all $u, v \in D^*$:*

(1) $\mathscr{T}_M(u, v) = (T_M(u_1, v_1), S_M(u_2, v_2), S_M(u_3, v_3))$;
(2) $\mathscr{T}_P(u, v) = (T_P(u_1, v_1), S_P(u_2, v_2), S_P(u_3, v_3))$;
(3) $\mathscr{T}_{LK}(u, v) = (T_{LK}(u_1, v_1), S_{LK}(u_2, v_2), S_{LK}(u_3, v_3))$;

(4) $\mathscr{T}_D(u,v) = (T_D(u_1,v_1), S_D(u_2,v_2), S_D(u_3,v_3))$;
(5) $\mathscr{T}_{nM}(u,v) = (T_{nM}(u_1,v_1), S_{nM}(u_2,v_2), S_{nM}(u_3,v_3))$;
(6) $\mathscr{T}(u,v) = (T_P(u_1,v_1), S_{LK}(u_2,v_2), S_D(u_3,v_3))$;
(7) $\mathscr{T}(u,v) = (T_D(u_1,v_1), S_P(u_2,v_2), S_P(u_3,v_3))$;
(8) $\mathscr{T}(u,v) = (T_{LK}(u_1,v_1), S_{LK}(u_2,v_2), S_M(u_3,v_3))$;
(9) $\mathscr{T}(u,v) = (T_M(u_1,v_1), S_P(u_2,v_2), S_M(u_3,v_3))$.

Example 4. *Some neutrosophic t-conorms are defined by, for all $u, v \in D^*$:*

(1) $\mathscr{S}_M(u,v) = (S_M(u_1,v_1), T_M(u_2,v_2), T_M(u_3,v_3))$;
(2) $\mathscr{S}_P(u,v) = (S_P(u_1,v_1), T_P(u_2,v_2), T_P(u_3,v_3))$;
(3) $\mathscr{S}_{LK}(u,v) = (S_{LK}(u_1,v_1), T_{LK}(u_2,v_2), T_{LK}(u_3,v_3))$;
(4) $\mathscr{S}_D(u,v) = (S_D(u_1,v_1), T_D(u_2,v_2), T_D(u_3,v_3))$;
(5) $\mathscr{S}_{nM}(u,v) = (S_{nM}(u_1,v_1), T_{nM}(u_2,v_2), T_{nM}(u_3,v_3))$;
(6) $\mathscr{S}(u,v) = (S_M(u_1,v_1), T_P(u_2,v_2), T_{LK}(u_3,v_3))$;
(7) $\mathscr{S}(u,v) = (S_P(u_1,v_1), T_{LK}(u_2,v_2), T_{LK}(u_3,v_3))$;
(8) $\mathscr{S}(u,v) = (S_{LK}(u_1,v_1), T_{LK}(u_2,v_2), T_M(u_3,v_3))$;
(9) $\mathscr{S}(u,v) = (S_D(u_1,v_1), T_P(u_2,v_2), T_{nM}(u_3,v_3))$.

Furthermore, the representation theorems of neutrosophic t-norms (t-conorms) are proposed as follows:

Theorem 1. *Let $\mathscr{T}$ be a binary operation on D^*. Then, for all $u, v \in D^*$,*

$$\mathscr{T}(u,v) = (T(u_1,v_1), S_1(u_2,v_2), S_2(u_3,v_3))$$

is a neutrosophic t-norm, where S_1, S_2 are t-conorms, T is a t-norm on $[0,1]$.

Proof. (NT1) Let S_1, S_2 be two t-conorms, T is a t-norm on $[0,1]$. Since $T(u_1,v_1) = T(v_1,u_1)$, $S_1(u_2,v_2) = S_1(v_2,u_2)$, $S_2(u_3,v_3) = S_2(v_3,u_3)$, $\mathscr{T}(u,v) = \mathscr{T}(v,u)$, for all $u, v \in D^*$.

(NT2) $\mathscr{T}(1_{D^*},u) = (T(1,u_1), S_1(0,u_2), S_2(0,u_3)) = (u_1,u_2,u_3) = u$, for all $u \in D^*$.

(NT3) For all $u, u', v, v' \in D^*$ with the condition $u \leq_1 u'$, $v \leq_1 v'$, we have $T(u_1,v_1) \leq T(u_1',v_1')$, $S_1(u_2,v_2) \geq S_1(u_2',v_2')$, $S_2(u_3,v_3) \geq S_2(u_3',v_3')$. Therefore, $\mathscr{T}(u,v) \leq_1 \mathscr{T}(u',v')$.

(NT4) $\mathscr{T}(u,\mathscr{T}(v,w)) = \mathscr{T}(u,(T(v_1,w_1), S_1(v_2,w_2), S_2(v_3,w_3))) = (T(u_1,T(v_1,w_1)), S_1(u_2, S_1(v_2,w_2)), S_2(u_3,S_2(v_3,w_3))) = (T(v_1,T(u_1,w_1)), S_1(v_2,S_1(u_2,w_2)), S_2(v_3, S_2(u_3,w_3))) = \mathscr{T}(v, \mathscr{T}(u,w))$, for all $u, v, w \in D^*$.

Hence, $\mathscr{T}(u,v)$ is a neutrosophic t-norm. □

Theorem 2. *Let $\mathscr{S}$: $(D^*)^2 \to D^*$ be a mapping. Then, for all $u, v \in D^*$,*

$$\mathscr{S}(u,v) = (S(u_1,v_1), T_1(u_2,v_2), T_2(u_3,v_3))$$

is a neutrosophic t-conorm, where S is a t-conorm, T_1, T_2 are t-norms on $[0,1]$.

Proof. The proof is similar to that of Theorem 1. □

Theorem 1 proposes a way to construct neutrosophic t-norms on D^* with t-norms and t-conorms which are defined on $[0,1]$. Unfortunately, the converse is not always true. It is not always possible to find two t-conorms S_1, S_2, a t-norm T on $[0,1]$ such that $\mathscr{T} = (T, S_1, S_2)$.

To distinguish these two kinds of neutrosophic t-norms, we introduce the notions of representable neutrosophic t-norms.

Definition 9. *A neutrosophic t-norm $\mathscr{T}$ is called representable, if and only if, there exist two t-conorms S_1, S_2 and a t-norm T on $[0,1]$ satisfying, for any $u, v \in D^*$,*

$$\mathscr{T}(u,v) = (T(u_1,v_1), S_1(u_2,v_2), S_2(u_3,v_3)).$$

Definition 10. *A neutrosophic t-norm $\mathscr{T}$ is called standard representable, if and only if, there exists a t-norm T and a t-conorm S on $[0,1]$ satisfying, for any $u,v \in D^*$,*

$$\mathscr{T}(u,v) = (T(u_1,v_1), S(u_2,v_2), S(u_3,v_3)).$$

Definition 11. *A N-dual representable neutrosophic t-norm $\mathscr{T}$ defined by, for any $u,v \in D^*$,*

$$\mathscr{T}(u,v) = (T(u_1,v_1), S(u_2,v_2), S(u_3,v_3)).$$

where T is a t-norm on $[0,1]$ and S is the N-dual t-conorm of T, that is, $T(u,v) = 1 - S(1-u, 1-v)$.

Definition 12. *A first N-dual representable neutrosophic t-norm $\mathscr{T}$ defined by, for any $u,v \in D^*$,*

$$\mathscr{T}(u,v) = (T(u_1,v_1), S_1(u_2,v_2), S_2(u_3,v_3)).$$

where T is a t-norm on $[0,1]$ and S_1 is the N-dual t-conorm of T, S_2 is a t-conorm on $[0,1]$.

Definition 13. *A second N-dual representable neutrosophic t-norm $\mathscr{T}$ defined by, for any $u,v \in D^*$,*

$$\mathscr{T}(u,v) = (T(u_1,v_1), S_1(u_2,v_2), S_2(u_3,v_3))$$

where T is a t-norm on $[0,1]$ and S_2 is the N-dual t-conorm of T, S_1 is a t-conorm on $[0,1]$.

Notice that the N-dual representable neutrosophic t-norms are not only the standard representable neutrosophic t-norms, but also the first N-dual representable neutrosophic t-norms and the second N-dual representable neutrosophic t-norms. Those neutrosophic t-norms presented in Example 3 are all representable neutrosophic t-norms, and (1)–(5) are N-dual representable neutrosophic t-norms, (8) is a first N-dual representable neutrosophic t-norm, (9) is a second N-dual representable neutrosophic t-norm.

Definition 14. *A neutrosophic t-conorm $\mathscr{S}$ is called representable, if and only if, there exists a t-conorm S and two t-norms T_1, T_2 on $[0,1]$ satisfying, for any $u,v \in D^*$,*

$$\mathscr{S}(u,v) = (S(u_1,v_1), T_1(u_2,v_2), T_2(u_3,v_3)).$$

For neutrosophic t-conorms, the rest of the related concepts can be obtained by contrasting with Definitions 10–13 of neutrosophic t-norms above.

The following propositions present a method for constructing new representable neutrosophic t-norms (t-conorms) with intuitionistic fuzzy t-norms (t-conorms).

Proposition 7. *Let $T(x,y)$ be a representable intuitionistic fuzzy t-norm: $T(x,y) = (t(x_1,y_1), s_2(x_3,y_3))$, for all $x = (x_1,x_3)$, $y = (y_1,y_3) \in L^*$, where t is a t-norm, s_2 is a t-conorm on $[0,1]$. Assume that s_1 is a t-conorm on $[0,1]$, satisfying, $0 \leq t(u_1,v_1) + s_1(u_2,v_2) + s_2(u_3,v_3) \leq 3$. Then $\mathscr{T}(u,v) = (t(u_1,v_1), s_1(u_2,v_2), s_2(u_3,v_3))$ is a representable neutrosophic t-norm, for any $u,v \in D^*$.*

Proposition 8. *Let $S(x,y)$ be a representable intuitionistic fuzzy t-conorm: $S(x,y) = (s(x_1,y_1), t_2(x_3,y_3))$, for all $u = (x_1,x_3)$, $v = (y_1,y_3) \in L^*$, where s is a t-conorm, t_2 is a t-norm on $[0,1]$. Suppose that t_1 is a t-norm on $[0,1]$, satisfying, $0 \leq s(u_1,v_1) + t_1(u_2,v_2) + t_2(u_3,v_3) \leq 3$. Then $\mathscr{S}(u,v) = (s(u_1,v_1), t_1(u_2,v_2), t_2(u_3,v_3))$ is a representable neutrosophic t-conorm, for all $u,v \in D^*$.*

De Morgan triple is the perfect combination of a fuzzy t-norm, a fuzzy t-conorm and a fuzzy negator because it describes the duality of a fuzzy t-norm and a fuzzy t-conorm with respect to a fuzzy negator. Thus, it is necessary to discuss De Morgan neutrosophic triples. First of all, neutrosophic negators as the extension of fuzzy negators, as well as intuitionistic negators can be defined as follows:

Definition 15. *A neutrosophic negator is a function $\mathcal{N}: D^* \to D^*$ that satisfies the following conditions:*

(NN1) $\mathcal{N}(u) \geq_1 \mathcal{N}(v)$, *for all* $u, v \in D^*$ *such that* $u \leq_1 v$;
(NN2) $\mathcal{N}(0_{D^*}) = 1_{D^*}$;
(NN3) $\mathcal{N}(1_{D^*}) = 0_{D^*}$.

If $\mathcal{N}(\mathcal{N}(u)) = u$, for all $u \in D^*$, then $\mathcal{N}$ is called an involutive neutrosophic negator.
The mapping $\mathcal{N}_S: D^* \to D^*$ defined by, for all $(u_1, u_2, u_3) \in D^*$,

$$\mathcal{N}_S(u_1, u_2, u_3) = (u_3, 1 - u_2, u_1)$$

is an involutive neutrosophic negator. Then we call it the standard neutrosophic negator. Of course, $\mathcal{N}(u) = (u_3, 1 - u_3, u_1)$, $\mathcal{N}(u) = (u_3, u_1, u_1)$ are neutrosophic negators.

Definition 16. *Let $\mathcal{T}$ be a neutrosophic t-norm, $\mathcal{S}$ be a neutrosophic t-conorm, $\mathcal{N}$ be a neutrosophic negator. The triple $(\mathcal{T}, \mathcal{N}, \mathcal{S})$ satisfied the following conditions, for all $u, v \in D^*$,*

$$\mathcal{N}(\mathcal{S}(u,v)) = \mathcal{T}(\mathcal{N}(u), \mathcal{N}(v));$$
$$\mathcal{N}(\mathcal{T}(u,v)) = \mathcal{S}(\mathcal{N}(u), \mathcal{N}(v))$$

is called a De Morgan neutrosophic triple. Moreover, $\mathcal{T}$ and $\mathcal{S}$ are dual with respect to $\mathcal{N}$.

Theorem 3. *Let $\mathcal{N}$ be an involutive neutrosophic negator.*

(1) *If $\mathcal{S}$ is a neutrosophic t-conorm, then the operator $\mathcal{T}$ defined by*

$$\mathcal{T}(u,v) = \mathcal{N}(\mathcal{S}(\mathcal{N}(u), \mathcal{N}(v))),$$

is a neutrosophic t-norm. Furthermore, $(\mathcal{T}, \mathcal{N}, \mathcal{S})$ is a De Morgan neutrosophic triple.

(2) *If $\mathcal{T}$ is a neutrosophic t-norm, then the operator $\mathcal{S}$ defined by*

$$\mathcal{S}(u,v) = \mathcal{N}(\mathcal{T}(\mathcal{N}(u), \mathcal{N}(v))),$$

is a neutrosophic t-conorm. Furthermore, $(\mathcal{T}, \mathcal{N}, \mathcal{S})$ is a De Morgan neutrosophic triple.

Proof. (1) Let $\mathcal{N}$ be an involutive neutrosophic negator, $\mathcal{S}$ be a neutrosophic t-conorm.

(NT1) For any $u, v \in D^*$, $\mathcal{T}(u,v) = \mathcal{N}(\mathcal{S}(\mathcal{N}(u), \mathcal{N}(v))) = \mathcal{N}(\mathcal{S}(\mathcal{N}(v), \mathcal{N}(u))) = \mathcal{T}(v,u)$, because $\mathcal{S}$ is commutative. Thus, $\mathcal{T}$ is commutative.

(NT2) For any $u, v, w \in D^*$, $\mathcal{T}(u, \mathcal{T}(v,w)) = \mathcal{T}(u, \mathcal{N}(\mathcal{S}(\mathcal{N}(v), \mathcal{N}(w)))) = \mathcal{N}(\mathcal{S}(\mathcal{N}(u), \mathcal{N}(\mathcal{N}(\mathcal{S}(\mathcal{N}(v), \mathcal{N}(w)))))) = \mathcal{N}(\mathcal{S}(\mathcal{N}(u), \mathcal{S}(\mathcal{N}(v), \mathcal{N}(w)))) = \mathcal{N}(\mathcal{S}(\mathcal{N}(v), \mathcal{S}(\mathcal{N}(u), \mathcal{N}(w)))) = \mathcal{T}(v, \mathcal{T}(u,w))$, because $\mathcal{S}$ is associative and $\mathcal{N}$ is involutive. Thus, $\mathcal{T}$ is associative.

(NT3) Let $u, u', v, v' \in D^*$ with the condition $u \leq_1 u'$, $v \leq_1 v'$. Then $\mathcal{N}(u) \geq_1 \mathcal{N}(u')$, $\mathcal{N}(v) \geq_1 \mathcal{N}(v')$, because $\mathcal{N}$ is non-increasing. Since $\mathcal{S}$ is non-decreasing in its every variable, $\mathcal{S}(\mathcal{N}(u), \mathcal{N}(v)) \geq_1 \mathcal{S}(\mathcal{N}(u'), \mathcal{N}(v'))$. Thus, $\mathcal{N}(\mathcal{S}(\mathcal{N}(u), \mathcal{N}(v))) \leq_1 \mathcal{N}(\mathcal{S}(\mathcal{N}(u'), \mathcal{N}(v')))$, that is, $\mathcal{T}(u,v) \leq_1 \mathcal{T}(u',v')$. Hence, $\mathcal{T}$ is non-decreasing.

(NT4) For any $u \in D^*$, $\mathcal{T}(u, 1_{D^*}) = \mathcal{N}(\mathcal{S}(\mathcal{N}(u), 0_{D^*})) = \mathcal{N}(\mathcal{N}(u)) = u$.

Therefore, $\mathcal{T}$ is a neutrosophic t-norm.

Furthermore, $(\mathcal{T}, \mathcal{N}, \mathcal{S})$ is a De Morgan neutrosophic triple.

(2) Similarly, assume that $\mathcal{T}$ is a neutrosophic t-norm, $\mathcal{S}$ can be proved to be a neutrosophic t-conorm and $(\mathcal{T}, \mathcal{N}, \mathcal{S})$ will be a De Morgan neutrosophic triple. □

Proposition 9. *Suppose that $(\mathcal{T}, \mathcal{N}, \mathcal{S})$ is a De Morgan neutrosophic triple, $\mathcal{N}$ is a standard neutrosophic negator. Then, for all $u \in D^*$,*

(1) $\mathscr{T}(u, 1_{D^*}) = u$ *if and only if* $\mathscr{S}(u, 0_{D^*}) = u$.
(2) $\mathscr{T}(u, 1_{D^*}) = (u_1, 0, u_3)$ *if and only if* $\mathscr{S}(u, 0_{D^*}) = (u_1, 1, u_3)$.
(3) $\mathscr{T}(u, 1_{D^*}) = (u_1, 1, u_3)$ *if and only if* $\mathscr{S}(u, 0_{D^*}) = (u_1, 0, u_3)$.

Example 5. *Some neutrosophic t-norms and neutrosophic t-conorms are dual with respect to* $\mathscr{N}_S$.

(1) $\mathscr{T}_M(u, v) = (T_M(u_1, v_1), S_M(u_2, v_2), S_M(u_3, v_3))$, $\mathscr{S}_M(u, v) = (S_M(u_1, v_1), T_M(u_2, v_2), T_M(u_3, v_3))$.

Indeed, $\mathscr{T}_M(\mathscr{N}(u), \mathscr{N}(v)) = \mathscr{T}_M((u_3, 1-u_2, u_1), (v_3, 1-v_2, v_1)) = (T_M(u_3, v_3), S_M(1-u_2, 1-v_2), S_M(u_1, v_1))$, *then* $\mathscr{N}(\mathscr{T}_M(\mathscr{N}(u), \mathscr{N}(v))) = (S_M(u_1, v_1), 1 - S_M(1-u_2, 1-v_2), T_M(u_3, v_3)) = (S_M(u_1, v_1), T_M(u_2, v_2), T_M(u_3, v_3)) = \mathscr{S}_M(u, v)$. *Thus,* $\mathscr{T}_M$ *and* $\mathscr{S}_M$ *are dual with respect to* $\mathscr{N}_S$.

(2) $\mathscr{T}_P(u, v) = (T_P(u_1, v_1), S_P(u_2, v_2), S_P(u_3, v_3))$, $\mathscr{S}_P(u, v) = (S_P(u_1, v_1), T_P(u_2, v_2), T_P(u_3, v_3))$.
(3) $\mathscr{T}(u, v) = (T_{LK}(u_1, v_1), S_P(u_2, v_2), S_P(u_3, v_3))$, $\mathscr{S}(u, v) = (S_P(u_1, v_1), T_P(u_2, v_2), T_{LK}(u_3, v_3))$.
(4) $\mathscr{T}(u, v) = (\frac{1}{2}(u_1 + v_1 - 1 + u_1 \cdot v_1) \vee 0, S_M(u_2, v_2), S_M(u_3, v_3))$, $\mathscr{S}(u, v) = (S_M(u_1, v_1), T_M(u_2, v_2), \frac{1}{2}(u_3 + v_3 - 1 + u_3 \cdot v_3) \vee 0)$.
(5) $\mathscr{T}(u, v) = (T_P(u_1, v_1), S_M(u_2, v_2), S_M(u_3, v_3))$, $\mathscr{S}(u, v) = (S_M(u_1, v_1), T_M(u_2, v_2), T_P(u_3, v_3))$.
(6) $\mathscr{T}(u, v) = (T_P(u_1, v_1), S_{LK}(u_2, v_2), S_{LK}(u_3, v_3))$, $\mathscr{S}(u, v) = (S_{LK}(u_1, v_1), T_{LK}(u_2, v_2), T_P(u_3, v_3))$.

Representable neutrosophic t-norms are mainly analyzed and discussed above. As for non-representable neutrosophic t-norms, we give the following theorem:

Theorem 4. *Let* $\mathscr{T}\colon (D^*)^2 \to D^*$ *be a mapping. Then, for all* $u, v \in D^*$,

$$\mathscr{T}(u, v) = \begin{cases} u & \text{if } v = 1_{D^*}, \\ v & \text{if } u = 1_{D^*}, \\ (\min(u_1, v_1), \max(1-u_1, 1-v_1), \max(u_3, v_3)) & \text{otherwise.} \end{cases}$$

is a non-representable neutrosophic t-norm.

Proof. Firstly, $\mathscr{T}$ is a neutrosophic t-norm. In fact,

(NT1) Obviously, $\mathscr{T}(u, v) = \mathscr{T}(v, u)$, for all $u, v \in D^*$.

(NT2) If $u = 1_{D^*}$ or $v = 1_{D^*}$, we can easily prove that $\mathscr{T}(u, \mathscr{T}(v, w)) = \mathscr{T}(\mathscr{T}(u, v), w)$. If $u \neq 1_{D^*}$ and $v \neq 1_{D^*}$, $\mathscr{T}(u, \mathscr{T}(v, w)) = (\min(u_1, \min(v_1, w_1)), \max(1-u_1, 1-\min(v_1, w_1)), \max(u_3, \max(v_3, w_3))) = (\min(u_1, v_1, w_1), \max(1-u_1, 1-v_1, 1-w_1), \max(u_3, v_3, w_3)) = (\min(\min(u_1, v_1), w_1), \max(1-\min(u_1, v_1), 1-w_1), \max(\max(u_3, v_3), w_3)) = \mathscr{T}(\mathscr{T}(u, v), w)$.

(NT3) $\mathscr{T}(1_{D^*}, u) = u$.

(NT4) If $u = 1_{D^*}$ or $v = 1_{D^*}$, we can easily prove that $\mathscr{T}$ is non-decreasing in every variable. If $u \neq 1_{D^*}$ and $v \neq 1_{D^*}$, let $u, u', v, v' \in D^*$ with the condition $u \leq_1 u'$, $v \leq_1 v'$. Then $u_1 \leq u'_1$, $v_1 \leq v'_1$, $u_3 \geq u'_3$, $v_3 \geq v'_3$. Thus, $\min(u_1, v_1) \leq \min(u'_1, v'_1)$, $\max(1-u_1, 1-v_1) \geq \max(1-u'_1, 1-v'_1)$, $\max(u_3, v_3) \geq \max(u'_3, v'_3)$. That is, $\mathscr{T}(u, v) \leq_1 \mathscr{T}(u', v')$. Therefore, $\mathscr{T}$ is a neutrosophic t-norm.

Secondly, for a representable neutrosophic t-norm $\mathscr{T}$, there exists a t-norm T and two t-conorms S_1, S_2 on $[0,1]$ such that, for all $u = (u_1, u_2, u_3), v = (v_1, v_2, v_3) \in D^*$, $\mathscr{T}(u, v) = (T(u_1, v_1), S_1(u_2, v_2), S_2(u_3, v_3))$. Let $u = (0.2, 0.5, 0.7)$, $u' = (0.3, 0.5, 0.7)$, $v = (0.4, 0.5, 0.9)$. From $\mathscr{T}(u, v) = (0.2, 0.8, 0.9)$ and $\mathscr{T}(u', v) = (0.3, 0.7, 0.9)$, we get $S_1(u_2, v_2) = 0.8$ and $S_1(u'_2, v_2) = 0.7$, so $S_1(u_2, v_2) \neq S_1(u'_2, v_2)$. Hence $S_1(u, v)$ is not independent from u_1, thus $\mathscr{T}$ is not representable. □

Furthermore, the dual neutrosophic t-conorm of $\mathscr{T}$ with respect to the standard neutrosophic negator $\mathscr{N}_S$ is $\mathscr{S}$ defined by, for all $u, v \in D^*$,

$$\mathscr{S}(u,v) = \begin{cases} u & if\ v = 0_{D^*}, \\ v & if\ u = 0_{D^*}, \\ (\max(u_1,v_1), \min(u_3,v_3), \min(u_3,v_3)) & otherwise. \end{cases}$$

Then, $\mathscr{S}$ is not representable.

Remark 1. *Let $\mathscr{T}$ be a non-representable neutrosophic t-norm on D^*, $\mathscr{S}$ be a neutrosophic t-conorm which is dual to $\mathscr{T}$ with respect to the standard neutrosophic negator $\mathscr{N}_s$. Then, $\mathscr{S}$ is not representable. Conversely, the dual neutrosophic t-norm with respect to an involutive neutrosophic negator $\mathscr{N}$ on D^* of a non-representable neutrosophic t-conorm is not representable.*

Example 6. *Let $\mathscr{T}$: $(D^*)^2 \to D^*$ be a mapping. Then, for all $u, v \in D^*$,*

$$\mathscr{T}(u,v) = \begin{cases} u & if\ v = 1_{D^*}, \\ v & if\ u = 1_{D^*}, \\ (\min(u_1,v_1), \max(1-u_1, 1-v_1), \max(1-u_1, 1-v_1)) & otherwise. \end{cases}$$

is a non-representable neutrosophic t-norm.

Meanwhile, the dual neutrosophic t-conorm $\mathscr{S}$ of $\mathscr{T}$ with respect to $\mathscr{N}_s$ is presented by, for all $u, v \in D^*$,

$$\mathscr{S}(u,v) = \begin{cases} u & if\ v = 0_{D^*}, \\ v & if\ u = 0_{D^*}, \\ (\max(1-u_3, 1-v_3), \min(u_3,v_3), \min(u_3,v_3)) & otherwise. \end{cases}$$

Then, $\mathscr{S}$ is not representable, too.

Example 7. *Let $\mathscr{S}$ be a mapping: $(D^*)^2 \to D^*$. Then, for all $u, v \in D^*$,*

$$\mathscr{S}(u,v) = \begin{cases} u & if\ v = 0_{D^*}, \\ v & if\ u = 0_{D^*}, \\ (\max(1-u_3, 1-v_3), \min(u_2,v_2), \min(u_3,v_3)) & otherwise. \end{cases}$$

is a non-representable neutrosophic t-conorm.

Meanwhile, the dual neutrosophic t-norm $\mathscr{T}$ of $\mathscr{S}$ with respect to $\mathscr{N}_s$ is presented by, for all $u, v \in D^*$,

$$\mathscr{T}(u,v) = \begin{cases} u & if\ v = 1_{D^*}, \\ v & if\ u = 1_{D^*}, \\ (\min(u_1,v_1), \max(u_2,v_2), \max(1-u_1, 1-v_1)) & otherwise. \end{cases}$$

Then, $\mathscr{T}$ is not representable, too.

5. Neutrosophic Residual Implications of Neutrosophic t-Norms

This section will introduce the notions of neutrosophic residual implications on the complete lattice D^*, investigate basic properties of neutrosophic residual implications, and give some important conclusions between neutrosophic t-norms and neutrosophic residual implications after proving that residual neutrosophic t-norms are $\vee$-distributive. Firstly, we give the notions of neutrosophic implications on D^*.

Definition 17. *A neutrosophic implication is a function $\mathscr{I}$: $(D^*)^2 \to D^*$ that satisfies the following conditions,*

(NI1) $\mathscr{I}$ is non-increasing with respect to $\leq_1$ in its first variable, that is, $\mathscr{I}(u,v) \geq_1 \mathscr{I}(u',v)$, where $u, u', v \in D^*$ and $u \leq_1 u'$;

(NI2) $\mathscr{I}$ is non-decreasing with respect to $\leq_1$ in its second variable, that is, $\mathscr{I}(u,v) \leq_1 \mathscr{I}(u,v')$, where $u, v, v' \in D^*$ and $v \leq_1 v'$;

(NI3) $\mathscr{I}(0_{D^*}, 0_{D^*}) = 1_{D^*}$;

(NI4) $\mathscr{I}(1_{D^*}, 1_{D^*}) = 1_{D^*}$;
(NI5) $\mathscr{I}(1_{D^*}, 0_{D^*}) = 0_{D^*}$.

Definition 18. *A function $\mathscr{I}\colon (D^*)^2 \to D^*$ is called a neutrosophic residual implication, if there exits a neutrosophic t-norm $\mathscr{T}$ such that*

$$\mathscr{I}(u,v) = \sup\{w | w \in D^*, \mathscr{T}(u,w) \leq_1 v\}.$$

If $\mathscr{I}$ is a neutrosophic residual implication generated from a neutrosophic t-norm $\mathscr{T}$, then it will be denoted by $\mathscr{I}_{\mathscr{T}}$.

Furthermore, a neutrosophic t-norm $\mathscr{T}$ satisfies the residual principle if and only if, for all $u, v, w \in D^*$,

$$\mathscr{T}(u,w) \leq_1 v \text{ if and only if } w \leq_1 \mathscr{I}_{\mathscr{T}}(u,v).$$

Similarly, we can get the definitions of neutrosophic co-implications:

Definition 19. *A neutrosophic co-implication is a function $\mathscr{J}\colon (D^*)^2 \to D^*$ that satisfies the following conditions:*

(NJ1) *$\mathscr{J}$ is non-increasing with respect to $\leq_1$ in its first variable, that is, $\mathscr{J}(u,v) \geq_1 \mathscr{J}(u',v)$, where $u, u', v \in D^*$ and $u \leq_1 u'$;*
(NJ2) *$\mathscr{J}$ is non-decreasing with respect to $\leq_1$ in its second variable, that is, $\mathscr{J}(u,v) \leq_1 \mathscr{J}(u,v')$, where $u, v, v' \in D^*$ and $v \leq_1 v'$;*
(NJ3) $\mathscr{J}(0_{D^*}, 0_{D^*}) = 0_{D^*}$;
(NJ4) $\mathscr{J}(1_{D^*}, 1_{D^*}) = 0_{D^*}$;
(NJ5) $\mathscr{J}(0_{D^*}, 1_{D^*}) = 1_{D^*}$.

Definition 20. *A function $\mathscr{J}\colon (D^*)^2 \to D^*$ is called a neutrosophic residual co-implication, if there exits a neutrosophic t-conorm $\mathscr{S}$ such that*

$$\mathscr{J}(u,v) = \inf\{w | w \in D^*, \mathscr{S}(u,w) \geq_1 v\}.$$

If $\mathscr{J}$ is a neutrosophic residual co-implication generated from a neutrosophic t-conorm $\mathscr{S}$, then it will be denoted by $\mathscr{J}_{\mathscr{S}}$.

Furthermore, a neutrosophic t-conorm $\mathscr{S}$ satisfies the residual principle if and only if, for all $u, v, w \in D^*$,

$$\mathscr{S}(u,w) \geq_1 v \text{ if and only if } w \geq_1 \mathscr{I}_{\mathscr{S}}(u,v).$$

Using the description of the above definitions, we can easily obtain the neutrosophic residual implications of neutrosophic t-norms discussed in Section 4.

Example 8. *The neutrosophic residual implications of the representable neutrosophic t-norms of Example 3 are given by, for all $u, v \in D^*$,*

(1) $\mathscr{I}_{\mathscr{T}_M}(u,v) = (I_{GD}(u_1,v_1), J_{GD}(u_2,v_2), J_{GD}(u_3,v_3))$;
(2) $\mathscr{I}_{\mathscr{T}_P}(u,v) = (I_{GG}(u_1,v_1), J_{GG}(u_2,v_2), J_{GG}(u_3,v_3))$;
(3) $\mathscr{I}_{\mathscr{T}_{LK}}(u,v) = (I_{LK}(u_1,v_1), J_{LK}(u_2,v_2), J_{LK}(u_3,v_3))$;
(4) $\mathscr{I}_{\mathscr{T}_D}(u,v) = (I_{WB}(u_1,v_1), J_{WB}(u_2,v_2), J_{WB}(u_3,v_3))$;
(5) $\mathscr{I}_{\mathscr{T}_{nM}}(u,v) = (I_{FD}(u_1,v_1), J_{FD}(u_2,v_2), J_{FD}(u_3,v_3))$;
(6) $\mathscr{I}_{\mathscr{T}}(u,v) = (I_{GG}(u_1,v_1), J_{LK}(u_2,v_2), J_{WB}(u_3,v_3))$;
(7) $\mathscr{I}_{\mathscr{T}}(u,v) = (I_{WB}(u_1,v_1), J_{GG}(u_2,v_2), J_{GG}(u_3,v_3))$;
(8) $\mathscr{I}_{\mathscr{T}}(u,v) = (I_{LK}(u_1,v_1), J_{LK}(u_2,v_2), J_{GD}(u_3,v_3))$;

(9) $\mathscr{I}_{\mathscr{T}}(u,v) = (I_{GD}(u_1,v_1), J_{GG}(u_2,v_2), J_{GD}(u_3,v_3))$.

Example 9. *The neutrosophic residual co-implications of the representable neutrosophic t-conorms of Example 4 are given by, for all* $u,v \in D^*$,

(1) $\mathscr{J}_{\mathscr{S}_M}(u,v) = (J_{GD}(u_1,v_1), I_{GD}(u_2,v_2), I_{GD}(u_3,v_3))$;
(2) $\mathscr{J}_{\mathscr{S}_P}(u,v) = (J_{GG}(u_1,v_1), I_{GG}(u_2,v_2), I_{GG}(u_3,v_3))$;
(3) $\mathscr{J}_{\mathscr{S}_{LK}}(u,v) = (J_{LK}(u_1,v_1), I_{LK}(u_2,v_2), I_{LK}(u_3,v_3))$;
(4) $\mathscr{J}_{\mathscr{S}_D}(u,v) = (J_{WB}(u_1,v_1), I_{WB}(u_2,v_2), I_{WB}(u_3,v_3))$;
(5) $\mathscr{I}_{\mathscr{S}_{nM}}(u,v) = (J_{FD}(u_1,v_1), I_{FD}(u_2,v_2), I_{FD}(u_3,v_3))$;
(6) $\mathscr{J}_{\mathscr{S}}(u,v) = (J_{GD}(u_1,v_1), I_{GG}(u_2,v_2), I_{LK}(u_3,v_3))$;
(7) $\mathscr{J}_{\mathscr{S}}(u,v) = (J_{GG}(u_1,v_1), I_{LK}(u_2,v_2), I_{LK}(u_3,v_3))$;
(8) $\mathscr{J}_{\mathscr{S}}(u,v) = (J_{LK}(u_1,v_1), I_{LK}(u_2,v_2), I_{GD}(u_3,v_3))$;
(9) $\mathscr{J}_{\mathscr{S}}(u,v) = (J_{WB}(u_1,v_1), I_{GG}(u_2,v_2), I_{FD}(u_3,v_3))$.

As we all know, t-conorms are dual operators of t-norms on $[0,1]$, in the same way, residual co-implications are dual operators of residual implications on $[0,1]$, with respect to $N(u) = 1-u$. Neutrosophic residual co-implications of neutrosophic t-conorms are dual operators of neutrosophic residual implications of neutrosophic t-norms, just as that neutrosophic t-conorms are dual operators of neutrosophic t-norms with respect to $\mathscr{N}_S$. As Examples 8 and 9 above show, if $\mathscr{S}$ is the dual neutrosophic t-conorm of a neutrosophic t-norm $\mathscr{T}$, then the neutrosophic residual co-implication $\mathscr{J}_{\mathscr{S}}$ is the dual operator of the neutrosophic residual implication of $\mathscr{I}_{\mathscr{T}}$.

Next, we will introduce the most important theorem in this section, which gives the sufficient condition that the residual operators induced by neutrosophic t-norms must be neutrosophic implications.

Theorem 5. *Let* $\mathscr{T}$ *be a neutrosophic t-norm on* D^* *with the neutral element* 1_{D^*}. *Then, for all* $u,v \in D^*$,

$$\mathscr{I}_{\mathscr{T}}(u,v) = \sup\{w | w \in D^*, \mathscr{T}(u,w) \leq_1 v\}$$

is a neutrosophic implication.

Proof. From Definition 18, $\mathscr{I}_{\mathscr{T}}(u,1_{D^*}) = \sup\{w|w \in D^*, \mathscr{T}(u,w) \leq_1 1_{D^*}\} = 1_{D^*}$, for all $u \in D^*$. Therefore, $\mathscr{I}_{\mathscr{T}}(1_{D^*},1_{D^*}) = 1_{D^*}$. Since $\mathscr{T}$ is non-decreasing, $\mathscr{I}_{\mathscr{T}}(1_{D^*},0_{D^*}) = \sup\{w|w \in D^*, \mathscr{T}(w,1_{D^*}) \leq_1 0_{D^*}\} = \sup\{w|w \in D^*, w \leq_1 0_{D^*}\} = 0_{D^*}$. $\mathscr{I}_{\mathscr{T}}(0_{D^*},0_{D^*}) = \sup\{w|w \in D^*, \mathscr{T}(w,0_{D^*}) \leq_1 0_{D^*}\} = 1_{D^*}$. Let $u,u' \in D^*$ with the condition $u \leq_1 u'$. Since the non-decreasingness of $\mathscr{T}$, $\{w|w \in D^*, \mathscr{T}(u,w) \leq_1 v\} \supseteq_1 \{w|w \in D^*, \mathscr{T}(u',w) \leq_1 v\}$, then $\sup\{w|w \in D^*, \mathscr{T}(u,w) \leq_1 v\} \geq_1 \sup\{w|w \in D^*, \mathscr{T}(u',w) \leq_1 v\}$. Thus, $\mathscr{I}_{\mathscr{T}}(u,v) \geq_1 \mathscr{I}_{\mathscr{T}}(u',v)$. That is $\mathscr{I}_{\mathscr{T}}$ is non-increasing with respect to $\leq_1$ in its first variable. Let $v,v' \in D^*$ with the condition $v \leq_1 v'$. Since the non-decreasingness of $\mathscr{T}$, $\{w|w \in D^*, \mathscr{T}(u,w) \leq_1 v\} \subseteq_1 \{w|w \in D^*, \mathscr{T}(u,w) \leq_1 v'\}$, then $\sup\{w|w \in D^*, \mathscr{T}(u,w) \leq_1 v\} \leq_1 \sup\{w|w \in D^*, \mathscr{T}(u,w) \leq_1 v'\}$. Thus, $\mathscr{I}_{\mathscr{T}}(u,v) \leq_1 \mathscr{I}_{\mathscr{T}}(u,v')$. That is $\mathscr{I}_{\mathscr{T}}$ is non-decreasing with respect to $\leq_1$ in its second variable. □

For neutrosophic residual implications, there are several important properties as follows:

Theorem 6. *Suppose that* $\mathscr{T}$ *is a neutrosophic t-norm on* D^* *with the neutral element* 1_{D^*}, $\mathscr{I}_{\mathscr{T}}$ *is a neutrosophic residual implication. Then, for all* $u,v,w \in D^*$,

(1) $\mathscr{I}_{\mathscr{T}}(0_{D^*},v) = 1_{D^*}$;
(2) $\mathscr{I}_{\mathscr{T}}(u,1_{D^*}) = 1_{D^*}$;
(3) $\mathscr{I}_{\mathscr{T}}(u,u) = 1_{D^*}$;
(4) $\mathscr{I}_{\mathscr{T}}(1_{D^*},v) = v$;
(5) $\mathscr{I}_{\mathscr{T}}(u,v) \geq_1 v$;

(6) $\mathscr{I}_{\mathscr{T}}(u,v) = 1_{D^*}$ *if and only if* $u \leq_1 v$;
(7) $u \leq_1 \mathscr{I}_{\mathscr{T}}(v,w)$ *if and only if* $v \leq_1 \mathscr{I}_{\mathscr{T}}(u,w)$;
(8) $u \leq_1 \mathscr{I}_{\mathscr{T}}(v, \mathscr{T}(u,v))$;
(9) $\mathscr{I}_{\mathscr{T}}(\mathscr{T}(u,v), \mathscr{T}(u,w)) \geq_1 \mathscr{I}_{\mathscr{T}}(v,w)$.

Proof. For all $u, v \in D^*$,

The proofs of (1)–(4) can be directly obtained by Definition 18.

(5) Since $\mathscr{I}$ is non-increasing with respect to $\leq_1$ in its first variable, $\mathscr{I}_{\mathscr{T}}(u,v) \geq_1 \mathscr{I}_{\mathscr{T}}(1_{D^*},v) = v$.

(6) On the one hand, since $u \leq_1 v$, $\mathscr{T}(1_{D^*},u) \leq_1 v$. Thus, $\mathscr{I}_{\mathscr{T}}(u,v) \geq_1 1_{D^*}$, that is $\mathscr{I}_{\mathscr{T}}(u,v) = 1_{D^*}$. On the other hand, if $\mathscr{I}_{\mathscr{T}}(u,v) = 1_{D^*}$, then $\mathscr{T}(1_{D^*},u) \leq_1 v$. Thus, $u \leq_1 v$.

(7) Since $u \leq_1 \mathscr{I}_{\mathscr{T}}(v,w)$, $\mathscr{T}(v,u) \leq_1 w$. Thus, $v \leq_1 \mathscr{I}_{\mathscr{T}}(u,w)$. Similarly, it follows from $v \leq_1 \mathscr{I}_{\mathscr{T}}(u,w)$ that $u \leq_1 \mathscr{I}_{\mathscr{T}}(v,w)$.

(8) Since $\mathscr{T}(u,v) \leq_1 \mathscr{T}(u,v)$, $u \leq_1 \mathscr{I}_{\mathscr{T}}(v, \mathscr{T}(u,v))$.

(9) $\mathscr{I}_{\mathscr{T}}(\mathscr{T}(u,v), \mathscr{T}(u,w)) = \sup\{t \mid t \in D^*, \mathscr{T}(\mathscr{T}(u,v),t) \leq_1 \mathscr{T}(u,w)\} = \sup\{t \mid t \in D^*, \mathscr{T}(u, \mathscr{T}(v,t)) \leq_1 \mathscr{T}(u,w)\} \geq_1 \sup\{t \mid t \in D^*, \mathscr{T}(v,t) \leq_1 w\} = \mathscr{I}_{\mathscr{T}}(v,w)$. □

Example 10. *Example 8 shows some neutrosophic residual implications of representable neutrosophic t-norms, furthermore, it is easy to verify that neutrosophic residual implications of representable neutrosophic t-norms satisfy the properties described in Theorem 6.*

For non-representable neutrosophic t-norms, take the neutrosophic t-norm $\mathscr{T}$ presented in Theorem 4 for example, then, for all $u, v \in D^*$,

$$\mathscr{I}_{\mathscr{T}}(u,v) = \begin{cases} 1_{D^*} & \text{if } v = 1_{D^*}, \\ v & \text{if } u = 1_{D^*}, \\ \left(I_{GD}(u_1,v_1), \begin{cases} u_1 & \text{if } u_1 \leq 1 - v_2, \\ 0 & \text{otherwise} \end{cases}, J_{GD}(u_3,v_3)\right) & \text{otherwise} \end{cases}.$$

is a neutrosophic implication and satisfies the properties given in Theorem 6.

Similarly, we have the following two important theorems of neutrosophic t-conorm on D^*:

Theorem 7. *Assume that $\mathscr{S}$ is a neutrosophic t-conorm on D^* with the neutral element 0_{D^*}. Then, for all $u, v \in D^*$,*

$$\mathscr{J}_{\mathscr{S}}(u,v) = \inf\{w | w \in D^*, \mathscr{S}(u,w) \geq_1 v\}.$$

is a neutrosophic co-implication.

Proof. From Definition 20, we can prove it using the proven ways of Theorem 5. □

Theorem 8. *Assume that $\mathscr{S}$ is a neutrosophic t-conorm on D^* with the neutral element 0_{D^*}, $\mathscr{J}_{\mathscr{S}}$ is a neutrosophic residual co-implication. Then, for all $u, v, w \in D^*$,*

(1) $\mathscr{J}_{\mathscr{S}}(1_{D^*},v) = 0_{D^*}$;
(2) $\mathscr{J}_{\mathscr{S}}(u,0_{D^*}) = 0_{D^*}$;
(3) $\mathscr{J}_{\mathscr{S}}(u,u) = 0_{D^*}$;
(4) $\mathscr{J}_{\mathscr{S}}(0_{D^*},v) = v$;
(5) $\mathscr{J}_{\mathscr{S}}(u,v) \leq_1 v$;
(6) $\mathscr{J}_{\mathscr{S}}(u,v) = 0_{D^*}$ *if and only if* $u \geq_1 v$;
(7) $u \geq_1 \mathscr{J}_{\mathscr{S}}(v,w)$ *if and only if* $v \geq_1 \mathscr{J}_{\mathscr{S}}(u,w)$;
(8) $u \geq_1 \mathscr{J}_{\mathscr{S}}(v, \mathscr{S}(u,v))$;
(9) $\mathscr{J}_{\mathscr{S}}(\mathscr{S}(u,v), \mathscr{S}(u,w)) \leq_1 \mathscr{J}_{\mathscr{S}}(v,w)$.

Example 11. *Example 9 shows some neutrosophic residual co-implications of representable neutrosophic t-conorms; furthermore, it is easy to verify that neutrosophic residual co-implications of representable neutrosophic t-conorms satisfy the properties described in Theorem 8.*

For non-representable neutrosophic t-conorms, take the neutrosophic t-conorm $\mathscr{S}$ presented in Theorem 4 for example, then, for all $u, v \in D^*$,

$$\mathscr{I}_{\mathscr{T}}(u,v) = \begin{cases} 0_{D^*} & \text{if } v = 0_{D^*}, \\ v & \text{if } u = 0_{D^*}, \\ (J_{GD}(u_1,v_1), \begin{cases} 1 & \text{if } u_3 \leq v_2, \\ v_2 & \text{otherwise} \end{cases}, I_{GD}(u_3,v_3)) & \text{otherwise} \end{cases}.$$

is a neutrosophic co-implication and it satisfies the properties given in Theorem 8.

In Definition 18, $\mathscr{I}_{\mathscr{T}}$ is called the neutrosophic residual implication. At the same time, $\mathscr{T}$ is called the residual neutrosophic t-norm. Then, some important properties of the residual neutrosophic t-norm will be discussed below.

Definition 21. *[45] A binary operation H on a complete lattice L is called left (right) infinitely $\vee$-distributive, if for all $u \in L$,*

$$H(\sup_{w \in W} w, v) = \sup_{w \in W} H(w,v) \ (H(u, \sup_{w \in W} w) = \sup_{w \in W} H(u,w));$$

H is called left (right) infinitely $\wedge$-distributive, if for all $u \in L$,

$$H(\inf_{w \in W} w, v) = \inf_{w \in W} H(w,v) \ (H(u, \inf_{w \in W} w) = \inf_{w \in W} H(u,w)),$$

where $W \subseteq L$. H is called infinitely $\vee$-distributive ($\wedge$-distributive) on L, if H is both left and right infinitely $\vee$-distributive ($\wedge$-distributive).

Theorem 9. *Assume that $\mathscr{T}$ is a residual neutrosophic t-norm on D^* with the neutral element 1_{D^*}. Then $\mathscr{T}$ is infinitely $\vee$-distributive on D^*.*

Proof. Let $W \subseteq D^*$. If $W = \emptyset$, then $\mathscr{T}(\sup_{w \in W} w, v) = \mathscr{T}(0_{D^*}, v) = 0_{D^*} = \sup_{w \in W} \mathscr{T}(w,v)$, for all $v \in D^*$. If $W \neq \emptyset$, since $\mathscr{T}$ is non-decreasing, $\mathscr{T}(\sup_{w \in W} w, v) \geq_1 \sup_{w \in W} \mathscr{T}(w,v)$, for any $v \in D^*$. Suppose $m = \sup_{w \in W} \mathscr{T}(w,v)$, then $\mathscr{T}(w,v) \leq_1 m$, now we have $w \in \{t \mid t \in D^*, \mathscr{T}(t,v) \leq_1 m\}$. By Definition 18, $w \leq_1 \mathscr{I}_{\mathscr{T}}(v,m)$, for all $w \in W$. Thus, $\sup_{w \in W} w \leq_1 \mathscr{I}_{\mathscr{T}}(v,m)$. Since $\mathscr{T}$ is non-decreasing, $\mathscr{T}(\sup_{w \in W} w, v) \leq_1 \mathscr{T}(\mathscr{I}_{\mathscr{T}}(v,m), v)$. Since $\mathscr{T}(u,w) \leq_1 v$ if and only if $w \leq_1 \mathscr{I}_{\mathscr{T}}(u,v)$, and $\mathscr{I}_{\mathscr{T}}(v,m) \leq_1 \mathscr{I}_{\mathscr{T}}(v,m)$, $\mathscr{T}(\mathscr{I}_{\mathscr{T}}(v,m), v) \leq_1 m = \sup_{w \in W} \mathscr{T}(w,v)$. Therefore, $\mathscr{T}(\sup_{w \in W} w, v) \leq_1 \sup_{w \in W} \mathscr{T}(w,v)$. □

Theorem 10. *Assume that $\mathscr{T}$ is a residual neutrosophic t-norm on D^* with the neutral element 1_{D^*}. Then, for all $u, v \in D^*$,*

(1) $\mathscr{T}(\mathscr{I}_{\mathscr{T}}(u,v), u) \leq_1 v$. *In particularly,* $\mathscr{T}(\mathscr{I}_{\mathscr{T}}(u,u), u) = u$, $\mathscr{T}(\mathscr{I}_{\mathscr{T}}(u, 0_{D^*}), u) = 0_{D^*}$;
(2) $\mathscr{I}_{\mathscr{T}}(\mathscr{T}(u,v), w) = \mathscr{I}_{\mathscr{T}}(v, \mathscr{I}_{\mathscr{T}}(u,w))$;
(3) $\mathscr{I}_{\mathscr{T}}(\sup_{u \in U} u, v) = \inf_{u \in U} \mathscr{I}_{\mathscr{T}}(u,v)$;
(4) $\mathscr{T}(\mathscr{I}_{\mathscr{T}}(u,v), \mathscr{I}_{\mathscr{T}}(v,w)) \leq_1 \mathscr{I}_{\mathscr{T}}(u,w)$;
(5) $\mathscr{T}(\mathscr{I}_{\mathscr{T}}(w, 1_{D^*}), \mathscr{I}_{\mathscr{T}}(u,v)) \leq_1 \mathscr{I}_{\mathscr{T}}(\mathscr{T}(u,w), v)$;
(6) $\mathscr{I}_{\mathscr{T}}(u, \mathscr{I}_{\mathscr{T}}(v,w)) = \mathscr{I}_{\mathscr{T}}(v, \mathscr{I}_{\mathscr{T}}(u,w))$;
(7) $\mathscr{T}(\mathscr{I}_{\mathscr{T}}(u, \mathscr{T}(v,u)), u) = \mathscr{T}(u,v)$;

(8) $\mathscr{I}_{\mathscr{T}}(u, \mathscr{T}(\mathscr{I}_{\mathscr{T}}(u,v),u)) = \mathscr{I}_{\mathscr{T}}(u,v)$.

Proof. We can directly prove (1)–(6) directly by the method of Theorems 4.3 and 4.6–4.9 in [46]; The proofs of (7) and (8) can be obtained directly from Theorem 3.5 in [48]. □

Naturally, we can prove that a residual neutrosophic t-conorm is infinitely $\wedge$-distributive, and then we can get some important properties of a residual neutrosophic t-conorm on D^*.

Theorem 11. *Assume that $\mathscr{S}$ is a residual neutrosophic t-conorm on D^* with the neutral element 0_{D^*}. Then $\mathscr{S}$ is infinitely $\wedge$-distributive on D^*.*

Theorem 12. *Assume that $\mathscr{S}$ is a residual neutrosophic t-conorm on D^* with the neutral element 0_{D^*}. Then, for all $u, v \in D^*$,*

(1) $\mathscr{S}(\mathscr{J}_{\mathscr{S}}(u,v),u) \geq_1 v$. *In particularly,* $\mathscr{S}(\mathscr{J}_{\mathscr{S}}(u,u),u) = u$, $\mathscr{S}(\mathscr{J}_{\mathscr{S}}(u,1_{D^*}),u) = 1_{D^*}$;
(2) $\mathscr{J}_{\mathscr{S}}(\mathscr{S}(u,v),w) = \mathscr{J}_{\mathscr{S}}(v, \mathscr{J}_{\mathscr{S}}(u,w))$;
(3) $\mathscr{J}_{\mathscr{S}}(\inf_{u\in U} u,v) = \sup_{u\in U} \mathscr{J}_{\mathscr{S}}(u,v)$;
(4) $\mathscr{S}(\mathscr{J}_{\mathscr{S}}(u,v), \mathscr{J}_{\mathscr{S}}(v,w)) \geq_1 \mathscr{J}_{\mathscr{S}}(u,w)$;
(5) $\mathscr{S}(\mathscr{J}_{\mathscr{S}}(w,1_{D^*}), \mathscr{J}_{\mathscr{S}}(u,v)) \geq_1 \mathscr{J}_{\mathscr{S}}(\mathscr{S}(u,w),v)$;
(6) $\mathscr{J}_{\mathscr{S}}(u, \mathscr{J}_{\mathscr{S}}(v,w)) = \mathscr{J}_{\mathscr{S}}(v, \mathscr{J}_{\mathscr{S}}(u,w))$;
(7) $\mathscr{S}(\mathscr{J}_{\mathscr{S}}(u,\mathscr{S}(v,u)),u) = \mathscr{S}(u,v)$;
(8) $\mathscr{J}_{\mathscr{S}}(u,\mathscr{S}(\mathscr{J}_{\mathscr{S}}(u,v),u)) = \mathscr{J}_{\mathscr{S}}(u,v)$.

Proof. The proofs of (1)–(6) can be obtained directly from Theorems 3.2 and 3.5–3.8 in [47]; the proofs of (7) and (8) can be obtained directly from Theorem 3.5 in [48]. □

6. Neutrosophic t-Norms Induced by Neutrosophic Implications on D^*

From Theorem 5, we know that neutrosophic implications can be induced by neutrosophic t-norms. In this section, the dual situation will be considered. Then, residuated lattices can be constructed on the basis of neutrosophic t-norms and their corresponding neutrosophic residual implications.

Definition 22. *Let $\mathscr{I}\colon (D^*)^2 \to D^*$ be a neutrosophic implication. The induced operator $\mathscr{T}_{\mathscr{I}}$ by $\mathscr{I}$ is defined as follows:*

$$\mathscr{T}_{\mathscr{I}}(u,v) = \inf\{w \mid w \in D^*, v \leq_1 \mathscr{I}(u,w)\}, \text{for all } u,v \in D^*.$$

Remark 2. *(1) $\mathscr{T}_{\mathscr{I}}(u,v)$ is a non-empty set, since $\mathscr{I}(u,1_{D^*}) = 1_{D^*}$, for all $u \in D^*$.*
(2) $\mathscr{T}_{\mathscr{I}}$ defined above is not always a neutrosophic t-norm. For example, for all $u, v \in D^*$,

$$\mathscr{I}(u,v) = (1 - u_1 + u_1 v_1, \begin{cases} 0 & \text{if } u_2 \geq v_2, \\ v_2 & \text{otherwise} \end{cases}, \begin{cases} 0 & \text{if } u_3 \geq v_3, \\ v_3 & \text{otherwise} \end{cases})$$

is a neutrosophic implication. However, $\mathscr{T}_{\mathscr{I}}$ is not a neutrosophic t-norm, because $\mathscr{T}_{\mathscr{I}}(1_{D^*},v) = (1, v_2, v_3) \neq v$.

Theorem 13. *Let $\mathscr{I}$ be a neutrosophic implication on D^*. The induced operator $\mathscr{T}_{\mathscr{I}}$ by $\mathscr{I}$:*

$$\mathscr{T}_{\mathscr{I}}(u,v) = \inf\{w \mid w \in D^*, v \leq_1 \mathscr{I}(u,w)\}$$

is a neutrosophic t-norm if $\mathscr{I}$ satisfies the following conditions, for all $u, v, w \in D^*$:

(1) $u \leq_1 \mathscr{I}(v,w)$ *if and only if* $v \leq_1 \mathscr{I}(u,w)$;
(2) $\mathscr{I}(\mathscr{I}(u,v),w) = \mathscr{I}(u,\mathscr{I}(v,w))$;
(3) $\mathscr{I}(u,v) = 1_{D^*}$ *if and only if* $u \leq_1 v$;

(4) $\mathscr{I}(1_{D^*}, u) = u$.

Proof. Firstly, we prove that $\mathscr{T}_{\mathscr{I}}$ is a neutrosophic t-norm.
(NT1) From (1), we can directly get $\mathscr{T}_{\mathscr{I}}(u,v) = \mathscr{T}_{\mathscr{I}}(v,u)$, for all $u, v \in D^*$.

(NT2) From (1) and (2), $\mathscr{T}_{\mathscr{I}}(\mathscr{T}_{\mathscr{I}}(u,v),w) = \mathscr{T}_{\mathscr{I}}(w,\mathscr{T}_{\mathscr{I}}(u,v)) = \inf\{t \mid t \in D^*, \mathscr{T}_{\mathscr{I}}(u,v) \leq_1 \mathscr{I}(w,t)\} = \inf\{t \mid t \in D^*, u \leq_1 \mathscr{I}(v,\mathscr{I}(w,t))\} = \inf\{t \mid t \in D^*, v \leq_1 \mathscr{I}(u,\mathscr{I}(w,t))\} = \inf\{t \mid t \in D^*, v \leq_1 \mathscr{I}(w,\mathscr{I}(u,t))\} = \inf\{t \mid t \in D^*, \mathscr{T}_{\mathscr{I}}(v,w) \leq_1 \mathscr{I}(u,t)\} = \mathscr{T}_{\mathscr{I}}(\mathscr{T}_{\mathscr{I}}(v,w),u) = \mathscr{T}_{\mathscr{I}}(u,\mathscr{T}_{\mathscr{I}}(v,w))$.

(NT3) Since $\mathscr{T}_{\mathscr{I}}(u,1_{D^*}) = \mathscr{T}_{\mathscr{I}}(1_{D^*},u) = \inf\{t \mid t \in D^*, u \leq_1 \mathscr{I}(1_{D^*},t)\} = \inf\{t \mid t \in D^*, u \leq_1 t\} = u$, $\mathscr{T}_{\mathscr{I}}(u,1_{D^*}) = \mathscr{T}_{\mathscr{I}}(1_{D^*},u) = u$.

(NT4) Assume $u, u', v, v' \in D^*$ with the condition $u \leq_1 u'$, $v \leq_1 v'$. Since $\mathscr{I}$ is a neutrosophic implication, $\mathscr{I}(u',t) \leq_1 \mathscr{I}(u,t)$, for all $t \in D^*$. For any $t_0 \in \{t \mid t \in D^*, v' \leq_1 \mathscr{I}(u',t)\}$, it follows that $v' \leq_1 \mathscr{I}(u',t_0)$. Since $v \leq_1 v'$, and $\mathscr{I}(u',t_0) \leq_1 \mathscr{I}(u,t_0)$, $v \leq_1 \mathscr{I}(u,t_0)$, that is, $t_0 \in \{t \mid t \in D^*, v \leq_1 \mathscr{I}(u,t)\}$. Thus, $\{t \mid t \in D^*, v' \leq_1 \mathscr{I}(u',t)\} \subseteq_1 \{t \mid t \in D^*, v \leq_1 \mathscr{I}(u,t)\}$. Hence, $\inf\{t \mid t \in D^*, v \leq_1 \mathscr{I}(u,t)\} \leq_1 \inf\{t \mid t \in D^*, v' \leq_1 \mathscr{I}(u',t)\}$, that is, $\mathscr{T}_{\mathscr{I}}(u,v) \leq_1 \mathscr{T}_{\mathscr{I}}(u',v')$.

Therefore, $\mathscr{T}_{\mathscr{I}}$ is a neutrosophic t-norm. □

Theorem 13 describes the conditions that an induced operator $\mathscr{T}_{\mathscr{I}}$ by $\mathscr{I}$ is a neutrosophic t-norm. Moreover, we can construct neutrosophic t-norms with neutrosophic implications according to these conditions.

Next, some important properties of the residual neutrosophic implication on D^* will be discussed.

Theorem 14. *Let $\mathscr{I}$ be a residual neutrosophic implication on D^*. Then $\mathscr{I}(u, \inf_{w\in W} w) = \inf_{w\in W} \mathscr{I}(u,w)$, for all $u \in D^*$, $W \subseteq_1 D^*$.*

Proof. Let $W \subseteq_1 D^*$. If $W = \emptyset$, then $\mathscr{I}(u, \inf_{w\in W} w) = \mathscr{I}(u, 1_{D^*}) = 1_{D^*} = \inf_{w\in W} \mathscr{I}(u,w)$, for any $u \in D^*$. If $W \neq \emptyset$, since $\mathscr{I}$ is non-decreasingness in its second variable, $\mathscr{I}(u, \inf_{w\in W} w) \leq_1 \inf_{w\in W} \mathscr{I}(u,w)$, for all $u \in D^*$. Suppose $n = \inf_{w\in W} \mathscr{I}(u,w)$, then $n \leq_1 \mathscr{I}(u,w)$, now we have $w \in \{t \mid t \in D^*, n \leq_1 \mathscr{I}(u,t)\}$. By Definition 22, $\mathscr{T}_{\mathscr{I}}(n,u) \leq_1 w$, for all $w \in W$. Thus, $\mathscr{T}_{\mathscr{I}}(n,u) \leq_1 \inf_{w\in W} w$. Since $\mathscr{I}$ is non-decreasingness in its second variable, $\mathscr{I}(u, \inf_{w\in W} w) \geq_1 \mathscr{I}(u, \mathscr{T}_{\mathscr{I}}(n,u))$. Since $u \leq_1 \mathscr{I}(v,w)$ if and only if $w \geq_1 \mathscr{T}_{\mathscr{I}}(u,v)$, and $\mathscr{T}_{\mathscr{I}}(n,u) \geq_1 \mathscr{T}_{\mathscr{I}}(n,u)$, $\mathscr{I}(u, \mathscr{T}_{\mathscr{I}}(n,u)) \geq_1 n = \inf_{w\in W} \mathscr{I}(u,w)$. Therefore, $\mathscr{I}(u, \inf_{w\in W} w) \geq_1 \inf_{w\in W} \mathscr{I}(u,w)$. □

From Theorem 14, we know that a residual neutrosophic implication satisfies infinitively $\wedge$-distributive in its second variable.

Theorem 15. *Assume that $\mathscr{I}$ is a residual neutrosophic implication on D^*. Then, for all $u, v \in D^*$,*

(1) $\mathscr{T}_{\mathscr{I}}(u, \mathscr{I}(u, \mathscr{T}_{\mathscr{I}}(u,v))) = \mathscr{T}_{\mathscr{I}}(u,v)$;
(2) $\mathscr{I}(u, \mathscr{T}_{\mathscr{I}}(u, \mathscr{I}(u,v))) = \mathscr{I}(u,v)$.

Proof. Let $u, v \in D^*$.

(1) $\mathscr{T}_{\mathscr{I}}(u, \mathscr{I}(u, \mathscr{T}_{\mathscr{I}}(u,v))) = \inf\{t \mid t \in D^*, \mathscr{I}(u, \inf\{t \mid t \in D^*, v \leq_1 \mathscr{I}(u,t)\}) \leq_1 \mathscr{I}(u,t)\} = \inf\{t \mid t \in D^*, \inf\{\mathscr{I}(u,t) \mid t \in D^*, v \leq_1 \mathscr{I}(u,t)\} \leq_1 \mathscr{I}(u,t)\} = \inf\{t \mid t \in D^*, v \leq_1 \mathscr{I}(u,t)\} = \mathscr{T}_{\mathscr{I}}(u,v)$.
(2) $\mathscr{I}(u, \mathscr{T}_{u,\mathscr{I}}(\mathscr{I}(u,v))) = \mathscr{I}(u, \inf\{t \mid t \in D^*, \mathscr{I}(u,v) \leq_1 \mathscr{I}(u,t)\}) = \inf\{\mathscr{I}(u,t) \mid t \in D^*, \mathscr{I}(u,t) \geq_1 \mathscr{I}(u,v)\} = \mathscr{I}(u,v)$.

□

Summarizing the results in Theorem 5 and 13, we get the following theorem.

Theorem 16. *(1) Assume that $\mathscr{T}$ is a neutrosophic t-norm on D^*. Then $\mathscr{I}_{\mathscr{T}}(u, \inf_{v\in V} v) = \inf_{v\in V} \mathscr{I}_{\mathscr{T}}(u,v)$ and $\mathscr{T} = \mathscr{T}_{\mathscr{I}_{\mathscr{T}}}$;*
(2) Let $\mathscr{I}$ be a neutrosophic implication on D^*. Then $\mathscr{T}_{\mathscr{I}}$ which satisfies the conditions presented in Theorem 13 is a infinitely $\vee$-distributive neutrosophic t-norm, and $\mathscr{I} = \mathscr{I}_{\mathscr{T}_{\mathscr{I}}}$.

Proof. (1) From Theorem 5, $\mathscr{I}_{\mathscr{T}}$ is a neutrosophic implication. Next, we prove $\mathscr{I}_{\mathscr{T}}(u, \inf_{v\in V} v) = \inf_{v\in V} \mathscr{I}_{\mathscr{T}}(u,v)$, for all $u \in D^*$, $V \subseteq_1 D^*$. Suppose $V \subseteq_1 D^*$. If $V = \emptyset$, then $\mathscr{I}_{\mathscr{T}}(u, \inf_{v\in V} v) = \mathscr{I}_{\mathscr{T}}(u, 1_{D^*}) = 1_{D^*} = \inf_{v\in V} \mathscr{I}_{\mathscr{T}}(u,v)$, for all $u \in D^*$. If $V \neq \emptyset$, then $\mathscr{I}_{\mathscr{T}}(u, \inf_{v\in V} v) = \sup\{t \mid t \in D^*, \mathscr{T}(t,u) \leq_1 \inf_{v\in V} v\} = \sup\{t \mid t \in D^*, \forall v \in V, \mathscr{T}(t,u) \leq_1 v\} = \sup\{t \in D^* \mid \forall v \in V, t \leq_1 \mathscr{I}_{\mathscr{T}}(u,v)\} = \sup\{t \in D^* \mid t \leq_1 \inf_{v\in V} \mathscr{I}_{\mathscr{T}}(u,v)\} = \inf_{v\in V} \mathscr{I}_{\mathscr{T}}(u,v)$. Finally, from Definitions 18 and 22, we get $\mathscr{T}_{\mathscr{I}_{\mathscr{T}}}(u,v) = \inf\{t \mid t \in D^*, v \leq_1 \mathscr{I}_{\mathscr{T}}(u,t)\} = \inf\{t \mid t \in D^*, \mathscr{T}(u,v) \leq_1 t\} = \mathscr{T}(u,v)$, for all $u,v \in D^*$. Thus, $\mathscr{T} = \mathscr{T}_{\mathscr{I}_{\mathscr{T}}}$.
(2) From Definition 22 and Theorem 13, $\mathscr{T}_{\mathscr{I}}$ is a neutrosophic t-norm. Next, we prove $\mathscr{T}_{\mathscr{I}}(\sup_{u\in U} u, v) = \sup_{u\in U} \mathscr{T}_{\mathscr{I}}(u,v)$, for all $v \in D^*$, $U \subseteq_1 D^*$. Suppose $U \subseteq_1 D^*$. If $U = \emptyset$, then $\mathscr{T}_{\mathscr{I}}(\sup_{u\in U} u, v) = \mathscr{T}_{\mathscr{I}}(0_{D^*}, v) = 0_{D^*} = \sup_{u\in U} \mathscr{T}_{\mathscr{I}}(u,v)$, for all $v \in D^*$. If $U \neq \emptyset$, then $\mathscr{T}_{\mathscr{I}}(\sup_{u\in U} u, v) = \inf\{t \mid t \in D^*, v \leq_1 \mathscr{I}(\sup_{u\in U} u, t)\} = \inf\{t \mid t \in D^*, \forall u \in U, v \leq_1 \mathscr{I}(u,t)\} = \inf\{t \mid t \in D^*, \forall u \in U, t \geq_1 \mathscr{T}_{\mathscr{I}}(u,v)\} = \inf\{t \mid t \in D^*, t \geq_1 \sup_{u\in U} \mathscr{T}_{\mathscr{I}}(u,v)\} = \sup_{u\in U} \mathscr{T}_{\mathscr{I}}(u,v)$. Since $\mathscr{T}_{\mathscr{I}}$ satisfies the commutative law, $\mathscr{T}_{\mathscr{I}}(u, \sup_{v\in V} v) = \sup_{v\in V} \mathscr{T}_{\mathscr{I}}(u,v)$. Hence, $\mathscr{T}_{\mathscr{I}}$ is infinitely $\vee$-distributive. At last, from Definitions 18 and 22, we get $\mathscr{I}_{\mathscr{T}_{\mathscr{I}}}(u,v) = \sup\{t \mid t \in D^*, \mathscr{T}_{\mathscr{I}}(t,u) \leq_1 v\} = \sup\{t \mid t \in D^*, t \leq_1 \mathscr{I}(u,v)\} = \mathscr{I}(u,v)$, for all $u,v \in D^*$. Thus, $\mathscr{I} = \mathscr{I}_{\mathscr{T}_{\mathscr{I}}}$. □

Sections 4 and 5 mainly discuss neutrosophic t-norms and their residual implications, then, we can get a residuated lattice by using these two neutrosophic logic operators as follows:

Theorem 17. *Let $\mathscr{T}$ be a neutrosophic t-norm on D^*. Suppose $(D^*; \vee_1, \wedge_1, ^c, 0_{D^*}, 1_{D^*})$ is a system on D^*. For all $u,v \in D^*$, we define:*

$$u \otimes v = \mathscr{T}_{\mathscr{I}_{\mathscr{T}}}(u,v); \; u \to v = \mathscr{I}_{\mathscr{T}}(u,v).$$

Then, $(D^*; \vee_1, \wedge_1, \otimes, \to, 0_{D^*}, 1_{D^*})$ is a residuated lattice.

Proof. Firstly, from Proposition 3, we know that $(D^*; \vee_1, \wedge_1, 0_{D^*}, 1_{D^*})$ is a bounded lattice.
Then, we prove that $(D^*; \otimes, 1_{D^*})$ is a commutative monoid. (1) For any $u \in D^*$, $1_{D^*} \otimes u = \inf\{t \mid t \in D^*, 1_{D^*} \leq_1 \mathscr{I}_{\mathscr{T}}(u,t)\} = \inf\{t \mid t \in D^*, \mathscr{I}_{\mathscr{T}}(u,t) = 1_{D^*}\} = \inf\{t \mid t \in D^*, u \leq_1 t\} = u$, $u \otimes 1_{D^*} = \inf\{t \mid t \in D^*, u \leq_1 \mathscr{I}_{\mathscr{T}}(1_{D^*}, t)\} = \inf\{t \mid t \in D^*, u \leq_1 t\} = u$. Thus, $1_{D^*} \otimes u = u \otimes 1_{D^*} = u$. (2) Theorem 16 proves that $\mathscr{T}_{\mathscr{I}_{\mathscr{T}}} = \mathscr{T}$ is a neutrosophic t-norm. Thus, $\mathscr{T}$ satisfies the commutative law, that is, $u \otimes v = v \otimes u$. (3) Similarly, $\mathscr{T}$ satisfies the associative law, that is, $u \otimes (v \otimes w) = (u \otimes v) \otimes w$.

Finally, we prove that $\otimes$ is a binary operation for which the equivalence

$$u \otimes v \leq_1 w \text{ if and only if } v \leq_1 u \to w$$

holds for all $u,v,w \in D^*$. On the one hand, by the definition of $\otimes$, we have $u \otimes v = \inf\{w \mid w \in D^*, v \leq_1 u \to w\}$, then $u \otimes v \leq_1 w$. Thus, $v \leq_1 u \to w$. On the other hand, from the definition of $\to$, we have $u \to w = \sup\{v \mid v \in D^*, u \otimes v \leq_1 w\}$. Thus, $u \otimes v \leq_1 w$.
Therefore, $(D^*; \vee_1, \wedge_1, \otimes, \to, 0_{D^*}, 1_{D^*})$ is a residuated lattice. □

Example 12. *Suppose $(D^*; \vee_1, \wedge_1, ^c, 0_{D^*}, 1_{D^*})$ is a system on D^*. For all $u,v \in D^*$, we define:*

$$u \otimes v = \mathscr{T}_{\mathscr{I}_{\mathscr{T}}}(u,v);$$
$$u \to v = \mathscr{I}_{\mathscr{T}}(u,v).$$

where $\mathscr{T}_{\mathscr{I}_{\mathscr{T}}}(u,v)$ *is that presented in Theorem 4,* $\mathscr{I}_{\mathscr{T}}(u,v)$ *is that presented in Example 10.*

Then, $(D^*; \vee_1, \wedge_1, \otimes, \to, 0_{D^*}, 1_{D^*})$ *is a residuated lattice.*

Proof. Firstly, from Proposition 3, we know that $(D^*; \vee_1, \wedge_1, 0_{D^*}, 1_{D^*})$ is a bounded lattice. Then, we prove that $(D^*; \otimes, 1_{D^*})$ is a commutative monoid. (1) For any $u \in D^*$, by the definition of $\otimes$, we get $1_{D^*} \otimes u = u \otimes 1_{D^*} = u$. (2) Obviously, $u \otimes v = v \otimes u$. (3) Suppose $u, v, w \in D^*$. If at least one of them is equal to 1_{D^*}, then $u \otimes (v \otimes w) = (u \otimes v) \otimes w$. Otherwise, $u \otimes (v \otimes w) = (\min(u_1, \min(v_1, w_1)), \max(1 - u_1, \max(1 - v_1, 1 - w_1)), \max(u_3, \max(v_3, w_3))) = (\min(u_1, v_1, w_1), \max(1 - u_1, 1 - v_1, 1 - w_1), \max(u_3, v_3, w_3)) = (\min(v_1, \min(u_1, w_1)), \max(1 - v_1, \max(1 - u_1, 1 - w_1)), \max(v_3, \max(u_3, w_3))) = (u \otimes v) \otimes w$.

Finally, we will prove

$$u \otimes v \leq_1 w \Leftrightarrow v \leq_1 u \to w, \text{ for all } u, v, w \in D^*.$$

If $u = 1_{D^*}$, $u \otimes v = v \leq_1 w$ if and only if $v \leq_1 u \to w \Rightarrow v \leq_1 1_{D^*} \to w \Rightarrow v \leq_1 w$; If $v = 1_{D^*}$, $u \otimes v = u \otimes 1_{D^*} = u \leq_1 w \Rightarrow u \leq_1 1_{D^*}$ if and only if $v \leq_1 u \to w \Rightarrow 1_{D^*} \leq 1_{D^*}$; If $u \neq 1_{D^*}$ and $v \neq 1_{D^*}$, $u \otimes v = (\min(u_1, \min(v_1, w_1)), \max(1 - u_1, \max(1 - v_1, 1 - w_1)), \max(u_3, \max(v_3, w_3))) \leq_1 w$ if and only if

$$v \leq_1 \begin{cases} 1_{D^*} & \text{if } w = 1_{D^*}, \\ w & \text{if } u = 1_{D^*}, \\ (I_{GD}(u_1, w_1), \begin{cases} u_1 & \text{if } u_1 \leq 1 - w_2, \\ 0 & \text{otherwise} \end{cases}, J_{GD}(u_3, w_3)) & \text{otherwise} \end{cases}$$

that is, $v \leq_1 u \to w$.

Therefore, $(D^*; \vee_1, \wedge_1, \otimes, \to, 0_{D^*}, 1_{D^*})$ is a residuated lattice. □

7. Conclusions

Neutrosophic logic plays a vital role in neutrosophic set theory. Neutrosophic t-norms, t-conorms, negators and implications are very important neutrosophic logic operators. In this paper, under the first type inclusion relation, the lattice structure of neutrosophic sets is discussed, $(D^*; \leq_1)$ and $(D^*; \vee_1, \wedge_1, ^c, 0_{D^*}, 1_{D^*})$ are proved to be a complete lattice and De Morgan algebra, respectively. On the complete lattice $(D^*; \leq_1)$, we introduce the definitions of neutrosophic t-norms, t-conorms, negators and their operations. Furthermore, De Morgan neutrosophic triples are defined, which describe that neutrosophic t-norms and t-conorms are dual with respect to the standard neutrosophic negator. Then, we introduce neutrosophic residual implications (co-implications) on the complete lattice $(D^*; \leq_1)$, propose a theorem which shows that residual operations induced by neutrosophic t-norms are neutrosophic implications, investigate basic properties for neutrosophic residual implications (co-implications), prove that residual neutrosophic t-norms are infinitely $\vee$-distributive, and give some important results for residual neutrosophic t-norms and neutrosophic residual implications. Finally, we introduce neutrosophic operations produced by neutrosophic implications, discuss the conditions that the neutrosophic operations are neutrosophic t-norms, and then construct residuated lattices. Based on these results, we will consider their applications in neutrosophic inference systems in the future.

Author Contributions: This paper is written through contributions of all authors. The individual contributions and responsibilities of all authors can be described as follows: The idea of this whole thesis was put forward by X.Z., he also provided valuable and helpful suggestions throughout the writing of the article. Q.H. investigated the work and completed the paper.

Funding: This work was supported by the National Natural Science Foundation of China (Grant no. 61573240).

Conflicts of Interest: The authors declare no conflict of interest.

References

1. Smarandache, F. *Neutrosophy: Neutrosophic Probability, Set, and Logic: Analytic Synthesis and Synthetic Analysis;* American Research Press: Santa Fe, NM, USA, 1998; pp. 25–38.
2. Wang, H.; Smarandache, F.; Zhang, Y.; Sunderraman, R. Single valued neutrosophic sets. *Multispace Multistruct. Neutrosophic Transdiscipl.* **2010**, *4*, 410–413.
3. Ye, J.; Smarandache, F. Similarity measure of refined single-valued neutrosophic sets and its multicriteria decision making method. *Neutrosophic Sets Syst.* **2016**, *12*, 41–44.
4. Ye, J. Vector similarity measures of simplified neutrosophic sets and their application in multicriteria decision making. *Int. J. Fuzzy Syst.* **2014**, *16*, 204–211.
5. Liu, P. The aggregation operators based on archimedean t-conorm and t-norm for single-valued neutrosophic numbers and their application to decision making. *Int. J. Fuzzy Syst.* **2016**, *18*, 849–863. [CrossRef]
6. Ye, J. Multicriteria decision-making method using the correlation coefficient under single-valued neutrosophic environment. *Int. J. Gen. Syst.* **2013**, *42*, 386–394. [CrossRef]
7. Hu, Q.; Zhang, X. New similarity measures of single-valued neutrosophic multisets based on the decomposition theorem and its application in medical diagnosis. *Symmetry* **2018**, *10*, 466. [CrossRef]
8. Wang, J.; Zhang, X. Two types of single-valued neutrosophic covering rough sets and an application to decision making. *Symmetry* **2018**, *10*, 710. [CrossRef]
9. Zhang, X.; Mao, X.; Wu, Y.; Zhai, X. Neutrosophic filters in pseudo-BCI algebras. *Int. J. Uncertain Quantif.* **2018**, *8*, 511–526. [CrossRef]
10. Zhang, X.; Hu, Q.; Smarandache, F.; An, X. On neutrosophic triplet groups: Basic properties, NT-subgroups and some notes. *Symmetry* **2018**, *10*, 289. [CrossRef]
11. Ma, Y.; Zhang, X.; Yang, X.; Zhou, X. Generalized neutrosophic extended triplet group. *Symmetry* **2019**, *11*, 327. [CrossRef]
12. Wu, X.; Zhang, X. The decomposition theorems of AG-neutrosophic extended triplet loops and strong AG-(1, 1)-loops. *Mathematics* **2019**, *7*, 268. [CrossRef]
13. Cuong, B.C.; Kreinovich, V. Picture fuzzy sets-a new concept for computational intelligence problems. In Proceedings of the IEEE Third World Congress on Information and Communication Technologies, Hanoi, Vietnam, 15–18 December 2013; pp. 1–6.
14. Zhang, X.; Bo, C.; Smarandache, F.; Dai, J. New inclusion relation of neutrosophic sets with applications and related lattice structure. *Int. J. Mach. Learn. Cybern.* **2018**, *9*, 1753–1763. [CrossRef]
15. Klement, E.P.; Mesiar, R.; Stupňanová, A. Picture fuzzy sets and 3-fuzzy sets. In Proceedings of the IEEE International Conference on Fuzzy Systems, Rio de Janeiro, Brazil, 8–13 July 2018.
16. Cuong, B.C.; Phong, P.H.; Smarandache, F. Standard neutrosophic soft theory: Some first results. *Neutrosophic Sets Syst.* **2016**, *12*, 80–91.
17. Cuong, B.C.; Ngan, R.T.; Ngoc, L.C. Some Algebraic Properties of Picture Fuzzy t-Norms and Picture Fuzzy t—Conorms on Standard Neutrosophic Sets. 2017. Available online: http://vixra.org/abs/1701.0144 (accessed on 20 May 2019).
18. Smarandache, F. Neutrosophic set—A generialization of the intuitionistics fuzzy sets. *J. Def. Manag.* **2006**, *1*, 38–42.
19. Smarandache, F. A unifying field in logics: Neutrosophic logic. *Multiple-Valued Log.* **2002**, *8*, 385–438.
20. Menger, K. Statistical metrics. *Natl. Acad. Sci.* **1942**, *28*, 535–537. [CrossRef]
21. Schweizer, B.; Sklar, A. Associative functions and statistical triangle inequalities. *Publ. Math.* **1961**, *8*, 169–186.
22. Schweizer, B.; Sklar, A. *Probabilistic Metric Spaces;* North-Holland: New York, NY, USA, 1983.
23. Simon, D. Fuzzy sets and fuzzy logic: Theory and applications. *Control Eng. Pract.* **1996**, *4*, 1332–1333. [CrossRef]
24. Klement, E.P.; Mesiar, R.; Pap, E. Triangular norms. Position paper I: Basic analytical and algebraic properties. *Fuzzy Sets Syst.* **2004**, *143*, 5–26. [CrossRef]
25. Li, D. Type-2 triangular norms and their residual operators. *Inf. Sci.* **2015**, *317*, 259–277. [CrossRef]
26. Zhang, B. Notes on type-2 triangular norms and their residual operators. *Inf. Sci.* **2016**, *346–347*, 338–350. [CrossRef]
27. Baczyński, M.; Jayaram, B. Fuzzy implications. *Stud. Fuzziness Soft Comput.* **2008**, *231*, 41–105.
28. Zhang, X.; Borzooei, R.A.; Jun, Y.B. Q-filters of quantum B-algebras and basic implication algebras. *Symmetry* **2018**, *10*, 573. [CrossRef]

29. Hu, B.; Kwong, C.K. On type-2 fuzzy sets and their t-norm operations. *Inf. Sci.* **2014**, *255*, 58–81. [CrossRef]
30. Wang, C.; Hu, B. On fuzzy-valued operations and fuzzy-valued fuzzy sets. *Fuzzy Sets Syst.* **2015**, *268*, 72–92. [CrossRef]
31. Fodor, J.C. Contrapositive symmetry of fuzzy implications. *Fuzzy Sets Syst.* **1995**, *69*, 141–156. [CrossRef]
32. Klement, E.P., Navara, M. A survey on different triangular norm-based fuzzy logics. *Fuzzy Sets Syst.* **1999**, *101*, 241–251. [CrossRef]
33. Zhou, H. *Probabilistically Quantitative Logic and Its Application*; Science Press: Beijing, China, 2015; pp. 219–235.
34. Deschrijver, G. Implication functions in interval-valued fuzzy set theory. *Stud. Fuzziness Soft Comput.* **2013**, *300*, 73–99.
35. Deschrijver, G.; Cornelis, C.; Kerre, E.E. On the representation of intuitionistic fuzzy t-norms and t-conorms. *IEEE Trans. Fuzzy Syst.* **2004**, *12*, 45–61. [CrossRef]
36. Deschrijver, G.; Kerre, E.E. Classes of intuitionistic fuzzy t-norms satisfying the residuation principle. *Int. J. Univ. Fuzziness Knowl.-Based Syst.* **2003**, *11*, 691–709. [CrossRef]
37. Goguen, J.A. L-fuzzy sets. *J. Math. Anal. Appl.* **1967**, *18*, 145–174. [CrossRef]
38. Cuong, B.C. Picture fuzzy sets. *J. Comput. Sci. Cybern.* **2014**, *30*, 409–420.
39. Cuong, B.C.; Hai, P.V. Some fuzzy logic operators for picture fuzzy sets. In Proceedings of the IEEE Seventh International Conference on Knowledge and Systems Engineering, Ho Chi Minh City, Vietnam, 8–10 October 2015; pp. 132–137.
40. Cuong, B.C.; Kreinovich, V.; Ngan, R.T. A classification of representable t-norm operators for picture fuzzy sets. In Proceedings of the IEEE Eighth International Conference on Knowledge and Systems Engineering (KSE), Hanoi, Vietnam, 6–8 October 2016; pp. 19–24.
41. Cuong, B.C.; Ngan, R.T.; Hai, B.D. An involutive picture fuzzy negator on picture fuzzy sets and some De Morgan triples. In Proceedings of the IEEE Seventh International Conference on Knowledge and Systems Engineering (KSE), Ho Chi Minh City, Vietnam, 8–10 October 2015; pp. 126–131.
42. Cuong, B.C.; Ngan, R.T.; Long, L.B. Some new De Morgan picture operator triples in picture fuzzy logic. *J. Comput. Sci. Cybern.* **2017**, *33*, 143–164.
43. Son, L.H.; Viet, P.V.; Hai, P.V. Picture inference system: A new fuzzy inference system on picture fuzzy set. *Appl. Intell.* **2016**, *46*, 1–18. [CrossRef]
44. Baets, B.D.; Fodor, J. Residual operators of uninorms. *Soft Comput.* **1999**, *3*, 89–100. [CrossRef]
45. Wang, Z.; Yu, Y. Pseudo-t-norms and implication operators on a complete Brouwerian lattice. *Fuzzy Sets Syst.* **2002**, *132*, 113–124. [CrossRef]
46. Wang, Z.; Fang, J. Residual operations of left and right uninorms on a complete lattice. *Fuzzy Sets Syst.* **2009**, *160*, 22–31. [CrossRef]
47. Wang, Z.; Fang, J. Residual coimplicators of left and right uninorms on a complete lattice. *Fuzzy Sets Syst.* **2008**, *160*, 2086–2096. [CrossRef]
48. Liu, H. Semi-uninorms and implications on a complete lattice. *Fuzzy Sets Syst.* **2012**, *191*, 72–82. [CrossRef]
49. Alkhazaleh, S. More on neutrosophic norms and conforms. *Neutrosophic Sets Syst.* **2015**, *9*, 23–30.
50. Smarandache, F. N-valued refined neutrosophic logic and its applications to physics. *Neutrosophic Theory Appl. Progress Phys.* **2013**, *4*, 36–44.
51. Smarandache, F. N-norm and n-conorm in neutrosophic logic and set, and the neutrosophic topologies. In *A Unifying Field in Logics: Neutrosophic Logic. Neutrosophy, Neutrosophic Set, Neutrosophic Probability*, 4th ed.; American Research Press: Santa Fe, NM, USA, 2005.
52. Atanassov, K. Intuitionistic fuzzy sets. *Fuzzy Sets Syst.* **1986**, *20*, 87–96. [CrossRef]
53. Smarandache, F. *Neutrosophic Perspectives: Triplets, Duplets, Multisets, Hybrid Operators, Modal Logic, Hedge Algebras. And Applications*; Infinite Study: Conshohocken, PA, USA, 2017.
54. Zhang, X.; Bo, C.; Smarandache, F.; Park, C. New operations of tatally dependent-neutrosophic sets and tatally dependent-neutrosophic soft sets. *Symmetry* **2018**, *10*, 187. [CrossRef]
55. Sevcovic, D. Free non-distributive morgan-stone algebras. *N. Z. J. Math.* **1996**, *25*, 85–94.

symmetry

MDPI

Article

Introduction to Non-Standard Neutrosophic Topology

Mohammed A. Al Shumrani [1] and Florentin Smarandache [2,*]

[1] Department of Mathematics, King Abdulaziz University, Jeddah 21589, Saudi Arabia; maalshmrani1@kau.edu.sa
[2] Mathematics Department, University of New Mexico, Gallup, NM 87301, USA
* Correspondence: smarand@unm.edu

Received: 10 May 2019; Accepted: 20 May 2019; Published: 23 May 2019

Abstract: For the first time we introduce non-standard neutrosophic topology on the extended non-standard analysis space, called non-standard real monad space, which is closed under neutrosophic non-standard infimum and supremum. Many classical topological concepts are extended to the non-standard neutrosophic topology, several theorems and properties about them are proven, and many examples are presented.

Keywords: non-standard analysis; extended non-standard analysis; monad; binad; left monad closed to the right; right monad closed to the left; pierced binad; unpierced binad; non-standard neutrosophic mobinad set; neutrosophic topology; non-standard neutrosophic topology

---

1. Introduction to Non-Standard Analysis

The purpose of this study is to initiate for the first time a new field of research, called non-standard neutrosophic algebraic structures, and we start with non-standard neutrosophic topology (NNT) in this paper. Being constructed on the set of hyperreals, that includes the infinitesimals, NNT can further be utilized in neutrosophic calculus applications.

As a branch of mathematical logic, non-standard analysis [1] deals with *hyperreal numbers*, which include *infinitesimals* and *infinities*.

The introduction of infinitesimals in calculus has been debated philosophically in the history of mathematics since the time of G. W. Leibniz, with pros and cons. Many mathematicians prefer the *epsilon-delta* use in calculus concepts' definitions and theorems' proofs.

By the 1960s Abraham Robinson had developed non-standard analysis [2] in a more rigorous way.

Besides calculus, non-standard analysis found applications in mathematical physics, mathematical economics, and in probability theory.

In 1998, Smarandache [3] used non-standard analysis in philosophy and in neutrosophic logic, in order to differentiate between *absolute truth* (which is truth in all possible worlds, according to Leibniz), and *relative truth* (which is, according to the same Leibniz, truth in at least one world). Let T represent the neutrosophic truth value, I the neutrosophic indeterminacy value, and F the neutrosophic falsehood value, with $T, I, F \in [^{-}0, 1^{+}]$. Then T (*absolute truth*) $= 1^{+} = \mu\ (1^{+})$, while T (*relative truth*) $=$ 1. This is analogously for *absolute falsehood* vs. *relative falsehood*, and *absolute indeterminacy* vs. *relative indeterminacy*.

Then he extended [3] the use of non-standard analysis to neutrosophic set (*absolute membership/indeterminacy/nonmembership* vs. *relative membership/indeterminacy/nonmembership* respectively) and to neutrosophic probability (*absolute occurrence/indeterminate occurrence/nonoccurence of an event* vs. *relative occurrence/indeterminate occurrence/nonoccurence of an event*, respectively).

We next recall several notions and results from classical non-standard analysis [2] that are needed to defining and developing the non-standard neutrosophic topology.

The set R^* of *nonstandard reals* (or *hyperreals*) is the generalization of the real numbers (R).

The *transfer principle* states that first-order statements that are valid in R are also valid in R^*.

$$R^* \text{ includes the infinites and the infinitesimals, which on the } \textit{hyperreal number line} \text{ may be represented as } 1/\varepsilon = \omega/1. \quad (1)$$

An *infinite* (or infinite number) (ω) is a number that is greater than anything:

$$1 + 1 + 1 + \ldots + 1 \text{ (for any number of finite terms)} \quad (2)$$

The infinitesimals are reciprocals of infinites.

An *infinitesimal* (or infinitesimal number) (ε) is a number ε such that $|\varepsilon| < 1/n$, for any non-null positive integer n.

An infinitesimal is so small that it cannot be measured, and it is very close to zero.

The infinitesimal in absolute value, is a number smaller than anything nonzero positive number.

In calculus one uses the infinitesimals.

$$\text{By } R_+{}^* \text{ we denote the set of positive non-zero hyperreal numbers.} \quad (3)$$

Left Monad {for simplicity, denoted [2] by $({}^-a)$ or only ${}^-a$} was defined as:

$$\mu({}^-a) = ({}^-a) = {}^-a = \overline{a} = \{a - x, x \in R_+{}^* | x \text{ is } infinitesimal\} \quad (4)$$

Right Monad {for simplicity, denoted [2] by (a^+) or only by a^+} was defined as:

$$\mu(a^+) = (a^+) = a^+ = \overset{+}{a} = \{a + x, x \in R_+{}^* | x \text{ is } infinitesimal\} \quad (5)$$

μ (a) is a *monad* (*halo*) of an element $a \in R^*$, which is formed by a subset of numbers infinitesimally close (to the left-hand side, or right-hand side) to a.

1.1. Non-Standard Analysis's First Extension

In 1998, Smarandache [3] introduced the pierced binad.

Pierced binad {for simplicity, denoted by $({}^-a^+)$ or only ${}^-a^+$} was defined as:

$$\begin{aligned} &\mu({}^-a^+) = ({}^-a^+) = {}^-a^+ = \overset{-+}{a} = \\ &= \{a - x, x \in R_+{}^* | x \text{ is } infinitesimal\} \cup \{a + x, x \in R_+{}^* | x \text{ is } infinitesimal\} \\ &= \{a \pm x, x \in R_+{}^* | x \text{ is } infinitesimal\} \end{aligned} \quad (6)$$

This extension was needed in order to be able to do union aggregations of non-standard neutrosophic sets, where a left monad μ $({}^-\text{a})$ had to be united with a right monad μ (a^+), as such producing a pierced binad: $\mu({}^-\text{a}) \cup \mu(\text{a}^+) =_N \mu({}^-\text{a}^+)$. Without this pierced binad we would not have been able to define the non-standard neutrosophic operators.

1.2. Non-Standard Analysis's Second Extension

Smarandache [4,5] introduced at the beginning of 2019 for the first time, the left monad closed to the right, the right monad closed to the left, and unpierced binad, defined as below:

Left Monad Closed to the Right

$$\begin{aligned} &\mu(\overset{-0}{a}) = (\overset{-0}{a}) = \overset{-0}{a} = \{a - x | x = 0, or\ x \in R_+{}^* \text{ and } x \text{ is } infinitesimal\} = \mu({}^-a) \cup \{a\} = ({}^-a) \cup \{a\} \\ &= {}^-a \cup \{a\} \end{aligned} \quad (7)$$

Right Monad Closed to the Left

$$\mu(\overset{0+}{a}) = (\overset{0+}{a}) = \overset{0+}{a} = \{a + x | x = 0, or\ x \in R_+{}^* \ and\ x\ is\ infinitesimal\} = \mu(a^+) \cup \{a\} = (a^+) \cup \{a\} \\ = a^+ \cup \{a\} \tag{8}$$

Unpierced Binad

$$\mu(\overset{-0+}{a}) = (\overset{-0+}{a}) = \overset{-0+}{a} = \{a - x | \ x \in R_+{}^* \ and\ x\ is\ infinitesimal\} \cup \{a + x | \ x \in R_+{}^* \ and\ x\ is\ infinitesimal\} \cup \{a\} \\ = \{a \pm x | x = 0, or\ x \in R_+{}^* \ and\ x\ is\ infinitesimal\} \\ = \mu({}^-a^+) \cup \{a\} = ({}^-a^+) \cup \{a\} = \ {}^-a^+ \cup \{a\} \tag{9}$$

Therefore, as seen, the element $\{a\}$ has been included in both the left and right monads, and also in the pierced binad respectively.

All monads and binads are subsets of R^*.

This second extension was done in order to be able to compute the non-standard aggregation operators (negation, conjunction, disjunction, implication, equivalence) in non-standard neutrosophic logic, set, and probability, and now we need them in non-standard neutrosophic topology.

1.3. The Best Notations for Monads and Binads

For any standard real number $a \in R$, we employ the following notations for monads and binads:

$$\overset{m}{a} \in \{a, \overset{-}{a}, \overset{-0}{a}, \overset{+}{a}, \overset{0+}{a}, \overset{-+}{a}, \overset{-0+}{a}\} \text{ and by convention } \overset{0}{a} = a; \tag{10}$$

where

$$m \in \{{}^-, {}^{-0}, {}^+, {}^{+0}, {}^{-+}, {}^{-0+}\} = \{{}^0, {}^-, {}^{-0}, {}^+, {}^{+0}, {}^{-+}, {}^{-0+}\}; \tag{11}$$

thus "m" written above the standard real number "a" means: a standard real number (^{0}, or nothing above), or a left monad (${}^-$), or a left monad closed to the right (${}^{-0}$), or a right monad (${}^+$), or a right monad closed to the left (${}^{0+}$), or a pierced binad (${}^{-+}$), or a unpierced binad (${}^{-0+}$) respectively.

Neutrosophic notations will have an index $_N$ associated to each symbol, for example: the classical symbol < (less than), becomes $<_N$ (neutrosophically less than, i.e., some indeterminacy is involved, especially with respect to infinitesimals, monads and binads).

Similarly for: $\cap$ and $\cap_N$, $\wedge$ and $\wedge_N$ etc.

1.4. Non-Standard Neutrosophic Inequalities

We have the following *neutrosophic non-standard inequalities* (taking into account the definitions of infinitesimals, monads and binads):

$$({}^-a) <_N a <_N (a^+) \tag{12}$$

because

$$\forall x \in R_+^*, a - x < a < a + x \tag{13}$$

where x is a (nonzero) positive infinitesimal.

The converse also is true:

$$(a^+) >_N \mathrm{a} >_N (-a) \tag{14}$$

Similarly:

$$({}^-a) \le_N({}^-a^+) \le_N(a^+) \tag{15}$$

To prove it, we rely on the fact that $({}^-a^+) = ({}^-a) \cup (a^+)$ and the number a is in between the subsets (on the real number line) ${}^-a = (a - \varepsilon, a)$ and $a^+ = (a, a + \varepsilon)$, so:

$$({}^-a) \le_N({}^-a) \cup (a^+) \ge_N(a^+) \tag{16}$$

Conversely, it is neutrosophically true too:

$$(a^{+}) \geq_N ({}^{-}a) \cup (a^{+}) \geq_N ({}^{-}a) \tag{17}$$

$$\text{Also, } \overset{-}{a} \leq_N \overset{-0}{a} \leq_N a \leq_N \overset{0+}{a} \leq_N \overset{+}{a} \text{ and } \overset{-}{a} \leq_N \overset{-+}{a} \leq_N \overset{-0+}{a} \leq_N \overset{+}{a} \tag{18}$$

Conversely, they are also neutrosophically true:

$$\overset{+}{a} \geq_N \overset{0+}{a} \geq_N a \geq_N \overset{-0}{a} \geq_N \overset{-}{a} \text{ and } \overset{+}{a} \geq_N \overset{-0+}{a} \geq_N \overset{-+}{a} \geq_N \overset{-}{a} \text{ respectively.} \tag{19}$$

Let a, b be two standard real numbers. If $a > b$, which is (standard) classical real inequality, then we have:

$$a >_N ({}^{-}b), a >_N (b^{+}), a >_N ({}^{-}b^{+}), a >_N \overset{-0}{b}, a >_N \overset{0+}{b}, a >_N \overset{-0+}{b}; \tag{20}$$

$$({}^{-}a) >_N b, ({}^{-}a) >_N ({}^{-}b), ({}^{-}a) >_N (b^{+}), ({}^{-}a) >_N ({}^{-}b^{+}), \overset{-}{a} >_N \overset{-0}{b}, \overset{-}{a} >_N \overset{0+}{b}, \overset{-}{a} >_N \overset{-0+}{b}; \tag{21}$$

$$(a^{+}) >_N b, (a^{+}) >_N b({}^{-}b), (a^{+}) >_N b(b^{+}), (a^{+}) >_N b({}^{-}b^{+}), \overset{+}{a} >_N \overset{-0}{b}, \overset{+}{a} >_N \overset{0+}{b}, \overset{+}{a} >_N \overset{-0+}{b}; \tag{22}$$

$$({}^{-}a^{+}) >_N b, ({}^{-}a^{+}) >_N ({}^{-}b), ({}^{-}a^{+}) >_N (b^{+}), ({}^{-}a^{+}) >_N ({}^{-}b^{+}), \text{ etc.} \tag{23}$$

No non-standard order relationship between a and $({}^{-}a^{+})$,

nor between a and $({}^{-0}a^{+})$. (24)

1.5. Neutrosophic Infimum and Neutrosophic Supremum

1.5.1. Neutrosophic Infimum

Let $(S, <_N)$ be a set, which is neutrosophically partially ordered, and let M be a subset of S.

The neutrosophic infimum of M, denoted by $inf_N(M)$, is the neutrosophically greatest element in S, which is neutrosophically less than or equal to all elements of M.

1.5.2. Neutrosophic Supremum

Let $(S, <_N)$ be a set, which is neutrosophically partially ordered, and let M be a subset of S.

The neutrosophic supremum of M, denoted by $sup_N(M)$, is the neutrosophically smallest element in S, which is neutrosophically greater than or equal to all elements of M.

The neutrosophic infimum and supremum are both extensions of the classical infimum and supremum respectively, using the *transfer principle* from the real set R to the neutrosophic real *MoBiNad* set NR_{MB} defined below.

1.5.3. Property

If $\overset{m_1}{a}, \overset{m_2}{b}$ are left monads, right monads, pierced binads, or unpierced monads, then both $\inf_N\{\overset{m_1}{a}, \overset{m_2}{b}\}$ and $\sup_N\{\overset{m_1}{a}, \overset{m_2}{b}\}$ are left monads or right monads. (25)

1.6. Non-Standard Real MoBiNad Set

MoBiNad [3] etymologically comes from **mo***nad* + **bi***nad*.

Let R and R^* be the set of standard real numbers, and respectively the set of hyper-reals (or non-standard reals) that contains the infinitesimals and infinites.

The Non-standard Real MoBiNad Set [2] is built as follows:

$$NR_{MB} = {}_N\left\{ \begin{array}{c} \varepsilon, \omega, a, (^-a), (^-a^0), (a^+), (^0a^+), (^-a^+), (^-a^{0+}) \text{ |where } \varepsilon \text{ are infinitesimals,} \\ \text{with } \varepsilon \in \mathbb{R}^*; \omega = 1/\varepsilon \text{ are infinites, with } \omega \in \mathbb{R}^*; \text{and } a \text{ are real numbers, with } a \in \mathbb{R} \end{array} \right\} \quad (26)$$

or,

$$NR_{MB} = {}_N\left\{ \varepsilon, \omega, \overset{m}{a} \middle| \begin{array}{c} \text{where } \varepsilon, \omega \in \mathbb{R}^*, \varepsilon \text{ are infinitesimals, } \omega = \frac{1}{\varepsilon} \text{ are infinitesimals;} \\ a \in \mathbb{R}; \text{ and } m \in \{^-, ^{-0}, ^+, ^{+0}, ^{-+}, ^{-0+}\} \end{array} \right\} \quad (27)$$

As a set, NR_{MB} is closed under addition, subtraction, multiplication, division [except division by $\overset{m}{a}$, with $a = 0$ and $m \in \{^-, ^{-0}, ^+, ^{0+}, ^{-+}, ^{-0+}\}$], and power

$$\{\left(\overset{m_1}{a}\right)^{\left(\overset{m_2}{b}\right)} \text{ with : either } a > 0, \text{ or } a \leq 0 \text{ but } b = \tfrac{p}{r} \text{ (irreducible fraction) and } p, r \text{ are} \quad (28)$$

positive integers with r an odd number}.

1.7. Remark

The neutrosophic infimum and neutrosophic maximum are well-defined on the Non-standard Real MoBiNad Set NR_{MB}, in the sense that we can compute inf_N and sup_N of any subset of NR_{MB}.

1.8. Non-Standard Real Open Monad Unit Interval

Since there is no relationship of order between a and $^-a^+$, not between a and $(^{-0}a^+)$, and we need a total order relationship on the set of non-standard real numbers, we remove all binads and keep only the open left monads and open right monads [we also remove the monads closed to one side].

$$]^-0,\ 1^+[_M = \{a,\ \varepsilon, ^-a, a^+ | a \in [0,1], \varepsilon \in R^*, \varepsilon > 0\}. \quad (29)$$

where a is subunitary real number, and ε is an infinitesimal number.

The non-standard neutrosophic unit interval $]^-0, 1^+[_M$ includes the previously defined $]^-0, 1^+[$ as follows:

$$]^-0, 1^+[=_N (^-0) \cup [0,1] \cup (1^+) \subset_N]^-0, 1^+[_M \quad (30)$$

where the index $_M$ means that the interval includes all open monads and infinitesimals between $^-0$ and 1^+.

2. General Monad Neutrosophic Set

Let U be a universe of discourse, and S ⊂ U be a subset. Then, a *Neutrosophic Set* is a set for which each element x from S has a degree of membership (T), a degree of indeterminacy (I), and a degree of non-membership (F), with T, I, F standard or non-standard real monad subsets or infinitesimals, neutrosophically included in or equal to the nonstandard real monad unit interval $]^-, ^+[_M$, or

$$T, I, F \subseteq_N]^-0, 1^+[_M \quad (31)$$

where

$$^-0 \leq_N inf_N T + inf_N I + inf_N F \leq_N sup_N T + sup_N I + sup_N F \leq 3^+. \quad (32)$$

2.1. Non-Standard Neutrosophic Set

Let us consider the above general definition of general neutrosophic set, and assume that at least one of T, I, or F (the neutrosophic components) is a non-standard real monad subset or infinitesimal, neutrosophically included in or equal to $]^-0, 1^+[_M$, where

$$^-0 \leq_N inf_N T + inf_N I + inf_N F \leq_N sup_N T + sup_N I + sup_N F \leq 3^+, \quad (33)$$

we have a non-standard neutrosophic set.

2.2. Non-Standard Fuzzy t-Norm and Fuzzy t-Conorm

Let T_1, and T_2, $\in]^{-}0, 1^{+}[_M$, be nonstandard real numbers (infinitesimals, or open monads), or standard (classical) real numbers, such that at least one of them is a non-standard real number. T_1 and T_2 are non-standard fuzzy degrees of membership. Then one has:

The non-standard fuzzy t-norms:

$$T_1 \wedge_F T_2 = inf_N \{T_1, T_2\} \tag{34}$$

The non-standard fuzzy t-conorms:

$$T_1 \vee_F T_2 = sup_N \{T_1, T_2\} \tag{35}$$

2.3. Aggregation Operators on Non-Standard Neutrosophic Set

Let T_1, I_1, F_1 and $T_2, I_2, F_2 \in]^{-}0, 1^{+}[_{MB}$, be nonstandard real numbers (infinitesimals, or monads), or standard (classical) real numbers, such that at least one of them is a non-standard real number.

Non-Standard Neutrosophic Conjunction

$$\begin{gathered}(T_1, I_1, F_1) \wedge_N (T_2, I_2, F_2) = (T_1 \wedge_F T_2, I_1 \vee_F I_2, F_1 \vee_F F_2) = \\ (inf_N (T_1, T_2), sup_N (I_1, I_2), sup_N (F_1, F_2))\end{gathered} \tag{36}$$

Non-Standard Neutrosophic Disjunctions

$$\begin{gathered}(T_1, I_1, F_1) \vee_N (T_2, I_2, F_2) = (T_1 \vee_F T_2, I_1 \wedge_F I_2, F_1 \wedge_F F_2) = \\ (sup_N (T_1, T_2), inf_N (I_1, I_2), inf_N (F_1, F_2))\end{gathered} \tag{37}$$

Non-Standard Neutrosophic Complement/Negation

We may use the notations C_N or $\neg_N$ for the neutrosophic complement.

$$C_N(T_1, I_1, F_1) = {}_{N}\neg_N(T_1, I_1, F_1) =_N (F_1, I_1, T_1). \tag{38}$$

Non-Standard Neutrosophic Inclusion/Inequality

$$(T_1, I_1, F_1) \leq_N (T_2, I_2, F_2) \ iff \ T_1 \leq_N T_2, I_1 \geq_N I_2, F_1 \geq_N F_2. \tag{39}$$

Let $A, B \in P(X)$, if $A \subseteq_N B$ then B is called a *neutrosophic superset* of A.

Non-standard Neutrosophic Equality

$$(T_1, I_1, F_1) =_N (T_2, I_2, F_2) \ \textit{iff} \ (T_1, I_1, F_1) \leq_N (T_2, I_2, F_2) \ \textit{and} \ (T_2, I_2, F_2) \leq_N (T_1, I_1, F_1). \tag{40}$$

Non-Standard Monad Neutrosophic Universe of Discourse

We now introduce for the first time the non-standard neutrosophic universe.

Definition 1. *A general set U, defined such that each element $x \in U$ has neutrosophic coordinates of the form $x(T_x, I_x, F_x)$, such that T_x represents the degree of truth-membership of the element x with respect to set U, I_x represents the degree of indeterminate-membership of the element x with respect to the set U, and F_x represents the degree of false-membership of the element x with respect to the set U; where T_x, I_x, and F_x are non-standard or standard subsets of the neutrosophic real monad set NR_M, but at least one of all of them is non-standard (i.e., contains infinitesimals, or open monads).*

Single-Valued Non-Standard Neutrosophic Topology

Let U be a single-valued non-standard neutrosophic universe of discourse, i.e., for all $x \in U$, their neutrosophic components T_x, I_x, F_x are single-values (either real numbers, or infinitesimals, or open monads) belonging to $]^-0, 1^+[$

Definition 2. *Let X be a non-standard neutrosophic subset of U. The neutrosophic empty-set, denoted by* $\mathbf{0}_N = (^-0, 1^+, 1^+)$, *is a set* $\Phi_N \subset X$ *whose all elements have the non-standard neutrosophic components equal to* $(^-0, 1^+, 1^+)$. *The whole set, denoted by* $1_N = (1^+, ^-0, ^-0)$, *is a set* $W_N \subset X$ *whose all elements have the non-standard neutrosophic components equal to* $(1^+, ^-0, ^-0)$.

Definition 3. *Let X be a non-standard neutrosophic set. Let* $A = (T_1, I_1, F_1)$ *and* $B = (T_2, I_2, F_2)$ *be non-standard neutrosophic numbers. Then:*

$$A \cap B = (\inf_N (T_1, T_2), \sup_N (I_1, I_2), \sup_N (F_1, F_2)) \tag{41}$$

$$A \cup B = (\sup_N (T_1, T_2), \inf_N (I_1, I_2), \inf_N (F_1, F_2)) \tag{42}$$

$$C_N A = (F_1, I_1, T_1) \tag{43}$$

Definition 4. *Let X be a non-standard neutrosophic set. Let A(X) be the family of all non-standard neutrosophic sets in X. Let* $\tau \subseteq A(X)$ *be a family of non-standard neutrosophic sets in X. Then* τ *is called a Non-standard Neutrosophic Topology on X, if it satisfies the following axioms:*

(i) $\mathbf{0}_N$ *and* $\mathbf{1}_N$ *are in* τ.
(ii) *The intersection of the elements of any finite subcollection of* τ *is in* τ.
(iii) *The union of the elements of any subcollection of* τ *is in* τ.

The pair (X, τ) *is called a non-standard neutrosophic topological space. All members of* τ *are called non-standard neutrosophic open sets in X.*

Example 1. *Let X be a non-standard neutrosophic set. Let* τ *be the set consisting of* $\mathbf{0}_N$ *and* $\mathbf{1}_N$. *Then* τ *is a topology on X. It is called the non-standard neutrosophic trivial topology.*

Example 2. *Let X be a non-standard neutrosophic set. Let A be a non-standard neutrosophic set in X. Let* $\tau = \{\mathbf{0}_N, \mathbf{1}_N, A\}$. *Then it can be easily shown that* τ *is a topology on X.*

Example 3. *Let X be a non-standard neutrosophic set. Let A and B be non-standard neutrosophic sets in X such that A is a neutrosophic superset of B. Let* $\tau = \{\mathbf{0}_N, \mathbf{1}_N, A, B\}$. *Then since* $A \cap B = B$ *and* $A \cup B = A$ *we deduce that* τ *is a topology on X.*

Example 4. *Let X be a non-standard neutrosophic set. Suppose we have a nested sequence*

$$A_1 \subseteq A_2 \subseteq A_3 \subseteq \ldots \subseteq A_{n-1} \subseteq A_n \subseteq \tag{44}$$

of non-standard neutrosophic sets in X such that each A_n *is a neutrosophic superset of* A_{n-1} *for each*

$$n \in \{1, 2, 3, \ldots\}.$$

Let $\tau = \{\mathbf{0}_N, \mathbf{1}_N, A_n : n \in N\}$. *Then since* $A_i \cap_N A_j = A_i$ *and* $A_i \cup_N A_j = A_j$ *for each i less than j, we deduce that* τ *is a topology on X.*

Example 5. *Let X be a non-standard neutrosophic infinite set:*

$$X = \left(x_{m,n,p} \left(\left(\overset{+}{0.7} \right)^m, (0.2)^n, \left(\overset{-}{0.6} \right)^p \right), x_{m,n,p} \in X; m, n, p \in \{1, 2, \ldots\} \right) \tag{45}$$

Let M_{100} be a family of subsets of X, such that each member $A_{m,n,p}$ of the family has:

$$m, n, p \in \{1, 2, \ldots, 100\}. \tag{46}$$

Then $\tau = \{\mathbf{0}_N, \mathbf{1}_N, M_{100}\}$ is a non-standard neutrosophic topology.

Proof. Any monad $\left(\overset{m}{a}\right)$ raised to the integer power $k > 0$, is equal to the monad of a^k:

$$\left(\overset{m}{a}\right)^k = \left(\overset{m}{a^k}\right) \tag{47}$$

Let's consider two non-standard neutrosophic elements from X:

$$x_{m_1,n_1,p_1}\left(\left(\overset{+}{0.7}\right)^{m_1}, (0.2)^{n_1}, \left(\overset{-}{0.6}\right)^{p_1}\right) \text{ and } x_{m_2,n_2,p_2}\left(\left(\overset{+}{0.7}\right)^{m_2}, (0.2)^{n_2}, \left(\overset{-}{0.6}\right)^{p_2}\right) \tag{48}$$

where

$$m_1, n_1, p_1, m_2, n_2, p_2 \in \{1, 2, \ldots, 100\}. \tag{49}$$

It is sufficient to prove that their non-standard neutrosophic finite intersection and the random union of elements from M_{100} are in M_{100}.

$$\begin{aligned} &x_{m_1,n_1,p_1} \cap_N x_{m_2,n_2,p_2} =_N \Big(\inf_N\{\left(\overset{+}{0.7}\right)^{m_1}, \left(\overset{+}{0.7}\right)^{m_2}\}, \\ &\mathrm{SUP}_N\{(0.2)^{n_1}, (0.2)^{n_2}\}, \mathrm{SUP}_N\{\left(\overset{-}{0.6}\right)^{p_1}, \left(\overset{-}{0.6}\right)^{p_2}\}\Big) \\ &= \left(\left(\overset{+}{0.7}\right)^{\max\{m_1,m_2\}}, (0.2)^{\min\{n_1,n_2\}}, \left(\overset{-}{0.6}\right)^{\min\{p_1,p_2\}}\right) \in M_{100} \end{aligned} \tag{50}$$

$$\text{because also } max\{m_1, m_2\}, min\{n_1, n_2\}, min\{p_1, p_2\} \in M_{100}. \tag{51}$$

$$\begin{aligned} &\bigcup_{m,n,p \in (\psi_1,\psi_2,\psi_3) \subseteq \{1,2,\ldots,100\}^3} \{x_{m,n,p}\left(\left(\overset{+}{0.7}\right)^{m}, (0.2)^{n}, \left(\overset{-}{0.6}\right)^{p}\right)\} \\ &= \left(\left(\overset{+}{0.7}\right)^{\min\{m, m \in \psi_1\}}, (0.2)^{\max\{n, n \in \psi_2\}}, \left(\overset{-}{0.6}\right)^{\max\{p, p \in \psi_3\}}\right) \in M_{100} \end{aligned} \tag{52}$$

□

Definition 5. *Let X be a non-standard neutrosophic set. Suppose that τ and τ′ are two topologies on X such that τ ⊂ τ′. Then we say that τ′ is finer than τ.*

Example 6. *Let X be a non-standard neutrosophic set. Let A and B be non-standard neutrosophic sets in X such that A is a neutrosophic superset of B. Let $\tau = \{\mathbf{0}_N, \mathbf{1}_N, A\}$ and $\tau' = \{\mathbf{0}_N, \mathbf{1}_N, B\}$.*
Then τ′ is finer than τ.

Example 7. *Let's consider the above Example 5. In addition to M_{100}, let's define L_{100} as follows:*

$$L_{100} = \{x_{m,n,p}\left(\left(\overset{+}{0.7}\right)^{m}, (0.2)^{n}, \left(\overset{-}{0.6}\right)^{p}\right), x_{m,n,p} \in X; m, n, p \in \{2, 4, 6, \ldots, 100\}\} \tag{53}$$

The non-standard neutrosophic topology $\tau = \{\mathbf{0}_N, \mathbf{1}_N, M_{100}\}$ is a finer non-standard neutrosophic topology than the non-standard neutrosophic topology $\tau' = \{\mathbf{0}_N, \mathbf{1}_N, L_{100}\}$.

Definition 6. *The subset Z of a non-standard neutrosophic topological space X is called a non-standard neutrosophic closed set if its complement C_N (Z) is open in X.*

Example 8. *Let Y be a non-standard neutrosophic infinite set*

$$Y = \{y_{m,n}\left(\left(\overset{+}{0.5}\right)^m, \left(\overset{-}{0.1}\right)^n, \left(\overset{+}{0.5}\right)^m\right), y_{m,n} \in Y; m, n \in \{1, 2, \ldots\}\} \tag{54}$$

and P (Y) the power set of Y.

Let $\tau \subseteq P(Y)$ be a non-standard neutrosophic topology.

Each non-standard neutrosophic set $A \in \tau$ is a non-standard neutrosophic open set and closed set in the same time, because its non-standard neutrosophic complement $C_N(A) = A$.

Proof. For any $y_{m,n} \in Y$ one has:

$$C_N(y_{m,n}) = C_n\left(\left(\overset{+}{0.5}\right)^m, \left(\overset{-}{0.1}\right)^n, \left(\overset{+}{0.5}\right)^m\right) = \left(\left(\overset{+}{0.5}\right)^m, \left(\overset{-}{0.1}\right)^n, \left(\overset{+}{0.5}\right)^m\right) = y_{m,n} \tag{55}$$

□

Theorem 1. *Unlike in classical topology, the non-standard neutrosophic empty-set $\mathbf{0}_N$ and the non-standard neutrosophic whole set $\mathbf{1}_N$ are not necessarily closed, since they are not the non-standard neutrosophic complement of each other.*

Proof.

$$C_N\,(^-0, 1^+, 1^+) =_N (1^+, 1^+, {}^-0) \neq (1^+, {}^-0, {}^-0), \text{ and reciprocally:} \tag{56}$$

$$C_N\,(1^+, {}^-0, {}^-0) =_N ({}^-0, {}^-0, 1^+) \neq ({}^-0, 1^+, 1^+). \tag{57}$$

Theorem 2. *In a non-stardard neutrosophic topology there may be non-standard neutrosophic sets which are both open and closed set.*

Proof. See the above Example 8. □

Theorem 3. *Unlike in classical topology, the intersection of two non-standard neutrosophic closed sets is not necessarily a non-standard neutrosophic closed set. Moreover, the union of two non-standard neutrosophic closed sets is not necessarily a non-standard neutrosophic closed set.*

Proof. Consider Example 3 above.

$$\text{Let } A = (T_2, I_2, F_2) \text{ and } B = (T_1, I_1, F_1). \text{ Note that } C_N A = (F_2, I_2, T_2) \text{ and } C_N B = (F_1, I_1, T_1). \tag{58}$$

$$\text{Then } C_N A \cap_N C_N B = (F_2, I_1, T_2). \tag{59}$$

$$\text{Since } C_N\,(C_N A \cap_N C_N B) = (T_2, I_1, F_2) \tag{60}$$

is not non-standard neutrosophic open set in X, we have that $C_N A \cap_N C_N B$ is not a non-standard neutrosophic closed set in X. Also,

$$C_N A \cap_N C_N B = (F_1, I_2, T_1). \tag{61}$$

$$\text{Since } C_N\,(C_N A \cap_N C_N B) = (T_1, I_2, F_1) \tag{62}$$

is not non-standard neutrosophic open set in X, we have that $C_N A \cap_N C_N B$ is not a non-standard neutrosophic closed set in X. □

General Remark 1. *Since the non-standard neutrosophic aggregation operators (conjunction, disjunction, complement) needed in non-standard neutrosophic topology, are defined by* ***classes of operators*** *(not by exact unique operators) respectively, the classical topological space theorems and properties extended (by the transfer principle) to the non-standard neutrosophic topological space may be valid for some non-standard neutrosophic operators, but invalid for other classes of neutrosophic aggregation operators.*

Even worth, due to the fact that non-standard neutrosophic conjunction/disjunction/complement are, in addition, based on fuzzy t-norms and fuzzy t-conorms, which are not fixed either, but characterized by classes!

{Similarly for fuzzy and intuitionistic fuzzy aggregation operators.}

For example, the neutrosophic intersection/$\backslash_N$ can be defined in 2 ways:

$$(T_1, I_1, F_1) /\backslash_N (T_2, I_2, F_2) = (T_1/\backslash_F T_2, I_1/\backslash_F I_2, F_1/\backslash_F F_2) \tag{63}$$

And

$$(T_1, I_1, F_1) /\backslash_N (T_2, I_2, F_2) = (T_1/\backslash_F T_2, I_1 \backslash/_F I_2, F_1/\backslash_F F_2). \tag{64}$$

In turn, the fuzzy t-norms ($/\backslash_F$) and fuzzy t-conorm ($\backslash/_F$) are also defined in many ways; for example I know at least 3 types of fuzzy t-norms:

$$a/\backslash_F\, b = min\ \{a, b\} \tag{65}$$

$$a/\backslash_F\, b = ab \tag{66}$$

$$a/\backslash_F\, b = max\ \{a + b - 1, 0\} \tag{67}$$

and 3 types of fuzzy t-conorms:

$$a\backslash/_F\, b = max\ \{a, b\} \tag{68}$$

$$a/\backslash_F\, b = a + b - ab \tag{69}$$

$$a/\backslash_F\, b = min\ \{a + b, 1\} \tag{70}$$

therefore there exist at least 2·3·3 = 18 possibilities to define the neutrosophic *t-norm* ($/\backslash_N$).

There exist at least the same number 18 of possibilities of defining the neutrosophic *t-conorm* ($\backslash/_N$).

From these 18 possibilities of defining/$\backslash_N$ and $\backslash/_N$ for some of them the classical topological theorems extended to non-standard neutrosophic topology may be valid, for others invalid.

Definition 7. *Let (X, τ) be a nonstandard neutrosophic topological space. Let A be a non-standard neutrosophic set in X. Then the Non-standard Neutrosophic Closure of A is the intersection of all non-standard neutrosophic closed supersets of A, and we denote it by cl_N (A). The Non-standard Neutrosophic Closure of A is the smallest nonstandard neutrosophic closed set in X that neutrosophically includes A.*

Example 9. *Let X be a non-standard neutrosophic set:*

$$X = \{x_1(\overset{-}{0.4}, \overset{+}{0.1}, \overset{-}{0.5}), x_2(\overset{-}{0.5}, \overset{+}{0.1}, \overset{-}{0.4}), x_3(\overset{-}{0.5}, \overset{+}{0.1}, \overset{-}{0.5})\} \tag{71}$$

and the following non-standard neutrosophic topology:

$$\tau = \{\Phi_N, 1_N, A_1\{x_1(\overset{-}{0.4}, \overset{+}{0.1}, \overset{-}{0.5}), A_2\{x_2(\overset{-}{0.5}, \overset{+}{0.1}, \overset{-}{0.4}), A_3\{x_3(\overset{-}{0.5}, \overset{+}{0.1}, \overset{-}{0.5})\}\} \tag{72}$$

where

$$\Phi_N = \{x_1(\overset{-}{0},\overset{+}{1},\overset{+}{1}), x_2(\overset{-}{0},\overset{+}{1},\overset{+}{1}), x_3(\overset{-}{0},\overset{+}{1},\overset{+}{1}), 1_N = x_1(\overset{+}{1},\overset{-}{0},\overset{-}{0}), x_1(\overset{+}{1},\overset{-}{0},\overset{-}{0}), x_1(\overset{+}{1},\overset{-}{0},\overset{-}{0})\} \quad (73)$$

Proof. τ is a non-standard neutrosophic topology because:

$$A_1 \cap_N A_2 = A_1, A_1 \cap_N A_3 = A_1, A_2 \cap_N A_3 = A_3 \quad (74)$$

$$A_1 \cup_N A_2 = A_2, A_1 \cup_N A_3 = A_3, A_2 \cup_N A_3 = A_2, A_1 \cup_N A_2 \cup_N A_3 = A_2. \quad (75)$$

(X, τ) is a non-standard neutrosophic topological space.

The non-standard neutrosophic sets A_1, A_2, A_3 are open sets since they belong to τ.

A_2 is the non-standard neutrosophic complement of A_1, or $C_N(A_2) = A_1$, therefore A_2 is a non-standard neutrosophic closed set in X.

A_3 is the non-standard neutrosophic complement of A_3 (itself), or $C_N(A_3) = A_3$, therefore A_3 is also a non-standard neutrosophic closed set in X.

A_2 and A_3 are nonstandard neutrosophic supersets of A_1, since $A_1 \subset A_2$ and $A_1 \subset A_3$.

Whence, the *Non-standard Neutrosophic Closure* of A_1 is the intersection of its non-standard neutrosophic closed supersets A_2 and A_3, or

$$cl_N(A_1) =_N A_2 \cap_N A_3 =_N A_3 \quad (76)$$

□

Definition 8. *The Non-standard Neutrosophic Interior of A is the union of all non-standard neutrosophic open subsets of A that are contained in A, and we denote it by* $int_N(A)$*.*

The Non-standard Neutrosophic Interior of A is the largest non-standard neutrosophic open set in X that is neutrosophically included into A.

Example 10. *Into the previous Example 9, let's compute* $int_N(A_2)$*.*

A_1 and A_3 are non-standard neutrosophic open sets in X, with $A_1 \subset_N A_2$ and $A_3 \subset_N A_2$ (77)

Whence

$$\mathrm{int}_N(A_2) = A_1 \cup_N A_3 = A_3. \quad (78)$$

Definition 9. *Let* (X, τ) *be a non-standard neutrosophic topological space, and let* $Y \subseteq_N X$ *be a non-standard neutrosophic subset of X. Then the collection* $\tau_Y = \{O \cap_N Y, O \in \tau\}$ *is a topology on Y. It is called the non-standard neutrosophic subspace topology and Y is called a non-standard neutrosophic subspace of X.*

Example 11. *In the same previous Example 9, let's take* $Y = A_3 \subset X$*, and the non-standard neutrosophic subspace topology*

$$\tau_Y = \{\Phi_N, 1_N, A_3, \{(\overset{-}{0.5}, \overset{+}{0.1}, \overset{-}{0.5})\}\} \quad (79)$$

Then Y is a non-standard neutrosophic topological subspace of X.

Definition 10. *Let X and Y be two non-standard neutrosophic topological spaces. A map f:*

$$X \to Y \quad (80)$$

is said to be non-standard neutrosophic continuous map if for each non-standard neutrosophic open set A in Y, the set $f^{-1}(A)$ *is a non-standard neutrosophic open set in X.*

Example 12. *Let X be a non-standard neutrosophic space. Let Y be a non-standard neutrosophic subspace of X. Then the inclusion map i:* Y → X *is non-standard neutrosophic continuous.*

Example 13. *Let X be a non-standard neutrosophic set. Suppose that* τ *and* τ' *are two non-standard neutrosophic topologies on X such that* τ' *is finer than* τ*. Then the identity map id:* $(X, \tau') \rightarrow (X, \tau)$ *is obviously non-standard neutrosophic continuous.*

Definition 11. *Let* (X_1, τ_1) *and* (X_2, τ_2) *be two non-standard neutrosophic topological spaces. Then* $\tau_1 \times \tau_2 =_N \{U \times V : U \in \tau_1, V \in \tau_2\}$ *defines a topology on the product*

$$X_1 \times X_2 \tag{81}$$

The topology $\tau_1 \times \tau_2$ *is called non-standard neutrosophic product topology.*

3. Development of Neutrosophic Topologies

Since the first definition of neutrosophic topology and neutrosophic topological space [3] in 1998, the neutrosophic topology has been developed tremendously in multiple directions and has added new topological concepts such as: neutrosophic crisp topological [6–9], neutrosophic crisp α-topological spaces [10], neutrosophic soft topological k-algebras [11–13], neutrosophic nano ideal topological structure [14], neutrosophic soft cubic set in topological spaces [15], neutrosophic alpha m-closed sets [16], neutrosophic crisp bi-topological spaces [17], ordered neutrosophic bi-topological space [18], neutrosophic frontier and neutrosophic semi-frontier [19], neutrosophic topological functions [20], neutrosophic topological manifold [21], restricted interval valued neutrosophic topological spaces [22], smooth neutrosophic topological spaces [23], nω–closed sets in neutrosophic topological spaces [24], and other topological properties [25,26], arriving now to the neutrosophic topology extended to the non-standard analysis space.

4. Conclusions

We have introduced for the first time the non-standard neutrosophic topology, non-standard neutrosophic toplogical space and subspace constructed on the non-standard unit interval$]{-0}, 1{+}[_M$ that is formed by real numbers and positive infinitesimals and open monads, together with several concepts related to them, such as: non-standard neutrosophic open/closed sets, non-standard neutrosophic closure and interior of a given set, and non-standard neutrosophic product topology. Several theorems were proven and non-standard neutrosophic examples were presented.

Non-standard neutrosophic topology (NNT) is initiated now for the first time. It is a neutrosophic topology defined on the *set of hyperreals*, while the previous neutrosophic topologies were initiated and developed on the *set of reals*.

The novelty of NNT is its possibility to be used in calculus due to the involvement of infinitesimals, while the previous neutrosophic topologies could not be used due to lack of infinitesimals.

Thus, the paper has contributed to the foundation of a new field of study, called non-standard neutrosophic topology.

As future work, we intend to study more non-standard neutrosophic algebraic structures, such as: non-standard neutrosophic group, non-standard neutroosphic ring and field, non-standard neutrosophic vector space and so on.

Author Contributions: The authors contributed in the following way: methodology by M.A.A.S.; and conceptualization by F.S.

Funding: This project was funded by the Deanship of Scientific Research (DSR) at King Abdulaziz University, Jeddah, under grant No. KEP-34-130-38. The authors, therefore, acknowledge with thanks DSR for technical and financial support.

Conflicts of Interest: The authors declare no conflict of interest.

References

1. Insall, M.; Weisstein, E.W. Nonstandard Analysis. From MathWorld—A Wolfram Web Resource. Available online: http://mathworld.wolfram.com/NonstandardAnalysis.html (accessed on 10 May 2019).
2. Robinson, A. *Non-Standard Analysis*; Princeton University Press: Princeton, NJ, USA, 1996.
3. Smarandache, F. A Unifying Field in Logics: Neutrosophic Logic Neutrosophy, Neutrosophic Set, Neutrosophic Probability and Statistics. 1998. Available online: http://fs.unm.edu/eBook-Neutrosophics6.pdf (accessed on 21 April 2019).
4. Smarandache, F. *Extended Nonstandard Neutrosophic Logic., Set, and Probability Based on Extended Nonstandard Analysis*; 31 pages, arXiv; Cornell University: Ithaca, NY, USA, 11 March 2019.
5. Smarandache, F. Extended Nonstandard Neutrosophic Logic, Set, and Probability Based on Extended Nonstandard Analysis. *arXiv* **2019**, arXiv:1903.04558.
6. AL-Nafee, A.B.; Al-Hamido, R.K.; Smarandache, F. Separation Axioms in Neutrosophic Crisp Topological Spaces. *Neutrosophic Sets Syst.* **2019**, *25*, 25–32. [CrossRef]
7. Al-Omeri, W. Neutrosophic crisp Sets via Neutrosophic crisp Topological Spaces. *Neutrosophic Sets Syst.* **2016**, *13*, 96–104. [CrossRef]
8. Salama, A.A.; Smarandache, F.; ALblowi, S.A. New Neutrosophic Crisp Topological Concepts. *Neutrosophic Sets Syst.* **2014**, *4*, 50–54. [CrossRef]
9. Salama, A.A.; Smarandache, F.; Kroumov, V. Neutrosophic Crisp Sets & Neutrosophic Crisp Topological Spaces. *Neutrosophic Sets Syst.* **2014**, *2*, 25–30. [CrossRef]
10. Salama, A.A.; Hanafy, I.M.; Elghawalby, H.; Dabash, M.S. Neutrosophic Crisp α-Topological Spaces. *Neutrosophic Sets Syst.* **2016**, *12*, 92–96. [CrossRef]
11. Akram, M.; Gulzar, H.; Smarandache, F. Neutrosophic Soft Topological K-Algebras. *Neutrosophic Sets Syst.* **2019**, *25*, 104–124. [CrossRef]
12. Mukherjee, A.; Datta, M.; Smarandache, F. Interval Valued Neutrosophic Soft Topological Spaces. *Neutrosophic Sets Syst.* **2014**, *6*, 18–27. [CrossRef]
13. Bera, T.; Mahapatra, N.K. On Neutrosophic Soft Topological Space. *Neutrosophic Sets Syst.* **2018**, *19*, 3–15. [CrossRef]
14. Parimala, M.; Karthika, M.; Jafari, S.; Smarandache, F.; Udhayakumar, R. Neutrosophic Nano ideal topological structure. *Neutrosophic Sets Syst.* **2019**, *24*, 70–76. [CrossRef]
15. Cruz, R.A.; Irudayam, F.N. Neutrosophic Soft Cubic Set in Topological Spaces. *Neutrosophic Sets Syst.* **2018**, *23*, 23–44. [CrossRef]
16. Mohammed, F.M.; Matar, S.F. Fuzzy Neutrosophic Alpha ^{m}-Closed Sets in Fuzzy Neutrosophic Topological Spaces. *Neutrosophic Sets Syst.* **2018**, *21*, 56–65. [CrossRef]
17. Hamido, R.K.; Gharibah, T. Neutrosophic Crisp Bi-Topological Spaces. *Neutrosophic Sets Syst.* **2018**, *21*, 66–73. [CrossRef]
18. Devi, R.N.; Dhavaseelan, R.; Jafari, S. On Separation Axioms in an Ordered Neutrosophic Bitopological Space. *Neutrosophic Sets Syst.* **2017**, *18*, 27–36. [CrossRef]
19. Iswarya, P.; Bageerathi, K. A Study on Neutrosophic Frontier and Neutrosophic Semi-frontier in Neutrosophic Topological Spaces. *Neutrosophic Sets Syst.* **2017**, *16*, 6–15. [CrossRef]
20. Arokiarani, I.; Dhavaseelan, R.; Jafari, S.; Parimala, M. On Some New Notions and Functions in Neutrosophic Topological Spaces. *Neutrosophic Sets Syst.* **2017**, *16*, 16–19. [CrossRef]
21. Salama, A.A.; El Ghawalby, H.; Ali, S.F. Topological Manifold Space via Neutrosophic Crisp Set Theory. *Neutrosophic Sets Syst.* **2017**, *15*, 18–21. [CrossRef]
22. Mukherjee, A.; Datta, M.; Sarkar, S. Restricted Interval Valued Neutrosophic Sets and Restricted Interval Valued Neutrosophic Topological Spaces. *Neutrosophic Sets Syst.* **2016**, *12*, 45–53. [CrossRef]
23. El-Gayyar, M.K. Smooth Neutrosophic Topological Spaces. *Neutrosophic Sets Syst.* **2016**, *12*, 65–72. [CrossRef]
24. Santhi, R.; Udhayarani, N. Nω –Closed Sets in Neutrosophic Topological Spaces. *Neutrosophic Sets Syst.* **2016**, *12*, 114–117. [CrossRef]

25. Thao, N.X.; Smarandache, F. Standard neutrosophic rough set and its topologies properties. *Neutrosophic Sets Syst.* **2016**, *14*, 65–70. [CrossRef]
26. Sweety, C.A.C.; Arockiarani, I. Topological structures of fuzzy neutrosophic rough sets. *Neutrosophic Sets Syst.* **2015**, *9*, 50–57. [CrossRef]

symmetry

MDPI

Article

Application of the Bipolar Neutrosophic Hamacher Averaging Aggregation Operators to Group Decision Making: An Illustrative Example

Muhammad Jamil [1], Saleem Abdullah [2,*], Muhammad Yaqub Khan [1], Florentin Smarandache [3] and Fazal Ghani [2]

1 Department of Mathematics and Statistics, Riphah International University, Islamabad 44000, Pakistan; 10616@students.riu.edu.pk (M.J.); muhammad.yaqub@riphah.edu.pk (M.Y.K.)

2 Department of Mathematics, Abdul Wali Khan University, Mardan 23200, Pakistan; fazalghanimaths@gmail.com

3 Department of Mathematics and Sciences, University of New Mexico, 705 Gurley Ave., Gallup, NM 87301, USA; fsmarandache@gmail.com

* Correspondence: saleemabdullah@awkum.edu.pk

Received: 5 April 2019; Accepted: 29 April 2019; Published: 21 May 2019

Abstract: The present study aims to introduce the notion of bipolar neutrosophic Hamacher aggregation operators and to also provide its application in real life. Then neutrosophic set (NS) can elaborate the incomplete, inconsistent, and indeterminate information, Hamacher aggregation operators, and extended Einstein aggregation operators to the arithmetic and geometric aggregation operators. First, we give the fundamental definition and operations of the neutrosophic set and the bipolar neutrosophic set. Our main focus is on the Hamacher aggregation operators of bipolar neutrosophic, namely, bipolar neutrosophic Hamacher weighted averaging (BNHWA), bipolar neutrosophic Hamacher ordered weighted averaging (BNHOWA), and bipolar neutrosophic Hamacher hybrid averaging (BNHHA) along with their desirable properties. The prime gain of utilizing the suggested methods is that these operators progressively provide total perspective on the issue necessary for the decision makers. These tools provide generalized, increasingly exact, and precise outcomes when compared to the current methods. Finally, as an application, we propose new methods for the multi-criteria group decision-making issues by using the various kinds of bipolar neutrosophic operators with a numerical model. This demonstrates the usefulness and practicality of this proposed approach in real life.

Keywords: BNHWA aggregation operator; BNHOWA aggregation operator; BNHHA aggregation operator; score function; accuracy function; certainty function; group decision making

1. Introduction

In the recent era of decision making, there is often incomplete, indeterminate, and inconsistent information. Zadeh introduced the notion of fuzzy set [1], which deals with uncertainty and can be applied in many fields. However, it has a shortcoming, i.e., it only expresses membership value and is unable to express non-membership value. At that point, Atanassov [2] introduced the idea of intuitionistic fuzzy set (IFS) to address issues with the fuzzy set. Every component in IFS is shown by a structured pair, and every pair is portrayed by a membership value (truth-membership) $\zeta_A(p)$ and a non-membership value (falsity-membership) $\text{Ю}_A(p)$ that satisfy the conditions $\zeta(p), \text{Ю}(p) \in [0,1]$ and $0 \leq \zeta(p), \text{Ю}(p) \leq 1$. IFSs can deal with incomplete data but cannot deal with the indeterminate and inconsistent data. Smarandache [3] developed the neutrosophic set (NS) by including an indeterminacy membership value $\text{Ґ}(p)$, which is a generality of IFS. NS can deal with information very effectively, i.e., incomplete, indeterminate, and inconsistent. When $\zeta(p) + \text{Ґ}(p) + \text{Ю}(p) < 1$, it shows that this information is indeterminate and when $\zeta(p) + \text{Ґ}(p) + \text{Ю}(p) > 1$, it is inconsistent information.

Single valued neutrosophic set (SVNS), as suggested by Wang et al. [4], was applied to decision making with the conditions $\zeta(p), \acute{\Gamma}(p), \text{ꝊO}(p) \in [0,1]$ and $0 \leq \zeta(p) + \acute{\Gamma}(p) + \text{ꝊO}(p) \leq 3$. Ye [5] introduced the correlation coefficient and also proposed the comparison method for SVNSs. Wang et al. [6] proposed the interval valued SVNSs to extend the truth, indeterminacy, and false membership to interval values. Ye [7] defined the similarity measures between interval valued neutrosophic sets on the basis of the Hamming and Euclidean distances and also proposed a multiple attribute decision-making method.

Aggregation operators are the important research areas, claiming the attention of today's researchers. Since the proposed theory of IFS, many scientists [8–15] have made essential contributions to the advancement of IFS theory. XU and Yager [12] developed the notion of aggregation operators based on IFS. They also applied these aggregation operators to decision making. Wang and Liu [16], proposed the idea of Einstein aggregation operators. Zhao and Wei [17] built up some of the Einstein hybrid aggregation operators. Fahmi et al. [18–21] developed aggregation operators based on triangular and trapezoidal cubic fuzzy numbers with applications to decision making. Rahman et al. [22–25] proposed aggregation operators on different extension of fuzzy numbers. The bipolar fuzzy set (BFS) [26–28] uses a substitute method to deal with uncertainty in decision making. The bipolar fuzzy set consists of a positive as well as negative membership degree. The membership degree of the bipolar fuzzy set ranges from −1 to 1. BFSs have been useful in various research domains and set theory decision analysis and organizational modeling [29], quantum computing [30], physics and philosophy [31], and graph theory [32]. Bipolar averaging and geometric fuzzy aggregations operators were defined by Gul [33]. Irfan et al. [34] presented the bipolar neutrosophic set with basic operations. They also proposed the comparison method for bipolar neutrosophic sets. Fan et al. [35] developed Heronian mean operators in a bipolar neutrosophic environment. Irfan et al. [36] presented the interval valued bipolar neutrosophic set with applications to pattern recognition. Irfan et al. [37] proposed the interval valued neutrosophic soft set with applications to decision making. Zhan et al. [38] proposed Schweizer-Sklar Muirhead mean aggregation operators based on single-valued neutrosophic set. Ashraf et al. introduced some logarithmic aggregation operators on neutrosophic sets [39].

Hamacher t-norm and t-conorm [40], which are the generalization of algebraic and Einstein t-norm and t-conorm, are more general and flexible. There are many researchers who extended the Hamacher operations to solve multiple attribute decision-making problems combined with other fuzzy environments, such as intuitionistic [41], interval valued intuitionistic [42], hesitant fuzzy [43], hesitant Pythagorean fuzzy [44], bipolar fuzzy numbers [45] and neutrosophic numbers [46,47]. Since the development of this field, there has been no significant research on Hamacher operations and its applicability to bipolar neutrosophic numbers. Here, we extended the Hamacher operations to bipolar neutrosophic numbers to develop bipolar neutrosophic Hamacher aggregation operators for multiple attribute decision-making problems.

Decision making plays a key role in present day management. Necessarily, sound decision making is a basic part of administration. Consciously or unconsciously, a manager makes a decision, or decisions, as it is his or her responsibility as a manager. Decision making has a consequential role as organizational and managerial activities are linked with that decision. A decision is explained as a sequence of actions, intentionally taken from a set of alternates, to accomplish managerial or organizational targets. The decision-making process is an incessant and obligatory part of organization management or activities that are carried out in business. Decisions are made to address the events of all activities related to business and organizations. Decision and economic theories are interconnected with the assumption that field experts make a decision to make the best use out of their personal interest and reasonableness. This, though, doesn't take into consideration the probabilities of intervening factors that make decision making dependent upon the situation. The aforementioned factors play a key role in normalizing decision making for the manager to achieve optimal targets.

Although there is much literature available regarding the present study, the points given below, connected to the bipolar neutrosophic set and its aggregation operator, motivated the researcher to construct a detail and deep inquiry in the present study. Our main rationalizations and are as follows:

(1) Single valued neutrosophic sets (SVNSs) help in dealing with uncertain information in a more reliable way. It is a generalization of classical set, fuzzy set, and intuitionistic fuzzy set etc., and adaptable to the framework in comparison with pre-existing fuzzy sets and its versions.
(2) To deal with uncertain real-life problems, bipolar fuzzy sets are of great value and prove to be helpful in dealing with the positive as well as the negative membership values.
(3) We tried to merge these ideas with the Hamacher aggregation operator and strive to develop a more effective tool to deal with uncertainty in the form of bipolar neutrosophic Haymaker averaging aggregation operators.
(4) The objective of the study was to propose bipolar neutrosophic Hamacher operators and also study its properties. Furthermore, we proposed three aggregation operators, namely bipolar neutrosophic Hamacher weighted averaging operators (BNHWA), and bipolar neutrosophic Hamacher ordered weighted averaging operators (BNHOWA) and bipolar neutrosophic Hamacher hybrid averaging operators (BNHHA). Multi-attribute decision making (MADM) program approach is established based on bipolar neutrosophic numbers
(5) So as to affirm the effectiveness of the proposed method, we applied bipolar neutrosophic numbers to the decision-making problem.
(6) The initial decision matrices were composed of bipolar neutrosophic numbers and transformed into a collective bipolar neutrosophic decision matrix.
(7) The proposed operators probably completely elaborate the vagueness of bipolar neutrosophic Hamacher aggregation operator.

The rest of the study is organized as follows:

- Section 2 comprises the fundamental definitions and their related properties, which are required later in paper.
- In Section 3 we introduce the BNHWA operator, BNHWOA operator and BNHHA operator.
- In Section 4, the new aggregation operators are applied to group decision making and we propose a numerical problem.
- In Section 5, there is a comparison of our method in relation to other methods and concluding remarks are also given.

2. Preliminaries

Fundamental definitions of neutrosophic set are provided in the current section for bipolar fuzzy set, bipolar neutrosophic set, score function, accuracy function, certainty function and Hamacher operations.

Definition 1. [3] *Let P be any fixed set. Then a neutrosophic set (NS) is as follows:*

$$B = \left\{ \left(p, \zeta(p), \Gamma'(p), \text{Ю}(p) \right) \middle| p \in P \right\}, \quad (1)$$

where the truth-membership is $\zeta : H \rightarrow E$, *the indeterminacy-membership is* $\Gamma' : H \rightarrow E$, *and the falsity-membership is* $\text{Ю} : H \rightarrow E$, *where* $E =]0^-, 1^+[$.$\zeta(p), \Gamma'(p)$ *and* $\text{Ю}(p)$ *are real standard or non-standards subsets of* $]0^-, 1^+[$, *which was proposed by Abraham Robinson in 1966 [48]. There is no restriction on the sum of* $\zeta(p), \Gamma'(p)$ *and* $\text{Ю}(p)$, *so* $0^- \leq \zeta(p) + \Gamma'(p) + \text{Ю}(p) \leq 3^+$.

As it is difficult to apply NS in real scientific and engineering areas, Wang et al. [4] proposed the concept of a single valued neutrosophic set (SVNS), which follows.

Definition 2. [4] *Let P be a non-empty set, with an element in P denoted by p, and then the single valued neutrosophic set (SVNS) of A in H is as follows:*

$$A_{NS} = \{(p, \zeta(p), \Gamma'(p), \text{Ю}(p)) | p \in P\}, \tag{2}$$

where the truth-membership is $\zeta : H \to N$, *the indeterminacy-membership is* $\Gamma' : H \to N$, *and the falsity-membership is* $\text{Ю} : H \to N$, *where* $N = [0, 1]$. *There is one condition, i.e.,*

$$0 \le \zeta(p) + \Gamma'(p) + \text{Ю}(p) \le 3.$$

The basic operations for two SVNSs are:

$$A_{NS} = \{(p, \zeta_A(p), \Gamma'_A(p), \text{Ю}_A(p)) | p \in P\},\ \mathrm{B}_{NS} = \{(p, \zeta_B(p), \Gamma'_B(p), \text{Ю}_B(p)) | p \in P\},$$

and are given as follows:

i. The subset $A_{NS} \subseteq B_{NS}$ if, and only if,

$$\zeta_A(p) \le \zeta_B(p), \Gamma'_A(p) \ge \Gamma'_B(p), \text{Ю}_A(p) \ge \text{Ю}_B(p).$$

ii. $A_{NS} = B_{NS}$ if, and only if,

$$\zeta_A(p) = \zeta_B(p), \Gamma'_A(p) = \Gamma'_B(p), \text{Ю}_A(p) = \text{Ю}_B(p).$$

iii. the complement A'_{NS} is

$$A'_{NS} = \{(p, \text{Ю}_A(p), 1 - \Gamma'_A(p), \zeta_A(p)) | p \in P\}.$$

iv. the intersection is defined by

$$A_{NS} \cap B_{NS} = \{(p, \min\{\zeta_A(p), \zeta_B(p)\}, \max\{\Gamma'_A(p), \Gamma'_B(p)\}, \max\{\text{Ю}_A(p), \text{Ю}_B(p)\}) | p \in P\}, \text{ and}$$

v. the union is defined by

$$A_{NS} \cup B_{NS} = \{(p, \max\{\zeta_A(p), \zeta_B(p)\}, \min\{\Gamma'_A(p), \Gamma'_B(p)\}, \min\{\text{Ю}_A(p), \text{Ю}_B(p)\}) | p \in P\},$$

Definition 3. [49] *Let* $u_1 = (\zeta_1, \Gamma'_1, \text{Ю}_1)$ *and* $u_2 = (\zeta_2, \Gamma'_2, \text{Ю}_2)$ *be two single-value neutrosophic numbers (SVNNs). Then, the operations for the SVNNs are as follows:*

i. $u_1 + u_2 = (\zeta_1 + \zeta_2 - \zeta_1\zeta_2, \Gamma'_1\Gamma'_2, \text{Ю}_1\text{Ю}_2)$,
ii. $u_1 \cdot u_2 = (\zeta_1 + \zeta_2, \Gamma'_1 + \Gamma'_2 - \Gamma'_1\Gamma'_2, \text{Ю}_1 + \text{Ю}_2 - \text{Ю}_1\text{Ю}_2)$,
iii. $\lambda(u_1) = \left(1 - (1 - \zeta_1)^\lambda, (\Gamma'_1)^\lambda, (\text{Ю}_1)^\lambda\right)$ *and*
iv. $(u_1)^\lambda = \left((\zeta_1)^\lambda, 1 - (1 - \Gamma'_1)^\lambda, 1 - (1 - \text{Ю}_1)^\lambda\right)$,

where $\lambda > 0$.

Definition 4. [49] *Let* $u_1 = (\zeta_1, \Gamma'_1, \text{Ю}_1)$ *be a SVNN. Then, the score function* $s(u_1)$ *is as follows:*

$$s(u_1) = \frac{(\zeta_1 + 1 - \Gamma'_1 + 1 - \text{Ю}_1)}{3}.$$

Definition 5. [49] *Let* $u_1 = (\zeta_1, \Gamma'_1, \text{IO}_1)$ *be a SVNN. Then, the accuracy function* $a(u_1)$ *is as follows:*

$$a(u_1) = (\zeta_1 - \text{IO}_1).$$

Definition 6. [49] *Let* $u_1 = (\zeta_1, \Gamma'_1, \text{IO}_1)$ *be a SVNN. Then, the certainty function* $c(u_1)$ *is as follows:*

$$c(u_1) = \zeta_1 .$$

Definition 7. [49] *Let* $u_1 = (\zeta_1, \Gamma'_1, \text{IO}_1)$ *and* $u_2 = (\zeta_2, \Gamma'_2, \text{IO}_2)$ *be two SVNNs. Then, the comparison is as follows:*

i. *if* $s(u_1) > s(u_2)$, *then* u_1 *is greater than* u_2, *denoted by* $u_1 > u_2$,
ii. *if* $s(u_1) = s(u_2)$ *and* $a(u_1) > a(u_2)$, *then* u_1 *is greater than* u_2, *denoted by* $u_1 > u_2$,
iii. *if* $s(u_1) = s(u_2)$, $a(u_1) = a(u_2)$ *and* $c(u_1) > c(u_2)$, *then* u_1 *is greater than* u_2, *denoted by* $u_1 > u_2$ *and*
iv. *if* $s(u_1) = s(u_2)$, $a(u_1) = a(u_2)$ *and* $c(u_1) = c(u_2)$, *then* u_1 *is equal to* u_2, *denoted by* $u_1 = u_2$.

Definition 8. [28] *Let P be a fixed set, and the bipolar fuzzy set is as follows:*

$$F = \{\langle p, \mu_F^+(p), \eta_F^-(p)\rangle | p \in P\}, \quad (3)$$

where the positive degree of membership is $\mu_F^+(p) : H \to N$ *and the negative degree of membership is* $\eta_F^-(p) : H \to M$, *where* $N = [1, 0]$ *and* $M = [-1, 0]$.

Definition 9. [34] *A bipolar neutrosophic set (BNS), A in P, is as follows:*

$$A = \{(p, \zeta^+(p), \Gamma'^+(p), \text{IO}^+(p), \zeta^-(p), \Gamma'^-(p), \text{IO}^-(p)) | p \in P\}, \quad (4)$$

Let $\zeta^+(p), \Gamma'^+(p), \text{IO}^+(p) = BN^+$ and $\zeta^-(p), \Gamma'^-(p), \text{IO}^-(p) = BN^-$, where, $\zeta^+(p), \Gamma'^+(p), \text{IO}^+(p)$ is the positive degree of truth, the indeterminate and false membership of $p \in P$ and $\zeta^-(p), \Gamma'^-(p), \text{IO}^-(p)$ is the negative degree of truth, the indeterminate and false membership of $p \in P$. Then $BN^+ : H \to N$ and $BN^- : H \to M$, where $N = [1, 0]$ and $M = [-1, 0]$. There are conditions where $0 \le \zeta^+(p) + \Gamma'^+(p) + \text{IO}^+(p) + \zeta^-(p) + \Gamma'^-(p) + \text{IO}^-(p) \le 6$.

Example 1. *Let* $P = \{p_1, p_2, p_3\}$, *then*

$$A = \left\{ \begin{array}{c} (p_1, 0.1, 0.5, 0.3, -0.4, -0.5, -0.6), \\ (p_2, 0.2, 0.7, 0.4, -0.3, -0.6, -0.2), \\ (p_3, 0.4, 0.6, 0.7, -0.2, -0.3, -0.1) \end{array} \right\}$$

is a bipolar neutrosophic subset of P.

Basic operations [34], for two bipolar neutrosophic sets (BNSs), are as follows:
Let

$$A_1 = \{(p, \zeta_1^+(p), \Gamma_1'^+(p), \text{IO}_1^+(p), \zeta_1^-(p), \Gamma_1'^-(p), \text{IO}_1^-(p)) | p \in P\}$$
$$\text{and } A_2 = \{(p, \zeta_2^+(p), \Gamma_2'^+(p), \text{IO}_2^+(p), \zeta_2^-(p), \Gamma_2'^-(p), \text{IO}_2^-(p)) | p \in P\}$$

be two BNSs, then:

i. $A_1 \subseteq A_2$ if, and only if,

$$\zeta_1^+(p) \le \zeta_2^+(p), \text{Ґ}_1^+(p) \le \text{Ґ}_2^+(p), \text{Ю}_1^+(p) \ge \text{Ю}_2^+(p)$$

and

$$\zeta_1^-(p) \ge \zeta_2^-(p), \text{Ґ}_1^-(p) \ge \text{Ґ}_2^-(p), \text{Ю}_1^-(p) \le \text{Ю}_2^-(p),$$

ii. $A_1 = A_2$ if, and only if,

$$\zeta_1^+(p) = \zeta_2^+(p), \text{Ґ}_1^+(p) = \text{Ґ}_2^+(p), \text{Ю}_1^+(p) = \text{Ю}_2^+(p)$$

and

$$\zeta_1^-(p) = \zeta_2^-(p), \text{Ґ}_1^-(p) = \text{Ґ}_2^-(p), \text{Ю}_1^-(p) = \text{Ю}_2^-(p),$$

iii. The union is defined as:

$$(A_1 \cup A_2) = \left\{\left(\max\left(\zeta_1^+(p), \zeta_2^+(p)\right), \frac{\text{Ґ}_1^+(p)+\text{Ґ}_2^+(p)}{2}, \min\left(\text{Ю}_1^+(p), \text{Ю}_2^+(p)\right), \min\left(\zeta_1^-(p), \zeta_2^-(p)\right), \frac{\text{Ґ}_1^-(p)+\text{Ґ}_2^-(p)}{2}, \max\left(\text{Ю}_1^-(p), \text{Ю}_2^-(p)\right)\right)\right\}, \text{ and}$$

iv. The intersection is defined as:

$$(A_1 \cap A_2) = \left\{\left(\min\left(\zeta_1^+(p), \zeta_2^+(p)\right), \frac{\text{Ґ}_1^+(p)+\text{Ґ}_2^+(p)}{2}, \max\left(\text{Ю}_1^+(p), \text{Ю}_2^+(p)\right), \max\left(\zeta_1^-(p), \zeta_2^-(p)\right), \frac{\text{Ґ}_1^-(p)+\text{Ґ}_2^-(p)}{2}, \min\left(\text{Ю}_1^-(p), \text{Ю}_2^-(p)\right)\right)\right\}.$$

Let $A = \left\{\left(p, \zeta^+(p), \text{Ґ}^+(p), \text{Ю}^+(p), \zeta^-(p), \text{Ґ}^-(p), \text{Ю}^-(p)\right) \middle| p \in P\right\}$ and be a BNS. Then the complement A^c is defined as:

$$\zeta_{A^c}^+(p) = \left\{1^+\right\} - \zeta_A^+(p), \text{Ґ}_{A^c}^+(p) = \left\{1^+\right\} - \text{Ґ}_A^+(p), \text{Ю}_{A^c}^+(p) = \left\{1^+\right\} - \text{Ю}_A^+(p)$$

and

$$\zeta_{A^c}^-(p) = \{1^-\} - \zeta_A^-(p), \text{Ґ}_{A^c}^-(p) = \{1^-\} - \text{Ґ}_A^-(p), \text{Ю}_{A^c}^-(p) = \{1^-\} - \text{Ю}_A^-(p).$$

Definition 10. *[34] Let* $u_1 = \left(\zeta_1^+, \text{Ґ}_1^+, \text{Ю}_1^+, \zeta_1^-, \text{Ґ}_1^-, \text{Ю}_1^-\right)$ *and* $u_2 = \left(\zeta_2^+, \text{Ґ}_2^+, \text{Ю}_2^+, \zeta_2^-, \text{Ґ}_2^-, \text{Ю}_2^-\right)$ *be two bipolar neutrosophic numbers (BNNs). Then, the operations for the BNNs are as follows:*

$$u_1 + u_2 = \left(\zeta_1^+ + \zeta_2^+ - \zeta_1^+\zeta_2^+, \text{Ґ}_1^+\text{Ґ}_2^+, \text{Ю}_1^+\text{Ю}_2^+, -\zeta_1^-\zeta_2^-, -\left(-\text{Ґ}_1^- - \text{Ґ}_2^- - \text{Ґ}_1^-\text{Ґ}_2^-\right), -\left(-\text{Ю}_1^- - \text{Ю}_2^- - \text{Ю}_1^-\text{Ю}_2^-\right)\right),$$

$$u_1 \cdot u_2 = \left(\zeta_1^+\zeta_2^+, \text{Ґ}_1^+ + \text{Ґ}_2^+ - \text{Ґ}_1^+\text{Ґ}_2^+, \text{Ю}_1^+ + \text{Ю}_2^+ - \text{Ю}_1^+\text{Ю}_2^+, -\left(-\zeta_1^- - \zeta_2^- - \zeta_1^-\zeta_2^-\right), -\text{Ґ}_1^-\text{Ґ}_2^-, -\text{Ю}_1^-\text{Ю}_2^-\right),$$

$$\lambda(u_1) = \left(1 - \left(1 - \zeta_1^+\right)^\lambda, \left(\text{Ґ}_1^+\right)^\lambda, \left(\text{Ю}_1^+\right)^\lambda, -\left(-\zeta_1^-\right)^\lambda, -\left(-\text{Ґ}_1^-\right)^\lambda, -\left(1 - \left(1 - \left(-\text{Ю}_1^-\right)\right)^\lambda\right)\right),$$

$$(u_1)^\lambda = \left(\left(\zeta_1^+\right)^\lambda, 1 - \left(1 - \text{Ґ}_1^+\right)^\lambda, 1 - \left(1 - \text{Ю}_1^+\right)^\lambda, -\left(1 - \left(1 - \left(-\zeta_1^-\right)\right)^\lambda\right), -\left(-\text{Ґ}_1^-\right)^\lambda, -\left(-\text{Ю}_1^-\right)^\lambda\right).$$

where $\lambda > 0$.

Definition 11. *[34] Let* $u = \left(\zeta^+, \text{Ґ}^+, \text{Ю}^+, \zeta^-, \text{Ґ}^-, \text{Ю}^-\right)$ *be a bipolar neutrosophic number (BNN), then the score function of u is as follows:*

$$S(u) = \frac{1}{6}\left(\zeta^+ + 1 - \text{Ґ}^+ + 1 - \text{Ю}^+ + 1 + \zeta^- - \text{Ґ}^- - \text{Ю}^-\right). \tag{5}$$

Definition 12. *[34] Let $u = \left(\zeta^{+}, \Gamma^{+}, Ю^{+}, \zeta^{-}, \Gamma^{-}, Ю^{-}\right)$ be a BNN, then the accuracy function of u is as follows:*

$$a(u) = \zeta^{+} - Ю^{+} + \zeta^{-} - Ю^{-}. \tag{6}$$

Definition 13. *[34] Let $u = \left(\zeta^{+}, \Gamma^{+}, Ю^{+}, \zeta^{-}, \Gamma^{-}, Ю^{-}\right)$ be a bipolar neutrosophic value, then the certainty function of u is as follows:*

$$c(u) = \zeta^{+} - Ю^{-}. \tag{7}$$

Definition 14. *[34] Let $u_1 = \left(\zeta_1^{+}, \Gamma_1^{+}, Ю_1^{+}, \zeta_1^{-}, \Gamma_1^{-}, Ю_1^{-}\right)$ and $u_2 = \left(\zeta_2^{+}, \Gamma_2^{+}, Ю_2^{+}, \zeta_2^{-}, \Gamma_2^{-}, Ю_2^{-}\right)$ be two BNNs, then the comparison method is as follows:*

(i) *If $S(u_1) > S(u_2)$, then u_1 is greater than u_2, denoted by $u_1 > u_2$,*
(ii) *If $S(u_1) = S(u_2)$, and $a(u_1) > a(u_2)$, then u_1 is superior to u_2, denoted by $u_1 > u_2$,*
(iii) *If $S(u_1) = S(u_2)$, $a(u_1) = a(u_2)$ and $c(u_1) > c(u_2)$, then u_1 is greater than u_2, denoted by $u_1 > u_2$ and*
(iv) *If $S(u_1) = S(u_2)$, $a(u_1) = a(u_2)$ and $c(u_1) = c(u_2)$, then u_1 is equal to u_2, denoted by $u_1 = u_2$*

Hamacher [40] proposed a more generalized t-norm and t-conorm. The Hamacher product, ⊗, is a t-norm and the Hamacher sum, ⊕, is a t-conorm, where:

$$T(a,b) = a \otimes b = \frac{ab}{\ddot{y} + (1-\ddot{y})(a+b-ab)}, \ddot{y} > 0$$
$$T*(a,b) = a \oplus b = \frac{a+b-ab-(1-\ddot{y})ab}{1-(1-\ddot{y})ab}, \ddot{y} > 0$$

when $\ddot{y} = 1$, the Hamacher t-norm and t-conorm will be reduced to algebraic t-norm and *t*-conorm, respectively:

$$T(a,b) = a \otimes b = ab$$
$$T*(a,b) = a \oplus b = a + b - ab$$

when $\ddot{y} = 2$, the Hamacher t-norm and t-conorm will be reduced to the Einstein t-norm and t-conorm, respectively [16]:

$$T(a,b) = a \otimes b = \frac{ab}{1+(1-a)(1-b)}$$
$$T*(a,b) = a \oplus b = \frac{a+b}{1+ab}$$

The following definitions introduce the Hamacher operations of bipolar neutrosophic set, as the notion of the bipolar neutrosophic Hamacher sum, product, scalar multiple and exponential operations are defined.

Definition 15. *Let $u = \left(\zeta^{+}, \Gamma^{+}, Ю^{+}, \zeta^{-}, \Gamma^{-}, Ю^{-}\right)$, $u_1 = \left(\zeta_1^{+}, \Gamma_1^{+}, Ю_1^{+}, \zeta_1^{-}, \Gamma_1^{-}, Ю_1^{-}\right)$ and $u_2 = \left(\zeta_2^{+}, \Gamma_2^{+}, Ю_2^{+}, \zeta_2^{-}, \Gamma_2^{-}, Ю_2^{-}\right)$ be three BNNs values, and $\lambda > 0$ be any real number, then we define basic Hamacher operators with $\ddot{y} > 0$.*

$$u_1 \oplus u_2 = \left(\begin{array}{c} \frac{\zeta_1^{+}+\zeta_2^{+}-\zeta_1^{+}\zeta_2^{+}-(1-\ddot{y})\zeta_1^{+}\zeta_2^{+}}{1-(1-\ddot{y})\zeta_1^{+}\zeta_2^{+}}, \frac{\Gamma_1^{+}\Gamma_2^{+}}{\ddot{y}+(1-\ddot{y})\left(\Gamma_1^{+}+\Gamma_2^{+}-\Gamma_1^{+}\Gamma_2^{+}\right)}, \frac{Ю_1^{+}Ю_2^{+}}{\ddot{y}+(1-\ddot{y})\left(Ю_1^{+}+Ю_2^{+}-Ю_1^{+}Ю_2^{+}\right)}, \\ \frac{-\zeta_1^{-}\zeta_2^{-}}{\ddot{y}+(1-\ddot{y})\left(\zeta_1^{-}+\zeta_2^{-}-\zeta_1^{-}\zeta_2^{-}\right)}, \frac{-\left(-\Gamma_1^{-}-\Gamma_2^{-}-\Gamma_1^{-}\Gamma_2^{-}-(1-\ddot{y})\Gamma_1^{-}\Gamma_2^{-}\right)}{1-(1-\ddot{y})\Gamma_1^{-}\Gamma_2^{-}}, \frac{-\left(-Ю_1^{-}-Ю_2^{-}-Ю_1^{-}Ю_2^{-}-(1-\ddot{y})Ю_1^{-}Ю_2^{-}\right)}{1-(1-\ddot{y})Ю_1^{-}Ю_2^{-}} \end{array} \right) \tag{8}$$

$$u_1 \otimes u_2 = \left(\begin{array}{c} \frac{\zeta_1^{+}\zeta_2^{+}}{\ddot{y}+(1-\ddot{y})\left(\zeta_1^{+}+\zeta_2^{+}-\zeta_1^{+}\zeta_2^{+}\right)}, \frac{\Gamma_1^{+}+\Gamma_2^{+}-\Gamma_1^{+}\Gamma_2^{+}-(1-\ddot{y})\Gamma_1^{+}\Gamma_2^{+}}{1-(1-\ddot{y})\Gamma_1^{+}\Gamma_2^{+}}, \frac{Ю_1^{+}+Ю_2^{+}-Ю_1^{+}Ю_2^{+}-(1-\ddot{y})Ю_1^{+}Ю_2^{+}}{1-(1-\ddot{y})Ю_1^{+}Ю_2^{+}}, \\ \frac{-\left(-\zeta_1^{-}-\zeta_2^{-}-\zeta_1^{-}\zeta_2^{-}-(1-\ddot{y})\zeta_1^{-}\zeta_2^{-}\right)}{1-(1-\ddot{y})\zeta_1^{-}\zeta_2^{-}}, \frac{-\Gamma_1^{-}\Gamma_2^{-}}{\ddot{y}+(1-\ddot{y})\left(\Gamma_1^{-}+\Gamma_2^{-}-\Gamma_1^{-}\Gamma_2^{-}\right)}, \frac{-Ю_1^{-}Ю_2^{-}}{\ddot{y}+(1-\ddot{y})\left(Ю_1^{-}+Ю_2^{-}-Ю_1^{-}Ю_2^{-}\right)} \end{array} \right) \tag{9}$$

$$\lambda(u)=\begin{pmatrix}\frac{(1+(\ddot{y}-1)\zeta^{+})^{\lambda}-(1-\zeta^{+})^{\lambda}}{(1+(\ddot{y}-1)\zeta^{+})^{\lambda}+(\ddot{y}-1)(1-\zeta^{+})^{\lambda}},\frac{\ddot{y}(\Gamma^{+})^{\lambda}}{(1+(\ddot{y}-1)(1-\Gamma^{+}))^{\lambda}+(\ddot{y}-1)(\Gamma^{+})^{\lambda}},\frac{\ddot{y}(\text{Ю}^{+})^{\lambda}}{(1+(\ddot{y}-1)(1-\text{Ю}^{+}))^{\lambda}+(\ddot{y}-1)(\text{Ю}^{+})^{\lambda}},\\ \frac{-\ddot{y}|\zeta^{-}|^{\lambda}}{(1+(\ddot{y}-1)(1+\zeta^{-}))^{\lambda}+(\ddot{y}-1)|\zeta^{-}|^{\lambda}},\frac{-\ddot{y}|\Gamma^{-}|^{\lambda}}{(1+(\ddot{y}-1)(1+\Gamma^{-}))^{\lambda}+(\ddot{y}-1)|\Gamma^{-}|^{\lambda}},\frac{-((1+(\ddot{y}-1)|\text{Ю}^{-}|)^{\lambda}-(1+\text{Ю}^{-})^{\lambda})}{(1+(\ddot{y}-1)|\text{Ю}^{-}|)^{\lambda}+(\ddot{y}-1)(1+\text{Ю}^{-})^{\lambda}}\end{pmatrix} \tag{10}$$

$$(u)^{\lambda}=\begin{pmatrix}\frac{(\zeta^{+})^{\lambda}}{(1+(\ddot{y}-1)(1-\zeta^{+}))^{\lambda}+(\ddot{y}-1)(\zeta^{+})^{\lambda}},\frac{(1+(\ddot{y}-1)\Gamma^{+})^{\lambda}-(1-\Gamma^{+})^{\lambda}}{(1+(\ddot{y}-1)\Gamma^{+})^{\lambda}+(\ddot{y}-1)(1-\Gamma^{+})^{\lambda}},\frac{(1+(\ddot{y}-1)\text{Ю}^{+})^{\lambda}-(1-\text{Ю}^{+})^{\lambda}}{(1+(\ddot{y}-1)\text{Ю}^{+})^{\lambda}+(\ddot{y}-1)(1-\text{Ю}^{+})^{\lambda}},\\ \frac{-((1+(\ddot{y}-1)|\zeta^{-}|)^{\lambda}-(1+\zeta^{-})^{\lambda})}{(1+(\ddot{y}-1)\zeta^{-})^{\lambda}+(\gamma-1)(1+\zeta^{-})^{\lambda}},\frac{-\ddot{y}|\Gamma^{-}|^{\lambda}}{(1+(\ddot{y}-1)(1+\Gamma^{-}))^{\lambda}+(\ddot{y}-1)|\Gamma^{-}|^{\lambda}},\frac{-\ddot{y}|\text{Ю}^{-}|^{\lambda}}{(1+(\ddot{y}-1)(1+\text{Ю}^{-}))^{\lambda}+(\ddot{y}-1)|\text{Ю}^{-}|^{\lambda}}\end{pmatrix} \tag{11}$$

3. Bipolar Neutrosophic Hamacher Aggregation Operators

We propose some properties of the Hamacher aggregation operators in this part of the paper for bipolar neutrosophic Hamacher weighted averaging (BNHWA), bipolar neutrosophic Hamacher ordered weighted averaging (BNHOWA) and bipolar neutrosophic Hamacher hybrid averaging (BNHHA).

Let $u_\ell=\left(\zeta_\ell^{+},\Gamma_\ell^{+},\text{Ю}_\ell^{+},\zeta_\ell^{-},\Gamma_\ell^{-},\text{Ю}_\ell^{-}\right)$ be a family of BNNs, where $\ell\in Z$ and $Z=\{1,2,3,\ldots,n\}$.

3.1. Bipolar Neutrosophic HamacherWeighted Averaging Aggregation Operator

Definition 16. *The bipolar neutrosophic Hamacher weighted averaging (BNHWA) operator can be defined as follows:*

$$BNHWA_{v}(u_1,u_2,\ldots..,u_n)=\overset{n}{\underset{\ell=1}{\oplus}}(v_\ell u_\ell)=v_1u_1\oplus v_2u_2\oplus\ldots\ldots..\oplus v_nu_n \tag{12}$$

where $v=(v_1,v_2,\ldots,v_n)^T$ is the weighted vector of u_ℓ, such that $v_\ell>0$ and $\sum_{\ell=1}^{n}v_\ell=1,\ddot{y}>0$.

Theorem 1. *The (BNHWA) operator gives a bipolar neutrosophic value when:*

$$BNHWA_{v}(u_1,u_2,\ldots..,u_n)=$$
$$\begin{pmatrix}\frac{\prod_{\ell=1}^{n}(1+(\ddot{y}-1)\zeta_\ell^{+})^{v_\ell}-\prod_{\ell=1}^{n}(1-\zeta_\ell^{+})^{v_\ell}}{\prod_{\ell=1}^{n}(1+(\ddot{y}-1)\zeta_\ell^{+})^{v_\ell}+(\ddot{y}-1)\prod_{\ell=1}^{n}(1-\zeta_\ell^{+})^{v_\ell}},\frac{\ddot{y}\prod_{\ell=1}^{n}(\Gamma_\ell^{+})^{v_\ell}}{\prod_{\ell=1}^{n}(1+(\ddot{y}-1)(1-\Gamma_\ell^{+}))^{v_\ell}+(\ddot{y}-1)\prod_{\ell=1}^{n}(\Gamma_\ell^{+})^{v_\ell}},\frac{\ddot{y}\prod_{\ell=1}^{n}(\text{Ю}_\ell^{+})^{v_\ell}}{\prod_{\ell=1}^{n}(1+(\ddot{y}-1)(1-\text{Ю}_\ell^{+}))^{v_\ell}+(\ddot{y}-1)\prod_{\ell=1}^{n}(\text{Ю}_\ell^{+})^{v_\ell}},\\ \frac{-\ddot{y}\prod_{\ell=1}^{n}|\zeta_\ell^{-}|^{v_\ell}}{\prod_{\ell=1}^{n}(1+(\ddot{y}-1)(1+\zeta_\ell^{-}))^{v_\ell}+(\ddot{y}-1)\prod_{\ell=1}^{n}|\zeta_\ell^{-}|^{v_\ell}},-\frac{\prod_{\ell=1}^{n}(1+(\ddot{y}-1)|\Gamma_\ell^{-}|)^{v_\ell}-\prod_{\ell=1}^{n}(1+\Gamma_\ell^{-})^{v_\ell}}{\prod_{\ell=1}^{n}(1+(\ddot{y}-1)|\Gamma_\ell^{-}|)^{v_\ell}+(\ddot{y}-1)\prod_{\ell=1}^{n}(1+\Gamma_\ell^{-})^{v_\ell}},-\frac{\prod_{\ell=1}^{n}(1+(\ddot{y}-1)|\text{Ю}_\ell^{-}|)^{v_\ell}-\prod_{\ell=1}^{n}(1+\text{Ю}_\ell^{-})^{v_\ell}}{\prod_{\ell=1}^{n}(1+(\ddot{y}-1)|\text{Ю}_\ell^{-}|)^{v_\ell}+(\ddot{y}-1)\prod_{\ell=1}^{n}(1+\text{Ю}_\ell^{-})^{v_\ell}}\end{pmatrix} \tag{13}$$

where $v=(v_1,v_2,\ldots,v_n)^T$ *is the weighted vector of* u_ℓ, *such that* $v_\ell>0$ *and* $\sum_{\ell=1}^{n}v_\ell=1,\ddot{y}>0$.

Proof. This theorem can be proved by mathematical induction as follows:

When $n=2$

$$v_1u_1=\begin{pmatrix}\frac{(1+(\ddot{y}-1)\zeta_1^{+})^{v_1}-(1-\zeta_1^{+})^{v_1}}{(1+(\ddot{y}-1)\zeta_1^{+})^{v_1}+(\ddot{y}-1)(1-\zeta_1^{+})^{v_1}},\frac{\ddot{y}(\Gamma_1^{+})^{v_1}}{(1+(\ddot{y}-1)(1-\Gamma_1^{+}))^{v_1}+(\ddot{y}-1)(\Gamma_1^{+})^{v_1}},\frac{\ddot{y}(\text{Ю}_1^{+})^{v_1}}{(1+(\ddot{y}-1)(1-\text{Ю}_1^{+}))^{v_1}+(\ddot{y}-1)(\text{Ю}_1^{+})^{v_1}},\\ \frac{-\ddot{y}|\zeta_1^{-}|^{v_1}}{(1+(\ddot{y}-1)(1+\zeta_1^{-}))^{v_1}+(\ddot{y}-1)|\zeta_1^{-}|^{v_1}},\frac{-\ddot{y}|\Gamma_1^{-}|^{v_1}}{(1+(\ddot{y}-1)(1+\Gamma_1^{-}))^{v_1}+(\ddot{y}-1)|\Gamma_1^{-}|^{v_1}},\frac{-((1+((\ddot{y}-1)|\text{Ю}_1^{-}|))^{v_1}-(1+\text{Ю}_1^{-})^{v_1})}{(1+(\ddot{y}-1)|\text{Ю}_1^{-}|)^{v_1}+(\ddot{y}-1)(1+\text{Ю}_1^{-})^{v_1}}\end{pmatrix}$$

$$v_2u_2=\begin{pmatrix}\frac{(1+(\ddot{y}-1)\zeta_2^{+})^{v_2}-(1-\zeta_2^{+})^{v_2}}{(1+(\ddot{y}-1)\zeta_2^{+})^{v_2}+(\ddot{y}-1)(1-\zeta_2^{+})^{v_2}},\frac{\ddot{y}(\Gamma_2^{+})^{v_2}}{(1+(\ddot{y}-1)(1-\Gamma_2^{+}))^{v_2}+(\ddot{y}-1)(\Gamma_2^{+})^{v_2}},\frac{\ddot{y}(\text{Ю}_2^{+})^{v_2}}{(1+(\ddot{y}-1)(1-\text{Ю}_2^{+}))^{v_2}+(\ddot{y}-1)(\text{Ю}_2^{+})^{v_2}},\\ \frac{-\ddot{y}|\zeta_2^{-}|^{v_2}}{(1+(\ddot{y}-1)(1+\zeta_2^{-}))^{v_2}+(\ddot{y}-1)|\zeta_2^{-}|^{v_2}},\frac{-\ddot{y}|\Gamma_2^{-}|^{v_2}}{(1+(\ddot{y}-1)(1+\Gamma_2^{-}))^{v_2}+(\ddot{y}-1)|\Gamma_2^{-}|^{v_2}},\frac{-((1+((\ddot{y}-1)|\text{Ю}_2^{-}|))^{v_2}-(1+\text{Ю}_2^{-})^{v_2})}{(1+(\ddot{y}-1)|\text{Ю}_2^{-}|)^{v_2}+(\ddot{y}-1)(1+\text{Ю}_2^{-})^{v_2}}\end{pmatrix}$$

and for

$$v_1u_1=\left(\begin{array}{c}\frac{(1+(\ddot{y}-1)\zeta_1^+)-(1-\zeta_1^+)}{(1+(\ddot{y}-1)\zeta_1^+)+(\ddot{y}-1)(1-\zeta_1^+)},\frac{\ddot{y}(\Gamma_1^+)}{(1+(\ddot{y}-1)(1-\Gamma_1^+))+(\ddot{y}-1)(\Gamma_1^+)},\frac{\ddot{y}(\text{Ю}_1^+)}{(1+(\ddot{y}-1)(1-\text{Ю}_1^+))+(\ddot{y}-1)(\text{Ю}_1^+)},\\ \frac{-\ddot{y}|\zeta_1^-|}{(1+(\ddot{y}-1)(1+\zeta_1^-))+(\ddot{y}-1)|\zeta_1^-|},-\frac{(1+((\ddot{y}-1)|\Gamma_1^-|)-(1+\Gamma_1^-)}{(1+(\ddot{y}-1)|\Gamma_1^-|)+(\ddot{y}-1)(1+\Gamma_1^-)},-\frac{(1+((\ddot{y}-1)|\text{Ю}_1^-|)-(1+\text{Ю}_1^-)}{(1+(\ddot{y}-1)|\text{Ю}_1^-|)+(\ddot{y}-1)(1+\text{Ю}_1^-)}\end{array}\right) \tag{14}$$

So it is proved for $n=1$.

Now when $n=r$ in Equation (14), then

$$BNHWA_v(u_1,u_2,\ldots,u_n)=$$
$$\left(\begin{array}{c}\frac{\prod_{\ell=1}^r(1+(\ddot{y}-1)\zeta_\ell^+)^{v_\ell}-\prod_{\ell=1}^r(1-\zeta_\ell^+)^{v_\ell}}{\prod_{\ell=1}^r(1+(\ddot{y}-1)\zeta_\ell^+)^{v_\ell}+(\ddot{y}-1)\prod_{\ell=1}^r(1-\zeta_\ell^+)^{v_\ell}},\frac{\ddot{y}\prod_{\ell=1}^r(\Gamma_\ell^+)^{v_\ell}}{\prod_{\ell=1}^r(1+(\ddot{y}-1)(1-\Gamma_\ell^+))^{v_\ell}+(\ddot{y}-1)\prod_{\ell=1}^r(\Gamma_\ell^+)^{v_\ell}},\frac{\ddot{y}\prod_{\ell=1}^r(\text{Ю}_\ell^+)^{v_\ell}}{\prod_{\ell=1}^r(1+(\ddot{y}-1)(1-\text{Ю}_\ell^+))^{v_\ell}+(\ddot{y}-1)\prod_{\ell=1}^r(\text{Ю}_\ell^+)^{v_\ell}},\\ \frac{-\ddot{y}\prod_{\ell=1}^r|\zeta_\ell^-|^{v_\ell}}{\prod_{\ell=1}^r(1+(\ddot{y}-1)(1+\zeta_\ell^-))^{v_\ell}+(\ddot{y}-1)\prod_{\ell=1}^r|\zeta_\ell^-|^{v_\ell}},\frac{-\ddot{y}\prod_{\ell=1}^r|\Gamma_\ell^-|}{\prod_{\ell=1}^r(1+(\ddot{y}-1)(1+\Gamma_\ell^-))^{v_\ell}+(\ddot{y}-1)\prod_{\ell=1}^r|\Gamma_\ell^-|^{v_\ell}},\frac{-\prod_{j=1}^k(1-(\ddot{y}-1)|\text{Ю}_\ell^-|)^{v_\ell}-\prod_{\ell=1}^r(1-\text{Ю}_\ell^-)^{v_\ell}}{\prod_{\ell=1}^r(1-(\ddot{y}-1)|\text{Ю}_\ell^-|)^{v_\ell}+(\ddot{y}-1)\prod_{\ell=1}^r(1+\text{Ю}_\ell^-)^{v_\ell}}\end{array}\right)$$

This proves that it is true for $n=r$

When $n=r+1$, then

$$BNHWA_v(u_1,u_2,\ldots,u_n)$$
$$=\left(\begin{array}{c}\frac{\prod_{\ell=1}^r(1+(\ddot{y}-1)\zeta_\ell^+)^{v_\ell}-\prod_{\ell=1}^r(1-\zeta_\ell^+)^{v_\ell}}{\prod_{\ell=1}^r(1+(\ddot{y}-1)\zeta_\ell^+)^{v_\ell}+(\ddot{y}-1)\prod_{\ell=1}^r(1-\zeta_\ell^+)^{v_\ell}},\frac{\ddot{y}\prod_{\ell=1}^r(\Gamma_\ell^+)^{v_\ell}}{\prod_{\ell=1}^r(1+(\ddot{y}-1)(1-\Gamma_\ell^+))^{v_\ell}+(\ddot{y}-1)\prod_{\ell=1}^r(\Gamma_\ell^+)^{v_\ell}},\\ \frac{\ddot{y}\prod_{\ell=1}^r(\text{Ю}_\ell^+)^{v_\ell}}{\prod_{\ell=1}^r(1+(\ddot{y}-1)(1-\text{Ю}_\ell^+))^{v_\ell}+(\ddot{y}-1)\prod_{\ell=1}^r(\text{Ю}_\ell^+)^{v_\ell}},\frac{-\ddot{y}\prod_{\ell=1}^r(|\zeta_\ell^-|)^{v_\ell}}{\prod_{\ell=1}^r(1+(\ddot{y}-1)(1+\zeta_\ell^-))^{v_\ell}+(\ddot{y}-1)\prod_{\ell=1}^r|\zeta_\ell^-|^{v_\ell}},\\ \frac{-\ddot{y}\prod_{\ell=1}^r|\Gamma_\ell^-|}{\prod_{\ell=1}^r(1+(\ddot{y}-1)(1+\Gamma_\ell^-))^{v_\ell}+(\ddot{y}-1)\prod_{\ell=1}^r|\Gamma_\ell^-|^{v_\ell}},-\frac{\prod_{\ell=1}^r(1+(\ddot{y}-1)|\text{Ю}_\ell^-|)^{v_\ell}-\prod_{\ell=1}^r(1+\text{Ю}_\ell^-)^{v_\ell}}{\prod_{\ell=1}^r(1+(\ddot{y}-1)|\text{Ю}_\ell^-|)^{v_\ell}+(\ddot{y}-1)\prod_{\ell=1}^r(1+\text{Ю}_\ell^-)^{v_\ell}}\end{array}\right)$$
$$\oplus\left(\begin{array}{c}\frac{(1+(\ddot{y}-1)\zeta_{r+1}^+)^{v_{r+1}}-(1-\zeta_{r+1}^+)^{v_{r+1}}}{(1+(\ddot{y}-1)\zeta_{r+1}^+)^{v_{r+1}}+(\ddot{y}-1)(1-\zeta_{r+1}^+)^{v_{r+1}}},\frac{\ddot{y}(\Gamma_{r+1}^+)^{v_{r+1}}}{(1+(\ddot{y}-1)(1-\Gamma_{r+1}^+))^{v_{r+1}}+(\ddot{y}-1)(\Gamma_{r+1}^+)^{v_{r+1}}},\\ \frac{\ddot{y}(\text{Ю}_{r+1}^+)^{v_{r+1}}}{(1+(\ddot{y}-1)(1-\text{Ю}_{r+1}^+))^{v_{r+1}}+(\ddot{y}-1)(\text{Ю}_{r+1}^+)^{v_{r+1}}},\frac{-\ddot{y}|\zeta_{r+1}^-|^{v_{r+1}}}{(1+(\ddot{y}-1)(1+\zeta_{r+1}^-))^{v_{r+1}}+(\ddot{y}-1)|\zeta_{r+1}^-|^{v_{r+1}}},\\ \frac{-\ddot{y}|\Gamma_{r+1}^-|^{v_{r+1}}}{(1+(\ddot{y}-1)(1+\Gamma_{r+1}^-))^{v_{r+1}}+(\ddot{y}-1)|\Gamma_{r+1}^-|^{v_{r+1}}},-\frac{(1-(1+(\ddot{y}-1)|\text{Ю}_{r+1}^-|))^{v_{r+1}}-(1+\text{Ю}_{r+1}^-)^{v_{r+1}})}{(1+(\ddot{y}-1)|\text{Ю}_{r+1}^-|)^{v_{r+1}}+(\ddot{y}-1)(1+\text{Ю}_{r+1}^-)^{v_{r+1}}}\end{array}\right)$$
$$=\left(\begin{array}{c}\frac{\prod_{\ell=1}^{r+1}(1+(\ddot{y}-1)\zeta_\ell^+)^{v_\ell}-\prod_{\ell=1}^{r+1}(1-\zeta_\ell^+)^{v_\ell}}{\prod_{\ell=1}^{r+1}(1+(\ddot{y}-1)\zeta_\ell^+)^{v_\ell}+(\ddot{y}-1)\prod_{\ell=1}^{r+1}(1-\zeta_\ell^+)^{v_\ell}},\frac{\ddot{y}\prod_{\ell=1}^{r+1}(\Gamma_\ell^+)^{v_\ell}}{\prod_{\ell=1}^{r+1}(1+(\ddot{y}-1)(1-\Gamma_\ell^+))^{v_\ell}+(\ddot{y}-1)\prod_{\ell=1}^{r+1}(\Gamma_\ell^+)^{v_\ell}},\\ \frac{\ddot{y}\prod_{\ell=1}^{r+1}(\text{Ю}_\ell^+)^{v_\ell}}{\prod_{\ell=1}^{r+1}(1+(\ddot{y}-1)(1-\text{Ю}_\ell^+))^{v_\ell}+(\ddot{y}-1)\prod_{\ell=1}^{r+1}(\text{Ю}_\ell^+)^{v_\ell}},\frac{-\ddot{y}\prod_{\ell=1}^{r+1}(|\zeta_\ell^-|)^{v_\ell}}{\prod_{\ell=1}^{r+1}(1+(\ddot{y}-1)(1+\zeta_\ell^-))^{v_\ell}+(\ddot{y}-1)\prod_{\ell=1}^{r+1}|\zeta_\ell^-|^{v_\ell}},\\ \frac{-\ddot{y}\prod_{\ell=1}^{r+1}|\Gamma_\ell^-|^{v_\ell}}{\prod_{\ell=1}^{r+1}(1+(\ddot{y}-1)(1+\Gamma_\ell^-))^{v_\ell}+(\ddot{y}-1)\prod_{\ell=1}^{r+1}|\Gamma_\ell^-|^{v_\ell}},-\frac{\prod_{\ell=1}^{r+1}(1+(\ddot{y}-1)|\text{Ю}_\ell^-|)^{v_\ell}-\prod_{\ell=1}^{r+1}(1+\text{Ю}_\ell^-)^{v_\ell}}{\prod_{\ell=1}^{r+1}(1+(\ddot{y}-1)|\text{Ю}_\ell^-|)^{v_\ell}+(\ddot{y}-1)\prod_{\ell=1}^{r+1}(1+\text{Ю}_\ell^-)^{v_\ell}}\end{array}\right).$$

Thus, Equation (14) is true for $n=r+1$, which proves Theorem 1. □

Theorem 2. *(Idempotency) Let* $u_\ell = \left(\zeta_\ell^+, \Gamma_\ell^+, \text{Ю}_\ell^+, \zeta_\ell^-, \Gamma_\ell^-, \text{Ю}_\ell^-\right)$*, where* $\ell \in Z$ *and* $Z = \{1, 2, 3, \ldots, n\}$ *be a collection of BNNs are equal, i.e.,* $u_\ell = u$ *for all* ℓ*, then:*

$$BNHWA_v(u_1, u_2, \ldots, u_n) = u.$$

Theorem 3. *(Boundedness) Let* $u^- = \min_\ell u_\ell$*,* $u^+ = \max_\ell u_\ell$*, then:*

$$u^- \leq BNHWA_v(u_1, u_2, \ldots, u_n) \leq u^+.$$

Theorem 4. *(Monotonicity) Let* $u_\ell = \left(\zeta_\ell^+, \Gamma_\ell^+, \text{Ю}_\ell^+, \zeta_\ell^-, \Gamma_\ell^-, \text{Ю}_\ell^-\right)$*, where* $\ell \in Z$ *and* $Z = \{1, 2, 3, \ldots, n\}$*, and* $u'_\ell = \left(\zeta'^+_\ell, \Gamma'^+_\ell, \text{Ю}'^+_\ell, \zeta'^-_\ell, \Gamma'^-_\ell, \text{Ю}'^-_\ell\right)$*, where* $\ell \in Z$ *and* $Z = \{1, 2, 3, \ldots, n\}$ *are two BNNs. If* $u_\ell \leq u'_\ell$*, for all* ℓ*, then:*

$$BNHWA_v(u_1, u_2, \ldots, u_n) \leq BNHWA_v\left(u'_1, u'_2, \ldots, u'_n\right)$$

A discussion of two cases of BNHWA operator follows.

- If $\ddot{\gamma} = 1$, then the BNHWA is converted to the bipolar neutrosophic weighted average (BNWA):

$$BNWA_v(u_1, u_2, \ldots, u_n) = \overset{n}{\underset{\ell=1}{\oplus}} (v_\ell u_\ell) = \left(\begin{array}{c} 1 - \prod_{\ell=1}^{n} \left(1 - \zeta_\ell^+\right)^{v_\ell}, \prod_{\ell=1}^{n} \left(\Gamma_\ell^+\right)^{v_\ell}, \prod_{\ell=1}^{n} \left(\text{Ю}_\ell^+\right)^{v_\ell}, \\ -\prod_{\ell=1}^{n} \left|\zeta_\ell^-\right|^{v_\ell}, -\left(1 - \left(\prod_{\ell=1}^{n} \left(1 + \Gamma_\ell^-\right)^{v_\ell}\right)\right), -\left(1 - \left(\prod_{\ell=1}^{n} \left(1 + \text{Ю}_\ell^-\right)^{v_\ell}\right)\right) \end{array} \right).$$

- If $\ddot{\gamma} = 2$, then the BNHWA is converted to the bipolar neutrosophic Einstein weighted average (BNEWA):

$$BNEWA_v(u_1, u_2, \ldots, u_n) = \overset{n}{\underset{\ell=1}{\oplus}} (v_\ell u_\ell) =$$

$$\left(\begin{array}{c} \frac{\prod_{\ell=1}^{n} (1+\zeta_\ell^+)^{v_\ell} - \prod_{\ell=1}^{n} (1-\zeta_\ell^+)^{v_\ell}}{\prod_{\ell=1}^{n} (1+\zeta_\ell^+)^{v_\ell} + \prod_{\ell=1}^{n} (1-\zeta_\ell^+)^{v_\ell}}, \frac{\prod_{\ell=1}^{n} (\Gamma_\ell^+)^{v_\ell}}{\prod_{\ell=1}^{n} (2-\Gamma_\ell^+)^{v_\ell} + \prod_{\ell=1}^{n} (\Gamma_\ell^+)^{v_\ell}}, \frac{\prod_{\ell=1}^{n} (\text{Ю}_\ell^+)^{v_\ell}}{\prod_{\ell=1}^{n} (2-\text{Ю}_\ell^+)^{v_\ell} + \prod_{\ell=1}^{n} (\text{Ю}_\ell^+)^{v_\ell}}, \\ \frac{-2\prod_{\ell=1}^{n} |\zeta_\ell^-|^{v_\ell}}{\prod_{\ell=1}^{n} (2+\zeta_\ell^-)^{v_\ell} + \prod_{\ell=1}^{n} |\zeta_\ell^-|^{v_\ell}}, -\frac{\prod_{\ell=1}^{n} \left(1+\left|\Gamma_\ell^-\right|\right)^{v_\ell} - \prod_{\ell=1}^{n} (1+\Gamma_\ell^-)^{v_\ell}}{\prod_{\ell=1}^{n} \left(1+\left|\Gamma_\ell^-\right|\right)^{v_\ell} + \prod_{\ell=1}^{n} (1+\Gamma_\ell^-)^{v_\ell}}, -\frac{\prod_{\ell=1}^{n} (1+|\text{Ю}_\ell^-|)^{v_\ell} - \prod_{\ell=1}^{n} (1+\text{Ю}_\ell^-)^{v_\ell}}{\prod_{\ell=1}^{n} (1+|\text{Ю}_\ell^-|)^{v_\ell} + \prod_{\ell=1}^{n} (1+\text{Ю}_\ell^-)^{v_\ell}} \end{array} \right).$$

3.2. Bipolar Neutrosophic Hamacher OrderedWeighted Averaging Aggregation Operator

Definition 17. *The bipolar neutrosophic Hamacher ordered weighted averaging (BNHOWA) operator can be defined as follows:*

$$BNHOWA_v(u_1, u_2, \ldots, u_n) = \overset{n}{\underset{\ell=1}{\oplus}} \left(v_\ell u_{\rho(\ell)}\right) = v_1 u_{\rho(1)} \oplus v_2 u_{\rho(2)} \oplus v_3 u_{\breve{\sigma}(3)} \oplus \ldots \oplus v_n u_{\rho(n)}, \quad (15)$$

where $(\rho(1), \rho(2), \ldots, \rho(n))$ *is a permutation with* $u_{\rho(\ell-1)} \geq u_{\rho(\ell)}$*,* $\forall \ell \in Z$*,* $Z = \{1, 2, 3, \ldots, n\}$*, and* $v = (v_1, v_2, \ldots, v_n)^T$ *is the weighted vector of* u_ℓ *such that* $v_\ell > 0$ *and* $\sum_{\ell=1}^{n} v_\ell = 1, \ddot{\gamma} > 0$.

Theorem 5. *The (BNHOWA) operator gives a bipolar neutrosophic value:*

$$BNHOWA_{v}(u_1,u_2,\ldots..,u_n)=\left(\begin{array}{c}\frac{\prod_{\ell=1}^{n}\left(1+(\ddot{\gamma}-1)\zeta^{+}_{\rho(\ell)}\right)^{v_\ell}-\prod_{\ell=1}^{n}\left(1-\zeta^{+}_{\rho(\ell)}\right)^{v_\ell}}{\prod_{\ell=1}^{n}\left(1+(\ddot{\gamma}-1)\zeta^{+}_{\rho(\ell)}\right)^{v_\ell}+(\ddot{\gamma}-1)\prod_{\ell=1}^{n}\left(1-\zeta^{+}_{\rho(\ell)}\right)^{v_\ell}},\frac{\ddot{\gamma}\prod_{\ell=1}^{n}\left(\Gamma^{+}_{\rho(\ell)}\right)^{v_\ell}}{\prod_{\ell=1}^{n}\left(1+(\ddot{\gamma}-1)(1-\Gamma^{+}_{\rho(\ell)})\right)^{v_\ell}+(\ddot{\gamma}-1)\prod_{\ell=1}^{n}\left(\Gamma^{+}_{\rho(\ell)}\right)^{v_\ell}},\\ \frac{\ddot{\gamma}\prod_{\ell=1}^{n}\left(\text{Ю}^{+}_{\rho(\ell)}\right)^{v_\ell}}{\prod_{\ell=1}^{n}\left(1+(\ddot{\gamma}-1)\left(1-\text{Ю}^{+}_{\rho(\ell)}\right)\right)^{v_\ell}+(\ddot{\gamma}-1)\prod_{\ell=1}^{n}\left(\text{Ю}^{+}_{\rho(\ell)}\right)^{v_\ell}},\frac{-\ddot{\gamma}\prod_{\ell=1}^{n}\left|\zeta^{-}_{\rho(\ell)}\right|^{v_\ell}}{\prod_{\ell=1}^{n}\left(1+(\ddot{\gamma}-1)\left(1+\zeta^{-}_{\rho(\ell)}\right)\right)^{v_\ell}+(\ddot{\gamma}-1)\prod_{\ell=1}^{n}\left|\zeta^{-}_{\rho(\ell)}\right|^{v_\ell}},\\ -\frac{\prod_{\ell=1}^{n}\left(1+(\ddot{\gamma}-1)\left|\Gamma^{-}_{\rho(\ell)}\right|\right)^{v_\ell}-\prod_{\ell=1}^{n}\left(1+\Gamma^{-}_{\rho(\ell)}\right)^{v_\ell}}{\prod_{\ell=1}^{n}\left(1+(\ddot{\gamma}-1)\left|\Gamma^{-}_{\rho(\ell)}\right|\right)^{v_\ell}+(\ddot{\gamma}-1)\prod_{\ell=1}^{n}\left(1+\Gamma^{-}_{\rho(\ell)}\right)^{v_\ell}},-\frac{\prod_{\ell=1}^{n}\left(1+(\ddot{\gamma}-1)\left|\text{Ю}^{-}_{\rho(\ell)}\right|\right)^{v_\ell}-\prod_{\ell=1}^{n}\left(1+\text{Ю}^{-}_{\rho(\ell)}\right)^{v_\ell}}{\prod_{\ell=1}^{n}\left(1+(\ddot{\gamma}-1)\left|\text{Ю}^{-}_{\rho(\ell)}\right|\right)^{v_\ell}+(\ddot{\gamma}-1)\prod_{\ell=1}^{n}\left(1+\text{Ю}^{-}_{\rho(\ell)}\right)^{v_\ell}}\end{array}\right) \quad (16)$$

where $(\rho(1),\rho(2),\ldots\ldots\ldots,\rho(n))$ *is a permutation with* $u_{\rho(\ell-1)}\geq u_{\rho(\ell)}$, $\forall\ell\in Z$, $Z=\{1,2,3,\ldots.,n\}$ *and* $v=(v_1,v_2,\ldots,v_n)^T$ *is the weighted vector of* u_ℓ *such that* $v_\ell>0$ *and* $\sum_{\ell=1}^{n}v_\ell=1,\ddot{\gamma}>0$.

Proof. The theorem is straightforward. □

Theorem 6. *(Idempotency) Let* $u_\ell=\left(\zeta^{+}_{\ell},\Gamma^{+}_{\ell},\text{Ю}^{+}_{\ell},\zeta^{-}_{\ell},\Gamma^{-}_{\ell},\text{Ю}^{-}_{\ell}\right)$, *where* $\ell\in Z$ *and* $Z=\{1,2,3,\ldots.,n\}$ *are a collection of equal BNNs, i.e.,* $u_\ell=u$ *for all* ℓ, *then:*

$$BNHOWA_{v}(u_1,u_2,\ldots..,u_n)=u.$$

Theorem 7. *(Boundedness) Let* $u^-=\underset{\ell}{min}u_\ell$, $u^+=\underset{\ell}{max}u_\ell$, *then:*

$$u^-\leq BNHOWA_{v}(u_1,u_2,\ldots..,u_n)\leq u^+.$$

Theorem 8. *(Monotonicity) Let* $u_\ell=\left(\zeta^{+}_{\ell},\Gamma^{+}_{\ell},\text{Ю}^{+}_{\ell},\zeta^{-}_{\ell},\Gamma^{-}_{\ell},\text{Ю}^{-}_{\ell}\right)$, *where* $\ell\in Z$ *and* $Z=\{1,2,3,\ldots.,n\}$ *and* $u'_\ell=\left(\zeta'^{+}_{\ell},\Gamma'^{+}_{\ell},\text{Ю}'^{+}_{\ell},\zeta'^{-}_{\ell},\Gamma'^{-}_{\ell},\text{Ю}'^{-}_{\ell}\right)$, *where* $\ell\in Z$ *and* $Z=\{1,2,3,\ldots.,n\}$ *are two BNNs. If* $u_\ell\leq u'_\ell$, *for all* ℓ, *then:*

$$BNHOWA_{\varpi}(u_1,u_2,\ldots..,u_n)\leq BNHOWA_{\varpi}\left(u'_1,u'_2,\ldots..,u'_n\right).$$

Now, we discuss two cases of the BNHOWA operator:

- If $\ddot{\gamma}=1$, the BNHOWA is converted to the bipolar neutrosophic ordered weighted average (BNOWA):

$$BNOWA_{v}(u_1,u_2,\ldots..,u_n)=\overset{n}{\underset{\ell=1}{\oplus}}\left(v_\ell u_{\rho(\ell)}\right)=\left(\begin{array}{c}1-\prod_{\ell=1}^{n}\left(1-\zeta^{+}_{\rho(\ell)}\right)^{v_\ell},\prod_{\ell=1}^{n}\left(\Gamma^{+}_{\rho(\ell)}\right)^{v_\ell},\prod_{\ell=1}^{n}\left(\text{Ю}^{+}_{\rho(\ell)}\right)^{v_\ell},\\ -\prod_{\ell=1}^{n}\left|\zeta^{-}_{\rho(\ell)}\right|^{v_\ell},-\left(1-\left(\prod_{\ell=1}^{n}\left(1+\Gamma^{-}_{\rho(\ell)}\right)^{v_\ell}\right)\right),-\left(1-\left(\prod_{\ell=1}^{n}\left(1+\text{Ю}^{-}_{\rho(\ell)}\right)^{v_\ell}\right)\right)\end{array}\right).$$

- If $\ddot{\gamma} = 2$, the BNHOWA is converted to the bipolar neutrosophic Einstein ordered weighted average (BNEOWA):

$$BNEOWA_v(u_1, u_2, \ldots, u_n) = \mathop{\oplus}_{\ell=1}^{n} \left(v_\ell u_{\rho(\ell)}\right)$$

$$= \left(\begin{array}{c} \frac{\prod_{\ell=1}^{n}\left(1+\zeta^{+}_{\rho(\ell)}\right)^{v_\ell} - \prod_{\ell=1}^{n}\left(1-\zeta^{+}_{\rho(\ell)}\right)^{v_\ell}}{\prod_{\ell=1}^{n}\left(1+\zeta^{+}_{\rho(\ell)}\right)^{v_\ell} + \prod_{\ell=1}^{n}\left(1-\zeta^{+}_{\rho(\ell)}\right)^{v_\ell}}, \frac{2\prod_{\ell=1}^{n}\left(\Gamma^{+}_{\rho(\ell)}\right)^{v_\ell}}{\prod_{\ell=1}^{n}\left(2-\Gamma^{+}_{\rho(\ell)}\right)^{v_\ell} + \prod_{\ell=1}^{n}\left(\Gamma^{+}_{\rho(\ell)}\right)^{v_\ell}}, \frac{2\prod_{\ell=1}^{n}\left(Ю^{+}_{\rho(\ell)}\right)^{v_\ell}}{\prod_{\ell=1}^{n}\left(2-Ю^{+}_{\rho(\ell)}\right)^{v_\ell} + \prod_{\ell=1}^{n}\left(Ю^{+}_{\rho(\ell)}\right)^{v_\ell}}, \\ \frac{-2\prod_{\ell=1}^{n}\left|\zeta^{-}_{\rho(\ell)}\right|^{v_\ell}}{\prod_{\ell=1}^{n}\left(2+\zeta^{-}_{\rho(\ell)}\right)^{v_\ell} + \prod_{\ell=1}^{n}\left|\zeta^{-}_{\rho(\ell)}\right|^{v_\ell}}, -\frac{\prod_{\ell=1}^{n}\left(1+\left|\Gamma^{-}_{\rho(\ell)}\right|\right)^{v_\ell} - \prod_{\ell=1}^{n}\left(1+\Gamma^{-}_{\rho(\ell)}\right)^{v_\ell}}{\prod_{\ell=1}^{n}\left(1+\left|\Gamma^{-}_{\rho(\ell)}\right|\right)^{v_\ell} + \prod_{\ell=1}^{n}\left(1+\Gamma^{-}_{\rho(\ell)}\right)^{v_\ell}}, -\frac{\prod_{\ell=1}^{n}\left(1+\left|Ю^{-}_{\rho(\ell)}\right|\right)^{v_\ell} - \prod_{\ell=1}^{n}\left(1+Ю^{-}_{\rho(\ell)}\right)^{v_\ell}}{\prod_{\ell=1}^{n}\left(1+\left|Ю^{-}_{\rho(\ell)}\right|\right)^{v_\ell} + \prod_{\ell=1}^{n}\left(1+Ю^{-}_{\rho(\ell)}\right)^{v_\ell}} \end{array} \right).$$

3.3. Bipolar Neutrosophic Hamacher HybridAveraging Aggregation Operator

Definition 18. *The bipolar neutrosophic Hamacher hybrid averaging (BNHHA) operator can be defined as follows:*

$$BNHHA_{w,v}(u_1, u_2, \ldots, u_n) = \mathop{\oplus}_{\ell=1}^{n}\left(v_\ell \dot{u}_{\rho(\ell)}\right) = v_1\dot{u}_{\rho(1)} \oplus v_2\dot{u}_{\rho(2)} \oplus v_3\dot{u}_{\rho(3)} \oplus \ldots \oplus v_n\dot{u}_{\rho(n)} \quad (17)$$

where $w = (w_1, w_2, \ldots, w_n)$ *is a weighting vector of* $u_\ell (\ell \in Z)$, $Z = \{1, 2, 3, \ldots, n\}$, *such that* $w_\ell \in [0, 1]$, $\sum_{\ell=1}^{n} w_\ell = 1$ *and* $\dot{u}_{\rho(\ell)}$ *is the* ℓ*-th largest element of the bipolar neutrosophic argument*, $\dot{u}_\ell\left(\dot{u}_\ell = (nv_\ell)u_\ell, \ell = 1, 2, \ldots, n\right)$ *and also* $v = (v_1, v_2, \ldots, v_n)$ *are weighting vectors of bipolar neutrosophic arguments* $u_\ell(\ell \in Z)$, $Z = \{1, 2, 3, \ldots, n\}$, *such that* $v_\ell \in [0, 1]$, $\sum_{\ell=1}^{n} v_\ell = 1$, *where n is the balancing coefficient. Note that BNHHA reduces to BNHWA if* $w = \left(\frac{1}{n}, \frac{1}{n}, \ldots, \frac{1}{n}\right)^T$ *and BNHOWA operator if:*

$$v = \left(\frac{1}{n}, \frac{1}{n}, \ldots, \frac{1}{n}\right).$$

Theorem 9. *The (BNHHA) operator returns a bipolar neutrosophic value when:*

$$BNHHA_{w,v}(u_1, u_2, \ldots, u_n) = \left(\begin{array}{c} \frac{\prod_{\ell=1}^{n}\left(1+(\ddot{\gamma}-1)\dot{\zeta}^{+}_{\rho(\ell)}\right)^{v_\ell} - \prod_{\ell=1}^{n}\left(1-\dot{\zeta}^{+}_{\rho(\ell)}\right)^{v_\ell}}{\prod_{\ell=1}^{n}\left(1+(\ddot{\gamma}-1)\dot{\zeta}^{+}_{\rho(\ell)}\right)^{v_\ell} + (\ddot{\gamma}-1)\prod_{\ell=1}^{n}\left(1-\dot{\zeta}^{+}_{\rho(\ell)}\right)^{v_\ell}}, \frac{\ddot{\gamma}\prod_{\ell=1}^{n}\left(\dot{\Gamma}^{+}_{\rho(\ell)}\right)^{v_\ell}}{\prod_{\ell=1}^{n}\left(1+(\ddot{\gamma}-1)\left(1-\dot{\Gamma}^{+}_{\rho(\ell)}\right)\right)^{v_\ell} + (\ddot{\gamma}-1)\prod_{\ell=1}^{n}\left(\dot{\Gamma}^{+}_{\rho(\ell)}\right)^{v_\ell}}, \\ \frac{\ddot{\gamma}\prod_{\ell=1}^{n}\left(\dot{Ю}^{+}_{\rho(\ell)}\right)^{v_\ell}}{\prod_{\ell=1}^{n}\left(1+(\ddot{\gamma}-1)\left(1-\dot{Ю}^{+}_{\rho(\ell)}\right)\right)^{v_\ell} + (\ddot{\gamma}-1)\prod_{\ell=1}^{n}\left(\dot{Ю}^{+}_{\rho(\ell)}\right)^{v_\ell}}, \frac{-\ddot{\gamma}\prod_{\ell=1}^{n}\left|\dot{\zeta}^{-}_{\rho(\ell)}\right|^{v_\ell}}{\prod_{\ell=1}^{n}\left(1+(\ddot{\gamma}-1)\left(1+\dot{\zeta}^{-}_{\rho(\ell)}\right)\right)^{v_\ell} + (\ddot{\gamma}-1)\prod_{\ell=1}^{n}\left|\dot{\zeta}^{-}_{\rho(\ell)}\right|^{v_\ell}}, \\ -\frac{\prod_{\ell=1}^{n}\left(1+(\ddot{\gamma}-1)\left|\dot{\Gamma}^{-}_{\rho(\ell)}\right|\right)^{v_\ell} - \prod_{\ell=1}^{n}\left(1+\dot{\Gamma}^{-}_{\rho(\ell)}\right)^{v_\ell}}{\prod_{\ell=1}^{n}\left(1+(\ddot{\gamma}-1)\left|\dot{\Gamma}^{-}_{\rho(\ell)}\right|\right)^{v_\ell} + (\ddot{\gamma}-1)\prod_{\ell=1}^{n}\left(1+\dot{\Gamma}^{-}_{\rho(\ell)}\right)^{v_\ell}}, -\frac{\prod_{\ell=1}^{n}\left(1+(\ddot{\gamma}-1)\left|\dot{Ю}^{-}_{\rho(\ell)}\right|\right)^{v_\ell} - \prod_{\ell=1}^{n}\left(1+\dot{Ю}^{-}_{\rho(\ell)}\right)^{v_\ell}}{\prod_{\ell=1}^{n}\left(1+(\ddot{\gamma}-1)\left|\dot{Ю}^{-}_{\rho(\ell)}\right|\right)^{v_\ell} + (\ddot{\gamma}-1)\prod_{\ell=1}^{n}\left(1+\dot{Ю}^{-}_{\rho(\ell)}\right)^{v_\ell}} \end{array} \right) \quad (18)$$

where $w = (w_1, w_2, \ldots, w_n)$ *is the weighting vector of* $u_\ell(\ell \in Z)$, $Z = \{1, 2, 3, \ldots, n\}$, *such that* $w_\ell \in [0, 1]$, $\sum_{\ell=1}^{n} w_\ell = 1$ *and* $\dot{u}_{\rho(\ell)}$ *is the* ℓ *-th largest element of the bipolar neutrosophic arguments*, $\dot{u}_\ell\left(\dot{u}_\ell = (nv_\ell)u_\ell, \ell = 1, 2, \ldots, n\right)$ *and also* $v = (v_1, v_2, \ldots, v_n)$ *are the weighting vector of bipolar neutrosophic arguments* $u_\ell(\ell \in Z)$, $Z = \{1, 2, 3, \ldots, n\}$, *such that* $v_\ell \in [0, 1]$, $\sum_{\ell=1}^{n} v_\ell = 1$, *where n is the balancing coefficient*, $\ddot{\gamma} > 0$.

Proof. The theorem is straightforward. □

Now, we discuss two cases of BNHOWA:

- If $\ddot{\gamma} = 1$, the BNHHA is converted to the bipolar neutrosophic hybrid averaging (BNHA):

$$BNHA_{w,v}(u_1, u_2, \ldots, u_n) = \overset{n}{\underset{\ell=1}{\oplus}}\left(v_\ell \dot{u}_{\rho(\ell)}\right) = \left(\begin{array}{l} 1 - \prod_{\ell=1}^{n}\left(1 - \dot{\zeta}^{+}_{\rho(\ell)}\right)^{v_\ell}, \prod_{\ell=1}^{n}\left(\dot{\Gamma}^{+}_{\rho(\ell)}\right)^{v_\ell}, \prod_{\ell=1}^{n}\left(\dot{\text{Ю}}^{+}_{\rho(\ell)}\right)^{v_\ell}, \\ -\prod_{\ell=1}^{n}\left|\dot{\zeta}^{-}_{\rho(\ell)}\right|^{v_\ell}, -\left(1 - \left(\prod_{\ell=1}^{n}\left(1 + \dot{\Gamma}^{-}_{\rho(\ell)}\right)^{v_\ell}\right)\right), -\left(1 - \left(\prod_{\ell=1}^{n}\left(1 + \dot{\text{Ю}}^{-}_{\rho(\ell)}\right)^{v_\ell}\right)\right) \end{array}\right).$$

- If $\ddot{\gamma} = 2$, the BNHHA is converted to the bipolar neutrosophic Einstein hybrid averaging (BNEHA):

$$BNEHA_{w,v}(u_1, u_2, \ldots, u_n) = \overset{n}{\underset{\ell=1}{\oplus}}\left(v_\ell \dot{u}_{\rho(\ell)}\right)$$

$$= \left(\begin{array}{l} \frac{\prod_{\ell=1}^{n}\left(1+\dot{\zeta}^{+}_{\rho(\ell)}\right)^{v_\ell} - \prod_{\ell=1}^{n}\left(1-\dot{\zeta}^{+}_{\rho(\ell)}\right)^{v_\ell}}{\prod_{\ell=1}^{n}\left(1+\dot{\zeta}^{+}_{\rho(\ell)}\right)^{v_\ell} + \prod_{\ell=1}^{n}\left(1-\dot{\zeta}^{+}_{\rho(\ell)}\right)^{v_\ell}}, \frac{2\prod_{\ell=1}^{n}\left(\dot{\Gamma}^{+}_{\rho(\ell)}\right)^{v_\ell}}{\prod_{\ell=1}^{n}\left(2-\dot{\Gamma}^{+}_{\rho(\ell)}\right)^{v_\ell} + \prod_{\ell=1}^{n}\left(\dot{\Gamma}^{+}_{\rho(\ell)}\right)^{v_\ell}}, \frac{2\prod_{\ell=1}^{n}\left(\dot{\text{Ю}}^{+}_{\rho(\ell)}\right)^{v_\ell}}{\prod_{\ell=1}^{n}\left(2-\dot{\text{Ю}}^{+}_{\rho(\ell)}\right)^{v_\ell} + \prod_{\ell=1}^{n}\left(\dot{\text{Ю}}^{+}_{\rho(\ell)}\right)^{v_\ell}}, \\ \frac{-2\prod_{\ell=1}^{n}\left|\dot{\zeta}^{-}_{\rho(\ell)}\right|^{v_\ell}}{\prod_{\ell=1}^{n}\left(2+\dot{\zeta}^{-}_{\rho(\ell)}\right)^{v_\ell} + \prod_{\ell=1}^{n}\left|\dot{\zeta}^{-}_{\rho(\ell)}\right|^{v_\ell}}, \frac{\prod_{\ell=1}^{n}\left(1+\left|\dot{\Gamma}^{-}_{\rho(\ell)}\right|\right)^{v_\ell} - \prod_{\ell=1}^{n}\left(1+\dot{\Gamma}^{-}_{\rho(\ell)}\right)^{v_\ell}}{\prod_{\ell=1}^{n}\left(1+\left|\dot{\Gamma}^{-}_{\rho(\ell)}\right|\right)^{v_\ell} + \prod_{\ell=1}^{n}\left(1+\dot{\Gamma}^{-}_{\rho(\ell)}\right)^{v_\ell}}, \frac{\prod_{\ell=1}^{n}\left(1+\left|\dot{\text{Ю}}^{-}_{\rho(\ell)}\right|\right)^{v_\ell} - \prod_{\ell=1}^{n}\left(1+\dot{\text{Ю}}^{-}_{\rho(\ell)}\right)^{v_\ell}}{\prod_{\ell=1}^{n}\left(1+\left|\dot{\text{Ю}}^{-}_{\rho(\ell)}\right|\right)^{v_\ell} + \prod_{\ell=1}^{n}\left(1+\dot{\text{Ю}}^{-}_{\rho(\ell)}\right)^{v_\ell}} \end{array}\right).$$

4. An Application of the Bipolar Neutrosophic Hamacher Averaging Aggregation Operators to Group Decision Making

In this section, we apply the bipolar neutrosophic Hamacher averaging aggregation operators to the multiple attribute group decision-making problems in which the attribute weights take the form of crisp numbers and the attribute values take the form of BNNs.

Algorithm 1: Bipolar Neutrosophic Group Decision Making Problems

Let $G = \{G_1, G_2, \ldots, G_m\}$ be the set of m alternatives, $L = \{L_1, L_2, \ldots, L_n\}$ be the set of n attributes or criterions, and $D = \{D_1, D_2, \ldots, D_k\}$ be the finite k decision makers. Let $v = (v_1, v_2, \ldots, v_n)^T$ be the weighted vector of the decision makers $\overline{D^s}(s = 1, 2, \ldots, k)$, such that $v_\ell \in [0, 1]$ and $\sum_{\ell=1}^{n} v_\ell = 1$. Let $w = (w_1, w_2, \ldots, w_n)^T$ be the weighted vector of the attribute set $L = \{L_1, L_2, \ldots, L_n\}$ such that $w_\ell \in [0, 1]$ and $\sum_{\ell=1}^{n} w_\ell = 1$. An alternative of the criterion is assessed by the decision maker and the values are represented by bipolar neutrosophic values, where $u^{(s)}_{\widehat{i}\widehat{j}} = \left[\left(\zeta^{+}_{\widehat{i}\widehat{j}}, \Gamma^{+}_{\widehat{i}\widehat{j}}, \text{Ю}^{+}_{\widehat{i}\widehat{j}}, \zeta^{-}_{\widehat{i}\widehat{j}}, \Gamma^{-}_{\widehat{i}\widehat{j}}, \text{Ю}^{-}_{\widehat{i}\widehat{j}}\right)\right]_{m \times n}$ is the decision matrix provided by the decision maker (Tables 1–3) and $u^{(s)}_{\widehat{i}\widehat{j}}$ is a bipolar neutrosophic number (Table 4) for alternative $G_{\widehat{i}}$, associated with criterion $L_{\widehat{i}}$.

The condition $\zeta^{+}_{\widehat{i}\widehat{j}}, \Gamma^{+}_{\widehat{i}\widehat{j}}, \text{Ю}^{+}_{\widehat{i}\widehat{j}}, \zeta^{-}_{\widehat{i}\widehat{j}}, \Gamma^{-}_{\widehat{i}\widehat{j}}$ and $\text{Ю}^{-}_{\widehat{i}\widehat{j}} \in [0, 1]$ is such that

$0 \le \zeta^{+}_{\widehat{i}\widehat{j}} + \Gamma^{+}_{\widehat{i}\widehat{j}} + \text{Ю}^{+}_{\widehat{i}\widehat{j}} + \zeta^{-}_{\widehat{i}\widehat{j}} + \Gamma^{-}_{\widehat{i}\widehat{j}} + \text{Ю}^{-}_{\widehat{i}\widehat{j}} \le 6$ for $\widehat{i} = 1, 2, \ldots, m$ and $\widehat{j} = 1, 2, \ldots, n$.

Step 1: Construct the decision matrix $\overline{D^s} = \left[u^{(s)}_{\widehat{i}\widehat{j}}\right]_{m \times n}$ $(s = 1, 2, \ldots, k)$ for the decision.

Step 2: Compute $BNHWA_v\left(r_{\widehat{i}1}, r_{\widehat{i}2}, \ldots, r_{\widehat{i}n}\right)$ for each $\widehat{i} = 1, 2, 3, \ldots, m$:

$$r_{\widehat{i}} = \left(\zeta^{+}_{\widehat{i}}, \Gamma^{+}_{\widehat{i}}, \text{Ю}^{+}_{\widehat{i}}, \zeta^{-}_{\widehat{i}}, \Gamma^{-}_{\widehat{i}}, \text{Ю}^{-}_{\widehat{i}}\right) =$$

$$BNHWA_v\left(r_{\widehat{i}1}, r_{\widehat{i}2}, \ldots, r_{\widehat{i}n}\right) = \overset{n}{\underset{\ell=1}{\oplus}}\left(v_\ell r_{i\ell}\right)$$

$$= \left(\begin{array}{l} \frac{\prod_{\ell=1}^{n}(1+(\ddot{\gamma}-1)\zeta^{+}_{\ell})^{v_\ell} - \prod_{\ell=1}^{n}(1-\zeta^{+}_{\ell})^{v_\ell}}{\prod_{\ell=1}^{n}(1+(\ddot{\gamma}-1)\zeta^{+}_{\ell})^{v_\ell} + (\ddot{\gamma}-1)\prod_{\ell=1}^{n}(1-\zeta^{+}_{\ell})^{v_\ell}}, \frac{\ddot{\gamma}\prod_{\ell=1}^{n}(\Gamma^{+}_{\ell})^{v_\ell}}{\prod_{\ell=1}^{n}(1+(\ddot{\gamma}-1)(1-\Gamma^{+}_{\ell}))^{v_\ell} + (\ddot{\gamma}-1)\prod_{\ell=1}^{n}(\Gamma^{+}_{\ell})^{v_\ell}}, \frac{\ddot{\gamma}\prod_{\ell=1}^{n}(\text{Ю}^{+}_{\ell})^{v_\ell}}{\prod_{\ell=1}^{n}(1+(\ddot{\gamma}-1)(1-\text{Ю}^{+}_{\ell}))^{v_\ell} + (\ddot{\gamma}-1)\prod_{\ell=1}^{n}(\text{Ю}^{+}_{\ell})^{v_\ell}}, \\ \frac{-\ddot{\gamma}\prod_{\ell=1}^{n}|\zeta^{-}_{\ell}|^{v_\ell}}{\prod_{\ell=1}^{n}(1+(\ddot{\gamma}-1)(1+\zeta^{-}_{\ell}))^{v_\ell} + (\ddot{\gamma}-1)\prod_{\ell=1}^{n}|\zeta^{-}_{\ell}|^{v_\ell}}, -\frac{\prod_{\ell=1}^{n}(1+(\ddot{\gamma}-1)|\Gamma^{-}_{\ell}|)^{v_\ell} - \prod_{\ell=1}^{n}(1+\Gamma^{-}_{\ell})^{v_\ell}}{\prod_{\ell=1}^{n}(1+(\ddot{\gamma}-1)|\Gamma^{-}_{\ell}|)^{v_\ell} + (\ddot{\gamma}-1)\prod_{\ell=1}^{n}(1+\Gamma^{-}_{\ell})^{v_\ell}}, -\frac{\prod_{\ell=1}^{n}(1+(\ddot{\gamma}-1)|\text{Ю}^{-}_{\ell}|)^{v_\ell} - \prod_{\ell=1}^{n}(1+\text{Ю}^{-}_{\ell})^{v_\ell}}{\prod_{\ell=1}^{n}(1+(\ddot{\gamma}-1)|\text{Ю}^{-}_{\ell}|)^{v_\ell} + (\ddot{\gamma}-1)\prod_{\ell=1}^{n}(1+\text{Ю}^{-}_{\ell})^{v_\ell}} \end{array}\right)$$

Step 3: Calculate the scores of $S\left(r_{\widehat{i}}\right)$ for the $\left(\widehat{i} = 1, 2, 3, \ldots, m\right)$.

Step 4: Rank all the software systems of $BNHWA_v\left(u_{\widehat{i}1}, u_{\widehat{i}2}, \ldots, u_{\widehat{i}n}\right)$ according to the scores values.

Step 5: Select the best alternative(s).

Illustrative Example

We considered an issue, taken from Deli [29], as an application for the proposed method in the present paper. The issue given is that an investment company wants to make some investments in the best possible options. There are four types of companies $G_{\widehat{i}}\left(\widehat{i} = 1,2,3\ldots\ldots\ldots m\right)$ that are available as alternatives, namely G_1: computer company, G_2: food company, G_3: car company, and G_4: arms company, to invest money. The investment company takes into account four attributes to evaluate the alternatives: L_1: risk, L_2: growth, L_3: environmental impact, and L_4: performance. We utilized the bipolar neutrosophic numbers to assess the four possible alternatives of $G_{\widehat{i}}\left(\widehat{i} = 1,2,3,4\right)$ under the four criteria. The weight vector of the attributes is $v = \left(\frac{1}{4}, \frac{1}{5}, \frac{3}{10}, \frac{1}{4}\right)^T$. There are three experts, i.e., $\overline{D^s}(s = 1,2,3)$, from a group of decision makers, whose weight vector is $v = \left(\frac{3}{10}, \frac{3}{10}, \frac{2}{5}\right)^T$. The expert opion about the companies based on atrribute are given in Tables 1–3.

Step 1: Decision matrices

Table 1. Bipolar Neutrosophic Decision Matrix, D_1.

	L_1	L_2	L_3	L_4
G_1	(0.5,0.4,0.3, −0.7, −0.5, −0.6)	(0.1, 0.5, 0.4, −0.2, −0.4, −0.7)	(0.3, 0.7, 0.4−0.6, −0.4, −0.6)	(0.2, 0.4, 0.7, −0.3, −0.5, −0.1)
G_2	(0.2, 0.6, 0.4, −0.3, −0.6, −0.8)	(0.5, 0.7, 0.6, −0.3, −0.4, −0.5)	(0.5, 0.5, 0.1, −0.7, −0.4, −0.8)	(0.6, 0.5, 0.4, −0.5, −0.4, −0.6)
G_3	(0.4, 0.5, 0.3, −0.5, −0.6, −0.7)	(0.8, 0.9, 0.2, −0.7, −0.4, −0.6)	(0.2, 0.6, 0.5, −0.5, −0.4, −0.7)	(0.5, 0.7, 0.3, −0.5, −0.4, −0.2)
G_4	(0.7, 0.6, 0.5, −0.6, −0.5, −0.4)	(0.5, 0.7, 0.6, −0.6, −0.3, −0.5)	(0.3, 0.1, 0.8, −0.9, −0.5, −0.6)	(0.2, 0.5, 0.7, −0.4, −0.5, −0.8)

Table 2. Bipolar Neutrosophic Decision Matrix, D_2.

	L_1	L_2	L_3	L_4
G_1	(0.2, 0.5,0.3, −0.4, −0.6, −0.5)	(0.4, 0.3, 0.7, −0.5, −0.4, −0.6)	(0.5, 0.7, 0.3, −0.4, −0.7, −0.6)	(0.1, 0.4, 0.6, −0.3, −0.4, −0.2)
G_2	(0.5, 0.6,0.4, −0.2, −0.4, −0.5)	(0.5, 0.1, 0.6, −0.6, −0.4, −0.2)	(0.3, 0.5, 0.4, −0.1, −0.4, −0.6)	(0.5, 0.3, 0.4−0.7, −0.4, −0.5)
G_3	(0.7, 0.4, 0.5, −0.4, −0.5, −0.6)	(0.7, 0.2, 0.4, −0.3, −0.5, −0.1)	(0.1, 0.7, 0.5, −0.4, −0.3, −0.8)	(0.4, 0.3, 0.5, −0.7, −0.4, −0.3)
G_4	(0.3, 0.4, 0.5, −0.7, −0.1, −0.3)	(0.8, 0.2, 0.1, −0.5, −0.3, −0.4)	(0.5, 0.2, 0.4, −0.1, −0.4, −0.7)	(0.4, 0.3, 0.7, −0.5, −0.2, −0.6)

Table 3. Bipolar Neutrosophic Decision Matrix, D_3.

	L_1	L_2	L_3	L_4
G_1	(0.4, 0.5, 0.1, −0.6, −0.4, −0.5)	(0.4, 0.3, 0.6, −0.1, −0.6, −0.3)	(0.2, 0.6, 0.3, −0.7, −0.2, −0.5)	(0.2, 0.1, 0.8, −0.9, −0.2, −0.3)
G_2	(0.5, 0.3, 0.2, −0.4, −0.1, −0.6)	(0.4, 0.5, 0.3, −0.7, −0.2, −0.3)	(0.5, 0.3, 0.4, −0.5, −0.4, −0.6)	(0.8, 0.6, 0.2, −0.1, −0.5, −0.4)
G_3	(0.3, 0.4, 0.6, −0.7, −0.2, −0.4)	(0.2, 0.9, 0.1, −0.4, −0.5, −0.6)	(0.3, 0.7, 0.2, −0.5, −0.3, −0.4)	(0.4, 0.1, 0.6, −0.3, −0.4, −0.1)
G_4	(0.5, 0.3, 0.6, −0.6, −0.3, −0.5)	(0.5, 0.6, 0.2, −0.5, −0.3, −0.6)	(0.4, 0.3, 0.7, −0.8, −0.4, −0.7)	(0.7, 0.4, 0.5, −0.5, −0.3, −0.4)

Step 2: We computed $BNHWA_v\left(r_{\widehat{i}1}, r_{\widehat{i}2}, \ldots.., r_{\widehat{i}n}\right)$ for $\ddot{\gamma} = 2$: The collective bipolar neutrosophic decision matrix is given in Table 4.

$$r_{\widehat{i}} = \left(\zeta^+_{\widehat{i}}, \Gamma^+_{\widehat{i}}, \text{Ю}^+_{\widehat{i}}, \zeta^-_{\widehat{i}}, \Gamma^-_{\widehat{i}}, \text{Ю}^-_{\widehat{i}}\right) =$$

$$BNHWA_v\left(r_{\widehat{i}1}, r_{\widehat{i}2}, \ldots.., r_{\widehat{i}n}\right) = \bigoplus_{\ell=1}^{n} (v_\ell r_{i\ell})$$

$$= \left(\begin{array}{l} \frac{\prod_{\ell=1}^{n}(1+(\ddot{\gamma}-1)\zeta^+_\ell)^{v_\ell} - \prod_{\ell=1}^{n}(1-\zeta^+_\ell)^{v_\ell}}{\prod_{\ell=1}^{n}(1+(\ddot{\gamma}-1)\zeta^+_\ell)^{v_\ell} + (\ddot{\gamma}-1)\prod_{\ell=1}^{n}(1-\zeta^+_\ell)^{v_\ell}}, \frac{\ddot{\gamma}\prod_{\ell=1}^{n}(\Gamma^+_\ell)^{v_\ell}}{\prod_{\ell=1}^{n}\left(1+(\ddot{\gamma}-1)(1-\Gamma^+_\ell)\right)^{v_\ell} + (\ddot{\gamma}-1)\prod_{\ell=1}^{n}(\Gamma^+_\ell)^{v_\ell}}, \frac{\ddot{\gamma}\prod_{\ell=1}^{n}(\text{Ю}^+_\ell)^{v_\ell}}{\prod_{\ell=1}^{n}(1+(\ddot{\gamma}-1)(1-\text{Ю}^+_\ell))^{v_\ell} + (\ddot{\gamma}-1)\prod_{\ell=1}^{n}(\text{Ю}^+_\ell)^{v_\ell}}, \\ \frac{-\ddot{\gamma}\prod_{\ell=1}^{n}|\zeta^-_\ell|^{v_\ell}}{\prod_{\ell=1}^{n}(1+(\ddot{\gamma}-1)(1+\zeta^-_\ell))^{v_\ell} + (\ddot{\gamma}-1)\prod_{\ell=1}^{n}|\zeta^-_\ell|^{v_\ell}}, -\frac{\prod_{\ell=1}^{n}\left(1+(\ddot{\gamma}-1)\left|\Gamma^-_\ell\right|\right)^{v_\ell} - \prod_{\ell=1}^{n}(1+\Gamma^-_\ell)^{v_\ell}}{\prod_{\ell=1}^{n}\left(1+(\ddot{\gamma}-1)\left|\Gamma^-_\ell\right|\right)^{v_\ell} + (\ddot{\gamma}-1)\prod_{\ell=1}^{n}(1+\Gamma^-_\ell)^{v_\ell}}, -\frac{\prod_{\ell=1}^{n}(1+(\ddot{\gamma}-1)|\text{Ю}^-_\ell|)^{v_\ell} - \prod_{\ell=1}^{n}(1+\text{Ю}^-_\ell)^{v_\ell}}{\prod_{\ell=1}^{n}(1+(\ddot{\gamma}-1)|\text{Ю}^-_\ell|)^{v_\ell} + (\ddot{\gamma}-1)\prod_{\ell=1}^{n}(1+\text{Ю}^-_\ell)^{v_\ell}} \end{array}\right)$$

Step 3: We calculated the score function, for $\ddot{\gamma} = 2$:

$$r_1 = (0.2993, 0.3560, 0.4013, -0.4544, -0.4369, -0.4585),$$
$$r_2 = (0.5060, 0.4265, 0.3217, -0.3607, -0.3854, -0.5668),$$
$$r_3 = (0.4150, 0.4765, 0.3629, -0.4780, -0.4010, -0.4920),$$
$$r_4 = (0.4989, 0.3360, 0.5028, -0.5272, -0.3517, -0.5683),$$

$$S\left(r_{\widetilde{i}}\right) = \frac{1}{6}\left(\zeta^{+} + 1 - \Gamma^{+} + 1 - \text{Ю}^{+} + 1 + \zeta^{-} - \Gamma^{-} - \text{Ю}^{-}\right)\text{and}$$

$$S(r_1) = 0.4972, S(r_2) = 0.5582, S(r_3) = 0.4984, S(r_4) = 0.5088.$$

Step 4: We calculated the scores for $\ddot{\gamma} = 2$, which gave:

$$G_2 \succ G_4 \succ G_3 \succ G_1$$

Step 5: Thus, the best option was G_2

Table 4. Collective Bipolar Neutrosophic Decision Matrix R.

	L_1	L_2
G_1	(0.3757,0.4683,0.1962, −0.5615, −0.4946, −0.5317)	(0.3155,0.3520,0.5616, −0.2056, −0.4865, −0.5313)
G_2	(0.4181,0.4622,0.3062, −0.3001, −0.3585, −0.6479)	(0.4614,0.3623,0.4622, −0.5292, −0.3233, −0.3359)
G_3	(0.4708,0.4283,0.4669, −0.5406, −0.4250, −0.5632)	(0.5855,0.6202,0.1902, −0.4417, −0.4712, −0.4741)
G_4	(0.5174,0.4072,0.5386, −0.6291, −0.3083, −0.4134)	(0.6132,0.4687,0.2350, −0.5288, −0.3000, −0.5147)
	L_3	L_4
G_1	(0.3264,0.3931,0.3276, −0.5709, −0.4174, −0.5255)	(0.1703,0.2364,0.7077, −0.4916, −0.3566, −0.2115)
G_2	(0.4441,0.4108,0.2709, −0.3623, −0.4000, −0.6722)	(0.6708,0.4669,0.3061, −0.3107, −0.4413, −0.4946)
G_3	(0.2115,0.6692,0.3536, −0.4683, −0.3308, −0.6406)	(0.4312,0.2639,0.4669, −0.4597, −0.4000, −0.1914)
G_4	(0.4029,0.1931,0.6264, −0.4973, −0.4312, −0.6724)	(0.4891,0.3942,0.6154, −0.4683, −0.3359, −0.6088)

5. Comparison with the Different Methods

There are various tools utilized by researchers so far in decision making. Chen et al. [50] utilized FSs, and, later on, Atanassov [2] utilized intuitionistic FSs, Dubois et al. [51] utilized BFSs, Zavadskas et al. [52] utilized NSs, Ali et al. [53] utilized bipolar neutrosophic soft sets, and Irfan et al. [34] utilized BNSs and so many others have studied decision making. In this paper, we applied the bipolarity to the neutrosophic sets via Hamacher operators. If $\ddot{\gamma} = 1$, then our proposed model corresponded to the same BNSs of the decision making as Irfan et al. [34].

The advantage of our proposed methods was that the decision maker could choose different values of $\ddot{\gamma}$ in accordance with their preferences (Table 5). Generally, when the values of $\ddot{\gamma} = 1, 2,$ are used, they form algebraic aggregation operators and Einstein aggregation operators. The aggregation operators suggested in this paper were more general and flexible in accordance with the different values of $\ddot{\gamma}$, keeping in view the above computation and analysis, it is derived that although the overall rating values of the alternatives are varying by using different values of $\ddot{\gamma}$, the ranking orders of the alternatives are slightly contrastive (Table 5). However, the most desirable investment company is G_2.

Table 5. Ranking of the four alternatives for the different values of $\ddot{\gamma}$.

$\ddot{\gamma}$	Aggregation Operators	Ranking
1	*BNHWA*	$G_2 \succ G_4 \succ G_1 \succ G_3$
1.5	*BNHWA*	$G_2 \succ G_4 \succ G_1 \succ G_3$
2	*BNHWA*	$G_2 \succ G_4 \succ G_3 \succ G_1$
2.5	*BNHWA*	$G_2 \succ G_4 \succ G_3 \succ G_1$
3	*BNHWA*	$G_2 \succ G_4 \succ G_1 \succ G_3$

6. Conclusions

The purpose of this paper was to study the different bipolar neutrosophic aggregation operators, based on Hamacher t-norms and *t*-conorms, and their application to multiple criteria group decision making where the criteria are bipolar neutrosophic values. Motivated by the Hamacher operations, we have proposed bipolar neutrosophic Hamacher aggregation operators. Firstly, we have introduced bipolar neutrosophic Hamacher aggregation operators, as well as their desirable properties. These aggregation operators were bipolar neutrosophic Hamacher weighted averaging (BNHWA), bipolar neutrosophic Hamacher ordered weighted averaging (BNHOWA), and bipolar neutrosophic Hamacher hybrid averaging (BNHHA). When $\ddot{y} = 1$, the bipolar neutrosophic Hamacher averaging operators reduced to the bipolar neutrosophic averaging aggregation operator and for $\ddot{y} = 2$, the bipolar neutrosophic Hamacher averaging operators transformed to the bipolar neutrosophic Einstein averaging aggregation operators. Finally, we have introduced a method for multi-attribute group decision making. A descriptive example of opting for the best company or alternative to investing money was provided. The results in this paper showed that our proposed methods were more effective and practical in real life. In our future study, we are determined to extend the proposed models to other domains and applications, such as risk analysis, pattern recognition, and so on.

Author Contributions: There was equal contribution by all authors.

Funding: The present study was been carried out without any monetary support.

Conflicts of Interest: All authors have given their full consent unanimously.

References

1. Zadeh, L.A. Fuzzy sets. *Inf. Control* **1965**, *8*, 338–353. [CrossRef]
2. Atanassov, K.T. Intuitionistic fuzzy sets. *Fuzzy Sets Syst.* **1986**, *20*, 87–96. [CrossRef]
3. Smarandache, F. *A Unifying Field in Logics Neutrosophy and Neutrosophic Probability, Set and Logic*; American Research Press: Rehoboth, Delaware, 1999.
4. Wang, H.; Smarandache, F.; Zhan, Y.; Sunderraman, R. Single valued neutrosophic sets. In Proceedings of the 10th 476 International Conference on Fuzzy theory and Technology, Salt Lake City, UT, USA, 21–26 July 2005.
5. Ye, J. Multicriteria decision-making method using the correlation coefficient under single valued neutrosophic environment. *Int. J. Gen. Syst.* **2013**, *42*, 386–394. [CrossRef]
6. Wang, H.; Smarandache, F.; Zhan, Y.Q. *Interval Neutrosophic Sets and Logic: Theory and Applications in Computing*; Hexis: Phoenix, AZ, USA, 2005.
7. Ye, J. Similarity measures between interval neutrosophic sets and their applications in multicriteria decision-making. *J. Intell. Fuzzy Syst.* **2014**, *26*, 165–172.
8. Yu, D.J. Group decision making based on generalized intuitionistic fuzzy prioritized geometric operator. *Int. J. Intell. Syst.* **2012**, *27*, 635–661. [CrossRef]
9. Li, D.F. The GOWA operator based approach to multiattribute decision making using intuitionistic fuzzy sets. *Math. Comput. Model.* **2011**, *53*, 1182–1196. [CrossRef]
10. Wei, G.W. Gray relational analysis method for intuitionistic fuzzy multiple attribute decision making. *Expert Syst. Appl.* **2011**, *38*, 11671–11677. [CrossRef]
11. Wei, G.W.; Wang, H.J.; Lin, R. Application of correlation coefficient to interval-valued intuitionistic fuzzy multiple attribute decision making with incomplete weight information. *Knowl. Inf. Syst.* **2011**, *26*, 337–349. [CrossRef]
12. Xu, Z.S.; Yager, R.R. Some geometric aggregation operators based on intuitionistic fuzzy sets. *Int. J. Gen. Syst.* **2006**, *35*, 417–433. [CrossRef]
13. Wang, W.; Liu, X. Intuitionistic Fuzzy Geometric Aggregation Operators Based on Einstein Operations. *Int. J. Intell. Syst.* **2011**, *26*, 1049–1075. [CrossRef]
14. Xu, Z.S. Intuitionistic fuzzy aggregation operators. *IEEE Trans. Fuzzy Syst.* **2007**, *15*, 1179–1187.
15. Xu, Z.S. Multi-person multi-attribute decision making models under intuitionistic fuzzy environment. *Fuzzy Optim. Decis. Mak.* **2007**, *6*, 221–236. [CrossRef]
16. Wang, W.; Liu, X. Intuitionistic Fuzzy Information Aggregation Using Einstein Operations. *IEEE Trans. Fuzzy Syst.* **2012**, *20*, 923–938. [CrossRef]
17. Zhao, X.; Wei, G. Some intuitionistic fuzzy Einstein hybrid aggregation operators and their application to multiple attribute decision making. *Knowl. Based Syst.* **2013**, *37*, 472–479. [CrossRef]

18. Fahmi, A.; Abdullah, S.; Amin, F.; Siddique, N.; Ali, A. Aggregation operators on triangular cubic fuzzy numbers and its application to multi-criteria decision making problems. *J. Intell. Fuzzy Syst.* **2017**, *33*, 3323–3337. [CrossRef]
19. Fahmi, A.; Abdullah, S.; Amin, F.; Khan, M.S.A. Trapezoidal cubic fuzzy number Einstein hybrid weighted averaging operators and its application to decision making problems. *Soft Comput.* **2018**. [CrossRef]
20. Fahmi, A.; Amin, F.; Abdullah, S.; Ali, A. Cubic fuzzy Einstein aggregation operators and its application to decision-making. *Int. J. Syst. Sci.* **2018**, *49*, 2385–2397. [CrossRef]
21. Fahmi, A.; Abdullah, S.; Amin, F.; Ali, A.; Ahmad, R.; Shakeel, M. Trapezoidal cubic hesitant fuzzy aggregation operators and their application in group decision-making. *Int. J. Fuzzy Syst.* **2019**, *36*, 3619–3635. [CrossRef]
22. Rahman, K.; Abdullah, S.; Jamil, M.; Khan, M.Y. Some Generalized intuitionistic fuzzy Einstein hybrid aggregation operators and their application to multiple attribute group decision making. *Int. J. Fuzzy Syst.* **2018**, *20*, 1567–1575. [CrossRef]
23. Ali Khan, M.S.; Abdullah, S.; Ali, A. Multiattribute group decision-making based on Pythagorean fuzzy Einstein prioritized aggregation operators. *In. J. Intel. Syst.* **2019**, *34*, 1001–1033. [CrossRef]
24. Ashraf, S.; Abdullah, S. Spherical aggregation operators and their application in multiattribute group decision-making. *Int. J. Intell. Syst.* **2019**, *34*, 493–523. [CrossRef]
25. Khan, M.S.A.; Abdullah, S.; Ali, A.; Amin, F.; Hussain, F. Pythagorean hesitant fuzzy Choquet integral aggregation operators and their application to multi-attribute decision-making. *Soft Comput.* **2019**, *23*, 251–267. [CrossRef]
26. Zhang, W.R. Bipolar fuzzy sets and relations: A computational frame work for cognitive modeling and multiagent decision analysis. In Proceedings of the First International Joint Conference of The North American Fuzzy Information Processing Society Biannual Conference, San Antonio, TX, USA, 18–21 December 1994; pp. 305–309.
27. Zhang, W.R. Bipolar fuzzy sets. In Proceedings of the IEEE International Conference on Fuzzy Systems, IEEE World Congress on Computational Intelligence, Anchorage, AK, USA, 4–9 May 1998; pp. 835–840.
28. Zhang, W.R.; Zhang, L. Bipolar logic and Bipolar fuzzy logic. *Inf. Sci.* **2004**, *165*, 265–287. [CrossRef]
29. Li, P.P. The global implications of the indigenous epistemological system from the east: How to apply Yin-Yang balancing to paradox management. *Cross Cult. Strateg. Manag.* **2016**, *23*, 42–47. [CrossRef]
30. Zhang, W.R. Bipolar quantum logic gates and quantum cellular combinatorics-a logical extension to quantum entanglement. *J. Quant. Inf. Sci.* **2013**, *3*, 93–105. [CrossRef]
31. Zhang, W.R. G-CPT symmetry of quantum emergence and submergence-an information conservational multiagent cellular automata unification of CPT symmetry and CP violation for equilibrium-based many world causal analysis of quantum coherence and decoherence. *J. Quant. Inf. Sci.* **2016**, *6*, 62–97. [CrossRef]
32. Akram, M. Bipolar fuzzy graphs. *Inf. Sci.* **2011**, *181*, 5548–5564. [CrossRef]
33. Gul, Z. Some Bipolar Fuzzy Aggregations Operators and Their Applications in Multicriteria Group Decision Making. M. Phil. Thesis, Hazara University, Mansehra, Pakistan, 2015.
34. Deli, I.; Ali, M.; Smarandache, F. Bipolar neutrosophic sets and their application based on multi-criteria decision making problems. In Proceedings of the 2015 International Conference on Advanced Mechatronic System, Beijing, China, 22–24 August 2015.
35. Fan, C.; Ye, J.; Fen, S.; Fan, E.; Hu, K. Multi-criteria decision-making method using heronian mean operators under a bipolar neutrosophic environment. *Mathematics* **2019**, *7*, 97. [CrossRef]
36. Deli, I.; Subas, Y.; Smarandache, F.; Ali, M. Interval valued bipolar neutrosophic sets and their application in pattern recognition. Conference Paper. *arXiv* **2016**, arXiv:289587637.
37. Deli, I. Interval-valued neutrosophic soft sets and its decision making. *Int. J. Mach. Learn. Cybern.* **2017**. [CrossRef]
38. Zhang, H.; Wang, F.; Geng, Y. Multi-Criteria decision-making method based on single-valued neutrosophic Schweizer-Sklar Muirhead mean aggregation operators. *Symmetry* **2019**, *11*, 152. [CrossRef]
39. Ashraf, S.; Abdullah, S.; Smarandache, F. Logarithmic Hybrid Aggregation Operators Based on Single Valued Neutrosophic Sets and Their Applications in Decision Support Systems. *Symmetry* **2019**, *11*, 364. [CrossRef]
40. Hamacher, H. Uber logische verknunpfungenn unssharfer Aussagen und deren Zugenhorige Bewetungsfunktione. *Prog. Cybern. Syst. Res.* **1978**, *3*, 276–288.
41. Huang, J.Y. Intuitionistic fuzzy Hamacher aggregation operators and their application to multiple attribute decision making. *J. Intell. Fuzzy Syst.* **2014**, *27*, 505–513.

42. Liu, P. Some Hamacher aggregation operators based on the interval-valued intuitionistic fuzzy numbers and their application to group decision making. *IEEE Trans. Fuzzy Syst.* **2014**, *22*, 83–97. [CrossRef]
43. Zhou, L.; Zhao, X.; Wei, G. Hesitant fuzzy Hamacher aggregation operators and their application to multiple attribute decision making. *J. Intell. Fuzzy Syst.* **2014**, *26*, 2689–2699.
44. Lu, M.; Wei, G.; Alsaadi, F.E.; Hayat, T.; Alsaadi, A. Hesitant Pythagorean fuzzy Hamacher aggregation operators and their application to multiple attribute decision making. *J. Intell. Fuzzy Syst.* **2017**, *33*, 1105–1117. [CrossRef]
45. Wei, G.; Alsaadi, F.E.; Hayat, T.; Alsaedi, A. Bipolar fuzzy Hamacher aggregation operators in multiple attribute decision making. *Int. J. Fuzzy Syst.* **2017**, *20*, 1–12. [CrossRef]
46. Liu, P.; Chu, Y.; Li, Y.; Chen, Y. Some generalized neutrosophic number Hamacher aggregation operators and their application to group decision making. *Int. J. Fuzzy Syst.* **2014**, *16*, 242–255.
47. Qun, W.; Peng, W.; Ligang, Z.; Huayou, C.; Xianjun, G. Some new Hamacher aggregation operators under single-valued neutrosophic 2-tuple linguistic environment and their applications to multi-attribute group decision making. *Comput. Ind. Eng.* **2017**, *116*, 144–162.
48. Robinson, A. *Non-Standard Analysis*; North-Holland Pub. Co.: Amsterdam, The Netherlands, 1966.
49. Peng, J.J.; Wang, J.Q.; Wang, J.; Zhang, H.Y.; Chen, X.H. Simplified neutrosophic sets and their applicationsin multi-criteria group decision-making problems. *Int. J. Syst. Sci.* **2015**. [CrossRef]
50. Chen, S.M. A new approach to handling fuzzy decision-making problems. *IEEE Trans. Syst. Man Cybern.* **1998**, *18*, 1012–1016. [CrossRef]
51. Dubois, D.; Kaci, S.; Prade, H. Bipolarity in reasoning and decision, an introduction. *Inf. Process. Manag. Uncertain. IPMU* **2004**, *4*, 959–966.
52. Zavadskas, E.K.; Bausys, R.; Kaklauskas, A.; Ubarte, I.; Kuzminske, A.; Gudience, N. Sustainable market valuation of buildings by the single-valued neutrosophic MAMVA method. *Appl. Soft Comput.* **2017**, *57*, 74–87. [CrossRef]
53. Ali, M.; Son, L.H.; Deli, I.; Tian, N.D. Bipolar neutrosophic soft sets and applications in decision making. *J. Intell. Fuzzy Syst.* **2017**, *33*, 4077–4087. [CrossRef]

symmetry

MDPI

Article

Neutrosophic Extended Triplet Group Based on Neutrosophic Quadruple Numbers

Qiaoyan Li [1], Yingcang Ma [1,*], Xiaohong Zhang [2] and Juanjuan Zhang [1]

[1] School of Science, Xi'an Polytechnic University, Xi'an 710048, China; qiaoyan_li@126.con (Q.L.); zhang79jj@126.com (J.Z.)

[2] School of Arts and Sciences, Shaanxi University of Science & Technology, Xi'an 710021, China; zhangxiaohong@sust.edu.cn

* Correspondence: mayingcang@126.com

Received: 4 May 2019; Accepted: 17 May 2019; Published: 21 May 2019

Abstract: In this paper, we explore the algebra structure based on neutrosophic quadruple numbers. Moreover, two kinds of degradation algebra systems of neutrosophic quadruple numbers are introduced. In particular, the following results are strictly proved: (1) the set of neutrosophic quadruple numbers with a multiplication operation is a neutrosophic extended triplet group; (2) the neutral element of each neutrosophic quadruple number is unique and there are only sixteen different neutral elements in all of neutrosophic quadruple numbers; (3) the set which has same neutral element is closed with respect to the multiplication operator; (4) the union of the set which has same neutral element is a partition of four-dimensional space.

Keywords: neutrosophic extended triplet group; neutrosophic quadruple numbers; neutrosophic set

1. Introduction

The notion of a neutrosophic set is proposed by F. Smarandache [1] in order to solve real-world problems and some in-depth analysis and research have been carried out [2–5]. Recently, Smarandache and Ali in [6] proposed a new algebraic system, neutrosophic triplet group (NTG), which different from classical groups. From the original definition of NTG, the neutral element is different from the classical algebraic unit element. By removing this restriction, the neutrosophic extended triplet group (NETG) is proposed in [7,8] and the classical group is regarded as a special case of NETG.

As a new algebraic structure, NTG (NETG) immediately attracted the attention of scholars and conducted in-depth research. These studies are mainly carried out by the following three aspects. Firstly, the structure properties of NTG (NETG) have been studied deeply. For examples, paper [8] has conducted an in-depth analysis of the nature of NTG, and the properties and structural features of NTG are studied by using theoretical analysis and software calculations. In paper [9], the notion of the neutrosophic triplet coset and its relation with the classical coset are proposed and the properties of the neutrosophic triplet cosets are given. The neutrosophic duplet sets, neutrosophic duplet semi-groups, and cancellable neutrosophic triplet groups are proposed and the characterizations of cancellable weak neutrosophic duplet semi-groups are established in paper [10]. In order to explore the structure of the algebraic system $(Z_n, \otimes)$, where $\otimes$ is the classical mod multiplication, paper [11] reveals that for each $n \in Z^+, n \geq 2, (Z_n, \otimes)$ is a commutative NETG if and only if the factorization of n is a product of single factors. Moreover, the generalized neutrosophic extended triplet group (GNETG) is proposed in [11] and verify that for each $n \in Z^+, n \geq 2, (Z_n, \otimes)$ is a commutative GNETG. Secondly, it is the application research on the algebraic system NET. For example, In paper [12], the distinguishing features between an NTG and other algebraic structures are investigated and the first isomorphism theorem was established for NTGs, furthermore, applications of the results on NTG to management and sports are discussed. In paper [13], NTGs and their applications to mathematical models, such as fuzzy cognitive maps model, neutrosophic cognitive maps model and fuzzy relational maps model, are discussed. Thirdly, extend the idea of NTG(NETG) to another algebraic system. For example, in paper [14,15], the extend to Abel–Grassmann groupoid (AG-groupoid) is studied. The neutrosophic triplet ring and a neutrosophic triplet field are discussed in paper [16,17]. A notion of neutrosophic triplet metric space is given and properties of neutrosophic triplet metric spaces are studied in [18]. The notion of neutrosophic triplet v-generalized metric space are introduced in [19]. Paper [20] applies the neutrosophic set theory to pseudo-BCI algebras. The idea of a neutrosophic triplet set to non-associative semihypergroups is given in paper [21]. The above results enrich the research content of the algebraic system NTG (NETG).

In neutrosophic logic, each proposition is approximated to represent respectively the truth (T), the falsehood (F), and the indeterminacy (I), where T, I, F are standard or non-standard subsets of the non-standard unit interval $]0^-, 1^+[= 0^- \cup [0,1] \cup 1^+$. The notion of neutrosophic quadruple number, which is represented by a known part and an unknown part to describe a neutrosophic logic proposition, was introduced by Florentin Smarandache in [22]. The algebra system $(NQ, *)$ based on neutrosophic quadruple numbers are introduced and the properties have discussed [22,23]. In this paper, we will reveal that $(NQ, *)$ is a NETG and some properties are discussed.

The paper is organized as follows. Section 2 gives the basic concepts. In Section 3, $(NQ, *)$ be a NETG is proved and some properties are discussed. In Section 4, two kinds of degradation algebra systems of $(NQ, *)$ are introduced and studied. Finally, the summary and future work are presented in Section 5.

2. Basic Concepts

In this section, we will provide the related basic definitions and properties of NETG and neutrosophic quadruple numbers, the details can be seen in [7,8,22,23].

Definition 1 ([7,8]). *Let N be a non-empty set together with a binary operation $*$. Then, N is called a neutrosophic extended triplet set if for any $a \in N$, there exists a neutral of "a" (denote by $neut(a)$), and an opposite of "a"(denote by $anti(a)$), such that $neut(a) \in N$, $anti(a) \in N$ and:*

$$a * neut(a) = neut(a) * a = a, \quad a * anti(a) = anti(a) * a = neut(a).$$

The triplet $(a, neut(a), anti(a))$ is called a neutrosophic extended triplet.

Definition 2 ([7,8]). *Let $(N, *)$ be a neutrosophic extended triplet set. Then, N is called a neutrosophic extended triplet group (NETG), if the following conditions are satisfied:*

(1) $(N, *)$ is well-defined, i.e., for any $a, b \in N$, one has $a * b \in N$.

(2) $(N, *)$ is associative, i.e., $(a * b) * c = a * (b * c)$ for all $a, b, c \in N$.

A NETG N is called a commutative NETG if for all $a, b \in N, a * b = b * a$.

Proposition 1 ([8]). *Let $(N, *)$ be a NETG. We have:*

(1) $neut(a)$ is unique for any $a \in N$.

(2) $neut(a) * neut(a) = neut(a)$ for any $a \in N$.

(3) $neut(neut(a)) = neut(a)$ for any $a \in N$.

Definition 3 ([22,23]). *A neutrosophic quadruple number is a number of the form (a, bT, cI, dF), where T, I, F have their usual neutrosophic logic meanings and $a, b, c, d \in \mathbb{R}$ or $\mathbb{C}$. The set NQ, defined by*

$$NQ = \{(a, bT, cI, dF) : a, b, c, d \in \mathbb{R} \text{ or } \mathbb{C}\}. \tag{1}$$

is called a neutrosophic set of quadruple numbers. For a neutrosophic quadruple number (a, bT, cI, dF), a is called the known part and (bT, cI, dF) is called the unknown part.

Definition 4 ([22,23]). *Let N be a set, endowed with a total order $a \prec b$, named "a prevailed by b" or "a less stronger than b" or "a less preferred than b". We consider $a \preceq b$ as "a prevailed by or equal to b" "a less stronger than or equal to b", or "a less preferred than or equal to b".*

For any elements $a, b \in N$, with $a \preceq b$, one has the absorbance law:

$$a \cdot b = b \cdot a = \text{absorb}(a, b) = \max(a, b) = b, \tag{2}$$

which means that the bigger element absorbs the smaller element. Clearly,

$$a \cdot a = a^2 = \text{absorb}(a, a) = \max(a, a) = a. \tag{3}$$

and

$$a_1 \cdot a_2 \cdots a_n = \max(a_1, a_2, \cdots, a_n). \tag{4}$$

Analogously, we say that "$a \succ b$" and we read: "a prevails to b" or "a is stronger than b" or "a is preferred to b". Also, $a \succeq b$, and we read: "a prevails or is equal to b" "a is stronger than or equal to b", or "a is preferred or equal to b".

Definition 5 ([22,23]). *Consider the set $\{T, I, F\}$. Suppose in an optimistic way we consider the prevalence order $T \succ I \succ F$. Then we have: $TI = IT = \max(T, I) = T$, $TF = FT = \max(T, F) = T$, $IF = FI = \max(I, F) = I$, $TT = T^2 = T$, $II = I^2 = I$, $FF = F^2 = F$.*

Analogously, suppose in a pessimistic way we consider the prevalence order $T \prec I \prec F$. Then we have: $TI = IT = \max(T, I) = I$, $TF = FT = \max(T, F) = F$, $IF = FI = \max(I, F) = F$, $TT = T^2 = T$, $II = I^2 = I$, $FF = F^2 = F$.

Definition 6 ([22,23]). *Let $a = (a_1, a_2T, a_3I, a_4F), b = (b_1, b_2T, b_3I, b_4F) \in NQ$, Suppose in an pessimistic way, the neutrosophic expert considers the prevalence order $T \prec I \prec F$. Then the multiplication operation is defined as following:*

$$\begin{aligned} a * b = \; & (a_1, a_2T, a_3I, a_4F) * (b_1, b_2T, b_3I, b_4F) \\ & = (a_1b_1, (a_1b_2 + a_2b_1 + a_2b_2)T, (a_1b_3 + a_2b_3 + a_3b_1 + a_3b_2 + a_3b_3)I, \\ & \quad (a_1b_4 + a_2b_4 + a_3b_4 + a_4b_1 + a_4b_2 + a_4b_3 + a_4b_4)F). \end{aligned} \tag{5}$$

Suppose in an optimistic way the neutrosophic expert considers the prevalence order $T \succ I \succ F$. Then:

$$\begin{aligned} a \star b = \; & (a_1, a_2T, a_3I, a_4F) * (b_1, b_2T, b_3I, b_4F) \\ & = (a_1b_1, (a_1b_2 + a_2b_1 + a_2b_2 + a_3b_2 + a_4b_2 + a_2b_3 + a_2b_4)T, \\ & \quad (a_1b_3 + a_3b_1 + a_3b_3 + a_3b_4 + a_4b_3)I, (a_1b_4 + a_4b_1 + a_4b_4)F). \end{aligned} \tag{6}$$

Proposition 2 ([22,23]). *Let $NQ = \{(a, bT, cI, dF) : a, b, c, d \in \mathbb{R} \text{ or } \mathbb{C}\}$. We have:*
(1) $(NQ, *)$ is a commutative monoid.
(2) $(NQ, \star)$ is a commutative monoid.

3. Main Results

From Proposition 2, we can see that $(NQ, *)$ (or $(NQ, \star)$) be a commutative monoid. In these section, we will show that the algebra system $(NQ, *)$(or $(NQ, \star)$) is a NETG.

Theorem 1. *For the algebra system $(NQ, *)$, for every element $a \in NQ$, there exists the neutral element $neut(a)$ and opposite element $anti(a)$.*

Proof analysis: the proof of this theorem contains two aspects. Firstly, given an element $a \in NQ, a = (a_1, a_2T, a_3I, a_4F)$, $a_i \in \mathbb{R}, i \in \{1, 2, 3, 4\}$. Being a_i can select every element in $\mathbb{R}$, we should discuss from different cases and in each case $netu(a)$ and $anti(a)$ should given. Secondly, we should prove that all the cases discussed above include all the elements in NQ.

Proof. Let $a = (a_1, a_2T, a_3I, a_4F)$, we consider $a_i \in \mathbb{R}, i \in \{1,2,3,4\}$ and the same results can be gotten when $a_i \in \mathbb{C}$.

Set $neut(a) = (b_1, b_2T, b_3I, b_4F), b_i \in \mathbb{R}, i \in \{1,2,3,4\}$ and $anti(a) = (c_1, c_2T, c_3I, c_4F), c_i \in \mathbb{R}, i \in \{1,2,3,4\}$. From Definition 1 we can get $a * neut(a) = a$, that is $a_1b_1 = a_1$ should hold. So we discuss from two cases, $a_1 = 0$ or $a_1 \neq 0$.

Case A: when $a_1 = 0$.

In this case, we have $a = (0, a_2T, a_3I, a_4F)$. From Definition 1, $a * anti(a) = neut(a)$, that is $0 \cdot c_1 = b_1$, so we have $b_1 = 0$, i.e., $neut(a) = (0, b_2T, b_3I, b_4F)$. Moreover, from $a * neut(a) = a$, we have $(0, a_2T, a_3I, a_4F) * (0, b_2T, b_3I, b_4F) = (0, a_2T, a_3I, a_4F)$, so we have $a_2b_2 = a_2$. So we discuss from $a_2 = 0$ or $a_2 \neq 0$.

Case A1: $a_1 = 0, a_2 = 0$. That is, $a = (0, 0, a_3I, a_4F), netu(a) = (0, b_2T, b_3I, b_4F), anti(a) = (c_1, c_2T, c_3I, c_4F)$. From $a * anti(a) = neut(a)$, we have $0c_1 + 0(c_1 + c_2) = b_2$, so $b_2 = 0$, i.e., $netu(a) = (0, 0, b_3I, b_4F)$. From $(0, 0, a_3I, a_4F) * (0, 0, b_3I, b_4F) = (0, 0, a_3I, a_4F)$, we have $a_3b_3 = a_3$. So we discuss from $a_3 = 0$ or $a_3 \neq 0$.

Case A11: $a_1 = a_2 = a_3 = 0$, that is, $a = (0, 0, 0, a_4F), netu(a) = (0, 0, b_3I, b_4F), anti(a) = (c_1, c_2T, c_3I, c_4F)$. In the same way, from $a * anti(a) = neut(a)$, we have $b_3 = 0$, i.e., $netu(a) = (0, 0, 0, b_4F)$. From $(0, 0, 0, a_4F) * (0, 0, 0, b_4F) = (0, 0, 0, a_4F)$, we have $a_4b_4 = a_4$. So we discuss from $a_4 = 0$ or $a_4 \neq 0$.

Case A111: $a_1 = a_2 = a_3 = a_4 = 0$, that is, $a = (0, 0, 0, 0)$, in this case, we can easily get $neut(a) = (0, 0, 0, 0)$ and $anti(a) = (c_1, c_2T, c_3I, c_4F)$, c_i can be chosen arbitrarily in $\mathbb{R}$.

Case A112: $a_1 = a_2 = a_3 = 0, a_4 \neq 0$, being that $a_4b_4 = a_4$ and $a_4 \neq 0$, we have $b_4 = 1$, that is, $a = (0, 0, 0, a_4F), netu(a) = (0, 0, 0, F), anti(a) = (c_1, c_2T, c_3I, c_4F)$. From $(0, 0, 0, a_4F) * (c_1, c_2T, c_3I, c_4F) = (0, 0, 0, F)$, we have $a_4(c_1 + c_2 + c_3 + c_4) = 1$, so the opposite element of a should satisfy $c_1 + c_2 + c_3 + c_4 = \frac{1}{a_4}, c_i \in \mathbb{R}$.

Case A12: $a_1 = a_2 = 0, a_3 \neq 0$. From $a_3b_3 = a_3$ and $a_3 \neq 0$, we have $b_3 = 1$. That is $a = (0, 0, a_3I, a_4F), netu(a) = (0, 0, I, b_4F), anti(a) = (c_1, c_2T, c_3I, c_4F)$. From $(0, 0, a_3I, a_4F) * (0, 0, I, b_4F) = (0, 0, a_3I, a_4F)$, we have $0b_4 + 0b_4 + a_3b_4 + a_4(0 + 0 + 1 + b_4) = a_4$, so $(a_3 + a_4)b_4 = 0$. We discuss from $a_3 + a_4 = 0$ or $a_3 + a_4 \neq 0$.

Case A121: $a_1 = a_2 = 0$, $a_3 \neq 0, a_3 + a_4 = 0$, that is $a = (0, 0, a_3I, -a_3F), neut(a) = (0, 0, I, b_4F), anti(a) = (c_1, c_2T, c_3I, c_4F)$. From $a * anti(a) = neut(a)$, that is $(0, 0, a_3I, -a_3F) * (c_1, c_2T, c_3I, c_4F) = (0, 0, I, b_4F)$. So we have $a_3(c_1 + c_2 + c_3) = 1$ and $a_3c_4 - a_3(c_1 + c_2 + c_3 + c_4) = b_4$ i.e., $c_1 + c_2 + c_3 = \frac{1}{a_3}$ and $b_4 = 1$. Thus $neut(a) = (0, 0, I, -F)$, $anti(a) = (c_1, c_2T, c_3I, c_4F)$, where $c_1 + c_2 + c_3 = \frac{1}{a_3}$, c_4 can be chosen arbitrarily in $\mathbb{R}$.

Case A122: $a_1 = a_2 = 0, a_3 \neq 0, a_3 + a_4 \neq 0$. From $(a_3 + a_4)b_4 = 0$, we have $b_4 = 0$. that is $a = (0, 0, a_3I, a_4F), neut(a) = (0, 0, I, 0), anti(a) = (c_1, c_2T, c_3I, c_4F)$. From $a * anti(a) = neut(a)$, that is $(0, 0, a_3I, a_4F) * (c_1, c_2T, c_3I, c_4F) = (0, 0, I, 0)$. So we have $a_3(c_1 + c_2 + c_3) = 1$ and $a_3c_4 - a_3(c_1 + c_2 + c_3 + c_4) = 0$ i.e., $c_1 + c_2 + c_3 = \frac{1}{a_3}$ and $c_4 = -\frac{a_4}{a_3(a_3 + a - 4)}$. Thus $neut(a) = (0, 0, I, 0)$, $anti(a) = (c_1, c_2T, c_3I, c_4F)$, where $c_1 + c_2 + c_3 = \frac{1}{a_3}, c_4 = -\frac{a_4}{a_3(a_3 + a_4)}$.

Case A2: when $a_1 = 0, a_2 \neq 0$. From $a_2b_2 = a_2$, we have $b_2 = 1$, that is, $a = (0, 0, a_3I, a_4F), netu(a) = (0, T, b_3I, b_4F), anti(a) = (c_1, c_2T, c_3I, c_4F)$. In the same way, from $a * neut(a) = a$, we have $(a_2 + a_3)b_3 = 0$, so we discuss from $a_2 + a_3 = 0$ or $a_2 + a_3 \neq 0$.

Case A21: when $a_1 = 0, a_2 \neq 0, a_2 + a_3 = 0$. that is, $a = (0, a_2T, -a_2I, a_4F), netu(a) = (0, T, b_3I, b_4F), anti(a) = (c_1, c_2T, c_3I, c_4F)$. In the same way, from $a * neut(a) = a$, we have $a_4 + a_4(b_3 + b_4) = a_4$, that is $a_4(b_3 + b_4) = 0$, so we discuss from $a_4 = 0$ or $a_4 \neq 0$.

Case A211: when $a_1 = 0, a_2 \neq 0, a_2 + a_3 = 0, a_4 = 0$. that is, $a = (0, a_2T, -a_2I, 0), netu(a) = (0, T, b_3I, b_4F), anti(a) = (c_1, c_2T, c_3I, c_4F)$. From $(0, a_2T, -a_2I, 0) * (c_1, c_2T, c_3I, c_4F) = (0, T, b_3I, b_4F)$, so we have $a_2(c_1 + c_2) = 1$ and $-a_2(c_1 + c_2) = b_3$, that is $b_3 = -1$. In the same way, we can get

$b_4 = 0$. Thus $neut(a) = (0, T, -I, 0)$, $anti(a) = (c_1, c_2T, c_3I, c_4F)$, where $c_1 + c_2 = \frac{1}{a_2}$, c_3, c_4 can be chosen arbitrarily in $\mathbb{R}$.

Case A212: when $a_1 = 0, a_2 \neq 0, a_2 + a_3 = 0, a_4 \neq 0$, From $a_4(b_3 + b_4) = 0$, we have $b_3 + b_4 = 0$, that is, $a = (0, a_2T, -a_2I, a_4F)$, $netu(a) = (0, T, b_3I, -b_3F)$, $anti(a) = (c_1, c_2T, c_3I, c_4F)$. From $(0, a_2T, -a_2I, a_4) * (c_1, c_2T, c_3I, c_4F) = (0, T, b_3I, -b_3F)$, so we have $a_2(c_1 + c_2) = 1$ and $-a_2(c_1 + c_2) = b_3$, i.e., $b_3 = -1$. Thus $neut(a) = (0, T, -I, F)$, $anti(a) = (c_1, c_2T, c_3I, c_4F)$, where $c_1 + c_2 = \frac{1}{a_2}$, $c_3 + c_4 = \frac{1}{a_4} - \frac{1}{a_2}$.

Case A22: when $a_1 = 0, a_2 \neq 0, a_2 + a_3 \neq 0$. From $(a_2 + a_3)b_3 = 0$, we have $b_3 = 0$. that is, $a = (0, a_2T, -a_2I, a_4F), netu(a) = (0, T, 0, b_4F), anti(a) = (c_1, c_2T, c_3I, c_4F)$. From $a * neut(a) = a$, we have $(a_2 + a_3 + a_4)b_4 = 0$, so we discuss from $a_2 + a_3 + a_4 = 0$ or $a_2 + a_3 + a_4 \neq 0$.

Case A221: when $a_1 = 0, a_2 \neq 0, a_2 + a_3 \neq 0, a_2 + a_3 + a_4 = 0$. In this case $a = (0, a_2T, a_3I, a_4F), netu(a) = (0, T, 0, b_4F), anti(a) = (c_1, c_2T, c_3I, c_4F)$. From $(0, a_2T, a_3I, a_4F) * (c_1, c_2T, c_3I, c_4F) = (0, T, 0, b_4F)$, so we have $a_2(c_1 + c_2) = 1$, $c_3 = -\frac{a_3}{a_2(2+a_3)}$, $(a_2 + a_3 + a_4)b_4 + a_4(c_1 + c_2 + c_3) = b_4$, so we have $b_4 = -1$. Thus $neut(a) = (0, T, 0, -F)$, $anti(a) = (c_1, c_2T, c_3I, c_4F)$, where $c_1 + c_2 = \frac{1}{a_2}$, $c_3 = -\frac{a_3}{a_2(2+a_3)}$, c_4 can be chosen arbitrarily in $\mathbb{R}$.

Case A222: when $a_1 = 0, a_2 \neq 0, a_2 + a_3 \neq 0, a_2 + a_3 + a_4 \neq 0$. From $(a_2 + a_3 + a_4)b_4 = 0$, we have $b_4 = 0$. that is, $a = (0, a_2T, a_3I, a_4F), netu(a) = (0, T, 0, 0), anti(a) = (c_1, c_2T, c_3I, c_4F)$. From $(0, a_2T, a_3I, 0) * (c_1, c_2T, c_3I, c_4F) = (0, T, 0, 0)$, so we have $a_2(c_1 + c_2) = 1$, $c_3 = -\frac{a_3}{a_2(2+a_3)}$, $(a_2 + a_3 + a_4)b_4 + a_4(c_1 + c_2 + c_3) = 0$, Thus $neut(a) = (0, T, 0, 0)$, $anti(a) = (c_1, c_2T, c_3I, c_4F)$, where $c_1 + c_2 = \frac{1}{a_2}$, $c_3 = -\frac{a_3}{a_2(2+a_3)}$, $c_4 = -\frac{a_4}{(a_2+a_3)(a_2+a_3+a_4)}$.

Case B: when $a_1 \neq 0$.

In this case, from $a_1b_1 = a_1$ and $a_1 \neq 0$, we have $b_1 = 1$. That is $a = (a_1, a_2T, a_3I, a_4F), neut(a) = (1, b_2T, b_3I, b_4F), anti(a) = (c_1, c_2T, c_3I, c_4F)$. From Definition 1, $a * neut(a) = a$, that is $a_1b_2 + a_2 + a_2b_2 = a_2$, so $(a_1 + a_2)b_2 = 0$. So we discuss from $a_1 + a_2 = 0$ or $a_1 + a_2 \neq 0$.

Case B1: when $a_1 \neq 0, a_1 + a_2 = 0$. That is $a = (a_1, -a_1T, a_3I, a_4F), neut(a) = (1, b_2T, b_3I, b_4F), anti(a) = (c_1, c_2T, c_3I, c_4F)$. From $a * anti(a) = neut(a)$, we have $c_1 = \frac{1}{a_1}$, $a_1c_2 - a_1c_1 - a_1c_2 = b_2$, so $b_2 = -1$. From $a * neut(a) = a$, so we have $a_3 + a_3b_2 + a_3b_3 = a_3$, i.e., $a_3(b_2 + b_3) = 0$. So we discuss from $a_3 = 0$ or $a_3 \neq 0$.

Case B11: when $a_1 \neq 0, a_1 + a_2 = 0, a_3 = 0$. That is $a = (a_1, -a_1T, 0, a_4F), neut(a) = (1, -T, b_3I, b_4F), anti(a) = (c_1, c_2T, c_3I, c_4F)$. From $a * neut(a) = a$, we have $a_1b_4 - a_1b_4 - 0b_4 + a_4(1 - 1 + b_3 + b_4) = a_4$, i.e., $a_4(b_3 + b_4) = a_4$. So we discuss from $a_4 = 0$ or $a_4 \neq 0$.

Case B111: when $a_1 \neq 0, a_1 + a_2 = 0, a_3 = 0, a_4 = 0$. That is $a = (a_1, -a_1T, 0, 0), neut(a) = (1, -T, b_3I, b_4F), anti(a) = (c_1, c_2T, c_3I, c_4F)$. From $a * anti(a) = neut(a)$, i.e., $(a_1, -a_1T, 0, 0) * (c_1, c_2T, c_3I, c_4F) = (1, -T, b_3I, b_4F)$, we have $c_1 = \frac{1}{a_1}$, $b_3 = b_4 = 0$. Thus $neut(a) = (1, -T, 0, 0)$, $anti(a) = (c_1, c_2T, c_3I, c_4F)$, which satisfies $c_1 = \frac{1}{a_1}$ and c_2, c_3, c_4 can be chosen arbitrarily in $\mathbb{R}$.

Case B112: when $a_1 \neq 0, a_1 + a_2 = 0, a_3 = 0, a_4 \neq 0$. From $a_4(b_3 + b_4) = a_4$, we have $b_3 + b_4 = 1$. That is $a = (a_1, -a_1T, 0, a_4F), neut(a) = (1, -T, b_3I, b_4F), anti(a) = (c_1, c_2T, c_3I, c_4F)$. From $a * anti(a) = neut(a)$, i.e., $(a_1, -a_1T, 0, a_4F) * (c_1, c_2T, c_3I, c_4F) = (1, -T, b_3I, b_4F)$, we have $c_1 = \frac{1}{a_1}$, $b_3 = 0, b_4 = 1$. Thus $neut(a) = (1, -T, 0, F)$, $anti(a) = (c_1, c_2T, c_3I, c_4F)$, where $c_1 = \frac{1}{a_1}$ and $c_2 + c_3 + c_4 = \frac{1}{a_4} - \frac{1}{a_1}$.

Case B12: when $a_1 \neq 0, a_1 + a_2 = 0, a_3 \neq 0$. From $a_3(b_2 + b_3) = 0$ and $a_3 \neq 0$, we have $b_2 + b_3 = 0$, i.e., $b_3 = 1$. That is $a = (a_1, -a_1T, a_3I, a_4F), neut(a) = (1, -T, I, b_4F), anti(a) = (c_1, c_2T, c_3I, c_4F)$. From $a * neut(a) = a$, we have $a_3b_4 + a_4 + a_4b_4 = a_4$, i.e., $(a_3 + a_4)b_4 = 0$. So we discuss from $a_3 + a_4 = 0$ or $a_3 + a_4 \neq 0$.

Case B121: when $a_1 \neq 0, a_1 + a_2 = 0, a_3 \neq 0, a_3 + a_4 = 0$. That is $a = (a_1, -a_1T, a_3I, -a_3F)$, $neut(a) = (1, -T, I, b_4F)$, $anti(a) = (c_1, c_2T, c_3I, c_4F)$. From $a * anti(a) = neut(a)$, i.e., $(a_1, -a_1T, a_3I, -a_3F) * (c_1, c_2T, c_3I, c_4F) = (1, -T, I, b_4F)$, we have $c_1 = \frac{1}{a_1}$, $c_2 + c_3 = \frac{1}{a_3} - \frac{1}{a_1}$, $-a_3(c_1 + c_2 + c_3) = b_4$, i.e., $b_4 = -1$. Thus $neut(a) = (1, -T, I, -F)$, $anti(a) = (c_1, c_2T, c_3I, c_4F)$, where $c_1 = \frac{1}{a_1}$, $c_2 + c_3 = \frac{1}{a_3} - \frac{1}{a_2}$, c_4 can be chosen arbitrarily in $\mathbb{R}$.

Case B122: when $a_1 \neq 0, a_1+a_2=0, a_3 \neq 0, a_3+a_4 \neq 0$, from $(a_3+a_4)b_4=0$, we have $b_4=0$. That is $a=(a_1,-a_1T,a_3I,a_4F), neut(a)=(1,-T,I,0), anti(a)=(c_1,c_2T,c_3I,c_4F)$. From $a*anti(a)=neut(a)$, i.e., $(a_1,-a_1T,a_3I,a_4F)*(c_1,c_2T,c_3I,c_4F)=(1,-T,I,0)$, we have $c_1=\frac{1}{a_1}, c_2+c_3=\frac{1}{a_3}-\frac{1}{a_1}$, $c_4=-\frac{a_4}{a_3(a_3+a_4)}$. Thus $neut(a)=(1,-T,I,-F), anti(a)=(c_1,c_2T,c_3I,c_4F)$, where $c_1=\frac{1}{a_1}, c_2+c_3=\frac{1}{a_3}-\frac{1}{a_1}, c_4=-\frac{a_4}{a_3(a_3+a_4)}$.

Case B2: when $a_1 \neq 0, a_1+a_2 \neq 0$, from $(a_1+a_2)b_2=0$, we have $b_2=0$. That is $a=(a_1,-a_1T,a_3I,a_4F), neut(a)=(1,0,b_3I,b_4F), anti(a)=(c_1,c_2T,c_3I,c_4F)$. From $a*neut(a)=a$, so we have $a_1b_3+a_2b_3+a_3+a_3b_3=a_3$, i.e., $(a_1+a_2+a_3)b_3=0$. So we discuss from $a_1+a_2+a_3=0$ or $a_1+a_2+a_3 \neq 0$.

Case B21: when $a_1 \neq 0, a_1+a_2 \neq 0, a_1+a_2+a_3=0$. That is $a=(a_1,a_2T,a_3I,a_4F), neut(a)=(1,0,b_3I,b_4F), anti(a)=(c_1,c_2T,c_3I,c_4F)$. From $a*neut(a)=a$, so we have $(a_1+a_2+a_3)b_4+a_4+a_4b_3+a_4b_4=a_4$, i.e., $(b_3+b_4)a_4=0$. So we discuss from $a_4=0$ or $a_4 \neq 0$.

Case B211: when $a_1 \neq 0, a_1+a_2 \neq 0, a_1+a_2+a_3=0, a_4=0$. That is $a=(a_1,a_2T,a_3I,0), neut(a)=(1,0,b_3I,b_4F), anti(a)=(c_1,c_2T,c_3I,c_4F)$. From $a*anti(a)=neut(a)$, i.e., $(a_1,a_2T,a_3I,0)*(c_1,c_2T,c_3I,c_4F)=(1,0,b_3I,b_4F)$, we have $c_1=\frac{1}{a_1}, c_2=-\frac{a_2}{a_1(a_1+a_2)}$. $a_3(c_1+c_2)=b_3$, $(a_1+a_2+a_3)c_4+0(c_1+c_2+c_3+c_4)=0$, which means $b_3=-1, b_4=0$. Thus $neut(a)=(1,0,-I,0)$, $anti(a)=(c_1,c_2T,c_3I,c_4F)$, where $c_1=\frac{1}{a_1}, c_2=-\frac{a_2}{a_1(a_1+a_2)}$, c_3, c_4 can be chosen arbitrarily in $\mathbb{R}$.

Case B212: when $a_1 \neq 0, a_1+a_2 \neq 0, a_1+a_2+a_3=0, a_4 \neq 0$. From $(b_3+b_4)a_4=0$, we have $b_3+b_4=0$. That is $a=(a_1,a_2T,a_3I,a_4F), neut(a)=(1,0,b_3I,-b_3F), anti(a)=(c_1,c_2T,c_3I,c_4F)$. From $a*anti(a)=neut(a)$, i.e., $(a_1,a_2T,a_3I,a_4F)*(c_1,c_2T,c_3I,c_4F)=(1,0,b_3I,-b_3F)$, we have $c_1=\frac{1}{a_1}, c_2=-\frac{a_2}{a_1(a_1+a_2)}$. $a_3(c_1+c_2)=b_3$, i.e., $b_3=-1, b_4=1$. Thus $neut(a)=(1,0,-I,F)$, $anti(a)=(c_1,c_2T,c_3I,c_4F)$, where $c_1=\frac{1}{a_1}, c_2=-\frac{a_2}{a_1(a_1+a_2)}, c_3+c_4=\frac{1}{a_4}-\frac{1}{a_1+a_2}$.

Case B22: when $a_1 \neq 0, a_1+a_2 \neq 0, a_1+a_2+a_3 \neq 0$, from $(a_1+a_2+a_3)b_3=0$, we have $b_3=0$. That is $a=(a_1,a_2T,a_3I,a_4F), neut(a)=(1,0,0,b_4F), anti(a)=(c_1,c_2T,c_3I,c_4F)$. From $a*neut(a)=a$, so we have $(a_1+a_2+a_3+a_4)b_4+a_4=a_4$, i.e., $(a_1+a_2+a_3+a_4)b_4=0$. So we discuss from $a_1+a_2+a_3+a_4=0$ or $a_1+a_2+a_3+a_4 \neq 0$.

Case B221: when $a_1 \neq 0, a_1+a_2 \neq 0, a_1+a_2+a_3 \neq 0, a_1+a_2+a_3+a_4=0$, That is $a=(a_1,a_2T,a_3I,a_4F), neut(a)=(1,0,0,b_4F), anti(a)=(c_1,c_2T,c_3I,c_4F)$. From $a*anti(a)=neut(a)$, i.e., $(a_1,a_2T,a_3I,a_4F)*(c_1,c_2T,c_3I,c_4F)=(1,0,0,b_4F)$, we have $c_1=\frac{1}{a_1}, c_2=-\frac{a_2}{a_1(a_1+a_2)}$. $c_3=-\frac{a_3}{(a_1+a_2)(a_1+a_2+a_3)}$, $a_4(c_1+c_2+c_3)=b_4$, so $b_4=-1$. Thus $neut(a)=(1,0,0,-F)$, $anti(a)=(c_1,c_2T,c_3I,c_4F)$, where $c_1=\frac{1}{a_1}, c_2=-\frac{a_2}{a_1(a_1+a_2)}, c_3=-\frac{a_3}{(a_1+a_2)(a_1+a_2+a_3)}$, c_4 can be chosen arbitrarily in $\mathbb{R}$.

Case B222: when $a_1 \neq 0, a_1+a_2 \neq 0, a_1+a_2+a_3 \neq 0, a_1+a_2+a_3+a_4 \neq 0$. From $(a_1+a_2+a_3+a_4)b_4+a_4=a_4$, we have $b_4=0$. That is $a=(a_1,a_2T,a_3I,a_4F), neut(a)=(1,0,0,0), anti(a)=(c_1,c_2T,c_3I,c_4F)$. From $a*anti(a)=neut(a)$, i.e., $(a_1,a_2T,a_3I,a_4F)*(c_1,c_2T,c_3I,c_4F)=(1,0,0,0)$, we have $c_1=\frac{1}{a_1}, c_2=-\frac{a_2}{a_1(a_1+a_2)}, c_3=-\frac{a_3}{(a_1+a_2)(a_1+a_2+a_3)}$. $(a_1+a_2+a_3+a_4)c_4+a_4(c_1+c_2+c_3)=0$, i.e., $c_4=-\frac{a_4}{(a_1+a_2+a_3)(a_1+a_2+a_3+a_4)}$. Thus $neut(a)=(1,0,0,0)$, $anti(a)=(c_1,c_2T,c_3I,c_4F)$, where $c_1=\frac{1}{a_1}, c_2=-\frac{a_2}{a_1(a_1+a_2)}, c_3=-\frac{a_3}{(a_1+a_2)(a_1+a_2+a_3)}, c_4=-\frac{a_4}{(a_1+a_2+a_3)(a_1+a_2+a_3+a_4)}$.

Finally, we should show that all the above cases include each element $a \in NQ$, i.e., $a_i, i=1,2,3,4$ can take all the values on $\mathbb{R}$. It is obvious that a_1 can take all the values on $\mathbb{R}$ because $a_1=0$ according to case A and that $a_1 \neq 0$ according to case B. Moreover, for case A, a_2 can take all the values on $\mathbb{R}$ because case A1 according to $a_2=0$ and case A2 according to $a_2 \neq 0$. For case B, a_2 can take all the values on $\mathbb{R}$ because case B1 according to $a_1+a_2=0$ and case B2 according to $a_1+a_2 \neq 0$. That is for each element $a=(a_1,a_2,a_3,a_4) \in NQ$, a_1, a_2 can select all of value in $\mathbb{R}$. We will verify that a_3 and a_4 can take all the values on $\mathbb{R}$ when case A1 or A2 or B1 or B2 respectively.

For case A1, a_3 can take all the value in $\mathbb{R}$ because case A11 according to $a_3=0$ and case A12 according to $a_3 \neq 0$. Similarly, for case A11, a_4 can take all the value in $\mathbb{R}$ because case A111 according to $a_4=0$ and case A112 according to $a_4 \neq 0$. For case A12, a_4 can take all the value in $\mathbb{R}$ because case A121 according to $a_3 + a_4 = 0$ and case A122 according to $a_3 + a_4 = 0$. The top left subgraph of Figure 1 shows that the four cases A111, A112, A211 and A222. The unique □ point represents the case A111, the + points represent the case A112, the $*$ points represent the case A121 and the • points represent the case A122. This explain the that for case A1, a_3 and a_4 can take all the points on the plane. For case A2, B1 or B2, we can get that a_3 and a_4 can take all the points on the plane respectively. The top right subgraph of Figure 1 represents the case A2 if we select $a_1 = 0, a_2 = 1$, the bottom left subgraph of Figure 1 represents the case B1 if we select $a_1 = 1, a_2 = -1$ and bottom right subgraph of Figure 1 represents the case B2 if we select $a_1 = 1, a_2 = 0$. The figure intuitively illustrates that all the points $(a_1, a_2, a_3, a_4), a_i \in \mathbb{R}$ are included.

Through the above analysis, we can get that for each element $a \in NQ$, there exists the neutral element $neut(a)$ and opposite element $anti(a)$. □

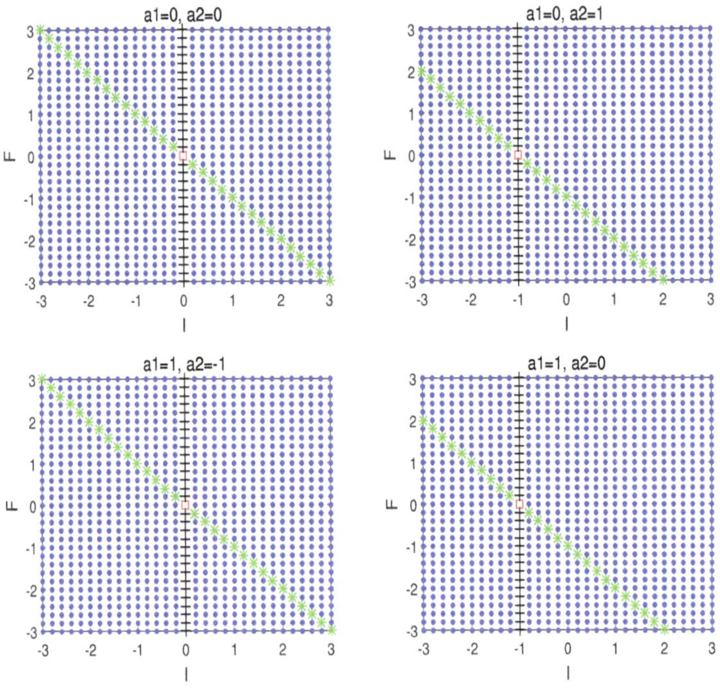

Figure 1. The demonstration figure shows that case A1 ($a_1 = a_2 = 0$, the top left subgraph) or A2 (select $a_1 = 0, 0 \neq a_2 = 1$, the top right subgraph) or B1 (Select $a_1 \neq 0, a_2 = -1$ which means $a_1 + a_2 = 0$, the bottom left subgraph) or B2 (select $a_1 = 1, a_2 = 0$, which means $a_1 + a_2 \neq 0$, the bottom right subgraph) can take all the values on the plane.

For algebra system $(NQ, *)$, Table 1 gives all the subset, which has the same neutral element, and the corresponding neutral element and opposite elements.

Table 1. The corresponding neutral element and opposite elements for $(NQ, *)$.

The Subset of NQ	**Neutral Elements**	**Opposite Element** (c_1, c_2T, c_3I, c_4F)
$\{(0,0,0,0)\}$	$(0,0,0,0)$	$c_i \in \mathbb{R}$
$\{(0,0,0,a_4F)\|a_4 \neq 0\}$	$(0,0,0,F)$	$c_1+c_2+c_3+c_4 = \frac{1}{a_4}$
$\{(0,0,a_3I,-a_3F)\|a_3 \neq 0\}$	$(0,0,I,-F)$	$c_1+c_2+c_3 = \frac{1}{a_3}, c_4 \in \mathbb{R}$
$\{(0,0,a_3I,a_4F)\|a_3 \neq 0, a_3+a_4 \neq 0\}$	$(0,0,I,0)$	$c_1+c_2+c_3 = \frac{1}{a_3}, c_4 = -\frac{a_4}{a_3(a_3+a_4)}$
$\{(0,a_2T,-a_2I,0)\|a_2 \neq 0\}$	$(0,T,-I,0)\}$	$c_1+c_2 = \frac{1}{a_2}, c_3, c_4 \in \mathbb{R}$
$\{(0,a_2T,-a_2I,a_4F)\|a_2 \neq 0, a_4 \neq 0\}$	$(0,T,-I,F)$	$c_1+c_2 = \frac{1}{a_2}, c_3+c_4 = \frac{1}{a_4} - \frac{1}{a_2}$
$\{(0,a_2T,a_3I,a_4F)\|a_2 \neq 0, a_2+a_3 \neq 0, a_2+a_3+a_4 = 0\}$	$(0,T,0,-F)$	$c_1+c_2 = \frac{1}{a_2}, c_3 = -\frac{a_3}{a_2(a_2+a_3)}, c_4 \in \mathbb{R}$
$\{(0,a_2T,a_3I,a_4F)\|a_2 \neq 0, a_2+a_3 \neq 0, a_2+a_3+a_4 \neq 0\}$	$(0,T,0,0)$	$c_1+c_2 = \frac{1}{a_2}, c_3 = -\frac{a_3}{a_2(a_2+a_3)}$, $c_4 = -\frac{a_4}{(a_2+a_3)(a_2+a_3+a_4)}$
$\{(a_1,-a_1T,0,0)\|a_1 \neq 0\}$	$(1,-T,0,0)\}$	$c_1 = \frac{1}{a_1}, c_2, c_3, c_4 \in \mathbb{R}$
$\{(a_1,-a_1T,0,a_4F)\|a_1 \neq 0, a_4 \neq 0\}$	$(1,-T,0,F)$	$c_1 = \frac{1}{a_1}, c_2+c_3+c_4 = \frac{1}{a_4} - \frac{1}{a_1}$
$\{(a_1,-a_1T,a_3I,-a_3F)\|a_1 \neq 0, a_3 \neq 0\}$	$(1,-T,I,-F)$	$c_1 = \frac{1}{a_1}, c_2+c_3 = \frac{1}{a_3} - \frac{1}{a_1}, c_4 \in \mathbb{R}$
$\{(a_1,-a_1T,a_3I,a_4F)\|a_1 \neq 0, a_3 \neq 0, a_3+a_4 \neq 0\}$	$(1,-T,I,0)$	$c_1 = \frac{1}{a_1}, c_2+c_3 = \frac{1}{a_3} - \frac{1}{a_1}$, $c_4 = -\frac{a_4}{a_3(a_3+a_4)}$
$\{(a_1,a_2T,a_3I,0)\|a_1 \neq 0, a_1+a_2 \neq 0, a_1+a_2+a_3 = 0\}$	$(1,0,-I,0)$	$c_1 = \frac{1}{a_1}, c_2 = -\frac{a_2}{a_1(a_1+a_2)}, c_3, c_4 \in \mathbb{R}$
$\{(a_1,a_2T,a_3I,a_4F)\|a_1 \neq 0, a_1+a_2 \neq 0, a_1+a_2+a_3 = 0, a_4 \neq 0\}$	$(1,0,-I,F)$	$c_1 = \frac{1}{a_1}, c_2 = -\frac{a_2}{a_1(a_1+a_2)}$, $c_3+c_4 = \frac{1}{a_4} - \frac{1}{a_1+a_2}$
$\{(a_1,a_2T,a_3I,a_4F)\|a_1 \neq 0, a_1+a_2 \neq 0, a_1+a_2+a_3 \neq 0, a_1+a_2+a_3+a_4 = 0\}$	$(1,0,0,-F)$	$c_1 = \frac{1}{a_1}, c_2 = -\frac{a_2}{a_1(a_1+a_2)}$, $c_3 = -\frac{a_3}{(a_1+a_2)(a_1+a_2+a_3)}, c_4 \in \mathbb{R}$
$\{(a_1,a_2T,a_3I,a_4F)\|a_1 \neq 0, a_1+a_2 \neq 0, a_1+a_2+a_3 \neq 0, a_1+a_2+a_3+a_4 \neq 0\}$	$(1,0,0,0)$	$c_1 = \frac{1}{a_1}, c_2 = -\frac{a_2}{a_1(a_1+a_2)}$, $c_3 = -\frac{a_3}{(a_1+a_2)(a_1+a_2+a_3)}$, $c_4 = -\frac{a_4}{(a_1+a_2+a_3)(a_1+a_2+a_3+a_4)}$

Example 1. *For the algebra system $(NQ, *)$, if $a = (a_1, a_2, a_3, a_4) = (1, -T, 3I, -F)$, i.e., $a_1 \neq 0, a_1 + a_2 = 0, a_3 \neq 0, a_3 + a_4 \neq 0$, then from Table 1, we can get $neut(a) = (1, -T, I, 0)$. Let $anti(a) = (c_1, c_2T, c_3I, c_4F)$, so $c_1 = \frac{1}{a_1} = 1, c_2 + c_3 = \frac{1}{a_3} - \frac{1}{a_1} = -\frac{2}{3}, c_4 = -\frac{a_4}{a_3(a_3+a_4)} = \frac{1}{6}$, so $anti(a) = (1, c_2T, c_3I, \frac{1}{6}F)$, where $c_2 + c_3 = -\frac{2}{3}$. Thus we can easily get the neutral element and opposite elements of each neutrosophic quadruple number. For more examples, see the following:*

1. *Let $b = (0, 0, I, -F)$, then $neut(b) = (0, 0, I, -F)$ and $anti(b) = (c_1, c_2T, c_3I, c_4F)$, where $c_1 + c_2 + c_3 = 1$, c_4 can be can be chosen arbitrarily in $\mathbb{R}$.*
2. *Let $c = (1, T, I, -F)$, then $neut(c) = (1, 0, 0, 0)$ and $anti(c) = (1, -\frac{1}{2}T, -\frac{1}{6}I, \frac{1}{6}F)$.*
3. *Let $d = (0, T, I, F)$, then $neut(d) = (0, T, 0, 0)$ and $anti(d) = (c_1, c_2T, -\frac{1}{2}I, -\frac{1}{6}F)$, where $c_1 + c_2 = 1$.*

In the following, we will discuss the algebra structure properties of $(NQ, *)$.

Proposition 3. *For algebra system $(NQ, *)$, let $NS = \{neut(a)|a \in NQ\}$, we have:*

(1) $NS = \{(1,0,0,0), (0,0,0,F), (0,0,I,-F), (0,0,I,0), (0,T,-I,0), (0,T,-I,F), (0,T,0,-F), (0,T,0,0), (1,-T,0,0), (1,-T,0,F), (1,-T,I,-F), (1,-T,I,0), (1,0,-I,0), (1,0,-I,F), (1,0,0,-F), (1,0,0,0)\}$.

(2) NS is closed with respect to operation $*$.

(3) Set $IS = \{a | a^2 = a, a \in NQ\}$*, which is all the set of idempotent elements of* $(NQ, *)$*, then* $NS = IS$*.*

Proof. (1) Obviously.

(2) If $c, d \in NS$, that is $neut(a) = c, neut(b) = d, a, b \in NQ$. From Proposition 1, $neut(a) * neut(b) = neut(a * b)$, i.e., $c * d = neut(a * b)$, then form Theorem 1, every element in NQ has neutral element, so $a * b$ also has neutral element, that is $neut(a * b) \in NS$, i.e., $c * d \in NS$, thus NS is closed with respect to operation $*$.

(3) From Proposition 1, $neut(a) * neut(a) = neut(a)$, so $neut(a)$ is a idempotent element and $NS \subseteq IS$. On the other hand if a is a idempotent element, so $a * a = a$, that is a exists the neutral element a and the opposite element a, so a is a neutral element, that is $IS \subseteq NS$. Thus $NS = IS$. □

Proposition 4. *For algebra system* $(NQ, *)$*, let* $V_c = \{a | a \in NQ \wedge neut(a) = c\}$*,* $V_{c*d} = \{a * b | a, b \in NQ \wedge neut(a) = c \wedge neut(b) = d\}$*, we have:*

(1) V_c *is closed with respect to operation* $*$*.*

(2) V_{c*d} *is closed with respect to operation* $*$*.*

Proof. (1) If $a, b \in V_c$, that is $neut(a) = neut(b) = c$. From Proposition 1, $neut(a) * neut(b) = neut(a * b)$, we can see that $neut(a * b) = neut(a) = c$, i.e., the neutral element of $a * b$ is the neutral element of a, so $a * b \in V_c$, that is V_c is closed with respect to operation $*$.

(2) If $a_1 * b_1, a_2 * b_2 \in V_{c*d}$, i.e., $neut(a_1) = neut(a_2) = c, neut(b_1) = neut(b_2) = d$. From Proposition 3(2), $a_1 * a_2 = a_3 \in V_c$, $b_1 * b_2 = b_3 \in V_d$, so $neut(a_3) = c, neut(b_3) = d$, from $(a_1 * b_1) * (a_2 * b_2) = a_3 * b_3$, so $neut(a_1 * a_2 * b_1 * b_2) = neut(a_3 * a_4)$, that is $a_3 * a_4 \in V_{c*d}$, that means $a_1 * a_2 * b_1 * b_2 \in V_{c*d}$. Thus V_{c*d} is closed with respect to operation $*$. □

Definition 7. *Assume that* $(N, *)$ *is a neutrosophic triplet group and* H *be a nonempty subset of* N*. Then* H *is called a neutrosophic triplet subgroup of* N *if;*

(1) $a * b \in H$ *for all* $a, b \in H$*;*

(2) there exists $anti(a) \in \{anti(a)\}$ *such that* $anti(a) \in H$ *for all* $a \in H$*, where* $\{anti(a)\}$ *is the set of opposite element of* a *in* $(N, *)$*.*

Theorem 2. *For algebra system* $(NQ, *)$*, let* $V_c = \{a | a \in NQ \wedge neut(a) = c\}$*,* $V_{c*d} = \{a * b | a, b \in NQ \wedge neut(a) = c, neut(b) = d\}$*, we have:*

(1) V_c *is a neutrosophic triplet subgroup of* NQ*.*

(2) V_{c*d} *is a neutrosophic triplet subgroup of* NQ*.*

Proof. (1) From Proposition 3, we can see that V_c is closed with respect to operation $*$. In the following, we will prove there exists $anti(a) \in \{anti(a)\}$ such that $anti(a) \in V_c$ for all $a \in V_c$.

Proof by contradiction.

Assume that $\{anti(a)\} \cap V_c = \emptyset$. From Proposition 1 we can see that $a * anti(a) = c$. On the other hand, $anti(a) \in NQ$, so $anti(a)$ exists neutral element, denoted by $neut(anti(a))$. Being $anti(a) \notin V_c$, so $neut(anti(a)) \neq c$.

From $a * anti(a) = c$, we have $a * anti(a) * neut(anti(a)) = c * neut(anti(a))$, being $anti(a) * neut(anti(a)) = anti(a)$ and $a * anti(a) = c$, we have $c * neut(anti(a)) = c$, and then we have $a * c * neut(anti(a)) = a * c = a$, that means $a * neut(anti(a)) = a$, so $neut(anti(a))$ is also a neutral element of a. This leads to the contradiction being the uniqueness of neutral element for each element. Therefore $\{anti(a)\} \cap V_c \neq \emptyset$. Thus from Definition 7, V_c is a neutrosophic triplet subgroup of NQ.

(2) The same way we can get V_{c*d} is a neutrosophic triplet subgroup of NQ. □

Theorem 3. *For algebra system* $(NQ, *)$*, let* $V_c = \{a | a \in NQ \wedge neut(a) = c\}$*, we have:*

(1) $V_c \cap V_d = \emptyset$ *if* $c \neq d$*.*

(2) $NQ = \cup_{c \in NS} V_c$. So $\cup_{c \in NS} V_c$ is a partition of NQ, where NS is a set, which contains all the neutral elements of $(NQ, *)$.

Proof. (1) Proof by contradiction.

Assume $V_c \cap V_d \neq \emptyset$ when $c \neq d$, so exist $a \in V_c \cap V_d$, such that a has two neutral elements c and d. This leads to the contradiction being the uniqueness of neutral element. So $V_c \cap V_d = \emptyset$ if $c \neq d$.

(2) From the proof of Theorem 1, we can get $NQ = \cup_{c \in NS} V_c$. So $\cup_{c \in NS} V_c$ is a partition of NQ. □

For the algebra system $(NQ, \star)$, we have the similar results. We describe as following and omit the proof.

Theorem 4. *For the algebra system $(NQ, \star)$, for every element $a \in NQ$, there exists the neutral element $neut(a)$ and opposite element $anti(a)$.*

For algebra system $(NQ, \star)$, Table 2 gives all the subset, which has the same neutral element, and the corresponding neutral element and opposite elements.

Table 2. The corresponding neutral element and opposite elements for $(NQ, \star)$.

The Subset of NQ	Neutral Elements	Opposite Element (c_1, c_2T, c_3I, c_4F)
$\{(0,0,0,0)\}$	$(0,0,0,0)$	$c_i \in \mathbb{R}$
$\{(0, a_2T, 0, 0) \mid a_2 \neq 0\}$	$(0, T, 0, 0)$	$c_1 + c_2 + c_3 + c_4 = \frac{1}{a_2}$
$\{(0, -a_3T, a_3I, 0) \mid a_3 \neq 0\}$	$(0, -T, I, 0)$	$c_1 + c_3 + c_4 = \frac{1}{a_3}, c_2 \in \mathbb{R}$
$\{(0, a_2T, a_3I, a_4F) \mid a_3 \neq 0, a_2 + a_3 \neq 0\}$	$(0, 0, I, 0)$	$c_1 + c_3 + c_4 = \frac{1}{a_3}$, $c_2 = -\frac{a_2}{a_3(a_2+a_3)}$
$\{(0, 0, -a_4I, a_4F) \mid a_4 \neq 0\}$	$(0, 0, -I, F)\}$	$c_1 + c_4 = \frac{1}{a_4}, c_2, c_3 \in \mathbb{R}$
$\{(0, a_2T, -a_4I, a_4F) \mid a_2 \neq 0, a_4 \neq 0\}$	$(0, T, -I, F)$	$c_1 + c_4 = \frac{1}{a_4}, c_2 + c_3 = \frac{1}{a_2} - \frac{1}{a_4}$
$\{(0, a_2T, a_3I, a_4F) \mid a_4 \neq 0, a_3 + a_4 \neq 0, a_2 + a_3 + a_4 = 0\}$	$(0, -T, 0, F)$	$c_1 + c_4 = \frac{1}{a_4}, c_3 = -\frac{a_3}{a_4(a_3+a_4)}, c_2 \in \mathbb{R}$
$\{(0, a_2T, a_3I, a_4F) \mid a_4 \neq 0, a_3 + a_4 \neq 0, a_2 + a_3 + a_4 \neq 0\}$	$(0, 0, 0, F)$	$c_1 + c_4 = \frac{1}{a_4}, c_3 = -\frac{a_3}{a_4(a_3+a_4)}$, $c_2 = -\frac{a_2}{(a_3+a_4)(a_2+a_3+a_4)}$
$\{(a_1, 0, 0, -a_1F) \mid a_1 \neq 0\}$	$(1, 0, 0, -F)\}$	$c_1 = -\frac{1}{a_1}, c_2, c_3, c_4 \in \mathbb{R}$
$\{(a_1, a_2T, 0, -a_1F) \mid a_1 \neq 0, a_2 \neq 0\}$	$(1, T, 0, -F)$	$c_1 = \frac{1}{a_1}, c_2 + c_3 + c_4 = \frac{1}{a_2} - \frac{1}{a_1}$
$\{(a_1, -a_3T, a_3I, -a_1F) \mid a_1 \neq 0, a_3 \neq 0\}$	$(1, -T, I, -F)$	$c_1 = \frac{1}{a_1}, c_3 + c_4 = \frac{1}{a_3} - \frac{1}{a_1}, c_4 \in \mathbb{R}$
$\{(a_1, a_2T, a_3I, -a_1F) \mid a_1 \neq 0, a_3 \neq 0, a_2 + a_3 \neq 0\}$	$(1, 0, I, -F)$	$c_1 = \frac{1}{a_1}, c_3 + c_4 = \frac{1}{a_3} - \frac{1}{a_1}$, $c_2 = -\frac{a_2}{a_3(a_2+a_3)}$
$\{(a_1, 0, a_3I, a_4F) \mid a_1 \neq 0, a_1 + a_4 \neq 0, a_1 + a_3 + a_4 = 0\}$	$(1, 0, -I, 0)$	$c_1 = \frac{1}{a_1}, c_4 = -\frac{a_4}{a_1(a_1+a_4)}, c_2, c_3 \in \mathbb{R}$
$\{(a_1, a_2T, a_3I, a_4F) \mid a_1 \neq 0, a_1 + a_4 \neq 0, a_1 + a_3 + a_4 = 0, a_2 \neq 0\}$	$(1, T, -I, 0)$	$c_1 = \frac{1}{a_1}, c_4 = -\frac{a_4}{a_1(a_1+a_4)}$, $c_2 + c_3 = \frac{1}{a_2} - \frac{1}{a_1+a_4}$
$\{(a_1, a_2T, a_3I, a_4F) \mid a_1 \neq 0, a_1 + a_4 \neq 0, a_1 + a_3 + a_4 \neq 0, a_1 + a_2 + a_3 + a_4 = 0\}$	$(1, -T, 0, 0)$	$c_1 = \frac{1}{a_1}, c_4 = -\frac{a_4}{a_1(a_1+a_4)}$, $c_3 = -\frac{a_3}{(a_1+a_4)(a_1+a_3+a_4)}, c_2 \in \mathbb{R}$
$\{(a_1, a_2T, a_3I, a_4F) \mid a_1 \neq 0, a_1 + a_4 \neq 0, a_1 + a_3 + a_4 \neq 0, a_1 + a_2 + a_3 + a_4 \neq 0\}$	$(1, 0, 0, 0)$	$c_1 = \frac{1}{a_1}, c_4 = -\frac{a_4}{a_1(a_1+a_4)}$, $c_3 = -\frac{a_3}{(a_1+a_4)(a_1+a_3+a_4)}$, $c_2 = -\frac{a_2}{(a_1+a_3+a_4)(a_1+a_2+a_3+a_4)}$

Theorem 5. *For an algebra system* $(NQ, \star)$, *let* $V_c = \{a|a \in NQ \wedge neut(a) = c\}$, $V_{c\star d} = \{a \star b|a, b \in NQ \wedge neut(a) = c, neut(b) = d\}$, *we have:*

(1) V_c *is a neutrosophic triplet subgroup of* NQ.

(2) $V_{c\star d}$ *is a neutrosophic triplet subgroup of* NQ.

Theorem 6. *For algebra system* $(NQ, \star)$, *Let* $V_c = \{a|a \in NQ \wedge neut(a) = c\}$, *we have:*

(1) $V_c \cap V_d = \emptyset$ *if* $c \neq d$.

(2) $NQ = \cup_{c\in NS} V_c$. *So* $\cup_{c\in NS} V_c$ *is a partition of* NQ, *where* NS *is a set, which contains all the neutral elements of* $(NQ, \star)$.

4. Two Kinds of Degenerate Systems of Neutrosophic Quadruple Numbers

The neutrosophic quadruple numbers consider (T, I, F) to solve real problems. In this section, we will explore two kinds of degenerate systems about neutrosophic quadruple numbers. The first system is only consider logical true, and the second system is only consider logical true and logical indeterminacy.

4.1. The Neutrosophic Binary Numbers

Definition 8. *A neutrosophic binary number is a number of the form* (a, bT), *where* T *have their usual neutrosophic logic true and* $a, b \in \mathbb{R}$ *or* $\mathbb{C}$. *The set* NB *defined by*

$$NB = \{(a, bT) : a, b \in \mathbb{R} \text{ or } \mathbb{C}\}. \quad (7)$$

is called a neutrosophic set of binary numbers. For a neutrosophic binary number (a, bT), a *is called the known part and* (bT) *is called the unknown part.*

Definition 9. *Let* $a = (a_1, a_2T), b = (b_1, b_2T) \in NB$, *the multiplication operation is defined as following:*

$$a * b = (a_1, a_2T) * (b_1, b_2T) = (a_1b_1, (a_1b_2 + a_2b_1 + a_2b_2)T). \quad (8)$$

We have the following results similar to $(NQ, *)$.

Theorem 7. *For the algebra system* $(NB, *)$, *for every element* $a \in NB$, *there exists the neutral element* $neut(a)$ *and opposite element* $anti(a)$.

For algebra system $(NB, *)$, Table 3 gives all the subset, which has the same neutral element, and the corresponding neutral element and opposite elements.

Table 3. The corresponding neutral element and opposite elements for $(NB, *)$.

The Subset	Neutral Elements	Opposite Element (c_1, c_2T)
$\{(0,0)\}$	$(0,0)$	$c_i \in \mathbb{R}$
$\{(0, a_2T)\|a_2 \neq 0\}$	$(0, T)$	$c_1 + c_2 = \frac{1}{a_2}$
$\{(a_1, -a_1T)\|a_1 \neq 0\}$	$(1, 0)$	$c_1 = \frac{1}{a_1}, c_2 \in \mathbb{R}$
$\{(a_1, a_2T)\|a_1 \neq 0$	$(1, -T)$	$c_1 = \frac{1}{a_1}, c_2 = -\frac{a_2}{a_1(a_1+a_2)}$

Theorem 8. *For algebra system* $(NB, *)$, *let* $V_c = \{a|a \in NB \wedge neut(a) = c\}$, $V_{c*d} = \{a * b|a, b \in NB \wedge neut(a) = c, neut(b) = d\}$, *we have:*

(1) V_c *is a neutrosophic triplet subgroup of* NB.

(2) V_{c*d} *is a neutrosophic triplet subgroup of* NB.

Theorem 9. *For an algebra system $(NB, *)$, let $V_c = \{a | a \in NB \wedge neut(a) = c\}$, we have:*
(1) $V_c \cap V_d = \emptyset$ if $c \neq d$.
(2) $NB = \cup_{c \in NS} V_c$. So $\cup_{c \in NS} V_c$ is a partition of NB, where NS is a set, which contains all the neutral elements of $(NB, *)$.

4.2. The Neutrosophic Triple Numbers

Definition 10. *A neutrosophic triple number is a number of the form (a, bT, cI), where T, I have their usual neutrosophic logic meanings and $a, b, c \in \mathbb{R}$ or $\mathbb{C}$. The set NT defined by*

$$NT = \{(a, bT, cI) : a, b, c \in \mathbb{R} \text{ or } \mathbb{C}\} \tag{9}$$

is called a neutrosophic set of triple numbers. For a neutrosophic triple number (a, bT, cI), a is called the known part and (bT, cI) is called the unknown part.

Definition 11. *Let $a = (a_1, a_2T, a_3I), b = (b_1, b_2T, b_3I) \in NT$, suppose in an pessimistic way, the neutrosophic expert considers the prevalence order $T \prec I$. Then the multiplication operation is defined as following:*

$$\begin{aligned} a * b = \ & (a_1, a_2T, a_3I) * (b_1, b_2T, b_3I) \\ & = (a_1b_1, (a_1b_2 + a_2b_1 + a_2b_2)T, (a_1b_3 + a_2b_3 + a_3b_1 + a_3b_2 + a_3b_3)I. \end{aligned} \tag{10}$$

Suppose in an optimistic way the neutrosophic expert considers the prevalence order $T \succ I$. Then:

$$\begin{aligned} a \star b = \ & (a_1, a_2T, a_3I) * (b_1, b_2T, b_3I) \\ & = (a_1b_1, (a_1b_2 + a_2b_1 + a_2b_2 + a_2b_2 + a_3b_2)T, (a_1b_3 + a_3b_1 + a_3b_3)I). \end{aligned} \tag{11}$$

Theorem 10. *For the algebra system $(NT, *)$, for every element $a \in NT$, there exists the neutral element $neut(a)$ and opposite element $anti(a)$.*

For algebra system $(NT, *)$, Table 4 gives all the subset, which has the same neutral element, and the corresponding neutral element and opposite elements.

Table 4. The corresponding neutral element and opposite elements for $(NT, *)$.

The Subset	Neutral Elements	Opposite Element (c_1, c_2T, c_3I)
$\{(0,0,0)\}$	$(0,0,0)$	$c_i \in \mathbb{R}$
$\{(0,0,a_3I) \| a_3 \neq 0\}$	$(0,0,I)$	$c_1 + c_2 + c_3 = \frac{1}{a_3}$
$\{(0,a_2T,-a_2I) \| a_2 \neq 0, a_2 + a_3 = 0\}$	$(0,T,-I)$	$c_1 + c_2 = \frac{1}{a_2}, c_3 \in \mathbb{R}$
$\{(0,a_2T,a_3I) \| a_2 \neq 0, a_2 + a_3 \neq 0\}$	$(0,T,0)$	$c_1 + c_2 = \frac{1}{a_2}, c_3 = -\frac{a_3}{a_2(a_2+a_3)}$
$\{(a_1,-a_1T,0) \| a_1 \neq 0\}$	$(1,-T,0)\}$	$c_1 = \frac{1}{a_1}, c_2, c_3 \in \mathbb{R}$
$\{(a_1,-a_1T,a_3I) \| a_1 \neq 0, a_3 \neq 0\}$	$(1,-T,I)$	$c_1 = \frac{1}{a_1}, c_2 + c_3 = \frac{1}{a_3} - \frac{1}{a_1}$
$\{(a_1,a_2T,a_3I) \| a_1 \neq 0, a_1 + a_2 \neq 0, a_1 + a_2 + a_3 = 0\}$	$(1,0,-I)$	$c_1 = \frac{1}{a_1}, c_2 = -\frac{a_2}{a_1(a_1+a_2)}, c_3 \in \mathbb{R}$
$\{(a_1,a_2T,a_3I) \| a_1 \neq 0, a_1 + a_2 \neq 0, a_1 + a_2 + a_3 \neq 0\}$	$(1,0,0)$	$c_1 = \frac{1}{a_1}, c_2 = -\frac{a_2}{a_1(a_1+a_2)}$, $c_3 = -\frac{a_3}{(a_1+a_2)(a_1+a_2+a_3)}$

Theorem 11. *For an algebra system $(NT, *)$, let $V_c = \{a|a \in NT \wedge neut(a) = c\}$, $V_{c*d} = \{a * b|a, b \in NT \wedge neut(a) = c, neut(b) = d\}$, we have:*
(1) V_c is a neutrosophic triplet subgroup of NT.
(2) V_{c*d} is a neutrosophic triplet subgroup of NT.

Theorem 12. *For an algebra system $(NT, *)$, let $V_c = \{a|a \in NT \wedge neut(a) = c\}$, we have:*
(1) $V_c \cap V_d = \emptyset$ if $c \neq d$.
(2) $NT = \cup_{c \in NS} V_c$. So $\cup_{c \in NS} V_c$ is a partition of NT, where NS is a set, which contains all the neutral elements of $(NT, *)$.

Theorem 13. *For the algebra system $(NT, \star)$, for every element $a \in NT$, there exists the neutral element $neut(a)$ and opposite element $anti(a)$.*

For an algebra system $(NT, \star)$, Table 5 gives all the subset, which has the same neutral element, and the corresponding neutral element and opposite elements.

Table 5. The corresponding neutral element and opposite elements for $(NT, \star)$.

The Subset	Neutral Elements	Opposite Element (c_1, c_2T, c_3I)
$\{(0,0,0)\}$	$(0,0,0)$	$c_i \in \mathbb{R}$
$\{(0, a_2T, 0)\|a_2 \neq 0\}$	$(0, T, 0)$	$c_1 + c_2 + c_3 = \frac{1}{a_2}$
$\{(0, a_3T, -a_3I)\|a_3 \neq 0, a_2 + a_3 = 0\}$	$(0, -T, I)$	$c_1 + c_3 = \frac{1}{a_3}, c_2 \in \mathbb{R}$
$\{(0, a_2T, a_3I)\|a_3 \neq 0, a_2 + a_3 \neq 0\}$	$(0, 0, I)$	$c_1 + c_3 = \frac{1}{a_3}, c_2 = -\frac{a_2}{a_3(a_2+a_3)}$
$\{(a_1, 0, -a_1I)\|a_1 \neq 0\}$	$(1, 0, -I)\}$	$c_1 = \frac{1}{a_1}, c_2, c_3 \in \mathbb{R}$
$\{(a_1, a_2T, -a_1I)\|a_1 \neq 0, a_2 \neq 0\}$	$(1, T, -I)$	$c_1 = \frac{1}{a_1}, c_2 + c_3 = \frac{1}{a_2} - \frac{1}{a_1}$
$\{(a_1, a_2T, a_3I)\|a_1 \neq 0, a_1 + a_3 \neq 0, a_1 + a_2 + a_3 = 0\}$	$(1, -T, 0)$	$c_1 = \frac{1}{a_1}, c_3 = -\frac{a_3}{a_1(a_1+a_3)}, c_2 \in \mathbb{R}$
$\{(a_1, a_2T, a_3I)\|a_1 \neq 0, a_1 + a_3 \neq 0, a_1 + a_2 + a_3 \neq 0\}$	$(1, 0, 0)$	$c_1 = \frac{1}{a_1}, c_3 = -\frac{a_3}{a_1(a_1+a_3)}$, $c_2 = -\frac{a_2}{(a_1+a_3)(a_1+a_2+a_3)}$

Theorem 14. *For algebra system $(NT, \star)$, let $V_c = \{a|a \in NT \wedge neut(a) = c\}$, $V_{c\star d} = \{a \star b|a, b \in NT \wedge neut(a) = c, neut(b) = d\}$, we have:*
(1) V_c is a neutrosophic triplet subgroup of NT.
(2) $V_{c\star d}$ is a neutrosophic triplet subgroup of NT.

Theorem 15. *For an algebra system $(NT, \star)$, let $V_c = \{a|a \in NT \wedge neut(a) = c\}$, we have:*
(1) $V_c \cap V_d = \emptyset$ if $c \neq d$.
(2) $NT = \cup_{c \in NS} V_c$. So $\cup_{c \in NS} V_c$ is a partition of NT, where NS is a set, which contains all the neutral elements of $(NT, \star)$.

5. Conclusions

In the paper, we prove that $(NQ, *)$(or $NQ, \star$) is a neutrosophic extended triplet group, and provide new examples of a neutrosophic extended triplet group. We also explore the algebra structure properties of neutrosophic quadruple numbers. Moreover, we discuss two kinds of degenerate systems of neutrosophic quadruple numbers. For neutrosophic quadruple numbers, the results in the paper can be extended to general fields. In the following, we will explore the relation of neutrosophic quadruple numbers and other algebra systems [24–26]. Moreover, on the one hand, we will discuss the neutrosophic quadruple numbers based on some particular ring which can form a neutrosophic extended triplet group, on the other hand, we will introduce a new operation $\circ$ in order to guarantee $(NQ, *, \circ)$ is a neutrosophic triplet ring.

Author Contributions: All authors have contributed equally to this paper.

Funding: This research was funded by the National Natural Science Foundation of China (Grant No. 11501435), Instructional Science and Technology Plan Projects of China National Textile and Apparel Council (no. 2016073) and Scientific Research Program Funded by Shaanxi Provincial Education Department (program no. 18JS042).

Acknowledgments: The authors would like to thank the reviewers for their many insightful comments and suggestions.

Conflicts of Interest: The authors declare no conflicts of interest.

References

1. Smarandache, F. *A Unifying Field in Logics. Neutrosophy: Neutrosophic Probability, Set and Logic*; American Research Press: Rehoboth, TX, USA, 1998.
2. Zhang, X.H.; Bo, C.X.; Smarandache, F.; Dai, J.H. New inclusion relation of neutrosophic sets with applications and related lattice structure. *Int. J. Mach. Learn. Cybern.* **2018**, *9*, 1753-1763. [CrossRef]
3. Zhang, X.H.; Bo, C.X.; Smarandache, F.; Park, C. New operations of totally dependent-neutrosophic sets and totally dependent-neutrosophic soft sets. *Symmetry* **2018**, *10*, 187. [CrossRef]
4. Ye, J. Improved correlation coefficients of single valued neutrosophic sets and interval neutrosophic sets for multiple attribute decision making. *J. Intell. Fuzzy Syst.* **2014**, *27*, 2453–2462.
5. Guo, Y.; Sengur, A. NCM: Neutrosophic c-means clustering algorithm. *Pattern Recognit.* **2015**, *48*, 2710–2724. [CrossRef]
6. Smarandache, F.; Ali, M. Neutrosophic triplet group. *Neural Comput. Appl.* **2018**, *29*, 595-601. [CrossRef]
7. Smarandache, F. *Neutrosophic Perspectives: Triplets, Duplets, Multisets, Hybrid Operators, Modal Logic, Hedge Algebras. And Applications*; Pons Publishing House: Brussels, Belgium, 2017.
8. Zhang, X.; Hu, Q.; Smarandache, F.; An, X. On Neutrosophic Triplet Groups: Basic Properties, NT-Subgroups, and Some Notes. *Symmetry* **2018**, *10*, 289. [CrossRef]
9. Bal, M.; Shalla, M.M.; Olgun, N. Neutrosophic Triplet Cosets and Quotient Groups. *Symmetry* **2017**, *10*, 126–139. [CrossRef]
10. Zhang, X.H.; Smarandache, F.; Liang, X.L. Neutrosophic Duplet Semi-Group and Cancellable Neutrosophic Triplet Groups. *Symmetry* **2017**, *9*, 275. [CrossRef]
11. Ma, Y.; Zhang, X.; Yang, X.; Zhou, X. Generalized Neutrosophic Extended Triplet Group. *Symmetry* **2019**, *11*, 327. [CrossRef]
12. Jaiyeola, T.G.; Smarandache, F. Some results on neutrosophic triplet group and their applications. *Symmetry* **2018**, *10*, 202. [CrossRef]
13. Vasantha, W.B.; Kandasamy, I.; Smarandache, F. *Neutrosophic Triplet Groups and Their Applications to Mathematical Modelling*; EuropaNova: Brussels, Belgium, 2017; ISBN 978-1-59973-533-7.
14. Wu, X.Y.; Zhang, X.H. The decomposition theorems of AG-neutrosophic extended triplet loops and strong AG-(l, l)-loops. *Mathematics* **2019**, *7*, 268. [CrossRef]
15. Zhang, X.H.; Wu, X.Y.; Mao, X.Y.; Smarandache, F.; Park, C. On Neutrosophic Extended Triplet Groups (Loops) and Abel-Grassmann's Groupoids (AG-Groupoids). *J. Intell. Fuzzy Syst.* **2019**, in press.
16. Smarandache, F. Hybrid Neutrosophic Triplet Ring in Physical Structures. *Bull. Am. Phys. Soc.* **2017**, *62*, 17.
17. Ali, M.; Smarandache, F.; Khan, M. Study on the development of neutrosophictriplet ring and neutrosophictriplet field. *Mathematics* **2018**, *6*, 46. [CrossRef]
18. Sahin, M.; Abdullah, K. Neutrosophic triplet normed space. *Open Phys.* **2017**, *15*, 697–704. [CrossRef]
19. Şahin, M.; Kargın, A. Neutrosophictriplet *v*-generalized metric space. *Axioms* **2018**, *7*, 67. [CrossRef]
20. Zhang, X.H.; Mao, X.Y.; Wu, Y.T.; Zhai, X.H. Neutrosophic filters in pseudo-BCI algebras. *Int. J. Uncertainty Quant.* **2018**, *8*, 511–526. [CrossRef]
21. Gulistan, M.; Nawaz, S.; Hassan, N. Neutrosophictriplet non-associative semihypergroups with application. *Symmetry* **2018**, *10*, 613. [CrossRef]

22. Smarandache, F. Neutrosophic Quadruple Numbers, Refined Neutrosophic Quadruple Numbers, Absorbance Law, and the Multiplication of Neutrosophic Quadruple Numbers. *Neutrosophic Sets Syst.* **2015**, *10*, 96–98.
23. Akinleye1, S.A.; Smarandache,F.; Agboola, A.A.A. On Neutrosophic Quadruple Algebraic Structures. *Neutrosophic Sets Syst.* **2016**, *12*, 122–126.
24. Zhang, X.H. Fuzzy anti-grouped filters and fuzzy normal filters in pseudo-BCI algebras. *J. Intell. Fuzzy Syst.* **2017**, *33*, 1767–1774. [CrossRef]
25. Zhang, X.H.; Borzooei, R.A. and Jun, Y.B. Q-filters of quantum B-algebras and basic implication algebras. *Symmetry* **2018**, *10*, 573. [CrossRef]
26. Zhang, X.H.; Park, C.; Wu, S.P. Soft set theoretical approach to pseudo-BCI algebras. *J. Intell. Fuzzy Syst.* **2018**, *34*, 559–568. [CrossRef]

symmetry

MDPI

Article

Multi-Attribute Decision Making Based on Probabilistic Neutrosophic Hesitant Fuzzy Choquet Aggregation Operators

Songtao Shao [1,†], Xiaohong Zhang [2,3,*] and Quan Zhao [1,†]

1 College of Information Engineering, Shanghai Maritime University, Shanghai 201306, China; 201740310005@stu.shmtu.edu.cn (S.S.); quanzhao@shmtu.edu.cn (Q.Z.)
2 Department of Mathematics, School of Arts and Sciences, Shaanxi University of Science & Technology, Xi'an 710021, China
3 Department of Mathematics, College of Arts and Sciences, Shanghai Maritime University, Shanghai 201306, China
* Correspondence: zhangxiaohong@sust.edu.cn or zhangxh@shmtu.edu.cn
† Current address: 1550, Haigang Main Road, Shanghai 201306, China.

Received: 21 March 2019; Accepted: 10 April 2019; Published: 3 May 2019

Abstract: Take the third-party logistics providers (3PLs) as an example, according to the characteristics of correlation between attributes in multi-attribute decision-making, two Choquet aggregation operators adoping probabilistic neutrosophic hesitation fuzzy elements (PNHFEs) are proposed to cope with the situations of correlation among criterions. This measure not only provides support for the correlation phenomenon between internal attributes, but also fully concerns the incidental uncertainty of the external space. Our goal is to make it easier for decision makers to cope with this uncertainty, thus we establish the notion of probabilistic neutrosophic hesitant fuzzy Choquet averaging (geometric) (PNHFCOA, PNHFCOG) operator. Based on this foundation, a method for aggregating decision makers' information is proposed, and then the optimal decision scheme is obtained. Finally, an example of selecting optimal 3PL is given to demonstrate the objectivity of the above-mentioned standpoint.

Keywords: probabilistic neutrosophic hesitant fuzzy set (PNHFS); decision-making; Choquet integral

1. Introduction

In the process of enterprise development, business leaders often encounter various multi-criteria decision-making (MADM) situations. In order for a company to survive in today's increasingly competitive real life, decision makers (DMs) must decide on the best solution when encountering MADM issues [1,2]. Therefore, how to effectively make optimal decisions in the MADM problems has become an emergency problem that global enterprises urgently need to solve. Establishing and perfecting research methods to suitable for MADM situations has attracted more and more attention from DMs [3]. The key to MADM is to choose the appropriate decision theory and computer software to aggregate the information of DMs and make the best decision in the decision process according to relevance of the information. In order to choose the optimal solution, different MADM schemes have been established to increase the development and competitiveness of enterprises. Since the most common feature in decision information is the ambiguity of the information. Therefore, some related researches based on fuzzy information have been proposed [4–12].

In reality, there is a correlation among attributes in the process of MADM. In addition, some researchers have begun to pay attention to this issue. For example, Brito et al. [13] proposed a new-type multi-criteria model based on the Choquet integral and epistemic mapping technique for evaluating eco-friendly cities. Krishnan et al. [14] introduced a new λ^0-measure authentication approach that expresses the mutual relation between the attributes. To more effectively highlight the correlation between attributes, Beg et al. [15] introduced the diminishing Chouqet hesitant 2-tuple average (DCH2TA) operator and established a MADM approach.

Compared with reality, the current research method does not consider the fuzzy information with possibility in the MADM problem. Thus, when there is a situation with possibility information, the existing methods will be invalid. To deal with this type of problem, there are two aspects that need to be solved: one is the representation of the PNHF information and the probability information of external environment. The other is the establishment of the MADM model considering the interrelationship among the attributes.

According to the analysis of the common fuzzy conditions in practice, fuzzy sets [16] have been used in many situations. Diversity based on fuzzy information, the fuzzy set theory has been continuously improved. For example, intuitionistic FS [17], hesitant FS [18], and dual hesitant FS [19]. When dealing with fuzzy data, DMs will often encounter the following three kinds of independent fuzzy information: truth fuzzy information, indeterminacy fuzzy information and falsity fuzzy information. The definition of neutrosophic set (NS) was constructed [20] to express the feature. For application to engineering projects, different types of generalized NSs are constructed such as single-valued NS [21], interval NS [22], and neutrosophic hesitant fuzzy (NHF) set [23]. These extended NS theories have been applied to medical diagnosis and other engineering fields [24–28]. Fuzzy set and neutrosophic set are also extended to the field of algebra [29–33]. In order to express three independent hesitant messages, NHFS was proposed and applied to MADM [34–36]. However, as can be seen from these references, those theories can only express information about cognitive uncertainty. Thus, in this article, we use PNHFS [3,37] to express the occasional uncertainty of information and the uncertainty of cognition. Under the MADM environment, due to the different cognitive situations of DMs and their own hesitation, the final evaluation value may not be unique. Depending on the complexity of the external environment, the probability of taking values also affects the evaluation values. Thus, probability plays a key role in interpreting the evaluation value, which avoid the loss of evaluation values, simultaneously. Not only subjective information of NHFS, but also objective probability information of each membership was clearly described.

The MADM problem under attribute correlation is one of the main research questions in this paper. As an important part of fuzzy mathematics, fuzzy integral can help DMs to better deal with MADM problems through modeling methods when attributes are associated with each other. The classic weight information is to satisfy the additivity condition, and the advantage of fuzzy integral is that it is not limited by the additivity condition, that is, the sum of the fuzzy integral may be greater than 1. When the Choquet integral (CI) [38] satisfies the additivity, the Choquet integral is converted into classic weight information. Therefore, the application range of Choquet integral is extensive, and it is more suitable for coping with MADM problems with uncertain information. For example, Khan et al. [39] proposed the (generalized) Pythagorean hesitant fuzzy Choquet averaging (geometric) operators under the MADM environment. Based on the hierarchical and interacting standards, Corrente et al. [40] Choquet integral preference model that can be robust ranking of universities evaluation. Labreuche et al. [41] developed two new Choquet models. Yager [42] used the standard Shapley value as an approximation of Choquet integra. Liu et al. [43] effectively investigate the MADM situations that the interrelationship between attributes, globally. Absolutely, above aggregation operators can not integrate PNHF information. Thus, we construct a new integration method by CI. Thus, we extend these operators to the PNHFSs and propose PNHF Choquet averaging (geometric) operators and establish a process to handle MADM situations.

Based on the above analysis: In Section 2, fundamental concepts are reviewed. In Section 3, the operators are included, the comparison approaches are described, and some basic properties are studied. In Section 4, an approach to MADM based on the PNHFCOA (PNHFCOG) operator is proposed. In Section 5, an illustrative situation is given to confirm the proposed method. In Section 6, our results are analyzed with the results of other methods. Finally, conclusions and future work are summarized.

2. Preliminaries

Some basic definitions can be referred to [3,23,37,38,44].

Definition 1. *A NHF set (NHFS) in a reference domain X set is depicted by:*

$$N = \{\langle x, \tilde{t}(x), \tilde{i}(x), \tilde{f}(x)\rangle | x \in X\}$$

in which $\tilde{t}(x)$ indicates that a set includes some truth-membership hesitant degrees of x, $\tilde{i}(x)$ indicates that a set includes some indeterminacy-membership hesitant degrees, and $\tilde{f}(x)$ indicates that a set includes some falsity-membership hesitant degrees. The following conditions are held: $\delta \in \tilde{t}(x), \gamma \in \tilde{i}(x), \eta \in \tilde{f}(x)$, $\delta, \gamma, \eta \in [0,1]$, $max\{\delta\} + max\{\gamma\} + max\{\eta\} \in [0,3]$.

Definition 2. *A PNHFS in a reference domain X is depicted by:*

$$N = \{\langle x, T(x)|P^T(x), I(x)|P^I(x), F(x)|P^F(x)\rangle | x \in X\}. \quad (1)$$

$T(x)|P^T(x)$, $I(x)|P^I(x)$ and $F(x)|P^F(x)$ describes three components of N, $T(x)$, $I(x)$ and $F(x)$ is depicted the three types hesitant degrees of x. $P^T(x)$, $P^I(x)$ and $P^F(x)$ describe the corresponding probability of $T(x)$, $I(x)$ and $F(x)$, The conditions that need to be held:

$$\alpha \in T(x), \beta \in I(x), \gamma \in F(x), \alpha, \beta, \gamma \in [0,1], max\{\alpha\} + max\{\beta\} + max\{\gamma\} \in [0,3];$$

$$P_a^T \in P^T, P_b^I \in P^I, P_c^F \in P^F. P_a^T, P_b^I, P_c^F \in [0,1]; \sum_{a=1}^{\#\tilde{t}} P_a^T \leq 1, \sum_{b=1}^{\#\tilde{i}} P_b^I \leq 1, \sum_{c=1}^{\#\tilde{f}} P_c^F \leq 1.$$

$\#\tilde{t}$, $\#\tilde{i}$ and $\#\tilde{f}$ describe the cardinal number of $T(x)|P^T(x)$, $I(x)|P^I(x)$, $F(x)|P^F(x)$, respectively.

Generally, $N = \{T|P^T, I|P^I, F|P^F\}$ is described a an PNHF number (PNHFE) of $N = \langle T(x)|P^T(x), I(x)|P^I(x), F(x)|P^F(x)\rangle$.

Definition 3. *A normalized PNHFE N satisfies the following conditions:*

$$\tilde{N} = \langle T(x)|\tilde{P}^T(x), I(x)|\tilde{P}^I(x), F(x)|\tilde{P}^F(x)\rangle, \quad (2)$$

where $\tilde{P}_a^T = \frac{P_a^T}{\sum P_a^T}$, $\tilde{P}_b^I = \frac{P_b^I}{\sum P_b^I}$, $\tilde{P}_c^F = \frac{P_c^F}{\sum P_c^F}$.

Definition 4. *Supposing that $N_1 = \{T_1|P^{T_1}, I_1|P^{I_1}, F_1|P^{F_1}\}$, $N_2 = \{T_2|P^{T_2}, I_2|P^{I_2}, F_2|P^{F_2}\}$ are two PNHFEs, some basic algorithms are defined by:*

(1) $(N_1)^c = \bigcup_{\alpha_1 \in T_1, \beta_1 \in I_1, \gamma_1 \in F_1} \{\gamma_1|P_1^{F_1}, 1 - \beta_1|P_1^{I_1}, \alpha_1|P_1^{T_1}\}$,

(2) $(N_1)^\lambda = \bigcup_{\alpha_1 \in T_1, \beta_1 \in I_1, \gamma_1 \in F_1} \{\{(\alpha_1)^\lambda|P_1^{T_1}\}, \{1 - (1 - \beta_1)^\lambda|P_1^{I_1}\}, \{1 - (1 - \gamma_1)^\lambda|P_1^{F_1}\}\}$,

(3) $\lambda(N_1) = \bigcup_{\alpha_1 \in T_1, \beta_1 \in I_1, \gamma_1 \in F_1} \{\{1 - (1 - \lambda_1)^\lambda|P_1^{T_1}\}, \{(\beta_1)^\lambda|P_1^{I_1}\}, \{(\gamma_1)^\lambda|P_1^{F_1}\}\}$,

(4) $N_1 \oplus N_2 = \bigcup_{\substack{\alpha_1 \in T_1, \beta_1 \in I_1, \gamma_1 \in F_1, \\ \eta_2 \in T_2, \pi_2 \in I_2, \mu_2 \in F_2}} \{\{\alpha_1 + \eta_2 - \alpha_2\eta_2|P_1^{T_1}P_2^{T_2}\}, \{\beta_1\pi_2|P_1^{I_1}P_2^{I_2}\}, \{\gamma_1\mu_2|P_1^{F_1}P_2^{F_2}\}\}$,

(5) $N_1 \otimes N_2 = \bigcup_{\substack{\alpha_1 \in T_1, \beta_1 \in I_1, \gamma_1 \in F_1, \\ \eta_2 \in T_2, \pi_2 \in I_2, \mu_2 \in F_2}} \{\{\alpha_1\eta_2|P_1^{T_1}P_2^{T_2}\}, \{\beta_1 + \pi_2 - \beta_1\pi_2|P_1^{I_1}P_2^{I_2}\}, \{\gamma_1 + \mu_2 - \gamma_1\mu_2|P_1^{F_1}P_2^{F_2}\}\}$,

where $P_1^{T_1}$; $P_1^{I_1}$ and $P_1^{F_1}$ are hesitant probabilities of $\alpha_1 \in T_1$, $\beta_1 \in I_1$ and $\gamma_1 \in F_1$, respectively. $P_2^{T_2}$; $P_2^{I_2}$ and $P_1^{F_2}$ are corresponding hesitant probabilities of $\eta_2 \in T_2$, $\pi_2 \in I_2$ and $\mu_2 \in F_2$.

Definition 5. *$P(Y)$ depicts the power set of a domain Y. Fuzzy measure μ: $P(Y) \to [0,1]$ satisfies conditions:*

(1) $\mu(\emptyset) = 0$, $\mu(Y) = 1$;
(2) $A \subseteq B$, *then* $\mu(A) \leq \mu(B), \forall A, B \subseteq P(Y)$;

Fuzzy measure μ satisfies property, $\forall Y_1, Y_2 \in P(Y)$, $A \cap B = \emptyset$

$$\mu(Y_1 \cup Y_2) = \mu(Y_1) + \mu(Y_2) + \lambda\mu(Y_1)\mu(Y_2), \lambda \in (-1, \infty).$$

Then, μ is described a λ-fuzzy measure.

Theorem 1. *A λ-fuzzy measure μ in a discourse $Y = \{y_1, y_2, \cdots, y_n\}$ satisfies the following formula:*

$$\mu(X) = \begin{cases} \frac{1}{\lambda}(\prod_{i=1}^{n}[1 + \lambda\mu(y_i)] - 1) & if \lambda \neq 0, \\ \sum_{i=1}^{n} \mu(y_i) & if \lambda = 0. \end{cases} \tag{3}$$

Since $\mu(Y) = 1$, parameter λ can be determined by

$$\lambda = \prod_{i=1}^{n}(1 + \lambda\mu(x_i)) - 1.$$

Definition 6. *f is a real function on $Y = \{y_1, y_2, \cdots, y_n\}$. The Choquet integral about fuzzy measure μ is depicted by:*

$$\int f d\mu = \sum_{a=1}^{n} f(y_{(a)})[\mu(A_{(a)}) - \mu(A_{(a+1)})]$$

in which $\{x_{\pi(1)}, x_{\pi(2)}, \cdots, x_{\pi(n)}\}$ is a new rank of Y, $f(y_{\pi(1)}) \leq f(y_{\pi(2)}) \leq \cdots \leq f(y_{\pi(n)})$, $A_{(a)} = \{y_{\pi(a)}, y_{\pi(a+1)}, \cdots, x_{\pi(n)}\}$ and $A_{(n+1)} = 0$.

3. PNHFSs and Aggregation Operators

The PNHFCOA and PNHFCOG operators are proposed in this section. Some basic properties are verified.

3.1. The Comparison Method of PNHFEs

When we describe decision information with PNHFS theory, a comparison method of PNHFEs is necessary. Thus, a approch of ranking PNHFEs is established.

Definition 7. *Supposing that $N = \{T|P^T, I|P^I, F|P^F\}$ is an PNHFE, then the score function of the PNHFE is expressed by the following formula:*

$$S(N) = \sum_{a=1}^{\#T} \alpha_a P_a^T + \sum_{b=1}^{\#I} (1 - \beta_b) P_b^I - \sum_{c=1}^{\#F} \gamma_c P_c^F. \tag{4}$$

Definition 8. *Supposing that $N = \{T|P^T, I|P^I, F|P^F\}$ is an PNHFE, then we can find the deviation function $D(N)$:*

$$D(N) = \sum_{a=1}^{\#T} (\alpha_a - S(N))^2 \cdot P_a^T + \sum_{b=1}^{\#I} (1 - \beta_b - S(N))^2 \cdot P_b^I + \sum_{c=1}^{\#F} (\gamma_c - S(N))^2 \cdot P_c^F. \tag{5}$$

The distance from the score valued in the PNHFE N is described by the deviation function. Thus, the deviation value is called a consistency indicator of the PNHFE N. The higher value of $D(N)$, the lower consistency of N. Based on Definitions 7 and 8, a method for ranking two PNHFEs is developed.

Definition 9. *For PNHFEs N_1 and N_2, the sort of N_1 and N_2 is determined by:*

(1) *If $S(N_1) > S(N_2)$, it indicates that PNHFE N_1 is superior to N_2;*
(2) *If $S(N_1) = S(N_2), D(N_1) > D(N_2)$, it indicates that PNHFE N_1 is inferior to N_2;*
(3) *If $S(N_1) = S(N_2), D(N_1) = D(N_2)$, it indicates that PNHFE N_1 is equal to N_2.*

3.2. The PNHFCOA Operator and PNHFCOG Operator

This section constructed the PNHFCOA operator and PNHFCOG operator under the probabilistic neutrosophic environment, and some basic properties are investigated. In this section, μ describes a fuzzy measure on a domain X, $k = 1, 2, \cdots, n$.

Definition 10. *Suppose that $N_k = \{T_k|P^{T_k}, I_k|P^{I_k}, F_k|P^{F_k}\}$ describes an PNHFE in a reference set X. Then, the PNHFCOA operator is expressed by the following formula:*

$$\begin{aligned}&PNHFCOA(N_1, N_2, \cdots, N_n) = \oplus_{\pi(k)=1}^{n}\mu_{\pi(k)}N_{\pi(k)}\\&= \bigcup_{\alpha_{\pi(k)}\in T_{\pi(k)},\beta_{\pi(k)}\in I_{\pi(k)},\gamma_{\pi(k)}\in F_{\pi(k)}} \{\{1-\prod_{\pi(k)=1}^{n}(1-\alpha_{\pi(k)})^{\mu_{\pi(k)}}|\prod_{\pi(k)=1}^{n}P_{\pi(k)}^{T_{\pi(k)}}\},\\&\{\prod_{\pi(k)=1}^{n}(\beta_{\pi(k)})^{\mu_{\pi(k)}}|\prod_{\pi(k)=1}^{n}P_{\pi(k)}^{I_{\pi(k)}}\},\{\prod_{\pi(k)=1}^{n}(\gamma_{\pi(k)})^{\mu_{\pi(k)}}|\prod_{\pi(k)=1}^{n}P_{\pi(k)}^{F_{\pi(k)}}\}\},\end{aligned} \tag{6}$$

where $P_{\pi(k)}^{T_{\pi(k)}}$, $P_{\pi(k)}^{I_{\pi(k)}}$ and $P_{\pi(k)}^{F_{\pi(k)}}$ are corresponding probability data of $\alpha_{\pi(k)}$, $\beta_{\pi(k)}$ and $\gamma_{\pi(k)}$. $\mu_{\pi(k)} = \mu(F_{\pi(k)} - F_{\pi(i+1)})$, $F_{\pi(k)} = \{x_{\pi(1)}, x_{\pi(2)}, \cdots, x_{\pi(k)}\}$ and $F_{\pi(0)} = 0$. $\{n_{\pi(k)}\}$ is a sequence such that $n_{\pi(1)} \leq n_{\pi(2)} \leq \cdots \geq n_{\pi(m)}$.

Theorem 2. *Supposing that N_k describes an PNHFE, PNHFCOA operator $PNHFCOA(N_1, N_2, \cdots, N_n)$ is still an PNHFE.*

Proof. The mathematical induction can be utilized.

(1) When $n = 1$, we have the following equation by Definition 10:

$$PNHFCOA(N_1) = \mu_{\pi(1)} \oplus N_{\pi(1)} = N_1. \tag{7}$$

Obviously, $PNHFCOA\{N_1\}$ is an PNHFE.

(2) When $n = 2$, we have

$$\begin{aligned}&PNHFCOA(N_1, N_2) = (\mu_{\pi(1)}N_{\pi(1)}) \oplus (\mu_{\pi(2)}N_{\pi(2)})\\&= \bigcup_{\alpha_{\pi(1)}\in T_{\pi(1)},\beta_{\pi(1)}\in I_{\pi(1)},\gamma_{\pi(1)}\in F_{\pi(1)}} \{1-(1-\alpha_{\pi(1)})^{\mu_{\pi(1)}}|P_{\pi(1)}^{T_{\pi(1)}}, \beta_{\pi(1)}^{\mu_{\pi(1)}}|P_{\pi(1)}^{I_{\pi(1)}}, \gamma_{\pi(1)}^{\mu_{\pi(1)}}|P_{\pi(1)}^{F_{\pi(1)}}\}\\&\oplus \bigcup_{\alpha_{\pi(2)}\in T_{\pi(2)},\beta_{\pi(2)}\in I_{\pi(2)},\gamma_{\pi(2)}\in F_{\pi(2)}} \{1-(1-\alpha_{\pi(2)})^{\mu_{\pi(2)}}|P_{\pi(2)}^{T_{\pi(2)}}, \beta_{\pi(2)}^{\mu_{\pi(2)}}|P_{\pi(2)}^{I_{\pi(2)}}, \gamma_{\pi(2)}^{\mu_{\pi(2)}}|P_{\pi(2)}^{F_{\pi(2)}}\}\\&= \bigcup_{\alpha_{\pi(k)}\in T_{\pi(k)},\beta_{\pi(k)}\in I_{\pi(k)},\gamma_{\pi(k)}\in F_{\pi(k)}} \{1-\prod_{\pi(k)=1}^{2}(1-\alpha_{\pi(k)})^{\mu_{\pi(k)}}|\prod_{\pi(k)=1}^{2}P_{\pi(1)}^{T_{\pi(k)}},\\&\prod_{\pi(k)=1}^{2}\beta_{\pi(k)}^{\mu_{\pi(k)}}|\prod_{\pi(k)=1}^{2}P_{\pi(k)}^{I_{\pi(k)}}, \prod_{\pi(k)=1}^{2}\gamma_{\pi(k)}^{\mu_{\pi(k)}}|\prod_{\pi(k)=1}^{2}P_{\pi(k)}^{F_{\pi(k)}}\}.\end{aligned}$$

Thus, we know $PNHFCOA\{N_1, N_2\}$ is an PNHFE.

(3) When $n = k$, Equation (9) is true, and we have

$$PNHFCOA(N_1, N_2, \cdots, N_k) = \oplus_{\pi(k)=1}^{k} \mu_{\pi(k)} N_{\pi(k)}$$
$$= \bigcup_{\alpha_{\pi(k)} \in T_{\pi(k)}, \beta_{\pi(k)} \in I_{\pi(k)}, \gamma_{\pi(k)} \in F_{\pi(k)}} \{\{1 - \prod_{\pi(k)=1}^{k} (1 - \alpha_{\pi(k)})^{\mu_{\pi(k)}} | \prod_{\pi(k)=1}^{k} P_{\pi(k)}^{T_{\pi(k)}}\},$$
$$\{\prod_{\pi(k)=1}^{k} (\beta_{\pi(k)})^{\mu_{\pi(k)}} | \prod_{\pi(k)=1}^{k} P_{\pi(k)}^{I_{\pi(k)}}\}, \{\prod_{\pi(k)=1}^{k} (\gamma_{\pi(k)})^{\mu_{\pi(k)}} | \prod_{\pi(k)=1}^{k} P_{\pi(k)}^{F_{\pi(k)}}\}\}.$$

Thus, the next formula is obtained, $n = k + 1$,

$$PNHFCOA(N_1, N_2, \cdots, N_k, N_{k+1}) = (\oplus_{\pi(k)=1}^{k} \mu_{\pi(k)} N_{\pi(k)}) \oplus (\mu_{\pi(k+1)} N_{\pi(k+1)})$$
$$= \bigcup_{\alpha_{\pi(k)} \in T_{\pi(k)}, \beta_{\pi(k)} \in I_{\pi(k)}, \gamma_{\pi(k)} \in F_{\pi(k)}} \{\{1 - \prod_{\pi(k)=1}^{k} (1 - \alpha_{\pi(k)})^{\mu_{\pi(k)}} | \prod_{\pi(k)=1}^{k} P_{\pi(k)}^{T_{\pi(k)}}\},$$
$$\{\prod_{\pi(k)=1}^{k} (\beta_{\pi(k)})^{\mu_{\pi(k)}} | \prod_{\pi(k)=1}^{k} P_{\pi(k)}^{I_{\pi(k)}}\}, \{\prod_{\pi(k)=1}^{k} (\gamma_{\pi(k)})^{\mu_{\pi(k)}} | \prod_{\pi(k)=1}^{k} P_{\pi(k)}^{F_{\pi(k)}}\}\}$$
$$\oplus \bigcup_{\alpha_{\pi(k+1)} \in T_{\pi(k+1)}, \beta_{\pi(k+1)} \in I_{\pi(k+1)}, \gamma_{\pi(k+1)} \in F_{\pi(k+1)}} \{1 - (1 - \alpha_{\pi(k+1)})^{\mu_{\pi(k+1)}} | P_{\pi(k+1)}^{T_{\pi(k+1)}},$$
$$\beta_{\pi(k+1)}^{\mu_{\pi(k+1)}} | P_{\pi(k+1)}^{I_{\pi(k+1)}}, \gamma_{\pi(k+1)}^{\mu_{\pi(k+1)}} | P_{\pi(k+1)}^{F_{\pi(k+1)}}\}$$
$$= \bigcup_{\alpha_{\pi(k)} \in T_{\pi(k)}, \beta_{\pi(k)} \in I_{\pi(k)}, \gamma_{\pi(k)} \in F_{\pi(k)}} \{\{1 - \prod_{\pi(k)=1}^{k+1} (1 - \alpha_{\pi(k)})^{\mu_{\pi(k)}} | \prod_{\pi(k)=1}^{k+1} P_{\pi(k)}^{T_{\pi(k)}}\},$$
$$\{\prod_{\pi(k)=1}^{k+1} (\beta_{\pi(k)})^{\mu_{\pi(k)}} | \prod_{\pi(k)=1}^{k+1} P_{\pi(k)}^{I_{\pi(k)}}\}, \{\prod_{\pi(k)=1}^{k+1} (\gamma_{\pi(k)})^{\mu_{\pi(k)}} | \prod_{\pi(k)=1}^{k+1} P_{\pi(k)}^{F_{\pi(k)}}\}\}.$$

Thus, for any n, the conclusion is right.

□

Next, when the fuzzy measure satisfies different conditions, different types of PNHFCOA can be obtained.

(1) Assume $\mu(F) = 1$, then

$$PNHFCOA(N_1, N_2, \cdots, N_n) = max\{N_1, N_2, \cdots, N_n\}.$$

(2) Assume $\mu(F) = 0$, then

$$PNHFCOA(N_1, N_2, \cdots, N_n) = min\{N_1, N_2, \cdots, N_n\}.$$

(3) Assume the condition $\mu(x_{\pi(k)}) = \mu(F_{\pi(k)} - F_{\pi(k-1)})$ is independent, the PNHFCOA operator is described an PNHFWA operator,

$$PNHFWA(N_1, N_2, \cdots, N_n) = \oplus_{k=1}^{n} \mu(x_k) N_k$$
$$= \bigcup_{\alpha_k \in T_k, \beta_k \in I_k, \gamma_k \in F_k} \{\{1 - \prod_{k=1}^{n} (1 - \alpha_k)^{\mu(x_k)} | \prod_{k=1}^{n} P_k^{T_k}\}, \{\prod_{k=1}^{n} (\beta_k)^{\mu(x_k)} | \prod_{k=1}^{n} P_k^{I_k}\}, \{\prod_{k=1}^{n} (\gamma_k)^{\mu(x_k)} | \prod_{k=1}^{n} P_k^{F_k}\}\}.$$

(4) Assume the condition $\mu(x_{\pi(k)}) = \frac{1}{n}$, the PNHFCOA operator and PNHFWA operator reduce to the PNHFA operator,

$$PNHFWA(N_1, N_2, \cdots, N_n) = \oplus_{k=1}^{n}\mu(x_k)N_k$$
$$= \bigcup_{\alpha_k\in T_k,\beta_k\in I_k,\gamma_k\in F_k} \{\{1-\prod_{k=1}^{n}(1-\alpha_k)^{\frac{1}{n}}|\prod_{k=1}^{n}P_k^{T_k}\},\{\prod_{k=1}^{n}(\beta_k)^{\frac{1}{n}}|\prod_{k=1}^{n}P_k^{I_k}\},\{\prod_{k=1}^{n}(\gamma_k)^{\frac{1}{n}}|\prod_{k=1}^{n}P_k^{F_k}\}\}.$$

Theorem 3. *(Monotonicity) Suppose* $N_k = \{\{\alpha_k|P_k^{T_k}\},\{\beta_k|P_k^{I_k}\},\{\gamma_k|P_k^{F_k}\}\}$ *and* $\tilde{N}_k = \{\{\tilde{\alpha}_k|P_k^{\tilde{T}_k}\},\{\tilde{\beta}_k|P_k^{\tilde{I}_k}\},\{\tilde{\gamma}_k|P_k^{\tilde{F}_k}\}\}$ *indicate two PNHFEs. The factor* $\pi(k)$ *satisfies* $N_{\pi(1)} \geq N_{\pi(2)} \geq \cdots \geq N_{\pi(n)}$ *and* $\tilde{N}_{\pi(1)} \geq \tilde{N}_{\pi(2)} \geq \cdots \geq \tilde{N}_{\pi(n)}$*. For any* $N_{\pi(k)}, \tilde{N}_{\pi(k)}$*, there are* $\alpha_{\pi(k)} \leq \tilde{\alpha}_{\pi(k)}$*,* $\beta_{\pi(k)} \geq \tilde{\beta}_{\pi(k)}$*,* $\gamma_{\pi(k)} \geq \tilde{\gamma}_{\pi(k)}$ *and* $P_{\pi(k)}^{T_{\pi(k)}} = P_{\pi(k)}^{\tilde{T}_{\pi(k)}}, P_{\pi(k)}^{I_{\pi(k)}} = P_{\pi(k)}^{\tilde{I}_{\pi(k)}}, P_{\pi(k)}^{F_{\pi(k)}} = P_{\pi(k)}^{\tilde{F}_{\pi(k)}}$*. Then,*

$$PNHFCOA\{N_1, N_2, \cdots, N_n\} \leq PNHFCOA\{\tilde{N}_1, \tilde{N}_2, \cdots, \tilde{N}_n\}.$$

Proof. By Definition 7, $N_{\pi(k)} \leq \tilde{N}_{\pi(k)}$. By Definition 10, the following inequality is obtained:

$$(1-\prod(1-\alpha_{\pi(k)})^{\mu_{\pi(k)}})\prod P_{\pi(k)}^{T_{\pi(k)}} + (1-\prod(\beta_{\pi(k)})^{\mu_{\pi(k)}}\prod P_{\pi(k)}^{I_{\pi(k)}}) - \prod(\gamma_{\pi(k)})^{\mu_{\pi(k)}}\prod P_j^{F_{\pi(k)}}\} \leq$$
$$(1-\prod(1-\tilde{\alpha}_{\pi(k)})^{\mu_{\pi(k)}})\prod P_{\pi(k)}^{T_{\pi(k)}} + (1-\prod(\tilde{\beta}_{\pi(k)})^{\mu_{\pi(k)}}\prod P_{\pi(k)}^{I_{\pi(k)}}) - \prod(\tilde{\gamma}_{\pi(k)})^{\mu_{\pi(k)}}\prod P_j^{F_{\pi(k)}}\}.$$

Then, by Definitions 7 and 9, the result is proved

$$PNHFCOA(N_1, N_2, \cdots, N_n) \leq PNHFCOA(\tilde{N}_1, \tilde{N}_2, \cdots, \tilde{N}_n).$$

□

Theorem 4. *(Boundedness) Suppose* $N_k = \{\{\alpha_k|P_k^{T_k}\},\{\beta_k|P^{I_k}\},\{\gamma_k|P_k^{F_k}\}\}$ *indicate an PNHFE,*

$$N^- = \{\{min\{\alpha_k\}|min\{P_k^{T_k}\}\},\{max\{\beta_k\}|max\{P_k^{I_k}\}\},\{max\{\gamma_k\}|max\{P_k^{F_k}\}\}\},$$
$$N^+ = \{\{max\{\alpha_k\}|max\{P_k^{T_k}\}\},\{min\{\beta_k\}|min\{P_k^{I_k}\}\},\{min\{\gamma_k\}|min\{P_k^{F_k}\}\}\}.$$

Then,

$$PNHFCOA(N^-) \leq PNHFCOA(N_1, N_2, \cdots, N_n) \leq PNHFCOA(N^+).$$

Proof. $\forall N_k$, we know:

$$min\{\alpha_k\} \leq \alpha_{\pi(k)} \leq max\{\alpha_k\},\ min\{\beta_k\} \leq \beta_{\pi(k)} \leq max\{\beta_k\},\ min\{\gamma_k\} \leq \gamma_{\pi(k)} \leq max\{\gamma_k\};$$
$$min\{P_k^{T_k}\} \leq P_{\pi(k)}^{T_{\pi(k)}} \leq max\{P_k^{T_k}\},\ min\{P_k^{I_k}\} \leq P_{\pi(k)}^{I_{\pi(k)}} \leq max\{P_k^{I_k}\},\ min\{P_k^{F_k}\} \leq P_{\pi(k)}^{F_{\pi(k)}} \leq max\{P_k^{F_k}\}.$$

Thus,

$$\begin{aligned}
1-\prod(1-\alpha_{\pi(k)})^{(\mu(F_{\pi(k)})-\mu(F_{\pi(k-1)}))} &\geq 1-\prod(1-min\{\alpha_i\})^{(\mu(F_{\pi(k)})-\mu(F_{\pi(k-1)}))} \\
&= 1-(1-min\{\alpha_k\})^{\sum(\mu(F_{\pi(k)})-\mu(F_{\pi(k-1)}))} \\
&= min\{\alpha_k\} = min\{\alpha_{\pi(k)}\}, \\
\prod(\beta_{\pi_{(k)}})^{(\mu(F_{\pi(k)})-\mu(F_{\pi(k-1)}))} &\leq \prod(max\{\beta_k\})^{(\mu(F_{\pi(k)})-\mu(F_{\pi(k-1)}))} \\
&= (max\{\beta_k\})^{\sum(\mu(F_{\pi(k)})-\mu(F_{\pi(k-1)}))} \\
&= max\{\beta_k\} = max\{\beta_{\pi(k)}\},
\end{aligned}$$

$$\prod(\gamma_{\pi(k)})^{(\mu(F_{\pi(k)})-\mu(F_{\pi(k-1)}))} \leq \prod(max\{\gamma_k\})^{(\mu(F_{\pi(k)})-\mu(F_{\pi(k-1)}))}$$
$$= (max\{\gamma_k\})^{\sum(\mu(F_{\pi(k)})-\mu(F_{\pi(k-1)}))}$$
$$= max\{\gamma_k\} = max\{\gamma_{\pi(k)}\}.$$

For the probabilities, it is easy to get

$$\prod min\{P_k^{T_k}\} = min\{P_{\pi(k)}^{T_{\pi(k)}}\} \leq \prod P_{\pi(k)}^{T_{\pi(k)}},$$
$$\prod P_{\pi(k)}^{I_{\pi(k)}} \leq \prod max\{P_{\pi(k)}^{I_{\pi(k)}}\} = \prod max\{P_i^{I_i}\},$$
$$\prod P_{\pi(k)}^{F_{\pi(k)}} \leq \prod max\{P_{\pi(k)}^{F_{\pi(k)}}\} = \prod max\{P_i^{F_i}\}.$$

Therefore, we have:

$$PNHFCOA(N^-) = \bigcup\{\{min\{\alpha_{\pi(k)}\}|\prod min\{P_k^{T_k}\}\}, \{max\{\beta_{\pi(k)}\}|\prod max\{P_{\pi(k)}^{I_{\pi(k)}}\}\}, \{max\{\gamma_{\pi(k)}\}|\prod max\{P_{\pi(k)}^{F_{\pi(k)}}\}\}\}.$$

By Definitions 7 and 9,

$$PNHFCOA(N^-) \leq PNHFCOA(N_1, N_2, \cdots, N_n).$$

Similarly, we can get

$$PNHFCOA(N_1, N_2, \cdots, N_n) \leq PNHFCOA(N^+).$$

□

Theorem 5. *(Idempotency) Supposing $N_i = \{\{\alpha|P_1\}, \{\beta|P_2\}, \{\gamma]|P_3\}\}$ ($i = 1, 2, \cdots, n$) is a normalized PNHFE, μ is a fuzzy measure on X; then,*

$$PNHFCOA(N_1, N_2, \cdots, N_X) = \{\{\alpha|P_1\}, \{\beta|P_2\}, \{\gamma|P_3\}\}. \quad (8)$$

Proof. When $P_1, P_2, P_3 \in \{1\}$, we have the following equation:

$$\prod P_j = P_j \ \ (j = 1, 2, 3).$$

Based on Definition 10, it is expressed by the following formula:

$$PNHFCOA(N_1, N_2, \cdots, N_n)$$
$$= \{\{1 - \prod(1-\alpha)^{\mu_{\pi(k)}}|\prod P_1\}, \{\prod(\beta)^{\mu_{\pi(k)}}|\prod P_2\}, \{\prod(\gamma)^{\mu_{\pi(k)}}|\prod P_3\}\}$$
$$= \{\{1 - (1-\alpha)^{\Sigma(\mu(F_{\pi(k)})-\mu(F_{\pi(k)}))}|P_1\}, \{(\beta)^{\Sigma(\mu(F_{\pi(k)})-\mu(F_{\pi(k)}))}|P_2\}, \{(\gamma)^{\Sigma(\mu(F_{\pi(k)})-\mu(F_{\pi(k)}))}|P_3\}\}$$
$$= \{\{\alpha|P_1\}, \{\beta|P_2\}, \{\gamma|P_3\}\}.$$

□

Theorem 6. *(Commutativity) Suppose $A = \{N_1, N_2, \cdots, N_n\}$ and $B = \{N_{\lambda(1)}, N_{\lambda(2)}, \cdots, N_{\lambda(n)}\}$ are two finite sets. If the position of the element in $\{N_{\pi(1)}, N_{\pi(2)}, \cdots, N_{\pi(n)}\}$ is changed arbitrarily to get $\{N_1, N_2, \cdots, N_n\}$. Then,*

$$PNHFCOA(N_1, N_2, \cdots, N_n) = PNHFCOA\{N_{\lambda(1)}, N_{\lambda(2)}, \cdots, N_{\lambda(n)}\}.$$

Proof. Based on Definition 10, the result is easy to get. □

Definition 11. *Suppose $N_k = \{T_k|P^{T_k}, I_k|P^{I_k}, F_k|P^{F_k}\}$ indicates an PNHFE. Then, the PNHFCOG operator is described by the following formula:*

$$\begin{aligned} &PNHFCOG(N_1, N_2, \cdots, N_n) = \otimes_{\pi(k)=1}^{n} \mu_{\pi(k)} N_{\pi(k)} \\ &= \bigcup_{\alpha_{\pi(k)} \in T_{\pi(k)}, \beta_{\pi(k)} \in I_{\pi(k)}, \gamma_{\pi(k)} \in F_{\pi(k)}} \{\{\prod_{\pi(k)=1}^{n} (\alpha_{\pi(k)})^{\mu_{\pi(k)}} | \prod_{\pi(k)=1}^{n} P_{\pi(k)}^{T_{\pi(k)}}\}, \\ &\{1 - \prod_{\pi(k)=1}^{n} (1 - \beta_{\pi(k)})^{\mu_{\pi(k)}} | \prod_{\pi(k)=1}^{n} P_{\pi(k)}^{I_{\pi(k)}}\}, \{1 - \prod_{\pi(k)=1}^{n} (1 - \gamma_{\pi(k)})^{\mu_{\pi(k)}} | \prod_{\pi(k)=1}^{n} P_{\pi(k)}^{F_{\pi(k)}}\}\}, \end{aligned} \tag{9}$$

where $P_{\pi(k)}^{T_{\pi(k)}}$, $P_{\pi(k)}^{I_{\pi(k)}}$ and $P_{\pi(k)}^{F_{\pi(k)}}$ are the corresponding probabilities of $\alpha_{\pi(k)}$, $\beta_{\pi(k)}$ and $\gamma_{\pi(k)}$. $\mu_{\pi(k)} = \mu(F_{\pi(k)} - F_{\pi(k-1)})$, $F_{\pi(k)} = \{x_{\pi(1)}, x_{\pi(2)}, \cdots, x_{\pi(k)}\}$ and $F_{\pi(0)} = 0$. The factor $n_{\pi(k)}$ hold $n_{\pi(1)} \geq n_{\pi(2)} \geq \cdots \geq n_{\pi(n)}$.

Theorem 7. *Suppose that N_k indicates an PNHFE, $PNHFCOG(N_1, N_2, \cdots, N_n)$ is still an PNHFE.*

Similarly, the fuzzy measure satisfies different conditions, different types of PNHFCOA can be obtained.

(1) Assume $\mu(F) = 1$, then

$$PNHFCOG(N_1, N_2, \cdots, N_n) = max\{N_1, N_2, \cdots, N_n\}.$$

(2) Assume $\mu(F) = 0$, then

$$PNHFCOG(N_1, N_2, \cdots, N_n) = min\{N_1, N_2, \cdots, N_n\}.$$

(3) Assume the prerequisite $\mu(x_{\pi(k)}) = \mu(F_{\pi(k)} - F_{\pi}(i-1))$ is independent, the PNHFCOG operator indicates an PNHFWG operator:

$$\begin{aligned} &PNHFWG(N_1, N_2, \cdots, N_n) = \otimes_{k=1}^{n} \mu(x_k) N_k \\ &= \bigcup_{\alpha_k \in T_k, \beta_k \in I_k, \gamma_k \in F_k} \{\{\prod_{k=1}^{n} (\alpha_k)^{\mu(x_k)} | \prod_{k=1}^{n} P_k^{T_k}\}, \{1 - \prod_{k=1}^{n} (1 - \beta_k)^{\mu(x_k)} | \prod_{k=1}^{n} P_k^{I_k}\}, \{1 - \prod_{k=1}^{n} (1 - \gamma_k)^{\mu(x_k)} | \prod_{k=1}^{n} P_k^{F_k}\}\}. \end{aligned}$$

(4) Assume the precondition $\mu(x_{\pi(k)}) = \frac{1}{n}$, the PNHFFCG operator and PNHFWG operator reduce to the PNHFG operator:

$$\begin{aligned} &PNHFWG(N_1, N_2, \cdots, N_n) = \otimes_{k=1}^{n} \mu(x_k) N_k \\ &= \bigcup_{\alpha_k \in T_k, \beta_k \in I_k, \gamma_k \in F_k} \{\{\prod_{k=1}^{n} (\alpha_k)^{\frac{1}{n}} | \prod_{k=1}^{n} P_k^{T_k}\}, \{1 - \prod_{k=1}^{n} (1 - \beta_k)^{\frac{1}{n}} | \prod_{k=1}^{n} P_k^{I_k}\}, \{1 - \prod_{k=1}^{n} (1 - \gamma_k)^{\frac{1}{n}} | \prod_{k=1}^{n} P_k^{F_k}\}\}. \end{aligned}$$

Theorem 8. *By analyzing the PNHFCOA operator, we can obtain the following theorems, obviously.*

(1) (Monotonicity) Assume $N_k = \{\{\alpha_k|P_k^{T_k}\}, \{\beta_k|P_k^{I_k}\}, \{\gamma_k|P_k^{F_k}\}\}$ and $\tilde{N}_k = \{\{\tilde{\alpha}_k|P_k^{\tilde{T}_k}\}, \{\tilde{\beta}_k|P_k^{\tilde{I}_k}\}, \{\tilde{\gamma}_k|P_k^{\tilde{F}_k}\}\}$ indicate two PNHFEs. The factor $\pi(k)$ satisfies condition $N_{\pi(1)} \geq N_{\pi(2)} \geq \cdots \geq N_{\pi(n)}$ and $\tilde{N}_{\pi(1)} \geq \tilde{N}_{\pi(2)} \geq \cdots \geq \tilde{N}_{\pi(n)}$. With $\forall N_{\pi(k)}$ and $\forall \tilde{N}_{\pi(k)}$, there are $\alpha_{\pi(k)} \leq \tilde{\alpha}_{\pi(k)}$, $\beta_{\pi(k)} \geq \tilde{\beta}_{\pi(k)}$, $\gamma_{\pi(k)} \geq \tilde{\gamma}_{\pi(k)}$ and $P_{\pi(k)}^{T_{\pi(k)}} = P_{\pi(k)}^{\tilde{T}_{\pi(k)}}$, $P_{\pi(k)}^{I_{\pi(k)}} = P_{\pi(k)}^{\tilde{I}_{\pi(k)}}$, $P_{\pi(k)}^{F_{\pi(k)}} = P_{\pi(k)}^{\tilde{F}_{\pi(k)}}$. Then,

$$PNHFCOG\{N_1, N_2, \cdots, N_n\} \leq PNHFCOG\{\tilde{N}_1, \tilde{N}_2, \cdots, \tilde{N}_n\}.$$

(2) (Boundedness) Assume $N_k = \{\{\alpha_k|P_k^{T_k}\}, \{\beta_k|P^{I_k}\}, \{\gamma_k|P_k^{F_k}\}\}$ indicates an PNHFE,

$$N^- = \{\{min\{\alpha_k\}|min\{P_k^{T_k}\}\}, \{max\{\beta_k\}|max\{P_k^{I_k}\}\}, \{max\{\gamma_k\}|max\{P_k^{F_k}\}\}\},$$
$$N^+ = \{\{max\{\alpha_k\}|max\{P_k^{T_k}\}\}, \{min\{\beta_k\}|min\{P_k^{I_k}\}\}, \{min\{\gamma_k\}|min\{P_k^{F_k}\}\}\}.$$

Then,

$$PNHFCOG(N^-) \leq PNHFCOA(N_1, N_2, \cdots, N_n) \leq PNHFCOG(N^+).$$

(3) *(Idempotency) Assume* $N_k = \{\{\alpha|P_1\}, \{\beta|P_2\}, \{\gamma]|P_3\}\}$ *is a normalized PNHFE, then*

$$PNHFCOG(N_1, N_2, \cdots, N_X) = \{\{\alpha|P_1\}, \{\beta|P_2\}, \{\gamma|P_3\}\}. \tag{10}$$

(4) *(Commutativity) Assume* $A = \{N_1, N_2, \cdots, N_n\}$ *and* $B = \{N_{\lambda(1)}, N_{\lambda(2)}, \cdots, N_{\lambda(n)}\}$ *are two finite sets. If the position of the element in* $\{N_{\pi(1)}, N_{\pi(2)}, \cdots, N_{\pi(n)}\}$ *is changed arbitrarily to get* $\{N_1, N_2, \cdots, N_n\}$*, then:*

$$PNHFCOG(N_1, N_2, \cdots, N_n) = PNHFCOG\{N_{\lambda(1)}, N_{\lambda(2)}, \cdots, N_{\lambda(n)}\}.$$

Lemma 1. *By reference [37], if* $x_k \geq 0, w_k \geq 0, \sum_{k=1}^{n}$*, we have*

$$\prod_{k=1}^{n}(x_k)^{w_k} \leq \sum_{k=1}^{n} w_k x_k.$$

Theorem 9. *Suppose* $N_k = \{\{\alpha_k|P_k^{T_k}\}, \{\beta_k|P_k^{I_k}\}, \{\gamma_k|P_k^{F_k}\}\}$ *indicates an PNHFE, we have*

$$PNHFCOG(N_1, N_2, \ldots, N_n) \leq PNHFCOA(N_1, N_2, \ldots, N_n). \tag{11}$$

Proof. Based on Lemma 1, Definitions 10 and 11, the following formula is obtained:

$$\prod(\alpha_{\pi(k)})^{\mu_{\pi(k)}} \leq \sum(\mu_{\pi(k)})\alpha_{\pi(k)} = 1 - \sum \mu_{\pi(k)}(1-\alpha_{\pi(k)}) \leq 1 - \prod(1-\alpha_{\pi(k)})^{\mu_{\pi(k)}}.$$

Obviously,

$$\prod(\alpha_{\pi(k)})^{\mu_{\pi(k)}} \prod P_{\pi(k)}^{T_{\pi(k)}} \leq (1 - \prod(1-\alpha_{\pi(k)})^{\mu_{\pi(k)}})P_{\pi(k)}^{T_{\pi(k)}}.$$

Similarly, we know

$$\prod(\beta_{\pi(k)})^{\mu_{\pi(k)}} \prod P_{\pi(k)}^{T_{\pi(k)}} \leq (1 - \prod(1-\beta_{\pi(k)})^{\mu_{\pi(k)}})P_{\pi(k)}^{T_{\pi(k)}},$$
$$\prod(\gamma_{\pi(k)})^{\mu_{\pi(k)}} \prod P_{\pi(k)}^{T_{\pi(k)}} \leq (1 - \prod(1-\gamma_{\pi(k)})^{\mu_{\pi(k)}})P_{\pi(k)}^{T_{\pi(k)}}.$$

Thus, based on Equation (4) and Definition 9,

$$PNHFCOG(N_1, N_2, \ldots, N_n) \leq PNHFCOA(N_1, N_2, \ldots, N_n).$$

□

4. A MADM method in PNHF Environment

For a MAMD problem under the PNHF environment, assume $Z_1, Z_2, \cdots, Z_n$ indicates all the alternatives, $D_1, D_2, \cdots, D_k$ indicates all the attributes. The evaluated information of Z_i with D_j is indicated by PNHFE $N_{ij} = \{\{T_{ij}|P_{ij}^{T_{ij}}\}, \{I_{ij}|P_{ij}^{I_{ij}}\}, \{F_{ij}|P_{ij}^{F_{ij}}\}\}$.

Based on these necessary prerequisites, we elicit specific steps.

- Step 1. Construct a PNHF decision matrix (PNHFDM) $E = (N_{ij})_{m \times k}$.

Rank all PNHFEs from smallest to largest according to Definition 9. Then, the reorder decision matrix can be obtained such that $N_{i\pi(1)} \leq N_{i\pi(2)} \leq \cdots \leq N_{i\pi(n)}$, $\{\pi(1), \pi(2), \cdots, \pi(n)\}$ indicates a new arrangement.

- Step 2. Calculate μ_j of criterion D_j.

In order to consider more interrelationships among criteria, by taking the g_λ fuzzy measure, the measure μ of each criterion could be determined.

- Step 3. Based on the goal, select a PNHFCOA (PNHFCOG) operator to aggregate all PNHFEs Z_i $(i = 1, 2, \cdots, n)$.

When we need to consider the group's major points, the PNHFCOA operator should be utilized. However, the individual major points could be emphasized based on the PNHFCOG operator. Thus, different types of operators can be used based on the different demands.

- Step 4. Reorder the alternatives

By the valued of $S(N_i)$, $D(N_i)$ and Definition 9, all the Z_i are ranked in decreasing order, DM picks an optimal option Z_i.

5. The Program of the Proposed Approach

Choosing the suitable 3PL plays a key role in business development, like improving efficiency and reducing costs, improving market share and service quality. ABC Machinery Manufacturing Company Limited as an automotive manufacturing company. The decision maker needs to select an optimal third part logistics supplier. There are four possible logistics suppliers that are denoted as Z_1, Z_2, Z_3, Z_4. The decision maker selects the following four attributes to access these alternatives: D_1: equipment system; D_2: effectiveness; D_3: safety; D_4: Correlation. The PNHFDM E is obtained, depicted in Table 1. Suppose that fuzzy measures of D_j are $\mu(D_1) = 0.3$, $\mu(D_2) = 0.3$, $\mu(D_3) = 0.3$, $\mu(D_4) = 0.2$, respectively.

- Step 1. Calculate the score values of all Z_i. The results are depicted in Table 2.

Thus, we can get reordered PNHFDM E'. Because of space constraints, the decision matrix E' is omitted.

- Step 2. Since the information of fuzzy measure is $\mu(C_1) = 0.3$, $\mu(C_2) = 0.3$, $\mu(C_3) = 0.3$, $\mu(C_4) = 0.2$, respectively. By Equation (3), we get $\lambda = -0.2317$. Thus, taking Z_1 as an example, we can get

$$\mu_{\pi(1)} = 0.2477, \mu_{\pi(2)} = 0.1732, \mu_{\pi(3)} = 0.2791, \mu_{\pi(4)} = 0.3.$$

- Step 3. Utilizing the PNHFCOA operator, by Equation (9), we can get

$$S(Z_1) = 0.6466, S(Z_2) = 0.6436, S(Z_3) = 0.5822, S(Z_4) = 0.6950.$$

- Step 4. Rank the PNHFEs by Definition 9,

$$Z_4 > Z_1 > Z_2 > Z_3.$$

The 3PL Company Z_1 is an optimal option.

Next, suppose that the PNHFCOG operator is utilized to solve this problem. Similarly, the score value of alternative A_i is obtained:

$$S(Z_1) = 0.6181, S(Z_2) = 0.6167, S(Z_3) = 0.5639, S(Z_4) = 0.6686.$$

Thus, the final ranking of alternatives is determined, as follows:

$$Z_4 > Z_1 > Z_2 > Z_3.$$

The 3PL Company A_1 is an optimal option.

Table 1. A PNHFDM Information E.

	C_1
A_1	$\{\{0.5\|0.3, 0.57\|0.22, 0.58\|0.27, 0.64\|0.21\}, \{0.43\|0.25, 0.48\|0.2, 0.49\|0.30, 0.55\|0.25\}, \{0.41\|0.27, 0.47\|0.23, 0.52\|0.23, 0.46\|0.27\}\}$
A_2	$\{\{0.44\|0.27, 0.49\|0.24, 0.48\|0.26, 0.52\|0.23\}, \{0.46\|0.47, 0.53\|0.53\}, \{0.29\|0.18, 0.33\|0.14, 0.36\|0.20, 0.41\|0.16, 0.41\|0.18, 0.47\|0.14\}\}$
A_3	$\{\{0.41\|0.30, 0.48\|0.22, 0.47\|0.27, 0.53\|0.21\}, \{0.46\|0.23, 0.49\|0.26, 0.49\|0.24, 0.53\|0.27\}, \{0.39\|0.24, 0.41\|0.25, 0.48\|0.26, 0.45\|0.25\}\}$
A_4	$\{\{0.47\|0.25, 0.51\|0.24, 0.50\|0.26, 0.53\|0.25\}, \{0.34\|0.33, 0.43\|0.35, 0.5\|0.32\}, \{0.42\|0.28, 0.45\|0.21, 0.53\|0.29, 0.56\|0.22\}\}$
	C_2
A_1	$\{\{0.40\|0.26, 0.51\|0.25, 0.49\|0.25, 0.58\|0.24\}, \{0.56\|0.27, 0.59\|0.24, 0.60\|0.26, 0.63\|0.23\}, \{0.39\|0.23, 0.43\|0.29, 0.42\|0.21, 0.47\|0.27\}\}$
A_2	$\{\{0.51\|0.53, 0.54\|0.47\}, \{0.49\|0.25, 0.52\|0.22, 0.57\|0:28, 0.60\|0.25\}, \{0.43\|0.18, 0.46\|0.18, 0.48\|0.17, 0.50\|0.16, 0.53\|0.16, 0.55\|0.15\}\}$
A_3	$\{\{0.54\|0.26, 0.60\|0.25, 0.63\|0.25, 0.68\|0.24\}, \{0.50\|0.48, 0.56\|0.52\}, \{0.43\|0.26, 0.46\|0.24, 0.46\|0.26T, 0.50\|0.24\}\}$
A_4	$\{\{0.61\|0.54, 0.67\|0.46\}, \{0.43\|0.27, 0.50\|0.26, 0.46\|0.24, 0.53\|0.23\}, \{0.42\|0.23, 0.50\|0.24T, 0.45\|0.26, 0.53\|0.27\}\}$
	C_3
A_1	$\{\{0.56\|0.24, 0.62\|0.24, 0.59\|0.26, 0.64\|0.26\}, \{0.33\|0.25, 0.36\|0.24, 0.37\|0.26, 0.41\|0.25\}, \{0.36\|0.33, 0.42\|0.36, 0.45\|0.31\}\}$
A_2	$\{\{0.65\|0.24T, 0.69\|0.27, 0.67\|0.23, 0.71\|0.26\}, \{0.43\|0.31, 0.52\|0.23, 0.46\|0.27, 0.55\|0.19\}, \{0.43\|0.26, 0.46\|0.25, 0.50\|0.25, 0.53\|0.24\}\}$
A_3	$\{\{0.51\|0.26, 0.54\|0.26, 0.57\|0.24, 0.60\|0.24\}, \{0.43\|0.26, 0.46\|0.24, 0.48\|0.26, 0.52\|0.24\}, \{0.49\|0.25, 0.54\|0.26, 0.57\|0.24, 0.62\|0.25\}\}$
A_4	$\{\{0.57\|0.24, 0.66\|0.28, 0.66\|0.22, 0.73\|0.26\}, \{0.43\|0.54, 0.49\|0.46\}, \{0.47\|0.16, 0.53\|0.17, 0.56\|0.16, 0.50\|0.17, 0.57\|0.18, 0.59\|0.17\}\}$
	C_4
A_1	$\{\{0.48\|0.47, 0.57\|0.53\}, \{0.40\|0.51, 0.47\|0.49\}, \{0.47\|0.16, 0.50\|0.15, 0.53\|0.15, 0.49\|0.19, 0.54\|0.18, 0.56\|0.17\}\}$
A_2	$\{\{0.51\|0.27, 0.62\|0.26, 0.54\|0.24, 0.64\|0.23\}, \{0.40\|0.25, 0.46\|0.28, 0.46\|0.22, 0.53\|0.25\}, \{0.39\|0.33, 0.42\|0.37, 0.45\|0.30\}\}$
A_3	$\{\{0.48\|0.28, 0.58\|0.23, 0.51\|0.26, 0.61\|0:23\}, \{0.42\|0.25, 0.45\|0.24, 0.47\|0.26, 0.50\|0.25\}, \{0.42\|0.27, 0.50\|0.26, 0.45\|0.24, 0.53\|0.23\}\}$
A_4	$\{\{0.66\|0.27, 0.73\|0.24, 0.71\|0:26, 0.77\|0.23\}, \{0.43\|0.38, 0.49\|0.33, 0.54\|0.29\}, \{0.36\|0.27, 0.41\|0.24, 0.39\|0.26T, 0.45\|0.23\}\}$

Table 2. The score values of PNHFE N_{ij}.

	D_1	D_2	D_3	D_4
Z_1	0.6185	0.4700	0.8259	0.5782
Z_2	0.6081	0.4885	0.7204	0.6941
Z_3	0.5395	0.6181	0.5273	0.6072
Z_4	0.5907	0.6825	0.6562	0.8329

6. Comparison with Other Approaches

Based on the same problem background, the comparison results are described.

Wu's method emphasizes the individual (group's) major points, which utilize the MADM problems with a simplified neutrosophic environment.

Peng put forward the TOPSIS-based QUALIFIEX method and the cross-entropy measurement to manage MADM situations with probability multi-valued neutrosophic elements. Then, the effectiveness of this method is demonstrated by an illustrative example.

However, in many actual problems, attributes are not independent. But by comparing the above methods, the association between attributes is not considered. However, attribute correlation is considered in our proposed method. The decision results are more reasonable and effective. The final results by different approaches are indicated in Table 3.

Table 3. Choice of optimal result.

Method	Sort of Results	Optimal Alternative	Worst Alternative
TOPSIS-based QUALIFLEX method [3]	$Z_4 > Z_2 > Z_1 > Z_3$	Z_4	Z_3
SNNPWA operator [45]	$Z_3 > Z_1 > Z_2 > Z_4$	Z_3	Z_4
SNNPWG operator [45]	$Z_3 > Z_2 > Z_1 > Z_4$	Z_2	Z_1
PNHFCOA operator	$Z_4 > Z_1 > Z_2 > Z_3$	Z_4	Z_3
PNHFCOG operator	$Z_4 > Z_1 > Z_2 > Z_3$.	Z_4	Z_3

Through the above analysis, the DMs' evaluation information is represented by PNHFEs. PNHFEs are more flexible in describing the hesitant MADM information and reporting the probabilistic values of all hesitant values. The Choquet integral and aggregation operators are integrated in our method. Next, the alternatives' information is aggregated and ranked. In this model, the interrelationship between attributes are involved by Choquet integral. The MADM problems are effectively resolved by utilizing interdependencies or interactions between attributes. The results are closer to the actual situations.

7. Conclusions

Firstly, our goal is to aggregate the PNHF information by the notion of PNHFS. By applying the Choquet integral, the weight information is extended, more information about the correlation between attributes is mined. The advantage of PNHFS is that it can explain the preferences of DMs without information loss. By investigating, we found both of the PNHFCOA and PNHFCOG operators satisfying the following characteristics: monotonity, boundedness, idempotency and commutativity. Those operators can reduce to some common aggregation operators. Moreover, those aggregation operators were used to an PNHF background, where fuzzy measure of attributes are recognized. All alternatives are reordered and choose an optimal option. Next, we present an illustrative situation to explain the objectivity of our method. The result received by the PNHFCOA and PNHFCOG operators are effective and flexible. The results are more suitable for the actual situations, because more information can be considered based on our method. Thus, when some uncertain problems or inconsistent and indeterminate information needs to be resolved, our proposed approach shows great advantages. In regard to the next jobs, more types of aggregation operators are investigated and applied in other practical situations, like medical diagnoses, group decision-making, risk evaluations, and fractal-wavelet modeling [46–50].

Author Contributions: All authors have contributed equally to this paper. The idea of the this paper is described by X.Z. The framework of the paper is constructed and analyzed by S.S. and Q.Z.

Funding: This research was funded by the National Natural Science Foundation of China (Grant No. 61573240).

Conflicts of Interest: The authors declare no conflict of interest.

References

1. Konwar, N.; Davvaz, B.; Debnath, P. Results on generalized intuitionistic fuzzy hypergroupoids. *J. Intell. Fuzzy Syst.* **2019**, *36*, 2571–2580. [CrossRef]
2. Fu, C.; Chang, W.; Xue, M.; Yang, S. Multiple criteria group decision-making with belief distributions and distributed preference relations. *Eur. J. Oper. Res.* **2019**, *273*, 623–633. [CrossRef]
3. Peng, H.G.; Zhang, H.Y.; Wang, J.Q. Probability multi-valued neutrosophic sets and its application in multi-criteria group decision-making problems. *Neural Comput. Appl.* **2017**, *20*, 563–583. [CrossRef]
4. Xue, J.; Wu, C.; Chen, Z.; Van Gelder, P.H.; Yan, X. Modeling human-like decision-making for inbound smart ships based on fuzzy decision trees. *Expert Syst. Appl.* **2019**, *115*, 172–188. [CrossRef]
5. Spizzichino, F.L. On the probabilistic meaning of copula-based extensions of fuzzy measures. Applications to target-based utilities and multi-state reliability systems. *Fuzzy Sets Syst.* **2019**, *354*, 1–19. [CrossRef]
6. Ma Y.C.; Zhang, X.H.; Yang, X.F.; Zhou X. Generalized neutrosophic extended triplet group. *Symmetry* **2019**, *11*, 327. [CrossRef]
7. Li, L.; Jin, Q.; Hu, K. Lattice-valued convergence associated with CNS spaces. *Fuzzy Sets Syst.* **2018**. [CrossRef]
8. Jin, Q.; Li, L.; Lv, Y.; Zhao, F.; Zou, J Connectedness for lattice-valued subsets in lattice-valued convergence spaces. *Quaest. Math.* **2019**, *42*, 135–150. [CrossRef]
9. Li, L.Q. *p*-Topologicalness–A Relative Topologicalness in ⊤-Convergence Spaces. *Mathematics* **2019**, *7*, 228. [CrossRef]
10. Zhang, X.H.; Park, C.; Wu, S.P. Soft set theoretical approach to pseudo-BCI algebras. *J. Intell. Fuzzy Syst.* **2018**, *34*, 559–568. [CrossRef]
11. Wu, X.Y.; Zhang, X.H. The decomposition theorems of AG-neutrosophic extended triplet loops and strong AG-(l, l)-loops. *Mathematics* **2019**, *7*, 268. [CrossRef]
12. Zhang, X.H. Fuzzy anti-grouped filters and fuzzy normal filters in pseudo-BCI algebras. *J. Intell. Fuzzy Syst.* **2017**, *33*, 1767–1774. [CrossRef]
13. Brito, V.T.F.; Ferreira, F.A.F.; Pérez-Gladish, B.; Govindan, K.; et.al. Developing a green city assessment system using cognitive maps and the Choquet integral. *J. Clean. Prod.* **2019**, *218*, 486–497. [CrossRef]
14. Krishnan, A.; Wahab, S.; Kasim, M.; Bakar, E. An alternate method to determine λ^0-measure values prior to applying Choquet integral in a multi-attribute decision-making environment. *Decis. Sci. Lett.* **2019**, *8*, 193–210. [CrossRef]
15. Beg I.; Jamil R N.; Rashid T. Diminishing Choquet Hesitant 2-Tuple Linguistic Aggregation Operator for Multiple Attributes Group Decision Making. *Int. J. Anal. Appl.* **2019**, *17*, 76–104.
16. Zadeh, L.A. Fuzzy sets. *Inf. Control.* **1965**, *8*, 338–353. [CrossRef]
17. Castillo, O.; Atanassov, K. Comments on Fuzzy Sets, Interval Type-2 Fuzzy Sets, General Type-2 Fuzzy Sets and Intuitionistic Fuzzy Sets. In *Recent Advances in Intuitionistic Fuzzy Logic Systems*; Springer: Cham, Switzerland, 2019; pp. 35–43.
18. Song, C.; Zhao, H.; Xu, Z.S.; Hao, Z. Interval probabilistic hesitant fuzzy set and its application in the Arctic geopolitical risk evaluation. *Int. J. Intell. Syst.* **2019**, *34*, 627–651. [CrossRef]
19. Farhadinia, B.; Xu, Z.S. *Distance Measures for Hesitant Fuzzy Sets and Their Extensions*; Springer: Singapore, 2019; pp. 31–58.
20. Smarandache, F. A unifying field in logics: Neutrosophic logic. *Multi-Valued Log.* **1999**, *8*, 489–503.
21. Wei, G.; Zhang, Z. Some single-valued neutrosophic Bonferroni power aggregation operators in multiple attribute decision-making. *J. Ambient Intell. Humaniz. Comput.* **2019**, *10*, 863–882. [CrossRef]
22. Thong, N.T.; Dat, L.Q.; Hoa, N.D.; Ali, M.; Smarandache, F. Dynamic interval valued neutrosophic set: Modeling decision-making in dynamic environments. *Comput. Ind.* **2019**, *108*, 45–52. [CrossRef]
23. Sun, R.; Hu, J.; Chen, X. Novel single-valued neutrosophic decision-making approaches based on prospect theory and their applications in physician selection. *Soft Comput.* **2019**, *23*, 211–225. [CrossRef]
24. Yong, R.; Ye, J. Multiple Attribute Projection Methods with Neutrosophic Sets. In *Fuzzy Multi-Criteria Decision-Making Using Neutrosophic Sets*; Springer: Cham, Switzerland, 2019; pp. 603–622.
25. Ye, J.; Du, S. Some distances, similarity and entropy measures for interval-valued neutrosophic sets and their relationship. *Int. J. Mach. Learn. Cybern.* **2019**, *10*, 347–355. [CrossRef]
26. Sahin, R.; Liu, P.D. Maximizing deviation method for neutrosophic multiple attribute decision-making with incomplete weight information. *Neural Comput. Appl.* **2015**, *27*, 2017–2029. [CrossRef]

27. Ye, J. Improved cosine similarity measures of simplified neutrosophic sets for medical diagnoses. *Artif. Intell. Med.* **2015**, *63*, 171–179. [CrossRef]
28. Tian, Z.P.; Zhang, H.Y.; Wang, J. Wang, J.Q.; Chen, X.H. Multi-criteria decision-making method based on a cross-entropy with interval neutrosophic sets. *Int. J. Syst. Sci.* **2016**, *47*, 3598–3608. [CrossRef]
29. Zhang, X.H.; Borzooei, R.A. and Jun, Y.B. Q-filters of quantum B-algebras and basic implication algebras. *Symmetry.* **2018**, *10*, 573. [CrossRef]
30. Zhang, X.H.; Wang, X.J.; Smarandache, F.; Jaíyéolá, T.G.; Lian, T. Singular neutrosophic extended triplet groups and generalized groups. *Cogn. Syst. Res.* **2018**. [CrossRef]
31. Zhang, X.H.; Bo, C.X.; Smarandache, F.; Park, C. New operations of totally dependent-neutrosophic sets and totally dependent-neutrosophic soft sets. *Symmetry* **2018**, *10*, 187. [CrossRef]
32. Zhang, X.H.; Mao, X.Y.; Wu, Y.T.; Zhai, X.H. Neutrosophic filters in pseudo-BCI algebras. *Int. J. Uncertainty Quant.* **2018**, *8*, 511–526. [CrossRef]
33. Smith, P. Exploring public transport sustainability with neutrosophic logic. *Transp. Plan. Technol.* **2019**, 1–17. [CrossRef]
34. Biswas, P.; Pramanik, S.; Giri, B.C. Neutrosophic TOPSIS with group decision-making. In *Fuzzy Multi-Criteria Decision-Making Using Neutrosophic Sets*; Springer: Cham, Switzerland, 2019; pp. 543–585.
35. Nirmal, N.P.; Bhatt, M.G. Development of Fuzzy-Single Valued Neutrosophic MADM Technique to Improve Performance in Manufacturing and Supply Chain Functions. In *Fuzzy Multi-Criteria Decision-Making Using Neutrosophic Sets*; Springer, Cham, Switzerland, 2019; pp. 711–729.
36. Li, G.; Niu, C.; Zhang, C. Multi-criteria decision-making approach using the fuzzy measures for environmental improvement under neutrosophic environment. *Ekoloji* **2019**, *28*, 1605–1615.
37. Shao, S.T.; Zhang, X.H.; Li, Yu.; Bo, C.X. Probabilistic single-valued (interval) neutrosophic hesitant fuzzy set and its application in multi-attribute decision-making *Symmetry* **2018**, *10*, 419. [CrossRef]
38. Choquet, G. Theory of capacities. *Ann. Inst. Fourier* **1953**, *5*, 131–295. [CrossRef]
39. Khan, M.S.A.; Abdullah, S.; Ali, A.; Amin, F.; Hussain, F. Pythagorean hesitant fuzzy Choquet integral aggregation operators and their application to multi-attribute decision-making. *Soft Comput.* **2019**, *23*, 251–267. [CrossRef]
40. Corrente, S.; Greco, S.; Słowiński, R. Robust ranking of universities evaluated by hierarchical and interacting criteria. In *Multiple Criteria Decision Making and Aiding*; Springer: Cham, Switzerland, 2019; pp. 145–192.
41. Labreuche, C.; Grabisch, M. Using multiple reference levels in Multi-Criteria Decision aid: The Generalized-Additive Independence model and the Choquet integral approaches. *Eur. J. Oper. Res.* **2018**, *267*, 598–611. [CrossRef]
42. Yager, R.R. On Using the Shapley Value to Approximate the Choquet Integral in Cases of Uncertain Arguments. *IEEE Trans. Fuzzy Syst.* **2018**, *26*, 1303–1310. [CrossRef]
43. Liu, P.; Tang, G. Some generalized Shapely interval-valued dual hesitant fuzzy uncertain linguistic Choquet geometric operators and their application to multiple attribute decision-making. *J. Intell. Fuzzy Syst.* **2019**, *36*, 557–574. [CrossRef]
44. Sugeno, M. *Theory of Fuzzy Integrals and Its Applications.* Ph.D. Thesis, Tokyo Institute of Technology, Tokyo, Japan, 1974.
45. Wu, X.H.; Wang, J.; Peng, J.J.; Chen, X.H. Cross-entropy and prioritized aggregation operator with simplified neutrosophic setsand their application in multi-criteria decision-making problems.*Int. J. Fuzzy Syst.* **2016**, *18*, 1104–1116. [CrossRef]
46. Guariglia, E. Primality, Fractality and Image Analysis. *Entropy* **2019**, *21*, 304. [CrossRef]
47. Guido, R.C.; Addison, P.; Walker, J. Introducing wavelets and time-frequency analysis. *IEEE Eng. Biol. Med. Mag.* **2009**, *28*, 13. [CrossRef]
48. Guido, R.C. Practical and useful tips on discrete wavelet transforms. *IEEE Signal Process. Mag.* **2015**, *32*, C162–C166 . [CrossRef]
49. Guariglia, E. Entropy and Fractal Antennas. *Entropy* **2016**, *18*, 84. [CrossRef]
50. Guariglia, E. Harmonic Sierpinski Gasket and Applications. *Entropy* **2018**, *20*, 714. [CrossRef]

symmetry

MDPI

Article

IoT and Its Impact on the Electronics Market: A Powerful Decision Support System for Helping Customers in Choosing the Best Product

Mohamed Abdel-Basset [1,*], Mai Mohamed [1], Victor Chang [2] and Florentin Smarandache [3]

[1] Department of Operations Research, Faculty of Computers and Informatics, Zagazig University, Sharqiyah 44519, Egypt; mmgaafar@zu.edu.eg
[2] Independent Researcher, Southampton SO16 3QX, UK; victorchang.research@gmail.com
[3] Math & Science Department, University of New Mexico, Gallup, NM 87301, USA; smarand@unm.edu
* Correspondence: analyst_mohamed@zu.edu.eg

Received: 25 March 2019; Accepted: 15 April 2019; Published: 1 May 2019

Abstract: Many companies have observed the significant benefits they can get via using internet. Since then, large companies have been able to develop business transactions with customers at anytime, anywhere, and in relation to anything, so that we now need a more comprehensive concept than the internet. This concept is the Internet of Things (IoT). IoT will influence decision making style in various phases of selling, buying and marketing process. Therefore, every individual and company should know precisely what IoT is, and how and why they should incorporate it in their operations. This motivated us to propose a smart system based on IoT to help companies and marketers make a powerful marketing strategy via utilizing obtained data from IoT devices. Not only this, but the proposed system can also solve the problems which face companies and customers in online shopping. Since there are different types of the same product, and also different criteria for purchasing which can be different between individuals, customers will need a decision support system to recommend them with the best selection. This motivates us to also propose a neutrsophic technique to deal with unclear and conflicting information which exists usually in the purchasing process. Therefore, the smart system and neutrosophic technique is considered as a comprehensive system which links between customers, companies, marketers to achieve satisfaction for each of them.

Keywords: e-marketing; Internet of Things; neutrsophic set; multi-criteria decision making techniques

1. Introduction

Internet of Things (IoT) was presented as a concept in 1999. It has provided a platform to connect to different hardware and mobile devices, so that different people can be connected to each other. The networks can be on the local wide area networks subscribed to each organization, or wireless networks, or both. IoT can also collect data via wireless sensors, and then connect to its central servers for processing and storage. Similarly, it enables people to connect to the internet and other people's mobile devices via central servers and/or wireless sensors. The efficient use of IoT can improve operational efficiency due to its capability to gather and explicate big data, as well as automate connections among machines [1].

The IoT can be applied in several areas such as smart cities, smart homes, education, agriculture, health, wearables, and industrial automation [2]. It provides enormous benefits to the society as a whole. We can see the effect of IoT in cars with built-in sensors, health-monitoring systems, biochip transponders which are used for farm animals, search and deliverance devices, smart washer/dryers which use Wi-Fi for remote monitoring, etc. There will be almost twenty billion devices on the IoT by 2020 according to Gartner [3].

Although more companies and retail stores have adopted IoT, many consumers are still unaware of IoT services.

Using IoT can significantly make users' day to day activities more convenient since many services can be accessed on their mobile devices. It also improves inventory management, tracks product usage, monitors selling rates and locations. Also, the IoT can improve the customer services to allow real-time communications. Additionally, it can allow businesses to forecast possible customers' concerns and cases, and proactively provide solutions [4]. By doing so, it can achieve a better customer satisfaction. As a result, IoT can also save time, reduce costs and also human errors.

Due to the significant role of IoT in enhancing services quality, managing customer demands, and achieving customer satisfaction and loyalty, some studies are presented that highlight this role. Jie et al. [5] illustrated in their study how e-retailers who deal with innovative products in the era of the Internet of Things (IoT) select product delivery service providers to ensure timely and efficient delivery to customers. Additionally, Desai [6] could model IoT services on the basis of service quality dimensions in the electricity distribution center of Bangalore Electricity Supply Company. The researchers in [7] have determined IoT solutions to improve the effectiveness of the service product.

Several research papers are presented to demonstrate discovering the capabilities for IoT adoption in the organization and also studying its effect on customers' experience. The way in which the IoT changes customers' experience while shopping in a retailing context is presented by Balaji and Roy [8]. The theoretical understanding of consumers' adoption and continued use of wearable technology for advanced health and fitness purposes is illustrated by Canhoto and Arp [9]. Wu, Chen, and Dou [10] gave insights into how companies can improve their brand building through the use of IoT technologies. A better understanding of the underlying causes of consumer resistance to smart and related products was developed by Mani and Chouk [11]. Woodside and Sood [12] proposed substantial revisions of the dominant logic of service because of the next take-off phase to adopt new radical innovations in the Internet of Things. Ehret and Wirtz [13] illustrated that the industrial IoT offers new opportunities and harbors threats that companies are not able to address with presented business models. Additionally, a smart framework for a shopping mall based on IoT technologies was presented by Pathan et al. [14].

IoT also has a huge impact on marketers since it provides them with the access to accurate big data. Marketers can track and record products, estimate the number of customers daily, analyze purchasing behaviors and understand the individual uses of products [4]. The analyzed outputs may eliminate the need for surveys or the collection of costly and time-consuming data, where ideas can be collected from the actual use of connected products and related data. It can also improve direct and hyper-local marketing, where personal messages can be sent via a number of connected mediums, for monitoring response and comments from customers. This can directly benefit all forms of marketing, since the information is the first-hand and the dissemination can reach more individuals regardless of their demographic, psychographic, or geographic generalizations. The growth of IoT will influence all marketing companies, particularly those focused on big data analytics. With more big data from consumers and businesses which became quickly available to marketers, analysts can turn raw data into useful insights, recommendations and predicted outcomes.

For understanding how IoT constitutes to marketing, few studies are presented. Balmer and Yen [15] proposed the appearance of what they call, 'The Corporate Marketing Internet Revolution' which calls for a radical rethinking of marketing practice and scholarship. The influence of the IoT on marketing practices was considered by De Cremer, Nguyen, and Simkin [16] via addressing the overlooked area of the dark side of the IoT. A smart marketing system based on IoT was proposed by Rajabi and Hakim [17] to help customers and shopping centers to interact with each other. The vision and challenges of advertising in the Internet of Things era was presented by Aksu et al. [18]. Celik [19], illustrated how IoT will be a great source of future marketing tools. An intelligent retail 4.0 Internet of Things consumer retailer model for smart and strategic marketing of in-store products was presented by Jayaram [20]. The potential applications of Internet of Things technologies and solutions for effective marketing at retail was presented by Bogdanovic [21]. Also, a precise positioning of marketing and behavior intentions of location-based mobile commerce in the Internet of things was presented by Tsai et al. [22].

The previous studies motivated us to design a smart and comprehensive framework for presenting the impact of IoT on customers, companies, and also on constituting marketing strategies. In this smart framework, customers, companies and marketers interact with each other smartly and achieve desired goals easily.

In addition to the proposed framework, we also proposed a novel neutrosophic multi-criteria decision making technique based on The Technique for Order of Preference by Similarity to Ideal Solution (TOPSIS) for supporting customers in selecting the best service or product from among several types. The TOPSIS technique is used for evaluating the performance of IoT in organizations [23]. Similarly, an unclear multicriteria group decision making algorithm based on the TOPSIS approach and the concept of similarity measures was developed by Wibowo and Grandhi [24] for evaluating the overall performance of IoT-based supply chains. We used a neutrosophic set in our technique since it considers the truth, indeterminacy and falsity membership degrees, so that it forms the best representation of reality rather than unclear and intuitionistic uncertain results [25].

The residual parts of this paper are presented as follows: Section 2 describes introductory concepts that includes the e-marketing concept and Internet of Things (IoT). A smart e-marketing model for aiding customers, companies, and marketers is suggested in Section 3. Section 4 presents a neutrosophic technique for aiding customers in selecting the best available product or service, and also the case study and experimental results of our suggested technique are presented. Section 5 concludes and identifies future trends of this paper.

2. Foundations of E-Marketing and Internet of Things (IoT)

The conventional marketing is very costly and takes more time to promote products. The development in technology that comprises the internet media and other digital media has led to the emergence of new marketing concepts.

E-marketing (also known as online marketing or internet marketing) refers to any marketing activities which are presented and serviced online via internet technologies. It includes not only advertising that is shown on websites, but also other types of online activities such as email and social networking.

E-marketing has extended and offered more opportunities for companies to reach out their customers and make direct requests served [26]. The popular media to introduce services and products of companies is on websites to blend information and social media. The three cornerstone principles of e-marketing are immediacy, personalization, and relevance.

The popular e-marketing methods are as follows:

- Search Engine Market (SEM): There are three major search engine marketing activities, which correspond to search engine optimization (SEO), pay-per-click (PPC) and trusted feed [27].
- Online Partnerships: It has three types, which include:
    - Link building: Which is a structured activity to comprise high-quality hyperlinks to your website from pertinent sites with a good page rank.
    - Affiliate marketing: Given rewards of affiliate by business as soon as the customer purchases a product through the own marketing efforts of affiliate, and it is a zero-risk advertising strategy since the merchant does not have to pay any fee until the products are purchased.
    - Online sponsorship: It links to a brand with associated content or context to create brand consciousness. The aim is to enhance the attractiveness of brand.
- Interactive advertising: This means placing ad banners on other websites, and in some respects, it is completely similar to a pay-per-click search engine.
- Email marketing: It is divided into two categories which are inbound and outbound email marketing.

- Online PR: Public relations (PR) means maximizing constructive mentions of an organization, its brands, products or websites on websites of third-party which are probably going to be visited by its target crowd.
- Viral marketing: Viral marketing uses e-mail to send a promotional message to another probable customer [28]. Offline campaigns:
- The use of communications tools like advertising and public relations which are delivered by traditional media for directing visitors to an online attendance.

Nowadays, many people depend on e-marketing websites and services for buying products and services. There is a stable increase in the number of buyers who make purchasing decisions over Google or social network searches or the comments of preceding customers regarding the quality and price of the product. This is due to permanent sales of 24 h/7 days/365 days for customers and businesses, access to customers in distant geographical areas, minimum costs, presenting the right products to the right customers, sustained relationships of customers in the future, and free advertising of businesses, products or services. Hence, e-marketing is an important way to build strong relationships with customers.

Despite all these advantages of e-marketing, there are still many disadvantages which are as follows:

(a) If the infrastructure of e-marketing is weak, users will not have many opportunities to access the internet, learn information online, buy online, and participate in online auctions, and so forth.
(b) If the content control is not good, it can easily to affect the brand image.
(c) It is hard to control the target audience due to the diverse methods of e-marketing.
(d) There is a need to synchronize good information, otherwise, it will lead to information disruption in management.
(e) Customer trust.
(f) Security and privacy issues.

Since the main marketing pertinent element is the fact that consumers anticipate businesses to perform transactions with them at anytime, anywhere, and in relation to anything, we need to apply a more widespread concept than the internet. This concept is based on IoT, since it has become a base for connecting things, sensors, actuators, and other smart technologies.

There are many technical solutions for IoT: Radio Frequency Identification (RFID), Near Field Communication (NFC), Bluetooth Low Energy (BLE), Wi-Fi, Z-wave and others [29]. Protocols like RFID and NFC have been used in retail practice for inventory tracking or payments. BLE is a protocol that has attracted the attention of retailers and marketers in recent years. As it has become the standard in most current smart phones, it presents real-time, contextual, personalized communication and activation at or close to point of purchase, identifying micro location. BLE is a modulation to the standard Bluetooth protocol for allowing short range, low bandwidth, low latency, and efficient communication.

IoT systems contain application, network and perception layers and comprise of a number of component modules.

By its nature, IoT generates an enormous amount of data. For making this data generate useful information and create value to the user, they should be connected and enabled via cloud services and big data analytics, ensuring compliance with security and privacy requirements.

Now the key question here is how will IoT affect e-marketing?

One of the primary goals of marketers is marketing data. The obtained data from IoT devices will help marketers to analyze buying behavior of customers and then determine customers' preferences. The IoT will help marketers to target their audience and then make more relevant advertisements. It will also save time on gathering and analyzing data. Real-time data which is obtained from IoT devices will help marketers to respond to their customers quickly and then achieve customer satisfaction. Every smart product will help marketers to connect with their customers and then increase customer engagement.

Then, we can define IoT in e-marketing as "the interconnection of our digital devices which introduces endless chances for brands to listen and react to the requirements of their customers–with the proper message, at the proper time, and on the proper device".

Despite the great benefits of using IoT in e-marketing, there exists some challenges. The major challenge is security issues. Although gathering data is a very important characteristic of IoT according to marketer's views, it is a very critical part from point of view of customers. The critical part in this process is that the retrieved data from IoT devices about customers, is personal, numerous and includes not only computers and mobile devices but other kinds of house equipment, wearables, etc. Thus, users' behaviors can be tracked at any time and everywhere. This can lead to hackers gaining unauthorized access to customer databases and physical objects which can pose threats to human lives. For example, cars which are based on IoT technology are susceptible to hacker attacks. The hacking of databases is not only dangerous from the point of view of customers but also from the company's view. For example, hacking companies' databases which contain personal information about customers may make the customers sue the company.

3. Model of IoT Application in E-Marketing

In this section we propose a smart model based on IoT technologies which helps customers and companies to interact with each other and meet their needs in the best way. It also helps marketers make the best marketing strategies for their companies through utilizing the significant data which is obtained from IoT devices. By applying this model, we can also solve the most popular problems which exist in e-marketing models.

In this model we focused only on food commodities as example of products, but this model has the potential to be applied on all kinds of products in our lives.

Before we begin to explain the main parts of the proposed model, let us ask ourselves some questions. What if we could know the food commodities that we need to buy, when we are out of the house or at work? For instance, if you run out of milk, a refrigerator can connect to the internet and decide your needs and present a message on its screen or your phone. Additionally, what if you buy these products with the highest quality and lowest prices? How about knowing all the information about products that you need to buy, and the best recommendations according to your purchase criteria, in only one click? What if you can get these products as soon as possible? How about achieving the highest degree of security when buying these products online? These questions are from the point of view of users.

But, from the point of view of companies the important question is: What if companies can achieve online identity verification of consumers, maintain customers' loyalty, solve the problem of product return and refund, and achieve data security?

Finally, from the point of view of marketers the most important question is: What will happen if marketers can get all data and information about customers' behaviors, habits and preferences?

Our proposed model answers the previous questions, and can help customers, companies and marketers to achieve their goals efficiently and effectively.

3.1. Knowing the Amount of Food Commodities in Our Kitchens When We Are Away from Home or at Work

The first part of the proposed model is to know the amount of current food commodities in our kitchens when we are away from home, at work or in our car. This helps people to know their needs and buy it quickly.

In order to do this, we first need to design a smart kitchen based on IoT technologies, but in this part we focus only on specific parts of the kitchen which can help us to know the stock of food commodities. These parts of kitchen are the refrigerator and some shelves in the kitchen that contains some goods like rice, cooking oil, coffee, tea, sugar, etc.

The smart refrigerator should contain an IP address that might sustain functions such as control units, sensors, communication modules, but the most important technologies that will help us in our smart system are: (1) Bluetooth/Wi-Fi. (2) RFID technology: a micro-chip in a label used for transmitting data when the label is exposed to radio waves. The RFID will maintain an updated list of the products in the fridge. Now, all items are tagged with RFID cards when entering the fridge for the first time, and every time items are placed in or removed from the fridge, the RFID antenna which installed inside the fridge recognizes the items and registers them as either in or out of the stock.

So using a smart refrigerator with RFID technology and Wi-Fi connectivity can help us know what is inside the fridge as well as what is consumed and what we need to buy by sending an electronic report to the owner's phone.

The second place that contains food commodities in our kitchen are kitchen shelves. We will design these shelves from keen glass with RFID technology and also connect them to Wi-Fi. The RFID reader in these shelves will determine the quantity of items required, and automatically send alert emails to its owner's phone if the product is less than the threshold as seen on the smart refrigerator.

3.2. Smart E-Marketing Application

In this sub-section, we suggest a smart app for Android and Apple iOS to help customers make their online shopping in an easy, simple, attractive and secure procedure.

This application saves time for customers who are searching for their needs in various websites. Some websites can deceive customers by not sending products with the required quality, difficulty in shipping and retrieval. But this app supports and advertises trusted sites of companies that are subject to certain specifications.

We will store this application in App Store and Google Play to enable various users to download it easily. Once the program is downloaded and installed, users can register in the application as in Figure 1. In this registration each user will fill his/her details and obtain an ID which will be unique for future prospects. This unique ID will store in our database and through it the user will receive messages on his/her phone as well as emails with all offers and discounts on existing products.

This app supports customers, vendors, and also affiliates. It also helps companies to build their websites and market their products smartly.

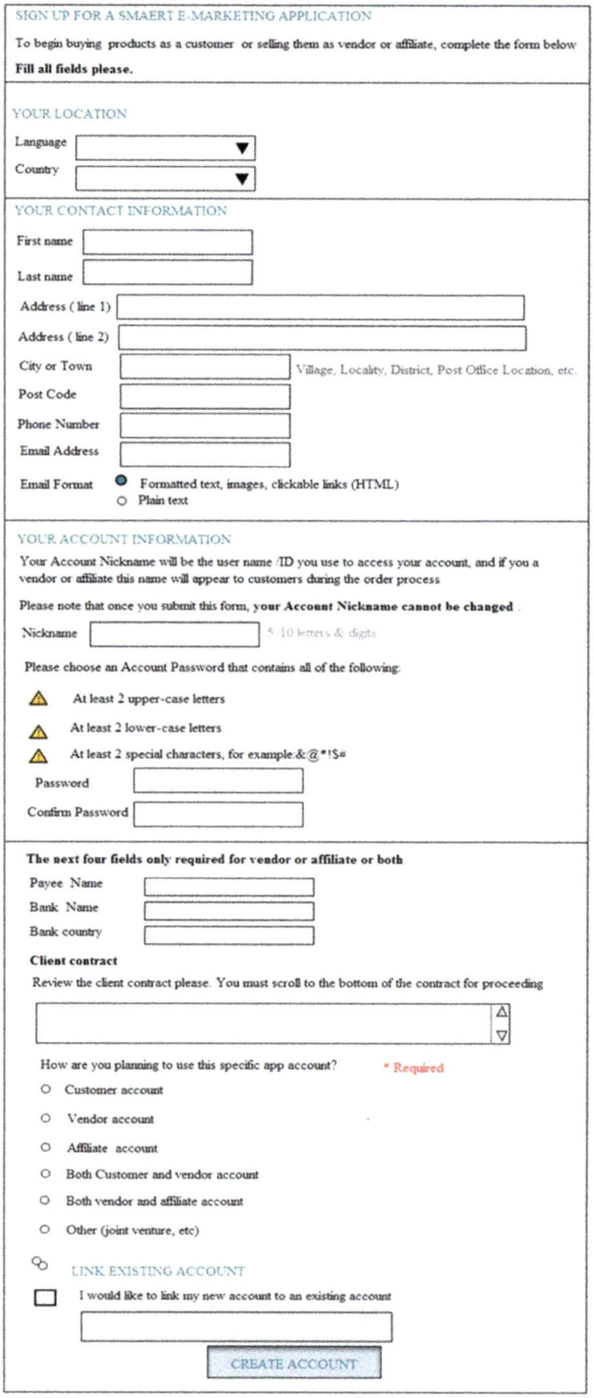

Figure 1. Registration process in app.

3.2.1. Customer Registration on the Application

In the first part of this section we illustrate how any person will be able to know his/her kitchen's stock of food commodities using IoT technology in the refrigerator and kitchen shelves. The RFID reader in the refrigerator and kitchen shelves will determine the quantity of items in each take, and automatically send alert emails if the product is less than the threshold. Hence, if such circumstances occurred one message will be sent to the owner's phone, then the owner will pass his/her needs to the proposed app.

The first step that the customer must do is the registration process in the proposed app as in Figure 1. After finishing the registration process of user (customer or shopper), this application can perform various processes as follows:

- Firstly, view videos in a unique way to allow the customer to see a full description of the app, its features and benefits.
- Enable the customer to select a category of product that he/she plans to buy either by selecting the category from the app directly or via customer voice since this app supports voice recognition technology.
- After determining products that the customer decides to buy, this app begins to compare prices between various websites and recommend the cheapest and highest quality products to the customer. The app deals only with companies which have an Secure Sockets Layer (SSL) certificate. Usually, SSL is used for securing credit card transactions, transferring data and logins, and more recently, is becoming the norm when securely browsing social media sites [30]. The customer has to choose the suitable product according to his/her purchasing criteria. We will illustrate this part with detail in next section via proposing a new neutrsophic technique for helping the customer in the selection process.
- Customers can also buy products from vendors who have registered in this app, since the vendor in this case is able to advertise his/her product using this app. We illustrate this part with details in the next subsection.
- This application also allows customers to get the best description of products by telling customers about products, clarifying why it is for them, characterizing how the product feels, how it can fix problems, save time, or make them happier, and can complete requests in text with photos, graphics and videos. Entertainment is not just a notification. This helps customers to assess products properly and obtain all the information they need, so that they feel comfortable about their purchasing products.
- If a customer buys his/her products continuously via the app, he/she will obtain large discounts.
- After completing the buying process, all information about customers and products that he/she bought and also the websites that he/she bought from it, will store in our cloud database. This helps marketers capture interactions, conversion metrics, and consumer behavior predictions and link them to purchase-intent data.
- The app also supports direct contacts between customers and vendors via chat, video calls, and customer service. This will help in solving all existing problems, increase customer confidence and satisfaction.
- The app will inform users continuously with special offers via sending messages and e-mails to their smart phones.

This app enables customers to make the payment process as in Figure 2. The development of the international payments market made the payment process very simple [30].

Figure 2. Available options of payment process.

3.2.2. Registering as a Vendor on the Application

This app not only supports customers, but it also supports sellers and affiliates. The seller in this app may be the person who plans to sell a product or a company that aims to advertise its products using this app and already has its own website.

Previously we explained how anyone can remotely access the stock of food commodities in his/her kitchen using IoT technology, and then we enabled them to buy the product easily through the proposed app if his/her stock's state is low.

But, if there exists a large amount of food commodities in our shelves that are about to expire or to dispose, the owner can sell them with special offers by using this app as follows:

- Make the registration process as in Figure 1.
- Next, the app verifies the user's identity in a streamlined manner and then converts it to digital data if it is validated. The verification process of vendor/seller consists of three steps:
    - Registration of vendor identity documents such as identification card, driving license, passports, vital standards as fingerprint and facial recognition.
    - Validate documents in addition to their holder.
    - Create a reliable digital ID. This can combat the fraud, strengthen compliance processes and enhance sensitive security services such as money transfer, etc.
- If the previous step can prove that he/she is an invalid user, then the app will automatically reject his/her registration process and automatically block him/her.
- But if he/she is a valid user, then the app will complete registration and give the vendor a special link. This link will help the vendor in dealing with customers, tracking his/her product status, and also receiving money.
- The vendor then begins to upload the product that he/she wants to sell and supports it with the detailed information, images, videos that explain the product status with detail.
- The product will submit for approval before the vendor is able start selling it within the app.

3.2.3. Registration as an Affiliate on the Application

As we know, the vendor is the person who sells their own products, and the affiliate is the person who promotes products for vendors in order to earn a commission. This smart app also supports the affiliate.

In the first step, the affiliate person must fill in a registration form exactly as the vendor did. A verification process will be performed for validating the affiliate as we explained previously with the vendor case.

After the verification process, the affiliate will have a formal registration in our app and receive special link for helping him/her in the promotion of products and receiving commissions.

After that, the app will recommend various products for the affiliate to select from for the promotion process. Not only this, but the app will also classify products into 'most popular', 'most gravity', and a 'new products' category.

By using the proposed app, any affiliate can obtain advice on how to achieve the greatest earning through the promotion process of products. It also generates a report with the most famous sites that support affiliation. Since, in marketing language, the new products usually achieve the highest earned value of affiliate, then our app will continuously send e-mails and messages to the affiliate with any new product either sold in the app or any site supported by the affiliate.

3.2.4. Use of the App by Companies That Have Their Own Inventory

This app is also designed for supporting companies to create their own website via providing the chance to create and market it smartly.

Before creating the website, the company must use IoT technologies in their inventory's management. Subsequently, products should have either an RFID tag or barcode label for offering the visibility of inventory levels, dates of expiration, item location, and product demand.

Using IoT, it will increase the capability to track and communicate with products. For instance, RFID tags will load information about an object, and communicate with an inventory system. Also, built-in RFID tags can drive information about an object's temperature, weather, damage, and traffic, etc. Moreover, built-in GPS locations permit the vendor company to know precisely where every item is. Every object will have its own unique identifier. As a result, the vendor company will be able to pin-point each and every item or piece of equipment. This will then effectively minimize stock reduction, shortages, and overstocks. The vendor company can identify precisely which areas are efficient and which are not. Therefore, inefficiencies and problems that were not exposed before will become simple to spot with recommendations for further actions.

The vendor company should also insert IoT technologies in their products' shipping services for customers. Our lives depend on transportation, since it is important to travel for work and leisure, as well as the delivery of food and goods to destinations. The growing use of sensors attached to both products and the enclosures (that move them from point A to point B) opens a new window into real-time discovery of actual conditions, with clear ramifications for cost control and accountability. Thus, by adding IoT technologies to shipping services we can track shipments, optimize delivery and shipping routes, minimize costs associated with inefficiencies in logistics, and raise our expectations for goods and services. Additionally, merging data from weather meters and road closure notes makes operations run easily. It can also inform stakeholders of real-time operations - a major win in an era of instant gratification.

After using IoT in inventory management, the shipping service of products and having explained its effective and efficient role, the next step is the creation of a company website for marketing and promotion of their products and services.

By using the proposed app the company can select the creation option of the website which is a free feature in our app. The required steps for creating the company's website are as follows:

- First the vendor company should complete a registration process as in Figure 1, after then, the company must produce all identity documents.
- The validation process of company documents has to be performed.
- For finding new customers and boosting their business, this app enables the company to build a high-quality website with the following prosperities:

- ○ A high-quality building platform.
- ○ Simple and attractive designs. The most important thing is to build trust for a business or company and ensure customers can find the content or sales information they are looking for.
- ○ It can map out company content.
- ○ It works on most browsers such as Firefox, Chrome, Safari, etc.
- ○ It is almost effortless to read on mobile platforms.
- ○ It is quick to load.
- ○ It supports the use of online social media such as LinkedIn, Facebook.
- ○ Use of offline channels such as press releases or groups you belong to.
- ○ Secure: using high level usernames and passwords, up-to date of platform software and any plugging/modules, and considering an external security monitoring software.
- ○ To assist people with physical and visual disabilities, this site will also provide voice support next to screens where the customer can make the entire purchase process of products by using the voice service.
- ○ The app also enables companies to create an account with best ePayment gateway.

- After creating the website of a vendor company, this app gives a full update on the related events and content on your sites, as well as customers' average time spent on your site, page views per visitor, percentage of reiterate visitors, and visitors' countries of origins.
- It also compares your large success stories with your less successful endeavors; it is simple to distinguish where your effort should be concentrated to enhance site page rank and draw in more traffic. As soon as you know where to direct your efforts, your expenditures and time can be used more effectively.

Now, let us ask ourselves a question: what is the relation between adding IoT in inventory management and a shipping service with the marketing process of company products via their site?

Large companies spend huge amounts of money on marketing their products, and money that can be used to produce a better product rather than being spent on reaching to the widest possible audience. Thus, why is marketing so expensive? This is because marketing agencies need to gather quite a lot of information to determine their target audience. Once they know their audience a campaign targeted towards them can be created. Since the IoT enables companies to obtain all information about their products, and then the marketer can create their marketing strategy easily and effectively. Moreover, this information can help to understand which products have reached the expiration of its validity and then the marketers can make a marketing strategy for selling this product with various offers and discounts.

By using the proposed app the marketers can obtain a huge amount of data about customers. This data includes customer location, time of buying, a list of purchases, and customer demographics as stored in our cloud database of customers. According to stored information about customers, the marketers can extract and analyze customers' preferences and habits, and build more attractive marketing plans.

In order to avoid any type of risks and misunderstanding, the marketers should inform consumers that their private data is stored and will be used for commercial purposes. The high level of transparency will help companies to minimize or eliminate consumers' dissatisfaction.

The general framework of proposed model is shown in Figure 3.

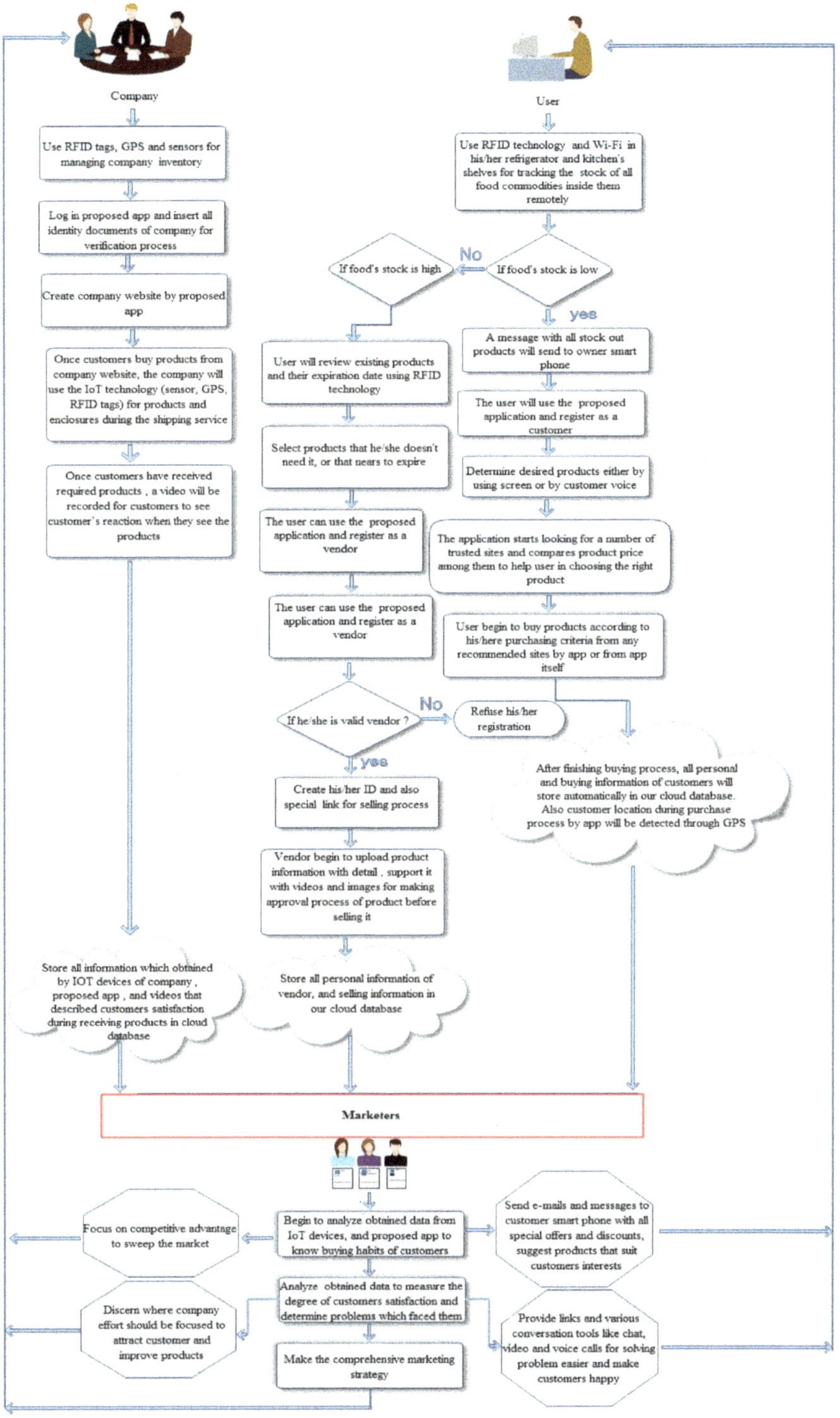

Figure 3. The proposed model.

4. Neutrosophic Technique for Helping the Customer to Select the Best Service or Product from Several Types

To determine the best choice between all of the possible alternatives, the multiple criteria decision making (MCDM) techniques are used widely. The problem of product or service selection on which decision maker has a typically vague and inaccurate result, which is a representative example of an MCDM problem. The traditional techniques have not been very effective for solving MCDM problems due to the inaccurate or unclear nature of the linguistic assessments. Finding the exact values for MCDM problems is complex and not possible in more real world cases. So, it is more rational to consider the values of alternatives regarding to the criteria as neutrosophic numbers (NNs). This part deals with The Technique for Order of Preference by Similarity to Ideal Solution (TOPSIS) method and expands the TOPSIS approach to the MCDM problem with single valued neutrosophic information. Here, the weights of criteria are calculated using the rank order centroids (ROCs) method, and the evaluation matrix for comparing alternatives relating to existing criteria is represented by using triangular neutrosophic numbers (TNNs).

So, in this section, we will explain how the proposed technique can help customers choose the best product among several types.

4.1. Proposed Neutrosophic Technique

As explained in the previous section, the customer will have the opportunity to choose from several products recommended by the proposed application. Next, the customer should choose the best product that suits his/her needs. But the assessment process of existing products is a multifaceted problem owing to several mismatched criteria. These include the interests of different customers, the diversity of products, and the breakdown of dealing with unclear and conflicting information which exists frequently during the selection processes. Therefore, we proposed a neutrosophic technique for helping customers to select the best product or service.

For determining the appropriate product or service, according to the several purchasing criteria of customer like price of product, delivery time to customer, quality of product, etc., let $C = \{C1, C2, \cdots, Cn\}$ be a given set of finite criteria for product or service, and $P = \{P1, P2, \cdots, Pm)$ be given set of finite alternatives (products).

The detailed steps for selecting the best possible product are as follows:

Step 1: Let the customer determine his/her selection criteria and rank them according to their needs.

Step 2: After determining the rank of purchase criteria by customer, the weight of these criteria must be calculated. Here we used the rank order centroids (ROCs) for assigning weights to these criteria [31]. For a set of ranks of order N, the ROC value which corresponds to the kth rank is given by:

$$r_k = \left(\sum\nolimits_{i=k}^{N} \left(\frac{1}{i}\right)\right)/N \tag{1}$$

For example, if we have a set of three ranks, associated ROCs values are:

$$r_1 = \left(1 + \frac{1}{2} + \frac{1}{3}\right)/3 = 0.61,$$

$$r_2 = \left(0 + \frac{1}{2} + \frac{1}{3}\right)/3 = 0.28,$$

$$r_3 = \left(0 + 0 + \frac{1}{3}\right)/3 = 0.11.$$

Step 3: After assigning relative weights to the purchase criteria, we begin to build the estimation matrix which consists of m alternatives and n criteria using the following linguistic variables as in Table 1. The crossing of every alternative and criteria indicated as x_{ij}. Then, we have $(x_{ij})_{m\times n}$ matrix.

Table 1. Linguistic variables for comparison matrices.

Linguistic Variables	Neutrosophic Numbers
Very Low/ Bad (VL/VB)	$\langle 0,1,2;0.10,0.85,0.90\rangle$
Low/ Bad (L/B)	$\langle 1,2,3;0.20,0.75,0.80\rangle$
Medium Low/Bad (ML/MB)	$\langle 2,3,4;0.35,0.65,0.60\rangle$
Medium/Fair (M/F)	$\langle 3,4,5;0.50,0.50,0.50\rangle$
Medium High/ Good (MH/MG)	$\langle 4,5,6;0.60,0.35,0.30\rangle$
High/Good (H/G)	$\langle 5,6,7;0.80,0.20,0.15\rangle$
Very High/Good (VH/VG)	$\langle 6,7,8;0.90,0.10,0.05\rangle$
Extremely High/Good (EH/EG)	$\langle 7,8,9;1.00,0.00,0.00\rangle$

Each value in Table 1 is a single valued triangular neutrosophic number which is a special case of single valued neutrosophic set:

Definition 1. *A single valued neutrosophic set A over X, is an object taking the form* $A = \{\langle x, T_A(x), I_A(x), F_A(x)\rangle : x \in X\}$, *where* $T_A(x)$: $X \rightarrow [0, 1]$, $I_A(x)$: $X \rightarrow [0, 1]$ *and* $F_A(x)$: $X \rightarrow [0, 1]$ *with* $0 \leq T_A(x) + I_A(x) + F_A(x) \leq 3$ *for all* $x \in X$. *The intervals* $T_A(x)$, $I_A(x)$ *and* $F_A(x)$ *represent the truth-membership degree, the indeterminacy-membership degree and the falsity membership degree of x to A, respectively and X is a universe of discourse. For convenience, a SVN number is represented by* $A = (a, b, c)$, *where* $a, b, c \in [0, 1]$ *and* $a + b + c \leq 3$ *[32].*

Definition 2. *A single valued triangular neutrosophic number* $\tilde{a} = \langle (a_1, a_2, a_3); T_a, I_a, F_a\rangle$ *is a special neutrosophic set on the real line set R, whose truth-membership, indeterminacy-membership and falsity-membership functions are defined as follows [32]:*

$$T_a(x) = \begin{cases} T_a\left(\frac{x-a_1}{a_2-a_1}\right) & (a_1 \leq x < a_2) \\ T_a & (x = a_2) \\ T_a\left(\frac{a_3-x}{a_3-a_2}\right) & (a_2 < x \leq a_3) \\ 0 & \text{otherwise,} \end{cases} \tag{2}$$

$$I_a(x) = \begin{cases} \frac{(a_2-x+I_a(x-a_1))}{(a_2-a_1)} & (a_1 \leq x < a_2) \\ I_a & (x = a_2) \\ \frac{(x-a_2+I_a(a_3-x))}{(a_3-a_2)} & (a_2 < x \leq a_3) \\ 1 & \text{otherwise,} \end{cases} \tag{3}$$

$$F_a(x) = \begin{cases} \frac{(a_2-x+F_a(x-a_1))}{(a_2-a_1)} & (a_1 \leq x < a_2) \\ F_a & (x = a_2) \\ \frac{(x-a_2+F_a\,(a_3-x))}{(a_3-a_2)} & (a_2 < x \leq a_3) \\ 1 & \text{otherwise.} \end{cases} \tag{4}$$

where T_a, I_a *and* $F_a(x)$, *represent the maximum truth-membership degree, minimum indeterminacy-membership degree and minimum falsity-membership degree respectively. A single valued triangular neutrosophic number* $\tilde{a} = (a_1, a_2, a_3); T_a, I_a, F_a$ *may express an ill-defined quantity about a, which is approximately equal to a.*

Definition 3. *Let* $\tilde{a} = \langle (a_1, a_2, a_3); T_a, I_a, F_a\rangle$ *and* $\tilde{b} = \langle (b_1, b_2, b_3); T_b, I_b, F_b\rangle$ *be two single valued triangular neutrosophic numbers and* $\gamma \neq 0$ *be any real number [32]. Then,*

1. *Addition of two triangular neutrosophic numbers*

$$\tilde{a} + \tilde{b} = \langle (a_1 + b_1, a_2 + b_2, a_3 + b_3); T_a \wedge T_b, I_a \vee I_b, F_a \vee F_b\rangle$$

2. *Subtraction of two triangular neutrosophic numbers*

$$\widetilde{a} - \widetilde{b} = \langle (a_1 - b_3, a_2 - b_2, a_3 - b_1); T_a \wedge T_b, I_a \vee I_b, F_a \vee F_b \rangle$$

3. *Inverse of a triangular neutrosophic number*

$$\widetilde{a}^{-1} = \langle \left(\frac{1}{a_3}, \frac{1}{a_2}, \frac{1}{a_1} \right); T_a, I_a, F_a \rangle, \text{ where } (\widetilde{a} \neq 0)$$

4. *Multiplication of triangular neutrosophic number by constant value*

$$\gamma\widetilde{a} = \begin{cases} \langle (\gamma a_1, \gamma a_2, \gamma a_3); T_a, I_a, F_a \rangle \text{ if } (\gamma > 0) \\ \langle (\gamma a_3, \gamma a_2, \gamma a_1); T_a, I_a, F_a \rangle \text{ if } (\gamma < 0) \end{cases}$$

5. *Division of triangular neutrosophic number by constant value*

$$\frac{\widetilde{a}}{\gamma} = \begin{cases} \langle \left(\frac{a_1}{\gamma}, \frac{a_2}{\gamma}, \frac{a_3}{\gamma} \right); T_a, I_a, F_a \rangle \text{ if } (\gamma > 0) \\ \langle \left(\frac{a_3}{\gamma}, \frac{a_2}{\gamma}, \frac{a_1}{\gamma} \right); T_a, I_a, F_a \rangle \text{ if } (\gamma < 0) \end{cases}$$

6. *Division of two triangular neutrosophic numbers*

$$\frac{\widetilde{a}}{\widetilde{b}} = \begin{cases} \langle \left(\frac{a_1}{b_3}, \frac{a_2}{b_2}, \frac{a_3}{b_1} \right); T_a \wedge T_b, I_a \vee I_b, F_a \vee F_b \rangle & \text{if}(a_3 > 0, b_3 > 0) \\ \langle \left(\frac{a_3}{b_3}, \frac{a_2}{b_2}, \frac{a_1}{b_1} \right); T_a \wedge T_b, I_a \vee I_b, F_a \vee F_b \rangle & \text{if}(a_3 < 0, b_3 > 0) \\ \langle \left(\frac{a_3}{b_1}, \frac{a_2}{b_2}, \frac{a_1}{b_3} \right); T_a \wedge T_b, I_a \vee I_b, F_a \vee F_b \rangle & \text{if}(a_3 < 0, b_3 < 0) \end{cases}$$

7. *Multiplication of two triangular neutrosophic numbers*

$$\widetilde{ab} = \begin{cases} \langle (a_1 b_1, a_2 b_2, a_3 b_3); T_a \wedge T_b, I_a \vee I_b, F_a \vee F_b \rangle & \text{if } (a_3 > 0, b_3 > 0) \\ \langle (a_1 b_1, a_2 b_2, a_3 b_3); T_a \wedge T_b, I_a \vee I_b, F_a \vee F_b \rangle & \text{if } (a_3 < 0, b_3 > 0) \\ \langle (a_3 b_3, a_2 b_2, a_1 b_1); T_a \wedge T_b, I_a \vee I_b, F_a \vee F_b \rangle & \text{if } (a_3 < 0, b_3 < 0) \end{cases}$$

So, the evaluation matrix of alternatives with respect to criteria will take the following form:

$$E = \langle e_{ij} \rangle_{m \times n} = \langle L_{ij}, M_{ij}, U_{ij}; T_{ij}, I_{ij}, F_{ij} \rangle_{m \times n} =$$

$$\begin{array}{c} \\ A_1 \\ A_2 \\ \vdots \\ A_m \end{array} \begin{array}{cccc} C_1 & C_2 & \cdots & C_n \\ \left(\begin{array}{c} \langle L_{11}, M_{11}, U_{11}; T_{11}, I_{11}, F_{11} \rangle \\ \langle L_{21}, M_{21}, U_{21}; T_{21}, I_{21}, F_{21} \rangle \\ \vdots \\ \langle L_{m1}, M_{m1}, U_{m1}; T_{m1}, I_{m1}, F_{m1} \rangle \end{array} \right. & \begin{array}{c} \langle L_{12}, M_{12}, U_{12}; T_{12}, I_{12}, F_{12} \rangle \\ \langle L_{22}, M_{22}, U_{22}; T_{22}, I_{22}, F_{22} \rangle \\ \vdots \\ \langle L_{m2}, M_{m2}, U_{m2}; T_{m2}, I_{m2}, F_{m2} \rangle \end{array} & \begin{array}{c} \cdots \\ \cdots \\ \ddots \\ \cdots \end{array} & \left. \begin{array}{c} \langle L_{1n}, M_{1n}, U_{1n}; T_{1n}, I_{1n}, F_{1n} \rangle \\ \langle L_{2n}, M_{2n}, U_{2n}; T_{2n}, I_{2n}, F_{2n} \rangle \\ \vdots \\ \langle L_{mn}, M_{mn}, U_{mn}; T_{mn}, I_{mn}, F_{mn} \rangle \end{array} \right) \end{array} \quad (5)$$

where, $e_{ij} = \langle L_{ij}, M, U_{ij}; T_{ij}, I_{ij}, F_{ij} \rangle$ is the triangular neutrosophic element of evaluation matrix E for $i = 1; 2; \ldots; m$ and $j = 1; 2; \ldots; n$. Since L, M, U are the lower, median and upper value of triangular neutrosophic number, and T, I, F are the truth, indeterminacy and falsity degrees of this triangular neutrosophic number.

Step 4: Calculate the neutrosophic weighted evaluation matrix as follows:

$$E^w = E \times w = \langle e_{ij}{}^{wj} \rangle_{m \times n} = \langle L_{ij}{}^{wj}, M_{ij}{}^{wj}, U_{ij}{}^{wj}; T_{ij}^{wj}, I_{ij}{}^{wj}, F_{ij}{}^{wj} \rangle_{m \times n} =$$

$$\begin{array}{c} \\ A_1 \\ A_2 \\ \vdots \\ A_m \end{array} \begin{array}{c} \begin{array}{cccc} C_1 & C_2 & \cdots & C_n \end{array} \\ \begin{pmatrix} \langle L_{11}^{w1}, M_{11}^{w1}, U_{11}^{w1}; T_{11}^{w1}, I_{11}^{w1}, F_{11}^{w1} \rangle & \langle L_{12}^{w2}, M_{12}^{w2}, U_{12}^{w2}; T_{12}^{w2}, I_{12}^{w2}, F_{12}^{w2} \rangle & \cdots & \langle L_{1n}^{wn}, M_{1n}^{wn}, U_{1n}^{wn}; T_{1n}^{wn}, I_{1n}^{wn}, F_{1n}^{wn} \rangle \\ \langle L_{21}^{w1}, M_{21}^{w1}, U_{21}^{w1}; T_{21}^{w1}, I_{21}^{w1}, F_{21}^{w1} \rangle & \langle L_{22}^{w2}, M_{22}^{w2}, U_{22}^{w2}; T_{22}^{w2}, I_{22}^{w2}, F_{22}^{w2} \rangle & \cdots & \langle L_{2n}^{wn}, M_{2n}^{wn}, U_{2n}^{wn}; T_{2n}^{wn}, I_{2n}^{wn}, F_{2n}^{wn} \rangle \\ \vdots & \vdots & \ddots & \vdots \\ \langle L_{m1}^{w1}, M_{m1}^{w1}, U_{m1}^{w1}; T_{m1}^{w1}, I_{m1}^{w1}, F_{m1}^{w1} \rangle & \langle L_{m2}^{w2}, M_{m2}^{w2}, U_{m2}^{w2}; T_{m2}^{w2}, I_{m2}^{w2}, F_{m2}^{w2} \rangle & \cdots & \langle L_{mn}^{wn}, M_{mn}^{wn}, U_{mn}^{wn}; T_{mn}^{wn}, I_{mn}^{wn}, F_{mn}^{wn} \rangle \end{pmatrix} \end{array} \quad (6)$$

Here, $\langle e_{ij}{}^{wj} \rangle_{m \times n} = \langle L_{ij}{}^{wj}, M_{ij}{}^{wj}, U_{ij}{}^{wj}; T_{ij}{}^{wj}, I_{ij}{}^{wj}, F_{ij}{}^{wj} \rangle_{m \times n}$ is an element of the weighted neutrosophic evaluation matrix E^w for $i = 1; 2; \cdots; m$ and $j = 1; 2; \cdots; n$.

Step 5: Define the neutrosophic positive and negative ideal solution NPIS and NNIS:

Their exists two types of attributes B_1 and B_2, which are the benefit and cost-type attribute respectively. So, v^+ is the neutrosophic positive ideal solution (NPIS), v^- is the neutrosophic negative ideal solution (NNIS) and have the following formula:

$$v^+ = \left[e_1{}^{w+}, \; e_2{}^{w+}, \cdots, e_n{}^{w+} \right]. \quad (7)$$

where $e_j{}^{w+} = \langle L_j{}^{w+}, M_j{}^{w+}, U_j{}^{w+}; T_j{}^{w+}, I_j{}^{w+}, F_j{}^{w+} \rangle$ for $j = 1, 2, \cdots, n$.

$$L_j{}^{w+} = \left\{ \left(\max{}_i \left\{ L_{ij}^{wj} \right\} \middle| j \in B_1 \right), \left(\min{}_i \left\{ L_{ij}^{wj} \right\} \middle| j \in B_2 \right) \right\}, \quad (8)$$

$$M_j{}^{w+} = \left\{ \left(\max{}_i \left\{ M_{ij}^{wj} \right\} \middle| j \in B_1 \right), \left(\min{}_i \left\{ M_{ij}^{wj} \right\} \middle| j \in B_2 \right) \right\}, \quad (9)$$

$$U_j{}^{w+} = \left\{ \left(\max{}_i \left\{ U_{ij}^{wj} \right\} \middle| j \in B_1 \right), \left(\min{}_i \left\{ U_{ij}^{wj} \right\} \middle| j \in B_2 \right) \right\}, \quad (10)$$

$$T_j{}^{w+} = \left\{ \left(\max{}_i \left\{ T_{ij}^{wj} \right\} \middle| j \in B_1 \right), \left(\min{}_i \left\{ T_{ij}^{wj} \right\} \middle| j \in B_2 \right) \right\}, \quad (11)$$

$$I_j{}^{w+} = \left\{ \left(\min{}_i \left\{ I_{ij}^{wj} \right\} \middle| j \in B_1 \right), \left(\max{}_i \left\{ I_{ij}^{wj} \right\} \middle| j \in B_2 \right) \right\}, \quad (12)$$

$$F_j{}^{w+} = \left\{ \left(\min{}_i \left\{ F_{ij}^{wj} \right\} \middle| j \in B_1 \right), \left(\max{}_i \left\{ F_{ij}^{wj} \right\} \middle| j \in B_2 \right) \right\}. \quad (13)$$

Also,

$$v^- = [e_1{}^{w-}, \; e_2{}^{w-}, \cdots, e_n{}^{w-}], \quad (14)$$

where $\langle e_j{}^{w-} = L_j{}^{w-}, M_j{}^{w-}, U_j{}^{w-}; T_j{}^{w-}, I_j{}^{w-}, F_j{}^{w-} \rangle$ for $j = 1, 2, \cdots, n$.

$$L_j{}^{w-} = \left\{ \left(\min{}_i \left\{ L_{ij}^{wj} \right\} \middle| j \in B_1 \right), \left(\max{}_i \left\{ L_{ij}^{wj} \right\} \middle| j \in B_2 \right) \right\}, \quad (15)$$

$$M_j{}^{w-} = \left\{ \left(\min{}_i \left\{ M_{ij}^{wj} \right\} \middle| j \in B_1 \right), \left(\max{}_i \left\{ M_{ij}^{wj} \right\} \middle| j \in B_2 \right) \right\}, \quad (16)$$

$$U_j{}^{w-} = \left\{ \left(\min{}_i \left\{ U_{ij}^{wj} \right\} \middle| j \in B_1 \right), \left(\max{}_i \left\{ U_{ij}^{wj} \right\} \middle| j \in B_2 \right) \right\}, \quad (17)$$

$$T_j{}^{w-} = \left\{ \left(\min{}_i \left\{ T_{ij}^{wj} \right\} \middle| j \in B_1 \right), \left(\max{}_i \left\{ T_{ij}^{wj} \right\} \middle| j \in B_2 \right) \right\}, \quad (18)$$

$$I_j{}^{w-} = \left\{ \left(\max{}_i \left\{ I_{ij}^{wj} \right\} \middle| j \in B_1 \right), \left(\min{}_i \left\{ I_{ij}^{wj} \right\} \middle| j \in B_2 \right) \right\}, \quad (19)$$

$$F_j{}^{w-} = \left\{ \left(\max{}_i \left\{ F_{ij}^{wj} \right\} \middle| j \in B_1 \right), \left(\min{}_i \left\{ F_{ij}^{wj} \right\} \middle| j \in B_2 \right) \right\}. \quad (20)$$

Step 6: Measure the Euclidian distance of each alternative from the NPIS and NNIS:

The normalized Euclidian distance of each alternative $\langle L_{ij}{}^{wj}, M_{ij}{}^{wj}, U_{ij}{}^{wj}; T_{ij}{}^{wj}, I_{ij}{}^{wj}, F_{ij}{}^{wj}\rangle$ from the neutrosophic positive ideal solution $\langle L_j{}^{w+}, M_j{}^{w+}, U_j{}^{w+}; T_j{}^{w+}, I_j{}^{w+}, F_j{}^{w+}\rangle$ for $i = 1, 2, \cdots, m$ and $j = 1, 2, \cdots, n$ written as follows:

$$D_N(A_i, \text{NPIS}) = D^{i+}{}_N(e_{ij}{}^{wj}, e_j{}^{w+}) = \sqrt{\frac{1}{6n}\sum_n^j \left\{ \begin{array}{c} \left(L_{ij}{}^{wj}(x_j) - L_j{}^{w+}(x_j)\right)^2 + \left(M_{ij}{}^{wj}(x_j) - M_j{}^{w+}(x_j)\right)^2 + \left(U_{ij}{}^{wj}(x_j) - U_j{}^{w+}(x_j)\right)^2 + \\ \left(T_{ij}{}^{wj}(x_j) - T_j{}^{w+}(x_j)\right)^2 + \left(I_{ij}{}^{wj}(x_j) - I_j{}^{w+}(x_j)\right)^2 + \left(F_{ij}{}^{wj}(x_j) - F_j{}^{w+}(x_j)\right)^2 \end{array} \right\}}. \tag{21}$$

Also, the normalized Euclidian distance of each alternative $\langle L_{ij}{}^{wj}, M_{ij}{}^{wj}, U_{ij}{}^{wj}; T_{ij}{}^{wj}, I_{ij}{}^{wj}, F_{ij}{}^{wj}\rangle$ from the neutrosophic negative ideal solution $\langle L_j{}^{w-}, M_j{}^{w-}, U_j{}^{w-}; T_j{}^{w-}, I_j{}^{w-}, F_j{}^{w-}\rangle$ for $i = 1, 2, \cdots, m$ and $j = 1, 2, \cdots, n$ written as follows:

$$D_N(A_i, \text{NNIS}) = D^{i-}{}_N\left(e_{ij}{}^{wj}, e_j{}^{w-}\right) = \sqrt{\frac{1}{6n}\sum_n^j \left\{ \begin{array}{c} \left(L_{ij}{}^{wj}(x_j) - L_j{}^{w-}(x_j)\right)^2 + \left(M_{ij}{}^{wj}(x_j) - M_j{}^{w-}(x_j)\right)^2 + \left(U_{ij}{}^{wj}(x_j) - U_j{}^{w-}(x_j)\right)^2 + \\ \left(T_{ij}{}^{wj}(x_j) - T_j{}^{w-}(x_j)\right)^2 + \left(I_{ij}{}^{wj}(x_j) - I_j{}^{w-}(x_j)\right)^2 + \left(F_{ij}{}^{wj}(x_j) - F_j{}^{w-}(x_j)\right)^2 \end{array} \right\}}. \tag{22}$$

Step 7: Calculate the closeness coefficient of each alternative according to the NPIS:

$$c_i^* = \frac{D^{i-}{}_N\left(e_{ij}{}^{wj}, e_j{}^{w-}\right)}{D^{i+}{}_N\left(e_{ij}{}^{wj}, e_j{}^{w+}\right) + D^{i-}{}_N\left(e_{ij}{}^{wj}, e_j{}^{w-}\right)} \text{ where } 0 \le c_i^* \le 1. \tag{23}$$

Step 8: Rank alternatives according to the largest values of c_i^*.

4.2. A Numerical Example

If a customer plans to buy a specific product using the proposed app, then the app will search for that product from various websites. After that, the app will return various products with different prices and different qualities. The customer should have to choose the best one according to his/her needs, so the decision in his/her hands. Since the selection of the best product is the customer mission (and it is a complex problem because of vague and incomplete information) we will apply the ROCs method and neutrosophic TOPSIS for the selection process as follows:

Step 1: Ask customer to insert his/her purchase criteria and rank them from the most to the least important. Here the customer ranked the purchase criteria as follows:

1. Quality,
2. Price,
3. Delivery Time.

Step 2: After determining the rank of purchase criteria by the customer, the weight of these criteria must be calculated using the ROCs method. Since the rank of purchase criteria according to customer needs are: Quality, Price and Time respectively. Then, by using the ROCs method, the weight of criteria will be as follows:

$$W_1(\text{Quality}) = 0.61,\ W_2(\text{Price}) = 0.28,\ W_3(\text{Time}) = 0.11.$$

Step 3: Assuming that the customer should choose one from the four initially selected products p_1, p_2, p_3, p_4 with respect to three criteria which determined previously, the decision maker will compare all alternatives according to criteria using the linguistic scale which was presented in Table 1. Since there is no absolute truth and the truth is always relative, the single valued triangular neutrosophic numbers have been used for handling unclear, imperfect and conflicting information which usually exists in actuality.

By comparing the four products with respect to existing criteria, the estimation matrix is as in Table 2.

Step 4: Construct the weighted decision matrix via multiplying weights of criteria by the estimation matrix as in Table 3.

Step 5: Define the neutrosophic positive and negative ideal solution, NPIS and NNIS by using equations from Equation (7) to Equation (20) according to attribute type (i.e., benefit or cost).

Table 2. Estimation matrix of alternatives according to criteria.

P	C_1	C_2	C_3
P_1	G	VH	B
P_2	EG	H	G
P_3	G	L	VG
P_4	VG	H	VG

Table 3. Weighted decision matrix of alternatives.

P	C_1	C_2	C_3
P_1	$\langle 3,4,4;0.80,0.20,0.15\rangle$	$\langle 2,2,2;0.90,0.10,0.05\rangle$	$\langle 0,0,0;0.20,0.75,0.80\rangle$
P_2	$\langle 4,5,5;1.00,0.00,0.00\rangle$	$\langle 1,2,2;0.80,0.20,0.15\rangle$	$\langle 0,1,1;0.80,0.20,0.15\rangle$
P_3	$\langle 3,4,4;0.80,0.20,0.15\rangle$	$\langle 0,1,1;0.20,0.75,0.80\rangle$	$\langle 1,1,1;0.90,0.10,0.05\rangle$
P_4	$\langle 4,4,5;0.90,0.10,0.05\rangle$	$\langle 1,2,2;0.80,0.20,0.15\rangle$	$\langle 0,1,1;0.90,0.10,0.05\rangle$

The NPIS and NNIS are given by:

$$\text{NPIS} = (\langle 4,4,4;1.00,0.00,0.00\rangle, \langle 0,1,1;0.2,0.75,0.80\rangle, \langle 1,0,0;0.9,0.1,0.05\rangle),$$

$$\text{NNIS} = (\langle 3,4,4;0.8,0.20,0.15\rangle, \langle 2,2,2;0.90,0.10,0.05\rangle, \langle 0,0,0;0.20,0.75,0.80\rangle).$$

If obtained values of lower, median, and/or upper are out of order, then reorder them and follow the work.

Step 6: Calculate the normalized Euclidian distance of each alternative from the NPIS as follows:

$$D(P_1, \text{NPIS}) = 0.78,\ D(P_2, \text{NPIS}) = 0.71,$$

$$D(P_3, \text{NPIS}) = 0.41,\ D(P_4, \text{NPIS}) = 0.67.$$

Step 7: Calculate the normalized Euclidian distance of each alternative from the neutrosophic negative ideal solution as follows:

$$D(P_1, \text{NNIS}) = 0.0,\ D(P_2, \text{NNIS}) = 0.63,$$

$$D(P_3, \text{NNIS}) = 0.81,\ D(P_4, \text{NNIS}) = 0.60.$$

Step 8: Calculate the closeness coefficient of each alternative according to the NPIS using Equation (23):

$$c_1^* = 0.00,\ c_2^* = 0.470,\ c_3^* = 0.663,\ c_4^* = 0.472.$$

Step 9: Rank alternatives according to the largest values of c_i^*:

$$P_3 > P_4 > P_2 > P_1$$

Hence, the best product which suits customer needs is P_3 as appears in Figure 4.

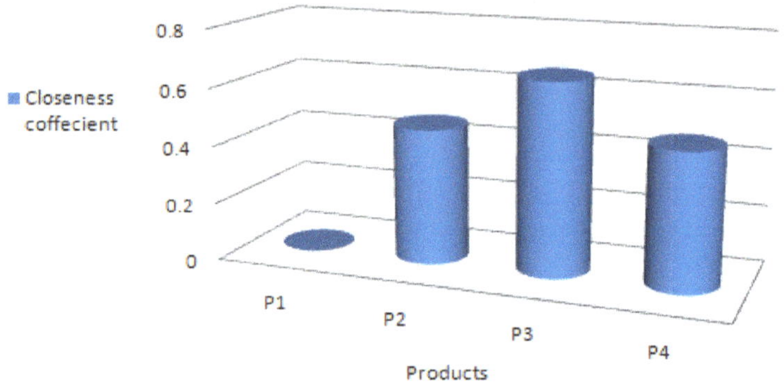

Figure 4. Products and their closeness coefficients.

5. Conclusion and Future Directions

In this paper we proposed a smart e-marketing system based on IoT. Using this system enables marketers to meet customers' expectations for products and services, and then achieve a high degree of satisfaction. It also enables marketers to have new streams of data, and discover consumer preferences and habits. Therefore, it enables marketers to provide customers with what they need with better matches.

By using the proposed system by customers and companies, all problems which face them in online shopping could be solved easily. It also helps customers, sellers, affiliate, companies, and marketers to achieve their goals with a high degree of accuracy.

In this system we focused only on food commodities as an example of products, but this system has the potential to be applied to all kinds of other products in our daily lives.

Although the proposed system can avoid different threats and hacking process, more involvements from governments to create legal basis is very significant and will make the proposed system better.

Since there are different types of the same product and also different criteria of buying, and the main problem of product or service selection is the vague and inaccurate knowledge of the decision maker, we presented a multi-criteria decision making technique based on neutrosophic TOPSIS to deal with unclear and conflicting information. In this technique we calculated the weights of criteria by using the rank order centroids (ROCs) method, and the evaluation matrix for comparing alternatives regarding to existing criteria is represented by using triangular neutrosophic numbers (TNNs). This technique will support customers in selecting the best product or service.

For our future work, we will expand our IoT research outputs, applications and services, aiming to apply them in different domains in agriculture, health, and industry. Furthermore, we will apply the proposed neutrosophic technique in various situations, not only for supporting customers, but also supporting marketers and companies.

Author Contributions: Conceptualization, M.A.-B.; methodology, M.A.-B., M.M. and F.S.; validation, M.A.-B., V.C. and F.S.; formal analysis, M.A.-B.; investigation, V.C.; resources, V.C.; writing—original draft, M.A.-B. and M.M.; writing—review and editing, V.C. and F.S.; visualization, M.M. and F.S.; project administration, M.A.-B.

Funding: This research received no external funding.

Conflicts of Interest: The authors declare no conflict of interest.

References

1. Abdel-Basset, M.; Manogaran, G.; Mohamed, M. Internet of Things (IoT) and its impact on supply chain: A framework for building smart, secure and efficient systems. *Future Gener. Comput. Syst.* **2018**. [CrossRef]

2. Abdel-Basset, M.; Manogaran, G.; Mohamed, M.; Rushdy, E. Internet of things in smart education environment: Supportive framework in the decision-making process. *Concurr. Comput. Pract. Exp.* **2018**, e4515. [CrossRef]
3. Gartner. Gartner Says 6.4 Billion Connected Things Will Be in Use in 2016, up 30 percent from 2015. Press Release, STAMFORD, Conn., 10 November 2015. Retrieved 8 September, 2016. Available online: http://www.gartner.com/newsroom/id/3165317 (accessed on 2 November 2018).
4. Nguyen, B.; Simkin, L. The internet of things (IoT) and marketing: The state of play, future trends and the implications for marketing. *J. Mark. Maneg.* **2017**, *33*, 1–6. [CrossRef]
5. Jie, Y.U.; Subramanian, N.; Ning, K.; Edwards, D. Product delivery service provider selection and customer satisfaction in the era of internet of things: A Chinese e-retailers' perspective. *Int. J. Prod. Econ.* **2015**, *159*, 104–116.
6. Desai, K.; Mahalakshmi, S. Impressions of service quality dimensions on customer's intention to use IoT at Bangalore Electricity Supply Company (BESCOM). *J. Adv. Manag. Res.* **2018**, *6*, 74–82.
7. Yerpude, S.; Singhal, T. Customer service enhancement through on-road vehicle assistance enabled with internet of things (IoT) solutions and frameworks: A futuristic perspective. *Int. J. Appl. Bus. Econ. Res.* **2018**, *15*, 551–565.
8. Balaji, M.S.; Roy, S.K. Value co-creation with Internet of things technology in the retail industry. *J. Mark. Manag.* **2017**, *33*, 7–31. [CrossRef]
9. Canhoto, A.I.; Arp, S. Exploring the factors that support adoption and sustained use of health and fitness wearable's. *J. Mark. Manag.* **2017**, *33*, 32–60. [CrossRef]
10. Wu, J.; Chen, J.; Dou, W. The internet of things and interaction style: The effect of smart interaction on brand attachment. *J. Mark. Manag.* **2017**, *33*, 61–75. [CrossRef]
11. Mani, Z.; Chouk, I. Drivers of consumers' resistance to smart products. *J. Mark. Manag.* **2017**, *33*, 76–97. [CrossRef]
12. Woodside, A.G.; Sood, S. Vignettes in the two-step arrival of the internet of things and its reshaping of marketing management's service-dominant logic. *J. Mark. Manag.* **2017**, *33*, 98–110. [CrossRef]
13. Ehret, M.; Wirtz, J. Unlocking value from machines: Business models and the industrial internet of things. *J. Mark. Manag.* **2017**, *33*, 111–130. [CrossRef]
14. Pathan, A.; Kokate, R.; Mutha, A.; Pingale, P.; Gadakh, P. Digital India: IoT based intelligent interactive super market framework for shopping mall. *Eng. Sci.* **2016**, *1*, 1–5.
15. Balmer, J.M.; Yen, D.A. The Internet of total corporate communications, quaternary corporate communications and the corporate marketing Internet revolution. *J. Mark. Manag.* **2017**, *33*, 131–144. [CrossRef]
16. De Cremer, D.; Nguyen, B.; Simkin, L. The integrity challenge of the internet-of-things (IoT): On understanding its dark side. *J. Mark. Manag.* **2017**, *33*, 145–158. [CrossRef]
17. Rajabi, N.; Hakim, A. An intelligent interactive marketing system based-on Internet of Things (IoT). In Proceedings of the Knowledge-Based Engineering and Innovation (KBEI), Tehran, Iran, 5–6 November 2015; pp. 243–247.
18. Aksu, H.; Babun, L.; Conti, M.; Tolomei, G.; Uluagac, A.S. Advertising in the IoT era: Vision and challenges. *IEEE Commun. Mag.* **2018**. [CrossRef]
19. Celik, F. Internet of Things as a Source of Future Marketing Tools. Bachelor's Thesis, University of Twente, Enschede, The Netherlands, 2016.
20. Jayaram, A. Smart Retail 4.0 IoT Consumer retailer model for retail intelligence and strategic marketing of in-store products. In Proceedings of the 17th International Business Horizon-INBUSH ERA-2017, Noida, India, 9 February 2017.
21. Đurđević, N.; Labus, A.; Bogdanović, Z.; Despotović-Zrakić, M. Internet of things in marketing and retail. *Int. J. Adv. Comput. Sci. Appl.* **2017**, *6*, 7–11.
22. Tsai, Y.T.; Wang, S.C.; Yan, K.Q.; Chang, C.M. Precise positioning of marketing and behavior intentions of location-based mobile commerce in the internet of things. *Symmetry* **2017**, *9*, 139. [CrossRef]
23. Nallakaruppan, M.K.; Kumaran, U.S. Quick fix for obstacles emerging in management recruitment measure using IOT-based candidate selection. *Serv. Oriented Comput. Appl.* **2018**, *12*, 275–284. [CrossRef]
24. Wibowo, S.; Grandhi, S. Fuzzy multicriteria analysis for performance evaluation of internet-of-things-based supply chains. *Symmetry* **2018**, *10*, 603. [CrossRef]
25. Abdel-Basset, M.; Gunasekaran, M.; Mohamed, M.; Chilamkurti, N. Three-way decisions based on neutrosophic sets and AHP-QFD framework for supplier selection problem. *Future Gener. Comput. Syst.* **2018**, *89*, 19–30. [CrossRef]

26. Ziemba, P.; Jankowski, J.; Wątróbski, J. Dynamic decision support in the internet marketing management. In *Transactions on Computational Collective Intelligence*; Springer: Cham, Switzerland, 2018; pp. 39–68.
27. Chaffey, D.; Ellis-Chadwick, F.; Mayer, R.; Johnston, K. *Internet Marketing: Strategy, Implementation and Practice*; Pearson Education: London, UK, 2009.
28. Karczmarczyk, A.; Jankowski, J.; Wątróbski, J. Multi-criteria decision support for planning and evaluation of performance of viral marketing campaigns in social networks. *PLoS ONE* **2018**, *13*. [CrossRef] [PubMed]
29. Al-Fuqaha, A.; Guizani, M.; Mohammadi, M.; Aledhari, M.; Ayyash, M. Internet of things: A survey on enabling technologies, protocols, and applications. *IEEE Commun. Surv. Tutor.* **2015**, *17*, 2347–2376. [CrossRef]
30. Strzelecki, A. Key Features of e-tailer shops in adaptation to cross-border e-commerce in the EU. *Sustainability* **2019**, *11*, 1589. [CrossRef]
31. McCaffrey, J.D. Using the multi-attribute global inference of quality (MAGIQ) technique for software testing. In Proceedings of the Information Technology: New Generations, Las Vegas, NV, USA, 27–29 April 2009.
32. Abdel-Basset, M.; Mohamed, M.; Chang, V. NMCDA: A framework for evaluating cloud computing services. *Future Gener. Comput. Syst.* **2018**, *86*, 12–29. [CrossRef]

symmetry

MDPI

Article

Neutrosophic Optimization Model and Computational Algorithm for Optimal Shale Gas Water Management under Uncertainty

Firoz Ahmad [1,†,‡], Ahmad Yusuf Adhami [1,‡] and Florentin Smarandache [2,*]

1 Department of Statistics and Operations Research, Aligarh Muslim University, Aligarh-202002, India; firoz.ahmad02@gmail.com (F.A.); yusufstats@gmail.com (A.Y.A.)
2 Mathematics Department, University of New Mexico, 705 Gurley Ave., Gallup, NM 87301, USA
* Correspondence: fsmarandache@gmail.com
† Current address: Department of Statistics and Operations Research, Aligarh Muslim University.
‡ These authors contributed equally to this work.

Received: 17 March 2019; Accepted: 11 April 2019; Published: 15 April 2019

Abstract: Shale gas energy is the most prominent and dominating source of power across the globe. The processes for the extraction of shale gas from shale rocks are very complex. In this study, a multiobjective optimization framework is presented for an overall water management system that includes the allocation of freshwater for hydraulic fracturing and optimal management of the resulting wastewater with different techniques. The generated wastewater from the shale fracking process contains highly toxic chemicals. The optimal control of a massive amount of contaminated water is quite a challenging task. Therefore, an on-site treatment plant, underground disposal facility, and treatment plant with expansion capacity were designed to overcome environmental issues. A multiobjective trade-off between socio-economic and environmental concerns was established under a set of conflicting constraints. A solution method—the neutrosophic goal programming approach—is suggested, inspired by independent, neutral/indeterminacy thoughts of the decision-maker(s). A theoretical computational study is presented to show the validity and applicability of the proposed multiobjective shale gas water management optimization model and solution procedure. The obtained results and conclusions, along with the significant contributions, are discussed in the context of shale gas supply chain planning policies over different time horizons.

Keywords: intuitionistic fuzzy parameters; uncertainty modeling; neutrosophic goal programming approach; shale gas water management system

1. Introduction

Energy sources play a dynamic role in the development, nourishment, and enrichment reputation of any country. Presently, many conventional sources of energy are being used for energy production, but shale gas energy is booming among different energy sources [1–3]. Apart from conventional sources of energy, shale gas—which is located within shale rocks—is the most promising source of natural gas. Recently, shale gas has become an emerging source of natural gas across the world [4,5]. The United States is the second-richest country after China in terms of the abundance of shale gas resources. Since the start of this century, significant interest has been shown in the potential extraction of shale gas across the world [6–8]. In 2000, only 1% of the US natural gas production was contributed by shale gas energy; by 2010, it was more than 20%, and according to predictions of the US government's Energy Information Administration (EIA), by 2035 more than 46% of the US' natural gas supply will be from shale gas [9]. The first extraction of shale gas from shale rocks was done in Fredonia (New York) in 1821 by using shallow and low-pressure fractures. However, horizontal drilling started in the

1930s, and the first well was fractured in the US in 1947. Presently, shale gas potential extraction and enriched abundance in many nations are being investigated. According to Sieminski et al. [10], in 2013, only a few countries (e.g., the US, Canada, and China) have sufficient shale gas enrichment, and future production is planned at commercial scale [10]. China has an apparent strategy to dramatically grow its shale gas production and investment, which has been restricted by its insufficient approach to water, land, and the latest technology. Shale hosts rocks trapping potential shale gas quantities that have numerous common properties, namely, being composed of organic material, a mature petroleum source, containing a high amount of natural gas in the thermogenic gas window spread inside the Earth's crust where there is high heat and pressure being applied to convert petroleum into natural gas. Most commonly, hydraulic fracturing (also known as *fracking*) and horizontal drilling are two dominant methods that are being used in the process of shale gas extraction across the world. The high concentration of released toxic and contaminated wastewater from the extraction and use of shale gas affects the environment. A challenge in the shale gas extraction process is preventing environmental pollution. This depends on drilling wells and their capacity, which varies with shale use. Water cannot be reused until a well is fractured and the water starts to withdraw from the well. A study was published by Kerr [11] in May 2011 that strongly suggested shale gas wells contain a rigorous abundance of toxic surface groundwater flows with flammable methane in North-Eastern Pennsylvania. Although the presented study was confined to the contamination of water, the impact in other areas that would be dug out for shale extraction purposes was not discussed.

Over the past few years, various research works have been published suggesting, in the context of optimal production policies, a selection area, supply chain network, and socio-economic balancing during shale gas extraction at the commercial level. Lutz et al. [12] presented a theoretical overview of shale gas development in the context of a more prominent resource-producing country such as the United States. The quantification of shale gas energy and wastewater generation throughout Pennsylvania was revealed with consolidated data obtained from 2189 wells. The concluding remarks were contrary to the current perception regarding the shale gas extraction-to-wastewater evulsion ratio, transportation, disposal facilities, treatment strategies, and the associated factors in the shale gas extraction processes. Yang et al. [13] presented the optimum usage of the water life cycle for drilled well-peds through a discontinuous-time bi-stage stochastic mixed-integer linear programming (SMILP) optimization framework under uncertainty. The model was optimized with a set of long-term historical data. The discussed approach was applied to two Marcellus shale gas uses, which showed the effectiveness of the addressed study. Yang et al. [14] discussed the optimal usage of water in the fracking and drilling mechanism during shale gas extraction processes at commercial scale, and also formulated a new mixed-integer linear programming (MILP) problem that inherently optimizes the capital investment for an optimal shale gas yielding scheme. A case study was implemented in the proposed optimization scenario. Li and Peng [15] investigated a new solution scheme based on interval-valued hesitant fuzzy information for the selection of promising shale gas areas, and discussed the applicability of the proposed approach by selecting the shale gas areas using multi-criteria decision making (MCDM). Although shale gas extraction has been done for over 100 years in two different prominent basins of the United States (i.e., the Appalachian Basin and the Illinois Basin), the wells seldom result in profitable production. The current shale gas extraction process, consisting of horizontal drilling and hydraulic fracturing, has made shale gas synthesis more advantageous.

In April 2012 [16], the cost of extraction incurred over shale gas in different coastal parts of the UK was approximated to be much higher than $200 per barrel, which was compared to oil prices of about $120 per barrel in the UK North Sea. North America has emerged as one of the potential leaders in developing and producing shale energy. In the US and Canada, after the successful economic accomplishment of the Barnett Shale use in Texas, the exploration of promising new sources of shale gas is being made. Gao and You [17] designed an active water cycle configuration for the shale gas extraction process and re-formulated it as a mixed-integer linear fractional programming (MILFP) optimization model under different objectives and sets of constraints. The models were globally optimized using various approaches such as the parametric method, a reformulation linearization approach, the branch and bound method, and the Charnes–Cooper transformation technique. The addressed mathematical models were applied to two case studies based on Marcellus shale play, in which on-site treatment techniques of wastewater gained importance in generating freshwater storage. Sang et al. [18] discussed a numerical optimization model of desorption and adsorption processes for hydraulic fracturing stimulations that was optimized by assuming polar co-ordinate and balance space, respectively. To estimate the receptacle volume of drilled horizontal shale well reservoirs, Gao and You [19] addressed a practical framework for the optimal flow of shale energy networks. The designed configuration comprises various coherent components such as freshwater, shale energy, wastewater management, transportation, and disposal facility with treatment plant options. The formulated models were built in the form of a mixed-integer nonlinear programming (MINLP) problem. The obtained results revealed the trade-off between economic and environmental objectives. Furthermore, Guerra et al. [20] also discussed the mathematical formulation and implementation of a comprehensive shale gas production framework with the integration of the water supply chain management system. The proposed optimization framework was illustrated with two case studies with different leading components of the shale gas production systems. Bartholomew and Mauter [21] also developed a multiobjective mixed-integer linear programming framework to highlight the trade-off between financial cost and human health & environment (HHE) costs in the overall shale gas water management system. The system's objectives were defined effectively, inherently representing different financial aspects of the shale gas production planning problem. Zhang et al. [22] presented a specific study on shale gas wastewater management systems under uncertainty. The presented optimization framework for shale gas wastewater management system corresponds to the disposal and treatment facilities under the expansion of treatment capacity. The proposed model has been designed by considering fuzzy and stochastic parameters with feasibility degree and probability distribution function at the different significance level. The concluding remarks revealed the optimal wastewater management in cases where underground disposal capacity is scarce. The uncertainty involved in the parameter reduced the reliability risk factor in shale gas production. In the present competitive epoch, different shale-gas-producing countries have motivated the wholesome and challenging study of shale gas production policies and the optimal supply chain network configuration. Lira-Barragán et al. [23] investigated a mathematical programming formulation for integrating water networks consumed for hydraulic fracturing processes in shale gas extraction. The proposed uncertainty pertained to the use of water for a different purpose and highlighted probabilistic aspects. Moreover, the developed models also cover the scheduling problem associated with the whole modeling framework for shale gas extraction. The different expected objective functions were incorporated, which led to the existence of uncertainty in the modeling approach. Interested researchers can find recent publications on shale gas development and future research scope in Chen et al. [24], Knee and Masker [25], Lan et al. [26], Ren et al. [27], Zhang et al. [28], Denham et al. [29], Al-Aboosi and El-Halwagi [30], Jin et al. [31], Ren and Zhang [32], and Wang et al. [33].

Research Gaps and Contribution

Shale gas extraction planning models and optimal strategic implementation inherently depend on various parametric factors that are actively indulged in the decision-making process. The requirement of a tremendous amount of freshwater for hydraulic fracturing (i.e., between 7000 and 40,000 m^3 per well) becomes challenging. The assessment of different freshwater sources is somewhat uneconomic, but other extractions can fulfill freshwater demand. The produced wastewater management system is also an indispensable issue and is very important in shale gas production planning models.

Many recent publications, such as Guo et al. [34], Gao et al. [35], Chebeir et al. [36], Chen et al. [37], Drouven and Grossmann [38], He et al. [39], and Wang et al. [40] have discussed different optimal modeling approaches for shale gas water management systems with socio-economic and environmental concerns. All of the above studies are confined to only uncertain modeling approaches, and have not discussed uncertainty among parameters' values; however, Zhang et al. [22] incorporated vagueness among parameters and represented it by fuzzy and stochastic quantification methodology. However, the study proposed by Zhang et al. [22] is lagging in two more practical aspects. First, it may not always be possible to have historical data for to the stochastic technique can be applied; additionally, due to some hesitation regarding imprecise parameters, the fuzzy number may not be an appropriate representative of uncertain parameters. Hence, better representation of the degree of hesitation under vagueness or imprecision can be made by using the intuitionistic fuzzy number, which considers the degree of belongingness as well as non-belongingness of the element in the possible set. Second, Zhang et al. [22] only designed the optimization framework for the optimal management of wastewater throughout the shale gas extraction processes, and did not consider the management of freshwater, which is also an integrated part of the whole shale gas extraction over time horizons. Thus, in this study we propose the unification of the two aspects discussed above. The proposed multiobjective shale gas water management system optimization model was designed after considering the most critical aspects of overall water management planning and optimization epoch. Furthermore, the concept of a neutrosophic goal programming approach is new and has not yet been applied in the field of such an emerging source of energy. The proposed model also ensures the trade-off between the socio-economic and environmental effects of shale gas production policies more realistically. The proposed shale gas optimization model also provides an opportunity to adopt the available on-site treatment technology along with the option of expanding the treatment plant, which would be beneficial for Pennsylvania because underground disposal facilities are scarce and most often wastewater is supplied to nearby cities in Ohio. The rest of the paper is summarized as follows:

In Section 2, the methodologies and technical definitions regarding intuitionistic fuzzy parameters and the neutrosophic goal programming approach (NGPA) are discussed, while Section 3 represents the multiobjective shale gas water management optimization model and implementation of the NGPA algorithm. A hypothetical case study is examined in Section 4 that shows the applicability and validity of the proposed approach. Finally, concluding remarks and findings are highlighted based on the present work in Section 5.

2. Methodology

The shale gas optimization and modeling framework discussed in this paper enviably involve significant work-flow procedures. The involvement of various critical terminological aspects in the proposed modeling and computational approach makes the shale gas optimization problem more pervasive. In order to represent these aspects, we have used some technical terminology which is able to define each and every aspect of the proposed model effectively and efficiently. The mathematical technical terminologies used in this study are intuitionistic fuzzy parameters [41–43] and those from multiobjective optimization problems [44–46] and the neutrosophic goal programming approach, which is based on the neutrosophic decision set (see [47–49]). On the basis of these mathematical technical terminologies, we developed an effective modeling and optimization framework for a shale gas water management system that dynamically characterizes the freshwater requirement and the dispensation of the generated wastewater from shale gas wells. The proposed model for shale gas water management systems contemplates different kinds of cost parameters (e.g., acquisition cost, transportation cost, treatment cost, disposal cost, and capital investment) involved in the accumulation process of freshwater and the dispensation of the generated wastewater from the shale gas extraction process. Apart from the cost, different parameters such as the freshwater storage capacity, underground injection disposal capacity of wastewater, wastewater treatment capacity, and the capacity for the expansion of wastewater treatment plants were considered in this study. Moreover, these parameters are not always in deterministic/crisp form, despite containing some kind of ambiguity and vagueness. This ambiguousness and vagueness can be represented by different uncertain parameters, such as fuzzy Zhang et al. [22], intuitionistic fuzzy, stochastic Zhang et al. [22], and other uncertain forms. The fuzzy parameters are only concerned with the maximization of membership degree (belongingness), whereas an intuitionistic fuzzy set is based on more intuition than a fuzzy set, as it deals with the maximization of membership (belongingness) and the minimization of non-membership degree (non-belongingness) of the element in the set. A stochastic parameter involves a probability distribution function with known mean and variances based on the randomly occurring events.

Furthermore, the proposed modeling approach was designed and incorporated with socio-economic and environmental facts. The potential production and distribution of shale gas energy at the commercial level is not a boon unless and until the proper pertinent initiatives are undertaken in order to overcome the by-products released by the shale gas extraction processes. Therefore, the proposed modeling and optimization approach inherently involves more than one objective (known as a multiobjective optimization problem), which is sufficient to justify the trade-off among different critical socio-economic and environmental aspects of shale gas energy. The mathematical formulation of multiple objectives ensures the economic and environmental aspects of shale gas extraction procedures. To deal with the proposed multiobjective shale gas water management optimization model, a neutrosophic goal programming approach was developed that reveals the actual situation more appropriately. The proposed NGPA considers the independent indeterminacy/neutral degree, which is the area of incognizance of a proposition's value. The selection of the proposed NGPA technique is quite effective, explanatory, and a good representative of real-life situations.

2.1. Intuitionistic Fuzzy Set

Definition 1 ([50]). *(Intuitionistic fuzzy set (IFS)) Let there be a universal set Y; then, an IFS $\tilde{W}$ in Y is given by the ordered triplets as follows:*

$$\tilde{W} = \{y,\ \mu_{\tilde{W}}(y),\ \nu_{\tilde{W}}(y) |\ y \in\ Y\},$$

where

$$\mu_{\tilde{W}}(y) : Y\ \rightarrow\ [0,\ 1];\ \ \nu_{\tilde{W}}(y) : Y\ \rightarrow\ [0,\ 1],$$

with conditions

$$0\ \leq\ \mu_{\tilde{W}}(y) + \nu_{\tilde{W}}(y)\ \leq 1,$$

where $\mu_{\tilde{W}}(y)$ and $\nu_{\tilde{W}}(y)$ denote the membership and non-membership functions of the element $y \in Y$ into the set $\tilde{W}$.

Definition 2 ([51] (Intuitionistic fuzzy number)). *An intuitionistic fuzzy set $\tilde{W} = \{y,\ \mu_{\tilde{W}}(y),\ \nu_{\tilde{W}}(y) |\ y\ \in\ Y\}$ of the real number R is said to be an intuitionistic fuzzy number if the following condition holds:*

(i) $\tilde{W}$ should be intuitionistic fuzzy normal and convex.
(ii) $\mu_{\tilde{W}}(y)$ and $\nu_{\tilde{W}}(y)$ should be upper and lower semi-continuous functions.
(iii) Supp $\tilde{W} = \{y \in R;\ \nu_{W}(y) < 1\}$ should be bounded.

Definition 3 ([43]). *(Triangular intuitionistic fuzzy number) An intuitionistic fuzzy number $\tilde{W}$ is said to be a triangular intuitionistic fuzzy number if the membership function $\mu_{\tilde{W}}(y)$ and non-membership function $\nu_{\tilde{W}}(y)$ are given by:*

$$\mu_{\tilde{W}}(y) = \begin{cases} \dfrac{y - a_1}{b - a_1}, & \text{if } a_1 \leq y \leq b, \\ 1, & \text{if } y = b, \\ \dfrac{a_2 - y}{a_2 - b}, & \text{if } b \leq y \leq a_2 \end{cases}$$

and

$$\nu_{\tilde{W}}(y) = \begin{cases} \dfrac{b-y}{b-a_3}, & \text{if } a_3 \le y \le b, \\ 0, & \text{if } y = b, \\ \dfrac{y-b}{a_4-b}, & \text{if } b \le y \le a_4, \end{cases}$$

where $a_3 \le a_1 \le b \le a_2 \le a_4$ *and is denoted by* $\tilde{W} = \{(a_1, b, a_2; \mu_{\tilde{W}}), (a_3, b, a_4; \nu_{\tilde{W}})\}$.

Definition 4 ([42]). *(Expected interval for intuitionistic fuzzy number) Let us consider that there exists an intuitionistic fuzzy number* $\tilde{W}$ *which belongs to the set of real numbers R with* $(a_1, a_2, a_3, a_4; b_1, b_2, b_3, b_4) \in \mathbb{R}$ *such that* $a_1 \le a_2 \le a_3 \le a_4 \le b_1 \le b_2 \le b_3 \le b_4$. *The four functions* $f_{\tilde{W}}(y), g_{\tilde{W}}(y), h_{\tilde{W}}(y), k_{\tilde{W}}(y) : \mathbb{R} \to [0,1]$ *such that* $f_{\tilde{W}}(y)$ *and* $g_{\tilde{W}}(y)$ *are non-decreasing and* $h_{\tilde{W}}(y)$ *and* $k_{\tilde{A}}(y)$ *are non-increasing functions, then the intuitionistic fuzzy number* $\tilde{W} = \{y,\ \mu_{\tilde{W}}(y),\ \nu_{\tilde{W}}(y) : y \in Y\}$ *can be represented by membership and non-membership functions stated as follows:*

$$\mu_{\tilde{W}}(y) = \begin{cases} 0, & \text{if } y \le a_1 \text{ or } y \ge a_4, \\ f_{\tilde{W}}(y), & \text{if } a_1 \le y \le a_2, \\ g_{\tilde{W}}(y), & \text{if } a_3 \le y \le a_4, \\ 1, & \text{if } a_2 \le y \le a_3 \end{cases}$$

and

$$\nu_{\tilde{W}}(y) = \begin{cases} 1, & \text{if } y \le b_1 \text{ or } y \ge b_4, \\ h_{\tilde{W}}(y), & \text{if } b_1 \le y \le b_2, \\ k_{\tilde{W}}(y), & \text{if } b_3 \le y \le b_4, \\ 0, & \text{if } b_2 \le y \le b_3. \end{cases}$$

Furthermore, Grzegrorzewski [52] discussed the expected interval for the intuitionistic fuzzy number $\tilde{W} = \{a_1, a_2, a_3, a_4; b_1, b_2, b_3, b_4\}$ *as a crisp interval and presented it as follows:*

$$EI(\tilde{W}) = [E_1(\tilde{W}), E_2(\tilde{W})]. \tag{1}$$

The lower and upper values of the expected interval for the intuitionistic fuzzy number $\tilde{W}$ *is defined as given below:*

$$E_1(\tilde{W}) = \frac{b_1 + a_2}{2} + \int_{b_1}^{b_2} h_{\tilde{W}}(y) - \int_{a_1}^{a_2} f_{\tilde{W}}(y),$$

$$E_2(\tilde{W}) = \frac{a_3 + b_4}{2} + \int_{a_3}^{a_4} g_{\tilde{W}}(y) - \int_{b_3}^{b_4} k_{\tilde{W}}(y),$$

where

$$h_{\tilde{W}}(y) = \frac{y - b_1}{b_2 - b_1}, \quad f_{\tilde{W}}(y) = \frac{y - a_1}{a_2 - a_1},$$

$$k_{\tilde{W}}(y) = \frac{y - b_4}{b_3 - b_4}, \quad g_{\tilde{W}}(y) = \frac{y - a_4}{a_3 - a_4}.$$

Definition 5 ([42]). *(Expected interval and value for triangular intuitionistic fuzzy number) Suppose that* $\tilde{W} = \{(a_1, b, a_2; \mu_{\tilde{W}}), (a_3, b, a_4; \nu_{\tilde{W}})\}$ *is a triangular intuitionistic fuzzy number with membership and non-membership functions* $\mu_{\tilde{W}}(y)$ *and* $\nu_{\tilde{W}}(y)$; *then, the expected interval of the triangular intuitionistic fuzzy number by using the above definition can be obtained as follows:*

$$E_1(\tilde{W}) = \frac{b_1 + a_2}{2} + \int_{b_1}^{b_2} h_{\tilde{W}}(y) - \int_{a_1}^{a_2} f_{\tilde{W}}(y) = \frac{3a + b_1 + (a - b_1)\nu_{\tilde{W}} - (a - a_1)\mu_{\tilde{W}}}{4} \tag{2}$$

and

$$E_2(\tilde{W}) = \frac{a_3+b_4}{2} + \int_{a_3}^{a_4} g_{\tilde{W}}(y) - \int_{b_3}^{b_4} k_{\tilde{W}}(y) = \frac{3a+b_2+(a_2-a)\mu_{\tilde{W}}+(a-b_2)\nu_{\tilde{W}}}{2}. \tag{3}$$

Grzegrorzewski [52] also suggested the expected value for the intuitionistic fuzzy number with the help of lower and upper values of the expected interval. Therefore, the expected value for the triangular intuitionistic fuzzy number is obtained as follows:

$$EV(\tilde{W}) = \left[\frac{E_1(\tilde{W})+E_2(\tilde{W})}{2}\right]. \tag{4}$$

Definition 6. *The general mathematical programming formulation of a multiobjective optimization problem with k objectives, j constraints, and q variables can be stated as follows:*

$$\begin{aligned} &Optimize\ (Z_1, Z_2, \cdots, Z_k)\ \ k = 1,2,\cdots,K \\ &s.t.\ g_j(x) \le d_j,\ \ j = 1,2,\cdots,j_1; \\ &\quad g_j(x) \ge d_j,\ \ j = j_1+1, j_1+2,\cdots,j_2; \\ &\quad g_j(x) = d_j,\ \ j = j_2+1, j_2+2,\cdots,J; \\ &\quad x_q \ge 0,\qquad q = 1,2,3,\cdots,Q;\ \ x_q \in X, \end{aligned} \tag{5}$$

where Z_k are a set of k different conflicting objectives, g_j are real-valued functions, and d_j are real numbers. x_q is a q-dimensional decision variable vector and X is a feasible solution set.

2.2. Neutrosophic Goal Programming Approach (NGPA)

In the past few decades, the extended version of the fuzzy set (FS) and intuitionistic fuzzy set (IFS) have been introduced. In order to reflect the insightful concept of indeterminacy or neutral thoughts in decision making, a new set called the *neutrosophic set* was introduced by Smarandache [47]. The technical erm *neutrosophic* holds two different words, which are *neutre* derived from French and meaning "neutral", and *sophia*, adopted from Greek and meaning "skill" or "wisdom". Therefore, the word "neutrosophic" concretely means "knowledge of neutral thoughts". The FS is mainly concerned with the maximization of the degree of belongingness (membership function) of an element in the set, whereas the IFS deals with two aspects, namely, the degree of belongingness (membership function) and the degree of non-belongingness (non-membership function) of the element in the set. The incorporation of the independent neutral/indeterminacy concept in the neutrosophic set differentiates itself from FS and IFS, providing more strength to decision-making processes.

Moreover, many real-life circumstances may not be easy to tackle with only the degree of belongingness and non-belongingness of the element in the set. However, the degree up to some level of belongingness and non-belongingness would be a significant touchstone in the decision-making process. For example, if we take the opinion about the victory of team X in a cricket match, and supposing they have the possible chance of winning equalling 0.8, the chance team X has of losing would be 0.4 and the chance that match would be a tie would be 0.5 (see [53]). All the possibilities are independent of each other and can take any value between 0 and 1. Therefore, this sort of decision-making problem is outside of the domain of FS and IFS, and consequently beyond the periphery of fuzzy programming and intuitionistic fuzzy programming approaches, respectively. Hence, independent indeterminacy conditions under the uncertainty domain are a more technical perspective in real-life optimization problems (see [48,49,53]).

An efficient approach called the neutrosophic goal programming approach (NGPA) based on the neutrosophic decision set [47] was designed in order to reach the best compromise solution of multiobjective optimization problems. The NGPA inherently comprises three membership functions, namely, the maximization of truth and indeterminacy degrees and the minimization of the falsity degree present in any optimization problem. It permits policymakers to manifest independent neutral inferences about decision-making processes and provides an opportunity to effectively reach goals using the NGPA technique.

Definition 7 ([47] (Neutrosophic Set)). *Let there be a universal discourse Y such that* $y \in Y$*, then a neutrosophic set W in Y is defined by three membership functions, namely, truth* $T_W(y)$*, indeterminacy* $I_W(y)$*, and falsity* $F_W(y)$*, denoted by the following form:*

$$W = \{< y, T_W(y), I_W(y), F_W(y) > | y \in Y\},$$

where $T_W(y), I_W(y)$*, and* $F_W(y)$ *are real standard or non-standard subsets belonging to* $]0^-, 1^+[$*, also given as,* $T_W(y) : y \rightarrow]0^-, 1^+[$*,* $I_W(y) : Y \rightarrow]0^-, 1^+[$*, and* $F_W(y) : Y \rightarrow]0^-, 1^+[$*. There is no restriction on the sum of* $T_W(y), I_W(y)$ *and* $F_W(y)$*, so we have*

$$0^- \leq sup\ T_W(y) + I_W(y) + sup\ F_W(y) \leq 3^+.$$

Definition 8 ([47]). *Let there be two single-valued neutrosophic sets A and B, then* $C = (A \cup B)$ *with truth* $T_C(y)$*, indeterminacy* $I_C(y)$*, and falsity* $F_C(y)$ *membership functions are given by:*

$T_C(y) = max\ (T_A(y), T_B(y))$,
$I_C(y) = min\ (I_A(y), I_B(y))$,
$F_C(y) = min\ (F_A(y), F_B(y))$ *for each* $y \in Y$.

Definition 9 ([47]). *Let there be two single-valued neutrosophic sets A and B, then* $C = (A \cap B)$ *with truth* $T_C(y)$*, indeterminacy* $I_C(y)$*, and falsity* $F_C(y)$ *membership functions are given by*

$T_C(y) = min\ (T_A(y), T_B(y))$,
$I_C(y) = max\ (I_A(y), I_B(y))$,
$F_C(y) = max\ (F_A(y), F_B(y))$ *for each* $y \in Y$.

The concept of fuzzy decision (D), fuzzy goal (G), and fuzzy constraint (C) was first discussed by Bellman and Zadeh [44] and extensively used in many real-life decision-making problems under fuzziness. Therefore, a fuzzy decision set can be defined as follows:

$$D = G \cap C.$$

Equivalently, the neutrosophic decision set D_N, with the set of neutrosophic goals and constraints, can be defined as:

$$D_N = (\cap_{k=1}^{K} G_k)(\cap_{j=1}^{J} C_j) = (y,\ T_D(y),\ I_D(y),\ F_D(y)\ ),$$

where

$$T_D(y) = min \left\{ \begin{array}{c} T_{G_1}(y), T_{G_2}(y), ..., T_{G_K}(y) \\ T_{C_1}(y), T_{C_2}(y), ..., T_{C_J}(y) \end{array} \right\} \forall\ y \in Y,$$

$$I_D(y) = max \left\{ \begin{array}{c} I_{G_1}(y), I_{G_2}(y), ..., I_{G_K}(y) \\ I_{C_1}(y), I_{C_2}(y), ..., I_{C_J}(y) \end{array} \right\} \forall\ y \in Y,$$

$$F_D(y) = max \left\{ \begin{array}{c} F_{G_1}(y), F_{G_2}(y), ..., F_{G_K}(y) \\ F_{C_1}(y), F_{C_2}(y), ..., F_{C_J}(y) \end{array} \right\} \forall\ y \in Y,$$

where $T_D(y), I_D(y)$, and $F_D(y)$ are the truth, indeterminacy, and falsity membership functions of neutrosophic decision set D_N, respectively.

In order to formulate the different membership functions for multiobjective optimization problems (MOOPs), we defined the bounds for each objective function. The lower and upper bounds for each objective function are represented by L_k and U_k which can be obtained as follows:

First, we solved each objective function as a single objective under the given constraints of the problem. After solving k objectives individually, we have the k solutions set, $X^1, X^2, ..., X^k$. After that, the obtained solutions are substituted for each objective function to provide the lower and upper bounds for each objective, as given below:

$$U_k = \max\left[Z_k(X^k)\right] \text{ and } L_k = \min\left[Z_k(X^k)\right] \quad \forall\, k = 1,2,3,...,K. \tag{6}$$

The bounds for k objective functions under the neutrosophic environment can be obtained as follows:

$$U_k^T = U_k, \quad L_k^T = L_k \quad \text{for truth membership,}$$
$$U_k^I = L_k^T + s_k, \quad L_k^I = L_k^T \quad \text{for indeterminacy membership,}$$
$$U_k^F = U_k^T, \quad L_k^F = L_k^T + t_k \quad \text{for falsity membership,}$$

where s_k and $t_k \in (0,1)$ are predetermined real numbers assigned by decision maker(s) (DM(s)). By using the above lower and upper bounds, we defined the linear membership functions under a neutrosophic environment:

$$T_k(Z_k(x)) = \begin{cases} 1 & if\ Z_k(x) < L_k^T \\ \frac{U_k^T - Z_k(x)}{U_k^T - L_k^T} & if\ L_k^T \leq Z_k(x) \leq U_k^T \\ 0 & if\ Z_k(x) > U_k^T, \end{cases} \tag{7}$$

$$I_k(Z_k(x)) = \begin{cases} 1 & if\ Z_k(x) < L_k^I \\ \frac{U_k^I - Z_k(x)}{U_k^I - L_k^I} & if\ L_k^I \leq Z_k(x) \leq U_k^I \\ 0 & if\ Z_k(x) > U_k^I, \end{cases} \tag{8}$$

$$F_k(Z_k(x)) = \begin{cases} 1 & if\ Z_k(x) > U_k^F \\ \frac{Z_k(x) - L_k^F}{U_k^F - L_k^F} & if\ L_k^F \leq Z_k(x) \leq U_k^F \\ 0 & if\ Z_k(x) < L_k^F. \end{cases} \tag{9}$$

In the above case, $L_k^{(.)} \neq U_k^{(.)}$ for all k objective functions. If for any membership $L_k^{(.)} = U_k^{(.)}$, then the value of these memberships will be equal to 1. The diagrammatic representation of the objective function with different components of membership functions under a neutrosophic environment is shown in Figure 1.

Moreover all the above three discussed membership degrees can be transformed into membership goals according to their respective degrees of attainment. The highest degree of truth membership function that can be achieved is unity (1), the indeterminacy membership function is neutral and independent with the highest attainment degree half (0.5), and the falsity membership function can be achieved with the highest attainment degree zero (0). Now the transformed membership goals under a neutrosophic environment can be expressed as follows:

$$T_k(Z_k(x)) + d_{kT}^- - d_{kT}^+ = 1, \tag{10}$$

$$I_k(Z_k(x)) + d_{kI}^- - d_{kI}^+ = 0.5, \tag{11}$$

$$F_k(Z_k(x)) + d_{kF}^- - d_{kF}^+ = 0, \tag{12}$$

where $d_{kT}^-, d_{kT}^+, d_{kI}^-, d_{kI}^+, d_{kF}^-$, and d_{kF}^+ are the over and under deviations such that $d_{kT}^-.d_{kT}^+ = 0$, $d_{kF}^-.d_{kF}^+ = 0$, and $d_{kF}^-.d_{kF}^+ = 0$ for truth membership, indeterminacy membership, and falsity membership goals under a neutrosophic environment.

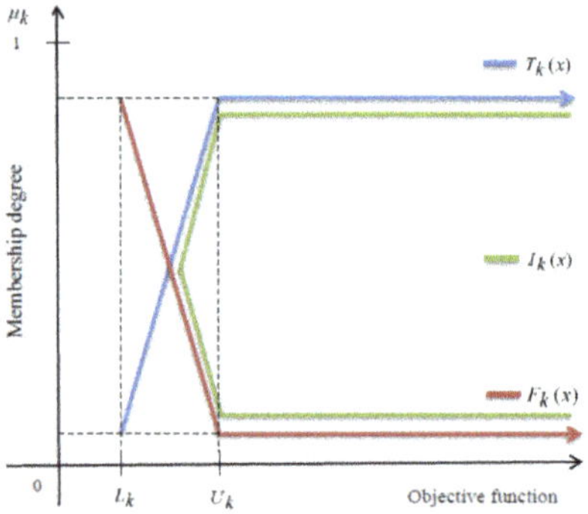

Figure 1. Diagrammatic representation of truth, indeterminacy, and falsity membership degree for the objective function.

Intuitionally, the aims are to maximize the truth and indeterminacy membership degrees of neutrosophic objectives and constraints, and minimize the falsity membership degree of neutrosophic objectives and constraints. The general formulation of the neutrosophic goal programming (NGP) model for multiobjective optimization problem (5) is represented as follows:

$$\begin{aligned}
&\text{Minimize } Z = \sum_{k=1}^{K} w_{kT}.d_{kT}^- + \sum_{k=1}^{K} w_{kI}.d_{kI}^- + \sum_{k=1}^{K} w_{kF}.d_{kF}^+,\\
&\text{subject to}\\
&T_k(Z_k(x)) + d_{kT}^- - d_{kT}^+ \geq 1;\\
&I_k(Z_k(x)) + d_{kI}^- - d_{kI}^+ \geq 0.5;\\
&F_k(Z_k(x)) + d_{kF}^- - d_{kF}^+ \leq 0;\\
&T_k(Z_k(x)) \geq I_k(Z_k(x));\\
&T_k(Z_k(x)) \geq F_k(Z_k(x));\\
&F_k(Z_k(x)) \geq 0,\ d_{kT}^-.d_{kT}^+ = 0;\\
&d_{kI}^-.d_{kI}^+ = 0,\ d_{kF}^-.d_{kF}^+ = 0;\\
&g_j(x) \leq d_j,\ j = 1,2,\cdots,m_1;\\
&g_j(x) \geq d_j,\ j = m_1+1, m_1+2,\cdots,m_2;\\
&g_j(x) = d_j,\ j = m_2+1, m_2+2,\cdots,m;\\
&x_i \geq 0,\quad i = 1,2,3,\cdots,q;\ x_i \in X;\\
&d_{kT}^-,\ d_{kT}^+,\ d_{kI}^-,\ d_{kI}^+\ d_{kF}^-,\ d_{kF}^+ \geq 0\ \forall\, k,
\end{aligned} \tag{13}$$

where w_{kT}, w_{kI}, and w_{kF} are the weights assigned to deviations of the truth, indeterminacy, and falsity membership goals of each objective function, respectively. Now the assignment of corresponding weighting schemes of different weights can be obtained as follows:
$w_{kT} = \dfrac{1}{U_k^T - L_k^T}$, $w_{kI} = \dfrac{1}{U_k^I - L_k^I}$, and $w_{kF} = \dfrac{1}{U_k^F - L_k^F}$.

Hence, the optimum evaluation of multiobjective optimization problems by using the NGP approach is a very useful technique as it involves the degree of indeterminacy, which is independent and certainly ensures the achievement of marginal evaluation of each membership goal by reducing the deviational values under a neutrosophic environment.

3. Shale Gas Water Management System: Modeling and Optimization under Uncertainty

Shale gas is a rapidly emerging and unconventional source of energy found trapped in shale rocks. The extraction process at the wholesale level is very complicated. Since shales usually possess low permeability to permit significant fluid inflow to a well-bore, most shale wells are not adequate sources of natural gas for commercial production. Other sources of natural gases include coal bed methane, methane hydrates, and tight sandstones. Most commonly, the area in which shale gases are trapped are known as resource plays. Shale has comparatively low matrix permeability, which affects gas production at the commercial level and requires the fracturing process to supply permeability. In the past few decades, shale gas has been produced from shale rocks with natural fractures. Shale gas production seems to be booming in recent years due to the latest potential technology in hydraulic fracturing (fracking), which has led in the direction of pervasive artificial fractures around good bores. Horizontal drilling is often used in the shale gas extraction process. Lateral lengths up to 10,000 feet (3000 m) within shale wells are dug out to create maximum borehole surface area in contact with the shale. While injecting water with high pressure into shale rocks, chemicals are added to facilitate the underground fracturing process, which releases natural gas. The fracturing fluid is primarily water and contains approximately 0.5% chemical additives that are fully dissolved into the water. Depending on the size of the area, millions of liters of water are used for fracking, which signifies that thousands of liters of chemicals are injected into the subsurface.

The massive amount of contaminated surface water and groundwater with fracking fluids has emerged as a problematic issue. Generally, accrued shale gas is usually trapped several thousand feet below ground. Different challenging environmental concerns are often observed. For example, methane migration, improper treatment of produced wastewater, and lack of an underground injection disposal site. About 50% to 70% of the injected volume of contaminated water is generated after fracking, and sufficient storage capacity for wastewater management is required. The remaining volume of water remains in the subsurface. The hydraulic fracturing process leads to the perception that it can lead to the contamination of groundwater aquifers. However, foul odors and very toxic local water supply above-ground are also unavoidable truths about shale gas. Acid mine wastewater can be released into groundwater, but it might cause significant contamination of underground freshwater. Usually, the harmful impact and water pollution associated with wastewater and coal production can be reduced to a certain extent in shale gas production. Apart from using water and industrial chemicals, it may also be feasible to frack shale gas with only liquefied propane gas. This extraction option simultaneously reduces water and environmental degradation. It can be implemented in regions like Pennsylvania that have experienced a marginal increment in the freshwater requirement for energy production. More explicitly, shale gas development in the United States represents less than half a percent of total domestic freshwater consumption, although this quantity can reach as high as 25% in particularly arid regions.

Therefore, the proposed shale gas water management strategy has been designed to optimize the allocation of water requirement for different purposes. The designed water supply chain network configuration contains various components, such as the acquisition of freshwater and its transportation, on-site treatment with different technology, underground injection disposal sites, and treatment plants for wastewater with an option for expansion with a to and fro transportation network. Different potential objectives addressing the project planning strategy were also considered in this present study. A well-defined set of dynamic constraints were imposed to represent the modeling approach more realistically. The integrated water flow supply chain network within shale gas planning periods is shown in Figure 2. In Figure 2, the different echelons are presented to highlight the proposed shale gas water management design. The flow of freshwater initiates from different freshwater sources S and is then shipped to various shale sites I. After fracking processes, a possible part of the generated toxic water would be treated by on-site treatment technology O, and the remaining wastewater would be used to dispatch for further management to different treatment plants or disposal sites J, respectively. To enhance the treatment capacity of sewage treatment plants, an opportunity to adopt the different expansion options M was incorporated along with the associated capital investment. Hence, the treated wastewater can be reused for household purposes and in turn can yield significant revenues from the reuse of water. The whole integrated water cycle continues to flow over different time horizons T. Therefore, to assure the optimal flow of water among different echelons, the proposed water flow network captures the actual behavior of flow-back and produced water during shale gas extraction processes. The shale gas project planning model explicitly includes different indices' set, decision variables, and values of parameters shown in Table 1, which presents the significant characteristic features during the shale gas synthesis process.

The proposed shale gas water flow network configuration is based on the following assumptions:

1. There is no scope for the transportation of water using pipelines throughout the planning horizons.
2. The expansion options of underground injection disposal sites have not been considered due to the financial crisis or uneconomic aspects throughout the planning horizons.
3. The expansion of the treatment plant has been considered in order to avoid excess wastewater at the subsurface level of underground water during all the planning periods.
4. An absolute option of on-site treatment technology has been included that enables the reuse of wastewater within the shale sites throughout the planning horizons.
5. The restrictive margin was designed for the minimum and maximum capacity of wastewater treatment by using different on-site treatment technologies throughout the planning horizons.
6. The overall produced wastewater volume was successfully managed by the proposed system during all the planning horizons.

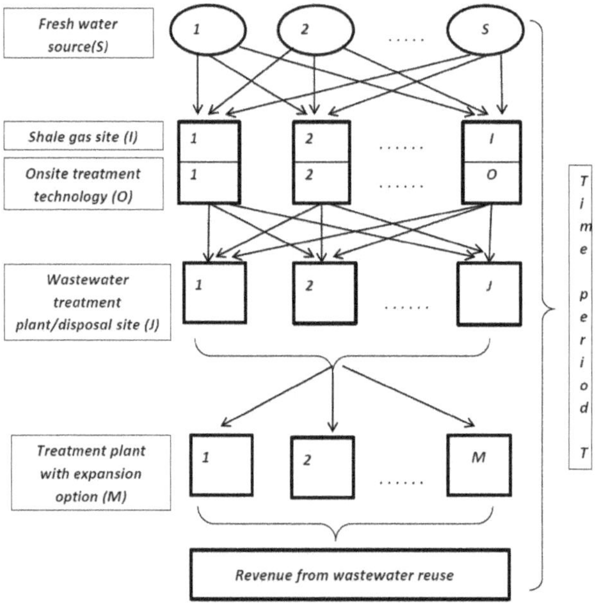

Figure 2. Representation of shale gas integrated water flow optimization network over time.

Table 1. Notions and descriptions.

Indices	Descriptions
i	Denotes the number of shale sites
j	Represents the number of disposal sites and treatment plants
m	Denotes the available options for the expansion capacity of the treatment plant
o	Denotes the on-site treatment technologies
t	Represents the time period
s	Denotes the source of freshwater
Decision variables	
$FW_{s,i,t}$	Amount of freshwater acquired from source s at shale site i in time period t
$WTO_{i,o,t}$	Amount of wastewater treated by on-site treatment technology o at shale site i in time period t
$WW_{i,j,t}$	Total amount of wastewater generated at shale site i and received by disposal site and treatment plant j in time period t
$WWD_{i,j,t}$	Amount of wastewater generated at shale site i and received by disposal site j in time period t
$WWT_{i,j,t}$	Amount of wastewater generated at shale site i and received by treatment plant j in time period t
$Y_{j,m,t}$	Binary variable representing the expansion capacity of the disposal site and treatment plant j by expansion option m in time period t
$YO_{i,o}$	Binary variable representing that on-site technology o is applied at shale site i
Parameters	
lo_o	Recovery factor for treating wastewater with on-site treatment technology o
$fdw_{i,t}$	Freshwater demand at shale site i in time period t
$fca_{s,t}$	Freshwater supply capacity at source s in time period t
rf_o	Ratio of freshwater to wastewater required for blending after treatment with on-site treatment technology o
$wwds_{j,t}$	Capacity for wastewater at disposal site j in time period t
$wwtp_{j,t}$	Capacity for wastewater at treatment plant j in time period t
$wdw_{j,t}$	Total wastewater capacity at disposal site and treatment plant j in time period t
$eo_{j,m,t}$	Represents increased treatment capacity of wastewater treatment plant j by using available expansion option m in time period t
$caq_{s,t}$	Denotes the unit acquisition cost of freshwater at source s in time period t
$ctf_{s,i,t}$	Denotes the unit transportation cost of freshwater from source s to shale site i in time period t
$ctw_{i,j,t}$	Denotes the unit transportation cost of wastewater from shale site i to disposal site and treatment plant j in time period t
$ctr_{j,t}$	Denotes the unit treatment cost of wastewater at treatment plant j in time period t
$cd_{j,t}$	Denotes the unit disposal cost of wastewater at disposal site j in time period t
$re_{j,t}$	Denotes the revenues from wastewater reuse from treatment plant j in time period t
$rr_{j,t}$	Denotes the reuse rate from wastewater treatment plant j in time period t
$cex_{j,m,t}$	Represents the investment cost of expanding the disposal site and treatment plant j by expansion option m in time period t
ocl_o	Denotes the minimum capacity for the on-site treatment of wastewater
ocu_o	Denotes the maximum capacity for the on-site treatment of wastewater

3.1. Objective Function

The first objective function is concerned with a different kind of cost incurred over the freshwater. It is quite a challenging task to collect the optimal amount of freshwater directly from natural freshwater sources; however, the option exists to acquire the freshwater from nearby the shale gas plant, which results in a lower acquisition cost. The transportation of freshwater is also required, which again appears as a transportation cost from source s to shale site i over period t, and both are of minimization type. Therefore, the cost function (14) related to freshwater can be furnished as follows:

$$Minimize\ Z_1 = \sum_{s=1}^{S}\sum_{i=1}^{I}\sum_{t=1}^{T}(caq_{s,t} + ctf_{s,i,t})FW_{s,i,t}. \quad (14)$$

The second objective mainly focuses on a different kinds of cost levied over the wastewater. It is crucial to manage the huge amount of contaminated or toxic wastewater released during the shale gas energy generation process. The produced amount of wastewater can be handled by either sending it to the treatment plants or by dumping into underground wastewater disposal sites. Both techniques are associated with some cost known as treatment and disposal facility costs, respectively. The transportation of wastewater from shale sites to different treatment plants and disposal sites results in additional transportation costs associated with the wastewater. The total revenues from the reuse of wastewater with some reuse rate are also associated with wastewater from shale site i to disposal and treatment plant j over period t. Therefore, the cost function (15) related to wastewater can be presented as follows:

$$Minimize\ Z_2 = \sum_{i=1}^{I}\sum_{j=1}^{J}\sum_{t=1}^{T}(ctr_{j,t} + cd_{j,t} + ctw_{i,j,t} - rr_{j,t}.re_{j,t})WW_{i,j,t}. \quad (15)$$

The third objective function provides the facility of proliferation at treatment plants and underground disposal sites with some predetermined expansion option. The different expansion options require capital investment, which is to be minimized with binary variable taking value 1 if the expansion option m is adopted at treatment plant j over time period t; otherwise 0. Therefore the total capital investment (16) for the expansion of wastewater treatment plant capacity can be summarized as follows:

$$Minimize\ Z_3 = \sum_{j=1}^{J}\sum_{m=1}^{M}\sum_{t=1}^{T}(cex_{j,m,t}).Y_{j,m,t}. \quad (16)$$

3.2. Constraints

The constraint given by (17) is related to freshwater demand at shale sites:

At each shale site, a certain quantity of freshwater is required for the hydraulic fracturing process. The total amount of freshwater obtained from different sources is not sufficient to meet the demand at shale sites, but it is indispensable to build up the other sources or by developing other techniques to obtain the freshwater. Therefore on-site treatment technology with a recovery factor for treating wastewater is also an important option to fulfill the demand of such a tremendous amount of freshwater. Hence, the sum of the total amount freshwater acquired from different freshwater sources s and freshwater obtained from various on-site treatment technologies o with the recovery factor for treating wastewater must be greater than or equal to its total requirement at each shale site i over period t:

$$\sum_{s=1}^{S} FW_{s,i,t} + \sum_{o=1}^{O} lo_o * WTO_{i,o,t} \geq fdw_{i,t} \quad \forall\ i,t. \quad (17)$$

The constraint has given in (18) is related to the freshwater capacity at each source:

The total amount of freshwater obtained from different sources has some limitations in terms of storage capacity at different sources. The optimal stock of freshwater at different sources may differ marginally. It is necessary to ensure that the total amount of freshwater can be obtained without substantially affecting the storage capacity of each freshwater source. Therefore, the total amount of freshwater acquired from different sources s with the consumption at each shale site i must be less than or equal to its storage capacity at source s over period t:

$$\sum_{i=1}^{I} FW_{s,i,t} \leq fca_{s,t} \quad \forall\ s,t. \quad (18)$$

The constraint given in (19) is related to wastewater capacity at underground disposal sites:

The amount of wastewater generated after the fracturing process contains various toxic chemicals dissolved in it. A proper disposal system with its associated available capacity must be built to overcome fatal environmental effects. Therefore, it must be assured that the amount of wastewater received at different disposal sites can be fully tackled. Thus, the amount of wastewater released from shale site i and received at each disposal site j must be less than or equal to the presumable capacity of each disposal site j over period t:

$$\sum_{i=1}^{I} WWD_{i,j,t} \leq wwds_{j,t} \quad \forall\ j,t. \tag{19}$$

The constraint given in (20) is related to the wastewater capacity at each treatment plant with its prevalence:

The wastewater treatment facility leads to the option of reusing wastewater. The amount of wastewater liberated from different shale sites restrains the tremendous amount of harmful chemicals that must be treated at the water treatment plant to ensure its reuse for different household purposes. Thus, the amount of wastewater released from different shale sites i and dispatched to different treatment plants j must be less than or equal to the sum of the total capacity of each treatment plant with its several expansion options m over period t:

$$\sum_{i=1}^{I} WWT_{i,j,t} \leq wwtp_{j,t} + \sum_{m=1}^{M} eo_{j,m,t}.Y_{j,m,t} \quad \forall\ j,t. \tag{20}$$

The constraint given in (21) is related to the overall wastewater capacity at the treatment plant and disposal site:

The total amount of wastewater generated during the shale gas extraction process must be confronted with proper cautionary measures. The option of the treatment plant and disposal site for dealing with wastewater must be sufficient to conquer its harmful effects. Therefore it must be ensured that the total amount of wastewater generated from the hydraulic fracturing process at shale site i is less than or equal to its total capacity at the treatment plant and disposal site j over period t:

$$\sum_{i=1}^{I} WW_{i,j,t} \leq wdw_{j,t} \quad \forall\ j,t. \tag{21}$$

The constraint given in (22) is related to different wastewater capacities at the treatment plant and disposal site:

This constraint ensures that regardless of what the excess amount of wastewater released from shale sites is, it must be fully managed by expanding the treatment plant capacity. Therefore, the different treatment plants have a potential storage capacity increment option within the investment costs. Thus, the sum of total wastewater capacity enhanced by expanding treatment plant j with expansion option m, the total capacity of underground disposal and treatment plant j must be less than or equal to the assorted capacity of disposal site and treatment plant j over time period t:

$$\sum_{m=1}^{M} eo_{j,m,t}.Y_{j,m,t} + wwds_{j,t} + wwtp_{j,t} \leq wdw_{j,t} \quad \forall\ i,j,t. \tag{22}$$

The constraint given in (23) is related to the different wastewater capacities at the treatment plant and disposal site:

The necessity and utilization of a huge amount of freshwater in the whole process of shale gas extraction requires thought regarding its acquisition. Various techniques are used to recycle freshwater. Therefore, one of the most trending techniques is on-site treatment with different technologies. Thus, the reuse specification for hydraulic fracturing with the blending ratio of freshwater to wastewater after the treatment of on-site treatment technology o must be less than or equal to the total amount of freshwater acquired at source s transported to shale site i over period t:

$$\sum_{o=1}^{O} rf_o.lo_o.WTO_{i,o,t} \leq FW_{s,i,t} \quad \forall\ s,i,t. \tag{23}$$

The constraint given in (24) is related to the minimum capacity of the ton-site treatment of wastewater:

This restriction was imposed with the fact that a minimal amount of freshwater must be obtained by using on-site treatment technology. The capital investment towards the setup of on-site treatment plant steers the utilization of on-site treatment technology. Thus, the minimum capacity of on-site wastewater treatment with technology o along with the binary variable taking value one if the certain technology is used (and otherwise 0) at shale site i must be less than or equal to the amount of wastewater treated by on-site treatment technology o over period t:

$$\sum_{o=1}^{O} ocl_o.YO_{i,o} \leq WTO_{i,o,t} \quad \forall\ i,t. \tag{24}$$

The constraint given in (25) is related to the maximum capacity of on-site wastewater treatment:

This restriction ensures that the maximal amount of freshwater is acquired by using on-site treatment technology. The upper limit for the on-site treatment of wastewater restricts the excessive holding of wastewater at the on-site treatment plant. Thus, this constraint provides the surety that the minimum capacity of on-site treatment of wastewater with technology o along with the binary variable taking value 1 if the certain technology is used (otherwise 0) at shale site i is greater than or equal to the amount of wastewater treated by on-site treatment technology o over time period t:

$$\sum_{o=1}^{O} ocu_o.YO_{i,o} \geq WTO_{i,o,t} \quad \forall\ i,t. \tag{25}$$

The constraint given in (26) is related to the total wastewater produced during the shale gas extraction process:

It must be ensured that the total amount of wastewater generated during the fracking procedures is strictly equal to the sum of different amounts of wastewater distributed to the treatment plant and disposal site. Therefore, the sum of the total amount of wastewater at the treatment plant and disposal site j dispatched from shale site i must be equal to the assorted wastewater capacity at disposal site and treatment plant j over time period t:

$$\sum_{i=1}^{I}\sum_{j=1}^{J}\sum_{t=1}^{T} WWD_{i,j,t} + \sum_{i=1}^{I}\sum_{j=1}^{J}\sum_{t=1}^{T} WWT_{i,j,t} = \sum_{i=1}^{I}\sum_{j=1}^{J}\sum_{t=1}^{T} WW_{i,j,t} \quad \forall\ i,j,t. \tag{26}$$

The proposed multiobjective shale gas optimization model under uncertainty is presented in model $\mathbf{M}_1$ with the fact that parameter values inherently contain vagueness and ambiguousness in the real-life decision-making process. The decision maker(s) or policy maker(s) is(are) not very sure about the exact parameter values due to a lack of proper information, relatively little experience, environmental issues, and other humanitarian logical perception. To overcome these issues, the settings are taken as the triangular intuitionistic fuzzy number and are more elaborately discussed in Section 3.3.

$$\mathbf{M}_1 : Minimize\ Z_1 = \sum_{s=1}^{S}\sum_{i=1}^{I}\sum_{t=1}^{T}\{c\tilde{a}q_{s,t} + c\tilde{t}f_{s,i,t}\}FW_{s,i,t}$$

$$Minimize\ Z_2 = \sum_{i=1}^{I}\sum_{j=1}^{J}\sum_{t=1}^{T}\{c\tilde{t}r_{j,t} + c\tilde{d}_{j,t} + c\tilde{t}w_{i,j,t} - rr_{j,t}.re_{j,t}\}WW_{i,j,t}$$

$$Minimize\ Z_3 = \sum_{j=1}^{J}\sum_{m=1}^{M}\sum_{t=1}^{T}\{c\tilde{e}x_{j,m,t}\}.Y_{j,m,t}$$

subject to:

$$\sum_{s=1}^{S} FW_{s,i,t} + \sum_{o=1}^{O} lo_o.WTO_{i,o,t} \geq f\tilde{d}w_{i,t} \quad \forall\ i,t$$

$$\sum_{i=1}^{I} FW_{s,i,t} \leq f\tilde{c}a_{s,t} \quad \forall\ s,t$$

$$\sum_{i=1}^{I} WWD_{i,j,t} \leq w\tilde{w}ds_{j,t} \quad \forall\ j,t$$

$$\sum_{i=1}^{I} WWT_{i,j,t} \leq w\tilde{w}tp_{j,t} + \sum_{m=1}^{M} eo_{j,m,t}.Y_{j,m,t} \quad \forall\ j,t$$

$$\sum_{i=1}^{I} WW_{i,j,t} \leq w\tilde{d}w_{j,t} \quad \forall\ j,t$$

$$\sum_{m=1}^{M} eo_{j,m,t}.Y_{j,m,t} + w\tilde{w}ds_{j,t} + w\tilde{w}tp_{j,t} \leq w\tilde{d}w_{j,t} \quad \forall\ i,j,t$$

$$\sum_{o=1}^{O} rf_o.lo_o.WTO_{i,o,t} \leq FW_{s,i,t} \quad \forall\ s,i,t$$

$$\sum_{o=1}^{O} ocl_o.YO_{i,o} \leq WTO_{i,o,t} \quad \forall\ i,t$$

$$\sum_{o=1}^{O} ocu_o.YO_{i,o} \geq WTO_{i,o,t} \quad \forall\ i,t$$

$$\sum_{i=1}^{I}\sum_{j=1}^{J}\sum_{t=1}^{T} WWD_{i,j,t} + \sum_{i=1}^{I}\sum_{j=1}^{J}\sum_{t=1}^{T} WWT_{i,j,t} = \sum_{i=1}^{I}\sum_{j=1}^{J}\sum_{t=1}^{T} WW_{i,j,t} \quad \forall\ i,j,t$$

$$FW_{s,i,t} \geq 0,\ \ WW_{i,j,t} \geq 0 \quad \forall\ s,i,j,t$$

$$WTO_{i,o,t} \geq 0,\ \ WWD_{i,j,t} \geq 0,\ \ WWT_{i,j,t} \geq 0 \quad \forall\ i,o,j\ and\ t$$

$$0 \leq Y_{j,m,t}\ ,YO_{i,o} \leq 1,\ \ Y_{j,m,t}\ and\ YO_{i,o} = integer, \quad \forall\ i,o,j,m\ and\ t,$$

where the notation ($\tilde{.}$) over different parameters represents the triangular intuitionistic fuzzy number for the indices' whole set.

3.3. *Intuitionistic Fuzzy Parameters*

The proposed multiobjective shale gas water management optimization model discussed in Section 3 inherently involves uncertainty or impreciseness. The existence of ambiguity among parameters makes it uncertain. It is not always feasible for decision maker(s) or project manager(s) to assign crisp/exact parameter values. Actual perceptions behind the uncertainty involve a lack of proper information, environmental conditions, the condition of roads, natural calamities, abrupt changes in the prices of fuel, different routes of transportation, shortages of freshwater on sunny days, etc. In such cases, only some vague and inconsistent pieces of information are available regarding the parameter values. Therefore, uncertainty can take different forms, such as fuzzy numbers, stochastic random variables, and other forms of change. Based on this confluent information, one may assume imprecise parameters and easily overcome uncertainty by applying the different techniques to obtain the best estimates of the parameters. In brief, we may distinguish between stochastic and fuzzy methods while dealing with the uncertain dataset. The uncertainty involved in the data due to randomness can be handled with a stochastic programming approach while it can be dealt with using fuzzy techniques due to vagueness or ambiguousness. In the present study, all the parameters were assumed to be triangular intuitionistic fuzzy numbers, which is more realistic as compared to fuzzy numbers as it simultaneously reveals both the degree of belongingness and the degree of non-belongingness. The defuzzification/ranking method of triangular intuitionistic fuzzy parameters is based on the expected interval and expected values of a lower and upper member of the set. Imprecise parameters involved in the different objective functions were converted to their crisp forms by using expected values, whereas uncertain parameters present in constraints were transformed into their deterministic forms using expected intervals. All the pieces of information regarding triangular intuitionistic fuzzy settings used in THE shale gas optimization model are summarized in Table 2.

Table 2. Information regarding the triangular intuitionistic fuzzy parameters of the shale gas model.

Intuitionistic Fuzzy Parameters	Triangular Intuitionistic Fuzzy Number	$EI(.) = [E_1(.), E_2(.)]$	$EV(.)$
$\tilde{caq}_{s,t}$	$\{(caq^{(1)}_{s,t}, caq_{s,t}, caq^{(2)}_{s,t}); (caq^{(3)}_{s,t}, caq_{s,t}, caq^{(4)}_{s,t})\}$	$\{\frac{3.caq + caq^{(3)} + (caq - caq^{(3)})\nu_{\tilde{caq}} - (caq - caq^{(1)})\mu_{\tilde{caq}}}{4}, \frac{3.caq + caq^{(4)} + (caq^{(2)} - caq)\mu_{\tilde{caq}} + (caq - caq^{(4)})\nu_{\tilde{caq}}}{2}\}$	$\frac{E_1(\tilde{caq}) + E_2(\tilde{caq})}{2}$
$\tilde{ctf}_{s,i,t}$	$\{(ctf^{(1)}_{s,i,t}, ctf_{s,i,t}, ctf^{(2)}_{s,i,t}); (ctf^{(3)}_{s,i,t}, ctf_{s,i,t}, ctf^{(4)}_{s,i,t})\}$	$\{\frac{3.ctf + ctf^{(3)} + (ctf - ctf^{(3)})\nu_{\tilde{ctf}} - (ctf - ctf^{(1)})\mu_{\tilde{ctf}}}{4}, \frac{3.ctf + ctf^{(4)} + (ctf^{(2)} - ctf)\mu_{\tilde{ctf}} + (ctf - ctf^{(4)})\nu_{\tilde{ctf}}}{2}\}$	$\frac{E_1(\tilde{ctf}) + E_2(\tilde{ctf})}{2}$
$\tilde{ctr}_{j,t}$	$\{(ctr^{(1)}_{j,t}, ctr_{j,t}, ctr^{(2)}_{j,t}); (ctr^{(3)}_{j,t}, ctr_{j,t}, ctr^{(4)}_{j,t})\}$	$\{\frac{3.ctr + ctr^{(3)} + (ctr - ctr^{(3)})\nu_{\tilde{ctr}} - (ctr - ctr^{(1)})\mu_{\tilde{ctr}}}{4}, \frac{3.ctr + ctr^{(4)} + (ctr^{(2)} - ctr)\mu_{\tilde{ctr}} + (ctr - ctr^{(4)})\nu_{\tilde{ctr}}}{2}\}$	$\frac{E_1(\tilde{ctr}) + E_2(\tilde{ctr})}{2}$
$\tilde{ctw}_{i,j,t}$	$\{(ctw^{(1)}_{i,j,t}, ctw_{i,j,t}, ctw^{(2)}_{i,j,t}); (ctw^{(3)}_{i,j,t}, ctw_{i,j,t}, ctw^{(4)}_{i,j,t})\}$	$\{\frac{3.ctw + ctw^{(3)} + (ctw - ctw^{(3)})\nu_{\tilde{ctw}} - (ctw - ctw^{(1)})\mu_{\tilde{ctw}}}{4}, \frac{3.ctw + ctw^{(4)} + (ctw^{(2)} - ctw)\mu_{\tilde{ctw}} + (ctw - ctw^{(4)})\nu_{\tilde{ctw}}}{2}\}$	$\frac{E_1(\tilde{ctw}) + E_2(\tilde{ctw})}{2}$
$\tilde{cd}_{j,t}$	$\{(cd^{(1)}_{j,t}, cd_{j,t}, cd^{(2)}_{j,t}); (cd^{(3)}_{j,t}, cd_{j,t}, cd^{(4)}_{j,t})\}$	$\{\frac{3.cd + cd^{(3)} + (cd - cd^{(3)})\nu_{\tilde{cd}} - (cd - cd^{(1)})\mu_{\tilde{cd}}}{4}, \frac{3.cd + cd^{(4)} + (cd^{(2)} - cd)\mu_{\tilde{cd}} + (cd - cd^{(4)})\nu_{\tilde{cd}}}{2}\}$	$\frac{E_1(\tilde{cd}) + E_2(\tilde{cd})}{2}$
$\tilde{cex}_{j,m,t}$	$\{(cex^{(1)}_{j,m,t}, cex_{j,m,t}, cex^{(2)}_{j,m,t}); (cex^{(3)}_{j,m,t}, cex_{j,m,t}, cex^{(4)}_{j,m,t})\}$	$\{\frac{3.cex + cex^{(3)} + (cex - cex^{(3)})\nu_{\tilde{cex}} - (cex - cex^{(1)})\mu_{\tilde{cex}}}{4}, \frac{3.cex + cex^{(4)} + (cex^{(2)} - cex)\mu_{\tilde{cex}} + (cex - cex^{(4)})\nu_{\tilde{cex}}}{2}\}$	$\frac{E_1(\tilde{cex}) + E_2(\tilde{cex})}{2}$
$\tilde{fdw}_{i,t}$	$\{(fdw^{(1)}_{i,t}, fdw_{i,t}, fdw^{(2)}_{i,t}); (fdw^{(3)}_{i,t}, fdw_{i,t}, fdw^{(4)}_{i,t})\}$	$\{\frac{3.fdw + fdw^{(3)} + (fdw - fdw^{(3)})\nu_{\tilde{fdw}} - (fdw - fdw^{(1)})\mu_{\tilde{fdw}}}{4}, \frac{3.fdw + fdw^{(4)} + (fdw^{(2)} - fdw)\mu_{\tilde{fdw}} + (fdw - fdw^{(4)})\nu_{\tilde{fdw}}}{2}\}$	$\frac{E_1(\tilde{fdw}) + E_2(\tilde{fdw})}{2}$
$\tilde{fca}_{s,t}$	$\{(fca^{(1)}_{s,t}, fca_{s,t}, fca^{(2)}_{s,t}); (fca^{(3)}_{s,t}, fca_{s,t}, fca^{(4)}_{s,t})\}$	$\{\frac{3.fca + fca^{(3)} + (fca - fca^{(3)})\nu_{\tilde{fca}} - (fca - fca^{(1)})\mu_{\tilde{fca}}}{4}, \frac{3.fca + fca^{(4)} + (fca^{(2)} - fca)\mu_{\tilde{fca}} + (fca - fca^{(4)})\nu_{\tilde{fca}}}{2}\}$	$\frac{E_1(\tilde{fca}) + E_2(\tilde{fca})}{2}$
$\tilde{wwds}_{j,t}$	$\{(wwds^{(1)}_{j,t}, wwds_{j,t}, wwds^{(2)}_{j,t}); (wwds^{(3)}_{j,t}, wwds_{j,t}, wwds^{(4)}_{j,t})\}$	$\{\frac{3.wwds + wwds^{(3)} + (wwds - wwds^{(3)})\nu_{\tilde{wwds}} - (wwds - wwds^{(1)})\mu_{\tilde{wwds}}}{4}, \frac{3.wwds + wwds^{(4)} + (wwds^{(2)} - wwds)\mu_{\tilde{wwds}} + (wwds - wwds^{(4)})\nu_{\tilde{wwds}}}{2}\}$	$\frac{E_1(\tilde{wwds}) + E_2(\tilde{wwds})}{2}$
$\tilde{wwtp}_{j,t}$	$\{(wwtp^{(1)}_{j,t}, wwtp_{j,t}, wwtp^{(2)}_{j,t}); (wwtp^{(3)}_{j,t}, wwtp_{j,t}, wwtp^{(4)}_{j,t})\}$	$\{\frac{3.wwtp + wwtp^{(3)} + (wwtp - wwtp^{(3)})\nu_{\tilde{wwtp}} - (wwtp - wwtp^{(1)})\mu_{\tilde{wwtp}}}{4}, \frac{3.wwtp + wwtp^{(4)} + (wwtp^{(2)} - wwtp)\mu_{\tilde{wwtp}} + (wwtp - wwtp^{(4)})\nu_{\tilde{wwtp}}}{2}\}$	$\frac{E_1(\tilde{wwtp}) + E_2(\tilde{wwtp})}{2}$
$\tilde{wdw}_{j,t}$	$\{(wdw^{(1)}_{j,t}, wdw_{j,t}, wdw^{(2)}_{j,t}); (wdw^{(3)}_{j,t}, wdw_{j,t}, wdw^{(4)}_{j,t})\}$	$\{\frac{3.wdw + wdw^{(3)} + (wdw - wdw^{(3)})\nu_{\tilde{wdw}} - (wdw - wdw^{(1)})\mu_{\tilde{wdw}}}{4}, \frac{3.wdw + wdw^{(4)} + (wdw^{(2)} - wdw)\mu_{\tilde{wdw}} + (wdw - wdw^{(4)})\nu_{\tilde{wdw}}}{2}\}$	$\frac{E_1(\tilde{wdw}) + E_2(\tilde{wdw})}{2}$

Therefore, the crisp/deterministic version of the proposed multiobjective shale gas water management system optimization model $\mathbf{M}_1$ based on the different crisp values of the parameters can be represented in the model $\mathbf{M}_2$ as follows:

$$\mathbf{M}_2 : Minimize\ Z_1 = \sum_{s=1}^{S}\sum_{i=1}^{I}\sum_{t=1}^{T}\{EV(c\tilde{a}q_{s,t}) + EV(c\tilde{t}f_{s,i,t})\}FW_{s,i,t},$$

$$Minimize\ Z_2 = \sum_{i=1}^{I}\sum_{j=1}^{J}\sum_{t=1}^{T}\{EV(c\tilde{t}r_{j,t}) + EV(c\tilde{d}_{j,t}) + EV(c\tilde{t}w_{i,j,t}) - rr_{j,t}.re_{j,t}\}WW_{i,j,t},$$

$$Minimize\ Z_3 = \sum_{j=1}^{J}\sum_{m=1}^{M}\sum_{t=1}^{T}\{EV(c\tilde{e}x_{j,m,t})\}Y_{j,m,t},$$

subject to:

$$\sum_{s=1}^{S} FW_{s,i,t} + \sum_{o=1}^{O} lo_o.WTO_{i,o,t} \geq E_1^{fdw_{i,t}} \quad \forall\ i,t$$

$$\sum_{i=1}^{I} FW_{s,i,t} \leq E_2^{fca_{s,t}} \quad \forall\ s,t$$

$$\sum_{i=1}^{I} WWD_{i,j,t} \leq E_2^{wwds_{j,t}} \quad \forall\ j,t$$

$$\sum_{i=1}^{I} WWT_{i,j,t} \leq E_2^{wwtp_{j,t}} + \sum_{m=1}^{M} eo_{j,m,t}.Y_{j,m,t} \quad \forall\ j,t$$

$$\sum_{i=1}^{I} WW_{i,j,t} \leq E_2^{wdw_{j,t}} \quad \forall\ j,t$$

$$\sum_{m=1}^{M} eo_{j,m,t}.Y_{j,m,t} + E_1^{wwds_{j,t}} + E_1^{wwtp_{j,t}} \leq E_2^{wdw_{j,t}} \quad \forall\ j,t$$

$$\sum_{o=1}^{O} rf_o.lo_o.WTO_{i,o,t} \leq FW_{s,i,t} \quad \forall\ s,i,t$$

$$\sum_{o=1}^{O} ocl_o.YO_{i,o} \leq WTO_{i,o,t} \quad \forall\ i,t$$

$$\sum_{o=1}^{O} ocu_o.YO_{i,o} \geq WTO_{i,o,t} \quad \forall\ i,t$$

$$\sum_{i=1}^{I}\sum_{j=1}^{J}\sum_{t=1}^{T} WWD_{i,j,t} + \sum_{i=1}^{I}\sum_{j=1}^{J}\sum_{t=1}^{T} WWT_{i,j,t} = \sum_{i=1}^{I}\sum_{j=1}^{J}\sum_{t=1}^{T} WW_{i,j,t} \quad \forall\ i,j,t$$

$$FW_{s,i,t} \geq 0,\ WW_{i,j,t} \geq 0 \quad \forall\ s,i,j,t$$

$$WTO_{i,o,t} \geq 0,\ WWD_{i,j,t} \geq 0,\ WWT_{i,j,t} \geq 0 \quad \forall\ i,o,j\ and\ t$$

$$0 \leq Y_{j,m,t}\ ,YO_{i,o} \leq 1,\ \ Y_{j,m,t}\ and\ YO_{i,o} = integer, \quad \forall\ i,o,j,m\ and\ t,$$

where $EV(.)$, $E_1^{(.)}$, and $E_2^{(.)}$ are the expected value and lower and upper intervals of triangular intuitionistic fuzzy numbers for the entire indices' set, respectively.

The discussed solution technique (i.e., the neutrosophic goal programming approach (NGPA)) is based on the neutrosophic decision set, which ensures the efficient implementation of the independent neutral thoughts of the decision maker(s). The obtained crisp model $\mathbf{M}_2$ can be transformed into $\mathbf{M}_3$ to achieve the globally optimal solution of the proposed multiobjective shale gas water management system optimization model.

$$\mathbf{M}_3: \text{Minimize } Z = (w_{1T}.d_{1T}^- + w_{2T}.d_{2T}^- + w_{3T}.d_{3T}^-) + (w_{1I}.d_{1I}^- + w_{2I}.d_{2I}^- + w_{3I}.d_{3I}^-) + (w_{1F}.d_{1F}^+ + w_{2F}.d_{2F}^+ + w_{3F}.d_{3F}^+)$$

subject to

$$\frac{U_1^T - \sum_{s=1}^S \sum_{i=1}^I \sum_{t=1}^T \{EV(c\tilde{a}q_{s,t}) + EV(c\tilde{t}f_{s,i,t})\}FW_{s,i,t}}{U_1^T - L_1^T} + d_{1T}^- - d_{1T}^+ = 1$$

$$\frac{U_1^I - \sum_{s=1}^S \sum_{i=1}^I \sum_{t=1}^T \{EV(c\tilde{a}q_{s,t}) + EV(c\tilde{t}f_{s,i,t})\}FW_{s,i,t}}{U_1^I - L_1^I} + d_{1I}^- - d_{1I}^+ = 0.5$$

$$\frac{\sum_{s=1}^S \sum_{i=1}^I \sum_{t=1}^T \{EV(c\tilde{a}q_{s,t}) + EV(c\tilde{t}f_{s,i,t})\}FW_{s,i,t} - L_1^F}{U_1^F - L_1^F} + d_{1F}^- - d_{1F}^+ = 0$$

$$\frac{U_2^T - \sum_{i=1}^I \sum_{j=1}^J \sum_{t=1}^T \{EV(c\tilde{t}r_{j,t}) + EV(\tilde{cd}_{j,t}) + EV(c\tilde{f}w_{i,j,t}) - rr_{j,t}.re_{j,t}\}WW_{i,j,t}}{U_2^T - L_2^T} + d_{2T}^- - d_{2T}^+ = 1$$

$$\frac{U_2^I - \sum_{i=1}^I \sum_{j=1}^J \sum_{t=1}^T \{EV(c\tilde{t}r_{j,t}) + EV(\tilde{cd}_{j,t}) + EV(c\tilde{f}w_{i,j,t}) - rr_{j,t}.re_{j,t}\}WW_{i,j,t}}{U_2^I - L_2^I} + d_{2I}^- - d_{2I}^+ = 0.5$$

$$\frac{\sum_{i=1}^I \sum_{j=1}^J \sum_{t=1}^T \{EV(c\tilde{t}r_{j,t}) + EV(\tilde{cd}_{j,t}) + EV(c\tilde{f}w_{i,j,t}) - rr_{j,t}.re_{j,t}\}WW_{i,j,t} - L_2^F}{U_2^F - L_2^F} + d_{2F}^- - d_{2F}^+ = 0$$

$$\frac{U_3^T - \sum_{j=1}^J \sum_{m=1}^M \sum_{t=1}^T \{EV(c\tilde{e}x_{j,m,t})\}Y_{j,m,t}}{U_3^T - L_3^T} + d_{3T}^- - d_{3T}^+ = 1$$

$$\frac{U_3^I - \sum_{j=1}^J \sum_{m=1}^M \sum_{t=1}^T \{EV(c\tilde{e}x_{j,m,t})\}Y_{j,m,t}}{U_3^I - L_3^I} + d_{3I}^- - d_{3I}^+ = 0.5$$

$$\frac{\sum_{j=1}^J \sum_{m=1}^M \sum_{t=1}^T \{EV(c\tilde{e}x_{j,m,t})\}Y_{j,m,t} - L_3^F}{U_3^F - L_3^F} + d_{3F}^- - d_{3F}^+ = 0$$

$$\sum_{s=1}^S FW_{s,i,t} + \sum_{o=1}^O lo_o.WTO_{i,o,t} \geq E_1^{fdw_{i,t}} \quad \forall\ i,t$$

$$\sum_{i=1}^I FW_{s,i,t} \leq E_2^{fca_{s,t}} \quad \forall\ s,t$$

$$\sum_{i=1}^I WWD_{i,j,t} \leq E_2^{wwds_{j,t}} \quad \forall\ j,t$$

$$\sum_{i=1}^I WWT_{i,j,t} \leq E_2^{wwtp_{j,t}} + \sum_{m=1}^M eo_{j,m,t}.Y_{j,m,t} \quad \forall\ j,t$$

$$\sum_{i=1}^I WW_{i,j,t} \leq E_2^{wdw_{j,t}} \quad \forall\ j,t$$

$$\sum_{m=1}^M eo_{j,m,t}.Y_{j,m,t} + E_1^{wwds_{j,t}} + E_1^{wwtp_{j,t}} \leq E_2^{wdw_{j,t}} \quad \forall\ j,t$$

$$\sum_{o=1}^O rf_o.lo_o.WTO_{i,o,t} \leq FW_{s,i,t} \quad \forall\ s,i,t$$

$$\sum_{o=1}^O ocl_o.YO_{i,o} \leq WTO_{i,o,t} \quad \forall\ i,t$$

$$\sum_{o=1}^O ocu_o.YO_{i,o} \geq WTO_{i,o,t} \quad \forall\ i,t$$

$$\sum_{i=1}^I \sum_{j=1}^J \sum_{t=1}^T WWD_{i,j,t} + \sum_{i=1}^I \sum_{j=1}^J \sum_{t=1}^T WWT_{i,j,t} = \sum_{i=1}^I \sum_{j=1}^J \sum_{t=1}^T WW_{i,j,t} \quad \forall\ i,j,t$$

$$FW_{s,i,t} \geq 0, \;\; WW_{i,j,t} \geq 0 \;\;\; \forall \; s,i,j,t$$
$$WTO_{i,o,t} \geq 0, \;\; WWD_{i,j,t} \geq 0, \;\; WWT_{i,j,t} \geq 0 \;\;\; \forall \; i,o,j \; and \; t$$
$$0 \leq Y_{j,m,t} \;, YO_{i,o} \leq 1, \;\; Y_{j,m,t} \; and \; YO_{i,o} = integer, \;\;\; \forall \; i,o,j,m \; and \; t$$
$$d_{kT}^{-}.d_{kT}^{+} = 0, \;\; d_{kI}^{-}.d_{kI}^{+} = 0, \;\; d_{kF}^{-}.d_{kF}^{+} = 0,$$

where $w_{1T}, w_{1I}, w_{1F}, w_{2T}, w_{2I}, w_{2F}, w_{3T}, w_{3I}$, and w_{3F} are the parameter weights assigned to different deviational variables of the neutrosophic membership goals.

3.4. Solution Algorithm

To reformulate the shale gas water management optimization model into the neutrosophic goal programming model, one needs to solve each objective function individually and has to determine the maximum and minimum values of each objective. With the help of these values, the upper and lower bounds for each membership function under a neutrosophic environment were obtained. Then, the truth, indeterminacy, and falsity membership functions for each objective were constructed. The transformation of membership functions into membership goals can be done by using the different deviational variables. The weighting scheme of each aim was designed based on the difference between the best and worst values of the respective objective function. The developed framework for the optimal shale gas water management computational model was transmuted under a neutrosophic environment. The stepwise solution procedures for the proposed neutrosophic goal programming approach can be summarized as follows:

Step 1. Design the proposed multiobjective shale gas water management optimization model as given in $\mathbf{M}_1$.

Step 2. Convert each intuitionistic fuzzy parameter involved in model $\mathbf{M}_1$ into its crisp form by using the expected interval and values method as given in Equations (2)–(4) or presented in Table 2.

Step 3. Modify model $\mathbf{M}_1$ into $\mathbf{M}_2$ and solve model $\mathbf{M}_2$ for each objective function individually in order to obtain the best and worst solutions.

Step 4. Determine the upper and lower bounds for each objective function by using Equation (6). Using U_k and L_k, define the upper and lower bounds for truth, indeterminacy, and falsity membership as given in Equations (7)–(9).

Step 5. Transform the truth, indeterminacy, and falsity membership degrees into their respective membership goals and deviational variables as defined in Equations (10)–(12).

Step 6. Formulate the neutrosophic goal programming model defined in $\mathbf{M}_3$ and solve the multiobjective shale gas water management optimization model in order to obtain the compromise solution using suitable techniques or some optimization software packages.

4. A Computational Study

The integrated framework representative of the multiobjective shale gas water management optimization model is presented based on a real-life scenario, hypothetical proposition, data, information, and a quick review of the published research (Lutz et al. [12], Rahm and Riha [54], Rahm et al. [55], Zhang et al. [22], Alawattegama [56]). The unified optimal shale gas water planning model was structured to manifest the real-life scenario in the current and future characteristic features of shale gas extraction processes. The proposed model includes the optimal acquisition of freshwater, on-site treatment of wastewater, expansion of treatment plant facility, underground injection disposal site, treatment plant facility, and primary socio-economic concerns and environmental issues with the technical and potential aspects in major shale gas plays in the United States. The acquisition of freshwater from different sources and the inventory holding of freshwater to a certain level for the smooth operation of the shale gas extraction processes is quite a challenging task. Therefore, the acquisition of freshwater is allowed some predetermined budget allocation at the different freshwater sources. The flow-back-produced water from shale play is a matter of grave concern.

The privilege of an on-site wastewater treatment facility for reuse purposes at a moderate scale is also feasible and laid down as a base of future technologies. Various burning socio-environmental issues are being raised against the contaminated wastewater generated from shale wells after fracturing processes at the national and international political levels. To overcome these issues, the orientation of wastewater underground disposal sites and treatment facilities with their expansion options have been taken under consideration. There is no scope for pipelines to any extent throughout the shale gas extraction process. All sorts of to and fro flow of freshwater and wastewater have been depicted with roadways. The planning periods are designed in such a way that shale gas production turnover results in an economically profitable scenario.

In this study, the shale gas water management system optimization model comprises one freshwater source, five shale sites with one drilled well at each shale site, and three on-site wastewater treatment facilities. The toxic wastewater management system includes one wastewater underground injection disposal site, two wastewater treatment plants with three expansion options for each treatment plant facility over three planning periods of 5 years each which are capable of representing the whole shale gas production process more realisticallyy. All the summarized parameters were assumed to be a triangular intuitionistic fuzzy number, and their defuzzified version can be obtained from Table 2. The acquisition costs (in \$/bbl) of freshwater at source and transportation cost (in \$/bbl) of freshwater by road over three planning periods are presented in Table 3. The various costs incurred on account of wastewater, such as transportation cost (in \$/bbl) from different shale sites to disposal site and treatment plants, underground injection disposal cost (in \$/bbl), and wastewater operational cost at different treatment plants (in \$/bbl) over three planning periods are summarized in Table 4. The capital investment costs(in \$/bbl) for alternative options for the expansion of treatment plant capacity with the respective enhanced potential volume (in bbl/day) over three planning periods are presented in Table 4. The crisp parameters which include revenues/profits from the reuse of wastewater (in \$), reuse rate (in bbl/day), recovery factor for treating wastewater with different treatment technology, and the required ratio of freshwater to sewer for blending after on-site treatment technology, along with the minimum and maximum capacities for on-site treatment with conflicting technology over three time periods are summarized in Table 5. The different restrictive intuitionistic fuzzy parameters (for freshwater and wastewater) were introduced for the optimal allocation of freshwater and wastewater according to their speculated destination. The freshwater acquisition capacity at the source, the requirement of freshwater at different shale sites, the underground wastewater disposal capacity, the wastewater treatment plant capacity, and the overall generated wastewater permitted for managerial purposes are summarized in Table 5. Throughout the project planning scheme, the decision maker(s) or project manager(s) intend to adopt the certainly feasible strategy that ensures the optimal allocation of freshwater and wastewater to their predetermined consumption points. However, during the whole planning periods, the decision maker(s) are confronted with the different multiple conflicting objectives which are to be optimized in order to achieve the global benefits from the production of shale gas energy as well as its commercial distribution. Hence, the proposed multiobjective shale gas water management optimization model experiments with these hypothetical datasets and was applied to tackle the project planning scheme.

Table 3. Acquisition and transportation costs of freshwater (\$/bbl).

Freshwater Acquisition Cost at Source $(\tilde{caq})$	**Time Period** $t = 1$	$t = 2$	$t = 3$
Source	**(1.9,2.1,2.3;1.8,2.1,2.4)**	**(1.6,1.8,2;1.5,1.8,2.1)**	**(0.9,1.2,1.5;0.8,1.2,1.6)**
Transportation costs of freshwater from source to shale site $(\tilde{ctf})$			
Source to shale site 1	(1.2,1.4,1.6;1.1,1.4,1.7)	(4.2,4.4,4.6;4.1,4.4,4.7)	(4.1,4.3,4.5;4.0,4.3,4.6)
Source to shale site 2	(2.1,2.3,2.5;1.9,2.3,2.7)	(3.2,3.4,3.6;3.1,3.4,3.7)	(3.2,3.4,3.6;3.0,3.4,3.8)
Source to shale site 3	(3.4,3.6,3.8;3.2,3.6,4.0)	(2.2,2.4.2.6;2.1,2.4,2.7)	(2.2,2.4,2.6;2.0,2.4,2.8)
Source to shale site 4	(2.2,2.4,2.6;2.1,2.4,2.7)	(1.5,1.8,1.9;1.4,1.8,2.1)	(1.5,1.7,1.9;1.4,1.7,2.0)
Source to shale site 5	(1.4,1.6,1.8;1.2,1.6,2)	(1.8,2,2.2;1.8,2,2.2)	(2.6,2.8,3.0;2.5,2.8,3.1)

Table 4. Different costs related to wastewater and capital investment for treatment plant expansions ($/bbl).

Transportation Cost From Shale Site to Facility ($c\tilde{f}w$)		**Time Period**		
Source	**Treatment and Disposal Facility**	$t = 1$	$t = 2$	$t = 3$
Shale site 1	Disposal site	(1.4,2.4,3.4)	(2,3,4)	(3.4,3.6,3.8)
Shale site 1	Treatment plant 1	(3.0,3.2,3.4)	(3.4,3.6,3.8)	(3.8,4.0,4.2)
Shale site 1	Treatment plant 2	(5.2,5.6,6.0)	(6.0,6.3,6.6)	(6.6,6.7,6.8)
Shale site 2	Disposal site	(6.0,6.5,7.0)	(6.6,6.9,7.2)	(7.1,7.4,7.7)
Shale site 2	Treatment plant 1	(2.8,2.9,3.0)	(3.5,3.7,3.9)	(4.2,4.4,4.6)
Shale site 2	Treatment plant 2	(3.2,3.4,3.6)	(3.5,3.9,4.3)	(4.1,4.2,4.3)
Shale site 3	Disposal site	(4.0,4.2,4.4)	(4.5,4.8,5.1)	(5.0,5.5,6.0)
Shale site 3	Treatment plant 1	(4.4,4.8,5.2)	(5.0,5.3,5.6)	(5.5,5.9,6.3)
Shale site 3	Treatment plant 2	(5.0,5.1,5.2)	(5.0,5.5,6.0)	(6.0,6.3,6.6)
Shale site 4	Disposal site	(2.5,2.7,2.9)	(3.0,3.2,3.4)	(3.5,3.9,7.3)
Shale site 4	Treatment plant 1	(5.5,6.0,6.5)	(6.5,6.7,6.9)	(7.1,7.3,7.5)
Shale site 4	Treatment plant 2	(3.3,3.6,3.9)	(4.0,4.3,4.6)	(4.4,4.9,5.4)
Shale site 5	Disposal site	(6.8,7.1,7.4)	(7.3,7.5,7.7)	(7.8,7.9,8.0)
Shale site 5	Treatment plant 1	(3.0,3.2,3.4)	(3.4,3.6,3.8)	(3.8,3.9,4.0)
Shale site 5	Treatment plant 2	(2.8,3.1,3.4)	(3.6,3.8,4.0)	(4.0,4.3,4.6)
Operational costs of treatment facility ($c\tilde{t}r$) and disposal facility ($c\tilde{d}$)	Disposal site	(0.5,0.7,0.9;0.4,0.7,1.0)	(0.4,0.6,0.8;0.3,0.6,0.9)	(2.1,2.3,2.6;2.1,2.3,2.6)
	Treatment plant 1	(3.6,3.8,4.0;3.5,3.8,4.1)	(0.5,0.7,0.9;0.4,0.7,1.0)	(1.4,1.6,1.8;1.2,1.6,2.0)
	Treatment plant 2	(2.5,2.7,2.9;2.4,2.7,3.0)	(1.5,1.7,1.9;1.4,1.7,2.0)	(1.5,1.7,1.9;1.4,1.7,2.0)
			Time period	
Capital cost of expanding treatment plant ($c\tilde{e}x$)	Expansion option m	$t = 1$	$t = 2$	$t = 3$
Treatment plant 1	1	(15.6,15.8,16.0;15.4,15.8,16.2)	(17.2,17.4,17.6;17.1,17.4,17.7)	(14.3,14.6,14.9;14.2,14.6,15.0)
Treatment plant 1	2	(09.6,09.8,10.0;09.5,09.8,10.1)	(16.2,16.4,16.6;16.1,16.4,16.7)	(12.2,12.4,12.6;12.1,12.4,12.7)
Treatment plant 1	3	(12.2,12.4,12.6;12.0,12.4,12.8)	(13.3,13.5,13.7;13.2,13.5,13.8)	(11.2,11.4,11.6;11.1,11.4,11.7)
Treatment plant 2	1	(14.2,14.4,14.6;14.0,14.4,14.8)	(12.1,12.3,12.5;12.0,12.3,12.6)	(13.1,13.3,13.5;13.0,13.3,13.6)
Treatment plant 2	2	(13.2,13.4,13.6;13.0,13.4,13.8)	(11.2,11.4,11.6;11.1,11.4,11.7)	(16.2,16.4,16.6;16.1,16.4,16.7)
Treatment plant 2	3	(12.2,12.5,12.8;12.1,12.5,12.9)	(11.3,11.5,11.7;11.2,11.5,11.8)	(17.2,17.4,17.6;17.1,17.4,17.7)
Increased treatment capacity (eo)				
Treatment plant 1	1	600	600	600
Treatment plant 1	2	750	750	750
Treatment plant 1	3	850	850	850
Treatment plant 2	1	550	550	550
Treatment plant 2	2	650	650	650
Treatment plant 2	3	800	800	800

Table 5. Capacity restrictions on freshwater and wastewater (bbl/day).

Freshwater Acquisition Capacity at Source $(\tilde{fca})$		**Time Period**	
	$t = 1$ **(200, 300, 400; 100, 300, 500)**	$t = 2$ **(300, 500, 700; 200, 500, 800)**	$t = 3$ **(500, 600, 700; 400, 600, 800)**
Freshwater demand at shale site $(\tilde{fdw})$			
Shale site 1	(300,000, 500,000, 700,000, 900,000)	(500,000, 700,000, 900,000, 1,100,000)	(1,300,000, 1,400,000, 1,500,000, 1,600,000)
Shale site 2	(500,000, 600,000, 700,000, 800,000)	(600,000, 700,000, 800,000, 900,000)	(1,000,000, 1,100,000, 1,200 000, 1,300,000)
Shale site 3	(700,000, 900,000, 1,100,000, 1,300,000)	(300,000, 400,000, 500,000, 600,000)	(600,000, 800,000, 1,000,000, 1,200,000)
Shale site 4	(800,000, 900,000, 1,000,000, 1,100,000)	(1,000,000, 1,100,000, 1,200,000, 1,300,000)	(1,000,000, 1,200,000, 1,400,000, 1,600,000)
Shale site 5	(600,000, 800,000, 1,000,000, 1,200,000)	(1,000,000, 1,200,000, 1,400,000, 1,600,000)	(1,000,000, 1,500,000, 2,000,000, 2,500,000)
Wastewater capacity at disposal site $(\tilde{wwds})$			
Disposal site	(200, 300, 400; 100, 300, 500)	(600, 800, 1000; 500, 800, 1100)	(400, 600, 800; 300, 600, 900)
Wastewater capacity at treatment plant $(\tilde{wwtp})$			
Treatment plant 1	(100,000, 200,000, 300,000, 400,000)	(200,000, 300,000, 400,000, 500,000)	(1,000,000, 1,200,000, 1,400,000, 1,600,000)
Treatment plant 2	(200,000, 400,000, 600,000, 800,000)	(1,300,000, 1,600,000, 1,800,000, 2,200,000)	(3,000,000, 3,200,000, 3,400,000, 3,600,000)
Overall wastewater capacity $(\tilde{wdw})$			
Disposal site	(620,000, 630,000, 640,000, 650,000)	(2,473,000, 2,474,000, 2,475,000, 2,476,000)	(4,460,000, 4,460,000, 4,470,000, 4,480,000)
Treatment plant 1	(600,000, 700,000, 800,000, 900,000)	(2,000,000, 3,000,000, 4,000,000, 5,000,000)	(4,070,000, 4,080,000, 4,090,000, 4,500,000)
Treatment plant 2	(3,002,000, 3,004,000, 3,006,000, 3,008,000)	(4,010,000, 4,020,000, 4,030,000, 4,040,000)	(5,100,000, 5,200 000, 5,300,000, 5,400,000)
Revenues from wastewater reuse (re)			
Treatment plant 1	1.20	1.30	1.50
Treatment plant 2	1.00	1.20	1.40
Reuse rate (rr)			
Treatment plant 1	0.75	0.85	0.95
Treatment plant 2	0.70	0.80	0.90
		Onsite treatment technology o	
	1	2	3
Recovery factor (lo)	0.15	0.45	0.65
Ratio of freshwater to wastewater for blending (rf)	0.43	0.40	0.38
Minimum capacity for on-site treatment (ocl)	150	200	300
Maximum capacity for on-site treatment (ocu)	5000	8000	9000

Results Analyses

The multiobjective shale gas water management optimization model was written in the AMPL language and solved using the BARON solver through NEOS server version 5.0 in the on-line facility provided by Wisconsin Institutes for Discovery at the University of Wisconsin in Madison for solving optimization problems, see Dolan [57], Drud [58], Server [59], and Gropp, W. Moré [60]. The technical description of the problem is presented as follows: The final multiobjective shale gas water management optimization model along with a set of well-defined multiple objectives comprised 219 variables including 42 binary variables, 27 non-linear variables, 150 linear variables, and 336 constraints, including 15 non-linear constraints and 321 linear constraints, 66 equality, and 270 inequality constraints. The total computational time for obtaining the final solution was 0.095 s (CPU time). The proposed multiobjective shale gas water management optimization model was solved with three weight parameters assigned to deviational variables of each membership goal with respect to their marginal membership degree. The first weight parameter w_{kT} was assigned to the truth deviational variable of each membership goal. The second weight parameter w_{kI} was assigned to the indeterminacy deviational variable of each membership goal, and the third weight parameter w_{kF} was assigned to the falsity deviational variable of each membership goal included in all three objective functions. The obtained optimal results were categorized into five main parts: (i) the optimal acquisition of freshwater from various sources to different shale sites in order to ensure smooth operation of the shale gas energy generation system; (ii) prominent emerging technologies for the on-site treatment of wastewater; (iii) the optimal wastewater management system strategy, which is challenging from the environmental point of view; (iv) the optimal expansion plan to enhance the treatment plant capacity; and (v) the optimal values of different conflicting objectives with their corresponding assigned weights. The optimal amounts of freshwater from source to different shale sites are summarized in Table 6. In planning period 1, the amount of freshwater requirements from source to five shale sites were 700.000, 186.765, 700.000, 300.480, and 74.100 bbl/day, respectively. In planning period 2, the requirements of freshwater at each shale site were obtained as 1125.000, 1125.000, 654.419, 131.542 and 212.553 bbl/day, respectively. In planning period 3, the consumption of freshwater at each shale site was 1275.000, 187.613, 528.153, 131.542, and 212.553 bbl/day in order to ensure smooth operation of the shale gas extraction processes. However, with the exception of shale sites 1 and 5, the requirements for freshwater increased for each planning horizon. The maximum requirement of freshwater was in shale site 5 with a volume 1275.000 bbl/day, whereas the minimum freshwater requirement was observed at shale site 1 during planning period 3, with 74.100 bbl/day due to the low and high cost of acquisition and transportation incurred over the amount of freshwater, respectively.

The most promising characteristic features of on-site wastewater treatment are the different technologies which are being used to reutilize the wastewater within candidate shale sites. The optimal allocation of wastewater for on-site treatment is summarized in Table 7. The on-site treatment of wastewater by different technologies are emerging options for generating freshwater, which was included in the proposed modeling and optimization framework. At shale site 1, the amount of freshwater after treatment by technology 1 was 150 bbl/day in all three planning horizons; the generation of freshwater after treatment by using technology 2 was 200 bbl/day in each planning period; and by applying technology 3 the values were 1551.71, 2401.65, and 2701.62 bbl/day, which was consistently increasing and ensuring the reuse of wastewater in these three planning horizons. At shale site 2, the amount of freshwater generated by the on-site treatment facility using technology 1 was 127.352 bbl/day in each planning period; the generation of freshwater after treatment using technology 2 was 200, 6250, and 200 bbl/day each; and by applying on-site treatment technology 3 they were 525.313, 4554.66, and 527.01 bbl/day in each planning horizon, respectively. At shale site 3, the amount of freshwater after treatment by technology 1 was 150 bbl/day in all three planning horizons; the generation of freshwater after treatment by using technology 2 was 200 bbl/day in the first and second planning periods, which were the same as shale site 1; whereas it was 2865.32 bbl/day in the third planning slot, and unlike by applying technology 3 the obtained amounts were 1551.71, 2060.51, and 2208.04 in all planning horizons, resulting in a significant increase in the freshwater generation pattern by on-site treatment. At shale site 4, the generation of freshwater using on-site treatment technology 1 was 147.779, 2023.26, and 147.779 bbl/day; by implementing on-site treatment technology 2 they were 200, 272.266, and 730.788 bbl/day, revealing the significant increment in the regenerated wastewater volumes in three planning slots. The amount of freshwater by using on-site treatment technology 3 was 751.275, 300.000, and 300.000 bbl/day in each planning horizon respectively. At shale site 5, the generation of freshwater using on-site treatment technology 1 was 150 bbl/day in each planning slot; by applying on-site treatment technology 2 it was 342.801, 861.73, and 861.73 bbl/day; and after implementing on-site treatment technology 3 it was 300, 860.539, and 860.539 bbl/day in each planning horizon, respectively. Therefore, the optimal regeneration of freshwater at each shale sites was effectively designed by implementing the on-site treatment technology component in the proposed shale gas water management study and could be potentially achieved using these technologies in an efficient manner under many adverse circumstances, especially where wastewater managerial issues are often encountered at the political level.

The presented wastewater managerial study includes one underground injection disposal site and two treatment plants with its three expansion options which are capable of representing the wastewater management system for the shale plays. Optimal distribution of total wastewater for underground injection disposal and treatment facility is summarized in Table 7. The toxic wastewater produced at shale site 1 was 6.75, 17.25, and 13.25 bbl/day, which is directly transported to the underground injection disposal site; whereas the total volume shipped to the treatment plant was 645, 842.50, and 0 bbl/day in all three time horizons. At shale site 2, the whole volume of wastewater was directly sent to the underground injection disposal site and it was not feasible to facilitate the usage of a treatment plant facility. At shale site 3, a certain volume of wastewater was delivered to an underground injection disposal site and treatment plant 2 without allocating any volume to treatment plant 1. The amount of wastewater shipped to the underground injection disposal site was 6.75, 17.25, and 13.25 bbl/day, and the optimal allocations to treatment plant 1 were 137.71, 675, and 850 bbl/day in the three planning periods. At shale sites 4 and 5, the overall volume of produced wastewater that would be delivered from both shale sites were the same and found to be 6.75, 17.25, and 13.25 bbl/day for underground injection disposal purposes: 645, 842.50, and 937.50 bbl/day towards treatment plant 1 whereas the optimal shipment volumes of wastewater from both shale sites to treatment plant 2 were 137.71, 675, and 850 bbl/day in all three planning horizons, respectively. The optimal allocation strategy for the total wastewater volumes was described in such a fashion that the optimal contribution of each wastewater management system components had equal significance. At all five shale sites, the generated amount of wastewater sent from each shale site to the underground injection disposal site were 6.75, 17.25, and 13.25 bbl/day over the three planning horizons, respectively, revealing the maximum permitted amount at the underground injection disposal site and restraining the subsurface water for a certain period. More elaborately, it could be concluded that during the various time horizons it was not found optimal and feasible to flow the wastewater towards the underground injection disposal site due to the significant cost of transportation and the underground injection disposal facility. At shale site 1, the amount of wastewater that would be shipped to treatment plant 1 was 645 and 842.50 bbl/day for planning horizons 1 and 2, respectively. The shipment of wastewater from shale site 1 to treatment plant 2 was not found to be feasible due to the significant increase in the transportation cost incurred over wastewater. At shale site 2, the allocation of any wastewater amount to treatment plants 1 and 2 was not found to be justified in all three planning periods. At shale site 3, it was not feasible to deliver any amount of wastewater to treatment plant 1 during all three planning periods, although the amount of wastewater that would be shipped to treatment plant 2 was 137.71, 675, and 850 bbl/day in the three planning horizons, respectively. At shale sites 4 and 5, the volume of wastewater that would be delivered from both shale sites were the same and found to be 645, 842.50, and 937.5 bbl/day towards treatment plant 1, whereas the optimal shipment volume of wastewater from both shale sites to treatment plant 2 were 137.71, 675, and 850 bbl/day in all three planning horizons respectively.

During all three planning horizons, treatment plant expansion options played a significant role in dealing with the excess volume of wastewater produced at different shale sites. The vital dominant characteristic of treatment plant expansion was mainly due to limited and rare existence of underground injection disposal site facilities in some places. The limitations imposed on underground injection disposal sites enabled the expanded scope of treatment plant expansions. The optimal strategy for the expansion of treatment plants is presented in Table 7. The optimal expansion results of treatment plant 1 during planning periods 1 and 3 by using expansion option 1 were 600 bbl/day each. By using expansion option 2 in planning periods 1 and 3, the optimal capacity was 750 bbl/day each; and by using expansion option 3 in planning periods 1 and 3, the optimal capacity was 850 bbl/day each. There was no need to expand the treatment capacity of treatment plant 1 in planning period 2. Moreover, the optimal expansion strategy for treatment plant 2 by using all three expansion options during planning horizon 1 were obtained as 550, 650, and 850 bbl/day, whereas in planning period 2, only expansion option 2 was suggested to enhance the treatment capacity. There was no more optimal strategy indicated for the rest of the expansion options. The compromise solution results obtained by solving the proposed multiobjective shale gas water management model are summarized in Table 6. The minimum total cost of acquisition and transportation of freshwater at the source and from different sources to shale sites was USD $525126.00, whereas the net cost incurred over the entire amount of wastewater management during the three planning periods was obtained as USD $4025940.00. The optimal strategy to expand the treatment plant capacity with the predetermined expansion option was presented efficiently and the total capital investment levied on the expansion of the wastewater treatment plant was USD $5548.97, which reveals that there is still adequate opportunity to expand the capacity of the treatment plant. Shale gas water management systems play an important role in the whole process of generating shale gas energy. The acquisition of a huge amount of freshwater for the fracturing process is a challenging task. The wastewater released from shale sites is toxic in nature and contains various harmful dissolved elements. Therefore, a well-organized wastewater management system includes disposal sites (underground injections) and the establishment of different treatment plants with expansion options.

The overall shale gas water modeling approach was presented, inevitably revealing more practical aspects of decision-making scenarios. Uncertainty among parameters due to vagueness and hesitation were addressed with the triangular intuitionistic fuzzy number, which complies over the degree of acceptance and non-acceptance simultaneously. For example, if the decision maker intends to quantify the value of freshwater requirement with some estimated value, such as each shale site requires approximately 54,800 bbl/day for fracking and horizontal drilling purposes, then the most likely estimated interval would be 54,750–54,850 bbl/day, along with some hesitation degree that may be given as 54,700–54900 bbl/day, which ensures less violation of risks with degree of acceptance and non-acceptance. The representation of different constraints imposed over various parameters also reflects the real scenario of Pennsylvania. In Pennsylvania, underground disposal facilities are very rare and most often wastewater is shipped to nearby cities in Ohio. The solution results have shown a similar situation, and less sewage has been allocated to a different underground disposal facility. Furthermore, the scope for on-site treatment technology and expansion capacity options of treatment plants have been optimally utilized. The resulting optimal allocation of wastewater for on-site treatment at different shale sites shows another advantage by reducing the transportation cost incurred over the treatment and disposal facilities. The opportunity for the expansion capacity option of the treatment plant—if needed—was propounded, and results show that some expansion option was adopted due to the lesser capital investment. The determination of the wastewater reuse rate at the treatment plant also yielded a significant amount of freshwater generation and ensured a lesser burden on the underground disposal facility, which again exhibits substantial characteristic features of the shale gas modeling approach of Pennsylvania. Thus, the proposed shale gas water management model can be easily applied to shale gas energy project planning problems that inherently involve uncertain parameters. The decision maker(s) or project manager(s) can conclusively determine the optimal allocation of each water component with a set of multiple conflicting objectives along with a profitable and economic strategy.

Table 6. Optimal amount of freshwater and value of objective functions.

	Amount of Freshwater $FW_{s,i,t}$
1 1 1	700.000
1 1 2	1125.000
1 1 3	1275.000
1 2 1	186.765
1 2 2	1125.000
1 2 3	187.613
1 3 1	700.000
1 3 2	654.419
1 3 3	528.153
1 4 1	300.480
1 4 2	131.542
1 4 3	131.542
1 5 1	74.100
1 5 2	212.553
1 5 3	212.553
Optimal objective values	
Minimum Z_1	525,126.00
Minimum Z_2	4,025,940.00
Minimum Z_3	5548.97

Table 7. Optimal amount of wastewater allocation and treatment plant expansion strategy.

	Total Amount of wastewater $WW_{i,j,t}$	Amount of Wastewater at Disposal Site $WWD_{i,j,t}$	Amount of Wastewater at Treatment Plant $WWT_{i,j,t}$	Amount of Wastewater for on-Site Treatment $WTO_{i,o,t}$
1 1 1	6.75	6.75	0	150
1 1 2	17.25	17.25	0	150
1 1 3	13.25	13.25	0	150
1 2 1	645	0	645	200
1 2 2	842.5	0	842.5	200
1 2 3	0	0	0	200
1 3 1	0	0	0	1551.71
1 3 2	0	0	0	2401.65
1 3 3	0	0	0	2701.62
2 1 1	6.75	6.75	0	127.352
2 1 2	17.25	17.25	0	127.352
2 1 3	13.25	13.25	0	127.352
2 2 1	0	0	0	200
2 2 2	0	0	0	6250
2 2 3	0	0	0	200
2 3 1	0	0	0	525.313
2 3 2	0	0	0	4554.66
2 3 3	0	0	0	527.01
3 1 1	6.75	6.75	0	150
3 1 2	17.25	17.25	0	150
3 1 3	13.25	13.25	0	150
3 2 1	0	0	0	200
3 2 2	0	0	0	200
3 2 3	0	0	0	2865.32
3 3 1	137.71	0	137.71	1551.32
3 3 2	675	0	675	2060.51
3 3 3	850	0	850	2208.04
4 1 1	6.75	6.75	0	147.779
4 1 2	17.25	17.25	0	2023.26
4 1 3	13.25	13.25	0	147.779
4 2 1	645	0	645	200
4 2 2	842.5	0	842.5	272.266
4 2 3	937.5	0	937.5	730.788
4 3 1	137.71	0	137.71	751.275
4 3 2	675	0	675	300
4 3 3	850	0	850	300
5 1 1	6.75	6.75	0	150
5 1 2	17.25	17.25	0	150
5 1 3	13.25	13.25	0	150
5 2 1	645	0	645	342.801
5 2 2	842.5	0	842.5	861.73
5 2 3	937.5	0	937.5	861.73
5 3 1	137.71	0	137.71	300
5 3 2	675	0	675	360.539
5 3 3	850	0	850	360.539
Increased treatment plant capacity (eo)	Expansion option (m)	Time period $t = 1$	$t = 2$	$t = 3$
Treatment plant 1	1	600	-	600
Treatment plant 1	2	750	-	750
Treatment plant 1	3	850	-	850
Treatment plant 2	1	550	550	-
Treatment plant 2	2	650	-	-
Treatment plant 2	3	800	-	-

5. Conclusions

The multiobjective shale gas water management optimization model addressed within synthesizes the optimum allocation of water resources for shale gas extraction processes. It assures the optimal distribution of freshwater and wastewater, which are the complementary components of shale gas energy production problems. The proposed shale gas modeling outlook is reliable and provides a helpful tool to investigate and analyze the trade-off between socio-economic and environmental concerns globally. The different costs incurred over freshwater, charges levied on wastewater, and capital investment of expanding treatment plant capacity along with the set of shale gas water management system constraints were optimized simultaneously. Uncertainty measures were incorporated among different parameters to demonstrate the actual situations encountered in real-life shale gas optimization frameworks. The accumulation of freshwater from various sources is a crucial task to fulfill commercial needs. However, alternate options were suggested for the generation of freshwater by using on-site treatment technology, which simultaneously reduced the transportation costs for freshwater. Underground injection disposal sites and treatment plant facilities are two major consumption points of generated wastewater from shale sites. A critical factor in the reuse of water in shale gas is the detailed coordination of activities. For greater convenience, auxiliary options have also been introduced to tackle the excess amount of wastewater in the form of on-site treatment technology and different potential expansions of treatment plant capacity at each shale site during each planning horizon. Unlike the various existing conventional solution techniques, the neutrosophic goal programming approach was suggested, which also considers the independent neutral thoughts of decision makers in the decision-making process. Since the proposed approach was applied to a small-scale shale gas extraction process (see Figure 2), it resulted in the globally optimal solution for all objectives simultaneously. However, it may not always be possible to have a globally optimal solution when dealing with large-scale dataset problems. The discussed approach cannot capture the stochastic nature of parameters, which consequently cannot be applied to stochastic optimization problems.

The significant contributions of the proposed multiobjective shale gas water management system are summarized as follows:

- The proposed study considers the overall shale gas water management system which consists of freshwater acquisition at sources, on-site wastewater treatment facilities at each shale site, underground injection disposal sewage facilities, different treatment plant options for the reuse of wastewater and the total wastewater capacity which are feasible to handle without affecting the environmental issues. The decision maker(s) or project manager(s) may adopt the presented shale gas modeling framework, which has a magnetic orientation concerning the overall water management system. However, pipeline facilities have not been included throughout the shale gas energy extraction due to their uneconomic aspect.
- Uncertainty among the parameter values is commonly known in the decision-making process. In this shale gas optimization model, the different parameters (e.g., acquisition cost, transportation cost, treatment cost, disposal cost, and capital investment) are taken as the triangular intuitionistic fuzzy number, which is based on more intuition and leads to more realistic uncertainty modeling texture. It also ensures that the system costs the reliability of each component (costs related to freshwater and wastewater) more realistically. The crisp versions of uncertain parameters were determined in terms of expected interval and expected values.
- A neutrosophic-based computational decision-making algorithm for such a complex and dynamic multiobjective shale gas water management optimization model provides benefits while obtaining globally optimal solutions. The indeterminacy/neutral thought is the region of the propositions' value uncertainly and originates from the independent and impartial thoughts. Therefore, the proposed NGPA is a dominating and suitable conventional optimization technique that is preferred over others due to the existence of its independent indeterminacy degree.

- The multiobjective shale gas project planning model was implemented with a possible dataset and the obtained optimal results were analyzed for each component of the shale gas system in a well-organized and efficient manner. Hence, it was concluded that the proposed optimal strategy for shale gas production could be adopted for more sophisticated and quite typical Marcellus shale plays in large-scale long-term scenarios.

Due to manuscript drafting constraints and space limitations, some important aspects remain untouched and may be explored as a future research scope. The presented shale gas water management modeling approach could be extended by considering different essential aspects such as the to and fro movement of water through the pipeline which was not considered in this paper. The presented computational study was demonstrated for small-scale shale-plays, which could be further explored for large-scale and long-term time horizons by enhancing the number of shale sites and different sources of freshwater and various destinations for wastewater.

Flow-back water does not exit instantaneously, but follows a decline curve. Most of the water exits in the first 3–4 weeks, but there is small and a continuous flow of produced water during all the shale-sites life. Therefore the presented modeling approach may be extended by capturing the above-discussed behavior of flow-back produced water. On-site treatment technology exerts less pressure on the underground disposal of wastewater and provides an opportunity to reuse the treated wastewater for fracking purposes within the shale sites itself. If there are no NORMs (normally occurring radioactive materials), the most costly part of water treatment is desalination. Therefore, the sort of on-site treatment technologies may be specified along with their actual cost, and the possibility of being used for on-site treatment purposes may be explored as a future study. Most of the water management system (e.g., the water treated in municipal wastewater treatment facilities that are usually not prepared to deal with hypersaline water) are presently forbidden and may be implemented and executed by including them under a good practices scheme in future work. From the decision-making point of view, hierarchical decision-making processes could be adopted, ensuring a decentralized decision-making scenario and providing more flexibility compared to multiobjective optimization techniques with a single decision maker. Apart from conventional solution techniques, some metaheuristic algorithms could be applied to solve such shale gas water management planning problems. Furthermore, the propounded neutrosophic modeling approach could be applied to real-life dataset problems such as supplier selection problems, inventory control problems, supply chain management, humanitarian logistic problems, etc. The proposed approach could be further extended by incorporating the multi-choice and stochastic parameters along with bi-level and multi-level decision-making scenarios.

Author Contributions: conceptualization, F.A. and A.Y.A.; methodology, F.A.; formal analysis, A.Y.A.; writing—original draft preparation, F.A.; writing—review and editing, F.A.; supervision, A.Y.A.; project administration, F.S.; funding acquisition, F.S.

Funding: This research received no external funding.

Acknowledgments: All authors are deeply thankful to potential reviewers and editors for their valuable suggestions to improve the quality and readability of the manuscript.

Conflicts of Interest: The authors declare no conflict of interest.

References

1. Zoback, M.D.; Arent, D.J. Shale gas: Development opportunities and challenges. In Proceedings of the 5th Asian Mining Congress, Kolkata, India, 13–15 February 2014; Volume 44.
2. Moniz, E.J.; Jacoby, H.D.; Meggs, A.J.; Armtrong, R.; Cohn, D.; Connors, S.; Deutch, J.; Ejaz, Q.; Hezir, J.; Kaufman, G. *The Future of Natural Gas*; Massachusetts Institute of Technology: Cambridge, MA, USA, 2011.
3. Shaffer, D.L.; Arias Chavez, L.H.; Ben-Sasson, M.; Romero-Vargas Castrillon, S.; Yip, N.Y.; Elimelech, M. Desalination and reuse of high-salinity shale gas produced water: drivers, technologies, and future directions. *Environ. Sci. Technol.* **2013**, *47*, 9569–9583. [CrossRef]

4. Asche, F.; Oglend, A.; Osmundsen, P. Gas versus oil prices the impact of shale gas. *Energy Policy* **2012**, *47*, 117–124. [CrossRef]
5. Wenzhi, Z.; Dazhong, D.; Jianzhong, L.; Guosheng, Z. The resource potential and future status in natural gas development of shale gas in China. *Eng. Sci.* **2012**, *14*, 46–52.
6. Vengosh, A.; Jackson, R.B.; Warner, N.; Darrah, T.H.; Kondash, A. A critical review of the risks to water resources from unconventional shale gas development and hydraulic fracturing in the United States. *Environ. Sci. Technol.* **2014**, *48*, 8334–8348. [CrossRef]
7. Vidic, R.D.; Brantley, S.L.; Vandenbossche, J.M.; Yoxtheimer, D.; Abad, J.D. Impact of shale gas development on regional water quality. *Science* **2013**, *340*, 1235009. [CrossRef]
8. Warner, N.R.; Christie, C.A.; Jackson, R.B.; Vengosh, A. Impacts of shale gas wastewater disposal on water quality in western Pennsylvania. *Environ. Sci. Technol.* **2013**, *47*, 11849–11857. [CrossRef]
9. Stevens, P. *The Shale Gas Revolution: Developments and Changes*; Chatham House: London, UK, 2012.
10. Sieminski, A. International energy outlook. In Proceedings of the Deloitte Oil and Gas Conference, Houston, TX, USA, 18 November 2014.
11. Kerr, R.A. Study: High-Tech Gas Drilling Is Fouling Drinking Water. *Science* **2011**, *332*, 775.
12. Lutz, B.D.; Lewis, A.N.; Doyle, M.W. Generation, transport, and disposal of wastewater associated with Marcellus Shale gas development. *Water Resour. Res.* **2013**, *49*, 647–656. [CrossRef]
13. Yang, L.; Grossmann, I.E.; Manno, J. Optimization models for shale gas water management. *AIChE J.* **2014**, *60*, 3490–3501. [CrossRef]
14. Yang, L.; Grossmann, I.E.; Mauter, M.S.; Dilmore, R.M. Investment optimization model for freshwater acquisition and wastewater handling in shale gas production. *AIChE J.* **2015**, *61*, 1770–1782. [CrossRef]
15. Li, L.G.; Peng, D.H. Interval-valued hesitant fuzzy Hamacher synergetic weighted aggregation operators and their application to shale gas areas selection. *Math. Probl. Eng.* **2014**, *2014*, 181050. [CrossRef]
16. Gloyston, H.; Johnstone, C. UK Has Vast Shale Gas Reserves, Geologists Say. 2012. Available online: https://www.birmingham.ac.uk/Documents/research/SocialSciences/NuclearEnergyFullReport.pdf (accessed on 16 March 2019).
17. Gao, J.; You, F. Optimal design and operations of supply chain networks for water management in shale gas production: MILFP model and algorithms for the water-energy nexus. *AIChE J.* **2015**, *61*, 1184–1208. [CrossRef]
18. Sang, Y.; Chen, H.; Yang, S.; Guo, X.; Zhou, C.; Fang, B.; Zhou, F.; Yang, J. A new mathematical model considering adsorption and desorption process for productivity prediction of volume fractured horizontal wells in shale gas reservoirs. *J. Nat. Gas Sci. Eng.* **2014**, *19*, 228–236. [CrossRef]
19. Gao, J.; You, F. Shale gas supply chain design and operations toward better economic and life cycle environmental performance: MINLP model and global optimization algorithm. *ACS Sustain. Chem. Eng.* **2015**, *3*, 1282–1291. [CrossRef]
20. Guerra, O.J.; Calderón, A.J.; Papageorgiou, L.G.; Siirola, J.J.; Reklaitis, G.V. An optimization framework for the integration of water management and shale gas supply chain design. *Comput. Chem. Eng.* **2016**, *92*, 230–255. [CrossRef]
21. Bartholomew, T.V.; Mauter, M.S. Multiobjective optimization model for minimizing cost and environmental impact in shale gas water and wastewater management. *ACS Sustain. Chem. Eng.* **2016**, *4*, 3728–3735. [CrossRef]
22. Zhang, X.; Sun, A.Y.; Duncan, I.J. Shale gas wastewater management under uncertainty. *J. Environ. Manag.* **2016**, *165*, 188–198. [CrossRef] [PubMed]
23. Lira-Barragán, L.F.; Ponce-Ortega, J.M.; Serna-González, M.; El-Halwagi, M.M. Optimal reuse of flowback wastewater in hydraulic fracturing including seasonal and environmental constraints. *AIChE J.* **2016**, *62*, 1634–1645. [CrossRef]
24. Chen, Y.; He, L.; Guan, Y.; Lu, H.; Li, J. Life cycle assessment of greenhouse gas emissions and water-energy optimization for shale gas supply chain planning based on multi-level approach: Case study in Barnett, Marcellus, Fayetteville, and Haynesville shales. *Energy Convers. Manag.* **2017**, *134*, 382–398. [CrossRef]
25. Knee, K.L.; Masker, A.E. Association between unconventional oil and gas (UOG) development and water quality in small streams overlying the Marcellus Shale. *Freshw. Sci.* **2019**, *38*, 113–130. [CrossRef]
26. Lan, Y.; Yang, Z.; Wang, P.; Yan, Y.; Zhang, L.; Ran, J. A review of microscopic seepage mechanism for shale gas extracted by supercritical CO_2 flooding. *Fuel* **2019**, *238*, 412–424. [CrossRef]

27. Ren, K.; Tang, X.; Jin, Y.; Wang, J.; Feng, C.; Höök, M. Bi-objective optimization of water management in shale gas exploration with uncertainty: A case study from Sichuan, China. *Resour. Conserv. Recycl.* **2019**, *143*, 226–235. [CrossRef]
28. Zhang, Y.; Clark, A.; Rupp, J.A.; Graham, J.D. How do incentives influence local public support for the siting of shale gas projects in China? *J. Risk Res.* **2019**. [CrossRef]
29. Denham, A.; Willis, M.; Zavez, A.; Hill, E. Unconventional natural gas development and hospitalizations: Evidence from Pennsylvania, United States, 2003–2014. *Public Health* **2019**, *168*, 17–25. [CrossRef]
30. Al-Aboosi, F.; El-Halwagi, M. An integrated approach to water-energy nexus in shale-gas production. *Processes* **2018**, *6*, 52. [CrossRef]
31. Jin, L.; Fu, H.; Kim, Y.; Wang, L.; Cheng, H.; Huang, G. The α-Representation Inexact T2 Fuzzy Sets Programming Model for Water Resources Management of the Southern Min River Basin under Uncertainty. *Symmetry* **2018**, *10*, 579. [CrossRef]
32. Ren, C.; Zhang, H. A Fuzzy Max–Min Decision Bi-Level Fuzzy Programming Model for Water Resources Optimization Allocation under Uncertainty. *Water* **2018**, *10*, 488. [CrossRef]
33. Wang, H.; Zhang, C.; Guo, P. An Interval Quadratic Fuzzy Dependent-Chance Programming Model for Optimal Irrigation Water Allocation under Uncertainty. *Water* **2018**, *10*, 684. [CrossRef]
34. Guo, M.; Lu, X.; Nielsen, C.P.; McElroy, M.B.; Shi, W.; Chen, Y.; Xu, Y. Prospects for shale gas production in China: Implications for water demand. *Renew. Sustain. Energy Rev.* **2016**, *66*, 742–750. [CrossRef]
35. Gao, J.; He, C.; You, F. Shale Gas Process and Supply Chain Optimization. In *Advances in Energy Systems Engineering*; Springer: Cham, Switzerland, 2017; pp. 21–46.
36. Chebeir, J.; Geraili, A.; Romagnoli, J. Development of Shale Gas Supply Chain Network under Market Uncertainties. *Energies* **2017**, *10*, 246. [CrossRef]
37. Chen, Y.; He, L.; Li, J.; Zhang, S. Multi-criteria design of shale-gas-water supply chains and production systems towards optimal life cycle economics and greenhouse gas emissions under uncertainty. *Comput. Chem. Eng.* **2018**, *109*, 216–235. [CrossRef]
38. Drouven, M.G.; Grossmann, I.E. Mixed-integer programming models for line pressure optimization in shale gas gathering systems. *J. Pet. Sci. Eng.* **2017**, *157*, 1021–1032. [CrossRef]
39. He, L.; Chen, Y.; Li, J. A three-level framework for balancing the tradeoffs among the energy, water, and air-emission implications within the life-cycle shale gas supply chains. *Resour. Conserv. Recycl.* **2018**, *133*, 206–228. [CrossRef]
40. Wang, J.; Liu, M.; Bentley, Y.; Feng, L.; Zhang, C. Water use for shale gas extraction in the Sichuan Basin, China. *J. Environ. Manag.* **2018**, *226*, 13–21. [CrossRef] [PubMed]
41. Ye, J. Expected value method for intuitionistic trapezoidal fuzzy multicriteria decision-making problems. *Expert Syst. Appl.* **2011**, *38*, 11730–11734. [CrossRef]
42. Nishad, A.K.; Singh, S.R. Solving multi-objective decision making problem in intuitionistic fuzzy environment. *Int. J. Syst. Assur. Eng. Manag.* **2015**, *6*, 206–215. [CrossRef]
43. Singh, S.K.; Yadav, S.P. Intuitionistic fuzzy non linear programming problem: Modeling and optimization in manufacturing systems. *J. Intell. Fuzzy Syst.* **2015**, *28*, 1421–1433.
44. Bellman, R.E.; Zadeh, L.A. Decision-Making in a Fuzzy Environment. *Manag. Sci.* **1970**, *17*, B-141–B-164. doi:10.1287/mnsc.17.4.B141. [CrossRef]
45. Zadeh, L. Fuzzy sets. *Inf. Control* **1965**, *8*, 338–353, doi:10.1016/S0019-9958(65)90241-X. [CrossRef]
46. Zimmermann, H.J. Description and optimization of fuzzy systems. *Int. J. Gen. Syst.* **1976**, *2*, 209–215. [CrossRef]
47. Smarandache, F. A Unifying Field in Logics: Neutrosophic Logic. In *Philosophy*; American Research Press(APP): Rehoboth, NM, USA, 1999; pp. 1–141.
48. Ahmad, F.; Adhami, A.Y. Neutrosophic programming approach to multiobjective nonlinear transportation problem with fuzzy parameters. *Int. J. Manag. Sci. Eng. Manag.* **2018**. [CrossRef]
49. Ahmad, F.; Adhami, A.Y.; Smarandache, F. Single Valued Neutrosophic Hesitant Fuzzy Computational Algorithm for Multiobjective Nonlinear Optimization Problem. *Neutrosophic Sets Syst.* **2018**, *22*. doi:10.5281/zenodo.2160357. [CrossRef]
50. Atanassov, K.T. Intuitionistic fuzzy sets. *Fuzzy Sets Syst.* **1986**, *20*, 87–96. doi:10.1016/S0165-0114(86)80034-3. [CrossRef]

51. Ebrahimnejad, A.; Verdegay, J.L. A new approach for solving fully intuitionistic fuzzy transportation problems. *Fuzzy Optim. Decis. Mak.* **2018**, *17*, 447–474. [CrossRef]
52. Grzegrorzewski, P. The hamming distance between intuitionistic fuzzy sets. In Proceedings of the 10th IFSA world congress, Istanbul, Turkey, 30 June–2 July 2003; Volume 30, pp. 35–38.
53. Rizk-Allah, R.M.; Hassanien, A.E.; Elhoseny, M. A multi-objective transportation model under neutrosophic environment. *Comput. Electr. Eng.* **2018**, *69*, 705–719. [CrossRef]
54. Rahm, B.G.; Riha, S.J. Toward strategic management of shale gas development: Regional, collective impacts on water resources. *Environ. Sci. Policy* **2012**, *17*, 12–23. [CrossRef]
55. Rahm, B.G.; Bates, J.T.; Bertoia, L.R.; Galford, A.E.; Yoxtheimer, D.A.; Riha, S.J. Wastewater management and Marcellus Shale gas development: Trends, drivers, and planning implications. *J. Environ. Manag.* **2013**, *120*, 105–113. [CrossRef]
56. Alawattegama, S.K. Survey of Well Water Contamination in a Rural Southwestern Pennsylvania Community with Unconventional Shale Gas Drilling. Ph.D. Thesis, Duquesne University, Pittsburgh, PA, USA, 2013.
57. Dolan, E. *The Neos Server 4.0 Administrative Guide*; Technical Report; Memorandum ANL/MCS-TM-250; Mathematics and Computer Science Division, Argonne National Laboratory: Argonne, IL, USA, 2001.
58. Drud, A.S. CONOPT—A large-scale GRG code. *ORSA J. Comput.* **1994**, *6*, 207–216. [CrossRef]
59. Server, N. *State-of-the-Art Solvers for Numerical Optimization*. 2016. Available online: https://neos-server.org/neos/ (accessed on 16 March 2019).
60. Gropp, W.; Moré, J. Optimization environments and the NEOS server. In *Approximation Theory and Optimization*; Cambridge University Press: Cambridge, UK, 1997; pp. 167–182.

symmetry

MDPI

Article

Extended Nonstandard Neutrosophic Logic, Set, and Probability Based on Extended Nonstandard Analysis

Florentin Smarandache

Mathematics Department, University of New Mexico, Gallup, NM 87301, USA; smarand@unm.edu

Received: 4 March 2019; Accepted: 27 March 2019; Published: 10 April 2019

Abstract: We extend for the second time the nonstandard analysis by adding the left monad closed to the right, and right monad closed to the left, while besides the pierced binad (we introduced in 1998) we add now the unpierced binad—all these in order to close the newly extended nonstandard space under nonstandard addition, nonstandard subtraction, nonstandard multiplication, nonstandard division, and nonstandard power operations. Then, we extend the Nonstandard Neutrosophic Logic, Nonstandard Neutrosophic Set, and Nonstandard Probability on this Extended Nonstandard Analysis space, and we prove that it is a nonstandard neutrosophic lattice of first type (endowed with a nonstandard neutrosophic partial order) as well as a nonstandard neutrosophic lattice of second type (as algebraic structure, endowed with two binary neutrosophic laws: inf_N and sup_N). Many theorems, new terms introduced, better notations for monads and binads, and examples of nonstandard neutrosophic operations are given.

Keywords: nonstandard analysis; extended nonstandard analysis; open and closed monads to the left/right; pierced and unpierced binads; MoBiNad set; infinitesimals; infinities; nonstandard reals; standard reals; nonstandard neutrosophic lattices of first type (as poset) and second type (as algebraic structure), nonstandard neutrosophic logic; extended nonstandard neutrosophic logic; nonstandard arithmetic operations; nonstandard unit interval; nonstandard neutrosophic infimum; nonstandard neutrosophic supremum

1. Short Introduction

In order to more accurately situate and fit the neutrosophic logic into the framework of extended nonstandard analysis [1–3], we present the nonstandard neutrosophic inequalities, nonstandard neutrosophic equality, nonstandard neutrosophic infimum and supremum, and nonstandard neutrosophic intervals, including the cases when the neutrosophic logic standard and nonstandard components T, I, F get values outside of the classical unit interval [0, 1], and a brief evolution of neutrosophic operators [4].

2. Theoretical Reason for the Nonstandard Form of Neutrosophic Logic

The only reason we have added the nonstandard form to neutrosophic logic (and similarly to neutrosophic set and probability) was in order to make a distinction between Relative Truth (which is truth in some Worlds, according to Leibniz) and Absolute Truth (which is truth in all possible Words, according to Leibniz as well) that occur in philosophy.

Another possible reason may be when the neutrosophic degrees of truth, indeterminacy, or falsehood are infinitesimally determined, for example a value infinitesimally bigger than 0.8 (or 0.8^+), or infinitesimally smaller than 0.8 (or $^-0.8$). But these can easily be overcome by roughly using interval neutrosophic values, for example (0.80, 0.81) and (0.79, 0.80), respectively.

3. Why the Sum of Neutrosophic Components Is Up to 3

We was more prudent when we presented the sum of single valued standard neutrosophic components [5–9], saying

$$\text{Let } T, I, F \text{ be single valued numbers, } T, I, F \in [0, 1], \text{ such that } 0 \leq T + I + F \leq 3. \quad (1)$$

The sum of the single-valued neutrosophic components, $T + I + F$ is up to 3 since they are considered completely (100%) independent of each other. But if the two components T and F are completely (100%) dependent, then $T + F \leq 1$ (as in fuzzy and intuitionistic fuzzy logics), and let us assume the neutrosophic middle component I is completely (100%) independent from T and F, then $I \leq 1$, whence $T + F + I \leq 1 + 1 = 2$.

But the degree of dependence/independence [10] between T, I, F all together, or taken two by two, may be, in general, any number between 0 and 1.

4. Neutrosophic Components outside the Unit Interval [0, 1]

Thinking out of box, inspired from the real world, was the first intent, i.e., allowing neutrosophic components (truth/indeterminacy/falsehood) values be outside of the classical (standard) unit real interval [0, 1] used in all previous (Boolean, multivalued, etc.) logics if needed in applications, so neutrosophic component values < 0 and > 1 had to occurs due to the Relative/Absolute stuff, with

$$^{-}0 <_N 0 \text{ and } 1^{+} >_N 1 \quad (2)$$

Later on, in 2007, I found plenty of cases and real applications in Standard Neutrosophic Logic and Set (therefore, not using the Nonstandard Neutrosophic Logic, Set, and Probability), and it was thus possible the extension of the neutrosophic set to Neutrosophic Overset (when some neutrosophic component is > 1), and to Neutrosophic Underset (when some neutrosophic component is < 0), and to Neutrosophic Offset (when some neutrosophic components are off the interval [0, 1], i.e., some neutrosophic component > 1 and some neutrosophic component < 0). Then, similar extensions to Neutrosophic Over/Under/Off Logic, Measure, Probability, Statistics, etc., [11–14], extending the unit interval [0, 1] to

$$[\Psi, \Omega], \text{ with } \Psi \leq 0 < 1 \leq \Omega, \quad (3)$$

where Ψ, Ω are standard (or nonstandard) real numbers.

5. Refined Neutrosophic Logic, Set, and Probability

We wanted to get the neutrosophic logic as general as possible [15], extending all previous logics (Boolean, fuzzy, intuitionistic fuzzy logic, intuitionistic logic, paraconsistent logic, and dialethism), and to have it able to deal with all kind of logical propositions (including paradoxes, nonsensical propositions, etc.).

That is why in 2013 we extended the Neutrosophic Logic to Refined Neutrosophic Logic / Set / Probability (from generalizations of 2-valued Boolean logic to fuzzy logic, also from the Kleene's and Lukasiewicz's and Bochvar's 3-symbol valued logics or Belnap's 4-symbol valued logic, to the most general n-symbol or n-numerical valued refined neutrosophic logic, for any integer $n \geq 1$), the largest ever so far, when some or all neutrosophic components T, I, F were split/refined into neutrosophic subcomponents $T_1, T_2, \ldots; I_1, I_2, \ldots; F_1, F_2, \ldots$, which were deduced from our everyday life [16].

6. From Paradoxism Movement to Neutrosophy Branch of Philosophy and then to Neutrosophic Logic

We started first from Paradoxism (that we founded in the 1980s in Romania as a movement based on antitheses, antinomies, paradoxes, contradictions in literature, arts, and sciences), then we introduced the Neutrosophy (as generalization of Dialectics of Hegel and Marx, which is actually the ancient YinYang Chinese philosophy), neutrosophy is a branch of philosophy studying the dynamics of triads, inspired from our everyday life, triads that have the form

$$\langle A\rangle, \text{ its opposite } \langle antiA\rangle, \text{ and their neutrals } \langle neutA\rangle, \tag{4}$$

where <A> is any item or entity [17]. (Of course, we take into consideration only those triads that make sense in our real and scientific world.)

The Relative Truth neutrosophic value was marked as *1*, while the Absolute Truth neutrosophic value was marked as 1^+ (a tinny bigger than the Relative Truth's value): $1^+ >_N 1$, where $>_N$ is a neutrosophic inequality, meaning 1^+ is neutrosophically bigger than *1*.

Similarly for Relative Falsehood/Indeterminacy (which is falsehood/indeterminacy in some Worlds) and Absolute Falsehood/Indeterminacy (which is falsehood/indeterminacy in all possible worlds).

7. Introduction to Nonstandard Analysis

An *infinitesimal* (or infinitesimal number) (ε) is a number ε, such that $|\varepsilon| < 1/n$, for any non-null positive integer n. An infinitesimal is close to zero, and so small that it cannot be measured.

The infinitesimal is a number smaller, in absolute value, than anything positive nonzero.

Infinitesimals are used in calculus.

An *infinite* (or infinite number) (ω) is a number greater than anything:

$$1 + 1 + 1 + \ldots + 1 \text{ (for any finite number terms)} \tag{5}$$

The infinites are reciprocals of infinitesimals.

The set of *hyperreals* (or *nonstandard reals*), denoted as R^*, is the extension of set of the real numbers, denoted as R, and it comprises the infinitesimals and the infinites, that may be represented on the *hyperreal number line*:

$$1/\varepsilon = \omega/1. \tag{6}$$

The set of hyperreals satisfies the *transfer principle*, which states that the statements of first order in R are valid in R^* as well.

A *monad* (*halo*) of an element $a \in R^*$, denoted by $\mu(a)$, is a subset of numbers infinitesimally close to a.

8. First Extension of Nonstandard Analysis

Let us denote by $R_+{}^*$ the set of positive nonzero hyperreal numbers.

We consider the left monad and right monad, and the (*pierced*) *binad* that we have introduced as extension in 1998 [5]:

Left Monad {that we denote, for simplicity, by (^-a) or only ^-a} is defined as:

$$\mu(^-a) = (^-a) = {}^-a = \overset{-}{a} = \{a - x,\ x \in R_+{}^* \mid x \text{ is } infinitesimal\}. \tag{7}$$

Right Monad {that we denote, for simplicity, by (a^+) or only by a^+} is defined as:

$$\mu(a^+) = (a^+) = a^+ = \overset{+}{a} = \{a + x,\ x \in R_+{}^* \mid x \text{ is } infinitesimal\}. \tag{8}$$

Pierced Binad {that we denote, for simplicity, by $(^-a^+)$ or only $^-a^+$} is defined as:

$$\begin{aligned}\mu(^-a^+) = (^-a^+) = {}^-a^+ = \overset{-+}{a} &= \{a - x,\ x \in R_+{}^* \mid x \text{ is } infinitesimal\} \cup \{a + x,\ x \in R_+{}^* \mid x \text{ is } infinitesimal\} \\ &= \{a \pm x,\ x \in R_+{}^* \mid x \text{ is } infinitesimal\}.\end{aligned} \tag{9}$$

The left monad, right monad, and the pierced binad are subsets of R^*.

9. Second Extension of Nonstandard Analysis

For the necessity of doing calculations that will be used in nonstandard neutrosophic logic in order to calculate the nonstandard neutrosophic logic operators (conjunction, disjunction, negation, implication, and equivalence) and in order to have the Nonstandard Real MoBiNad Set closed under arithmetic operations, we extend, for the time being, the left monad to the Left Monad Closed to the Right, the right monad to the Right Monad Closed to the Left, and the Pierced Binad to the Unpierced Binad, defined as follows [18–21].

Left Monad Closed to the Right

$$\mu\left(\overset{-0}{a}\right) = \left(\overset{-0}{a}\right) = \overset{-0}{a} = \{a - x \mid x = 0,\ or\ x \in R_+{}^*\ and\ x\ is\ infinitesimal\} = \mu({}^-a) \cup \{a\} = ({}^-a)\ 0 \cup \{a\} = {}^-a \cup \{a\}. \tag{10}$$

Right Monad Closed to the Left

$$\mu\left(\overset{0+}{a}\right) = \left(\overset{0+}{a}\right) = \overset{0+}{a} = \{a + x \mid x = 0,\ or\ x \in R_+{}^*\ and\ x\ is\ infinitesimal\} = \mu(a^+) \cup \{a\} = (a^+)\ 0 \cup \{a\} = a^+ \cup \{a\}. \tag{11}$$

Unpierced Binad

$$\mu\left(\overset{-0+}{a}\right) = \left(\overset{-0+}{a}\right) = \overset{-0+}{a} = \{a - x | x \in R_+{}^* andxisinfinitesimal\} \cup \{a + x | x \in R_+{}^* and\ xisinfinitesimal\} \cup \{a\} = \{a \pm x | x = 0, orx \in R_+{}^* andxisinfinitesimal\} = \mu({}^-a^+) \cup \{a\} = ({}^-a^+) \cup \{a\} = {}^- a^+ \cup \{a\} \tag{12}$$

The element $\{a\}$ has been included into the left monad, right monad, and pierced binad respectively.

10. Nonstandard Neutrosophic Function

In order to be able to define equalities and inequalities in the sets of monads, and in the sets of binads, we construct a *nonstandard neutrosophic function* that approximates the monads and binads to tiny open (or half open and half closed respectively) standard real intervals as below. It is called 'neutrosophic' since it deals with indeterminacy: unclear, vague monads and binads, and the function approximates them with some tiny real subsets.

Taking an arbitrary infinitesimal

$$\varepsilon_1 > 0, \text{ and writing } {}^-a = a - \varepsilon_1,\ a^+ = a + \varepsilon_1, \text{ and } {}^-a^+ = a \pm \varepsilon_1, \tag{13}$$

or taking an arbitrary infinitesimal $\varepsilon_2 \geq 0$, and writing

$$\overset{-0}{a} = (a - \varepsilon_2, a],\ \overset{0+}{a} = [a, a + \varepsilon_2),\ \overset{-0+}{a} = (a - \varepsilon_2, a + \varepsilon_2) \tag{14}$$

We meant to actually pick up a representative from each class of the monads and of the binads.

Representations of the monads and binads by intervals is not quite accurate from a classical point of view, but it is an approximation that helps in finding a partial order and computing nonstandard arithmetic operations on the elements of the nonstandard set NR_{MB}.

Let ε be a generic positive infinitesimal, while a be a generic standard real number.

Let $P(R)$ be the power set of the real number set R.

$$\mu_N : NR_{MB} \to P(R) \tag{15}$$

For any $a \in R$, the set of real numbers, one has

$$\mu_N((^{-}a)) =_N (a - \varepsilon, a), \tag{16}$$

$$\mu_N((a^{+})) =_N (a, a + \varepsilon), \tag{17}$$

$$\mu_N((^{-}a^{+})) =_N (a - \varepsilon, a) \cup (a, a + \varepsilon), \tag{18}$$

$$\mu_N\left(\left(\overset{-0}{a}\right)\right) =_N (a - \varepsilon, a], \tag{19}$$

$$\mu_N\left(\left(\overset{0+}{a}\right)\right) =_N [a, a + \varepsilon), \tag{20}$$

$$\mu_N\left(\left(\overset{-0+}{a}\right)\right) =_N (a - \varepsilon, a + \varepsilon), \tag{21}$$

$$\mu_N\left(\left(\overset{0}{a}\right)\right) =_N \mu_N(a) =_N a = [a, a], \tag{22}$$

in order to set it as real interval too.

11. General Notations for Monads and Binads

Let $a \in R$ be a standard real number. We use the following general notation for monads and binads.

$$\overset{m}{a} \in \left\{a, \overset{-}{a}, \overset{-0}{a}, \overset{+}{a}, \overset{0+}{a}, \overset{-+}{a}, \overset{-0+}{a}\right\} \text{ and by convention } \overset{0}{a} = a; \tag{23}$$

or

$$m \in \{\,, ^{-}, ^{-0}, ^{+}, ^{+0}, ^{-+}, ^{-0+}\} = \{^{0}, ^{-}, ^{-0}, ^{+}, ^{+0}, ^{-+}, ^{-0+}\}; \tag{24}$$

therefore "m" above a standard real number "a" may mean anything: a standard real number (^{0}, or nothing above), a left monad ($^{-}$), a left monad closed to the right ($^{-0}$), a right monad ($^{+}$), a right monad closed to the left ($^{0+}$), a pierced binad ($^{-+}$), or a unpierced binad ($^{-0+}$), respectively.

The notations of monad's and binad's diacritics above (not laterally) the number a as

$$\overset{-}{a}, \overset{-0}{a}, \overset{+}{a}, \overset{0+}{a}, \overset{-+}{a}, \overset{-0+}{a} \tag{25}$$

are the best, since they also are designed to avoid confusion for the case when the real number a is negative.

For example, if $a = {}^{-}2$, then the corresponding monads and binads are respectively represented as:

$$\overset{-}{^{-}2}, \overset{-0}{^{-}2}, \overset{+}{^{-}2}, \overset{0+}{^{-}2}, \overset{-+}{^{-}2}, \overset{-0+}{^{-}2} \tag{26}$$

Classical and Neutrosophic Notations

Classical notations on the set of real numbers:

$$<, \leq, >, \geq, \wedge, \vee, \rightarrow, \leftrightarrow, \cap, \cup, \subset, \supset, \subseteq, \supseteq, =, \in, \\ +, -, \times, \div, \hat{}, * \tag{27}$$

Operations with real subsets:

$$\circledast \tag{28}$$

Neutrosophic notations on nonstandard sets (that involve indeterminacies, approximations, and vague boundaries):

$$<_N, \leq_N, >_N, \geq_N, \wedge_N, \vee_N, \rightarrow_N, \leftrightarrow_N, \cap_N, \cup_N, \subset_N, \supset_N, \subseteq_N, \supseteq_N, =_N, \in_N +_N, -_N, \times_N, \div_N, \hat{}_N, *_N \tag{29}$$

12. Neutrosophic Strict Inequalities

We recall the neutrosophic strict inequality which is needed for the inequalities of nonstandard numbers.

Let α and β be elements in a partially ordered set M.

We have defined the neutrosophic strict inequality

$$\alpha >_N \beta \tag{30}$$

and read as

"α is neutrosophically greater than β"

if α in general is greater than β, or α is approximately greater than β, or *subject to some indeterminacy* (unknown or unclear ordering relationship between α and β) *or subject to some contradiction* (situation when α is smaller than or equal to β) *α is greater than β*.

It means that in most of the cases, on the set M, *α is greater than β*.

And similarly for the opposite neutrosophic strict inequality

$$\alpha <_N \beta \tag{31}$$

13. Neutrosophic Equality

We have defined the neutrosophic inequality

$$\alpha =_N \beta \tag{32}$$

and read as

"α is neutrosophically equal to β"

if α in general is equal to β, or α is approximately equal to β, or *subject to some indeterminacy* (unknown or unclear ordering relationship between α and β) *or subject to some contradiction* (situation when α is not equal to β) *α is equal to β*.

It means that in most of the cases, on the set M, *α is equal to β*.

14. Neutrosophic (Nonstrict) Inequalities

Combining the neutrosophic strict inequalities with neutrosophic equality, we get the $\geq_N$ and $\leq_N$ neutrosophic inequalities.

Let α and β be elements in a partially ordered set M.

The neutrosophic (nonstrict) inequality

$$\alpha \geq_N \beta \tag{33}$$

and read as

"α is neutrosophically greater than or equal to β"

if α in general is greater than or equal to β, or α is approximately greater than or equal to β, or subject to some indeterminacy (unknown or unclear ordering relationship between α and β) or subject to some contradiction (situation when α is smaller than β) α is greater than or equal to β.

It means that in most of the cases, on the set M, *α is greater than or equal to β*.

And similarly for the opposite neutrosophic (nonstrict) inequality

$$\alpha \leq_N \beta. \tag{34}$$

15. Neutrosophically Ordered Set

Let M be a set. $(M, <_N)$ is called a neutrosophically ordered set if

$$\forall \alpha, \beta \in M, \text{onehas : either} \alpha <_N \beta, \text{or} \alpha =_N \beta, \text{or} \alpha >_N \beta. \tag{35}$$

16. Neutrosophic Infimum and Neutrosophic Supremum

As an extension of the classical infimum and classical supremum, and using the neutrosophic inequalities and neutrosophic equalities, we define the neutrosophic infimum (denoted as inf_N) and the neutrosophic supremum (denoted as sup_N).

Neutrosophic Infimum.

Let $(S, <_N)$ be a set that is neutrosophically partially ordered, and M a subset of S.

The neutrosophic infimum of M, denoted as $inf_N(M)$ is the neutrosophically greatest element in S that is neutrosophically less than or equal to all elements of M:

Neutrosophic Supremum.

Let $(S, <_N)$ be a set that is neutrosophically partially ordered and M a subset of S.

The neutrosophic supremum of M, denoted as $sup_N(M)$ is the neutrosophically smallest element in S that is neutrosophically greater than or equal to all elements of M.

17. Definition of Nonstandard Real MoBiNad Set

Let $\mathbb{R}$ be the set of standard real numbers, and $\mathbb{R}^*$ be the set of hyper-reals (or nonstandard reals) that consists of infinitesimals and infinites.

The Nonstandard Real MoBiNad Set is now defined for the first time as follows

$$NR_{MB} =_N \left\{ \begin{array}{c} \varepsilon,\ \omega,\ a, \left(^-a\right), \left(^-a^0\right),\ \left(a^+\right), \left(^0a^+\right), \left(^-a^+\right), \left(^-a^{0+}\right) | \text{where } \varepsilon \text{ are infinitesimals,} \\ \text{with } \varepsilon \in \mathbb{R}^*;\ \omega = 1/\varepsilon \text{ are infinites, with } \omega \in \mathbb{R}^*; \text{ and } a \text{ are real numbers, with } a \in \mathbb{R} \end{array} \right\} \tag{36}$$

Therefore

$$NR_{MB} =_N \mathbb{R}^* \cup \mathbb{R} \cup \mu\left(^-\mathbb{R}\right) \cup \mu\left(^-\mathbb{R}^0\right) \cup \mu\left(\mathbb{R}^+\right) \cup \mu\left(^0\mathbb{R}^+\right) \cup \mu\left(^-\mathbb{R}^+\right) \cup \mu\left(^-\mathbb{R}^{0\,+}\right), \tag{37}$$

where

$\mu\left(^-\mathbb{R}\right)$ is the set of all real left monads,
$\mu\left(^-\mathbb{R}^0\right)$ is the set of all real left monads closed to the right,
$\mu\left(\mathbb{R}^+\right)$ is the set of all real right monads,
$\mu\left(^0\mathbb{R}^+\right)$ is the set of all real right monads closed to the left,
$\mu\left(^-\mathbb{R}^+\right)$ is the set of all real pierced binads,
and $\mu\left(^-\mathbb{R}^{0\,+}\right)$ is the set of all real unpierced binads.

Also,

$$NR_{MB} =_N \left\{ \varepsilon,\ \omega,\ \overset{m}{a} \middle| \begin{array}{c} \text{where } \varepsilon,\ \omega \in \mathbb{R}^*, \varepsilon \text{ are infinitesimals, } \omega = \frac{1}{\varepsilon} \text{ are infinities;} \\ a \in \mathbb{R}; \text{and } m \in \left\{ ,^-,^{-\,0},^+,^{+\,0},^{-\,+},^{-\,0\,+} \right\} \end{array} \right\} \tag{38}$$

NR_{MB} is closed under addition, subtraction, multiplication, division (except division by $\overset{m}{a}$, with $a = 0$ and $m \in \left\{ ,^-,^{-\,0},^+,^{0+},^{-\,+},^{-\,0\,+} \right\}$), and power

$\left\{ \left(\overset{m_1}{a}\right)^{\left(\overset{m_2}{b}\right)} \right.$ with: either $a > 0$, or $a = 0$ and $m \in \left\{ ,^+,^{0+} \right\}$ and $b > 0$, or $a < 0$ but $b = \frac{p}{r}$ (irreducible fraction) and p, r are integers with r an odd positive integer $r \in \{1, 3, 5, \ldots\}\}$.

These mobinad (nonstandard) above operations are reduced to set operations, using Set Analysis and Neutrosophic Analysis (both introduced by the author [22] (page 11), which are generalizations of Interval Analysis), and they deal with sets that have indeterminacies.

18. Etymology of MoBiNad

MoBiNad comes from **mo***nad* + **bi***nad*, introduced now for the first time.

19. Definition of Nonstandard Complex MoBiNad Set

The Nonstandard Complex MoBiNad Set, introduced here for the first time, is defined as

$$NC_{MB} =_N \left\{\alpha + \beta i | \text{ where } i = \sqrt{-1};\ \alpha,\ \beta \in NR_{MB}\right\} \tag{39}$$

20. Definition of Nonstandard Neutrosophic Real MoBiNad Set

The Nonstandard Neutrosophic Real MoBiNad Set, introduced now for the first time, is defined as

$$NNR_{MB} =_N \left\{\alpha + \beta I | \text{ where } I = \text{literal indeterminacy},\ I^2 = I;\ \alpha,\ \beta \in NR_{MB}\right\}. \tag{40}$$

21. Definition of Nonstandard Neutrosophic Complex MoBiNad Set

The Nonstandard Neutrosophic Complex MoBiNad Set, introduced now for the first time, is defined as

$$NNC_{MB} =_N \left\{\alpha + \beta I | \text{ where } I = \text{literal indeterminacy},\ I^2 = I;\ \alpha,\ \beta \in NC_{MB}\right\} \tag{41}$$

22. Properties of the Nonstandard Neutrosophic Real Mobinad Set

Since in nonstandard neutrosophic logic we use only the nonstandard neutrosophic real mobinad set, we study some properties of it.

Theorem 1. *The nonstandard real mobinad set (NR_{MB}, $\leq_N$), endowed with the nonstandard neutrosophic inequality is a lattice of first type [as partially ordered set (poset)].*

Proof. The set NR_{MB} is partially ordered, because (except the two-element subsets of the form $\left\{a, \overset{-+}{a}\right\}$, and $\left\{a, \overset{-0+}{a}\right\}$, with $a \in \mathbb{R}$, beetwen which there is no order) all other elements are ordered:

If $a < b$, where $a,\ b \in \mathbb{R}$, then: $\overset{m_1}{a} <_N \overset{m_2}{b}$, for any monads or binads

$$m_1,\ m_2 \in_N \left\{{}^-, {}^{-0}, {}^+, {}^{0+}, {}^{-+}, {}^{-0+}\right\}. \tag{42}$$

If $a = b$, one has:

$${}^-a <_N a, \tag{43}$$

$$a^- <_N a^+ \tag{44}$$

$$a <_N a^+ \tag{45}$$

$${}^-a \leq_N {}^-a^+, \tag{46}$$

$${}^-a \leq_N {}^-a^+, \tag{47}$$

and there is no neutrosophic ordering relationship between a and ${}^-a^+$,

nor between a and $\overset{-0+}{a}$ (that is why $\leq_N$ on NR_{MB} is a partial ordering set). (48)

$$\text{If } a > b, \text{ then} : \overset{m_1}{a} >_N \overset{m_2}{b}, \text{ for any monads or binads } m_1, m_2. \tag{49}$$

□

Any two-element set $\{\alpha, \beta\} \subset_N NR_{MB}$ has a neutrosophic nonstandard infimum (meet, or greatest lower bound) that we denote by $\inf_N$, and a neutrosophic nonstandard supremum (joint, or least upper bound) that we denote by $\sup_N$, where both

$$\inf_N\{\alpha, \beta\} \text{ and } \sup_N\{\alpha, \beta\} \in NR_{MB}. \tag{50}$$

For the nonordered elements a and ${}^-a^+$:

$$\inf_N\{a, {}^-a^+\} =_N {}^-a \in_N NR_{MB}, \tag{51}$$

$$\sup_N\{a, {}^-a^+\} =_N a^+ \in_N NR_{MB} \tag{52}$$

And similarly for nonordered elements a and ${}^-a^{0\,+}$:

$$\inf_N\{a, {}^-a^{0\,+}\} =_N {}^-a \in_N NR_{MB}, \tag{53}$$

$$\sup_N\{a, {}^-a^{0\,+}\} =_N a^+ \in_N NR_{MB}. \tag{54}$$

Dealing with monads and binads which neutrosophically are real subsets with indeterminate borders, and similarly $a = [a, a]$ can be treated as a subset, we may compute $\inf_N$ and $\sup_N$ of each of them.

$$\inf_N({}^-a) =_N {}^-a \text{ and } \sup_N({}^-a) =_N {}^-a \tag{55}$$

$$\inf_N(a^+) =_N a^+ \text{ and } \sup_N(a^+) =_N a^+; \tag{56}$$

$$\inf_N({}^-a^+) =_N {}^-a \text{ and } \sup_N({}^-a^+) =_N a^+; \tag{57}$$

$$\inf_N({}^-a^{0\,+}) =_N {}^-a \text{ and } \sup_N({}^-a^{0\,+}) =_N a^+. \tag{58}$$

Also,

$$\inf_N(a) =_N a \text{ and } \sup_N(a) =_N a. \tag{59}$$

If $a < b$, then $\overset{m_1}{a} <_N \overset{m_2}{b}$, hence

$$\inf_N\left\{\overset{m_1}{a}, \overset{m_2}{b}\right\} =_N \inf_N\left(\overset{m_1}{a}\right) \text{ and } \sup_N\left\{\overset{m_1}{a}, \overset{m_2}{b}\right\} =_N \sup_N \overset{m_2}{b}, \tag{60}$$

which are computed as above.

Similarly, if

$$a > b, \text{ with } \overset{m_1}{a} <_N \overset{m_2}{b}. \tag{61}$$

If $a = b$, then: $\inf_N\left\{\overset{m_1}{a}, \overset{m_2}{a}\right\} =_N$ the neutrosophically smallest ($<_N$) element among

$$\inf_N\left\{\overset{m_1}{a}\right\} \text{ and } \inf_N\left\{\overset{m_2}{a}\right\}. \tag{62}$$

While $\sup_N\left\{\overset{m_1}{a}, \overset{m_2}{a}\right\} =_N$ the neutrosophically greatest ($>_N$) element among

$$\sup_N\left\{\overset{m_1}{a}\right\} \text{ and } \sup_N\left\{\overset{m_2}{a}\right\}. \tag{63}$$

Examples:

$$\inf_N\left({}^-a,\ a^+\right) =_N {}^-a \text{ and } \sup_N\left({}^-a,\ a^+\right) =_N a^+; \tag{64}$$

$$\inf_N\left({}^-a,{}^-a^+\right) =_N {}^-a \text{ and } \sup_N\left({}^-a,{}^-a^+\right) =_N a^+; \tag{65}$$

$$\inf_N\left({}^-a^+,\ a^+\right) =_N {}^-a \text{ and } \sup_N\left({}^-a^+,\ a^+\right) =_N a^+. \tag{66}$$

Therefore, $(NR_{MB}, \leq_N)$ is a nonstandard real mobinad lattice of first type (as partially ordered set).

Consequence

If we remove all pierced and unpierced binads from NR_{MB} and we denote the new set by $NR_M = \left\{\varepsilon,\ \omega,\ a,{}^-a,{}^-a^0,\ a^+,{}^0a^+,\ \text{where } \varepsilon \text{ are infinitesimals, } \omega \text{ are infinites, and } a \in \mathbb{R}\right\}$ we obtain a totally neutrosophically ordered set.

Theorem 2. *Any finite non-empty subset L of* $(NR_{MB}, \leq_N)$ *is also a sublattice of first type.*

Proof. It is a consequence of any classical lattice of first order (as partially ordered set). □

Theorem 3. *$(NR_{MB}, \leq_N)$ is bounded neither to the left nor to the right, since it does not have a minimum (bottom, or least element), or a maximum (top, or greatest element).*

Proof. Straightforward, since NR_{MB} includes the set of real number $R = (-\infty,\ +\infty)$ which is clearly unbounded to the left and right-hand sides. □

Theorem 4. *$(NR_{MB},\ inf_N,\ sup_N)$, where inf_N and sup_N are two binary operations, dual to each other, defined before as a lattice of second type (as an algebraic structure).*

Proof. We have to show that the two laws inf_N and sup_N are commutative, associative, and verify the absorption laws.

Let $\alpha,\ \beta,\ \gamma \in NR_{MB}$ be two arbitrary elements.

Commutativity Laws

(i)

$$\inf_N\{\alpha,\ \beta\} =_N \inf_N\{\beta,\ \alpha\} \tag{67}$$

(ii)

$$\sup_N\{\alpha,\ \beta\} =_N \sup_N\{\beta,\ \alpha\} \tag{68}$$

Their proofs are straightforward.

Associativity Laws

(i)

$$\inf_N\{\alpha,\ \inf_N\{\beta,\gamma\}\} =_N \inf_N\{\inf_N\{\alpha,\ \beta\},\ \gamma\}. \tag{69}$$

□

Proof.

$$\inf_N\{\alpha,\ \inf_N\{\beta,\gamma\}\} =_N \inf_N\{\alpha,\ \beta,\ \gamma\}, \tag{70}$$

and

$$\inf_N\{\inf_N\{\alpha,\ \beta\},\ \gamma\} =_N \inf_N\{\alpha,\ \beta,\ \gamma\}, \tag{71}$$

where we have extended the binary operation $\inf_N$ to a trinary operation $\inf_N$.

(ii)

$$\sup_N\left\{\alpha,\ \sup_N\{\beta,\gamma\}\right\} =_N \sup_N\left\{\sup_N\{\alpha,\ \beta\},\ \gamma\right\} \tag{72}$$

□

Proof.

$$\sup_N\{\alpha,\ \sup_N\{\beta,\gamma\}\} =_N \sup_N\{\alpha,\ \beta,\ \gamma\}, \tag{73}$$

and

$$\sup_N\{\alpha,\ \sup_N\{\beta,\gamma\}\} =_N \sup_N\{\alpha,\ \beta,\ \gamma\}, \tag{74}$$

where similarly we have extended the binary operation $\sup_N$ to a trinary operation $\sup_N$.

Absorption Laws (as peculiar axioms to the theory of lattice)

(i) We need to prove that

$$\inf_N\{\alpha,\ \sup_N\{\alpha,\ \beta\}\} =_N \alpha. \tag{75}$$

Let $\alpha \leq_N \beta$, then

$$\inf_N\{\alpha,\ \sup_N\{\alpha,\ \beta\}\} =_N \inf_N\{\alpha,\alpha\} =_N \alpha. \tag{76}$$

Let $\alpha >_N \beta$, then

$$\inf_N\{\alpha,\ \sup_N\{\alpha,\ \beta\}\} =_N \inf_N\{\alpha,\alpha\} =_N \alpha. \tag{77}$$

(ii) Now, we need to prove that

$$\sup_N\{\alpha,\ \inf_N\{\alpha,\ \beta\}\} =_N \alpha. \tag{78}$$

Let $\alpha \leq_N \beta$, then

$$\sup_N\{\alpha,\ \inf_N\{\alpha,\ \beta\}\} =_N \sup_N\{\alpha,\ \alpha\} =_N \alpha. \tag{79}$$

Let $\alpha >_N \beta$, then

$$\sup_N\{\alpha,\ \inf_N\{\alpha,\ \beta\}\} =_N \sup_N\{\alpha,\beta\} =_N \alpha. \tag{80}$$

Consequence

The binary operations $\inf_N$ and $\sup_N$ also satisfy the idempotent laws:

$$\inf_N\{\alpha,\ \alpha\} =_N \alpha, \tag{81}$$

$$\sup_N\{\alpha,\ \alpha\} =_N \alpha. \tag{82}$$

□

Proof. The axioms of idempotency follow directly from the axioms of absorption proved above. □

Thus, we have proved that (NR_{MB}, inf_N, sup_N) is a lattice of second type (as algebraic structure).

23. Definition of General Nonstandard Real MoBiNad Interval

Let $a,\ b \in \mathbb{R}$, with

$$-\infty < a \leq b < \infty, \tag{83}$$

$$]^-a,\ b^+[_{MB}= \{x \in NR_{MB}, ^-a \leq_N x \leq_N b^+\}. \tag{84}$$

As particular edge cases:

$$]^-a,\ a^+[_{MB}=_N \{^-a,\ a, ^-a^+, a^+\}, \tag{85}$$

a discrete nonstandard real set of cardinality 4.

$$]^-a,\ ^-a[_{MB}=_N \{^-a\}; \tag{86}$$

$$]a^+,\ a^+[_{MB} =_N \{a^+\} \tag{87}$$

$$]a,\ a^+[_{MB} =_N \{a,\ a^+\} \tag{88}$$

$$]^-a,\ a[_{MB}=_N \{^-a,\ a\} \tag{89}$$

$$]^{-}a,\ ^{-}a^{+}[_{MB}=_N\left\{^{-}a,^{-}a^{+},\ a^{+}\right\}, \tag{90}$$

where $a\notin]^{-}a,^{-}a^{+}[_{MB}$ since $a\not\leq_N {}^{-}a^{+}$ (there is no relation of order between a and $^{-}a^{+}$);

$$]^{-}a^{+},\ a^{+}[_{MB}=_N\left\{^{-}a^{+},\ a^{+}\right\}. \tag{91}$$

Theorem 5.

$$\left(\ ]^{-}a,\ b^{+}[,\ \leq_N\right)\ \textit{is a nonstandard real mobinad sublattice of first type (poset)}. \tag{92}$$

Proof. Straightforward since $]^{-}a,\ b^{+}[$ is a sublattice of the lattice of first type NR_{MB}. □

Theorem 6.

$$\left(\ ]^{-}a,\ b^{+}[,\ \inf_N,\ \sup_N,^{-}a,\ b^{+}\right)\ \textit{is a nonstandard bounded real mobinad sublattice of second type (as algebraic structure)}. \tag{93}$$

Proof. $]^{-}a,\ b^{+}[_{MB}$ as a nonstandard subset of NR_{MB} is also a poset, and for any two-element subset

$$\{\alpha,\ \beta\}\subset_N\ ]^{-}0,\ 1^{+}[_{MB} \tag{94}$$

one obviously has the triple neutrosophic nonstandard inequality:

$$^{-}a\leq_N \inf_N\{\alpha,\ \beta\}\leq_N \sup_N\{\alpha,\ \beta\}\leq_N b^{+} \tag{95}$$

hence $(\ ]^{-}a,\ b^{+}[_{MB}\leq_N)$ is a nonstandard real mobinad sublattice of first type (poset), or sublattice of NR_{MB}.

Further on, $]^{-}a,\ b^{+}[$, endowed with two binary operations $\inf_N$ and $\sup_N$, is also a sublattice of the lattice NR_{MB}, since the lattice axioms (Commutative Laws, Associative Laws, Absorption Laws, and Idempotent Laws) are clearly verified on $]^{-}a,\ b^{+}[$.

The nonstandard neutrosophic modinad Identity Join Element (Bottom) is ^{-}a, and the nonstandard neutrosophic modinad Identity Meet Element (Top) is b^{+}, or

$$\inf_N]^{-}a,\ b^{+}[=_N{}^{-}\text{ and }\sup_N]^{-}a,\ b^{+}[=_N b^{+}. \tag{96}$$

The sublattice Identity Laws are verified below.

$$\text{Let }\alpha\in_N]^{-}a,\ b^{+}[,\text{ whence }^{-}a\leq_N\alpha\leq_N b^{+}. \tag{97}$$

Then:

$$\inf_N\left\{\alpha,\ b^{+}\right\}=_N\alpha,\text{ and }\sup_N\{\alpha,^{-}a\}=_N\alpha. \tag{98}$$

□

24. Definition of Nonstandard Real MoBiNad Unit Interval

$$\begin{aligned}&]^{-}0,\ 1^{+}[_{MB}=_N\{x\in NR_{MB},^{-}0\leq_N x\leq_N 1^{+}\}\\ &=_N\left\{\begin{array}{c}\varepsilon,a,\overset{-}{a},\overset{-0}{a},\overset{+}{a},\overset{0+}{a},\overset{-+}{a},\overset{-0+}{a}\ \Big|\ \textit{where }\varepsilon\textit{ are infinitesimals,}\\ \varepsilon\in\mathbb{R}^{*},\textit{ with }\varepsilon>0,\textit{ and }a\ \in[0,\ 1]\end{array}\right\}\end{aligned} \tag{99}$$

This is an extension of the previous definition (1998) [5] of nonstandard unit interval

$$]^{-}0,\ 1^{+}[=_N (^{-}0) \cup [0,\ 1] \cup \left(1^{+}\right) \quad (100)$$

Associated to the first published definitions of neutrosophic set, logic, and probability was used. One has

$$]^{-}0,\ 1^{+}[\subset_N\,]^{-}0,\ 1^{+}[_{MB} \quad (101)$$

where the index $_{MB}$ means: all monads and binads included in $]^{-}0,\ 1^{+}[$, for example,

$$(^{-}0.2),\ (^{-}0.3^{0}),\ (0.5^{+}),\ (^{-}0.7^{+}),\ (^{-}0.8^{0+})\ \text{etc.} \quad (102)$$

or, using the top diacritics notation, respectively,

$$\overset{-}{0.2},\overset{-0}{0.3},\overset{+}{0.5},\overset{-+}{0.7},\ \overset{-0+}{0.8}\ \text{etc.} \quad (103)$$

Theorem 7. *The Nonstandard Real MoBiNad Unit Interval $]^{-}0,\ 1^{+}[_{MB}$ is a partially ordered set (poset) with respect to $\leq_N$, and any of its two elements have an inf_N and sup_N hence $]^{-}0,\ 1^{+}[_{MB}$ is a nonstandard neutrosophic lattice of first type (as poset).*

Proof. Straightforward. □

Theorem 8. *The Nonstandard Real MoBiNad Unit Interval $]^{-}0,\ 1^{+}[_{MB}$, endowed with two binary operations inf_N and sup_N, is also a nonstandard neutrosophic lattice of second type (as an algebraic structure).*

Proof. Replace $a = 0$ and $b = 1$ into the general nonstandard real mobinad interval $]^{-}a,\ b^{+}[$. □

25. Definition of Extended General Neutrosophic Logic

We extend and present in a clearer way our 1995 definition (published in 1998) of neutrosophic logic.

Let $\mathcal{U}$ be a universe of discourse of propositions and $P \in \mathcal{U}$ be a generic proposition.

A General Neutrosophic Logic is a multivalued logic in which each proposition P has a degree of truth (T), a degree of indeterminacy (I), and a degree of falsehood (F), and where T, I, and F are standard real subsets or nonstandard real mobinad subsets of the nonstandard real mobinat unit interval $]^{-}0,\ 1^{+}[_{MB}$,

With

$$T,\ I,\ F \subseteq_N]^{-}0,\ 1^{+}[_{MB} \quad (104)$$

where

$$^{-}0 \leq_N \text{inf}_N T + \text{inf}_N I + \text{inf}_N F \leq_N \text{sup}_N T + \text{sup}_N I + \text{sup}_N F \leq 3^{+}. \quad (105)$$

26. Definition of Standard Neutrosophic Logic

If in the above definition of general neutrosophic logic all neutrosophic components, T, I, and F are standard real subsets, included in or equal to the standard real unit interval, $T,\ I,\ F \subseteq [0,\ 1]$, where

$$0 \leq \text{inf}T + \text{inf}I + \text{inf}F \leq \text{sup}T + \text{sup}I + \text{sup}F \leq 3 \quad (106)$$

we have a standard neutrosophic logic.

27. Definition of Extended Nonstandard Neutrosophic Logic

If in the above definition of general neutrosophic logic at least one of the neutrosophic components T, I, or F is a nonstandard real mobinad subset, neutrosophically included in or equal to the nonstandard real mobinad unit interval $]^{-}0,\ 1^{+}[_{MB}$, where

$$^{-}0 \leq_N \mathrm{inf}_N T + \mathrm{inf}_N I + \mathrm{inf}_N F \leq_N \mathrm{sup}_N T + \mathrm{sup}_N I + \mathrm{sup}_N F \leq 3^{+}, \tag{107}$$

we have an extended nonstandard neutrosophic logic.

Theorem 9. *If M is a standard real set, $M \subset \mathbb{R}$, then*

$$inf_N(M) = inf(M)\ and\ sup_N(M) = sup(M). \tag{108}$$

Proof. The neutrosophic infimum and supremum coincide with the classical infimum and supremum since there is no indeterminacy on the set M, meaning M contains no nonstandard numbers. □

28. Definition of Extended General Neutrosophic Set

We extend and present in a clearer way our 1995 definition of neutrosophic set.

Let $\mathcal{U}$ be a universe of discourse of elements and $S \in \mathcal{U}$ a subset.

A Neutrosophic Set is a set such that each element x from S has a degree of membership (T), a degree of indeterminacy (I), and a degree of nonmembership (F), where T, I, and F are standard real subsets or nonstandard real mobinad subsets, neutrosophically included in or equal to the nonstandard real mobinat unit interval

$$]^{-}0,\ 1^{+}[_{MB}, \text{with } T,\ I,\ F \subseteq_N]^{-}0,\ 1^{+}[_{MB}, \tag{109}$$

where

$$^{-}0 \leq_N \mathrm{inf}_N T + \mathrm{inf}_N I + \mathrm{inf}_N F \leq_N \mathrm{sup}_N T + \mathrm{sup}_N I + \mathrm{sup}_N F \leq 3^{+}. \tag{110}$$

29. Definition of Standard Neutrosophic Set

If in the above general definition of neutrosophic set all neutrosophic components, T, I, and F, are standard real subsets included in or equal to the classical real unit interval, $T,\ I,\ F \subseteq [0,\ 1]$, where

$$0 \leq \mathrm{inf}T + \mathrm{inf}I + \mathrm{inf}F \leq \mathrm{sup}T + \mathrm{sup}I + \mathrm{sup}F \leq 3, \tag{111}$$

we have a standard neutrosophic set.

30. Definition of Extended Nonstandard Neutrosophic Set

If in the above general definition of neutrosophic set at least one of the neutrosophic components T, I, or F is a nonstandard real mobinad subsets, neutrosophically included in or equal to $]^{-}0,\ 1^{+}[_{MB}$, where

$$^{-}0 \leq_N \mathrm{inf}_N T + \mathrm{inf}_N I + \mathrm{inf}_N F \leq_N \mathrm{sup}_N T + \mathrm{sup}_N I + \mathrm{sup}_N F \leq 3^{+}, \tag{112}$$

we have a nonstandard neutrosophic set.

31. Definition of Extended General Neutrosophic Probability

We extend and present in a clearer way our 1995 definition of neutrosophic probability.

Let $\mathcal{U}$ be a universe of discourse of events, and $E \in \mathcal{U}$ be an event.

A Neutrosophic Probability is a multivalued probability such that each event E has a chance of occuring (T), an indeterminate (unclear) chance of occuring or not occuring (I), and a chance of not occuring (F), and where T, I, and F are standard or nonstandard real mobinad subsets, neutrosophically included in or equal to the nonstandard real mobinat unit interval

$$]^{-}0,\ 1^{+}[_{MB},\ T,\ I,\ F \subseteq_N]^{-}0,\ 1^{+}[_{MB},\ \text{where } ^{-}0 \leq_N \inf_N T + \inf_N I + \inf_N F \leq_N \sup_N T + \sup_N I + \sup_N F \leq 3^{+}. \tag{113}$$

32. Definition of Standard Neutrosophic Probability

If in the above general definition of neutrosophic probability all neutrosophic components, T, I, and F are standard real subsets, included in or equal to the standard unit interval $T,\ I,\ F \subseteq [0,\ 1]$, where

$$0 \leq \inf T + \inf I + \inf F \leq \sup T + \sup I + \sup F \leq 3, \tag{114}$$

we have a standard neutrosophic probability.

33. Definition of Extended Nonstandard Neutrosophic Probability

If in the above general definition of neutrosophic probability at least one of the neutrosophic components T, I, F is a nonstandard real mobinad subsets, neutrosophically included in or equal to $]^{-}0,\ 1^{+}[_{MB}$, where

$$^{-}0 \leq_N \inf_N T + \inf_N I + \inf_N F \leq_N \sup_N T + \sup_N I + \sup_N F \leq 3^{+}, \tag{115}$$

we have a nonstandard neutrosophic probability.

34. Classical Operations with Real Sets

Let $A,\ B \subseteq \mathbb{R}$ be two real subsets. Let $\circledast$ and $*$ denote any of the real subset classical operations and real number classical operations respectively: addition (+), subtraction (−), multiplication (×), division (÷), and power (^).

Then,

$$A \circledast B = \{a * b,\ \text{where } a \in A \text{ and } b \in B\} \tag{116}$$

Thus

$$A \oplus B = \{a + b | a \in A, b \in B\} \tag{117}$$

$$A \ominus B = \{a - b | a \in A, b \in B\} \tag{118}$$

$$A \otimes B = \{a \times b | a \in A, b \in B\} \tag{119}$$

$$A \oslash B = \{a \div b | a \in A; b \in B, b \neq 0\} \tag{120}$$

$$A^B = \left\{a^b \middle| a \in A, a > 0; b \in B\right\} \tag{121}$$

For the division (÷), of course, we consider $b \neq 0$. While for the power (^), we consider $a > 0$.

35. Operations on the Nonstandard Real MoBiNad Set (NR_{MB})

For all nonstandard (addition, subtraction, multiplication, division, and power) operations

$$\alpha, \beta \in_N NR_{MB},\ \alpha *_N \beta =_N \mu_N(\alpha) \circledast \mu_N(\beta) \tag{122}$$

where $*_N$ is any neutrosophic arithmetic operations with neutrosophic numbers ($+_N$, $-_N$, $\times_N$, $\div_N$, $\hat{}_N$), while the corresponding $\circledast$ is an arithmetic operation with real subsets.

So, we approximate the nonstandard operations by standard operations of real subsets.

We sink the nonstandard neutrosophic real mobinad operations into the standard real subset operations, then we resurface the last ones back to the nonstandard neutrosophic real mobinad set.

Let ε_1 and ε_2 be two non-null positive infinitesimals. We present below some particular cases, all others should be deduced analogously.

Nonstandard Addition

First Method

$$({}^{-}a) + ({}^{-}b) =_N (a - \varepsilon_1,\ a) + (b - \varepsilon_2,\ b) =_N (a + b - \varepsilon_1 - \varepsilon_2,\ a + b) =_N (a + b - \varepsilon,\ a + b) =_N {}^{-}(a + b), \quad (123)$$

where we denoted $\varepsilon_1 + \varepsilon_2 = \varepsilon$ (the addition of two infinitesimals is also an infinitesimal).

Second Method

$$({}^{-}a) + ({}^{-}b) =_N (a - \varepsilon_1) + (b - \varepsilon_2) =_N (a + b - \varepsilon_1 - \varepsilon_2) =_N {}^{-}(a + b) \quad (124)$$

Adding two left monads, one also gets a left monad.

Nonstandard Subtraction

First Method

$$\begin{aligned}({}^{-}a) - ({}^{-}b) =_N\ & (a - \varepsilon_1, a) \\ & -(b - \varepsilon_2, b) =_N (a - \varepsilon_1 - b,\ a - b + \varepsilon_2) =_N (a - b - \varepsilon_1,\ a - b \\ & +\varepsilon_2) =_N \begin{pmatrix} - & 0 & + \\ & a-b & \end{pmatrix}\end{aligned} \quad (125)$$

Second Method

$$({}^{-}a) - ({}^{-}b) =_N (a - \varepsilon_1) - (b - \varepsilon_2) =_N a - b - \varepsilon_1 + \varepsilon_2, \quad (126)$$

since ε_1 and ε_2 may be any positive infinitesimals,

$$=_N \begin{cases} -(a-b), \text{ when } \varepsilon_1 > \varepsilon_2; \\ \begin{pmatrix} 0 \\ a-b \end{pmatrix}, \text{ when } \varepsilon_1 = \varepsilon_2 \\ (a-b)^{+}, \text{ when } \varepsilon_1 < \varepsilon_2. \end{cases} =_N \begin{pmatrix} 0 \\ a-b \end{pmatrix} =_N a - b; \quad (127)$$

Subtracting two left monads, one obtains an unpierced binad (that is why the unpierced binad had to be introduced).

Nonstandard Division

Let $a, b > 0$.

$$({}^{-}a) \div ({}^{-}b) =_N (a - \varepsilon_1,\ a) \div (b - \varepsilon_2,\ b) =_N \left(\frac{a - \varepsilon_1}{b}, \frac{a}{b - \varepsilon_2}\right) \quad (128)$$

Since

$$\varepsilon_1 > 0 \text{ and } \varepsilon_2 > 0,\ \frac{a - \varepsilon_1}{b} < \frac{a}{b} \text{ and } \frac{a}{b - \varepsilon_2} > \frac{a}{b}, \quad (129)$$

while between $\frac{a-\varepsilon_1}{b}$ and $\frac{a}{b-\varepsilon_2}$ there is a continuum whence there are some infinitesimals ε_1^0 and ε_2^0 such that $\frac{a-\varepsilon_1^0}{b-\varepsilon_2^0} = \frac{a}{b}$, or $ab - b\varepsilon_1^0 = ab - a\varepsilon_2^0$, and for a given ε_1^0 there exists an

$$\varepsilon_2^0 = \varepsilon_1^0 \cdot \frac{b}{a} \quad (130)$$

Hence

$$\frac{({}^{-}a)}{({}^{-}b)} =_N \begin{pmatrix} - & 0 & + \\ & \frac{a}{b} & \end{pmatrix} \quad (131)$$

For a or/and b negative numbers, it is similar but it is needed to compute the inf_N and sup_N of the products of intervals.

Dividing two left monads, one obtains an unpierced binad.

Nonstandard Multiplication

Let $a,\ b \geq 0$.

$$\begin{array}{ll}\left({}^{-}a^{0}\right)\times\left({}^{-}b^{0\,+}\right) & =_N (a-\varepsilon_1,\ a] \\ & \times(b-\varepsilon_2,\ b+\varepsilon_2) =_N ((a-\varepsilon_1)\cdot(b-\varepsilon_2), a\cdot(b+\varepsilon_2)) =_N \left({}^{-}ab^{0\,+}\right)\end{array} \tag{132}$$

Since

$$(a-\varepsilon_1)\cdot(b-\varepsilon_2) < a\cdot b \text{ and } a\cdot(b+\varepsilon_2) > a\cdot b. \tag{133}$$

For a or/and b negative numbers, it is similar but it is needed to compute the inf_N and sup_N of the products of intervals.

Multiplying a positive left monad closed to the right, with a positive unpierced binad, one obtains an unpierced binad.

Nonstandard Power

Let $a,\ b > 1$.

$$\left({}^{0}a^{+}\right)^{\left({}^{-}b^{0}\right)} =_N [a,\ a+\varepsilon_1)^{(b-\varepsilon_2,\ b]} =_N (a^{b-\varepsilon_2},\ (a+\varepsilon_1)^b) =_N \begin{pmatrix} - & 0 & + \\ & a^b & \end{pmatrix} \tag{134}$$

$$\text{since } a^{b-\varepsilon_1} < a^b \text{ and } (a+\varepsilon_1)^b > a^b. \tag{135}$$

Raising a right monad closed to the left to a power equal to a left monad closed to the right, for both monads above 1, the result is an unpierced binad.

Consequence

In general, when doing arithmetic operations on nonstandard real monads and binads, the result may be a different type of monad or binad.

That is why is was imperious to extend the monads to closed monads, and the pierced binad to unpierced binad, in order to have the whole nonstandard neutrosophic real mobinad set closed under arithmetic operations.

36. Conditions of Neutrosophic Nonstandard Inequalities

Let NR_{MB} be the Nonstandard Real MoBiNad. Let's endow $(NR_{MB}, <_N)$ with a neutrosophic inequality.

Let $\alpha, \beta \in NR_{MB}$, where α, β may be real numbers, monads, or binads.

And let

$$\begin{array}{c}\begin{pmatrix} - \\ a \end{pmatrix}, \begin{pmatrix} -0 \\ a \end{pmatrix}, \begin{pmatrix} + \\ a \end{pmatrix}, \begin{pmatrix} 0+ \\ a \end{pmatrix}, \begin{pmatrix} -+ \\ a \end{pmatrix}, \begin{pmatrix} -0+ \\ a \end{pmatrix} \in NR_{MB}, \text{ and} \\ \begin{pmatrix} - \\ b \end{pmatrix}, \begin{pmatrix} -0 \\ b \end{pmatrix}, \begin{pmatrix} + \\ b \end{pmatrix}, \begin{pmatrix} 0+ \\ b \end{pmatrix}, \begin{pmatrix} -+ \\ b \end{pmatrix}, \begin{pmatrix} -0+ \\ b \end{pmatrix} \in NR_{MB},\end{array} \tag{136}$$

be the left monads, left monads closed to the right, right monads, right monads closed to the left, and binads, and binads nor pierced of the elements (standard real numbers) a and b, respectively. Since all monads and binads are real subsets, we may treat the single real numbers

$$a = [a, a] \text{ and } b = [b, b] \text{ as real subsets too} \tag{137}$$

as real subsets too.

NR_{MB} is a set of subsets, and thus we deal with neutrosophic inequalities between subsets.

(i) If the subset α has many of its elements above all elements of the subset β,

(ii) then $\alpha >_N \beta$ (partially).

(iii) If the subset α has many of its elements below all elements of the subset β,

(iv) then $\alpha <_N \beta$ (partially).

(*v*) If the subset α has many of its elements equal with elements of the subset β,
(*vi*) then $\alpha =_N \beta$ (partially).

If the subset α verifies (*i*) and (*iii*) with respect to subset β, then $\alpha \geq_N \beta$.
If the subset α verifies (*ii*) and (*iii*) with respect to subset β, then $\alpha \leq_N \beta$.
If the subset α verifies (*i*) and (*ii*) with respect to subset β, then there is no neutrosophic order (inequality) between α and β.

For example, between a and $(^-a^+)$ there is no neutrosophic order, similarly between a and $\overset{-0+}{a}$.

Similarly, if the subset α verifies (*i*), (*ii*) and (*iii*) with respect to subset β, then there is no neutrosophic order (inequality) between α and β.

37. Open Neutrosophic Research

The quantity or measure of "many of its elements" of the above (*i*), (*ii*), or (*iii*) conditions depends on each *neutrosophic application* and on its *neutrosophic experts*.

An approach would be to employ the *Neutrosophic Measure* [23,24], that handles indeterminacy, which may be adjusted and used in these cases.

In general, we do not try in purpose to validate or invalidate an existing scientific result, but to investigate how an existing scientific result behaves in a new environment (that may contain indeterminacy), or in a new application, or in a new interpretation.

38. Nonstandard Neutrosophic Inequalities

For the neutrosophic nonstandard inequalities, we propose, based on the previous six neutrosophic equalities, the following.

$$(^-\mathrm{a}) <_N, \mathrm{a} <_N (a^+) \tag{138}$$

Since the standard real interval $(a - \varepsilon, a)$ is below a, and a is below the standard real interval $(a, a + \varepsilon)$ by using the approximation provided by the nonstandard neutrosophic function μ, or because

$$\forall x \in R_+^*, a - x < a < a + x \tag{139}$$

where x is of course a (nonzero) positive infinitesimal (the above double neutrosophic inequality actually becomes a double classical standard real inequality for each fixed positive infinitesimal).

The converse double neutrosophic inequality is also neutrosophically true:

$$(a^+) >_N, \mathrm{a} >_N (^-a) \tag{140}$$

Another nonstandard neutrosophic double inequality:

$$(^-\mathrm{a}) \leq_N (^-a^+) \leq_N (a^+) \tag{141}$$

This double neutrosophic inequality may be justified since $(^-a^+) = (^-a) \cup (a^+)$ and, geometrically, on the Real Number Line, the number a is in between the subsets $^-a = (a - \varepsilon, a)$ and $a^+ = (a, a + \varepsilon)$, so

$$(^-a) \leq_N (^-a) \cup (a^+) \leq_N (a^+) \tag{142}$$

Hence the left side of the inequality's middle term coincides with the inequality first term, while the right side of the inequality middle term coincides with the third inequality term.

Conversely, it is neutrosophically true as well:

$$(a^+) \geq_N (^-a) \cup (a^+) \geq_N (^-a) \tag{143}$$

Also,

$$\bar{a} \leq_N \overset{-0}{a} \leq_N a \leq_N \overset{0+}{a} \leq_N \overset{+}{a} \text{ and } \bar{a} \leq_N \overset{-+}{a} \leq_N \overset{-0+}{a} \leq_N \overset{+}{a} \quad (144)$$

Conversely, they are also neutrosophically true:

$$\overset{+}{a} \geq_N \overset{0+}{a} \geq_N a \geq_N \overset{-0}{a} \geq_N \bar{a} \text{ and } \overset{+}{a} \geq_N \overset{-0+}{a} \geq_N \overset{-+}{a} \geq_N \bar{a} \text{ respectively.} \quad (145)$$

If $a > b$, which is a (standard) classical real inequality, then we have the following neutrosophic nonstandard inequalities.

$$a >_N ({}^{-}b), a >_N \left(b^{+}\right), a >_N \left({}^{-}b^{+}\right), a >_N \overset{-0}{b}, a >_N \overset{0+}{b}, a >_N \overset{-0+}{b}; \quad (146)$$

$$({}^{-}a) >_N b, ({}^{-}a) >_N ({}^{-}b), ({}^{-}a) >_N \left(b^{+}\right), ({}^{-}a) >_N \left({}^{-}b^{+}\right), \bar{a} >_N \overset{-0}{b}, \bar{a} >_N \overset{0+}{b}, \bar{a} >_N \overset{-0+}{b}; \quad (147)$$

$$\left(a^{+}\right) >_N b, \left(a^{+}\right) >_N ({}^{-}b), \left(a^{+}\right) >_N \left(b^{+}\right), \left(a^{+}\right) >_N \left({}^{-}b^{+}\right), \overset{+}{a} >_N \overset{-0}{b}, \overset{+}{a} >_N \overset{0+}{b}, \overset{+}{a} >_N \overset{-0+}{b}; \quad (148)$$

$$({}^{-}a^{+}) >_N b, ({}^{-}a^{+}) >_N ({}^{-}b), ({}^{-}a^{+}) >_N (b^{+}), ({}^{-}a^{+}) >_N ({}^{-}b^{+}), \text{etc.} \quad (149)$$

No Ordering Relationships

For any standard real number a, there is no relationship of order between the elements a and $({}^{-}a^{+})$, or between the elements a and

$$\left(\overset{-0+}{a}\right) \quad (150)$$

Therefore, NR_{MB} is a neutrosophically partially order set.

If one removes all binads from NR_{MB}, then $(NR_{MB}, \leq_N)$ is neutrosophically totally ordered. (151)

Theorem 10. *Using the nonstandard general notation one has:*

If $a > b$, which is a (standard) classical real inequality, then

$$\overset{m_1}{a} >_N \overset{m_2}{b} \text{ for any } m_1, m_2 \in \left\{, {}^{-}, {}^{-0}, {}^{+}, {}^{+0}, {}^{-0}, {}^{-0+}\right\}. \quad (152)$$

Conversely, if $a < b$, which is a (standard) classical real inequality, then

$$\overset{m_1}{a} <_N \overset{m_2}{b} \text{ for any } m_1, m_2 \in \left\{, {}^{-}, {}^{-0}, {}^{+}, {}^{+0}, {}^{-0}, {}^{-0+}\right\}. \quad (153)$$

39. Nonstandard Neutrosophic Equalities

Let a, b be standard real numbers; if $a = b$ that is a (classical) standard equality, then

$$({}^{-}a) =_N ({}^{-}b), (a^{+}) =_N (b^{+}), ({}^{-}a^{+}) =_N ({}^{-}b^{+}), \quad (154)$$

$$\left(\overset{-0}{a}\right) =_N \left(\overset{-0}{b}\right), \left(\overset{0+}{a}\right) =_N \left(\overset{0+}{b}\right), \left(\overset{-0+}{a}\right) =_N \left(\overset{-0+}{b}\right) \quad (155)$$

40. Nonstandard Neutrosophic Belongingness

On the nonstandard real set NR_{MB}, we say that

$$\overset{m}{c} \in_N \left]\overset{m_1}{a}, \overset{m_2}{b}\right[\text{ iff } \overset{m_1}{a} \leq_N \overset{m}{c} \leq_N \overset{m_2}{b}, \quad (156)$$

where

$$m_1, m_2, m \in \{, ^{-}, ^{-0}, ^{+}, ^{+0}, ^{-+}, ^{-0+}\}. \tag{157}$$

We use the previous nonstandard neutrosophic inequalities.

41. Nonstandard Hesitant Sets

Nonstandard Hesitant sets are sets of the form:

$$A = \{a_1, a_2, \ldots, a_n\},\ 2 \le n < \infty,\ A \subset_N NR_{MB}, \tag{158}$$

where at least one element $a_{i_0}, 1 \le i_0 \le n$, is an infinitesimal, a monad, or a binad (of any type); while other elements may be standard real numbers, infinitesimals, or also monads or binads (of any type).

If the neutrosophic components T, I, and F are nonstandard hesitant sets, then one has a Nonstandard Hesitant Neutrosophic Logic/Set/Probability.

42. Nonstandard Neutrosophic Strict Interval Inclusion

On the nonstandard real set NR_{MB},

$$]\overset{m_1}{a}, \overset{m_2}{b}[\subset_N]\overset{m_3}{c}, \overset{m_4}{d}[\tag{159}$$

iff

$$\overset{m_3}{c} \le_N \overset{m_1}{a} <_N \overset{m_2}{b} <_N \overset{m_4}{d} \text{ or } \overset{m_3}{c} <_N \overset{m_1}{a} <_N \overset{m_2}{b} \le_N \overset{m_4}{d} \text{ or } \overset{m_3}{c} <_N \overset{m_1}{a} <_N \overset{m_2}{b} <_N \overset{m_4}{d} \tag{160}$$

43. Nonstandard Neutrosophic (Nonstrict) Interval Inclusion

On the nonstandard real set NR_{MB},

$$]\overset{m_1}{a}, \overset{m_2}{b}[\subseteq_N]\overset{m_3}{c}, \overset{m_4}{d}[\text{ iff} \tag{161}$$

$$\overset{m_3}{c} \le_N \overset{m_1}{a} <_N \overset{m_2}{b} \le_N \overset{m_4}{d}. \tag{162}$$

44. Nonstandard Neutrosophic Strict Set Inclusion

The nonstandard set A is neutrosophically strictly included in the nonstandard set B, $A \subset_N B$, if:

$$\forall x \in_N A, x \in_N B, \text{ and } \exists y \in_N B : y \notin_N A. \tag{163}$$

45. Nonstandard Neutrosophic (Nonstrict) Set Inclusion

The nonstandard set A is neutrosophically not strictly included in the nonstandard set B,

$$A \subseteq_N B, \text{ iff:} \tag{164}$$

$$\forall x \in_N A, x \in_N B. \tag{165}$$

46. Nonstandard Neutrosophic Set Equality

The nonstandard sets A and B are neutrosophically equal,

$$A =_N B, \text{ iff:} \tag{166}$$

$$A \subseteq_N B \text{ and } B \subseteq_N A. \tag{167}$$

47. The Fuzzy, Neutrosophic, and Plithogenic Logical Connectives $\wedge$, $\vee$, $\rightarrow$

All fuzzy, intuitionistic fuzzy, and neutrosophic logic operators are *inferential approximations*, not written in stone. They are improved from application to application.

Let's denote:

$$\wedge_F, \wedge_N, \wedge_P \text{ representing respectively the fuzzy conjunction, neutrosophic conjunction, and plithogenic conjunction;} \tag{168}$$

similarly

$$\vee_F, \vee_N, \vee_P \text{ representing respectively the fuzzy disjunction, neutrosophic disjunction, and plithogenic disjunction,} \tag{169}$$

and

$$\rightarrow_F, \rightarrow_N, \rightarrow_P \text{ representing respectively the fuzzy implication, neutrosophic implication, and plithogenic implication.} \tag{170}$$

I agree that my beginning neutrosophic operators (when I applied the same *fuzzy t-norm*, or the same *fuzzy t-conorm*, to all neutrosophic components T, I, F) were less accurate than others developed later by the neutrosophic community researchers. This was pointed out in 2002 by Ashbacher [25] and confirmed in 2008 by Rivieccio [26]. They observed that if on T_1 and T_2 one applies a *fuzzy t-norm*, for their opposites F_1 and F_2, one needs to apply the *fuzzy t-conorm* (the opposite of fuzzy t-norm), and reciprocally.

About inferring I_1 and I_2, some researchers combined them in the same directions as T_1 and T_2. Then,

$$(T_1, I_1, F_1) \wedge_N (T_2, I_2, F_2) = (T_1 \wedge_F T_2, I_1 \wedge_F I_2, F_1 \vee_F F_2), \tag{171}$$

$$(T_1, I_1, F_1) \vee_N (T_2, I_2, F_2) = (T_1 \vee_F T_2, I_1 \vee_F I_2, F_1 \wedge_F F_2), \tag{172}$$

$$(T_1, I_1, F_1) \rightarrow_N (T_2, I_2, F_2) = (F_1, I_1, T_1) \vee_N (T_2, I_2, F_2) = (F_1 \vee_F T_2, I_1 \vee_F I_2, T_1 \wedge_F F_2). \tag{173}$$

others combined I_1 and I_2 in the same direction as F_1 and F_2 (since both I and F are negatively qualitative neutrosophic components, while F is qualitatively positive neutrosophic component), the most used one is as follows.

$$(T_1, I_1, F_1) \wedge_N (T_2, I_2, F_2) = (T_1 \wedge_F T_2, I_1 \vee_F I_2, F_1 \vee_F F_2), \tag{174}$$

$$(T_1, I_1, F_1) \vee_N (T_2, I_2, F_2) = (T_1 \vee_F T_2, I_1 \wedge_F I_2, F_1 \wedge_F F_2), \tag{175}$$

$$(T_1, I_1, F_1) \rightarrow_N (T_2, I_2, F_2) = (F_1, I_1, T_1) \vee_N (T_2, I_2, F_2) = (F_1 \vee_F T_2, I_1 \wedge_F I_2, T_1 \wedge_F F_2). \tag{176}$$

Even more, recently, in an extension of neutrosophic set to *plithogenic set* [27] (which is a set whose each element is characterized by many attribute values), the *degrees of contradiction* $c(,)$ between the neutrosophic components T, I, and F have been defined (in order to facilitate the design of the aggregation operators), as follows:

$$c(T, F) = 1 \text{ (or 100\%, because they are totally opposite)}, c(T, I) = c(F, I) = 0.5 \text{ (or 50\%, because they are only half opposite).} \tag{177}$$

Then,

$$(T_1, I_1, F_1) \wedge_P (T_2, I_2, F_2) = (T_1 \wedge_F T_2, 0.5(I_1 \wedge_F I_2) + 0.5(I_1 \vee_F I_2), F_1 \vee_F F_2), \tag{178}$$

$$(T_1, I_1, F_1) \vee_P (T_2, I_2, F_2) = (T_1 \vee_F T_2, 0.5(I_1 \vee_F I_2) + 0.5(I_1 \wedge_F I_2), F_1 \wedge_F F_2), \tag{179}$$

$$(T_1, I_1, F_1) \rightarrow_N (T_2, I_2, F_2) = (F_1, I_1, T_1) \vee_N (T_2, I_2, F_2) = (F_1 \vee_F T_2, 0.5(I_1 \vee_F I_2) + 0.5(I_1 \wedge_F I_2), T_1 \wedge_F F_2). \tag{180}$$

48. Fuzzy t-norms and Fuzzy t-conorms

The most used $\wedge_F$ (Fuzzy t-norms), and $\vee_F$ (Fuzzy t-conorms) are as follows.

Let

$$a, b \in [0, 1]. \tag{181}$$

Fuzzy t-norms (fuzzy conjunctions, or fuzzy intersections):

$$a \wedge_F b = \min\{a, b\}; \tag{182}$$

$$a \wedge_F b = ab; \tag{183}$$

$$a \wedge_F b = \max\{a + b - 1, 0\}. \tag{184}$$

Fuzzy t-conorms (fuzzy disjunctions, or fuzzy unions):

$$a \vee_F b = \max\{a, b\}; \tag{185}$$

$$a \vee_F b = a + b \text{ - } ab; \tag{186}$$

$$a \vee_F b = \min\{a + b, 1\} \tag{187}$$

49. Nonstandard Neutrosophic Operators

Nonstandard Neutrosophic Conjunctions

$$(T_1, I_1, F_1) \wedge_N (T_2, I_2, F_2) = (T_1 \wedge_F T_2, I_1 \vee_F I_2, F_1 \vee_F F_2) = (inf_N(T_1, T_2), sup_N(I_1, I_2), sup_N(F_1, F_2)) \tag{188}$$

$$(T_1, I_1, F_1) \wedge_N (T_2, I_2, F_2) = (T_1 \wedge_F T_2, I_1 \vee_F I_2, F_1 \vee_F F_2) = (T_1 \times_N T_2, I_1 +_N I_2 -_N I_1 \times_N I_2, F_1 +_N F_2 -_N F_1 \times_N F_2) \tag{189}$$

Nonstandard Neutrosophic Disjunctions

$$(T_1, I_1, F_1) \vee_N (T_2, I_2, F_2) = (T_1 \vee_F T_2, I_1 \wedge_F I_2, F_1 \wedge_F F_2) = (sup_N(T_1, T_2), inf_N(I_1, I_2), inf_N(F_1, F_2)) \tag{190}$$

$$(T_1, I_1, F_1) \vee_N (T_2, I_2, F_2) = (T_1 \vee_F T_2, I_1 \wedge_F I_2, F_1 \wedge_F F_2) = (T_1 +_N T_2 -_N T_1 \times_N T_2, I_1 \times_N I_2, F_1 \times_N F_2) \tag{191}$$

Nonstandard Neutrosophic Negations

$$\neg(T_1, I_1, F_1) = (F_1, I_1, T_1) \tag{192}$$

$$\neg(T_1, I_1, F_1) = (F_1, (1^+) \text{ -}_N I_1, T_1) \tag{193}$$

Nonstandard Neutrosophic Implications

$$(T_1, I_1, F_1) \rightarrow_N (T_2, I_2, F_2) = (F_1, I_1, T_1) \vee_N (T_2, I_2, F_2) = (F_1 \vee_F T_2, I_1 \wedge_F I_2, T_1 \wedge_F F_2) = (F_1 +_N T_2 -_N F_1 \times_N T_2, I_1 \times_N I_2, T_1 \times_N F_2) \tag{194}$$

$$(T_1, I_1, F_1) \rightarrow_N (T_2, I_2, F_2) = (F_1, (1^+) -_N I_1, T_1) \vee_N (T_2, I_2, F_2) = (F_1 \vee_F T_2, ((1^+) -_N I_1) \wedge_F I_2, T_1 \wedge_F F_2) = (F_1 +_N T_2 -_N F_1 \times_N T_2, ((1^+) -_N I_1) \times_N I_2, T_1 \times_N F_2) \tag{195}$$

Let $P_1(T_1, I_1, F_1)$ and $P_2(T_2, I_2, F_2)$ be two nonstandard neutrosophic logical propositions, whose nonstandard neutrosophic components are, respectively,

$$T_1, I_1, F_1, T_2, I_2, F_2 \in_N NR_{MB}. \tag{196}$$

50. Numerical Examples of Nonstandard Neutrosophic Operators

Let us take a particular numeric example, where

$$P_1 =_N (\overset{0+}{0.3}, \overset{-+}{0.2}, 0.4), P_2 =_N (\overset{-0}{0.6}, \overset{-0+}{0.1}, \overset{+}{0.5}) \tag{197}$$

are two nonstandard neutrosophic logical propositions.

We use the nonstandard arithmetic operations previously defined *Numerical Example of Nonstandard Neutrosophic Conjunction*

$$\overset{0+}{0.3} \times \overset{-0}{0.6} =_N [0.3, 0.3+\varepsilon_1) \times (0.6-\varepsilon_2, 0.6) = (0.18 - 0.3\varepsilon_2, 0.18 + 0.6\varepsilon_1) =_N \overset{-0+}{0.18} \tag{198}$$

$$\begin{aligned}
&\overset{-+}{0.2} +_N \overset{-0+}{0.1} -_N \overset{-+}{0.2} \times_N \overset{-0+}{0.1} =_N [(0.2-\varepsilon_1, 0.2) \cup (0.2, 0.2+\varepsilon_1)] + (0.1-\varepsilon_2, 0.1+\varepsilon_2) \\
&-[(0.2-\varepsilon_1, 0.2) \cup (0.2, 0.2+\varepsilon_1)] \times (0.1-\varepsilon_2, 0.1+\varepsilon_2) \\
&= [(0.3-\varepsilon_1-\varepsilon_2, 0.3+\varepsilon_2) \cup (0.3-\varepsilon_2, 0.3+\varepsilon_1+\varepsilon_2)] \\
&-[(0.2-\varepsilon_1) \times (0.1-\varepsilon_2), (0.02+0.2\varepsilon_2)] \cup [(0.02-0.2\varepsilon_2), (0.2+\varepsilon_1) \times (0.1+\varepsilon_2)] \\
&= [\overset{-0+}{0.3} \cup \overset{-0+}{0.3}] - [\overset{-0+}{0.02} \cup \overset{-0+}{0.02}] = [\overset{-0+}{0.3}] - [\overset{-0+}{0.02}] = 0.3 \overset{-0+}{-} 0.02 =_N \overset{-0+}{0.28}
\end{aligned} \tag{199}$$

$$\begin{aligned}
&0.4 +_N \overset{+}{0.5} =_N [0.4, 0.4] + (0.5, 0.5+\varepsilon_1) - [0.4, 0.4] \times (0.5, 0.5+\varepsilon_1) \\
&= (0.4+0.5, 0.4+0.5+\varepsilon_1) - (0.4 \times 0.5, 0.4 \times 0.5 + 0.4\varepsilon_1) \\
&= (0.9, 0.9+\varepsilon_1) - (0.2, 0.2+0.4\varepsilon_1) \\
&= (0.9-0.2-0.4\varepsilon_1, 0.9+\varepsilon_1-0.2) = (0.7-0.4\varepsilon_1, 0.7+\varepsilon_1) =_N \overset{-0+}{0.70}
\end{aligned} \tag{200}$$

Hence

$$P_1 \wedge P_2 =_N (\overset{-0+}{0.18}, \overset{-0+}{0.28}, \overset{-0+}{0.70}) \tag{201}$$

Numerical Example of Nonstandard Neutrosophic Disjunction

$$\begin{aligned}
&\overset{0+}{0.3} +_N \overset{-0}{0.6} - \overset{0+}{0.3} \times_N \overset{-0}{0.6} =_N \{[0.3, 0.3+\varepsilon_1) + (0.6-\varepsilon_1, 0.6]\} - \{[0.3, 0.3+\varepsilon_1) \times (0.6-\varepsilon_1, 0.6]\} \\
&= (0.9-\varepsilon_1, 0.9+\varepsilon_1) - (0.18-0.3\varepsilon_1, 0.18+0.6\varepsilon_1) = (0.72-1.6\varepsilon_1, 0.72+1.3\varepsilon_1) =_N \overset{-0+}{0.72}
\end{aligned} \tag{202}$$

$$\overset{-+}{0.2} \times_N \overset{-0+}{0.1} =_N \left(0.2 \overset{-0+}{\times} 0.1\right) =_N \overset{-0+}{0.02} \tag{203}$$

$$0.4 \times_N \overset{+}{0.5} =_N \left(0.4 \overset{+}{\times} 0.5\right) =_N \overset{+}{0.20} \tag{204}$$

Hence

$$P_1 \vee_N P_2 =_N (\overset{-0+}{0.72}, \overset{-0+}{0.02}, \overset{+}{0.20}) \tag{205}$$

Numerical Example of Nonstandard Neutrosophic Negation

$$\neg_N P_1 =_N \neg_N (\overset{0+}{0.3}, \overset{-+}{0.2}, 0.4) =_N (0.4, \overset{-+}{0.2}, \overset{0+}{0.3}) \tag{206}$$

Numerical Example of Nonstandard Neutrosophic Implication

$$(P_1 \to {}_N P_2) \Leftrightarrow {}_N(\neg_N P_1 \vee_N P_2) =_N (0.4, \overset{-+}{0.2}, \overset{0+}{0.3}) \vee_N (\overset{-0}{0.6}, \overset{-0+}{0.1}, \overset{+}{0.5}) \tag{207}$$

Afterwards,

$$0.4 +_N \overset{-0}{0.6} - 0.4 \times_N \overset{-0}{0.6} =_N \left(0.4 \overset{-0}{+} 0.6\right) -_N \left(0.4 \overset{-0}{\times} 0.6\right) =_N \overset{-0}{1.0} -_N \overset{-0}{0.24} =_N \overset{-0+}{0.76} \tag{208}$$

$$\overset{-+}{0.2} \times_N \overset{-0+}{0.1} =_N \overset{-0+}{0.02} \tag{209}$$

$$\overset{0+}{0.3} \times \overset{+}{0.5} =_N \overset{+}{0.15} \tag{210}$$

whence

$$\neg_N P_1 =_N (\overset{-0+}{0.76}, \overset{-0+}{0.02}, \overset{+}{0.15}) \tag{211}$$

Therefore, we have showed above how to do nonstandard neutrosophic arithmetic operations on some concrete examples.

51. Conclusions

In the history of mathematics, critics on nonstandard analysis, in general, have been made by Paul Halmos, Errett Bishop, Alain Connes, and others.

That's why we have extended in 1998 for the first time the monads to pierced binad, and then in 2019 for the second time we extended the left monad to left monad closed to the right, the right monad to right monad closed to the left, and the pierced binad to unpierced binad. These were necessary in order to construct a general nonstandard neutrosophic real mobinad space, which is closed under the nonstandard neutrosophic arithmetic operations (such as addition, subtraction, multiplication, division, and power), which are needed in order to be able to define the nonstandard neutrosophic operators (such as conjunction, disjunction, negation, implication, and equivalence) on this space, and to transform the newly constructed nonstandard neutrosophic real mobinad space into a lattice of first order (as partially ordered nonstandard set, under the neutrosophic inequality $\leq_N$) and a lattice of second type (as algebraic structure, endowed with two binary laws: neutrosophic infimum (*infN*) and neutrosophic supremum (sup_N)).

As a consequence of extending the nonstandard analysis, we also extended the nonstandard neutrosophic logic, set, measure, probability and statistics.

As future research it would be to introduce the nonstandard neutrosophic measure, and to find applications of extended nonstandard neutrosophic logic, set, probability into calculus, since in calculus one deals with infinitesimals and their aggregation operators, due to the tremendous number of applications of the neutrosophic theories [28].

Funding: This research received no external funding.

Conflicts of Interest: The author declares no conflict of interest.

References

1. Imamura, T. Note on the Definition of Neutrosophic Logic. *arXiv* **2018**, arXiv:1811.02961.
2. Smarandache, F. Answers to Imamura Note on the Definition of Neutrosophic Logic. *arXiv* **1812**.
3. Smarandache, F. *About Nonstandard Neutrosophic Logic (Answers to Imamura's 'Note on the Definition of Neutrosophic Logic')*; Cornell University: New York City, NY, USA, 2019.
4. Smarandache, F. *Extended Nonstandard Neutrosophic Logic, Set, and Probability Based on Extended Nonstandard Analysis*; Cornell University: Ithaca, NY, USA, 2019; p. 31.
5. Smarandache, F. *Neutrosophy: Neutrosophic Probability, Set, and Logic*; ProQuest Information & Learning: Ann Arbor, MI, USA, 1998; p. 105.
6. Smarandache, F. Definition of neutrosophic logic—A generalization of the intuitionistic fuzzy logic. In Proceedings of the 3rd Conference of the European Society for Fuzzy Logic and Technology, Zittau, Germany, 10–12 September 2003; pp. 141–146.

7. Thao, N.X.; Smarandache, F. (I, T)-Standard neutrosophic rough set and its topologies properties. *Neutrosophic Sets Syst.* **2016**, *14*, 65–70. [CrossRef]
8. Thao, N.X.; Cuong, B.C.; Smarandache, F. Rough Standard Neutrosophic Sets: An Application on Standard Neutrosophic Information Systems. *Neutrosophic Sets Syst.* **2016**, *14*, 80–92. [CrossRef]
9. Cuong, B.C.; Phong, P.H.; Smarandache, F. Standard Neutrosophic Soft Theory—Some First Results. *Neutrosophic Sets Syst.* **2016**, *12*, 80–91. [CrossRef]
10. Smarandache, F. Degree of Dependence and Independence of the (Sub)Components of Fuzzy Set and Neutrosophic Set. *Neutrosophic Sets Syst.* **2016**, *11*, 95–97.
11. Smarandache, F.; Overset, N.; Underset, N.; Offset, N. *Similarly for Neutrosophic Over-/Under-/Off-Logic, Probability, and Statistics*; Pons Publishing House: Brussels, Belgium, 2016; 168p.
12. Smarandache, F. *Applications of Neutrosophic Sets in Image Identification, Medical Diagnosis, Fingerprints and Face Recognition and Neutrosophic Overset/Underset/Offset*; COMSATS Institute of Information Technology: Abbottabad, Pakistan, 26 December 2017.
13. Smarandache, F. Interval-Valued Neutrosophic Oversets, Neutrosophic Understes, and Neutrosophic Offsets. *Int. J. Sci. Eng. Investig.* **2016**, *5*, 1–4.
14. Smarandache, F. Operators on Single-Valued Neutrosophic Oversets, Neutrosophic Undersets, and Neutrosophic Offsets. *J. Math. Inform.* **2016**, *5*, 63–67. [CrossRef]
15. Smarandache, F. A Unifying Field in Logics: Neutrosophic Logic. *Mult. Valued Logic* **2002**, *8*, 385–438.
16. Smarandache, F. n-Valued Refined Neutrosophic Logic and Its Applications in Physics. *Prog. Phys.* **2013**, *4*, 143–146.
17. Smarandache, F. Neutrosophy, A New Branch of Philosophy. *Mult. Valued Logic* **2002**, *8*, 297–384.
18. Robinson, A. *Non-Standard Analysis*; Princeton University Press: Princeton, NJ, USA, 1996.
19. Loeb, P.A.; Wolff, M. (Eds.) Nonstandard analysis for the working mathematician. In *Mathematics and Its Applications*; Kluwer Academic Publishers: Dordrecht, The Netherlands, 2000; p. 311.
20. Insall, M.; Weisstein, E.W.; Nonstandard Analysis. From MathWorld—A Wolfram Web Resource. Available online: http://mathworld.wolfram.com/NonstandardAnalysis.html (accessed on 1 April 2019).
21. Insall, M.; Transfer Principle. From MathWorld—A Wolfram Web Resource, Created by Eric W. *Weisstein.* Available online: http://mathworld.wolfram.com/TransferPrinciple.html (accessed on 1 April 2019).
22. Smarandache, F. *Neutrosophic Precalculus and Neutrosophic Calculus*; EuropaNova: Brussels, Belgium, 2015; p. 154.
23. Smarandache, F. *Introduction to Neutrosophic Measure, Neutrosophic Integral, and Neutrosophic Probability*; Sitech & Educational: Craiova, Romania; Columbus, OH, USA, 2013; 140p.
24. Smarandache, F. An Introduction to Neutrosophic Measure. In Proceedings of the 2014 American Physical Society April Meeting, Savannah, Georgia, 5–8 April 2014; Volume 59.
25. Ashbacher, C. *Introduction to Neutrosophic Logic*; ProQuest Information & Learning: Ann Arbor, MI, USA, 2002.
26. Rivieccio, U. Neutrosophic logics: Prospects and problems. *Fuzzy Sets Syst.* **2008**, *159*, 1860–1868. [CrossRef]
27. Smarandache, F. *Plithogeny, Plithogenic Set, Logic, Probability, and Statistics*; Pons Publishing House: Brussels, Belgium, 2017; p. 141.
28. Peng, X.; Dai, J. A bibliometric analysis of neutrosophic set: Two decades review from 1998 to 2017. In *Artificial Intelligence Review*; Springer: Amsterdam, The Netherlands, 2018.

symmetry

MDPI

Article

Simplified Neutrosophic Exponential Similarity Measures for Evaluation of Smart Port Development

Jihong Chen [1,2], Kai Xue [1], Jun Ye [3,*], Tiancun Huang [1,*], Yan Tian [2], Chengying Hua [1] and Yuhua Zhu [4]

1 College of Transport and Communications, Shanghai Maritime University, Shanghai 201306, China; cxjh2004@163.com (J.C.); rookiex23@163.com (K.X.); 201830610136@stu.shmtu.edu.cn (C.H.)
2 School of Business, Beijing Institute of Technology, Zhuhai, Zhuhai 519088, China; Tandty@163.com
3 Department of Electrical and Information Engineering, Shaoxing University, Shaoxing 312000, China
4 School of Management, Fudan University, Shanghai 200433, China; yhzhu@shmtu.edu.cn
* Correspondence: yehjun@aliyun.com (J.Y.); huangtiancun1563@163.com (T.H.); Tel.: +86-575-8832-7323 (J.Y.); +86-21-3828-2334 (T.H.)

Received: 30 January 2019; Accepted: 29 March 2019; Published: 3 April 2019

Abstract: Smart ports represent the current trend of port development. Intelligent operations reduce the daily production cost of ports, facilitate efficient production, strengthen the risk mitigation ability and comply with the requirements for long-term development. However, a systematic and scientific smart port evaluation method is missing to nail down the evaluation indicators of a smart port and enable accurate evaluation of a port's degree of intelligence. This paper analyzes the concept of the smart port, establishes a set of smart port evaluation indicator systems, and applies a single-valued neutrosophic exponential similarity measure to port evaluation to enable quantitative evaluation of port integrity. This evaluation method is capable of decision-making in the event of incomplete, uncertain, and inconsistent information during general evaluation, opening up a new method for smart port evaluation, and acting as a helpful tool for ports to carry out improvements during actual application.

Keywords: smart port; simplified neutrosophic set; single-valued neutrosophic set; exponential similarity measure; port evaluation

1. Introduction

More and more opportunities for international trade cooperation are emerging as the world economy integrates. Ports, as a key link in global transportation, play an important role in world economic and trade development. However, currently port enterprises are battling sluggish growth of revenue and cutthroat competition due to homogenization. The international community is paying increasing attention to environmental protection issues, demonstrating higher sensitivity to climate change issues. All these problems have been pushing ports toward upgrading and transformation. In recent years, smart ports have become a dominant mode for port development, representing the highest level of modern port development. Smart port is based on systematic, strategic and social thinking, featuring integrated application of cloud computing, big data, Internet of Things, mobile internet, intelligent sensing and other next-generation information technologies to achieve all-round perception, ubiquitous interconnection, intelligent integration, deep computing, and coordinated operation and promoting organic connection and sharing of various resource elements and related parties in the port organization ecosystem, so as to eventually form a modern port that is smarter, safer, more efficient, more flexible, greener, and with strong cultural presence.

To materialize an operation mode for the smart port, scholars have conducted a ton of research on port optimization, including improving logistics supply efficiency, enhancing port service functions, reducing environmental pollution at ports, and cutting energy consumption of equipment. These studies have provided practical and effective steps to port organizations and improved their shortcomings. However, the current evaluation of smart port performance remains imperfect, lacking an objective and scientific evaluation method, and clear indicators for the evaluation, which is not conducive to discovering problems in ports. For the purpose of establishing a systematic and scientific smart port evaluation approach, this paper builds up a system of indicators for smart port evaluation, namely by using single-valued neutrosophic exponential similarity measure to quantify the degree of port intelligence. The research results can offer a theoretical basis for the port industry and stakeholders to launch smart port construction.

The rest of the paper is organized as follows. The literature review in Section 2 summarizes the current research programs on port efficiency evaluation with the aim to draw lessons from them. Section 3 determines the smart port evaluation indicators. Section 4 proposes the research approach in this paper, and introduces the neutrosophic exponential similarity measure. Section 5 conducts an empirical study to evaluate smart port based on the indicators in a simplified neutrosophic environment. Section 6 presents the conclusion of this paper.

2. Literature Review

Development efficiency evaluation helps ports to identify their own shortcomings, and facilitates better designing of plans and policies tailored to distinct situations of different ports, which is of constructive significance for port development. At present, parametric analysis and nonparametric analysis are two dominant approaches for port efficiency evaluation both domestically and internationally. Specifically, data envelopment analysis (DEA) is often used in the nonparametric analysis. Scholars select different input and output indicators from different angles according to their own research needs, and construct corresponding port operation efficiency evaluation models. Wu et al. used the DEA model to test the sensitivity of individual input and output decision-making units, finding that berth count and capital investment are the most sensitive factors affecting the throughput of a container port [1]. Tongzon studied the operational efficiency of international ports with the DEA model, and compared the operational advantages of several international ports, discovering that the relationship between the efficiency of an international port and its size is not clear [2,3]. Cullinane et al. took into account time-varying factors in port efficiency evaluation modeling, and established the DEA time window analysis model to study the relative efficiencies of world's major container ports, finding that the evaluated efficiency of a container port fluctuates over time [4]. Cullinane et al. studied the advantages and disadvantages of port privatization, and used the DEA model to conduct an empirical study on the relationship between privatization and container port efficiency [5]. Rajasekar and Deo studied the size effect and its efficiency of Indian Major Ports using DEA-Additive models, discovering that there is no significant difference between size and its efficiency of the port [6]. Wang and Han used the traditional DEA, into fuzzy DEA using fuzzy number characteristics in order to measure the efficiency of twelve international container ports in Taiwan and surrounding areas without having to consider weighting values of inputs and outputs. This approach allows objective and easy measurement of international container port efficiency. By the fuzzy DEA computation, it was found that the results of judging under input orientation and output orientation were consistent [7]. Cullinane and Wang studied the fundamentals of DEA and demonstrated how DEA can be applied to measure the efficiency of container ports. As a benchmarking approach to study efficiency, DEA enables a port to evaluate its performance vis-a-vis its peers. In so doing, the possible waste of resources and the industry best practice can be identified [8]. Gamassa and Chen used the DEA model to measured and analyzed the East and West African major ports efficiency over time, and the findings demonstrate that though West African ports have bigger ports size and have a higher Container throughput TEUs compared to East African ports [9]. In addition, Chin et al. used the DEA model to study the efficiencies of ports in Singapore, Greece, ASEAN, etc., and proposed enhanced measures targeting their shortcomings [10–12].

The parametric approach for port efficiency evaluation primarily uses the so-called stochastic frontier analysis. The stochastic frontier analysis refers to calculating the deviation degree between the sample port and frontier ports in terms of efficiency to evaluate the efficiency of the sample port. There are no definite frontier ports in terms of efficiency, they are just the most efficient ports relative to other ports, and vary with different port sample sets [13]. Coto-Millan et al. used a stochastic frontier cost function to evaluate the economic efficiency of Spanish ports using panel data, and found that the port operation model has a significant impact on economic efficiency, but the port size has no relationship with port economic efficiency [14]. Cullinane et al. used the cross-section and panel data versions of the "stochastic frontier model" to evaluate the relative efficiency of major container ports in Asia, and found that port size is closely related to port efficiency [15]. Notteboom and Winkelmans used the Bayesian stochastic frontier model to evaluate the port efficiency in Asia and Europe, and found that port efficiency has nothing to do with privatization but is positively related to port size [16]. In addition, many scholars used the analytical hierarchy process to evaluate port performance [17–19], the fuzzy analytical network process (FANP), and other multiple attribute decision-making methods were also chosen to evaluate port efficiency [20,21]. Such evaluation methods are easier for practical application, but feature international indicators of the evaluation system and hence a high degree of interference.

Founded by Smarandache in 1980, neutrosophy studies the neutrosophic origin, nature and scope, as well as the roles of different ideologies [22]. In recent years, neutrosophic set theory has been widely used in decision-making and evaluation research in many industries. Neutrosophy boasts wide application, such as in the fields of engineering, medicine, military science, cybernetics and physics, for logical deduction, aggregation, and probability statistics. Fu and Ye proposed new exponential similarity measures (ESMs) between simplified neutrosophic sets (SNSs), including single-valued neutrosophic ESMs and interval neutrosophic ESMs, and their initial evaluation/diagnosis method of the BPH symptoms with simplified neutrosophic information [23]. Peng et al. proposed simplified neutrosophic sets for multi-criteria decision-making problems [24]. Sahin and Liu used two new operational laws in which the bases are positive real numbers and interval numbers, respectively and the exponents are SNNs, and discussed some of their desired properties [25]. Şahin and Küçük established a useful method dealing with subsethood similarity measure between two SVNSs [26]. Biswas et al. applied the TOPSIS method in single-valued neutrosophic information [27]. Şahin and Liu introduced a maximizing deviation method under neutrosophic environment and utilized it for solving a numerical example with incomplete weight information [28]. Akram and Shahzadi defined the notion of the interval valued neutrosophic soft set (ivn-soft sets), which is a combination of an interval valued neutrosophic set and a soft set to investigate the decision making based on ivn-soft sets by level soft sets [29]. Smarandache and Ali introduced the notion of neutrosophic triplet, which is a group of three elements that satisfy certain properties with some binary operation [30]. Rizk-Allah et al. developed a new compromise algorithm for multi-objective transportation problem (MO-TP), which is inspired by Zimmermann's fuzzy programming and the neutrosophic set terminology [31]. Liu and Teng introduced the definition, the properties, the score function, the accuracy function, and the operational laws of the normal neutrosophic numbers (NNNs), and used an illustrative example to demonstrate the practicality and effectiveness of the proposed method [32]. Abdel-Basset and Mohamed proposed a general framework for smart city evaluation with imperfect and incomplete information through using single valued neutrosophic and rough set theories [33]. Thong et al. developed a new Technique for Order of Preference by Similarity to Ideal Solution (TOPSIS) method based on the proposed Dynamic Interval-valued Neutrosophic Set(DIVNS) theory [34]. However, neutrosophy in the field of transportation application is rarely mentioned, so this paper proposes to apply a single-valued neutrosophic exponential similarity measure to evaluate smart port in a simplified neutral environment.

Most of the current research on port efficiency evaluation adopts the aforementioned methods, or makes further improvements based on the aforesaid studies. In recent years, some new non-classical nature-inspired evaluation and optimization methods have also been developed and applied [35–37].

However, port evaluation indicators are uncertain in practical application, no matter whether the parametric analysis or the nonparametric analysis approach is used for port performance evaluation. Besides, the evaluation indicators for different ports are not consistent and applying different evaluation methods for the same port will also produce different results. To avoid this, we should nail down the indicators related to smart port evaluation and then use the single-valued neutrosophic exponential similarity measure to evaluate smart port on that basis.

3. Evaluating Indicators

Port is the junction of water and land transport, an important base for industrial activities, a comprehensive logistics center and a new growth point for urban economic development. To facilitate more efficient, safer and greener port operation, experts never ceased the search of new information technologies, and the concept of the smart port came into being.

Currently, the smart port in the broad sense is a result of digital technologies, business model innovation and resource value innovation. The smart port features excellent port operation, an open ecosystem, and active expansion in sustainable innovative businesses.

Smart ports attempt to apply advanced information technology as well as automated and intelligent mechanical equipment to the daily production and operation management of ports, realizing the automation of port production and operation, the whole process of port logistics supply chain services, the facilitation of port financial trade and the rationalization of port energy saving and emission reduction. Smart ports enable seamless connection and synergy between vehicles, ships, people, cargoes and various systems of the port, improving its daily operational efficiency and amplifying its advantages. By referring to literature in port and related fields, this paper singles out specific evaluation indicators for the smart port, as shown in Table 1. The smart port mentioned in this paper refers to a comprehensive conceptual port, which renders intelligence and advancedness to the port in terms of production and operation systems, logistics supply chain systems, financial and trade service technologies, and energy conservation and emission reduction capacities. This enables a safe, efficient, convenient, green and sustainable development form of the port to improve the comprehensive competitiveness of ports.

Specifically, the evaluation indicators of the port production and operation systems include the application of emerging information technologies, such as port production dispatching automation, Internet of Things and cloud computing, and emergency response capabilities. These indicators emphasize the application of intelligent technology in port operation, which is the breakthrough sign of the development of information and intelligence in many ports in the world. Smart Ports use intelligent technology to automate production scheduling, reduce manual work, and enhance the handling capacity of port emergency events.

The evaluation indicators of the port logistics supply chain systems include the intelligent level of door-to-door full-course services of port logistics, the electronic processing of logistics documents, the standardization level of operations and the "Internet +" logistics supply chain services of the port. These indicators mainly consider the ability of ports to develop door-to-door supply chain services, especially require ports to realize the intellectualization, standardization, and convenience of logistics services through intelligent technology. Ports can provide efficient, fast, and convenient integrated logistics services.

The evaluation indicators of the port financial and trade service technologies include port integration and facilitation as well as customs clearance efficiency, and sharing of financial service resources in the port supply chain. These indicators consider the service expansion capability of ports in the context of intellectualization, and require ports to use intelligent technology to achieve convenient customs clearance environment, goods trade, and supply chain financial service extension.

The evaluation indicators of the port energy conservation and emission reduction capacities include the application status of green energy sources at the port and the emission control and governance capacities over port pollutants. These indicators mainly consider how to use intelligent technology to achieve energy reduction and emission reduction, and how to build and develop green ports. Green port is also a new concept of port development in the world. It is a sustainable port with a good balance between environmental protection and economic interests. Guided by the concept of green development, green ports achieve environmental health, energy consumption reduction, and pollution reduction through intelligent technologies and emission reduction measures.

Table 1. Classification of Smart Port Evaluation Indicators.

Area	Evaluation Indicator	Reference
Port Production and Operation Systems	Production dispatching automation	[38–40]
	Application of emerging information technologies at ports such as the Internet of Things and cloud computing	[39,41,42]
	Emergency response capabilities	[38,41]
Port Logistics Supply Chain System	Intelligent level of door-to-door full-course services of port logistics	[40–43]
	Electronic processing of logistics documents, the standardization level of operations	[40–43]
	"Internet +" logistics supply chain services	[40–43]
Port Financial and Trade Service Technologies	Port integration and facilitation as well as customs clearance efficiency	[38,44,45]
	Sharing of financial service resources in the port supply chain	[44,45]
Port Energy Conservation and Emission Reduction Capacities	Application status of green energy sources at the port	[38,46,47]
	Emission control and governance capacities over port pollutants	[46–49]

4. Research Methodology

4.1. Basic Concepts of SNSs

Simplified neutrosophic set (SNS) is a powerful tool that attracts the attention of many scholars in dealing with uncertainty and vagueness [24]. Ye proposed the Simplified Neutrosophic Set (SNS), a subset of the neutrosophic set which is more suitable for processing issues that contain many incomplete, uncertain, and inconsistent information to apply neutrosophy to science and engineering better [23,26]. SNS can be defined as follows.

Let X be a space of points (objects), with a generic element in X denoted by x. An SNS N in X is characterized by a truth-membership function $T_N(x)$, an indeterminacy-membership function $I_N(x)$ and a falsity-membership function $F_N(x)$. Then, an *SNS N* can be expressed as $N = \{<x, T_N(x), I_N(x), F_N(x)>|x \in X\}$, where the sum of $T_N(x), I_N(x), F_N(x) \subseteq [0, 1]$ satisfies the condition $0 \leq \sup T_N(x) + \sup I_N(x) + \sup F_N(x) \leq 3$ for each point x in X. Then, SNS is a subclass of the neutrosophic set and includes the concepts of single valued neutrosophic set (SVNS).

Assume that $A = \{<x, T_A(x), I_A(x), F_A(x)>|x \in X\}$ and $B = \{<x, T_B(x), I_B(x), F_B(x)>|x \in X\}$ are two SNSs, where $T_A(x), I_A(x), F_A(x) \in [0, 1]$, $0 \leq T_A(x) + I_A(x) + F_A(x) \leq 3$ for each point x in X, i.e., A and B are two Simplified Neutrosophic Sets (SNSs). The SNS is a effective generalization of the fuzzy set that is designed for some situations in which each element has different truth membership function, indeterminacy membership function and falsity membership function. Then, the inclusion, equation, and complement for SNSs A and B are defined, respectively, as follows:

(1) $B \subseteq A$ if and only if $T_A(x) \leq T_B(x), I_A(x) \geq I_B(x), F_A(x) \geq F_B(x)$ for any x in X,
(2) $A = B$ if and only if $A \subseteq B$ and $B \subseteq A$,
(3) $A^c = \{<x, F_A(x), 1 - I_A(x), T_A(x)>|x \in X\}$ and $B^c = \{<x, F_B(x), 1 - I_B(x), T_B(x)>|x \in X\}$.

Assume that $A = \{\langle x, T_A(x), I_A(x), F_A(x)\rangle | x \in X\}$ and $B = \{\langle x, T_B(x), I_B(x), F_B(x)\rangle | x \in X\}$ are two SNSs in X. If $T_A(x), I_A(x), F_A(x) \subseteq [0, 1]$, $0 \le \sup T_A(x) + \sup I_A(x) + \sup F_A(x) \le 3$, $T_B(x), I_B(x), F_B(x) \subseteq [0, 1]$, and $0 \le \sup T_B(x) + \sup I_B(x) + \sup F_B(x) \le 3$ for each point x in X, then A and B are reduced to two interval neutrosophic sets (INSs) [50]. Thus, the inclusion, equation, and complement for SNSs B and A are defined, respectively, as follows:

(4) $B \subseteq A$ if and only if $\inf T_B(x) \le \inf T_A(x)$, $\inf I_B(x) \ge \inf I_A(x)$, $\inf F_B(x) \ge \inf F_A(x)$, $\sup T_B(x) \le \sup T_A(x)$, $\sup I_B(x) \ge \sup I_A(x)$, $\sup F_B(x) \ge \sup F_A(x)$ for any x in X;

(5) $B = A$ if and only if $B \subseteq A$ and $A \subseteq B$;

(6) $A^c = \{\langle x, [\inf F_A(x), \sup F_A(x)], [1 - \sup I_A(x), 1 - \inf I_A(x)], [\inf T_A(x), \sup T_A(x)]\rangle | x \in X\}$ and $B^c = \{\langle x, [\inf F_B(x), \sup F_B(x)], [1 - \sup I_B(x), 1 - \inf I_B(x)], [\inf T_B(x), \sup T_B(x)]\rangle | x \in X\}$

Especially when the upper and lower ends of the interval numbers $T_A(x), I_A(x), F_A(x)$ in A and $T_B(x), I_B(x), F_B(x)$ in B are equal, the INSs A and B are reduced to the single valued neutrosophic sets (SVNSs) A and B. Therefore, SVNSs are the special cases of INSs, and also SVNSs and INSs are also the special cases of SNSs.

4.2. Exponential Similarity Measures of SVNS

The single valued neutrosophic set (SVNS) is a generalization of classic set, fuzzy set, interval valued fuzzy set, intuitionistic fuzzy set, and para-consistent set [23,26]. This section describes the steps of applying the exponential similarity measuring method, as detailed below.

Step 1. Determine the decision goal. This paper aims to obtain the evaluation approach of smart port, consulting relevant literature to determine the specific indicators of the evaluation system. I_m represents the m-th indicator.

Step 2. Develop criteria for indicators. According to different evaluation environments, different degrees of decision-making statuses are selected, and D_n is used to denote the n-th degree. Second, these degrees are represented by their respective SVNS information.

Step 3. Conduct a preliminary evaluation. t industry experts are invited to evaluate the indicators. Truth, Indeterminacy, and Falsity represent the degree of recognition, from high to low, of the indicator performance. Specific SVNS value calculation is as follows:

Assume a experts select Truth, b experts choose Indeterminacy, and c experts choose Falsity. Then the SNS value of this indicator is $<a/t, b/t, c/t>$.

Step 4. Exponential similarity measure. Set the standard to be $A = \{<x_j, T_A(x_j), I_A(x_j), F_A(x_j)> | x_j \in X\}$, and the preliminary evaluation is $B = \{<x_j, T_B(x_j,), I_B(x_j,), F_B(x_j)> | x \in X\}$, which is any two SVNSs in the range of $X = \{x_1, x_2, \ldots, x_m\}$. Based on the exponential function, the exponential similarity measure of the standard and the preliminary evaluation is defined as follows:

$$E_i(A, B) = \frac{1}{m}\sum_{j=1}^{m} \frac{\exp\{-\frac{1}{3}[|T_A(x_j) - T_B(x_j)| + |I_A(x_j) - I_B(x_j)| + |F_A(x_j) - F_B(x_j)|]\} - \exp(-1)}{1 - \exp(-1)} \quad (1)$$

When the weights of the indicators are different, a weight coefficient w_j can be added. $w_j \in [0, 1]$ and the sum of them is 1. The specific expression is as follows:

$$W_i(A, B) = \sum_{j=1}^{m} w_j \frac{\exp\{-\frac{1}{3}[|T_A(x_j) - T_B(x_j)| + |I_A(x_j) - I_B(x_j)| + |F_A(x_j) - F_B(x_j)|]\} - \exp(-1)}{1 - \exp(-1)} \quad (2)$$

Step 5. Make the calculation using MATLAB. The above content is coded and calculated in MATLAB to obtain the maximum similarity measure to indicate the most appropriate evaluation.

5. Application Method and Results

In this section, the aforementioned methods will apply. First, follow Step 1 to organize the smart port evaluation indicators listed above into the Table 2.

Table 2. Evaluation by 5 experts for a port P_k.

Indicators	Truth	Indeterminacy	Falsity
	(T)	(I)	(F)
I_1 Port production scheduling of fully automated			
I_2 The application of the Internet of things, cloud computing and other emerging information technologies in ports			
I_3 The ability of the port to deal with emergencies			
I_4 Intelligent level of port logistics door-to-door service			
I_5 Port logistics documents, data processing and other links electronic, standardized operation level			
I_6 Port "Internet +" logistics supply chain service			
I_7 Integration and facilitation of ports and customs clearance efficiency			
I_8 Port supply chain financial service resource sharing			
I_9 Application of green energy in ports			
I_{10} The ability to control and control the discharge of pollutants from ports			

Follow Step 2 to preliminarily classify smart port into five degrees: strong, relatively strong, average, relatively weak, and weak, to quantify the initially evaluated degrees of the smart port. See Table 3 for details.

Table 3. Five types of smart port degree with simplified neutrosophic information.

Indicators	D_1	D_2	D_3	D_4	D_5
	(Strong)	(Relatively Strong)	(Average)	(Relatively Weak)	(Weak)
I_1	<1.0,0.0,0.0>	<0.8,0.2,0.0>	<0.6,0.4,0.0>	<0.4,0.4,0.2>	<0.2,0.4,0.4>
I_2	<1.0,0.0,0.0>	<0.8,0.2,0.0>	<0.6,0.4,0.0>	<0.4,0.4,0.2>	<0.2,0.4,0.4>
I_3	<1.0,0.0,0.0>	<0.8,0.2,0.0>	<0.6,0.4,0.0>	<0.4,0.4,0.2>	<0.2,0.4,0.4>
I_4	<1.0,0.0,0.0>	<0.8,0.2,0.0>	<0.6,0.4,0.0>	<0.4,0.4,0.2>	<0.2,0.4,0.4>
I_5	<1.0,0.0,0.0>	<0.8,0.2,0.0>	<0.6,0.4,0.0>	<0.4,0.4,0.2>	<0.2,0.4,0.4>
I_6	<1.0,0.0,0.0>	<0.8,0.2,0.0>	<0.6,0.4,0.0>	<0.4,0.4,0.2>	<0.2,0.4,0.4>
I_7	<1.0,0.0,0.0>	<0.8,0.2,0.0>	<0.6,0.4,0.0>	<0.4,0.4,0.2>	<0.2,0.4,0.4>
I_8	<1.0,0.0,0.0>	<0.8,0.2,0.0>	<0.6,0.4,0.0>	<0.4,0.4,0.2>	<0.2,0.4,0.4>
I_9	<1.0,0.0,0.0>	<0.8,0.2,0.0>	<0.6,0.4,0.0>	<0.4,0.4,0.2>	<0.2,0.4,0.4>
I_{10}	<1.0,0.0,0.0>	<0.8,0.2,0.0>	<0.6,0.4,0.0>	<0.4,0.4,0.2>	<0.2,0.4,0.4>

From Table 3 we can see that the evaluated degrees of smart port indicators correspond to the following SVNS information.

$S_1 = \{<I_1,1.0,0.0,0.0>, <I_2,1.0,0.0,0.0>, <I_3,1.0,0.0,0.0>, <I_4,1.0,0.0,0.0>, <I_5,1.0,0.0,0.0>, <I_6,1.0,0.0,0.0>, <I_7,1.0,0.0,0.0>, <I_8,1.0,0.0,0.0>, <I_9,1.0,0.0,0.0>, <I_{10},1.0,0.0,0.0>\}$,

$S_2 = \{<I_1,0.8,0.2,0.0>, <I_2,0.8,0.2,0.0>, <I_3,0.8,0.2,0.0>, <I_4,0.8,0.2,0.0>, <I_5,0.8,0.2,0.0>, <I_6,0.8,0.2,0.0>, <I_7,0.8,0.2,0.0>, <I_8,0.8,0.2,0.0>, <I_9,0.8,0.2,0.0>, <I_{10},0.8,0.2,0.0>\}$,

$S_3 = \{<I_1,0.6,0.4,0.0>, <I_2,0.6,0.4,0.0>, <I_3,0.6,0.4,0.0>, <I_4,0.6,0.4,0.0>, <I_5,0.6,0.4,0.0>, <I_6,0.6,0.4,0.0>, <I_7,0.6,0.4,0.0>, <I_8,0.6,0.4,0.0>, <I_9,0.6,0.4,0.0>, <I_{10},0.6,0.4,0.0>\}$,

$S_4 = \{<I_1,0.4,0.4,0.2>, <I_2,0.4,0.4,0.2>, <I_3,0.4,0.4,0.2>, <I_4,0.4,0.4,0.2>, <I_5,0.4,0.4,0.2>, <I_6,0.4,0.4,0.2>, <I_7,0.4,0.4,0.2>, <I_8,0.4,0.4,0.2>, <I_9,0.4,0.4,0.2>, <I_{10},0.4,0.4,0.2>\}$,

$S_5 = \{<I_1,0.2,0.4,0.4>, <I_2,0.2,0.4,0.4>, <I_3,0.2,0.4,0.4>, <I_4,0.2,0.4,0.4>, <I_5,0.2,0.4,0.4>, <I_6,0.2,0.4,0.4>, <I_7,0.2,0.4,0.4>, <I_8,0.2,0.4,0.4>, <I_9,0.2,0.4,0.4>, <I_{10},0.2,0.4,0.4>\}$.

Follow Step 3 to provide Table 2 to five experts for parallel preliminary evaluation of the three ports. The evaluation results are shown in Table 4.

Table 4. Five experts' evaluation of three smart ports.

Indicators	Port A			Port B			Port C		
	T	I	F	T	I	F	T	I	F
I_1	5/5	0/5	0/5	4/5	0/5	1/5	3/5	1/5	1/5
I_2	4/5	1/5	0/5	3/5	2/5	0/5	3/5	0/5	2/5
I_3	4/5	0/5	1/5	2/5	0/5	3/5	1/5	1/5	3/5
I_4	3/5	2/5	0/5	4/5	1/5	0/5	4/5	0/5	1/5
I_5	4/5	1/5	0/5	3/5	1/5	1/5	2/5	1/5	2/5
I_6	5/5	0/5	0/5	3/5	1/5	1/5	3/5	1/5	1/5
I_7	3/5	0/5	2/5	4/5	0/5	1/5	2/5	2/5	1/5
I_8	4/5	1/5	0/5	3/5	1/5	1/5	4/5	1/5	0/5
I_9	5/5	0/5	0/5	3/5	0/5	2/5	2/5	3/5	0/5
I_{10}	3/5	2/5	0/5	2/5	2/5	1/5	3/5	2/5	0/5

From Table 4, the indicator degrees of Port P_k (k = 1, 2, 3) can be expressed with the following SVNS information:

$P_1 = \{\langle I_1,1.0,0.0,0.0\rangle, \langle I_2,0.8,0.2,0.0\rangle, \langle I_3,0.8,0.0,0.2\rangle, \langle I_4,0.6,0.4,0.0\rangle, \langle I_5,0.8,0.2,0.0\rangle, \langle I_6,1.0,0.0,0.0\rangle, \langle I_7,0.6,0.0,0.4\rangle, \langle I_8,0.8,0.2,0.0\rangle, \langle I_9,1.0,0.0,0.0\rangle, \langle I_{10},0.6,0.0,4.0\rangle\}$,

$P_2 = \{\langle I_1,0.8,0.0,0.2\rangle, \langle I_2,0.6,0.4,0.0\rangle, \langle I_3,0.4,0.0,0.6\rangle, \langle I_4,0.8,0.2,0.0\rangle, \langle I_5,0.6,0.2,0.2\rangle, \langle I_6,0.6,0.2,0.2\rangle, \langle I_7,0.8,0.0,0.2\rangle, \langle I_8,0.6,0.2,0.2\rangle, \langle I_9,0.6,0.0,0.4\rangle, \langle I_{10},0.4,0.4,0.2\rangle\}$,

$P_3 = \{\langle I_1,0.6,0.2,0.2\rangle, \langle I_2,0.6,0.0,0.4\rangle, \langle I_3,0.2,0.2,0.6\rangle, \langle I_4,0.8,0.0,0.2\rangle, \langle I_5,0.4,0.2,0.4\rangle, \langle I_6,0.6,0.2,0.2\rangle, \langle I_7,0.4,0.4,0.2\rangle, \langle I_8,0.8,0.2,0.0\rangle, \langle I_9,0.4,0.6,0.0\rangle, \langle I_{10},0.6,0.4,0.0\rangle\}$.

According to Step 4, assume that the weight of each element I_j is $w_j = 1/10$ for $j = 1, 2, \ldots, 10$. Then, by using MTALAB, we can get the results of the similarity measure between the port P_k (k = 1, 2, 3) and the indicator degree D_i (i = 1, 2, 3, 4, 5), as shown in Table 5.

Table 5. Similarity measure values of between P_k and D_i with SVNSs.

	S_1	S_2	S_3	S_4	S_5
$W_1(P_1,D_i)$	0.8099	0.8445	0.7556	0.6189	0.4841
$W_1(P_2,D_i)$	0.6340	0.7356	0.7380	0.7531	0.6167
$W_1(P_3,D_i)$	0.5905	0.7207	0.7380	0.7531	0.6513

In Table 5, the maximum similarity measure indicates the most proper evaluation. In the three smart ports, the result of port P_1 is "Relative Strong", that of port P_2 is "Average", and that of port P_3 is "Relative Weak". The difference in evaluation results for different ports is obvious.

In order to compare our method with other methods, we neglect the indeterminacy and falsity situations. When only the truth situation is considered, the neutrosophic sets degenerate into the traditional fuzzy sets. Therefore, under the fuzzy set framework, the indicator degrees of Port P_k (k = 1, 2, 3) can be expressed with the following SVNS information:

$P_1 = \{\langle I_1,1.0,0.0,0.0\rangle, \langle I_2,0.8,0.0,0.0\rangle, \langle I_3,0.8,0.0,0.0\rangle, \langle I_4,0.6,0.0,0.0\rangle, \langle I_5,0.8,0.0,0.0\rangle, \langle I_6,1.0,0.0,0.0\rangle, \langle I_7,0.6,0.0,0.0\rangle, \langle I_8,0.8,0.0,0.0\rangle, \langle I_9,1.0,0.0,0.0\rangle, \langle I_{10},0.6,0.0,0.0\rangle\}$,

$P_2 = \{\langle I_1,0.8,0.00.0\rangle, \langle I_2,0.6,0.0,0.0\rangle, \langle I_3,0.4,0.0,0.0\rangle, \langle I_4,0.8,0.0,0.0\rangle, \langle I_5,0.6,0.0,0.0\rangle, \langle I_6,0.6,0.0,0.0\rangle, \langle I_7,0.8,0.0,0.0\rangle, \langle I_8,0.6,0.0,0.0\rangle, \langle I_9,0.6,0.0,0.0\rangle, \langle I_{10},0.4,0.0,0.0\rangle\}$,

$P_3 = \{\langle I_1,0.6,0.0,0.0\rangle, \langle I_2,0.6,0.0,0.4\rangle, \langle I_3,0.2,0.2,0.6\rangle, \langle I_4,0.8,0.0,0.2\rangle, \langle I_5,0.4,0.0,0.0\rangle, \langle I_6,0.6,0.0,0.0\rangle, \langle I_7,0.4,0.0,0.0\rangle, \langle I_8,0.8,0.0,0.0\rangle, \langle I_9,0.4,0.0,0.0\rangle, \langle I_{10},0.6,0.0,0.0\rangle\}$.

Similarly, we can get the results of the similarity measure between the port P_k ($k = 1, 2, 3$) and the indicator degree D_i ($i = 1, 2, 3, 4, 5$) under the traditional fuzzy set framework, as shown in Table 6.

Table 6. Similarity measure values of between P_k and D_i with Fuzzy Set.

	S_1	S_2	S_3	S_4	S_5
$W_1(P_1,D_i)$	0.8999	0.8407	0.7150	0.5531	0.4114
$W_1(P_2,D_i)$	0.8038	0.8038	0.7668	0.6308	0.4794
$W_1(P_3,D_i)$	0.7775	0.7775	0.7406	0.6391	0.5018

From the data in Table 6, we can see that there is a big difference between the evaluation results using fuzzy sets and our method (see Table 5). From the evaluation results of fuzzy sets, the result of port P_1 is "Strong", that of port P_2 is "Strong" or "Relative Strong", and that of port P_3 is also "Strong" or "Relative Strong" (see Table 6). In this way, it is difficult to distinguish the differences among the evaluation results of the three smart ports. Therefore, our evaluation method is much more effective and reasonable than the traditional fuzzy set method. From the case study in this paper, simplified neutrosophic exponential similarity measures can be well used in the evaluation of smart port development and get more reasonable evaluation results. However, from the point of view of the specific research process, this method has not added the influence of decision experts' weight. In the future research, we can consider the influence of decision experts' weight on the evaluation results comprehensively, and further overcome the subjective limitations of expert evaluation, so as to make the evaluation results of smart ports more reasonable and flexible.

6. Conclusions and Future Directions

The intelligent operation is an imperative development direction of advanced ports in the future. Securing all-round and sustainable development is the key to enhancing the competitiveness of ports. To this end, it is of more practical significance to study and analyze smart port evaluation. Based on exponential functions, this paper proposes to apply single-valued neutrosophic exponential similarity measure to evaluate smart port in a simplified neutral environment. This evaluation approach is advantageous over other existing port evaluation methods in that it has a more complete evaluation system to render a clearly quantitative evaluation result. Besides, it addresses the decision-making in the context of incomplete, uncertain and inconsistent information for smart port evaluation, making its evaluation results more scientific and rigorous. The contribution of this study is threefold. First, this study makes an initiative for the assessment of world smart port development. Second, this research provides an effective method for the evaluation system of amert ports. Third, the achievements of this study can provide decision-making basis and practical tool for international organizations, relevant governments or policy makers to formulate reasonable and effective governance strategy of global port industry and smart port development.

Using single-valued neutrosophic exponential similarity measure to analyze and evaluate smart port is an innovative attempt. This paper still has limitations. In the future, further research can be carried out focusing on the following three aspects. First, more smart port types and orientations can be taken into consideration in future to further tap to the evaluation of the smart ports using next-generation information technologies, while including different types of data in the evaluation scope to build more accurate evaluation indicators. Second, the evaluation indicators can be further subdivided. In the future, we can further look at the four aspects of smart port, namely the daily production and operation, the logistics supply chain system, the financial and trade services, and the energy conservation and emission reduction, for in-depth research and establishment of a more practical evaluation system that better complies with the actual situations. Third, the probabilistic approach can be introduced into the single-valued neutrosophic exponential similarity measure of this paper to give the evaluation model a certain predictive ability for the future development direction of smart ports, helping the port to locate find more room for improvement, so as to elevate the comprehensive competitiveness and put forward more accurate and effective suggestions for port building.

Author Contributions: J.C. and K.X. conceived the research idea, and co-wrote the paper. J.Y. proposed exponential similarity measures (ESMs) between simplified neutrosophic sets (SNSs), including single valued neutrosophic ESMs and interval neutrosophic ESMs. T.H. and Y.T. performed data analysis, established the mathematical models and revised the paper. C.H. and Y.Z. collected the data and revised the paper.

Funding: This paper was supported by the National Natural Science Foundation of China (Grant No. 51879156 and 51409157), Shanghai Pujiang Program (17PJC053), and the Program of Humanities and Social Science of the Ministry of Education of China (14YJC630008).

Conflicts of Interest: The authors declare no conflict of interest.

References

1. Wu, J.; Yan, H.; Liu, J. DEA models for identifying sensitive performance measures in container port evaluation. *Marit. Econ. Logist.* **2010**, *12*, 215–236. [CrossRef]
2. Tongzon, J. Efficiency measurement of selected Australian and other international ports using data envelopment analysis. *Transp. Res. Part A Policy A Pract.* **2001**, *35*, 107–122. [CrossRef]
3. Tongzon, J.L. Systematizing international benchmarking for ports. *Marit. Manag.* **1995**, *22*, 171–177. [CrossRef]
4. Kevin, C.; Dong-Wook, S.; Ping, J.; Teng-Fei, W. An application of DEA windows analysis to container port production efficiency. *Rev. Netw. Econ.* **2004**, *3*, 1–23.
5. Cullinane, K.; Ji, P.; Wang, T.F. The Relationship between privatization and DEA estimates of efficiency in the container port industry. *J. Econ. Bus.* **2005**, *57*, 433–462. [CrossRef]
6. Rajasekar, T.; Deo, M. The size effect of Indian major ports on its efficiency using Dea-Additive models. *Int. J. Adv. Manag. Econ.* **2018**, *1*, 12–18.
7. Wang, Y.J.; Han, T.C. Efficiency measurement for international container ports of Taiwan and surrounding areas by fuzzy data envelopment analysis. *J. Mar. Sci. Technol.-Taiwan* **2018**, *26*, 185–193.
8. Cullinane, K.; Wang, T.-F. Chapter 23 Data Envelopment Analysis (DEA) and Improving Container Port Efficiency. *Res. Transp. Econ.* **2006**, *17*, 517–566. [CrossRef]
9. Gamassa, P.K.P.; Chen, Y. Comparison of port efficiency between Eastern and Western African ports using DEA Window Analysis. In Proceedings of the 2017 14th International Conference on Service Systems and Service Management (ICSSSM), Dalian, China, 16–18 June 2017.
10. Chin, A.; Tongzon, J. Maintaining singapore as a major shipping and air transport hub. In *Competitiveness of The Singapore Economy: A Strategic Perspective*; Yam, T.K., Heng, T.M., Eds.; World Scientific: Singapore, 1998; pp. 83–114.
11. Barros, C.P.; Athanassiou, M. Efficiency in european seaports with DEA: Evidence from Greece and Portugal. *Marit. Econ. Logist.* **2004**, *6*, 122–140. [CrossRef]
12. Tongzon, J.L.; Ganesalingam, S. An evaluation of ASEAN port performance and efficiency. *Asian Econ. J.* **1994**, *8*, 317–330. [CrossRef]
13. Chen, J.; Wan, Z.; Zhang, F.; Park, N.K.; He, X.; Yin, W. Operational efficiency evaluation of iron ore logistics at the ports of Bohai Bay in China: Based on the PCA-DEA Model. *Math. Probl. Eng.* **2016**, *2016*. [CrossRef]
14. Coto-Millan, P.; Baños-Pino, J.; Rodríguez-Alvarez, A. Economic efficiency in Spanish ports: Some empirical evidence. *Marit. Manag.* **2000**, *27*, 169–174. [CrossRef]
15. Cullinane, K.; Song, D.W.; Gray, R. A stochastic frontier model of the efficiency of major container terminals in Asia: Assessing the influence of administrative and ownership structures. *Transp. Res. Part A Policy Pract.* **2002**, *36*, 743–762. [CrossRef]
16. Notteboom, T.E.; Winkelmans, W. Structural changes in logistics: How will port authorities face the challenge? *Marit. Manag.* **2001**, *28*, 71–89. [CrossRef]
17. Yuhling, S.U.; Liang, G.S.; Chin-Feng, L.I.U.; Tsung-Yu, C.H.O.U. A study on integrated port performance comparison based on the concept of balanced scorecard. *J. Eastern Asia Soc. Transp. Stud.* **2003**, *5*, 609–624.
18. Lirn, T.C.; Thanopoulou, H.A.; Beresford, A.K.C. Transhipment port selection and decision-making behaviour: Analysing the Taiwanese case. *Int. J. Logist. Res. Appl.* **2003**, *6*, 229–244. [CrossRef]

19. Chiu, R.-H.; Lin, L.-H.; Ting, S.-C. Evaluation of Green Port Factors and Performance: A Fuzzy AHP Analysis. *Math. Probl. Eng.* **2014**, *2014*, 1–12. [CrossRef]
20. Onut, S.; Tuzkaya, U.R.; Torun, E. Selecting container port via a fuzzy ANP-based approach: A case study in the Marmara Region, Turkey. *Transp. Policy* **2011**, *18*, 182–193. [CrossRef]
21. Brooks, M.R.; Schellinck, T.; Pallis, A.A. A systematic approach for evaluating port effectiveness. *Marit. Manag.* **2011**, *38*, 315–334. [CrossRef]
22. Smarandache, F. A unifying field in logics: Neutrosophic logic. *Multiple-Valued Logic* **1999**, *8*, 489–503.
23. Fu, J.; Ye, J. Simplified Neutrosophic Exponential Similarity Measures for the Initial Evaluation/Diagnosis of Benign Prostatic Hyperplasia Symptoms. *Symmetry* **2017**, *9*, 154. [CrossRef]
24. Peng, J.-J.; Wang, J.-Q.; Zhang, H.-Y.; Chen, X.-H. An outranking approach for multi-criteria decision-making problems with simplified neutrosophic sets. *Appl. Soft Comput.* **2014**, *25*, 336–346. [CrossRef]
25. Şahin, R.; Liu, P. Some approaches to multi criteria decision making based on exponential operations of simplified neutrosophic numbers. *J. Intell. Syst.* **2017**, *32*, 2083–2099. [CrossRef]
26. Şahin, R.; Küçük, A. Subsethood measure for single valued neutrosophic sets. *J. Intell. Syst.* **2015**, *29*, 525–530. [CrossRef]
27. Biswas, P.; Pramanik, S.; Giri, B.C. TOPSIS method for multi-attribute group decision-making under single-valued neutrosophic environment. *Neural Comput. Appl.* **2016**, *27*, 727–737. [CrossRef]
28. Şahin, R.; Liu, P. Maximizing deviation method for neutrosophic multiple attribute decision making with incomplete weight information. *Neural Comput. Appl.* **2016**, *27*, 2017–2029. [CrossRef]
29. Akram, M.; Shahzadi, S. Neutrosophic soft graphs with application. *J. Intell. Fuzzy Syst.* **2017**, *32*, 841–858. [CrossRef]
30. Smarandache, F.; Ali, M. Neutrosophic triplet group. *Neural Comput. Appl.* **2018**, *29*, 595–601. [CrossRef]
31. Rizk-Allah, R.M.; Hassanien, A.E.; Elhoseny, M. A multi-objective transportation model under neutrosophic environment. *Comput. Electr. Eng.* **2018**, *69*, 705–719. [CrossRef]
32. Liu, P.; Teng, F. Multiple attribute decision making method based on normal neutrosophic generalized weighted power averaging operator. *Int. J. Mach. Learn. Cybern.* **2018**, *9*, 281–293. [CrossRef]
33. Abdel-Basset, M.; Mohamed, M. The role of single valued neutrosophic sets and rough sets in smart city: Imperfect and incomplete information systems. *Measurement* **2018**, *124*, 47–55. [CrossRef]
34. Thong, N.T.; Dat, L.Q.; Son, L.H.; Hoa, N.D.; Ali, M.; Smarandache, F. Dynamic interval valued neutrosophic set: Modeling decision making in dynamic environments. *Comput. Ind.* **2019**, *108*, 45–52. [CrossRef]
35. López, L.F.D.M.; Blas, N.G.; Arteta, A. The optimal combination: Grammatical swarm, particle swarm optimization and neural networks. *J. Comput. Sci.* **2012**, *3*, 46–55. [CrossRef]
36. Albert, A.A.; Blas, N.G.; de Mingo López, L.F. Natural combination to trade in the stock market. *Soft Comput.* **2016**, *20*, 2433–2450. [CrossRef]
37. López, L.F.M.; Blas, N.G.; Albert, A.A. Multidimensional knapsack problem optimization using a binary particle swarm model with genetic operations. *Soft Comput.* **2018**, *22*, 2567–2582. [CrossRef]
38. Angeloudis, P.; Bell, M.G. An uncertainty-aware AGV assignment algorithm for automated container terminals. *Transp. Res. E: Logist. Transp. Rev.* **2010**, *46*, 354–366. [CrossRef]
39. Luo, S.; Ren, B. *The Monitoring and Managing Application of Cloud Computing Based on Internet of Things*; Elsevier North-Holland, Inc.: New York, NY, USA, 2016.
40. Gu, S.; Hua, J.; Lv, T. Evaluation of customer satisfaction of "Door-to-Door" whole-process logistic service with interval-valued intuitionistic fuzzy information. *J. Intell. Fuzzy Syst.* **2016**, *30*, 2487–2495. [CrossRef]
41. Zhang, Y.; Zou, D.; Zheng, J.; Fang, X.; Luo, H. Formation mechanism of quick emergency response capability for urban rail transit: Inter-organizational collaboration perspective. *Adv. Mech. Eng.* **2016**, *8*. [CrossRef]
42. Cho, H.; Choi, H.; Lee, W.; Jung, Y.; Baek, Y. LITeTag: Design and implementation of an RFID system for IT-based port logistics. *J. Commun.* **2006**, *1*, 48–57. [CrossRef]
43. Wang, L. Study on Port Logistics Marketing under the Environment of Supply Chain. *Int. J. Bus. Manag.* **2011**, *6*, 267. [CrossRef]
44. Rudjanakanoknad, J.; Suksirivoraboot, W.; Sukdanont, S. Evaluation of International Ports in Thailand through Trade Facilitation Indices from Freight Forwarders. *Procedia—Soc. Behav. Sci.* **2014**, *111*, 1073–1082. [CrossRef]
45. Lai, G.; Debo, L.G.; Sycara, K. Sharing inventory risk in supply chain: The implication of financial constraint. *Omega* **2009**, *37*, 811–825. [CrossRef]

46. Peris-Mora, E.; Diez Orejas, J.M.; Subirats, A.; Ibáñez, S.; Alvarez, P. Development of a system of indicators for sustainable port management. *Mar. Pollut. Bull.* **2005**, *50*, 1649–1660. [CrossRef] [PubMed]
47. Zhao, M.; Zhang, Y.; Ma, W.; Fu, Q.; Yang, X.; Li, C.; Zhou, B.; Yu, Q.; Chen, L. Characteristics and ship traffic source identification of air pollutants in China's largest port. *Atmos. Environ.* **2013**, *64*, 277–286. [CrossRef]
48. Chen, J.; Wan, Z.; Zhang, F.; Park, N.-K.; Zheng, A.; Zhao, J. Evaluation and comparison of the development performances of typical free trade port zones in China. *Transp. Res. Part A Policy Pract.* **2018**, *118*, 506–526. [CrossRef]
49. Chen, J.; Huang, T.; Xie, X.; Lee, P.T.-W.; Hua, C. Constructing Governance Framework of a Green and Smart Port. *J. Mar. Sci. Eng.* **2019**, *7*, 83. [CrossRef]
50. Wang, H.; Smarandache, F.; Zhang, Y.-Q.; Sunderraman, R. *Interval Neutrosophic Sets and Logic: Theory and Applications in Computing*; Hexis: Frontignan, France, 2005.

symmetry

MDPI

Article

A Refined Approach for Forecasting Based on Neutrosophic Time Series

Mohamed Abdel-Basset [1,*], Victor Chang [2], Mai Mohamed [1] and Florentin Smarandache [3]

1 Department of Operations Research, Faculty of Computers and Informatics, Zagazig University, Sharqiyah 44519, Egypt; mmgaafar@zu.edu.eg
2 School of Electronics and Computer Science, University of Southampton, Southampton SO17 1BJ, UK; victorchang.research@gmail.com
3 Mathematics Department, University of New Mexico, Gallup, NM 87301, USA; fsmarandache@gmail.com
* Correspondence: analyst_mohamed@yahoo.com

Received: 3 February 2019; Accepted: 26 March 2019; Published: 1 April 2019

Abstract: This research introduces a neutrosophic forecasting approach based on neutrosophic time series (NTS). Historical data can be transformed into neutrosophic time series data to determine their truth, indeterminacy and falsity functions. The basis for the neutrosophication process is the score and accuracy functions of historical data. In addition, neutrosophic logical relationship groups (NLRGs) are determined and a deneutrosophication method for NTS is presented. The objective of this research is to suggest an idea of first-and high-order NTS. By comparing our approach with other approaches, we conclude that the suggested approach of forecasting gets better results compared to the other existing approaches of fuzzy, intuitionistic fuzzy, and neutrosophic time series.

Keywords: neutrosophic time series; triangular neutrosophic number; neutrosophic logical relationship; neutrosophic logical relationship groups

1. Introduction

There are different methods in the literature on fuzzy and intuitionistic fuzzy time series methods to forecast future values. The major difference between traditional and fuzzy time series is that the values of traditional time series are presented in numbers, whereas the values in fuzzy time series are fuzzy sets or linguistic values with real meanings. In intuitionistic fuzzy time series, the values are intuitionistic fuzzy sets or linguistic values. The first method in literature for forecasting future values based on fuzzy time series was introduced by Song and Chissom [1]. They also applied time-variant and time-invariant models for forecasting the enrollment data at the University of Alabama [1,2]. The identification of fuzzy relationship and the defuzzification process in both models were the main steps for calculating forecasted values. In time variant fuzzy time series it is proposed that autocorrelation is dependent due to the time, while in time invariant it is proposed that autocorrelation is independent due to the time.

The term "fuzzy relationship" means a collection of fuzzy sets which are caused only by other sets. In addition, the "defuzzification" process means converting the fuzzy values into crisp ones. Furthermore, a straightforward approach for time series forecasting was presented by Chen [3] by using uncomplicated arithmetic computations. To enhance the accuracy of forecasted outputs, some papers suggested various methods on fuzzy time series (FTS) forecasting [4–7]. A high-order FTS method was also presented by Chen [8] and Singh [9], and a method of bivariate fuzzy time series analysis for the forecasting of a stock index was introduced by Hsu et al. [10]. Furthermore, a framework developed for evaluation and forecasting based on the fuzzy NEAT F-PROMETHEE method was presented by Ziemba and Becker [11] for taking into account the uncertainty of input data, which is particularly burdened with the forecast values of the information and communication technologies development indicators.

The concept of fuzzy set was introduced by Zadeh [12], and it was generalized by Atanassov [13] to intuitionistic fuzzy set (IFS) to make it more suitable to handle ambiguity. The IFS considers both the membership (truth) and non-membership (falsity) degrees. However, the fuzzy set considers only the membership degree. Recently, the IFS was used for handling the fuzzy time series forecasting by Gangwar and Kumar [14] and Wang et al. [15]. In addition, the notion of intuitionistic fuzzy time series (IFTS)was employed in forecasting, as in [16–18]. Several researchers [19,20] proposed forecasting models using a genetic algorithm, or suggested a method of forecasting based on aggregated FTS and particle swarm optimization [21]. A novel method of forecasting based on hesitant fuzzy set was proposed by Bisht and Kumar [22], and fuzzy descriptor models for earthquakes was introduced by Bahrami and Shafiee [23]. A heuristic adaptive-order IFTS forecasting model was presented by Wang et al. [24]. Subsequently, Abhishekh et al. [25,26] presented a weighted type 2 FTS and score function-based IFTS forecasting approach. Moreover, Abhishekh and Kumar [27] suggested an approach for forecasting rice production in the area of FTS.

Since the accuracy rates of forecasting in the previous approaches are not good enough in the field of fuzzy and intuitionistic fuzzy time series, we introduce the notion of first- and high-order neutrosophic time series data for this research. Additionally, with the growing need to represent vague and random information, neutrosophic set (NS) theory [28] is an effective extension of fuzzy and intuitionistic fuzzy set theories. Smarandache [29] suggested NSs, which consist of truth membership function, indeterminacy membership function, and falsity membership function, as a better representation of reality. Neutrosophic sets received wide attention, as well as benefitting from various practical applications in diverse fields [30–39]. However, there are only two recent research papers published in the forecasting field (e.g., for stock market analysis). Guan et al. [40] proposed a new forecasting model based on multi-valued neutrosophic sets and two-factor third-order fuzzy logical relationships to forecast the stock market. Subsequently, Guan et al. [41] proposed a new forecasting method based on high-order fluctuation trends and information entropy.

The aim of this research is to enhance accuracy rates of forecasting in the area of fuzzy, intuitionistic fuzzy, and neutrosophic time series (NTS). In this research, we present the notion of forecasting based on first-and high-order NTS data by determining the suitable length of neutrosophic numbers that influence on expected values. We also suggest a neutrosophication of historical time series data, based on the biggest score function (i.e., the maximum value of score function), and define neutrosophic logical relationship groups (NLRGs) for obtaining forecasted outputs. The suggested approach of neutrosophic time series forecasting has been validated and compared with different existing models for showing its superiority.

The remaining parts of this research are organized as follows. The essential concepts of neutrosophic set and neutrosophic time series are briefly presented in Section 2. Section 3 presents the proposed neutrosophic time series method for the forecasting process. Section 4 validates the proposed method by applying it to two numerical examples for showing its effectiveness; a comparison with other existing methods is presented. Finally, Section 5 concludes the research and determines future trends.

2. Some Basic Definitions of Neutrosophic Set and Neutrosophic Time Series

Neutrosophic time series is a concept for solving forecasting problems using neutrosophic concepts. In this section, we present the basic concepts of the neutrosophic set and of the neutrosophic time series (NTS).

Definition 1. *Let X be a finite universal set. A neutrosophic set N in X is an object having the following form:* $N = \{\langle x, T_N(x), I_N(x), F_N(x)\rangle | x \in X\}$, *where* $T_N(x) : X \to [0,1]$ *determines the degree of truth membership function,*

$I_N(x): X \to [0,1]$ *determines the degree of indeterminacy, and function* $F_N(x): X \to [0,1]$ *determines the degree of non-membership or falsity function. For every* $x \in X, 0^- \le T_N(x) + I_N(x) + F_N(x) \le 3^+$ [29].

Definition 2. *A single valued triangular neutrosophic number* $\widetilde{N} = \langle (n_1, n_2, n_3); T_{\widetilde{N}}, I_{\widetilde{N}}, F_{\widetilde{N}} \rangle$ *is a special neutrosophic set on the real number set R whose truth (membership), indeterminacy, and falsity (non-membership) degrees are as follows* [29]:

$$T_{\widetilde{N}}(x) = \begin{cases} T_{\widetilde{N}}\left(\frac{x-n_1}{n_2-n_1}\right) & (n_1 \le x \le n_2) \\ T_{\widetilde{N}} & (x = n_2) \\ T_{\widetilde{N}}\left(\frac{n_3-x}{n_3-n_2}\right) & (n_2 < x \le n_3) \\ 0 & \textbf{otherwise}, \end{cases} \tag{1}$$

$$I_{\widetilde{N}}(x) = \begin{cases} \frac{(n_2-x+I_{\widetilde{N}}(x-n_1))}{(n_2-n_1)} & (n_1 \le x \le n_2\ ) \\ I_{\widetilde{N}} & (x = n_2) \\ \frac{(x-n_2+I_{\widetilde{N}}(n_3-x))}{(n_3-n_2)} & (n_2 < x \le n_3) \\ 1 & \textit{otherwise}, \end{cases} \tag{2}$$

$$F_{\widetilde{N}}(x) = \begin{cases} \frac{(n_2-x+F_{\widetilde{N}}(x-n_1))}{(n_2-n_1)} & (n_1 \le x \le n_2\ ) \\ F_{\widetilde{N}} & (x = n_2) \\ \frac{(x-n_2+F_{\widetilde{N}}\ (n_3-x))}{(n_3-n_2)} & (n_2 < x \le n_3) \\ 1 & \text{otherwise}, \end{cases} \tag{3}$$

where $0 \le T_{\widetilde{N}} \le 1, 0 \le I_{\widetilde{N}} \le 1, 0 \le F_{\widetilde{N}} \le 1, 0 \le T_{\widetilde{N}} + I_{\widetilde{N}} + F_{\widetilde{N}} \le 3$, $n_1, n_2, n_3 \in R$, and being the lower, median, and upper values of the triangular neutrosophic number.

Definition 3. *Let X and Y be two finite universal sets. A neutrosophic relation R from X to Y is a neutrosophic set in the direct product space X to Y:*

$$R = \{\langle (x,y), T_N(x,y), I_N(x,y), F_N(x) \rangle | (x,y) \in X \times Y\}$$

where $0^- \le T_N(x,y) + I_N(x,y) + F_N(x,y) \le 3^+, \forall (x,y) \in X \times Y$ for $T_N(x,y) \to [0,1]$, $I_N(x,y) \to [0,1]$, and $F_N(x,y) \to [0,1] : X \times Y \to [0,1]$.

Definition 4. *Let* $X(t)(t = 1,2,\ldots,)$*, a subset of* $\mathbf{R}$*, be the universe of discourse on which neutrosophic sets* $f_i(t) = \langle T_N(x,y), I_N(x,y), F_N(x) \rangle (i = 1,2,\ldots)$ *are defined.* $F(t) = \{f_1(x), f_2(x), \ldots\}$ *is a collection of* $f_i(t)$ *and it defines a neutrosophic time series on* $X(t)(t = 0,1,2,\ldots)$.

Definition 5. *If there exists a neutrosophic relationship* $R(t-1,t)$*, such that* $F(t) = F(t-1) \times R(t-1,t)$*, where '×' represents an operator, then* $F(t)$ *is said to be caused by* $F(t-1)$*. The relationship between* $F(t)$ *and* $F(t-1)$ *is symbolized by* $F(t-1) \to F(t)$.

Definition 6. *Let* $F(t)$ *caused by* $F(t-1)$ *only and symbolized by* $F(t-1) \to F(t)$*; consequently, a neutrosophic relationship exists between* $F(t)$ *and* $F(t-1)$ *that is denoted as* $F(t) = F(t-1) \times R(t-1,t)$*, since* R *is a first-order model of* $F(t)$*. The* $F(t)$ *is a time-invariant neutrosophic time series if* $R(t-1,t)$ *is independent of time* t, $R(t,t-1) = R(t-1,t-2) \forall\ t$*. Otherwise,* $F(t)$ *is called a time-variant neutrosophic time series.*

Definition 7. Let $F(t-1) = \widetilde{N}_i$ and $(t) = \widetilde{N}_j$; a neutrosophic logical relationship (NLR) can be defined as $\widetilde{N}_i \to \widetilde{N}_j$, where $\widetilde{N}_i, \widetilde{N}_j$ are the current and next state of NLR. Since $F(t)$ is occurred by more than one neutrosophic set $F(t-n), F(t-n+1), \ldots F(t-1)$, then the neutrosophic relationship is represented by

$\widetilde{N}_{i1}, \widetilde{N}_{i2}, \ldots, \widetilde{N}_{in} \rightarrow \widetilde{N}_j$, where $F(t-n) = \widetilde{N}_{i1}, F(t-n+1) = \widetilde{N}_{i2}$. The relationship is called high-order neutrosophic time series model.

3. Neutrosophic Time Series Forecasting Algorithm

Because a neutrosophic set plays a significant role in decision-making and data analysis problems by handling vague, inconsistent, and incomplete information [30–39], we propose in this section an enhanced approach of forecasting using the concept of neutrosophic time series (NTS).

The stepwise method of the suggested algorithm of neutrosophic time series forecasting is dependent on historical time series data.

3.1. The Proposed Method of Forecasting Based on First-Order NTS Data

Step 1: By depending on the range of the existing data set, determine the universe of discourse U as follows:

- Select the largest D_l and the smallest D_s from all available data D_v, then

$$U = [D_s - D_1,\ D_l + D_2] \tag{4}$$

where D_1 and D_2 are two proper positive numbers assigned by experts in the problem domain. So, we can define D_1, D_2 as the values by which the range of the universe of discourse is less than the specified value of D_s for the first (i.e., D_1) or greater than the specified value of D_l for the latter (i.e., D_2).

Step 2: Create a partition of the universe of discourse, to m triangular neutrosophic numbers as follows:

- Decide the suitable length (Le) of available time series data:
  - ○ Among the value D_{v-1}, D_v, calculate all absolute differences and take the average of these differences.
  - ○ Consider half the average as the initial length.
  - ○ According to the obtained result, use the base mapping table [42] to determine the base for the length of intervals.
  - ○ Round the result to determine the appropriate length of neutrosophic numbers.
  - ○ For example: if we have these time series data $30, 50, 80, 120, 100, 70$, then the absolute differences will be $20, 30, 40, 20, 30$, and the average of these values $= 28$. Then, half of the average will be 14 and this is the initial value of length. By using the base mapping table [42], the base for length $=$ 10 because 14 locates in the range $[11 - 100]$ and by rounding the length 14 by the base ten, the result will equal 10. Here, the appropriate length of neutrosophic numbers equals 10.
- Compute the number of triangular neutrosophic numbers (m) as follows:

$$m = \frac{D_l + D_2 - D_s + D_1}{le} \tag{5}$$

Step 3: According to the numbers of triangular neutrosophic numbers on the universe of discourse and determined length (le), begin to construct the triangular neutrosophic numbers. The triangular neutrosophic numbers are $\widetilde{N}_1, \widetilde{N}_2, \ldots, \widetilde{N}_m$.

As we illustrated in Definition 2, each triangular neutrosophic number consists of two parts which are the value of the triangular neutrosophic number (lower, median, upper) and the degree of confirmation (truth/membership degree T, indeterminacy degree I, falsity/non-membership degree F). The initial value of T, I, F must be determined by experts according to the existing problem.

Step 4: Make a neutrosophication process of the existing data:

For $i, j = 1, 2, \ldots, v$ (the end of data):

Rule 1: Use this equation to calculate the score degree, and if the score degree of two neutrosophic numbers is not equal for any data, then choose the maximum value of the score degree:

$$SC_{\widetilde{N}_j}(x_i) = 2 + T_{\widetilde{N}_j}(x_i) - I_{\widetilde{N}_j}(x_i) - F(x_i) \tag{6}$$

Then, select $SC_{\widetilde{N}_k} = max\ (SC_{\widetilde{N}_k}, SC_{\widetilde{N}_k}, \ldots, SC_{\widetilde{N}_k})$ for $x_i, i = 1, 2 \ldots, n, 1 \leq k \leq n$, and assign the neutrosophic number $\widetilde{N}_k$ to x_i.

Rule 2: If two neutrosophic numbers have the same score degree, then use the following equation to calculate the score degree, and select the minimum accuracy degree:

$$AC_{\widetilde{N}_j}(x_i) = 2 + T_{\widetilde{N}_j}(x_i) - I_{\widetilde{N}_j}(x_i) + F(x_i) \tag{7}$$

Furthermore, $AC_{\widetilde{N}_k} = min\ (AC_{\widetilde{N}_k}, AC_{\widetilde{N}_k}, \ldots, AC_{\widetilde{N}_k})$ for x_i, $i = 1, 2 \ldots, n, 1 \leq k \leq n$; assign the neutrosophic number $\widetilde{N}_k$ to x_i.

Step 5: Construct the neutrosophic logical relationships (NLRs) as follows:

If $\widetilde{N}_j$,$\widetilde{N}_k$ are the neutrosophication values of year n and year $n+1$, respectively, then the NLR is symbolized as $\widetilde{N}_j \rightarrow \widetilde{N}_k$.

Step 6: Based on the NLR, begin to establish the neutrosophic logical relationship groups (NLRGs).

Step 7: Calculate the forecasted values as follows:

Rule 1: If the neutrosophication value of $data_i$ is $\widetilde{N}_k$ and it is not caused by any other neutrosophication values and, by looking at the NLRG of this value, you cannot find the value which it depends on (i.e., $\neq\rightarrow \widetilde{N}_k$), then the forecasted value in this case will equal—(i.e., leave it empty). The $\neq$ symbol means no value.

Rule 2: If the neutrosophication value of $data_i$ is $\widetilde{N}_k$ and it is caused by $\widetilde{N}_j$ ($\widetilde{N}_j \rightarrow \widetilde{N}_k$), then look at NLRG of $\widetilde{N}_j$, and

- If NLRG of $\widetilde{N}_j$ is empty (i.e., $\widetilde{N}_j \rightarrow \varnothing$, or $\widetilde{N}_j \rightarrow \widetilde{N}_j$), then the forecasted value is the middle value of $\widetilde{N}_j$.
- If NLRG of $\widetilde{N}_j$ is one-to-one (i.e., $\widetilde{N}_j \rightarrow \widetilde{N}_k$), then the forecasted value is the middle value of $\widetilde{N}_k$.
- If NLRG of $\widetilde{N}_j$ is one-to-many (i.e., $\widetilde{N}_j \rightarrow \widetilde{N}_{k1}, \widetilde{N}_{k2}, \ldots, \widetilde{N}_{kn}$), then the forecasted value is the average of the middle values of $\widetilde{N}_{k1}, \widetilde{N}_{k2}, \ldots, \widetilde{N}_{kn}$.

Step 8: Use the following equations to calculate the forecasting error:

$$\text{Root mean square error (RMSE)} = \sqrt{\frac{\sum_{i=1}^{n} (\text{Forecast}_i - \text{Actual}_i)^2}{n}}, \tag{8}$$

$$\text{Forecasting error} = \frac{|\text{ Forecast} - \text{Actual}|}{Actual} \times 100, \tag{9}$$

$$\text{Average forecasting error (AFE) (\%)} = \frac{\text{Sum of forecasting error}}{number\ of\ errors} \times 100. \tag{10}$$

3.2. *The Proposed Method of Forecasting Based on High-Order NTS Data*

We can also apply the proposed method of forecasting based on high-order NTS data:

- All steps from 1 to 4 are the same as previously, but in step 5 we begin to construct the neutrosophic logical relationships (NLRs) of the nth order NTS, where $n \geq 2$.
- Based on the NLR of the nth order, NTS begin to establish the neutrosophic logical relationship groups (NLRGs).
- Calculate the forecasted values as follows:

- Rule 1: If the neutrosophication values of $data_i$ is $\widetilde{N}_l$ and it is not caused by any other neutrosophication values and, by looking at the NLRG of this value, you cannot find the values which it depends on (i.e., $\neq\rightarrow \widetilde{N}_l$), then the forecasted value in this case will equal—(i.e., leave it empty). The $\neq$ symbol means no value.
- Rule 2: If the neutrosophication value of $data_i$ is $\widetilde{N}_l$ and it is caused by $\widetilde{N}_{in}, \widetilde{N}_{i(n-1)}, \ldots,$ $\widetilde{N}_{ik}$ (i.e., $\widetilde{N}_{in}, \widetilde{N}_{i(n-1)}, \ldots, \widetilde{N}_{ik} \rightarrow \widetilde{N}_l$), then look at the NLRG of $\widetilde{N}_{in}, \widetilde{N}_{i(n-1)}, \ldots, \widetilde{N}_{ik}$, and

  - If $\widetilde{N}_{in}, \widetilde{N}_{i(n-1)}, \ldots, \widetilde{N}_{ik} \rightarrow \varnothing$, then the forecasted value at this year is the average of the middle value of $\widetilde{N}_{in}, \widetilde{N}_{i(n-1)}, \ldots, \widetilde{N}_{ik}$.
  - If $\widetilde{N}_{in}, \widetilde{N}_{i(n-1)}, \ldots, \widetilde{N}_{ik} \rightarrow \widetilde{N}_j$, then the forecasted value at this year is the middle value of $\widetilde{N}_j$.
  - If $\widetilde{N}_{in}, \widetilde{N}_{i(n-1)}, \ldots, \widetilde{N}_{ik} \rightarrow \widetilde{N}_j, \widetilde{N}_{j1}, \widetilde{N}_{j2}$, then the forecasted value at this year is the average of the middle value of $\widetilde{N}_j, \widetilde{N}_{j1}, \widetilde{N}_{j2}$.

4. Numerical Examples

In this section, we solve two numerical examples and compare outputs with other existing methods for verifying the applicability and superiority of the suggested method.

4.1. Numerical Example 1

In this example, the suggested approach is implemented on the benchmarking time series data of student enrollments at the University of Alabama from year 1971 to 1992 adopted from [26]. The steps are as follows:

Step 1: Let the two proper positive numbers D_1 and D_2 be 5 and 13, determined by the expert. By selecting the largest and the smallest observation from all available data which are presented in Table 1, then $D_l = 19,337$ and $D_s = 13,055$, respectively. Consequently, the universe of discourse U = $[13,055 - 5,\ 19,337 + 13] = [13,050,\ 19,350]$.

Step 2: Create a partition of the universe of discourse, to m triangular neutrosophic numbers, as follows:

- Determine the suitable length (Le) of available time series data:
  - From Table 1, the average of absolute differences = 510.3.
  - The initial length $= \frac{510.3}{2} = 255.15$.
  - By using the base mapping table [42], the base for length of intervals = 100, since it is located in the range $[101, 1000]$.
  - By rounding 255.15 with regard to base 100, then the appropriate length of neutrosophic numbers = 300.
- Compute the number of triangular neutrosophic numbers (m) as follows:

$$m = \frac{19350 - 13050}{300} = 21.$$

Then, we can partition U into 21 triangular neutrosophic numbers with length $=$ 300.

Step 3: According to the number of triangular neutrosophic numbers on the universe of discourse and determined length (*le*), begin to construct the triangular neutrosophic numbers as follows:

$$\begin{aligned}
\widetilde{N}_1 &= \langle 13050, 13350, 13650; 0.90, 0.10, 0.10\rangle,\\
\widetilde{N}_2 &= \langle 13350, 13650, 13950; 0.80, 0.20, 0.10\rangle,\\
\widetilde{N}_3 &= \langle 13650, 13950, 14250; 0.90, 0.20, 0.10\rangle,\\
\widetilde{N}_4 &= \langle 13950, 14250, 14550; 0.85, 0.15, 0.10\rangle,\\
\widetilde{N}_5 &= \langle 14250, 14550, 14850; 0.75, 0.10, 0.30\rangle,\\
\widetilde{N}_6 &= \langle 14550, 14850, 15150; 0.90, 0.10, 0.10\rangle,\\
\widetilde{N}_7 &= \langle 14850, 15150, 15450; 0.60, 0.30, 0.40\rangle,\\
\widetilde{N}_8 &= \langle 15150, 15450, 15750; 0.80, 0.20, 0.20\rangle,\\
\widetilde{N}_9 &= \langle 15450, 15750, 16050; 0.70, 0.20, 0.30\rangle,\\
\widetilde{N}_{10} &= \langle 15750, 16050, 16350; 0.90, 0.10, 0.30\rangle,\\
\widetilde{N}_{11} &= \langle 16050, 16350, 16650; 0.85, 0.10, 0.15\rangle,\\
\widetilde{N}_{12} &= \langle 16350, 16650, 16950; 0.80, 0.20, 0.20\rangle,\\
\widetilde{N}_{13} &= \langle 16650, 16950, 17250; 0.90, 0.10, 0.30\rangle,\\
\widetilde{N}_{14} &= \langle 16950, 17250, 17550; 0.90, 0.10, 0.30\rangle,\\
\widetilde{N}_{15} &= \langle 17250, 17550, 17850; 0.75, 0.10, 0.30\rangle,\\
\widetilde{N}_{16} &= \langle 17550, 17850, 18150; 0.65, 0.20, 0.35\rangle,\\
\widetilde{N}_{17} &= \langle 17850, 18150, 18450; 0.90, 0.10, 0.10\rangle,\\
\widetilde{N}_{18} &= \langle 18150, 18450, 18750; 0.90, 0.10, 0.10\rangle,\\
\widetilde{N}_{19} &= \langle 18450, 18750, 19050; 0.60, 0.20, 0.30\rangle,\\
\widetilde{N}_{20} &= \langle 18750, 19050, 19350; 0.90, 0.10, 0.10\rangle,\\
\widetilde{N}_{21} &= \langle 19050, 19350, 19350; 0.90, 0.10, 0.10\rangle.
\end{aligned}$$

Step 4: Make a neutrosophication of the available time series data:

The first value of actual enrollments is 13,055 which is located only in the range of triangular neutrosophic number $\widetilde{N}_1$, then the neutrosophication value of 13,055 is $\widetilde{N}_1$ as in Table 1.

Also, the second value of actual enrollments (i.e., 13,563) locates in the range of triangular neutrosophic numbers $\widetilde{N}_1 = \langle 13050, 13350, 13650; 0.90, 0.10, 0.10\rangle$ and $\widetilde{N}_2 = \langle 13350, 13650, 13950; 0.80, 0.20, 0.10\rangle$.

Then, we must select the highest score degree of 13,563 as follows:

The membership, indeterminacy, and non-membership degrees of this value are calculated by using Equations (1)–(3) as follows:

$$T_{\widetilde{N}_1}(13563) = 0.261,\ \mathrm{I}_{\widetilde{N}_1}(13563) = 0.739, \mathrm{F}_{\widetilde{N}_1}(13563) = 0.739.$$

We must also calculate membership, indeterminacy, and non-membership degrees of 13,563 according to $\widetilde{N}_2 = \langle 13350, 13650, 13950; 0.80, 0.20, 0.10\rangle$ as follows:

$$T_{\widetilde{N}_2}(13563) = 0.568, \mathrm{I}_{\widetilde{N}_2}(13563) = 0.432,\ \mathrm{F}_{\widetilde{N}_2}(13563) = 0.361.$$

In this case, we must calculate the score degree of 13563 in both $\widetilde{\mathrm{N}}_1$ and $\widetilde{\mathrm{N}}_2$ and select the maximum value.

$$SC_{\widetilde{N}_1}(13563) = 2 + 0.262 - 0.739 - 0.739 = 0.783,$$
$$\text{and } SC_{\widetilde{N}_2}(13563) = 2 + 0.568 - 0.432 - 0.361 = 1.775.$$

Since the score degree of 13563 in $\widetilde{N}_2$ is greater than $\widetilde{N}_1$, then the neutrosophication value of 13563 is $\widetilde{N}_2$, as in Table 1.

We will apply the previous steps on the remaining data as follows:

The value 13867 locates in the range of $\widetilde{N}_2 = \langle 13350, 13650, 13950; 0.80, 0.20, 0.10\rangle$, and $\widetilde{N}_3 = \langle 13650, 13950, 14250; 0.90, 0.20, 0.10\rangle$.

Then

$$\begin{gathered} T_{\widetilde{N}_2}(13867) = 0.221, I_{\widetilde{N}_2}(13867) = 0.156, F_{\widetilde{N}_2}(13867) = 0.751. \\ T_{\widetilde{N}_3}(13867) = 0.651, I_{\widetilde{N}_3}(13867) = 0.421, F_{\widetilde{N}_3}(13867) = 0.349, \\ SC_{\widetilde{N}_2}(13867) = 2 + 0.221 - 0.156 - 0.751 = 1.314, \\ \text{and } SC_{\widetilde{N}_3}(13867) = 2 + 0.651 - 0.421 - 0.349 = 1.881. \end{gathered}$$

So, the neutrosophication value of 13867 is $\widetilde{N}_3$.

Also, the value of 14,696 locates in the range of $\widetilde{N}_5 = \langle 14250, 14550, 14850; 0.75, 0.10, 0.30\rangle$, $\widetilde{N}_6 = \langle 14550, 14850, 15150; 0.90, 0.10, 0.10\rangle$, then

$$\begin{gathered} T_{\widetilde{N}_5}(14696) = 0.385, I_{\widetilde{N}_5}(14696) = 0.538, F_{\widetilde{N}_5}(14696) = 0.641. \\ T_{\widetilde{N}_6}(14696) = 0.438, I_{\widetilde{N}_6}(14696) = 0.562, F_{\widetilde{N}_6}(14696) = 0.562. \\ SC_{\widetilde{N}_5}(14696) = 2 + 0.385 - 0.538 - 0.641 = 1.206, \\ SC_{\widetilde{N}_6}(14696) = 2 + 0.438 - 0.562 - 0.562 = 1.314. \end{gathered}$$

So, the neutrosophication value of 14,696 is $\widetilde{N}_6$.

The value 15,460 locates in the range of $\widetilde{N}_8 = \langle 15150, 15450, 15750; 0.80, 0.20, 0.20\rangle$, and $\widetilde{N}_9 = \langle 15450, 15750, 16050; 0.70, 0.20, 0.30\rangle$, then

$$\begin{gathered} T_{\widetilde{N}_8}(15460) = 0.773, I_{\widetilde{N}_8}(15460) = 0.226, F_{\widetilde{N}_8}(15460) = 0.226. \\ T_{\widetilde{N}_9}(15460) = 0.023, I_{\widetilde{N}_9}(15460) = 0.973, F_{\widetilde{N}_9}(15460) = 0.973. \\ SC_{\widetilde{N}_8}(15460) = 2 + 0.773 - 0.226 - 0.226 = 2.321, \\ \text{and } SC_{\widetilde{N}_9}(15460) = 2 + 0.023 - 0.973 - 0.976 = 0.074. \end{gathered}$$

So, the neutrosophication value of $15,460$ is $\widetilde{N}_8$.

The value of 15,311 locates in the range of $\widetilde{N}_7 = \langle 14850, 15150, 15450; 0.60, 0.30, 0.40\rangle$ and $\widetilde{N}_8 = \langle 15150, 15450, 15750; 0.80, 0.20, 0.20\rangle$, then

$$\begin{gathered} T_{\widetilde{N}_7}(15311) = 0.278, I_{\widetilde{N}_7}(15311) = 0.675, F_{\widetilde{N}_7}(15311) = 0.722. \\ SC_{\widetilde{N}_7}(15311) = 2 + 0.278 - 0.675 - 0.722 = 0.881. \\ T_{\widetilde{N}_8}(15311) = 0.429, I_{\widetilde{N}_8}(15311) = 0.570, F_{\widetilde{N}_8}(15311) = 0.570. \\ SC_{\widetilde{N}_8}(15311) = 2 + 0.429 - 0.570 - 0.570 = 1.289. \end{gathered}$$

So, the neutrosophication value of $15,311$ is $\widetilde{N}_8$.

The value of 15,603 locates in the range of $\widetilde{N}_8 = \langle 15150, 15450, 15750; 0.80, 0.20, 0.20\rangle$ and $\widetilde{N}_9 = \langle 15450, 15750, 16050; 0.70, 0.20, 0.30\rangle$, then

$$\begin{gathered} T_{\widetilde{N}_8}(15603) = 0.392, I_{\widetilde{N}_8}(15603) = 0.608, F_{\widetilde{N}_8}(15603) = 0.608. \\ SC_{\widetilde{N}_8}(15603) = 2 + 0.392 - 0.608 - 0.608 = 1.176. \\ T_{\widetilde{N}_9}(15603) = 0.357, I_{\widetilde{N}_9}(15603) = 0.592, F_{\widetilde{N}_9}(15603) = 0.643. \\ SC_{\widetilde{N}_9}(15603) = 2 + 0.357 - 0.592 - 0.643 = 1.122. \end{gathered}$$

So, the neutrosophication value of $15,603$ is $\widetilde{N}_8$.

The value of 15,861 locates in the range of $\widetilde{N}_9 = \langle 15450, 15750, 16050; 0.70, 0.20, 0.30\rangle$, and $\widetilde{N}_{10} = \langle 15750, 16050, 16350; 0.90, 0.10, 0.30\rangle$, then

$$\begin{aligned}
T_{\widetilde{N}_9}(15861) &= 0.441, I_{\widetilde{N}_9}(15861) = 0.496, F_{\widetilde{N}_9}(15861) = 0.559. \\
SC_{\widetilde{N}_9}(15861) &= 2 + 0.441 - 0.496 - 0.559 = 1.386. \\
T_{\widetilde{N}_{10}}(15861) &= 0.333, I_{\widetilde{N}_{10}}(15861) = 0.667, F_{\widetilde{N}_{10}}(15861) = 0.741. \\
SC_{\widetilde{N}_{10}}(15861) &= 2 + 0.333 - 0.667 - 0.741 = 0.925.
\end{aligned}$$

So, the neutrosophication value of 15861 is $\widetilde{N}_9$.

The value of 16,807 locates in the range of $\widetilde{N}_{12} = \langle 16350, 16650, 16950; 0.80, 0.20, 0.20\rangle$, $\widetilde{N}_{13} = \langle 16650, 16950, 17250; 0.90, 0.10, 0.30\rangle$ then,

$$\begin{aligned}
T_{\widetilde{N}_{12}}(16807) &= 0.381, I_{\widetilde{N}_{12}}(16807) = 0.618, F_{\widetilde{N}_{12}}(16807) = 0.618. \\
SC_{\widetilde{N}_{12}}(16807) &= 2 + 0.381 - 0.618 - 0.618 = 1.145. \\
T_{\widetilde{N}_{13}}(16807) &= 0.471, I_{\widetilde{N}_{13}}(16807) = 0.529, F_{\widetilde{N}_{13}}(16807) = 0.634. \\
SC_{\widetilde{N}_{13}}(16807) &= 2 + 0.471 - 0.529 - 0.634 = 1.308.
\end{aligned}$$

So, the neutrosophication value of 16807 is $\widetilde{N}_{13}$.

The value of 16919 locates in the range of $\widetilde{N}_{12} = \langle 16350, 16650, 16950; 0.80, 0.20, 0.20\rangle$, $\widetilde{N}_{13} = \langle 16650, 16950, 17250; 0.90, 0.10, 0.30\rangle$, then

$$\begin{aligned}
T_{\widetilde{N}_{12}}(16919) &= 0.063, I_{\widetilde{N}_{12}}(16919) = 0.917, F_{\widetilde{N}_{12}}(16919) = 0.917. \\
SC_{\widetilde{N}_{12}}(16919) &= 2 + 0.063 - 0.917 - 0.917 = 0.229. \\
T_{\widetilde{N}_{13}}(16919) &= 0.807, I_{\widetilde{N}_{13}}(16919) = 0.193, F_{\widetilde{N}_{13}}(16919) = 0.372. \\
SC_{\widetilde{N}_{13}}(16919) &= 2 + 0.807 - 0.193 - 0.372 = 2.24.
\end{aligned}$$

So, the neutrosophication value of 16919 is $\widetilde{N}_{13}$.

The value of 16388 locates in the range of $\widetilde{N}_{11} = \langle 16050, 16350, 16650; 0.85, 0.10, 0.15\rangle$, $\widetilde{N}_{12} = \langle 16350, 16650, 16950; 0.80, 0.20, 0.20\rangle$, then

$$\begin{aligned}
T_{\widetilde{N}_{11}}(16388) &= 0.742, I_{\widetilde{N}_{11}}(16388) = 0.214, F_{\widetilde{N}_{11}}(16388) = 0.257. \\
SC_{\widetilde{N}_{11}}(16388) &= 2 + 0.742 - 0.214 - 0.257 = 2.271. \\
T_{\widetilde{N}_{12}}(16388) &= 0.101, I_{\widetilde{N}_{12}}(16388) = 0.898, F_{\widetilde{N}_{12}}(16388) = 0.898. \\
SC_{\widetilde{N}_{12}}(16388) &= 2 + 0.101 - 0.898 - 0.898 = 0.305.
\end{aligned}$$

So, the neutrosophication value of 16388 is $\widetilde{N}_{11}$.

The value of 15433 locates in the range of $\widetilde{N}_7 = \langle 14850, 15150, 15450; 0.60, 0.30, 0.40\rangle$, and $\widetilde{N}_8 = \langle 15150, 15450, 15750; 0.80, 0.20, 0.20\rangle$, then

$$\begin{aligned}
T_{\widetilde{N}_7}(15433) &= 0.034, I_{\widetilde{N}_7}(15433) = 0.960, F_{\widetilde{N}_7}(15433) = 0.966. \\
SC_{\widetilde{N}_7}(15433) &= 2 + 0.034 - 0.960 - 0.966 = 0.108. \\
T_{\widetilde{N}_8}(15433) &= 0.754, I_{\widetilde{N}_8}(15433) = 0.245, F_{\widetilde{N}_8}(15433) = 0.245. \\
SC_{\widetilde{N}_8}(15433) &= 2 + 0.754 - 0.245 - 0.245 = 2.264.
\end{aligned}$$

So, the neutrosophication value of 15433 is $\widetilde{N}_8$.

The value of 15497 locates in the range of $\widetilde{N}_8 = \langle 15150, 15450, 15750; 0.80, 0.20, 0.20\rangle$ and $\widetilde{N}_9 = \langle 15450, 15750, 16050; 0.70, 0.20, 0.30\rangle$ then,

$$\begin{aligned}
T_{\widetilde{N}_8}(15497) &= 0.674,\ I_{\widetilde{N}_8}(15497) = 0.325,\ F_{\widetilde{N}_8}(15497) = 0.325. \\
SC_{\widetilde{N}_8}(15497) &= 2.024.
\end{aligned}$$

Also,

$$T_{\widetilde{N}_9}(15497) = 0.109, I_{\widetilde{N}_9}(15497) = 0.874,\ F_{\widetilde{N}_9}(15497) = 0.890.$$
$$SC_{\widetilde{N}_9}(15497) = 0.345.$$

So, the neutrosophication value of 15, 433 is $\widetilde{N}_8$.

The value of 15,145 locates in the range of $\widetilde{N}_6 = \langle 14550, 14850, 15150; 0.90, 0.10, 0.10 \rangle$, and $\widetilde{N}_7 = \langle 14850, 15150, 15450; 0.60, 0.30, 0.40 \rangle$, then

$$T_{\widetilde{N}_6}(15145) = 0.015,\ I_{\widetilde{N}_6}(15145) = 0.985,\ F_{\widetilde{N}_6}(15145) = 0.985.$$
$$SC_{\widetilde{N}_6}(15145) = 0.045.$$

Also,

$$T_{\widetilde{N}_7}(15145) = 0.59, I_{\widetilde{N}_7}(15145) = 0.311,\ F_{\widetilde{N}_7}(15145) = 0.41.$$
$$SC_{\widetilde{N}_7}(15145) = 1.869.$$

So, the neutrosophication value of 15,145 is $\widetilde{N}_7$.

The value of 15, 163 locates in the range of $\widetilde{N}_7 = \langle 14850, 15150, 15450; 0.60, 0.30, 0.40 \rangle$, $\widetilde{N}_8 = \langle 15150, 15450, 15750; 0.80, 0.20, 0.20 \rangle$, then

$$T_{\widetilde{N}_7}(15163) = 0.6,\ I_{\widetilde{N}_7}(15163) = 0.330,\ F_{\widetilde{N}_7}(15163) = 0.426.$$
$$SC_{\widetilde{N}_7}(15163) = 1.844.$$

Also

$$T_{\widetilde{N}_8}(15163) = 0.034,\ I_{\widetilde{N}_8}(15163) = 0.965,\ F_{\widetilde{N}_8}(15163) = 0.965.$$
$$SC_{\widetilde{N}_8}(15163) = 0.104.$$

So, the neutrosophication value of 15, 163 is $\widetilde{N}_7$.

The value of 15, 984 locates in the range of $\widetilde{N}_9 = \langle 15450, 15750, 16050; 0.70, 0.20, 0.30 \rangle$, $\widetilde{N}_{10} = \langle 15750, 16050, 16350; 0.90, 0.10, 0.30 \rangle$, then

$$T_{\widetilde{N}_9}(15984) = 0.154, I_{\widetilde{N}_9}(15984) = 0.824, F_{\widetilde{N}_9}(15984) = 0.846.$$
$$SC_{\widetilde{N}_9}(15984) = 0.484.$$

Also,

$$T_{\widetilde{N}_{10}}(15984) = 0.702, I_{\widetilde{N}_{10}}(15984) = 0.298, F_{\widetilde{N}_{10}}(15984) = 0.454,$$
$$SC_{\widetilde{N}_{10}}(15984) = 1.95.$$

So, the neutrosophication value of 15984 is $\widetilde{N}_{10}$.

The value of 16859 locates in the range of $\widetilde{N}_{12} = \langle 16350, 16650, 16950; 0.80, 0.20, 0.20 \rangle$, $\widetilde{N}_{13} = \langle 16650, 16950, 17250; 0.90, 0.10, 0.30 \rangle$, then

$$T_{\widetilde{N}_{12}}(16859) = 0.242, I_{\widetilde{N}_{12}}(16859) = 0.757, \quad F_{\widetilde{N}_{12}}(16859) = 0.757,$$
$$SC_{\widetilde{N}_{12}}(16859) = 0.728.$$

Also,

$$T_{\widetilde{N}_{13}}(16859) = 0.627,\ I_{\widetilde{N}_{13}}(16859) = 0.373, \quad F_{\widetilde{N}_{13}}(16859) = 0.512,$$
$$SC_{\widetilde{N}_{13}}(16859) = 1.442.$$

So, the neutrosophication value of 16859 is $\widetilde{N}_{13}$.

The value of 18150 locates in the range of $\widetilde{N}_{16} = \langle 17550, 17850, 18150; 0.65, 0.20, 0.35\rangle$, $\widetilde{N}_{17} = \langle 17850, 18150, 18450; 0.90, 0.10, 0.10\rangle$, then

$$T_{\widetilde{N}_{16}}(18150) = 0,\ I_{\widetilde{N}_{16}}(18150) = 1, \quad F_{\widetilde{N}_{16}}(18150) = 1,$$
$$SC_{\widetilde{N}_{16}}(18150) = 0.$$

Also,

$$T_{\widetilde{N}_{17}}(18150) = 0.90,\ I_{\widetilde{N}_{17}}(18150) = 0.1,\ F_{\widetilde{N}_{17}}(18150) = 0.1,$$
$$SC_{\widetilde{N}_{17}}(18150) = 2.7.$$

So, the neutrosophication value of 18150 is $\widetilde{N}_{17}$.

The value of 18970 locates in the range of $\widetilde{N}_{19} = \langle 18450, 18750, 19050; 0.60, 0.20, 0.30\rangle$, $\widetilde{N}_{20} = \langle 18750, 19050, 19350; 0.90, 0.10, 0.10\rangle$, then

$$T_{\widetilde{N}_{19}}(18970) = 0.16,\ I_{\widetilde{N}_{19}}(18970) = 0.786, \quad F_{\widetilde{N}_{19}}(18970) = 0.813.$$
$$SC_{\widetilde{N}_{19}}(18970) = 0.561.$$

Also,

$$T_{\widetilde{N}_{20}}(18970) = 0.66, I_{\widetilde{N}_{20}}(18970) = 0.34, \quad F_{\widetilde{N}_{20}}(18970) = 0.34.$$
$$SC_{\widetilde{N}_{20}}(18970) = 1.98.$$

So, the neutrosophication value of 18, 970 is $\widetilde{N}_{20}$.

The value of 19, 328 locates in the range of $\widetilde{N}_{20} = \langle 18750, 19050, 19350; 0.90, 0.10, 0.10\rangle$, $\widetilde{N}_{21} = \langle 19050, 19350, 19; 0.90, 0.10, 0.10\rangle$, then

$$T_{\widetilde{N}_{20}}(19328) = 0.066, I_{\widetilde{N}_{20}}(19328) = 0.992, \quad F_{\widetilde{N}_{20}}(19328) = 0.992.$$
$$SC_{\widetilde{N}_{20}}(19328) = 0.082.$$

Also,

$$T_{\widetilde{N}_{21}}(19328) = 0.834,\ I_{\widetilde{N}_{21}}(19328) = 0.166, \quad F_{\widetilde{N}_{21}}(19328) = 0.166.$$
$$SC_{\widetilde{N}_{21}}(19328) = 2.502.$$

So, the neutrosophication value of 19, 328 is $\widetilde{N}_{21}$.

The value of 19, 337 locates in the range of $\widetilde{N}_{20} = \langle 18750, 19050, 19350; 0.90, 0.10, 0.10\rangle$, $\widetilde{N}_{21} = \langle 19050, 19350, 19; 0.90, 0.10, 0.10\rangle$, then

$$T_{\widetilde{N}_{20}}(19337) = 0.039, I_{\widetilde{N}_{20}}(19337) = 0.961, \quad F_{\widetilde{N}_{20}}(19337) = 0.961.$$
$$SC_{\widetilde{N}_{20}}(19337) = 0.117.$$

Also,

$$T_{\widetilde{N}_{21}}(19337) = 0.861, I_{\widetilde{N}_{21}}(19337) = 0.139, \quad F_{\widetilde{N}_{21}}(19337) = 0.139.$$
$$SC_{\widetilde{N}_{21}}(19337) = 2.583.$$

So, the neutrosophication value of 19, 337 is $\widetilde{N}_{21}$.

Finally, the value of 18, 876 locates in the range of $\widetilde{N}_{19} = \langle 18450, 18750, 19050; 0.60, 0.20, 0.30\rangle$, $\widetilde{N}_{20} = \langle 18750, 19050, 19350; 0.90, 0.10, 0.10\rangle$, then

$$T_{\widetilde{N}_{19}}(18876) = 0.348, I_{\widetilde{N}_{19}}(18876) = 0.536, \quad F_{\widetilde{N}_{19}}(18876) = 0.594.$$
$$SC_{\widetilde{N}_{19}}(18876) = 1.218.$$

Also,

$$T_{\widetilde{N}_{20}}(18876) = 0.378,\ I_{\widetilde{N}_{20}}(18876) = 0.622, \quad F_{\widetilde{N}_{20}}(18876) = 0.622.$$
$$SC_{\widetilde{N}_{20}}(18876) = 1.134.$$

So, the neutrosophication value of 18, 876 is $\widetilde{N}_{19}$.

Table 1. Actual and neutrosophication values of student enrollments.

Years	Actual Enrollments	Neutrosophication Values of Enrollments $\widetilde{N}$
1971	13,055	$\widetilde{N}_1$
1972	13,563	$\widetilde{N}_2$
1973	13,867	$\widetilde{N}_3$
1974	14,696	$\widetilde{N}_6$
1975	15,460	$\widetilde{N}_8$
1976	15,311	$\widetilde{N}_8$
1977	15,603	$\widetilde{N}_8$
1978	15,861	$\widetilde{N}_9$
1979	16,807	$\widetilde{N}_{13}$
1980	16,919	$\widetilde{N}_{13}$
1981	16,388	$\widetilde{N}_{11}$
1982	15,433	$\widetilde{N}_8$
1983	15,497	$\widetilde{N}_8$
1984	15,145	$\widetilde{N}_7$
1985	15,163	$\widetilde{N}_7$
1986	15,984	$\widetilde{N}_{10}$
1987	16,859	$\widetilde{N}_{13}$
1988	18,150	$\widetilde{N}_{17}$
1989	18,970	$\widetilde{N}_{20}$
1990	19,328	$\widetilde{N}_{21}$
1991	19,337	$\widetilde{N}_{21}$
1992	18,876	$\widetilde{N}_{19}$

Step 5: Construct the neutrosophic logical relationships (NLRs) as in Table 2:

Table 2. Neutrosophic logical relationships.

$\widetilde{N}_1 \rightarrow \widetilde{N}_2$	$\widetilde{N}_2 \rightarrow \widetilde{N}_3$	$\widetilde{N}_3 \rightarrow \widetilde{N}_6$	$\widetilde{N}_6 \rightarrow \widetilde{N}_8$	$\widetilde{N}_8 \rightarrow \widetilde{N}_8$
$\widetilde{N}_8 \rightarrow \widetilde{N}_9$	$\widetilde{N}_9 \rightarrow \widetilde{N}_{13}$	$\widetilde{N}_{13} \rightarrow \widetilde{N}_{13}$	$\widetilde{N}_{13} \rightarrow \widetilde{N}_{11}$	$\widetilde{N}_{11} \rightarrow \widetilde{N}_8$
$\widetilde{N}_8 \rightarrow \widetilde{N}_7$	$\widetilde{N}_7 \rightarrow \widetilde{N}_7$	$\widetilde{N}_7 \rightarrow \widetilde{N}_{10}$	$\widetilde{N}_{10} \rightarrow \widetilde{N}_{13}$	$\widetilde{N}_{13} \rightarrow \widetilde{N}_{17}$
$\widetilde{N}_{17} \rightarrow \widetilde{N}_{20}$	$\widetilde{N}_{20} \rightarrow \widetilde{N}_{21}$	$\widetilde{N}_{21} \rightarrow \widetilde{N}_{21}$	$\widetilde{N}_{21} \rightarrow \widetilde{N}_{19}$	

Step 6: Based on NLR, begin to establish the neutrosophic logical relationship groups (NLRGs) as in Table 3.

Table 3. Neutrosophic logical relationship groups (NLRGs) of enrollments.

$\widetilde{N}_1 \rightarrow \widetilde{N}_2$		
$\widetilde{N}_2 \rightarrow \widetilde{N}_3$		
$\widetilde{N}_3 \rightarrow \widetilde{N}_6$		
$\widetilde{N}_6 \rightarrow \widetilde{N}_8$		
$\widetilde{N}_7 \rightarrow \widetilde{N}_7$	$\widetilde{N}_7 \rightarrow \widetilde{N}_{10}$	
$\widetilde{N}_8 \rightarrow \widetilde{N}_7$	$\widetilde{N}_8 \rightarrow \widetilde{N}_8$	$\widetilde{N}_8 \rightarrow \widetilde{N}_9$
$\widetilde{N}_9 \rightarrow \widetilde{N}_{13}$		
$\widetilde{N}_{10} \rightarrow \widetilde{N}_{13}$		
$\widetilde{N}_{11} \rightarrow \widetilde{N}_8$		
$\widetilde{N}_{13} \rightarrow \widetilde{N}_{11}$	$\widetilde{N}_{13} \rightarrow \widetilde{N}_{13}$	$\widetilde{N}_{13} \rightarrow \widetilde{N}_{17}$
$\widetilde{N}_{17} \rightarrow \widetilde{N}_{20}$		
$\widetilde{N}_{20} \rightarrow \widetilde{N}_{21}$		
$\widetilde{N}_{21} \rightarrow \widetilde{N}_{19}$	$\widetilde{N}_{21} \rightarrow \widetilde{N}_{21}$	

Step 7: Calculate the forecasted values as in Table 4:

Table 4. Actual and forecasted values of enrollments.

Years	Actual Enrollments	Forecasted Values of Enrollments
1971	13,055	–
1972	13,563	13, 650
1973	13,867	13,950
1974	14,696	14, 850
1975	15,460	15, 450
1976	15,311	15, 450
1977	15,603	15, 450
1978	15,861	15, 450
1979	16,807	16, 950
1980	16,919	17, 150
1981	16,388	17, 150
1982	15,433	15,450
1983	15,497	15, 450
1984	15,145	15, 450
1985	15,163	15, 600
1986	15,984	15, 600
1987	16,859	16, 950
1988	18,150	17, 150
1989	18,970	19, 050
1990	19,328	19, 350
1991	19,337	19, 050
1992	18,876	19, 050

To calculate the forecasted value of 13,055 in year 1971, do the following:

- Look at the neutrosophication value of 13055 in year 1971 which is $\widetilde{N}_1$ as it appears in Table 1.
- Go to NLRG which is presented in Table 3, and because $\widetilde{N}_1$ is the first neutrosophication value of data, then it is not caused by any other value (i.e., $\neq\rightarrow \widetilde{N}_1$) as in Table 3.

Therefore, the forecasted value of 13,055 is—Which means leaving it empty, as we illustrated in Step 7, Rule 1 of the proposed algorithm.

Also, to calculate the forecasted value of 13,563 in year 1972, do the following:

- Look at the neutrosophication value of 13,563 in year 1972 which is $\widetilde{N}_2$ as it appears in Table 1, and because $\widetilde{N}_2$ is caused by $\widetilde{N}_1$ (i.e., $\widetilde{N}_1 \rightarrow \widetilde{N}_2$), then
- Go to Table 3, and look at the NLRG which starts with $\widetilde{N}_1$, and we noted that it is $\widetilde{N}_1 \rightarrow \widetilde{N}_2$. Then the forecasted value of 13,563 is the middle value of $\widetilde{N}_2$.

Another illustrating example for calculating the forecasted value of 18,876 in year 1992:

- Look at the neutrosophication value of 18,876 in year 1992 which is $\widetilde{N}_{19}$ as it appears in Table 1. Since $\widetilde{N}_{19}$ is caused by $\widetilde{N}_{21}$, then
- Go to Table 3, and look at the NLRG which starts with $\widetilde{N}_{21}$ (i.e., $\widetilde{N}_{21} \rightarrow \widetilde{N}_{19}, \widetilde{N}_{21} \rightarrow \widetilde{N}_{21}$). Then the forecasted value of 18876 is the average of the middle values of $\widetilde{N}_{19}$, $\widetilde{N}_{21}$, and it will equal 19,050.

The other forecasted values are calculated in the same manner.

The actual and forecasted values of enrollments appear in Figure 1.

Figure 1. Forecasted and actual enrollments.

The forecasted enrollment data obtained with the suggested method, along with the forecasted data obtained with the models in [14,17,43–46], are presented in Table 5.

Table 5. Forecasted values by suggested method and other methods.

Years	Actual Values	Forecasted Values						
		Proposed	[43]	[44]	[45]	[46]	[14]	[17]
1971	13,055	–	–	–	–	–	–	–
1972	13,563	13,650	14,242.0	14,025	13,250	14,031.35	14,586	13,693
1973	13,867	13,950	14,242.0	14,568	13,750	14,795.36	14,586	13,693
1974	14,696	14,850	14,242.0	14,568	13,750	14,795.36	15,363	14,867
1975	15,460	15,450	15,774.3	15,654	14,500	14,795.36	15,363	15,287
1976	15,311	15,450	15,774.3	15,654	15,375	16,406.57	15,442	15,376
1977	15,603	15,450	15,774.3	15,654	15,375	16,406.57	15,442	15,376
1978	15,861	15,450	15,774.3	15,654	15,625	16,406.57	15,442	15,376
1979	16,807	16,950	16,146.5	16,197	15,875	16,406.57	15,442	16,523
1980	16,919	17,150	16,988.3	17,283	16,833	17,315.29	17,064	16,606
1981	16,388	17,150	16,988.3	17,283	16,833	17,315.29	17,064	17,519
1982	15,433	15,450	16,146.5	16,197	16,500	17,315.29	15,438	16,606
1983	15,497	15,450	15,474.3	15,654	15,500	16,406.57	15,442	15,376
1984	15,145	15,450	15,474.3	15,654	15,500	16,406.57	15,442	15,376
1985	15,163	15,600	15,474.3	15,654	15,125	16,406.57	15,363	15,287
1986	15,984	15,600	15,474.3	15,654	15,125	16,406.57	15,363	15,287
1987	16,859	16,950	16,146.5	15,654	16,833	16,406.57	15,438	16,523
1988	18,150	17,150	16,988.3	16,197	16,667	17,315.29	17,064	17,519
1989	18,970	19,050	19,144.0	17,283	18,125	19,132.79	19,356	19,500
1990	19,328	19,350	19,144.0	18,369	18,750	19,132.79	19,356	19,000
1991	19,337	19,050	19,144.0	19,454	19,500	19,132.79	19,356	19,500
1992	18,876	19,050	19,144.0	19,454	19,500	19,132.79	19,356	19,500

By comparing the proposed method with other existing methods in Table 5, the RMSE and AFE tools confirm that the suggested method is better than others, as shown in Table 6.

Table 6. Error measures.

Tool	Proposed	[43]	[44]	[45]	[46]	[14]	[17]
RMSE	342.68	478.45	781.47	646.67	805.17	642.68	493.56
AFE (%)	1.44	2.39	3.61	2.98	4.28	2.96	2.33

We combined forecasted values with respect to all methods in Figure 2.

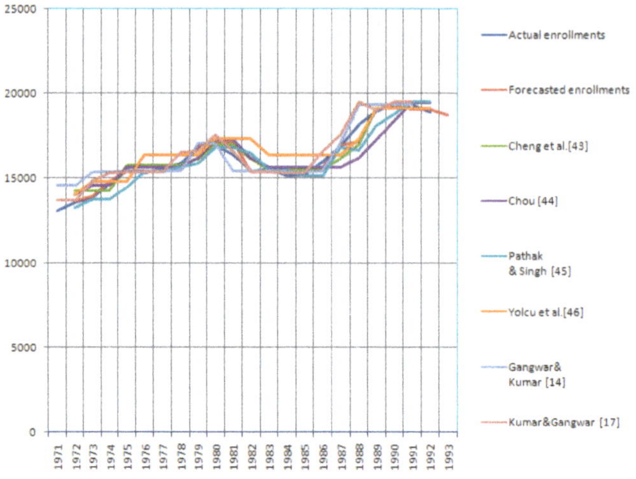

Figure 2. Comparison figures between all forecasted values.

If we plan to find the second-order neutrosophic logical relationships of the previous example by applying the proposed method of forecasting based on the second-order NTS, they are as shown in Table 7.

Table 7. Second-order NLR.

$\widetilde{N}_1, \widetilde{N}_2 \rightarrow \widetilde{N}_3$
$\widetilde{N}_2, \widetilde{N}_3 \rightarrow \widetilde{N}_6$
$\widetilde{N}_3, \widetilde{N}_6 \rightarrow \widetilde{N}_8$
$\widetilde{N}_6, \widetilde{N}_8 \rightarrow \widetilde{N}_8$
$\widetilde{N}_8, \widetilde{N}_8 \rightarrow \widetilde{N}_8$
$\widetilde{N}_8, \widetilde{N}_8 \rightarrow \widetilde{N}_9$
$\widetilde{N}_8, \widetilde{N}_9 \rightarrow \widetilde{N}_{13}$
$\widetilde{N}_9, \widetilde{N}_{13} \rightarrow \widetilde{N}_{13}$
$\widetilde{N}_{13}, \widetilde{N}_{13} \rightarrow \widetilde{N}_{11}$ $\widetilde{N}_{13}, \widetilde{N}_{11} \rightarrow \widetilde{N}_8$
$\widetilde{N}_{11}, \widetilde{N}_8 \rightarrow \widetilde{N}_8$
$\widetilde{N}_8, \widetilde{N}_8 \rightarrow \widetilde{N}_7$
$\widetilde{N}_8, \widetilde{N}_7 \rightarrow \widetilde{N}_7$
$\widetilde{N}_7, \widetilde{N}_7 \rightarrow \widetilde{N}_{10}$
$\widetilde{N}_7, \widetilde{N}_{10} \rightarrow \widetilde{N}_{13}$
$\widetilde{N}_{10}, \widetilde{N}_{13} \rightarrow \widetilde{N}_{17}$
$\widetilde{N}_{13}, \widetilde{N}_{17} \rightarrow \widetilde{N}_{20}$
$\widetilde{N}_{17}, \widetilde{N}_{20} \rightarrow \widetilde{N}_{21}$
$\widetilde{N}_{20}, \widetilde{N}_{21} \rightarrow \widetilde{N}_{21}$
$\widetilde{N}_{21}, \widetilde{N}_{21} \rightarrow \widetilde{N}_{19}$

The second-order neutrosophic logical relationship groups of the previous example are as shown in Table 8.

We compared forecasted values of enrollments based on the second order of neutrosophic logical relationship groups of the proposed method with the method of second order presented by Gautam and Singh [47]. The results are shown in Table 9.

Table 8. Second-order NLRGs.

$\widetilde{N}_1, \widetilde{N}_2 \rightarrow \widetilde{N}_3$		
$\widetilde{N}_2, \widetilde{N}_3 \rightarrow \widetilde{N}_6$		
$\widetilde{N}_3, \widetilde{N}_6 \rightarrow \widetilde{N}_8$		
$\widetilde{N}_6, \widetilde{N}_8 \rightarrow \widetilde{N}_8$		
$\widetilde{N}_8, \widetilde{N}_8 \rightarrow \widetilde{N}_8$	$\widetilde{N}_8, \widetilde{N}_8 \rightarrow \widetilde{N}_9$	$\widetilde{N}_8, \widetilde{N}_8 \rightarrow \widetilde{N}_7$
$\widetilde{N}_8, \widetilde{N}_9 \rightarrow \widetilde{N}_{13}$		
$\widetilde{N}_9, \widetilde{N}_{13} \rightarrow \widetilde{N}_{13}$		
$\widetilde{N}_{13}, \widetilde{N}_{13} \rightarrow \widetilde{N}_{11}$		
$\widetilde{N}_{13}, \widetilde{N}_{11} \rightarrow \widetilde{N}_8$		
$\widetilde{N}_{11}, \widetilde{N}_8 \rightarrow \widetilde{N}_8$		
$\widetilde{N}_8, \widetilde{N}_7 \rightarrow \widetilde{N}_7$		
$\widetilde{N}_7, \widetilde{N}_7 \rightarrow \widetilde{N}_{10}$		
$\widetilde{N}_7, \widetilde{N}_{10} \rightarrow \widetilde{N}_{13}$		
$\widetilde{N}_{10}, \widetilde{N}_{13} \rightarrow \widetilde{N}_{17}$		
$\widetilde{N}_{13}, \widetilde{N}_{17} \rightarrow \widetilde{N}_{20}$		
$\widetilde{N}_{17}, \widetilde{N}_{20} \rightarrow \widetilde{N}_{21}$		
$\widetilde{N}_{20}, \widetilde{N}_{21} \rightarrow \widetilde{N}_{21}$		
$\widetilde{N}_{21}, \widetilde{N}_{21} \rightarrow \widetilde{N}_{19}$		

Table 9. Actual and forecasted values of enrollments based on the second order of the proposed method vs. the Gautam and Singh [47] method.

Years	Actual Enrollments	Second-Order Forecasted Values of the Proposed Method	Forecasted Values in [47]
1971	13,055	–	–
1972	13,563	–	–
1973	13,867	13,950	13,800
1974	14,696	14,850	14,400
1975	15,460	15,450	15,300
1976	15,311	15,450	15,300
1977	15,603	15,450	15,600
1978	15,861	15,450	15,600
1979	16,807	16,950	16,800
1980	16,919	16,950	16,800
1981	16,388	16,350	16,200
1982	15,433	15,450	15,300
1983	15,497	15,450	15,300
1984	15,145	15,450	15,000
1985	15,163	15,150	15,000
1986	15,984	16,050	15,900
1987	16,859	16,950	16,800
1988	18,150	18,150	18,000
1989	18,970	19,050	18,900
1990	19,328	19,350	19,200
1991	19,337	19,350	19,200
1992	18,876	18,750	18,600

The MSE and AFE of the two methods are presented in Table 10.

Table 10. Error measures of the proposed method and the Gautam and Singh method [47].

Tool	Proposed	[47]
MSE	19,823.4	24,443.4
AFE (%)	0.60	0.81

From Table 10, it appears that our proposed method of second order is also better than the proposed method of second order presented by Gautam and Singh [47].

In addition, the third-order neutrosophic logical relationship groups of the previous example are constructed and shown in Table 11.

Table 11. Third-order NLRGs.

$\widetilde{N}_1, \widetilde{N}_2, \widetilde{N}_3 \rightarrow \widetilde{N}_6$
$\widetilde{N}_2, \widetilde{N}_3, \widetilde{N}_6 \rightarrow \widetilde{N}_8$
$\widetilde{N}_3, \widetilde{N}_6, \widetilde{N}_8 \rightarrow \widetilde{N}_8$
$\widetilde{N}_6, \widetilde{N}_8, \widetilde{N}_8 \rightarrow \widetilde{N}_8$
$\widetilde{N}_8, \widetilde{N}_8, \widetilde{N}_8 \rightarrow \widetilde{N}_9$
$\widetilde{N}_8, \widetilde{N}_8, \widetilde{N}_9 \rightarrow \widetilde{N}_{13}$
$\widetilde{N}_8, \widetilde{N}_9, \widetilde{N}_{13} \rightarrow \widetilde{N}_{13}$
$\widetilde{N}_9, \widetilde{N}_{13}, \widetilde{N}_{13} \rightarrow \widetilde{N}_{11}$
$\widetilde{N}_{13}, \widetilde{N}_{13}, \widetilde{N}_{11} \rightarrow \widetilde{N}_8$
$\widetilde{N}_{13}, \widetilde{N}_{11}, \widetilde{N}_8 \rightarrow \widetilde{N}_8$
$\widetilde{N}_{11}, \widetilde{N}_8, \widetilde{N}_8 \rightarrow \widetilde{N}_7$
$\widetilde{N}_8, \widetilde{N}_8, \widetilde{N}_7 \rightarrow \widetilde{N}_7$
$\widetilde{N}_8, \widetilde{N}_7, \widetilde{N}_7 \rightarrow \widetilde{N}_{10}$
$\widetilde{N}_7, \widetilde{N}_7, \widetilde{N}_{10} \rightarrow \widetilde{N}_{13}$
$\widetilde{N}_7, \widetilde{N}_{10}, \widetilde{N}_{13} \rightarrow \widetilde{N}_{17}$
$\widetilde{N}_{10}, \widetilde{N}_{13}, \widetilde{N}_{17} \rightarrow \widetilde{N}_{20}$
$\widetilde{N}_{13}, \widetilde{N}_{17}, \widetilde{N}_{20} \rightarrow \widetilde{N}_{21}$
$\widetilde{N}_{17}, \widetilde{N}_{20}, \widetilde{N}_{21} \rightarrow \widetilde{N}_{21}$
$\widetilde{N}_{20}, \widetilde{N}_{21}, \widetilde{N}_{21} \rightarrow \widetilde{N}_{19}$

We also compared the forecasted values of enrollments based on the third order of neutrosophic logical relationship groups of the proposed method with the proposed methods of third order presented by [8,9,47], and the results are shown in Table 12.

Table 12. Actual and forecasted values of enrollments based on the third order of the proposed method vs. the methods presented by [8,9,47].

Years	Actual Enrollments	Third-Order Forecasted Values of the Proposed Method	Forecasted Values in [47]	Forecasted Values in [8]	Forecasted Values in [9]
1971	13,055	–	–	–	–
1972	13,563	–	–	–	–
1973	13,867	–	–	–	–
1974	14,696	14,850	14,400	14,500	14,750
1975	15,460	15,450	15,300	15,500	15,750
1976	15,311	15,450	15,300	15,500	15,500
1977	15,603	15,450	15,600	15,500	15,500
1978	15,861	15,750	15,600	15,500	15,500
1979	16,807	16,950	16,800	16,500	16,500
1980	16,919	16,950	16,800	16,500	16,500
1981	16,388	16,350	16,200	16,500	16,500
1982	15,433	15,450	15,300	15,500	15,500
1983	15,497	15,450	15,300	15,500	15,500
1984	15,145	15,150	15,000	15,500	15,250
1985	15,163	15,150	15,000	15,500	15,500
1986	15,984	16,050	15,900	15,500	15,500
1987	16,859	16,950	16,800	16,500	16,500
1988	18,150	18,150	18,000	18,500	18,500
1989	18,970	19,050	18,900	18,500	18,500
1990	19,328	19,350	19,200	19,500	19,500
1991	19,337	19,350	19,200	19,500	19,500
1992	18,876	18,750	18,600	18,500	18,750

The MSE and AFE of the methods are presented in Table 13.

Table 13. Error measures of the proposed method and the [8,9,47] methods.

Tool	Proposed	[47]	[8]	[9]
MSE	7367.316	25,493.6	86,694	76,509
AFE (%)	0.40	0.82	1.52	1.40

4.2. Numerical Example 2

We verified the proposed method by solving the TAIEX2004 example [40], and by putting D_1 and D_2 equal 56 and 61, respectively, then U = [5600.17, 6200.69]. Also we calculated the suitable length as we illustrated previously and found that it is equal to 40. Therefore, the number of triangular neutrosophic numbers is equal to 12. For these neutrosophic numbers, the decision makers determined the truth, indeterminacy, and falsity degrees equal to $0.9, 0.1, 0.1$, respectively. The actual and forecasted values of the TAIEX2004 example are presented in Table 14 and Figure 3.

Table 14. Actual and forecasted values of TAIEX2004.

Dates	Actual Values	Forecasted Values of the Proposed Method
01/11/2004	5656.17	–
02/11/2004	5759.61	5760.17
03/11/2004	5862.85	5813.5
04/11/2004	5860.73	5900.17
05/11/2004	5931.31	5900.17
08/11/2004	5937.46	5903.02
09/11/2004	5945.2	5903.02
10/11/2004	5948.49	5940.17
11/11/2004	5874.52	5940.17
12/11/2004	5917.16	5903.02
15/11/2004	5906.69	5903.02
16/11/2004	5910.85	5903.02
17/11/2004	6028.68	5940.17
18/11/2004	6049.49	5940.17
19/11/2004	6026.55	5940.17
22/11/2004	5838.42	5830.17
23/11/2004	5851.1	5830.17
24/11/2004	5911.31	5903.02
25/11/2004	5855.24	5830.17
26/11/2004	5778.65	5813.5
29/11/2004	5785.26	5813.5
30/11/2004	5844.76	5860.17
1/12/2004	5798.62	5830.17
02/12/2004	5867.95	5860.17
03/12/2004	5893.27	5900.17
06/12/2004	5919.17	5900.17
07/12/2004	5925.28	5903.02
08/12/2004	5892.51	5903.02
09/12/2004	5913.97	5900.17
10/12/2004	5911.63	5903.02
13/12/2004	5878.89	5903.02
14/12/2004	5909.65	5900.17
15/12/2004	6002.58	5903.02
16/12/2004	6019.23	6040.17
17/12/2004	6009.32	6040.17
20/12/2004	5985.94	6040.17
21/12/2004	5987.85	6040.17
22/12/2004	6001.52	6040.17

Table 14. *Cont.*

Dates	Actual Values	Forecasted Values of the Proposed Method
23/11/2004	5851.1	5830.17
24/11/2004	5911.31	5903.02
25/11/2004	5855.24	5830.17
26/11/2004	5778.65	5813.5
29/11/2004	5785.26	5813.5
30/11/2004	5844.76	5860.17
1/12/2004	5798.62	5830.17
02/12/2004	5867.95	5860.17
03/12/2004	5893.27	5900.17
06/12/2004	5919.17	5900.17
07/12/2004	5925.28	5903.02
08/12/2004	5892.51	5903.02
09/12/2004	5913.97	5900.17
10/12/2004	5911.63	5903.02
13/12/2004	5878.89	5903.02
14/12/2004	5909.65	5900.17
15/12/2004	6002.58	5903.02
16/12/2004	6019.23	6040.17
17/12/2004	6009.32	6040.17
20/12/2004	5985.94	6040.17
21/12/2004	5987.85	6040.17
22/12/2004	6001.52	6040.17
23/12/2004	5997.67	6040.17
24/12/2004	6019.42	6040.17
27/12/2004	5985.94	6040.17
28/12/2004	6000.57	6040.17
29/12/2004	6088.49	6040.17
30/12/2004	6100.86	6080.17
31/12/2004	6139.69	6080.17

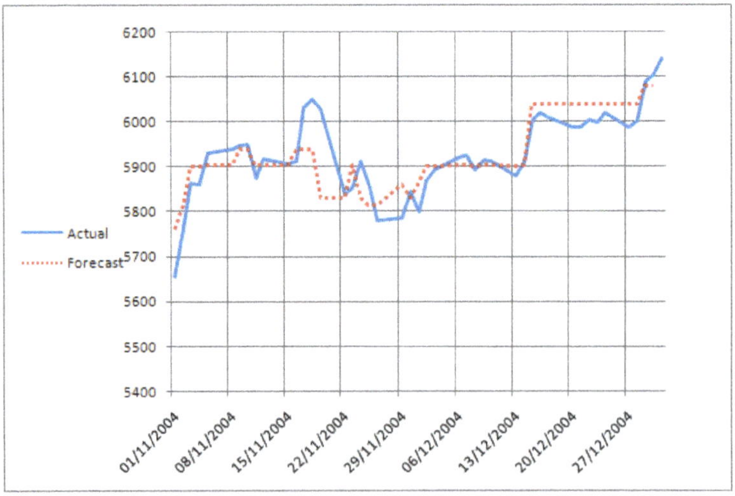

Figure 3. Actual and forecasted values of TAIEX2004.

The RMSE and AFE of the proposed method are presented in Table 15.

Table 15. Error measures of the proposed method.

Tool	Proposed
RMSE	42.05
AFE (%)	0.005

To confirm the performance of the suggested method, we compared it with other existing methods and the results are shown in Table 16 and Figure 4.

Table 16. Error measures of the proposed method and other existing methods which solved the TAIEX2004 example.

Methods	RMSE
Guan et al.'s method [40]	53.01
Huarng et al.'s method [48]	73.57
Chen and Kao's method [49]	58.17
Cheng et al.'s method [50]	54.24
Chen et al.'s method [51]	56.16
Chen and Chang's method [52]	60.48
Chen and Chen's method [53]	61.94
Yu and Huarng's method [54]	55.91
Proposed method	42.05

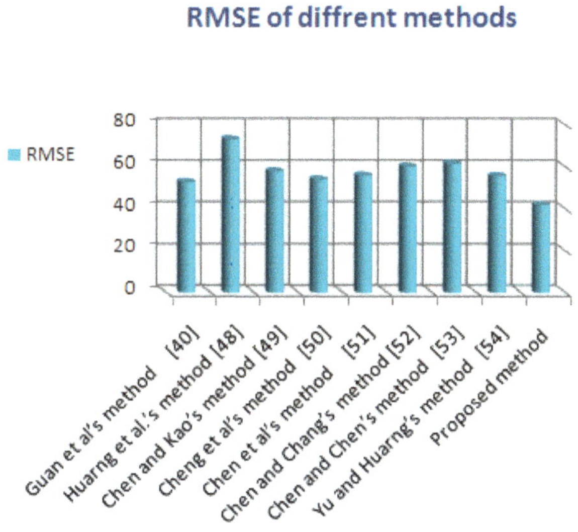

Figure 4. The RMSE of different methods that solved the TAIEX2004 example.

TAIEX2004 is used as a baseline to compare our method with other competitive methods, to compare and identify how all the methods can manage error reduction. The RMSE is a common approach used in financial analysis [55]. Compared with the existing methods as shown in Table 16, our proposed method can offer the least presence of errors since it has the most minimized RMSE. In other words, our method appears to be performing the best in reducing errors and ensuring all our analyses are accurate with insights. This may provide a new insight for business intelligence with artificial intelligence, cloud computing, and neutrosophic research.

5. Conclusion and Future Directions

The objective of this research was to enhance the accuracy rates of forecasting, since the forecasting accuracy rates in the existing approaches of fuzzy and intuitionistic fuzzy time series were not accurate enough. Thus, in this research we introduced the notion of first-and high-order neutrosophic time series data by defining the fitting length of intervals and proposing a novel method for calculating forecasted values. In order to obtain truth, indeterminacy, and falsity membership degrees of historical data, we defined triangular neutrosophic numbers. The neutrosophication process of historical time series data depends on the biggest score function of the triangular neutrosophic numbers. For the deneutrosophication process of first- and high-order NTS, we used simple arithmetic computations. The suggested approach of first- and high-order neutrosophic time series proved its superiority against other existing methods in the field of fuzzy, intuitionistic fuzzy, and neutrosophic time series. In the future, we plan to apply meta-heuristic optimization techniques for improving the accuracy of the suggested method. We will apply this model for predicting other time series, such as demand forecasting, electricity consumption, etc. Furthermore, we may consider using other approaches for comparing similarities of historical data, like information entropy.

Author Contributions: All authors contributed equally to this research. The individual responsibilities and contribution of all authors can be described as follows: the idea of this entire paper was put forward by M.A.-B. and M.M.; V.C. completed the preparatory work of the paper. F.S. analyzed the existing work. The revision and submission of this paper was completed by F.S. and M.A.-B.

Funding: This research received no external funding.

Acknowledgments: The authors are very thankful to the reviewers for their constructive suggestion to improve the quality of this research.

Conflicts of Interest: The authors declare no conflict of interest.

References

1. Song, Q.; Chissom, B.S. Forecasting enrollments with fuzzy time series—Part I. *Fuzzy Sets Syst.* **1993**, *54*, 1–9. [CrossRef]
2. Song, Q.; Chissom, B.S. Forecasting enrollments with fuzzy time series—Part II. *Fuzzy Sets Syst.* **1994**, *62*, 1–8. [CrossRef]
3. Chen, S.M. Forecasting enrollments based on fuzzy time series. *Fuzzy Sets Syst.* **1996**, *81*, 311–319. [CrossRef]
4. Lee, H.S.; Chou, M.T. Fuzzy forecasting based on fuzzy time series. *Int. J. Comput. Math.* **2004**, *81*, 781–789. [CrossRef]
5. Singh, S.R. A simple method of forecasting based on fuzzy time series. *Appl. Math. Comput.* **2007**, *186*, 330–339. [CrossRef]
6. Singh, S.R. A robust method of forecasting based on fuzzy time series. *Appl. Math. Comput.* **2007**, *188*, 472–484. [CrossRef]
7. Tsaur, R.C.; Yang, J.C.O.; Wang, H.F. Fuzzy relation analysis in fuzzy time series model. *Comput. Math. Appl.* **2005**, *49*, 539–548. [CrossRef]
8. Chen, S.M. Forecasting enrollments based on high-order fuzzy time series. *Cybern. Syst.* **2002**, *33*, 1–16. [CrossRef]
9. Singh, S.R. A simple time variant method for fuzzy time series forecasting. *Cybern. Syst. Int. J.* **2007**, *38*, 305–321. [CrossRef]
10. Hsu, Y.Y.; Tse, S.M.; Wu, B. A new approach of bivariate fuzzy time series analysis to the forecasting of a stock index. *Int. J. Uncertain. Fuzziness Knowl.-Based Syst.* **2003**, *11*, 671–690. [CrossRef]
11. Ziemba, P.; Becker, J. Analysis of the Digital Divide Using Fuzzy Forecasting. *Symmetry* **2019**, *11*, 166. [CrossRef]
12. Zadeh, L.A. On fuzzy algorithms. In *Fuzzy Sets, Fuzzy Logic, and Fuzzy Systems: Selected Papers by Lotfi A Zadeh*; World Scientific: Singapore, 1996; pp. 127–147.
13. Atanassov, K.T. Intuitionistic fuzzy sets. *Fuzzy Sets Syst.* **1986**, *20*, 87–96. [CrossRef]

14. Gangwar, S.S.; Kumar, S. Probabilistic and intuitionistic fuzzy sets—Based method for fuzzy time series forecasting. *Cybern. Syst.* **2014**, *45*, 349–361. [CrossRef]
15. Wang, Y.N.; Lei, Y.; Fan, X.; Wang, Y. Intuitionistic fuzzy time series forecasting model based on intuitionistic fuzzy reasoning. *Math. Prob. Eng.* **2016**, *2016*, 5035160. [CrossRef]
16. Joshi, B.P.; Kumar, S. Intuitionistic fuzzy sets based method for fuzzy time series forecasting. *Cybern. Syst.* **2012**, *43*, 34–47. [CrossRef]
17. Kumar, S.; Gangwar, S.S. Intuitionistic fuzzy time series: An approach for handling non-determinism in time series forecasting. *IEEE Trans. Fuzzy Syst.* **2016**, *24*, 1270–1281. [CrossRef]
18. Kumar, S.; Gangwar, S.S. A fuzzy time series forecasting method induced by intuitionistic fuzzy sets. *Int. J. Model. Simul. Sci. Comput.* **2015**, *6*, 1550041. [CrossRef]
19. Eğrioglu, E.; Aladag, C.H.; Yolcu, U.; Dalar, A.Z. A hybrid high order fuzzy time series forecasting approach based on PSO and ANNS methods. *Am. J. Intell. Syst.* **2016**, *6*, 22–29.
20. Cai, Q.; Zhang, D.; Wu, B.; Leung, S.C. A novel stock forecasting model based on fuzzy time series and genetic algorithm. *Procedia Comput. Sci.* **2013**, *18*, 1155–1162. [CrossRef]
21. Huang, Y.L.; Horng, S.J.; He, M.; Fan, P.; Kao, T.W.; Khan, M.K.; Kuo, I.H. A hybrid forecasting model for enrollments based on aggregated fuzzy time series and particle swarm optimization. *Expert Syst. Appl.* **2011**, *38*, 8014–8023. [CrossRef]
22. Bisht, K.; Kumar, S. Fuzzy time series forecasting method based on hesitant fuzzy sets. *Expert Syst. Appl.* **2016**, *64*, 557–568. [CrossRef]
23. Bahrami, B.; Shafiee, M. Fuzzy descriptor models for earthquake time prediction using seismic time series. *Int. J. Uncertain. Fuzziness Knowl.-Based Syst.* **2015**, *23*, 505–519. [CrossRef]
24. Wang, Y.; Lei, Y.; Wang, Y.; Zheng, K. A heuristic adaptive-order intuitionistic fuzzy time series forecasting model. *J. Electron. Inf. Technol.* **2016**, *38*, 2795–2802.
25. Abhishekh, S.S.G.; Singh, S.R. A refined weighted method for forecasting based on type 2 fuzzy time series. *Int. J. Model. Simul.* **2018**, *38*, 180–188. [CrossRef]
26. Abhishekh, S.S.G.; Singh, S.R. A score function-based method of forecasting using intuitionistic fuzzy time series. *New Math. Nat. Comput.* **2018**, *14*, 91–111. [CrossRef]
27. Abhishekh, S.K. A computational method for rice production forecasting based on high-order fuzzy time series. *Int. J. Fuzzy Math. Arch.* **2017**, *13*, 145–157.
28. Abdel-Basset, M.; Mohamed, M.; Chang, V. NMCDA: A framework for evaluating cloud computing services. *Future Gener. Comput. Syst.* **2018**, *86*, 12–29. [CrossRef]
29. Smarandache, F. A Unifying Field in Logics: Neutrosophic Logic. In *Neutrosophy, Neutrosophic Set, Neutrosophic Probability*, 3rd ed.; American Research Press: Rehoboth, DE, USA, 1999; Volume 8, pp. 489–503.
30. Abdel-Basset, M.; Mohamed, M. The role of single valued neutrosophic sets and rough sets in smart city: Imperfect and incomplete information systems. *Measurement* **2018**, *124*, 47–55. [CrossRef]
31. Abdel-Basset, M.; Mohamed, M.; Smarandache, F.; Chang, V. Neutrosophic Association Rule Mining Algorithm for Big Data Analysis. *Symmetry* **2018**, *10*, 106. [CrossRef]
32. Abdel-Basset, M.; Mohamed, M.; Zhou, Y.; Hezam, I. Multi-criteria group decision making based on neutrosophic analytic hierarchy process. *J. Intell. Fuzzy Syst.* **2017**, *33*, 4055–4066. [CrossRef]
33. Abdel-Basset, M.; Gunasekaran, M.; Mohamed, M.; Smarandache, F. A novel method for solving the fully neutrosophic linear programming problems. *Neural Comput. Appl.* **2018**, 1–11. [CrossRef]
34. Abdel-Basset, M.; Mohamed, M.; Sangaiah, A.K. Neutrosophic AHP-Delphi Group decision making model based on trapezoidal neutrosophic numbers. *J. Ambient Intell. Hum. Comput.* **2018**, *9*, 1427–1443. [CrossRef]
35. Abdel-Basset, M.; Mohamed, M.; Hussien, A.N.; Sangaiah, A.K. A novel group decision-making model based on triangular neutrosophic numbers. *Soft Comput.* **2018**, *22*, 6629–6643. [CrossRef]
36. Abdel-Basset, M.; Zhou, Y.; Mohamed, M.; Chang, V. A group decision making framework based on neutrosophic VIKOR approach for e-government website evaluation. *J. Intell. Fuzzy Syst.* **2018**, *34*, 4213–4224. [CrossRef]
37. Abdel-Basset, M.; Gunasekaran, M.; Mohamed, M.; Chilamkurti, N. Three-way decisions based on neutrosophic sets and AHP-QFD framework for supplier selection problem. *Futur. Gener. Comput. Syst.* **2018**, *89*, 19–30. [CrossRef]
38. Mohamed, M.; Abdel-Baset, M.; Smarandache, F.; Zhou, Y. *A Critical Path Problem in Neutrosophic Environment*; Infinite Study: El Segundo, CA, USA, 2017.

39. Abdel-Basset, M.; Mohamed, M.; Smarandache, F. A hybrid neutrosophic group ANP-TOPSIS framework for supplier selection problems. *Symmetry* **2018**, *10*, 226. [CrossRef]
40. Guan, H.; He, J.; Zhao, A.; Dai, Z.; Guan, S. A Forecasting Model Based on Multi-Valued Neutrosophic Sets and Two-Factor, Third-Order Fuzzy Fluctuation Logical Relationships. *Symmetry* **2018**, *10*, 245. [CrossRef]
41. Guan, H.; Dai, Z.; Guan, S.; Zhao, A. A Forecasting Model Based on High-Order Fluctuation Trends and Information Entropy. *Entropy* **2018**, *20*, 669. [CrossRef]
42. Huarng, K. Effective lengths of intervals to improve forecasting in fuzzy time series. *Fuzzy Sets Syst.* **2001**, *123*, 387–394. [CrossRef]
43. Cheng, C.H.; Cheng, G.W.; Wang, J.W. Multi-attribute fuzzy time series method based on fuzzy clustering. *Expert Syst. Appl.* **2008**, *34*, 1235–1242. [CrossRef]
44. Chou, M.T. Long-term predictive value interval with the fuzzy time series. *J. Mar. Sci. Technol.* **2011**, *19*, 509–513.
45. Pathak, H.K.; Singh, P. A new bandwidth interval based forecasting method for enrollments using fuzzy time series. *Appl. Math.* **2011**, *2*, 504–507. [CrossRef]
46. Yolcu, U.; Egrioglu, E.; Uslu, V.R.; Basaran, M.A.; Aladag, C.H. A new approach for determining the length of intervals for fuzzy time series. *Appl. Soft Comput.* **2009**, *9*, 647–651. [CrossRef]
47. Gautam, S.S.; Singh, S.R. A refined method of forecasting based on high-order intuitionistic fuzzy time series data. *Prog. Artif. Intell.* **2018**, *7*, 339–350.
48. Huarng, K.H.; Yu, T.H.K.; Hsu, Y.W. A multivariate heuristic model for fuzzy time-series forecasting. *IEEE Trans. Syst. Man Cybern. Part B (Cybern.)* **2007**, *37*, 836–846. [CrossRef]
49. Chen, S.M.; Kao, P.Y. TAIEX forecasting based on fuzzy time series, particle swarm optimization techniques and support vector machines. *Inf. Sci.* **2013**, *247*, 62–71. [CrossRef]
50. Cheng, S.H.; Chen, S.M.; Jian, W.S. Fuzzy time series forecasting based on fuzzy logical relationships and similarity measures. *Inf. Sci.* **2016**, *327*, 272–287. [CrossRef]
51. Chen, S.M.; Manalu, G.M.T.; Pan, J.S.; Liu, H.C. Fuzzy forecasting based on two-factors second-order fuzzy-trend logical relationship groups and particle swarm optimization techniques. *IEEE Trans. Cybern.* **2013**, *43*, 1102–1117. [CrossRef] [PubMed]
52. Chen, S.M.; Chang, Y.C. Multi-variable fuzzy forecasting based on fuzzy clustering and fuzzy rule interpolation techniques. *Inf. Sci.* **2010**, *180*, 4772–4783. [CrossRef]
53. Chen, S.M.; Chen, C.D. TAIEX forecasting based on fuzzy time series and fuzzy variation groups. *IEEE Trans. Fuzzy Syst.* **2011**, *19*, 1–12. [CrossRef]
54. Yu, T.H.K.; Huarng, K.H. A neural network-based fuzzy time series model to improve forecasting. *Expert Syst. Appl.* **2010**, *37*, 3366–3372. [CrossRef]
55. Chang, V. The business intelligence as a service in the cloud. *Futur. Gener. Comput. Syst.* **2014**, *37*, 512–534. [CrossRef]

symmetry MDPI

Article

A Novel Approach to Neutrosophic Soft Rough Set under Uncertainty

Ashraf Al-Quran [1], Nasruddin Hassan [2,*] and Emad Marei [3,†]

1 Department of Mathematics, Faculty of Science, Jerash University, Jerash 26150, Jordan; a.quraan@jpu.edu.jo
2 Faculty of Science and Technology, School of Mathematical Sciences, Universiti Kebangsaan Malaysia, UKM Bangi Selangor 43600, DE, Malaysia
3 Department of Mathematics, Faculty of Science and Art, Shaqra University, Shaqra 11961, Saudi Arabia; awajan@su.edu.sa
* Correspondence: nas@ukm.edu.my; Tel.: +60-3-8921-3710
† Deceased 13 July 2018.

Received: 5 January 2019; Accepted: 25 February 2019; Published: 15 March 2019

Abstract: To handle indeterminate and incomplete data, neutrosophic logic/set/probability were established. The neutrosophic truth, falsehood and indeterminacy components exhibit symmetry as the truth and the falsehood look the same and behave in a symmetrical way with respect to the indeterminacy component which serves as a line of the symmetry. Soft set is a generic mathematical tool for dealing with uncertainty. Rough set is a new mathematical tool for dealing with vague, imprecise, inconsistent and uncertain knowledge in information systems. This paper introduces a new rough set model based on neutrosophic soft set to exploit simultaneously the advantages of rough sets and neutrosophic soft sets in order to handle all types of uncertainty in data. The idea of neutrosophic right neighborhood is utilised to define the concepts of neutrosophic soft rough (NSR) lower and upper approximations. Properties of suggested approximations are proposed and subsequently proven. Some of the NSR set concepts such as NSR-definability, NSR-relations and NSR-membership functions are suggested and illustrated with examples. Further, we demonstrate the feasibility of the newly rough set model with decision making problems involving neutrosophic soft set. Finally, a discussion on the features and limitations of the proposed model is provided.

Keywords: decision making; membership function; neutrosophic set; neutrosophic soft rough; relations; rough set approximation; soft set

---

1. Introduction

The limitation of deterministic research is currently recognized in areas of management, social sciences, operations research and economics. Uncertain theories such as probability, fuzzy sets [1], intuitionistic fuzzy sets [2], vague sets [3] and theory of interval mathematics [4] are applied in realms which are ambiguous and uncertain.

Rough set theory, initiated by Pawlak [5], is an effective mathematical tool to the vague and imperfect knowledge. Rough set expresses vagueness by bounded region of a set, which can be interpreted as using the vagueness of Frege's idea. Pawlak argued that any vague concept can be replaced by the lower and upper approximations of precise sets using an equivalence relation. In the real application, the equivalence relation is a very stringent condition which limits the applications of rough sets in the real world. For this reason, by replacing the equivalence relation with covering, similarity, tolerance, preference, dominance relations, and different neighborhood operators, various kinds of rough set generalizations model were proposed [6–11].

A soft set is a set-valued map defined by Molodtsov [12], to approximately describe objects using several parameters. Maji et al. [13] applied the theory of soft set to solve decision making problems with the help of rough mathematics. The adequate parametrization capabilities of soft set theory and the lack of such capabilities in the fuzzy set served as the motivation to introduce the fuzzy soft sets [14].

Feng et al. [15] proved properties of soft rough set model. Smarandache [16,17] proposed neutrosophic set to handle problems containing imprecise, incomplete, uncertain and indeterminate data. Neutrosophic sets progressed rapidly to neutrosophic oversets, neutrosophic undersets, neutrosophic offsets [18], neutrosophic cubic sets [19], neutrosophic and generalised neutrosophic soft sets [20–22], neutrosophic rough sets [23–26], neutrosophic vague sets [27,28] and complex neutrosophic sets [29–32].

We will propose an approach to neutrosophic soft rough set and show that the traditional rough approach is a special case of our approach. Furthermore, we will study the neutrosophic soft rough approximations and apply them to decision making. The paper is organized into seven sections. Section 2 provides literature review. In Section 3, the concept of neutrosophic right neighborhood is defined. This section further defines neutrosophic soft rough set approximations. Properties of NSR-lower and NSR-upper approximations are included along with supported proofs and illustrated examples. Section 4 delves into neutrosophic soft rough set and generalization of rough concepts. NSR-set concepts include neutrosophic soft rough (NSR) definability, neutrosophic soft rough (NSR)-membership function, neutrosophic soft rough (NSR)-membership relations, neutrosophic soft rough (NSR)-inclusion relations and neutrosophic soft rough (NSR)-equality relations. Properties of these concepts are proven and examples provided. Section 5 provides an application of the proposed neutrosophic soft rough model on decision making. In Section 6, we conduct a discussion about the features and limitations of the proposed model by making a comparison with the existing models. In the final section, we outline future work and draw conclusions to this work.

2. Preliminaries

We start by reviewing the concepts of rough set, neutrosophic set and soft set.

Pawlak considered the set X with the equivalence relation E and called the pair (X, E) as a Pawlak approximation space. Then he assigned two operations (lower and upper approximations) to any vague subset $M \subseteq X$. These operations process some information connected with the relationship E and they are analogous with Kuratowski's operations, which are generated by the closure and the completion.

The lower, upper and boundary approximations are defined as follows.

Definition 1 ([5]). *Let E be an equivalence relation on a universe X and $M \subseteq X$. Then the pair (X, E) is referred to as a Pawlak approximation space. The lower, upper and boundary approximations of M are defined as follows.*

$$\underline{E}(M) = \cup\{[x]_E : [x]_E \subseteq M\},$$

$$\overline{E}(M) = \cup\{[x]_E : [x]_E \cap M \neq \phi\},$$

$$BND_E(M) = \overline{E}(M) - \underline{E}(M).$$

where $[x]_E = \{x' \in X : E(x) = E(x')\}$.

Definition 2 ([5]). *Let $A = (X, E)$ be an approximation space and let $M \subset X$. By the accuracy of approximation of M in A we mean the number*

$$\alpha_E(M) = \frac{\mid \underline{E}(M) \mid}{\mid \overline{E}(M) \mid}, \overline{E}(M) \neq \emptyset.$$

Obviously, $0 \leq \alpha_E(M) \leq 1$. If $\overline{E}(M) = \underline{E}(M)$, then M is crisp (exact) set, with respect to E, otherwise M is rough set.

The following proposition lists the properties of Pawlak's approximations.

Proposition 1 ([5]). *For every $M, Y \subset X$ and every approximation space $A = (X, E)$ the following properties hold:*
(i) $\underline{E}(M) \subseteq M \subseteq \overline{E}(M)$.
(ii) $\underline{E}(\phi) = \phi = \overline{E}(\phi)$ *and* $\underline{E}(X) = X = \overline{E}(X)$.
(iii) $\overline{E}(M \cup Y) = \overline{E}(M) \cup \overline{E}(Y)$.
(iv) $\underline{E}(M \cap Y) = \underline{E}(M) \cap \underline{E}(Y)$.
(v) $M \subseteq Y$, *then* $\underline{E}(M) \subseteq \underline{E}(Y)$ *and* $\overline{E}(M) \subseteq \overline{E}(Y)$.
(vi) $\underline{E}(M \cup Y) \supseteq \underline{E}(M) \cup \underline{E}(Y)$.
(vii) $\overline{E}(M \cap Y) \subseteq \overline{E}(M) \cap \overline{E}(Y)$.
(viii) $\underline{E}(M^c) = [\overline{E}(M)]^c$, *where* M^c *is the complement of* M.
(ix) $\overline{E}(M^c) = [\underline{E}(M)]^c$.
(x) $\underline{E}(\underline{E}(M)) = \overline{E}(\underline{E}(M)) = \underline{E}(M)$.
(xi) $\overline{E}(\overline{E}(M)) = \underline{E}(\overline{E}(M)) = \overline{E}(M)$.

Definition 3 ([33]). *An information system is a quadruple $IS = (U, A, V, f)$, where U is a non-empty finite set of objects, A is a non-empty finite set of attributes, $V = \cup\{V_e, e \in A\}$, V_e is the set of values of attribute e, and $f : U \times A \to V$, is called an information (knowledge) function.*

Definition 4 ([12]). *Let X be an initial universe set, E be a set of parameters, $A \subseteq E$ and let $P(X)$ denote the power set of X. Then, a pair $K = (L, A)$ is called a soft set over X, where L is a mapping given by $L : A \to P(X)$. In other words, a soft set over X is a parameterized family of subsets of X. For $a \in A$, $L(a)$ may be considered as the set of a-approximate elements of K.*

The neutrosophic set was defined by Smarandache below.

Definition 5 ([17]). *A neutrosophic set N on the universe of discourse X is defined as*

$$N = \{\langle n, T_N(n), I_N(n), F_N(n)\rangle : n \in X\}, where$$

$$^{-}0 \leq T_N(n) + I_N(n) + F_N(n) \leq 3^{+}, where$$

$$T, I, F : X \longrightarrow]^{-}0; 1^{+}[.$$

3. Neutrosophic Soft Rough Set Approximations (NSR-Set Approximations)

In this section, we give a definition of neutrosophic soft set (NSS in short) with an illustrative example. We will introduce and provide examples of NSR-lower and NSR-upper approximations.

Definition 6. *Let a universe X, E the parameter set and $A \subseteq E$. A neutrosophic soft set H over X is a neutrosophic set valued function from A to $P(X)$. It can be written as*

$$H = \{(a, < x, T_{H(a)}(x), I_{H(a)}(x), F_{H(a)}(x) >: x \in X) : a \in E\},$$

where $P(X)$ denotes the power neutrosophic set of X.

In other words, the neutrosophic soft set H is a parameterized family of neutrosophic subsets of X. For any parameter a, $H(a)$ is referred as the neutrosophic value set of parameter a.

The example below will convey the meaning of neutrosophic soft set.

Example 1. *Let X be a set of houses and E be a set of parameters (or qualities). Consider $E = \{$cheap, beautiful, green surrounding, spacious$\}$. To define an (NSS) means to point out cheap houses, beautiful houses and so* *on. If there are five houses in X, where,* $X = \{x_1, x_2, x_3, x_4, x_5\}$ *and the set of parameters* $A = \{a_1, a_2, a_3, a_4\}$*, where* $A \subset E$*, and each* a_i *is a specific property for houses:* a_1 *stands for (cheap),* a_2 *stands for (beautiful),* a_3 *stands for (green surrounding),* a_4 *stands for (spacious).*

An (NSS) can be represented as in Table 1, such that the entries are h_{ij} corresponding to the house x_i and the parameter a_j, where h_{ij} = (true membership value of x_i, indeterminacy-membership value of x_i, falsity membership value of x_i) in $H(a_j)$. Table 1, represents the (NSS) (H, A) as follows.

Table 1. Tabular representation of the neutrosophic soft set (NSS) (H, A).

X	a_1	a_2	a_3	a_4
x_1	(0.6, 0.6, 0.2)	(0.8, 0.4, 0.3)	(0.7, 0.4, 0.3)	(0.8, 0.6, 0.4)
x_2	(0.4, 0.6, 0.6)	(0.6, 0.2, 0.4)	(0.6, 0.4, 0.3)	(0.7, 0.6, 0.6)
x_3	(0.6, 0.4, 0.2)	(0.8, 0.1, 0.3)	(0.7, 0.2, 0.5)	(0.7, 0.6, 0.4)
x_4	(0.6, 0.3, 0.3)	(0.8, 0.2, 0.2)	(0.5, 0.2, 0.6)	(0.7, 0.5, 0.6)
x_5	(0.8, 0.2, 0.3)	(0.8, 0.3, 0.2)	(0.7, 0.3, 0.4)	(0.9, 0.5, 0.7)

In the following, we define the concept of the neutrosophic right neighborhood.

Definition 7. *Let X be a universal set and* Γ *be the power set of X. Let* (H, A) *be an (NSS) on X, and* $\omega = A \times X$*. Let S be a mapping given by*

$$S : \omega \to \Gamma,$$

where $S_a(x) = S(a, x) = \{x_i \in X : T_a(x_i) \geq T_a(x) \text{ and } I_a(x_i) \geq I_a(x) \text{ and } F_a(x_i) \leq F_a(x)\}$*.*

Then for any element $x \in X$*,* $S_a(x)$ *is called a neutrosophic right neighborhood, with respect to* $a \in A$*.*

Definition 8. *Let X be a universal set and* (H, A) *be an (NSS) on X. Then for all* $x \in X$ *and* $a \in A$*, the family of all neutrosophic right neighborhoods is defined as follows.*

$$\psi = \{S_a(x) : x \in X, a \in A\}.$$

The example below conveys the meaning of neutrosophic right neighborhoods.

Example 2. *We can deduce the statements below from Example 1.*
$S_{a_1}(x_1) = S_{a_2}(x_1) = S_{a_3}(x_1) = S_{a_4}(x_1) = \{x_1\}$,
$S_{a_1}(x_2) = S_{a_3}(x_2) = \{x_1, x_2\}$, $S_{a_2}(x_2) = \{x_1, x_2, x_4, x_5\}$, $S_{a_4}(x_2) = \{x_1, x_2, x_3\}$,
$S_{a_1}(x_3) = S_{a_4}(x_3) = \{x_1, x_3\}$, $S_{a_2}(x_3) = \{x_1, x_3, x_4, x_5\}$, $S_{a_3}(x_3) = \{x_1, x_3, x_5\}$,
$S_{a_1}(x_4) = \{x_1, x_3, x_4\}$, $S_{a_2}(x_4) = \{x_4, x_5\}$, $S_{a_3}(x_4) = X$, $S_{a_4}(x_4) = \{x_1, x_2, x_3, x_4\}$,
$S_{a_1}(x_5) = S_{a_2}(x_5) = S_{a_4}(x_5) = \{x_5\}$, $S_{a_3}(x_5) = \{x_1, x_5\}$.

It follows that, $\psi = \{ \{x_1\}, \{x_5\}, \{x_1, x_2\}, \{x_1, x_3\}, \{x_1, x_5\}, \{x_4, x_5\}, \{x_1, x_2, x_3\}, \{x_1, x_3, x_4\}, \{x_1, x_3, x_5\}, \{x_1, x_2, x_3, x_4\}, \{x_1, x_2, x_4, x_5\}, \{x_1, x_3, x_4, x_5\}, X \}$.

Proposition 2. *Let* (H, A) *be an (NSS) on a universe X,* ψ *is the parameterised family of all neutrosophic right neighborhoods and* $R_a : X \to \psi, R_a(x) = S_a(x)$*. Then the statements below hold.*
(i) R_a *is reflexive relation.*
(ii) R_a *is transitive relation.*

Proof. Let $\langle x_1, T_a(x_1), I_a(x_1), F_a(x_1)\rangle$, $\langle x_2, T_a(x_2), I_a(x_2), F_a(x_2)\rangle$ and $\langle x_3, T_a(x_3), I_a(x_3), F_a(x_3)\rangle \in (H, A)$. Then,
(i) Obviously, for all $i = 1,2,3$, $T_a(x_i) \geq T_a(x_i)$, $I_a(x_i) \geq I_a(x_i)$, $F_a(x_i) \leq F_a(x_i)$ Hence, for every $a \in A$, $x_i \in S_a(x_i)$ and $x_i R_a x_i$ and thus R_a is reflexive relation.

(ii) Let $x_1 R_a x_2$ and $x_2 R_a x_3$. Then, $x_2 \in S_a(x_1)$ and $x_3 \in S_a(x_2)$. Hence, $T_a(x_2) \geq T_a(x_1)$, $I_a(x_2) \geq I_a(x_1)$, $F_a(x_2) \leq F_a(x_1)$, $T_a(x_3) \geq T_a(x_2)$, $I_a(x_3) \geq I_a(x_2)$ and $F_a(x_3) \leq F_a(x_2)$. Consequently, wa have $T_a(x_3) \geq T_a(x_1)$, $I_a(x_3) \geq I_a(x_1)$ and $F_a(x_3) \leq F_a(x_1)$. It follows that $x_3 \in S_a(x_1)$ and $x_1 R_a x_3$ and thus R_a is transitive relation. □

Note that R_a in Proposition 2 may not necessarily be symmetric as shown below.

Example 3. *From Example 2, we have,* $S_{a_1}(x_1) = \{x_1\}$ *and* $S_{a_1}(x_3) = \{x_1, x_3\}$. *Hence,* $(x_3, x_1) \in R_{a_1}$ *but* $(x_1, x_3) \notin R_{a_1}$. *Thus* R_a *is not symmetric relation.*

We define the neutrosophic soft rough lower and upper approximations below.

Definition 9. *Let* (H, A) *be an (NSS) on a universe X, with* ψ *being the family of all neutrosophic right neighborhoods. The neutrosophic soft lower and neutrosophic soft upper approximations of any subset M based on* ψ*, respectively, are*

$$NR_* M = \cup\{Y \in \psi : Y \subseteq M\},$$

$$NR^* M = \cup\{Y \in \psi : Y \cap M \neq \emptyset\}.$$

$NR_* M$ *and* $NR^* M$ *can be referred as neutrosophic soft rough approximations of M (NSR-set approximations) with respect to A.*

Remark 1. *For any considered set M in an (NSS)* (H, A)*, the sets* $Pos_{NR} M = NR_* M$, $Neg_{NR} M = [NR^* M]^c$, $b_{NR} M = NR^* M - NR_* M$ *are called the NSR-positive, NSR-negative and NSR-boundary regions of a considered set M, respectively. The meaning of* $Pos_{NR} M$ *is the set of all elements, which are surely belonging to M,* $Neg_{NR} M$ *is the set of all elements, which do not belong to M and* $b_{NR} M$ *is the elements of M, not determined by* (H, A).

The proposition below lists the properties of neutrosophic soft rough approximations.

Proposition 3. *Let* (H, A) *be an (NSS) on a universe X, and let* $M, Z \subseteq X$. *Then the following properties hold.*
(i) $NR_* M \subseteq M \subseteq NR^* M$.
(ii) $NR_* \emptyset = NR^* \emptyset = \emptyset$.
(iii) $NR_* X = NR^* X = X$.
(iv) $M \subseteq Z \Rightarrow NR_* M \subseteq NR_* Z$.
(v) $M \subseteq Z \Rightarrow NR^* M \subseteq NR^* Z$.
(vi) $NR_*(M \cap Z) \subseteq NR_* M \cap NR_* Z$.
(vii) $NR_*(M \cup Z) \supseteq NR_* M \cup NR_* Z$.
(viii) $NR^*(M \cap Z) \subseteq NR^* M \cap NR^* Z$.
(ix) $NR^*(M \cup Z) = NR^* M \cup NR^* Z$.

Proof. (i) From Definition 9, we can deduce that $NR_* M \subseteq M$. In addition, let $x \in M$, but R_a defined in Proposition 2 is reflexive relation. For all $a \in A$, there exists $S_a(x)$ such that $x \in S_a(x)$ and there exists $Y \in \psi$ such that $Y \cap M \neq \emptyset$. Hence, $x \in NR^* M$. Thus $NR_* M \subseteq M \subseteq NR^* M$.

(ii) Proof of (ii) follows directly from Definition 9.

(iii) From property (i), we have $X \subseteq NR^* X$. Since X is the universe set $NR^* X = X$. From Definition 9, we have $NR_* X = \cup\{Y \in \psi : Y \subseteq X\}$, but for all $x \in X$, there exists $S_a(x) \in \psi$ such that $x \in S_a(x) \subseteq X$. Hence, $NR_* X = X$. Thus $NR_* X = NR^* X = X$.

(iv) Let $M \subseteq Z$ and $x \in NR_* M$. There exists $Y \in \psi$ such that $x \in Y \subseteq M$. However, $M \subseteq Z$, thus $x \in Y \subseteq Z$. Hence, $x \in NR_* Z$. Consequently, $NR_* M \subseteq NR_* Z$.

(v) Let $M \subseteq Z$ and $x \in NR^* M$. There exists $Y \in \psi$ such that $x \in Y$, $Y \cap M \neq \emptyset$. However, $M \subseteq Z$, thus $Y \cap Z \neq \emptyset$. Hence, $x \in NR^* Z$. Thus $NR^* M \subseteq NR^* Z$.

(vii) Let $x \in NR_*(M \cap Z) = \cup\{Y \in \psi : Y \subseteq (M \cap Z)\}$. There exists $Y \in \psi$ such that $x \in Y \subseteq (M \cap Z)$, $x \in Y \subseteq M$ and $x \in Y \subseteq Z$. Consequently, $x \in NR_* M$ and $x \in NR_* Z$, implying $x \in NR_* M \cap NR_* Z$. Thus $NR_*(M \cap Z) \subseteq NR_* M \cap NR_* Z$.

(viii) Let $x \notin NR_*(M \cup Z) = \cup\{Y \in \psi : Y \subseteq M \cup Z\}$. For all $a \in M$, $x \in Y$, we have $Y \not\subseteq M \cup Z$, thus for all $a \in A$, $x \in Y$, we have $Y \not\subseteq M$ and $Y \not\subseteq Z$. Consequently, $x \notin NR_* M$ and $x \notin NR_* Z$, implying $x \notin NR_* M \cup NR_* Z$. Thus $NR_*(M \cup Z) \supseteq NR_* M \cup NR_* Z$.

(ix) Let $x \in NR^*(M \cap Z) = \cup\{Y \in \psi : Y \cap (M \cap Z) \neq \emptyset\}$. There exists $Y \in \psi$ such that $x \in Y$, $Y \cap (M \cap Z) \neq \emptyset$, $Y \cap M \neq \emptyset$ and $Y \cap Z \neq \emptyset$. Consequently, $x \in NR^* M$ and $x \in NR^* Z$, implying $x \in NR^* M \cap NR^* Z$. Thus $NR^*(M \cap Z) \subseteq NR^* M \cap NR^* Z$.

(x) Let $x \notin NR^*(M \cup Z) = \cup\{Y \in \psi : Y \cap (M \cup Z) \neq \emptyset\}$. For all $a \in A$, $x \in Y$, we have $Y \cap (M \cup Z) = \emptyset$. For all $a \in A$, $x \in Y$, we have $Y \cap M = \emptyset$ and $Y \cap Z = \emptyset$. Consequently, $x \notin NR^* M$ and $x \notin NR^* Z$, implying $x \notin NR^* M \cup NR^* Z$. Therefore, $NR^*(M \cup Z) \supseteq NR^* M \cup NR^* Z$. In addition, let $x \in NR^*(M \cup Z) = \cup\{Y \in \psi : Y \cap (M \cup Z) \neq \emptyset\}$, and thus, there exists $Y \in \psi$ such that $x \in Y$, $Y \cap (M \cup Z) \neq \emptyset$. It follows that, $Y \cap M \neq \emptyset$ or $Y \cap Z \neq \emptyset$. Consequently, $x \in NR^* M$ or $x \in NR^* Z$. Hence, $x \in NR^* M \cup NR^* Z$, and $NR^* M \cup NR^* Z \supseteq NR^*(M \cup Z)$. Thus $NR^* M \cup NR^* Z = NR^*(M \cup Z)$. □

The converse of property (i) in Proposition 3 does not hold, as shown below.

Example 4. *From Example 1, if* $M = \{x_1, x_4\}$, *then* $NR_* M = \{x_1\}$ *and* $NR^* M = X$. *Hence,* $NR_* M \neq M$ *and* $M \neq NR^* M$.

The converse of property (iv) in Proposition 3 does not hold, as shown below.

Example 5. *From Example 1, if* $M = \{x_2\}$ *and* $Z = \{x_1, x_2\}$, *then* $NR_* M = \emptyset$, $NR_* Z = \{x_1, x_2\}$. *Thus* $NR_* M \neq NR_* Z$.

The converse of property (v) in Proposition 3 does not hold, as shown below.

Example 6. *According to Example 1. Let* $A = \{a_1\}$, *then* $\psi = \{\{x_1\}, \{x_5\}, \{x_1, x_2\}, \{x_1, x_3\}, \{x_1, x_3, x_4\}\}$. *If* $M = \{x_2\}$ *and* $Z = \{x_1, x_2\}$, *then* $NR^* M = \{x_1, x_2\}$ *and* $NR^* Z = \{x_1, x_2, x_3, x_4\}$. *Hence,* $NR^* M \neq NR^* Z$.

The converse of property (vi) in Proposition 3 does not hold, as shown below.

Example 7. *From Example 1, if* $M = \{x_1, x_3, x_4\}$ *and* $Z = \{x_1, x_4, x_5\}$, *then* $NR_* M = \{x_1, x_3, x_4\}$, $NR_* Z = \{x_1, x_4, x_5\}$ *and* $NR_*(M \cap Z) = \{x_1\}$. *Hence,* $NR_*(M \cap Z) \neq NR_* M \cap NR_* Z$.

The converse of property (vii) in Proposition 3 does not hold, as shown below.

Example 8. *From Example 1, if* $M = \{x_1\}$ *and* $Z = \{x_2\}$, *then* $NR_* M = \{x_1\}$, $NR_* Z = \emptyset$ *and* $NR_*(M \cup Z) = \{x_1, x_2\}$. *Hence,* $NR_*(M \cup Z) \neq NR_* M \cup NR_* Z$.

The converse of property (viii) in Proposition 3 does not hold, as shown below.

Example 9. *From Example 6, if* $M = \{x_2, x_5\}$ *and* $Z = \{x_1, x_3, x_5\}$, *then* $NR^* M = \{x_1, x_2, x_5\}$, $NR^* Z = X$ *and* $NR^*(M \cap Z) = \{x_5\}$. *Hence,* $NR^*(M \cap Z) \neq NR^* M \cap NR^* Z$.

Proposition 4. *Let (H, A) be an (NSS) on a universe X, and let $M, Z \subseteq X$. Then the properties below hold.*
(i) $NR_* \ NR_* M = NR_* M$.
(ii) $NR^* \ NR^* M \supseteq NR^* M$.
(iii) $NR_* \ NR^* M = NR^* M$.
(iv) $NR^* \ NR_* M \supseteq NR_* M$.
(v) $NR_* M^c \supseteq [NR^* M]^c$.
(vi) $NR^* M^c \supseteq [NR_* M]^c$.

Proof. (i) Let $W = NR_* M$ and $x \in W = \cup\{Y \in \psi : Y \subseteq M\}$. Then, for some $a \in A$, $x \in Y \subseteq W$. So, $x \in NR_* W$. Therefore, $W \subseteq NR_* W$. Hence $NR_* M \subseteq NR_* \ NR_* M$. From property (i) of Proposition 3, $NR_* M \subseteq M$ and using property (iv) of Proposition 3, we obtain $NR_* \ NR_* M \subseteq NR_* M$. Subsequently, $NR_* \ NR_* M = NR_* M$.

(ii) Let $W = NR^* M$. Using property (i) of Proposition 3, we get $W \subseteq NR^* W$. Hence $NR^* NR^* M \supseteq NR^* M$.

(iii) Let $W = NR^* M$. Using property (i) of Proposition 3, we get $NR_* W \subseteq W$. Let $x \in W = \cup\{Y \in \psi : Y \cap M \neq \emptyset\}$, thus there exists $Y \in \psi$ where $x \in Y \subseteq W$ such that $x \in NR_* W$. Subsequently, $W \subseteq NR_* W$, with $W = NR_* W$, and $W = NR^* M$. Therefore, $NR_* NR^* M = NR^* M$.

(iv) Let $W = NR_* M$. Using property (i) of Proposition 3, we get $W \subseteq NR^* W$. Hence $NR^* NR_* M \supseteq NR_* M$.

(v) Let $x \notin NR_* M^c$. For all $Y \in \psi$ such that $x \in Y$, we have $Y \not\subset M^c$ and $Y \cap M^c = \emptyset$. Thus $Y \cap M \neq \emptyset$, where $x \in NR^* M$ but $x \notin [NR^* M]^c$. Therefore, $NR_* M^c \supseteq [NR^* M]^c$.

(vi) From property (v) of Proposition 4, we get $NR_* M^c \supseteq [NR^* M]^c$.

Therefore, $NR_* M \supseteq [NR^* M^c]^c$ meaning that $NR^* M^c \supseteq [NR_* M]^c$. □

The converse of property (ii) in Proposition 4 does not hold, as shown below.

Example 10. *From Example 6, if $X = \{x_2\}$, we will have $NR^* X = \{x_1, x_2\}$ and $NR^* \ NR^* X = \{x_1, x_2, x_3, x_4\}$. Therefore, $NR^* \ NR^* X \neq NR^* X$.*

The converse of property (iv) in Proposition 4 does not hold, as shown below.

Example 11. *From Example 6, if $M = \{x_1, x_4\}$, then $NR_* M = \{x_1\}$ and $NR^* \ NR_* M = \{x_1, x_2, x_3, x_4\}$. Hence, $NR^* \ NR_* M \neq NR_* M$.*

The converse of property (v) in Proposition 4 does not hold, as shown below.

Example 12. *From Example 6, if $M = \{x_3\}$, then $NR_* M^c = \{x_1, x_2, x_5\}$ and $[NR^* M]^c = \{x_2, x_5\}$. Hence, $NR_* M^c \neq [NR^* M]^c$.*

The converse of property (vi) in Proposition 4 does not hold, as shown below.

Example 13. *From Example 6, if $M = \{x_1, x_2, x_4, x_5\}$, then $[NR_* M]^c = \{x_3, x_4\}$ and $NR^* M^c = \{x_1, x_3, x_4\}$. Hence, $[NR_* M]^c \neq NR^* M^c$.*

Proposition 5. *Let (H, A) be an (NSS) on a universe X, and let $M, Z \subseteq X$. Then,*

$$NR_* (M - Z) \subseteq NR_* M - NR_* Z.$$

Proof. Let $u \in NR_* (M - Z) = \cup\{Y \in \psi : Y \subseteq (M - Z)\}$. There exists $Y \in \psi$ where $u \in Y \subseteq (M - Z)$, $u \in Y \subseteq M$ and $u \in Y \not\subseteq Z$. Subsequently, $u \in NR_* M$ but $u \notin NR_* Z$, hence $u \in NR_* M - NR_* Z$. Thus, $NR_* (M - Z) \subseteq NR_* M - NR_* Z$. □

The converse of Proposition 5 does not hold, as shown below.

Example 14. *From Example 1, if* $M = \{x_1, x_3, x_5\}$ *and* $Z = \{x_1, x_5\}$*, then* $NR_* M = \{x_1, x_3, x_5\}$*,* $NR_* Z = \{x_1, x_5\}$*,* $NR_*(M - Z) = \emptyset$ *and* $NR_* M - NR_* Z = \{x_3\}$*. Hence,* $NR_* M - NR_* Z \neq NR_*(M - Z)$*.*

Proposition 6. *Let* (H, A) *be an (NSS) on a universe X, and let* $M, Z \subseteq X$*. Then, the property below holds.*

$$NR^*(M - Z) \neq NR^* M - NR^* Z.$$

Example 15. *From Example 6, if* $M = \{x_1, x_3, x_5\}$ *and* $Z = \{x_1, x_5\}$*, then* $NR^* M = X$*,* $NR^* Z = X$*,* $NR^*(M - Z) = \{x_1, x_3, x_4\}$ *and* $NR^* M - NR^* Z = \emptyset$*. Hence,* $NR^* M - NR^* Z \neq NR^*(M - Z)$*.*

4. The Concepts of Neutrosophic Soft Rough Set

We will now define the neutrosophic soft rough concepts as a generalization of rough concepts, illustrated by examples.

Definition 10. *Let* (H, A) *be an (NSS) on a universe X and let* $M \subseteq X$*. A subset* $M \subseteq X$ *is called*
(i) NSR-definable (NSR-exact) set, if $NR_* M = NR^* M = M$*.*
(ii) Internally NSR-definable set, if $NR_* M = M$ *and* $NR^* M \neq M$*.*
(iii) Externally NSR-definable set, if $NR_* M \neq M$ *and* $NR^* M = M$*.*
(iv) NSR-rough set, if $NR_* M \neq M$ *and* $NR^* M \neq M$*.*

Example 16. *From Example 6, we have* $\{x_1, x_2, x_3, x_4\}$ *is NSR-definable set, whereas* $\{x_1\}$*,* $\{x_5\}$*,* $\{x_1, x_2\}$*,* $\{x_1, x_3\}$*,* $\{x_1, x_5\}$*,* $\{x_1, x_3, x_4\}$*,* $\{x_1, x_3, x_5\}$*,* $\{x_1, x_2, x_3, x_5\}$*,* $\{x_1, x_3, x_4, x_5\}$ *are internally NSR-definable sets, whereas the rest of the subsets of X are NSR-rough sets.*

The degree of NSR-crispness (exactness) of any subset $M \subseteq X$, can be determined by using NSR_p-accuracy measure denoted by $C_{NSR} M$, which is defined as follows.

Definition 11. *Let* (H, A) *be an (NSS) on a universe X and let* $M \subseteq X$*. Then,*

$$C_{NSR} M = \frac{|NR_* M|}{|NR^* M|},$$

where $M \neq \phi$ *and* $\mid M \mid$ *denotes the cardinality of sets.*

Remark 2. *Let* (H, A) *be an (NSS) on a universe X. A subset* $M \subseteq X$ *is NSR-definable, if and only if,* $C_{NSR} M = 1$*.*

Neutrosophic soft rough (NSR)-membership function is defined below.

Definition 12. *Let* (H, A) *be an (NSS) on a universe X and let* $M \subseteq X$*.*
NSR-membership function of an element m to a set M denoted by $N_M(m)$ *is defined as follows.*

$$N_M(m) = \frac{|m_A \cap M|}{|m_A|},$$

where $S_A(m) = \cap\{S_a(m) : a \in A\}$ *and* $S_a(m)$ *is a neutrosophic right neighborhood defined in Definition 7.*

Proposition 7. *Let* (H, A) *be an (NSS) on a universe X,* $M \subseteq X$ *and let* $N_M(m)$ *be the membership function defined in Definition 12. Then the properties below holds:*

$$N_M(m) \in [0, 1]$$

Proof. From Definition 12, we have $\phi \subseteq S_A(m) \cap M \subseteq S_A(m)$, then $0 \leq |S_A(m) \cap M| \leq |S_A(m)|$ and $0 \leq \frac{|S_A(m) \cap M|}{|S_A(m)|} \leq 1$, thus $N_M(m) \in [0,1]$. □

Proposition 8. *Let (H, A) be an (NSS) on a universe X and let $M \subseteq X$. Then,*

$$m \in M \Leftrightarrow N_M(m) = 1$$

Proof. Let $N_M(m) = 1$, if and only if, $\frac{|S_A(m) \cap M|}{|S_A(m)|} = 1$, if and only if, $|S_A(m) \cap M| = |S_A(m)|$, if and only if, $S_A(m) \subseteq M$. However, from Proposition 2, we have R_a is a reflexive relation for all $a \in A$. Hence $m \in S_a(m)$, $\forall a \in A$. It follows that $m \in S_A(m)$. Hence $m \in M$, if and only if, $N_M(m) = 1$. □

Proposition 9. *Let (H, A) be an (NSS) on a universe X and let $M \subseteq X$. If $M_1 \subseteq M_2$, then the properties below hold:*
(i) $N_{M_1}(m) \leq N_{M_2}(m)$
(ii) $N_{NR_*M_1}(m) \leq N_{NR_*M_2}(m)$
(iii) $N_{NR^*M_1}(m) \leq N_{NR^*M_2}(m)$

Proof. (i) If $M_1 \subseteq M_2$, it follows that $S_A(m) \cap M_1 \subseteq S_A(m) \cap M_2$, then $|S_A(m) \cap M_1| \leq |S_A(m) \cap M_2|$ and $\frac{|S_A(m) \cap M_1|}{|S_A(m)|} \leq \frac{|S_A(m) \cap M_2|}{|S_A(m)|}$, thus $N_{M_1}(m) \leq N_{M_2}(m)$.
(ii) We get the proof directly from property (i) of Proposition 9 and property (iv) of Proposition 3.
(iii) We get the proof directly from property (ii) of Proposition 9 and property (v) of Proposition 3. □

Proposition 10. *Let (H, A) be an (NSS) on a universe X and let $M \subseteq X$, then the following properties hold:*
(i) $N_{NR_*M}(m) \leq N_M(m)$
(ii) $N_M(m) \leq N_{NR^*M}(m)$
(iii) $N_{NR_*M}(m) \leq N_{NR^*M}(m)$

Proof. The proof of properties (i), (ii) and (iii) can be obtained directly from Propositions 3 and property (i) of Proposition 9. □

Definition 13. *Let (H, A) be a (NSS) on a universe X and let $m \in X$, $M \subseteq X$. NSR-membership relations, denoted by $\underline{\in}_{NSR}$ and $\overline{\in}_{NSR}$, are defined below.*

$$m \underline{\in}_{NSR} M, \ \ if \ \ m \in NR_*M,$$

$$m \overline{\in}_{NSR} M, \ \ if \ \ m \in NR^*M.$$

Proposition 11. *Let (H, A) be an (NSS) on a universe X and let $m \in X$, $M \subseteq X$. Then,*

$$(i) m \underline{\in}_{NSR} M \longrightarrow m \in M$$

$$(ii) m \ \overline{\notin}_{NSR} M \longrightarrow m \notin M$$

Proof. Proof of (i) and (ii) follows directly from Definition 13 and Proposition 3. □

The following example illustrates that the converse of properties (i) and (ii) in Proposition 11 do not hold.

Example 17. *In Example 1, if $M = \{x_2, x_5\}$, then $NR_*M = \{x_5\}$ and $NR^*M = X$. Hence $x_2 \underline{\notin}_{NSR} M$, although $x_2 \in M$ and $x_3 \notin M$, although $x_3 \overline{\in}_{NSR} M$.*

Proposition 12. *Let (H, A) be an (NSS) on a universe X, and let $M \subseteq X$. Then the properties below hold:*

$$(i) m \underline{\in}_{NSR} M \longrightarrow N_M(m) = 1$$

$$(ii) N_M(m) = 1 \longrightarrow m \overline{\in}_{NSR} M$$

Proof. The proof of properties (i) and (ii) can be obtained directly from Definition 13 and Propositions 11. □

The converse of property (ii) in Proposition 12 does not hold, as shown below.

Example 18. *In Example 1, if $M = \{x_1, x_4\}$, then $NR_* M = \{x_1\}$ and $S_A(x_4) = \{x_4\}$, it follows that $N_M(x_4) = \frac{|S_A(x_4) \cap M|}{|S_A(x_4)|} = \frac{|\{x_4\}|}{|\{x_4\}|} = 1$, although $x_4 \underline{\notin}_{NSR} M$.*

The converse of Proposition 13 does not hold, as shown below.

Example 19. *In Example 1, if $M = \{x_1, x_4\}$, then $NR^* M = X$ and $S_A(x_3) = \{x_1, x_3\}$, it follows that x_3 $\overline{\in}_{NSR} M$, although $N_M(x_3) = \frac{|\{x_1\}|}{|\{x_1, x_3\}|} = \frac{1}{2} \neq 1$.*

Proposition 13. *Let (H, A) be an (NSS) on a universe X and let $M \subseteq X$. Then,*

$$m \overline{\notin}_{NSR} M \longrightarrow N_M(m) = 0$$

Proof. Let $m \overline{\notin}_{NSR} M$, then $m \notin NR^* M$, also from Definition 9, we conclude that $S_a(m) \cap M = \phi, \forall a \in A$, but $S_A(m) = \cap\{S_a(m) : a \in A\}$. Thus $S_A(m) \cap M = \phi$ and $|S_A(m) \cap M| = 0$, and hence $N_M(m) = 0$. □

The following example illustrates that the converse of Proposition 13 does not hold.

Example 20. *In Example 1, if $M = \{x_2\}$, then $NR^* M = \{x_1, x_2\}$ and $S_A(x_1) = \{x_1\}$. It follows that $x_1 \overline{\in}_{NSR} M$, although $N_M(x_1) = \frac{|\{x_1\} \cap \{x_2\}|}{|\{x_1\}|} = 0$.*

Proposition 14. *Let (H, A) be an (NSS) on a universe X and let $M \subseteq X$. Then,*
(i) $N_M(m) = 0 \longrightarrow m \notin M$
(ii) $N_M(m) = 0 \longrightarrow m \underline{\notin}_{NSR} M$

Proof. The proof of properties (i) and (ii) are straightforward and therefore are omitted. □

The converse of property (i) in Proposition 14 does not hold, as shown below.

Example 21. *In Example 1, if $M = \{x_1, x_3, x_4\}$, then $S_A(x_2) = \{x_1, x_2\}$ and $N_M(x_2) = \frac{|\{x_1\}|}{|\{x_1, x_2\}|} = \frac{1}{2} \neq 0$, although $x_2 \notin M$.*

The converse of property (ii) in Proposition 14 does not hold, as shown below.

Example 22. *In Example 1, if $M = \{x_1, x_4, x_5\}$, then $NR_* M = \{x_1, x_4, x_5\}$ and $x_{2_A} = \{x_1, x_2\}$, it follows that $N_M(x_2) = \frac{|\{x_1\}|}{|\{x_1, x_2\}|} = \frac{1}{2} \neq 0$, although $x_2 \underline{\notin}_{NSR} M$.*

Definition 14. *Let (H, A) be an (NSS) on a universe X and let $M, Z \subseteq X$. NSR-inclusion relations, denoted by $\underset{\rightarrow}{\subset}_{NSR}$ and $\overset{\rightarrow}{\subset}_{NSR}$, are defined as follows.*

$$M \underset{\rightarrow}{\subset}_{NSR} Z, \text{ if } NR_* M \subseteq NR_* Z,$$

$$M \overrightarrow{\subset}_{NSR} Z, \text{ if } NR^{*}M \subseteq NR^{*}Z.$$

Proposition 15. *Let (H, A) be an (NSS) on a universe X and let $M, Z \subseteq X$. Then,*

$$M \subseteq Z \longrightarrow M \underset{\rightarrow NSR}{\subset} Z \text{ and } M \overrightarrow{\subset}_{NSR} Z.$$

Proof. It can be directly obtained from Proposition 3. □

The inverse of Proposition 15 does not hold, as shown below.

Example 23. *In Example 6, if $M = \{x_1, x_4\}$ and $Z = \{x_1, x_5\}$, then $NR_*M = \{x_1\}$, $NR_*Z = \{x_1, x_5\}$, $NR^*M = \{x_1, x_2, x_3, x_4\}$ and $NR^*Z = X$. Hence, $M \underset{\rightarrow NSR}{\subset} Z$ and $M \overrightarrow{\subset}_{NSR} Z$, although $M \not\subseteq Z$.*

Proposition 16. *Let (H, A) be an (NSS) on a universe X and let $M, Z \subseteq X$. If $M \underset{\rightarrow NSR}{\subset} Z$, then the following properties hold:*
(i) $N_{NR_*M}(m) \le N_{NR_*Z}(m)$
(ii) $N_{NR_*M}(m) \le N_Z(m)$
(iii) $N_{NR_*M}(m) \le N_{NR^*Z}(m)$

Proof. The proof can be directly obtained from Definition 14 and Proposition 9. □

Proposition 17. *Let (H, A) be an (NSS) on a universe X and let $M, Z \subseteq X$. If $M \overrightarrow{\subset}_{NSR} Z$, then the properties below hold:*
(i) $N_{NR^*M}(m) \le N_{NR^*Z}(m)$
(ii) $N_M(m) \le N_{NR^*Z}(m)$
(iii) $N_{NR_*M}(m) \le N_{NR^*Z}(m)$

Proof. It can be directly obtained from Definition 14 and Proposition 9. □

Definition 15. *Let (H, A) be an (NSS) on a universe X and let $M, Z \subseteq X$. NSR-equality relations are defined as follows.*

$$M \eqsim_{NSR} Z, \text{ if } NR_*M = NR_*Z,$$
$$M \simeq_{NSR} Z, \text{ if } NR^*M = NR^*Z,$$
$$M \approx_{NSR} Z, \text{ if } M \eqsim_{NSR} Z \text{ and } M \simeq_{NSR} Z.$$

The example below illustrates Definition 15.

Example 24. *In Example 6, suppose $M_1 = \{x_2\}$, $M_2 = \{x_3\}$, $M_3 = \{x_1, x_2\}$, $M_4 = \{x_1, x_4\}$, $M_5 = \{x_3, x_5\}$ and $M_6 = \{x_4, x_5\}$. Then, $NR_*M_1 = NR_*M_2 = \emptyset$, $NR^*M_3 = NR^*M_4 = \{x_1, x_2, x_3, x_4\}$, $NR_*M_5 = NR_*M_6 = \{x_5\}$ and $NR^*M_5 = NR^*M_6 = \{x_1, x_3, x_4, x_5\}$. Consequently, $M_1 \eqsim_{NSR} M_2$, $M_3 \simeq_{NSR} M_4$ and $M_5 \approx_{NSR} M_6$.*

Proposition 18. *Let (H, A) be an (NSS) on a universe X and let $M, Z \subseteq X$. Then,*
(i) $M \eqsim_{NSR} NR_*M$
(ii) $M = Z \longrightarrow M \approx_{NSR} Z$
(iii) $M \subseteq Z, Z \eqsim_{NSR} \emptyset \longrightarrow M \eqsim_{NSR} \emptyset$
(iv) $M \subseteq Z, M \eqsim_{NSR} X \longrightarrow Z \eqsim_{NSR} X$
(v) $M \subseteq Z, Z \simeq_{NSR} \emptyset \longrightarrow M \simeq_{NSR} \emptyset$
(vi) $M \subseteq Z, M \simeq_{NSR} X \longrightarrow Z \simeq_{NSR} X$

Proof. It can be directly obtained from Propositions 3 and 4. □

Proposition 19. *Let (H, A) be an (NSS) on a universe X and let $M, Z \subseteq X$. If $M \eqsim_{NSR} Z$, then the following properties hold:*
(i) $N_{NR_*M}(m) = N_{NR_*Z}(m)$
(ii) $N_{NR_*M}(m) \leq N_Z(m)$
(iii) $N_{NR_*M}(m) \leq N_{NR^*Z}(m)$

Proof. The proof of properties (i), (ii) and (iii) can be obtained directly from Definition 15 and Proposition 9. □

Proposition 20. *Let (H, A) be an (NSS) on a universe X and let $M, Z \subseteq X$. If $M \simeq_{NSR} Z$, then the following properties hold:*
(i) $N_{NR^*M}(m) = N_{NR^*Z}(m)$
(ii) $N_Z(m) \leq N_{NR^*M}(m)$
(iii) $N_{NR_*Z}(m) \leq N_{NR^*M}(m)$

Proof. The proof of properties (i), (ii) and (iii) can be obtained directly from Definition 15 and Proposition 9. □

5. Application of the Proposed Neutrosophic Soft Rough Model in Decision Making

This section presents an employment of the suggested neutrosophic soft rough approximations to the multi attribute decision making problems.

Consider Example 1 and suppose that we are requested to make a decision about the most desirable house based on the given attributes. To solve this problem, we apply the following decision steps.

Step 1: Input the NSS (H, A).
Step 2: Compute the accuracy measure to each alternative (house) in the given NSS (H, A), separately.
Step 3: Choose the (element) alternative which has the highest accuracy measure as the optimal solution. If there is more than one alternative with highest accuracy measure, we do the following steps.
Step 4: Consider the alternatives that have the highest accuracy measure and create a new NSS $(\widehat{H}, A)$, which consists of the selected alternatives x_i and the corresponding parameters a_j.
Step 5: Find the values of $S_{ij} = T_{a_j}(x_i) + I_{a_j}(x_i) - F_{a_j}(x_i)$, where T, I and F represent, respectively the truth, indeterminacy and falsity membership functions of the NSS $(\widehat{H}, A)$.
Step 6: Compute the score $C(x_i) = \sum_{j=1}^{m} S_{ij}$ of each element of the selected alternatives, where m is the number of the parameters.
Step 7: Determine the value of the highest score. Then the decision is to choose the alternative with the highest score. If more than one alternative has the maximum score, then any one of those alternatives can be the optimal solution.

Table 2 gives the accuracy measure to all alternatives.

Table 2. Accuracy measures of the alternatives.

Alternatives	x_1	x_2	x_3	x_4	x_5
$C_{NSR}M$	$\frac{1}{2}$	0	0	0	$\frac{1}{2}$

From Table 2, it is clear that there are two alternatives (houses) with the highest accuracy measure which are house 1 and house 5. Thus, we proceed to the next steps and create the NSS $(\widehat{H}, A)$ of the considered alternatives as in Table 3.

Table 3. Tabular representation of $(\widehat{H}, A)$.

X	a_1	a_2	a_3	a_4
x_1	(0.6, 0.6, 0.2)	(0.8, 0.4, 0.3)	(0.7, 0.4, 0.3)	(0.8, 0.6, 0.4)
x_5	(0.8, 0.2, 0.3)	(0.8, 0.3, 0.2)	(0.7, 0.3, 0.4)	(0.9, 0.5, 0.7)

Now we calculate the values of S_{ij} to the considered alternatives (houses) as in the Table 4.

Table 4. Values of S_{ij} to the considered houses.

X	a_1	a_2	a_3	a_4
x_1	1	0.9	0.8	1
x_5	0.7	0.9	0.6	0.7

The scores $C(x_i)$ of the considered houses can be shown as in Table 5.

Table 5. Scores of the considered houses.

X	**Score**
x_1	3.7
x_5	2.9

From Table 5, it is clear that the house x_1 gets the highest score which is 3.7. Thus the decision is to choose house 1 as the appropriate solution under the parameter set A.

6. Discussion

We will discuss the features and limitations of our model by conducting a comparison with the existing models. Discussion will begin on the features of the proposed model before moving on to its limitations.

To illustrate the features of our model, we compare it with traditional rough approach [5,33], neutrosophic rough set approaches [10,23–25], and fuzzy and intuitionistic fuzzy rough soft approaches [34–36].

We begin by making a comparison between the proposed neutrosophic soft rough approach and the traditional rough approach. The following Table 6 shows the properties of both traditional rough and the proposed neutrosophic soft rough approaches.

Table 6. Properties of traditional rough and neutrosophic soft rough.

Traditional Rough Properties	**Neutrosophic Soft Rough Properties**
$\underline{E}(M \cap Z) = \underline{E}(M) \cap \underline{E}(Z)$	$NR_*(M \cap Z) \subseteq NR_* M \cap NR_* Z$
$\overline{E}(\overline{E}(M)) = \overline{E}(M)$	$NR^* NR^* M \supseteq NR^* M$
$\overline{E}(\underline{E}(M)) = \underline{E}(M)$	$NR^* NR_* M \supseteq NR_* M$
$\underline{E}(M^c) = [\overline{E}(M)]^c$	$NR_* M^c \supseteq [NR^* M]^c$
$\overline{E}(M^c) = [\underline{E}(M)]^c$	$NR^* M^c \supseteq [NR_* M]^c$

In the proposed neutrosophic soft rough approach, let us consider the NSS (H, A) on the universe X, where $x \in X$ and $M \subseteq X$. If we consider the case where $T_a(x_i) > 0.5$, then $a(x) = 1$, otherwise $a(x) = 0$. Thus, the neutrosophic right neighborhood of an element x is replaced by the following equivalence class $[x] = \{x_i \in X : a(x_i) = a(x), a \in A\}$. Subsequently, the neutrosophic soft rough set approximates to that of Pawlak, i.e., the lower and upper approximations of the proposed model will be $NR_*M = \{x \in X : [x] \subseteq M\}$ and $NR^*M = \{x \in X : [x] \cap M \neq \emptyset\}$. Therefore, all properties of traditional rough set approximations will be satisfied.

We continue our discussion by comparing the proposed neutrosophic soft rough approach with other approaches which combine rough set to neutrosophic set [10,23–25]. It can be seen that these approaches have the inadequacy of the parametrization tool to facilitate the representation of parameters, while the soft set in the proposed model can represent the problem parameters in a more complete manner. This feature makes the proposed model superior to these models and other models that do not incorporate soft sets into their structures.

Now, we compare the proposed model to the fuzzy and intuitionistic fuzzy soft rough approaches [34–36]. The proposed approach combines rough set to neutrosophic soft set which is a generalisation of fuzzy and intuitionistic fuzzy soft set. Neutrosophic soft sets consider three membership functions instead of two as in the intuitionistic fuzzy soft set and one as in the fuzzy soft set. Fuzzy sets handle the uncertainty in data, intuitionistic fuzzy sets deal with ambiguous and incomplete data, while neutrosophic sets hold the features of all of the aforementioned sets in addition to its ability to handle the indeterminacy in data. Thus, combining neutrosophic soft sets to the rough sets provides the opportunity to deal with complicated data that cannot be handled by other models. From Example 1, it can be seen that fuzzy soft set and intuitionistic fuzzy soft set cannot describe the data presented by the neutrosophic soft set, which makes these models incapable to be applied directly on decision making problems with neutrosophic soft information. Conversely, the newly proposed model can directly address fuzzy and intuitionistic fuzzy soft rough set based decision making, since the intuitionistic fuzzy soft set is a special case of neutrosophic soft set and can be easily represented in the form of neutrosophic soft set. For example, the intuitionistic fuzzy soft value $(0.4, 0.5)$ can be represented as $(0.4, 0.1, 0.5)$ by means of neutrosophic soft set, since the sum of the degrees of membership, nonmembership and indeterminacy of an intuitionistic fuzzy value equals to 1. Note that the indeterminacy degree in intuitionistic fuzzy set is provided by default and cannot be defined alone unlike the neutrosophic set where the indeterminacy is defined independently and quantified explicitly.

Then we enlarge the discussion by presenting two limitations of the proposed model: (1) It cannot be used to solve multi attribute group decision making problems which incorporate the opinions of more than one expert. For more illustration, if we consider Example 1 and suppose that there are three experts who are requested to provide their opinions on each house under each (attribute) parameter, then we need a mechanism to incorporate the opinions of the three experts in one model (neutrosophic soft set), otherwise, we have to construct three neutrosophic soft sets and this increases the amount of both mathematical calculations and investigation of several operators in incorporating three neutrosophic soft sets to find out the optimal solution; (2) There exist some neutrosophic soft set based decision making problems in which the proposed algorithm is likely to get an empty decision (optimum) set. Consider Example 1, and consider the NSS (G, A) as in Table 7.

Table 7. Tabular representation of the NSS (G, A).

X	a_1	a_2	a_3	a_4
x_1	(0.6, 0.6, 0.2)	(0.8, 0.4, 0.3)	(0.7, 0.4, 0.3)	(0.8, 0.6, 0.4)
x_2	(0.4, 0.6, 0.6)	(0.6, 0.2, 0.4)	(0.6, 0.4, 0.3)	(0.7, 0.6, 0.6)
x_3	(0.6, 0.4, 0.2)	(0.8, 0.2, 0.2)	(0.7, 0.4, 0.3)	(0.7, 0.6, 0.4)
x_4	(0.6, 0.3, 0.3)	(0.8, 0.2, 0.2)	(0.5, 0.2, 0.6)	(0.7, 0.5, 0.6)
x_5	(0.6, 0.6, 0.2)	(0.8, 0.4, 0.3)	(0.7, 0.3, 0.4)	(0.8, 0.6, 0.4)

We then obtain the family of all neutrosophic right neighborhood $\psi = \{ \{x_1, x_3\}, \{x_1, x_5\}, \{x_1, x_2, x_3\}, \{x_1, x_3, x_5\}, \{x_1, x_2, x_3, x_5\}, \{x_1, x_3, x_4, x_5\}, \{x_3, x_4\}, X \}$.

As a result, the accuracy measure table is as follows.

From Table 8, it can be seen that the accuracy measure to each alternative (house) equals zero, which means that none of the houses can be selected as a candidate to be an optimal solution. Thus the proposed approach fails to handle this case.

Table 8. Accuracy measures of the alternatives in the NSS (G, A).

Alternatives	x_1	x_2	x_3	x_4	x_5
$C_{NSR}M$	0	0	0	0	0

7. Conclusions

We proposed a novel approach to rough sets based on neutrosophic soft sets and deduced that the traditional rough approach is a special case of the proposed approach. The lower and upper neutrosophic soft rough approximations are defined and their properties are verified. We have further defined some essential neutrosophic soft rough concepts such as neutrosophic soft rough (NSR) definability, neutrosophic soft rough (NSR)-membership relations and functions. Properties of these concepts are deduced, proven and shown by several examples. In addition, we provided an algorithm based on the proposed neutrosophic soft rough sets approximations. Finally, we have made a comparative analysis and a discussion to reveal the features and limitations of the proposed model. For the future prospects, we will extend this model by using topological structures and commit to exploring the application of the proposed model to data mining and attribute reduction.

Author Contributions: E.M. proposed the concept of neutrosophic right neighborhood; A.A.-Q. and N.H. studied the neutrosophic soft rough set approximations and their properties; A.A.-Q. introduced some concepts on neutrosophic soft rough set and provided some illustrated examples; N.H. edited the paper and all authors wrote it.

Funding: This research was funded by Universiti Kebangsaan Malaysia GUP-2017-105.

Acknowledgments: We are indebted to Universiti Kebangsaan Malaysia for providing financial support and facilities for this research under the grant GUP-2017-105.

Conflicts of Interest: Authors declare that they have no conflict of interest.

References

1. Zadeh, L.A. Fuzzy sets. *Inf. Control* **1965**, *8*, 338–353. [CrossRef]
2. Atanassov, K. Intuitionistic fuzzy sets. *Fuzzy Sets Syst.* **1986**, *20*, 87–96. [CrossRef]
3. Gau, L.W.; Buehrer, D.J. Vague sets. *IEEE Trans Syst Man Cybern.* **1993**, *23*, 610–614. [CrossRef]
4. Gorzalzany, M.B. A method of inference in approximate reasoning based on interval-valued fuzzy sets. *Fuzzy Sets Syst.* **1987**, *21*, 1–17. [CrossRef]
5. Pawlak, Z. Rough sets. *Int. J. Comput. Inf. Sci.* **1982**, *11*, 341–356. [CrossRef]
6. Slowinski, R.; Vanderpooten, D. A generalized definition of rough approximations based on similarity. *IEEE Trans. Knowl. Data Eng.* **2000**, *12*, 331–336. [CrossRef]
7. Greco, S.; Matarazzo, B.; Slowinski, R. Rough approximation of a preference relation by dominance relations. *Eur. J. Oper. Res.* **1999**, *117*, 63–83. [CrossRef]
8. Skowron, A.; Stepaniuk, J. Tolerance approximation spaces. *Fundamen. Inform.* **1996**, *27*, 245–253.
9. Greco, S.; Matarazzo, B.; Slowinski, R. Rough approximation by dominance relation. *Int. J. Intell. Syst.* **2002**, *17*, 153–171. [CrossRef]
10. Wang, J.; Zhang, X. Two types of single valued neutrosophic covering rough sets and an application to decision making. *Symmetry* **2018**, *10*, 710. [CrossRef]
11. Yao, Y.Y. Relational interpretation of neighborhood operators and rough set approximation operator. *Inf. Sci.* **1998**, *111*, 239–259. [CrossRef]
12. Molodtsov, D. Soft set theory: First results. *Comput. Math. Appl.* **1999**, *37*, 19–31. [CrossRef]
13. Maji, P.K.; Roy, A.R.; Biswas, R. An application of soft sets in a decision making problem. *Comput. Math. Appl.* **2002**, *44*, 1077–1083. [CrossRef]
14. Maji, P.K.; Biswas, R.; Roy, A.R. Fuzzy soft set theory. *J. Fuzzy Math.* **2001**, *3*, 589–602.

15. Feng, F.; Liu, X.; Fotea, V.L.; Jun, Y.B. Soft sets and soft rough sets. *Inf. Sci.* **2011**, *181*, 1125–1137. [CrossRef]
16. Smarandache, F. *Neutrosophy: Neutrosophic Probability, Set, and Logic*; American Research Press: Rehoboth, IL, USA, 1998.
17. Smarandache, F. Neutrosophic set—A generalisation of the intuitionistic fuzzy sets. *Int. J. Pure Appl. Math.* **2005**, *24*, 287–297.
18. Smarandache, F. *Neutrosophic Overset, Neutrosophic Underset, and Neutrosophic Offset. Similarly for Neutrosophic over-/under-/off- Logic, Probability, and Statistics*; Pons Editions: Brussels, Belgium, 2016.
19. Jun, Y.B.; Smarandache, F.; Kim, C.S. Neutrosophic cubic sets. *New Math. Nat. Comput.* **2017**, *13*, 41–54. [CrossRef]
20. Maji, P.K. Neutrosophic soft set. *Ann. Fuzzy Math. Inform.* **2013**, *5*, 157–168.
21. Abu Qamar, M.; Hassan, N. Generalized Q-neutrosophic soft expert set for decision under uncertainty. *Symmetry* **2018**, *10*, 621. [CrossRef]
22. Ulucay, V.; Sahin, M.; Hassan, N. Generalized neutrosophic soft expert set for multiple-criteria decision-making. *Symmetry* **2018**, *10*, 437. [CrossRef]
23. Broumi, S.; Smarandache, F.; Dhar, M. Rough neutrosophic sets. *Neutrosophic Sets Syst.* **2014**, *3*, 62–67.
24. Bo, C.; Zhang, X.; Shao, S.; Smarandache, F. Multi-granulation neutrosophic rough sets on a single domain and dual domains with applications. *Symmetry* **2018**, *10*, 296. [CrossRef]
25. Bo, C.; Zhang, X.; Shao, S.; Smarandache, F. New multigranulation neutrosophic rough set with applications. *Symmetry* **2018**, *10*, 578. [CrossRef]
26. Broumi, S.; Smarandache, F. Interval-valued neutrosophic soft rough sets. *Int. J. Comput. Math.* **2015**, *2015*, 232919 . [CrossRef]
27. Al-Quran, A.; Hassan, N. Neutrosophic vague soft set and its applications. *Malays. J. Math. Sci.* **2017**, *11*, 141–163.
28. Al-Quran, A.; Hassan, N. Neutrosophic vague soft multiset for decision under uncertainty. *Songklanakarin J. Sci. Technol.* **2018**, *40*, 290–305.
29. Ali, M.; Smarandache, F. Complex neutrosophic set. *Neural Comput. Appl.* **2017**, *28*, 1817–1834. [CrossRef]
30. Al-Quran, A.; Alkhazaleh, S. Relations between the complex neutrosophic sets with their applications in decision making. *Axioms* **2018**, *7*, 64. [CrossRef]
31. Al-Quran, A.; Hassan, N. The complex neutrosophic soft expert relation and its multiple attribute decision-making method. *Entropy* **2018**, *20*, 101. [CrossRef]
32. Quek, S.G.; Broumi, S.; Selvachandran, G.; Bakali, A.; Talea, M.; Smarandache, F. Some results on the graph theory for complex neutrosophic sets. *Symmetry* **2018**, *10*, 190. [CrossRef]
33. Pawlak, Z.; Skowron, A. Rudiments of rough sets. *Inf. Sci.* **2007**, *177*, 3–27. [CrossRef]
34. Zhang, Z. A rough set approach to intuitionistic fuzzy soft set based decision making. *Appl. Math. Model.* **2012**, *36*, 4605–4633. [CrossRef]
35. Zhang, H.; Shu, L.; Liao, S. Intuitionistic fuzzy soft rough set and its application in decision making. *Abstr. Appl. Anal.* **2014**, *2014*, 287314. [CrossRef]
36. Meng, D.; Zhang, X.; Qin, K. Soft rough fuzzy sets and soft fuzzy rough sets. *Comput. Math. Appl.* **2011**, *26*, 4635–4645. [CrossRef]

symmetry

MDPI

Article

Fuzzy Parameterized Complex Neutrosophic Soft Expert Set for Decision under Uncertainty

Ashraf Al-Quran [1], Nasruddin Hassan [2,*] and Shawkat Alkhazaleh [3]

1 Department of Mathematics, Faculty of Science, Jerash University, Jerash 26150, Jordan; a.quraan@jpu.edu.jo
2 School of Mathematical Sciences, Faculty of Science and Technology, Universiti Kebangsaan Malaysia, 43600 UKM Bangi Selangor DE, Malaysia
3 Department of Mathematics, Faculty of Science, Zarqa University, Zarqa 13132, Jordan; shmk79@gmail.com
* Correspondence: nas@ukm.edu.my; Tel.: +603-8921-3710

Received: 5 January 2019; Accepted: 26 February 2019; Published: 15 March 2019

Abstract: In the definition of the complex neutrosophic soft expert set (CNSES), parameters set is a classical set, and the parameters have the same degree of importance, which is considered as 1. This poses a limitation in modeling of some problems. This paper introduces the concept of fuzzy parameterized complex neutrosophic soft expert set (FP-CNSES) to handle this issue by assigning a degree of importance to each of the problem parameters. We further develop FP-CNSES by establishing the concept of weighted fuzzy parameterized complex neutrosophic soft expert set (WFP-CNSES) based on the idea that each expert has a relative weight. These new mathematical frameworks reduce the chance of unfairness in the decision making process. Some essential operations with their properties and relevant laws related to the notion of FP-CNSES are defined and verified. The notation of mapping on fuzzy parameterized complex neutrosophic soft expert classes is defined and some properties of fuzzy parameterized complex neutrosophic soft expert images and inverse images was investigated. FP-CNSES is used to put forth an algorithm on decision-making by converting it from complex state to real state and subsequently provided the detailed decision steps. Then, we provide the comparison of FP-CNSES to the current methods to show the ascendancy of our proposed method.

Keywords: complex neutrosophic set; complex neutrosophic soft expert set; fuzzy parameterized single valued neutrosophic soft expert set; single valued neutrosophic set; soft expert set

1. Introduction

In a world where not everything is certain, the need to represent uncertain data was successfully fulfilled by Zadeh [1] by introducing the concept of fuzzy sets. Intuitionistic fuzzy sets were introduced by Atanassov [2] as an extension of Zadeh's notion of fuzzy set, which proved to be a better model of uncertainty. The words "neutrosophy" and "neutrosophic" were introduced for the first time by Smarandache [3]. Then, neutrosophic set [4] was defined as a more general platform, which extends the concepts of the fuzzy set and intuitionistic fuzzy set. To apply neutrosophic set to real- life problems, its operators need to be specified. Thus, single valued neutrosophic set and its basic operations were defined [5] as a special case of neutrosophic set. Molodtsov [6] proposed the concept of soft set, to bring a topological flavor to the models of uncertainty and associate family of subsets of universe to parameters. Soft set was then extended to the soft expert set [7], which was further developed to fuzzy soft expert set (FSES) [8], intuitionistic fuzzy soft expert sets (IFSES) [9] and single valued neutrosophic soft expert sets (SVNSES) [10]. At the same time, there have been some practical applications in soft expert set theory and its extensions that are used in decision making [11–13].

Among the significant milestones in the development of soft sets and soft expert sets and their generalizations is the introduction of the fuzzy parameterized aspect. This new aspect has further improved the theories of soft and soft expert sets and made them better suited to be used in solving decision making problems, especially when used with the more accurate generalizations of soft and soft expert sets such as FSESs, IFSESs and other hybrid models mentioned above. The fuzzy parameterized aspect was firstly established by Cagman et al. [14] who introduced the concept of fuzzy parameterized soft set by giving a degree of importance to each element in the set of parameters. Subsequently, this aspect was attached and/or added to the existing generalizations of the soft sets, soft expert sets and fuzzy sets. Bashir and Salleh [15] introduced the notion of fuzzy parameterized soft expert sets (FP-SES), while Hazaymeh et al. [16] the notion of fuzzy parameterized fuzzy soft expert sets (FP-FSES), followed by Selvachandran and Salleh [17] who introduced the notion of fuzzy parameterized intuitionistic fuzzy soft expert set (FP-IFSES) as a generalization of the work by Hazaymeh et al. [16]. However, these sets can only handle incomplete and uncertainty information but not indeterminate and inconsistent information, which usually exists in real situations. To treat this deficiency, Al-Quran and Hassan [18] defined the fuzzy parameterized single valued neutrosophic soft expert set (FP-SVNSES), which proves superior to these models with three independent membership functions. The FP-SVNSES model is also significantly more advantageous compared to SVNSES, as it has added advantages to SVNSES by virtue of the fuzzy parameterized feature, which provides more information enhancing the quality of the information presented by the SVNSES, which in turn, increases the accuracy of the final decision.

The development of the uncertainty sets that have been mentioned above are not limited to the real field but extended to the complex field. The introduction of fuzzy sets was followed by their extension to the complex fuzzy set [19,20]. Alkouri and Salleh [21] introduced the concept of complex intuitionistic fuzzy set (CIFS) to represent information that happens repeatedly over a period of time. To handle imprecise, indeterminate, inconsistent, and incomplete information that has periodic nature, Ali and Smarandache [22] introduced complex neutrosophic set, where each membership function associates with a phase term. This feature gives wave-like properties that could be used to describe constructive and destructive interference depending on the phase value of an element, as well as its ability to deal with indeterminacy. Inspired by this, Al-Quran and Hassan [23] generalized the CNSES from the definitions of the complex neutrosophic set and soft expert set on the basis of the SVNSES. In depth, the rationales of introducing the complex neutrosophic set and the soft expert set are considered as a potent motivation to the introduction of the concept of CNSES. CNSES is actually an extension of the SVNSES to the complex space, which makes it superior to all of the aforementioned uncertainty sets. Subsequently, Al-Quran and Hassan [24] studied the CNSESs further by establishing a novel structure of relation between two CNSESs, called complex neutrosophic soft expert relation, to evaluate the degree of interaction between the CNSESs, which is defined as a subset of the Cartesian product of the CNSESs.

Over the years, many techniques and methods have been proposed as tools to be used to find the solutions of problems that are nonlinear or vague in nature, with every method introduced superior to its predecessors. Following in this direction, we extend the studies on CNSESs [23] and FP-SVNSES [18] through the establishment of the notion of FP-CNSES to keep the advantages of CNSES while holding the FP-SVNSES features. On the one hand, the novelty of CNSES appears in its ability to provide a succinct, elegant and comprehensive representation of two-dimensional neutrosophic information as well as the adequate parameterization and the opinions of the experts, all in a single set. This two-dimensional information is presented by amplitude and phase terms simultaneously where the phase terms give neutrosophic information that may interfere, constructively or destructively, with the neutrosophic information presented by the associated amplitude terms, thus making this model highly suitable for use in decision making problems to select the best alternative. On the other hand, FP-SVNSES has the fuzzy parameterized feature, which gives a degree of importance to each parameter in the domain of the SVNSES. All of these features together are contained in the proposed FP-CNSES.

To facilitate our discussion, we first review some background on complex neutrosophic set and FP-SVNSES in Section 2. In Section 3, we introduce the concept of FP-CNSES and give its theoretic operations. In Section 4, we study a mapping on fuzzy parameterized complex neutrosophic soft expert classes and its properties. In Section 5 we discuss an application of the FP-CNSES in decision making. Section 6 provides a comprehensive comparison among FP-CNSES and other recent approaches to manifest the dominance of our proposed method. In Section 7, we define the concept of the WFP-CNSES where experts' relative weights are considered and applied to solve a decision making problem. Section 8 outlines the conclusion of this paper.

2. Preliminaries

In this section, a summary of the literature on complex neutrosophic set and FP-SVNSES relevant to this paper is presented.

We begin by recalling the definition of complex neutrosophic set and its basic operations in the following two definitions.

Definition 1. [22] *Let X be a universe of discourse; a complex neutrosophic set S in X is characterized by a truth membership function $T_S(x)$, an indeterminacy membership function $I_S(x)$, and a falsity membership function $F_S(x)$ that assign an element $x \in X$ a complex-valued grade of $T_S(x)$, $I_S(x)$, and $F_S(x)$ in S. By definition, the values $T_S(x)$, $I_S(x)$, and $F_S(x)$ and their sum may all be within the unit circle in the complex plane and are of the form, $T_S(x) = p_S(x).e^{j\mu_S(x)}$, $I_S(x) = q_S(x).e^{j\nu_S(x)}$ and $F_S(x) = r_S(x).e^{j\omega_S(x)}$; $p_S(x), q_S(x), r_S(x)$ and $\mu_S(x), \nu_S(x), \omega_S(x)$ are, respectively, real valued and $p_S(x), q_S(x), r_S(x) \in [0,1]$ such that $0^- \leq P_S(x) + q_S(x) + r_S(x) \leq 3^+$.*

Definition 2. [22] *Let A and B be two complex neutrosophic sets, where A is characterized by a truth membership function $T_A(x) = p_A(x).e^{j\mu_A(x)}$, an indeterminacy membership function $I_A(x) = q_A(x).e^{j\nu_A(x)}$ and a falsity membership function $F_A(x) = r_A(x).e^{j\omega_A(x)}$ and B is characterized by a truth membership function $T_B(x) = p_B(x).e^{j\mu_B(x)}$, an indeterminacy membership function $I_B(x) = q_B(x).e^{j\nu_B(x)}$ and a falsity membership function $F_B(x) = r_B(x).e^{j\omega_B(x)}$.*

We define the the complement, subset, union and intersection operations as follows.

1. *The complement of A, denoted as $\tilde{c}(A)$, is specified by functions:*
   $T_{\tilde{c}(A)}(u) = p_{\tilde{c}(A)}(u).e^{j\mu_{\tilde{c}(A)}(u)} = r_A(u).e^{j(2\pi-\mu_A(u))}$,
   $I_{\tilde{c}(A)}(u) = q_{\tilde{c}(A)}(u).e^{j\nu_{\tilde{c}(A)}(u)} = (1-q_A(u)).e^{j(2\pi-\nu_A(u))}$, *and*
   $F_{\tilde{c}(A)}(u) = r_{\tilde{c}(A)}(u).e^{j\omega_{\tilde{c}(A)}(u)} = p_A(u).e^{j(2\pi-\omega_A(u))}$.

2. *A is said to be complex neutrosophic subset of B ($A \subseteq B$) if and only if the following conditions are satisfied:*
   (a) $T_A(u) \leq T_B(u)$ such that $p_A(u) \leq p_B(u)$ and $\mu_A(u) \leq \mu_B(u)$.
   (b) $I_A(u) \geq I_B(u)$ such that $q_A(u) \geq q_B(u)$ and $\nu_A(u) \geq \nu_B(u)$.
   (c) $F_A(u) \geq F_B(u)$ such that $r_A(u) \geq r_B(u)$ and $\omega_A(u) \geq \omega_B(u)$.

3. *The union(intersection) of A and B, denoted as $A \cup (\cap) B$ and the truth membership function $T_{A\cup(\cap)B}(u)$, the indeterminacy membership function $I_{A\cup(\cap)B}(u)$, and the falsity membership function $F_{A\cup(\cap)B}(u)$ are defined as:*
   $T_{A\cup(\cap)B}(u) = [(p_A(u) \vee (\wedge) p_B(u))].e^{j(\mu_A(u)\vee(\wedge)\mu_B(u))}$,
   $I_{A\cup(\cap)B}(u) = [(q_A(u) \wedge (\vee) q_B(u))].e^{j(\nu_A(u)\wedge(\vee)\nu_B(u))}$ *and*
   $F_{A\cup(\cap)B}(u) = [(r_A(u) \wedge (\vee) r_B(u))].e^{j(\omega_A(u)\wedge(\vee)\omega_B(u))}$,
   where $\vee$ = max and $\wedge$ = min .

Al-Quran and Hassan [18] defined the FP-SVNSES, agree FP-SVNSES and disagree FP-SVNSES as follows.

Definition 3. *[18] Let U be a universe set, E be a set of parameters, I^E denote all fuzzy subsets of E, X be a set of experts, and $O = \{1 = agree,\ 0 = disagree\}$ a set of opinions. Let $Z = \Psi \times X \times O$ and $A \subseteq Z$, where $\Psi \subset I^E$. A pair $(f, A)_\Psi$ is called a fuzzy parameterized single valued neutrosophic soft expert set $(FPSVNSES)$ over U, where F is a mapping given by*

$$f_\Psi : A \to SVN(U),$$

and $SVN(U)$ denotes the set of all single valued neutrosophic subsets of U.

Definition 4. *An agree FP-SVNSES $(f, A)_{\Psi_1}$ over U is a FP-SVNSE subset of $(f, A)_\Psi$ where the opinions of all experts are "agree" and is defined as follows:*

$$(f, A)_{\Psi_1} = \{F_\Psi(\epsilon) : \epsilon \in \Psi \times X \times \{1\}\}.$$

A disagree FP-SVNSES $(f, A)_{\Psi_0}$ over U is a FP-SVNSE subset of $(f, A)_\Psi$ where the opinions of all experts are "disagree" and is defined as follows:

$$(f, A)_{\Psi_0} = \{F_\Psi(\epsilon) : \epsilon \in \Psi \times X \times \{0\}\}.$$

3. Fuzzy Parameterized Complex Neutrosophic Soft Expert Set

In this section, we introduce the definition of FP-CNSES, which is a generalization of the concept of FP-SVNSES. We define some operations on this concept, namely subset, equality, complement, union and intersection. We also give some properties on these operations.

We begin by proposing the definition of FP-CNSES, and give an illustrative example of it.

Definition 5. *Let U be a universe set, E be a set of parameters, $FZ(E)$ denote all fuzzy subsets of E, X be a set of experts, and $O = \{1 = agree,\ 0 = disagree\}$ a set of opinions. Let $Y = \Gamma \times X \times O$ and $A \subseteq Y$ where $\Gamma \subset FZ(E)$. A pair $(H, A)_\Gamma$ is called a fuzzy parameterized complex neutrosophic soft expert set (FP-CNSES) over U, where H is a mapping given by*

$$H_\Gamma : A \to CN^U,$$

and CN^U denotes the power complex neutrosophic set of U.

The FP-CNSES $(H, A)_\Gamma$ can be written as the following set of ordered pairs:

$$(H, A)_\Gamma = \Big\{(a, \{\frac{u}{H_\Gamma(a)(u)} : u \in U\}) : a \in A\Big\},$$

where $A \subseteq \Gamma \times X \times O = \Big\{(\frac{e}{\mu_\Gamma(e)}, x, o) : e \in E, x \in X \text{and } o \in O\Big\}$, such that $\mu_\Gamma(e)$ is the corresponding membership function of the fuzzy set Γ and $\forall u \in U, \forall a \in A$, $H_\Gamma(a)(u) = \langle T_{H_\Gamma(a)}(u), I_{H_\Gamma(a)}(u), F_{H_\Gamma(a)}(u)\rangle$, where $T_{H_\Gamma(a)}(u) = p_{H_\Gamma(a)}(u).e^{j\mu_{H_\Gamma(a)}(u)}, I_{H(a)}(u) = q_{H_\Gamma(a)}(u).e^{j\nu_{H_\Gamma(a)}(u)}$ and $F_{H_\Gamma(a)}(u) = r_{H_\Gamma(a)}(u).e^{j\omega_{H_\Gamma(a)}(u)}$ with $T_{H_\Gamma(a)}(u), I_{H_\Gamma(a)}(u)$ and $F_{H_\Gamma(a)}(u)$ representing the complex-valued truth membership function, complex-valued indeterminacy membership function and complex-valued falsity membership function, respectively, for the FP-CNSES $(H, A)_\Gamma$. The values $T_{H_\Gamma(a)}(u), I_{H_\Gamma(a)}(u), F_{H_\Gamma(a)}(u)$ are within the unit circle in the complex plane and both the amplitude terms $p_{H_\Gamma(a)}(u), q_{H_\Gamma(a)}(u), r_{H_\Gamma(a)}(u)$ and the phase terms $\mu_{H_\Gamma(a)}(u), \nu_{H_\Gamma(a)}(u), \omega_{H_\Gamma(a)}(u)$ are real valued such that $p_{H_\Gamma(a)}(u), q_{H_\Gamma(a)}(u)$, $r_{H_\Gamma(a)}(u) \in [0,1]$ and $0 \leq p_{H_\Gamma(a)}(u) + q_{H_\Gamma(a)}(u) + r_{H_\Gamma(a)}(u) \leq 3$.

Example 1. *Suppose that a car company produces two different models of cars and wish to take the opinion of its team of experts concerning these two models of cars before and after testing these cars. These two models of cars form the universe of elements,* $U = \{u_1, u_2\}$*. The team of experts is represented by the set* $X = \{x_1, x_2\}$*. Suppose that the team of experts consider a set of parameters,* $E = \{e_1, e_2, e_3\}$*, where* $e_i (i = 1, 2, 3)$ *denotes the decisions "reliability", "comfortability" and "durability", respectively, and suppose* $\Gamma = \left\{ \frac{e_1}{0.4}, \frac{e_2}{0.6}, \frac{e_3}{0.5} \right\}$ *is a fuzzy subset of* $FZ(E)$*. It is to be noted that the parameters may be affected and altered after the cars are tested. By applying the FP-CNSES and considering the opinions of the experts in process one (before testing the car) as amplitude terms of membership, non-membership and indeterminate membership, and setting the opinions of the experts in the second process (after testing the car) as phase terms of membership, non-membership, and indeterminacy, the first and second processes form a FP-CNSES as a whole, which is shown below:*

$$
\begin{aligned}
&(H, A)_\Gamma = \\
&\left\{ \left\{ \left(\tfrac{e_1}{0.4}, x_1, 1 \right), \left\{ \frac{u_1}{\langle 0.6e^{2\pi(0.5)}, 0.1e^{2\pi(0)}, 0.2e^{2\pi(0.4)} \rangle}, \frac{u_2}{\langle 0.5e^{2\pi(0.8)}, 0.4e^{2\pi(0.7)}, 0.3e^{2\pi(1)} \rangle} \right\} \right\}, \right. \\
&\left\{ \left(\tfrac{e_1}{0.4}, x_2, 1 \right), \left\{ \frac{u_1}{\langle 0.5e^{2\pi(0.4)}, 0.7e^{2\pi(0.5)}, 0.1e^{2\pi(0.2)} \rangle}, \frac{u_2}{\langle 0.8e^{2\pi(0.7)}, 0.7e^{2\pi(0.5)}, 0.4e^{2\pi(0.3)} \rangle} \right\} \right\}, \\
&\left\{ \left(\tfrac{e_2}{0.6}, x_1, 1 \right), \left\{ \frac{u_1}{\langle 0.3e^{2\pi(0.2)}, 0.5e^{2\pi(0.6)}, 0.4e^{2\pi(0.5)} \rangle}, \frac{u_2}{\langle 0.6e^{2\pi(0.5)}, 0.7e^{2\pi(0.6)}, 0.1e^{2\pi(0.2)} \rangle} \right\} \right\}, \\
&\left\{ \left(\tfrac{e_2}{0.6}, x_2, 1 \right), \left\{ \frac{u_1}{\langle 0.9e^{2\pi(0.8)}, 0.5e^{2\pi(0.3)}, 0.7e^{2\pi(0.4)} \rangle}, \frac{u_2}{\langle 0.5e^{2\pi(0.4)}, 0.8e^{2\pi(0.9)}, 0.1e^{2\pi(0.2)} \rangle} \right\} \right\}, \\
&\left\{ \left(\tfrac{e_3}{0.5}, x_1, 1 \right), \left\{ \frac{u_1}{\langle 0.4e^{2\pi(0.2)}, 0.5e^{2\pi(0.3)}, 0.7e^{2\pi(0.6)} \rangle}, \frac{u_2}{\langle 0.5e^{2\pi(0.2)}, 0.7e^{2\pi(0.2)}, 0.8e^{2\pi(0.2)} \rangle} \right\} \right\}, \\
&\left\{ \left(\tfrac{e_3}{0.5}, x_2, 1 \right), \left\{ \frac{u_1}{\langle 0.8e^{2\pi(0.1)}, 0.3e^{2\pi(0.2)}, 0.7e^{9\pi(0.2)} \rangle}, \frac{u_2}{\langle 0.1e^{2\pi(0.6)}, 0.2e^{2\pi(0.4)}, 0.6e^{2\pi(0.8)} \rangle} \right\} \right\}, \\
&\left\{ \left(\tfrac{e_1}{0.4}, x_1, 0 \right), \left\{ \frac{u_1}{\langle 0.6e^{2\pi(0.5)}, 0.7e^{2\pi(0.4)}, 0.9e^{2\pi(0.3)} \rangle}, \frac{u_2}{\langle 0.9e^{2\pi(0.8)}, 0.8e^{2\pi(0.8)}, 0.7e^{2\pi(0.7)} \rangle} \right\} \right\}, \\
&\left\{ \left(\tfrac{e_1}{0.4}, x_2, 0 \right), \left\{ \frac{u_1}{\langle 0.4e^{2\pi(0.2)}, 0.6e^{2\pi(0.4)}, 0.7e^{2\pi(0.8)} \rangle}, \frac{u_2}{\langle 0.2e^{2\pi(0.4)}, 0.5e^{2\pi(0.4)}, 0.6e^{2\pi(0.5)} \rangle} \right\} \right\}, \\
&\left\{ \left(\tfrac{e_2}{0.6}, x_1, 0 \right), \left\{ \frac{u_1}{\langle 0.3e^{2\pi(0.2)}, 0.6e^{2\pi(0.2)}, 0.9e^{2\pi(0.2)} \rangle}, \frac{u_2}{\langle 0.6e^{2\pi(0.4)}, 0.5e^{2\pi(0.6)}, 0.7e^{2\pi(0.8)} \rangle} \right\} \right\}, \\
&\left\{ \left(\tfrac{e_2}{0.6}, x_2, 0 \right), \left\{ \frac{u_1}{\langle 0.5e^{2\pi(0.8)}, 0.3e^{2\pi(0.1)}, 0.8e^{2\pi(0.9)} \rangle}, \frac{u_2}{\langle 0.3e^{2\pi(0.7)}, 0.8e^{2\pi(0.7)}, 0.7e^{2\pi(0.6)} \rangle} \right\} \right\}, \\
&\left\{ \left(\tfrac{e_3}{0.5}, x_1, 0 \right), \left\{ \frac{u_1}{\langle 0.4e^{2\pi(0.7)}, 0.9e^{2\pi(0.8)}, 0.1e^{2\pi(0.3)} \rangle}, \frac{u_2}{\langle 0.8e^{2\pi(0.2)}, 0.3e^{2\pi(0.2)}, 0.5e^{2\pi(0.2)} \rangle} \right\} \right\}, \\
&\left. \left\{ \left(\tfrac{e_3}{0.5}, x_2, 0 \right), \left\{ \frac{u_1}{\langle 0.5e^{2\pi(0.4)}, 0.6e^{2\pi(0.7)}, 0.1e^{2\pi(0.9)} \rangle}, \frac{u_2}{\langle 0.1e^{2\pi(0)}, 0.4e^{2\pi(0.3)}, 0.1e^{2\pi(0.1)} \rangle} \right\} \right\} \right\}.
\end{aligned}
$$

In the FP-CNSES $(H, A)_\Gamma$ *above, the amplitude term of the membership in the first process and the phase term in the second process form a complex-valued truth membership function. Similarly, the amplitude term of non-membership in process one and the phase term of non-membership in the second process form a complex-valued falsity membership function. In addition, the amplitude term of undecidedness in the first process and the phase term of indeterminacy in the second process form the complex-valued indeterminate membership function.*

Now, we put forward the definition of an agree FP-CNSES and the definition of a disagree FP-CNSES.

Definition 6. *An agree FP-CNSES* $(H, A)_{\Gamma_1}$ *over U is a FP-CNSE subset of* $(H, A)_\Gamma$ *where the opinions of all experts are "agree" and is defined as follows:*

$$(H, A)_{\Gamma_1} = \left\{ \left(a, \left\{ \frac{u}{H_\Gamma(a)(u)} : u \in U \right\} \right) : a \in A \subseteq \Gamma \times X \times \{1\} \right\}$$

Definition 7. *A disagree FP-CNSES* $(H, A)_{\Gamma_0}$ *over U is a FP-CNSE subset of* $(H, A)_\Gamma$ *where the opinions of all experts are "disagree" and is defined as follows:*

$$(H, A)_{\Gamma_0} = \left\{ \left(a, \left\{ \frac{u}{H_\Gamma(a)(u)} : u \in U \right\} \right) : a \in A \subseteq \Gamma \times X \times \{0\} \right\}$$

In the following, we give some basic definitions and set theoretic operations of FP-CNSESs.

We begin by proposing the definition of the subset of two FP-CNSESs and the equality of two FP-CNSESs.

Definition 8. *Let $(H,A)_\Gamma$ and $(G,B)_\Delta$ be two FP-CNSESs over a universe U, we say that $(H,A)_\Gamma$ is a fuzzy parameterized complex neutrosophic soft expert subset of $(G,B)_\Delta$ denoted by $(H,A)_\Gamma \subseteq (G,B)_\Delta$ if and only if* 1. $A \subseteq B$, *and* 2. $\forall a \in A, H_\Gamma(a)$ *is complex neutrosophic subset of* $G_\Delta(a)$.

Definition 9. *For two FP-CNSESs $(H,A)_\Gamma$ and $(G,B)_\Gamma$ over a universe U, we say that $(H,A)_\Gamma$ is equal to $(G,B)_\Gamma$ and we write $(H,A)_\Gamma = (G,B)_\Gamma$ if $(H,A)_\Gamma \subseteq (G,B)_\Gamma$ and $(G,B)_\Gamma \subseteq (H,A)_\Gamma$.*

In the following, we propose the definition of the complement of a FP-CNSES along with an illustrative example and give a proposition on the complement of a FP-CNSES.

Let U be a universe of discourse and $(H,A)_\Gamma$ be a FP-CNSES on U, which is as defined below:

$$(H,A)_\Gamma = \Big\{\big(a, \{\frac{u}{H_\Gamma(a)(u)} : u \in U\}\big) : a \in A\Big\}.$$

Definition 10. *The complement of $(H,A)_\Gamma$ is denoted by $(H,A)_\Gamma^c$ and is defined by*

$$(H,A)_\Gamma^c = \Big\{\big(a, \{\frac{u}{H_\Gamma^c(a)(u)} : u \in U\}\big) : a \in \neg A\Big\},$$

where

$$H_\Gamma^c(a)(u) = \langle T_{H_\Gamma^c(a)}(u), I_{H_\Gamma^c(a)}(u), F_{H_\Gamma^c(a)}(u)\rangle,$$

such that:

$$\begin{aligned}
T_{H_\Gamma^c(a)}(u) &= p_{H_\Gamma^c(a)}(u).e^{j\mu_{H_\Gamma^c(a)}(u)} = r_{H_\Gamma(a)}(u).e^{j(2\pi-\mu_{H_\Gamma(a)}(u))},\\
I_{H_\Gamma^c(a)}(u) &= q_{H_\Gamma^c(a)}(u).e^{j\nu_{H_\Gamma^c(a)}(u)} = (1-q_{H_\Gamma(a)}(u)).e^{j(2\pi-\nu_{H_\Gamma(a)}(u))},\\
F_{H_\Gamma^c(a)}(u) &= r_{H_\Gamma^c(a)}(u).e^{j\omega_{H_\Gamma^c(a)}(u)} = p_{H_\Gamma(a)}(u).e^{j(2\pi-\omega_{H_\Gamma(a)}(u))},
\end{aligned}$$

and $\neg A \subseteq \Gamma^{\tilde{c}} \times X \times O$, where $\tilde{c}$ is the fuzzy complement.

Example 2. *Consider the approximation given in Example 1, where*

$$H_\Gamma(\frac{e_1}{0.4}, x_1, 1) = \{\frac{u_1}{\langle 0.6e^{j2\Pi(0.5)},\ 0.1e^{j2\Pi(0)},\ 0.2e^{j2\Pi(0.4)}\rangle}, \frac{u_2}{\langle 0.5e^{j2\Pi(0.8)},\ 0.4e^{j2\Pi(0.7)},\ 0.3e^{j2\Pi(1)}\rangle}\}.$$

By using the complex neutrosophic complement and the fuzzy complement, we obtain the complement of the approximation given by

$$H_\Gamma(\frac{e_1}{0.6}, x_1, 1) = \{\frac{u_1}{\langle 0.2e^{j2\Pi(0.5)},\ 0.9e^{j2\Pi(1)},\ 0.6e^{j2\Pi(0.6)}\rangle}, \frac{u_2}{\langle 0.3e^{j2\Pi(0.2)},\ 0.6e^{j2\Pi(0.3)},\ 0.5e^{j2\Pi(0)}\rangle}\}.$$

Proposition 1. *If $(H,A)_\Gamma$ is a FP-CNSES over U, then $\big((H,A)_\Gamma^c\big)^c = (H,A)_\Gamma$.*

Proof. From Definition 10, we have $(H,A)_\Gamma^c = \Big\{\big(a, \{\frac{u}{H_\Gamma^c(a)(u)} : u \in U\}\big) : a \in \neg A\Big\}$, where, $H_\Gamma^c(a)(u)$ $= \langle T_{H_\Gamma^c(a)}(u), I_{H_\Gamma^c(a)}(u), F_{H_\Gamma^c(a)}(u)\rangle = \langle p_{H_\Gamma^c(a)}(u).e^{j\mu_{H_\Gamma^c(a)}(u)}, q_{H_\Gamma^c(a)}(u).e^{j\nu_{H_\Gamma^c(a)}(u)}, r_{H_\Gamma^c(a)}(u).e^{j\omega_{H_\Gamma^c(a)}(u)}\rangle$, $= \langle r_{H_\Gamma(a)}(u).e^{j(2\pi-\mu_{H_\Gamma(a)}(u))}, (1-q_{H_\Gamma(a)}(u)).e^{j(2\pi-\nu_{H_\Gamma(a)}(u))}, p_{H_\Gamma(a)}(u).e^{j(2\pi-\omega_{H_\Gamma(a)}(u))}\rangle$, and $\neg A \subseteq \Gamma^{\tilde{c}} \times X \times O$ Thus, $((H,A)_\Gamma^c)^c = \Big\{\big(a, \{\frac{u}{(H_\Gamma^c)^c(a)(u)} : u \in U\}\big) : a \in \neg(\neg A)\Big\}$, where $(H_\Gamma^c)^c(a)(u) = \langle r_{H_\Gamma^c(a)}(u).e^{j(2\pi-\mu_{H_\Gamma^c(a)}(u))}, (1-q_{H_\Gamma^c(a)}(u)).e^{j(2\pi-\nu_{H_\Gamma^c(a)}(u))}, p_{H_\Gamma^c(a)}(u)\ .e^{j(2\pi-\omega_{H_\Gamma^c(a)}(u))}\rangle$,

$= \langle p_{H_\Gamma(a)}(u).e^{j\left(2\pi-(2\pi-\mu_{H_\Gamma(a)}(u))\right)}, (1-(1-q_{H_\Gamma(a)}(u)))\ .e^{j\left(2\pi-(2\pi-\nu_{H_\Gamma(a)}(u))\right)},$
$r_{H_\Gamma(a)}(u).e^{j\left(2\pi-(2\pi-\omega_{H_\Gamma(a)}(u))\right)} \rangle = \langle p_{H_\Gamma(a)}(u).e^{j\mu_{H_\Gamma(a)}(u)}, q_{H_\Gamma(a)}(u).e^{j\nu_{H_\Gamma(a)}(u)}, r_{H_\Gamma(a)}(u).e^{j\omega_{H_\Gamma(a)}(u)} \rangle,$
$= \langle T_{H_\Gamma(a)}(u), I_{H_\Gamma(a)}(u), F_{H_\Gamma(a)}(u) \rangle, = H_\Gamma(a)(u)$, and $\neg(\neg A) \subseteq (\Gamma^{\tilde{c}})^{\tilde{c}} \times X \times O$ and since $(\Gamma^{\tilde{c}})^{\tilde{c}} = \Gamma$, this completes the proof. □

We introduce the definition of union and intersection operations of two FP-CNSESs along with an illustrative example and some propositions on these two operations.

Let $(H,A)_\Gamma$ and $(G,B)_\Delta$ be two FP-CNSESs over a universe U, where $(H,A)_\Gamma = \Big\{(a, \{\frac{u}{H_\Gamma(a)(u)} : u \in U\}) : a \in A\Big\}$ and $(G,B)_\Delta = \Big\{(b, \{\frac{u}{G_\Delta(b)(u)} : u \in U\}) : b \in B\Big\}$.

Definition 11. *The union of $(H,A)_\Gamma$ and $(G,B)_\Delta$, denoted by $(H,A)_\Gamma \tilde{\cup} (G,B)_\Delta$, is the FP-CNSES $(K,C)_\Theta$ such that $C_\Theta = A_\Gamma \cup B_\Delta$ and $\Theta = \Gamma \hat{\cup} \Delta$, $\hat{\cup}$ is the fuzzy union, and $\forall \epsilon \in C_\Theta, \forall\, u \in U$,*

$$T_{K_\Theta(\epsilon)}(u) = \begin{cases} p_{H_\Gamma(\epsilon)}(u).e^{j\mu_{H_\Gamma(\epsilon)}(u)} & , if\ \epsilon \in A_\Gamma - B_\Delta \\ p_{G_\Delta(\epsilon)}(u).e^{j\mu_{G_\Delta(\epsilon)}(u)} & , if\ \epsilon \in B_\Delta - A_\Gamma \\ (p_{H_\Gamma(\epsilon)}(u) \vee p_{G_\Delta(\epsilon)}(u)).e^{j(\mu_{H_\Gamma(\epsilon)}(u) \vee \mu_{G_\Delta(\epsilon)}(u))} & , if\ \epsilon \in A_\Gamma \cap B_\Delta, \end{cases}$$

$$I_{K_\Theta(\epsilon)}(u) = \begin{cases} q_{H_\Gamma(\epsilon)}(u).e^{j\nu_{H_\Gamma(\epsilon)}(u)} & , if\ \epsilon \in A_\Gamma - B_\Delta \\ q_{G_\Delta(\epsilon)}(u).e^{j\nu_{G_\Delta(\epsilon)}(u)} & , if\ \epsilon \in B_\Delta - A_\Gamma \\ (q_{H_\Gamma(\epsilon)}(u) \wedge q_{G_\Delta(\epsilon)}(u)).e^{j(\nu_{H_\Gamma(\epsilon)}(u) \wedge \nu_{G_\Delta(\epsilon)}(u))} & , if\ \epsilon \in A_\Gamma \cap B_\Delta, \end{cases}$$

$$F_{K_\Theta(\epsilon)}(u) = \begin{cases} r_{H_\Gamma(\epsilon)}(u).e^{j\omega_{H_\Gamma(\epsilon)}(u)} & , if\ \epsilon \in A_\Gamma - B_\Delta \\ r_{G_\Delta(\epsilon)}(u).e^{j\omega_{G_\Delta(\epsilon)}(u)} & , if\ \epsilon \in B_\Delta - A_\Gamma \\ (r_{H_\Gamma(\epsilon)}(u) \wedge r_{G_\Delta(\epsilon)}(u)).e^{j(\omega_{H_\Gamma(\epsilon)}(u) \wedge \omega_{G_\Delta(\epsilon)}(u))} & , if\ \epsilon \in A_\Gamma \cap B_\Delta, \end{cases}$$

where $\vee = max$, and $\wedge = min$.

Definition 12. *The intersection of $(H,A)_\Gamma$ and $(G,B)_\Delta$, denoted by $(H,A)_\Gamma \tilde{\cap} (G,B)_\Delta$, is the FP-CNSES $(K,C)_\Theta$ such that $C_\Theta = A_\Gamma \cup B_\Delta$ and $\Theta = \Gamma \hat{\cap} \Delta$, $\hat{\cap}$ is the fuzzy intersection, and $\forall \epsilon \in C_\Theta, \forall\, u \in U$,*

$$T_{K_\Theta(\epsilon)}(u) = \begin{cases} p_{H_\Gamma(\epsilon)}(u).e^{j\mu_{H_\Gamma(\epsilon)}(u)} & , if\ \epsilon \in A_\Gamma - B_\Delta \\ p_{G_\Delta(\epsilon)}(u).e^{j\mu_{G_\Delta(\epsilon)}(u)} & , if\ \epsilon \in B_\Delta - A_\Gamma \\ (p_{H_\Gamma(\epsilon)}(u) \wedge p_{G_\Delta(\epsilon)}(u)).e^{j(\mu_{H_\Gamma(\epsilon)}(u) \wedge \mu_{G_\Delta(\epsilon)}(u))} & , if\ \epsilon \in A_\Gamma \cap B_\Delta, \end{cases}$$

$$I_{K_\Theta(\epsilon)}(u) = \begin{cases} q_{H_\Gamma(\epsilon)}(u).e^{j\nu_{H_\Gamma(\epsilon)}(u)} & , if\ \epsilon \in A_\Gamma - B_\Delta \\ q_{G_\Delta(\epsilon)}(u).e^{j\nu_{G_\Delta(\epsilon)}(u)} & , if\ \epsilon \in B_\Delta - A_\Gamma \\ (q_{H_\Gamma(\epsilon)}(u) \vee q_{G_\Delta(\epsilon)}(u)).e^{j(\nu_{H_\Gamma(\epsilon)}(u) \vee \nu_{G_\Delta(\epsilon)}(u))} & , if\ \epsilon \in A_\Gamma \cap B_\Delta, \end{cases}$$

$$F_{K_\Theta(\epsilon)}(u) = \begin{cases} r_{H_\Gamma(\epsilon)}(u).e^{j\omega_{H_\Gamma(\epsilon)}(u)} & , if\ \epsilon \in A_\Gamma - B_\Delta \\ r_{G_\Delta(\epsilon)}(u).e^{j\omega_{G_\Delta(\epsilon)}(u)} & , if\ \epsilon \in B_\Delta - A_\Gamma \\ (r_{H_\Gamma(\epsilon)}(u) \vee r_{G_\Delta(\epsilon)}(u)).e^{j(\omega_{H_\Gamma(\epsilon)}(u) \vee \omega_{G_\Delta(\epsilon)}(u))} & , if\ \epsilon \in A_\Gamma \cap B_\Delta, \end{cases}$$

where $\vee = max$, and $\wedge = min$.

Example 3. *Consider Example 1. Let* $\Gamma = \left\{\frac{e_1}{0.3}, \frac{e_2}{0.1}, \frac{e_3}{0.9}\right\}$ *be a fuzzy subset of E, and* $\Delta = \left\{\frac{e_1}{0.4}, \frac{e_2}{0.8}, \frac{e_3}{0.2}\right\}$ *be another fuzzy subset over E.*

$$A_\Gamma = \left\{\left(\frac{e_1}{0.3}, x_1, 1\right), \left(\frac{e_2}{0.1}, x_2, 1\right), \left(\frac{e_3}{0.9}, x_2, 0\right)\right\},$$

$$B_Y = \left\{\left(\frac{e_1}{0.4}, x_2, 0\right), \left(\frac{e_2}{0.8}, x_2, 1\right), \left(\frac{e_3}{0.2}, x_1, 0\right)\right\}.$$

Suppose $(H,A)_\Gamma$ *and* $(G,B)_\Delta$ *are two FP-CNSESs over the same U given by*

$$\begin{aligned}
&(H,A)_\Gamma = \\
&\left\{\left\{\left(\tfrac{e_1}{0.3}, x_1, 1\right), \left\{\frac{u_1}{\langle 0.4e^{2\pi(0.5)}, 0.5e^{2\pi(0.6)}, 0.8e^{2\pi(0.3)}\rangle}, \frac{u_2}{\langle 0.3e^{2\pi(0.8)}, 0.7e^{2\pi(0.1)}, 0.6e^{2\pi(0.7)}\rangle}\right\}\right\},\right. \\
&\left\{\left(\tfrac{e_2}{0.1}, x_2, 1\right), \left\{\frac{u_1}{\langle 0.7e^{2\pi(0.4)}, 0.9e^{2\pi(0.3)}, 0.4e^{2\pi(0.1)}\rangle}, \frac{u_2}{\langle 0.6e^{2\pi(0.3)}, 0.2e^{2\pi(0.6)}, 0.4e^{2\pi(0.7)}\rangle}\right\}\right\}, \\
&\left.\left\{\left(\tfrac{e_3}{0.9}, x_2, 0\right), \left\{\frac{u_1}{\langle 0.4e^{2\pi(0.1)}, 0.3e^{2\pi(0.4)}, 0.5e^{2\pi(0.9)}\rangle}, \frac{u_2}{\langle 0.6e^{2\pi(0.2)}, 0.9e^{2\pi(0.8)}, 0.1e^{2\pi(0.3)}\rangle}\right\}\right\}\right\},
\end{aligned}$$

and

$$\begin{aligned}
&(G,B)_\Delta = \\
&\left\{\left\{\left(\tfrac{e_1}{0.4}, x_2, 0\right), \left\{\frac{u_1}{\langle 0.9e^{2\pi(0.5)}, 0.5e^{2\pi(0.8)}, 0.4e^{2\pi(0.4)}\rangle}, \frac{u_2}{\langle 0.1e^{2\pi(0.9)}, 0.7e^{2\pi(0.2)}, 0.6e^{2\pi(0.7)}\rangle}\right\}\right\},\right. \\
&\left\{\left(\tfrac{e_2}{0.8}, x_2, 1\right), \left\{\frac{u_1}{\langle 0.7e^{2\pi(0.6)}, 0.4e^{2\pi(0.3)}, 0.6e^{2\pi(0.4)}\rangle}, \frac{u_2}{\langle 0.7e^{2\pi(0.9)}, 0.2e^{2\pi(0.4)}, 0.1e^{2\pi(0.6)}\rangle}\right\}\right\}, \\
&\left.\left\{\left(\tfrac{e_3}{0.2}, x_1, 0\right), \left\{\frac{u_1}{\langle 0.7e^{2\pi(0.4)}, 0.2e^{2\pi(0.5)}, 0.9e^{2\pi(0.3)}\rangle}, \frac{u_2}{\langle 0.1e^{2\pi(0.5)}, 0.3e^{2\pi(0.7)}, 0.9e^{2\pi(0.8)}\rangle}\right\}\right\}\right\}.
\end{aligned}$$

By using the complex neutrosophic union and the fuzzy union (maximum), we have

$$\begin{aligned}
&(H,A)_\Gamma \,\widetilde{\cup}\, (G,B)_\Delta = \\
&\left\{\left\{\left(\tfrac{e_1}{0.3}, x_1, 1\right), \left\{\frac{u_1}{\langle 0.4e^{2\pi(0.5)}, 0.5e^{2\pi(0.6)}, 0.8e^{2\pi(0.3)}\rangle}, \frac{u_2}{\langle 0.3e^{2\pi(0.8)}, 0.7e^{2\pi(0.1)}, 0.6e^{2\pi(0.7)}\rangle}\right\}\right\},\right. \\
&\left\{\left(\tfrac{e_2}{0.8}, x_2, 1\right), \left\{\frac{u_1}{\langle 0.7e^{2\pi(0.6)}, 0.4e^{2\pi(0.3)}, 0.4e^{2\pi(0.1)}\rangle}, \frac{u_2}{\langle 0.7e^{2\pi(0.9)}, 0.2e^{2\pi(0.4)}, 0.1e^{2\pi(0.6)}\rangle}\right\}\right\}, \\
&\left\{\left(\tfrac{e_3}{0.9}, x_2, 0\right), \left\{\frac{u_1}{\langle 0.4e^{2\pi(0.1)}, 0.3e^{2\pi(0.4)}, 0.5e^{2\pi(0.9)}\rangle}, \frac{u_2}{\langle 0.6e^{2\pi(0.2)}, 0.9e^{2\pi(0.8)}, 0.1e^{2\pi(0.3)}\rangle}\right\}\right\}, \\
&\left\{\left(\tfrac{e_1}{0.4}, x_2, 0\right), \left\{\frac{u_1}{\langle 0.9e^{2\pi(0.5)}, 0.5e^{2\pi(0.8)}, 0.4e^{2\pi(0.4)}\rangle}, \frac{u_2}{\langle 0.1e^{2\pi(0.9)}, 0.7e^{2\pi(0.2)}, 0.6e^{2\pi(0.7)}\rangle}\right\}\right\}, \\
&\left.\left\{\left(\tfrac{e_3}{0.2}, x_1, 0\right), \left\{\frac{u_1}{\langle 0.7e^{2\pi(0.4)}, 0.2e^{2\pi(0.5)}, 0.9e^{2\pi(0.3)}\rangle}, \frac{u_2}{\langle 0.1e^{2\pi(0.5)}, 0.3e^{2\pi(0.7)}, 0.9e^{2\pi(0.8)}\rangle}\right\}\right\}\right\}.
\end{aligned}$$

By using the complex neutrosophic intersection and the fuzzy intersection (minimum), we have

$$\begin{aligned}
&(H,A)_\Gamma \,\widetilde{\cap}\, (G,B)_\Delta = \\
&\left\{\left\{\left(\tfrac{e_1}{0.3}, x_1, 1\right), \left\{\frac{u_1}{\langle 0.4e^{2\pi(0.5)}, 0.5e^{2\pi(0.6)}, 0.8e^{2\pi(0.3)}\rangle}, \frac{u_2}{\langle 0.3e^{2\pi(0.8)}, 0.7e^{2\pi(0.1)}, 0.6e^{2\pi(0.7)}\rangle}\right\}\right\},\right. \\
&\left\{\left(\tfrac{e_2}{0.8}, x_2, 1\right), \left\{\frac{u_1}{\langle 0.7e^{2\pi(0.4)}, 0.9e^{2\pi(0.3)}, 0.6e^{2\pi(0.4)}\rangle}, \frac{u_2}{\langle 0.6e^{2\pi(0.3)}, 0.2e^{2\pi(0.6)}, 0.4e^{2\pi(0.7)}\rangle}\right\}\right\}, \\
&\left\{\left(\tfrac{e_3}{0.9}, x_2, 0\right), \left\{\frac{u_1}{\langle 0.4e^{2\pi(0.1)}, 0.3e^{2\pi(0.4)}, 0.5e^{2\pi(0.9)}\rangle}, \frac{u_2}{\langle 0.6e^{2\pi(0.2)}, 0.9e^{2\pi(0.8)}, 0.1e^{2\pi(0.3)}\rangle}\right\}\right\}, \\
&\left\{\left(\tfrac{e_1}{0.4}, x_2, 0\right), \left\{\frac{u_1}{\langle 0.9e^{2\pi(0.5)}, 0.5e^{2\pi(0.8)}, 0.4e^{2\pi(0.4)}\rangle}, \frac{u_2}{\langle 0.1e^{2\pi(0.9)}, 0.7e^{2\pi(0.2)}, 0.6e^{2\pi(0.7)}\rangle}\right\}\right\}, \\
&\left.\left\{\left(\tfrac{e_3}{0.2}, x_1, 0\right), \left\{\frac{u_1}{\langle 0.7e^{2\pi(0.4)}, 0.2e^{2\pi(0.5)}, 0.9e^{2\pi(0.3)}\rangle}, \frac{u_2}{\langle 0.1e^{2\pi(0.5)}, 0.3e^{2\pi(0.7)}, 0.9e^{2\pi(0.8)}\rangle}\right\}\right\}\right\}.
\end{aligned}$$

Proposition 2. *Let* $(H,A)_\Gamma$, $(G,B)_\Delta$ *and* $(S,W)_\Theta$ *be any three FP-CNSESs over a universe U. Then,*

1. $((H,A)_\Gamma \,\widetilde{\cup}\, (G,B)_\Delta)\widetilde{\cap}(S,W)_\Theta = ((H,A)_\Gamma \,\widetilde{\cap}\, (S,W)_\Theta)\widetilde{\cup}((G,B)_\Delta\widetilde{\cap}(S,W)_\Theta)$;
2. $((H,A)_\Gamma \,\widetilde{\cap}\, (G,B)_\Delta)\widetilde{\cup}(S,W)_\Theta = ((H,A)_\Gamma \,\widetilde{\cup}\, (S,W)_\Theta)\widetilde{\cap}((G,B)_\Delta\widetilde{\cup}(S,W)_\Theta)$.

Proof.
(1) Assume that $((H,A)_\Gamma \,\widetilde{\cup}\, (G,B)_\Delta) = (Q,R)_\Pi$, where $R_\Pi = A_\Gamma \cup B_\Delta$ and $\Pi = \Gamma \widehat{\cup} \Delta$, $(L,D)_\Phi = (Q,R)_\Pi \widetilde{\cap} (S,W)_\Theta$ where $D_\Phi = R_\Pi \cup W_\Theta$ and $\Phi = \Pi \widehat{\cap} \Theta$. Thus, $\Phi = (\Gamma \widehat{\cup} \Delta) \widehat{\cap} \Theta = (\Gamma \widehat{\cap} \Theta) \widehat{\cup} (\Delta \widehat{\cap} \Theta)$, since the distributive property is valid for fuzzy sets.

Let $(L,D)_{\Phi} = \left\{\left(\epsilon,\left\{\frac{u}{L_{\Phi}(\epsilon)(u)} : u \in U\right\}\right) : \epsilon \in D\right\}$, where $L_{\Phi}(\epsilon)(u) = \langle T_{L_{\Phi}(\epsilon)}(u), I_{L_{\Phi}(\epsilon)}(u), F_{L_{\Phi}(\epsilon)}(u)\rangle$. We consider the case when $\epsilon \in A_{\Gamma} \cap B_{\Delta} \cap W_{\Theta}$ as the other cases are trivial. Then, we have,

$$
\begin{aligned}
&T_{L_{\Phi}(\epsilon)}(u) = T_{(Q_{\Pi}(\epsilon)\tilde{\cap}S_{\Theta}(\epsilon))}(u) = T_{(H_{\Gamma}(\epsilon)\tilde{\cup}G_{\Delta}(\epsilon))\tilde{\cap}S_{\Theta}(\epsilon)}(u),\\
&= min(max(p_{H_{\Gamma}(\epsilon)}(u), p_{G_{\Delta}(\epsilon)}(u)), p_{S_{\Theta}(\epsilon)}(u)).e^{jmin(max(\mu_{H_{\Gamma}(\epsilon)}(u),\mu_{G_{\Delta}(\epsilon)}(u)),\mu_{S_{\Theta}(\epsilon)}(u))},\\
&= max(min(p_{H_{\Gamma}(\epsilon)}(u), p_{S_{\Theta}(\epsilon)}(u)), min(p_{G_{\Delta}(\epsilon)}(u), p_{S_{\Theta}(\epsilon)}(u))).e^{jmax(min(\mu_{H_{\Gamma}(\epsilon)}(u),\mu_{S_{\Theta}(\epsilon)}(u)),min(\mu_{G_{\Delta}(\epsilon)}(u),\mu_{S_{\Theta}(\epsilon)}(u)))},\\
&= T_{(H_{\Gamma}(\epsilon)\tilde{\cap}S_{\Theta}(\epsilon))\tilde{\cup}(G_{\Delta}(\epsilon)\tilde{\cap}S_{\Theta}(\epsilon))}(u),
\end{aligned}
$$

which implies that $T_{(H_{\Gamma}(\epsilon)\tilde{\cup}G_{\Delta}(\epsilon))\tilde{\cap}S_{\Theta}(\epsilon)}(u) = T_{(H_{\Gamma}(\epsilon)\tilde{\cap}S_{\Theta}(\epsilon))\tilde{\cup}(G_{\Delta}(\epsilon)\tilde{\cap}S_{\Theta}(\epsilon))}(u)$.

The proofs for the identity and falsity terms follow similarly. Therefore,

$$((H,A)_{\Gamma} \tilde{\cup} (G,B)_{\Delta})\tilde{\cap}(S,W)_{\Theta} = ((H,A)_{\Gamma} \tilde{\cap} (S,W)_{\Theta})\tilde{\cup}((G,B)_{\Delta}\tilde{\cap}(S,W)_{\Theta}).$$

(2) The proof is similar to that in Part (1) and therefore is omitted. □

4. MAPPING ON FP-CNSESs

In this section, we introduce the notion of a mapping on fuzzy parameterized complex neutrosophic soft expert classes. fuzzy parameterized complex neutrosophic soft expert classes are collections of FP-CNSESs. We define the fuzzy parameterized complex neutrosophic soft expert images and fuzzy parameterized complex neutrosophic soft expert inverse images of FP-CNSESs. We give some operations and properties related with this concept.

Definition 13. *Let U be a universe set, E be a set of parameters, $FZ(E)$ denote all fuzzy subsets of E, where $\Gamma \subset FZ(E)$, X be a set of experts, and $O = \{1 = agree,\ 0 = disagree\}$ a set of opinions. Let $Y = \Gamma \times X \times O$. Then, the collection of all FP-CNSESs over U with parameters from Y is called a fuzzy parameterized complex neutrosophic soft expert class and is denoted as $\widetilde{(U,Y)}$.*

Definition 14. *Let $\widetilde{(U,Y)}$ and $\widetilde{(V,Y')}$ be fuzzy parameterized complex neutrosophic soft expert classes. Let $r : U \to V$ and $s : Y \to Y'$ be mappings. Then, a mapping $F : \widetilde{(U,Y)} \to \widetilde{(V,Y')}$ is defined as follows:*

For a FP-CNSES $(H,A)_{\Gamma}$ in $\widetilde{(U,Y)}$, $(F(H,A)_{\Gamma}, M)$, where $M = s(Y) \subseteq Y'$ is a FP-CNSES in $\widetilde{(V,Y')}$ obtained as follows:

$$F(H,A)_{\Gamma}(\beta)(v) = \begin{cases} \bigcup\limits_{u \in r^{-1}(v)} \left(\bigcup\limits_{\alpha \in s^{-1}(\beta)\cap A} H_{\Gamma}(\alpha)\right)(u) & \text{if } r^{-1}(v) \text{ and } s^{-1}(\beta)\cap A \neq \emptyset, \\ (0,1,1) & \text{otherwise.} \end{cases}$$

For $u \in r^{-1}(v)$, $\beta \in M \subseteq Y'$, $v \in V$ and $\forall \alpha \in s^{-1}(\beta) \cap A$, $(F(H,A)_{\Gamma}, M)$ is called a fuzzy parameterized complex neutrosophic soft expert image of the FP-CNSES $(H,A)_{\Gamma}$. If $M = Y'$, then we shall write $(F(H,A)_{\Gamma}, M)$ as $F(H,A)_{\Gamma}$.

Definition 15. *Let $\widetilde{(U,Y)}$ and $\widetilde{(V,Y')}$ be two fuzzy parameterized complex neutrosophic soft expert classes. Let $r : U \to V$ and $s : Y \to Y'$ be mappings. Then, a mapping $F^{-1} : \widetilde{(V,Y')} \to \widetilde{(U,Y)}$ is defined as follows:*

For a FP-CNSES $(G,B)_{\Delta}$ in $\widetilde{(V,Y')}$, $(F^{-1}(G,B)_{\Delta}, N)$, where $N = s^{-1}(B)$, is a FP-CNSES in $\widetilde{(U,Y)}$ obtained as follows:

$$F^{-1}(G,B)_{\Delta}(\alpha)(u) = \begin{cases} G_{\Delta}(s(\alpha))(r(u)) & \text{if } s(\alpha) \in B, \\ (0,1,1) & \text{otherwise,} \end{cases}$$

For $\alpha \in N \subseteq Y$ and $u \in U$. $(F^{-1}(G,B)_{\Delta}, N)$ is called a fuzzy parameterized complex neutrosophic soft expert inverse image of the FP-CNSES $(G,B)_{\Delta}$. If $N = Y$, we write $(F^{-1}(G,B)_{\Delta}, N)$ as $F^{-1}(G,B)_{\Delta}$.

Definition 16. *Let* $F : (\widetilde{U,Y}) \to (\widetilde{V,Y'})$ *be a mapping and* $(H,A)_\Gamma$, $(H',A')_\Delta$ *be FP-CNSESs in* $(\widetilde{U,Y})$. *Then, for* $\beta \in Y'$, $v \in V$, *the union and intersection of the fuzzy parameterized complex neutrosophic soft expert images* $F(H,A)_\Gamma$ *and* $F(G,B)_\Delta$ *are defined as follows.*

$$\left(F(H,A)_\Gamma \tilde{\cup} F(H',A')_\Delta\right)(\beta)(v) = F(H,A)_\Gamma(\beta)(v) \cup F(H',A')_\Delta(\beta)(v),$$

$$\left(F(H,A)_\Gamma \tilde{\cap} F(H',A')_\Delta\right)(\beta)(v) = F(H,A)_\Gamma(\beta)(v) \cap F(H',A')_\Delta(\beta)(v),$$

where $\tilde{\cup}$ *and* $\tilde{\cap}$ *denote fuzzy parameterized complex neutrosophic soft expert union and intersection of fuzzy parameterized complex neutrosophic soft expert images in* $(\widetilde{V,Y'})$.

Definition 17. *Let* $F : (\widetilde{U,Y}) \to (\widetilde{V,Y'})$ *be a mapping and* $(G,B)_\Sigma$, $(G',B')_\Omega$ *FP-CNSESs in* $(\widetilde{V,Y'})$. *Then, for* $\alpha \in Y$, $u \in U$, *the union and intersection of the fuzzy parameterized complex neutrosophic soft expert inverse images* $F^{-1}(G,B)_\Sigma$ *and* $F^{-1}(G',B')_\Omega$ *are defined as follows.*

$$\left(F^{-1}(G,B)_\Sigma \tilde{\cup} F^{-1}(G',B')_\Omega\right)(\alpha)(u) = F^{-1}(G,B)_\Sigma(\alpha)(u) \cup F^{-1}(G',B')_\Omega(\alpha)(u),$$

$$(F^{-1}(G,B)_\Sigma \tilde{\cap} F^{-1}(G',B')_\Omega)(\alpha)(u) = F^{-1}(G,B)_\Sigma(\alpha)(u) \cap F^{-1}(G',B')_\Omega(\alpha)(u),$$

where $\tilde{\cup}$ *and* $\tilde{\cap}$ *denote fuzzy parameterized complex neutrosophic soft expert union and intersection of fuzzy parameterized complex neutrosophic soft expert inverse images in* $(\widetilde{U,Y})$.

Proposition 3. *Let* $F : (\widetilde{U,Y}) \to (\widetilde{V,Y'})$ *be a mapping. Then, for FP-CNSESs* $(H,A)_\Gamma$ *and* $(H',A')_\Delta$ *in the fuzzy parameterized complex neutrosophic soft expert class* $(\widetilde{U,Y})$, *we have:*

1. $F(\tilde{\phi}) = \tilde{\phi}$.
2. $F(\tilde{\psi}) = \tilde{\psi}$.
3. $F\left((H,A)_\Gamma \tilde{\cup} (H',A')_\Delta\right) = F(H,A)_\Gamma \tilde{\cup} F(H',A')_\Delta$.
4. $F\left((H,A)_\Gamma \tilde{\cap} (H',A')_\Delta\right) \subseteq F(H,A)_\Gamma \tilde{\cap} F(H',A')_\Delta$.
5. *If* $(H,A)_\Gamma \subseteq (H',A')_\Delta$, *then* $F(H,A)_\Gamma \subseteq F(H',A')_\Delta$.

Proof. The proof is straightforward by Definitions 14 and 16. □

Proposition 4. *Let* $F : (\widetilde{U,Y}) \to (\widetilde{V,Y'})$ *be a mapping. Then, for FP-CNSESs* $(G,B)_\Sigma$ *and* $(G',B')_\Omega$ *in the fuzzy parameterized complex neutrosophic soft expert class* $(\widetilde{V,Y'})$, *we have:*

1. $F^{-1}(\tilde{\phi}) = \tilde{\phi}$.
2. $F^{-1}(\tilde{\psi}) = \tilde{\psi}$.
3. $F^{-1}\left((G,B)_\Sigma \tilde{\cup} (G',B')_\Omega\right) = F^{-1}(G,B)_\Sigma \tilde{\cup} F^{-1}(G',B')_\Omega$.
4. $F^{-1}\left((G,B)_\Sigma \tilde{\cap} (G',B')_\Omega\right) = F^{-1}(G,B)_\Sigma \tilde{\cap} F^{-1}(G',B')_\Omega$.
5. *If* $(G,B)_\Sigma \subseteq (G',B')_\Omega$, *then* $F^{-1}(G,B)_\Sigma \subseteq F^{-1}(G',B')_\Omega$.

Proof. The proof is straightforward by Definitions 15 and 17. □

5. An Application of Fuzzy Parameterized Complex Neutrosophic Soft Expert Set

In this section, we present an application of FP-CNSES in a decision making problem by considering the following example.

Example 4. *Suppose that an engineering company wishes to evaluate two kinds of a certain product from a manufacturer and choose the most suitable one. Suppose that the company takes the opinion of its experts concerning these two kinds of product on two phases: once before using the products and again after trying a sample of each of the two kinds of the product. Suppose that* $U = \{u_1, u_2\}$ *is the universe consisting of the two alternatives (the two kinds of the product) and* $E = \{e_1, e_2, e_3\}$ *is the attributes set, where* e_1 *stands for "easy to use",* e_2 *stands for " functional " and* e_3 *stands for "durable". The attributes* e_1, e_2 *and* e_3 *are important* *with degrees 0.3, 0.6 and 0.8, respectively. That is, the fuzzy subset of parameters is* $\Gamma = \left\{\frac{e_1}{0.3}, \frac{e_2}{0.6}, \frac{e_3}{0.8}\right\}$. *Let* $X = \{x_1, x_2\}$ *be a set of experts. Now, the team of experts are requested to make a decision about the most desirable alternative based on the goals and the constraints according to a chosen subset* Γ *of* $FZ(E)$ *to construct a FP-CNSES.*

$$
\begin{aligned}
&(H,A)_\Gamma = \\
&\left\{\left\{\left(\tfrac{e_1}{0.3}, x_1, 1\right), \left\{\frac{u_1}{\langle 0.5e^{2\pi(0.4)}, 0.2e^{2\pi(0.4)}, 0.1e^{2\pi(0.2)}\rangle}, \frac{u_2}{\langle 0.6e^{2\pi(0.7)}, 0.5e^{2\pi(0.4)}, 0.1e^{2\pi(1)}\rangle}\right\}\right\},\right. \\
&\left\{\left(\tfrac{e_1}{0.3}, x_2, 1\right), \left\{\frac{u_1}{\langle 0.6e^{2\pi(0.5)}, 0.6e^{2\pi(0.4)}, 0.1e^{2\pi(0.3)}\rangle}, \frac{u_2}{\langle 0.9e^{2\pi(0.8)}, 0.5e^{2\pi(0.6)}, 0.5e^{2\pi(0.4)}\rangle}\right\}\right\}, \\
&\left\{\left(\tfrac{e_2}{0.6}, x_1, 1\right), \left\{\frac{u_1}{\langle 0.4e^{2\pi(0.3)}, 0.7e^{2\pi(0.8)}, 0.8e^{2\pi(0.7)}\rangle}, \frac{u_2}{\langle 0.4e^{2\pi(0.9)}, 0.2e^{2\pi(0.5)}, 0.1e^{2\pi(0.2)}\rangle}\right\}\right\}, \\
&\left\{\left(\tfrac{e_2}{0.6}, x_2, 1\right), \left\{\frac{u_1}{\langle 0.8e^{2\pi(0.9)}, 0.4e^{2\pi(0.3)}, 0.7e^{2\pi(0.4)}\rangle}, \frac{u_2}{\langle 0.5e^{2\pi(0.4)}, 0.8e^{2\pi(0.6)}, 0.6e^{2\pi(0.2)}\rangle}\right\}\right\}, \\
&\left\{\left(\tfrac{e_3}{0.8}, x_1, 1\right), \left\{\frac{u_1}{\langle 0.2e^{2\pi(0.2)}, 0.3e^{2\pi(0.3)}, 0.7e^{2\pi(0.6)}\rangle}, \frac{u_2}{\langle 0.5e^{2\pi(0.2)}, 0.7e^{2\pi(0.2)}, 0.8e^{2\pi(0.2)}\rangle}\right\}\right\}, \\
&\left\{\left(\tfrac{e_3}{0.8}, x_2, 1\right), \left\{\frac{u_1}{\langle 0.8e^{2\pi(0.1)}, 0.3e^{2\pi(0.2)}, 0.7e^{9\pi(0.2)}\rangle}, \frac{u_2}{\langle 0.1e^{2\pi(0.6)}, 0.2e^{2\pi(0.4)}, 0.6e^{2\pi(0.8)}\rangle}\right\}\right\}, \\
&\left\{\left(\tfrac{e_1}{0.3}, x_1, 0\right), \left\{\frac{u_1}{\langle 0.5e^{2\pi(0.5)}, 0.5e^{2\pi(0.4)}, 0.9e^{2\pi(0.3)}\rangle}, \frac{u_2}{\langle 0.9e^{2\pi(0.8)}, 0.7e^{2\pi(0.8)}, 0.5e^{2\pi(0.7)}\rangle}\right\}\right\}, \\
&\left\{\left(\tfrac{e_1}{0.3}, x_2, 0\right), \left\{\frac{u_1}{\langle 0.4e^{2\pi(0.6)}, 0.7e^{2\pi(0.4)}, 0.7e^{2\pi(0.8)}\rangle}, \frac{u_2}{\langle 0.5e^{2\pi(0.4)}, 0.5e^{2\pi(0.4)}, 0.3e^{2\pi(0.5)}\rangle}\right\}\right\}, \\
&\left\{\left(\tfrac{e_2}{0.6}, x_1, 0\right), \left\{\frac{u_1}{\langle 0.3e^{2\pi(0.2)}, 0.6e^{2\pi(0.2)}, 0.9e^{2\pi(0.2)}\rangle}, \frac{u_2}{\langle 0.7e^{2\pi(0.4)}, 0.5e^{2\pi(0.8)}, 0.6e^{2\pi(0.8)}\rangle}\right\}\right\}, \\
&\left\{\left(\tfrac{e_2}{0.6}, x_2, 0\right), \left\{\frac{u_1}{\langle 0.3e^{2\pi(0.5)}, 0.3e^{2\pi(0.1)}, 0.9e^{2\pi(0.3)}\rangle}, \frac{u_2}{\langle 0.3e^{2\pi(0.7)}, 0.8e^{2\pi(0.7)}, 0.7e^{2\pi(0.6)}\rangle}\right\}\right\}, \\
&\left\{\left(\tfrac{e_3}{0.8}, x_1, 0\right), \left\{\frac{u_1}{\langle 0.5e^{2\pi(0.4)}, 0.4e^{2\pi(0.5)}, 0.4e^{2\pi(0.3)}\rangle}, \frac{u_2}{\langle 0.8e^{2\pi(0.2)}, 0.3e^{2\pi(0.2)}, 0.5e^{2\pi(0.2)}\rangle}\right\}\right\}, \\
&\left.\left\{\left(\tfrac{e_3}{0.8}, x_2, 0\right), \left\{\frac{u_1}{\langle 0.4e^{2\pi(0.6)}, 0.2e^{2\pi(0.3)}, 0.1e^{2\pi(0.6)}\rangle}, \frac{u_2}{\langle 0.3e^{2\pi(0)}, 0.6e^{2\pi(0.4)}, 0.3e^{2\pi(0.1)}\rangle}\right\}\right\}\right\}.
\end{aligned}
$$

In the FP-CNSES $(H, A)_\Gamma$ above, the amplitude terms of membership, non-membership and indeterminate membership represent the opinions of the experts in phase one (before using the products), while the phase terms of membership, non-membership, and indeterminacy represent the opinions of the experts in the second phase (after trying a sample of each of the two kinds of the product). Thus, the amplitude term of the membership in the first phase and the phase term of the membership in the second phase form a complex-valued truth membership function of the FP-CNSES $(H, A)_\Gamma$. Similarly, the amplitude term of non-membership in phase one and the phase term of non-membership in the second phase form a complex-valued falsity membership function. In addition, the amplitude term of undecidedness in the first phase and the phase term of indeterminacy in the second phase form the complex-valued indeterminate membership function. Now, our problem is to select the most desirable kind of the product to the engineering company.

To solve this decision making problem, we use the FP-CNSES $(H, A)_\Gamma$ together with a generalized algorithm. This algorithm converts the fuzzy parameterized complex neutrosophic soft expert values (FP-CNSEVs) to fuzzy parameterized single-valued neutrosophic soft expert values (FP-SVNSEVs) using a practical formula which give a decision-making with a simple computational process without the need to carry out directed operations on complex numbers. In this formula, we give a weight to the amplitude terms (a weight to the experts' opinions before using the products) by multiplying the weight vector to each amplitude term. Similarly we give a weight to the phase terms (a weight to the experts' opinions after using the products) by multiplying the weight vector to each phase term. Then, we combine the values of the weighted amplitude terms and phase terms to obtain the FP-SVNSEVs which represent the experts' opinions on both phases together. Thus, after doing these simple calculations to all membership functions of the FP-CNSES $(H, A)_\Gamma$, we proceed to the final decision using the fuzzy parameterized single valued neutrosophic soft expert method (FPSVNSEM) [18].

The generalized algorithm is as follows.

Algorithm 1: Fuzzy parameterized complex neutrosophic soft expert method (FP-CNSEM).

1. Input the FP-CNSES $(H, A)_\Gamma$
2. Convert the FP-CNSES $(H, A)_\Gamma$ to the FPSVNSES $(\widehat{H}, A)_\Gamma$ by obtaining the weighted aggregation values of $T_{\widehat{H}_\Gamma(a_i)}(u_j), I_{\widehat{H}_\Gamma(a_i)}(u_j)$ and $F_{\widehat{H}_\Gamma(a_i)}(u_j), \forall a_i \in A$ and $\forall u_j \in U$ as the following formulas:
$$T_{\widehat{H}_\Gamma(a_i)}(u_j) = w_1 p_{H_\Gamma(a_i)}(u_j) + w_2(1/2\pi)\mu_{H_\Gamma(a_i)}(u_j),$$
$$I_{\widehat{H}_\Gamma(a_i)}(u_j) = w_1 q_{H_\Gamma(a_i)}(u_j) + w_2(1/2\pi)\nu_{H_\Gamma(a_i)}(u_j),$$
$$F_{\widehat{H}_\Gamma(a_i)}(u_j) = w_1 r_{H_\Gamma(a_i)}(u_j) + w_2(1/2\pi)\omega_{H_\Gamma(a_i)}(u_j),$$
where $p_{H_\Gamma(a_i)}(u_j), q_{H_\Gamma(a_i)}(u_j), r_{H_\Gamma(a_i)}(u_j)$ and $\mu_{H_\Gamma(a_i)}(u_j), \nu_{H_\Gamma(a_i)}(u_j), \omega_{H_\Gamma(a_i)}(u_j)$ are the amplitude and phase terms in the FP-CNSES $(H, A)_\Gamma$, respectively. $T_{\widehat{H}_\Gamma(a_i)}(u_j), I_{\widehat{H}_\Gamma(a_i)}(u_j)$ and $F_{\widehat{H}_\Gamma(a_i)}(u_j)$ are the truth-membership function, indeterminacy-membership function and falsity-membership function in the FP-SVNSES $(\widehat{H}, A)_\Gamma$, respectively, and w_1, w_2 are the weights for the amplitude terms (the first decision process) and the phase terms (the second decision process), respectively, where w_1 and $w_2 \in [0, 1]$ and $w_1 + w_2 = 1$.
3. Find the values of $c_{ij} = T_{\widehat{H}_\Gamma(a_i)}(u_j) - I_{\widehat{H}_\Gamma(a_i)}(u_j) - F_{\widehat{H}_\Gamma(a_i)}(u_j), \forall u_j \in U$ and $\forall a_i \in A$.
4. Compute the score of each element $u_j \in U$ by the following formulas :
$$K_j = \sum_{x\in X}\sum_{i=1}^{n} c_{ij}(\mu_\Gamma(e_i)),\ S_j = \sum_{x\in X}\sum_{i=1}^{n} c_{ij}(\mu_\Gamma(e_i))$$
for the agree FP-SVNSES and disagree FP-SVNSES, where $\mu_\Gamma(e_i)$ is the corresponding membership function of the fuzzy set Γ, X is the set of the experts and n is the number of the parameters (attributes).
5. Find the values of the score $r_j = K_j - S_j$ for each element $u_j \in U$.
6. Determine the value of the highest score $m = max_{u_j \in U}\{r_j\}$. Then, the decision is to choose element u_j as the optimal solution to the problem. If there are more than one elements with the highest r_j score, then any one of those elements can be chosen as the optimal solution.

Then, we can conclude that the optimal choice for the team of the experts is to select the kind u_j as the most desirable kind of the product to the company.

Now, to convert the FP-CNSES $(H, A)_\Gamma$ to the FP-SVNSES $(\widehat{H}, A)_\Gamma$, suppose that the weight vector for the amplitude terms is $w_1 = 0.3$ and the weight vector for the phase terms is $w_2 = 0.7$ and obtain the weighted aggregation values of $T_{\widehat{H}_\Gamma(a_i)}(u_j), I_{\widehat{H}_\Gamma(a_i)}(u_j)$ and $F_{\widehat{H}_\Gamma(a_i)}(u_j), \forall a_i \in A$ and $\forall u_j \in U$. To illustrate this step, we calculate $T_{\widehat{H}_\Gamma(a_1)}(u_1), I_{\widehat{H}_\Gamma(a_1)}(u_1)$ and $F_{\widehat{H}_\Gamma(a_1)}(u_1)$, such that $a_1 = (\frac{e_1}{0.3}, x_1, 1)$ as shown below:

$$\begin{aligned} T_{\widehat{H}_\Gamma(\frac{e_1}{0.3},x_1,1)}(u_1) &= w_1 p_{H_\Gamma(\frac{e_1}{0.3},x_1,1)}(u_1) + w_2(1/2\pi)\mu_{H_\Gamma(\frac{e_1}{0.3},x_1,1)}(u_1) \\ &= 0.3(0.5) + 0.7(1/2\pi)(2\pi)(0.4) \\ &= 0.43 \end{aligned}$$

$$\begin{aligned} I_{\widehat{H}_\Gamma(\frac{e_1}{0.3},x_1,1)}(u_1) &= w_1 q_{H_\Gamma(\frac{e_1}{0.3},x_1,1)}(u_1) + w_2(1/2\pi)\nu_{H_\Gamma(\frac{e_1}{0.3},x_1,1)}(u_1) \\ &= 0.3(0.2) + 0.7(1/2\pi)(2\pi)(0.4) \\ &= 0.34 \end{aligned}$$

$$\begin{aligned} F_{\widehat{H}_\Gamma(\frac{e_1}{0.3},x_1,1)}(u_1) &= w_1 r_{H_\Gamma(\frac{e_1}{0.3},x_1,1)}(u_1) + w_2(1/2\pi)\omega_{H_\Gamma(\frac{e_1}{0.3},x_1,1)}(u_1) \\ &= 0.3(0.1) + 0.7(1/2\pi)(2\pi)(0.2) \\ &= 0.17. \end{aligned}$$

Then, the FP-SVNSEV

$$(T_{\widehat{H}_\Gamma(a_1)}(u_1), I_{\widehat{H}_\Gamma(a_1)}(u_1), F_{\widehat{H}_\Gamma(a_1)}(u_1)) = (0.43, 0.34, 0.17).$$

In the same manner, we calculate the others FP-SVNSEVs, $\forall a \in A$ and $\forall u \in U$ as in Table 1.

Table 1. Values of $(\widehat{H}, A)_\Gamma$.

	u_1	u_2
$(\frac{e_1}{0.3}, x_1, 1)$	$\langle 0.43, 0.34, 0.17\rangle$	$\langle 0.67, 0.43, 0.73\rangle$
$(\frac{e_1}{0.3}, x_2, 1)$	$\langle 0.53, 0.46, 0.24\rangle$	$\langle 0.83, 0.57, 0.53\rangle$
$(\frac{e_2}{0.6}, x_1, 1)$	$\langle 0.33, 0.77, 0.73\rangle$	$\langle 0.75, 0.41, 0.71\rangle$
$(\frac{e_2}{0.6}, x_2, 1)$	$\langle 0.87, 0.33, 0.49\rangle$	$\langle 0.43, 0.66, 0.32\rangle$
$(\frac{e_3}{0.8}, x_1, 1)$	$\langle 0.20, 0.30, 0.63\rangle$	$\langle 0.29, 0.35, 0.38\rangle$
$(\frac{e_3}{0.8}, x_2, 1)$	$\langle 0.31, 0.23, 0.35\rangle$	$\langle 0.45, 0.34, 0.74\rangle$
$(\frac{e_1}{0.3}, x_1, 0)$	$\langle 0.50, 0.43, 0.48\rangle$	$\langle 0.83, 0.77, 0.64\rangle$
$(\frac{e_1}{0.3}, x_2, 0)$	$\langle 0.54, 0.49, 0.77\rangle$	$\langle 0.43, 0.43, 0.44\rangle$
$(\frac{e_2}{0.6}, x_1, 0)$	$\langle 0.23, 0.32, 0.41\rangle$	$\langle 0.49, 0.71, 0.74\rangle$
$(\frac{e_2}{0.6}, x_2, 0)$	$\langle 0.44, 0.16, 0.48\rangle$	$\langle 0.58, 0.73, 0.63\rangle$
$(\frac{e_3}{0.8}, x_1, 0)$	$\langle 0.43, 0.47, 0.33\rangle$	$\langle 0.38, 0.23, 0.29\rangle$
$(\frac{e_3}{0.8}, x_2, 0)$	$\langle 0.54, 0.27, 0.45\rangle$	$\langle 0.09, 0.46, 0.16\rangle$

Tables 2 and 3 give the values of $c_{ij} = T_{\widehat{H}_\Gamma(a_i)}(u_j) - I_{\widehat{H}_\Gamma(a_i)}(u_j) - F_{\widehat{H}_\Gamma(a_i)}(u_j)$ and the score of each element $u_j \in U$ for agree FP-SVNSES and disagree FP-SVNSES, respectively.

Table 2. Numerical degree of agree FP-SVNSES.

U	u_1	u_2
$(\frac{e_1}{0.3}, x_1)$	−0.08	−0.49
$(\frac{e_1}{0.3}, x_2)$	−0.17	−0.27
$(\frac{e_2}{0.6}, x_1)$	−1.17	−0.37
$(\frac{e_2}{0.6}, x_2)$	0.05	−0.55
$(\frac{e_3}{0.8}, x_1)$	−0.73	−0.44
$(\frac{e_3}{0.8}, x_2)$	−0.27	−0.63
$K_j = \sum_{x\in X}\sum_{i=1}^{3} c_{ij}(\mu_\Gamma(e_i))$	$K_1 = -1.547$	$K_2 = -1.636$

Table 3. Numerical degree for disagree FP-SVNSES.

U	u_1	u_2
$(\frac{e_1}{0.3}, x_1)$	−0.41	−0.58
$(\frac{e_1}{0.3}, x_2)$	−0.72	−0.44
$(\frac{e_2}{0.6}, x_1)$	−0.50	−0.96
$(\frac{e_2}{0.6}, x_2)$	−0.20	−0.78
$(\frac{e_3}{0.8}, x_1)$	−0.37	−0.14
$(\frac{e_3}{0.8}, x_2)$	−0.18	−0.53
$S_j = \sum_{x\in X}\sum_{i=1}^{3} c_{ij}(\mu_\Gamma(e_i))$	$S_1 = -1.199$	$S_2 = -1.886$

Let K_j and S_j represent the score of each numerical degree for the agree FP-SVNSES and disagree FP-SVNSES, respectively. These values are given in Table 4.

Table 4. The score $r_j = K_j - S_j$.

$K_j = \sum_{x \in X} \sum_{i=1}^{3} c_{ij}(\mu_\Gamma(e_i))$	$S_j = \sum_{x \in X} \sum_{i=1}^{3} c_{ij}(\mu_\Gamma(e_i))$	$r_j = K_j - S_j$
$K_1 = -1.547$	$S_1 = -1.199$	-0.348
$K_2 = -1.636$	$S_2 = -1.886$	0.25

Thus, $m = max_{u_j \in U}\{r_j\} = r_2$. Therefore, the team of experts advise selecting the kind u_2 of this product as a desirable alternative.

6. Comparison between FP-CNSES and Other Existing Methods

In this section, we compare our proposed FP-CNSES model to FP-SVNSES [18] and FP-IFSES [17], which are generalizations of FP-SES [15] and FP-FSES [16].

To reveal the significance of our proposed FP-CNSES compared to FP-SVNSES and FP-IFSES, let us consider Example 4 above. In this example, we apply the FP-CNSES to handle a decision-making problem, which consists of two decision processes. Our proposed model is applied to both the decision processes by considering the opinions in process one as amplitude terms of truth membership, falsity membership and indeterminacy membership, and setting the second decision process as phase terms of truth membership, falsity membership, and indeterminacy. Thus, both decision processes form a FP-CNSES as whole.

On the other hand, when we apply the FP-IFSES to both processes, it tells us only about the truth membership and falsity membership in the first decision process, but cannot tell anything about the undecidedness. The situation is similar in the second decision process. Thus, FP-IFSES fails to handle this situation. Now, when we apply the FP-SVNSES, it tells about the truth membership, falsity membership and indeterminacy membership in the first round, and, similarly, it tells about the second round but it cannot describe both decision processes simultaneously.

Thus, both FP-SVNSES and FP-IFSES cannot directly solve such a decision-making problem with fuzzy parameterized complex neutrosophic soft expert information. In contrast, the FP-CNSES can directly address the fuzzy parameterized single valued neutrosophic soft expert problem, since the FP-SVNSES is a special case of FP-CNSES and can be easily represented in the form of FP-CNSES. In other words, the FP-SVNSES is a FP-CNSES with phase terms equal zeros. For example, the FP-SVNSEV (0.2, 0.4, 0.6) can be represented as $(0.2e^{j2\Pi(0)}, 0.4e^{j2\Pi(0)}, 0.6e^{j2\Pi(0)})$ by means of FP-CNSES. Furthermore, our method is applicable for fuzzy parameterized intuitionistic fuzzy soft expert problem, since FP-IFSES is a special case of FP-SVNSES and consequently of FP-CNSES. For example, the fuzzy parameterized intuitionistic fuzzy soft expert value (0.1, 0.6) can be (0.1, 0.3, 0.6) by means of FP-SVNSES and hence can be $(0.1e^{j2\Pi(0)}, 0.3e^{j2\Pi(0)}, 0.6e^{j2\Pi(0)})$ by means of FP-CNSES, since the sum of the degrees of membership, nonmembership and indeterminacy of an intuitionistic fuzzy value equal to 1. Note that the indeterminacy degree in intuitionistic fuzzy set is provided by default and cannot be defined alone unlike the SVNS where the indeterminacy is defined independently and quantified explicitly.

The advantage of using FP-CNSES manifests in representing information of two dimensions for one object in the same time (i.e., the expert's time can be saved by constructing/building one set (FP-CNSES) instead of two sets (FP-SVNSESs), not to mention reducing the amount of both mathematical calculations and investigation of several operators in incorporating two FP-SVNSESs to find the optimal solution). A practical formula is employed to convert the FP-CNSES to the FP-SVNSES, which gives a decision-making with a simple computational process without the need to carry out directed operations on complex numbers.

7. Weighted Fuzzy Parameterized Complex Neutrosophic Soft Expert Set

Experts opinions are vital for any decision making process as most real-life situations deal with elements and parameters that are subjective, biased and have the potential to be distorted and interpreted differently by different parties. Each expert might have his/her own thought, which differs from others in various aspects but all of the experts should have a common goal to reach the final destination. Moreover, as the domains of expertise of different experts are different, one expert may be more confident on his/her opinion than the other on the same set of attributes. For this type of environment, equal weights assignment to different experts may lead to improper and biased solution. Feeling the need of prioritizing different experts motivates us to develop a new idea for assigning relative weight to each expert by establishing a novel notion called weighted fuzzy parameterized complex neutrosophic soft expert set (WFP-CNSES). The relative weight is assigned to each of the experts where the choice of the experts may not be of equal importance.

This new mathematical framework reduces the chance of unfairness in the decision making process and brings more credibility to the final decision.

We begin this section by first proposing the concept of the weighted fuzzy parameterized single-valued neutrosophic soft expert set (WFP-SVNSES).

Definition 18. *Let U be a universe set, E be a set of parameters, I^E denote all fuzzy subsets of E, X be a set of experts, I^X denotes all fuzzy subsets of X and $O = \{1 = agree,\ 0 = disagree\}$ a set of opinions. Let $Z = \Psi \times W \times O$ and $A \subseteq Z$ where $\Psi \subset I^E$ and $W \subset I^X$. A pair $(f, A,)_{\Psi,W}$ is called a weighted fuzzy parameterized single-valued neutrosophic soft expert set over U, where f is a mapping given by*

$$f_{\Psi,W} : A \to SVN(U),$$

and $SVN(U)$ denotes the set of all single-valued neutrosophic subsets of U.

We then generalize the WFP-SVNSES by establishing the concept of the weighted fuzzy parameterized complex neutrosophic soft expert set (WFP-CNSES), which is actually an extended version of WFP-SVNSES, on the complex space.

Definition 19. *Let U be a universe set, E be a set of parameters, $FZ(E)$ denote all fuzzy subsets of E, X be a set of experts, $FZ(X)$ denotes all fuzzy subsets of X and $O = \{1 = agree,\ 0 = disagree\}$ a set of opinions. Let $Y = \Gamma \times W \times O$ and $A \subseteq Y$ where $\Gamma \subset FZ(E)$ and $W \subset FZ(X)$. A pair $(H, A,)_{\Gamma,W}$ is called a weighted fuzzy parameterized complex neutrosophic soft expert set $(WFP - CNSES)$ over U, where H is a mapping given by*

$$H_{\Gamma,W} : A \to CN^U,$$

and CN^U denotes the power complex neutrosophic set of U.

The WFP-CNSES $(H, A)_{\Gamma,W}$ can be written as the following set of ordered pairs: $(H, A)_{\Gamma,W} = \left\{(a, \{\frac{u}{H_{\Gamma,W(a)(u)}} : u \in U\}) : a \in A\right\}$, where $A \subseteq \Gamma \times W \times O =$ $\left\{(\frac{e}{\mu_\Gamma(e)}, \frac{x}{\mu_W(x)}, o) : e \in E, x \in X \text{and } o \in O\right\}$, such that $\mu_\Gamma(e)$ and $\mu_W(x)$ are the corresponding membership functions of the fuzzy sets Γ and W, respectively.

From Definition 19, it is clear that the expert set in the WFP-CNSES $(H, A)_{\Gamma,W}$ is a fuzzy set, where the membership function $\mu_W(x)$ of the fuzzy set W represents weight of the expert x, and $\mu_W(x) \in [0, 1], \forall x \in X$.

For more illustration, we consider the following example.

Example 5. *Consider Example 4. Suppose that the weights for the experts x_1 and x_2 are 0.5 and 0.8, respectively. Then, the fuzzy subset of experts is $W = \left\{ \frac{x_1}{0.5}, \frac{x_2}{0.8} \right\}$ and the FP-CNSES $(H, A)_\Gamma$ in Example 4 is converted to the WFP-CNSES $(H, A)_{\Gamma,W}$ where,*

$$
\begin{aligned}
&(H,A)_{\Gamma,W} = \\
&\Big\{ \Big\{ \Big(\tfrac{e_1}{0.3}, \tfrac{x_1}{0.5}, 1\Big), \Big\{ \tfrac{u_1}{\langle 0.5e^{2\pi(0.4)}, 0.2e^{2\pi(0.4)}, 0.1e^{2\pi(0.2)} \rangle}, \tfrac{u_2}{\langle 0.6e^{2\pi(0.7)}, 0.5e^{2\pi(0.4)}, 0.1e^{2\pi(1)} \rangle} \Big\} \Big\}, \\
&\Big\{ \Big(\tfrac{e_1}{0.3}, \tfrac{x_2}{0.8}, 1\Big), \Big\{ \tfrac{u_1}{\langle 0.6e^{2\pi(0.5)}, 0.6e^{2\pi(0.4)}, 0.1e^{2\pi(0.3)} \rangle}, \tfrac{u_2}{\langle 0.9e^{2\pi(0.8)}, 0.5e^{2\pi(0.6)}, 0.5e^{2\pi(0.4)} \rangle} \Big\} \Big\}, \\
&\Big\{ \Big(\tfrac{e_2}{0.6}, \tfrac{x_1}{0.5}, 1\Big), \Big\{ \tfrac{u_1}{\langle 0.4e^{2\pi(0.3)}, 0.7e^{2\pi(0.8)}, 0.8e^{2\pi(0.7)} \rangle}, \tfrac{u_2}{\langle 0.4e^{2\pi(0.9)}, 0.2e^{2\pi(0.5)}, 0.1e^{2\pi(0.2)} \rangle} \Big\} \Big\}, \\
&\Big\{ \Big(\tfrac{e_2}{0.6}, \tfrac{x_2}{0.8}, 1\Big), \Big\{ \tfrac{u_1}{\langle 0.8e^{2\pi(0.9)}, 0.4e^{2\pi(0.3)}, 0.7e^{2\pi(0.4)} \rangle}, \tfrac{u_2}{\langle 0.5e^{2\pi(0.4)}, 0.8e^{2\pi(0.6)}, 0.6e^{2\pi(0.2)} \rangle} \Big\} \Big\}, \\
&\Big\{ \Big(\tfrac{e_3}{0.8}, \tfrac{x_1}{0.5}, 1\Big), \Big\{ \tfrac{u_1}{\langle 0.2e^{2\pi(0.2)}, 0.3e^{2\pi(0.3)}, 0.7e^{2\pi(0.6)} \rangle}, \tfrac{u_2}{\langle 0.5e^{2\pi(0.2)}, 0.7e^{2\pi(0.2)}, 0.8e^{2\pi(0.2)} \rangle} \Big\} \Big\}, \\
&\Big\{ \Big(\tfrac{e_3}{0.8}, \tfrac{x_2}{0.8}, 1\Big), \Big\{ \tfrac{u_1}{\langle 0.8e^{2\pi(0.1)}, 0.3e^{2\pi(0.2)}, 0.7e^{9\pi(0.2)} \rangle}, \tfrac{u_2}{\langle 0.1e^{2\pi(0.6)}, 0.2e^{2\pi(0.4)}, 0.6e^{2\pi(0.8)} \rangle} \Big\} \Big\}, \\
&\Big\{ \Big(\tfrac{e_1}{0.3}, \tfrac{x_1}{0.5}, 0\Big), \Big\{ \tfrac{u_1}{\langle 0.5e^{2\pi(0.5)}, 0.5e^{2\pi(0.4)}, 0.9e^{2\pi(0.3)} \rangle}, \tfrac{u_2}{\langle 0.9e^{2\pi(0.8)}, 0.7e^{2\pi(0.8)}, 0.5e^{2\pi(0.7)} \rangle} \Big\} \Big\}, \\
&\Big\{ \Big(\tfrac{e_1}{0.3}, \tfrac{x_2}{0.8}, 0\Big), \Big\{ \tfrac{u_1}{\langle 0.4e^{2\pi(0.6)}, 0.7e^{2\pi(0.4)}, 0.7e^{2\pi(0.8)} \rangle}, \tfrac{u_2}{\langle 0.5e^{2\pi(0.4)}, 0.5e^{2\pi(0.4)}, 0.3e^{2\pi(0.5)} \rangle} \Big\} \Big\}, \\
&\Big\{ \Big(\tfrac{e_2}{0.6}, \tfrac{x_1}{0.5}, 0\Big), \Big\{ \tfrac{u_1}{\langle 0.3e^{2\pi(0.2)}, 0.6e^{2\pi(0.2)}, 0.9e^{2\pi(0.2)} \rangle}, \tfrac{u_2}{\langle 0.7e^{2\pi(0.4)}, 0.5e^{2\pi(0.8)}, 0.6e^{2\pi(0.8)} \rangle} \Big\} \Big\}, \\
&\Big\{ \Big(\tfrac{e_2}{0.6}, \tfrac{x_2}{0.8}, 0\Big), \Big\{ \tfrac{u_1}{\langle 0.3e^{2\pi(0.5)}, 0.3e^{2\pi(0.1)}, 0.9e^{2\pi(0.3)} \rangle}, \tfrac{u_2}{\langle 0.3e^{2\pi(0.7)}, 0.8e^{2\pi(0.7)}, 0.7e^{2\pi(0.6)} \rangle} \Big\} \Big\}, \\
&\Big\{ \Big(\tfrac{e_3}{0.8}, \tfrac{x_1}{0.5}, 0\Big), \Big\{ \tfrac{u_1}{\langle 0.5e^{2\pi(0.4)}, 0.4e^{2\pi(0.5)}, 0.4e^{2\pi(0.3)} \rangle}, \tfrac{u_2}{\langle 0.8e^{2\pi(0.2)}, 0.3e^{2\pi(0.2)}, 0.5e^{2\pi(0.2)} \rangle} \Big\} \Big\}, \\
&\Big\{ \Big(\tfrac{e_3}{0.8}, \tfrac{x_2}{0.8}, 0\Big), \Big\{ \tfrac{u_1}{\langle 0.4e^{2\pi(0.6)}, 0.2e^{2\pi(0.3)}, 0.1e^{2\pi(0.6)} \rangle}, \tfrac{u_2}{\langle 0.3e^{2\pi(0)}, 0.6e^{2\pi(0.4)}, 0.3e^{2\pi(0.1)} \rangle} \Big\} \Big\} \Big\}.
\end{aligned}
$$

Suppose we are interested in solving the decision making problem in Example 4, where the data are represented by the means of the WFP-CNSES $(H, A)_{\Gamma,W}$ above, where each expert has his/her own weight. Then, we may use the following algorithm, which is a generalization of Algorithm 1.

Now, we use Algorithm 2 to select the most desirable kind of the product to the engineering company.

Algorithm 2: Weighted fuzzy parameterized complex neutrosophic soft expert method (WFP-CNSEM).

---

1. Input the WFP-CNSES $(H, A)_{\Gamma,W}$.

2. Convert the WFP-CNSES $(H, A)_{\Gamma,W}$ to the WFP-SVNSES $(\widehat{H}, A)_{\Gamma,W}$ as it was illustrated in step 2 of Algorithm 1. Note that the WFP-CNSES $(H, A)_{\Gamma,W}$ and the FP-CNSES $(H, A)_\Gamma$ has the same evaluation information and the difference between them lies in the structure of the expert set which does not affect the conversion process.

3. Find the values of c_{ij} for agree WFP-SVNSES and disagree WFP-SVNSES respectively, where $c_{ij} = T_{\widehat{H}_{\Gamma,W}(a_i)}(u_j) - I_{\widehat{H}_{\Gamma,W}(a_i)}(u_j) - F_{\widehat{H}_{\Gamma,W}(a_i)}(u_j), \forall u_j \in U$ and $\forall a_i \in A$.

4. Compute the score of each element $u_j \in U$ by the following formulas:
$K_j = \sum_{x \in X} \sum_i c_{ij}(\mu_\Gamma(e))(\mu_W(x))$, $S_j = \sum_{x \in X} \sum_i c_{ij}(\mu_\Gamma(e))(\mu_W(x))$, for the agree WFP-SVNSES and disagree WFP-SVNSES, where $\mu_\Gamma(e)$ and $\mu_W(x)$ are the corresponding membership functions of the fuzzy sets Γ and W, respectively.

5. Find the values of the score $r_j = K_j - S_j$ for each element $u_j \in U$.

6. Determine the value of the highest score $m = \max_{u_j \in U}\{r_j\}$. Then, the decision is to choose element u_j as the optimal solution to the problem.

Tables 5 and 6 give the numerical degree of agree WFP-SVNSES and disagree WFP-SVNSES, respectively.

Table 5. Numerical degree of agree WFP-SVNSES.

U	u_1	u_2
$\left(\frac{e_1}{0.3}, \frac{x_1}{0.5}\right)$	−0.08	−0.49
$\left(\frac{e_1}{0.3}, \frac{x_2}{0.8}\right)$	−0.17	−0.27
$\left(\frac{e_2}{0.6}, \frac{x_1}{0.5}\right)$	−1.17	−0.37
$\left(\frac{e_2}{0.6}, \frac{x_2}{0.8}\right)$	0.05	−0.55
$\left(\frac{e_3}{0.8}, \frac{x_1}{0.5}\right)$	−0.73	−0.44
$\left(\frac{e_3}{0.8}, \frac{x_2}{0.8}\right)$	−0.27	−0.63
$K_j = \sum_{x \in X} \sum_i c_{ij}(\mu_\Gamma(e))(\mu_W(x))$	$K_1 = -0.845$	$K_2 = -1.093$

Table 6. Numerical degree of disagree WFP-SVNSES.

U	u_1	u_2
$\left(\frac{e_1}{0.3}, \frac{x_1}{0.5}\right)$	−0.41	−0.58
$\left(\frac{e_1}{0.3}, \frac{x_2}{0.8}\right)$	−0.72	−0.44
$\left(\frac{e_2}{0.6}, \frac{x_1}{0.5}\right)$	−0.50	−0.96
$\left(\frac{e_2}{0.6}, \frac{x_2}{0.8}\right)$	−0.20	−0.78
$\left(\frac{e_3}{0.8}, \frac{x_1}{0.5}\right)$	−0.37	−0.14
$\left(\frac{e_3}{0.8}, \frac{x_2}{0.8}\right)$	−0.18	−0.53
$S_j = \sum_{x \in X} \sum_i c_{ij}(\mu_\Gamma(e))(\mu_W(x))$	$S_1 = -0.744$	$S_2 = -1.250$

Let K_j and S_j represent the score of each numerical degree of agree WFP-SVNSES and disagree WFP-SVNSES, respectively. These values are given in Table 7.

Table 7. The score $r_j = K_j - S_j$.

$K_j = \sum_{x \in X} \sum_i c_{ij}(\mu_\Gamma(e))(\mu_W(x))$	$S_j = \sum_{x \in X} \sum_i c_{ij}(\mu_\Gamma(e))(\mu_W(x))$	$r_j = K_j - S_j$
$K_1 = -0.845$	$S_1 = -0.744$	−0.101
$K_2 = -1.093$	$S_2 = -1.250$	0.157

Thus, $m = \max_{u_j \in U}\{r_j\} = r_2$. Therefore, the team of experts advise selecting the kind u_2 of this product as a desirable alternative.

Using Algorithms 1 and 2, we obtained the same results. However, it is clear that giving more consideration to the expert (weight) reduces the chance of unfairness in the decision making process and as a result adds more credibility to the final decision.

To illustrate the significance of the adjustable approach using WFP-CNSES as compared to that of FP-CNSES, let us consider Example 5 above.

It can be seen that WFP-CNSES is basically a FP-CNSES with weighted experts and every FP-CNSES can be considered as a WFP-CNSES with a relative weight equal to 1 assigned to all of the experts' opinions.

For example, the approximation

$$\left\{\left(\frac{e_1}{0.3}, \frac{x_1}{0.5}, 1\right), \left\{\frac{u_1}{\langle 0.5e^{2\pi(0.4)}, 0.2e^{2\pi(0.4)}, 0.1e^{2\pi(0.2)}\rangle}, \frac{u_2}{\langle 0.6e^{2\pi(0.7)}, 0.5e^{2\pi(0.4)}, 0.1e^{2\pi(1)}\rangle}\right\}\right\}$$

can be represented as

$$\left\{\left(\frac{e_1}{0.3}, \frac{x_1}{1}, 1\right), \left\{\frac{u_1}{\langle 0.5e^{2\pi(0.4)}, 0.2e^{2\pi(0.4)}, 0.1e^{2\pi(0.2)}\rangle}, \frac{u_2}{\langle 0.6e^{2\pi(0.7)}, 0.5e^{2\pi(0.4)}, 0.1e^{2\pi(1)}\rangle}\right\}\right\}$$

by means of FP-CNSES.

In other words, the expert set in the FP-CNSES is a classical set, where the opinions of the experts have the same degree of importance, which is considered as 1. This limitation may lead to improper and biased solution. In contrast, the expert set in the WFP-CNSES is a fuzzy set, i.e., a membership degree that represents an importance degree (a relative weight) between zero and one is assigned to each of the experts. This makes the decision process more realistic and reduces the biased information given by the experts.

8. Conclusions

We introduced the concept of FP-CNSES, which is a FP-SVNSES defined in a complex setting. We defined FP-CNSES operations and their properties. We then studied a mapping on fuzzy parameterized complex neutrosophic soft expert classes and its properties. An adjustable approach to decision making problems based on FP-CNSES was also introduced. A comparison of the FP-IFSES and FP-SVNSES to FP-CNSES was made and the preferability of FP-CNSES was revealed. Finally, we defined the notion of WFP-CNSES where experts' relative weights were considered and applied it to solve a decision making problem. This adjustable approach to decision making can be helpful to deal with more room of uncertainty compared to using FP-CNSES on the same problem. Both the newly proposed approaches efficiently capture the incomplete, indeterminate, and inconsistent information and extend existing decision-making methods to provide a more comprehensive outlook for decision-makers. WFP-CNSES seems to be a promising new concept, paving the way toward numerous possibilities for future research. We intend to investigate this concept further by studying its operations and their properties to develop some real applications.

Author Contributions: A.A.-Q. proposed the concept of the fuzzy parameterized complex neutrosophic soft expert set; N.H. provided a concrete example on the decision making problem; A.A.-Q. and S.A. proposed the FP-CNSES decision making method; and N.H. edited the manuscript. All authors wrote the manuscript.

Funding: Universiti Kebangsaan Malaysia GUP-2017-105.

Acknowledgments: We are indebted to Universiti Kebangsaan Malaysia for providing financial support and facilities for this research under the grant GUP-2017-105.

Conflicts of Interest: Authors declare that they have no conflict of interest.

References

1. Zadeh, L.A. Fuzzy sets. *Inf. Control.* **1965**, *8*, 338–353. [CrossRef]
2. Atanassov, K. Intuitionistic fuzzy sets. *Fuzzy Sets Syst.* **1986**, *20*, 87–96. [CrossRef]
3. Smarandache, F. *Neutrosophy: Neutrosophic Probability, Set, and Logic;* American Research Press: Rehoboth, IL, USA, 1998.
4. Smarandache, F. Neutrosophic set—A generalisation of the intuitionistic fuzzy sets. *Int. J. Pure. Appl. Math.* **2005**, *24*, 287–297.
5. Wang, H.; Smarandache, F.; Zhang, Y.; Sunderraman, R. Single valued neutrosophic sets. *Multisp. Multistruct.* **2010**, *4*, 410–413.
6. Molodtsov, D. Soft set theory: First results. *Comput. Math. Appl.* **1999**, *37*, 19–31. [CrossRef]
7. Alkhazaleh, S.; Salleh, A.R. Soft expert sets. *Adv. Decis. Sci.* **2011**, *2011*, 757868. [CrossRef]

8. Alkhazaleh, S.; Salleh, A.R. Fuzzy soft expert set and its application. *Appl. Math.* **2014**, *5*, 1349–1368. [CrossRef]
9. Broumi, S.; Smarandache, F. Intuitionistic fuzzy soft expert sets and its application in decision making. *J. New Theory* **2015**, *1*, 89–105.
10. Broumi, S.; Smarandache, F. Single valued neutrosophic soft expert sets and their application in decision making. *J. New Theory* **2015**, *3*, 67–88.
11. Al-Quran, A.; Hassan, N. Neutrosophic vague soft expert set theory. *J. Intell. Fuzzy Syst.* **2016**, *30*, 3691–3702. [CrossRef]
12. Abu Qamar, M.; Hassan, N. Generalized Q-neutrosophic soft expert set for decision under uncertainty. *Symmetry* **2018**, *10*, 621. [CrossRef]
13. Ulucay, V.; Sahin, M.; Hassan, N. Generalized neutrosophic soft expert set for multiple-criteria decision-making. *Symmetry* **2018**, *10*, 437. [CrossRef]
14. Cagman, N.; Citak, F.; Enginoglu, S. FP-soft set theory and its applications. *Ann. Fuzzy Math. Inform.* **2011**, *2*, 219–226.
15. Bashir, M.; Salleh, A.R. Fuzzy parameterized soft expert set. *Abstr. Appl. Anal.* **2012**, *2012*, 258361. [CrossRef]
16. Hazaymeh, A.; Abdullah, I.B.; Balkhi, Z.; Ibrahim, R. Fuzzy parameterized fuzzy soft expert set. *Appl. Math. Sci.* **2012**, *6*, 5547–5564.
17. Selvachandran, G.; Salleh, A.R. Fuzzy parameterized intuitionistic fuzzy soft expert set theory and its application in decision making. *Int. J. Soft Comput.* **2016**, *11*, 52–63.
18. Al-Quran, A.; Hassan, N. Fuzzy parameterised single valued neutrosophic soft expert set theory and its application in decision making. *Int. J. Appl. Decis. Sci.* **2016**, *9*, 212–227.
19. Ramot, D.; Milo, R.; Friedman, M.; Kandel, A. Complex fuzzy sets. *IEEE Trans. Fuzzy Syst.* **2002**, *10*, 171–186. [CrossRef]
20. Hu, B.; Bi, L.; Dai, S. The orthogonality between complex fuzzy sets and its application to signal detection. *Symmetry* **2017**, *9*, 175. [CrossRef]
21. Alkouri, A.; Salleh, A.R. Complex intuitionistic fuzzy sets. *AIP Conf. Proc.* **2012**, *1482*, 464–470.
22. Ali, M.; Smarandache, F. Complex neutrosophic set. *Neural Comput. Appl.* **2017**, *28*, 1817–1834. [CrossRef]
23. Al-Quran, A.; Hassan, N. The complex neutrosophic soft expert set and its application in decision making. *J. Intell. Fuzzy. Syst.* **2018**, *34*, 569–582. [CrossRef]
24. Al-Quran, A.; Hassan, N. The complex neutrosophic soft expert relation and its multiple attribute decision-making method. *Entropy* **2018**, *20*, 101. [CrossRef]

symmetry

MDPI

Article

Logarithmic Hybrid Aggregation Operators Based on Single Valued Neutrosophic Sets and Their Applications in Decision Support Systems

Shahzaib Ashraf [1], Saleem Abdullah [1,*], Florentin Smarandache [2] and Noor ul Amin [3]

[1] Department of Mathematics, Abdul Wali Khan Uniersity, Mardan 23200, Pakistan; shahzaibashraf@awkum.edu.pk
[2] Department of Mathematics and Sciences, University of New Mexico, 705 Gurley Ave., Gallup, NM 87301, USA; fsmarandache@gmail.com
[3] Department of Information Technology, Hazara University, Mansehra 21300, Pakistan; namin@hu.edu.pk
* Correspondence: saleemabdullah@awkum.edu.pk

Received: 17 January 2019; Accepted: 7 March 2019; Published: 11 March 2019

Abstract: Recently, neutrosophic sets are found to be more general and useful to express incomplete, indeterminate and inconsistent information. The purpose of this paper is to introduce new aggregation operators based on logarithmic operations and to develop a multi-criteria decision-making approach to study the interaction between the input argument under the single valued neutrosophic (SVN) environment. The main advantage of the proposed operator is that it can deal with the situations of the positive interaction, negative interaction or non-interaction among the criteria, during decision-making process. In this paper, we also defined some logarithmic operational rules on SVN sets, then we propose the single valued neutrosophic hybrid aggregation operators as a tool for multi-criteria decision-making (MCDM) under the neutrosophic environment and discussd some properties. Finally, the detailed decision-making steps for the single valued neutrosophic MCDM problems were developed, and a practical case was given to check the created approach and to illustrate its validity and superiority. Besides this, a systematic comparison analysis with other existent methods is conducted to reveal the advantages of our proposed method. Results indicate that the proposed method is suitable and effective for decision process to evaluate their best alternative.

Keywords: single valued neutrosophic sets; logarithmic operational laws; logarithmic aggregation operators; MCGDM problems

1. Introduction

The information involves, in most of the real-life decision-making problems are often incomplete, indeterminate and inconsistent. Fuzzy set theory introduced by Zadeh [1] deals with imprecise, inconsistent information. Although fuzzy set information proved to be very handy but it cannot express the information about rejection. Atanassov [2] introduced the intuitionistic fuzzy set (IFS) to bring in non-membership. Non membership function represents degree of rejection. To incorporate indeterminate and inconsistent information, in addition to incomplete information, the concept of neutrosophic set (NS) proposed by Smarandache [3]. A NS generalizes the notion of the classic set, fuzzy set (FS) [1], IFS [2], paraconsistent set [4], dialetheist set, paradoxist set [4], and tautological set [4] to name a few. In NS, indeterminacy is quantified explicitly, and truth, indeterminacy, and falsity memberships are expressed independently. The NS generalizes different types of non-crisp sets but in real scientific and engineering applications the NS and the set-theoretic operators require to be specified. For a detailed study on NS we refer to [5–17].

Related Work

Most of the weighted aggregation operators consider situations in which criteria and preferences of experts are independent, which means that additivity is a main property of these operators. However, in real life decision-making problems, the criteria of the problems are often interdependent or interactive.

Most of the weighted average operators are based on the basic algebraic product and algebraic sum of single valued neutrosophic numbers (SVNNs) which are not the only operations available to model the intersection and union of SVNNs. The logarithmic algebraic product and sum are two good alternatives of algebraic operations which can be used the model intersection and union of SVNNs. Moreover, it is observed that in the literature there is little investigation on aggregation operators utilizing the logarithmic operations on SVNNs. For a detailed review on the applications of logarithmic operations, we refer to [10]. As already mentioned that the single valued neutrosophic set (SVNS) is an effective tool to describe the uncertain, incomplete and indeterminate information. The logarithmic single valued neutrosophic hybrid and logarithmic generalized single valued neutrosophic algebraic operators have the ability to express interactions among the criteria and it can replace the weighted average to aggregate dependent criteria for obtaining more accurate results. Motivated by these, we find it interesting to develop the logarithmic single valued neutrosophic hybrid aggregation operators for decision-making with neutrosophic information.

Also, we proposed the possibility of a degree-ranking technique for SVNNs from the probability point of view, since the ranking of SVNNs is very important for decision-making under the SVN environment. Furthermore, we proposed a multi-criteria decision-making model based on the logarithmic single valued neutrosophic hybrid weighted operators. Forstudy the multi-criteria decision-making models, we refer [18–31].

The aim of writing this paper is to introduce a decision-making method for MCDM problems in which there exist interrelationships among the criteria. The contributions of this research are:

(1) A novel logarithmic operations for neutrosophic information is defined, which can overcome the weaknesses of algebraic operations and obtain the relationship between various SVNNs.

(2) Logarithmic operators for IFSs are extended to logarithmic single-valued neutrosophic hybrid operators and logarithmic generalized single-valued neutrosophic operators, namely, logarithmic single valued neutrosophic hybrid weighted averaging (L-SVNHWA), logarithmic single valued neutrosophic hybrid weighted geometric (L-SVNHWG), logarithmic generalized single-valued neutrosophic weighted averaging (L-GSVNWA) and logarithmic single-valued neutrosophic weighted geometric (L-GSVNWG) to SVNSs, which can overcome the algebraic operators drawbacks.

(3) A decision-making approach to handle the MCDM problems under the neutrosophic informations is introduced.

To attain our research goals which are stated above, the arrangement of the paper is offered as: Section 2 concentrates on basic definitions and operations of existing extensions of fuzzy set theories. In Section 3, some novel logarithmic operational laws of SVNSs are presented. Section 4 defines the logarithmic hybrid aggregation operators for SVNNs. In Section 5, an algorithm for handling the neutrosophic MCDM problem based on the developed logarithmic operators is presented. In Section 5.1, an application to verify the novel method is given and Section 5.2 presents the comparison study about algebraic and logarithmic aggregation operators. Section 6 consists of the conclusion of the study.

2. Preliminaries

This section includes the concepts and basic operations of existing extensions of fuzzy sets to make the study self contained.

Definition 1. [2] *For a set $\Re$, by an intuitionistic fuzzy set in $\Re$, we have a structure*

$$\zeta = \{\langle P_\sigma(r), N_\sigma(r)\rangle \,|r \in \Re\}, \tag{1}$$

in which $P_\sigma : \Re \to \Theta$ and $N_\sigma : \Re \to \Theta$ indicate the membership and non-membership grades in $\Re$, $\Theta = [0,1]$ be the unit interval. Also the following condition is satisfied by P_σ and N_σ, $0 \le P_\sigma(r) + N_\sigma(r) \le 1; \forall\, r \in \Re$. Then ζ is said to be intuitionistic fuzzy set in $\Re$.

Definition 2. [32] *For a set $\Re$, by a neutrosophic set in $\Re$, we have a structure*

$$\zeta = \{\langle P_\sigma(r), I_\sigma(r), N_\sigma(r)\rangle \,|r \in \Re\}, \tag{2}$$

in which $P_\sigma : \Re \to \Theta$, $I_\sigma : \Re \to \Theta$ and $N_\sigma : \Re \to \Theta$ indicate the truth, indeterminacy and falsity memberships in $\Re$, $\Theta =]0^-, 1^+[$. Also the following condition is satisfied by P_σ, I_σ and N_σ, $0^- \le P_\sigma(r) + I_\sigma(r) + N_\sigma(r) \le 3^+; \forall\, r \in \Re$. Then, ζ is said to be neutrosophic set in $\Re$.

Definition 3. [33] *For a set $\Re$, by a single valued neutrosophic set in $\Re$, we mean a structure*

$$\zeta = \{\langle P_\sigma(r), I_\sigma(r), N_\sigma(r)\rangle \,|r \in \Re\}, \tag{3}$$

in which $P_\sigma : \Re \to \Theta$, $I_\sigma : \Re \to \Theta$ and $N_\sigma : \Re \to \Theta$ indicate the truth, indeterminacy and falsity memberships in $\Re$, $\Theta = [0,1]$. Also the following condition is satisfied by P_σ, I_σ and N_σ, $0 \le P_\sigma(r) + I_\sigma(r) + N_\sigma(r) \le 3$; $\forall r \in \Re$. Then, ζ is said to be a single valued neutrosophic set in $\Re$. We denote this triplet $\zeta = \langle P_\sigma(r), I_\sigma(r), N_\sigma(r)\rangle$, in whole study called SVNN.

Ye [14], Wang et al. [33] and [34] proposed the basic operations of SVNNs, which are as follows:

Definition 4. [34] *For any two SVNNs $\zeta_p = \left\langle P_{\xi_p}(r), I_{\xi_p}(r), N_{\xi_p}(r)\right\rangle$ and $\zeta_q = \left\langle P_{\xi_q}(r), I_{\xi_q}(r), N_{\xi_q}(r)\right\rangle$ in $\Re$. The union, intersection and compliment are proposed as:*
(1) $\zeta_p \subseteq \zeta_q$ iff $\forall r \in \Re$, $P_{\xi_p}(r) \le P_{\xi_q}(r)$, $I_{\xi_p}(r) \ge I_{\xi_q}(r)$ and $N_{\xi_p}(r) \ge N_{\xi_q}(r)$;
(2) $\zeta_p = \zeta_q$ iff $\zeta_p \subseteq \zeta_q$ and $\zeta_q \subseteq \zeta_p$;
(3) $\zeta_p \cup \zeta_q = \left\langle \max\left(P_{\xi_p}, P_{\xi_q}\right),\ \min\left(I_{\xi_p}, I_{\xi_q}\right),\ \min\left(N_{\xi_p}, N_{\xi_q}\right)\right\rangle$;
(4) $\zeta_p \cap \zeta_q = \left\langle \min\left(P_{\xi_p}, P_{\xi_q}\right),\ \max\left(I_{\xi_p}, I_{\xi_q}\right),\ \max\left(N_{\xi_p}, N_{\xi_q}\right)\right\rangle$;
(5) $\zeta_p^c = \left\langle N_{\xi_p}, I_{\xi_p}, P_{\xi_p}\right\rangle$.

Definition 5. [13,15,33] *For any two SVNNs $\zeta_p = \left\langle P_{\xi_p}(r), I_{\xi_p}(r), N_{\xi_p}(r)\right\rangle$ and $\zeta_q = \left\langle P_{\xi_q}(r), I_{\xi_q}(r), N_{\xi_q}(r)\right\rangle$ in $\Re$ and $\beta \ge 0$.Then the operations of SVNNs are proposed as:*
(1) $\zeta_p \oplus \zeta_q = \left\{P_{\xi_p} + P_{\xi_q} - P_{\xi_p} \cdot P_{\xi_q},\ I_{\xi_p} \cdot I_{\xi_q},\ N_{\xi_p} \cdot N_{\xi_q}\right\}$;
(2) $\beta \cdot \zeta_p = \left\{1 - (1 - P_{\xi_p})^\beta,\ (I_{\xi_p})^\beta,\ (N_{\xi_p})^\beta\right\}$;
(3) $\zeta_p \otimes \zeta_q = \left\{P_{\xi_p} \cdot P_{\xi_q},\ I_{\xi_p} + I_{\xi_q} - I_{\xi_p} \cdot I_{\xi_q}, N_{\xi_p} + N_{\xi_q} - N_{\xi_p} \cdot N_{\xi_q}\right\}$;
(4) $\zeta_p^\beta = \left\{(P_{\xi_p})^\beta,\ 1 - (1 - I_{\xi_p})^\beta, 1 - (1 - N_{\xi_p})^\beta\right\}$.
(5) $$\beta^{\zeta_p} = \begin{cases} \left(\beta^{1-P_{\xi_p}}, 1 - \beta^{I_{\xi_p}}, 1 - \beta^{N_{\xi_p}}\right) & if\ \beta \in (0,1) \\ \left(\left(\frac{1}{\beta}\right)^{1-P_{\xi_p}}, 1 - \left(\frac{1}{\beta}\right)^{I_{\xi_p}}, 1 - \left(\frac{1}{\beta}\right)^{N_{\xi_p}}\right) & if\ \beta \ge 1 \end{cases}$$

Definition 6. [33] *For any three SVNNs $\zeta_p = \left\langle P_{\xi_p}(r), I_{\xi_p}(r), N_{\xi_p}(r)\right\rangle$, $\zeta_q = \left\langle P_{\xi_q}(r), I_{\xi_q}(r), N_{\xi_q}(r)\right\rangle$ and $\zeta_l = \langle P_{\sigma_l}(r), I_{\sigma_l}(r), N_{\sigma_l}(r)\rangle$ in $\Re$ and $\beta_1, \beta_2 \ge 0$. Then, we have*

(1) $\zeta_p \oplus \zeta_q = \zeta_q \oplus \zeta_p$;
(2) $\zeta_p \otimes \zeta_q = \zeta_q \otimes \zeta_p$;
(3) $\beta_1(\zeta_p \oplus \zeta_q) = \beta_1\zeta_p \oplus \beta_1\zeta_q,\ \beta_1 > 0$;
(4) $(\zeta_p \otimes \zeta_q)^{\beta_1} = \zeta_p^{\beta_1} \otimes \zeta_q^{\beta_1},\ \beta_1 > 0$;
(5) $\beta_1\zeta_p \oplus \beta_2\zeta_p = (\beta_1 + \beta_2)\zeta_p, \beta_1 > 0, \beta_2 > 0$;
(6) $\zeta_p^{\beta_1} \otimes \zeta_p^{\beta_2} = \zeta_p^{\beta_1+\beta_2}, \beta_1 > 0, \beta_2 > 0$;
(7) $(\zeta_p \oplus \zeta_q) \oplus \zeta_l = \zeta_p \oplus (\zeta_q \oplus \zeta_l)$;
(8) $(\zeta_p \otimes \zeta_q) \otimes \zeta_l = \zeta_p \otimes (\zeta_q \otimes \zeta_l)$.

Definition 7. *[33] For any SVNN $\zeta_p = \left\langle P_{\xi_p}(r), I_{\xi_p}(r), N_{\xi_p}(r) \right\rangle$ in $\Re$. Then score and accuracy values are defined as:*
(1) $\widetilde{S}(\zeta_p) = P_{\xi_p} - I_{\xi_p} - N_{\xi_p}$
(2) $\widetilde{A}(\zeta_p) = P_{\xi_p} + I_{\xi_p} + N_{\xi_p}$

The above definitions of score and accuracy funtions suggest which SVNN is greater than other SVNNs. The comparison technique is defined in following definition.

Definition 8. *[33] For any SVNNs $\zeta_p = \left\langle P_{\xi_p}(r), I_{\xi_p}(r), N_{\xi_p}(r) \right\rangle (p = 1, 2)$ in $\Re$.*
Then comparison techniques are proposed as:
(1) If $\widetilde{S}(\zeta_1) < \widetilde{S}(\zeta_2)$, then $\zeta_1 < \zeta_2$,
(2) If $\widetilde{S}(\zeta_1) > \widetilde{S}(\zeta_2)$, then $\zeta_1 > \zeta_2$,
(3) If $\widetilde{S}(\zeta_1) = \widetilde{S}(\zeta_2)$, and
(a) $\widetilde{A}(\zeta_1) < \widetilde{A}(\zeta_2)$, then $\zeta_1 < \zeta_2$,
(b) $\widetilde{A}(\zeta_1) > \widetilde{A}(\zeta_2)$, then $\zeta_1 > \zeta_2$,
(c) $\widetilde{A}(\zeta_1) = \widetilde{A}(\zeta_2)$, then $\zeta_1 \approx \zeta_2$.

Garg and Nancy [10] proposed some logarithmic-based aggregation operators, which are as follows:

Definition 9. *[10] For any collection of SVNNs $\zeta_p = \left\langle P_{\xi_p}(r), I_{\xi_p}(r), N_{\xi_p}(r) \right\rangle (p = 1, 2, ..., n)$ in $\Re$, with $0 < \sigma_p \leq \min\left\{P_{\xi_p}, 1 - I_{\xi_p}, 1 - N_{\xi_p}\right\} < 1, \sigma \neq 1$. Then, the structure of logarithmic single valued neutrosophic weighted averaging (L-SVNWA) operator is defined as:*

$$L - SVNWA(\zeta_1, \zeta_2, ..., \zeta_n) = \begin{pmatrix} 1 - \prod_{p=1}^{n} \left(\ell og_{\sigma_p} P_{\xi_p}\right)^{\beta_p}, \\ \prod_{p=1}^{n} \left(\ell og_{\sigma_p}\left(1 - I_{\xi_p}\right)\right)^{\beta_p}, \\ \prod_{p=1}^{n} \left(\ell og_{\sigma_p}\left(1 - N_{\xi_p}\right)\right)^{\beta_p} \end{pmatrix}, \tag{4}$$

where $\beta_p\ (p = 1, 2, ..., n)$ are weight vectors with $\beta_p \geq 0$ and $\sum_{p=1}^{n} \beta_p = 1$.

Definition 10. [10] *For any collection of SVNNs* $\zeta_p = \left\langle P_{\xi_p}(r), I_{\xi_p}(r), N_{\xi_p}(r) \right\rangle$ $(p = 1, 2, ..., n)$ *in* $\Re$, *with* $0 < \sigma_p \leq \min\left\{P_{\xi_p}, 1 - I_{\xi_p}, 1 - N_{\xi_p}\right\} < 1, \sigma \neq 1$. *Then, the structure of the logarithmic single-valued neutrosophic-ordered weighted averaging (L-SVNOWA) operator is defined as:*

$$L - SVNOWA(\zeta_1, \zeta_2, ..., \zeta_n) = \begin{pmatrix} 1 - \prod_{p=1}^{n} \left(\ell og_{\sigma_p} P_{\zeta_{\eta(p)}}\right)^{\beta_p}, \\ \prod_{p=1}^{n} \left(\ell og_{\sigma_p}\left(1 - I_{\zeta_{\eta(p)}}\right)\right)^{\beta_p}, \\ \prod_{p=1}^{n} \left(\ell og_{\sigma_p}\left(1 - N_{\zeta_{\eta(p)}}\right)\right)^{\beta_p} \end{pmatrix}, \tag{5}$$

where β_p $(p = 1, 2, ..., n)$ *are weighting vector with* $\beta_p \geq 0$, $\sum_{p=1}^{n} \beta_p = 1$ *and pth largest weighted value is* $\zeta_{\eta(p)}$ *consequently by total order* $\zeta_{\eta(1)} \geq \zeta_{\eta(2)} \geq ... \geq \zeta_{\eta(n)}$.

Definition 11. [10] *For any collection of SVNNs* $\zeta_p = \left\langle P_{\xi_p}(r), I_{\xi_p}(r), N_{\xi_p}(r) \right\rangle$ $(p = 1, 2, ..., n)$ *in* $\Re$, *with* $0 < \sigma_p \leq \min\left\{P_{\xi_p}, 1 - I_{\xi_p}, 1 - N_{\xi_p}\right\} < 1, \sigma \neq 1$. *Then, the structure of logarithmic single-valued neutrosophic-weighted geometric (L-SVNWG) operator is defined as:*

$$L - SVNWG(\zeta_1, \zeta_2, ..., \zeta_n) = \begin{pmatrix} \prod_{p=1}^{n} \left(1 - \ell og_{\sigma_p} P_{\xi_p}\right)^{\beta_p}, \\ 1 - \prod_{p=1}^{n} \left(1 - \ell og_{\sigma_p}\left(1 - I_{\xi_p}\right)\right)^{\beta_p}, \\ 1 - \prod_{p=1}^{n} \left(1 - \ell og_{\sigma_p}\left(1 - N_{\xi_p}\right)\right)^{\beta_p} \end{pmatrix}, \tag{6}$$

where β_p $(p = 1, 2, ..., n)$ *are weight vectors with* $\beta_p \geq 0$ *and* $\sum_{p=1}^{n} \beta_p = 1$.

Definition 12. [10] *For any collection of SVNNs* $\zeta_p = \left\langle P_{\xi_p}(r), I_{\xi_p}(r), N_{\xi_p}(r) \right\rangle$ $(p = 1, 2, ..., n)$ *in* $\Re$, *with* $0 < \sigma_p \leq \min\left\{P_{\xi_p}, 1 - I_{\xi_p}, 1 - N_{\xi_p}\right\} < 1, \sigma \neq 1$. *Then, the structure of logarithmic single valued neutrosophic ordered weighted geometric (L-SVNOWG) operator is defined as:*

$$L - SVNOWG(\zeta_1, \zeta_2, ..., \zeta_n) = \begin{pmatrix} \prod_{p=1}^{n} \left(1 - \ell og_{\sigma_p} P_{\xi_{\eta(p)}}\right)^{\beta_p}, \\ 1 - \prod_{p=1}^{n} \left(1 - \ell og_{\sigma_p}\left(1 - I_{\xi_{\eta(p)}}\right)\right)^{\beta_p}, \\ 1 - \prod_{p=1}^{n} \left(1 - \ell og_{\sigma_p}\left(1 - N_{\xi_{\eta(p)}}\right)\right)^{\beta_p} \end{pmatrix}, \tag{7}$$

where β_p $(p = 1, 2, ..., n)$ *are weighting vector with* $\beta_p \geq 0$ *and* $\sum_{p=1}^{n} \beta_p = 1$ *and pth are the largest weighted value is* $\zeta_{\eta(p)}$ *consequently by total order* $\zeta_{\eta(1)} \geq \zeta_{\eta(2)} \geq ... \geq \zeta_{\eta(n)}$.

3. Logarithmic Operational Laws

Motivated by the well growing concept of SVNSs, we introduce some novel logarithmic operational laws for single valued neutrosophic numbers. As in real number systems $\ell og_\sigma 0$ is meaningless and $\ell og_\sigma 1$ is not defined therefore, in our study we take non-empty SVNSs and $\sigma \neq 1$, where σ is any real number.

Definition 13. *For any SVNN* $\zeta_p = \left\langle P_{\xi_p}(r), I_{\xi_p}(r), N_{\xi_p}(r) \right\rangle$ *in* $\Re$. *The logarithmic SVNN is defined as:*

$$\ell og_\sigma \zeta_p = \left\{\left\langle 1 - \left(\ell og_\sigma P_{\xi_p}(r)\right), \ell og_\sigma\left(1 - I_{\xi_p}(r)\right), \ell og_\sigma\left(1 - N_{\xi_p}(r)\right)\right\rangle | r \in \Re\right\}, \tag{8}$$

in which $P_\sigma : \Re \to \Theta$, $I_\sigma : \Re \to \Theta$ *and* $N_\sigma : \Re \to \Theta$ *are indicated the truth, indeterminacy and falsity memberships in* $\Re$, $\Theta = [0,1]$ *be the unit interval. Also following condition is satisfied by* P_σ, I_σ *and* N_σ, $0 \le P_\sigma(r) + I_\sigma(r) + N_\sigma(r) \le 3; \forall\, r \in \Re$. *Therefore the truth membership grade is*

$$1-\left(\ell og_\sigma P_{\xi_p}(r)\right) : \Re \to \Theta, \text{ such that } 0 \le 1-\left(\ell og_\sigma P_{\xi_p}(r)\right) \le 1, \text{ for all } r \in \Re$$

the indeterminacy membership is

$$\ell og_\sigma\left(1-I_{\xi_p}(r)\right) : \Re \to \Theta, \text{ such that } 0 \le \ell og_\sigma\left(1-I_{\xi_p}(r)\right) \le 1, \text{ for all } r \in \Re$$

and falsity membership is

$$\ell og_\sigma\left(1-N_{\xi_p}(r)\right) : \Re \to \Theta, \text{such that } 0 \le \ell og_\sigma\left(1-N_{\xi_p}(r)\right) \le 1, \text{ for all } r \in \Re.$$

Therefore

$$\begin{aligned} \ell og_\sigma \zeta_p &= \left\{\left\langle 1-\left(\ell og_\sigma P_{\xi_p}(r)\right), \ell og_\sigma\left(1-I_{\xi_p}(r)\right), \ell og_\sigma\left(1-N_{\xi_p}(r)\right)\right\rangle | r \in \Re\right\} \\ 0 &< \sigma \le \min\left\{P_{\xi_p}, 1-I_{\xi_p}, 1-N_{\xi_p}\right\} \le 1, \sigma \ne 1 \end{aligned}$$

is SVNS.

Definition 14. *For any SVNN* $\zeta_p = \left\langle P_{\xi_p}(r), I_{\xi_p}(r), N_{\xi_p}(r)\right\rangle$ *in* $\Re$. *If*

$$\ell og_\sigma \zeta_p = \begin{cases} \begin{pmatrix} 1-\left(\ell og_\sigma P_{\xi_p}(r)\right), \\ \ell og_\sigma\left(1-I_{\xi_p}(r)\right), \\ \ell og_\sigma\left(1-N_{\xi_p}(r)\right) \end{pmatrix} & 0 < \sigma \le \min\left\{P_{\xi_p}, 1-I_{\xi_p}, 1-N_{\xi_p}\right\} < 1 \\ \begin{pmatrix} 1-\left(\ell og_{\frac{1}{\sigma}} P_{\xi_p}(r)\right), \\ \ell og_{\frac{1}{\sigma}}\left(1-I_{\xi_p}(r)\right), \\ \ell og_{\frac{1}{\sigma}}\left(1-N_{\xi_p}(r)\right) \end{pmatrix} & \begin{matrix} 0 < \frac{1}{\sigma} \le \min\left\{P_{\xi_p}, 1-I_{\xi_p}, 1-N_{\xi_p}\right\} < 1, \\ \sigma \ne 1 \end{matrix} \end{cases} \tag{9}$$

then the function $\ell og_\sigma \zeta_p$ *is known to be a logarithmic operator for SVNS, and its value is said to be logarithmic SVNN (L-SVNN). Here, we take* $\ell og_\sigma 0 = 0, \sigma > 0, \sigma \ne 1$.

Theorem 1. *[10] For any SVNN* $\zeta_p = \left\langle P_{\xi_p}(r), I_{\xi_p}(r), N_{\xi_p}(r)\right\rangle$ *in* $\Re$, *then* $\ell og_\sigma \zeta_p$ *is also be SVNN.*

Now, we give some discussion on the basic properties of the L-SVNN.

Definition 15. *For any two L-SVNNs* $\ell og_\sigma \zeta_p = \begin{pmatrix} 1-\left(\ell og_\sigma P_{\xi_p}(r)\right), \\ \ell og_\sigma\left(1-I_{\xi_p}(r)\right), \\ \ell og_\sigma\left(1-N_{\xi_p}(r)\right) \end{pmatrix}$ *and* $\ell og_\sigma \zeta_q = \begin{pmatrix} 1-\left(\ell og_\sigma P_{\xi_q}(r)\right), \\ \ell og_\sigma\left(1-I_{\xi_q}(r)\right), \\ \ell og_\sigma\left(1-N_{\xi_q}(r)\right) \end{pmatrix}$ *in* $\Re$ *and* $\beta \ge 0$.*Then the logarithmic operations of L-SVNNs are propose as*

$$(1)\ \ell og_\sigma \zeta_p \oplus \ell og_\sigma \zeta_q = \left\{ \begin{matrix} 1-\left(\ell og_\sigma P_{\xi_p}(r)\right)\cdot\left(\ell og_\sigma P_{\xi_q}(r)\right), \\ \ell og_\sigma\left(1-I_{\xi_p}(r)\right)\cdot \ell og_\sigma\left(1-I_{\xi_q}(r)\right), \\ \ell og_\sigma\left(1-N_{\xi_p}(r)\right)\cdot \ell og_\sigma\left(1-N_{\xi_q}(r)\right) \end{matrix} \right\};$$

$$(2)\ \beta\cdot \ell og_{\sigma}\zeta_p=\left\{\begin{array}{c} 1-\left(\ell og_{\sigma}P_{\xi_p}(r)\right)^{\beta},\\ \left(\ell og_{\sigma}\left(1-I_{\xi_p}(r)\right)\right)^{\beta},\\ \left(\ell og_{\sigma}\left(1-N_{\xi_p}(r)\right)\right)^{\beta}\end{array}\right\};$$

$$(3)\ \ell og_{\sigma}\zeta_p\otimes \ell og_{\sigma}\zeta_q=\left\{\begin{array}{c} 1-\left(\ell og_{\sigma}P_{\xi_p}(r)\right)\cdot 1-\left(\ell og_{\sigma}P_{\xi_q}(r)\right),\\ 1-\left(1-\ell og_{\sigma}\left(1-I_{\xi_p}(r)\right)\right)\cdot\left(1-\ell og_{\sigma}\left(1-I_{\xi_q}(r)\right)\right),\\ 1-\left(1-\ell og_{\sigma}\left(1-N_{\xi_p}(r)\right)\right)\cdot\left(1-\ell og_{\sigma}\left(1-N_{\xi_q}(r)\right)\right)\end{array}\right\};$$

$$(4)\ \left(\ell og_{\sigma}\zeta_p\right)^{\beta}=\left\{\begin{array}{c} \left(1-\left(\ell og_{\sigma}P_{\xi_p}(r)\right)\right)^{\beta},\\ 1-\left(1-\ell og_{\sigma}\left(1-I_{\xi_p}(r)\right)\right)^{\beta},\\ 1-\left(1-\ell og_{\sigma}\left(1-N_{\xi_p}(r)\right)\right)^{\beta}\end{array}\right\}.$$

Theorem 2. *[10] For any two L-SVNNs* $\ell og_{\sigma}\zeta_p=\left(\begin{array}{c} 1-\left(\ell og_{\sigma}P_{\xi_p}(r)\right),\\ \ell og_{\sigma}\left(1-I_{\xi_p}(r)\right),\\ \ell og_{\sigma}\left(1-N_{\xi_p}(r)\right)\end{array}\right)$ $(p=1,2)$ *in* $\Re$, *with* $0<\sigma\leq \min\left\{P_{\xi_p},1-I_{\xi_p},1-N_{\xi_p}\right\}<1,\sigma\neq 1,\beta,\beta_1,\beta_2>0$ *be any real numbers. Then*
(1) $\beta\left(\ell og_{\sigma}\zeta_1\oplus \ell og_{\sigma}\zeta_2\right)=\beta\ell og_{\sigma}\zeta_1\oplus \beta\ell og_{\sigma}\zeta_2$;
(2) $\left(\ell og_{\sigma}\zeta_1\otimes \ell og_{\sigma}\zeta_2\right)^{\beta}=\left(\ell og_{\sigma}\zeta_1\right)^{\beta}\otimes\left(\ell og_{\sigma}\zeta_2\right)^{\beta}$;
(3) $\beta_1\ell og_{\sigma}\zeta_1\oplus \beta_2\ell og_{\sigma}\zeta_1=\left(\beta_1+\beta_2\right)\ell og_{\sigma}\zeta_1$;
(4) $\left(\ell og_{\sigma}\zeta_1\right)^{\beta_1}\otimes\left(\ell og_{\sigma}\zeta_1\right)^{\beta_2}=\left(\ell og_{\sigma}\zeta_1\right)^{(\beta_1+\beta_2)}$;
(5) $\left(\left(\ell og_{\sigma}\zeta_1\right)^{\beta_1}\right)^{\beta_2}=\left(\ell og_{\sigma}\zeta_1\right)^{\beta_1\beta_2}$.

Comparison Technique for L-SVNNs

Definition 16. *[10] For any L-SVNN* $\ell og_{\sigma}\zeta_p=\left(\begin{array}{c} 1-\left(\ell og_{\sigma}P_{\xi_p}(r)\right),\\ \ell og_{\sigma}\left(1-I_{\xi_p}(r)\right),\\ \ell og_{\sigma}\left(1-N_{\xi_p}(r)\right)\end{array}\right)$ *in* $\Re$. *Then score and accuracy values are define as*
(1) $\widetilde{S}(\ell og_{\sigma}\zeta_p)=1-\left(\ell og_{\sigma}P_{\xi_p}(r)\right)-\ell og_{\sigma}\left(1-I_{\xi_p}(r)\right)-\left(\ell og_{\sigma}\left(1-N_{\xi_p}(r)\right)\right)$
(2) $\widetilde{A}(\ell og_{\sigma}\zeta_p)=1-\left(\ell og_{\sigma}P_{\xi_p}(r)\right)+\ell og_{\sigma}\left(1-I_{\xi_p}(r)\right)+\left(\ell og_{\sigma}\left(1-N_{\xi_p}(r)\right)\right)$

The above defined score and accuracy values suggest which L-SVNN are greater than other L-SVNNs. The comparison technique is defined in the following definition.

Definition 17. *For any L-SVNNs* $\ell og_{\sigma}\zeta_p=\left(\begin{array}{c} 1-\left(\ell og_{\sigma}P_{\xi_p}(r)\right),\\ \ell og_{\sigma}\left(1-I_{\xi_p}(r)\right),\\ \ell og_{\sigma}\left(1-N_{\xi_p}(r)\right)\end{array}\right)$ $(p=1,2)$ *in* $\Re$. *Then, comparison technique is proposed as:*
(1) *If* $\widetilde{S}(\ell og_{\sigma}\zeta_1)<\widetilde{S}(\ell og_{\sigma}\zeta_2)$ *then* $\ell og_{\sigma}\zeta_1<\ell og_{\sigma}\zeta_2$,
(2) *If* $\widetilde{S}(\ell og_{\sigma}\zeta_1)>\widetilde{S}(\ell og_{\sigma}\zeta_2)$ *then* $\ell og_{\sigma}\zeta_1>\ell og_{\sigma}\zeta_2$,
(3) *If* $\widetilde{S}(\ell og_{\sigma}\zeta_1)=\widetilde{S}(\ell og_{\sigma}\zeta_2)$ *then*
(a) $\widetilde{A}(\ell og_{\sigma}\zeta_1)<\widetilde{A}(\ell og_{\sigma}\zeta_2)$ *then* $\ell og_{\sigma}\zeta_1<\ell og_{\sigma}\zeta_2$,
(b) $\widetilde{A}(\ell og_{\sigma}\zeta_1)>\widetilde{A}(\ell og_{\sigma}\zeta_2)$ *then* $\ell og_{\sigma}\zeta_1>\ell og_{\sigma}\zeta_2$,
(c) $\widetilde{A}(\ell og_{\sigma}\zeta_1)=\widetilde{A}(\ell og_{\sigma}\zeta_2)$ *then* $\ell og_{\sigma}\zeta_1\approx \ell og_{\sigma}\zeta_2$.

4. Logarithmic Aggregation Operators for L-SVNNs

Now, we propose novel logarithmic hybrid aggregation operators for L-SVNNs based on logarithmic operations laws as follows:

4.1. Logarithmic Hybrid Averaging Operator

Definition 18. *For any collection of SVNNs* $\zeta_p = \left\langle P_{\xi_p}(r), I_{\xi_p}(r), N_{\xi_p}(r) \right\rangle$ $(p = 1, ..., n)$ *in* $\Re$, *with* $0 < \sigma_p \leq \min\left\{P_{\xi_p}, 1 - I_{\xi_p}, 1 - N_{\xi_p}\right\} < 1, \sigma \neq 1$. *The structure of logarithmic single valued neutrosophic hybrid weighted averaging (L-SVNHWA) operator is*

$$L - SVNHWA(\zeta_1, \zeta_2, ..., \zeta_n) = \sum_{p=1}^{n} \omega_p \ell og_{\sigma_p} \zeta^*_{\eta(p)}, \tag{10}$$

where β_p $(p = 1, ..., n)$ *is the weighting vector with* $\beta_p \geq 0$ *and* $\sum_{p=1}^{n} \beta_p = 1$ *and pth biggest weighted value is* $\zeta^*_{\eta(p)}$ $\left(\zeta^*_{\eta(p)} = n\beta_p \zeta_{\eta(p)}, \ P \in N\right)$ *consequently by total order* $\zeta^*_{\eta(1)} \geq \zeta^*_{\eta(2)} \geq ... \geq \zeta^*_{\eta(n)}$. *Also, the associated weights are* $\omega = (\omega_1, \omega_2, ..., \omega_n)$ *with* $\omega_p \geq 0$, $\Sigma_{p=1}^{n} \omega_p = 1$.

Theorem 3. *For any collection of SVNNs* $\zeta_p = \left\langle P_{\xi_p}(r), I_{\xi_p}(r), N_{\xi_p}(r) \right\rangle$ $(p = 1, ..., n)$ *in* $\Re$, *with* $0 < \sigma_p \leq \min\left\{P_{\xi_p}, 1 - I_{\xi_p}, 1 - N_{\xi_p}\right\} < 1, \sigma \neq 1$. *Then by using logarithmic operations and Definition 18,* $L - SVNHWA$ *is defined as*

$$L - SVNHWA(\zeta_1, \zeta_2, ..., \zeta_n) = \begin{cases} \begin{pmatrix} 1 - \prod_{p=1}^{n} \left(\ell og_{\sigma_p} P^*_{\xi_{\eta(p)}}\right)^{\omega_p}, \\ \prod_{p=1}^{n} \left(\ell og_{\sigma_p}\left(1 - I^*_{\xi_{\eta(p)}}\right)\right)^{\omega_p}, \\ \prod_{p=1}^{n} \left(\ell og_{\sigma_p}\left(1 - N^*_{\xi_{\eta(p)}}\right)\right)^{\omega_p} \end{pmatrix} & 0 < \sigma_p \leq \min\left\{\begin{array}{c} P_{\xi_p}, \\ 1 - I_{\xi_p}, \\ 1 - N_{\xi_p} \end{array}\right\} < 1 \\ \begin{pmatrix} 1 - \prod_{p=1}^{n} \left(\ell og_{\frac{1}{\sigma_p}} P^*_{\xi_{\eta(p)}}\right)^{\omega_p}, \\ \prod_{p=1}^{n} \left(\ell og_{\frac{1}{\sigma_p}}\left(1 - I^*_{\xi_{\eta(p)}}\right)\right)^{\omega_p}, \\ \prod_{p=1}^{n} \left(\ell og_{\frac{1}{\sigma_p}}\left(1 - N^*_{\xi_{\eta(p)}}\right)\right)^{\omega_p} \end{pmatrix} & \begin{array}{c} 0 < \frac{1}{\sigma_p} \leq \min\left\{\begin{array}{c} P_{\xi_p}, \\ 1 - I_{\xi_p}, \\ 1 - N_{\xi_p} \end{array}\right\} < 1, \\ \sigma \neq 1 \end{array} \end{cases} \tag{11}$$

where β_p $(p = 1, ..., n)$ *are weighting vector with* $\beta_p \geq 0$ *and* $\sum_{p=1}^{n} \beta_p = 1$ *and pth biggest weighted value is* $\zeta^*_{\eta(p)}$ $\left(\zeta^*_{\eta(p)} = n\beta_p \zeta_{\eta(p)}, \ P \in N\right)$ *consequently by total order* $\zeta^*_{\eta(1)} \geq \zeta^*_{\eta(2)} \geq ... \geq \zeta^*_{\eta(n)}$. *Also the associated weights are* $\omega = (\omega_1, \omega_2, ..., \omega_n)$ *with* $\omega_p \geq 0$, $\Sigma_{p=1}^{n} \omega_p = 1$.

Proof. Using mathematical induction to prove Equation (3), we proceed as:

(a) For $n = 2$, since

$$\omega_1 \ell og_{\sigma_1} \zeta^*_{\eta(1)} = \begin{pmatrix} 1 - \left(\ell og_{\sigma_1} P^*_{\xi_{\eta(1)}}\right)^{\omega_1}, \\ \left(\ell og_{\sigma_1}\left(1 - I^*_{\xi_{\eta(1)}}\right)\right)^{\omega_1}, \\ \left(\ell og_{\sigma_1}\left(1 - N^*_{\xi_{\eta(1)}}\right)\right)^{\omega_1} \end{pmatrix}$$

and

$$\omega_2 \ell og_{\sigma_2} \zeta^*_{\eta(2)} = \begin{pmatrix} 1 - \left(\ell og_{\sigma_2} P^*_{\xi_{\eta(2)}}\right)^{\omega_2}, \\ \left(\ell og_{\sigma_2}\left(1 - I^*_{\xi_{\eta(2)}}\right)\right)^{\omega_2}, \\ \left(\ell og_{\sigma_2}\left(1 - N^*_{\xi_{\eta(2)}}\right)\right)^{\omega_2} \end{pmatrix}$$

Then

$$\begin{aligned} L - SVNHWA(\zeta_1, \zeta_2) &= \omega_1 \ell og_{\sigma_1} \zeta^*_{\eta(1)} \oplus \omega_2 \ell og_{\sigma_2} \zeta^*_{\eta(2)} \\ &= \begin{pmatrix} 1 - \left(\ell og_{\sigma_1} P^*_{\xi_{\eta(1)}}\right)^{\omega_1}, \\ \left(\ell og_{\sigma_1}\left(1 - I^*_{\xi_{\eta(1)}}\right)\right)^{\omega_1}, \\ \left(\ell og_{\sigma_1}\left(1 - N^*_{\xi_{\eta(1)}}\right)\right)^{\omega_1} \end{pmatrix} \oplus \begin{pmatrix} 1 - \left(\ell og_{\sigma_2} P^*_{\xi_{\eta(2)}}\right)^{\omega_2}, \\ \left(\ell og_{\sigma_2}\left(1 - I^*_{\xi_{\eta(2)}}\right)\right)^{\omega_2}, \\ \left(\ell og_{\sigma_2}\left(1 - N^*_{\xi_{\eta(2)}}\right)\right)^{\omega_2} \end{pmatrix} \\ &= \begin{pmatrix} 1 - \left(\ell og_{\sigma_1} P^*_{\xi_{\eta(1)}}\right)^{\omega_1} \cdot \left(\ell og_{\sigma_2} P^*_{\xi_{\eta(2)}}\right)^{\omega_2}, \\ \left(\ell og_{\sigma_1}\left(1 - I^*_{\xi_{\eta(1)}}\right)\right)^{\omega_1} \cdot \left(\ell og_{\sigma_2}\left(1 - I^*_{\xi_{\eta(2)}}\right)\right)^{\omega_2}, \\ \left(\ell og_{\sigma_1}\left(1 - N^*_{\xi_{\eta(1)}}\right)\right)^{\omega_1} \cdot \left(\ell og_{\sigma_2}\left(1 - N^*_{\xi_{\eta(2)}}\right)\right)^{\omega_2} \end{pmatrix} \\ &= \begin{pmatrix} 1 - \prod_{p=1}^{2} \left(\ell og_{\sigma_p} P^*_{\xi_{\eta(p)}}\right)^{\omega_p}, \\ \prod_{p=1}^{n} \left(\ell og_{\sigma_p}\left(1 - I^*_{\xi_{\eta(p)}}\right)\right)^{\omega_p}, \\ \prod_{p=1}^{n} \left(\ell og_{\sigma_p}\left(1 - N^*_{\xi_{\eta(p)}}\right)\right)^{\omega_p} \end{pmatrix}. \end{aligned}$$

(b) Now Equation (3) is true for $n = k$,

$$L - SVNHWA(\zeta_1, \zeta_2, ..., \zeta_k) = \begin{pmatrix} 1 - \prod_{p=1}^{k} \left(\ell og_{\sigma_p} P^*_{\xi_{\eta(p)}}\right)^{\omega_p}, \\ \prod_{p=1}^{k} \left(\ell og_{\sigma_p}\left(1 - I^*_{\xi_{\eta(p)}}\right)\right)^{\omega_p}, \\ \prod_{p=1}^{k} \left(\ell og_{\sigma_p}\left(1 - N^*_{\xi_{\eta(p)}}\right)\right)^{\omega_p} \end{pmatrix}.$$

(c) Now, we prove that Equation (3) for $n = k + 1$, that is

$$L - SVNHWA(\zeta_1, \zeta_2, ..., \zeta_k) = \sum_{p=1}^{k} \omega_p \ell og_{\sigma_p} \zeta^*_{\eta(p)} + \omega_{k+1} \ell og_{\sigma_{k+1}} \zeta^*_{\eta(k+1)}$$

$$L-SVNHWA\left(\zeta_1,\zeta_2,...,\zeta_k\right)$$

$$=\begin{pmatrix} 1-\prod_{p=1}^{k}\left(\ell og_{\sigma_p}P^*_{\xi_{\eta(p)}}\right)^{\omega_p}, \\ \prod_{p=1}^{k}\left(\ell og_{\sigma_p}\left(1-I^*_{\xi_{\eta(p)}}\right)\right)^{\omega_p}, \\ \prod_{p=1}^{k}\left(\ell og_{\sigma_p}\left(1-N^*_{\xi_{\eta(p)}}\right)\right)^{\omega_p} \end{pmatrix} \oplus \begin{pmatrix} 1-\left(\ell og_{\sigma_{k+1}}P^*_{\xi_{\eta(k+1)}}\right)^{\omega_{k+1}}, \\ \left(\ell og_{\sigma_{k+1}}\left(1-I^*_{\xi_{\eta(k+1)}}\right)\right)^{\omega_{k+1}}, \\ \left(\ell og_{\sigma_{k+1}}\left(1-N^*_{\xi_{\eta(k+1)}}\right)\right)^{\omega_{k+1}} \end{pmatrix}$$

$$=\begin{pmatrix} 1-\prod_{p=1}^{k+1}\left(\ell og_{\sigma_p}P^*_{\xi_{\eta(p)}}\right)^{\omega_p}, \\ \prod_{p=1}^{k+1}\left(\ell og_{\sigma_p}\left(1-I^*_{\xi_{\eta(p)}}\right)\right)^{\omega_p}, \\ \prod_{p=1}^{k+1}\left(\ell og_{\sigma_p}\left(1-N^*_{\xi_{\eta(p)}}\right)\right)^{\omega_p} \end{pmatrix}$$

Thus Equation (3) is true for $n = z+1$. Hence its satisfies for whole n. Therefore

$$L-SVNHWA\left(\zeta_1,\zeta_2,...,\zeta_n\right)=\begin{pmatrix} 1-\prod_{p=1}^{n}\left(\ell og_{\sigma_p}P^*_{\xi_{\eta(p)}}\right)^{\omega_p}, \\ \prod_{p=1}^{n}\left(\ell og_{\sigma_p}\left(1-I^*_{\xi_{\eta(p)}}\right)\right)^{\omega_p}, \\ \prod_{p=1}^{n}\left(\ell og_{\sigma_p}\left(1-N^*_{\xi_{\eta(p)}}\right)\right)^{\omega_p} \end{pmatrix}.$$

In a similarly way, if $0<\frac{1}{\sigma_p}\leq \min\left\{P_{\xi_p},1-I_{\xi_p},1-N_{\xi_p}\right\}<1, \sigma\neq 1$, we can also obtain

$$L-SVNHWA\left(\zeta_1,\zeta_2,...,\zeta_n\right)=\begin{pmatrix} 1-\prod_{p=1}^{n}\left(\ell og_{\frac{1}{\sigma_p}}P^*_{\xi_{\eta(p)}}\right)^{\omega_p}, \\ \prod_{p=1}^{n}\left(\ell og_{\frac{1}{\sigma_p}}\left(1-I^*_{\xi_{\eta(p)}}\right)\right)^{\omega_p}, \\ \prod_{p=1}^{n}\left(\ell og_{\frac{1}{\sigma_p}}\left(1-N^*_{\xi_{\eta(p)}}\right)\right)^{\omega_p} \end{pmatrix}$$

which completes the proof. □

Remark 1. *If $\sigma_1=\sigma_2=\sigma_3=...=\sigma_n=\sigma$, that is $0<\sigma\leq \min\left\{P_{\xi_p},1-I_{\xi_p},1-N_{\xi_p}\right\}<1, \sigma\neq 1$, then $L-SVNHWA$ operator is reduced as follows*

$$L-SVNHWA\left(\zeta_1,\zeta_2,...,\zeta_n\right)=\begin{pmatrix} 1-\prod_{p=1}^{n}\left(\ell og_{\sigma}P^*_{\xi_{\eta(p)}}\right)^{\omega_p}, \\ \prod_{p=1}^{n}\left(\ell og_{\sigma}\left(1-I^*_{\xi_{\eta(p)}}\right)\right)^{\omega_p}, \\ \prod_{p=1}^{n}\left(\ell og_{\sigma}\left(1-N^*_{\xi_{\eta(p)}}\right)\right)^{\omega_p} \end{pmatrix}. \quad (12)$$

Properties

$L-SVNHWA$ operator satisfies some properties are enlist below;

(1) Idempotency: For any collection of SVNNs $\zeta_p=\left\langle P_{\xi_p}(r),I_{\xi_p}(r),N_{\xi_p}(r)\right\rangle$ $(p=1,...,n)$ in $\Re$. Then, if collection of SVNNs $\zeta_p=\left\langle P_{\xi_p}(r),I_{\xi_p}(r),N_{\xi_p}(r)\right\rangle$ $(p=1,...,n)$ are identical, that is

$$L-SVNHWA\left(\zeta_1,\zeta_2,...,\zeta_n\right)=\zeta. \quad (13)$$

(2) Boundedness: for any collection of SVNNs $\zeta_p = \left\langle P_{\xi_p}(r), I_{\xi_p}(r), N_{\xi_p}(r)\right\rangle$ $(p = 1, ..., n)$ in $\Re$. $\zeta_p^- = \left\langle \min_p P^*_{\xi_p}, \max_p I^*_{\xi_p}, \max_p N^*_{\xi_p}\right\rangle$ and $\zeta_p^+ = \left\langle \max_p P^*_{\xi_p}, \min_p I^*_{\xi_p}, \min_p N^*_{\xi_p}\right\rangle$ $(p = 1, ..., n)$ in $\Re$, therefore

$$\zeta_{p}^{-} \subseteq L - SVNHWA\left(\zeta_1, \zeta_2, ..., \zeta_n\right) \subseteq \zeta_p^{+}. \tag{14}$$

(3) Monotonically: for any collection of SVNNs $\zeta_p = \left\langle P_{\xi_p}(r), I_{\xi_p}(r), N_{\xi_p}(r)\right\rangle$ $(p = 1, ..., n)$ in $\Re$. If $\zeta_{\eta(p)} \subseteq \zeta^*_{\eta(p)}$ for $(p = 1, ..., n)$, then

$$L - SVNHWA\left(\zeta_1, \zeta_2, ..., \zeta_n\right) \subseteq L - SVNHWA\left(\zeta_1^*, \zeta_2^*, ..., \zeta_n^*\right). \tag{15}$$

4.2. Logarithmic Hybrid Geometric Operators

Definition 19. *For any collection of SVNNs* $\zeta_p = \left\langle P_{\xi_p}(r), I_{\xi_p}(r), N_{\xi_p}(r)\right\rangle$ $(p = 1, ..., n)$ *in* $\Re$, *with* $0 < \sigma_p \leq \min\left\{P_{\xi_p}, 1 - I_{\xi_p}, 1 - N_{\xi_p}\right\} < 1, \sigma \neq 1$. *The structure of logarithmic single valued neutrosophic hybrid weighted geometric (L-SVNHWG) operator is*

$$L - SVNHWG\left(\zeta_1, \zeta_2, ..., \zeta_n\right) = \prod_{p=1}^{n}\left(\ell og_{\sigma_p} \zeta^*_{\eta(p)}\right)^{\omega_p} \tag{16}$$

where $\beta_p\,(p = 1, ..., n)$ *are weight vectors with* $\beta_p \geq 0$ *and* $\sum_{p=1}^{n} \beta_p = 1$ *and pth biggest weighted value is* $\zeta^*_{\eta(p)}\left(\zeta^*_{\eta(p)} = \left(\zeta_{\eta(p)}\right)^{n\beta_p}, P \in N\right)$ *consequently by total order* $\zeta^*_{\eta(1)} \geq \zeta^*_{\eta(2)} \geq ... \geq \zeta^*_{\eta(n)}$. *Also associated weights are* $\omega = (\omega_1, \omega_2, ..., \omega_n)$ *with* $\omega_p \geq 0$, $\Sigma_{p=1}^{n}\omega_p = 1$.

Theorem 4. *For any collection of SVNNs* $\zeta_p = \left\langle P_{\xi_p}(r), I_{\xi_p}(r), N_{\xi_p}(r)\right\rangle$ $(p = 1, ..., n)$ *in* $\Re$, *with* $0 < \sigma_p \leq \min\left\{P_{\xi_p}, 1 - I_{\xi_p}, 1 - N_{\xi_p}\right\} < 1, \sigma \neq 1$. *Then by using logarithmic operations and Definition 19, L−SVNHWG define as*

$$L - SVNHWG\left(\zeta_1, \zeta_2, ..., \zeta_n\right) = \begin{cases} \begin{pmatrix} \prod_{p=1}^{n}\left(1 - \ell og_{\sigma_p} P^*_{\xi_{\eta(p)}}\right)^{\beta_p} \\ 1 - \prod_{p=1}^{n}\left(1 - \left(\ell og_{\sigma_p}\left(1 - I^*_{\xi_{\eta(p)}}\right)\right)\right)^{\beta_p} \\ 1 - \prod_{p=1}^{n}\left(1 - \left(\ell og_{\sigma_p}\left(1 - n^*_{\xi_{\eta(p)}}\right)\right)\right)^{\beta_p} \end{pmatrix} & 0 < \sigma_p \leq \min\left\{\begin{array}{c} P_{\xi_p}, \\ 1 - I_{\xi_p}, \\ 1 - N_{\xi_p} \end{array}\right\} < 1 \\ \begin{pmatrix} \prod_{p=1}^{n}\left(1 - \ell og_{\frac{1}{\sigma_p}} P^*_{\xi_{\eta(p)}}\right)^{\beta_p} \\ 1 - \prod_{p=1}^{n}\left(1 - \left(\ell og_{\frac{1}{\sigma_p}}\left(1 - I^*_{\xi_{\eta(p)}}\right)\right)\right)^{\beta_p} \\ 1 - \prod_{p=1}^{n}\left(1 - \left(\ell og_{\frac{1}{\sigma_p}}\left(1 - n^*_{\xi_{\eta(p)}}\right)\right)\right)^{\beta_p} \end{pmatrix} & 0 < \frac{1}{\sigma_p} \leq \min\left\{\begin{array}{c} P_{\xi_p}, \\ 1 - I_{\xi_p}, \\ 1 - N_{\xi_p} \end{array}\right\} < 1, \\ & \sigma \neq 1 \end{cases} \tag{17}$$

where $\beta_p\,(p = 1, ..., n)$ *are weight vectors with* $\beta_p \geq 0$ *and* $\sum_{p=1}^{n} \beta_p = 1$ *and pth biggest weighted value is* $\zeta^*_{\eta(p)}\left(\zeta^*_{\eta(p)} = \left(\zeta_{\eta(p)}\right)^{n\beta_p}, P \in N\right)$ *consequently by total order* $\zeta^*_{\eta(1)} \geq \zeta^*_{\eta(2)} \geq ... \geq \zeta^*_{\eta(n)}$. *Also associated weights are* $\omega = (\omega_1, \omega_2, ..., \omega_n)$ *with* $\omega_p \geq 0$, $\Sigma_{p=1}^{n}\omega_p = 1$.

Proof. Using mathematical induction to prove Equation (4), we proceed as:

(a) For $n = 2$, since

$$(\ell og_{\sigma_1}\zeta_1^*)^{\omega_1} = \begin{pmatrix} \left(1-\ell og_{\sigma_1}P^*_{\xi_1}\right)^{\omega_1} \\ 1-\left(1-\left(\ell og_{\sigma_1}\left(1-I^*_{\xi_1}\right)\right)\right)^{\omega_1} \\ 1-\left(1-\left(\ell og_{\sigma_1}\left(1-N^*_{\xi_1}\right)\right)\right)^{\omega_1} \end{pmatrix}$$

and

$$(\ell og_{\sigma_2}\zeta_2^*)^{\omega_2} = \begin{pmatrix} \left(1-\ell og_{\sigma_2}P^*_{\xi_2}\right)^{\omega_2} \\ 1-\left(1-\left(\ell og_{\sigma_2}\left(1-I^*_{\xi_2}\right)\right)\right)^{\omega_2} \\ 1-\left(1-\left(\ell og_{\sigma_2}\left(1-N^*_{\xi_2}\right)\right)\right)^{\omega_2} \end{pmatrix}.$$

Then

$$\begin{aligned} & L-SVNHWG(\zeta_1,\zeta_2) = (\ell og_{\sigma_1}\zeta_1^*)^{\omega_1} \otimes (\ell og_{\sigma_2}\zeta_2^*)^{\omega_2} \\ = & \begin{pmatrix} \left(1-\ell og_{\sigma_1}P^*_{\xi_1}\right)^{\omega_1} \\ 1-\left(1-\left(\ell og_{\sigma_1}\left(1-I^*_{\xi_1}\right)\right)\right)^{\omega_1} \\ 1-\left(1-\left(\ell og_{\sigma_1}\left(1-N^*_{\xi_1}\right)\right)\right)^{\omega_1} \end{pmatrix} \otimes \begin{pmatrix} \left(1-\ell og_{\sigma_2}P^*_{\xi_2}\right)^{\omega_2} \\ 1-\left(1-\left(\ell og_{\sigma_2}\left(1-I^*_{\xi_2}\right)\right)\right)^{\omega_2} \\ 1-\left(1-\left(\ell og_{\sigma_2}\left(1-N^*_{\xi_2}\right)\right)\right)^{\omega_2} \end{pmatrix} \\ = & \begin{Bmatrix} \left(1-\ell og_{\sigma_1}P^*_{\xi_1}\right)^{\omega_1} \cdot \left(1-\ell og_{\sigma_2}P^*_{\xi_2}\right)^{\omega_2} \\ 1-\left(1-\left(\ell og_{\sigma_1}\left(1-I^*_{\xi_1}\right)\right)\right)^{\omega_1} \cdot \left(1-\left(\ell og_{\sigma_2}\left(1-I^*_{\xi_2}\right)\right)\right)^{\omega_2} \\ 1-\left(1-\left(\ell og_{\sigma_1}\left(1-N^*_{\xi_1}\right)\right)\right)^{\omega_1} \cdot \left(1-\left(\ell og_{\sigma_2}\left(1-N^*_{\xi_2}\right)\right)\right)^{\omega_2} \end{Bmatrix} \\ = & \begin{pmatrix} \prod_{p=1}^{2}\left(1-\ell og_{\sigma_p}P^*_{\xi_p}\right)^{\omega_p}, \\ 1-\prod_{p=1}^{2}\left(1-\left(\ell og_{\sigma_p}\left(1-I^*_{\xi_p}\right)\right)\right)^{\omega_p} \\ 1-\prod_{p=1}^{2}\left(1-\left(\ell og_{\sigma_p}\left(1-N^*_{\xi_p}\right)\right)\right)^{\omega_p} \end{pmatrix}. \end{aligned}$$

(b) Now Equation (4) is true for $n = k$,

$$L-SVNHWG(\zeta_1,\zeta_2,...,\zeta_k) = \begin{pmatrix} \prod_{p=1}^{k}\left(1-\ell og_{\sigma_p}P^*_{\xi_p}\right)^{\omega_p} \\ 1-\prod_{p=1}^{k}\left(1-\left(\ell og_{\sigma_p}\left(1-I^*_{\xi_p}\right)\right)\right)^{\omega_p} \\ 1-\prod_{p=1}^{k}\left(1-\left(\ell og_{\sigma_p}\left(1-N^*_{\xi_p}\right)\right)\right)^{\omega_p} \end{pmatrix},$$

(c) Now, we prove that Equation (4) for $n = k+1$, that is

$$L-SVNHWG(\zeta_1,\zeta_2,...,\zeta_k,\zeta_{k+1}) = \prod_{p=1}^{k}\left(\ell og_{\sigma_p}\zeta_p\right)^{\omega_p} \otimes \left(\ell og_{\sigma_{k+1}}\zeta_{k+1}\right)^{\omega_{k+1}}$$

$$L-SVNHWG\left(\zeta_1,\zeta_2,...,\zeta_k,\zeta_{k+1}\right)$$
$$= \begin{pmatrix} \prod_{p=1}^{k}\left(1-\ell og_{\sigma_p}P^*_{\xi_p}\right)^{\omega_p} \\ 1-\prod_{p=1}^{k}\left(1-\left(\ell og_{\sigma_p}\left(1-I^*_{\xi_p}\right)\right)\right)^{\omega_p} \\ 1-\prod_{p=1}^{k}\left(1-\left(\ell og_{\sigma_p}\left(1-N^*_{\xi_p}\right)\right)\right)^{\omega_p} \end{pmatrix} \otimes \begin{pmatrix} \left(1-\ell og_{\sigma_p}P^*_{\xi_{k+1}}\right)^{\omega_{k+1}} \\ 1-\left(1-\left(\ell og_{\sigma_p}\left(1-I^*_{\xi_{k+1}}\right)\right)\right)^{\omega_{k+1}} \\ 1-\left(1-\left(\ell og_{\sigma_p}\left(1-N^*_{\xi_{k+1}}\right)\right)\right)^{\omega_{k+1}} \end{pmatrix}$$
$$= \begin{pmatrix} \prod_{p=1}^{k+1}\left(1-\ell og_{\sigma_p}P^*_{\xi_p}\right)^{\omega_p} \\ 1-\prod_{p=1}^{k+1}\left(1-\left(\ell og_{\sigma_p}\left(1-I^*_{\xi_p}\right)\right)\right)^{\omega_p} \\ 1-\prod_{p=1}^{k+1}\left(1-\left(\ell og_{\sigma_p}\left(1-N^*_{\xi_p}\right)\right)\right)^{\omega_p} \end{pmatrix}$$

Thus Equation (4) is true for $n = z+1$. Hence it is satisfied for all n. Therefore

$$L-SVNHWG\left(\zeta_1,\zeta_2,...,\zeta_n\right) = \begin{pmatrix} \prod_{p=1}^{n}\left(1-\ell og_{\sigma_p}P^*_{\xi_p}\right)^{\omega_p} \\ 1-\prod_{p=1}^{n}\left(1-\left(\ell og_{\sigma_p}\left(1-I^*_{\xi_p}\right)\right)\right)^{\omega_p} \\ 1-\prod_{p=1}^{n}\left(1-\left(\ell og_{\sigma_p}\left(1-N^*_{\xi_p}\right)\right)\right)^{\omega_p} \end{pmatrix}.$$

In a similar way, if $0 < \frac{1}{\sigma_p} \leq \min\left\{P_{\xi_p}, 1-I_{\xi_p}, 1-N_{\xi_p}\right\} < 1, \sigma \neq 1$, we can also obtain

$$L-SVNHWG\left(\zeta_1,\zeta_2,...,\zeta_n\right) = \begin{pmatrix} \prod_{p=1}^{n}\left(1-\ell og_{\frac{1}{\sigma_p}}P^*_{\xi_p}\right)^{\omega_p} \\ 1-\prod_{p=1}^{n}\left(1-\left(\ell og_{\frac{1}{\sigma_p}}\left(1-I^*_{\xi_p}\right)\right)\right)^{\omega_p} \\ 1-\prod_{p=1}^{n}\left(1-\left(\ell og_{\frac{1}{\sigma_p}}\left(1-N^*_{\xi_p}\right)\right)\right)^{\omega_p} \end{pmatrix}$$

which completes the proof. □

Remark 2. *If $\sigma_1 = \sigma_2 = \sigma_3 = ... = \sigma_n = \sigma$, that is $0 < \sigma \leq \min\left\{P_{\xi_p}, 1-I_{\xi_p}, 1-N_{\xi_p}\right\} < 1, \sigma \neq 1$, then $L-SVNHWG$ operator reduced as follows*

$$L-SVNHWG\left(\zeta_1,\zeta_2,...,\zeta_n\right) = \begin{pmatrix} \prod_{p=1}^{n}\left(1-\ell og_{\sigma}P^*_{\xi_p}\right)^{\omega_p} \\ 1-\prod_{p=1}^{n}\left(1-\left(\ell og_{\sigma}\left(1-I^*_{\xi_p}\right)\right)\right)^{\omega_p} \\ 1-\prod_{p=1}^{n}\left(1-\left(\ell og_{\sigma}\left(1-N^*_{\xi_p}\right)\right)\right)^{\omega_p} \end{pmatrix}. \tag{18}$$

Properties

$L-SVNHWG$ operator satisfies some properties are enlist below;

(1) Idempotency: for any collection of SVNNs $\zeta_p = \left\langle P_{\xi_p}(r), I_{\xi_p}(r), N_{\xi_p}(r)\right\rangle$ $(p = 1,...,n)$ in $\Re$. Then, if collection of SVNNs $\zeta_p = \left\langle P_{\xi_p}(r), I_{\xi_p}(r), N_{\xi_p}(r)\right\rangle$ $(p = 1,...,n)$ are identical, that is

$$L-SVNHWG\left(\zeta_1,\zeta_2,...,\zeta_n\right) = \zeta. \tag{19}$$

(2) Boundedness: for any collection of SVNNs $\zeta_p = \left\langle P_{\xi_p}(r), I_{\xi_p}(r), N_{\xi_p}(r) \right\rangle$ $(p = 1, ..., n)$ in $\Re$. $\zeta_p^- = \left\langle \min_p P_{\xi_p}, \max_p I_{\xi_p}, \max_p N_{\xi_p} \right\rangle$ and $\zeta_p^+ = \left\langle \max_p P_{\xi_p}, \min_p I_{\xi_p}, \min_p N_{\xi_p} \right\rangle$ $(p = 1, ..., n)$ in $\Re$, therefore

$$\zeta_p^- \subseteq L - SVNHWG(\zeta_1, \zeta_2, ..., \zeta_n) \subseteq \zeta_p^+. \tag{20}$$

(3) Monotonically: for any collection of SVNNs $\zeta_p = \left\langle P_{\xi_p}(r), I_{\xi_p}(r), N_{\xi_p}(r) \right\rangle$ $(p = 1, ..., n)$ in $\Re$. If $\zeta_p \subseteq \zeta_p^*$ for $(p = 1, ..., n)$, then

$$L - SVNHWG(\zeta_1, \zeta_2, ..., \zeta_n) \subseteq L - SVNHWG(\zeta_1^*, \zeta_2^*, ..., \zeta_n^*). \tag{21}$$

4.3. Generalized Logarithmic Averaging Operator

Definition 20. *For any collection of SVNNs $\zeta_p = \left\langle P_{\xi_p}(r), I_{\xi_p}(r), N_{\xi_p}(r) \right\rangle$ $(p = 1, ..., n)$ in $\Re$, with $0 < \sigma_p \leq \min\left\{P_{\xi_p}, 1 - I_{\xi_p}, 1 - N_{\xi_p}\right\} < 1$, $\sigma \neq 1$. The structure of logarithmic generalized single-valued neutrosophic weighted averaging (L-GSVNWA) operator is*

$$L - GSVNWA(\zeta_1, \zeta_2, ..., \zeta_n) = \left(\sum_{p=1}^{n} \beta_p \ell og_{\sigma_p}(\zeta_p)^{\gamma} \right)^{\frac{1}{\gamma}} \tag{22}$$

where β_p $(p = 1, ..., n)$ are weighting vector with $\beta_p \geq 0$ and $\sum_{p=1}^{n} \beta_p = 1$.

Theorem 5. *For any collection of SVNNs $\zeta_p = \left\langle P_{\xi_p}(r), I_{\xi_p}(r), N_{\xi_p}(r) \right\rangle$ $(p = 1, ..., n)$ in $\Re$, with $0 < \sigma_p \leq \min\left\{P_{\xi_p}, 1 - I_{\xi_p}, 1 - N_{\xi_p}\right\} < 1, \sigma \neq 1, \gamma \geq 1$. Then by using logarithmic operations and Definition 20, $L - GSVNWA$ define as*

$$L - GSVNWA(\zeta_1, \zeta_2, ..., \zeta_n)$$

$$\begin{cases} \begin{pmatrix} \left(1 - \prod_{p=1}^{n}\left(1 - \left(1 - \left(\ell og_{\sigma_p} P_{\xi_p}\right)\right)^{\gamma}\right)^{\beta_p}\right)^{\frac{1}{\gamma}}, \\ 1 - \left[1 - \prod_{p=1}^{n}\left(1 - \left(1 - \ell og_{\sigma_p}\left(1 - I_{\xi_p}\right)\right)^{\gamma}\right)^{\beta_p}\right]^{\frac{1}{\gamma}} \\ 1 - \left[1 - \prod_{p=1}^{n}\left(1 - \left(1 - \ell og_{\sigma_p}\left(1 - N_{\xi_p}\right)\right)^{\gamma}\right)^{\beta_p}\right]^{\frac{1}{\gamma}} \end{pmatrix} & 0 < \sigma_p \leq \min\left\{\begin{array}{c} P_{\xi_p}, \\ 1 - I_{\xi_p}, \\ 1 - N_{\xi_p} \end{array}\right\} < 1 \\ \begin{pmatrix} \left(1 - \prod_{p=1}^{n}\left(1 - \left(1 - \left(\ell og_{\frac{1}{\sigma_p}} P_{\xi_p}\right)\right)^{\gamma}\right)^{\beta_p}\right)^{\frac{1}{\gamma}}, \\ 1 - \left[1 - \prod_{p=1}^{n}\left(1 - \left(1 - \ell og_{\frac{1}{\sigma_p}}\left(1 - I_{\xi_p}\right)\right)^{\gamma}\right)^{\beta_p}\right]^{\frac{1}{\gamma}} \\ 1 - \left[1 - \prod_{p=1}^{n}\left(1 - \left(1 - \ell og_{\frac{1}{\sigma_p}}\left(1 - N_{\xi_p}\right)\right)^{\gamma}\right)^{\beta_p}\right]^{\frac{1}{\gamma}} \end{pmatrix} & \begin{array}{c} 0 < \frac{1}{\sigma_p} \leq \min\left\{\begin{array}{c} P_{\xi_p}, \\ 1 - I_{\xi_p}, \\ 1 - N_{\xi_p} \end{array}\right\} < 1, \\ \sigma \neq 1 \end{array} \end{cases} \tag{23}$$

where β_p $(p = 1, ..., n)$ are weighting vector with $\beta_p \geq 0$ and $\sum_{p=1}^{n} \beta_p = 1$.

Apparently, if we use $\gamma = 1$, then the $L - GSVNWA$ operator is becomes into $L - SVNWA$ operator.

Proof. Theorem 5 take the form by utilized the technique of mathematical induction and procedure is eliminate here. □

Remark 3. *If* $\sigma_1 = \sigma_2 = \sigma_3 = ... = \sigma_n = \sigma$, *that is* $0 < \sigma \leq \min\left\{P_{\tilde{\zeta}_p}, 1 - I_{\tilde{\zeta}_p}, 1 - N_{\tilde{\zeta}_p}\right\} < 1, \sigma \neq 1$, *then* $L - GSVNWA$ *operator reduced as follows*

$$L - GSVNWA\left(\zeta_1, \zeta_2, ..., \zeta_n\right) = \begin{pmatrix} \left(1 - \prod_{p=1}^{n}\left(1 - \left(1 - \left(\ell og_{\sigma} P_{\tilde{\zeta}_p}\right)\right)^{\gamma}\right)^{\beta_p}\right)^{\frac{1}{\gamma}}, \\ 1 - \left[1 - \prod_{p=1}^{n}\left(1 - \left(1 - \ell og_{\sigma}\left(1 - I_{\tilde{\zeta}_p}\right)\right)^{\gamma}\right)^{\beta_p}\right]^{\frac{1}{\gamma}} \\ 1 - \left[1 - \prod_{p=1}^{n}\left(1 - \left(1 - \ell og_{\sigma}\left(1 - N_{\tilde{\zeta}_p}\right)\right)^{\gamma}\right)^{\beta_p}\right]^{\frac{1}{\gamma}} \end{pmatrix}. \tag{24}$$

Properties

$L - GSVNWA$ operator satisfies some properties are enlist below;

(1) Idempotency: For any collection of SVNNs $\zeta_p = \left\langle P_{\tilde{\zeta}_p}(r), I_{\tilde{\zeta}_p}(r), N_{\tilde{\zeta}_p}(r)\right\rangle$ $(p = 1, ..., n)$ in $\Re$. Then, if collection of SVNNs $\zeta_p = \left\langle P_{\tilde{\zeta}_p}(r), I_{\tilde{\zeta}_p}(r), N_{\tilde{\zeta}_p}(r)\right\rangle$ $(p = 1, ..., n)$ are identical, that is

$$L - GSVNWA\left(\zeta_1, \zeta_2, ..., \zeta_n\right) = \zeta. \tag{25}$$

(2) Boundedness: for any collection of SVNNs $\zeta_p = \left\langle P_{\tilde{\zeta}_p}(r), I_{\tilde{\zeta}_p}(r), N_{\tilde{\zeta}_p}(r)\right\rangle$ $(p = 1, ..., n)$ in $\Re$. $\zeta_p^- = \left\langle \min_p P_{\tilde{\zeta}_p}, \max_p I_{\tilde{\zeta}_p}, \max_p N_{\tilde{\zeta}_p}\right\rangle$ and $\zeta_p^+ = \left\langle \max_p P_{\tilde{\zeta}_p}, \min_p I_{\tilde{\zeta}_p}, \min_p N_{\tilde{\zeta}_p}\right\rangle$ $(p = 1, ..., n)$ in $\Re$, therefore

$$\zeta_p^- \subseteq L - GSVNWA\left(\zeta_1, \zeta_2, ..., \zeta_n\right) \subseteq \zeta_p^+. \tag{26}$$

(3) Monotonically: for any collection of SVNNs $\zeta_p = \left\langle P_{\tilde{\zeta}_p}(r), I_{\tilde{\zeta}_p}(r), N_{\tilde{\zeta}_p}(r)\right\rangle$ $(p = 1, ..., n)$ in $\Re$. If $\zeta_p \subseteq \zeta_p^*$ for $(p = 1, ..., n)$, then

$$L - GSVNWA\left(\zeta_1, \zeta_2, ..., \zeta_n\right) \subseteq L - GSVNWA\left(\zeta_1^*, \zeta_2^*, ..., \zeta_n^*\right). \tag{27}$$

4.4. Generalized Logarithmic Geometric Operator

Definition 21. *For any collection of SVNNs* $\zeta_p = \left\langle P_{\tilde{\zeta}_p}(r), I_{\tilde{\zeta}_p}(r), N_{\tilde{\zeta}_p}(r)\right\rangle$ $(p = 1, ..., n)$ *in* $\Re$, *with* $0 < \sigma_p \leq \min\left\{P_{\tilde{\zeta}_p}, 1 - I_{\tilde{\zeta}_p}, 1 - N_{\tilde{\zeta}_p}\right\} < 1, \sigma \neq 1$. *The structure of logarithmic generalized single valued neutrosophic weighted geometric (L-GSVNWG) operator is*

$$L - GSVNWG\left(\zeta_1, \zeta_2, ..., \zeta_n\right) = \left(\sum_{p=1}^{n}\left(\ell og_{\sigma_p}\left(\zeta_p\right)^{\gamma}\right)^{\beta_p}\right)^{\frac{1}{\gamma}} \tag{28}$$

where $\beta_p\,(p = 1, ..., n)$ *are weighting vector with* $\beta_p \geq 0$ *and* $\sum_{p=1}^{n} \beta_p = 1$.

Theorem 6. *For any collection of SVNNs* $\zeta_p = \left\langle P_{\xi_p}(r), I_{\xi_p}(r), N_{\xi_p}(r) \right\rangle$ $(p = 1, ..., n)$ *in* $\Re$, *with* $0 < \sigma_p \leq \min\left\{P_{\xi_p}, 1 - I_{\xi_p}, 1 - N_{\xi_p}\right\} < 1, \sigma \neq 1, \gamma \geq 1$. *Then by using logarithmic operations and definition (21),* $L - GSVNWG$ *define as*

$$L - GSVNWG(\zeta_1, \zeta_2, ..., \zeta_n) \left\{ \begin{array}{ll} \left(\begin{array}{c} 1 - \left[1 - \prod_{p=1}^{n} \left(1 - \left(\ell og_{\sigma_p} P_{\xi_p}\right)^{\gamma}\right)^{\beta_p}\right]^{\frac{1}{\gamma}}, \\ \left(1 - \prod_{p=1}^{n} \left(1 - \left(\ell og_{\sigma_p}\left(1 - I_{\xi_p}\right)\right)^{\gamma}\right)^{\beta_p}\right)^{\frac{1}{\gamma}} \\ \left(1 - \prod_{p=1}^{n} \left(1 - \left(\ell og_{\sigma_p}\left(1 - N_{\xi_p}\right)\right)^{\gamma}\right)^{\beta_p}\right)^{\frac{1}{\gamma}} \end{array} \right) & 0 < \sigma_p \leq \min\left\{ \begin{array}{c} P_{\xi_p}, \\ 1 - I_{\xi_p}, \\ 1 - N_{\xi_p} \end{array} \right\} < 1 \\ \left(\begin{array}{c} 1 - \left[1 - \prod_{p=1}^{n} \left(1 - \left(\ell og_{\frac{1}{\sigma_p}} P_{\xi_p}\right)^{\gamma}\right)^{\beta_p}\right]^{\frac{1}{\gamma}}, \\ \left(1 - \prod_{p=1}^{n} \left(1 - \left(\ell og_{\frac{1}{\sigma_p}}\left(1 - I_{\xi_p}\right)\right)^{\gamma}\right)^{\beta_p}\right)^{\frac{1}{\gamma}} \\ \left(1 - \prod_{p=1}^{n} \left(1 - \left(\ell og_{\frac{1}{\sigma_p}}\left(1 - N_{\xi_p}\right)\right)^{\gamma}\right)^{\beta_p}\right)^{\frac{1}{\gamma}} \end{array} \right) & \begin{array}{c} 0 < \frac{1}{\sigma_p} \leq \min\left\{ \begin{array}{c} P_{\xi_p}, \\ 1 - I_{\xi_p}, \\ 1 - N_{\xi_p} \end{array} \right\} < 1, \\ \sigma \neq 1 \end{array} \end{array} \right. \tag{29}$$

where β_p $(p = 1, ..., n)$ *is the weighting vector with* $\beta_p \geq 0$ *and* $\sum_{p=1}^{n} \beta_p = 1$.

Apparently, if we use $\gamma = 1$, then the $L - GSVNWG$ operator is becomes into $L - SVNWG$ operator.

Proof. Theorem 6 takes the form by utilizing the technique of mathematical induction and the procedure is eliminated here. □

Remark 4. *If* $\sigma_1 = \sigma_2 = \sigma_3 = ... = \sigma_n = \sigma$, *that is* $0 < \sigma \leq \min\left\{P_{\xi_p}, 1 - I_{\xi_p}, 1 - N_{\xi_p}\right\} < 1, \sigma \neq 1$, *then* $L - GSVNWG$ *operator reduced as follows*

$$L - GSVNWG(\zeta_1, \zeta_2, ..., \zeta_n) = \left(\begin{array}{c} 1 - \left[1 - \prod_{p=1}^{n} \left(1 - \left(\ell og_{\sigma} P_{\xi_p}\right)^{\gamma}\right)^{\beta_p}\right]^{\frac{1}{\gamma}}, \\ \left(1 - \prod_{p=1}^{n} \left(1 - \left(\ell og_{\sigma}\left(1 - I_{\xi_p}\right)\right)^{\gamma}\right)^{\beta_p}\right)^{\frac{1}{\gamma}} \\ \left(1 - \prod_{p=1}^{n} \left(1 - \left(\ell og_{\sigma}\left(1 - N_{\xi_p}\right)\right)^{\gamma}\right)^{\beta_p}\right)^{\frac{1}{\gamma}} \end{array} \right). \tag{30}$$

Properties

$L - GSVNWG$ operator satisfies some properties are enlist below;

(1) Idempotency: For any collection of SVNNs $\zeta_p = \left\langle P_{\xi_p}(r), I_{\xi_p}(r), N_{\xi_p}(r) \right\rangle$ $(p = 1, ..., n)$ in $\Re$. Then, if collection of SVNNs $\zeta_p = \left\langle P_{\xi_p}(r), I_{\xi_p}(r), N_{\xi_p}(r) \right\rangle$ $(p = 1, ..., n)$ are identical, that is

$$L - GSVNWG(\zeta_1, \zeta_2, ..., \zeta_n) = \zeta. \tag{31}$$

(2) Boundedness: for any collection of SVNNs $\zeta_p = \left\langle P_{\xi_p}(r), I_{\xi_p}(r), N_{\xi_p}(r) \right\rangle$ $(p = 1, ..., n)$ in $\Re$. $\zeta_p^- = \left\langle \min_p P_{\xi_p}, \max_p I_{\xi_p}, \max_p N_{\xi_p} \right\rangle$ and $\zeta_p^+ = \left\langle \max_p P_{\xi_p}, \min_p I_{\xi_p}, \min_p N_{\xi_p} \right\rangle$ $(p = 1, ..., n)$ in $\Re$, therefore

$$\zeta_p^- \subseteq L - GSVNWG(\zeta_1, \zeta_2, ..., \zeta_n) \subseteq \zeta_p^+. \tag{32}$$

(3) Monotonically: for any collection of SVNNs $\zeta_p = \left\langle P_{\xi_p}(r), I_{\xi_p}(r), N_{\xi_p}(r) \right\rangle$ $(p = 1, ..., n)$ in $\Re$. If $\zeta_p \subseteq \zeta_p^*$ for $(p = 1, ..., n)$, then

$$L - GSVNWG(\zeta_1, \zeta_2, ..., \zeta_n) \subseteq L - GSVNWG(\zeta_1^*, \zeta_2^*, ..., \zeta_n^*). \tag{33}$$

5. Proposed Technique for Solving Decision-Making Problems

This section includes the new approach to decision-making based on the single-valued neutrosophic sets, and we will propose a decision-making matrix as indicated below.

Let $H = (h_1, h_2, ..., h_m)$ be a distinct collection of m probable alternatives and $Y = (y_1, y_2, ..., y_n)$ be a finite collection of n criteria, where h_i indicate the i-th alternatives and y_j indicate the j-th criteria. Let $D = (d_1, d_2, ..., d_t)$ be a finite set of t experts, where d_k indicate the k-th expert. The expert d_k supply her appraisal of an alternative h_i on an attribute y_j as a SVNNs $(i = 1, ..., m; j = 1, ..., n)$. The expert's information is represented by the SVNS decision-making matrix $D^s = \left[E_{ip}^{(s)}\right]_{m \times n}$. Assume that $\beta_p(p = 1, ..., m)$ is the weight vector of the attribute y_j, where $0 \leq \beta_p \leq 1$, $\sum_{p=1}^{n} \beta_p = 1$ and $\psi = (\psi_1, \psi_2, ..., \psi_m)$ be the weights of the decision makers d_k such that $\psi_k \leq 1$, $\sum_{k=1}^{n} \psi_k = 1$.

When we construct the SVNS decision-making matrices, $D^s = \left[E_{ip}^{(s)}\right]_{m \times n}$ for decision. Basically, criteria have two types, one is benefit criteria and other one is cost criteria. If the SVNS decision matrices have cost-type criteria metrics$D^s = \left[E_{ip}^s\right]_{m \times n}$ can be converted into the normalized SVNS decision matrices, $R^s = \left[r_{ip}^{(s)}\right]_{m \times n}$, where $r_{ip}^s = \begin{cases} E_{ip}^s, & \text{for benefit criteria } A_p \\ \overline{E}_{ip}^s, & \text{for cost criteria } A_p, \end{cases}$ $j = 1, ..., n$, and $\overline{E}_{ip}^s$ is the complement of E_{ip}^s. The normalization is not required, if the criteria have the same type.

Step 1: In this step, we get the neutrosophic information, using the all proposed logarithmic aggregation operators to evolute the alternative preference values with associated weights, which are $\omega = (\omega_1, \omega_2, ..., \omega_n)$ with $\omega_p \geq 0$, $\Sigma_{p=1}^n \omega_p = 1$.

Step 2: We find the score value $\widetilde{S}(\ell og_\sigma \zeta_p)$ and the accuracy value $\widetilde{A}(\ell og_\sigma \zeta_p)$ of the cumulative total preference value h_i $(i = 1, ..., m)$.

Step 3: By definition, we give ranking to the alternatives h_i $(i = 1, ..., m)$ and choose the best alternative which has the maximum score value.

5.1. Numerical Example

Assume that there is a committee which selects five applicable emerging technology enterprises $H_g (g = 1, ..., 5)$, which are given as follows.
(1) Augmented reality (H_1),
(2) Personalized medicine (H_2),
(3) Artificial intelligence (H_3),
(4) Gene drive (H_4) and
(5) Quantum computing (H_5).

They assess the possible rising technology enterprises according to the five attributes, which are
(1) Advancement (D_1),
(2) Market risk (D_2),
(3) Financial investments (D_3),

(4) Progress of science and technology (D_4) and
(5) Designs (D_5).

To avoid the conflict between them, the decision makers take the attribute weights as $\beta = (0.15, 0.28, 0.20, 0.22, 0.15)^T$. They construct the SVNS decision-making matrix given in Table 1.

Table 1. Emerging Technology Enterprises D^1.

	D_1	D_2	D_3	D_4	D_5
H_1	$(0.5, 0.3, 0.4)$	$(0.3, 0.2, 0.5)$	$(0.2, 0.2, 0.6)$	$(0.4, 0.2, 0.3)$	$(0.3, 0.3, 0.4)$
H_2	$(0.7, 0.1, 0.3)$	$(0.3, 0.2, 0.7)$	$(0.6, 0.3, 0.2)$	$(0.2, 0.4, 0.6)$	$(0.7, 0.1, 0.2)$
H_3	$(0.5, 0.3, 0.4)$	$(0.4, 0.2, 0.6)$	$(0.6, 0.1, 0.2)$	$(0.3, 0.1, 0.5)$	$(0.6, 0.4, 0.3)$
H_4	$(0.7, 0.3, 0.2)$	$(0.2, 0.2, 0.7)$	$(0.4, 0.5, 0.2)$	$(0.2, 0.2, 0.5)$	$(0.4, 0.5, 0.4)$
H_5	$(0.4, 0.1, 0.3)$	$(0.2, 0.1, 0.5)$	$(0.4, 0.1, 0.5)$	$(0.6, 0.3, 0.4)$	$(0.3, 0.2, 0.4)$

Since D_1, D_3 are benefit-type criteria and D_2, D_4 is cost type criteria, the normalization is required for these decision matrices. Normalized decision matrices are shown in Table 2.

Table 2. Emerging Technology Enterprises R^1.

	D_1	D_2	D_3	D_4	D_5
H_1	$(0.5, 0.3, 0.4)$	$(0.5, 0.2, 0.3)$	$(0.2, 0.2, 0.6)$	$(0.3, 0.2, 0.4)$	$(0.3, 0.3, 0.4)$
H_2	$(0.7, 0.1, 0.3)$	$(0.7, 0.2, 0.3)$	$(0.6, 0.3, 0.2)$	$(0.6, 0.4, 0.2)$	$(0.7, 0.1, 0.2)$
H_3	$(0.5, 0.3, 0.4)$	$(0.6, 0.2, 0.4)$	$(0.6, 0.1, 0.2)$	$(0.5, 0.1, 0.3)$	$(0.6, 0.4, 0.3)$
H_4	$(0.7, 0.3, 0.2)$	$(0.7, 0.2, 0.2)$	$(0.4, 0.5, 0.2)$	$(0.5, 0.2, 0.2)$	$(0.4, 0.5, 0.4)$
H_5	$(0.4, 0.1, 0.3)$	$(0.5, 0.1, 0.2)$	$(0.4, 0.1, 0.5)$	$(0.4, 0.3, 0.6)$	$(0.3, 0.2, 0.4)$

Step 1: Now, we apply all the proposed logarithmic aggregation operators to collective neutrosophic information as follows.

Case 1: Using logarithmic single-valued neutrosophic hybrid weighted averaging aggregation operator, we obtained the results shown in Table 3.

Table 3. Aggregated information using the logarithmic single valued neutrosophic hybrid weighted averaging (L-SVNHWA) operator for $\sigma = 0.3$.

H_1	$(0.17624, 0.23432, 0.43885)$
H_2	$(0.66164, 0.16229, 0.21840)$
H_3	$(0.52788, 0.18347, 0.32224)$
H_4	$(0.49410, 0.30962, 0.20985)$
H_5	$(0.22496, 0.12393, 0.39318)$

Case 2: Using Logarithmic single valued neutrosophic hybrid weighted geometric aggregation operator, we obtainedthe results shown in Table 4.

Table 4. Aggregated information using logarithmic single valued neutrosophic hybrid weighted geometric (L-SVNHWG) operator for $\sigma = 0.1$.

H_1	$(0.52472, 0.12638, 0.24189)$
H_2	$(0.81968, 0.10633, 0.11764)$
H_3	$(0.74946, 0.11782, 0.17620)$
H_4	$(0.70685, 0.18942, 0.11685)$
H_5	$(0.58497, 0.07427, 0.23305)$

Step 2: We find the score index $\widetilde{S}(\ell og_{\sigma}\zeta_p)$ and the accuracy index $\widetilde{A}(\ell og_{\sigma}\zeta_p)$ of the cumulative overall preference value h_i $(i = 1, 2, 3, 4, 5)$.

Case 1: Using the score of aggregated information for L-SVNHWA operator, we obtained the results shown in Table 5.

Table 5. Score of aggregated information for L-SVNHWA operator.

$\tilde{S}(\ell og_{0.3}H_1)$	-1.14345	$\tilde{A}(\ell og_{0.3}H_1)$	0.25985
$\tilde{S}(\ell og_{0.3}H_2)$	0.30519	$\tilde{A}(\ell og_{0.3}H_2)$	1.0087
$\tilde{S}(\ell og_{0.3}H_3)$	-0.02207	$\tilde{A}(\ell og_{0.3}H_3)$	0.96078
$\tilde{S}(\ell og_{0.3}H_4)$	-0.08895	$\tilde{A}(\ell og_{0.3}H_4)$	0.91781
$\tilde{S}(\ell og_{0.3}H_5)$	-0.76389	$\tilde{A}(\ell og_{0.3}H_5)$	0.28571

Case 2: Score of Aggregated information for L-SVNHWG Operator, we obtained the results shown in Table 6.

Table 6. Score of aggregated information for L-SVNHWG operator.

$\tilde{S}(\ell og_{0.1}H_1)$	0.540979	$\tilde{A}(\ell og_{0.1}H_1)$	0.89888
$\tilde{S}(\ell og_{0.1}H_2)$	0.810463	$\tilde{A}(\ell og_{0.1}H_2)$	1.01683
$\tilde{S}(\ell og_{0.1}H_3)$	0.736126	$\tilde{A}(\ell og_{0.1}H_3)$	1.01338
$\tilde{S}(\ell og_{0.1}H_4)$	0.704159	$\tilde{A}(\ell og_{0.1}H_4)$	0.994506
$\tilde{S}(\ell og_{0.1}H_5)$	0.618387	$\tilde{A}(\ell og_{0.1}H_5)$	0.903179

Step 3: We find the best (suitable) alternative which has the maximum score value from the set of alternatives h_i $(i = 1,2,3,4,5)$. Overall preference value and ranking of the alternatives are summarized in Table 7.

Table 7. Overall preference value and ranking of the alternatives.

	$\tilde{S}(H_1)$	$\tilde{S}(H_2)$	$\tilde{S}(H_3)$	$\tilde{S}(H_4)$	$\tilde{S}(H_5)$	**Ranking**
$L-SVNHWA$	-1.143	0.305	-0.022	-0.088	-0.763	$H_2 > H_3 > H_4 > H_5 > H_1$
$L-SVNHWG$	0.540	0.810	0.736	0.704	0.618	$H_2 > H_3 > H_4 > H_5 > H_1$

5.2. Comparison with Existing Methods

This section consists of the comparative analysis of several existing aggregation operators of neutrosophic information with the proposed logarithmic single valued hybrid weighted aggregation operators. Existing methods for aggregated neutrosophic information are shown in Table 8–11.

Table 8. Average aggregated SVN information.

	SVNWA [35]	**SVNOWA** [35]	**NWA** [14]
H_1	$(0.3779, 0.2259, 0.4002)$	$(0.3820, 0.2449, 0.4071)$	$(0.3779, 0.2314, 0.4223)$
H_2	$(0.6615, 0.2052, 0.2381)$	$(0.6663, 0.1801, 0.2430)$	$(0.6615, 0.2426, 0.2446)$
H_3	$(0.5656, 0.1763, 0.3131)$	$(0.5597, 0.1838, 0.3122)$	$(0.5656, 0.2109, 0.3272)$
H_4	$(0.5722, 0.2929, 0.2219)$	$(0.5706, 0.3145, 0.2219)$	$(0.5722, 0.3348, 0.2338)$
H_5	$(0.4165, 0.1413, 0.3607)$	$(0.3960, 0.1373, 0.3696)$	$(0.4165, 0.1633, 0.4131)$

Table 9. Average aggregated SVN information.

	SVNFWA [12]	**SVNHWA** [11] $\gamma = 2$
H_1	$(0.3755, 0.2262, 0.4018)$	$(0.3725, 0.2264, 0.4033)$
H_2	$(0.6611, 0.2072, 0.2385)$	$(0.6608, 0.2086, 0.2388)$
H_3	$(0.5652, 0.1779, 0.3141)$	$(0.5648, 0.1790, 0.3149)$
H_4	$(0.5692, 0.2956, 0.2225)$	$(0.5663, 0.2978, 0.2230)$
H_5	$(0.4159, 0.1422, 0.3646)$	$(0.4151, 0.1427, 0.3680)$

Table 10. Average aggregated SVN information.

	SVNHWA [11] $\gamma = 3$	L-SVNWA [10]
H_1	$(0.3693, 0.2266, 0.4048)$	$(0.3130, 0.1753, 0.3544)$
H_2	$(0.6604, 0.2099, 0.2390)$	$(0.6486, 0.1989, 0.2313)$
H_3	$(0.5645, 0.1800, 0.3157)$	$(0.4989, 0.1733, 0.3321)$
H_4	$(0.5635, 0.3000, 0.2234)$	$(0.5585, 0.2736, 0.1942)$
H_5	$(0.4143, 0.1432, 0.3714)$	$(0.2849, 0.1249, 0.3758)$

Table 11. Average aggregated SVN information.

	L-SVNOWA [10]
H_1	$(0.3229, 0.1926, 0.3607)$
H_2	$(0.6549, 0.1719, 0.2368)$
H_3	$(0.4896, 0.1823, 0.3303)$
H_4	$(0.5561, 0.2975, 0.1942)$
H_5	$(0.2442, 0.1209, 0.3834)$

Now, we analyze the ranking of the alternatives according to their aggregated information (in Table 12).

Table 12. Overall ranking of the alternatives.

Existing Operators	Ranking
NWA [14]	$H_2 > H_3 > H_4 > H_5 > H_1$
SVNWA [35]	$H_2 > H_3 > H_4 > H_5 > H_1$
SVNOWA [35]	$H_2 > H_3 > H_4 > H_5 > H_1$
SVNWG [35]	$H_2 > H_3 > H_4 > H_5 > H_1$
SVNOWG [35]	$H_2 > H_3 > H_4 > H_5 > H_1$
SVNFWA [12]	$H_2 > H_3 > H_4 > H_5 > H_1$
SVNHWA [11] $\gamma = 2$	$H_2 > H_3 > H_4 > H_5 > H_1$
SVNHWA [11] $\gamma = 3$	$H_2 > H_3 > H_4 > H_5 > H_1$
NWG [14]	$H_2 > H_3 > H_4 > H_5 > H_1$
SVNFWG [12]	$H_2 > H_3 > H_4 > H_5 > H_1$
SVNHWG [11] $\gamma = 2$	$H_2 > H_3 > H_4 > H_5 > H_1$
SVNHWG [11] $\gamma = 3$	$H_2 > H_3 > H_4 > H_5 > H_1$
SNWEA [15]	$H_2 > H_3 > H_5 > H_4 > H_1$
L-SVNWA [10]	$H_2 > H_4 > H_3 > H_5 > H_1$
L-SVNOWA [10]	$H_2 > H_4 > H_3 > H_5 > H_1$
L-SVNWG [10]	$H_2 > H_4 > H_3 > H_1 > H_5$
L-SVNOWG [10]	$H_2 > H_3 > H_4 > H_5 > H_1$
Proposed Operators	**Ranking**
L-SVNHWA	$H_2 > H_3 > H_4 > H_5 > H_1$
L-SVNHWG	$H_2 > H_3 > H_4 > H_5 > H_1$
L-GSVNWA	$H_2 > H_4 > H_3 > H_5 > H_1$
L-GSVNWG	$H_2 > H_4 > H_3 > H_1 > H_5$

The bast alternative was H_2. The obtained results utilizing logarithmic single valued neutrosophic hybrid weighted operators and logarithmic generalized single valued neutrosophic weighted operators were same as results shows existing methods. Hence, this study proposed novel logarithmic aggregation operators to aggregate the neutrosophic information more effectively and efficiently. Utilizing the proposed logarithmic aggregation operators, we sound the best alternative from a set of alternatives given by the decision maker. Hence the proposed MCDM technique based on logarithmic operators lets us find the best alternative as an applications in decision support systems.

6. Conclusions

In this work, an attempt has been made to present different kinds of logarithmic weighted averaging and geometric aggregation operators based on the single-valued neutrosophic set environment. Earlier, it has been observed that the various aggregation operators are defined under the SVNSs environment where the aggregation operators based on the algebraic or Einstein t-norm and t-conorm. In this paper, we proposed novel logarithmic hybrid aggregation operators and also logarithmic generalized averaging and geometric aggregation operators. Aggregation operators, namely L-SVNHWA, L-SVNHWG, L-GSVNWA and L-GSVNWA are developed under the SVNSs environment and we have studied their properties in detail. Further, depending on the standardization of the decision matrix and the proposed aggregation operators, a decision-making approach is presented to find the best alternative to the SVNSs environment. An illustrative example is taken for illustrating the developed approach, and their results are compared with some of the existing approaches of the SVNSs environment to show the validity of it. From the studies, we conclude that the proposed approach is more generic and suitable for solving the stated problem.

In the future, we shall link the proposed operators with some novel fuzzy sets, like as type 2 fuzzy sets, neutrosophic sets, and so on. Moreover, we may examine if our constructed approach can also be applied in different areas, such as personal evaluation, medical artificial intelligence, energy management and supplier selection evaluation.

Author Contributions: Conceptualization, S.A. (Shahzaib Ashraf) and S.A. (Saleem Abdullah); methodology, S.A. (Shahzaib Ashraf); software, S.A. (Shahzaib Ashraf); validation, S.A. (Shahzaib Ashraf), S.A. (Saleem Abdullah), F.S. and N.u.A.; formal analysis, F.S. and N.u.A.; investigation, S.A. (Shahzaib Ashraf); writing–original draft preparation, S.A.; writing–review and editing, S.A. (Shahzaib Ashraf); visualization, F.S.; supervision, S.A. (Saleem Abdullah); funding acquisition, F.S.

Funding: This research received no external funding.

Acknowledgments: The authors would like to thank the editor in chief, associate editor and the anonymous referees for detailed and valuable comments which helped to improve this manuscript.

Conflicts of Interest: The authors declare no conflict of interest.

References

1. Zadeh, L.A. Fuzzy sets. *Inf. Control* **1965**, *8*, 338–353. [CrossRef]
2. Atanassov, K.T. Intuitionistic fuzzy sets. *Fuzzy Sets Syst.* **1986**, *20*, 87–96. [CrossRef]
3. Smarandache, F. A unifying field in logics: Neutrosophic logic. *Am. Res. Press Rehoboth* **1999**, 1–141.
4. Soria Frisch, A.; Koppen, M.; Sy, T. Is she gonna like it? automated inspection system using fuzzy aggregation. In *Intelligent Systems for Information Processing*; Elsevier: Amsterdam, The Netherlands, 2003; pp. 465–476.
5. Broumi, S.; Bakal, A.; Talea, M.; Smarandache, F.; Vladareanu, L. Applying Dijkstra algorithm for solving neutrosophic shortest path problem. In Proceedings of the 2016 International Conference on Advanced Mechatronic Systems (ICAMechS), Melbourne, VIC, Australia, 30 November–3 December 2016; IEEE: Piscataway, NJ, USA, 2016; pp. 412–416.
6. Broumi, S.; Bakal, A.; Talea, M.; Smarandache, F.; ALi, M. Shortest path problem under bipolar neutrosphic setting. *Appl. Mech. Mater.* **2017**, *859*, 59–66. [CrossRef]
7. Broumi, S.; Bakal, A.; Talea, M.; Smarandache, F.; Vladareanu, L. Computation of shortest path problem in a network with sv-trapezoidal neutrosophic numbers. In Proceedings of the 2016 International Conference on Advanced Mechatronic Systems (ICAMechS), Melbourne, VIC, Australia, 30 November–3 December 2016; IEEE: Piscataway, NJ, USA, 2016; pp. 417–422.
8. Broumi, S.; Smarandache, F. Correlation coefficient of interval neutrosophic set. In *Applied Mechanics and Materials*; Trans Tech Publications: Zurich, Switzerland, 2013; Volume 436, pp. 511–517.

9. Broumi, S.; Talea, M.; Smarandache, F.; Bakali, A. Decision-making method based on the interval valued neutrosophic graph. In Proceedings of the Future Technologies Conference (FTC), San Francisco, CA, USA, 6–7 December 2016; pp. 44–50.
10. Garg, H. New logarithmic operational laws and their applications to multiattribute decision making for single-valued neutrosophic numbers. *Cognit. Syst. Res.* **2018**, *52*, 931–946. [CrossRef]
11. Liu, P.; Chu, Y.; Li, Y.; Chen, Y. Some generalized neutrosophic number hamacher aggregation operators and their application to group decision making. *Int. J. Fuzzy Syst.* **2014**, *16*, 242–255.
12. Nancy, G.H.; Garg, H. Novel single-valued neutrosophic decision making operators under frank norm operations and its application. *Int. J. Uncertain. Quant.* **2016**, *6*, 361–375. [CrossRef]
13. Ye, J. Single valued neutrosophic cross-entropy for multicriteria decision making problems. *Appl. Math. Model.* **2014**, *38*, 1170–1175. [CrossRef]
14. Ye, J. Subtraction and division operations of simplified neutrosophic sets. *Information* **2017**, *8*, 51. [CrossRef]
15. Ye, J. Exponential operations and aggregation operators of interval neutrosophic sets and their decision making methods. *SpringerPlus* **2016**, *5*, 1488. [CrossRef] [PubMed]
16. Garg, H.; Nancy, G.H. Multi-Criteria Decision-Making Method Based on Prioritized Muirhead Mean Aggregation Operator under Neutrosophic Set Environment. *Symmetry* **2018**, *10*, 280. [CrossRef]
17. Khan, Q.; Hassan, N.; Mahmood, T. Neutrosophic Cubic Power Muirhead Mean Operators with Uncertain Data for Multi-Attribute Decision-Making. *Symmetry* **2018**, *10*, 444. [CrossRef]
18. Adlassnig, K.P. Fuzzy set theory in medical diagnosis. *IEEE Trans. Syst. Man Cybern.* **1986**, *16*, 260–265. [CrossRef]
19. Ashraf, S.; Mahmood, T.; Abdullah, S.; Khan, Q. Different Approaches to Multi-Criteria Group Decision Making Problems for Picture Fuzzy Environment. *Bull. Braz. Math. Soc. New Ser.* **2018**, 1–25. [CrossRef]
20. Ashraf, S.; Mehmood, T.; Abdullah, S.; Khan, Q. Picture Fuzzy Linguistic Sets and Their Applications for Multi-Attribute Group. *Nucleus* **2018**, *55*, 66–73.
21. Zeng, S.; Asharf, S.; Arif, M.; Abdullah, S. Application of Exponential Jensen Picture Fuzzy Divergence Measure in Multi-Criteria Group Decision Making. *Mathematics* **2019**, *7*, 191. [CrossRef]
22. Ashraf, S.; Abdullah, S. Spherical aggregation operators and their application in multi attribute group decision making. *Int. J. Intell. Syst.* **2019**, *34*, 493–523. [CrossRef]
23. Ashraf, S.; Abdullah, S.; Mahmood, T.; Ghani, F.; Mahmood, T. Spherical fuzzy sets and their applications in multi-attribute decision making problems. *J. Intell. Fuzzy Syst.* **2019**, 1–16. [CrossRef]
24. Wang, R.; Wang, J.; Gao, H.; Wei, G. Methods for MADM with Picture Fuzzy Muirhead Mean Operators and Their Application for Evaluating the Financial Investment Risk. *Symmetry* **2019**, *11*, 6. [CrossRef]
25. Xu, Z. Intuitionistic fuzzy aggregation operators. *IEEE Trans. Fuzzy Syst.* **2007**, *15*, 1179–1187.
26. Zhang, H.; Zhang, R.; Huang, H.; Wang, J. Some Picture Fuzzy Dombi Heronian Mean Operators with Their Application to Multi-Attribute Decision-Making. *Symmetry* **2018**, *10*, 593. [CrossRef]
27. Ashraf, S.; Abdullah, S.; Mahmood, T. GRA method based on spherical linguistic fuzzy Choquet integral environment and its application in multi attribute decision making problems. *Math. Sci.* **2018**, *12*, 263–275. [CrossRef]
28. Cuong, B.C. *Picture Fuzzy Sets—First Results. Part 1, Seminar Neuro-Fuzzy Systems with Applications*; Technical Report; Institute of Mathematics: Hanoi, Vietnam, 2013.
29. Deschrijver, G.; Cornelis, C.; Kerre, E.E. On the representation of intuitionistic fuzzy t-norms and t-conorms. *IEEE Trans. Fuzzy Syst.* **2004**, *12*, 45–61. [CrossRef]
30. Li, L.; Zhang, R.; Wang, J.; Shang, X.; Bai, K. A Novel Approach to Multi-Attribute Group Decision-Making with q-Rung Picture Linguistic Information. *Symmetry* **2018**, *10*, 172. [CrossRef]
31. Xu, Z.; Yager, R.R. Some geometric aggregation operators based on intuitionistic fuzzy sets. *Int. J. Gen. Syst.* **2006**, *35*, 417–433. [CrossRef]
32. Smarandache, F. *Neutrosophy. Neutrosophic Probability, Set, and Logic*; ProQuest Information & Learning: Ann Arbor, MI, USA, 1998.
33. Wang, H.; Smarandache, F.; Zhang, Y.Q.; Sunderraman, R. Single valued neutrosophic sets. *Multispace Multistruct.* **2010**, *4*, 410–413.

34. Zhang, X.; Bo, C.; Smarandache, F.; Dai, J. New inclusion relation of neutrosophic sets with applications and related lattice structure. *Int. J. Mach. Learn. Cybern.* **2018**, *9*, 1753–1763. [CrossRef]
35. Peng, J.J.; Wang, J.Q.; Wang, J.; Zhang, H.Y.; Chen, Z.H. Simplified neutrosophic sets and their applications in multi-criteria group decision-making problems. *Int. J. Syst. Sci.* **2016**, *47*, 2342–2358. [CrossRef]

symmetry

MDPI

Article

Application of Neutrosophic Logic to Evaluate Correlation between Prostate Cancer Mortality and Dietary Fat Assumption

Muhammad Aslam * and **Mohammed Albassam**

Department of Statistics, Faculty of Science, King Abdulaziz University, Jeddah 21551, Saudi Arabia; malbassam@kau.edu.sa

* Correspondence: aslam_ravian@hotmail.com or magmuhammad@kau.edu.sa; Tel.: +966-59-3329841

Received: 3 January 2019; Accepted: 1 March 2019; Published: 6 March 2019

Abstract: This paper presents an epidemiological study on the dietary fat that causes prostate cancer in an uncertainty environment. To study this relationship under the indeterminate environment, data from 30 countries are selected for the prostate cancer death rate and dietary fat level in the food. The neutrosophic correlation and regression line are fitted on the data. We note from the neutrosophic analysis that the prostate cancer death rate increases as the dietary fat level in the people increases. The neutrosophic regression coefficient also confirms this claim. From this study, we conclude that neutrosophic regression is a more effective model under uncertainty than the regression model under classical statistics. We also found a statistical correlation between dietary fat and prostate cancer risk.

Keywords: prostate cancer; neutrosophic statistics; dietary fat level; neutrosophic regression; neutrosophic correlation

---

1. Introduction

In modern society, cancer is the most widespread disease in both developed and under-developing countries. The cancer disease, its symptoms and treatment impact the victim patients. Lin et al. [1] pointed out that high anxiety and depression is noted in cancer patients. According to Siegel et al. [2], cancer is the disease that is the leading cause of death in the USA. As mentioned by Rahib et al. [3], in 2030, it is expected the sufficient deaths because of cancer. In men, worldwide, prostate cancer is the second common type of cancer. Prostate cancer increases slowly and is not harmful during the initial stage. The main symptoms of prostate cancer are troubling in urine, blood during urination and pain in the bones. Cao et al. [4] mentioned, “Prostate cancer is the third highest cause of male mortality in the developed world”. Torre et al. [5] pointed out that more than 37,000 prostate cancer cases are recorded every year in the UK. According to Jemal et al. [6], prostate cancer increases with the increase in age. Lin et al. [1] described that prostate cancer is a common urologic malignancy and results in a high death rate. In Taiwan, it is the fifth most common cancer. Arnold et al. [7] studied the effect of smoking and weight on cancer. [8] studied the factors that contribute significantly to the cancer. Cao et al. [4] presented seven new biomarkers for the diagnosis of cancer. Applegate et al. [9] studied the relationship between soy food and prostate cancer. More information on cancer can be seen in Carter et al. [10].

Regression analysis, principle components analysis and partial least square analysis have been widely used for analyzing the data in a variety of fields. The regression analysis study is applied to see the effect of the independent variable (s) on the dependent variable (s). The regression analysis is a powerful statistical method to use for examining the relationship between one or more explanatory variables and dependent variables. This method is widely used for prediction and forecasting purposes.

The applications of these methods can be seen in Abdul-Wahab et al. [11], Cervigón et al. [12], Kumar and Chong [13] and Karamacoska et al. [14].

Fuzzy logic is applied when there is uncertainty in the observations or in the parameters. Fuzzy logic has been widely applied in epidemical studies. Saritas et al. [15] introduced the fuzzy system to analyze prostate cancer. Benecchi [16] used the fuzzy logic and artificial neural network to develop a system to predict prostate cancer. Saritas et al. [17] described the fuzzy approach to determine prostate cancer. Yuksel et al. [18] introduced the soft expert system to diagnose this cancer. Fu et al. [19] studied the risk evaluation of prostate cancer. More information of this topic can be read in Cosma et al. [20], Al-Dmour et al [21] and Ludwig et al. [22]. More details about the use of fractal geometry and wavelet analysis in the cancer study can be seen in Rodrigo et al. [23], Rodrigo Capobianco Guido [24], Guariglia [25], Guariglia [26] and Guariglia [27].

According to Smarandache [28], neutrosophic logic is the generalized form of the fuzzy logic. The neutrosophic logic also considered an indeterminacy interval for the analysis. Smarandache [29] introduced neutrosophic statistics, which is the extension of classical statistics and is applied for the analysis under the uncertainty environment. Neutrosophic regression is the extension of classical regression analysis. Neutrosophic regression is applied when some observations in the sample are uncertain. Details of neutrosophic regression can be seen in Smarandache [29]. The applications of the neutrosophic statistics can be seen in Chen et al. [30], Chen et al. [30], Aslam [31] and Aslam [32].

Many studies on prostate cancer in social and biological sciences are available using classical statistics. The existing studies are not helpful when the data is recorded in an indeterminate environment. In this paper, we will study the relationship between prostate cancer and dietary fat level using the neutrosophic interval method. By exploring the literature on prostate cancer and according to the best of our knowledge, there is no study on the relationship between the death rate due to prostate cancer and dietary fat level using neutrosophic statistics. In this paper, we aim to present the analysis for these variables using the data of 30 countries under neutrosophic statistics. We expect that the proposed method will be more adequate and effective in analyzing the relation between the prostate cancer death rate and the dietary fat level than the analysis under classical statistics.

2. Material Methods

In this section, we will give details about the source of data, material and methods under the neutrosophic statistical interval method.

2.1. Data Description

The epidemiological studies showed that the dietary fat level causes an increase in the prostate cancer rate across countries, see [33]. The dietary fat can measure through the fat by each ingredient in our daily food or by calculating the per percentage of calories using glucose machines through blood in our daily food. In the earlier case, it was not possible to keep the record of exact or determined dietary fat level, and in the latter case it is was also not determined. Measuring the diet through these methods is not a determined value and recorded in the interval rage. When the variables are expressed in an interval, the relationship between dietary level and prostate cancer cannot be studied using the classical regression model. Therefore, under an uncertain environment, when the dietary fat level and prostate cancer rate are in the interval range, the neutrosophic statistics can be applied to study the relationship between the dietary fat level and prostate cancer. By following [33], the data on dietary fat level and prostate cancer from the 30 countries under the uncertainty level is reported in Table 1. The variable prostrate death rate is expressed per 100,000 and dietary fat consumption is expressed in gram/day.

Table 1. Prostate cancer death rate of 30 countries.

D-Rate	Diet Fat	County No.	D-Rate	Diet Fat	County No.
[10.1,10.3]	[97,97]	16	[0.9,1.1]	[38,38]	1
[11.4,11.4]	[73,75]	17	[1.3,1.3]	[29,31]	2
[11.1,11.1]	[112,112]	18	[1.6,1.6]	[42,42]	3
[13.1,13.3]	[100,100]	19	[4.5,4.5]	[57,57]	4
[12.9,13.1]	[134,134]	20	[4.8,4.10]	[96,98]	5
[13.4,13.4]	[142,142]	21	[5.4,5.6]	[47,49]	6
[13.9,14.2]	[119,119]	22	[5.5,5.5]	[67,67]	7
[14.4,14.4]	[137,137]	23	[5.6,5.6]	[72,74]	8
[14.4,14.6]	[152,152]	24	[6.4,6.6]	[93,93]	9
[15.1,15.3]	[129,129]	25	[7.8,7.8]	[58,58]	10
[15.9,15.9]	[156,156]	26	[8.4,8.6]	[95,95]	11
[16.3,16.4]	[147,147]	27	[8.8,8.8]	[67,69]	12
[16.8,16.9]	[133,133]	28	[9,9]	[62,62]	13
[18.4,18.4]	[132,132]	29	[9.1,9.1]	[96,96]	14
[12.4,12.6]	[143,144]	30	[9.4,9.4]	[86,87]	15

2.2. Study Participants

[33] presented data on the prostate cancer death rate and consumption of fat in the 30 countries. The data presented in Table 1 is the extension of the data provided by [33] under the neutrosophic statistics. The data was collected from the 30 countries. The data given in Table 1 becomes the same as in [33] if no indeterminate observations are found in the dietary fat level and prostate cancer.

2.3. Study Outcomes

In this study, the neutrosophic dietary fat level is an explanatory variable and the neutrosophic death rate due to prostate cancer in the 30 countries is a response variable. The aim of this study is to see the effect on the prostate cancer death rate due to an increase in the dietary fat level under the neutrosophic statistical interval method. Therefore, the objective of the study is to measure the prostate cancer death rate. We defined the neutrosophic death due to prostate cancer may cause of increase in fat. The data given in Table 1 is the extension of the data provided by [33] in the neutrosophic interval rage.

2.4. Statistical Methods

We will study the effect of the death rate using the fat level as an explanatory variable. The relationship between the two variables is studied using the neutrosophic regression proposed by [29]. The neutrosophic regression is the extension of the classical regression used to see the relationship between variables when the data is taken from an incomplete, indeterminate or uncertain sample of the population. Let the neutrosophic variable $X_N \epsilon \{X_L, X_U\}$ represent the dietary fat level and neutrosophic variable $Y_{NN} \epsilon \{Y_L, Y_U\}$ denote the death rate due to prostate cancer. By following [29], the neutrosophic regression which expressed the relation between these two variables is given as

$$Y_N = a_N + b_N X_N;\ a_N \epsilon \{a_L, a_U\},\ b_N \epsilon \{b_L, b_U\} \quad (1)$$

where a_N and b_N are the neutrosophic intercept and rate of change per unit, respectively. The neutrosophic regression given in Equation (1) can be written as follows

$$D.rate = a_N + b_N Diet\ Fat;\ a_N \epsilon \{a_L, a_U\}\ b_N \epsilon \{b_L, b_U\} \quad (2)$$

The neutrosophic correlation coefficient $r_N \epsilon \{r_L, r_U\}$ between the two variables is defined as

$$r_N = \frac{n_N \sum xy - \sum x \sum y}{\sqrt{\left\{n_N \sum x^2 - \left(\sum x\right)^2\right\}\left\{n_N \sum y^2 - \left(\sum y\right)^2\right\}}}; \; n_N \epsilon \{n_L, n_U\} \tag{3}$$

We presented the scatter plot under the neutrosophic statistical interval methods. We presented the neutrosophic regression models to measure the prostate cancer death rate. We presented the fitted model and determined the residual sum of squares for using the neutrosophic regression model. We used a 5% level of significance to test the null hypothesis that the neutrosophic regression coefficient is zero. We also made a comparison between the neutrosophic regressions with a regression under the classical statistics. The data is analyzed using EXCEL and R.

3. Results

We plotted the death rate (D. rate) and dietary fat (Diet Fat) on the scatter diagram in Figure 1. We note from Figure 1 that there is an increasing trend in the D. rate as the fat level increases. Therefore, Figure 1 indicates that a relationship between the death rate due to prostate cancer and the dietary fat level exists. The neutrosophic correlation between two variables is $r_N \epsilon \{0.8811, 0.8851\}$. The fitted neutrosophic regression model of two variables is given as

$$Y_N = [28.36, 29.95] + [6.757, 6.911,] X_N \tag{4}$$

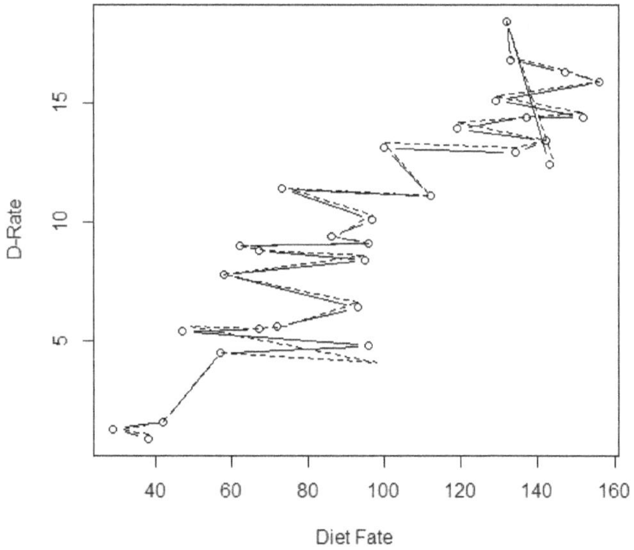

Figure 1. The scatter plot of two variables.

From Equation (3), we note that for a unit of change in dietary fat, the death rate due to prostate cancer increases between 6.75 to 6.911. We also observe from the model given in Equation (3) that when dietary fat is zero, the death rate due to prostate cancer can be expected to be between 28.38 to 29.95 per 100,000. The 95% neutrosophic confidence interval for the neutrosophic slope (rate of change) $b_N \epsilon \{b_L, b_U\}$ is $([5.35, 5.50], [8.16, 8.31])$. This confidence interval shows that one can expect that the minimum death rate per 10,000 will be between 5.35 to 5.50 and the maximum death rate per 10,000 will be between 8.16 to 8.31. The neutrosophic p-values [0.0001,0.0001]$\leq$ 0.05 show the significance of death due to prostate cancer.

4. Discussion

The relationship between two variables using the regression model under classical statistics is given as: $Y = 28.38 + 6.911X$. By comparing the neutrosophic regression with the classical regression, we note that the neutrosophic regression provides the parameters, confidence interval and p-values in the indeterminacy interval range. The regression model using the classical statistics provides the determined values of all parameters involved in the regression line. Therefore, under uncertainty, the proposed neutrosophic regression analysis will be more helpful to see the significance of death due to the increase in fate rate. Further, the forecasting of a death rate due to cancer can be expected in the indeterminacy interval rage rather than the exact or the determined values. The neutrosophic regression model is more flexible and adequate under the uncertainty environment than the classical regression model.

Study Limitations

The proposed neutrosophic regression model is the extension of the classical regression. The neutrosophic regression becomes the regression model in the classical statistics if no uncertain observations are in the sample or in the population. The proposed neutrosophic regression can be applied adequately in the uncertainty environment.

5. Conclusions

The present study showed that there is a strong relationship between the prostate cancer rate and dietary fat. We presented a neutrosophic regression model that can be used more effectively in the uncertainty environment than the classical regression model. The neutrosophic regression provides the parameters estimation in the indeterminacy interval rage. The neutrosophic regression shows a significant effect on the death rate due to the inverse in the fat. We also found a statistical correlation between dietary fat and prostate cancer risk. We conclude from this study that the use of the neutrosophic regression model for the estimation and forecasting of prostate cancer death will be helpful to make an adequate decision to control the death rate. We recommend that a health manager should adopt the neutrosophic regression model for better and more effective analysis under an uncertainty environment. The proposed study using multiple regression analysis, principal component analysis or partial least squares analysis can be extended for future research.

Author Contributions: Conceived and designed the experiments, M.A.; M.A.B.; Performed the experiments, M.A.; Analyzed the data, M.A.; Contributed reagents/materials/analysis tools, M.A.; Wrote the paper, M.A., M.A.B.

Funding: This article was funded by the Deanship of Scientific Research (DSR) at King Abdulaziz University, Jeddah. The authors, therefore, acknowledge with thanks DSR technical and financial support.

Acknowledgments: The authors are deeply thankful to the editor and reviewers for their valuable suggestions to improve the quality of this manuscript.

Conflicts of Interest: The authors declare no conflict of interest regarding this paper.

References

1. Lin, P.-H.; Liu, J.-M.; Hsu, R.-J.; Chuang, H.-C.; Chang, S.-W.; Pang, S.-T.; Chang, Y.-H.; Chuang, C.-K.; Lin, S.-K. Depression negatively impacts survival of patients with metastatic prostate cancer. *Int. J. Environ. Res. Public Health* **2018**, *15*, 2148. [CrossRef] [PubMed]
2. Siegel, R.L.; Miller, K.D.; Jemal, A. Cancer statistics, 2017. *CA Cancer J. Clin.* **2017**, *67*, 7–30. [CrossRef] [PubMed]

3. Rahib, L.; Smith, B.D.; Aizenberg, R.; Rosenzweig, A.B.; Fleshman, J.M.; Matrisian, L.M. Projecting cancer incidence and deaths to 2030: The unexpected burden of thyroid, liver, and pancreas cancers in the united states. *Cancer Res.* **2014**, *74*, 2913–2921. [CrossRef] [PubMed]
4. Cao, K.; Arthurs, C.; Atta-ul, A.; Millar, M.; Beltran, M.; Neuhaus, J.; Horn, L.-C.; Henrique, R.; Ahmed, A.; Thrasivoulou, C. Quantitative analysis of seven new prostate cancer biomarkers and the potential future of the 'biomarker laboratory'. *Diagnostics* **2018**, *8*, 49. [CrossRef] [PubMed]
5. Torre, L.A.; Bray, F.; Siegel, R.L.; Ferlay, J.; Lortet-Tieulent, J.; Jemal, A. Global cancer statistics, 2012. *CA Cancer J. Clin.* **2015**, *65*, 87–108. [CrossRef] [PubMed]
6. Jemal, A.; Siegel, R.; Ward, E.; Murray, T.; Xu, J.; Smigal, C.; Thun, M.J. Cancer statistics, 2006. *CA Cancer J. Clin.* **2006**, *56*, 106–130. [CrossRef] [PubMed]
7. Arnold, L.D.; Patel, A.V.; Yan, Y.; Jacobs, E.J.; Thun, M.J.; Calle, E.E.; Colditz, G.A. Are racial disparities in pancreatic cancer explained by smoking and overweight/obesity? *Cancer Epidemiol. Prev. Biomarkers* **2009**, *18*, 2397–2405. [CrossRef] [PubMed]
8. Scarton, L.; Yoon, S.; Oh, S.; Agyare, E.; Trevino, J.; Han, B.; Lee, E.; Setiawan, V.; Permuth, J.; Schmittgen, T. Pancreatic cancer related health disparities: A commentary. *Cancers* **2018**, *10*, 235. [CrossRef] [PubMed]
9. Applegate, C.; Rowles, J.; Ranard, K.; Jeon, S.; Erdman, J. Soy consumption and the risk of prostate cancer: An updated systematic review and meta-analysis. *Nutrients* **2018**, *10*, 40. [CrossRef] [PubMed]
10. Carter, H.B.; Albertsen, P.C.; Barry, M.J.; Etzioni, R.; Freedland, S.J.; Greene, K.L.; Holmberg, L.; Kantoff, P.; Konety, B.R.; Murad, M.H. Early detection of prostate cancer: Aua guideline. *J. Urol.* **2013**, *190*, 419–426. [CrossRef] [PubMed]
11. Abdul-Wahab, S.A.; Bakheit, C.S.; Al-Alawi, S.M. Principal component and multiple regression analysis in modelling of ground-level ozone and factors affecting its concentrations. *Environ. Model. Softw.* **2005**, *20*, 1263–1271. [CrossRef]
12. Cervigón, R.; Moreno, J.; Reilly, R.; Pérez-Villacastín, J.; Castells, F. Quantification of anaesthetic effects on atrial fibrillation rate by partial least-squares. *Physiol. Meas.* **2012**, *33*, 1757. [CrossRef] [PubMed]
13. Kumar, S.; Chong, I. Correlation analysis to identify the effective data in machine learning: Prediction of depressive disorder and emotion states. *Int. J. Environ. Res. Public Health* **2018**, *15*, 1907. [CrossRef] [PubMed]
14. Karamacoska, D.; Barry, R.J.; Steiner, G.Z. Using principal components analysis to examine resting state eeg in relation to task performance. *Psychophysiology* **2019**, e13327. [CrossRef] [PubMed]
15. Saritas, I.; Allahverdi, N.; Sert, I.U. A fuzzy expert system design for diagnosis of prostate cancer. *a a* **2003**, *1*, 50.
16. Benecchi, L. Neuro-fuzzy system for prostate cancer diagnosis. *Urology* **2006**, *68*, 357–361. [CrossRef] [PubMed]
17. Saritas, I.; Allahverdi, N.; Sert, I.U. A fuzzy approach for determination of prostate cancer. *Int. J. Intelligent Syst. Appl. Eng.* **2013**, *1*, 1–7.
18. Yuksel, S.; Dizman, T.; Yildizdan, G.; Sert, U. Application of soft sets to diagnose the prostate cancer risk. *J. Inequal. Appl.* **2013**, *2013*, 229. [CrossRef]
19. Fu, J.; Ye, J.; Cui, W. An evaluation method of risk grades for prostate cancer using similarity measure of cubic hesitant fuzzy sets. *J. Biomed. Inform.* **2018**, *87*, 131–137. [CrossRef] [PubMed]
20. Cosma, G.; McArdle, S.E.; Reeder, S.; Foulds, G.A.; Hood, S.; Khan, M.; Pockley, A.G. Identifying the presence of prostate cancer in individuals with psa levels< 20 ng ml− 1 using computational data extraction analysis of high dimensional peripheral blood flow cytometric phenotyping data. *Front. Immunol.* **2017**, *8*, 1771. [PubMed]
21. Al-Dmour, J.A.; Sagahyroon, A.; Al-Ali, A.; Abusnana, S. A fuzzy logic–based warning system for patients classification. *Health Inform. J.* **2017**. [CrossRef] [PubMed]
22. Ludwig, S.A.; Picek, S.; Jakobovic, D. Classification of cancer data: Analyzing gene expression data using a fuzzy decision tree algorithm. In *Operations Research Applications in Health Care Management*; Springer: New York, NY, USA, 2018; pp. 327–347.
23. Guido, R.C.; Addison, P.S.; Walker, J. Introducing wavelets and time–frequency analysis [introduction to the special issue]. *IEEE Eng. Med. Biol. Mag.* **2009**, *28*, 13. [CrossRef] [PubMed]
24. Guido, R.C. Practical and useful tips on discrete wavelet transforms [sp tips & tricks]. *IEEE Signal Process. Mag.* **2015**, *32*, 162–166.
25. Guariglia, E. Entropy and fractal antennas. *Entropy* **2016**, *18*, 84. [CrossRef]

26. Guariglia, E. Spectral Analysis of the Weierstrass-Mandelbrot Function. In Proceedings of the 2017 2nd International Multidisciplinary Conference on Computer and Energy Science (SpliTech), Split, Croatia, 12–14 July 2017; pp. 1–6.
27. Guariglia, E. Harmonic sierpinski gasket and applications. *Entropy* **2018**, *20*, 714. [CrossRef]
28. Smarandache, F. Neutrosophic logic-a generalization of the intuitionistic fuzzy logic. *Multispace Multistruct. Neutrosophic Transdiscip.* **2010**, *4*, 396.
29. Smarandache, F. *Introduction to Neutrosophic Statistics*. Infinite Study. 2014. Available online: https://arxiv.org/pdf/1406.2000 (accessed on 2 January 2019).
30. Chen, J.; Ye, J.; Du, S.; Yong, R. Expressions of rock joint roughness coefficient using neutrosophic interval statistical numbers. *Symmetry* **2017**, *9*, 123. [CrossRef]
31. Aslam, M. A new sampling plan using neutrosophic process loss consideration. *Symmetry* **2018**, *10*, 132. [CrossRef]
32. Aslam, M. Design of sampling plan for exponential distribution under neutrosophic statistical interval method. *IEEE Access* **2018**, *6*, 64153–64158. [CrossRef]
33. Triola, M.F. *Elementary Statistics*; Pearson/Addison-Wesley: Reading, MA, USA, 2006.

symmetry

MDPI

Article

Generalized Neutrosophic Extended Triplet Group

Yingcang Ma [1,*], Xiaohong Zhang [2], Xiaofei Yang [1] and Xin Zhou [1]

[1] School of Science, Xi'an Polytechnic University, Xi'an 710048, China; yangxiaofei2002@163.com (X.Y.); sxxzx1986@163.com (X.Z.)
[2] School of Arts and Sciences, Shaanxi University of Science & Technology, Xi'an 710021, China; zhangxiaohong@sust.edu.cn
* Correspondence: mayingcang@126.com

Received: 2 February 2019; Accepted: 25 February 2019; Published: 5 March 2019

Abstract: Neutrosophic extended triplet group is a new algebra structure and is different from the classical group. In this paper, the notion of generalized neutrosophic extended triplet group is proposed and some properties are discussed. In particular, the following conclusions are strictly proved: (1) an algebraic system is a generalized neutrosophic extended triplet group if and only if it is a quasi-completely regular semigroup; (2) an algebraic system is a weak commutative generalized neutrosophic extended triplet group if and only if it is a quasi-Clifford semigroup; (3) for each $n \in Z^+, n \geq 2, (Z_n, \otimes)$ is a commutative generalized neutrosophic extended triplet group; (4) for each $n \in Z^+, n \geq 2, (Z_n, \otimes)$ is a commutative neutrosophic extended triplet group if and only if $n = p_1 p_2 \cdots p_m$, i.e., the factorization of n has only single factor.

Keywords: neutrosophic extended triplet group; generalized neutrosophic extended triplet group; quasi-completely regular semigroup; clifford semigroup

1. Introduction

Groups are very important algebraic structures and have been applied in many fields, such as cryptology, engineering, physics, chemistry, etc. Recently, a new algebraic system, neutrosophic triplet group (NTG), is proposed by Smarandache and Ali in [1]. For an NTG $(N, *)$, every element $a \in N$ has its own neutral element (denote by $neut(a)$) satisfying condition $a * neut(a) = neut(a) * a = a$, and there exits at least one opposite element (denote by $anti(a)$) in N relative to $neut(a)$ such that $a * anti(a) = anti(a) * a = neut(a)$. Some further studies can be found in [2–11].

From the original definition of NTG, we can see that it is an extension of a group. However, it is different from a group. In NTG, the neutral element is different from the unit element of the classical algebraic system. By removing this restriction, an algebraic system, which is called neutrosophic extended triplet group (NETG), is proposed, and the classical group is regarded as a special case of NETG. Moreover, further studies of NETG can be found in [12–18] and some important results have been achieved.

For the algebraic system $(Z_n, \otimes)$, $\otimes$ is the classical mod multiplication, where $Z_n = \{[0], [1], \cdots, [n-1]\}$ and $n \geq 2$ is a positive integer. $(Z_n, \otimes)$ is only a commutative semigroup before NTG and NETG are introduced. Recently, properties of algebraic structure of $(Z_n, \otimes)$ in some special cases are studied in [19,20] by the point view of NTG. We can see that for some positive integers n, $(Z_n, \otimes)$ is a commutative NETG, but for some positive integers n, $(Z_n, \otimes)$ is not a commutative NETG. So, there are two problems, 1 What conditions can guarantee $(Z_n, \otimes)$ to be a NETG? 2 Is there a new algebraic system that makes $(Z_n, \otimes)$ is the algebraic system for each positive integer n? We will give positive answers for the two problems in this paper.

The paper is organized as follows. Section 2 gives the related work. In Section 3, the conditions which guarantee the algebraic system $(Z_n, \otimes)$ to be a NETG are deeply studied. In Section 4, the generalized neutrosophic extended triplet group (GNETG) is proposed and some properties and structure are discussed. Finally, the summary and future work are presented in Section 5.

2. Related Works

In this section, we will give the related research and results of the NETG. Some related notions are introduced at first. Let G be non-empty set, $*$ is a binary operation on G. If $\forall a, b \in G$, implies $a * b \in G$, then $(G, *)$ is called a groupoid. A groupoid G is called a semigroup if $*$ satisfies associative law, i.e., $(a * b) * c = a * (b * c)$ $(\forall a, b, c \in G)$. A semigroup G is called a group if there exists the unit element and for each element in G exists its inverse element.

Definition 1. *[11] Let N be a non-empty set together with a binary operation $*$. Then, N is called a neutrosophic extended triplet set if for any $a \in N$, there exists a neutral of "a" (denote by $neut(a)$), and an opposite of "a"(denote by $anti(a)$), such that $neut(a) \in N$, $anti(a) \in N$ and:*

$$a * neut(a) = neut(a) * a = a, \quad a * anti(a) = anti(a) * a = neut(a).$$

The triplet $(a, neut(a), anti(a))$ is called a neutrosophic extended triplet.

Definition 2. *[11,15] Let $(N, *)$ be a neutrosophic extended triplet set. Then, N is called a neutrosophic extended triplet group (NETG), if the following conditions are satisfied:*
(1) $(N, *)$ is well-defined, i.e., for any $a, b \in N$, one has $a * b \in N$.
(2) $(N, *)$ is associative, i.e., $(a * b) * c = a * (b * c)$ for all $a, b, c \in N$.
A NETG N is called a commutative NETG if for all $a, b \in N, a * b = b * a$.

From Definition 2, we can see that a NETG is an extension of the classical group. There are some important results about the properties and structure of a NETG. For examples, G is a NETG iff it is a completely regular semigroup, G is a singular NETG iff it is a generalized group, and so on. We introduce some related notions and results in the following. Figure 1 gives the relations of the NETG and other algebraic structures.

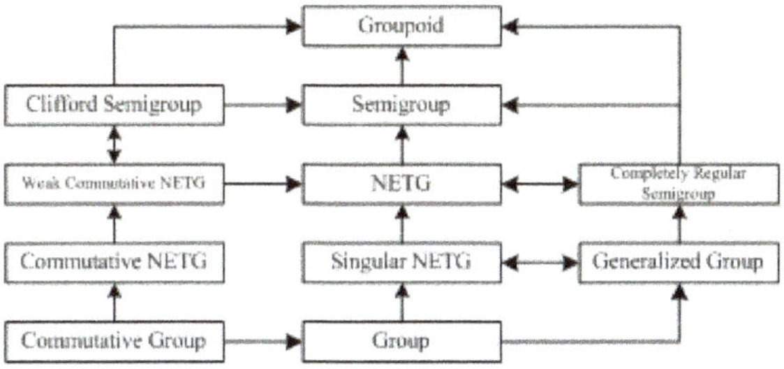

Figure 1. The relations of NETG and other algebraic structures.

Proposition 1. *[15] $(N, *)$ be a NETG. We have:*
(1) $neut(a)$ is unique for any $a \in N$.
(2) $neut(a) * neut(a) = neut(a)$ for any $a \in N$.
(3) $neut(neut(a)) = neut(a)$ for any $a \in N$.

Definition 3. *[17] A NETG $(N, *)$ is said to be singular, if $anti(a)$ is unique for any $a \in N$.*

Definition 4. *[21] A generalized group $(G, *)$ is a non-empty set admitting a binary operation $*$ called multiplication subject to the set of rules given below:*
(1) $(a * b) * c = a * (b * c), \forall\, a, b, c \in G$.

(2) For each $a \in G$, there exists a unique $e(a) \in G$ such that $a * e(a) = e(a) * a = a$.
(3) For each $a \in G$, there exists $a^{-1} \in G$ such that $a * a^{-1} = a^{-1} * a = e(a)$.

Theorem 1. *[17] $(N, *)$ is a singular NETG iff $(N, *)$ is a generalized group.*

Definition 5. *[15] Let $(N, *)$ be a NETG. Then N is called a weak commutative NETG if $a * neut(b) = neut(b) * a$ for all $a, b \in N$.*

Definition 6. *[22] A semigroup $(S, *)$ is called regular if for any $a \in S$, there exists a unary operation $a \to a^{-1}$ on S such that*

$$(a^{-1})^{-1} = a, \quad a * a^{-1} * a = a.$$

A semigroup S is called quasi-regular if for any $a \in S$, there exists a positive integer n such that a^n is regular.

Definition 7. *[22] A semigroup $(S, *)$ is called completely regular if for any $a \in S$, there exists a unary operation $a \to a^{-1}$ on S such that*

$$(a^{-1})^{-1} = a, \quad a * a^{-1} * a = a, \quad a * a^{-1} = a^{-1} * a.$$

A semigroup S is called quasi-completely regular if for any $a \in S$, there exists a positive integer n such that a^n is completely regular.

Proposition 2. *[22] Let $(S, *)$ be a semigroup. Then the following statements are equivalent:*
(1) S is completely regular;
(2) every element of S lies in a subgroup of S;
(3) every H-class in S is a group.

Definition 8. *[22] A semigroup $(S, *)$ is called Clifford semigroup, if it is completely regular and in which for any $a, b \in S$, such that*

$$(a * a^{-1}) * (b * b^{-1}) = (b * b^{-1}) * (a * a^{-1}).$$

Proposition 3. *[22] Let $(S, *)$ be a semigroup. Then the following statements are equivalent:*
(1) S is Clifford semigroup;
(2) S is a semilattice of groups;
(3) S is regular, and the idempotents of S are central.

Theorem 2. *[18] Let $(N, *)$ be a groupoid. Then N is a NETG iff it is a completely regular semigroup.*

Theorem 3. *[18] Let $(N, *)$ be a groupoid. Then N is a weak commutative NETG iff it is a Clifford semigroup.*

3. The Relations of $(Z_n, \otimes)$ and NETG

Lemma 1. *In an algebra system $(Z_n, \otimes)$, where Z is the integer set, we have for $a, b, c \in Z$, $ab \equiv c (\text{mod } n)$ holds iff $[a] \otimes [b] = [c]$ holds.*

Lemma 2. *For each positive integer $n \geq 2$, linear congruence equation $ax \equiv c (\text{mod } n)$ has a solution iff $\gcd(a, n) \mid c$, where $\gcd(a, b)$ is the greatest common divisor of a and b.*

Proposition 4. *The solution of x for equation $[a] \otimes [x] = [a]$ is $x = mn/\gcd(a, n) + 1$, where $m \in Z$.*

Proof. From Lemma 1, $[a] \otimes [x] = [a]$ has a solution iff $ax \equiv a (\text{mod } n)$ has a solution, and by Lemma 2, $ax \equiv a (\text{mod } n)$ must have a solution being $\gcd(a, n) \mid a$ holds obviously. We will prove the solution to the equation $[a] \otimes [x] = [a]$ is $mn/\gcd(a, n)$ in the following.

Let $\gcd(a,n) = d$, then there exist two integers a_1 and n_1, such that $a = a_1 * d, n = n_1 * d$ and $\gcd(a_1,n_1) = 1$. If x satisfies the equation $ax \equiv a(\text{mod } n)$, we can get that $n|(ax-a)$, thus $n_1|(a_1x-a_1)$, i.e., $a_1x - a_1 = m_1n_1, m_1 \in Z$, that is $a_1(x-1) = m_1n_1$. Since $\gcd(a_1,n_1) = 1$, we have $a_1|m_1$. Let $m_1 = ma_1, m \in Z$, then, $a_1(x-1) = ma_1n_1$, that is $x = mn_1 + 1, m \in Z$.

On the other hand, being $a(mn_1+1) - a = amn_1 = a_1mn$, that is $n|(a(mn_1+1)-a)$, i.e., $a(mn_1+1) \equiv a(\text{mod } n)$ holds, so $x = mn_1 + 1$ is the solution of equation $ax \equiv a(\text{mod } n)$. This tells us that the solution for equation $[a] \otimes [x] = [a]$ is $x = mn/\gcd(a,n) + 1$, where $m \in Z$. □

Remark: *From Proposition 4, we can see that the solution of* $[a] \otimes [x] = [a]$ *is just decided by the* $\gcd(a,n)$. *If* $\gcd(a,n) = \gcd(b,n)$, *then* $[a] \otimes [x] = [a]$ *and* $[b] \otimes [x] = [b]$ *have the same solution in* $(Z_n, \otimes)$.

Theorem 4. *In the algebra system* $(Z_n, \otimes)$, *for each* $[a] \in Z_n$, $[a]$ *has* $neut([a])$ *and* $anti([a])$ *iff* $\gcd(\gcd(a,n), n/\gcd(a,n)) = 1$.

Proof. Necessity: In general, let $[a] \in Z_n$ and the neutral elements of $[a]$ exists, that is there are $neut([a])$ and $anti([a])$ such that $[a] \otimes neut([a]) = [a]$ and $[a] \otimes anti([a]) = neut([a])$. That is the linear congruence equations $ax \equiv a(\text{mod } n)$ and $ay \equiv x(\text{mod } n)$ have solutions. Let $\gcd(a,n) = d$ and $a = a_1 * d, n = n_1 * d$. From Proposition 4, $x = mn_1 + 1, m \in Z$. From Lemma 2, y has solution for equation $ay \equiv x(\text{mod } n)$ iff $\gcd(a,n)|x$, that is $\gcd(a,n)|(mn_1+1)$, which means $\gcd(\gcd(a,n), n/\gcd(a,n)) = 1$.

Sufficiency: If $\gcd(\gcd(a,n), n/\gcd(a,n)) = 1$, we should verify that there is x and y, such that $ax \equiv a(\text{mod } n)$ and $ay \equiv x(\text{mod } n)$. From Proposition 4, $ax \equiv a(\text{mod } n)$ has a solution and $x = mn_1 + 1, m \in Z$. From the given condition $\gcd(d,n_1) = 1$, we can infer that there exist $p, q \in Z$, such that $pd + qn_1 = 1$, which means $d|((-q)n_1 + 1)$, that is $d|x$ when $m = -q$. So there exists y, such that $ay \equiv x(\text{mod } n)$, which means if $\gcd(\gcd(a,n), n/\gcd(a,n)) = 1$, thus for $[a] \in Z_n$, $[a]$ has $neut([a])$ and $anti([a])$. □

Example 1. *For* $(Z_8, \otimes)$, $n = 8 = 2^3$, *the operation table as following Table 1. From Definition 1, we have* $netu([0]) = [0], \{anti([0])\} = \{[0],[1],[2],[3],[4],[5],[6],[7]\}, netu([1]) = [1], anti([1]) = [1], netu([3]) = [1], anti([3]) = [3], netu([5]) = [1], anti([5]) = [5], netu([7]) = [1], anti([7]) = [7]$, *but* $[2], [4]$ *and* $[6]$ *have not the neutral element and opposite element.*

We can get the above results by Theorem 4. Being $\gcd(\gcd(1,8), 8/\gcd(1,8)) = \gcd(\gcd(3,8), 8/\gcd(3,8)) = \gcd(\gcd(5,8), 8/\gcd(5,8)) = \gcd(\gcd(7,8), 8/\gcd(7,8)) = 1$, *so from Theorem 4,* $[1], [3], [5]$ *and* $[7]$ *exist the neutral element and opposite element. In fact, from Proposition 4, being* $\gcd(1,8) = \gcd(3,8) = \gcd(5,8) = \gcd(7,8) = 1$, *i.e., they have the same greatest common divisor, so they have the same neutral element, that is* $[1]$. *In the same way,* $[0]$ *has its neutral element and opposite element. However,* $\gcd(\gcd(2,8), 8/\gcd(2,8)) = \gcd(\gcd(4,8), 8/\gcd(4,8)) = \gcd(\gcd(6,8), 8/\gcd(6,8)) = 2 \neq 1$, *so* $[2], [4]$ *and* $[6]$ *do not have the neutral element and opposite element.*

Table 1. The operation table of Z_8.

⊗	[0]	[1]	[2]	[3]	[4]	[5]	[6]	[7]
[0]	[0]	[0]	[0]	[0]	[0]	[0]	[0]	[0]
[1]	[0]	[1]	[2]	[3]	[4]	[5]	[6]	[7]
[2]	[0]	[2]	[4]	[6]	[0]	[2]	[4]	[6]
[3]	[0]	[3]	[6]	[1]	[4]	[7]	[2]	[5]
[4]	[0]	[4]	[0]	[4]	[0]	[4]	[0]	[4]
[5]	[0]	[5]	[2]	[7]	[4]	[1]	[6]	[3]
[6]	[0]	[6]	[4]	[2]	[0]	[6]	[4]	[2]
[7]	[0]	[7]	[6]	[5]	[4]	[3]	[2]	[1]

Proposition 5. *In the algebra system $(Z_n, \otimes)$, if the neutral element exists, then it is unique.*

Proof. In general, if $[a] \in Z_n$ and the neutral elements of $[a]$ exists, that is there are $neut([a])$ and $anti([a])$ such that $[a] \otimes neut([a]) = [a]$ and $[a] \otimes anti([a]) = neut([a])$. Let $\gcd(a, n) = d$ and there are two integers a_1 and n_1, such that $a = a_1 * d, n = n_1 * d$.

By Proposition 4, the neutral element of $[a]$ has form $n_1 m + 1$, where $m \in Z$, and from Theorem 4, there exists $m_0 \in Z$ such that $d|(n_1 m_0 + 1)$. Since $d|n$, we have $d|(n_1 m_0 + 1 \pm n)$, which means there must exist $m \in Z$ such that $0 \leq n_1 m + 1 < n$, so the neutral element of $[a]$ exists.

In the following, we will show that if there exist two integers m_1 and m_2, such that $d|(n_1 m_1 + 1)$ and $d|(n_1 m_2 + 1)$, then $n|n_1(m_1 - m_2)$.

From $d|(n_1 m_1 + 1)$ and $d|(n_1 m_2 + 1)$, we have $d|(n_1 m_1 + 1 - n_1 m_2 - 1)$, i.e., $d|(n_1(m_1 - m_2))$. Being $\gcd(d, n_1) = 1$, then $d|(m_1 - m_2)$, that is $n|n_1(m_1 - m_2)$.

From the above analysis, we can get that the neutral element of $[a]$ is unique in $(Z_n, \otimes)$ if it exists. □

Remark: *From Proposition 5, we can see that if the neutral element of $[a]$ exists, then it is unique. Let $neut([a])$ is the neutral element of $[a]$, then we have that $(neut([a], neut([a])), [1])$ and $(neut([a], neut([a])), neut([a]))$ are the neutrosophic extended triplets in $(Z_n, \otimes)$.*

Example 2. *For $(Z_{36}, \otimes)$, $n = 36 = 2^2 3^2$. From the above results we have:*

(1) $[2]$ *has not the neutral element and opposite element being* $\gcd(\gcd(2, 36), 36/\gcd(2, 36)) = 2 \neq 1$. *In fact,* $[2], [3], [6], [10], [12], [14], [15], [18], [21], [22], [24], [26], [30], [33]$ *and* $[34]$ *do not have the neutral element and opposite element by Theorem 4 for the same reason.*

(2) $[4]$ *and* $[8]$ *have the same neutral element being* $\gcd(4, 36) = \gcd(8, 36) = 4$. *In fact,* $[4], [8], [16], [20], [28]$ *and* $[32]$ *have the same neutral element, which is* $[28]$.

(3) $[9]$ *and* $[27]$ *have the same neutral element* $[9]$ *being* $\gcd(9, 36) = \gcd(27, 36) = 9$.

(4) $[1], [5], [7], [11], [13], [17], [19], [23], [25], [29], [31]$ *and* $[35]$ *have the same neutral element* $[1]$.

(5) $[0]$ *has the neutral element* $[0]$.

(6) We can see that the neutral element of each element is unique and explain that Proposition 5 is correct. Moreover, $([0], [0], [1]), ([0], [0], [0]), ([1], [1], [1]), ([9], [9], [1]), ([9], [9], [9]), ([28], [28], [1])$ *and* $([28], [28], [28])$ *are some neutrosophic extended triplets in* $(Z_{36}, \otimes)$, *which verify the above Remark.*

Theorem 5. *An algebra system $(Z_n, \otimes)$ is a NETG iff the factorization of n is a product of single factors, i.e., $n = p_1 p_2 \cdots p_k$, where $p_i (i = 1, 2, \cdots, k)$ is a prime number.*

Proof. Necessity: Suppose that $(Z_n, \otimes)$ is a NETG, we verify the factorization of n is the product of a single factor and proof by contradiction.

Assume, the factorization of n is $n = p_1^{k_1} p_2^{k_2} \cdots p_t^{k_t}$, and $\exists i$, such that $k_i \geq 2$. Without losing generality, let $k_1 \geq 2$. We will prove $neut([p_1])$ and $anti([p_1])$ do not exist.

Because $\gcd(\gcd(p_1, n), n/\gcd(p_1, n)) = p_1^{k_1 - 1}$ and $k_1 \geq 2$, then

$$\gcd(\gcd(p_1, n), n/\gcd(p_1, n)) \neq 1.$$

By Theorem 4, we have that $neut([p_1])$ and $anti([p_1])$ do not exist. This contradicts that $(Z_n, \otimes)$ is NETG. So, the factorization of n is a product of single factors if $(Z_n, \otimes)$ is a NETG.

Sufficiency: If the factorization of n is a product of some single factors, i.e., $n = p_1 p_2 \cdots p_k$, where $p_i (i = 1, 2, \cdots, k)$ is a prime number. We will prove for each $[a] \in Z_n$, $neut([a])$ and $anti([a])$ exist.

Let $d = \gcd(a, n)$, because $n = p_1 p_2 \cdots p_k$, so $\gcd(d, n/d) = 1$, i.e.,

$$\gcd(\gcd(a, n), n/\gcd(a, n)) = 1.$$

From Theorem 4, $[a]$ has the neutral element and opposite element. From the arbitrariness of a, we can know that $(Z_n, \otimes)$ is a NETG if the factorization of n is a product of some single factors. □

Remark: *For an algebra system $(Z_n, \otimes)$ and $n = p_1 p_2 \cdots p_k$, where $p_i (i = 1, 2, \cdots, k)$ is a prime number. From the above analysis, the number of different neutral elements in Z_n is 2^k.*

Example 3. *For $(Z_{30}, \otimes)$, $n = 30 = 2^1 3^1 5^1$, so each element in Z_{30} has its corresponding neutral element. From the above Remark, the number of different neutral elements in Z_{30} is $2^3 = 8$. In fact, being different factors of 30 are $1, 2, 3, 5, 6, 10, 15$, and 30. Thus, the different neutral elements are $[1], [16], [21], [25], [6], [10], [15]$ and $[0]$ respectively. In detail (just consider the neutral element):*

(1) $[1], [7], [11], [13], [17], [19], [23]$ *and* $[29]$ *have the same neutral element, which is* $[1]$.
(2) $[2], [4], [8], [14], [16], [22], [26]$ *and* $[28]$ *have the same neutral element, which is* $[16]$.
(3) $[3], [9], [21]$ *and* $[27]$ *have the same neutral element, which is* $[21]$.
(4) $[5]$ *and* $[25]$ *have the same neutral element, which is* $[25]$.
(5) $[6], [12], [18]$ *and* $[24]$ *have the same neutral element, which is* $[6]$.
(6) $[10]$ *and* $[20]$ *have the same neutral element, which is* $[10]$.
(7) $[15]$ *has neutral element* $[15]$.
(8) $[0]$ *has neutral element* $[0]$.

4. GNETG and Quasi-Completely Regular Semigroup

Definition 9. *Let N be a non-empty set together with a binary operation $*$. Then, N is called a generalized neutrosophic extended triplet set if for any $a \in N$, there exist at least a positive integer n, such that a^n exists neutral element, denoted by $neut(a^n)$, and opposite element, denoted by $anti(a^n)$. The triplet $(a, neut(a^n), anti(a^n))$ is called a generalized neutrosophic extended triplet with degree n.*

Definition 10. *Let $(N, *)$ is a generalized neutrosophic extended triplet set. Then, N is called a GNETG, if the following conditions are satisfied:*
(1) $(N, *)$ is well-defined, i.e., for any $a, b \in N$, $a * b \in N$.
(2) $(N, *)$ is associative, i.e., $(a * b) * c = a * (b * c)$ for all $a, b, c \in N$.
A GNETG N is called a commutative generalized neutrosophic extended triplet group if for all $a, b \in N$, $a * b = b * a$.

Remark: *From Definition 9, it is obviously that a neutrosophic extended triplet is a generalized neutrosophic extended triplet with degree 1, so a neutrosophic extended triplet set is a generalized neutrosophic extended triplet set. Moreover, a NETG is a GNETG, but a GNETG is not a NETG in general.*

Example 4. *Let $S = \{a, e, f, g\}$, an operation $*$ on S is defined as in Table 2. We can see that $(e, e, a), (e, e, e), (f, f, f)$ and (g, g, g) are neutrosophic extended triplets, but a does not exist the neutral element and opposite element. Thus, S is not a NETG. Moreover, $a^2 = e$ has the neutral element and opposite element, so $(S, *)$ is a GNETG. (a, e, a) and (a, e, e) are generalized neutrosophic extended triplets with degree 2. We can infer that $(S, *)$ is a GNETG but not a NETG. Moreover, it is not a commutative GNETG being $e * f \neq f * e$.*

Table 2. A GNETG of Example 4.

$*$	a	e	f	g
a	e	e	g	e
e	e	e	e	e
f	f	f	f	f
g	g	g	g	g

Example 5. *Consider* $(Z_{12}, \otimes)$, $12 = 2^2 3^1$. *From Theorem 4, we have:*

(1) $[0], [1], [3], [4], [5], [7], [8], [9]$ *and* $[11]$ *exist the neutral element and opposite element.*
(2) $[2]$ *does not exist the neutral element and opposite element, but we can see that* $[2]^2 = [4]$ *exists the neutral element and opposite element.*
(3) $[6]$ *does not exist the neutral element and opposite element, but we can see that* $[6]^2 = [0]$ *exists the neutral element and opposite element.*
(4) $[10]$ *does not exist the neutral element and opposite element, but we can see that* $[10]^2 = [4]$ *exists the neutral element and opposite element.*

So Z_{12} is a generalized neutrosophic extended triplet set, but it is not a neutrosophic extended triplet set. Moreover, $(Z_{12}, \otimes)$ is a commutative GNETG. The following theorem shows that for each positive integer $n \geq 2$, $(Z_n, \otimes)$ is a commutative GNETG.

Theorem 6. *For each* $[a] \in Z_n$, *there exists a positive integer* m, *such that* $[a]^m$ *has the neutral element and opposite element. That is,* $(Z_n, \otimes)$ *is commutative GNETG.*

Proof. If the factorization of n is a product of single factors, From Theorem 5, $(Z_n, \otimes)$ is a NETG, so $[a]^1 = [a]$ has the neutral element and opposite element.

If $n = p_1^{k_1} p_2^{k_2} \cdots p_t^{k_t}$, and $\exists i$, such that $k_i \geq 2$. For each $[a] \in Z_n$, it is easy to get that there is a positive integer m, such that

$$gcd((gcd(a^m, n)), n/(gcd(a^m, n))) = 1,$$

that is $[a^m] = [a]^m$ has the neutral element and opposite element from Theorem 4. So, for each $[a] \in Z_n$, there exists a positive integer m, such that $[a]^m$ has the neutral element and opposite element, i.e., $(Z_n, \otimes)$ is a commutative GNETG. □

Definition 11. *[22] Let* $(S, *)$ *be a semigroup, an element* x *in* S *is said to be periodic if there exists a positive integer* n *such that* $x^{n+1} = x$. S *is pointwise periodic if each* x *in* S *is periodic.*

Proposition 6. *Let* $(S, *)$ *be a pointwise periodic semigroup, then* $(S, *)$ *is a NETG, so it is a GNETG.*

Proof. Since $(S, *)$ is a pointwise periodic semigroup, so for each element x in a S, there exists a positive integer n such that $x^{n+1} = x$.

If $n = 1$, then $x^2 = x$, so (x, x, x) is a neutrosophic extended triplet.

If $n \geq 2$, so $x * x^n = x^n * x = x$ and $x * x^{n-1} = x^{n-1} * x = x^n$, that is, (x, x^n, x^{n-1}) is a neutrosophic extended triplet.

By the arbitrariness of x, we have $(S, *)$ is a NETG. Of course, it is a GNETG. □

Example 6. *Let* $S = \{a, b, c\}$, *an operation* $*$ *on* S *is defined as following Table 3. Because* $a^2 = a, b^3 = b, c^2 = c$, *so* $(S, *)$ *is a pointwise periodic semigroup. Moreover,* $(a, a, a), (b, a, b), (c, c, a), (c, c, b)$ *and* (c, c, c) *are neutrosophic extended triplets, that is* $(S, *)$ *is a NETG.*

Table 3. A pointwise periodic semigroup of Example 6.

$*$	a	b	c
a	a	b	c
b	b	a	c
c	c	c	c

From Example 6, we can also see that if $(S, *)$ has zero element (c is the zero element in Example 6), then the neutral element of zero element is itself and the opposite element of zero element is every element in S.

Proposition 7. *Let $(N,*)$ is a GNETG, $a \in N$ and $(a, neut(a^n), anti(a^n))$ is a generalized neutrosophic extended triplet with degree n. We have:*
(1) $neut(a^n)$ is unique.
(2) $neut(a^n) * neut(a^n) = neut(a^n)$.

Proof. Assume $c, d \in \{neut(a^n)\}$, so $a^n * c = c * a^n = a^n, a^n * d = d * a^n = a^n$, and there exists $x, y \in N$ such that

$$a^n * x = x * a^n = c, a^n * y = y * a^n = d.$$

We can obtain

$$c * d = (x * a^n) * d = x * (a^n * d) = x * a^n = c,$$

$$c * d = (a^n * x) * (a^n * y) = (a^n * (x * a^n)) * y = d.$$

We have $c = d = c * d$. So $neut(a^n)$ is unique and $neut(a^n) * neut(a^n) = neut(a^n)$. □

Proposition 8. *Let $(N,*)$ is a GNETG, $a \in N$ and $(a, neut(a^n), anti(a^n))$ is a generalized neutrosophic extended triplet with degree n. Then*
(1) $a^n * x * neut(a^n) = x * neut(a^n) * a^n = neut(a^n)$, for any $x \in \{anti(a^n)\}$.
(2) $a^n * neut(a^n) * x = neut(a^n) * x * a^n = neut(a^n)$ for any $x \in \{anti(a^n)\}$.
(3) $neut((neut(a^n))) = neut(x * neut(a^n)) = neut(a^n), x \in \{anti(a^n)\} \cup \{a^n\}$.
(4) $neut(a^n) * x = y * neut(a^n) = z * neut(a^n)$, for any $x, y, z \in \{anti(a^n)\}$.
(5) $anti(x * neut(a^n)) * neut(x * neut(a^n)) = a^n$, for any $x \in \{anti(a^n)\}$.

Proof. (1) For any $x \in \{anti(a^n)\}$, from Definition 9 and Proposition 7, we have

$$a^n * x * neut(a^n) = (a^n * x) * neut(a^n) = neut(a^n) * neut(a^n) = neut(a^n),$$

$$x * neut(a^n) * a^n = x * (neut(a^n) * a^n) = x * a^n = neut(a^n).$$

So $a^n * x * neut(a^n) = x * neut(a^n) * a^n = neut(a^n)$, for any $x \in \{anti(a^n)\}$.

(2) For any $x \in \{anti(a^n)\}$, From Definition 9 and Proposition 7, we have

$$a^n * neut(a^n) * x = (a^n * neut(a^n)) * x = a^n * x = neut(a^n),$$

$$neut(a^n) * x * a^n = neut(a^n) * (x * a^n) = neut(a^n) * neut(a^n) = neut(a^n).$$

So $a^n * neut(a^n) * x = neut(a^n) * x * a^n = neut(a^n)$ for any $x \in \{anti(a^n)\}$.

(3) We prove $neut(a^n) = neut((neut(a^n)))$ firstly.

For any $x \in \{anti(a^n)\}$ and $y \in \{anti(neut(a^n))\}$,

$$(y * x) * a^n = y * (x * a^n) = y * neut(a^n) = neut(neut(a^n)).$$

Moreover,

$$((y * x) * a^n) * neut(a^n) = (y * x) * (a^n * neut(a^n)) = (y * x) * a^n = neut(neut(a^n)).$$

Thus, $neut(a^n) = neut(neut(a^n)) * neut(a^n) = ((y * x) * a^n) * neut(a^n) = neut(neut(a^n))$.

In the following we will prove $neut(x * neut(a^n)) = neut(a^n), x \in \{anti(a^n)\} \cup \{a^n\}$.

It is obvious that $neut(x * neut(a^n)) = neut(a^n)$ when $x = a^n$. If $x \in \{anti(a^n)\}$, by Proposition 7 we have:

$$(x * neut(a^n)) * neut(a^n) = x * (neut(a^n) * neut(a^n)) = x * neut(a^n),$$

$$neut(a^n) * (x * neut(a^n)) = (neut(a^n) * x) * neut(a^n) = (x * neut(a^n)) * neut(a^n)$$
$$= x * (neut(a^n) * neut(a^n)) = x * neut(a^n).$$

Moreover, we can get

$$(x * neut(a^n)) * a^n = x * (neut(a^n) * a^n) = x * a^n = neut(a^n),$$

$$a^n * (x * neut(a^n)) = (a^n * x) * neut(a^n) = neut(a^n) * neut(a^n) = neut(a^n).$$

Thus, $neut(a^n) = neut(x * neut(a^n))$.

(4) For any $x, y, z \in \{anti(a^n)\}$, we have

$$neut(a^n) * x = (y * a^n) * x = y * (a^n * x) = y * neut(a^n).$$

Moreover,

$$y * neut(a^n) = y * (a^n * z) = (y * a^n) * z = neut(a^n) * z = (z * a^n) * z = z * (a^n * z) = z * neut(a^n).$$

Thus, (4) holds.

(5) Suppose $x \in \{anti(a^n)\}$, for each $y \in \{anti(x * neut(a^n))\}$, from (4) we know that $anti(x * neut(a^n)) * neut(x * neut(a^n))$ is unique. Applying (1), $a^n \in \{anti(x * neut(a^n))\}$, that is,

$$anti(x * neut(a^n)) * neut(x * neut(a^n)) = a^n neut(x * neut(a^n)).$$

By (3), $neut(x * neut(a^n)) = neut(a^n)$. Thus, $anti(x * neut(a^n)) * neut(x * neut(a^n)) = a^n * neut(x * neut(a^n)) = a^n * neut(a^n) = a^n$. □

Example 7. *Let $S = \{a, e, b, f, c, g\}$, an operation $*$ on S is defined as following Table 4. Since $neut(e) = e, \{anti(e)\} = \{a, e\}, neut(f) = f, \{anti(f)\} = \{b, f\}, neut(g) = g, \{anti(g)\} = \{a, e, b, f, c, g\}$ and $a^2 = e, b^2 = f, c^2 = g$, so $(S, *)$ is a GNETG. We can get that (Corresponding to the results of Proposition 8):*

Table 4. A GNETG of Example 7.

$*$	a	e	b	f	c	g
a	e	e	c	c	c	g
e	e	e	c	c	c	g
b	g	g	f	f	g	g
f	g	g	f	f	g	g
c	g	g	c	c	g	g
g	g	g	g	g	g	g

(1) *Being $a^2 * a * neut(a^2) = a * neut(a^2) * a^2 = neut(a^2)$, $a^2 * e * neut(a^2) = e * neut(a^2) * a^2 = neut(a^2)$, that is, for any $x \in \{anti(a^2)\}$, $a^2 * x * neut(a^2) = x * neut(a^2) * a^2 = neut(a^2)$.*

(2) *Being $b^2 * neut(b^2) * b = neut(b^2) * b * b^2 = neut(b^2)$, $b^2 * neut(b^2) * f = neut(b^2) * f * b^2 = neut(b^2)$, that is, for any $x \in \{anti(b^2)\}$, $b^2 * neut(b^2) * x = neut(b^2) * x * b^2 = neut(b^2)$.*

(3) *Being $neut((neut(a^2))) = neut(a * neut(a^2)) = neut(a^2)$, $neut((neut(a^2))) = neut(e * neut(a^2)) = neut(a^2)$, that is, for any $x \in \{anti(a^n)\} \cup \{a^n\}$, $neut((neut(a^n))) = neut(x * neut(a^n)) = neut(a^n)$.*

(4) *For each element $x, y, z \in S$, $neut(c^2) * x = y * neut(c^2) = z * neut(c^2) = g$, that is, for any $x, y, z \in \{anti(c^2)\}$, $neut(c^2) * x = y * neut(c^2) = z * neut(c^2)$.*

(5) *Being $b * f = b^2 = f, f * f = b^2 = f$, that is, for any $x \in \{anti(b^2)\}$, $anti(x * neut(b^2)) * neut(x * neut(b^2)) = b^2$.*

Proposition 9. *Let* $(N,*)$ *be a commutative GNETG, then* $\forall a, b \in N$, *there are two positive integers* m *andn such that the following hold:*
(1) $neut(a^n) * neut(b^m) = neut(a^n * b^m)$.
(2) $anti(a^n) * anti(b^m) \in \{anti(a^n * b^m)\}$.

Proof. Being $(N,*)$ be a commutative GNETG, then for $a \in N$, there is a positive integer n, such that a^n exists the neutral element and opposite element, denoted by $neut(a^n)$ and $anti(a^n)$ respectively. For $b \in N$, there is a positive integer m, such that b^m exists the and opposite element, denoted by $neut(b^m)$ and $anti(b^m)$ respectively. So

$$\begin{aligned}(neut(a^n) * neut(b^m)) * (a^n * b^m) &= ((neut(a^n) * neut(b^m)) * a^n) * b^m \\ &= ((neut(a^n) * a^n) * neut(b^m)) * b^m \\ &= (a^n * neut(b^m)) * b^m \\ &= a^n * (neut(b^m) * b^m) = a^n * b^m.\end{aligned}$$

in the same way, we have $(a^n * b^m) * (neut(a^n) * neut(b^m)) = a^n * b^m$. That is:

$$(a^n * b^m) * (neut(a^n) * neut(b^m)) = (neut(a^n) * neut(b^m)) * (a^n * b^m) = a^n * b^m.$$

Moreover, for any $anti(a^n) \in \{anti(a^n)\}$ and $anti(b^m) \in \{anti(b^m)\}$, we can get,

$$\begin{aligned}(anti(a^n) * anti(b^m)) * (a^n * b^m) &= ((anti(a^n) * anti(b^m)) * a^n) * b^m \\ &= ((anti(a^n) * a^n) * anti(b^m)) * b^m \\ &= (neut(a^n) * anti(b^m)) * b^m \\ &= neut(a^n) * (anti(b^m) * b^m) = neut(a^n) * neut(b^m).\end{aligned}$$

Similarly, we have $(a^n * b^m) * (anti(a^n) * anti(b^m)) = neut(a^n) * neut(b^m)$. That is:

$$(a^n * b^m) * (anti(a^n) * anti(b^m)) = (anti(a^n) * anti(b^m)) * (a^n * b^m) = neut(a^n) * neut(b^m).$$

Thus, we have

$$neut(a^n) * neut(b^m) \in \{neut(a^n * b^m)\}.$$

From this, by Proposition 7, we get: $neut(a^n) * neut(b^m) = neut(a^n * b^m)$. Therefore, we get $anti(a^n) * anti(b^m) \in \{anti(a^n * b^m)\}$. □

Example 8. *Consider* $(Z_{12}, \otimes)$, *which is a commutative GNETG from Example 5. Without losing generality, select* [2] *and* [4], *then exist two positive integers 2 and 1, such that* $neut([2]^2) * neut([4]^1) = neut([2]^2 * [4]^1) = [4]$, *so explain (1) of Proposition 9 is correct. Being* $\{anti[4]\} = \{[1], [4], [7]\}$ *and* [1], [4], [7] *is a sub algebra structure of* $(Z_{12}, \otimes)$ *by Table 5. So, explain (2) of Proposition 9 is correct.*

Table 5. The operator table of $\{[1], [4], [7]\}$ in $(Z_{12}, \otimes)$.

⊗	[1]	[4]	[7]
[1]	[1]	[4]	[7]
[4]	[4]	[4]	[4]
[7]	[1]	[4]	[1]

Example 9. *Apply the $(S,*)$ in Example 7, Since it is not a commutative GNETG, we can get that:*

(1) *Being for each $2 \leq n \in Z^+, a^n = e$, and for each $m \in Z^+, f^m = f$, so $neut(a^n) * neut(f^m) = e * f = c$. On the other hand, $neut(a^n * f^m) = neut(e * f) = neut(c)$, but $neut(c)$ does not exist, so the (1) of Proposition 9 does not hold.*
(2) *Being $\{anti(a^n)\} = \{a, e\}, \{anti(f^m)\} = \{f\}, a^n * f^m = c$, but $anti(c)$ does not exist, we can get that the (2) of Proposition 9 does not hold.*

Theorem 7. *Let $(N,*)$ be a groupoid. Then N is a GNETG if and only if it is a quasi-completely regular semigroup.*

Proof. Necessity: Suppose N is a GNETG, from Definition 10, for each $a \in N$, there is a positive integer n, such that a^n exists the neutral element and opposite element, denoted by $neut(a^n)$ and $anti(a^n)$ respectively. Set

$$(a^n)^{-1} = anti(a^n) * neut(a^n),$$

by Proposition 8, $(a^n)^{-1}$ is unique and we have

$$((a^n)^{-1})^{-1} = anti(anti(a^n) * neut(a^n)) * neut(anti(a^n) * neut(a^n)) = a^n,$$

$$a^n * (a^n)^{-1} * a^n = a^n * anti(a^n) * neut(a^n) * a^n = a^n,$$

$$a^n * anti(a^n) * neut(a^n) = anti(a^n) * neut(a^n) * a^n = neut(a^n), \text{i.e. } a^n * (a^n)^{-1} = (a^n)^{-1} * a^n$$

from Definition 7, N is a quasi-completely regular semigroup.

Sufficiency: If N is a completely quasi-regular semigroup. For any $a \in N$, there is a positive integer n and $(a^n)^{-1} \in N$, such that $a^n * (a^n)^{-1} * a^n = a^n$ and $a^n * (a^n)^{-1} = (a^n)^{-1} * a^n$, set

$$neut(a^n) = a^n * (a^n)^{-1},$$

then

$$neut(a^n) * a^n = a^n * (a^n)^{-1} * a^n = a^n,$$

$$a^n * neut(a^n) = a^n * a^n * (a^n)^{-1} = a^n * (a^n)^{-1} * a^n = a^n,$$

$$a^n * (a^n)^{-1} = (a^n)^{-1} * a^n = neut(a^n).$$

From Definition 10, we have that N is a GNETG and $(a^n)^{-1} \in \{anti(a^n)\}$. □

Example 10. *Apply $(S,*)$ in the Example 7, we know that it is a GNETG. We will show that it is a quasi-completely regular semigroup in the following.*

For e, there exists an inverse element $e^{-1} = e$, such that $(e^{-1})^{-1} = e, e * e^{-1} * e = e, e * e^{-1} = e^{-1} * e$, so e is completely regular. f and g are completely regular for the same reason. Moreover, being $a^2 = e, b^2 = f$ and $c^2 = g$, so a^2, b^2 and c^2 are completely regular, so $(S,*)$ is a quasi-completely regular semigroup by Definition 7.

Definition 12. *Let $(N,*)$ is a GNETG. N is called a weak commutative GNETG if there are two positive integers n, m, such that $a^n * neut(b^m) = neut(b^m) * a^n$ for all $a, b \in N$.*

Example 11. *Let $S = \{a,b,c,d,f\}$, an operation $*$ on S is defined as following Table 6. Since $(a,a,a), (b,a,b), (c,c,c)$ and (f,f,f) are neutrosophic extended triplets, but d does not exist the neutral element and opposite element. Thus, S is not a NETG. Moreover, $d^2 = c$ exists the neutral element and opposite element, so $(S,*)$ is GNETG and (d,c,c) is a generalized neutrosophic extended triplet with degree 2. We can infer that $(S,*)$ is a GNETG but not a NETG. Moreover, it is not a commutative GNETG being $b * d \neq d * b$, we can show that it is a weak commutative GNETG.*

For a, b, c, d, and f, there exist positive integers 1, 1, 1, 2, and 1 respectively, so $S' = \{a^1, b^1, c^1, d^2, f^1\} = \{a, b, c, f\}$ being $d^2 = c$. We know that $neut(a) = a, neut(b) = a, neut(c) = c, neut(f) = f$, so $\{neut(a), neut(b), neut(c), neut(f)\} \in S'$. From Table 6, we can get the sub algebra system $(S', *)$ of $(S, *)$ as following Table 7, and $(S', *)$ is commutative. Thus, $(S, *)$ is a weak commutative GNETG.

Table 6. The operation table of Example 11.

*	a	b	c	d	f
a	a	b	c	d	f
b	b	a	c	d	f
c	c	c	c	c	c
d	c	c	c	c	c
f	f	f	c	d	f

Table 7. The sub algebra system S' of S in Example 11.

*	a	b	c	f
a	a	b	c	f
b	b	a	c	f
c	c	c	c	c
f	f	f	c	f

Example 12. *Select the $(S, *)$ in Example 4, being $e^n = e, f^m = f$ for every $n, m \in Z^+$, $e^n * neut(f^m) = e * f = e$, $f^m * neut(e^n) = f * e = f$, so $e^n * neut(f^m) = e \neq f = neut(f^m) * e^n, \forall n, m \in Z^+$. Thus, the $(S, *)$ in Example 4 is not weak commutative GNETG.*

Proposition 10. *Let $(N, *)$ be a GNETG. Then $(N, *)$ is a weak commutative GNETG iff there are two positive integers n and m, such that N satisfies the following conditions:*
(1) $neut(a^n) * neut(b^m) = neut(b^m) * neut(a^n)$, for all $a, b \in N$.
(2) $neut(a^n) * neut(b^m) * a^n = a^n * neut(b^m)$, for all $a, b \in N$.

Proof. Necessity: If $(N, *)$ is a weak commutative GNETG, then there are two positive integers n, m, such that a^n and b^m exist the neutral element and opposite element. So, from Proposition 7 and Definition 12, we have

$$neut(a^n) * neut(b^m) = neut(b^m) * neut(a^n), \forall a, b \in N.$$

Moreover,

$$\begin{aligned} neut(a^n) * neut(b^m) * a^n &= neut(a^n) * (neut(b^m) * a^n) \\ &= neut(a^n) * (a^n * neut(b^m)) \\ &= (neut(a^n) * a^n) * neut(b^m) = a^n * neut(b^m). \end{aligned}$$

Sufficiency: Suppose that N satisfies the conditions (1) and (2) above. Then:

$$\begin{aligned} a^n * neut(b^m) &= (neut(a^n) * neut(b^m)) * a^n \\ &= (neut(b^m) * neut(a^n)) * a^n \\ &= neut(b^m) * (neut(a^n) * a^n) = neut(b^m) * a^n. \end{aligned}$$

From Definition 12 we know that $(N, *)$ is a weak commutative GNETG. □

Proposition 11. *Let $(N,*)$ be a weak commutative GNETG, then for $a,b \in N$, the following conditions are satisfied:*
(1) $neut(a^n) * neut(b^m) = neut(b^m * a^n)$;
(2) $anti(a^n) * anti(b^m) \in \{anti(b^m * a^n)\}$.

Proof. If $a,b \in N$, then there are two positive integers m,n, such that a^n and b^m exist the neutral element and opposite element. We have

$$\begin{aligned}(neut(a^n) * neut(b^m)) * (b^m * a^n) &= ((neut(a^n) * neut(b^m)) * b^m) * a^n \\ &= (neut(a^n) * (b^m * neut(b^m))) * a^n \\ &= (neut(a^n) * b^m) * a^n \\ &= (b^m * neut(a^n)) * a^n = b^m * (neut(a^n) * a^n) = b^m * a^n.\end{aligned}$$

Similarly, we have $(b^m * a^n) * (neut(a^n) * neut(b^m)) = (neut(a^n) * neut(b^m)) * (b^m * a^n) = b^m * a^n$. Moreover, for any $anti(a^n) \in \{anti(a^n)\}$ and $anti(b^m) \in \{anti(b^m)\}$, we have:

$$\begin{aligned}(anti(a^n) * anti(b^m)) * (b^m * a^n) &= ((anti(a^n) * anti(b^m)) * b^m) * a^n \\ &= (anti(a^n) * (anti(b^m) * b^m)) * a^n \\ &= (anti(a^n) * neut(b^m)) * a^n \\ &= anti(a^n) * (neut(b^m) * a^n) \\ &= anti(a^n) * (a^n * neut(b^m)) \\ &= (anti(a^n) * a^n) * neut(b^m) = neut(a^n) * neut(b^m).\end{aligned}$$

Similarly, we have $(b^m * a^n) * (anti(a^n) * anti(b^m)) = (anti(a^n) * anti(b^m)) * (b^m * a^n) = neut(a^n) * neut(b^m)$. So, we have $neut(a^n) * neut(b^m) \in \{neut(b^m * a^n)\}$. From this, we can get $neut(a^n) * neut(b^m) = neut(b^m * a^n)$. Thus, we have $anti(a^n) * anti(b^m) \in anti(b^m * a^n)$. □

Example 13. *Let $S = \{a,e,b,f,c,g\}$, an operation $*$ on S is defined as following Table 8. Being $netu(e) = e, \{anti(e)\} = \{a,e\}, netu(f) = f, \{anti(f)\} = \{b,f\}, netu(g) = g, \{anti(g)\} = \{a,e,b,f,c,g\}$ and $a^2 = e, b^2 = f, c^2 = g$, so $(S,*)$ is a GNETG. It is easy to verify $(S,*)$ is a weak commutative GNETG by Definition 12. We also can get:*

(1) Being $neut(b^2) * neut(e^1) = f * e = g$, and $neut(e^1 * b^2) = neut(g) = g$, so, $neut(b^2) * neut(e^1) = neut(e^1 * b^2)$.
(2) Being $\{anti(b^2)\} = \{b,f\}$, $\{anti(e^1)\} = \{a,e\}$, $\{anti(b^2 * e^1)\} = \{anti(g)\} = \{a,e,b,f,c,g\}$, thus, $anti(b^2) * anti(e^1) \in \{anti(b^2 * e^1)\}$.

Table 8. The operation table of Example 13.

$*$	a	e	b	f	c	g
a	e	e	c	c	g	g
e	e	e	g	g	g	g
b	c	g	f	f	c	g
f	c	g	f	f	c	g
c	g	g	c	c	g	g
g	g	g	g	g	g	g

Example 14. *Apply the $(S,*)$ in Example 4, Being it is not a weak commutative GNETG, we can get that:*

(1) Being $neut(a^2) * neut(f^1) = e * f = e$, but $neut(f^1 * a^2) = neut(f) = f$, so, $neut(a^2) * neut(f^1) \neq neut(f^1 * a^2)$.

(2) *Being* $\{anti(a^2)\} = \{a, e\}$, $\{anti(f^1)\} = \{f\}$, $\{anti(f^1 * a^2)\} = \{anti(f)\} = \{f\}$, *thus*, $anti(a^2) * anti(f^1) \notin \{anti(f^1 * a^2)\}$.

Definition 13. *A semigroup S is called a quasi-Clifford semigroup, if it is quasi-completely regular and for any* $a, b \in S$, *there are two positive integers* n, m *such that*

$$a^n * (b^m * (b^m)^{-1}) = (b^m * (b^m)^{-1}) * a^n.$$

Theorem 8. *Let* $(N, *)$ *be a groupoid. Then N is a weak commutative GNETG iff it is a quasi-Clifford semigroup.*

Proof. Necessity: Suppose that N is a weak commutative GNETG. By Theorem 7, we know that N is a quasi-completely regular semigroup, then there are two positive integers m, n, such that a^n and b^m exist the neutral element and opposite element. Set

$$(a^n)^{-1} = anti(a^n) * neut(a^n).$$

For any $a, b \in N$, so we have

$$a^n * (b^m * (b^m)^{-1}) = (b^m * (b^m)^{-1}) * a^n.$$

From Definition 13, we know that N is a quasi-Clifford semigroup.

Sufficiency: Assume that N is a quasi-Clifford semigroup, then there are two positive integers m and n, such that a^n and b^m are completely regular. Then there exists $(a^n)^{-1}$ and $(b^m)^{-1}$. Set

$$neut(a^n) = a^n * (a^n)^{-1}, neut(b^m) = b^m * (b^m)^{-1}.$$

Applying Definition 13, being $a^n * (b^m * (b^m)^{-1}) = (b^m * (b^m)^{-1}) * a^n$, we have $a^n * neut(b^m) = neut(b^m) * a^n$, we can get that N is a weak commutative GNETG by Definition 12. □

Example 15. *Apply the* $(S, *)$ *in Example 11, Being it is a weak commutative GNETG from Example 11. We show that it is a quasi-Clifford semigroup. From Theorem 7, we can see that* $(S, *)$ *is a quasi-completely regular semigroup, we just show for any* $x, y \in S$, *there are two positive integers n and m such that* $x^n * (y^m * (y^m)^{-1}) = (y^m * (y^m)^{-1}) * x^n$.

From Example 11, for a, b, c, d, *and* f, *there exist positive integers* $1, 1, 1, 2$, *and* 1 *respectively, and set* $a^{-1} = a, b^{-1} = b, c^{-1} = c, (d^2)^{-1} = c, f^{-1} = f$. *For any* $x, y \in \{a^1, b^1, c^1, d^2, f^1\}$, *without losing generality, let* $x = a, y = b$, *we can get* $a^1 * (b^1 * (b^1)^{-1}) = (b^1 * (b^1)^{-1}) * a^1 = b$. *We can verify other cases, thus* $(S, *)$ *is a quasi-Clifford semigroup.*

Example 16. *Apply the* $(S, *)$ *in Example 4, Being it is not a weak commutative GNETG from Example 12. We show that there exists* $x, y \in S$, *for any two positive integers n and m such that* $x^n * (y^m * (y^m)^{-1}) \neq (y^m * (y^m)^{-1}) * x^n$.

From Example 4, for any $n, m \in Z^+$, $e^n = e, f^m = f$ *and* $(e^n)^{-1} = e, (f^m)^{-1} = f$, *but* $e^n * (f^m * (f^m)^{-1}) = e \neq f = (f^m * (f^m)^{-1}) * e^n$. *That is for* $e, f \in S$, *there are not two positive integers* n, m *such that* $e^n * (f^m * (f^m)^{-1}) = (f^m * (f^m)^{-1}) * e^n$. *So* $(S, *)$ *is not a quasi-Clifford semigroup.*

5. Conclusions

In the paper, from the perspective of semigroup theory, we deeply studied the GNETG and obtained some important results. We proved that the GNETG is equal to the quasi-completely regular semigroup, and the weak commutative GNETG is equal to the quasi-Clifford semigroup. Moreover, we investigated the relationship between $(Z_n, \otimes)$ and a NETG. All these results are interesting for exploring the structure characterization of GNETG. As the next research topics, we will explore the structure of some special GNETG and their relationships with related logic algebras. Moreover, we will discuss the integration of the related topics, such as the combination of neutrosophic set, fuzzy set, soft set and algebra systems [23,24].

Author Contributions: All authors have contributed equally to this paper.

Funding: This research was funded by National Natural Science Foundation of China (Grant No. 11501435), Instructional Science and Technology Plan Projects of China National Textile and Apparel Council (No. 2016073) and Scientific Research Program Funded by Shaanxi Provincial Education Department (Program No. 18JS042).

Acknowledgments: The authors would like to thank the reviewers for their many insightful comments and suggestions.

Conflicts of Interest: The authors declare no conflicts of interest.

References

1. Smarandache, F.; Ali, M. Neutrosophic triplet group. *Neural Comput. Appl.* **2018**, *29*, 595–601. [CrossRef]
2. Vasantha, W.B.; Kandasamy, I.; Smarandache, F. *Neutrosophic Triplet Groups and Their Applications to Mathematical Modelling*; EuropaNova: Brussels, Belgium, 2017; ISBN 978-1-59973-533-7.
3. Smarandache, F. Hybrid Neutrosophic Triplet Ring in Physical Structures. *Bull. Am. Phys. Soc.* **2017**, *62*, 17.
4. Bal, M.; Shalla, M.M.; Olgun, N. Neutrosophic Triplet Cosets and Quotient Groups. *Symmetry* **2017**, *10*, 126. [CrossRef]
5. Smarandache, F.; Ali, M. Neutrosophic Triplet Ring and its Applications. In Proceedings of the 18th Annual Meeting of the APS Northwest Section, Pacific University, Forest Grove, OR, USA, 1–3 June 2017.
6. Sahin, M.; Abdullah, K. Neutrosophic triplet normed space. *Open Phys.* **2017**, *15*, 697–704. [CrossRef]
7. Jaiyeola, T.G.; Smarandache, F. Some results on neutrosophic triplet group and their applications. *Symmetry* **2018**, *10*, 202. [CrossRef]
8. Şahin, M.; Kargın, A. neutrosophic triplet v-generalized metric space. *Axioms* **2018**, *7*, 67. [CrossRef]
9. Ali, M.; Smarandache, F.; Khan, M. Study on the development of neutrosophic triplet ring and neutrosophic triplet field. *Mathematics* **2018**, *6*, 46. [CrossRef]
10. Gulistan, M.; Nawaz, S.; Hassan, N. Neutrosophic triplet non-associative semihypergroups with application. *Symmetry* **2018**, *10*, 613. [CrossRef]
11. Smarandache, F. *Neutrosophic Perspectives: Triplets, Duplets, Multisets, Hybrid Operators, Modal Logic, Hedge Algebras. And Applications*; Pons Publishing House: Brussels, Belgium, 2017.
12. Zhang, X.H.; Bo, C.X.; Smarandache, F.; Dai, J.H. New inclusion relation of neutrosophic sets with applications and related lattice structure. *Int. J. Mach. Learn. Cybern.* **2018**, *9*, 1753–1763. [CrossRef]
13. Zhang, X.H.; Bo, C.X.; Smarandache, F.; Park, C. New operations of totally dependent-neutrosophic sets and totally dependent-neutrosophic soft sets. *Symmetry* **2018**, *10*, 187. [CrossRef]
14. Zhang, X.H.; Wu, X.Y.; Smarandache, F.; Hu, M.H. Left (right)-quasi neutrosophic triplet loops (groups) and generalized BE-algebras. *Symmetry* **2018**, *10*, 241. [CrossRef]
15. Zhang, X.; Hu, Q.; Smarandache, F.; An, X. On Neutrosophic Triplet Groups: Basic Properties, NT-Subgroups, and Some Notes. *Symmetry* **2018**, *10*, 289. [CrossRef]
16. Zhang, X.H.; Smarandache, F.; Liang, X.L. Neutrosophic Duplet Semi-Group and Cancellable Neutrosophic Triplet Groups. *Symmetry* **2017**, *9*, 275. [CrossRef]
17. Zhang, X.H.; Wang, X.J.; Smarandache, F.; Jaíyéolá, T.G.; Liang, X.L. Singular neutrosophic extended triplet groups and generalized groups. *Cognit. Syst. Res.* **2018**, in press. [CrossRef]
18. Zhang, X.H.; Wu, X.Y.; Mao, X.Y.; Smarandache, F.; Park, C. On Neutrosophic Extended Triplet Groups (Loops) and Abel-Grassmann's Groupoids (AG-Groupoids). *J. Intell. Fuzzy Syst.* **2019**, in press.
19. Vasantha, W.B.; Kandasamy, I.; Smarandache, F. A Classical Group of Neutrosophic Triplet Groups Using $\{Z_{2p}, \times\}$. *Symmetry* **2018**, *10*, 193.
20. Vasantha, W.B.; Kandasamy, I.; Smarandache, F. Neutrosophic Duplets of $\{Z_{p^n}, \times\}$ and $\{Z_{pq}, \times\}$ and Their Properties. *Symmetry* **2018**, *10*, 344.
21. Akinmoyewa, J.T. A study of some properties of generalized groups. *Octogon Math. Mag.* **2009**, *17*, 599–626.
22. Howie, J.M. *Fundamentals of Semigroup Theory*; Oxford University Press: Oxford, UK, 1995.

23. Zhang, X.H. Fuzzy anti-grouped filters and fuzzy normal filters in pseudo-BCI algebras. *J. Intell. Fuzzy Syst.* **2017**, *33*, 1767–1774. [CrossRef]
24. Zhang, X.H.; Park, C.; Wu, S.P. Soft set theoretical approach to pseudo-BCI algebras. *J. Intell. Fuzzy Syst.* **2018**, *34*, 559–568. [CrossRef]

symmetry

MDPI

Article

A Generalized Approach towards Soft Expert Sets via Neutrosophic Cubic Sets with Applications in Games

Muhammad Gulistan [1] and Nasruddin Hassan [2,*]

1 Department of Mathematics & Statistics, Hazara University, Mansehra 21130, Pakistan; gulistanmath@hu.edu.pk or gulistanm21@yahoo.com

2 School of Mathematical Sciences, Faculty of Science and Technology, Universiti Kebangsaan Malaysia, Bangi 43600, Selangor, Malaysia

* Correspondence: nas@ukm.edu.my; Tel.: +60-1-9214-5750

Received: 28 January 2019; Accepted: 19 February 2019; Published: 22 February 2019

Abstract: Games are considered to be the most attractive and healthy event between nations and peoples. Soft expert sets are helpful for capturing uncertain and vague information. By contrast, neutrosophic set is a tri-component logic set, thus it can deal with uncertain, indeterminate, and incompatible information where the indeterminacy is quantified explicitly and truth membership, indeterminacy membership, and falsity membership independent of each other. Subsequently, we develop a combined approach and extend this concept further to introduce the notion of the neutrosophic cubic soft expert sets (NCSESs) by using the concept of neutrosophic cubic soft sets, which is a powerful tool for handling uncertain information in many problems and especially in games. Then we define and analyze the properties of internal neutrosophic cubic soft expert sets (INCSESs) and external neutrosophic cubic soft expert sets (ENCSESs), P-order, P-union, P-intersection, P-AND, P-OR and R-order, R-union, R-intersection, R-AND, and R-OR of NCSESs. The NCSESs satisfy the laws of commutativity, associativity, De Morgan, distributivity, idempotentency, and absorption. We derive some conditions for P-union and P-intersection of two INCSESs to be an INCSES. It is shown that P-union and P-intersection of ENCSESs need not be an ENCSES. The R-union and R-intersection of the INCSESs (resp., ENCSESs) need not be an INCSES (resp. ENCSES). Necessary conditions for the P-union, R-union and R-intersection of two ENCSESs to be an ENCSES are obtained. We also study the conditions for R-intersection and P-intersection of two NCSESs to be an INCSES and ENCSES. Finally, for its applications in games, we use the developed procedure to analyze the cricket series between Pakistan and India. It is shown that the proposed method is suitable to be used for decision-making, and as good as or better when compared to existing models.

Keywords: neutrosophic sets; cubic sets; soft sets; neutrosophic cubic soft sets; neutrosophic cubic soft expert system; multicriteria decision-making

1. Introduction

Researchers always try to discover methods to handle imprecise and vague information, which is not possible using classical set theory. In this regard, Zadeh gave the concept of fuzzy set [1], to cope with uncertainty. However, fuzzy sets were considered imperfect since it is not always easy to give an exact degree of membership to any element. To overcome this problem, the interval-valued fuzzy set was proposed by Turksen [2]. Atanassov [3] extended the notion of fuzzy sets to intuitionistic fuzzy sets by introducing the non-membership of an element with its membership in a set X, which were proven to be a better tool than fuzzy sets. Furthermore, the intuitionistic fuzzy sets are used in many directions [4]. Smarandache gave the notion of neutrosophic sets as a generalization of intuitionistic fuzzy sets and fuzzy sets [5]. The idea of neutrosophic sets are further expanded to different directions [6–9] by various researchers. Jun et al. [10] gave the idea of cubic set and it was characterized by interval-valued fuzzy set and fuzzy set, which is a more general tool to capture uncertainty and vagueness, since fuzzy set deals with single-value membership while interval-valued fuzzy set ranges the membership in the form of intervals. The hybrid platform provided by cubic set has the main advantage since it contains more information than a fuzzy set and interval-valued fuzzy set. By using this concept, different problems arising in several areas can be solved by choosing the finest choice by means of cubic sets as in the works of Abughazalah and Yaqoob [11], Rashid et al. [12], Gulistan et al. [13], Ma et al. [14], Naveed at al. [15], Gulistan et al. [16], Khan et al. [17,18], Yaqoob et al. [19], and Aslam et al. [20].

More recently, Jun et al. [21] gave the idea of neutrosophic cubic set and it was subsequently used in many areas by Khan et al. [22] and Gulistan et al. [23,24].

On the other hand, Molodtsov [25] introduced the concept of soft sets that can be seen as a new mathematical theory for dealing with uncertainty. It was applied to many different fields by Maji et al. [26] who later defined fuzzy soft set theory and some properties of fuzzy soft sets [27]. Hybrids of soft sets were further developed [28–32].

Alkhazaleh and Salleh in 2011 defined the concept of soft expert set in which the user could know the opinion of all the experts in one model and gave an application of this concept in the decision-making problem [33]. Arokia et al. [34] studied fuzzy parameterizations for decision-making in risk management systems via soft expert set. Arokia and Arockiarani [35] provided a fusion of soft expert set and matrix models. Alkhazaleh and Salleh [36] extended the concept of soft expert set in terms of fuzzy set and provided its application. Bashir and Salleh [37] provided the concept of fuzzy parameterized soft expert set. Bashir et al. [38] discussed possibility fuzzy soft expert set. Alhazaymeh et al. [39] provided the application of generalized vague soft expert set in decision-making. Broumi and Smarandache [40] extended the soft expert sets in terms of intuitionistic fuzzy sets. Abu Qamar and Hassan [41,42] presented the idea of Q-neutrosophic soft relation and its entropy measures of distance and similarity. Sahin et al. [43] gave the idea of neutrosophic soft expert sets while Uluçay et al. [44], introduced the concept of generalized neutrosophic soft expert set for multiple-criteria decision-making. Neutrosophic vague soft expert set theory was put forward by Al-Quran and Hassan [45] and developed it further to complex neutrosophic soft expert set [46,47]. Qayyum et al. [48] gave the idea of cubic soft expert sets for a more general approach. Ziemba and Becker [49] presented analysis of the digital divide using fuzzy forecasting, which is a new approach in decision-making.

Hence it is natural to extend the concept of expert sets to neutrosophic cubic soft expert sets for a more generalized approach. The major contribution of this paper is the development of neutrosophic cubic soft expert sets(NCSESs) by using the concept of neutrosophic cubic soft sets which generalizes the concept of fuzzy soft expert sets, intuitionistic soft expert sets, and cubic soft expert sets. We define and analyze the properties of internal neutrosophic cubic soft expert sets (INCSESs) and external neutrosophic cubic soft expert sets (ENCSESs), P-order, P-union, P-intersection, P-AND, P-OR, and R-order, R-union, R-intersection, R-AND, and R-OR of NCSESs. The NCSESs satisfy the laws of commutativity, associativity, De Morgan, distributivity, idempotentency, and absorption. We derive some conditions for P-union and P-intersection of two INCSESs to be an INCSES. It is shown that P-union and P-intersection of ENCSESs need not be an ENCSES. The R-union and R-intersection of the INCSESs (resp., ENCSESs) need not be an INCSES (resp. ENCSES). Necessary conditions for the P-union, R-union, and R-intersection of two ENCSESs to be an ENCSES are obtained. We also study the conditions for R-intersection and P-intersection of two NCSESs to be an INCSES and ENCSES. This paper is organized as follows. Section 2 will be on preliminaries, while Section 3 develops an approach to neutrosophic cubic soft expert set. We focus on the basic operations, namely P-order, R-order, P-containment, R-containment, P-union, P-intersection, R-union, R-intersection, complement, P-AND, P-OR, R-AND, and R-OR of NCSESs. Section 4 will present more results on NCSESs, followed by Section 5 on application in analyzing a cricket series. A comparison analysis will be discussed in Section 6 and a conclusion is drawn in Section 7.

2. Preliminaries

Here we recall some of the basic material from the literature to develop the new theory. For simplicity, the symbol (F_S^E, A) stands for the soft expert set, N stands for the neutrosophic set, I^N stands for the interval neutrosophic set and (NC) for the neutrosophic cubic sets.

In psychology, decision-making (also spelled decision-making) is regarded as the cognitive process resulting in the selection of a belief or a course of action among several alternative possibilities. Every decision-making process produces a final choice, which may or may not prompt action. Decision-making is the process of identifying and choosing alternatives based on the values, preferences, and beliefs of the decision-maker. Experts set is a technique used in decision-making problems, which is further extended to generalized forms, such as fuzzy experts set, intuitionistic fuzzy expert set, cubic expert sets, neutrosophic expert set and other hybrids. We begin by stating the definition of expert set.

Definition 1. [33] *Let U be a universe, E be a set of parameters, and X be a set of experts. Let $O = \{0 =$ disagree, $1 =$agree$\}$ be a set of two valued opinion, $Z = E \times X \times O$ and $A \subseteq Z$. A pair (F_S^E, A) is called a soft expert set over U, where F_S^E is a mapping given by $F_S^E : A \longrightarrow P(U)$ where $P(U)$ denotes the power set of U.*

Definition 2. [33] *Two soft expert sets (F_S^E, A) and (G_S^E, B) over U, $(F_S^E, A) \subseteq (G_S^E, B)$ if*

$$H_S^E(a) = \begin{cases} A \subseteq B \\ F_S^E(a) \subseteq G_S^E(a) \text{ for all } a \in A \end{cases}$$

and $(F_S^E, A) = (G_S^E, B)$ if and only if $(F_S^E, A) \subseteq (G_S^E, B)$ as well as $(G_S^E, B) \subseteq (F_S^E, A)$.

Definition 3. [33] *Let E be a set of parameters and X be a set of experts. The NOT set $\acute{I}Z$ of $Z = E \times X \times O$ is defined by*

$$\acute{I}Z = \{(\acute{I}e_i, x_i, o_k) . \acute{I}e_i \notin E, x_j \in X \text{ and } o_k \in O \,\forall\, i,\ j,\ k\}$$

Definition 4. [33] *The complement of a soft expert set (F_S^E, A) is denoted by $(F_S^E, A)^c = (F_S^{Ec}, \acute{I}A)$ where $F_S^{Ec} : \acute{I}A \longrightarrow P(U)$ is a mapping given by $F_S^{Ec}(a) = U - F_S^E(\acute{I}a)$, for all $a \in \acute{I}A$.*

Definition 5. [33] *If $Z = E \times X \times \{1\}$ in Definition 1 then (F_S^E, A) is called agree soft expert set over U and it is denoted by $(F_S^E, A)_1$.*

Definition 6. [33] *If $Z = E \times X \times \{0\}$ in Definition 1 then (F_S^E, A) is called disagree soft expert set over U and it is denoted by $(F_S^E, A)_0$*

Definition 7. [33] *The union of two soft expert sets (F_S^E, A) and (G_S^E, B) over U denoted by $(F_S^E, A) \cup (G_S^E, B)$, is the soft expert set (H_S^E, C) where $C = A \cup B$, and for all $a \in C$,*

$$H_S^E(a) = \begin{cases} F_S^E(a) \text{ if } a \in A - B \\ G_S^E(a) \text{ if } a \in B - A \\ F_S^E(a) \cup G_S^E(a) \text{ if } a \in A \cap B. \end{cases}$$

Definition 8. [33] *The intersection of two soft expert sets (F_S^E, A) and (G_S^E, B) over U denoted by $(F_S^E, A) \cap (G_S^E, B)$, is the soft expert set (H_S^E, C) where $C = A \cap B$, and for all $a \in C$,*

$$H_S^E(a) = \begin{cases} F_S^E(a) \text{ if } a \in A - B \\ G_S^E(a) \text{ if } a \in B - A \\ F_S^E(a) \cap G_S^E(a) \text{ if } a \in A \cap B. \end{cases}$$

Definition 9. *[33] If (F_S^E, A) and (G_S^E, B) are two soft expert sets over U then $(F_S^E, A) AND\ (G_S^E, B)$ denoted by $(F_S^E, A) \wedge (G_S^E, B)$, is defined by*

$$\left(F_S^E, A\right) \wedge (G_S^E, B) = (H_S^E, A \times B)$$

where $H_S^E(a,b) = F_S^E(a) \cap G_S^E(b)$, for all $(a,b) \in A \times B$.

Definition 10. *[33] If (F_S^E, A) and (G_S^E, B) are two soft expert sets then $(F_S^E, A) OR (G_S^E, B)$ denoted by $(F_S^E, A) \vee (G_S^E, B)$ is defined by*

$$\left(F_S^E, A\right) \vee (G_S^E, B) = (H_S^E, A \times B)$$

where $H_S^E(a,b) = F_S^E(a) \cup G_S^E(b)$, for all $(a,b) \in A \times B$.

Definition 11. *[5] A neutrosophic set in X is the structure of the form*

$$N := \{\langle x, T_N(x), I_N(x), F_N(x)\rangle : x \in X\}$$

where $T_N, I_N, F_N : X \to [0,1]$ such that $0 \leq T_N(x) + I_N(x) + F_N(x) \leq 3$.

Definition 12. *[8] An interval neutrosophic set in X is the structure of the form*

$$I^N := \left\{\left\langle x, T_{I^N}(x), I_{I^N}(x), F_{I^N}(x)\right\rangle : x \in X\right\}$$

where $T_N, I_N, F_N : X \to D[0,1]$ such that $[0,0] \preceq T_N(x) + I_N(x) + F_N(x) \preceq [3,3]$.

Definition 13. *[21] A neutrosophic cubic set in X is a pair $(NC) = (I^N, N)$ where*

$$I^N := \{\langle x, T_{I^N}(x), I_{I^N}(x), F_{I^N}(x)\rangle : x \in X\}$$

is an interval neutrosophic set in X where $T_{I^N}, I_{I^N}, F_{I^N} : X \to D[0,1]$ and

$$N := \{\langle x, T_N(x), I_N(x), F_N(x)\rangle : x \in X\}$$

is a neutrosophic set in X where $T_N, I_N, F_N : X \to [0,1]$.

3. Neutrosophic Cubic Soft Expert Set

In this section, we develop an approach to neutrosophic cubic soft expert set which is a more general approach for soft expert set theory. We focus on the basic operations namely, P-order, R-order, P-containment, R-containment, P-union, P-intersection, R-union, R-intersection, complement, P-AND, P-OR, R-AND, and R-OR of neutrosophic cubic soft expert sets. The symbol $((NC)_S^E, E, X)$ stands for the neutrosophic cubic soft expert set.

Definition 14. *Let U be a finite set containing n alternatives, E be a set of criteria, X be a set of experts. A triplet $((NC)_S^E, E, X)$ is called neutrosophic cubic soft expert set over U, if and only if $(NC)_S^E : E \times X \to NCP(U)$ is a mapping into the set of all neutrosophic cubic set in U and defined as*

$$((NC)_S^E, E, X) = \left\{(NC)_S^E(e,x) = \{\left\langle u, I^N_{(e,x)}(u), N_{(e,x)}(u)\right\rangle, u \in U\}, (e,x) \in E \times X\right\},$$

where

$$I^N_{(e,x)}(u) = \{(u, \widetilde{T}_{IN}(u), \widetilde{I}_{IN}(u), \widetilde{F}_{IN}(u))\},\ N_{(e,x)}(x) = \{(u, T_N(u), I_N(u), F_N(u)\},$$

such that

$$[0,0] \preceq \widetilde{T}_{IN}(u) + \widetilde{I}_{IN}(u) + \widetilde{F}_{IN}(u) \preceq [3,3],\ 0 \leq T_N(u) + I_N(u) + F_N(u) \leq 3.$$

Example 1. *Let $U = \{u_1 = India, u_2 = Pakistan\}$ be the set of countries playing a cricket series, $E = \{e_1 =$ playing conditions , $e_2 =$ historic record$\}$ be the set of factors affecting the series, $X = \{x_1, x_2, x_3\}$ be the set of experts giving their expert opinion. Let $E \times X = \{(e_1, x_1), (e_1, x_2), (e_2, x_1), (e_2, x_2)\}$. Then the neutrosophic cubic soft expert set $((NC)^E_S, E, X)$ is given by*

$$(NC)^E_S(e_1, x_1) = \left\{ \begin{array}{l} (u_1, [0.5, 0.6], [0.2, 0.3], [0.1, 0.2], 0.1, 0.4, 0.5), \\ (u_2, [0.6, 0.9], [0.6, 0.9], [0.6, 0.9], 0.9, 0.7, 0.6), \end{array} \right\}$$

$$(NC)^E_S(e_1, x_2) = \left\{ \begin{array}{l} (u_1, [0.5, 0.6], [0.2, 0.3], [0.1, 0.2], 0.1, 0.4, 0.5), \\ (u_2, [0.6, 0.9], [0.6, 0.9], [0.6, 0.9], 0.9, 0.7, 0.6), \end{array} \right\}$$

$$(NC)^E_S(e_1, x_3) = \left\{ \begin{array}{l} (u_1, [0.5, 0.6], [0.2, 0.3], [0.1, 0.2], 0.4, 0.3, 0.5), \\ (u_2, [0.6, 0.9], [0.6, 0.9], [0.6, 0.9], 0.9, 0.7, 0.6), \end{array} \right\}$$

$$(NC)^E_S(e_2, x_1) = \left\{ \begin{array}{l} (u_1, [0.6, 0.9], [0.6, 0.9], [0.6, 0.9], 0.9, 0.7, 0.6), \\ (u_2, [0.6, 0.9], [0.6, 0.9], [0.6, 0.9], 0.9, 0.7, 0.6), \end{array} \right\}$$

$$(NC)^E_S(e_2, x_2) = \left\{ \begin{array}{l} (u_1, [0.6, 0.9], [0.6, 0.9], [0.6, 0.9], 0.9, 0.7, 0.6), \\ (u_2, [0.6, 0.9], [0.6, 0.9], [0.6, 0.9], 0.9, 0.7, 0.6), \end{array} \right\}$$

$$(NC)^E_S(e_2, x_3) = \left\{ \begin{array}{l} (u_1, [0.6, 0.9], [0.6, 0.9], [0.6, 0.9], 0.9, 0.7, 0.6), \\ (u_2, [0.6, 0.9], [0.6, 0.9], [0.6, 0.9], 0.9, 0.7, 0.6), \end{array} \right\}$$

The function of the form $(\widetilde{T}_{IN}(u), T_N(u))$ denotes the range of values where the experts are sure to give certain membership to a certain element, $(\widetilde{I}_{IN}(u), I_N(u))$ denotes the range of values where the experts are hesitant and $(\widetilde{F}_{IN}(u), F_N(u))$ denotes the range of values where the experts are sure to give negative points to a certain element as a non-membership. Thus, experts have a wide range of scale to make their conclusion as compared to the previous defined versions of fuzzy sets. More specific in the current example is the function of the form $(\widetilde{T}_{IN}(u), T_N(u))$ which gives the expert opinion for the past performance of these two countries, $(\widetilde{I}_{IN}(u), I_N(u))$ gives the expert opinion for running series between these two countries and $(\widetilde{F}_{IN}(u), F_N(u))$ gives the expert opinion for the upcoming series between these two countries which is not to be held in the near future.

Definition 15. *A neutrosophic cubic soft expert set*

$$((NC)^E_S, E, X) = \left\{(NC)^E_S(e,x) = \{\left\langle u, I^N_{(e,x)}(u), N_{(e,x)}(u)\right\rangle, u \in U\}, (e,x) \in E \times X\right\}$$

over U is said to be:
(i) Internal truth neutrosophic cubic soft experts set (briefly, ITNCSESs) if for all $u \in U$, so that

$$T^-_{IN}(u) \leq T_N(u) \leq T^+_{IN}(u), \forall\ u \in U.$$

(ii) Internal indeterminacy neutrosophic cubic soft experts set (briefly, IINCSESs) if for all $u \in U$, so that

$$I^-_{IN}(u) \leq I_N(u) \leq I^+_{IN}(u), \forall\ u \in U.$$

(iii) Internal falsity neutrosophic cubic soft experts set (briefly, IFNCSESs) if for all $u \in U$, so that

$$F^-_{IN}(u) \leq F_N(u) \leq F^+_{IN}(u), \forall u \in U.$$

If a neutrosophic cubic soft expert set $((NC)_S^E, E, X)$ in X, satisfies (i), (ii), (iii), then it is known as internal neutrosophic cubic soft expert set in X, abbreviated as $(INCSESs)$.

Example 2. *Consider the Example 1. Then the internal neutrosophic cubic soft expert set is given by*

$$(NC)_S^E(e_1, x_1) = \left\{ \begin{array}{l} (u_1, [0.4, 0.6], [0.2, 0.5], [0.1, 0.5], 0.5, 0.4, 0.3), \\ (u_2, [0.6, 0.9], [0.5, 0.9], [0.6, 0.8], 0.7, 0.6, 0.7), \end{array} \right\}$$

$$(NC)_S^E(e_2, x_1) = \left\{ \begin{array}{l} (u_1, [0.5, 0.7], [0.2, 0.4], [0.1, 0.4], 0.6, 0.3, 0.2), \\ (u_2, [0.6, 0.9], [0.7, 0.9], [0.6, 0.8], 0.7, 0.8, 0.7), \end{array} \right\}$$

$$(NC)_S^E(e_1, x_2) = \left\{ \begin{array}{l} (u_1, [0.5, 0.8], [0.1, 0.3], [0.1, 0.4], 0.7, 0.2, 0.3), \\ (u_2, [0.5, 0.9], [0.6, 0.8], [0.4, 0.9], 0.6, 0.7, 0.6), \end{array} \right\}$$

$$(NC)_S^E(e_2, x_2) = \left\{ \begin{array}{l} (u_1, [0.6, 0.8], [0.3, 0.9], [0.6, 0.9], 0.7, 0.7, 0.8), \\ (u_2, [0.3, 0.9], [0.7, 0.9], [0.6, 0.8], 0.6, 0.8, 0.7) \end{array} \right\}$$

Remark 1. *We can draw the following conclusion from Example 2;*

(i) If the value of $N_{(e,x)}(u)$ lies in the interval $I^N_{(e,x)}(u)$, then it means that the respective team is going to maintain its progress in different time frames.

(ii) If the panel of experts consists of the internal panel (meaning that the experts are from the same country or same cricket board), then it is known as INCSESs.

Definition 16. *A neutrosophic cubic soft expert set*

$$((NC)_S^E, E, X) = \left\{ (NC)_S^E(e, x) = \{ \left\langle u, I^N_{(e,x)}(u), N_{(e,x)}(u) \right\rangle, u \in U\}, (e, x) \in E \times X \right\}$$

over U is said to be:
(i) External truth neutrosophic cubic soft expert set (briefly, $ETNCSESs$) if for all $u \in U$, we have

$$T_N(u) \notin (T^-_{IN}(u), T^+_{IN}(u)), \forall\, u \in U$$

(ii) External indeterminacy neutrosophic cubic soft expert set (briefly, $EINCSESs$) if for all $u \in U$, we have

$$I_N(x) \notin (I^-_{IN}(x), I^+_{IN}(x)), \forall\, u \in U$$

(iii) External falsity neutrosophic cubic soft expert set (briefly, $EFNCSESs$) if for all $u \in U$, we have

$$F_N(x) \notin (F^-_{IN}(x), F^+_{IN}(x)), \forall\, u \in U$$

If a neutrosophic cubic soft expert set $((NC)_S^E, E, X)$ over U, satisfies (i), (ii), (iii), then it is known as external neutrosophic cubic soft expert set in X, abbreviated as $(ENCSESs)$.

Example 3. *Let U be the set of countries playing a one-day international (ODI) triangular series provided in Example 1, then the external neutrosophic cubic soft expert set is given by;*

$$(NC)_S^E(e_1, x_1) = \left\{ \begin{array}{l} (u_{1,} [0.4, 0.6], [0.2, 0.5], [0.1, 0.5], 0.3, 0.1, 0.7), \\ (u_{2,} [0.6, 0.7], [0.5, 0.6], [0.6, 0.8], 0.8, 0.7, 0.9), \end{array} \right\}$$

$$(NC)_S^E(e_2, x_1) = \left\{ \begin{array}{l} (u_{1,} [0.5, 0.7], [0.2, 0.4], [0.1, 0.4], 0.8, 0.5, 0.6), \\ (u_{2,} [0.4, 0.6], [0.3, 0.5], [0.6, 0.8], 0.7, 0.2, 0.4), \end{array} \right\}$$

$$(NC)_S^E(e_1, x_2) = \left\{ \begin{array}{l} (u_{1,} [0.5, 0.8], [0.1, 0.3], [0.1, 0.4], 0.4, 0.4, 0.5), \\ (u_{2,} [0.1, 0.3], [0.6, 0.8], [0.4, 0.6], 0.6, 0.5, 0.3), \end{array} \right\}$$

$$(NC)_S^E(e_2, x_2) = \left\{ \begin{array}{l} (u_{1,} [0.6, 0.7], [0.3, 0.4], [0.6, 0.8], 0.8, 0.7, 0.5), \\ (u_{2,} [0.3, 0.5], [0.7, 0.8], [0.6, 0.7], 0.6, 0.5, 0.8), \end{array} \right\}$$

Remark 2. *We can draw the following conclusion from Example 3;*

(i) If the value of $N_{(e,x)}(u)$ does not lie in the interval $I_{(e,x)}^N(u)$, then it means the respective team is not maintaining its progress in different time frames.

(ii) If the panel of experts consists of the external panel (meaning that the experts are not from the same country or same cricket board), then it is known as ENCSESs.

Our next discussion is to define some basic operations on neutrosophic cubic soft expert sets to get more insight of neutrosophic cubic soft expert sets.

Definition 17. *A NCSESs $((NC)_{S1}^E, E_1, X_1)$ over U is said to be P-order contained in another NCSESs $((NC)_{S2}^E, E_2, X_2)$ over U, denoted by $((NC)_{S1}^E, E_1, X_1) \subseteq_P ((NC)_{S2}^E, E_2, X_2)$,*

if (i) $E_1 \subseteq E_2$,

(ii) $X_1 \subseteq X_2$,

(iii) $(NC)_{S1}^E(e, x) \subseteq_P (NC)_{S2}^E(e, x)$ for all $e \in E_1, x \in X_1$, where condition (iii) implies that

$$I_{1(NC)_S^E(e_1,x_1)}^N(x) \preceq I_{2(NC)_S^E(e_2,x_2)}^N(x), \ N_{1(NC)_S^E(e_1,x_1)}(x) \leq N_{2(NC)_S^E(e_2,x_2)}(x).$$

Definition 18. *A NCSESs $((NC)_{S1}^E, E_1, X_1)$ over U is said to be R-order contained in another NCSES $((NC)_{S2}^E, E_2, X_2)$ over U, denoted by $((NC)_{S1}^E, E_1, X_1) \subseteq_R ((NC)_{S2}^E, E_2, X_2)$,*

if (i) $E_1 \subseteq E_2$,

(ii) $X_1 \subseteq X_2$,

(iii) $(NC)_{S1}^E(e, x) \subseteq_R (NC)_{S2}^E(e, x)$ for all $e \in E_1, x \in X_1$,

where condition (iii) implies that

$$I_{1(NC)_S^E(e_1,x_1)}^N(x) \preceq I_{2(NC)_S^E(e_2,x_2)}^N(x), \ N_{1(NC)_S^E(e_1,x_1)}(x) \geq N_{2(NC)_S^E(e_2,x_2)}(x).$$

Definition 19. *Two NCSESs $((NC)_{S1}^E, E_1, X_1)$ and $((NC)_{S2}^E, E_2, X_2)$ over U is said to be equal which is denoted by $((NC)_{S1}^E, E_1, X_1) = ((NC)_{S2}^E, E_2, X_2)$,*

if (i) $A = B$,

(ii) $X_1 = X_2$,

(iii) $(NC)_{S1}^E(e) = (NC)_{S2}^E(e)$ for all $e \in E = A = B, x \in X = X_1 = X_2$,

where condition (iii) implies that

$$I_{1(NC)_S^E(e_1,x_1)}^N(x) = I_{2(NC)_S^E(e_2,x_2)}^N(x), \ N_{1(NC)_S^E(e_1,x_1)}(x) = N_{2(NC)_S^E(e_2,x_2)}(x).$$

Remark 3. *(a) We observe from Definitions 17–19, that for any two NCSESs $((NC)_{S1}^E, E_1, X_1)$ and $((NC)_{S2}^E, E_2, X_2)$ over U;*

(i) If $((NC)_{S1}^E, E_1, X_1) \subseteq_P ((NC)_{S2}^E, E_2, X_2)$ and $((NC)_{S2}^E, E_2, X_2) \subseteq_P ((NC)_{S1}^E, E_1, X_1)$,

then $((NC)^E_{S1}, E_1, X_1) = ((NC)^E_{S2}, E_2, X_2)$,
(ii) If $((NC)^E_{S1}, E_1, X_1) \subseteq_R ((NC)^E_{S2}, E_2, X_2)$ *and* $((NC)^E_{S2}, E_2, X_2) \subseteq_R ((NC)^E_{S1}, E_1, X_1)$,
then $((NC)^E_{S1}, E_1, X_1) = ((NC)^E_{S2}, E_2, X_2)$.

(b) Using Definitions 17–19, one can easily compare the performance of two cricket teams in different time frames.

Definition 20. *Let* $((NC)^E_{S1}, E_1, X_1)$ *and* $((NC)^E_{S2}, E_2, X_2)$ *be two NCSESs in U.*

Then we define (i) $((NC)^E_{S1}, E_1, X_1) \cup_P ((NC)^E_{S2}, E_2, X_2) = ((NC)^E_{S3}, E_3, X_3)$, *where* $E_3 = E_1 \cup E_2, X_3 = X_1 \cup X_2$

$$(NC)^E_{S3}(e_i) = \begin{cases} (NC)^E_{S1}(e_i) & \text{if } e_i \in E_1 - E_2 \\ (NC)^E_{S2}(e_i) & \text{if } e_i \in E_2 - E_1 \\ (NC)^E_{S1}(e_i) \vee_P (NC)^E_{S2}(e_i) & \text{if } e_i \in E_1 \cap E_2 \end{cases}$$

where

$$(NC)^E_{S1}(e_i) \vee_P (NC)^E_{S2}(e_i) = \left\{\left\langle u, I^N_{(NC)^E_{S1}(e_i)} \vee I^N_{(NC)^E_{S2}(e_i)}, N_{(NC)^E_{S1}(e_i)} \vee N_{(NC)^E_{S2}(e_i)} \right\rangle : u \in U\right\}$$
$$= \left\{\left\langle \begin{matrix} u, \begin{pmatrix} \widetilde{T}_{I^N_{(NC)^E_{S1}(e_i)}}(u) \vee \widetilde{T}_{I^N_{(NC)^E_{S2}(e_i)}}(u), \widetilde{I}_{I^N_{(NC)^E_{S1}(e_i)}}(u) \vee \widetilde{I}_{I^N_{(NC)^E_{S2}(e_i)}}(u) \\ , \widetilde{F}_{I^N_{(NC)^E_{S1}(e_i)}}(u) \wedge \widetilde{T}_{I^N_{(NC)^E_{S2}(e_i)}}(u) \end{pmatrix}, \\ \begin{pmatrix} T_{N_{(NC)^E_{S1}(e_i)}}(u) \vee T_{N_{(NC)^E_{S2}(e_i)}}(u), I_{N_{(NC)^E_{S1}(e_i)}}(u) \vee I_{N_{(NC)^E_{S2}(e_i)}}(u), \\ F_{N_{(NC)^E_{S1}(e_i)}}(u) \wedge F_{N_{(NC)^E_{S2}(e_i)}}(u) \end{pmatrix} \end{matrix} \right\rangle\right\}$$

(ii) $((NC)^E_{S1}, E_1, X_1) \cap_P ((NC)^E_{S2}, E_2, X_2) = ((NC)^E_{S3}, E_3, X_3)$, *where* $E_3 = E_1 \cap E_2, X_3 = X_1 \cap X_2$

$$(NC)^E_{S3}(e_i) = \left\{ (NC)^E_{S1}(e_i) \wedge_P (NC)^E_{S2}(e_i) \text{ if } e_i \in E_1 \cap E_2 \right.$$

where

$$(NC)^E_{S1}(e_i) \wedge_P (NC)^E_{S2}(e_i) = \left\{\left\langle u, I^N_{(NC)^E_{S1}(e_i)} \wedge I^N_{(NC)^E_{S2}(e_i)}, N_{(NC)^E_{S1}(e_i)} \wedge N_{(NC)^E_{S2}(e_i)} \right\rangle : u \in U\right\}$$
$$= \left\{\left\langle \begin{matrix} u, \begin{pmatrix} \widetilde{T}_{I^N_{(NC)^E_{S1}(e_i)}}(u) \wedge \widetilde{T}_{I^N_{(NC)^E_{S2}(e_i)}}(u), \widetilde{I}_{I^N_{(NC)^E_{S1}(e_i)}}(u) \wedge \widetilde{I}_{I^N_{(NC)^E_{S2}(e_i)}}(u), \\ \widetilde{F}_{I^N_{(NC)^E_{S1}(e_i)}}(u) \vee \widetilde{T}_{I^N_{(NC)^E_{S2}(e_i)}}(u) \end{pmatrix} \\ \begin{pmatrix} T_{N_{(NC)^E_{S1}(e_i)}}(u) \wedge T_{N_{(NC)^E_{S2}(e_i)}}(u), I_{N_{(NC)^E_{S1}(e_i)}}(u) \wedge I_{N_{(NC)^E_{S2}(e_i)}}(u), \\ F_{N_{(NC)^E_{S1}(e_i)}}(u) \vee F_{N_{(NC)^E_{S2}(e_i)}}(u) \end{pmatrix} \end{matrix} \right\rangle\right\}$$

(iii) $((NC)^E_{S1}, E_1, X_1) \cup_R ((NC)^E_{S2}, E_2, X_2) = ((NC)^E_{S3}, E_3, X_3)$, *where* $E_3 = E_1 \cup E_2, X_3 = X_1 \cup X_2$,

$$(NC)^E_{S3}(e_i) = \begin{cases} (NC)^E_{S1}(e_i) & \text{if } e_i \in E_1 - E_2 \\ (NC)^E_{S2}(e_i) & \text{if } e_i \in E_2 - E_1 \\ (NC)^E_{S1}(e_i) \vee_R (NC)^E_{S2}(e_i) & \text{if } e_i \in E_1 \cap E_2 \end{cases}$$

where

$$(NC)^E_{S1}(e_i) \vee_R (NC)^E_{S2}(e_i) = \left\{\left\langle u, I^N_{(NC)^E_{S1}(e_i)} \vee I^N_{(NC)^E_{S2}(e_i)}, N_{(NC)^E_{S1}(e_i)} \wedge N_{(NC)^E_{S2}(e_i)}\right\rangle : u \in U\right\}$$

$$= \{\left\langle u, \begin{pmatrix} \widetilde{T}_{I^N_{(NC)^E_{S1}(e_i)}}(u) \vee \widetilde{T}_{I^N_{(NC)^E_{S2}(e_i)}}(u), \widetilde{I}_{I^N_{(NC)^E_{S1}(e_i)}}(u) \vee \widetilde{I}_{I^N_{(NC)^E_{S2}(e_i)}}(u), \\ \widetilde{F}_{I^N_{(NC)^E_{S1}(e_i)}}(u) \wedge \widetilde{T}_{I^N_{(NC)^E_{S2}(e_i)}}(u) \end{pmatrix} \begin{pmatrix} T_{N_{(NC)^E_{S1}(e_i)}}(u) \wedge T_{N_{(NC)^E_{S2}(e_i)}}(u), I_{N_{(NC)^E_{S1}(e_i)}}(u) \wedge I_{N_{(NC)^E_{S2}(e_i)}}(u), \\ F_{N_{(NC)^E_{S1}(e_i)}}(u) \vee F_{N_{(NC)^E_{S2}(e_i)}}(u) \end{pmatrix} \right\rangle$$

(iv) $((NC)^E_{S1}, E_1, X_1) \cap_R ((NC)^E_{S2}, E_2, X_2) = ((NC)^E_{S3}, E_3, X_3)$, *where* $E_3 = E_1 \cap E_2, X_3 = X_1 \cap X_2$,

$$(NC)^E_{S3}(e_i) = \left\{ (NC)^E_{S1}(e_i) \wedge_R (NC)^E_{S2}(e_i) \text{ if } e_i \in E_1 \cap E_2 \right.$$

where

$$(NC)^E_{S1}(e_i) \wedge_R (NC)^E_{S2}(e_i) = \left\{\left\langle u, I^N_{(NC)^E_{S1}(e_i)} \wedge I^N_{(NC)^E_{S2}(e_i)}, N_{(NC)^E_{S1}(e_i)} \vee N_{(NC)^E_{S2}(e_i)}\right\rangle : u \in U\right\}$$

$$= \{\left\langle u, \begin{pmatrix} \widetilde{T}_{I^N_{(NC)^E_{S1}(e_i)}}(u) \wedge \widetilde{T}_{I^N_{(NC)^E_{S2}(e_i)}}(u), \widetilde{I}_{I^N_{(NC)^E_{S1}(e_i)}}(u) \wedge \widetilde{I}_{I^N_{(NC)^E_{S2}(e_i)}}(u), \\ \widetilde{F}_{I^N_{(NC)^E_{S1}(e_i)}}(u) \vee \widetilde{T}_{I^N_{(NC)^E_{S2}(e_i)}}(u) \end{pmatrix} \begin{pmatrix} T_{N_{(NC)^E_{S1}(e_i)}}(u) \vee T_{N_{(NC)^E_{S2}(e_i)}}(u), I_{N_{(NC)^E_{S1}(e_i)}}(u) \vee I_{N_{(NC)^E_{S2}(e_i)}}(u), \\ F_{N_{(NC)^E_{S1}(e_i)}}(u) \wedge F_{N_{(NC)^E_{S2}(e_i)}}(u) \end{pmatrix} \right\rangle\}$$

(v) The complement of a neutrosophic cubic soft expert set $((NC)^E_S, E, X)$ *denoted by*

$$\left((NC)^E_S, E, X\right)^c = \{(NC)^{Ec}_S(e_i) = \{\left\langle u, \widetilde{1} - I^N_{(NC)^E_S(e_i)}(u), 1 - N_{(NC)^E_S(e_i)}(u)\right\rangle, u \in U\}, e_i \in A\}.$$

Proposition 1. *Let* $((NC)^E_{S1}, E_1, X_1)$, $((NC)^E_{S2}, E_2, X_2)$, $((NC)^E_{S3}, E_3, X_3)$, $((NC)^E_{S4}, E_4, X_4)$ *be NCSESs in U. Then*

(i) If $((NC)^E_{S1}, E_1, X_1) \subseteq_P ((NC)^E_{S2}, E_2, X_2)$ *and* $((NC)^E_{S3}, E_3, X_3) \subseteq_P ((NC)^E_{S4}, E_4, X_4)$ *then* $((NC)^E_{S1}, E_1, X_1) \subseteq_P ((NC)^E_{S4}, E_4, X_4)$.

(ii) If $((NC)^E_{S1}, E_1, X_1) \subseteq_P ((NC)^E_{S2}, E_2, X_2)$ *then* $((NC)^E_{S2}, E_2, X_2)^c \subseteq_P ((NC)^E_{S1}, E_1, X_1)^c$.

(iii) If $((NC)^E_{S1}, E_1, X_1) \subseteq_P ((NC)^E_{S2}, E_2, X_2)$ *and* $((NC)^E_{S1}, E_1, X_1) \subseteq_P ((NC)^E_{S4}, E_4, X_4)$ *then* $((NC)^E_{S1}, E_1, X_1) \subseteq_P ((NC)^E_{S2}, E_2, X_2) \cap_P ((NC)^E_{S4}, E_4, X_4)$.

(iv) If $((NC)^E_{S1}, E_1, X_1) \subseteq_P ((NC)^E_{S2}, E_2, X_2)$ *and* $((NC)^E_{S3}, E_3, X_3) \subseteq_P ((NC)^E_{S2}, E_2, X_2)$ *then* $((NC)^E_{S1}, E_1, X_1) \cup_P ((NC)^E_{S3}, E_3, X_3) \subseteq_P ((NC)^E_{S2}, E_2, X_2)$.

(v) If $((NC)^E_{S1}, E_1, X_1) \subseteq_P ((NC)^E_{S2}, E_2, X_2)$ *and* $((NC)^E_{S3}, E_3, X_3) \subseteq_P ((NC)^E_{S4}, E_4, X_4)$ *then* $((NC)^E_{S1}, E_1, X_1) \cup_P ((NC)^E_{S3}, E_3, X_3) \subseteq_P ((NC)^E_{S2}, E_2, X_2) \cup_P ((NC)^E_{S4}, E_4, X_4)$, $((NC)^E_{S1}, E_1, X_1) \cap_P ((NC)^E_{S3}, E_3, X_3) \subseteq_P ((NC)^E_{S2}, E_2, X_2) \cap_P ((NC)^E_{S4}, E_4, X_4)$.

Proof. The proof is straightforward. □

Theorem 1. *For any two NCSESs* $((NC)^E_{S1}, E_1, X_1)$ *and* $((NC)^E_{S2}, E_2, X_2)$ *over U the following properties hold:*

(i) Idempotent law: $((NC)^E_{S1}, E_1, X_1) \cup_P ((NC)^E_{S1}, E_1, X_1) = ((NC)^E_{S1}, E_1, X_1)$, $((NC)^E_{S1}, E_1, X_1) \cap_P ((NC)^E_{S1}, E_1, X_1) = ((NC)^E_{S1}, E_1, X_1)$.

(ii) Commutative law: $((NC)^E_{S1}, E_1, X_1) \cup_P ((NC)^E_{S2}, E_2, X_2) = ((NC)^E_{S2}, E_2, X_2) \cup_P ((NC)^E_{S1}, E_1, X_1)$, $((NC)^E_{S1}, E_1, X_1) \cap_P ((NC)^E_{S2}, E_2, X_2) = ((NC)^E_{S2}, E_2, X_2) \cap_P ((NC)^E_{S1}, E_1, X_1)$.

(iii) Associative law: $(((NC)^E_{S1}, E_1, X_1) \cup_P ((NC)^E_{S2}, E_2, X_2)) \cup_P ((NC)^E_{S3}, E_3, X_3) = ((NC)^E_{S1}, E_1, X_1) \cup_P(((NC)^E_{S2}, E_2, X_2) \cup_P ((NC)^E_{S3}, E_3, X_3))$,

$(((NC)^E_{S1}, E_1, X_1) \cap_P ((NC)^E_{S2}, E_2, X_2)) \cap_P ((NC)^E_{S3}, E_3, X_3) = ((NC)^E_{S1}, E_1, X_1) \cap_P (((NC)^E_{S2}, E_2, X_2) \cap_P((NC)^E_{S3}, E_3, X_3))$.

(iv) Distributive and De Morgan's laws also true.

(v) Involution law: $(((NC)^E_{S1}, E_1, X_1)^c)^c = ((NC)^E_{S1}, E_1, X_1)$.

Proposition 2. *For any two NCSESs* $((NC)^E_{S1}, E_1, X_1)$ *and* $((NC)^E_{S2}, E_2, X_2)$ *over* U *the following properties are equivalent:*

(i) $((NC)^E_{S1}, E_1, X_1) \subseteq_P ((NC)^E_{S2}, E_2, X_2)$,

(ii) $((NC)^E_{S1}, E_1, X_1) \cap_P ((NC)^E_{S2}, E_2, X_2) = ((NC)^E_{S1}, E_1, X_1)$,

(iii) $((NC)^E_{S1}, E_1, X_1) \cup_P ((NC)^E_{S2}, E_2, X_2) = ((NC)^E_{S2}, E_2, X_2)$.

Proof. $(i) \Rightarrow (ii)$ By Definition 20, we have

$$\left((NC)^E_{S1}, E_1, X_1\right) \cap_P \left((NC)^E_{S2}, E_2, X_2\right) = ((NC)^E_{S1} \cap_P (NC)^E_{S2}, A \cap B) = ((NC)^E_{S1} \cap_P (NC)^E_{S2}, A)$$

as $A \subseteq B$ by hypothesis. Now for any $e \in E_1$, since $(NC)^E_{S1}(e) \subseteq_P (NC)^E_{S2}(e)$, using Definition 17, implies that $I^N_{(NC)^E_{S1}(e_i)}(u) \preceq I^N_{(NC)^E_{S2}(e_i)}(u)$ and $N_{(NC)^E_{S1}(e_i)}(u) \leq N_{(NC)^E_{S2}(e_i)}(u)$ for any $u \in U$, where $(NC)^E_{S1}(e_i) = \{\left\langle u, I^N_{(NC)^E_{S1}(e_i)}(u), N_{(NC)^E_{S1}(e_i)}(u)\right\rangle u \in U\}$. Since $I^{N-}_{(NC)^E_{S1}(e_i)}(u) \leq I^{N-}_{(NC)^E_{S2}(e_i)}(u)$ and $I^{N+}_{(NC)^E_{S1}(e_i)}(u) \leq I^{N+}_{(NC)^E_{S2}(e_i)}(u)$. Thus

$$\inf\{I^N_{(NC)^E_{S1}(e_i)}(u), I^N_{(NC)^E_{S2}(e_i)}(u)\} = [\inf\{I^{N-}_{(NC)^E_{S1}(e_i)}(u) \leq I^{N-}_{(NC)^E_{S2}(e_i)}(u)\},$$
$$\inf\{I^{N+}_{(NC)^E_{S1}(e_i)}(u) \leq I^{N+}_{(NC)^E_{S2}(e_i)}(u)\}] = [I^{N-}_{(NC)^E_{S1}(e_i)}(u) \leq I^{N+}_{(NC)^E_{S1}(e_i)}(u)]$$

and $\inf\{N_{(NC)^E_{S1}(e_i)}(u), N_{(NC)^E_{S2}(e_i)}(u)\} = N_{(NC)^E_{S1}(e_i)}(u)$. So It is ok.

$$(NC)^E_{S1}(e) \cap_P (NC)^E_{S2}(e) = \{\left\langle u, \inf\{I^N_{(NC)^E_{S1}(e_i)}(u), I^N_{(NC)^E_{S2}(e_i)}(u)\},\right.$$
$$\left.\inf\{N_{(NC)^E_{S1}(e_i)}(u), N_{(NC)^E_{S2}(e_i)}(u)\right\rangle : u \in U\} = \{\left\langle u, I^N_{(NC)^E_{S1}(e_i)}(u), N_{(NC)^E_{S1}(e_i)}(u)\right\rangle : u \in U\}$$
$$= (NC)^E_{S1}(e)$$

Hence $(NC)^E_{S1}(e) \cap_P (NC)^E_{S2}(e) = (NC)^E_S(e)$. □

(ii)⇒(iii) By Definition 20, we have

$$\left((NC)^E_{S1}, E_1, X_1\right) \cup_P \left((NC)^E_{S2}, E_2, X_2\right) = ((NC)^E_{S1} \cup_P (NC)^E_{S2}, A \cup B) = ((NC)^E_{S1} \cup_P (NC)^E_{S2}, A)$$

as $A \cap A = A$ and $(NC)^E_{S2} \cap (NC)^E_{S2} = (NC)^E_{S2}$, by hypothesis. Now for any $e \in E_1$, since $(NC)^E_{S1}(e) \cap_P (NC)^E_{S2}(e) = (NC)^E_{S1}(e)$, by Definition 20, we have

$$\inf\{I^N_{(NC)^E_{S1}(e_i)}(u), I^N_{(NC)^E_{S2}(e_i)}(u)\} = I^N_{(NC)^E_{S1}(e_i)}(u)$$
$$\text{and } \inf\{N_{(NC)^E_{S1}(e_i)}(u), N_{(NC)^E_{S2}(e_i)}(u)\} = N_{(NC)^E_{S1}(e_i)}(u)$$

this implies that

$$\sup\{I^N_{(NC)^E_{S1}(e_i)}(u), I^N_{(NC)^E_{S2}(e_i)}(u)\} = I^N_{(NC)^E_{S2}(e_i)}(u)$$
$$\text{and } \sup\{N_{(NC)^E_{S1}(e_i)}(u), N_{(NC)^E_{S2}(e_{i1})}(u)\} = N_{(NC)^E_{S2}(e_i)}(u)$$

Thus, we have

$$(NC)^E_{S1}(e) \cup_P (NC)^E_{S2}(e) = \{\left\langle \begin{array}{c} u, \sup\{I^N_{(NC)^E_{S1}(e_i)}(u), I^N_{(NC)^E_{S2}(e_i)}(u)\}, \\ \sup\{N_{(NC)^E_{S1}(e_i)}(u), N_{(NC)^E_{S2}(e_i)}(u) \end{array} \right\rangle : u \in U\}$$

$$= \{\left\langle u, I^N_{(NC)^E_{S2}(e_i)}(u), N_{(NC)^E_{S2}(e_i)}(u) \right\rangle : u \in U\} = (NC)^E_{S2}(e).$$

Hence $(NC)^E_{S1}(e) \cap_P (NC)^E_{S2}(e) = (NC)^E_{S2}(e)$.

(iii)⇒(i) By hypothesis we have

$$\left((NC)^E_{S1}, E_1, X_1\right) \cup_P \left((NC)^E_{S2}, E_2, X_2\right) = ((NC)^E_{S1} \cup_P (NC)^E_{S2}, A \cup B) = ((NC)^E_{S1} \cup_P (NC)^E_{S2}, A)$$

as $A \cup A = A$ and $(NC)^E_{S2} \cup (NC)^E_{S2} = (NC)^E_{S2} \Rightarrow A \subseteq A$ and $(NC)^E_{S2} \subseteq (NC)^E_{S2}$. Also

$$(NC)^E_{S1}(e) \cup_P (NC)^E_{S2}(e) = \{\left\langle \begin{array}{c} u, \sup\{I^N_{(NC)^E_{S1}(e_i)}(u), I^N_{(NC)^E_{S2}(e_i)}(u)\}, \\ \sup\{N_{(NC)^E_{S1}(e_i)}(u), N_{(NC)^E_{S2}(e_i)}(u) \end{array} \right\rangle : u \in U\}$$

$$= \{\left\langle u, I^N_{(NC)^E_{S2}(e_i)}(u), N_{(NC)^E_{S2}(e_i)}(u) \right\rangle : u \in U\} = (NC)^E_{S2}(e)$$

this implies that $I^N_{(NC)^E_{S1}(e_i)}(u) \preceq I^N_{(NC)^E_{S2}(e_i)}(u)$ and $N_{(NC)^E_{S1}(e_i)}(u) \leq N_{(NC)^E_{S2}(e_i)}(u)$ for any $u \in U$.

Hence $((NC)^E_{S1}, E_1, X_1) \subseteq_P ((NC)^E_{S2}, E_2, X_2)$.

Corollary 1. *If we take $X_1 = X_2 = X$ in the Proposition 2, then the following are equivalent:*
(i) $((NC)^E_{S1}, E_1, X) \subseteq_P ((NC)^E_{S2}, E_2, X)$,
(ii) $((NC)^E_{S1}, E_1, X) \cap_P ((NC)^E_{S2}, E_2, X) = ((NC)^E_{S1}, E_1, X)$,
(iii) $((NC)^E_{S1}, E_1, X) \cup_P ((NC)^E_{S2}, E_2, X) = ((NC)^E_{S2}, E_2, X)$.

Proof. The proof is straightforward. □

4. More on NCSESs

In this section, we discuss different types of union and intersection of the NCSESs and their related conditions.

1. The following example shows that R-Union of two $INCSESs$ in U need not be $INCSESs$ in U.

Example 4. *Let $((NC)^E_{S1}, E_1, X_1)$ and $((NC)^E_{S2}, E_2, X_2)$ be two $INCSESs$ in U, where*

$$\left((NC)^E_{S1}, E_1, X_1\right) = \{I^N_{(NC)^E_{S1}(e_i)} = ([0.1, 0.2], [0.4, 0.5], [0.5, 0.6]), N_{(NC)^E_{S1}(e_i)} = (0.2, 0.3, 0.4)\}$$

and

$$\left((NC)^E_{S2}, E_2, X_2\right) = \{I^N_{(NC)^E_{S2}(e_i)} = ([0.3, 0.4], [0.3, 0.5], [0.5, 0.7]), N_{(NC)^E_{S2}(e_i)} = (0.4, 0.6, 0.3)\}.$$

Now by Definition 20, we have $((NC)^E_{S1}, E_1, X_1) \cup_R ((NC)^E_{S2}, E_2, X_2) = ((NC)^E_{S3}, E_3, X_3)$

$$\left((NC)^E_{S3}, E_3, X_3\right) = \{I^N_{(NC)^E_{S3}(e_i)} = ([0.3, 0.4], [0.4, 0.5], [0.5, 0.7], N_{(NC)^E_{S3}(e_i)} = (0.2, 0.3, 0.4)\}$$

As $0.2 \notin [0.4, 0.5], 0.3 \notin [0.4, 0.5]$ *and* $0.4 \notin [0.5, 0.7]$.

Hence $((NC)^E_{S1}, E_1, X_1) \cup_R ((NC)^E_{S2}, E_2, X_2)$ is not a $INCSES$ in U.

The following theorem gives the condition under which R-union of two $INCSESs$ in U is also a $INCSES$ in U.

Theorem 2. *Let* $((NC)^E_{S1}, E_1, X_1)$ *and* $((NC)^E_{S2}, E_2, X_2)$ *be two INCSESs in U,*
where $((NC)^E_{S1}, E_1, X_1) = \left\{(NC)^E_{S1}(e,x) = \{\left\langle u, I^N_{1(e,x)}(u), N_{1(e,x)}(u)\right\rangle, u \in U\}, (e,x) \in E_1 \times X_1\right\}$
and $((NC)^E_{S2}, E_2, X_2) = \left\{(NC)^E_{S2}(f,y) = \{\left\langle u, I^N_{2(f,y)}(u), N_{2(f,y)}(u)\right\rangle, u \in U\}, (f,y) \in E_2 \times X_2\right\}$
such that

$$\sup\{I^{N-}_{(NC)^E_{S1}(e_i)}(u), I^{N-}_{(NC)^E_{S2}(e_i)}(u)\} \leq \left\{N_{(NC)^E_{S1}(e_i)}(u) \wedge N_{(NC)^E_{S2}(e_i)}(u)\right\}$$

for all $u \in U$ *and* $(g,z) \in (E_1 \cap E_2 \times X_1 \cap X_2)$. *Then* $((NC)^E_{S1}, E_1, X_1) \cup_R ((NC)^E_{S2}, E_2, X_2)$ *is INCSESs in U.*

Proof. By Definition 20, we know $((NC)^E_{S1}, E_1, X_1) \cup_R ((NC)^E_{S2}, E_2, X_2) = ((NC)^E_{S3}, E_3, X_3)$, where $E_3 = E_1 \cup E_2, X_3 = X_1 \cup X_2$,

$$(NC)^E_{S3}(e_3, x_3) = \begin{cases} (NC)^E_{S1}(e_3, x_3) & \text{if } (e_3, x_3) \in (E_1 \times X_1) - (E_2 \times X_2) \\ (NC)^E_{S2}(e_3, x_3) & \text{if } (e_3, x_3) \in (E_2 \times X_2) - (E_1 \times X_1) \\ (NC)^E_{S1}(e_3, x_3) \vee_R (NC)^E_{S2}(e_3, x_3) & \text{if } (e_3, x_3) \in (E_1 \times E_2) \cap (X_1 \times X_2) \end{cases}$$

where

$$(NC)^E_{S1}(e_3, x_3) \vee_R (NC)^E_{S2}(e_3, x_3) = \left\{\left\langle u, I^N_{(NC)^E_{S1}(e_3,x_3)} \vee I^N_{(NC)^E_{S2}(e_3,x_3)}, N_{(NC)^E_{S1}(e_3,x_3)} \wedge N_{(NC)^E_{S2}(e_3,x_3)}\right\rangle : u \in U\right\}$$

$$= \left\{\left\langle u, \begin{pmatrix} \widetilde{T}_{I^N_{(NC)^E_{S1}(e_3,x_3)}}(u) \vee \widetilde{T}_{I^N_{(NC)^E_{S2}(e_3,x_3)}}(u), \\ \widetilde{I}_{I^N_{(NC)^E_{S1}(e_3,x_3)}}(u) \vee \widetilde{I}_{I^N_{(NC)^E_{S2}(e_3,x_3)}}(u), \\ \widetilde{F}_{I^N_{(NC)^E_{S1}(e_3,x_3)}}(u) \wedge \widetilde{T}_{I^N_{(NC)^E_{S2}(e_3,x_3)}}(u) \end{pmatrix} \begin{pmatrix} T_{N_{(NC)^E_{S1}(e_3,x_3)}}(u) \wedge T_{N_{(NC)^E_{S2}(e_3,x_3)}}(u), \\ I_{N_{(NC)^E_{S1}(e_3,x_3)}}(u) \wedge I_{N_{(NC)^E_{S2}(e_3,x_3)}}(u), \\ F_{N_{(NC)^E_{S1}(e_3,x_3)}}(u) \vee F_{N_{(NC)^E_{S2}(e_3,x_3)}}(u) \end{pmatrix}\right\rangle\right.$$

If $(e_3, x_3) \in (E_1 \times X_1) - (E_2 \times X_2)$ or if $(e_3, x_3) \in (E_2 \times X_2) - (E_1 \times X_1)$ then the result is trivial. If $(e_3, x_3) \in (E_1 \cap E_2 \times X_1 \cap X_2)$, then

$$(NC)^E_{S3}(e_3, x_3) = \left\{\left\langle u, I^N_{(NC)^E_{S1}(e_3,x_3)} \vee I^N_{(NC)^E_{S2}(e_3,x_3)}, N_{(NC)^E_{S1}(e_3,x_3)} \wedge N_{(NC)^E_{S2}(e_3,x_3)}\right\rangle : u \in U\right\}. \quad \square$$

Since $((NC)^E_{S1}, E_1, X_1)$ and $((NC)^E_{S2}, E_2, X_2)$ are $INCSESs$ in U. So $I^{N-}_{(NC)^E_{S1}(e_3,x_3)}(u) \leq N_{(NC)^E_{S1}(e_3,x_3)}(u) \leq I^{N+}_{(NC)^E_{S1}(e_3,x_3)}(u)$ and $I^{N-}_{(NC)^E_{S2}(e_3,x_3)}(u) \leq N_{(NC)^E_{S2}(e_3,x_3)}(u) \leq I^{N+}_{(NC)^E_{S2}(e_3,x_3)}(u)$. Also

$$\begin{aligned} I^{N-}_{(NC)^E_{S1}(e_3,x_3)}(u) \vee I^{N-}_{(NC)^E_{S2}(e_3,x_3)}(u) &\leq N_{(NC)^E_{S1}(e_3,x_3)}(u) \wedge N_{(NC)^E_{S2}(e_3,x_3)}(u) \\ &\leq I^{N+}_{(NC)^E_{S1}(e_3,x_3)}(u) \vee I^{N+}_{(NC)^E_{S2}(e_3,x_3)}(u) \end{aligned}$$

for all $u \in U$ and $(e_3, x_3) \in (E_2 \cap E_2, X_1 \cap X_2)$. Hence $((NC)^E_{S1}, E_1, X_1) \cup_R ((NC)^E_{S2}, E_2, X_2)$ is $INCSESs$ in U.

2. The following example yields that R-intersection of two $INCSESs$ need not be a $INCSESs$.

Example 5. *Let* $((NC)^E_{S1}, E_1, X_1)$ *and* $((NC)^E_{S2}, E_2, X_2)$ *be two INCSESs in U, where*

$$\left((NC)^E_{S1}, E_1, X_1\right) = \{I^N_{(NC)^E_{S1}(e_i)} = ([0.1, 0.2], [0.3, 0.5], [0.3, 0.6]), N_{(NC)^E_{S1}(e_i)} = (0.2, 0.3, 0.4)\}$$

and

$$\left((NC)^E_{S2}, E_2, X_2\right) = \{I^N_{(NC)^E_{S2}(e_i)} = ([0.2, 0.6], [0.3, 0.6], [0.5, 0.7]), N_{(NC)^E_{S2}(e_i)} = (0.4, 0.6, 0.5)\}.$$

Now by Definition 20, we have $((NC)^E_{S1}, E_1, X_1) \cap_R ((NC)^E_{S2}, E_2, X_2) = ((NC)^E_{S3}, E_3, X_3)$

$$\left((NC)^E_{S2}, E_2, X_2\right) = \{I^N_{(NC)^E_{S3}(e_i)} = ([0.1, 0.2], [0.3, 0.5], [0.5, 0.7], N_{(NC)^E_{S\ni}(ei)} = (0.4, 0.6, 0.4)\}.$$

As $0.4 \notin [0.1, 0.2], 0.6 \notin [0.3, 0.5]$ *and* $0.4 \notin [0.5, 0.7]$.

Hence $((NC)^E_{S1}, E_1, X_1) \cap_R ((NC)^E_{S2}, E_2, X_2)$ is not a $INCSES$ in U.
The following theorem gives the condition that R-intersection of two $INCSESs$ is to be a $INCSES$.

Theorem 3. *Let* $((NC)^E_{S1}, E_1, X_1)$ *and* $((NC)^E_{S2}, E_2, X_2)$ *be two INCSESs in U,*
where $((NC)^E_{S1}, E_1, X_1) = \left\{(NC)^E_{S1}(e, x) = \{\left\langle u, I^N_{1(e,x)}(u), N_{1(e,x)}(u)\right\rangle, u \in U\}, (e, x) \in E_1 \times X_1\right\}$
and
$((NC)^E_{S2}, E_2, X_2) = \left\{(NC)^E_{S2}(f, y) = \{\left\langle u, I^N_{2(f,y)}(u), N_{2(f,y)}(u)\right\rangle, u \in U\}, (f, y) \in E_2 \times X_2\right\}$ *such that*

$$\inf\{I^{N+}_{(NC)^E_S(e_i)}(x), I^{N+}_{(NC)^E_{S2}(e_i)}(u)\} \geq \left\{N_{(NC)^E_S(e_i)}(u) \vee N_{(NC)^E_{S2}(e_i)}(u)\right\}$$

for all $u \in U$ *and* $(g, z) \in (E_1 \cap E_2 \times X_1 \cap X_2)$. *Then* $((NC)^E_{S1}, E_1, X_1) \cap_R ((NC)^E_{S2}, E_2, X_2)$ *is a INCSES in U.*

Proof. Similar to the proof of the Theorem 2. □

3. The following example yields that R-union of two $ENCSESs$ need not be an $ENCSESs$.

Example 6. *Let* $((NC)^E_{S1}, E_1, X_1)$ *and* $((NC)^E_{S2}, E_2, X_2)$ *be two ENCSESs in U, where*

$$\left((NC)^E_{S1}, E_1, X_1\right) = \{I^N_{(NC)^E_{S1}(e_i)} = ([0.3, 0.4], [0.4, 0.7], [0.3, 0.6]), N_{(NC)^E_S(e_i)} = (0.5, 0.3, 0.7)\}$$

and

$$\left((NC)^E_{S2}, E_2, X_2\right) = \{I^N_{(NC)^E_{S2}(e_i)} = ([0.2, 0.6], [0.3, 0.5], [0.5, 0.6]), N_{(NC)^E_{S2}(e_i)} = (0.1, 0.2, 0.4)\}.$$

Now by Definition 20, we have $((NC)^E_{S1}, E_1, X_1) \cup_R ((NC)^E_{S2}, E_2, X_2) = ((NC)^E_{S3}, E_3, X_3)$

$$\left((NC)^E_{S3}, E_3, X_3\right) = \{I^N_{(NC)^E_{S3}(e_i)} = ([0.3, 0.6], [0.4, 0.7], [0.5, 0.6], N_{(NC)^E_{S3}(ei)} = (0.1, 0.2, 0.7)\}.$$

As $0.1 \notin [0.3, 0.6], 0.2 \notin [0.4, 0.7]$ *and* $0.7 \notin [0.5, 0.6]$.

Hence R-union is not a $ENCSESs$ in U.
The following theorem gives the condition that R-union of two $ENCSESs$ to be a $ENCSESs$.

Theorem 4. *Let* $((NC)^E_{S1}, E_1, X_1)$ *and* $((NC)^E_{S2}, E_2, X_2)$ *be two ENCSESs in U,*
where $((NC)^E_{S1}, E_1, X_1) = \left\{(NC)^E_{S1}(e, x) = \{\left\langle u, I^N_{1(e,x)}(u), N_{1(e,x)}(u)\right\rangle, u \in U\}, (e, x) \in E_1 \times X_1\right\}$
and

$$\left((NC)^E_{S2}, E_2, X_2\right) = \left\{(NC)^E_{S2}(f,y) = \{\left\langle u, I^N_{2(f,y)}(u), N_{2(f,y)}(u)\right\rangle, u \in U\}, (f,y) \in E_2 \times X_2\right\}$$

such that

$$\begin{aligned}
&\{\inf\{\sup\{I^{N+}_{(NC)^E_{S1}(e_i)}(u), I^{N-}_{(NC)^E_{S2}(e_i)}(u)\}, \sup I^{N-}_{(NC)^E_{S1}(e_i)}(u), I^{N+}_{(NC)^E_{S2}(e_i)}(u)\}\} \\
&< \left\{N_{(NC)^E_{S1}(e_i)}(u) \wedge N_{(NC)^E_{S2}(e_i)}(u)\right\} \\
&\leq \{\sup\{I^{N+}_{(NC)^E_{S1}(e_i)}(u), I^{N-}_{(NC)^E_{S2}(e_i)}(u)\}, \sup I^{N-}_{(NC)^E_{S1}(e_i)}(u), I^{N+}_{(NC)^E_{S2}(e_i)}(u)\}\},
\end{aligned}$$

for all $u \in U$ and $(g,z) \in (E_1 \cap E_2 \times X_1 \cap X_2)$. Then $\left((NC)^E_{S1}, E_1, X_1\right) \cup_R \left((NC)^E_{S2}, E_2, X_2\right)$ is a ENCSES in U.

Proof. By Definition 20, we know $\left((NC)^E_{S1}, E_1, X_1\right) \cup_R \left((NC)^E_{S2}, E_2, X_2\right) = \left((NC)^E_{S3}, E_3, X_3\right)$, where $E_3 = E_1 \cup E_2, X_3 = X_1 \cup X_2$,

$$(NC)^E_{S3}(e_3,x_3) = \begin{cases} (NC)^E_{S1}(e_3,x_3) & \text{if } (e_3,x_3) \in (E_1 \times X_1) - (E_2 \times X_2) \\ (NC)^E_{S2}(e_3,x_3) & \text{if } (e_3,x_3) \in (E_2 \times X_2) - (E_1 \times X_1) \\ (NC)^E_{S1}(e_3,x_3) \vee_R (NC)^E_{S2}(e_3,x_3) & \text{if } (e_3,x_3) \in (E_1 \times E_2) \cap (X_1 \times X_2) \end{cases}$$

where

$$\begin{aligned}
(NC)^E_{S1}(e_3,x_3) \vee_R (NC)^E_{S2}(e_3,x_3) &= \left\{\left\langle \begin{array}{c} u, I^N_{(NC)^E_{S1}(e_3,x_3)} \vee I^N_{(NC)^E_{S2}(e_3,x_3)}, \\ N_{(NC)^E_{S1}(e_3,x_3)} \wedge N_{(NC)^E_{S2}(e_3,x_3)} \end{array}\right\rangle : u \in U\right\} \\
&= \{\left\langle \begin{array}{c} u, \begin{pmatrix} \widetilde{T}_{I^N_{(NC)^E_{S1}(e_3,x_3)}}(u) \vee \widetilde{T}_{I^N_{(NC)^E_{S2}(e_3,x_3)}}(u), \\ \widetilde{I}_{I^N_{(NC)^E_{S1}(e_3,x_3)}}(u) \vee \widetilde{I}_{I^N_{(NC)^E_{S2}(e_3,x_3)}}(u), \\ \widetilde{F}_{I^N_{(NC)^E_{S1}(e_3,x_3)}}(u) \wedge \widetilde{T}_{I^N_{(NC)^E_{S2}(e_3,x_3)}}(u) \end{pmatrix} \\ \begin{pmatrix} T_{N_{(NC)^E_{S1}(e_3,x_3)}}(u) \wedge T_{N_{(NC)^E_{S2}(e_3,x_3)}}(u), \\ I_{N_{(NC)^E_{S1}(e_3,x_3)}}(u) \wedge I_{N_{(NC)^E_{S2}(e_3,x_3)}}(u), \\ F_{N_{(NC)^E_{S1}(e_3,x_3)}}(u) \vee F_{N_{(NC)^E_{S2}(e_3,x_3)}}(u) \end{pmatrix} \end{array}\right\rangle
\end{aligned}$$

If $(e_3,x_3) \in (E_1 \cap E_2 \times X_1 \cap X_2)$, take

$$h = \begin{pmatrix} \{\inf\{\sup\{I^{N+}_{(NC)^E_{S1}(e_i)}(u), I^{N-}_{(NC)^E_{S2}(e_i)}(u)\}, \\ \sup\{I^{N-}_{(NC)^E_{S1}(e_i)}(u), I^{N+}_{(NC)^E_{S2}(e_i)}(u)\}\} \end{pmatrix}$$

and

$$k = \begin{pmatrix} \{\sup\{I^{N+}_{(NC)^E_{S1}(e_i)}(u), I^{N-}_{(NC)^E_{S2}(e_i)}(u)\}, \\ \sup I^{N-}_{(NC)^E_{S1}(e_i)}(u), I^{N+}_{(NC)^E_{S2}(e_i)}(u)\}\} \end{pmatrix}$$

Then h is one of

$$\left(I^{N+}_{(NC)^E_{S1}(e_i)}(u), I^{N-}_{(NC)^E_{S2}(e_i)}(u), I^{N-}_{(NC)^E_{S1}(e_i)}(u), I^{N+}_{(NC)^E_{S2}(e_i)}(u)\right).$$

If we choose $h = I^{N-}_{(NC)^E_{S2}(e_i)}(u)$ or $I^{N+}_{(NC)^E_{S2}(e_i)}(u)$, then

$$I^{N-}_{(NC)^E_{S1}(e_i)}(u) \leq I^{N+}_{(NC)^E_{S1}(e_i)}(u) \leq I^{N-}_{(NC)^E_{S2}(e_i)}(u) \leq I^{N+}_{(NC)^E_{S2}(e_i)}(u)$$

and so $k = I^{N+}_{(NC)^E_{S1}(e_i)}(u)$. Thus

$$\left(\begin{array}{c} \sup\{I^{N-}_{(NC)^E_{S1}(e_i)}(u), I^{N-}_{(NC)^E_{S2}(e_i)}(u)\} = I^{N-}_{(NC)^E_{S2}(e_i)}(u) \\ = h > \left\{N_{(NC)^E_{S1}(e_i)}(u) \wedge N_{(NC)^E_{S2}(e_i)}(u)\right\}. \end{array}\right).$$

Hence

$$\left(\begin{array}{c} \left\{N_{(NC)^E_{S1}(e_i)}(u) \wedge N_{(NC)^E_{S2}(e_i)}(u)\right\} \\ \notin (\sup\{I^{N-}_{(NC)^E_{S1}(e_i)}(u), I^{N-}_{(NC)^E_{S2}(e_i)}(u)\}, \\ \sup\{I^{N+}_{(NC)^E_{S1}(e_i)}(u), I^{N+}_{(NC)^E_{S2}(e_i)}(u)\}). \end{array}\right).$$

Now if $h = I^{N+}_{(NC)^E_{S2}(e_i)}(u)$ then $I^{N-}_{(NC)^E_{S1}(e_i)}(u) \leq I^{N+}_{(NC)^E_{S2}(e_i)}(u) \leq I^{N+}_{(NC)^E_{S1}(e_i)}(u)$ and so $\sup\{I^{N-}_{(NC)^E_{S1}(e_i)}(u), I^{N-}_{(NC)^E_{S2}(e_i)}(u)\}$. Assume $k = I^{N-}_{(NC)^E_{S1}(e_i)}(u)$, then we have

$$\begin{array}{c} I^{N-}_{(NC)^E_{S2}(e_i)}(u) \leq I^{N-}_{(NC)^E_{S1}(e_i)}(u) < \left\{N_{(NC)^E_{S1}(e_i)}(u) \wedge N_{(NC)^E_{S2}(e_i)}(u)\right\} \\ < I^{N+}_{(NC)^E_{S2}(e_i)}(u) \leq I^{N+}_{(NC)^E_{S1}(e_i)}(u). \end{array}$$

So, we can write

$$\left(\begin{array}{c} I^{N-}_{(NC)^E_{S2}(e_i)}(u) \leq I^{N-}_{(NC)^E_{S1}(e_i)}(u) < \\ \left\{N_{(NC)^E_{S1}(e_i)}(u) \wedge N_{(NC)^E_{S2}(e_i)}(u)\right\} \\ = I^{N+}_{(NC)^E_{S2}(e_i)}(u) \leq I^{N+}_{(NC)^E_{S1}(e_i)}(u) \end{array}\right).$$

For the case

$$\left(\begin{array}{c} I^{N-}_{(NC)^E_{S2}(e_i)}(u) \leq I^{N-}_{(NC)^E_{S1}(e_i)}(u) \\ < \left\{N_{(NC)^E_{S1}(e_i)}(u) \wedge N_{(NC)^E_{S2}(e_i)}(u)\right\} \\ = I^{N+}_{(NC)^E_{S2}(e_i)}(u) \leq I^{N+}_{(NC)^E_{S1}(e_i)}(u) \end{array}\right)$$

which contradicted the fact that $((NC)^E_{S1}, E_1, X_1)$ and $((NC)^E_{S2}, E_2, X_2)$ be two *ENCSESs* in *U*. For the case

$$\left(\begin{array}{c} I^{N-}_{(NC)^E_{S2}(e_i)}(u) \leq I^{N-}_{(NC)^E_{S1}(e_i)}(u) = \\ \left\{N_{(NC)^E_{S1}(e_i)}(u) \wedge N_{(NC)^E_{S2}(e_i)}(u)\right\} \\ < I^{N+}_{(NC)^E_{S2}(e_i)}(u) \leq I^{N+}_{(NC)^E_{S1}(e_i)}(u) \end{array}\right)$$

we have

$$\left(\begin{array}{c} \left\{N_{(NC)^E_{S1}(e_i)}(u) \wedge N_{(NC)^E_{S2}(e_i)}(u)\right\} \\ \notin (\sup\{I^{N-}_{(NC)^E_{S1}(e_i)}(u), I^{N-}_{(NC)^E_{S2}(e_i)}(u)\}) \\ = I^{N-}_{(NC)^E_{S1}(e_i)}(u) = \left\{N_{(NC)^E_{S1}(e_i)}(u) \wedge N_{(NC)^E_{S2}(e_i)}(u)\right\} \end{array}\right).$$

Again, assume that $k = I^{N-}_{(NC)^E_{S2}(e_i)}(u)$, then we have

$$\begin{array}{c} I^{N-}_{(NC)^E_{S1}(e_i)}(u) \leq I^{N-}_{(NC)^E_{S2}(e_i)}(u) < \left\{N_{(NC)^E_{S1}(e_i)}(u) \wedge N_{(NC)^E_{S2}(e_i)}(u)\right\} \\ < I^{N+}_{(NC)^E_{S2}(e_i)}(u) \leq I^{N+}_{(NC)^E_{S1}(e_i)}(u) \end{array}$$

or

$$I^{N-}_{(NC)^E_{S1}(e_i)}(u) \leq I^{N-}_{(NC)^E_{S2}(e_i)}(u) = \left\{N_{(NC)^E_{S1}(e_i)}(u) \wedge N_{(NC)^E_{S2}(e_i)}(u)\right\}$$
$$< I^{N+}_{(NC)^E_{S2}(e_i)}(u) \leq I^{N+}_{(NC)^E_{S1}(e_i)}(u).$$

For the case

$$I^{N-}_{(NC)^E_{S1}(e_i)}(u) \leq I^{N-}_{(NC)^E_{S2}(e_i)}(u) < \left\{N_{(NC)^E_{S1}(e_i)}(u) \wedge N_{(NC)^E_{S2}(e_i)}(u)\right\}$$
$$< I^{N+}_{(NC)^E_{S2}(e_i)}(u) \leq I^{N+}_{(NC)^E_{S1}(e_i)}(u)$$

which contradict $((NC)^E_{S1}, E_1, X_1)$ and $((NC)^E_{S2}, E_2, X_2)$ be two *ENCSESs* in U. For the case

$$I^{N-}_{(NC)^E_{S1}(e_i)}(u) \leq I^{N-}_{(NC)^E_{S2}(e_i)}(u) = \left\{N_{(NC)^E_{S1}(e_i)}(u) \wedge N_{(NC)^E_{S2}(e_i)}(u)\right\}$$
$$< I^{N+}_{(NC)^E_{S2}(e_i)}(u) \leq I^{N+}_{(NC)^E_{S1}(e_i)}(u)$$

we have

$$\left\{N_{(NC)^E_{S1}(e_i)}(u) \wedge N_{(NC)^E_{S2}(e_i)}(u)\right\} \notin \{((\sup\{I^{N-}_{(NC)^E_{S1}(e_i)}(u), I^{N-}_{(NC)^E_{S2}(e_i)}(u)\}),$$
$$(\sup\{I^{N+}_{(NC)^E_{S1}(e_i)}(u), I^{N+}_{(NC)^E_{S2}(e_i)}(u)\})\}$$

because

$$((\sup\{I^{N-}_{(NC)^E_{S1}(e_i)}(u), I^{N-}_{(NC)^E_{S2}(e_i)}(u)\}) =, I^{N-}_{(NC)^E_{S2}(e_i)}(u) = \left\{N_{(NC)^E_{S1}(e_i)}(u) \wedge N_{(NC)^E_{S2}(e_i)}(u)\right\}.$$

If $e_i = (e_3, x_3) \in (E_1 \times X_1) - (E_2 \times X_2)$ or if $e_i = (e_3, x_3) \in (E_2 \times X_2) - (E_1 \times X_1)$ then the result is trivial. Hence $((NC)^E_{S1}, E_1, X_1) \cup_R ((NC)^E_{S2}, E_2, X_2)$ is a *ENCSES* in U. □

4. The following example shows that *R*-intersection of two *ENCSESs* need not be *ENCSESs*.

Example 7. *Let $((NC)^E_{S1}, E_1, X_1)$ and $((NC)^E_{S2}, E_2, X_2)$ be two ENCSESs in U, where*

$$\left((NC)^E_{S1}, E_1, X_1\right) = \{I^N_{(NC)^E_{S1}(e_i)} = ([0.3, 0.4], [0.4, 0.7], [0.5, 0.6]), N_{(NC)^E_{S1}(e_i)} = (0.2, 0.3, 0.4)\}$$

and

$$\left((NC)^E_{S2}, E_2, X_2\right) = \{I^N_{(NC)^E_{S2}(e_i)} = ([0.2, 0.3], [0.3, 0.5], [0.6, 0.7]), N_{(NC)^E_{S2}(e_i)} = (0.4, 0.6, 0.5)\}.$$

Now by Definition 20, we have $((NC)^E_{S1}, E_1, X_1) \cap_R ((NC)^E_{S2}, E_2, X_2) = ((NC)^E_{S3}, E_3, X_3)$

$$\left((NC)^E_{S3}, E_3, X_3\right) = \{I^N_{(NC)^E_{S3}(e_i)} = ([0.2, 0.3], [0.3, 0.5], [0.6, 0.7], N_{(NC)^E_{S3}(ei)} = (0.4, 0.6, 0.4)\}.$$

As $0.4 \notin [0.2, 0.3], 0.6 \notin [0.3, 0.5]$ and $0.4 \notin [0.6, 0.7]$. Hence $((NC)^E_{S1}, E_1, X_1) \cap_R ((NC)^E_{S2}, E_2, X_2)$ is not a ENCSES in U.

The following theorem gives the condition that *R*-intersection of two *ENCSESs* is also a *ENCSES*.

Theorem 5. *Let $((NC)^E_{S1}, E_1, X_1)$ and $((NC)^E_{S2}, E_2, X_2)$ be two ENCSESs in U,*

where $((NC)^E_{S1}, E_1, X_1) = \left\{(NC)^E_{S1}(e,x) = \{\left\langle u, I^N_{1(e,x)}(u), N_{1(e,x)}(u)\right\rangle, u \in U\}, (e,x) \in E_1 \times X_1\right\}$
and
$((NC)^E_{S2}, E_2, X_2) \quad = \quad \left\{(NC)^E_{S2}(f,y) = \{\left\langle u, I^N_{2(f,y)}(u), N_{2(f,y)}(u)\right\rangle, u \in U\}, (f,y) \in E_2 \times X_2\right\}$
such that

$$\begin{aligned}
&\{\inf\{\sup\{I^{N+}_{(NC)^E_S(e_i)}(u), I^{N-}_{(NC)^E_{S2}(e_i)}(u)\}, \sup I^{N-}_{(NC)^E_S(e_i)}(u), I^{N+}_{(NC)^E_{S2}(e_i)}(u)\}\} \\
&< \left\{N_{(NC)^E_S(e_i)}(u) \vee N_{(NC)^E_{S2}(e_i)}(u)\right\} \\
&\leq \{\sup\{I^{N+}_{(NC)^E_S(e_i)}(u), I^{N-}_{(NC)^E_{S2}(e_i)}(u)\}, \sup I^{N-}_{(NC)^E_S(e_i)}(u), I^{N+}_{(NC)^E_{S2}(e_i)}(u)\}\}
\end{aligned}$$

for all $u \in U$ *and* $(g,z) \in (E_1 \cap E_2 \times X_1 \cap X_2)$. *Then* $((NC)^E_{S1}, E_1, X_1) \cap_R ((NC)^E_{S2}, E_2, X_2)$ *is ENCSESs in* U.

Proof. Similar to the proof of Theorem 4. □

5. The following example shows that *P*-union of two *ENCSESs* need not to be a *ENCSES*.

Example 8. *Let* $((NC)^E_{S1}, E_1, X_1)$ *and* $((NC)^E_{S2}, E_2, X_2)$ *be two ENCSESs in U, where*

$$\left((NC)^E_{S1}, E_1, X_1\right) = \{I^N_{(NC)^E_{S1}(e_i)} = ([0.3, 0.5], [0.4, 0.7], [0.5, 0.6]), N_{(NC)^E_{S1}(e_i)} = (0.6, 0.8, 0.4)\}$$

and

$$\left((NC)^E_{S2}, E_2, X_2\right) = \{I^N_{(NC)^E_{S2}(e_i)} = ([0.2, 0.3], [0.3, 0.5], [0.6, 0.7]), N_{(NC)^E_{S2}(e_i)} = (0.4, 0.6, 0.5)\}.$$

Now by Definition 20, we have

$$\left((NC)^E_{S3}, E_3, X_3\right) = \{I^N_{(NC)^E_{S3}(e_i)} = ([0.3, 0.5], [0.4, 0.7], [0.5, 0.6], N_{(NC)^E_{S3}(ei)} = (0.6, 0.8, 0.4)\}.$$

As $0.6 \notin [0.3, 0.5], 0.8 \notin [0.4, 0.7]$ *and* $0.4 \notin [0.5, 0.6]$.

Hence P-union of $((NC)^E_{S1}, E_1, X_1)$ and $((NC)^E_{S2}, E_2, X_2)$ is not *ENCSES* in U.
The following theorem gives the condition under which *P*-union of two *ENCSESs* is a *ENCSESs*.

Theorem 6. *Let* $((NC)^E_{S1}, E_1, X_1)$ *and* $((NC)^E_{S2}, E_2, X_2)$ *be two ENCSESs in U,*
where $((NC)^E_{S1}, E_1, X_1) = \left\{(NC)^E_{S1}(e,x) = \{\left\langle u, I^N_{1(e,x)}(u), N_{1(e,x)}(u)\right\rangle, u \in U\}, (e,x) \in E_1 \times X_1\right\}$
and
$((NC)^E_{S2}, E_2, X_2) \quad = \quad \left\{(NC)^E_{S2}(f,y) = \{\left\langle u, I^N_{2(f,y)}(u), N_{2(f,y)}(u)\right\rangle, u \in U\}, (f,y) \in E_2 \times X_2\right\}$
such that

$$\begin{aligned}
&\sup\{\inf\{I^{N+}_{(NC)^E_S(e_i)}(u), I^{N-}_{(NC)^E_{S2}(e_i)}(u)\}, \{I^{N-}_{(NC)^E_S(e_i)}(u), I^{N+}_{(NC)^E_{S2}(e_i)}(u)\}\} \\
&< \left\{N_{(NC)^E_S(e_i)}(u) \vee N_{(NC)^E_{S2}(e_i)}(u)\right\} \\
&\leq \inf\{\sup\{I^{N+}_{(NC)^E_S(e_i)}(u), I^{N-}_{(NC)^E_{S2}(e_i)}(u)\}, \sup I^{N-}_{(NC)^E_S(e_i)}(u), I^{N+}_{(NC)^E_{S2}(e_i)}(u)\}\}
\end{aligned}$$

for all $e \in (A \cap A \times B \cap B)$ *and* $u \in U$. *Then* $((NC)^E_{S1}, E_1, X_1) \cup_P ((NC)^E_{S2}, E_2, X_2)$ *is a ENCSES over* U.

Proof. Similar to the proof of Theorems 2 and 4. □

6. The following example shows that *P*-intersection of two *ENCSESs* need not be a *ENCSES*.

Example 9. *Let* $((NC)^E_{S1}, E_1, X_1)$ *and* $((NC)^E_{S2}, E_2, X_2)$ *be two ENCSESs in U, where*

$$\left((NC)^E_{S1}, E_1, X_1\right) = \{I^N_{(NC)^E_{S1}(e_i)} = ([0.3, 0.5], [0.4, 0.7], [0.5, 0.6]), N_{(NC)^E_{S1}(e_i)} = (0.6, 0.8, 0.4)\}$$

and

$$\left((NC)^E_{S2}, E_2, X_2\right) = \{I^N_{(NC)^E_{S2}(e_i)} = ([0.2, 0.3], [0.3, 0.5], [0.6, 0.7]), N_{(NC)^E_{S2}(e_i)} = (0.4, 0.6, 0.5)\}.$$

Now by Definition 20, we have

$$\left((NC)^E_{S3}, E_3, X_3\right) = \{I^N_{(NC)^E_{S3}(e_i)} = ([0.2, 0.3], [0.3, 0.5], [0.6, 0.7], N_{(NC)^E_{S3}(ei)} = (0.4, 0.6, 0.5)\}.$$

As $0.4 \notin [0.2, 0.3], 0.6 \notin [0.3, 0.5]$ *and* $0.5 \notin [0.6, 0.7]$.

Hence P-intersection of $((NC)^E_{S1}, E_1, X_1)$ and $((NC)^E_{S2}, E_2, X_2)$ is not $ENCSES$ in U.

Theorem 7. *Let* $\{(NC)^E_{Si}\}_{i\in I} = \{((NC)^E_{Si}, E_i, X_i)\}_{i\in I}$ *be a family of internal cubic soft expert set (INCSESs) sets over U,* $u \in U$ *for any* $e_i \in E_i$, $x \in X_i\}$. *Then* $\cup_{P_{i\in I}}\{(NC)^E_{Si}\}$ *and* $\cap_{P_{i\in I}}\{(NC)^E_{Si}\}$ *are internal cubic soft expert set (INCSESs) sets over U.*

Proof. As $\{(NC)^E_{Si}\}_{i\in I}$ be a family of internal generalized cubic soft expert set ($INCSESs$) over U so we have $I^{N-}_{(NC)^E_{Si}(e,x)}(u) \leq N_{(NC)^E_{Si}(e,x)}(u) \leq I^{N-}_{(NC)^E_{Si}(e,x)}(u)$ for each $i \in I$ this implies that

$$\cup_{P_{i\in I}} I^{N-}_{(NC)^E_{Si}(e,x)}(u) \leq \cup_{P_{i\in I}} N_{(NC)^E_{Si}(e,x)}(u) \leq \cup_{P_{i\in I}} I^{N-}_{(NC)^E_{Si}(e,x)}(u)$$

and

$$\cap_{P_{i\in I}} I^{N-}_{(NC)^E_{Si}(e,x)}(u) \leq \cap_{P_{i\in I}} N_{(NC)^E_{Si}(e,x)}(u) \leq \cap_{P_{i\in I}} I^{N-}_{(NC)^E_{Si}(e,x)}(u).$$

Hence $\cup_{\substack{p\\ i\in I}}\{(NC)^E_{Si}\}$ and $\cap_{\substack{p\\ i\in I}}\{(NC)^E_{Si}\}$ are ($INCSESs$) over U. □

Theorem 8. *Let* $((NC)^E_{S1}, E_1, X_1)$ *and* $((NC)^E_{S2}, E_2, X_2)$ *be two ENCSESs in U,*
where $((NC)^E_{S1}, E_1, X_1) = \left\{(NC)^E_{S1}(e,x) = \{\left\langle u, I^N_{1(e,x)}(u), N_{1(e,x)}(u)\right\rangle, u \in U\}, (e,x) \in E_1 \times X_1\right\}$
and $((NC)^E_{S2}, E_2, X_2) = \left\{(NC)^E_{S2}(f,y) = \{\left\langle u, I^N_{2(f,y)}(u), N_{2(f,y)}(u)\right\rangle, u \in U\}, (f,y) \in E_2 \times X_2\right\}$
such that

$$\begin{aligned}
&\sup\{\inf\{I^{N+}_{(NC)^E_S(e_i)}(u), I^{N-}_{(NC)^E_{S2}(e_i)}(u)\}, \{I^{N-}_{(NC)^E_S(e_i)}(u), I^{N+}_{(NC)^E_{S2}(e_i)}(u)\}\} \\
&< \left\{N_{(NC)^E_S(e_i)}(u) \wedge N_{(NC)^E_{S2}(e_i)}(u)\right\} \\
&\leq \inf\{\sup\{I^{N+}_{(NC)^E_S(e_i)}(u), I^{N-}_{(NC)^E_{S2}(e_i)}(u)\}, \sup I^{N-}_{(NC)^E_S(e_i)}(u), I^{N+}_{(NC)^E_{S2}(e_i)}(u)\}\}
\end{aligned}$$

for all $e \in (A \cap A \times B \cap B)$ *and* $u \in U$. *Then* $((NC)^E_{S1}, E_1, X_1) \cap_P ((NC)^E_{S2}, E_2, X_2)$ *is also an ENCSESs and INCSESs over U.*

Proof. Similar to the proof of Theorems 2 and 4. □

Theorem 9. *Let* $((NC)^E_{S1}, E_1, X_1)$ *and* $((NC)^E_{S2}, E_2, X_2)$ *be two ENCSESs in U,*
where $((NC)^E_{S1}, E_1, X_1) = \left\{(NC)^E_{S1}(e,x) = \{\left\langle u, I^N_{1(e,x)}(u), N_{1(e,x)}(u)\right\rangle, u \in U\}, (e,x) \in E_1 \times X_1\right\}$
and

$$((NC)^E_{S2}, E_2, X_2) = \left\{(NC)^E_{S2}(f,y) = \{\left\langle u, I^N_{2(f,y)}(u), N_{2(f,y)}(u)\right\rangle, u \in U\}, (f,y) \in E_2 \times X_2\right\}$$ *such that*

$$\left(N_{(NC)^E_S(e_i x)} \vee N_{(NC)^E_{S2}(e_i)}\right)(u) \in \begin{cases} \inf\{\sup\{I^{N+}_{(NC)^E_S(e_i)}(u), I^{N-}_{(NC)^E_{S2}(e_i)}(u)\}, \sup\{I^{N-}_{(NC)^E_S(e_i)}(u), I^{N+}_{(NC)^E_{S2}(e_i)}(u)\}\}, \\ \{\inf\{I^{N+}_{(NC)^E_S(e_i)}(u), I^{N-}_{(NC)^E_{S2}(e_i)}(u)\}, \inf\{I^{N-}_{(NC)^E_S(e_i)}(u), I^{N+}_{(NC)^E_{S2}(e_i)}(u)\}\} \end{cases}$$

for all $e \in (A \cap A \times B \cap B)$ *and* $u \in U$. *Then* $((NC)^E_{S1}, E_1, X_1) \cap_R ((NC)^E_{S2}, E_2, X_2)$ *is a ENCSES and INCSES over* U.

Proof. Similar to the proof of Theorems 2 and 4. □

Theorem 10. *Let* $((NC)^E_{S1}, E_1, X_1)$, *and* $((NC)^E_{S2}, E_2, X_2)$, *be any two NCSESs in* U. *Then*
(i) $((NC)^E_{S1}, E_1, X_1) \cup_P (((NC)^E_{S1}, E_1, X_1) \cap_P ((NC)^E_{S2}, E_2, X_2)) = ((NC)^E_{S1}, E_1, X_1)$,
(ii) $((NC)^E_{S1}, E_1, X_1) \cap_P (((NC)^E_{S2}, E_2, X_2) \cup_P ((NC)^E_{S2}, E_2, X_2)) = ((NC)^E_{S1}, E_1, X_1)$,
(iii) $(((NC)^E_{S1}, E_1, X_1) \cup_R (((NC)^E_{S2}, E_2, X_2)) \cap_R ((NC)^E_{S2}, E_2, X_2)) = ((NC)^E_{S1}, E_1, X_1)$,
(iv) $((NC)^E_{S1}, E_1, X_1) \cap_R (((NC)^E_{S2}, E_2, X_2) \cup_R ((NC)^E_{S2}, E_2, X_2)) = ((NC)^E_{S1}, E_1, X_1)$.

Proof. The proof is straightforward. □

Next we define some more operations on NCSESs.

Definition 21. *For two neutrosophic cubic soft expert sets (NCSESs)* $((NC)^E_{S1}, E_1, X_1)$ *and* $((NC)^E_{S2}, E_2, X_2)$ *over* U, *P-AND is denoted as*

$$\left((NC)^E_{S1}, E_1, X_1\right) \bigwedge_P \left((NC)^E_{S2}, E_2, X_2\right) = \left((NC)^E_{S3}, (E_1 \times E_2), (X_1 \times X_2)\right),$$

where

$$\begin{aligned}\left((NC)^E_{S3}, (E_1 \times E_2), (X_1 \times X_2)\right) &= (NC)^E_{S3}((e,f),(x,y)) \\ &= (NC)^E_{S1}(e,x) \cap_P (NC)^E_{S2}(f,y) \\ &= \{\left\langle u, I^N_{1(e,x)}(u), N_{1(e,x)}(u)\right\rangle, u \in U\}, (e,x) \in E_1 \times X_1\}\end{aligned}$$

and $\{\left\langle u, I^N_{2(e,x)}(u), N_{2(e,x)}(u)\right\rangle, u \in U\}, (e,x) \in E_1 \times X_1\}$.

Definition 22. *For any two NCSESs* $((NC)^E_{S1}, E_1, X_1)$ *and* $((NC)^E_{S2}, E_2, X_2)$ *over* U, *R-AND is denoted as*

$$\left((NC)^E_{S1}, E_1, X_1\right) \bigwedge_R \left((NC)^E_{S2}, E_2, X_2\right) = \left((NC)^E_{S3}, (E_1 \times E_2), (X_1 \times X_2)\right),$$

where

$$\begin{aligned}\left((NC)^E_{S3}, (E_1 \times E_2), (X_1 \times X_2)\right) &= (NC)^E_{S3}((e,f),(x,y)) = (NC)^E_{S1}(e,x) \cap_R (NC)^E_{S2}(f,y) \\ &= \{\left\langle u, I^N_{1(e,x)}(u), N_{1(e,x)}(u)\right\rangle, u \in U\}, (e,x) \in E_1 \times X_1\}\end{aligned}$$

and $\{\left\langle u, I^N_{2(e,x)}(u), N_{2(e,x)}(u)\right\rangle, u \in U\}, (e,x) \in E_1 \times X_1\}$.

Definition 23. *For any two NCSESs* $((NC)^E_{S1}, E_1, X_1)$ *and* $((NC)^E_{S2}, E_2, X_2)$ *over* U, *P-OR is denoted as*

$$\left((NC)^E_{S1}, E_1, X_1\right) \bigvee_P \left((NC)^E_{S2}, E_2, X_2\right) = \left((NC)^E_{S3}, (E_1 \times E_2), (X_1 \times X_2)\right),$$

where

$$\left((NC)^E_{S3}, (E_1 \times E_2), (X_1 \times X_2)\right) = (NC)^E_{S3}((e,f),(x,y)) = (NC)^E_{S1}(e,x) \cap_R (NC)^E_{S2}(f,y)$$
$$= \{\left\langle u, I^N_{1(e,x)}(u), N_{1(e,x)}(u)\right\rangle, u \in U\}, (e,x) \in E_1 \times X_1\}$$
$$\text{and } \{\left\langle u, I^N_{2(e,x)}(u), N_{2(e,x)}(u)\right\rangle, u \in U\}, (e,x) \in E_1 \times X_1\}.$$

Definition 24. *For any two NCSESs $((NC)^E_{S1}, E_1, X_1)$ and $((NC)^E_{S2}, E_2, X_2)$ over U, R-OR is denoted as*

$$\left((NC)^E_{S1}, E_1, X_1\right) \bigvee_R \left((NC)^E_{S2}, E_2, X_2\right) = \left((NC)^E_{S3}, (E_1 \times E_2), (X_1 \times X_2)\right),$$

where

$$\left((NC)^E_{S3}, (E_1 \times E_2), (X_1 \times X_2)\right) = (NC)^E_{S3}((e,f),(x,y)) = (NC)^E_{S1}(e,x) \cap_R (NC)^E_{S2}(f,y)$$
$$= \{\left\langle u, I^N_{1(e,x)}(u), N_{1(e,x)}(u)\right\rangle, u \in U\}, (e,x) \in E_1 \times X_1\}$$
$$\text{and } \{\left\langle u, I^N_{2(e,x)}(u), N_{2(e,x)}(u)\right\rangle, u \in U\}, (e,x) \in E_1 \times X_1\}.$$

Theorem 11. *Let $((NC)^E_{S1}, E_1, X_1)$, be a neutrosophic cubic soft expert sets in U. If $((NC)^E_{S1}, E_1, X_1)$, is an INCSESs (resp., ENCSESs). Then $((NC)^E_{S1}, E_1, X_1)^C$ is INCSESs (resp., ENCSESs) respectively.*

Proof. Since $((NC)^E_{S1}, E_1, X_1)$, is an $INCSESs$ (resp., $ENCSESs$) in U, so for any $e \in E_1$ we have $(NC)^E_{S1}(e) = \{\left\langle x, I^N_{(NC)^E_{S1}(e_i)}(x), N_{(NC)^E_{S1}(e_i)}(u)\right\rangle : u \in U\}$. As

$$I^{N-}_{(NC)^E_{S1}(e_i)}(u) \leq N_{(NC)^E_{S1}(e_i)}(u) \leq I^{N+}_{(NC)^E_{S1}(e_i)}(u)$$
$$\Rightarrow 1 - I^{N+}_{(NC)^E_{S1}(e_i)}(u) \leq 1 - N_{(NC)^E_{S1}(e_i)}(u) \leq 1 - I^{N-}_{(NC)^E_{S1}(e_i)}(u).$$

Hence $((NC)^E_{S1}, E_1, X_1)^C$ is an $INCSES$. Also

$$N_{(NC)^E_{S1}(e_i)}(u) \notin (I^{N-}_{(NC)^E_{S1}(e_i)}(u), I^{N+}_{(NC)^E_{S1}(e_i)}(u)) \ \forall\, u \in U$$
$$\Rightarrow (1 - N_{(NC)^E_{S1}(e_i)}(u) \notin (1 - I^{N+}_{(NC)^E_{S1}(e_i)}(u), 1 - I^{N-}_{(NC)^E_{S1}(e_i)}(u))$$

Hence $((NC)^E_{S1}, E_1, X_1)^C$ is an $ENCSES$. □

5. Applications

In this section, we use NCSESs to construct an algorithm and applied it to a decision-making problem. The series between Pakistan and India remains a hot cake for cricket lovers and this cricket rivalry existed between them at the start of partition. The first series between the two teams took place in 1951–52, when Pakistan made a tour of India. India made a tour of Pakistan for the first time in 1954–55. Between 1962 and 1977, no cricket was played between the two countries owing to two major wars in 1965 and 1971. The 1999 Kargil War and the 2008 Mumbai terrorist attacks have also interrupted the game of cricket between the two nations. The growth of large expatriate populations from India and Pakistan across the world led to neutral venues such as the United Arab Emirates and Canada hosting several bilateral and multilateral ODI series involving the two teams. Tickets for The India-Pakistan match in the 2015 World Cup in Australia sold out in 12 min at the ticket counters.

We try to use NCSESs in India–Pakistan cricket rivalry to conclude which country is suffering more from the above mentioned conflicts.

For this we first define the neutrosophic cubic soft expert weight average operator ($NCSEWAO$) and score function.

Definition 25. *Let* $((NC)^E_S, E, X) = \left\{(NC)^E_S(e,x) = \{\left\langle u, I^N_{(e,x)}(u), N_{(e,x)}(u)\right\rangle, u \in U\}, (e,x) \in E \times X\right\}$ *be a NCSESs. Then neutrosophic cubic soft expert weight average operator (NCSEWAO) is denoted and define as*

$$\mathbb{Q}_{w_i}(I^N_{(NC)^E_S(e_i)}(x), N_{(NC)^E_S(e_i)}(x)) = \left(\begin{array}{c} \left[\dfrac{\prod_{i=1}^{n}\left(1+I^{N-}_{(NC)^E_S(e_i)}(u)^{w_i}\right)-\prod_{i=1}^{n}\left(1-I^{N-}_{(NC)^E_S(e_i)}(u)^{w_i}\right)}{\prod_{i=1}^{n}\left(1+I^{N-}_{(NC)^E_S(e_i)}(u)^{w_i}\right)+\prod_{i=1}^{n}\left(1-I^{N-}_{(NC)^E_S(e_i)}(u)^{w_i}\right)}, \dfrac{\prod_{i=1}^{n}\left(1+I^{N+}_{(NC)^E_S(e_i)}(u)^{w_i}\right)-\prod_{i=1}^{n}\left(1-I^{N+}_{(NC)^E_S(e_i)}(u)^{w_i}\right)}{\prod_{i=1}^{n}\left(1+I^{N+}_{(NC)^E_S(e_i)}(u)^{w_i}\right)+\prod_{i=1}^{n}\left(1-I^{N+}_{(NC)^E_S(e_i)}(u)^{w_i}\right)} \right], \\ \prod_{i=1}^{n}\{N_{(NC)^E_S(e_i)}(u)^{w_i}\} \end{array} \right)$$

where w_i *is the weight of expert opinion,* $w_i \in [0,1]$ *and* $\sum_{i=1}^{n} w_i = 1$.

Definition 26. *Let* $(NC)^E_S = \left\langle \left[I^{N-}_{(NC)^E_S(e_i)}(u), I^{N+}_{(NC)^E_S(e_i)}(u)\right], N_{(NC)^E_S(e_i)}(u)\right\rangle$ *be neutrosophic cubic soft expert value. A score function Q of* $(NC)^E_S$ *is defined as* $Q((NC)^E_S) = \dfrac{I^{N-}_{(NC)^E_S(e_i)}(u)+I^{N+}_{(NC)^E_S(e_i)}(u)-N_{(NC)^E_S(e_i)}(u)}{9}$, *where* $Q((NC)^E_S) \in [-3,3]$.

Decision-making problems have been studied using fuzzy soft sets. Now, we are going to present multicriteria neutrosophic cubic soft set in decision-making along with weights and score function. For this we propose the following algorithmic steps as pictured in Figure 1.

Step 1: Define a decision problem by inputting the neutrosophic cubic soft expert set $(NC)^E_S(e_i, x_i)$.

Step 2: Use the opinions of experts in the form of $NCSESs$ to determine the opinions regarding given criteria. Make a separate table for the opinion of each expert.

Step 3: Assign weight to each expert according to their expertise.

Step 4: Apply neutrosophic cubic soft expert weighted average operator to each above table and find the neutrosophic cubic soft expert weighted average corresponding to each attribute.

Step 5: Calculate the $\bigvee_P$ of U_j.

Step 6 : Calculate the score of each U_j.

Step 7: Generate the non-increasing order of all the alternatives according to their scores.

Example 10. *Let* $U = \{u_1 = India, u_2 = Pakistan\}$ *be the set of countries playing a cricket series,* $E = \{e_1 =$ *Pakistan Cricket Board (PCB),* $e_2 =$ *The Board of Control for Cricket in India (BCCI),* $e_3 =$ *Bilateral relations between Pakistan and India*$\}$ *be the set of factors affecting the series,* $X = \{p, q, r\}$ *be the set of experts giving their expert opinion. The expert may consider the most burning parameter as "ICC's Future Tours Programme (FTP)" when they are giving their opinion by considering three times of frames as past, present, and future.*

Step 1: After a serious discussion, the committee constructed the following neutrosophic cubic soft expert set.

$$(NC)_S^E(e_1,p) = \left\{ \begin{array}{l} (u_1, [0.4,0.6], [0.2,0.5], [0.1,0.5], 0.3,0.1,0.7), \\ (u_2, [0.6,0.7], [0.5,0.6], [0.6,0.8], 0.8,0.7,0.9), \end{array} \right\}$$

$$(NC)_S^E(e_2,p) = \left\{ \begin{array}{l} (u_1, [0.5,0.7], [0.2,0.4], [0.1,0.4], 0.8,0.5,0.6), \\ (u_2, [0.4,0.6], [0.3,0.5], [0.6,0.8], 0.7,0.2,0.4), \end{array} \right\}$$

$$(NC)_S^E(e_3,p) = \left\{ \begin{array}{l} (u_1, [0.5,0.8], [0.1,0.3], [0.1,0.4], 0.4,0.4,0.5), \\ (u_2, [0.1,0.3], [0.6,0.8], [0.4,0.6], 0.6,0.5,0.3), \end{array} \right\}$$

$$(NC)_S^E(e_1,q) = \left\{ \begin{array}{l} (u_1, [0.5,0.8], [0.1,0.3], [0.1,0.4], 0.4,0.4,0.5), \\ (u_2, [0.1,0.3], [0.6,0.8], [0.4,0.6], 0.6,0.5,0.3), \end{array} \right\}$$

$$(NC)_S^E(e_2,q) = \left\{ \begin{array}{l} (u_1, [0.6,0.7], [0.3,0.4], [0.6,0.8], 0.8,0.7,0.5), \\ (u_2, [0.3,0.5], [0.7,0.8], [0.6,0.7], 0.6,0.5,0.8), \end{array} \right\}$$

$$(NC)_S^E(e_3,q) = \left\{ \begin{array}{l} (u_1, [0.4,0.6], [0.2,0.5], [0.1,0.5], 0.3,0.1,0.7), \\ (u_2, [0.6,0.7], [0.5,0.6], [0.6,0.8], 0.8,0.7,0.9), \end{array} \right\}$$

$$(NC)_S^E(e_1,r) = \left\{ \begin{array}{l} (u_1, [0.6,0.7], [0.3,0.4], [0.6,0.8], 0.8,0.7,0.5), \\ (u_2, [0.3,0.5], [0.7,0.8], [0.6,0.7], 0.6,0.5,0.8), \end{array} \right\}$$

$$(NC)_S^E(e_2,r) = \left\{ \begin{array}{l} (u_1, [0.4,0.6], [0.2,0.5], [0.1,0.5], 0.3,0.1,0.7), \\ (u_2, [0.6,0.7], [0.5,0.6], [0.6,0.8], 0.8,0.7,0.9), \end{array} \right\}$$

$$(NC)_S^E(e_3,r) = \left\{ \begin{array}{l} (u_1, [0.4,0.6], [0.2,0.5], [0.1,0.5], 0.3,0.1,0.7), \\ (u_2, [0.6,0.7], [0.5,0.6], [0.6,0.8], 0.8,0.7,0.9), \end{array} \right\}$$

Step 2: Opinion of expert p

$$(e_1,p) = \left\{ \begin{array}{l} (u_1, [0.4,0.6], [0.2,0.5], [0.1,0.5], 0.3,0.1,0.7), \\ (u_2, [0.6,0.7], [0.5,0.6], [0.6,0.8], 0.8,0.7,0.9), \end{array} \right\}$$

$$(e_2,p) = \left\{ \begin{array}{l} (u_1, [0.5,0.7], [0.2,0.4], [0.1,0.4], 0.8,0.5,0.6), \\ (u_2, [0.4,0.6], [0.3,0.5], [0.6,0.8], 0.7,0.2,0.4), \end{array} \right\}$$

$$(e_3,p) = \left\{ \begin{array}{l} (u_1, [0.5,0.8], [0.1,0.3], [0.1,0.4], 0.4,0.4,0.5), \\ (u_2, [0.1,0.3], [0.6,0.8], [0.4,0.6], 0.6,0.5,0.3), \end{array} \right\}$$

Opinion of expert q

$$(e_1,q) = \left\{ \begin{array}{l} (u_1, [0.5,0.8], [0.1,0.3], [0.1,0.4], 0.4,0.4,0.5), \\ (u_2, [0.1,0.3], [0.6,0.8], [0.4,0.6], 0.6,0.5,0.3), \end{array} \right\}$$

$$(e_2,q) = \left\{ \begin{array}{l} (u_1, [0.6,0.7], [0.3,0.4], [0.6,0.8], 0.8,0.7,0.5), \\ (u_2, [0.3,0.5], [0.7,0.8], [0.6,0.7], 0.6,0.5,0.8), \end{array} \right\}$$

$$(e_3,q) = \left\{ \begin{array}{l} (u_1, [0.4,0.6], [0.2,0.5], [0.1,0.5], 0.3,0.1,0.7), \\ (u_2, [0.6,0.7], [0.5,0.6], [0.6,0.8], 0.8,0.7,0.9), \end{array} \right\}$$

Opinion of expert r

$$(e_1, r) = \left\{ \begin{array}{l} (u_1, [0.6, 0.7], [0.3, 0.4], [0.6, 0.8], 0.8, 0.7, 0.5), \\ (u_2, [0.3, 0.5], [0.7, 0.8], [0.6, 0.7], 0.6, 0.5, 0.8), \end{array} \right\}$$

$$(e_2, r) = \left\{ \begin{array}{l} (u_1, [0.4, 0.6], [0.2, 0.5], [0.1, 0.5], 0.3, 0.1, 0.7), \\ (u_2, [0.6, 0.7], [0.5, 0.6], [0.6, 0.8], 0.8, 0.7, 0.9), \end{array} \right\}$$

$$(e_3, r) = \left\{ \begin{array}{l} (u_1, [0.4, 0.6], [0.2, 0.5], [0.1, 0.5], 0.3, 0.1, 0.7), \\ (u_2, [0.6, 0.7], [0.5, 0.6], [0.6, 0.8], 0.8, 0.7, 0.9), \end{array} \right\}$$

Step 3: Let $w = (0.5, 0.25, 0.25)$ be the weight vector assigned to the experts p, q and r respectively.

Step 4: The neutrosophic cubic soft expert weighted average (NCSEWA) of each attribute is displayed in Table 1.

Step 5: Calculate the $\bigvee_P$ of the first and second columns of Table 1 by using Definition 23. Thus we have

$$U_1 = \bigvee_{j=1}^{3} = ([0.66, 0.75], [0.80, 0.99], [0.61, 0.90], (0.95, 0.92, 0.91))$$

$$U_2 = \bigvee_{j=1}^{3} = ([0.97, 0.99], [0.98, 0.99], [0.98, 0.99], (0.68, 0.47, 0.51))$$

Step 6: Using Definition 26, we have $Q(u_1) = 0.2244$, $Q(u_2) = 0.4711$.

Step 7: The score of the $NCSESs$ values corresponding to $\bigvee_P$ of U_j implies the following order $u_2 > u_1$.

Thus, we can conclude that the country u_2 = Pakistan is affected more by the factors, e_1 = PCB, e_2 = BCCI, e_3 = Bilateral relations between Pakistan and India.

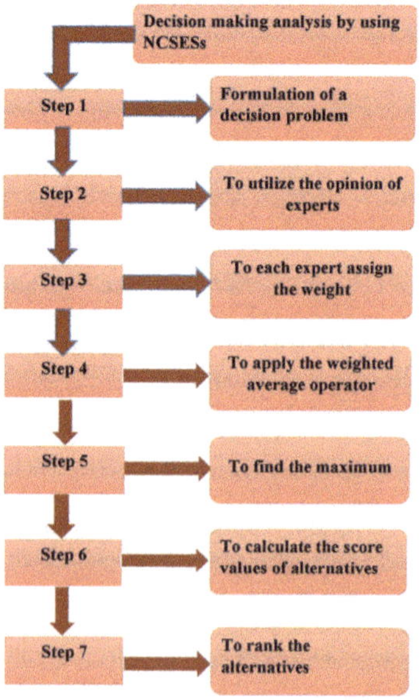

Figure 1. Flow chart of NCSESs based on MADM Problem.

Table 1. NCSEWA of each attribute.

	u_1	u_2
e_1	$\left(\begin{array}{c}([0.009, 0.051], [0.77, 0.87],\\ [0.61, 0.90]),\\ (0.41, 0.15, 0.58)\end{array}\right)$	$\left(\begin{array}{c}([0.97, 0.98], [0.98, 0.99],\\ [0.98, 0.99]),\\ (0.52, 0.14, 0.19)\end{array}\right)$
e_2	$\left(\begin{array}{c}([0.66, 0.75], [0.40, 0.49],\\ [0.54, 0.80]),\\ (0.46, 0.028, 0.048)\end{array}\right)$	$\left(\begin{array}{c}\left(\begin{array}{c}[0.78, 0.99], [0.57, 0.86],\\ [0.80, 0.99]\end{array}\right),\\ (0.68, 0.47, 0.34)\end{array}\right)$
e_3	$\left(\begin{array}{c}([0.34, 0.671], [0.80, 0.99],\\ [0.23, 0.71]),\\ (0.95, 0.92, 0.91)\end{array}\right)$	$\left(\begin{array}{c}([0.43, 0.67], [0.76, 0.98],\\ [0.34, 0.61]),\\ (0.25, 0.32, 0.51)\end{array}\right)$

6. Comparison Analysis

In this paper, we extend the concept of soft expert sets to neutrosophic cubic soft expert sets. This new idea generalizes the different types of expert sets presented by Alkhazaleh and Salleh [33,36], Broumi and Smarandache [40], Qayyum et al. [48] and Sahin et al. [43].

1. If we consider only truth part or indeterminacy part or falsity part, then our model and the model presented by Qayyum et al. [48] coincides with each other.

2. Since NCSESs consists of interval neutrosophic soft expert soft sets and neutrosophic soft expert sets and if we consider only the part containing the neutrosophic soft expert sets then our model reduces to the model presented by Sahin et al. [43].

3. If we consider the part containing the interval neutrosophic soft expert soft sets, we get a special class of soft expert sets.

4. Similarly imposing some extra conditions to our model it will reduce it to those models presented by Alkhazaleh and Salleh [33,36], Broumi and Smarandache [40].

7. Conclusions

The model of NCSESs can be used in many decision-making problems and it will produce more reliable results as compared to the previously defined versions of soft expert sets. This paper is of introductory nature where we establish this new generalized theory of soft expert sets with its basic properties and provide an application. In future we shall apply this model to other areas to illustrate its novelty. We have defined different operations of NCSESs including different versions of NCSESs. We have designed aggregation operators and score functions of neutrosophic cubic soft expert value. We have also constructed an algorithm based on this new concept and applied the developed approach to a numerical example.

The substantial impact of our outcome for the research field is that we gave a generalized form of soft expert system which certainly improves the decision-making theory due to its wide range of values in the form of truth, indeterminacy, and falsity. We have not used any software for the decision-making, since the preliminary research is on a small scale. We intend to expand the research to a larger scale on an applied problem so that we can develop a software and related interface. We also intend to apply the developed approach to a variety of games, signature theory and others, since the nature of NCSESs enables us to deal with vague and inconsistent data. In decision-making theory, we often deal with data which are inconsistent and vague. Thus, ex-ante decision-making problems can also be handled easily through NCSESs. In future, arguments and modern theories linked to behavioral patterns could strengthen the idea of NCSESs, especially using aggregation operators and methods that effectively deal with the uncertainty and inaccuracy of the input data [49].

Author Contributions: All authors contributed equally.

Funding: Universiti Kebangsaan Malaysia- Grant No. GUP-2017-105.

Acknowledgments: We are indebted to Universiti Kebangsaan Malaysia for providing financial support under the grant GUP-2017-105.

Conflicts of Interest: The authors declare no conflict of interest.

References

1. Zadeh, L.A. Fuzzy sets. *Inform. Control* **1965**, *8*, 338–353. [CrossRef]
2. Turksen, I.B. Interval-valued fuzzy sets based on normal forms. *Fuzzy Sets Syst.* **1986**, *20*, 191–210. [CrossRef]
3. Atanassov, K. Intuitionistic fuzzy sets. *Fuzzy Sets Syst.* **1986**, *20*, 87–96. [CrossRef]
4. Atanassov, K.; Pasi, G.; Yager, R. Intuitionistic fuzzy interpretations of multi-criteria multi-person and multi-measurement tool decision making. *Int. J. Syst. Sci.* **2005**, *36*, 859–868. [CrossRef]
5. Smarandache, F. *A Unifying Field in Logics: Neutrosophic Logic. Neutrosophy, Neutrosophic Set, Neutrosophic Probability*; American Reserch Press: Rehoboth, IL, USA, 1999.
6. Smarandache, F. Neutrosophic set-a generalization of the intuitionistic fuzzy set. *Int. J. Pure Appl. Math.* **2005**, *24*, 287–297.
7. De, S.K.; Beg, I. Triangular dense fuzzy neutrosophic sets. *Neutrosophic Sets Syst.* **2016**, *13*, 26–37.
8. Wang, H.; Smarandache, F.; Zhang, Y.Q.; Sunderraman, R. *Interval Neutrosophic Sets and Logic: Theory and Applications in Computing*; Hexis: Phoenix, AZ, USA, 2005.
9. Gulistan, M.; Nawaz, S.; Hassan, N. Neutrosophic triplet non-associative semihypergroups with application. *Symmetry* **2018**, *10*, 613. [CrossRef]
10. Jun, Y.B.; Kim, C.S.; Yang, K.O. Cubic Sets. *Ann. Fuzzy Math. Inform.* **2012**, *4*, 83–98.
11. Abughazalah, N.; Yaqoob, N. Applications of cubic structures to subsystems of finite state machines. *Symmetry* **2018**, *10*, 598. [CrossRef]
12. Rashid, S.; Yaqoob, N.; Akram, M.; Gulistan, M. Cubic graphs with application. *Int. J. Anal. Appl.* **2018**, *16*, 733–750.
13. Gulistan, M.; Yaqoob, N.; Vougiouklis, T.; Wahab, H.A. Extensions of cubic ideals in weak left almost semihypergroups. *J. Intell. Fuzzy Syst.* **2018**, *34*, 4161–4172. [CrossRef]
14. Ma, X.L.; Zhan, J.; Khan, M.; Gulistan, M.; Yaqoob, N. Generalized cubic relations in Hv-LA-semigroups. *J. Discret. Math. Sci. Cryptogr.* **2018**, *21*, 607–630. [CrossRef]
15. Yaqoob, N.; Gulistan, M.; Leoreanu-Fotea, V.; Hila, K. Cubic hyperideals in LA-semihypergroups. *J. Intell. Fuzzy Syst.* **2018**, *34*, 2707–2721. [CrossRef]
16. Gulistan, M.; Khan, M.; Yaqoob, N.; Shahzad, M. Structural properties of cubic sets in regular LA-semihypergroups. *Fuzzy Inf. Eng.* **2017**, *9*, 93–116. [CrossRef]
17. Khan, M.; Gulistan, M.; Yaqoob, N.; Hussain, F. General cubic hyperideals of LA-semihypergroups. *Afr. Mat.* **2016**, *27*, 731–751. [CrossRef]
18. Khan, M.; Jun, Y.B.; Gulistan, M.; Yaqoob, N. The generalized version of Jun's cubic sets in semigroups. *J. Intell. Fuzzy Syst.* **2015**, *28*, 947–960.
19. Yaqoob, N.; Mostafa, S.M.; Ansari, M.A. On cubic KU-ideals of KU-algebras. *ISRN Algebra* **2013**, *2013*, 935905. [CrossRef]
20. Aslam, M.; Aroob. T.; Yaqoob, N. On cubic Γ-hyperideals in left almost Γ-semihypergroups. *Ann. Fuzzy Math. Inform.* **2013**, *5*, 169–182.
21. Jun, Y.B. ; Smarandache, F.; Kim, C.S. Neutrosophic cubic sets. *New Math. Nat. Comput.* **2017**, *13*, 41–54. [CrossRef]
22. Khan, M.; Gulistan, M.; Yaqoob, N.; Shabir, M. Neutrosophic cubic (α,β)-ideals in semigroups with application. *J. Intell. Fuzzy Syst.* **2018**, *35*, 2469–2483. [CrossRef]
23. Gulistan, M.; Yaqoob, N.; Rashid, Z.; Smarandache, F.; Wahab, H.A. A study on neutrosophic cubic graphs with real life applications in industries. *Symmetry* **2018**, *10*, 203. [CrossRef]
24. Gulistan, M.; Khan, A.; Abdullah, A.; Yaqoob, N. Complex neutrosophic subsemigroups and ideals. *Int. J. Anal. Appl.* **2018** *16*, 97–116.
25. Molodtsov, D.A. Soft set theory-first results. *Comput. Math. Appl.* **1999**, *37*, 19–31. [CrossRef]
26. Maji, P.K.; Roy, A.R. Soft set theory. *Comput. Math. Appl.* **2003**, *45*, 555–562. [CrossRef]
27. Maji, P.K.; Biswas, R.; Roy, A.R. Fuzzy soft sets. *J. Fuzzy Math.* **2001**, *9*, 589–602.

28. Cagman, N.; Enginoglu, S.; Citak, F. Fuzzy soft set theory and its application. *Iran. J. Fuzzy Syst.* **2011**, *8*, 137–147.
29. Feng, F.; Jun, Y.B.; Liu, X.; Li. L. An adjustable approach to fuzzy soft set based decision making. *J. Comput. Appl. Math.* **2010**, *234*, 10–20. [CrossRef]
30. Gorzalczany, M.B. A method of inference in approximate reasoning based on interval valued fuzzy sets. *Fuzzy Sets Syst.* **1987**, *21*, 1–17. [CrossRef]
31. Kong, Z.; Gao, L.; Wang, L. Comment on "A fuzzy soft set theoretic approach to decision making problems". *J. Comput. Appl. Math.* **2009**, *223*, 540–542. [CrossRef]
32. Roy, A.R.; Maji, P.K. A fuzzy soft set theoretic approach to decision making problems. *J. Comput. Appl. Math.* **2007**, *203*, 412–418. [CrossRef]
33. Alkhazaleh, S.; Salleh, A.R. Soft expert sets. *Adv. Decis. Sci.* **2011**, *2011*, 757868. [CrossRef]
34. Arokia Lancy, A.; Tamilarasi, C.; Arockiarani, I. Fuzzy parameterization for decision making in risk management system via soft expert set. *Int. J. Innov. Res. Stud.* **2013**, *2*, 339–344.
35. Arokia Lancy, A.; Arockiarani, I. A fusion of soft expert set and matrix models. *Int. J. Res. Eng. Technol.* **2013**, *2*, 531–535.
36. Alkhazaleh, S.; Salleh, A.R. Fuzzy soft expert set and its application. *Appl. Math.* **2014**, *5*, 1349–1368. [CrossRef]
37. Bashir, M.; Salleh, A.R. Fuzzy parameterized soft expert set. *Abstr. Appl. Anal.* **2012**, *2012*, 25836. [CrossRef]
38. Bashir, M.; Salleh, A.R. Possibility fuzzy soft expert set. *Open J. Appl. Sci.* **2012**, *12*, 208–211. [CrossRef]
39. Alhazaymeh, K.; Hassan, N. Application of generalized vague soft expert set in decision making. *Int. J. Pure Appl. Math.* **2014**, *93*, 361–367. [CrossRef]
40. Broumi, S.; Smarandache, F. Intuitionistic fuzzy soft expert sets and its application in decision making. *J. New Theory* **2015**, *1*, 89–105.
41. Abu Qamar, M.; Hassan, N. Q-neutrosophic soft relation and its application in decision making. *Entropy* **2018**, *20*, 172. [CrossRef]
42. Abu Qamar, M.; Hassan, N. Entropy measures of distance and similarity of Q-neutrosophic soft sets and some applications. *Entropy* **2018**, *20*, 672. [CrossRef]
43. Sahin, M.; Alkhazaleh, S.; Ulucay, V. Neutrosophic soft expert sets. *Appl. Math.* **2015**, *6*, 116–127. [CrossRef]
44. Uluçay, V.; Sahin, M.; Hassan, N. Generalized neutrosophic soft expert set for multiple-criteria decision-making. *Symmetry* **2018**, *10*, 437. [CrossRef]
45. Al-Quran, A.; Hassan, N. Neutrosophic vague soft expert set theory. *J. Intell. Fuzzy Syst.* **2016**, *30*, 3691–3702. [CrossRef]
46. Al-Quran, A.; Hassan, N. The complex neutrosophic soft expert set and its application in decision making. *J. Intell. Fuzzy Syst.* **2018**, *34*, 569–582 [CrossRef]
47. Al-Quran, A.; Hassan, N. The complex neutrosophic soft expert relation and its multiple attribute decision-making method. *Entropy* **2018**, *20*, 101. [CrossRef]
48. Qayyum, A.; Abdullah, S.; Aslam, M. Cubic soft expert sets and their application in decision making. *J. Intell. Fuzzy Syst.* **2016**, *31*, 1585–1596. [CrossRef]
49. Ziemba, P.; Becker, J. Analysis of the digital divide using fuzzy forecasting. *Symmetry* **2019**, *11*, 166. [CrossRef]

symmetry

MDPI

Article

Hybrid Weighted Arithmetic and Geometric Aggregation Operator of Neutrosophic Cubic Sets for MADM

Lilian Shi * and Yue Yuan

Department of Electrical Automation, Shaoxing University, 508 Huancheng West Road, Shaoxing 312000, China; 15733719615@163.com

* Correspondence: cssll@usx.edu.cn

Received: 4 January 2019; Accepted: 19 February 2019; Published: 21 February 2019

Abstract: Neutrosophic cubic sets (NCSs) can express complex multi-attribute decision-making (MADM) problems with its interval and single-valued neutrosophic numbers simultaneously. The weighted arithmetic average (WAA) and geometric average (WGA) operators are common aggregation operators for handling MADM problems. However, the neutrosophic cubic weighted arithmetic average (NCWAA) and neutrosophic cubic geometric weighted average (NCWGA) operators may result in some unreasonable aggregated values in some cases. In order to overcome the drawbacks of the NCWAA and NCWGA, this paper developed a new neutrosophic cubic hybrid weighted arithmetic and geometric aggregation (NCHWAGA) operator and investigates its suitability and effectiveness. Then, we established a MADM method based on the NCHWAGA operator. Finally, a MADM problem with neutrosophic cubic information was provided to illustrate the application and effectiveness of the proposed method.

Keywords: weighted geometric operator; weighted average operator; neutrosophic cubic sets; multi-attribute decision-making (MADM); neutrosophic cubic hybrid weighted arithmetic and geometric aggregation operator (NCHWAGA)

1. Introduction

Zadeh [1] proposed the classic fuzzy set to describe fuzzy problems with the membership degree in the closed interval [0,1]. Atanassov [2] presented the concept of the intuitionistic fuzzy set (IFS) to express fuzzy problems by the membership function and non-membership function. Smarandache [3] defined neutrosophic logic and introduced neutrosophic sets (NSs) to describe fuzzy problems by the truth, falsity, and indeterminacy membership functions. For easy engineering applications, some subclasses of NSs are defined. Wang et al. developed interval neutrosophic sets (INSs) [4] and single-valued neutrosophic sets (SVNSs) [5]. Ye presented simplified neutrosophic sets (SNSs) [6]. Wang et al. also presented multi-valued neutrosophic sets (MVNSs) [7]. Since then, INSs, SVNSs, SNSs and MVNSs have been widely applied in decision-making [8,9], medical diagnoses [10,11], and fault diagnoses [12,13]. Furthermore, many scholars developed some extension forms of NSs by combining neutrosophic sets with other sets, such as refined single-valued neutrosophic sets [14], intuitionistic neutrosophic soft sets [15], single-valued neutrosophic hesitant fuzzy sets [16], and rough neutrosophic sets [17].

Recently, Ali et al. [18,19] also put forward the concepts of neutrosophic cubic sets (NCSs) by combining neutrosophic sets with cubic sets, and defined internal and external NCSs. NCSs are described by two parts simultaneously, where the truth, falsity, and indeterminacy membership functions can be expressed by an interval value and an exact value simultaneously. Obviously, an NCS can be combined by an INS and an SVNS, and it contains much more information than an INS or an SVNS. Thus, some researchers have applied NCSs in decision-making problems effectively. Lu et al. [20] studied cosine measures of NCSs to deal with multiple attribute decision-making (MADM) problems. Banerjee et al. [21] established a grey relational analysis (GRA) method for MADM in NCS environment. Pramanik et al. [22] investigated a multi-criteria group decision making (MCGDM) method based on the similarity measure of NCSs. Moreover, aggregation operators have been widely applied in many MADM problems [23–29], and some aggregation operators have been studied for MADM problems in an NCS environment. Shi et al. [30] developed Dombi aggregation operators of NCSs for MADM. Zhan et al. [31] also proposed the neutrosophic cubic weighted arithmetic average (NCWAA) and neutrosophic cubic geometric weighted average (NCWGA) operators by extending the WAA and WGA operators to NCSs. However, the aforementioned NCWAA and NCWGA operators may cause some unreasonable results in some cases. In order to overcome the shortcomings of the NCWAA and NCWGA operators, this paper developed a new neutrosophic cubic hybrid weighted arithmetic and geometric aggregation (NCHWAGA) operator and analyzed its effectiveness for MADM by numerical examples. The main advantage of the proposed NCHWAGA operator can overcome the shortcomings of the existing NCWAA and NCWGA operators in some situations and obtain the moderate aggregation values.

The rest of the paper is organized as follows. Section 2 briefly introduces some basic concepts of NCSs and analyzes the shortcomings of the NCWAA and NCWGA operators. Then, Section 3 presents the NCHWAGA operator and investigated its properties. We establish a MADM approach based on the NCHWAGA operator in Section 4. Subsequently, Section 5 provides numerical examples with neutrosophic cubic information to demonstrate the application and effectiveness of the developed approach. Finally, Section 6 presents conclusions and possible future research.

2. Preliminaries

Some basic concepts and ranking methods of NCSs were introduced in this section.

Definition 1. *[19,32] Let $\mathcal{Z}$ be a universal set. A NCS G in $\mathcal{Z}$ is denoted as follows:*

$$G = \{x, < T_c(x), V_c(x), F_c(x) >, < t_c(x), v_c(x), f_c(x)> \mid x \in \mathcal{Z}\},$$

where $< T_c(x), V_c(x), F_c(x) >$ is an INS [4] in $\mathcal{Z}$, and the intervals $T_c(x) = [T_c{}^L(x), T_c{}^U(x)] \subseteq [0,1]$, $V_c(x) = [V_c{}^L(x), V_c{}^U(x)] \subseteq [0,1]$, and $F_c(x) = [F_c{}^L(x), F_c{}^U(x)] \subseteq [0,1]$ for $x \in \mathcal{Z}$ represent respectively the truth, indeterminacy, and falsity membership functions; then $< t_c(x), v_c(x), f_c(x) >$ is an SVNS [3,5] in $\mathcal{Z}$, and $t_c(x)$, $v_c(x), f_c(x) \in [0,1]$ for $x \in \mathcal{Z}$ represent the truth, indeterminacy, and falsity membership functions, respectively.

Then, we called a basic element $(x, < T_c(x), V_c(x), F_c(x) >, < t_c(x), v_c(x), f_c(x) >)$ in an NCS G a neutrosophic cubic number (NCN) [20]; for convenience, we denoted it as $g = (< [T^L, T^U], [V^L, V^U], [F^L, F^U] >, < t, v, f >)$, where $t, v, f \in [0,1]$ and $[T^L, T^U], [V^L, V^U], [F^L, F^U] \subseteq [0,1]$ satisfy the condition $0 \le T^U + V^U + F^U \le 3$ and $0 \le t + v + f \le 3$.

Then, an NCS $G = \{x, < T_c(x), V_c(x), F_c(x) >, < t_c(x), v_c(x), f_c(x) > \mid x \in \mathcal{Z}\}$ is called an internal NCS if $T_c{}^L(x) \le t_c(x) \le T_c{}^U(x)$, $V_c{}^L(x) \le v_c(x) \le V_c{}^U(x)$, and $F_c{}^L(x) \le f_c(x) \le F_c{}^U(x)$ for $x \in \mathcal{Z}$; and an NCS G is called an external NCS if $t_c(x) \notin (T_c{}^L(x), T_c{}^U(x))$, $v_c(x) \notin (V_c{}^L(x), V_c{}^U(x))$, and $f_c(x) \notin (F_c{}^L(x), F_c{}^U(x))$ for $x \in \mathcal{Z}$ [18,19].

Let $g_1 = (< [T_1{}^L, T_1{}^U], [V_1{}^L, V_1{}^U], [F_1{}^L, F_1{}^U] >, < t_1, v_1, f_1 >)$ and $g_2 = (< [T_2{}^L, T_2{}^U], [V_2{}^L, V_2{}^U], [F_2{}^L, F_2{}^U] >, < t_2, v_2, f_2 >)$ be two NCNs, then there are following operational laws:

(1) $(g_1)^C = (< [F_1{}^L, F_1{}^U], [1 - V_1{}^U, 1 - V_1{}^L], [T_1{}^L, T_1{}^U] >, < f_1, 1 - v_1, t_1 >)$ (complement of g_1);

(2) $g_1 \oplus g_2 = (< [T_1{}^L + T_2{}^L - T_1{}^L T_2{}^L, T_1{}^U + T_2{}^U - T_1{}^U T_2{}^U], [V_1{}^L V_2{}^L, V_1{}^U V_2{}^U], [F_1{}^L F_1{}^L, F_1{}^U F_2{}^U] >, < t_1 + t_2 - t_1 t_2, v_1 v_2, f_1 f_2 >)$;

(3) $g_1 \otimes g_2 = (< [T_1{}^L T_2{}^L, T_1{}^U T_2{}^U], [V_1{}^L + V_2{}^L - V_1{}^L V_2{}^L, V_1{}^U + V_2{}^U - V_1{}^U V_2{}^U], [F_1{}^L + F_1{}^L - F_1{}^L F_1{}^L, F_1{}^U + F_2{}^U - F_1{}^U F_2{}^U] >, < t_1 t_2, v_1 + v_2 - v_1 v_2, f_1 + f_2 - f_1 f_2 >)$;

(4) $\lambda g_1 = (< [1-(1-T_1{}^L)^\lambda, 1-(1-T_1{}^U)^\lambda], [(V_1{}^L)^\lambda, (V_1{}^U)^\lambda], [(F_1{}^L)^\lambda, (F_1{}^U)^\lambda] >, < 1-(1-t_1)^\lambda, (v_1)^\lambda, (f_1)^\lambda >)$ for $\lambda > 0$;

(5) $(g_1)^\lambda = (< [(T_1{}^L)^\lambda, (T_1{}^U)^\lambda], [1-(1-V_1{}^L)^\lambda, 1-(1-V_1{}^U)^\lambda], 1-(1-F_1{}^L)^\lambda, 1-(1-F_1{}^U)^\lambda] >, < (t_1)^\lambda, 1-(1-v_1)^\lambda, 1-(1-f_1)^\lambda >)$ for $\lambda > 0$;

For any NCN $g = (< [T^L, T^U], [V^L, V^U], [F^L, F^U] >, < t, v, f >)$, its score and accuracy functions [33] can be defined as follows:

$$\Psi(x) = [(4 + T^L + T^U - V^L - V^U - F^L - F^U) + (2 + t - v - f)]/9 \quad (1)$$

$$\Gamma(x) = [(T^L + T^U - F^L - F^U) + (t - f)]/3 \quad (2)$$

Based on the functions $\Psi(x)$ and $\Gamma(x)$, two NCNs can be compared and ranked by definition as follows:

Definition 2. *[33] Let* $g_1 = (< [T_1{}^L, T_1{}^U], [V_1{}^L, V_1{}^U], [F_1{}^L, F_1{}^U] >, < t_1, v_1, f_1 >)$ *and* $g_2 = [T_2{}^L, T_2{}^U], [V_2{}^L, V_2{}^U], [F_2{}^L, F_2{}^U] >, < t_2, v_2, f_2 >)$ *be two NCNs, then their comparative relations are defined as follows:*

(i) *If* $\Psi(g_1) > \Psi(g_2)$, *then* $g_1 \succ g_2$;
(ii) *If* $\Psi(g_1) = \Psi(g_2)$ *and* $\Gamma(g_1) > \Gamma(g_2)$, *then* $g_1 \succ g_2$;
(iii) *If* $\Psi(g_1) = \Psi(g_2)$ *and* $\Gamma(g_1) = \Gamma(g_2)$, *then* $g_1 \sim g_2$.

Assume that $g_i = (< [T_i{}^L, T_i{}^U], [V_i{}^L, V_i{}^U], [F_i{}^L, F_i{}^U] >, < t_i, v_i, f_i >)$ $(i = 1, 2, \ldots, n)$ be a collection of NCNs. Then the NCWAA and NCWGA are provided, respectively, as follows [31]:

NCWAA $(g_1, g_2, \ldots, g_n)$

$$= \sum_{i=1}^{n} \zeta_i g_i = \left(\begin{array}{l} \left\langle \begin{array}{l} \left[1 - \prod_{i=1}^{n} (1 - T_i^L)^{\zeta_i}, 1 - \prod_{i=1}^{n} (1 - T_i^U)^{\zeta_i}\right], \left[\prod_{i=1}^{n} (V_i^L)^{\zeta_i}, \prod_{i=1}^{n} (V_i^U)^{\zeta_i}\right], \\ \left[\prod_{i=1}^{n} (F_i^L)^{\zeta_i}, \prod_{i=1}^{n} (F_i^U)^{\zeta_i}\right] \end{array} \right\rangle, \\ \left\langle 1 - \prod_{i=1}^{n} (1 - t_i)^{\zeta_i}, \prod_{i=1}^{n} (v_i)^{\zeta_i}, \prod_{i=1}^{n} (f_i)^{\zeta_i} \right\rangle \end{array} \right) \quad (3)$$

NCWGA $(g_1, g_2, \ldots, g_n)$

$$= \prod_{i=1}^{n} g_i^{\zeta_i} = \left(\begin{array}{l} \left\langle \begin{array}{l} \left[\prod_{i=1}^{n} (T_i^L)^{\zeta_i}, \prod_{i=1}^{n} (T_i^U)^{\zeta_i}\right], \left[1 - \prod_{i=1}^{n} (1 - V_i^L)^{\zeta_i}, 1 - \prod_{i=1}^{n} (1 - V_i^U)^{\zeta_i}\right], \\ \left[1 - \prod_{i=1}^{n} (1 - F_i^L)^{\zeta_i}, 1 - \prod_{i=1}^{n} (1 - F_i^U)^{\zeta_i}\right] \end{array} \right\rangle, \\ \left\langle \prod_{i=1}^{n} (t_i)^{\zeta_i}, 1 - \prod_{i=1}^{n} (1 - v_i)^{\zeta_i}, 1 - \prod_{i=1}^{n} (1 - f_i)^{\zeta_i} \right\rangle \end{array} \right) \quad (4)$$

where $\zeta_i \in (i = 1, 2, \ldots, n)$, satisfying $\sum_{i=1}^{n} \zeta_i = 1$.

Although the above-weighted average and geometric operators were used for multi-criteria decision making [31], some unreasonable results are implied in the following two cases.

Case 1. *Let* $g_1 = (< [0.001, 0.002], [0, 0], [0, 0] >, <0.001, 0, 0 >)$ *and* $g_2 = (< [0, 1], [0, 0], [0, 0] >, <1, 0, 0 >)$ *be two NCNs, with their weights* $\zeta_1 = 0.9$ *and* $\zeta_2 = 0.1$, *respectively.*

Then, by Equations (3) and (4), NCWAA $(g_1, g_2) = (< [0.001, 1], [0, 0], [0, 0] >, <1, 0, 0 >)$ and NCWGA $(g_1, g_2) = (< [0, 0.004], [0, 0], [0, 0] >, <0.002, 0, 0 >)$.

Case 2. *Also take two NCNs $g_1 = (< [0.001, 0.002], [0, 0], [0, 0] >, <0.001, 0, 0 >)$ and $g_2 = (< [0, 1], [0, 0], [0, 0] >, <1, 0, 0 >)$ with their weights $\zeta_1 = 0.1$ and $\zeta_2 = 0.9$, respectively.*

Then, by Equation (3) and (4), NCWAA $(g_1, g_2) = (< [0, 1], [0, 0], [0, 0] >, <1, 0, 0 >)$ and NCWGA $(g_1, g_2) = (< [0, 0.537], [0, 0], [0, 0] >, <0.501, 0, 0 >)$.

The above aggregated results indicate that the aggregated values of NCWAA (g_1, g_2) operator tend to the maximum value, while the aggregated results of NCWGA (g_1, g_2) operator tend to the maximum weight value. It is obvious that the NCWAA and NCWGA operators may cause unreasonable results of NCNs in some cases. In order to overcome the drawbacks, it is necessary to improve the NCWAA and NCWGA operators provided in [31]. Hence, in the next section, a new NCHWAGA is proposed by extending the hybrid arithmetic and geometric aggregation operators presented in [34,35].

3. Hybrid Arithmetic and Geometric Aggregation Operators of NCNs

In this section, we present the NCHWAGA operator and investigated its properties.

3.1. NCHWAGA Operator

Definition 3. *Let $g_i = (< [T_i^L, T_i^U], [V_i^L, V_i^U], [F_i^L, F_i^U] >, < t_i, v_i, f_i >)$ $(i = 1, 2, \ldots, n)$ be a collection of NCNs. Then, the NCHWAGA operator is defined by:*

$$NCHWAGA(g_1, g_2, \ldots, g_n) = \left(\sum_{i=1}^{n} \zeta_i g_i\right)^{\rho} \left(\prod_{i=1}^{n} g_i^{\zeta_i}\right)^{(1-\rho)} \tag{5}$$

where $\rho \in [0, 1]$; and ζ_i $(i = 1, 2, \ldots, n)$ is the weight of g_i $(i = 1, 2, \ldots, n)$, satisfying $\zeta_i \in [0, 1]$ and $\sum_{i=1}^{n} \zeta_i = 1$.

Theorem 1. *Let $g_i = (< [T_i^L, T_i^U], [V_i^L, V_i^U], [F_i^L, F_i^U] >, < t_i, v_i, f_i >)$ $(i = 1, 2, \ldots, n)$ be a collection of NCNs, and ζ_i $(i = 1, 2, \ldots, n)$ be the corresponding weight of g_i $(i = 1, 2, \ldots, n)$, satisfying $\zeta_i \in [0, 1]$ and $\sum_{i=1}^{n} \zeta_i = 1$. Then, the aggregated value of the NCHWAGA operator is also an NCN, which can be calculated by:*

$$\begin{aligned} NCHWAGA(g_1, g_2, \ldots, g_n) &= \left(\sum_{i=1}^{n} \zeta_i g_i\right)^{\rho} \left(\prod_{i=1}^{n} g_i^{\zeta_i}\right)^{(1-\rho)} \\ &= \left(\begin{array}{l} \left\langle \begin{array}{l} \left[\left(1-\prod_{i=1}^{n}(1-T_i^L)^{\zeta_i}\right)^{\rho}\left(\prod_{i=1}^{n}(T_i^L)^{\zeta_i}\right)^{(1-\rho)}, \left(1-\prod_{i=1}^{n}(1-T_i^U)^{\zeta_i}\right)^{\rho}\left(\prod_{i=1}^{n}(T_i^U)^{\zeta_i}\right)^{(1-\rho)}\right], \\ \left[1-\left(1-\prod_{i=1}^{n}(V_i^L)^{\zeta_i}\right)^{\rho}\left(\prod_{i=1}^{n}(1-V_i^L)^{\zeta_i}\right)^{(1-\rho)}, 1-\left(1-\prod_{i=1}^{n}(V_i^U)^{\zeta_i}\right)^{\rho}\left(\prod_{i=1}^{n}(1-V_i^U)^{\zeta_i}\right)^{(1-\rho)}\right], \\ \left[1-\left(1-\prod_{i=1}^{n}(F_i^L)^{\zeta_i}\right)^{\rho}\left(\prod_{i=1}^{n}(1-F_i^L)^{\zeta_i}\right)^{(1-\rho)}, 1-\left(1-\prod_{i=1}^{n}(F_i^U)^{\zeta_i}\right)^{\rho}\left(\prod_{i=1}^{n}(1-F_i^U)^{\zeta_i}\right)^{(1-\rho)}\right] \end{array} \right\rangle, \\ \left\langle \left(1-\prod_{i=1}^{n}(1-t_i)^{\zeta_i}\right)^{\rho}\left(\prod_{i=1}^{n}(t_i)^{\zeta_i}\right)^{(1-\rho)}, 1-\left(1-\prod_{i=1}^{n}(v_i)^{\zeta_i}\right)^{\rho}\left(\prod_{i=1}^{n}(1-v_i)^{\zeta_i}\right)^{(1-\rho)}, 1-\left(1-\prod_{i=1}^{n}(f_i)^{\zeta_i}\right)^{\rho}\left(\prod_{i=1}^{n}(1-f_i)^{\zeta_i}\right)^{(1-\rho)} \right\rangle \end{array} \right) \end{aligned} \tag{6}$$

Proof. Based on the operational laws of NCNs in Section 2 and the NCWAA and NCWGA, we can obtain the following result:

$$\begin{aligned} NCHWAGA(g_1, g_2, \ldots, g_n) &= \left(\sum_{i=1}^{n} \zeta_i g_i\right)^{\rho} \left(\prod_{i=1}^{n} g_i^{\zeta_i}\right)^{(1-\rho)} \\ &= \left(\begin{array}{l} \left\langle \begin{array}{l} \left[1-\prod_{i=1}^{n}(1-T_i^L)^{\zeta_i}, 1-\prod_{i=1}^{n}(1-T_i^U)^{\zeta_i}\right], \left[\prod_{i=1}^{n}(V_i^L)^{\zeta_i}, \prod_{i=1}^{n}(V_i^U)^{\zeta_i}\right], \\ \left[\prod_{i=1}^{n}(F_i^L)^{\zeta_i}, \prod_{i=1}^{n}(F_i^U)^{\zeta_i}\right] \end{array} \right\rangle, \\ \left\langle 1-\prod_{i=1}^{n}(1-t_i)^{\zeta_i}, \prod_{i=1}^{n}(v_i)^{\zeta_i}, \prod_{i=1}^{n}(f_i)^{\zeta_i} \right\rangle \end{array} \right)^{\rho} \\ &\times \left(\begin{array}{l} \left\langle \begin{array}{l} \left[\prod_{i=1}^{n}(T_i^L)^{\zeta_i}, \prod_{i=1}^{n}(T_i^U)^{\zeta_i}\right], \left[1-\prod_{i=1}^{n}(1-V_i^L)^{\zeta_i}, 1-\prod_{i=1}^{n}(1-V_i^U)^{\zeta_i}\right], \\ \left[1-\prod_{i=1}^{n}(1-F_i^L)^{\zeta_i}, 1-\prod_{i=1}^{n}(1-F_i^U)^{\zeta_i}\right] \end{array} \right\rangle, \\ \left\langle \prod_{i=1}^{n}(t_i)^{\zeta_i}, 1-\prod_{i=1}^{n}(1-v_i)^{\zeta_i}, 1-\prod_{i=1}^{n}(1-f_i)^{\zeta_i} \right\rangle \end{array} \right)^{(1-\rho)} \end{aligned}$$

$$
=\left(\begin{array}{l}
\left\langle\begin{array}{l}
\left[\left(1-\prod_{i=1}^{n}(1-T_i^{L})^{\xi_i}\right)^{\rho},\left(1-\prod_{i=1}^{n}(1-T_i^{U})^{\xi_i}\right)^{\rho}\right],\\
\left[1-\left(1-\prod_{i=1}^{n}(V_i^{L})^{\xi_i}\right)^{\rho},1-\left(1-\prod_{i=1}^{n}(V_i^{U})^{\xi_i}\right)^{\rho}\right],\\
\left[1-\left(1-\prod_{i=1}^{n}(F_i^{L})^{\xi_i}\right)^{\rho},1-\left(1-\prod_{i=1}^{n}(F_i^{U})^{\xi_i}\right)^{\rho}\right]
\end{array}\right\rangle,\\
\left\langle\left(1-\prod_{i=1}^{n}(1-t_i)^{\xi_i}\right)^{\rho},1-\left(1-\prod_{i=1}^{n}(v_i)^{\xi_i}\right)^{\rho},1-\left(1-\prod_{i=1}^{n}(f_i)^{\xi_i}\right)^{\rho}\right\rangle
\end{array}\right)
$$

$$
\times\left(\begin{array}{l}
\left\langle\begin{array}{l}
\left[\left(\prod_{i=1}^{n}(T_i^{L})^{\xi_i}\right)^{(1-\rho)},\left(\prod_{i=1}^{n}(T_i^{U})^{\xi_i}\right)^{(1-\rho)}\right],\\
\left[1-\left(\prod_{i=1}^{n}(1-V_i^{L})^{\xi_i}\right)^{(1-\rho)},1-\left(\prod_{i=1}^{n}(1-V_i^{U})^{\xi_i}\right)^{(1-\rho)}\right],\\
\left[1-\left(\prod_{i=1}^{n}(1-F_i^{L})^{\xi_i}\right)^{(1-\rho)},1-\left(\prod_{i=1}^{n}(1-F_i^{U})^{\xi_i}\right)^{(1-\rho)}\right]
\end{array}\right\rangle,\\
\left\langle\left(\prod_{i=1}^{n}(t_i)^{\xi_i}\right)^{(1-\rho)},1-\left(\prod_{i=1}^{n}(1-v_i)^{\xi_i}\right)^{(1-\rho)},1-\left(\prod_{i=1}^{n}(1-f_i)^{\xi_i}\right)^{(1-\rho)}\right\rangle
\end{array}\right)
$$

$$
=\left(\begin{array}{l}
\left\langle\begin{array}{l}
\left[\left(1-\prod_{i=1}^{n}(1-T_i^{L})^{\xi_i}\right)^{\rho}\left(\prod_{i=1}^{n}(T_i^{L})^{\xi_i}\right)^{(1-\rho)},\left(1-\prod_{i=1}^{n}(1-T_i^{U})^{\xi_i}\right)^{\rho}\left(\prod_{i=1}^{n}(T_i^{U})^{\xi_i}\right)^{(1-\rho)}\right],\\
\left[1-\left(1-\prod_{i=1}^{n}(V_i^{L})^{\xi_i}\right)^{\rho}+1-\left(\prod_{i=1}^{n}(1-V_i^{L})^{\xi_i}\right)^{(1-\rho)}-\left(1-\left(1-\prod_{i=1}^{n}(V_i^{L})^{\xi_i}\right)^{\rho}\right)\left(1-\left(\prod_{i=1}^{n}(1-V_i^{L})^{\xi_i}\right)^{(1-\rho)}\right),\right.\\
\left.1-\left(1-\prod_{i=1}^{n}(V_i^{U})^{\xi_i}\right)^{\rho}+1-\left(\prod_{i=1}^{n}(1-V_i^{U})^{\xi_i}\right)^{(1-\rho)}-\left(1-\left(1-\prod_{i=1}^{n}(V_i^{U})^{\xi_i}\right)^{\rho}\right)\left(1-\left(\prod_{i=1}^{n}(1-V_i^{U})^{\xi_i}\right)^{(1-\rho)}\right)\right],\\
\left[1-\left(1-\prod_{i=1}^{n}(F_i^{L})^{\xi_i}\right)^{\rho}+1-\left(\prod_{i=1}^{n}(1-F_i^{L})^{\xi_i}\right)^{(1-\rho)}-\left(1-\left(1-\prod_{i=1}^{n}(F_i^{L})^{\xi_i}\right)^{\rho}\right)\left(1-\left(\prod_{i=1}^{n}(1-F_i^{L})^{\xi_i}\right)^{(1-\rho)}\right),\right.\\
\left.1-\left(1-\prod_{i=1}^{n}(F_i^{U})^{\xi_i}\right)^{\rho}+1-\left(\prod_{i=1}^{n}(1-F_i^{U})^{\xi_i}\right)^{(1-\rho)}-\left(1-\left(1-\prod_{i=1}^{n}(F_i^{U})^{\xi_i}\right)^{\rho}\right)\left(1-\left(\prod_{i=1}^{n}(1-F_i^{U})^{\xi_i}\right)^{(1-\rho)}\right)\right]
\end{array}\right\rangle,\\
\left\langle\begin{array}{l}
\left(1-\prod_{i=1}^{n}(1-t_i)^{\xi_i}\right)^{\rho}\left(\prod_{i=1}^{n}(t_i)^{\xi_i}\right)^{(1-\rho)},1-\left(1-\prod_{i=1}^{n}(v_i)^{\xi_i}\right)^{\rho}+1-\left(\prod_{i=1}^{n}(1-v_i)^{\xi_i}\right)^{(1-\rho)}-\left(1-\left(1-\prod_{i=1}^{n}(v_i)^{\xi_i}\right)^{\rho}\right)\\
\times\left(1-\left(\prod_{i=1}^{n}(1-v_i)^{\xi_i}\right)^{(1-\rho)}\right),1-\left(1-\prod_{i=1}^{n}(f_i)^{\xi_i}\right)^{\rho}+1-\left(\prod_{i=1}^{n}(1-f_i)^{\xi_i}\right)^{(1-\rho)}-\left(1-\left(1-\prod_{i=1}^{n}(f_i)^{\xi_i}\right)^{\rho}\right)\\
\times\left(1-\left(\prod_{i=1}^{n}(1-f_i)^{\xi_i}\right)^{(1-\rho)}\right)
\end{array}\right\rangle
\end{array}\right)
$$

$$
=\left(\begin{array}{l}
\left\langle\begin{array}{l}
\left[\left(1-\prod_{i=1}^{n}(1-T_i^{L})^{\xi_i}\right)^{\rho}\left(\prod_{i=1}^{n}(T_i^{L})^{\xi_i}\right)^{(1-\rho)},\left(1-\prod_{i=1}^{n}(1-T_i^{U})^{\xi_i}\right)^{\rho}\left(\prod_{i=1}^{n}(T_i^{U})^{\xi_i}\right)^{(1-\rho)}\right],\\
\left[\left(\begin{array}{l}1-\left(1-\prod_{i=1}^{n}(V_i^{L})^{\xi_i}\right)^{\rho}+1-\left(\prod_{i=1}^{n}(1-V_i^{L})^{\xi_i}\right)^{(1-\rho)}-\left(1-\left(\prod_{i=1}^{n}(1-V_i^{L})^{\xi_i}\right)^{(1-\rho)}\right)\\+\left(1-\prod_{i=1}^{n}(V_i^{L})^{\xi_i}\right)^{\rho}-\left(1-\prod_{i=1}^{n}(V_i^{L})^{\xi_i}\right)^{\rho}\left(\prod_{i=1}^{n}(1-V_i^{L})^{\xi_i}\right)^{(1-\rho)}\end{array}\right),\right.\\
\left.\left(\begin{array}{l}1-\left(1-\prod_{i=1}^{n}(V_i^{U})^{\xi_i}\right)^{\rho}+1-\left(\prod_{i=1}^{n}(1-V_i^{U})^{\xi_i}\right)^{(1-\rho)}-\left(1-\left(\prod_{i=1}^{n}(1-V_i^{U})^{\xi_i}\right)^{(1-\rho)}\right)\\+\left(1-\prod_{i=1}^{n}(V_i^{U})^{\xi_i}\right)^{\rho}-\left(1-\prod_{i=1}^{n}(V_i^{U})^{\xi_i}\right)^{\rho}\left(\prod_{i=1}^{n}(1-V_i^{U})^{\xi_i}\right)^{(1-\rho)}\end{array}\right)\right],\\
\left[\left(\begin{array}{l}1-\left(1-\prod_{i=1}^{n}(F_i^{L})^{\xi_i}\right)^{\rho}+1-\left(\prod_{i=1}^{n}(1-F_i^{L})^{\xi_i}\right)^{(1-\rho)}-\left(1-\left(\prod_{i=1}^{n}(1-F_i^{L})^{\xi_i}\right)^{(1-\rho)}\right)\\+\left(1-\prod_{i=1}^{n}(F_i^{L})^{\xi_i}\right)^{\rho}-\left(1-\prod_{i=1}^{n}(F_i^{L})^{\xi_i}\right)^{\rho}\left(\prod_{i=1}^{n}(1-F_i^{L})^{\xi_i}\right)^{(1-\rho)}\end{array}\right),\right.\\
\left.\left(\begin{array}{l}1-\left(1-\prod_{i=1}^{n}(F_i^{U})^{\xi_i}\right)^{\rho}+1-\left(\prod_{i=1}^{n}(1-F_i^{U})^{\xi_i}\right)^{(1-\rho)}-\left(1-\left(\prod_{i=1}^{n}(1-F_i^{U})^{\xi_i}\right)^{(1-\rho)}\right)\\+\left(1-\prod_{i=1}^{n}(F_i^{U})^{\xi_i}\right)^{\rho}-\left(1-\prod_{i=1}^{n}(F_i^{U})^{\xi_i}\right)^{\rho}\left(\prod_{i=1}^{n}(1-F_i^{U})^{\xi_i}\right)^{(1-\rho)}\end{array}\right)\right]
\end{array}\right\rangle,\\
\left\langle\begin{array}{l}
\left(1-\prod_{i=1}^{n}(1-t_i)^{\xi_i}\right)^{\rho}\left(\prod_{i=1}^{n}(t_i)^{\xi_i}\right)^{(1-\rho)},1-\left(1-\prod_{i=1}^{n}(v_i)^{\xi_i}\right)^{\rho}+1-\left(\prod_{i=1}^{n}(1-v_i)^{\xi_i}\right)^{(1-\rho)}\\
-\left(1-\left(\prod_{i=1}^{n}(1-v_i)^{\xi_i}\right)^{(1-\rho)}-\left(1-\prod_{i=1}^{n}(v_i)^{\xi_i}\right)^{\rho}\right)-\left(1-\prod_{i=1}^{n}(v_i)^{\xi_i}\right)^{\rho}\left(\prod_{i=1}^{n}(1-v_i)^{\xi_i}\right)^{(1-\rho)},\\
1-\left(1-\prod_{i=1}^{n}(f_i)^{\xi_i}\right)^{\rho}+1-\left(\prod_{i=1}^{n}(1-f_i)^{\xi_i}\right)^{(1-\rho)}-\left(1-\left(\prod_{i=1}^{n}(1-f_i)^{\xi_i}\right)^{(1-\rho)}-\left(1-\prod_{i=1}^{n}(f_i)^{\xi_i}\right)^{\rho}\right)\\
-\left(1-\prod_{i=1}^{n}(f_i)^{\xi_i}\right)^{\rho}\left(\prod_{i=1}^{n}(1-f_i)^{\xi_i}\right)^{(1-\rho)}
\end{array}\right\rangle
\end{array}\right)
$$

$$
=\left(\begin{array}{l}
\left\langle\begin{array}{l}
\left[\left(1-\prod_{i=1}^{n}(1-T_i^{L})^{\xi_i}\right)^{\rho}\left(\prod_{i=1}^{n}(T_i^{L})^{\xi_i}\right)^{(1-\rho)},\left(1-\prod_{i=1}^{n}(1-T_i^{U})^{\xi_i}\right)^{\rho}\left(\prod_{i=1}^{n}(T_i^{U})^{\xi_i}\right)^{(1-\rho)}\right],\\
\left[1-\left(1-\prod_{i=1}^{n}(V_i^{L})^{\xi_i}\right)^{\rho}\left(\prod_{i=1}^{n}(1-V_i^{L})^{\xi_i}\right)^{(1-\rho)},1-\left(1-\prod_{i=1}^{n}(V_i^{U})^{\xi_i}\right)^{\rho}\left(\prod_{i=1}^{n}(1-V_i^{U})^{\xi_i}\right)^{(1-\rho)}\right],\\
\left[1-\left(1-\prod_{i=1}^{n}(F_i^{L})^{\xi_i}\right)^{\rho}\left(\prod_{i=1}^{n}(1-F_i^{L})^{\xi_i}\right)^{(1-\rho)},1-\left(1-\prod_{i=1}^{n}(F_i^{U})^{\xi_i}\right)^{\rho}\left(\prod_{i=1}^{n}(1-F_i^{U})^{\xi_i}\right)^{(1-\rho)}\right]
\end{array}\right\rangle,\\
\left\langle\left(1-\prod_{i=1}^{n}(1-t_i)^{\xi_i}\right)^{\rho}\left(\prod_{i=1}^{n}(t_i)^{\xi_i}\right)^{(1-\rho)},1-\left(1-\prod_{i=1}^{n}(v_i)^{\xi_i}\right)^{\rho}\left(\prod_{i=1}^{n}(1-v_i)^{\xi_i}\right)^{(1-\rho)},1-\left(1-\prod_{i=1}^{n}(f_i)^{\xi_i}\right)^{\rho}\left(\prod_{i=1}^{n}(1-f_i)^{\xi_i}\right)^{(1-\rho)}\right\rangle
\end{array}\right)
$$

The proof is finished. Hence, Theorem 1 is true. □

Let $g_i = (< [T_i^L, T_i^U], [V_i^L, V_i^U], [F_i^L, F_i^U] >, < t_i, v_i, f_i >)$ $(i = 1, 2, \ldots, n)$ be a collection of NCNs. Corresponding to the properties of the NCWAA and NCWGA operators [31], the NCNHWAGA operator also satisfies these properties:

(i) Idempotency: If $g_i = g$ for i = 1, 2, ... , n, then *NCHWAGA* ($g_1, g_2, \ldots, g_n$) = g.
(ii) Boundedness: If $g_{\min} = min$ ($g_1, g_2, \ldots, g_n$) and $g_{\max} = max$ ($g_1, g_2, \ldots, g_n$) for i = 1, 2, ... , n, then $g_{\min} \le$ *NCHWAGA* ($g_1, g_2, \ldots, g_n$) $\le g_{\max}$.
(iii) Monotonicity: If $g_i \le g_i^*$ for i = 1, 2, ... , n, then *NCHWAGA* ($g_1, g_2, \ldots, g_n$) $\le$ *NCHWAGA* ($g_1^*, g_2^*, \ldots, g_n^*$).

For different values of $\rho \in [0, 1]$, we can discuss the families of the NCHWAGA operator in some special cases as follows:

(i) The NCHWAGA operator reduces to the NCWAA operator [31] if ρ = 1.
(ii) The NCHWAGA operator reduces to the NCWGA operator [31] if ρ = 0.
(iii) The NCHWAGA operator is the mean of the NCWAA and NCWGA operators [31] if ρ = 0.5.

3.2. Numerical Example

We still consider the above two numerical examples in Section 2 to demonstrate the effectiveness of the presented NCHWAGA operator. Generally taking ρ = 0.5, we calculate aggregated values of the NCHWAGA operator.

For Case 1: *Let two NCNs* g_1 = (< [0.001, 0.002], [0, 0], [0, 0] >, <0.001, 0, 0 >) *and* g_2 = (< [0, 1], [0, 0], [0, 0] >, <1, 0, 0 >) *with their weights* ζ_1 = 0.9 *and* ζ_2 = 0.1, *by Equation (6), we obtain NCHWAGA* (g_1, g_2) = (< [0, 0.061], [0, 0], [0, 0] >, <0.045, 0, 0 >), *which is between NCWAA* (g_1, g_2) = (< [0.001, 1], [0, 0], [0, 0] >, <1, 0, 0 >) *and NCWGA* (g_1, g_2) = (< [0, 0.004], [0, 0], [0, 0] >, <0.002, 0, 0 >).

For Case 2: *Also take two NCNs* g_1 = (< [0.001, 0.002], [0, 0], [0, 0] >, <0.001, 0, 0 >) *and* g_2 = (< [0, 1], [0, 0], [0, 0] >, <1, 0, 0 >) *with their weights* ζ_1 = 0.1 *and* ζ_2 = 0.9, *then, by Equation (6), we get NCHWAGA* (g_1, g_2) = (< [0, 0.733], [0, 0], [0, 0] >, <0.708, 0, 0 >), *which is between NCWAA* (g_1, g_2) = (< [0, 1], [0, 0], [0, 0] >, <1, 0, 0 >) *and NCWGA* (g_1, g_2) = (< [0, 0.537], [0, 0], [0, 0] >, <0.501, 0, 0 >).

In the above two cases, we can obtain the moderate values by the NCHWAGA operator. Obviously, the NCHWAGA operator can overcome the drawbacks of the NCWAA and NCWGA provided in Reference [31].

4. MADM Method Using the NCHWAGA Operator

In this section, we provide a MADM method based on the NCHWAGA operator to deal with neutrosophic cubic information.

In a MADM problem, assume that $G = \{G_1, G_2, \ldots, G_k\}$ is a set of k alternatives and $P = \{P_1, P_2, \ldots, P_n\}$ is a set of attributes. Suppose that the weight vector of P is $\omega_P = \{\omega_{P_1}, \omega_{P_2}, \ldots, \omega_{P_n}\}$ with $\omega_{P_j} \in [0,1]$ and $\sum_{j=1}^{n} \omega_{P_j} = 1$. The evaluation value of an alternative G_i under an attribute P_j can be expressed using an NCN $g_{ij} = (< [T_{ij}^L, T_{Ij}^U], [V_{ij}^L, V_{ij}^U], [F_{ij}^L, F_{ij}^U] >, < t_{ij}, v_{ij}, f_{ij} >)$ (i = 1, 2, ... , k; j = 1, 2, ... , n), where $[T_{ij}^L, T_{ij}^U], [V_{ij}^L, V_{ij}^U], [F_{ij}^L, F_{ij}^U] \subseteq [0,1]$, and $t_{ij}, v_{ij}, f_{ij} \in [0,1]$. Then, we can construct a decision matrix $G = (g_{ij})_{k \times n}$ with the NCN information, and provide the following MADM procedures based on the proposed NCHWAGA operator:

Step 1. Calculate the aggregated value of g_i for each alternative G_i (i = 1, 2, ... , k) using the NCHWAGA operator:

$$
g_i = NCHWAGA(g_{i1}, g_{i2}, \ldots, g_{in}) = \left(\sum_{j=1}^{n} \omega_{p_j} g_{ij}\right)^{\rho} \left(\prod_{j=1}^{n} g_{ij}^{\omega_{p_j}}\right)^{(1-\rho)}
$$

$$
= \left(\begin{array}{c} \left\langle \begin{array}{l} \left[\left(1 - \prod_{j=1}^{n} \left(1 - T_{ij}^{L}\right)^{\omega_{p_j}}\right)^{\rho} \left(\prod_{j=1}^{n} \left(T_{ij}^{L}\right)^{\omega_{p_j}}\right)^{(1-\rho)}, \left(1 - \prod_{j=1}^{n} (1 - T_{i}^{U})^{\omega_{p_j}}\right)^{\rho} \left(\prod_{j=1}^{n} \left(T_{ij}^{U}\right)^{\omega_{p_j}}\right)^{(1-\rho)} \right], \\ \left[1 - \left(1 - \prod_{j=1}^{n} \left(V_{ij}^{L}\right)^{\omega_{p_j}}\right)^{\rho} \left(\prod_{j=1}^{n} \left(1 - V_{ij}^{L}\right)^{\omega_{p_j}}\right)^{(1-\rho)}, 1 - \left(1 - \prod_{j=1}^{n} \left(V_{ij}^{U}\right)^{\omega_{p_j}}\right)^{\rho} \left(\prod_{j=1}^{n} \left(1 - V_{ij}^{U}\right)^{\omega_{p_j}}\right)^{(1-\rho)} \right], \\ \left[1 - \left(1 - \prod_{j=1}^{n} \left(F_{ij}^{L}\right)^{\omega_{p_j}}\right)^{\rho} \left(\prod_{j=1}^{n} \left(1 - F_{ij}^{L}\right)^{\omega_{p_j}}\right)^{(1-\rho)}, 1 - \left(1 - \prod_{j=1}^{n} \left(F_{j}^{U}\right)^{\omega_{p_j}}\right)^{\rho} \left(\prod_{j=1}^{n} \left(1 - F_{ij}^{U}\right)^{\omega_{p_j}}\right)^{(1-\rho)} \right] \end{array} \right\rangle \\ \left\langle \left(1 - \prod_{j=1}^{n} (1 - t_{ij})^{\omega_{p_j}}\right)^{\rho} \left(\prod_{j=1}^{n} (t_{ij})^{\omega_{p_j}}\right)^{(1-\rho)}, 1 - \left(1 - \prod_{j=1}^{n} (v_{ij})^{\omega_{p_j}}\right)^{\rho} \left(\prod_{j=1}^{n} (1 - v_{ij})^{\omega_{p_j}}\right)^{(1-\rho)}, 1 - \left(1 - \prod_{j=1}^{n} (f_{ij})^{\omega_{p_j}}\right)^{\rho} \left(\prod_{j=1}^{n} (1 - f_{ij})^{\omega_{p_j}}\right)^{(1-\rho)} \right\rangle \end{array} \right) \quad (7)
$$

where $\omega_{P_j} \in [0,1]$ and $\sum_{j=1}^{n} \omega_{P_j} = 1$ for $j = 1, 2, \ldots, n$.

Step 2. Obtain the score values of $\Psi(x)$ (the accuracy degrees of $\Gamma(x)$ if necessary) of the collective NCN g_i ($i = 1, 2, \ldots, k$) by Equations (1) and (2).

Step 3. Rank all the alternatives corresponding to the values of $\Psi(x)$ and $\Gamma(x)$, and select the best one(s) based on the largest value.

Step 4. End.

5. Illustrative Example and Comparison Analysis

This section introduces an illustrative example adapted from Reference [20] to demonstrate the application of the above MADM method. A company wants to invest some money in one of the four possible alternatives G_i ($i = 1, 2, 3, 4$). G_1, G_2, G_3 and G_4 represent a textile company, an automobile company, a computer company, and a software company, respectively. The four alternatives need to be evaluated according to the three attributes P_j ($j = 1, 2, 3$). P_1, P_2 and P_3 represent respectively the risk, the growth, and the environmental impact. Corresponding to the three attributes, the weight vector is $\omega_P = (0.32,\ 0.38,\ 0.3)$. When the decision maker evaluates the four alternatives G_i ($i = 1, 2, 3, 4$) based on the three attributes P_j ($j = 1, 2, 3$) with the NCN information, the decision matrix can be established as shown in Table 1.

Table 1. The decision matrix with the neutrosophic cubic number (NCN) information.

Alternative	Attribute (P_1)	Attribute (P_2)	Attribute (P_3)
G_1	(< [0.5, 0.6], [0.1, 0.3], [0.2, 0.4]>, <0.2, 0.6, 0.3>)	(< [0.5, 0.6], [0.1, 0.3], [0.2, 0.4]>, <0.2, 0.6, 0.3>)	(< [0.6, 0.8], [0.2, 0.3], [0.1, 0.2]>, <0.7, 0.2, 0.1>)
G_2	(< [0.6, 0.8], [0.1, 0.2], [0.2, 0.3]>, <0.7, 0.1, 0.2>)	(< [0.6, 0.7], [0.1, 0.2], [0.2, 0.3]>, <0.6, 0.3, 0.4>)	(< [0.6, 0.7], [0.3, 0.4], [0.1, 0.2]>, <0.7, 0.4, 0.2>)
G_3	(< [0.4, 0.6], [0.2, 0.3], [0.1, 0.3]>, <0.6, 0.2, 0.2>)	(< [0.5, 0.6], [0.2, 0.3], [0.3, 0.4]>, <0.6, 0.3, 0.4>)	(< [0.5, 0.7], [0.2, 0.3], [0.3, 0.4]>, <0.6, 0.2, 0.3>)
G_4	(< [0.7, 0.8], [0.1, 0.2], [0.1, 0.2]>, <0.8, 0.1, 0.2>)	(< [0.6, 0.7], [0.1, 0.2], [0.1, 0.3]>, <0.7, 0.1, 0.2>)	(< [0.6, 0.7], [0.3, 0.4], [0.2, 0.3]>, <0.7, 0.3, 0.2>)

Then, we apply the NCHWAGA operator to handle the MADM problem as follows:

Step 1. By Equation (7) for $\rho = 0.5$, we calculate the aggregated value of the collective NCN g_i for the each alternative G_i ($i = 1, 2, 3, 4$) as follows:

g_1 = (< [0.5302, 0.6645], [0.1272, 0.3000], [0.1669, 0.3355] >, <0.3430, 0.4709, 0.2306>)
g_2 = (< [0.6000, 0.7335], [0.1523, 0.2563], [0.1669, 0.2685] >, <0.6628, 0.2525, 0.2346>)
g_3 = (< [0.4677, 0.6307], [0.2000, 0.3000], [0.2264, 0.3672] >, <0.6000, 0.2365, 0.3025>)
g_4 = (< [0.6328, 0.7335], [0.1523, 0.2563], [0.1272, 0.2665] >, <0.7335, 0.1523, 0.2000>)

Step 2. By Equation (1), we calculate the score values of $\Psi(g_i)$ for the alternatives G_i ($i = 1, 2, 3, 4$) as the follows:

$$\Psi(g_1) = 0.6563,\ \Psi(g_2) = 0.7405,\ \Psi(g_3) = 0.6740,\ \Psi(g_4) = 0.7717.$$

Step 3. According to $\Psi(g_4) > \Psi(g_2) > \Psi(g_3) > \Psi(g_1)$, the ranking of the alternatives is $G_4 \succ G_2 \succ G_3 \succ G_1$. So, the alternative G_4 is the best one.

Compared with the MADM method introduced in Reference [20], Table 2 lists the decision results based on the NCHWAGA operator and cosine similarity measures of the NCSs. Obviously, the best alternatives and the ranking orders based on the NCHWAGA operator proposed in this paper are the same as in Reference [20].

For further relative comparison, Table 3 lists the MADM results using the NCHWAGA operator proposed in this paper and the NCWAA and NCWGA operators provided in Reference [31], respectively. The results listed in Table 3 show that the aggregated values of the NCHWAGA operator tend to the moderate values between the aggregated results of the NCWAA and NCWGA operators. Then, the ranking orders based on the NCHWAGA operator have little difference with the NCWAA and NCWGA operators. However, the best alternative given in all the MADM methods is identical. Furthermore, the results in Table 3 also show that the aggregated values of the NCHWAGA operator tend to moderate values between the aggregated values of the NCWAA and NCWGA operators in [31]. Therefore, the NCHWAGA operator can overcome the drawbacks of the NCWAA and NCWGA operators, and it is more effective and more suitable than the NCWAA and NCWGA operators to handle MADM problems under a neutrosophic cubic environment in some cases.

Table 2. Decision results based on the neutrosophic cubic hybrid weighted arithmetic and geometric aggregation (NCHWAGA) operator and cosine similarity measures.

MADM Method	Score Values (Cosine Measures Value)	Ranking Order	The Best Alternative
NCHWAGA ($\rho = 0.5$)	0.6563, 0.7405, 0.6740, 0.7717	$G_4 \succ G_2 \succ G_1 \succ G_3$	G_4
Cosine Measure S_{w1} [20]	0.9564, 0.9855, 0.9596, 0.9945	$G_4 \succ G_2 \succ G_1 \succ G_3$	G_4
Cosine Measure S_{w2} [20]	0.9769, 0.9944, 0.9795, 0.9972	$G_4 \succ G_2 \succ G_1 \succ G_3$	G_4
Cosine Measure S_{w3} [20]	0.9892, 0.9959, 0.9897, 0.9989	$G_4 \succ G_2 \succ G_1 \succ G_3$	G_4

Table 3. Decision results based on different aggregation operators.

Aggregation Operator	Aggregated Result	Score Value	Ranking Order	The Best Alternative
NCHWAGA ($\rho = 0.5$)	$g_1 = (< [0.5302, 0.6645], [0.1272, 0.3000], [0.1669, 0.3355] >, < 0.3430, 0.4709, 0.2306 >)$	$\Psi(g_1) = 0.6563$	$G_4 \succ G_2 \succ G_1 \succ G_3$	G_4
	$g_2 = (< [0.6000, 0.7335], [0.1523, 0.2563], [0.1669, 0.2685] >, <0.6628, 0.2525, 0.2346>)$	$\Psi(g_2) = 0.7405$		
	$g_3 = (< [0.4677, 0.6307], [0.2000, 0.3000], [0.2264, 0.3672] >, <0.6000, 0.2365, 0.3025>)$	$\Psi(g_3) = 0.6740$		
	$g_4 = (< [0.6328, 0.7335], [0.1523, 0.2563], [0.1272, 0.2665] >, <0.7335, 0.1523, 0.2000>)$	$\Psi(g_4) = 0.7717$		
NCWAA [31]	$g_1 = (< [0.5324, 0.6751], [0.1231, 0.3000], [0.1625, 0.3249] >, < 0.4039, 0.4315, 0.2158 >),$	$\Psi(g_1) = 0.6726$	$G_4 \succ G_2 \succ G_3 \succ G_1$	G_4
	$g_2 = (< [0.6000, 0.7365], [0.1390, 0.2462], [0.1625, 0.2656] >, <0.6653, 0.2301, 0.2114>)$	$\Psi(g_2) = 0.7497$		
	$g_3 = (< [0.4700, 0.6331], [0.2000, 0.3000], [0.2111, 0.3648] >, <0.6000, 0.2333, 0.2939>)$	$\Psi(g_3) = 0.6778$		
	$g_4 = (< [0.6352, 0.7365], [0.1390, 0.2462], [0.1231, 0.2635] >, <0.7365, 0.1390, 0.2000>)$	$\Psi(g_4) = 0.7775$		
NCWGA [31]	$g_1 = (< [0.5281, 0.6541], [0.1312, 0.3000], [0.1712, 0.3459] >, < 0.2912, 0.5075, 0.2452 >)$	$\Psi(g_1) = 0.6414$	$G_4 \succ G_2 \succ G_3 \succ G_1$	G_4
	$g_2 = (< [0.6000, 0.7306], [0.1654, 0.2661], [0.1712, 0.2714] >, <0.6602, 0.2757, 0.2571>)$	$\Psi(g_2) = 0.7315$		
	$g_3 = (< [0.4655, 0.6284], [0.2000, 0.3000], [0.2414, 0.3697] >, <0.6000, 0.2396, 0.3110>)$	$\Psi(g_3) = 0.6703$		
	$g_4 = (< [0.6303, 0.7306], [0.1654, 0.2661], [0.1312, 0.2694] >, <0.7306, 0.1654, 0.2000>)$	$\Psi(g_4) = 0.7660$		

6. Conclusions

This paper developed the NCHWAGA operator of NCSs and investigated its properties. The main advantage of the proposed NCHWAGA operator can overcome the drawbacks implied by the existing NCWAA and NCWGA operators [31] in some cases and reach the moderate aggregated values. Then, the MADM method based on the NCHWAGA operator was established under an NCS environment. Finally, we provided an illustrative example to demonstrate the application of the established MADM method. By comparison, we found that the developed MADM method was more effective and more suitable to solve decision-making problems with neutrosophic cubic information in some cases. In the real world, a refined neutrosophic set [14] is very suitable to express complex problems of decision-making, since it can be described by its refined types of sub-truths, sub-indeterminacies, and sub-falsities. Therefore, we shall further extend the NCHWAGA operator to neutrosophic refined cubic sets for MADM by using the refined neutrosophic sets. In addition, the proposed method will be also extended to neutrosophic cubic oversets/undersets/offsets using the neutrosophic overset/underset/offset [36] in the future.

Author Contributions: L.S. investigated the NCHWAGA operator and its MADM method; Y.Y. analyzed the numerical examples; we wrote this paper together.

Funding: This research received no external funding.

Acknowledgments: This research is supposed by the Public Technology Research Project of Zhejiang Province (LGG18F030008).

Conflicts of Interest: The authors declare no conflicts of interests.

Abbreviations

GRA	Grey relational analysis
INSs	Interval neutrosophic sets
MADM	Multi-attribute decision-making
MCGDM	Multi-criteria group decision making
MVNSs	Multi-valued neutrosophic sets
NCHWAGA	Neutrosophic cubic hybrid weighted arithmetic and geometric aggregation
NCSs	Neutrosophic cubic sets
NCWAA	Neutrosophic cubic weighted arithmetic average
NCWGA	Neutrosophic cubic geometric weighted average
SNSs	Simplified neutrosophic sets
SVNSs	Single-valued neutrosophic sets
WAA	Weighted arithmetic average
WGA	Geometric average

References

1. Zadeh, L.A. The concept of a linguistic variable and its application to approximate reasoning Part I. *Inf. Sci.* **1975**, *8*, 199–249. [CrossRef]
2. Atanassov, K.T. Intuitionstic fuzzy sets. *Fuzzy Sets Syst.* **1986**, *20*, 87–96. [CrossRef]
3. Smarandache, F. *Neutrosophy: Neutrosophic Probability, Set, and Logic*; American Research Press: Rehoboth, DE, USA, 1998.
4. Wang, H.; Smarandache, F.; Zhang, Y.Q.; Sunderraman, R. *Interval Neutrosophic Sets and Logic: Theory and Applications in Computing*; Hexis: Phoenix, AZ, USA, 2005.
5. Wang, H.; Smarandache, F.; Zhang, Y.Q.; Sunderraman, R. Single valued neutrosophic sets. *Multisp. Multistruct.* **2010**, *4*, 410–413.
6. Ye, J. A multicriteria decision-making method using aggregation operators for simplified neutrosophic sets. *J. Intell. Fuzzy Syst.* **2014**, *26*, 2459–2466. [CrossRef]
7. Wang, J.Q.; Li, X.E. An application of the TODIM method with multi-valued neutrosophic set. *Control Decis.* **2015**, *30*, 1139–1142.
8. Peng, J.J.; Wang, J.Q. Multi-valued neutrosophic sets and its application in multi-criteria decision-making problems. *Neutrosophic Sets Syst.* **2015**, *10*, 3–17. [CrossRef]
9. Liu, P.D.; Tang, G.L. Some power generalized aggregation operators based on the interval neutrosophic numbers and their application to decision making. *J. Intell. Fuzzy Syst.* **2016**, *30*, 2517–2528. [CrossRef]
10. Ye, J. Improved cosine similarity measures of simplified neutrosophic sets for medical diagnoses. *Artif. Intell. Med.* **2015**, *63*, 171–179. [CrossRef]
11. Ye, J.; Fu, J. Multi-period medical diagnosis method using a single valued neutrosophic similarity measure based on tangent function. *Comput. Methods Programs Biomed.* **2016**, *123*, 142–149. [CrossRef]
12. Shi, L. Correlation Coefficient of simplified neutrosophic sets for bearing fault diagnosis. *Shock Vib.* **2016**. [CrossRef]

13. Ye, J. Single valued neutrosophic similarity measures based on cotangent function and their application in the fault diagnosis of steam turbine. *Soft Comput.* **2017**, *21*, 817–825. [CrossRef]
14. Smarandache, F. n-Valued refined neutrosophic logic and its applications in physics. *Prog. Phys.* **2013**, *4*, 143–146.
15. Broumi, S.; Smarandache, F. Intuitionistic neutrosophic soft set. *J. Inf. Comput. Sci.* **2013**, *8*, 130–140.
16. Ye, J. Multiple-attribute decision-making method under a single-valued neutrosophic hesitant fuzzy environment. *J. Intell. Fuzzy Syst.* **2015**, *24*, 23–36. [CrossRef]
17. Broum, S.; Smarandache, F.; Dhar, M. Rough neutrosophic sets. *Neutrosophic Sets Syst.* **2014**, *3*, 62–67.
18. Ali, M.; Deli, I.; Smarandache, F. The theory of neutrosophic cubic sets and their applications in pattern recognition. *J. Intell. Fuzzy Syst.* **2016**, *30*, 1957–1963. [CrossRef]
19. Jun, Y.B.; Smarandache, F.; Kim, C.S. Neutrosophic cubic sets. *New Math. Nat. Comput.* **2017**, *13*, 41–45. [CrossRef]
20. Lu, Z.; Ye, J. Cosine measures of neutrosophic cubic sets for multiple attribute decision-making. *Symmetry* **2017**, *9*, 121. [CrossRef]
21. Banerjee, D.; Giri, B.C.; Pramanik, S.; Smarandache, F. GRA for multi attribute decision making in neutrosophic cubic set environment. *Neutrosophic Sets Syst.* **2017**, *15*, 64–73.
22. Pramanik, S.; Dalapati, S.; Alam, S.; Roy, T.K.; Smarandache, F. Neutrosophic cubic MCGDM method based on similarity measure. *Neutrosophic Sets Syst.* **2017**, *16*, 44–56.
23. Xu, Z.S. A note on linguistic hybrid arithmetic averaging operator in multiple attribute group decision making with linguistic information. *Group Decis. Negot.* **2006**, *15*, 593–604. [CrossRef]
24. Liu, P. Multiple attribute decision-making methods based on normal intuitionistic fuzzy interaction aggregation operators. *Symmetry* **2017**, *9*, 261. [CrossRef]
25. Liu, P.; Wang, P. Some improved linguistic intuitionistic fuzzy aggregation operators and their applications to multiple-attribute decision making. *Int. J. Inf. Technol. Decis. Mak.* **2017**, *16*, 817–850. [CrossRef]
26. Wei, G.W.; Alsaadi, F.E.; Hayat, T.; Alsaedi, A. Picture 2-tuple linguistic aggregation operators in multiple attribute decision making. *Soft Comput.* **2018**, *22*, 989–1002. [CrossRef]
27. Wei, G.W. Pythagorean fuzzy interaction aggregation operators and their application to multiple attribute decision making. *J. Intell. Fuzzy Syst.* **2017**, *33*, 2119–2132. [CrossRef]
28. Luo, X.; Xu, Z.; Gou, X. Exponential operational laws and new aggregation operators of intuitionistic Fuzzy information based on Archimedean T-conorm and T-norm. *Int. J. Mach. Learn. Cybern.* **2018**, *9*, 1261–1269. [CrossRef]
29. Wang, J.; Tian, C.; Zhang, X.; Zhang, H.; Wang, T. Multi-criteria decision-making method based on simplified neutrosophic linguistic information with cloud model. *Symmetry* **2018**, *10*, 197. [CrossRef]
30. Shi, L.; Ye, J. Dombi aggregation operators of neutrosophic cubic sets for multiple attribute decision-making. *Algorithms* **2018**, *11*, 29. [CrossRef]
31. Zhan, J.; Khan, M.; Gulistan, M.; Ali, A. Applications of neutrosophic cubic sets in multi-criteria decision making. *Int. J. Uncertain. Quantif.* **2017**, *7*, 377–394. [CrossRef]
32. Jun, Y.B.; Smarandache, F.; Kim, C.S. P-union and P-intersection of neutrosophic cubic sets. *St. Univ. Ovidius Constanta* **2017**, *25*, 99–115. [CrossRef]
33. Ye, J. Linguistic neutrosophic cubic numbers and their multiple attribute decision-making method. *Information* **2017**, *8*, 110. [CrossRef]
34. Ye, J. Intuitionistic fuzzy hybrid arithmetic and geometric aggregation operators for the decision-making of mechanical design schemes. *Appl. Intell.* **2017**, *47*, 743–751. [CrossRef]
35. Lu, Z.; Ye, J. Single-valued neutrosophic hybrid arithmetic and geometric aggregation pperators and their decision-making method. *Information* **2017**, *8*, 84. [CrossRef]
36. Smarandache, F. Neutrosophic Overset, Neutrosophic Underset, and Neutrosophic Offset. Similarly for Neutrosophic Over-/Under-/Off-Logic, Probability, and Statistics. Pons Editions: Bruxelles, Belgique, 2016. Available online: https://arxiv.org/ftp/arxiv/papers/1607/1607.00234.pdf (accessed on 4 January 2019).

symmetry

Article

A Single-Valued Neutrosophic Linguistic Combined Weighted Distance Measure and Its Application in Multiple-Attribute Group Decision-Making

Chengdong Cao, Shouzhen Zeng * and Dandan Luo

School of Business, Ningbo University, Ningbo 315211, China; 156001316@nbu.edu.cn (C.C.); nbluodandan@163.com (D.L.)

* Correspondence: zengshouzhen@nbu.edu.cn; Tel.: +86-15867202316

Received: 19 January 2019; Accepted: 15 February 2019; Published: 21 February 2019

Abstract: The aim of this paper is to present a multiple-attribute group decision-making (MAGDM) framework based on a new single-valued neutrosophic linguistic (SVNL) distance measure. By unifying the idea of the weighted average and ordered weighted averaging into a single-valued neutrosophic linguistic distance, we first developed a new SVNL weighted distance measure, namely a SVNL combined and weighted distance (SVNLCWD) measure. The focal characteristics of the devised SVNLCWD are its ability to combine both the decision-makers' attitudes toward the importance, as well as the weights, of the arguments. Various desirable properties and families of the developed SVNLCWD were contemplated. Moreover, a MAGDM approach based on the SVNLCWD was formulated. Lastly, a real numerical example concerning a low-carbon supplier selection problem was used to describe the superiority and feasibility of the developed approach.

Keywords: single-valued neutrosophic linguistic set; distance measure; combined weighted average; MAGDM; low-carbon supplier selection

1. Introduction

Multiple-attribute group decision-making (MAGDM) is one of the most commonly used methods to rank and select potential alternatives based on the decision information of multiple decision-makers (or experts). In real MAGDM problems, the increasing uncertainties of objects make it increasingly difficult for people to precisely express judgments about their attributes during the process of decision-making. Indeed, this is related not only to the nature of the objects but also to the ambiguity of the underlying human intervention and cognitive thinking in general. Handling imprecision or vagueness effectively in these complex situations is a matter of great concern in MAGDM problems. Recently, a new tool for solving the uncertainty or inaccuracy of such information was introduced by Ye [1], namely the single-valued neutrosophic linguistic set (SVNLS). By unifying the features of single-valued neutrosophic sets (SVNS) [2,3] and linguistic terms [4], the SVNLS can eliminate both of their shortcomings, and has been proven to be suitable to measure a higher degree of uncertainty for subjective evaluations. As an effective extension of the linguistic terms and SVNS, the basic element of the SVNLS is the single-valued neutrosophic linguistic value (SVNLV), which makes it more effective for handling uncertain and imprecise information when contrasted with the existing fuzzy tools, such as the intuitionistic linguistic set [5] and the Pythagorean fuzzy set [6]. Following the latest research trend, the SVNLS has been widely applied to handle MAGDM problems under indeterminacy and complex environments. Ye [1] investigated the classic technique for order preference by similarity to an ideal solution (TOPSIS) method in SVNLS situation and studied its usefulness for decision-making problems. Ye [7] developed some neutrosophic linguistic operators and investigated their applications in selecting a flexible manufacturing system. Wang et al. [8] extended the Maclaurin symmetric mean operator to aggregate SVNL information. Chen et al. [9] developed a novel distance measure for SVNLS based on the ordered weighted viewpoint. Ji et al. [10] proposed a combined multi-attribute border approximation area comparison (MABAC) and the elimination and choice translating reality (ELECTRE) approach for SVNLS and studied its application in selecting outsourcing providers. Wu et al. [11] investigated the usefulness of the SVNLS in a 2-tuple environment of MAGDM analysis. Kazimieras et al. [12] developed a new SVN decision-making model by applying the weighted aggregated sum product assessment (WASPAS) method. Garg and Nancy [13] proposed several prioritized aggregation operators for SVNLS to handle the priority among the attributes.

Distance measurement is one of the most widely used tools in MAGDM, and can be used to measure the differences between the expected solutions and potential alternatives. Recently, a new distance measurement method based on the ordered weighted viewpoint, i.e., the ordered weighted averaging distance (OWAD) operator proposed by Merigó and Gil-Lafuente [14] has attracted increasing attention from researchers. The essence of this distance operator is that it enables decision-makers to incorporate their attitudinal bias into the decision-making process by imposing some weighting schemes to the individual distances. To date, several OWAD extensions and their subsequent applications in solving MAGDM problems have appeared in recent studies, such as the induced OWAD operator [15], intuitionistic fuzzy OWAD operator [16], hesitant fuzzy OWAD operator [17], probabilistic OWAD operator [18], Pythagorean fuzzy generalized OWAD operator [19], fuzzy linguistic induced Euclidean OWAD operator [20], continuous OWAD operator [21] and the intuitionistic fuzzy weighted induced OWAD operator [22]. More recently, Chen et al. [8] further presented a definition of the single-valued neutrosophic linguistic OWAD (SVNLOWAD) operator, on the basis of which a modified TOPSIS model was then proposed for MAGDM problems in a SVNL situation.

Although the OWAD operator and its numerous extensions, such as the SVNLOWAD operator, have shown their superiority in practical applications, they possess a defect in that they can integrate only the special interests of the experts, while ignoring the importance of the attributes in the outcome of a decision. To overcome this shortcoming, this study develops a combined weighted distance for SVNLSs, called the single-valued neutrosophic linguistic combined weighted distance (SVNLCWD) operator. The proposed combined weighted distance operator is superior in that it involves both subjective information on the importance of the ordered attributes and the importance of specific attributes. We further explored some of the key properties and particular cases of the proposed operator. Finally, we applied the SVNLCWD operator to a MAGDM problem concerning low-carbon supplier selection to verify its effectiveness and superiority.

2. Preliminaries

In this section, we will briefly review some of concepts we need to use in the following sections, including the definition of the SVNLS, the OWAD and the SVNLOWAD operator.

2.1. Linguistic Set

Let $S = \{s_\alpha | \alpha = 1, \ldots, l\}$ be a finitely ordered discrete term set, where s_α indicates a possible value for a linguistic variable (LV) and l is an odd number. For instance, taking $l = 7$, then a linguistic term set S could be specified $S = \{s_1, s_2, s_3, s_4, s_5, s_6, s_7\}$ = *{extremely poor, very poor, poor, fair, good, very good, extremely good}*. In this case, any two LVs s_i and s_j in S should satisfy rules (1)-(4) [23]:

(1) $Neg(s_i) = s_{-i}$;
(2) $s_i \le s_j \Leftrightarrow i \le j$;
(3) $\max(s_i, s_j) = s_j$, if $i \le j$;
(4) $\min(s_i, s_j) = s_i$, if $i \le j$.

To minimize information loss in the operational process, the discrete term set S shall be extended to a continuous set $\overline{S} = \{s_\alpha | \alpha \in R\}$. Any two LVs $s_\alpha, s_\beta \in \overline{S}$, satisfy the following operational rules [24]:

(1) $s_\alpha \oplus s_\beta = s_{\alpha+\beta}$;
(2) $\mu s_\alpha = s_{\mu\alpha}$, $\mu \geq 0$;
(3) $s_\alpha / s_\beta = s_{\alpha/\beta}$.

2.2. Single-Valued Neutrosophic Set (SVNS)

The neutrosophic set was introduced for the first time by Smarandache in 1998 [2], while Ye introduced the linguistic neutrosophic set in 2015 [1] and Ye developed the single-valued neutrosophic set (SVNS) in 2013 [25].

Definition 1. Let y be an element in a finite set Y. A SVNS P in Y can be defined as in (1):

$$P = \{ \langle y, T_P(y), I_P(y), F_P(y) \rangle | y \in Y \}, \tag{1}$$

where the truth-membership function $T_P(y)$, the indeterminacy-membership function $I_P(y)$, and the falsity-membership function $F_P(y)$ shall satisfy the following conditions:

$$0 \leq T_P(y), I_P(y), F_P(y) \leq 1, \; 0 \leq T_P(y) + I_P(y) + F_P(y) \leq 3. \tag{2}$$

For convenience of calculation, we call the triplet $(T_P(y), I_P(y), F_P(y))$ single-valued neutrosophic value (SVNV) and simply denote it as $y = (T_y, I_y, F_y)$. Let $y = (T_y, I_y, F_y)$ and $z = (T_z, I_z, F_z)$ be two SVNVs, their mathematical operational laws are defined as follows:

(1) $y \oplus z = (T_y + T_z - T_y * T_z, I_y * T_z, F_y * F_z)$;
(2) $\lambda y = (1 - (1 - T_y)^\lambda, (I_y)^\lambda, (F_y)^\lambda)$, $\lambda > 0$;
(3) $y^\lambda = ((T_y)^\lambda, 1 - (1 - I_y)^\lambda, 1 - (1 - F_y)^\lambda)$, $\lambda > 0$.

2.3. Single-Valued Neutrosophic Linguistic Set (SVNLS)

On the basis of the SVNS, Ye gave the definition and operational laws of the single-valued neutrosophic linguistic set (SVNLS), listed in the definitions 2–5.

Definition 2. Let Y be a finite universe set, a SVNLS Q in Y is defined as in (3):

$$Q = \left\{ \left\langle y, [s_{\theta(y)}, (T_P(y), I_P(y), F_P(y))] \right\rangle \middle| y \in Y \right\}, \tag{3}$$

where $s_{\theta(y)} \in \overline{S}$, the truth-membership function $T_q(y)$, the indeterminacy-membership function $I_q(y)$, and the falsity-membership function $F_q(y)$ satisfy condition (4):

$$0 \leq T_q(y), I_q(y), F_q(y) \leq 1, \; 0 \leq T_q(y) + I_q(y) + F_q(y) \leq 3. \tag{4}$$

For a SVNLS Q in Y, the SVNLV $\left\langle s_{\theta(y)}, (T_P(y), I_P(y), F_P(y)) \right\rangle$ is simply denoted as $y = \left\langle s_{\theta(y)}, (T_y, I_y, F_y) \right\rangle$ for computational convenience.

Definition 3. Let $y_i = \left\langle s_{\theta(y_i)}, (T_{y_i}, I_{y_i}, F_{y_i}) \right\rangle (i = 1, 2)$ be two SVNLVs, then

(1) $y_1 \oplus y_2 = \left\langle s_{\theta(y_1)+\theta(y_2)}, (T_{y_1} + T_{y_2} - T_{y_1} * T_{y_2}, I_{y_1} * T_{y_2}, F_{y_1} * F_{y_2}) \right\rangle$;
(2) $\lambda y_1 = \left\langle s_{\lambda\theta(y_1)}, (1 - (1 - T_{y_1})^\lambda, (I_{y_1})^\lambda, (F_{y_1})^\lambda) \right\rangle$, $\lambda > 0$;
(3) $y_1^\lambda = \left\langle s_{\theta^\lambda(y_1)}, ((T_{y_1})^\lambda, 1 - (1 - I_{y_1})^\lambda, 1 - (1 - F_{y_1})^\lambda) \right\rangle$, $\lambda > 0$.

Definition 4. The distance measure between the SVNLVs $y_i = \left\langle s_{\theta(y_i)}, (T_{y_i}, I_{y_i}, F_{y_i}) \right\rangle (i = 1, 2)$ is defined as in (5):

$$d(y_1, y_2) = \left[\left|\theta(y_1)T_{y_1} - \theta(y_2)T_{y_2}\right|^{\lambda} + \left|\theta(y_1)I_{y_1} - \theta(y_2)I_{y_2}\right|^{\lambda} + \left|\theta(y_1)F_{y_1} - \theta(y_2)F_{y_2}\right|^{\lambda} \right]^{1/\lambda}. \quad (5)$$

If we assign different weights to the individual distances of the SVNLVs, we get the single-valued neutrosophic linguistic weighted distance (SVNLWD) measure [8].

Definition 5. Let y_j, y_j' $(j = 1, \ldots, n)$ be the two collections of SVNLVs, a single-valued neutrosophic linguistic weighted distance measure of dimension n is a mapping SVNLWD: $\Omega^n \times \Omega^n \to R$, which has an associated weighting vector W with $w_j \in [0, 1]$ and $\sum_{j=1}^{n} w_j = 1$, such that:

$$SVNLWD((y_1, y_1'), \ldots, (y_n, y_n')) = \sum_{j=1}^{n} w_j d(y_j, y_j'), \quad (6)$$

The OWAD operator developed by Merigó and Gil-Lafuente [14] aims to aggregate individual distances as arguments on the basis of the ordered weighted averaging (OWA) operator [26]. Let $A = \{a_1, a_2, \ldots, a_n\}$ and $B = \{b_1, b_2, \ldots, b_n\}$ be two crisp sets, and the OWAD operator can be defined as follows.

Definition 6. An OWAD operator is defined as a mapping OWAD: $R^n \times R^n \to R$ with the weighting vector $W = \{ w_j | \sum_{i=1}^{n} w_j = 1, \ 0 \le w_j \le 1 \}$, such that:

$$OWAD(\langle a_1, b_1 \rangle, \ldots, \langle a_n, b_n \rangle) = \sum_{j=1}^{n} w_j d_j, \quad (7)$$

where d_j is the j-th largest number among $|a_i - b_i|$.

On the basis of the OWAD operator, Chen et al. [9] introduced the SVNLOWAD operator to aggregate SVNL information.

Definition 7. Let y_j, y_j' $(j = 1, \ldots, n)$ be the two collections of SVNLVs. If

$$SVNLOWAD((y_1, y_1'), \ldots, (y_n, y_n')) = \sum_{j=1}^{n} w_j d(y_j, y_j'), \quad (8)$$

then the SVNLOWAD is called the single-value neutrosophic linguistic OWAD, where $d(y_j, y_j')$ represents the j-th largest value among the individual distances $d(y_i, y_i')(i = 1, \ldots, n)$ defined in Equation (5). $w = (w_1, \ldots, w^n)^T$ is a weighting vector related to the SVNLOWAD operator, satisfying $\sum_{j=1}^{n} w_j = 1$ and $w_j \in [0, 1]$.

The properties of commutativity, monotonicity, boundedness and idempotency can easily be established for the SVNLOWAD operator. Based on the above analysis, we can find that, although the SVNLOWAD and SVNLWD operators have been widely used to solve MAGDM problems in SVNL environments, these two operators exhibit certain deficiencies. Next, we shall propose a combined weighted distance measure to alleviate these shortcomings.

3. SVNL Combined Weighted Distance (SVNLCWD) Operator

The SVNL combined weighted distance (SVNLCWD) operator unifies both the advantages of the SVNLWD and the SVNLOWAD operators in the same framework. Therefore, it is able to integrate the decision-makers' attitudes using ordered weighted arguments as well as embedding the importance of alternatives based on the weighted average method. Moreover, it allows decision-makers to adjust the allocation ratio of the SVNLOWAD and SVNLWD flexibly based on the needs of the particular problem or their interests. The SVNLCWD operator can be defined as follows.

Definition 8. Let y_j, y_j' $(j = 1, \ldots, n)$ be the two collections of SVNLVs. If

$$SVNLCWD((y_1, y_1'), \ldots, (y_n, y_n')) = \sum_{j=1}^{n} \overline{w}_j D_j, \tag{9}$$

then the SVNLCWD is called the single-value neutrosophic linguistic combined weighted distance operator, where D_j represents the j-th largest value among the individual distances $d(y_i, y_i')(i = 1, 2..., n)$ defined in Equation (5). There are two weights assigned to each distance D_j: ω_j, is the weight for weighted averaging (WA) with $\sum_{j=1}^{n} \omega_j = 1$ and $\omega_j \in [0,1]$, and w_j, is the weight for the OWA meeting $\sum_{j=1}^{n} w_j = 1$ and $w_j \in [0,1]$. The integrated weight $\overline{w}_j$ is defined as:

$$\overline{w}_j = \delta \omega_j + (1 - \delta) w_j, \tag{10}$$

where $\delta \in [0,1]$ and ω_j is indeed ω_i re-ordered to be associated to $d(y_i, y_i')(i = 1, \ldots, n)$.

Based on the basic operational laws (i.e., ordered weighted and weighted average), the SVNLCWD operator can be decomposed linearly into a combination of the SVNLOWAD and SVNLWD:

Definition 9. Let y_j, y_j' $(j = 1, \ldots, n)$ be the two collections of SVNLNs. If

$$SVNLCWD((y_1, y_1'), \ldots, (y_n, y_n')) = \delta \sum_{i=1}^{n} \omega_i d(y_i, y_i') + (1 - \delta) \sum_{j=1}^{n} w_j D_j, \tag{11}$$

where D_j represents the j-th largest value among the individual distances $d(y_i, y_i')(i = 1, \ldots, n)$ defined in Equation (5), and $\delta \in [0,1]$. Obviously, the SVNLCWD is reduced to the SVNLOWAD and SVNLWD, when $\delta = 0$ and $\delta = 1$, respectively.

Example 3.1. Let $Y = (y_1, y_2, y_3, y_4, y_5) = (\langle s_2, (0.5, 0.3, 0.4) \rangle, \langle s_5, (0.5, 0.2, 0.2) \rangle, \langle s_5, (0.3, 0.3, 0.6) \rangle, \langle s_2, (0.1, 0.4, 0.6) \rangle, \langle s_7, (0.5, 0.8, 0.2) \rangle)$ and $Y' = (y'_1, y'_2, y'_3, y'_4, y'_5) = (\langle s_5, (0.2, 0.9, 0) \rangle, \langle s_3, (0.5, 0.7, 0.2) \rangle, \langle s_5, (0.4, 0.4, 0.5) \rangle, \langle s_4, (0.5, 0.7, 0.2) \rangle, \langle s_3, (0.4, 0.2, 0.6) \rangle)$ be two SVNLSs defined in set $S = \{s_1, s_2, s_3, s_4, s_5, s_6, s_7\}$. Let $w = (0.15, 0.3, 0.2, 0.25, 0.1)^T$ be the weighting vector of SVNLCWD measure. Then, the aggregating process by the SVNLCWD can be displayed as follows:

(1) Compute the individual distances $d(y_i, y'_i)(i = 1, 2, \ldots, 5)$ (let $\lambda = 1$) according to Equation (5):

$$d(y_1, y'_1) = |2 \times 0.5 - 5 \times 0.2| + |2 \times 0.3 - 5 \times 0.9| + |2 \times 0.4 - 5 \times 0| = 4.7.$$

Similarly, we get

$$d(y_2, y'_2) = 2.4,\ d(y_3, y'_3) = 1.5,$$
$$d(y_4, y'_4) = 3.2,\ d(y_5, y'_5) = 7.7.$$

(2) Sort the $d(y_i, y'_i)(i = 1, 2, \ldots, 5)$ in decreasing order:

$$D_1 = d(y_5, y'_5) = 7.7,\ D_2 = d(y_1, y'_1) = 4.7,\ D_3 = d(y_4, y'_4) = 3.2,$$
$$D_4 = d(y_2, y'_2) = 2.4,\ D_5 = d(y_3, y'_3) = 1.5.$$

(3) Let the weighting vector $\omega = (0.1, 0.15, 0.2, 0.35, 0.2)^T$ and $\delta = 0.4$, calculate the integrated weights $\overline{w}_j$ according to Equation (10):

$$\overline{w}_1 = 0.4 \times 0.2 + (1 - 0.4) \times 0.15 = 0.17,\ \overline{w}_2 = 0.4 \times 0.1 + (1 - 0.4) \times 0.3 = 0.22,$$
$$\overline{w}_3 = 0.4 \times 0.35 + (1 - 0.4) \times 0.2 = 0.26,\ \hat{w}_4 = 0.4 \times 0.15 + (1 - 0.4) \times 0.25 = 0.21,$$
$$\overline{w}_5 = 0.4 \times 0.2 + (1 - 0.4) \times 0.1 = 0.14.$$

(4) Use the SVNLCWD measure defined in Equation (9) to perform the following aggregation:

$$\begin{gathered} SVNLCWD(Y, Y') \\ = 0.17 \times 7.7 + 0.22 \times 4.7 + 0.26 \times 3.2 + 0.21 \times 2.4 + 0.14 \times 1.5 \\ = 3.889 \end{gathered}$$

We can also perform the aggregation process of the SVNLCWD using Equation (11):

$$\begin{gathered} SVNLCWD(Y, Y') \\ = 0.4 \times SVNLWD + (1 - 0.4) \times SVNLOWAD \\ = 0.4 \times 3.79 + 0.6 \times 3.955 \\ = 3.889 \end{gathered}$$

Apparently, we obtain the same results using both methods. However, compared with the SVNLOWAD operator, the proposed SVNLCWD operator can not only incorporate decision-makers' interests and biases according to the ordered weights, but also highlights the importance of the input arguments based on the weighted average tool.

Furthermore, by setting varied weighting schemes on the SVNLCWD operator, we can obtain a series of SVNL weighted distance measures:

- If $w_1 = 1, w_2 = \cdots = w_n = 0$, then max-SVNLWD (SVNLMaxD) is formed.
- If $w_1 = \cdots = w_{n-1} = 0, w_n = 1$, then the min-SVNLWD (SVNLMinD) is obtained.
- The step-SVNLCWD operator is rendered by imposing $w_1 = \cdots = w_{k-1} = 0, w_k = 1$ and $w_{k+1} = \cdots = w_n = 0$.
- According to techniques used in the recent literature [27,28], we can create more special cases of the SVNLCWD, such as the Median-SVNLCWD, the Centered-SVNLCWD and the Olympic-SVNLCWD operators.

The SVNLCWD operator has the following desirable properties that all aggregation operators should ideally possess:

Theorem 1. (Commutativity–aggregation operator). Let $((x_1, x'_1), \ldots, (x_n, x'_n))$ be any permutation of the set of SVNLVs $((y_1, y'_1), \ldots, (y_n, y'_n))$, then

$$SVNLCWD((x_1, x'_1), \ldots, (x_n, x'_n)) = SVNLCWD((y_1, y'_1), \ldots, (y_n, y'_n)) \quad (12)$$

The property of commutativity can also be demonstrated from the perspective of distance measure:

$$SVNLCWD((y_1, y'_1), \ldots, (y_n, y'_n)) = SVNLCWD((y'_1, y_1), \ldots, (y'_n, y_n)) \tag{13}$$

Theorem 2. (Monotonicity). If $d(y_i, y'_i) \geq d(x_i, x'_i)$ for all i, the following property holds

$$SVNLCWD((y_1, y'_1), \ldots, (y_n, y'_n)) \geq SVNLCWD((x_1, x'_1), \ldots, (x_n, x'_n)) \tag{14}$$

Theorem 3. (Boundedness). This feature shows that the aggregation result lies between the minimum and maximum arguments (distances) to be aggregated:

$$\min_i(d(y_i, y'_i)) \leq SVNLCWD((y_1, y'_1), \ldots, (y_n, y'_n)) \leq \max_i(d(y_i, y'_i)) \tag{15}$$

Theorem 4. (Idempotency). If $d(y_i, y'_i) = D$ for all i, then

$$SVNLCWD((y_1, y'_1), \ldots, (y_n, y'_n)) = D \tag{16}$$

Theorem 5. (Nonnegativity). In case distances are aggregated, the result of aggregation is positive:

$$SVNLCWD((y_1, y'_1), \ldots, (y_n, y'_n)) \geq 0 \tag{17}$$

Theorem 6. (Reflexivity). In case the two vectors involved in the aggregation coincide, the resulting variable is zero:

$$SVNLCWD((y_1, y_1), \ldots, (y_n, y_n)) = 0 \tag{18}$$

4. New MAGDM Method Using the SVNLCWD Operator

The SVNLCWD operator can be used in a wide range of environments, such as data analysis, financial investment and engineering applications [29–32]. Subsequently, a new approach was developed for MAGDM problems in SVNL situations. Suppose that $C = \{C_1, C_2, \ldots, C_m\}$ is the set of schemes, and $A = \{A_1, A_2, \ldots, A_n\}$ is a set of finite attributes.

Step 1: Let each decision-maker (DM) $e_k(k = 1, 2, \ldots, t)$ (whose weight is ε_k, meeting $\varepsilon_k \geq 0$ and $\sum_{k=1}^{t} \varepsilon_k = 1$) provide his/her evaluation on the attributes expressed by the SVNLVs, and then form the individual matrix $Y^k = \left(y_{ij}^{(k)}\right)_{m \times n}$.

Step 2: Aggregate all evaluations of the individual DMs into a collective one, and then construct the group matrix:

$$Y = \left(y_{ij}\right)_{m \times n} = \begin{pmatrix} y_{11} & \cdots & y_{1n} \\ \vdots & \ddots & \vdots \\ y_{m1} & \cdots & y_{mn} \end{pmatrix}, \tag{19}$$

where the SVNLN $y_{ij} = \sum_{k=1}^{t} \varepsilon_k y_{ij}^{(k)}$.

Step 3: Construct the ideal levels for each attribute to establish the ideal scheme (see Table 1).

Table 1. Ideal scheme.

	A_1	A_2	$\cdots$	A_n
I	$\widetilde{y}_1$	$\widetilde{y}_2$	$\cdots$	$\widetilde{y}_n$

Step 4: Utilize the SVNLCWD to compute the distances between the ideal scheme I and the different alternatives $C_i(i = 1, 2, \ldots, m)$.

Step 5: Sort all alternatives and identify the best alternative(s) according to the results derived from Step 4.

5. An Illustrative Example: Low-Carbon Supplier Selection

We will focus on a numerical example of the low-carbon supplier selection problem provided by Chen et al. [9]. Three experts are invited to evaluate and prioritize a suitable low-carbon supplier as a manufacturer, with respect to the four potential suppliers C_i $(i = 1, 2, 3, 4)$ using the attributes: low-carbon technology (A_1), risk factor (A_2), cost (A_3) and capacity (A_4). The preference presented by the experts regarding these four attributes is formed into three individual SVNL decision matrices under the linguistic term set $S = \{s_1$ = extremely poor, s_2 = very poor, s_3 = poor, s_4 = fair, s_5 = good, s_6 = very good, s_7 = extremely good$\}$, as listed in Tables 2–4.

Table 2. SVNL decision matrix Y^1.

	A_1	A_2	A_3	A_4
C_1	$\left\langle s_5^{(1)}, (0.7, 0.0, 0.1) \right\rangle$	$\left\langle s_4^{(1)}, (0.6, 0.1, 0.2) \right\rangle$	$\left\langle s_3^{(1)}, (0.3, 0.1, 0.2) \right\rangle$	$\left\langle s_6^{(1)}, (0.6, 0.1, 0.2) \right\rangle$
C_2	$\left\langle s_6^{(1)}, (0.6, 0.1, 0.2) \right\rangle$	$\left\langle s_5^{(1)}, (0.6, 0.1, 0.2) \right\rangle$	$\left\langle s_4^{(1)}, (0.5, 0.2, 0.2) \right\rangle$	$\left\langle s_3^{(1)}, (0.6, 0.2, 0.4) \right\rangle$
C_3	$\left\langle s_4^{(1)}, (0.3, 0.2, 0.3) \right\rangle$	$\left\langle s_4^{(1)}, (0.5, 0.2, 0.3) \right\rangle$	$\left\langle s_3^{(1)}, (0.5, 0.3, 0.1) \right\rangle$	$\left\langle s_5^{(1)}, (0.3, 0.5, 0.2) \right\rangle$
C_4	$\left\langle s_5^{(1)}, (0.4, 0.2, 0.3) \right\rangle$	$\left\langle s_5^{(1)}, (0.4, 0.2, 0.3) \right\rangle$	$\left\langle s_3^{(1)}, (0.3, 0.2, 0.5) \right\rangle$	$\left\langle s_4^{(1)}, (0.5, 0.3, 0.3) \right\rangle$

Table 3. SVNL decision matrix Y^2.

	A_1	A_2	A_3	A_4
C_1	$\left\langle s_4^{(3)}, (0.6, 0.1, 0.2) \right\rangle$	$\left\langle s_4^{(3)}, (0.5, 0.2, 0.2) \right\rangle$	$\left\langle s_3^{(3)}, (0.4, 0.1, 0.1) \right\rangle$	$\left\langle s_5^{(3)}, (0.7, 0.2, 0.1) \right\rangle$
C_2	$\left\langle s_5^{(3)}, (0.5, 0.2, 0.3) \right\rangle$	$\left\langle s_4^{(3)}, (0.7, 0.2, 0.2) \right\rangle$	$\left\langle s_5^{(3)}, (0.7, 0.2, 0.1) \right\rangle$	$\left\langle s_6^{(3)}, (0.4, 0.6, 0.2) \right\rangle$
C_3	$\left\langle s_6^{(3)}, (0.5, 0.1, 0.3) \right\rangle$	$\left\langle s_5^{(3)}, (0.6, 0.1, 0.3) \right\rangle$	$\left\langle s_4^{(3)}, (0.6, 0.2, 0.1) \right\rangle$	$\left\langle s_4^{(3)}, (0.3, 0.6, 0.2) \right\rangle$
C_4	$\left\langle s_6^{(3)}, (0.5, 0.2, 0.3) \right\rangle$	$\left\langle s_6^{(3)}, (0.6, 0.2, 0.4) \right\rangle$	$\left\langle s_5^{(3)}, (0.2, 0.1, 0.6) \right\rangle$	$\left\langle s_4^{(3)}, (0.5, 0.2, 0.3) \right\rangle$

Table 4. SVNL decision matrix Y^3.

	A_1	A_2	A_3	A_4
C_1	$\left\langle s_4^{(2)}, (0.8, 0.1, 0.2) \right\rangle$	$\left\langle s_5^{(2)}, (0.7, 0.2, 0.3) \right\rangle$	$\left\langle s_4^{(2)}, (0.4, 0.2, 0.2) \right\rangle$	$\left\langle s_6^{(2)}, (0.6, 0.3, 0.3) \right\rangle$
C_2	$\left\langle s_6^{(2)}, (0.7, 0.2, 0.3) \right\rangle$	$\left\langle s_6^{(2)}, (0.7, 0.2, 0.3) \right\rangle$	$\left\langle s_5^{(2)}, (0.6, 0.2, 0.2) \right\rangle$	$\left\langle s_4^{(2)}, (0.5, 0.4, 0.2) \right\rangle$
C_3	$\left\langle s_6^{(2)}, (0.4, 0.2, 0.4) \right\rangle$	$\left\langle s_6^{(2)}, (0.6, 0.3, 0.4) \right\rangle$	$\left\langle s_4^{(2)}, (0.6, 0.1, 0.3) \right\rangle$	$\left\langle s_5^{(2)}, (0.4, 0.4, 0.1) \right\rangle$
C_4	$\left\langle s_5^{(2)}, (0.4, 0.3, 0.4) \right\rangle$	$\left\langle s_6^{(2)}, (0.5, 0.1, 0.2) \right\rangle$	$\left\langle s_5^{(2)}, (0.3, 0.1, 0.6) \right\rangle$	$\left\langle s_3^{(2)}, (0.7, 0.1, 0.1) \right\rangle$

Assume that the weights of the experts are $\varepsilon_1 = 0.37$, $\varepsilon_2 = 0.30$ and $\varepsilon_3 = 0.33$, respectively. Then we can aggregate the individual opinion and form the group SVNL decision matrix, which is listed in Table 5.

Table 5. Group SVNL decision matrix R.

	A_1	A_2	A_3	A_4
C_1	$\langle s_{4.37}, (0.714, 0.000, 0.155)\rangle$	$\langle s_{4.33}, (0.611, 0.155, 0.229)\rangle$	$\langle s_{3.67}, (0.365, 0.128, 0.163)\rangle$	$\langle s_{5.70}, (0.633, 0.180, 0.186)\rangle$
C_2	$\langle s_{5.70}, (0.611, 0.155, 0.258)\rangle$	$\langle s_{4.70}, (0.666, 0.155, 0.229)\rangle$	$\langle s_{2.37}, (0.602, 0.200, 0.162)\rangle$	$\langle s_{4.23}, (0.514, 0.350, 0.258)\rangle$
C_3	$\langle s_{5.26}, (0.399, 0.163, 0.330)\rangle$	$\langle s_{4.96}, (0.566, 0.186, 0.330)\rangle$	$\langle s_{3.37}, (0.566, 0.185, 0.144)\rangle$	$\langle s_{4.70}, (0.335, 0.491, 0.159)\rangle$
C_4	$\langle s_{5.30}, (0.432, 0.229, 0.330)\rangle$	$\langle s_{5.63}, (0.450, 0.159, 0.286)\rangle$	$\langle s_{2.37}, (0.271, 0.129, 0.561)\rangle$	$\langle s_{3.67}, (0.578, 0.185, 0.209)\rangle$

According to their objectives, the experts carry out a similar analysis to determine the ideal scheme, which represents the optimal results that a supplier should have. The resulting vector (Table 6) further serves as a reference point.

Table 6. Ideal scheme.

	A_1	A_2	A_3	A_4
I	$\langle s_7, (0.9, 0, 0)\rangle$	$\langle s_7, (1, 0, 0.1)\rangle$	$\langle s_7, (0.9, 0, 0.1)\rangle$	$\langle s_6, (0.9, 0.1, 0)\rangle$

Assume that the weight vectors of the attributes and the SVNLCWD are $\omega = (0.25, 0.40, 0.20, 0.15)^T$ and $w = (0.2, 0.15, 0.3, 0.35)^T$, respectively. Considering the available information, we can employ the developed SVNLCWD (without loss of generality, let $\delta = 0.5$) to compute the distances between the ideal scheme I and the different alternatives $C_i (i = 1, 2, 3, 4)$:

$$SVNLCWD(I, C_1) = 5.176,\ SVNLCWD(I, C_2) = 5.660,$$
$$SVNLCWD(I, C_3) = 6.544,\ SVNLCWD(I, C_4) = 6.641.$$

Note that smaller values of distances show preferable alternatives. Thus, the ranking of the alternatives through the values of $SVNLCWD(I, C_i)(i = 1, 2, 3, 4)$ yields:

$$A_1 \succ A_2 \succ A_3 \succ A_4.$$

The results show that A_1 had the smallest distance from the ideal scheme, which means it was the most desirable alternative.

To better reflect the superiority of the SVNLCWD, we used the SVNLWD and the SVNLOWAD to measure the relative performance of the ideal scheme to all alternatives. For the SVNLWD measure, we obtained:

$$SVNLWD(I, C_1) = 5.249,\ SVNLWD(I, C_2) = 5.669,$$
$$SVNLWD(I, C_3) = 6.621,\ SVNLWD(I, C_4) = 6.789.$$

For the SVNLOWAD operator, we obtained:

$$SVNLOWAD(I, C_1) = 5.103,\ SVNLOWAD(I, C_2) = 5.652,$$
$$SVNLOWAD(I, C_3) = 6.466,\ SVNLOWAD(I, C_4) = 6.492.$$

It is easy to see that the most desirable alternative was A_1 for both the SVNLWD and SVNLOWAD operators, which coincides with the results derived using the proposed SVNLCWD operator. Moreover, the comparison of the SVNLWD and SVNLOWAD operators indicates that the SVNLCWD operator was able to account for the degrees of pessimism or optimism of the attitudes of decision-makers, and the different values of importance assigned to the various criteria during the process of aggregation. Furthermore, this method has more flexibility as it can execute the selection procedure by assigning different parameter values for the operator.

6. Conclusions

In this paper, we proposed a new combined weighted distance measure for SVNLSs, i.e., the SVNL combined weighted distance operator, to overcome the drawbacks of the existing method. Given that the developed combined weighted distance measure for SVNLSs involves both the SVNL weighted average and SVNL ordered weighted models, it takes into account both the attitudes toward separate criteria, as well as toward positions in the ordered array. Moreover, the SVNLCWD operator generalizes different types of SVNL aggregation operators, such as the SVNLMaxD, the SVNLMinD, the SVNLOWAD and the step-SVNLCWD operators. Thus, it provides a further generalization of previous methods by presenting a more general model to deal with the complex environments in a more flexible and efficient manner.

The illustrative example dealt with a selection problem of a low-carbon supplier. We conducted the sensitivity analysis to verify the robustness of the results by means of the changes in the aggregation rules (implemented by switching to different aggregation operators) and the changes in the relative importance of the ordered weights and arithmetic weights. Therefore, the proposed methodology can simulate different degrees of pessimism or optimism displayed by the decision-makers and account for the relative importance imposed on the various criteria in the aggregation process.

In future research, we will propose some methodological extensions and applications of the SVNLCWD with other decision-making approaches, such as induced aggregation and moving averaging.

Author Contributions: S.Z. and D.L. revised the manuscript and conceived the MAGDM framework. C.C. drafted the initial manuscript and analyzed the data.

Funding: This paper was funded by the National Natural Science Foundation of China (No. 71671165), Zhejiang Province Natural Science Foundation (No. LY18G010007), Major Humanities and Social Sciences Research Projects in Zhejiang Universities (No. 2018QN058), Cooperation Project between Ningbo City and Chinese Academy of Social Sciences (No. NZKT201711), Key SRIP Project of Ningbo University (No. 2018SRIP0109) and K. C. Wong Magna Fund in Ningbo University.

Conflicts of Interest: The authors declare no conflict of interest.

References

1. Ye, J. An extended TOPSIS method for multiple attribute group decision-making based on single valued neutrosophic linguistic numbers. *J. Intell. Fuzzy Syst.* **2015**, *28*, 247–255.
2. Smarandache, F. *Neutrosophy: Neutrosophic Probability, Set, and Logic*; American Research Press: Rehoboth, DE, USA, 1998.
3. Wang, H.; Smarandache, F.; Zhang, Y.Q.; Sunderraman, R. Single valued neutrosophic sets. *Multispace Multistruct.* **2010**, *4*, 410–413.
4. Zadeh, L.A. Fuzzy sets. *Inf. Control* **1965**, *18*, 338–353. [CrossRef]
5. Wang, J.Q.; Li, H.B. Multi-criteria decision-making method based on aggregation operators for intuitionistic linguistic fuzzy numbers. *Control Decis.* **2010**, *25*, 1571–1574.
6. Yager, R.R. Pythagorean Membership Grades in Multicriteria Decision Making. *IEEE Trans. Fuzzy Syst.* **2014**, *22*, 958–965. [CrossRef]
7. Ye, J. Aggregation operators of neutrosophic linguistic numbers for multiple attribute group decision making. *SpringerPlus* **2016**, *5*, 67. [CrossRef]
8. Wang, J.Q.; Yang, Y.; Li, L. Multi-criteria decision-making method based on single-valued neutrosophic linguistic Maclaurin symmetric mean operators. *Neural Comput. Applic.* **2018**, *30*, 1529–1547. [CrossRef]
9. Chen, J.; Zeng, S.; Zhang, C. An OWA Distance-Based, Single-Valued Neutrosophic Linguistic TOPSIS Approach for Green Supplier Evaluation and Selection in Low-Carbon Supply Chains. *Int. J. Environ. Res. Public Health (IJERPH)* **2018**, *15*, 1439. [CrossRef]
10. Ji, P.; Zhang, H.Y.; Wang, J.Q. Selecting an outsourcing provider based on the combined MABAC–ELECTRE method using single-valued neutrosophic linguistic sets. *Comput. Ind. Eng.* **2018**, *120*, 429–441. [CrossRef]

11. Wu, Q.; Wu, P.; Zhou, L.; Chen, H.; Guan, X. Some new Hamacher aggregation operators under single-valued neutrosophic 2-tuple linguistic environment and their applications to multi-attribute group decision making. *Comput. Ind. Eng.* **2018**, *116*, 144–162. [CrossRef]
12. Zavadskas, E.K.; Baušys, R.; Lazauskas, M. Sustainable Assessment of Alternative Sites for the Construction of a Waste Incineration Plant by Applying WASPAS Method with Single-Valued Neutrosophic Set. *Sustainability* **2015**, *7*, 15923–15936. [CrossRef]
13. Garg, H. Nancy. Linguistic single-valued neutrosophic prioritized aggregation operators and their applications to multiple-attribute group decision-making. *J. Ambient Intell. Human. Comput.* **2018**, *9*, 1975–1997. [CrossRef]
14. Merigó, J.M.; Gil-Lafuente, A.M. New decision-making techniques and their application in the selection of financial products. *Inf. Sci.* **2010**, *180*, 2085–2094. [CrossRef]
15. Merigó, J.M.; Casanovas, M. Decision-making with distance measures and induced aggregation operators. *Comput. Ind. Eng.* **2011**, *60*, 66–76. [CrossRef]
16. Zeng, S.; Su, W. Intuitionistic fuzzy ordered weighted distance operator. *Knowl. Based Syst.* **2011**, *24*, 1224–1232. [CrossRef]
17. Xu, Z.; Xia, M. Distance and similarity measures for hesitant fuzzy sets. *Inf. Sci.* **2011**, *181*, 2128–2138. [CrossRef]
18. Zeng, S.; Merigó, J.M.; Su, W. The uncertain probabilistic OWA distance operator and its application in group decision making. *Appl. Math. Model.* **2013**, *37*, 6266–6275. [CrossRef]
19. Qin, Y.; Hong, Z.; Liu, Y. Multicriteria decision making method based on generalized Pythagorean fuzzy ordered weighted distance measures. *J. Intell. Fuzzy Syst.* **2017**, *33*, 3665–3675. [CrossRef]
20. Xian, S.D.; Sun, W.J. Fuzzy linguistic induced Euclidean OWA distance operator and its application in group linguistic decision making. *Int J Intell Syst.* **2014**, *29*, 478–491. [CrossRef]
21. Zhou, L.; Wu, J.; Chen, H. Linguistic continuous ordered weighted distance measure and its application to multiple attributes group decision making. *Appl. Soft Comput.* **2014**, *25*, 266–276. [CrossRef]
22. Li, Z.; Sun, D.; Zeng, S. Intuitionistic Fuzzy Multiple Attribute Decision-Making Model Based on Weighted Induced Distance Measure and Its Application to Investment Selection. *Symmetry* **2018**, *10*, 261. [CrossRef]
23. Herrera, F.; Herrera-Viedma, E. Linguistic decision analysis: Steps for solving decision problems under linguistic information. *Fuzzy Sets Syst.* **2000**, *115*, 67–82. [CrossRef]
24. Xu, Z. A Note on Linguistic Hybrid Arithmetic Averaging Operator in Multiple Attribute Group Decision Making with Linguistic Information. *Group Decis. Negot.* **2006**, *15*, 593–604. [CrossRef]
25. Ye, J. Multicriteria decision-making method using the correlation coefficient under single-valued neutrosophic environment. *Int. J. Gen. Syst.* **2013**, *42*, 386–394. [CrossRef]
26. Yager, R.R. On ordered weighted averaging aggregation operators in multi-criteria decision making. *IEEE Trans. Syst. Man Cybern. B* **1988**, *18*, 183–190. [CrossRef]
27. Zeng, S.; Xiao, Y. A METHOD BASED ON TOPSIS AND DISTANCE MEASURES FOR HESITANT FUZZY MULTIPLE ATTRIBUTE DECISION MAKING. *Technol. Econ. Dev. Econ.* **2018**, *24*, 969–983. [CrossRef]
28. Merigó, J.M.; Palacios-Marqués, D.; Soto-Acosta, P. Distance measures, weighted averages, OWA operators and Bonferroni means. *Appl. Soft Comput.* **2017**, *50*, 356–366. [CrossRef]
29. Li, S.; Chen, G.; Mou, X. ON THE DYNAMICAL DEGRADATION OF DIGITAL PIECEWISE LINEAR CHAOTIC MAPS. *Int. J. Bifurc. Chaos* **2005**, *15*, 3119–3151. [CrossRef]
30. Nie, X.; Coca, D. A matrix-based approach to solving the inverse Frobenius-Perron problem using sequences of density functions of stochastically perturbed dynamical systems. *Commun. Nonlinear Sci. Numer. Simul.* **2018**, *54*, 248–266. [CrossRef]
31. Guido, R.; Addison, P.; Walker, J. Introducing wavelets and time-frequency analysis [Introduction to the special issue. *IEEE Eng. Med. Biol. Mag.* **2009**, *28*, 13. [CrossRef]
32. Guariglia, E. Harmonic Sierpinski Gasket and Applications. *Entropy* **2018**, *20*, 714. [CrossRef]

symmetry

MDPI

Article

Multi-Attribute Decision Making Method Based on Aggregated Neutrosophic Set

Wen Jiang *, **Zihan Zhang** and **Xinyang Deng**

School of Electronics and Information, Northwestern Polytechnical University, Xi'an 710072, China; zhangzihanwpu@126.com (Z.Z.); xinyang.deng@nwpu.edu.cn (X.D.)
* Correspondence: jiangwen@nwpu.edu.cn; Tel.: +86-29-8843-1267

Received: 30 December 2018; Accepted: 17 February 2019; Published: 20 February 2019

Abstract: Multi-attribute decision-making refers to the decision-making problem of selecting the optimal alternative or sorting the scheme when considering multiple attributes, which is widely used in engineering design, economy, management and military, etc. But in real application, the attribute information of many objects is often inaccurate or uncertain, so it is very important for us to find a useful and efficient method to solve the problem. Neutrosophic set is proposed from philosophical point of view to handle inaccurate information efficiently, and a single-valued neutrosophic set (SVNS) is a special case of neutrosophic set, which is widely used in actual application fields. In this paper, a new method based on single-valued neutrosophic sets aggregation to solve multi-attribute decision making problem is proposed. Firstly, the neutrosophic decision matrix is obtained by expert assessment, a score function of single-valued neutrosophic sets (SVNSs) is defined to obtain the positive ideal solution (PIS) and the negative ideal solution (NIS). Then all alternatives are aggregated based on TOPSIS method to make decision. Finally numerical examples are given to verify the feasibility and rationality of the method.

Keywords: multi-attribute decision making; single-valued neutrosophic set; aggregation; TOPSIS

1. Introduction

With the rapid development of human society, the social environment has become more and more complex, which makes us have difficulties in making actual decision [1–9]. Therefore, in recent years, more and more attention has been paid to multi-attribute decision making [10–16]. In the real world, the attribute information of many objects is often inaccurate or uncertain [17–23], which makes the decision progress difficult. In view of this problem, Zadeh put forward the concept of fuzzy set (FS) in 1965 [24], which made up for the lack of flexibility of classical set theory to some extent. But the traditional FS can not describe the decision information well in some conditions, Atanassov proposed the intuitionistic fuzzy set (IFS) [25] on the basis of FS. An IFS is given by $A = \{\langle x, \mu_A(x), \nu_A(x)\rangle\}$ where $\mu_A(x), \nu_A(x) \in [0,1]$ denote the degree of membership and non-membership of x to A respectively. For instance, A is an IFS in $X = \{1,2,3\}$, $A = (0.6, 0.3)/1 + (0.7, 0.1)/2 + (0.5, 0.4)/3$. Because IFSs take into account the information of membership, non-membership and hesitation simultaneously, compared with FSs, IFSs can describe the fuzzy nature [26–29] of the objective world more precisely. IFS attracted attention of many scholars, rich achievements were made in the study of intuitionistic fuzzy sets [30–33]. Torra [34] proposed another generalized form of the fuzzy set, named the hesitant fuzzy set (HFS) in 2009. HFS allowed the membership of each element in the domain to belong to a certain set which was combined by a number of different values. The element of the HFS is named as the hesitant fuzzy element (HFE) by Xia et al. [35], the mathematical symbol of HFS is expressed by Xia as $A = \{\langle x, h_A(x)\rangle | x \in X\}$ where $h_A(x)$ is set of some values in $[0,1]$, denoting the possible membership degrees of element $x \in X$ to the set A. $h_A(x)$ is called a HFE. For example, B is a HFS in $X = \{1,2,3\}$

and $B = \{0.6, 07, 0.8\}/1 + \{0.2\}/2 + \{0.3, 0.9\}/3$. The addition and multiplication operations on the hesitant fuzzy element are also given by Xia. The hesitant intuitionistic fuzzy sets (HIFSs) is proposed by Zhou et al. [36]. Zhou proposed that the HIFSs here were the generalization of the IFSs. The group decision-making method under the condition of the uncertain intuitionistic fuzzy priority relation matrix and aggregation operator was also given by Zhou in that paper.

The IFS considers both the truth-membership $T_A(x)$ and the falsity membership $F_A(x)$ with $T_A(x), F_A(x) \in [0,1]$ and $0 \leq T_A(x) + F_A(x) \leq 1$. For IFSs, the indeterminacy is $1 - T_A(x) - F_A(x)$ by default. The IFS can handle incomplete information but can hardly process inaccurate information.

In order to better describe uncertain information, Smarandache introduced Neutrosophy in 1995 [37,38]. Neutrosophy is a kind of philosophy which studies the nature, scope, and origin of neutralities, as well as their joint parts with different ideational spectra [37]. Neutrosophic set (NS) was also propose by Smarandache. NS is a very powerful tool which generalizes the concept of the classic set, fuzzy set, interval-valued fuzzy set, IFS, interval-valued intuitionistic fuzzy set [39], dialetheist set, paraconsistent set, and tautological set [37]. A NS A is defined on a universe of discourse u. An element x in set A is expressed as $x = x(t, i, f) \in A$, where t is a truth-membership function, i is an indeterminacy-membership function, and f is a falsity-membership function, t, i, and f are the real standard or non-standard subsets of $]0^-, 1^+[$. For a NS, the indeterminacy is denoted explicitly in contrast of that of IFSs. The indeterminacy can be divided into more parts so as to better express the inaccurate information [37]. However neutrosophic sets are hard to use in actual occasions. So Wang et al. [40] proposed the concept of single-valued neutrosophic sets (SVNSs). For instance, $C = (0.7, 0.5, 0.2)$ is a SVNS, in which the truth-membership $t = 0.7$, the indeterminacy $i = 0.5$ and the falsity-membership $f = 0.2$. Because SVNS are easy to express the inaccurate information, SVNSs are widely used in actual situations, such as in medicine [41], image processing [42], multi-criteria decision-making [43–47], fault diagnosis [41,48,49], etc.

Neutrosophic set has many advantages in handling uncertain information. A lot of researches are conducted on it. In [50], SVNS is used to express the decision information, weighted average operator, TOPSIS method is adopted to propose the multi-criteria decision making method. Peng introduced relevant concepts about the interval neutrosophic set [51], and the multi-criteria decision making problem is also analyzed by combining the ranking method. Jiang proposed a new method to measure the similarity between SVNSs using Dempster-Shafer evidence theory [52]. Ye proposed a method to make decision based on the weighted correlation coefficient of SVNSs in [44], and proposed another method to make decision in multi-criteria environment based on single valued neutrosophic cross-entropy in [45]. Ye utilized cross entropy between the ideal solution and an alternative to get the rank of all alternatives according to the values of cross entropy and to choose the most suitable one(s). A decision method for the interval neutrosophic set is proposed based on cross entropy by Tian et al. [53]. Deli applied bipolar neutrosophic sets to multi-criteria decision situations [54]. The TOPSIS method can be effectively combined with SVNSs to accomplish multi-attribute decision making problems. By the way, the key point of TOPSIS is that the ideal alternative should have the shortest distance from the PIS and the farthest distance from the NIS. The standard TOPSIS to new multi-attribute decision-making called simplified-TOPSIS is proposed by Elhassouny [55], which simplifies the process of the classical TOPSIS and get the same result.

This paper mainly introduces a new multi-attribute decision-making method based on SVNSs. First, based on the given decision matrix, use score function to get the PIS and the NIS. Second, aggregate all alternatives to get aggregated neutrosophic set. Last, use TOPSIS method to rank all alternatives to make decision.

The paper is organized as follow. In Section 2, we present some preliminaries. Section 3 will focus on the proposed multi-criteria decision method based on aggregated neutrosophic set. Afterwards, two illustrative examples are introduced in Section 4. In the final section, conclusions are drawn.

2. Preliminaries

A brief review of some preliminaries will be given in the following part.

2.1. Neutrosophic Sets

Neutrosophic set is an efficient tool to deal with the uncertain information. In [37], Smarandache gave the definitions about a NS as follow:

Definition 1. *Y is a universal space of point (objects) with a generic element of Y denoted by y. A neutrosophic set $N \subset Y$ is consist of a truth-membership function $T_N(y)$, an indeterminacy-membership function $I_N(y)$ and a falsity-membership function $F_N(y)$. $T_A(y)$, These three components are real standard or real nonstandard subset of $[0^-, 1^+]$. So that all three components*

$$T_N(y) \to [0^-, 1^+]$$

$$I_N(y) \to [0^-, 1^+]$$

$$F_N(y) \to [0^-, 1^+]$$

$T_N(y), I_N(y)$ and $F_N(y)$ are related as follow:

$$0^- \leq \sup T_N(y) + \sup I_N(y) + \sup F_N(y) \leq 3^+.$$

Definition 2. *A neutrosophic set N has its' complement which is characterized by N^c and is characterized as*

$$T_N^c(y) = 1^+ - T_N(y)$$

$$I_N^c(y) = 1^+ - I_N(y)$$

$$F_N^c(y) = 1^+ - F_N(y)$$

for every y in Y.

2.2. SVNS

For NS is difficult to be used in practical occasions, Wang [40] proposed the concept of single-valued neutrosophic sets (SVNSs), which can be easily used in actual situations. The definition of SVNSs are introduced as follow:

Definition 3. *Let Y be a space of points (objects) which generic elements in Y denoted by y. A SVNS N is characterized by a truth-membership function T_N, an indeterminacy-membership function I_N, and a falsity-membership function F_N with $T_N, I_N, F_N \in [0, 1]$.*

When Y is continuous, an SVNS N can be expressed as:

$$N = \int_X \langle T_N(y), I_N(y), F_N(y) \rangle | y, \quad y \in Y. \tag{1}$$

When Y is discrete, an SVNS N can be characterized as:

$$N = \sum_y \langle T_N(y), I_N(y), F_N(y) \rangle | y, \quad y \in Y \tag{2}$$

For convenience, a SVNS is usually denoted by its' simplified symbol $N = \langle T_N(Y), I_N(Y), F_N(Y) \rangle$ for all $y \in Y$.

Definition 4. *A SVNS N has its' complement N^c which is characterized as*

$$T_N^c(y) = F_N(y)$$

$$I_N^c(y) = 1 - I_N(y)$$

$$F_N^c(y) = T_N(y)$$

for all y in Y.

Definition 5. *A SVNS N is contained in the other SVNS M, $N \subseteq M$, if and only if*

$$T_N(y) \leq T_M(y)$$

$$I_N(y) \geq I_M(y)$$

$$F_N(y) \geq F_M(y)$$

for all y in Y.

Definition 6. *Two SVNSs N and M are equal, written as $N = M$, if and only if $N \subseteq M$ and $N \subseteq M$.*

2.3. Score Function

Definition 7. *Assume $A = (T_A, I_A, F_A)$ be a single valued neutrosophic number, then the score function [56] is defined as*

$$S(A) = \frac{T_A - I_A - F_A}{3}. \tag{3}$$

Score function is a very useful tool to illustrate which neutrosophic number is better. For example, $A_1 = (0.5, 0.3, 0.4), A_2 = (0.6, 0.4, 0.5)$, use the score function and get $S(A_1) = -\frac{2}{15}, S(A_2) = -\frac{1}{6}$, $S(A_1) > S(A_2)$, this result is in line with intuition, in this case, A_1 is better than A_2.

2.4. Distance between Two Neutrosophic Sets

Assume that there two NSs, shows as follow:

$$M = \langle t_M(x_i), i_M(x_i), f_M(x_i) \rangle \quad i = 1, 2, \ldots, n \tag{4}$$

$$N = \langle t_N(x_i), i_N(x_i), f_N(x_i) \rangle \quad i = 1, 2, \ldots, n \tag{5}$$

Then the Hamming distance between M and N is defined as follow:

$$d_1(M, N) = \sum_{i=1}^{n} \{|t_M(x_i) - t_N(x_i)| + |i_M(x_i) - i_N(x_i)| + |f_M(x_i) - f_N(x_i)|\} \tag{6}$$

the standard Hamming distance between M and N is defined as follow:

$$d_2(M, N) = \frac{1}{3n} \sum_{i=1}^{n} \{|t_M(x_i) - t_N(x_i)| + |i_M(x_i) - i_N(x_i)| + |f_M(x_i) - f_N(x_i)|\} \tag{7}$$

the Euclid distance between M and N is defined as follow:

$$d_3(M, N) = \sqrt{\sum_{i=1}^{n} \{(t_M(x_i) - t_N(x_i))^2 + (i_M(x_i) - i_N(x_i))^2 + (f_M(x_i) - f_N(x_i))^2\}} \tag{8}$$

the standard Euclid distance between M and N is defined as follow:

$$d_4(M,N) = \sqrt{\frac{1}{3n}\sum_{i=1}^{n}\{(t_M(x_i) - t_N(x_i))^2 + (i_M(x_i) - i_N(x_i))^2 + (f_M(x_i) - f_N(x_i))^2\}} \tag{9}$$

3. An Improved Multi-Criteria Decision Making Method

Multi-criteria decision making often faces complex environments. At present, multi-criteria decision making methods mostly handle multiple attributes of an alternative separately for decision making. Few studies try to aggregate all neutrosophic sets of one alternative before decision making to simplify the decision progress. Based on the idea of information fusion, the NSs can be aggregated before using TOPSIS method to make a decision. In this way, decision progress can be simplified. A multi-attribute decision making method based on aggregated SVNSs is introduced in this part.

step1: Construct neutrosophic decision matrix.

The single-valued neutrosophic decision matrix is constructed based on expert assessments. For instance, when an expert is asked the opinion of an alternative A_1 with respect to a criterion C_1, the expert may say that possibility in which the alternative is good is 0.5 and false degree is 0.4 and the expert is not sure is 0.2. For the NS, it can be expressed as $d_{11} = \langle 0.5, 0.2, 0.4 \rangle$.

Assume there are m alternatives and n criteria. The neutrosophic decision matrix D is expressed as follow:

$$\begin{bmatrix} d_{11} & d_{12} & \dots & d_{1n} \\ d_{21} & d_{22} & \dots & d_{2n} \\ \vdots & \vdots & \ddots & \vdots \\ d_{m1} & d_{m2} & \dots & d_{mn} \end{bmatrix}$$

where $d_{ij} = (T_{ij}, I_{ij}, F_{ij})$, $1 \le i \le m$, $1 \le j \le n$, T_{ij}, I_{ij}, F_{ij} are the truth-membership degree, the indeterminacy-membership degree, the falsity-membership degree of alternative A_i with respect to criterion j.

step2: Determine the PIS and the NIS.

In this step, score function mentioned above is utilized to get the PIS A^* and the NIS A^{-*}.

Assume $A^* = (d_1^*, d_2^*, \dots, d_n^*)$, among them $S(d_j^*) = \max_i\{S(d_{ij})\}$, $j = 1, 2, \dots, n$, and $A^{-*} = (d_1^{-*}, d_2^{-*}, \dots, d_n^{-*})$ among them $S(d_j^{-*}) = \min_i\{S(d_{ij})\}$, $j = 1, 2, \dots, n$, the score function can be used to illustrate that how good or bad a neutrosophic set is. A simple example is given as follow:

Example 1. *Assume there is a decision matrix which is obtained from expert assessments showing as follow:*

$$\begin{bmatrix} (0.4, 0.2, 0.3) & (0.4, 0.2, 0.4) & (0.5, 0.2, 0.2) \\ (0.6, 0.1, 0.2) & (0.6, 0.1, 0.2) & (0.2, 0.2, 0.5) \\ (0.7, 0.1, 0.1) & (0.5, 0.3, 0.3) & (0.4, 0.3, 0.2) \end{bmatrix}$$

Specifically, let us consider the neutrosophic set $(0.4, 0.2, 0.3)$, $T_{11} = 0.4, I_{11} = 0.2, F_{11} = 0.3$, *use Equation (3) to obtain the score function as follows:*

$$S(d_{11}) = \frac{0.4 - 0.2 - 0.3}{3} = -\frac{1}{30}$$

In this way, scores of all neutrosophic sets can be obtained, neutrosophic sets which have the highest or the lowest scores will be choose to constitute the PIS(A^*) and the NIS(A^{-*}). For example, in the first column of the matrix:

$$S(d_{11}) = -\frac{1}{30}, S(d_{21}) = \frac{1}{10}, S(d_{31}) = \frac{1}{6}$$

In this case, d_{31} and d_{11} are chosen to constitute the first column of A^* and A^{-*} respectively. A^* and A^{-*} can be obtained as follow:

$$A^* = ((0.7, 0.1, 0.1), (0.6, 0.1, 0.2), (0.5, 0.2, 0.2))$$

$$A^{-*} = ((0.4, 0.2, 0.3), (0.4, 0.2, 0.4), (0.2, 0.2, 0.5))$$

step3: Aggregate the SVNSs.

In this step, all the attributes of a neutrosophic set are aggregated. Arithmetic average method is used to combine neutrosophic set. $A_j^a = (t_j^a, i_j^a, f_j^a)$ is the aggregated neutrosophic set, $1 \leq j \leq m$, where

$$t_j^a = \frac{t_{j1} + t_{j2} + \ldots + t_{jn}}{n}$$

$$i_j^a = \frac{i_{j1} + i_{j2} + \ldots + i_{jn}}{n}$$

$$f_j^a = \frac{f_{j1} + f_{j2} + \ldots + f_{jn}}{n}$$

n is the number of attributes.

This method is used to aggregate all the alternatives, the PIS and the NIS. So $A_j^a = (t_j^a, i_j^a, f_j^a)$, $1 \leq j \leq m$, $A^{*a} = (t^{*a}, i^{*a}, f^{*a})$ and $A^{-*a} = (t^{-*a}, i^{-*a}, f^{-*a})$ are obtained.

step4: Multi-criteria decision making based on TOPSIS method

In this step, TOPSIS method is utilized to finish a multi-attribute neutrosophic decision making. In step 3, the aggregated SVNSs, the aggregated PIS and NIS are obtained. Then the distance for each aggregated SVNS between A^{*a} and between A^{-*a} can be calculated by Euclid distance Equation (8). The distance between A_j^a and A^{*a} is calculated as follows:

$$D_j^* = \sqrt{(t_j^a - t^{*a})^2 + (i_j^a - i^{*a})^2 + (f_j^a - f^{*a})^2} \quad j = 1, 2, \ldots, m \tag{10}$$

The distance between A_j^a and A^{-*a} is got as follows:

$$D_j^{-*} = \sqrt{(t_j^a - t^{-*a})^2 + (i_j^a - i^{-*a})^2 + (f_j^a - f^{-*a})^2} \quad j = 1, 2, \ldots, m \tag{11}$$

The relative closeness T_j based on TOPSIS is got as follows:

$$T_j = \frac{D_j^{-*}}{(D_j^{-*} + D_j^*)} \quad j = 1, 2, \ldots, m \tag{12}$$

Obviously, the bigger the value of T_j is, the farther the alternative is from the NIS, and the closer the alternative is to the PIS, and vice versa. The ranked of all the alternatives is obtained in the descending order of the value of T_j.

4. Illustrative Example

In this section, two examples are used to demonstrate the application of the proposed method.

Example 2. *Let us consider the decision-making problem originated from [56]. There is an investment company which wants to invest some money in a schools. There are four possible schools to invest the money: $\{A_1, A_2, A_3, A_4\}$. The investment company is going to choose one school to invest with respect to the following four criteria: $\{C_1, C_2, C_3, C_4\}$.*

Step1: *Experts evaluate all the four possible schools according to the four attributes. According to the evaluation results, the following SVNS decision matrix is shown in Table 1:*

Table 1. Solution matrix for SVNS.

	C_1	C_2	C_3	C_4
A_1	$(0.6, 0.3, 0.5)$	$(0.5, 0.7, 0.6)$	$(0.7, 0.6, 0.5)$	$(0.5, 0.5, 0.6)$
A_2	$(0.6, 0.4, 0.5)$	$(0.4, 0.5, 0.6)$	$(0.3, 0.5, 0.6)$	$(0.4, 0.5, 0.6)$
A_3	$(0.5, 0.6, 0.7)$	$(0.7, 0.2, 0.8)$	$(0.7, 0.6, 0.3)$	$(0.4, 0.4, 0.5)$
A_4	$(0.4, 0.3, 0.2)$	$(0.5, 0.4, 0.3)$	$(0.6, 0.7, 0.2)$	$(0.4, 0.3, 0.2)$

Step2: *Then the score function is utilized to get A^* and A^{-*}. The calculation results are expressed in Table 2:*

Table 2. Scores matrix for SVNS.

	C_1	C_2	C_3	C_4
A_1	$-\frac{1}{15}$	$-\frac{4}{15}$	$-\frac{2}{15}$	$-\frac{1}{5}$
A_2	$-\frac{1}{10}$	$-\frac{7}{30}$	$-\frac{4}{15}$	$-\frac{7}{30}$
A_3	$-\frac{4}{15}$	$-\frac{1}{10}$	$-\frac{1}{15}$	$-\frac{1}{6}$
A_4	$-\frac{1}{30}$	$-\frac{1}{15}$	$-\frac{1}{10}$	$-\frac{1}{30}$

From Table 2, $S(d_{41}) = -\frac{1}{30}$ is the maximum in the first column, $S(d_{42})$, $S(d_{33})$, $S(d_{44})$ is the maximum in column $2, 3, 4$ respectively. The neutrosophic PIS A^* is composed of $d_{41}, d_{42}, d_{33}, d_{44}$. Similarly, the d_{ij} which has the minimum score in each column is chosen to constitute the neutrosophic NIS A^{-*}. The PIS A^* and the NIS A^{-*} can be obtained as follow:

$$A^* = ((0.4, 0.3, 0.2), (0.5, 0.4, 0.3), (0.7, 0.6, 0.3), (0.4, 0.3, 0.2))$$

$$A^{-*} = ((0.5, 0.6, 0.7), (0.5, 0.7, 0.6), (0.3, 0.5, 0.6), (0.4, 0.5, 0.6))$$

Step3: *Aggregate all the alternatives, A^* and A^{-*}. For example:*

$$A^* = ((0.4, 0.3, 0.2), (0.5, 0.4, 0.3), (0.7, 0.6, 0.3), (0.4, 0.3, 0.2))$$

The $T_{A^*}, I_{A^*}, F_{A^*}$ of A_c^* can be obtained as follow:

$$T_{A^*} = \frac{0.4 + 0.5 + 0.7 + 0.4}{4} = 0.500$$

$$I_{A^*} = \frac{0.3 + 0.4 + 0.6 + 0.3}{4} = 0.400$$

$$F_{A^*} = \frac{0.2 + 0.3 + 0.3 + 0.2}{4} = 0.250$$

$$A_c^* = (T_{A^*}, I_{A^*}, F_{A^*}) = (0.500, 0.400, 0.250)$$

All aggregated neutrosophic sets are obtained in the same way and shown as follow:

$$A_{1c} = (0.575, 0.525, 0.550)$$

$$A_{2c} = (0.425, 0.475, 0.575)$$

$$A_{3c} = (0.575, 0.450, 0.575)$$

$$A_{4c} = (0.475, 0.425, 0.225)$$

$$A_c^* = (0.500, 0.400, 0.250)$$

$$A_c^{-*} = (0.425, 0.575, 0.625)$$

Step4: *All alternatives are ranked by the TOPSIS method. Euclid distances between alternatives and the PIS are calculated as follow:*

$$d(A_{1c}, A_c^*) = \sqrt{(T_{A_{1c}} - T_{A_c^*})^2 + (I_{A_{1c}} - I_{A_c^*})^2 + (F_{A_{1c}} - F_{A_c^*})^2} = 0.334$$

$$d(A_{2c}, A_c^*) = \sqrt{(T_{A_{2c}} - T_{A_c^*})^2 + (I_{A_{2c}} - I_{A_c^*})^2 + (F_{A_{2c}} - F_{A_c^*})^2} = 0.342$$

$$d(A_{3c}, A_c^*) = \sqrt{(T_{A_{3c}} - T_{A_c^*})^2 + (I_{A_{3c}} - I_{A_c^*})^2 + (F_{A_{3c}} - F_{A_c^*})^2} = 0.337$$

$$d(A_{4c}, A_c^*) = \sqrt{(T_{A_{4c}} - T_{A_c^*})^2 + (I_{A_{4c}} - I_{A_c^*})^2 + (F_{A_{4c}} - F_{A_c^*})^2} = 0.043$$

Euclid distances between alternatives and the NIS are calculated as follow:

$$d(A_{1c}, A_c^{-*}) = \sqrt{(T_{A_{1c}} - T_{A_c^{-*}})^2 + (I_{A_{1c}} - I_{A_c^{-*}})^2 + (F_{A_{1c}} - F_{A_c^{-*}})^2} = 0.175$$

$$d(A_{2c}, A_c^{-*}) = \sqrt{(T_{A_{2c}} - T_{A_c^{-*}})^2 + (I_{A_{2c}} - I_{A_c^{-*}})^2 + (F_{A_{2c}} - F_{A_c^{-*}})^2} = 0.112$$

$$d(A_{3c}, A_c^{-*}) = \sqrt{(T_{A_{3c}} - T_{A_c^{-*}})^2 + (I_{A_{3c}} - I_{A_c^{-*}})^2 + (F_{A_{3c}} - F_{A_c^{-*}})^2} = 0.202$$

$$d(A_{4c}, A_c^{-*}) = \sqrt{(T_{A_{4c}} - T_{A_c^{-*}})^2 + (I_{A_{4c}} - I_{A_c^{-*}})^2 + (F_{A_{4c}} - F_{A_c^{-*}})^2} = 0.430$$

The relative closeness are calculated as follow:

$$T_1 = \frac{d(A_{1c}, A_c^{-*})}{d(A_{1c}, A_c^{-*}) + d(A_{1c}, A_c^*)} = 0.344$$

$$T_2 = \frac{d(A_{2c}, A_c^{-*})}{d(A_{2c}, A_c^{-*}) + d(A_{2c}, A_c^*)} = 0.247$$

$$T_3 = \frac{d(A_{3c}, A_c^{-*})}{d(A_{3c}, A_c^{-*}) + d(A_{3c}, A_c^*)} = 0.375$$

$$T_4 = \frac{d(A_{4c}, A_c^{-*})}{d(A_{4c}, A_c^{-*}) + d(A_{4c}, A_c^*)} = 0.909$$

Rank the T_i, $i = 1, 2, 3, 4$ in descending order, and get $T_4 > T_3 > T_1 > T_2$. In this case, A_4 is chosen as the ideal solution. This example shows that by using the proposed method decision results can be easily obtained.

Example 3. *For a further understanding and comparison of our approach, another problem is considered. The data in this example originated from [57]. Let us suppose that decision makers intend to select the most suitable tablet from the four chosen tablets (A_1, A_2, A_3, A_4) by considering six attributes namely: Features C_1, Hardware C_2, Display C_3, Communication C_4, Affordable Price C_5, Customer care C_6. By using the presented method, the problem is handled by the following steps.*

Step1: *Assume that $A_i(A_1, A_2, A_3, A_4)$ are alternatives with respect to six criteria $(C_1, C_2, C_3, C_4, C_5, C_6)$. Neutrosophic decision matrix are shown in Table 3:*

Table 3. Neutrosophic decision matrix.

	C_1	C_2	C_3
A_1	$(0.864, 0.136, 0.081)$	$(0.853, 0.147, 0.092)$	$(0.800, 0.200, 0.150)$
A_2	$(0.667, 0.333, 0.277)$	$(0.727, 0.273, 0.219)$	$(0.667, 0.333, 0.277)$
A_3	$(0.880, 0.120, 0.067)$	$(0.887, 0.113, 0.064)$	$(0.834, 0.166, 0.112)$
A_4	$(0.667, 0.333, 0.277)$	$(0.735, 0.265, 0.195)$	$(0.768, 0.232, 0.180)$
	C_4	C_5	C_6
A_1	$(0.704, 0.296, 0.241)$	$(0.823, 0.177, 0.123)$	$(0.864, 0.136, 0.081)$
A_2	$(0.744, 0.256, 0.204)$	$(0.652, 0.348, 0.293)$	$(0.608, 0.392, 0.336)$
A_3	$(0.779, 0.256, 0.204)$	$(0.811, 0.189, 0.109)$	$(0.850, 0.150, 0.092)$
A_4	$(0.727, 0.273, 0.221)$	$(0.791, 0.209, 0.148)$	$(0.808, 0.192, 0.127)$

Step2: *Score function are used to calculate and get the score matrix in Table 4.*

Table 4. Score matrix.

	C_1	C_2	C_3	C_4	C_5	C_6
A_1	0.216	0.205	0.150	0.056	0.174	0.216
A_2	0.019	0.078	0.019	0.095	0.004	-0.040
A_3	0.231	0.237	0.185	0.106	0.171	0.203
A_4	0.019	0.092	0.119	0.078	0.145	0.163

From Table 4, the neutrosophic PIS can be got as follow:

$$A^* = ((0.880, 0.120, 0.067), (0.887, 0.113, 0.064), (0.834, 0.166, 0.112),$$
$$(0.779, 0.256, 0.204), (0.823, 0.177, 0.123), (0.864, 0.136, 0.081))$$

and the neutrosophic NIS is shown as follow:

$$A^{-*} = ((0.667, 0.333, 0.277), (0.727, 0.273, 0.219), (0.667, 0.333, 0.277),$$
$$(0.704, 0.296, 0.241), (0.652, 0.348, 0.293), (0.608, 0.392, 0.336)).$$

Step3: *Neutrosophic sets aggregation is done in this step. After calculation, results are shown as follows:*

$$A_{1c} = (0.818, 0.182, 0.128)$$
$$A_{2c} = (0.677, 0.323, 0.268)$$
$$A_{3c} = (0.840, 0.166, 0.108)$$
$$A_{4c} = (0.749, 0.251, 0.191)$$
$$A_c^* = (0.844, 0.161, 0.109)$$
$$A_c^{-*} = (0.671, 0.329, 0.274)$$

Step4: *In the last step, Euclid distances between $A_{ic}, i = 1, 2, 3, 4$ and A_c^* and between $A_{ic}, i = 1, 2, 3, 4$ and A_c^{-*} are obtained, and the relative closeness of each alternative $T_i (i = 1, 2, 3, 4)$ is obtained. Euclid distances between alternatives and the neutrosophic PIS are as follow:*

$$d(A_{1c}, A_c^*) = 0.038 \quad d(A_{2c}, A_c^*) = 0.282$$
$$d(A_{3c}, A_c^*) = 0.007 \quad d(A_{4c}, A_c^*) = 0.154$$

Euclid distances between alternatives and the neutrosophic NIS are below:

$$d(A_{1c}, A^{-*}) = 0.254 \quad d(A_{2c}, A^{-*}) = 0.011$$

$$d(A_{3c}, A^{-*}) = 0.288 \quad d(A_{4c}, A^{-*}) = 0.138$$

So the relative closeness of each alternative to the ideal solution can be easily obtained as follow:

$$T_1 = 0.870 \quad T_2 = 0.038$$

$$T_3 = 0.976 \quad T_4 = 0.473$$

the rank of them can be obtained: $T_3 > T_1 > T_4 > T_2$*. Obviously,* A_3 *is the best solution.*

According to the results, method which uses aggregated SVNSs in this paper is more simplified than that proposed by Pramanik [57]. Additionally, the Pramanik's results are shown as follow:

$$T_1 = 0.8190 \quad T_2 = 0,1158$$

$$T_3 = 0.8605 \quad T_4 = 0.4801$$

The same rank of alternatives are got, but in our approach the difference between the data is greater and the results are clearer than that in Pramanik's. The presented method is different with Pramanik's method in handling the attributes. In Pramanik's method, attribute weight is used to revise the neutrosophic set. In the presented method neutrosophic sets of different attributes are aggregated to one neutrosophic set, which simplifies the decision-making process and makes the process more efficient. In this way, multi-criteria decision making can be easily made.

To verify the rationality and usefulness of the presented method, correlation coefficient method [44] and cross-entropy method [45] are used to calculate the same decision problem in Table 3. In Table 3, B_i *refers to the correlation between alternative* A_i *and the PIS* A^**, and* E_i *refers to the distance between* A_i *and the PIS* A^**. The comparison of presented method with these three methods are shown in Table 5.*

Table 5. Comparison with other methods.

Methods	Indexes for Decision Making	Rank of Indexes	The Chosen Alternative
Elhassouny's [57]	$T_1 = 0.8190, T_2 = 0,1158, T_3 = 0.8605, T_4 = 0.4801$	$T_3 > T_1 > T_4 > T_2$	A_3
correlation coefficient [44]	$B_1 = 0.959, B_2 = 0.846, B_3 = 0.970, B_4 = 0.917$	$B_3 > B_1 > B_4 > B_2$	A_3
cross-entropy [45]	$E_1 = 0.530, E_2 = 0.966, E_3 = 0.466, E_4 = 0.763$	$E_3 < E_1 < E_4 < E_2$	A_3
Presented method	$T_1 = 0.870, T_2 = 0.038, T_3 = 0.976, T_4 = 0.473$	$T_3 > T_1 > T_4 > T_2$	A_3

5. Conclusions

In this paper, a multi-attribute decision method used aggregated neutrosophic set and TOPSIS is proposed. In our method, arithmetic average method is used to aggregate neutrosophic sets, and aggregated sets can be used by TOPSIS method to get a final rank. The proposed approach could reduce the computation since it firstly aggregate the neutrosophic sets before ranking, so it is useful in real application with high requirements of real-time. Two examples have demonstrated the rationality and feasibility of our approach. However, this method also has some limitations, averaging the validity of the criteria may lead to the loss of some criteria information. If the criteria information is especially emphasized in the decision process, this presented method may be not so appropriate to be used. In the future, more work will be done on the score function to make the decision making process more accurate, and the geometric average method is also need to be checked. Continuous work in the application of complex decision-making problems such as group decision-making problems and other domains such as fuzzy system is also need to be done.

Author Contributions: W.J. and X.D. proposed the method; W.J., Z.Z. and X.D. analyzed the results of experiment; Z.Z. wrote the paper; W.J. and X.D. revised and improved the paper.

Funding: The work is partially supported by National Natural Science Foundation of China (Program No. 61671384, 61703338), Natural Science Basic Research Plan in Shaanxi Province of China (Program No. 2018JQ6085).

Conflicts of Interest: The authors declare that there is no conflict of interest regarding the publication of this paper.

References

1. Jiang, W. A correlation coefficient for belief functions. *Int. J. Approx. Reason.* **2018**, *103*, 94–106. [CrossRef]
2. Deng, X.; Jiang, W.; Wang, Z. Zero-sum polymatrix games with link uncertainty: A Dempster-Shafer theory solution. *Appl. Math. Comput.* **2019**, *340*, 101–112. [CrossRef]
3. Kahneman, D.; Tversky, A. Prospect Theory. An analysis of decision making under risk. *Econometrica* **1979**, *47*, 263–291. [CrossRef]
4. Han, Y.; Deng, Y. A hybrid intelligent model for Assessment of critical success factors in high risk emergency system. *J. Ambient Intell. Humaniz. Comput.* **2018**, *9*, 1933–1953. [CrossRef]
5. Bellman, R.E.; Zadeh, L.A. Decision-Making in a Fuzzy Environment. *Manag. Sci.* **1970**, *17*, B141–B164. [CrossRef]
6. Wendt, D.; Vlek, C. Utility, Probability, and Human Decision Making. *Soc. Sci. Electron. Publ.* **1977**, *185*, 1124–1131.
7. Xiao, F. Multi-sensor data fusion based on the belief divergence measure of evidences and the belief entropy. *Inf. Fusion* **2019**, *46*, 23–32. [CrossRef]
8. Zhang, L.; Wu, X.; Zhu, H.; AbouRizk, S.M. Perceiving safety risk of buildings adjacent to tunneling excavation: An information fusion approach. *Autom. Constr.* **2017**, *73*, 88–101. [CrossRef]
9. He, Z.; Jiang, W. An evidential Markov decision making model. *Inf. Sci.* **2018**, *467*, 357–372. [CrossRef]
10. He, Z.; Jiang, W. An evidential dynamical model to predict the interference effect of categorization on decision making. *Knowl.-Based Syst.* **2018**, *150*, 139–149. [CrossRef]
11. Liu, J.C.; Li, D.F. Corrections to "TOPSIS-based nonlinear-programming methodology for multi-attribute decision making with interval-valued intuitionistic fuzzy sets" [Apr 10 299-311]. *IEEE Trans. Fuzzy Syst.* **2018**, *26*, 391. [CrossRef]
12. Broumi, S.; Smarandache, F. Single valued neutrosophic trapezoid linguistic aggregation operators based multi-attribute decision making. *Bull. Pure Appl. Sci. Math. Stat.* **2017**, *33e*, 135. [CrossRef]
13. Deli, I.; Subas, Y. A ranking method of single valued neutrosophic numbers and its applications to multi-attribute decision making problems. *Int. J. Mach. Learn. Cybern.* **2017**, *8*, 1309–1322. [CrossRef]
14. Pramanik, S.; Dey, P.P.; Giri, B.C.; Smarandache, F. *An Extended TOPSIS for Multi-Attribute Decision Making Problems with Neutrosophic Cubic Information*; University of New Mexico: Albuquerque, NM, USA, 2017.
15. Figueira, J.; Greco, S.; Ehrogott, M. Multiple Criteria Decision Analysis: State of the Art Surveys. *International* **2018**, *142*, 192–202.
16. Siregar, D.; Arisandi, D.; Usman, A.; Irwan, D.; Rahim, R. Research of Simple Multi-Attribute Rating Technique for Decision Support. *J. Phys. Conf. Ser.* **2017**, *930*, 012015. [CrossRef]
17. Deng, X.; Jiang, W. Dependence assessment in human reliability analysis using an evidential network approach extended by belief rules and uncertainty measures. *Ann. Nuclear Energy* **2018**, *117*, 183–193. [CrossRef]
18. Deng, X. Analyzing the monotonicity of belief interval based uncertainty measures in belief function theory. *Int. J. Intell. Syst.* **2018**, *33*, 1869–1879. [CrossRef]
19. Li, Y.; Deng, Y. Generalized Ordered Propositions Fusion Based on Belief Entropy. *Int. J. Comput. Commun. Control* **2018**, *13*, 792–807. [CrossRef]
20. Deng, X.; Xiao, F.; Deng, Y. An improved distance-based total uncertainty measure in belief function theory. *Appl. Intell.* **2017**, *46*, 898–915. [CrossRef]
21. Zhang, X.; Mahadevan, S. Aircraft re-routing optimization and performance assessment under uncertainty. *Decis. Support Syst.* **2017**, *96*, 67–82. [CrossRef]
22. Huang, Z.; Yang, L.; Jiang, W. Uncertainty measurement with belief entropy on the interference effect in the quantum-like Bayesian Networks. *Appl. Math. Comput.* **2019**, *347*, 417–428. [CrossRef]
23. Jiang, W.; Hu, W. An improved soft likelihood function for Dempster-Shafer belief structures. *Int. J. Intell. Syst.* **2018**, *33*, 1264–1282. [CrossRef]
24. Zadeh, L.A. Fuzzy sets. *Inf. Control* **1965**, *8*, 338–353. [CrossRef]

25. Atanassov, K.T. Intuitionistic fuzzy sets. *Fuzzy Sets Syst.* **1986**, *20*, 87–96. [CrossRef]
26. Han, Y.; Deng, Y. An enhanced fuzzy evidential DEMATEL method with its application to identify critical success factors. *Soft Comput.* **2018**, *22*, 5073–5090. [CrossRef]
27. Kang, B.; Deng, Y.; Hewage, K.; Sadiq, R. A method of measuring uncertainty for Z-number. *IEEE Trans. Fuzzy Syst.* **2018**. [CrossRef]
28. Zhang, L.; Wu, X.; Qin, Y.; Skibniewski, M.J.; Liu, W. Towards a Fuzzy Bayesian Network Based Approach for Safety Risk Analysis of Tunnel-Induced Pipeline Damage. *Risk Anal.* **2016**, *36*, 278–301. [CrossRef] [PubMed]
29. Fei, L.; Wang, H.; Chen, L.; Deng, Y. A new vector valued similarity measure for intuitionistic fuzzy sets based on OWA operators. *Iranian J. Fuzzy Syst.* **2018**. [CrossRef]
30. Yager, R.R. *Some Aspects of Intuitionistic Fuzzy Sets*; Kluwer Academic Publishers: Dordrecht, The Netherlands, 2009; pp. 67–90.
31. Li, D.F. TOPSIS-Based nonlinear-programming methodology for multiattribute decision making with interval-valued intuitionistic fuzzy sets. *IEEE Trans. Fuzzy Syst.* **2018**, *26*, 391. [CrossRef]
32. Bustince, H.; Burillo, P. Vague sets are intuitionistic fuzzy sets. *Fuzzy Sets Syst.* **1996**, *79*, 403–405. [CrossRef]
33. Szmidt, E.; Kacprzyk, J. Distances between intuitionistic fuzzy sets. *Fuzzy Sets Syst.* **2000**, *114*, 505–518. [CrossRef]
34. Torra, V. Hesitant fuzzy sets. *Int. J. Intell. Syst.* **2010**, *25*, 529–539. [CrossRef]
35. Xia, M.; Xu, Z. Hesitant fuzzy information aggregation in decision making. *Int. J. Approx. Reason.* **2011**, *52*, 395–407. [CrossRef]
36. Zhou, W.; Xu, Z.; Chen, M. *Preference Relations Based on Hesitant-Intuitionistic Fuzzy Information and Their Application in Group Decision Making*; Pergamon Press, Inc.: Oxford, UK, 2015; pp. 163–175.
37. Smarandache, F. A unifying field in logics: Neutrosophic logic. *Multiple-Valued Logic* **1999**, *8*, 489–503.
38. Smarandache, F. *Neutrosophy: Neutrosophic Probability, Set, and Logic*; American Research Press: Ann Arbor, Michigan, USA, 1998; 105p.
39. Atanassov, K.T. *Interval Valued Intuitionistic Fuzzy Sets*; Elsevier North-Holland, Inc.: Amsterdam, The Netherlands, 1989; pp. 343–349.
40. Wang, H.; Smarandache, F.; Zhang, Y.; Sunderraman, R. Single valued neutrosophic sets. In Proceedings of the 8th Joint Conference on Information Sciences, Salt Lake, UT, USA, 21–26 July 2005; pp. 94–97.
41. Ma, Y.X.; Wang, J.Q.; Wang, J.; Wu, X.H. An interval neutrosophic linguistic multi-criteria group decision-making method and its application in selecting medical treatment options. *Neural Comput. Appl.* **2017**, *28*, 2745–2765. [CrossRef]
42. Guo, Y.; Cheng, H.D. *New Neutrosophic Approach to Image Segmentation*; Elsevier Science Inc.: Amsterdam, The Netherlands, 2009; pp. 587–595.
43. Ye, J. A multicriteria decision-making method using aggregation operators for simplified neutrosophic sets. *J. Intell. Fuzzy Syst.* **2014**, *26*, 2459–2466.
44. Ye, J. Multicriteria decision-making method using the correlation coefficient under single-valued neutrosophic environment. *Int. J. Gen. Syst.* **2013**, *42*, 386–394. [CrossRef]
45. Ye, J. Single valued neutrosophic cross-entropy for multicriteria decision making problems. *Appl. Math. Model.* **2014**, *38*, 1170–1175. [CrossRef]
46. Xiao, F. A novel multi-criteria decision making method for assessing health-care waste treatment technologies based on D numbers. *Eng. Appl. Artif. Intell.* **2018**, *71*, 216–225. [CrossRef]
47. Deng, X.; Jiang, W. D number theory based game-theoretic framework in adversarial decision making under a fuzzy environment. *Int. J. Approx. Reason.* **2019**, *106*, 194–213. [CrossRef]
48. Jiang, W.; Zhong, Y.; Deng, X. A Neutrosophic Set Based Fault Diagnosis Method Based on Multi-Stage Fault Template Data. *Symmetry* **2018**, *10*, 346. [CrossRef]
49. Zhou, D.; Al-Durra, A.; Zhang, K.; Ravey, A.; Gao, F. Online remaining useful lifetime prediction of proton exchange membrane fuel cells using a novel robust methodology. *J. Power Sources* **2018**, *399*, 314–328. [CrossRef]
50. Şahin, R.; Yiğider, M. A multi-criteria neutrosophic group decision making metod based TOPSIS for supplier selection. *arXiv* **2014**, arXiv:1412.5077.
51. Peng, J.J.; Wang, J.Q.; Zhang, H.Y.; Chen, X.H. An outranking approach for multi-criteria decision-making problems with simplified neutrosophic sets. *Appl. Soft Comput.* **2016**, *27*, 615–627. [CrossRef]

52. Jiang, W.; Shou, Y. A novel single-valued neutrosophic set similarity measure and its application in multicriteria decision-making. *Symmetry* **2017**, *9*, 127. [CrossRef]
53. Tian, Z.; Zhang, H.; Wang, J.; Wang, J.; Chen, X. Multi-criteria decision-making method based on a cross-entropy with interval neutrosophic sets. *Int. J. Syst. Sci.* **2015**, *47*, 3598–3608. [CrossRef]
54. Deli, I.; Ali, M.; Smarandache, F. Bipolar neutrosophic sets and their application based on multi-criteria decision making problems. In Proceedings of the 2015 International Conference on Advanced Mechatronic Systems, Beijing, China, 22–24 August 2015; pp. 249–254.
55. Elhassouny, A.; Smarandache, F. Neutrosophic-simplified-TOPSIS multi-criteria decision-making using combined simplified-TOPSIS method and neutrosophics. In Proceedings of the 2016 IEEE International Conference on Fuzzy Systems (FUZZ-IEEE), Vancouver, BC, Canada, 24–29 July 2016; pp. 2468–2474.
56. Chai, Q.Z.; Li, P.; Di, R.; Zhang, Y.W. TOPSIS method based on single-valued neutrosophic set. *J. Jiangsu Univ. Sci. Technol. (Nat. Sci. Ed.)* **2018**, *32*, 262–265. (In Chinese)
57. Pramanik, S.; Pramanik, S.; Giri, B.C. TOPSIS method for multi-attribute group decision-making under single-valued neutrosophic environment. *Neural Comput. Appl.* **2016**, *27*, 727–737.

symmetry

MDPI

Article

Neutrosophic Cubic Einstein Geometric Aggregation Operators with Application to Multi-Criteria Decision Making Method

Majid Khan [1], Muhammad Gulistan [1,*], Naveed Yaqoob [2], Madad Khan [3] and Florentin Smarandache [4]

1 Department of Mathematics and Statistics, Hazara University, Mansehra 21130, Pakistan; majid_swati@yahoo.com

2 Department of Mathematics, College of Science Al-Zulfi, Majmaah University, Al-Zulfi 11932, Saudi Arabia; na.yaqoob@mu.edu.sa or nayaqoob@ymail.com

3 Department of Mathematics, COMSATS University Iaslamabad, Abbottabad Campus, Abbottabad 22060, Pakistan; madadmath@yahoo.com

4 Department of Mathematics, University of New Mexico, Albuquerque, NM 87301, USA; fsmarandache@gmail.com

* Correspondence: gulistanmath@hu.edu.pk or gulistanm21@yahoo.com

Received: 26 December 2018; Accepted: 31 January 2019; Published: 16 February 2019

Abstract: Neutrosophic cubic sets (NCs) are amore generalized version of neutrosophic sets(Ns) and interval neutrosophic sets (INs). Neutrosophic cubic setsare better placed to express consistent, indeterminate and inconsistent information, which provides a better platform to deal with incomplete, inconsistent and vague data. Aggregation operators play a key role in daily life, and in relation to science and engineering problems. In this paper we defined the algebraic and Einstein sum, multiplication and scalar multiplication, score and accuracy functions. Using these operations we defined geometric aggregation operators and Einstein geometric aggregation operators. First, we defined the algebraic and Einstein operators of addition, multiplication and scalar multiplication. We defined score and accuracy function to compare neutrosophic cubic values. Then we definedthe neutrosophic cubic weighted geometric operator (NCWG), neutrosophic cubic ordered weighted geometric operator (NCOWG), neutrosophic cubic Einstein weighted geometric operator (NCEWG), and neutrosophic cubic Einstein ordered weighted geometric operator (NCEOWG) over neutrosophic cubic sets. A multi-criteria decision making method is developed as an application to these operators. This method is then applied to a daily life problem.

Keywords: neutrosophic cubic weighted geometric operator (NCWG); neutrosophic cubic ordered weighted geometric operator (NCOWG); neutrosophic cubic Einstein weighted geometric operator (NCEWG); neutrosophic cubic Einstein ordered weighted geometric operator (NCEOWG)

1. Introduction

The theory of fuzzy sets was introduced by Zadeh [1].Soon after, it attracted experts of sciences and engineering due to its possibilistic behavior. The applicability of fuzzy sets extended it to interval valued fuzzy sets(IVFs) [2,3]. In 1986, K. Atnassov developed the theory of intuitionistic fuzzy sets [4], which were further extended to interval valued intuitionistic fuzzy sets in 1989 [5]. In 2012, Y.B. Jun generalized the idea of fuzzy sets and intuitionistic fuzzy sets to form cubic sets [6]. Smarandache presented his theory regarding the inconsistent and indeterminate behavior of data in 1999, and named it the neutrosophic set [7]. Neutrosophic sets consist of three components:Truth, indeterminate and falsehood, which provides a more general platform to deal with vague and insufficient data. In 2005, Wang et al. [8] presented the idea of interval valued neutrosophic sets. Interval valued neutrosophic sets provide a range to experts which makes them more comfortable with making the choice. Jun et al. defined the neutrosophic cubic set [9,10]. Neutrosophic cubic sets are a generalization of neutrosophic sets and interval neutrosophic sets. They enable us to choose both interval values and single value membership. This characteristic of neutrosophic cubic sets enables us to deal with uncertain and vague data more efficiently.

Decision making is one of the most important factors in scienceand day-to-day life as well. Aggregation operators are an imperative part of modern decision making. A lack of data or information makes it difficult for decision makers to take an appropriatedecision. This uncertain situation can be minimized using the vague nature neutrosophic cubic set and its extensions. Neutrosophic cubic set (NCs) are a more generalized version of neutrosophic sets (Ns) and interval neutrosophic sets (INs). Neutrosophic cubic setsare better placed to express consistent, indeterminate, and inconsistent information, which provides a better platform to deal with incomplete, inconsistent, and vague data. Aggregation operators have a key role in daily life, science and engineering problems. Zhan et al. [11] in their workapplications of neutrosophic cubic sets in multi-criteria decision making in 2017. Banerjee et al. [12] usedgrey rational analysis in their workGRA for multi attribute decision making in neutrosophic cubic set environment in 2017.Lu and Ye [13] definedcosine measure for neutrosophic cubic sets for multiple attribte decision making in 2017. Pramanik et al. [14] defined neutrosophic cubic MCGDM method based on similarity measurein 2017. Shi and Ye [15] defined Dombi aggregation operators of neutrosophic cubic set for multiple attribute deicision makingin 2018. Baolin et al. [16] applied Einstein aggregations onneutrosophic sets in a novel generalized simplified neutrosophic number Einstein aggregation operator 2018. Alot of work has been done and is being done by different researchers in decision making using neutrosophic cubic sets.

In this paper, we define algebraic and Einstein sum, multiplication and scalar multiplication, score and accuracy functions. Using these operations, we define geometric aggregation operators and Einstein geometric aggregation operators. First, we define algebraic and Einstein operators of addition, multiplication and scalar multiplication. We then define score and accuracy functions to compare neutrosophic cubic values. Following this, we propose a neutrosophic cubic ordered weighted geometric operator (NCOWG), neutrosophic cubic Einstein weighted geometric operator (NCEWG), and a neutrosophic cubic Einstein ordered weighted geometric operator (NCEOWG) over neutrosophic cubic sets. A multi-criteria decision making method is then developed as an application for these operators. This method is then applied to a daily life problem.

2. Preliminaries

This section consists of two parts: Notations, which consists of notations with their descriptions and some previous definitions; and results. We recommend the reader to see [1–3,6–9,16].

2.1. Notations

This section consists of some notations with their descriptions, as shown in Table 1.

Table 1. Some notations with their descriptions.

S. No	Notation	Description
1	U	Ground set
2	u	Element of ground set (U).
3	ψ	Fuzzy set
4	$\widetilde{\Psi} = [\Psi^L, \Psi^U]$	Interval valued fuzzy set which is an interval of [0,1]. The left extreme ψ^L is referred as lower fuzzy and right extreme ψ^U is referred as upper fuzzy function.
5	(T_N, I_N, F_N)	components of neutrosophic sets each one is fuzzy sets.
6	$\left(\widetilde{T}_N, \widetilde{I}_N, \widetilde{F}_N\right)$	The components of interval neutrosophic each one is an interval valued fuzzy set.
7	$\left(\widetilde{T}_N, \widetilde{I}_N, \widetilde{F}_N, T_N, I_N, F_N\right)$	The components of neutrosophic cubic set. Referred to 5 and 6.
8	Γ^*, Γ	t-conorm, t-norm
9	$\oplus, \otimes$	Algebraic sum, product
10	$\oplus_E, \otimes_E$	Einstein sum, product

2.2. Pre-Defined Definitions

This section consists of some predefined definitions and results.

Definition 1 [1]. *A mapping $\psi: U \to [0,1]$ is called a fuzzy set, and $\psi(u)$ is called a membership function, simply denoted by ψ.*

Definition 2 [2,3]. *A mapping $\widetilde{\Psi}: U \to D[0,1]$, where $D[0,1]$ is the interval value of $[0,1]$, called the interval valued fuzzy set(IVF). For all $u \in U$ $\widetilde{\Psi}(u) = \{[\psi^L(u), \psi^U(u)] | \psi^L(u), \psi^U(u) \in [0,1]$ and $\psi^L(u) \le \psi^U(u)\}$ is membership degree of u in $\widetilde{\Psi}$. This is simply denoted by $\widetilde{\Psi} = [\Psi^L, \Psi^U]$.*

Definition 3 [6]. *A structure $C = \left\{\left(u, \widetilde{\Psi}(u), \Psi(u)\right) | u \in U\right\}$ is a cubic set in U in which $\widetilde{\Psi}(u)$ is IVF in U, that is, $\widetilde{\Psi} = [\Psi^L, \Psi^U]$ and Ψ is a fuzzy set in U. This can be simply denoted by $C = \left(\widetilde{\Psi}, \Psi\right)$. C^U denotes the collection of cubic sets in U.*

Definition 4 [7]. *A structure $N = \{(T_N(u), I_N(u), F_N(u)) | u \in U\}$ is a neutrosophic set(Ns), where $\{T_N(u), I_N(u), F_N(u) \in [0,1]\}$ are called truth, indeterminacy and falsity functions, respectively.This can be simply denoted by $N = (T_N, I_N, F_N)$.*

Definition 5 [8]. *An interval neutrosophic set (INs) in U is a structure $N = \left\{\left(\widetilde{T}_N(u), \widetilde{I}_N(u), \widetilde{F}_N(u)\right) | u \in U\right\}$, where $\left\{\widetilde{T}_N(u), \widetilde{I}_N(u), \widetilde{F}_N(u) \in D[0,1]\right\}$ is calledtruth, indeterminacy an falsity functionin U, respectively. This can be simply denoted by $N = \left(\widetilde{T}_N, \widetilde{I}_N, \widetilde{F}_N\right)$. For convenience, we denote $N = \left(\widetilde{T}_N, \widetilde{I}_N, \widetilde{F}_N\right)$ by $N = \left(\widetilde{T}_N = [T_N^L, T_N^U], \widetilde{I}_N = [I_N^L, I_N^U], \widetilde{F}_N = [F_N^L, F_N^U]\right)$.*

Definition 6 [9]. *A structure $N = \left\{\left(u, \widetilde{T}_N(u), \widetilde{I}_N(u), \widetilde{F}_N(u), T_N(u), I_N(u), F_N(u)\right) | u \in U\right\}$ is neutrosophic cubic set(NCs) in U, in which $\left(\widetilde{T}_N = [T_N^L, T_N^U], \widetilde{I}_N = [I_N^L, I_N^U], \widetilde{F}_N = [F_N^L, F_N^U]\right)$ is an interval neutrosophic set and (T_N, I_N, F_N) is a neutrosophic set in U. Simply denoted by $N = \left(\widetilde{T}_N, \widetilde{I}_N, \widetilde{F}_N, T_N, I_N, F_N\right)$, $[0,0] \le \widetilde{T}_N + \widetilde{I}_N + \widetilde{F}_N \le [3,3]$ and $0 \le T_N + I_N + F_N \le 3$. N^U denotes the collection of neutrosophic cubic sets in U. Simply denoted by $N = \left(\widetilde{T}_N, \widetilde{I}_N, \widetilde{F}_N, T_N, I_N, F_N\right)$.*

Definition 7 [16]. *The t-operators are basically union and intersection operators in the theory of fuzzy sets, which are denoted by t-conorm (Γ^*) and t-norm (Γ), respectively. The role of t-operators is very important in fuzzy theory and its applications.*

Definition 8 [16]. $\Gamma^* : [0,1] \times [0,1] \to [0,1]$ *is called t-conorm if it satisfies the following axioms:*

Axiom 1. $\Gamma^*(1,u) = 1$ *and* $\Gamma^*(0,u) = 0$;
Axiom 2. $\Gamma^*(u,v) = \Gamma^*(v,u)$ *for all a and b;*
Axiom 3. $\Gamma^*(u,\Gamma^*(v,w)) = \Gamma^*(\Gamma^*(u,v),w)$ *for all a, b and c;*
Axiom 4. *If* $u \le u'$ *and* $v \le v'$, *then* $\Gamma^*(u,v) \le \Gamma^*(u',v')$.

Most known t-conorms are as follows:

1. The default t-conorm: $\Gamma^*_{\max}(u,v) = \max(u,v)$.
2. The bounded t-conorm: $\Gamma^*_{bounded}(u,v) = \min(1, u+v)$.
3. The algebraic t-conorm: $\Gamma^*_{algebraic}(u,v) = u+v-uv$.

Definition 9 [16]. $\Gamma : [0,1] \times [0,1] \to [0,1]$ *is called t-norm if it satisfies the following axioms:*

Axiom 5. $\Gamma(1,u) = u$ *and* $\Gamma(0,u) = 0$;
Axiom 6. $\Gamma(u,v) = \Gamma(v,u)$ *for all a and b;*
Axiom 7. $\Gamma(u,\Gamma(v,w)) = \Gamma(\Gamma(u,v),w)$ *for all a, b and c;*
Axiom 8. *If* $u \le u'$ *and* $v \le v'$, *then* $\Gamma(u,v) \le \Gamma(u',v')$.

Most well known t-norms are as follows:

1. The default t-norm: $\Gamma_{\min}(u,v) = \min(u,v)$.
2. The bounded t-norm: $\Gamma_{bounded}(u,v) = \max(0, u+v-1)$.
3. The algebraic t-norm: $\Gamma_{algebraic}(u,v) = uv$.

If $\Gamma^*(u,v)$, $\Gamma(u,v)$ are continuous and $\Gamma^*(u,u) > u$, $\Gamma(u,u) < u$, then Γ^* and Γ are said to be Archimedes t-conorm and t-norm, respectively. Any pair of dual t-conorm (Γ^*) and t-norm (Γ) is used. It is known that t-norms and t-conorms operators satisfy the condition of conjunction and disjunction operators, respectively. However, the algebraic operations, like algebraic sum and product, are not unique and may correspond to union and intersection. The t-conorms and t-norms families have a vast range, which corresponds to unions and intersections. Among these, the Einstein sum and Einstein product are good choices since they give the smooth approximation like algebraic sum and algebraic product, respectively. Einstein sum $\oplus_E$ and Einstein product $\otimes_E$ are examples of t-conorm and t-norm, respectively:

$$\Gamma^*_E(u,v) = \frac{u+v}{1+uv}$$

$$\Gamma_E(u,v) = \frac{uv}{1+(1-u)(1-v)}$$

Group decision making is an important aspect of decision making theory. We are often in situationsin which we have to deal with more then one expert, attribute and alternative. Motivated by such situations, a multi-attribute decision making method for more then one expert is proposed on neutrosophic cubic aggregation operators.This whole work consisted of six sections. In Section 3, we define some algebraicEinstein operations and score and accuracy functions, along with some important results and examples. On the basis of these definitions and results, we define geometric and Einstein geometric aggregation operators on neutrosophic cubic sets in Section 4. In Section 5, an algorithm is proposed based on neutrosophic cubic geometric and Einstein geometric aggregation operators to deal with multi-attribute decision making problems. In the final section, a numerical example from daily life is presented as an application of the work.

3. Operations on Neutrosophic Cubic Sets

In this section, we introduce some new operations on neutrosophic cubic sets which are further used in the article.

3.1. Algebraic Addition, Multiplication and Scalar Multiplication

We introduce the algebraic addition, multiplication, and scalar multiplication on neutrosophic cubic sets(NCs). An important result of exponential multlipliction is established on the basis of these defintions, which provides the basis to define neutrosophic cubic geometric aggregation operators.

Definition 10. *The sum of two neutrosophic cubic sets(NCs),* $A = \left(\widetilde{T}_A, \widetilde{I}_A, \widetilde{F}_A, T_A, I_A, F_A\right)$, *where* $\widetilde{T}_A = [T_A^L, T_A^U], \widetilde{I}_A = [I_A^L, I_A^U], \widetilde{F}_A = [F_A^L, F_A^U]$, *and* $B = \left(\widetilde{T}_B, \widetilde{I}_B, \widetilde{F}_B, T_B, I_B, F_B\right)$, *where* $\widetilde{T}_B = [T_B^L, T_B^U], \widetilde{I}_B = [I_B^L, I_B^U], \widetilde{F}_B = [F_B^L, F_B^U]$ *is defined as*

$$A \oplus B = \begin{pmatrix} [T_A^L + T_B^L - T_A^L T_B^L, T_A^U + T_B^U - T_A^U T_B^U], \\ [I_A^L + I_B^L - I_A^L I_B^L, I_A^U + I_B^U - I_A^U I_B^U], \\ [F_A^L F_B^L, F_A^U F_B^U], \\ T_A T_B, I_A I_B, F_A + F_B - F_A F_B \end{pmatrix}$$

Definition 11. *The product between two neutrosophic cubic sets (NCs),* $A = \left(\widetilde{T}_A, \widetilde{I}_A, \widetilde{F}_A, T_A, I_A, F_A\right)$, *where* $\widetilde{T}_A = [T_A^L, T_A^U], \widetilde{I}_A = [I_A^L, I_A^U], \widetilde{F}_A = [F_A^L, F_A^U]$ *and* $B = \left(\widetilde{T}_B, \widetilde{I}_B, \widetilde{F}_B, T_B, I_B, F_B\right)$, *where* $\widetilde{T}_B = [T_B^L, T_B^U], \widetilde{I}_B = [I_B^L, I_B^U], \widetilde{F}_B = [F_B^L, F_B^U]$ *is defined as*

$$A \otimes B = \begin{pmatrix} [T_A^L T_B^L, T_A^U T_B^U], \\ [I_A^L I_B^L, I_A^U I_B^U], \\ [F_A^L + F_B^L - F_A^L F_B^L, F_A^U + F_B^U - F_A^U F_B^U], \\ T_A + T_B - T_A T_B, I_A + I_B - I_A I_B, F_A F_B \end{pmatrix}$$

Definition 12. *The scalar multiplication on a neutrosophic cubic set (NCs),* $A = \left(\widetilde{T}_A, \widetilde{I}_A, \widetilde{F}_A, T_A, I_A, F_A\right)$, *where* $\widetilde{T}_A = [T_A^L, T_A^U], \widetilde{I}_A = [I_A^L, I_A^U], \widetilde{F}_A = [F_A^L, F_A^U]$, *and a Scalar* k is defined as

$$kA = \begin{pmatrix} \left[1 - (1 - T_A^L)^k, 1 - (1 - T_A^U)^k\right], \\ \left[1 - (1 - I_A^L)^k, 1 - (1 - I_A^U)^k\right], \\ \left[(F_A^L)^k, (F_A^U)^k\right], \\ (T_A)^k, (I_A)^k, 1 - (1 - F_A)^k \end{pmatrix}$$

The following result is established to deal with the exponential multiplication on neutrosophic cubic values. This result enables us to define geometric aggregation operators along some important results on neutrosophic cubic sets.

Theorem 1. *Let* $A = \left(\widetilde{T}_A, \widetilde{I}_A, \widetilde{F}_A, T_A, I_A, F_A\right)$, *where* $\widetilde{T}_A = [T_A^L, T_A^U], \widetilde{I}_A = [I_A^L, I_A^U], \widetilde{F}_A = [F_A^L, F_A^U]$, *be a neutrosophic cubic value, then the exponential operation can be defined by*

$$A^k = \begin{pmatrix} \left[(T_A^L)^k, (T_A^U)^k\right], \\ \left[(I_A^L)^k, (I_A^U)^k\right], \\ \left[1 - (1 - F_A^L)^k, 1 - (1 - F_A^U)^k\right], \\ 1 - (1 - T_A)^k, 1 - (1 - I_A)^k, (F_A)^k \end{pmatrix}$$

where $A^k = A \otimes A \otimes, \ldots \otimes A(k - times)$, *and* A^k *is a neutrosophic cubic value for every positive value of* k.

Proof. We prove the theorem by mathematical induction, as the $k = 1$, $A^1 = A$ result holds.We assume that for $k = m$ the result is true:

$$A^m = \begin{pmatrix} \left[(T_A^L)^m, (T_A^U)^m\right], \\ \left[(I_A^L)^m, (I_A^U)^m\right], \\ \left[1-(1-F_A^L)^m, 1-(1-F_A^U)^m\right], \\ 1-(1-T_A)^m, 1-(1-I_A)^m, (F_A)^m \end{pmatrix}$$

That is A^m is neutrosophic cubic value. We prove that for $k = m+1$ is also neutrosophic cubic value.

Since

$$\begin{aligned} A^m \otimes A &= \begin{pmatrix} \left[(T_A^L)^m, (T_A^U)^m\right], \\ \left[(I_A^L)^m, (I_A^U)^m\right], \\ \left[1-(1-F_A^L)^m, 1-(1-F_A^U)^m\right], \\ 1-(1-T_A)^m, 1-(1-I_A)^m, (F_A)^m \end{pmatrix} \otimes \begin{pmatrix} [(T_A^L), (T_A^U)], \\ [(I_A^L), (I_A^U)], \\ [F_A^L, F_A^U], \\ T_A, I_A, F_A \end{pmatrix} \\ &= \begin{pmatrix} \left[(T_A^L)^{m+1}, (T_A^U)^{m+1}\right], \\ \left[(I_A^L)^{m+1}, (I_A^U)^{m+1}\right], \\ \left[1-(1-F_A^L)^m + F_A^L - \left(1-(1-F_A^L)^m\right)F_A^L, 1-(1-F_A^U)^m + F_A^U - \left(1-(1-F_A^U)^m\right)F_A^U\right], \\ 1-(1-T_A)^m + T_A - (1-(1-T_A)^m)T_A, 1-(1-I_A)^m + I_A - (1-(1-I_A)^m I_A), (F_A)^{m+1} \end{pmatrix} \\ &= \begin{pmatrix} \left[(T_A^L)^{m+1}, (T_A^U)^{m+1}\right], \\ \left[(I_A^L)^{m+1}, (I_A^U)^{m+1}\right], \\ \left[1-(1-F_A^L)^m + F_A^L - F_A^L + (1-F_A^L)^m F_A^L, 1-(1-F_A^U)^m + F_A^U - F_A^U + (1-F_A^U)^m F_A^U\right], \\ 1-(1-T_A)^m + T_A - T_A + (1-T_A)^m T_A, 1-(1-I_A)^m + I_A - I_A + (1-I_A)^m I_A, (F_A)^{m+1} \end{pmatrix} \\ &= \begin{pmatrix} \left[(T_A^L)^{m+1}, (T_A^U)^{m+1}\right], \\ \left[(I_A^L)^{m+1}, (I_A^U)^{m+1}\right], \\ \left[1-(1-F_A^L)^m + (1-F_A^L)^m F_A^L, 1-(1-F_A^U)^m + (1-F_A^U)^m F_A^U\right], \\ 1-(1-T_A)^m + (1-T_A)^m T_A, 1-(1-I_A)^m + (1-I_A)^m I_A, (F_A)^{m+1} \end{pmatrix} \\ &= \begin{pmatrix} \left[(T_A^L)^{m+1}, (T_A^U)^{m+1}\right], \\ \left[(I_A^L)^{m+1}, (I_A^U)^{m+1}\right], \\ \left[1-(1-F_A^L)^m(1-F_A^L), 1-(1-F_A^U)^m(1-F_A^U)\right], \\ 1-(1-T_A)^m(1-T_A), 1-(1-I_A)^m(1-I_A), (F_A)^{m+1} \end{pmatrix} \\ &= \begin{pmatrix} \left[(T_A^L)^{m+1}, (T_A^U)^{m+1}\right], \\ \left[(I_A^L)^{m+1}, (I_A^U)^{m+1}\right], \\ \left[1-(1-F_A^L)^{m+1}, 1-(1-F_A^U)^{m+1}\right], \\ 1-(1-T_A)^{m+1}, 1-(1-I_A)^{m+1}, (F_A)^{m+1} \end{pmatrix} \\ &= A^{m+1}. \end{aligned}$$

□

3.2. Einstein Addition, Multiplication and Scalar Multiplication

Taking into account the dual t-conorm (Γ^*) and t-norm (Γ), the Einstein operations of union, intersection, addition, multiplication and scalar multiplication are defined on the neutrosophic cubic sets.An important result of Einstein exponential multlipliction is established on the basis of these defintions, which provides the base with which to define neutrosophic cubic Einstein geometric aggregation operators.

Definition 13. *The Einstein union between two neutrosophic cubic sets (NCs),* $A = \left(\widetilde{T}_A, \widetilde{I}_A, \widetilde{F}_A, T_A, I_A, F_A\right)$ *where* $\widetilde{T}_A = [T_A^L, T_A^U], \widetilde{I}_A = [I_A^L, I_A^U], \widetilde{F}_A = [F_A^L, F_A^U]$*, and* $B = \left(\widetilde{T}_B, \widetilde{I}_B, \widetilde{F}_B, T_B, I_B, F_B\right)$ *where* $\widetilde{T}_B = [T_B^L, T_B^U], \widetilde{I}_B = [I_B^L, I_B^U], \widetilde{F}_B = [F_B^L, F_B^U]$ *is defined as*

$$A \vee B = \left(\Gamma\left\{\widetilde{T}_A, \widetilde{T}_B\right\}, \Gamma\left\{\widetilde{I}_A, \widetilde{I}_B\right\}, \Gamma^*\left\{\widetilde{F}_A, \widetilde{F}_B\right\}, \Gamma^*\{T_A, T_B\}, \Gamma^*\{I_A, I_B\}, \Gamma\{F_A, F_B\}\right)$$

Definition 14. *The Einstein intersection between two neutrosophic cubic sets(NCS),* $A = \left(\widetilde{T}_A, \widetilde{I}_A, \widetilde{F}_A, T_A, I_A, F_A\right)$*, where* $\widetilde{T}_A = [T_A^L, T_A^U], \widetilde{I}_A = [I_A^L, I_A^U], \widetilde{F}_A = [F_A^L, F_A^U]$ *and* $B = \left(\widetilde{T}_B, \widetilde{I}_B, \widetilde{F}_B, T_B, I_B, F_B\right)$*, where* $\widetilde{T}_B = [T_B^L, T_B^U], \widetilde{I}_B = [I_B^L, I_B^U], \widetilde{F}_B = [F_B^L, F_B^U]$ *is defined as*

$$A \wedge B = \left(\Gamma^*\left\{\widetilde{T}_A, \widetilde{T}_B\right\}, \Gamma^*\left\{\widetilde{I}_A, \widetilde{I}_B\right\}, \Gamma\left\{\widetilde{F}_A, \widetilde{F}_B\right\}, \Gamma\{T_A, T_B\}, \Gamma\{I_A, I_B\}, \Gamma^*\{F_A, F_B\}\right).$$

On the basis of Einstein union and intersection the Einstein sum and product is defined over neutrosophic cubic values.

Definition 15. *The Einstein sum between two neutrosophic cubic sets (NCS),* $A = \left(\widetilde{T}_A, \widetilde{I}_A, \widetilde{F}_A, T_A, I_A, F_A\right)$*, where* $\widetilde{T}_A = [T_A^L, T_A^U], \widetilde{I}_A = [I_A^L, I_A^U], \widetilde{F}_A = [F_A^L, F_A^U]$ *and* $B = \left(\widetilde{T}_B, \widetilde{I}_B, \widetilde{F}_B, T_B, I_B, F_B\right)$*, where* $\widetilde{T}_B = [T_B^L, T_B^U], \widetilde{I}_B = [I_B^L, I_B^U], \widetilde{F}_B = [F_B^L, F_B^U]$ *is defined as*

$$A \oplus_E B = \begin{pmatrix} \left[\frac{T_A^L + T_B^L}{1 + T_A^L T_B^L}, \frac{T_A^U + T_B^U}{1 + T_A^U T_B^U}\right], \\ \left[\frac{I_A^L + I_B^L}{1 + I_A^L I_B^L}, \frac{I_A^U + I_B^U}{1 + I_A^U I_B^U}\right], \\ \left[\frac{F_A^L F_B^L}{1 + (1 - F_A^L)(1 - F_B^L)}, \frac{F_A^U F_B^U}{1 + (1 - F_A^U)(1 - F_B^U)}\right] \\ \frac{T_A T_B}{1 + (1 - T_A)(1 - T_B)}, \frac{I_A I_B}{1 + (1 - I_A)(1 - I_B)}, \frac{F_A + F_B}{1 + F_A F_B} \end{pmatrix}$$

Definition 16. *The Einstein product between two neutrosophic cubic sets (NCS),* $A = \left(\widetilde{T}_A, \widetilde{I}_A, \widetilde{F}_A, T_A, I_A, F_A\right)$*, where* $\widetilde{T}_A = [T_A^L, T_A^U], \widetilde{I}_A = [I_A^L, I_A^U], \widetilde{F}_A = [F_A^L, F_A^U]$ *and* $B = \left(\widetilde{T}_B, \widetilde{I}_B, \widetilde{F}_B, T_B, I_B, F_B\right)$*, where* $\widetilde{T}_B = [T_B^L, T_B^U], \widetilde{I}_B = [I_B^L, I_B^U], \widetilde{F}_B = [F_B^L, F_B^U]$ *is defined as*

$$A \otimes_E B = \begin{pmatrix} \left[\frac{T_A^L T_B^L}{1 + (1 - T_A^L)(1 - T_B^L)}, \frac{T_A^U T_B^U}{1 + (1 - T_A^U)(1 - T_B^U)}\right], \\ \left[\frac{I_A^L I_B^L}{1 + (1 - I_A^L)(1 - I_B^L)}, \frac{I_A^U I_B^U}{1 + (1 - I_A^U)(1 - I_B^U)}\right], \\ \left[\frac{F_A^L + F_B^L}{1 + F_A^L F_B^L}, \frac{F_A^U + F_B^U}{1 + F_A^U F_B^U}\right] \\ \frac{T_A + T_B}{1 + T_A T_B}, \frac{I_A + I_B}{1 + I_A I_B}, \frac{F_A F_B}{1 + (1 - F_A)(1 - F_B)} \end{pmatrix}$$

Definition 17. *The scalar multiplication on a neutrosophic cubic set(NCS),* $A = \left(\widetilde{T}_A, \widetilde{I}_A, \widetilde{F}_A, T_A, I_A, F_A\right)$*, where* $\widetilde{T}_A = [T_A^L, T_A^U], \widetilde{I}_A = [I_A^L, I_A^U], \widetilde{F}_A = [F_A^L, F_A^U]$*, and scalar k is defined as*

$$k_E A = \begin{pmatrix} \left[\frac{(1+T_A^L)^k-(1-T_A^L)^k}{(1+T_A^L)^k+(1-T_A^L)^k}, \frac{(1+T_A^U)^k-(1-T_A^U)^k}{(1+T_A^U)^k+(1-T_A^U)^k}\right], \\ \left[\frac{(1+I_A^L)^k-(1-I_A^L)^k}{(1+I_A^L)^k+(1-I_A^L)^k}, \frac{(1+I_A^U)^k-(1-I_A^U)^k}{(1+I_A^U)^k+(1-I_A^U)^k}\right], \\ \left[\frac{2(F_A^L)^k}{(2-F_A^L)^k+(F_A^L)^k}, \frac{2(F_A^U)^k}{(2-F_A^U)^k+(F_A^U)^k}\right], \\ \frac{2(T_A)^k}{(2-T_A)^k+(T_A)^k}, \frac{2(I_A)^k}{(2-I_A)^k+(I_A)^k}, \frac{(1+F_A)^k-(1-F_A)^k}{(1+F_A)^k+(1-F_A)^k} \end{pmatrix}$$

After defining the scalar multiplication over the neutrosophic cubic set, we established the following result, which deals with the Einstein exponential multiplication on neutrosophic cubic values. This result enabled us to define Einstein geometric aggregation operators along with some important results on neutrosophic cubic sets.

Theorem 2. *Let $A = \left(\widetilde{T}_A, \widetilde{I}_A, \widetilde{F}_A, T_A, I_A, F_A\right)$, where $\widetilde{T}_A = [T_A^L, T_A^U], \widetilde{I}_A = [I_A^L, I_A^U], \widetilde{F}_A = [F_A^L, F_A^U]$, be a neutrosophic cubic value, then the exponential operation defined by*

$$A^{E^k} = \begin{pmatrix} \left[\frac{2(T_A^L)^k}{2(-T_A^L)^k+(T_A^L)^k}, \frac{2(T_A^U)^k}{(2-T_A^U)^k+(T_A^U)^k}\right], \\ \left[\frac{2(I_A^L)^k}{(2-I_A^L)^k+(I_A^L)^k}, \frac{2(I_A^U)^k}{(2-I_A^U)^k+(I_A^U)^k}\right], \\ \left[\frac{(1+F_A^L)^k-(1-F_A^L)^k}{(1+F_A^L)^k+(1-F_A^L)^k}, \frac{(1+F_A^U)^k-(1-F_A^U)^k}{(1+F_A^U)^k+(1-F_A^U)^k}\right], \\ \frac{(1+T_A)^k-(1-T_A)^k}{(1+T_A)^k+(1-T_A)^k}, \frac{(1+I_A)^k-(1-I_A)^k}{(1+I_A)^k+(1-I_A)^k}, \frac{2(F_A)^k}{(2-F_A)^k+(F_A)^k} \end{pmatrix}$$

where $A^{E^k} = A \otimes_E A \otimes_E \ldots \otimes_E A(k - times)$, moreover A^{E^k} is a neutrosophic cubic value for every positive value of k.

Proof. We prove the theorem by mathematical induction. For $k = 1$

$$A^E = \begin{pmatrix} \left[\frac{2(T_A^L)}{(2-T_A^L)+(T_A^L)}, \frac{2(T_A^U)}{(2-T_A^U)+(T_A^U)}\right], \\ \left[\frac{2(I_A^L)}{(2-I_A^L)+(I_A^L)}, \frac{2(I_A^U)}{(2-I_A^U)+(I_A^U)}\right], \\ \left[\frac{(1+F_A^L)-(1-F_A^L)}{(1+F_A^L)+(1-F_A^L)}, \frac{(1+F_A^U)-(1-F_A^U)}{(1+F_A^U)+(1-F_A^U)}\right], \\ \frac{(1+T_A)-(1-T_A)}{(1+T_A)+(1-T_A)}, \frac{(1+I_A)-(1-I_A)}{(1+I_A)+(1-I_A)}, \frac{2(F_A)}{(2-F_A)+(F_A)} \end{pmatrix}$$

We observe that the components $T_A^L, T_A^U, I_A^L, I_A^U, F_A$ are of the form $\frac{2x}{(2-x)+x}$, and F_A^L, F_A^U, T_A, I_A are of the form $\frac{(1+y)-(1-y)}{(1+y)+(1-y)}$,

For all $x, y \in [0,1]$, clearly $x = \frac{2x}{(2-x)+x}$ and $y = \frac{(1+y)-(1-y)}{(1+y)+(1-y)}$

Hence A^E is neutrosophic cubic value.

Assuming $k = m$ is a neutrosophic cubic value i.e.,

$$A^{E^m} = \begin{pmatrix} \left[\frac{2(T_A^L)^m}{(2-T_A^L)^m+(T_A^L)^m}, \frac{2(T_A^U)^m}{(2-T_A^U)^m+(T_A^U)^m}\right], \\ \left[\frac{2(I_A^L)^m}{(2-I_A^L)^m+(I_A^L)^m}, \frac{2(I_A^U)^m}{(2-I_A^U)^m+(I_A^U)^m}\right], \\ \left[\frac{(1+F_A^L)^m-(1-F_A^L)^m}{(1+F_A^L)^m+(1-F_A^L)^m}, \frac{(1+F_A^U)^m-(1-F_A^U)^m}{(1+F_A^U)^m+(1-F_A^U)^m}\right], \\ \frac{(1+T_A)^m-(1-T_A)^m}{(1+T_A)^m+(1-T_A)^m}, \frac{(1+I_A)^m-(1-I_A)^m}{(1+I_A)^m+(1-I_A)^m}, \frac{2(F_A)^m}{(2-F_A)^m+(F_A)^m} \end{pmatrix}$$

is a neutrosophic cubic value. Then we prove $A^{E^{k+1}}$ is neutrosophic cubic value.

Consider,

$$A^{E^m} \otimes_E A^E = \begin{pmatrix} \left[\frac{2(T_A^L)^m}{(2-T_A^L)^m+(T_A^L)^m}, \frac{2(T_A^U)^m}{(2-T_A^U)^m+(T_A^U)^m}\right], \\ \left[\frac{2(I_A^L)^m}{(2-I_A^L)^m+(I_A^L)^m}, \frac{2(I_A^U)^m}{(2-I_A^U)^m+(I_A^U)^m}\right], \\ \left[\frac{(1+F_A^L)^m-(1-F_A^L)^m}{(1+F_A^L)^m+(1-F_A^L)^m}, \frac{(1+F_A^U)^m-(1-F_A^U)^m}{(1+F_A^U)^m+(1-F_A^U)^m}\right], \\ \frac{(1+T_A)^m-(1-T_A)^m}{(1+T_A)^m+(1-T_A)^m}, \frac{(1+I_A)^m-(1-I_A)^m}{(1+I_A)^m+(1-I_A)^m}, \frac{2(F_A)^m}{(2-F_A)^m+(F_A)^m} \end{pmatrix} \otimes_E \begin{pmatrix} \left[\frac{2(T_A^L)}{(2-T_A^L)+(T_A^L)}, \frac{2(T_A^U)}{(2-T_A^U)+(T_A^U)}\right], \\ \left[\frac{2(I_A^L)^1}{(2-I_A^L)+(I_A^L)}, \frac{2(I_A^U)^1}{(2-I_A^U)+(I_A^U)}\right], \\ \left[\frac{(1+F_A^L)-(1-F_A^L)}{(1+F_A^L)+(1-F_A^L)}, \frac{(1+F_A^U)-(1-F_A^U)}{(1+F_A^U)+(1-F_A^U)}\right], \\ \frac{(1+T_A)-(1-T_A)}{(1+T_A)+(1-T_A)}, \frac{(1+I_A)-(1-I_A)}{(1+I_A)+(1-I_A)}, \frac{2(F_A)}{(2-F_A)+(F_A)} \end{pmatrix}$$

$$= \begin{pmatrix} \left[\frac{\frac{4\left(T_A^L\right)^{m+1}}{\left((2-T_A^L)^m+\left(T_A^L\right)^m\right)\left((2-T_A^L)+T_A^L\right)}}{1+\left(1-\frac{2\left(T_A^L\right)^m}{(2-T_A^L)^m+\left(T_A^L\right)^m}\right)\left(1-\frac{2T_A^L}{(2-T_A^L)+T_A^L}\right)}, \frac{\frac{4\left(T_A^U\right)^{m+1}}{\left((2-T_A^U)^m+\left(T_A^U\right)^m\right)\left((2-T_A^U)+T_A^U\right)}}{1+\left(1-\frac{2\left(T_A^U\right)^m}{(2-T_A^U)^m+\left(T_A^U\right)^m}\right)\left(1-\frac{2T_A^U}{(2-T_A^U)+T_A^U}\right)}\right], \\ \left[\frac{\frac{4\left(I_A^L\right)^{m+1}}{\left((2-I_A^L)^m+\left(I_A^L\right)^m\right)\left((2-I_A^L)+I_A^L\right)}}{1+\left(1-\frac{2\left(I_A^L\right)^m}{(2-I_A^L)^m+\left(I_A^L\right)^m}\right)\left(1-\frac{2I_A^L}{(2-I_A^L)+I_A^L}\right)}, \frac{\frac{4\left(I_A^U\right)^{m+1}}{\left((2-I_A^U)^m+\left(I_A^U\right)^m\right)\left((2-I_A^U)+I_A^U\right)}}{1+\left(1-\frac{2\left(I_A^U\right)^m}{(2-I_A^U)^m+\left(I_A^U\right)^m}\right)\left(1-\frac{2I_A^U}{(2-I_A^U)+I_A^U}\right)}\right], \\ \left[\frac{\left(\frac{(1+F_A^L)^m-(1-F_A^L)^m}{(1+F_A^L)^m+(1-F_A^L)^m}\right)+\left(\frac{(1+F_A^L)-(1-F_A^L)}{(1+F_A^L)+(1-F_A^L)}\right)}{1+\left(\frac{(1+F_A^L)^m-(1-F_A^L)^m}{(1+F_A^L)^m+(1-F_A^L)^m}\right)\left(\frac{(1+F_A^L)-(1-F_A^L)}{(1+F_A^L)+(1-F_A^L)}\right)}, \frac{\left(\frac{(1+F_A^U)^m-(1-F_A^U)^m}{(1+F_A^U)^m+(1-F_A^U)^m}\right)+\left(\frac{(1+F_A^U)-(1-F_A^U)}{(1+F_A^U)+(1-F_A^U)}\right)}{1+\left(\frac{(1+F_A^U)^m-(1-F_A^U)^m}{(1+F_A^U)^m+(1-F_A^U)^m}\right)\left(\frac{(1+F_A^U)-(1-F_A^U)}{(1+F_A^U)+(1-F_A^U)}\right)}\right], \\ \frac{\left(\frac{(1+T_A)^m-(1-T_A)^m}{(1+T_A)^m+(1-T_A)^m}\right)+\left(\frac{(1+T_A)-(1-T_A)}{(1+T_A)+(1-T_A)}\right)}{1+\left(\frac{(1+T_A)^m-(1-T_A)^m}{(1+T_A)^m+(1-T_A)^m}\right)\left(\frac{(1+T_A)-(1-T_A)}{(1+T_A)+(1-T_A)}\right)}, \frac{\left(\frac{(1+I_A)^m-(1-I_A)^m}{(1+I_A)^m+(1-I_A)^m}\right)+\left(\frac{(1+I_A)-(1-I_A)}{(1+I_A)+(1-I_A)}\right)}{1+\left(\frac{(1+I_A)^m-(1-I_A)^m}{(1+I_A)^m+(1-I_A)^m}\right)\left(\frac{(1+I_A)-(1-I_A)}{(1+I_A)+(1-I_A)}\right)}, \\ \frac{\frac{4\left(F_A\right)^{m+1}}{\left((2-F_A)^m+\left(F_A\right)^m\right)\left((2-F_A)+F_A\right)}}{1+\left(1-\frac{2\left(F_A\right)^m}{(2-F_A)^m+\left(F_A\right)^m}\right)\left(1-\frac{2F_A}{(2-F_A)+F_A}\right)} \end{pmatrix}$$

$$= \begin{pmatrix} \left[\frac{\frac{4\left(T_A^L\right)^{m+1}}{\left((2-T_A^L)^m+\left(T_A^L\right)^m\right)\left((2-T_A^L)+T_A^L\right)}}{1+\left(\frac{(2-T_A^L)^m+\left(T_A^L\right)^m-2\left(T_A^L\right)^m}{(2-T_A^L)^m+\left(T_A^L\right)^m}\right)\left(\frac{(2-T_A^L)+T_A^L-2T_A^L}{(2-T_A^L)+T_A^L}\right)}, \frac{\frac{4\left(T_A^U\right)^{m+1}}{\left((2-T_A^U)^m+\left(T_A^U\right)^m\right)\left((2-T_A^U)+T_A^U\right)}}{1+\left(\frac{(2-T_A^U)^m+\left(T_A^U\right)^m-2\left(T_A^U\right)^m}{(2-T_A^U)^m+\left(T_A^U\right)^m}\right)\left(\frac{(2-T_A^U)+T_A^U-2T_A^U}{(2-T_A^U)+T_A^U}\right)}\right], \\ \left[\frac{\frac{4\left(I_A^L\right)^{m+1}}{\left((2-I_A^L)^m+\left(I_A^L\right)^m\right)\left((2-I_A^L)+I_A^L\right)}}{1+\left(\frac{(2-I_A^L)^m+\left(I_A^L\right)^m-2\left(I_A^L\right)^m}{(2-I_A^L)^m+\left(I_A^L\right)^m}\right)\left(\frac{(2-I_A^L)+I_A^L-2I_A^L}{(2-I_A^L)+I_A^L}\right)}, \frac{\frac{4\left(I_A^U\right)^{m+1}}{\left((2-I_A^U)^m+\left(I_A^U\right)^m\right)\left((2-I_A^U)+I_A^U\right)}}{1+\left(\frac{(2-I_A^U)^m+\left(I_A^U\right)^m-2\left(I_A^U\right)^m}{(2-I_A^U)^m+\left(I_A^U\right)^m}\right)\left(\frac{(2-I_A^U)+I_A^U-2I_A^U}{(2-I_A^U)+I_A^U}\right)}\right], \\ \left[\begin{array}{c} \frac{\frac{\left(\left(1+F_A^L\right)^m-\left(1-F_A^L\right)^m\right)\left(\left(1+F_A^L\right)+\left(1-F_A^L\right)\right)+\left(\left(1+F_A^L\right)^m+\left(1-F_A^L\right)^m\right)\left((1+F_A^L)-(1-F_A^L)\right)}{\left((1+F_A^L)^m+(1-F_A^L)^m\right)\left((1+F_A^L)+(1-F_A^L)\right)}}{\frac{\left((1+F_A^L)^m+(1-F_A^L)^m\right)\left(\left(1+F_A^L\right)+\left(1-F_A^L\right)\right)+\left(1+F_A^L\right)^{m+1}-\left(1+F_A^L\right)^m\left(1-F_A^L\right)-\left(1-F_A^L\right)^m\left(1+F_A^L\right)+\left(1-F_A^L\right)^{m+1}}{\left((1+F_A^L)^m+(1-F_A^L)^m\right)\left((1+F_A^L)+(1-F_A^L)\right)}}, \\ \frac{\frac{\left(\left(1+F_A^U\right)^m-\left(1-F_A^U\right)^m\right)\left(\left(1+F_A^U\right)+\left(1-F_A^U\right)\right)+\left(\left(1+F_A^U\right)^m+\left(1-F_A^U\right)^m\right)\left((1+F_A^U)-(1-F_A^U)\right)}{\left((1+F_A^U)^m+(1-F_A^U)^m\right)\left((1+F_A^U)+(1-F_A^U)\right)}}{\frac{\left((1+F_A^U)^m+(1-F_A^U)^m\right)\left(\left(1+F_A^U\right)+\left(1-F_A^U\right)\right)+\left(1+F_A^U\right)^{m+1}-\left(1+F_A^U\right)^m\left(1-F_A^U\right)-\left(1-F_A^U\right)^m\left(1+F_A^U\right)+\left(1-F_A^U\right)^{m+1}}{\left((1+F_A^U)^m+(1-F_A^U)^m\right)\left((1+F_A^U)+(1-F_A^U)\right)}} \end{array}\right], \\ \frac{\frac{\left((1+T_A)^m-(1-T_A)^m\right)\left((1+T_A)+(1-T_A)\right)+\left((1+T_A)^m+(1-T_A)^m\right)\left((1+T_A)-(1-T_A)\right)}{\left((1+T_A)^m+(1-T_A)^m\right)\left((1+T_A)+(1-T_A)\right)}}{\frac{\left((1+T_A)^m+(1-T_A)^m\right)\left((1+T_A)+(1-T_A)\right)+(1+T_A)^{m+1}-(1+T_A)^m(1-T_A)-(1-T_A)^m(1+T_A)+(1-T_A)^{m+1}}{\left((1+T_A)^m+(1-T_A)^m\right)\left((1+T_A)+(1-T_A)\right)}}, \\ \frac{\frac{\left((1+I_A)^m-(1-I_A)^m\right)\left((1+I_A)+(1-I_A)\right)+\left((1+I_A)^m+(1-I_A)^m\right)\left((1+I_A)-(1-I_A)\right)}{\left((1+I_A)^m+(1-I_A)^m\right)\left((1+I_A)+(1-I_A)\right)}}{\frac{\left((1+I_A)^m+(1-I_A)^m\right)\left((1+I_A)+(1-I_A)\right)+(1+I_A)^{m+1}-(1+I_A)^m(1-I_A)-(1-I_A)^m(1+I_A)+(1-I_A)^{m+1}}{\left((1+I_A)^m+(1-I_A)^m\right)\left((1+I_A)+(1-I_A)\right)}}, \\ \frac{\frac{4\left(F_A\right)^{m+1}}{\left((2-F_A)^m+\left(F_A\right)^m\right)\left((2-F_A)+F_A\right)}}{1+\left(\frac{(2-F_A)^m+\left(F_A\right)^m-2\left(F_A\right)^m}{(2-F_A)^m+\left(F_A\right)^m}\right)\left(\frac{(2-F_A)+F_A-2F_A}{(2-F_A)+F_A}\right)} \end{pmatrix}$$

$$
= \left(\begin{array}{c}
\left[\frac{\frac{4(T_A^L)^{m+1}}{((2-T_A^L)^m+(T_A^L)^m)((2-T_A^L)+T_A^L)}}{\frac{(((2-T_A^L)^m+(T_A^L)^m)((2-T_A^L)+T_A^L))+(((2-T_A^L)^m-(T_A^L)^m)((2-T_A^L)-T_A^L))}{((2-T_A^L)^m+(T_A^L)^m)((2-T_A^L)+T_A^L)}}, \frac{\frac{4(T_A^U)^{m+1}}{((2-T_A^U)^m+(T_A^U)^m)((2-T_A^U)+T_A^U)}}{\frac{(((2-T_A^U)^m+(T_A^U)^m)((2-T_A^U)+T_A^U))+(((2-T_A^U)^m-(T_A^U)^m)((2-T_A^U)-T_A^U))}{((2-T_A^U)^m+(T_A^U)^m)((2-T_A^U)+T_A^U)}} \right], \\
\left[\frac{\frac{4(I_A^L)^{m+1}}{((2-I_A^L)^m+(I_A^L)^m)((2-I_A^L)+I_A^L)}}{\frac{(((2-I_A^L)^m+(I_A^L)^m)((2-I_A^L)+I_A^L))+(((2-I_A^L)^m-(I_A^L)^m)((2-I_A^L)-I_A^L))}{((2-I_A^L)^m+(I_A^L)^m)((2-I_A^L)+I_A^L)}}, \frac{\frac{4(I_A^U)^{m+1}}{((2-I_A^U)^m+(I_A^U)^m)((2-I_A^U)+I_A^U)}}{\frac{(((2-I_A^U)^m+(I_A^U)^m)((2-I_A^U)+I_A^U))+(((2-I_A^U)^m-(I_A^U)^m)((2-I_A^U)-I_A^U))}{((2-I_A^U)^m+(I_A^U)^m)((2-I_A^U)+I_A^U)}} \right], \\
\left[\frac{(1+F_A^L)^{m+1}+(1+F_A^L)^m(1-F_A^L)-(1-F_A^L)^m(1+F_A^L)-(1-F_A^L)^{m+1}+(1+F_A^L)^{m+1}-(1+F_A^L)^m(1-F_A^L)+(1-F_A^L)^m(1+F_A^L)-(1-F_A^L)^{m+1}}{(1+F_A^L)^{m+1}+(1+F_A^L)^m(1-F_A^L)-(1-F_A^L)^m(1+F_A^L)+(1-F_A^L)^{m+1}+(1+F_A^L)^{m+1}-(1+F_A^L)^m(1-F_A^L)+(1-F_A^L)^m(1+F_A^L)+(1-F_A^L)^{m+1}}, \frac{(1+F_A^U)^{m+1}+(1+F_A^U)^m(1-F_A^U)-(1-F_A^U)^m(1+F_A^U)-(1-F_A^U)^{m+1}+(1+F_A^U)^{m+1}-(1+F_A^U)^m(1-F_A^U)+(1-F_A^U)^m(1+F_A^U)-(1-F_A^U)^{m+1}}{(1+F_A^U)^{m+1}+(1+F_A^U)^m(1-F_A^U)-(1-F_A^U)^m(1+F_A^U)+(1-F_A^U)^{m+1}+(1+F_A^U)^{m+1}-(1+F_A^U)^m(1-F_A^U)+(1-F_A^U)^m(1+F_A^U)+(1-F_A^U)^{m+1}} \right], \\
\frac{(1+T_A)^{m+1}+(1+T_A)^m(1-T_A)-(1-T_A)^m(1+T_A)-(1-T_A)^{m+1}+(1+T_A)^{m+1}-(1+T_A)^m(1-T_A)+(1-T_A)^m(1+T_A)-(1-T_A)^{m+1}}{(1+T_A)^{m+1}+(1+T_A)^m(1-T_A)-(1-T_A)^m(1+T_A)+(1-T_A)^{m+1}+(1+T_A)^{m+1}-(1+T_A)^m(1-T_A)+(1-T_A)^m(1+T_A)+(1-T_A)^{m+1}}, \\
\frac{(1+I_A)^{m+1}+(1+I_A)^m(1-I_A)-(1-I_A)^m(1+I_A)-(1-I_A)^{m+1}+(1+I_A)^{m+1}-(1+I_A)^m(1-I_A)+(1-I_A)^m(1+I_A)-(1-I_A)^{m+1}}{(1+I_A)^{m+1}+(1+I_A)^m(1-I_A)-(1-I_A)^m(1+I_A)+(1-I_A)^{m+1}+(1+I_A)^{m+1}-(1+I_A)^m(1-I_A)+(1-I_A)^m(1+I_A)+(1-I_A)^{m+1}}, \\
\frac{\frac{4(F_A)^{m+1}}{((2-F_A)^m+(F_A)^m)((2-F_A)+F_A)}}{\frac{(((2-F_A)^m+(F_A)^m)((2-F_A)+F_A))+(((2-F_A)^m-(F_A)^m)((2-F_A)-F_A))}{((2-F_A)^m+(F_A)^m)((2-F_A)+F_A)}}
\end{array} \right)
$$

$$
= \left(\begin{array}{c}
\left[\frac{4(T_A^L)^{m+1}}{(2-T_A^L)^{m+1}+T_A^L(2-T_A^L)^m+(T_A^L)^{m+1}+T_A^{Lm}(2-T_A^L)+((2-T_A^L)^{m+1}-T_A^L(2-T_A^L)^m+(T_A^L)^{m+1}-(T_A^L)^m(2-T_A^L))}, \frac{4(T_A^U)^{m+1}}{(2-T_A^U)^{m+1}+T_A^U(2-T_A^U)^m+(T_A^U)^{m+1}+T_A^{Um}(2-T_A^U)+((2-T_A^U)^{m+1}-T_A^U(2-T_A^U)^m+(T_A^U)^{m+1}-(T_A^U)^m(2-T_A^U))} \right], \\
\left[\frac{4(I_A^L)^{m+1}}{(2-I_A^L)^{m+1}+I_A^L(2-I_A^L)^m+(I_A^L)^{m+1}+I_A^{Lm}(2-I_A^L)+((2-I_A^L)^{m+1}-I_A^L(2-I_A^L)^m+(I_A^L)^{m+1}-(I_A^L)^m(2-I_A^L))}, \frac{4(I_A^U)^{m+1}}{(2-I_A^U)^{m+1}+I_A^U(2-I_A^U)^m+(I_A^U)^{m+1}+I_A^{Um}(2-I_A^U)+((2-I_A^U)^{m+1}-I_A^U(2-I_A^U)^m+(I_A^U)^{m+1}-(I_A^U)^m(2-I_A^U))} \right], \\
\left[\frac{2((1+F_A^L)^{m+1}-(1-F_A^L)^{m+1})}{2((1+F_A^L)^{m+1}+(1-F_A^L)^{m+1})}, \frac{2((1+F_A^U)^{m+1}-(1-F_A^U)^{m+1})}{2((1+F_A^U)^{m+1}+(1-F_A^U)^{m+1})} \right], \\
\frac{2((1+T_A)^{m+1}-(1-T_A)^{m+1})}{2((1+T_A)^{m+1}+(1-T_A)^{m+1})}, \\
\frac{2((1+I_A)^{m+1}-(1-I_A)^{m+1})}{2((1+I_A)^{m+1}+(1-I_A)^{m+1})}, \\
\frac{4(F_A)^{m+1}}{(2-F_A)^{m+1}+F_A(2-F_A)^m+(F_A)^{m+1}+F_A^m(2-F_A)+((2-F_A)^{m+1}-F_A(2-F_A)^m+(F_A)^{m+1}-(F_A)^m(2-F_A))}
\end{array} \right)
$$

$$= \begin{pmatrix} \left[\frac{4(T_A^L)^{m+1}}{2\left((2-T_A^L)^{m+1}+(T_A^L)^{m+1}\right)}, \frac{4(T_A^U)^{m+1}}{2\left((2-T_A^U)^{m+1}+(T_A^U)^{m+1}\right)} \right], \\ \left[\frac{4(I_A^L)^{m+1}}{2\left((2-I_A^L)^{m+1}+(I_A^L)^{m+1}\right)}, \frac{4(I_A^U)^{m+1}}{2\left((2-I_A^U)^{m+1}+(I_A^U)^{m+1}\right)} \right], \\ \left[\frac{\left((1+F_A^L)^{m+1}-(1-F_A^L)^{m+1}\right)}{\left((1+F_A^L)^{m+1}+(1-F_A^L)^{m+1}\right)}, \frac{\left((1+F_A^U)^{m+1}-(1-F_A^U)^{m+1}\right)}{\left((1+F_A^U)^{m+1}+(1-F_A^U)^{m+1}\right)} \right], \\ \frac{\left((1+T_A)^{m+1}-(1-T_A)^{m+1}\right)}{\left((1+T_A)^{m+1}+(1-T_A)^{m+1}\right)}, \\ \frac{\left((1+I_A)^{m+1}-(1-I_A)^{m+1}\right)}{\left((1+I_A)^{m+1}+(1-I_A)^{m+1}\right)}, \\ \frac{4(F_A)^{m+1}}{2\left((2-F_A)^{m+1}+(F_A)^{m+1}\right)} \end{pmatrix}$$

$$= \begin{pmatrix} \left[\frac{2(T_A^L)^{m+1}}{\left((2-T_A^L)^{m+1}+(T_A^L)^{m+1}\right)}, \frac{2(T_A^U)^{m+1}}{\left((2-T_A^U)^{m+1}+(T_A^U)^{m+1}\right)} \right], \\ \left[\frac{2(I_A^L)^{m+1}}{\left((2-I_A^L)^{m+1}+(I_A^L)^{m+1}\right)}, \frac{2(I_A^U)^{m+1}}{\left((2-I_A^U)^{m+1}+(I_A^U)^{m+1}\right)} \right], \\ \left[\frac{\left((1+F_A^L)^{m+1}-(1-F_A^L)^{m+1}\right)}{\left((1+F_A^L)^{m+1}+(1-F_A^L)^{m+1}\right)}, \frac{\left((1+F_A^U)^{m+1}-(1-F_A^U)^{m+1}\right)}{\left((1+F_A^U)^{m+1}+(1-F_A^U)^{m+1}\right)} \right], \\ \frac{\left((1+T_A)^{m+1}-(1-T_A)^{m+1}\right)}{\left((1+T_A)^{m+1}+(1-T_A)^{m+1}\right)}, \\ \frac{\left((1+I_A)^{m+1}-(1-I_A)^{m+1}\right)}{\left((1+I_A)^{m+1}+(1-I_A)^{m+1}\right)}, \\ \frac{2(F_A)^{m+1}}{\left((2-F_A)^{m+1}+(F_A)^{m+1}\right)} \end{pmatrix}$$

Which shows that $k = m + 1$ is a neutrosophic cubic value. □

3.3. Score and Accuracy Function of Neutrosophic Cubic Set

For the comparison of two neutrosophic values, the score and accuracy function are defined. The score function is used tocompare two neutrosophic cubic values; sometimes the score of two neutrosophic cubic values becomes equal, although they have different components of truth, indeterminancy and falsity functions. This situation can be overcome by the help of an accuracy function. The following definition, along with examples, providesa better view of understanding to the reader.

Definition 18. *Let* $N = \left(\widetilde{T}_N, \widetilde{I}_N, \widetilde{F}_N, T_N, I_N, F_N\right)$, *where* $\widetilde{T}_N = [T_N^L, T_N^U], \widetilde{I}_N = [I_N^L, I_N^U], \widetilde{F}_N = [F_N^L, F_N^U]$, *be a neutrosophic cubic value and we define the score function as*

$$S(N) = \left[T_N^L - F_N^L + T_N^U - F_N^U + T_N - F_N\right]$$

Sometimes the situation arises that the score of two neutrosophic cubic values are equal. In such a situation, a comparison is made on the basis of an accuracy function.

Definition 19. *Let* $N = \left(\widetilde{T}_N, \widetilde{I}_N, \widetilde{F}_N, T_N, I_N, F_N\right)$, *where* $\widetilde{T}_N = [T_N^L, T_N^U], \widetilde{I}_N = [I_N^L, I_N^U], \widetilde{F}_N = [F_N^L, F_N^U]$, *be a neutrosophic cubic value, the accuracy function is defined as*

$$H(u) = \frac{1}{9}\left\{T_N^L + I_N^L + F_N^L + T_N^U + I_N^U + F_N^U + T_N + I_N + F_N\right\}$$

The following definition is accomplished for the comparison relation of the neutrosophic cubic values.

Definition 20. *Let N_1 and N_2 be two neutrosophic cubic values, where S_{N_1} and S_{N_2} are scores and H_{N_1} and H_{N_2} are accuracy functions of N_1 and N_2, respectively.*

1. *If* $S_{N_1} > S_{N_2} \Rightarrow N_1 > N_2$
2. *If* $S_{N_1} = S_{N_2}$ *and* $H_{N_1} > H_{N_2} \Rightarrow N_1 > N_2$ $H_{N_1} = H_{N_2} \Rightarrow N_1 = N_2$

Example 1. *Let* $N_1 = ([0.5, 0.9][0.6, 0.9][0.1, 0.4], 0.3, 0.4, 0.4)$ *and* $N_2 = ([0.2, 0.8][0.5, 0.9][0.4, 0.8], 0.4, 0.45, 0.8)$ *be two neutrosophic sets.*

Then

$$S_{N_1} = 0.8, and\, S_{N_2} = -0.6$$

$$S_{N_1} > S_{N_2} \Rightarrow N_1 > N_2$$

In the following example the score funtions are equal, so accuracy functions are used to compare neutrosophic cubic values.

Example 2. *Let* $N_1 = ([0.4, 0.9][0.5, 0.8][0.1, 0.7], 0.4, 0.5, 0.8)$ *and* $N_2 = ([0.4, 0.6][0.5, 0.9][0.6, 0.7], 0.7, 0.5, 0.3)$ *be two neutrosophic sets.*

$$S_{N_1} = 0.1, S_{N_2} = 0.1$$

$$S_{N_1} = S_{N_2} \Rightarrow N_1 = N_2$$

$$H_{N_1} = 0.566,\ H_{N_2} = 0.577$$

$$H_{N_1} < H_{N_2} \Rightarrow N_1 < N_2$$

4. Neutrosophic Cubic Geometric and Einstein Geometric Aggregation Operators

In this section, we introduce the concept of neutrosophic cubic geometric aggregation operators and neutrosophic cubic Einstein geometric aggregation operators.

This section consists of two sub-sections: In Section 4.1, the neutrosophic cubic geometric aggregation operators are defined on the basis of Section 3.1; and in Section 4.2, the neutrosophic cubic Einstein geometric aggregation operators are defined on the basis of Section 3.2.

4.1. Neutrosophic Cubic Weighted Geometric Aggregation Operator

We define neutrosophic cubic geometric aggregation operators using Section 3.1.

Definition 21. *We define the neutrosophic cubic weighted geometric operator(NCWG) as*

$$NCWG : R^m \to R \text{ defined by } NCWG_w(N_1, N_2, \ldots, N_m) = \bigotimes_{j=1}^{m} N_j^{w_j}$$

where the weight $W = (w_1, w_2, \ldots, w_m)^T$ of corresponding neutrosophic cubic values is such that each $w_j \in [0, 1]$ and $\sum_{j=1}^{m} w_j = 1$.

In NCWG, the neutrosophic cubic values are first weighted then aggregated.

Definition 22. *We define the neutrosophic cubic ordered weighted geometric operator(NCOWG) as*

$$NCOWG : R^m \to R \text{ defined by } NCOWG_w(N_1, N_2, \dots, N_m) = \overset{m}{\underset{j=1}{\otimes}} N_{(\gamma)_j}^{w_j}$$

where $N_{(\gamma)_j}$ *are descending ordered neutrosophic cubic values, and the weight* $W = (w_1, w_2, \dots, w_m)^T$ *of corresponding neutrosophic cubic values* $N_j (j = 1, 2, 3, \dots, m)$ *is such that each* $w_j \in [0, 1]$ *and* $\sum_{j=1}^{m} w_j = 1$.

In NCOWG, the neutrosophic cubic values are first arranged in decending order, weighted and then aggregated.

Theorem 3. *Let* $N_j = \left(\widetilde{T}_{N_j}, \widetilde{I}_{N_j}, \widetilde{F}_{N_j}, T_{N_j}, I_{N_j}, F_{N_j}\right)$, *where* $\widetilde{T}_{N_j} = \left[T_{N_j}^L, T_{N_j}^U\right], \widetilde{I}_{N_j} = \left[I_{N_j}^L, I_{N_j}^U\right], \widetilde{F}_{N_j} = \left[F_{N_j}^L, F_{N_j}^U\right] (j = 1, 2, \dots, n)$ *are a collection of neutrosophic cubic values, then neutrosophic cubic weighted geometric(NCWG) operator of* N_j *is also a neutrosophic cubic value and*

$$NCWG(N_j) = \begin{pmatrix} \left[\prod_{j=1}^{m}\left(T_{N_j}^L\right)^{w_j}, \prod_{j=1}^{m}\left(T_{N_j}^U\right)^{w_j}\right], \\ \left[\prod_{j=1}^{m}\left(I_{N_j}^L\right)^{w_j}, \prod_{j=1}^{m}\left(I_{N_j}^U\right)^{w_j}\right] \\ \left[1 - \prod_{j=1}^{m}(1 - F_{N_j}^L)^{w_j}, 1 - \prod_{j=1}^{m}(1 - F_{N_j}^U)^{w_j}\right] \\ 1 - \prod_{j=1}^{m}\left(1 - \left(T_{N_j}\right)\right)^{w_j}, 1 - \prod_{j=1}^{m}\left(1 - \left(I_{N_j}\right)\right)^{w_j}, \prod_{j=1}^{m}\left(F_{N_j}\right)^{w_j} \end{pmatrix}$$

where the weight $W = (w_1, w_2, \dots, w_m)^T$ *of* $N_j (j = 1, 2, 3, \dots, m)$ *such that* $w_j \in [0, 1]$ *and* $\sum_{j=1}^{m} w_j = 1$.

Proof. By mathematical induction for $m = 2$, using

$$\overset{2}{\underset{j=1}{\otimes}} N_j^{w_j} = N_1^{w_1} \otimes N_2^{w_2}$$

$$= \begin{pmatrix} \left[(T_{N_j}^L)^{w_1}, (T_{N_j}^U)^{w_1}\right], \\ \left[(I_{N_j}^L)^{w_1}, (I_{N_j}^U)^{w_1}\right], \\ \left[1 - \left(1 - F_{N_j}^L\right)^{w_1}, 1 - \left(1 - F_{N_j}^U\right)^{w_1}\right], \\ 1 - \left(1 - \left(T_{N_j}\right)\right)^{w_1}, 1 - \left(1 - \left(I_{N_j}\right)\right)^{w_1}, \left(F_{N_j}\right)^{w_1} \end{pmatrix} \otimes \begin{pmatrix} \left[(T_{N_j}^L)^{w_2}, (T_{N_j}^U)^{w_2}\right], \\ \left[(I_{N_j}^L)^{w_2}, (I_{N_j}^U)^{w_2}\right], \\ \left[1 - \left(1 - F_{N_j}^L\right)^{w_2}, 1 - \left(1 - F_{N_j}^U\right)^{w_2}\right], \\ 1 - \left(1 - \left(T_{N_j}\right)\right)^{w_2}, 1 - \left(1 - \left(I_{N_j}\right)\right)^{w_2}, (F_{N_j})^{w_2} \end{pmatrix}$$

$$= \begin{pmatrix} \left[\prod_{j=1}^{2} (T_{N_j}^L)^{w_j}, \prod_{j=1}^{2} (T_{N_j}^U)^{w_j}\right], \\ \left[\prod_{j=1}^{2} (I_{N_j}^L)^{w_j}, \prod_{j=1}^{2} (I_{N_j}^U)^{w_j}\right], \\ \left[1 - \prod_{j=1}^{2}\left(1 - F_{N_j}^L\right)^{w_j}, \prod_{j=1}^{2}\left(F_{N_j}^U\right)^{w_j}\right], \\ 1 - \prod_{j=1}^{2}\left(1 - T_{N_j}\right)^{w_j}, 1 - \prod_{j=1}^{2}\left(1 - I_{N_j}\right)^{w_j}, \prod_{j=1}^{2}\left(F_{N_j}\right)^{w_j} \end{pmatrix}$$

For $m = n$, we have

$$\underset{j=1}{\overset{n}{\otimes}} N_j^{w_j} = \begin{pmatrix} \left[\prod_{j=1}^{n} (T_{N_j}^L)^{w_j}, \prod_{j=1}^{n} (T_{N_j}^U)^{w_j}\right], \\ \left[\prod_{j=1}^{n} (I_{N_j}^L)^{w_j}, \prod_{j=1}^{n} (I_{N_j}^U)^{w_j}\right], \\ \left[1 - \prod_{j=1}^{n} \left(1 - F_{N_j}^L\right)^{w_j}, \prod_{j=1}^{n} \left(F_{N_j}^U\right)^{w_j}\right], \\ 1 - \prod_{j=1}^{n} \left(1 - T_{N_j}\right)^{w_j}, 1 - \prod_{j=1}^{n} \left(1 - I_{N_j}\right)^{w_j}, \prod_{j=1}^{n} \left(F_{N_j}\right)^{w_j} \end{pmatrix}$$

We prove the result holds for $m = n + 1$,

$$N_{n+1}^{w_{n+1}} = \begin{pmatrix} \left[(T_{N_{n+1}}^L)^{w_{n+1}}, (T_{N_{n+1}}^U)^{w_{n+1}}\right], \\ \left[(I_{N_{j+1}}^L)^{w_{j+1}}, (I_{N_{j+1}}^U)^{w_{j+1}}\right], \\ \left[1 - \left(1 - F_{N_{n+1}}^L\right)^{w_{n+1}}, 1 - \left(1 - F_{N_{n+1}}^U\right)^{w_{n+1}}\right], \\ 1 - (1 - T_{N_{n+1}})^{w_{n+1}}, 1 - (1 - I_{N_{n+1}})^{w_{n+1}}, (F_{N_{n+1}})^{w_{n+1}} \end{pmatrix}$$

$$\underset{j=1}{\overset{n}{\otimes}} N_j^{w_j} \oplus N_{n+1}^{w_{n+1}}$$

$$= \begin{pmatrix} \left[\prod_{j=1}^{n} (T_{N_j}^L)^{w_j}, \prod_{j=1}^{n} (T_{N_j}^U)^{w_j}\right], \\ \left[\prod_{j=1}^{n} (I_{N_j}^L)^{w_j}, \prod_{j=1}^{n} (I_{N_j}^U)^{w_j}\right], \\ \left[1 - \prod_{j=1}^{n} \left(1 - F_{N_j}^L\right)^{w_j}, \prod_{j=1}^{n} \left(F_{N_j}^U\right)^{w_j}\right], \\ 1 - \prod_{j=1}^{n} \left(1 - T_{N_j}\right)^{w_j}, 1 - \prod_{j=1}^{n} \left(1 - I_{N_j}\right)^{w_j}, \prod_{j=1}^{n} \left(F_{N_j}\right)^{w_j} \end{pmatrix} \oplus \begin{pmatrix} \left[(T_{N_{n+1}}^L)^{w_{n+1}}, (T_{N_{n+1}}^U)^{w_{n+1}}\right], \\ \left[(I_{N_{j+1}}^L)^{w_{j+1}}, (I_{N_{j+1}}^U)^{w_{j+1}}\right], \\ \left[1 - \left(1 - F_{N_{n+1}}^L\right)^{w_{n+1}}, 1 - \left(1 - F_{N_{n+1}}^U\right)^{w_{n+1}}\right], \\ 1 - (1 - T_{N_{n+1}})^{w_{n+1}}, 1 - (1 - I_{N_{n+1}})^{w_{n+1}}, (F_{N_{n+1}})^{w_{n+1}} \end{pmatrix}$$

$$\underset{j=1}{\overset{n+1}{\otimes}} N_j^{w_j} = \begin{pmatrix} \left[\prod_{j=1}^{n} \left(T_{N_j}^L\right)^{w_j} \left(T_{N_{m+1}}^L\right)^{w_{m+1}}, \prod_{j=1}^{n} \left(T_{N_j}^U\right)^{w_j} \left(T_{N_{m+1}}^U\right)^{w_{m+1}}\right], \\ \left[\prod_{j=1}^{n} \left(I_{N_j}^L\right)^{w_j} \left(I_{N_{m+1}}^L\right)^{w_{m+1}}, \prod_{j=1}^{n} \left(I_{N_j}^U\right)^{w_j} \left(I_{N_{m+1}}^U\right)^{w_{m+1}}\right], \\ \left[\begin{array}{c} 1 - \prod_{j=1}^{n} (1 - F_{N_j}^L)^{w_j} + 1 - (1 - F_{N_{m+1}}^L)^{w_{m+1}} - \\ \left(1 - \prod_{j=1}^{n} (1 - F_{N_j}^L)^{w_j}\right)\left(1 - (1 - F_{N_{m+1}}^L)^{w_{m+1}}\right), \\ 1 - \prod_{j=1}^{n} (1 - F_{N_j}^U)^{w_j} + 1 - (1 - F_{N_{m+1}}^U)^{w_{m+1}} - \\ \left(1 - \prod_{j=1}^{n} (1 - F_{N_j}^U)^{w_j}\right)\left(1 - (1 - F_{N_{m+1}}^U)^{w_{m+1}}\right), \end{array}\right], \\ 1 - \prod_{j=1}^{n} \left(1 - T_{N_j}\right)^{w_j} + 1 - (1 - T_{N_{m+1}})^{w_{m+1}} - \\ \left(1 - \prod_{j=1}^{n} \left(1 - T_{N_j}\right)^{w_j}\right)\left(1 - (1 - T_{N_{m+1}})^{w_{m+1}}\right) \\ 1 - \prod_{j=1}^{n} \left(1 - I_{N_j}\right)^{w_j} + 1 - (1 - I_{N_{m+1}})^{w_{m+1}} - \\ \left(1 - \prod_{j=1}^{n} \left(1 - I_{N_j}\right)^{w_j}\right)\left(1 - (1 - I_{N_{m+1}})^{w_{m+1}}\right) \\ \prod_{j=1}^{n} \left(F_{N_j}\right)^{w_j} (F_{N_{m+1}})^{w_{m+1}} \end{pmatrix}$$

$$= \left(\begin{array}{c} \left[\prod_{j=1}^{n+1} \left(T_{N_j}^L\right)^{w_j}, \prod_{j=1}^{n+1} \left(T_{N_j}^U\right)^{w_j} \right] \\ \left[\prod_{j=1}^{n+1} \left(I_{N_j}^L\right)^{w_j}, \prod_{j=1}^{n+1} \left(I_{N_j}^U\right)^{w_j} \right], \\ \left[\begin{array}{c} 2 - \prod_{j=1}^{n+1} (1 - F_{N_j}^L)^{w_j} - 1 + \prod_{j=1}^{n} (1 - F_{N_j}^L)^{w_j} + (1 - F_{N_{m+1}}^L)^{w_{m+1}} \\ - \left(\prod_{j=1}^{n} (1 - F_{N_j}^L)^{w_j} \right) (1 - F_{N_{m+1}}^L)^{w_{m+1}}, \\ 2 - \prod_{j=1}^{n+1} (1 - F_{N_j}^U)^{w_j} - 1 + \prod_{j=1}^{n+1} (1 - F_{N_j}^U)^{w_j} + (1 - F_{N_{m+1}}^U)^{w_{m+1}} \\ - \left(\prod_{j=1}^{n} (1 - F_{N_j}^U)^{w_j} \right) (1 - F_{N_{m+1}}^U)^{w_{m+1}} \end{array} \right], \\ 2 - \prod_{j=1}^{n+1} (1 - T_{N_j})^{w_j} - 1 + \prod_{j=1}^{n} (1 - T_{N_j})^{w_j} + (1 - T_{N_{m+1}})^{w_{m+1}} \\ - \left(\prod_{j=1}^{n} (1 - T_{N_j})^{w_j} \right) (1 - T_{N_{m+1}})^{w_{m+1}}, \\ 2 - \prod_{j=1}^{n+1} (1 - I_{N_j})^{w_j} - 1 + \prod_{j=1}^{n} (1 - I_{N_j})^{w_j} + (1 - I_{N_{m+1}})^{w_{m+1}} \\ - \left(\prod_{j=1}^{n} (1 - I_{N_j})^{w_j} \right) (1 - I_{N_{m+1}})^{w_{m+1}} \\ , \prod_{j=1}^{n+1} \left(F_{N_j}\right)^{w_j} \end{array} \right)$$

$$= \left(\begin{array}{c} \left[\prod_{J=1}^{n+1} \left(T_{N_j}^L\right)^{w_j}, \prod_{J=1}^{n+1} \left(T_{N_j}^U\right)^{w_j} \right] \\ \left[\prod_{J=1}^{n+1} \left(I_{N_j}^L\right)^{w_j}, \prod_{J=1}^{n+1} \left(I_{N_j}^U\right)^{w_j} \right], \\ \left[1 - \prod_{J=1}^{n+1} \left(1 - F_{N_j}^L\right)^{w_j}, 1 - \prod_{J=1}^{n+1} \left(1 - F_{N_j}^U\right)^{w_j} \right], \\ , 1 - \prod_{J=1}^{n+1} \left(1 - T_{N_j}\right)^{w_j}, 1 - \prod_{J=1}^{n+1} \left(1 - I_{N_j}\right)^{w_j}, \\ \prod_{J=1}^{n+1} \left(1 - F_{N_j}\right)^{w_j} \end{array} \right)$$

□

Theorem 4. *Let* $N_j = \left(\widetilde{T}_{N_j}, \widetilde{I}_{N_j}, \widetilde{F}_{N_j}, T_{N_j}, I_{N_j}, F_{N_j}\right)$, *where* $\widetilde{T}_{N_j} = \left[T_{N_j}^L, T_{N_j}^U\right], \widetilde{I}_{N_j} = \left[I_{N_j}^L, I_{N_j}^U\right], \widetilde{F}_{N_j} = \left[F_{N_j}^L, F_{N_j}^U\right]$, $(j = 1, 2, \ldots, m)$ *is a collection of neutrosophic cubic values The weight* $W = (w_1, w_2, \ldots, w_m)^T$ *of* $N_j (j = 1, 2, 3, \ldots, m)$, *be such that* $w_j \in [0, 1]$ *and* $\sum_{j=1}^{m} w_j = 1$.

1. ***Idempotency:*** *If for all* $N_j = \left(\widetilde{T}_{N_j}, \widetilde{I}_{N_j}, \widetilde{F}_{N_j}, T_{N_j}, I_{N_j}, F_{N_j}\right)$, *where* $\widetilde{T}_{N_j} = \left[T_{N_j}^L, T_{N_j}^U\right]$, $\widetilde{I}_{N_j} = \left[I_{N_j}^L, I_{N_j}^U\right], \widetilde{F}_{N_j} = \left[F_{N_j}^L, F_{N_j}^U\right]$, $(j = 1, 2, \ldots, m)$ *are equal, that is,* $N_j = N$ *for all k, then NCW* $G_w(N_1, N_2, \ldots, N_m) = N$
2. ***Monotonicity:*** *Let* $B_j = \left(\widetilde{T}_{B_j}, \widetilde{I}_{B_j}, \widetilde{F}_{B_j}, T_{B_j}, I_{B_j}, F_{B_j}\right)$ *where* $\widetilde{T}_{B_j} = \left[T_{B_j}^L, T_{B_j}^U\right]$, $\widetilde{I}_{B_j} = \left[I_{B_j}^L, I_{B_j}^U\right], \widetilde{F}_{B_j} = \left[F_{B_j}^L, F_{B_j}^U\right]$ $(j = 1, 2, \ldots, m)$ *is the collection of neutrosophic cubic values. If* $S_{B_j}(u) \geq S_{N_j}(u)$ *and* $B_j(u) \geq N_j(u)$ *then* $NCWG_w(N_1, N_2, \ldots, N_m) \leq NCWG_w(B_1, B_2, \ldots, B_m)$.

3. ***Boundary:*** $N^- \leq NCWG_w\{(N_1)_T, (N_2)_T, \ldots, (N_m)_T\} \leq N^+$, *where*

$$N^- = \left\{\min_j T^L_{N_j}, \min_j I^L_{N_j}, 1 - \max_j F^L_{N_j}, \min_j T_{N_j}, \min_j I_{N_j}, 1 - \max_j F^L_{N_j}, \min_j T_{N_j}, \min_j I_{N_j}, 1 - \max_j F^L_{N_j}\right\},$$

$$N^+ = \left\{\max_j T^U_{N_j}, \max_j I^U_{N_j}, 1 - \min_j F^U_{N_j}, \max_j T_{N_j}, \max_j I_{N_j}, 1 - \min_j F_{N_j}, \max_j T_{N_j}, \max_j I_{N_j}, 1 - \min_j F_{N_j}\right\}$$

Proof.

1. **Idempotent**: Since $N_j = N$, so

$$NCWG(N_j) = \begin{pmatrix} \left[\prod_{j=1}^m (T^L_N)^{w_j}, \prod_{j=1}^m (T^U_N)^{w_j}\right], \\ \left[\prod_{j=1}^m (I^L_N)^{w_j}, \prod_{j=1}^m (I^U_N)^{w_j}\right], \\ \left[1 - \prod_{j=1}^m (1 - F^L_N)^{w_j}, 1 - \prod_{j=1}^m (1 - F^L_N)^{w_j}\right], \\ 1 - \prod_{j=1}^m (1 - T_N)^{w_j}, 1 - \prod_{j=1}^m (1 - I_N)^{w_j}, \prod_{j=1}^m (F_N)^{w_j} \end{pmatrix}$$

$$= \begin{pmatrix} \left[(T^L_N)^{\sum_{j=1}^m w_j}, (T^U_N)^{\sum_{j=1}^m w_j}\right], \\ \left[(I^L_N)^{\sum_{j=1}^m w_j}, (I^U_N)^{\sum_{j=1}^m w_j}\right], \\ \left[1 - (1 - F^L_N)^{\sum_{j=1}^m w_j}, 1 - (1 - F^U_N)^{\sum_{j=1}^m w_j}\right], \\ 1 - (1 - T_N)^{\sum_{j=1}^m w_j}, 1 - (1 - I_N)^{\sum_{j=1}^m w_j}, (F_N)^{\sum_{j=1}^m w_j} \end{pmatrix}$$

$$= \left(\widetilde{T}_N, \widetilde{I}_N, \widetilde{F}_N, T_N, I_N, F_N\right)$$

2. **Monotonicity:** Since NCOWG is strictly monotone function.
3. **Boundary:** Let $u = \min N^-$ and $y = \max N^+$, then by monotonicity we have $u \leq NCOWA(N_j) \leq y \Rightarrow N^- \leq NCOWG(N_j) \leq N^+$.

□

Theorem 5. *Let* $N_j = \left(\widetilde{T}_{N_j}, \widetilde{I}_{N_j}, \widetilde{F}_{N_j}, T_{N_j}, I_{N_j}, F_{N_j}\right)$, *where* $\widetilde{T}_{N_j} = \left[T^L_{N_j}, T^U_{N_j}\right], \widetilde{I}_{N_j} = \left[I^L_{N_j}, I^U_{N_j}\right], \widetilde{F}_{N_j} = \left[F^L_{N_j}, F^U_{N_j}\right]$, $(j = 1, 2, \ldots, n)$ *be the collection of neutrosophic cubic values and* $W = (w_1, w_2, \ldots, w_n)^T$ *is the weight of the NCOWG, with* $w_j \in [0,1]$ *and* $\sum_{j=1}^m w_j = 1$.

1. *If* $W = (1, 0, \ldots, 0)^T$, *then* $NCOWG(N_1, N_2, \ldots, N_n) = \max N_j$
2. *If* $W = (0, 0, \ldots, 1)^T$, *then* $NCOWG(N_1, N_2, \ldots, N_n) = \min N_j$
3. *If* $w_j = 1$, $w_l = 0$, *and* $j \neq l$, *then* $NCOWG(N_1, N_2, \ldots, N_n) = N_j$

where N_j *is the jth largest of* $(N_1, N_2, \ldots, N_n)$.

Proof. Since in NCOWG the neutrosophic values are ordered in descending order. □

4.2. Neutrosophic Cubic Einstein Weighted Geometric Aggregation Operator

We define neutrosophic cubic Einstein geometric aggregation operators using Section 3.2.

Definition 23. *The neutrosophic cubic Einstein weighted geometric operator(NCEWA) is defined as* $NCEWG\colon R^m \to R$*, defined by* $NCEWG_w(N_1, N_2, \ldots, N_m) = \overset{m}{\underset{j=1}{\otimes}} \left(N_j^E\right)^{w_j}$

where, $W = (w_1, w_2, \ldots, w_m)^T$ *is the weight of* $N_j (j = 1, 2, 3, \ldots, m)$*, such that* $w_j \in [0, 1]$ *and* $\sum_{j=1}^{m} w_j = 1$.

That is, first all the neutrosophic values are weighted then aggregated using Einstein operations.

Definition 24. *Order neutrosophic cubic Einstein weighted geometric operator(NCEOWG) is defined as* $NCEOWG : R^m \to R$ *by* $NCEOWG_w(N_1, N_2, \ldots, N_m) = \overset{m}{\underset{j=1}{\otimes}} \left(B_j^E\right)^{w_j}$

where B_j *is the jth largest,* $W = (w_1, w_2, \ldots, w_m)^T$ *is the weight of* $N_j (j = 1, 2, 3, \ldots, m)$*, such that* $w_j \in [0, 1]$ *and* $\sum_{j=1}^{m} w_j = 1$.

That is, first all the neutrosophic values are ordered and then weighted, after ordering weighted values are aggregated using Einstein operations.The fundamental concept of ordered weighted operators is to rearrange the neutrosophic cubic values in descending order.

Theorem 6. *Let* $N_j = \left(\widetilde{T}_{N_j}, \widetilde{I}_{N_j}, \widetilde{F}_{N_j}, T_{N_j}, I_{N_j}, F_{N_j}\right)$*, where* $\widetilde{T}_{N_j} = \left[T^L_{N_j}, T^U_{N_j}\right], \widetilde{I}_{N_j} = \left[I^L_{N_j}, I^U_{N_j}\right]$*,* $\widetilde{F}_{N_j} = \left[F^L_{N_j}, F^U_{N_j}\right]$*,* $(j = 1, 2, \ldots, m)$ *is a collection of neutrosophic cubic values, then their Einstein weighted geometric aggregated value by NCEWG operator is also a neutrosophic cubic value, and*

$$NCEWG(N_j) = \left(\begin{array}{c} \left[\frac{2\prod_{j=1}^{m}\left(T^L_{N_j}\right)^{w_j}}{\prod_{j=1}^{m}\left(2-T^L_{N_j}\right)^{w_j}+\prod_{j=1}^{m}\left(T^L_{N_j}\right)^{w_j}}, \frac{2\prod_{j=1}^{m}\left(T^U_{N_j}\right)^{w_j}}{\prod_{j=1}^{m}\left(2-T^U_{N_j}\right)^{w_j}+\prod_{j=1}^{m}\left(T^U_{N_j}\right)^{w_j}} \right], \\ \left[\frac{2\prod_{j=1}^{m}\left(I^L_{N_j}\right)^{w_j}}{\prod_{j=1}^{m}\left(2-I^L_{N_j}\right)^{w_j}+\prod_{j=1}^{m}\left(I^L_{N_j}\right)^{w_j}}, \frac{2\prod_{j=1}^{m}\left(I^U_{N_j}\right)^{w_j}}{\prod_{j=1}^{m}\left(2-I^U_{N_j}\right)^{w_j}+\prod_{j=1}^{m}\left(I^U_{N_j}\right)^{w_j}} \right], \\ \left[\frac{\prod_{j=1}^{m}\left(1+F^L_{N_j}\right)^{w_j}-\prod_{j=1}^{m}\left(1-F^L_{N_j}\right)^{w_j}}{\prod_{j=1}^{m}\left(1+F^L_{N_j}\right)^{w_j}+\prod_{j=1}^{m}\left(1-F^L_{N_j}\right)^{w_j}}, \frac{\prod_{j=1}^{m}\left(1+F^U_{N_j}\right)^{w_j}-\prod_{j=1}^{m}\left(1-F^U_{N_j}\right)^{w_j}}{\prod_{j=1}^{m}\left(1+F^U_{N_j}\right)^{w_j}+\prod_{j=1}^{m}\left(1-F^U_{N_j}\right)^{w_j}} \right], \\ \frac{\prod_{j=1}^{m}\left(1+T_{N_j}\right)^{w_j}-\prod_{j=1}^{m}\left(1-T_{N_j}\right)^{w_j}}{\prod_{j=1}^{m}\left(1+T_{N_j}\right)^{w_j}+\prod_{j=1}^{m}\left(1-T_{N_j}\right)^{w_j}}, \\ \frac{\prod_{j=1}^{m}\left(1+I_{N_j}\right)^{w_j}-\prod_{j=1}^{m}\left(1-I_{N_j}\right)^{w_j}}{\prod_{j=1}^{m}\left(1+I_{N_j}\right)^{w_j}+\prod_{j=1}^{m}\left(1-I_{N_j}\right)^{w_j}}, \\ \frac{2\prod_{j=1}^{m}\left(F_{N_j}\right)^{w_j}}{\prod_{j=1}^{m}\left(2-F_{N_j}\right)^{w_j}+\prod_{j=1}^{m}\left(F_{N_j}\right)^{w_j}} \end{array} \right)$$

where $W = (w_1, w_2, \ldots, w_m)^T$ *is the weight vector of* $N_j (j = 1, 2, 3, \ldots, m)$*, such that* $w_j \in [0, 1]$ *and* $\sum_{j=1}^{m} w_j = 1$.

Proof. We use mathematical induction to prove this result, for $m = 2$, using definition (Einstein sum and Einstein scalar multiplication).

$$\left(N_1^E\right)^{w_1} = \left(\begin{array}{c} \left[\frac{2\left(T_{N_1}^L\right)^{w_1}}{\left(2-T_{N_1}^L\right)^{w_1}+T_{N_1}^L}, \frac{2\left(T_{N_1}^U\right)^{w_1}}{\left(2-T_{N_1}^U\right)^{w_1}+T_{N_1}^U} \right], \\ \left[\frac{2\left(I_{N_1}^L\right)^{w_1}}{\left(2-I_{N_1}^L\right)^{w_1}+I_{N_1}^L}, \frac{2\left(I_{N_1}^U\right)^{w_1}}{\left(2-I_{N_1}^U\right)^{w_1}+I_{N_1}^U} \right], \\ \left[\frac{(1+F_{N_1}^L)^{w_1}-(1-F_{N_1}^L)^{w_1}}{(1+F_{N_1}^L)^{w_1}+(1-F_{N_1}^L)^{w_1}}, \frac{(1+F_{N_1}^U)^{w_1}-(1-F_{N_1}^U)^{w_1}}{(1+F_{N_1}^U)^{w_1}+(1-F_{N_1}^U)^{w_1}} \right], \\ \frac{(1+T_{N_1})^{w_1}-(1-T_{N_1})^{w_1}}{(1+T_{N_1})^{w_1}+(1-T_{N_1})^{w_1}}, \\ \frac{(1+I_{N_1})^{w_1}-(1-I_{N_1})^{w_1}}{(1+I_{N_1})^{w_1}+(1-I_{N_1})^{w_1}}, \\ \frac{2\left(F_{N_1}\right)^{w_1}}{\left(2-F_{N_1}\right)^{w_1}+F_{N_1}} \end{array} \right)$$

$$\left(N_2^E\right)^{w_2} = \left(\begin{array}{c} \left[\frac{2\left(T_{N_2}^L\right)^{w_2}}{\left(2-T_{N_2}^L\right)^{w_2}+T_{N_2}^L}, \frac{2\left(T_{N_2}^U\right)^{w_2}}{\left(2-T_{N_2}^U\right)^{w_2}+T_{N_2}^U} \right], \\ \left[\frac{2\left(I_{N_2}^L\right)^{w_2}}{\left(2-I_{N_2}^L\right)^{w_2}+I_{N_2}^L}, \frac{2\left(I_{N_2}^U\right)^{w_2}}{\left(2-I_{N_2}^U\right)^{w_2}+I_{N_2}^U} \right], \\ \left[\frac{(1+F_{N_2}^L)^{w_2}-(1-F_{N_2}^L)^{w_2}}{(1+F_{N_2}^L)^{w_2}+(1-F_{N_2}^L)^{w_2}}, \frac{(1+F_{N_2}^U)^{w_2}-(1-F_{N_2}^U)^{w_2}}{(1+F_{N_2}^U)^{w_2}+(1-F_{N_2}^U)^{w_2}} \right], \\ \frac{(1+T_{N_2})^{w_2}-(1-T_{N_2})^{w_2}}{(1+T_{N_2})^{w_2}+(1-T_{N_2})^{w_2}}, \frac{(1+I_{N_2})^{w_2}-(1-I_{N_2})^{w_2}}{(1+I_{N_2})^{w_2}+(1-I_{N_2})^{w_2}}, \\ \frac{2\left(F_{N_2}\right)^{w_2}}{\left(2-F_{N_2}\right)^{w_2}+F_{N_2}} \end{array} \right)$$

$$\overset{2}{\underset{j=1}{\otimes}} \left(N_j^E\right)^{w_j} = \left(\begin{array}{c} \left[\frac{2\prod_{j=1}^{2}\left(T_{N_j}^L\right)^{w_j}}{\prod_{j=1}^{2}\left(2-T_{N_j}^L\right)^{w_j}+\prod_{j=1}^{2}\left(T_{N_j}^L\right)^{w_j}}, \frac{2\prod_{j=1}^{2}\left(T_{N_j}^U\right)^{w_j}}{\prod_{j=1}^{2}\left(2-T_{N_j}^U\right)^{w_j}+\prod_{j=1}^{2}\left(T_{N_j}^U\right)^{w_j}} \right], \\ \left[\frac{2\prod_{j=1}^{2}\left(I_{N_j}^L\right)^{w_j}}{\prod_{j=1}^{2}\left(2-I_{N_j}^L\right)^{w_j}+\prod_{j=1}^{2}\left(I_{N_j}^L\right)^{w_j}}, \frac{2\prod_{j=1}^{2}\left(I_{N_j}^U\right)^{w_j}}{\prod_{j=1}^{2}\left(2-I_{N_j}^U\right)^{w_j}+\prod_{j=1}^{2}\left(I_{N_j}^U\right)^{w_j}} \right], \\ \left[\frac{\prod_{j=1}^{2}\left(1+F_{N_j}^L\right)^{w_j}-\prod_{j=1}^{2}\left(1-F_{N_j}^L\right)^{w_j}}{\prod_{j=1}^{2}\left(1+F_{N_j}^L\right)^{w_j}+\prod_{j=1}^{2}\left(1-F_{N_j}^L\right)^{w_j}}, \frac{\prod_{j=1}^{2}\left(1+F_{N_j}^U\right)^{w_j}-\prod_{j=1}^{2}\left(1-F_{N_j}^U\right)^{w_j}}{\prod_{j=1}^{2}\left(1+F_{N_j}^U\right)^{w_j}+\prod_{j=1}^{2}\left(1-F_{N_j}^U\right)^{w_j}} \right], \\ \frac{\prod_{j=1}^{2}\left(1+T_{N_j}\right)^{w_j}-\prod_{j=1}^{2}\left(1-T_{N_j}\right)^{w_j}}{\prod_{j=1}^{2}\left(1+T_{N_j}\right)^{w_j}+\prod_{j=1}^{2}\left(1-T_{N_j}\right)^{w_j}}, \frac{\prod_{j=1}^{2}\left(1+I_{N_j}\right)^{w_j}-\prod_{j=1}^{2}\left(1-I_{N_j}\right)^{w_j}}{\prod_{j=1}^{2}\left(1+I_{N_j}\right)^{w_j}+\prod_{j=1}^{2}\left(1-I_{N_j}\right)^{w_j}}, \\ \frac{2\prod_{j=1}^{2}\left(F_{N_j}\right)^{w_j}}{\prod_{j=1}^{2}\left(2-F_{N_j}\right)^{w_j}+\prod_{j=1}^{2}\left(F_{N_j}\right)^{w_j}} \end{array} \right)$$

for $m = n$

$$\overset{n}{\underset{j=1}{\otimes}}\left(N_j^E\right)^{w_j} = \left(\begin{array}{c} \left[\frac{2\prod_{j=1}^{n}\left(T_{N_j}^L\right)^{w_j}}{\prod_{j=1}^{n}\left(2-T_{N_j}^L\right)^{w_j}+\prod_{j=1}^{n}\left(T_{N_j}^L\right)^{w_j}}, \frac{2\prod_{j=1}^{n}\left(T_{N_j}^U\right)^{w_j}}{\prod_{j=1}^{n}\left(2-T_{N_j}^U\right)^{w_j}+\prod_{j=1}^{n}\left(T_{N_j}^U\right)^{w_j}}\right], \\ \left[\frac{2\prod_{j=1}^{n}\left(I_{N_j}^L\right)^{w_j}}{\prod_{j=1}^{n}\left(2-I_{N_j}^L\right)^{w_j}+\prod_{j=1}^{n}\left(I_{N_j}^L\right)^{w_j}}, \frac{2\prod_{j=1}^{n}\left(I_{N_j}^U\right)^{w_j}}{\prod_{j=1}^{n}\left(2-I_{N_j}^U\right)^{w_j}+\prod_{j=1}^{n}\left(I_{N_j}^U\right)^{w_j}}\right], \\ \left[\frac{\prod_{j=1}^{n}\left(1+F_{N_j}^L\right)^{w_j}-\prod_{j=1}^{n}\left(1-F_{N_j}^L\right)^{w_j}}{\prod_{j=1}^{n}\left(1+F_{N_j}^L\right)^{w_j}+\prod_{j=1}^{n}\left(1-F_{N_j}^L\right)^{w_j}}, \frac{\prod_{j=1}^{n}\left(1+F_{N_j}^U\right)^{w_j}-\prod_{j=1}^{n}\left(1-F_{N_j}^U\right)^{w_j}}{\prod_{j=1}^{n}\left(1+F_{N_j}^U\right)^{w_j}+\prod_{j=1}^{n}\left(1-F_{N_j}^U\right)^{w_j}}\right], \\ \frac{\prod_{j=1}^{n}\left(1+T_{N_j}\right)^{w_j}-\prod_{j=1}^{n}\left(1-T_{N_j}\right)^{w_j}}{\prod_{j=1}^{n}\left(1+T_{N_j}\right)^{w_j}+\prod_{j=1}^{n}\left(1-T_{N_j}\right)^{w_j}}, \frac{\prod_{j=1}^{n}\left(1+I_{N_j}\right)^{w_j}-\prod_{j=1}^{n}\left(1-I_{N_j}\right)^{w_j}}{\prod_{j=1}^{n}\left(1+I_{N_j}\right)^{w_j}+\prod_{j=1}^{n}\left(1-I_{N_j}\right)^{w_j}}, \\ \frac{2\prod_{j=1}^{n}\left(F_{N_j}\right)^{w_j}}{\prod_{j=1}^{n}\left(2-F_{N_j}\right)^{w_j}+\prod_{j=1}^{n}\left(F_{N_j}\right)^{w_j}} \end{array}\right)$$

We prove the result holds for $m = n + 1$

$$as\left(N_{n+1}^E\right)^{w_{n+1}} = \left(\begin{array}{c} \left[\frac{2\left(T_{N_{n+1}}^L\right)^{w_{n+1}}}{\left(2-T_{N_{n+1}}^L\right)^{w_{n+1}}+\left(T_{N_{n+1}}^L\right)^{w_{n+1}}}, \frac{2\left(T_{N_{n+1}}^U\right)^{w_{n+1}}}{\left(2-T_{N_{n+1}}^U\right)^{w_{n+1}}+\left(T_{N_{n+1}}^U\right)^{w_{n+1}}}\right], \\ \left[\frac{2\left(I_{N_{n+1}}^L\right)^{w_{n+1}}}{\left(2-I_{N_{n+1}}^L\right)^{w_{n+1}}+\left(I_{N_{n+1}}^L\right)^{w_{n+1}}}, \frac{2\left(I_{N_{n+1}}^U\right)^{w_{n+1}}}{\left(2-I_{N_{n+1}}^U\right)^{w_{n+1}}+\left(I_{N_{n+1}}^U\right)^{w_{n+1}}}\right], \\ \left[\frac{(1+F_{N_{n+1}}^L)^{w_{n+1}}-(1-F_{N_{n+1}}^L)^{w_{n+1}}}{(1+F_{N_{n+1}}^L)^{w_{n+1}}+(1-F_{N_{n+1}}^L)^{w_{n+1}}}, \frac{(1+F_{N_{n+1}}^U)^{w_{n+1}}-(1-F_{N_{n+1}}^U)^{w_{n+1}}}{(1+F_{N_{n+1}}^U)^{w_{n+1}}+(1-F_{N_{n+1}}^U)^{w_{n+1}}}\right], \\ \frac{(1+T_{N_{n+1}})^{w_{n+1}}-(1-T_{N_{n+1}})^{w_{n+1}}}{(1+T_{N_{n+1}})^{w_{n+1}}+(1-T_{N_{n+1}})^{w_{n+1}}}, \frac{(1+I_{N_{n+1}})^{w_{n+1}}-(1-I_{N_{n+1}})^{w_{n+1}}}{(1+I_{N_{n+1}})^{w_{n+1}}+(1-I_{N_{n+1}})^{w_{n+1}}}, \\ \frac{2\left(F_{N_{n+1}}\right)^{w_{n+1}}}{\left(2-F_{N_{n+1}}\right)^{w_{n+1}}+\left(F_{N_{n+1}}\right)^{w_{n+1}}} \end{array}\right)$$

$$so\ \overset{n}{\underset{j=1}{\otimes}}\left(N_j^E\right)^{w_j} \otimes_E \left(N_{m+1}^E\right)^{w_{m+1}} =$$

$$\left(\begin{array}{c} \left[\frac{2\prod_{j=1}^{n}\left(T_{N_j}^L\right)^{w_j}}{\prod_{j=1}^{n}\left(2-T_{N_j}^L\right)^{w_j}+\prod_{j=1}^{n}\left(T_{N_j}^L\right)^{w_j}}, \frac{2\prod_{j=1}^{n}\left(T_{N_j}^U\right)^{w_j}}{\prod_{j=1}^{n}\left(2-T_{N_j}^U\right)^{w_j}+\prod_{j=1}^{n}\left(T_{N_j}^U\right)^{w_j}}\right], \\ \left[\frac{2\prod_{j=1}^{n}\left(I_{N_j}^L\right)^{w_j}}{\prod_{j=1}^{n}\left(2-I_{N_j}^L\right)^{w_j}+\prod_{j=1}^{n}\left(I_{N_j}^L\right)^{w_j}}, \frac{2\prod_{j=1}^{n}\left(I_{N_j}^U\right)^{w_j}}{\prod_{j=1}^{n}\left(2-I_{N_j}^U\right)^{w_j}+\prod_{j=1}^{n}\left(I_{N_j}^U\right)^{w_j}}\right], \\ \left[\frac{\prod_{j=1}^{n}\left(1+F_{N_j}^L\right)^{w_j}-\prod_{j=1}^{n}\left(1-F_{N_j}^L\right)^{w_j}}{\prod_{j=1}^{n}\left(1+F_{N_j}^L\right)^{w_j}+\prod_{j=1}^{n}\left(1-F_{N_j}^L\right)^{w_j}}, \frac{\prod_{j=1}^{n}\left(1+F_{N_j}^U\right)^{w_j}-\prod_{j=1}^{n}\left(1-F_{N_j}^U\right)^{w_j}}{\prod_{j=1}^{n}\left(1+F_{N_j}^U\right)^{w_j}+\prod_{j=1}^{n}\left(1-F_{N_j}^U\right)^{w_j}}\right], \\ \frac{\prod_{j=1}^{n}\left(1+T_{N_j}\right)^{w_j}-\prod_{j=1}^{n}\left(1-T_{N_j}\right)^{w_j}}{\prod_{j=1}^{n}\left(1+T_{N_j}\right)^{w_j}+\prod_{j=1}^{n}\left(1-T_{N_j}\right)^{w_j}}, \frac{\prod_{j=1}^{n}\left(1+I_{N_j}\right)^{w_j}-\prod_{j=1}^{n}\left(1-I_{N_j}\right)^{w_j}}{\prod_{j=1}^{n}\left(1+I_{N_j}\right)^{w_j}+\prod_{j=1}^{n}\left(1-I_{N_j}\right)^{w_j}}, \\ \frac{2\prod_{j=1}^{n}\left(F_{N_j}\right)^{w_j}}{\prod_{j=1}^{n}\left(2-F_{N_j}\right)^{w_j}+\prod_{j=1}^{n}\left(F_{N_j}\right)^{w_j}} \end{array}\right) \oplus_E \left(\begin{array}{c} \left[\frac{2\left(T_{N_{n+1}}^L\right)^{w_{n+1}}}{\left(2-T_{N_{n+1}}^L\right)^{w_{n+1}}+\left(T_{N_{n+1}}^L\right)^{w_{n+1}}}, \frac{2\left(T_{N_{n+1}}^U\right)^{w_{n+1}}}{\left(2-T_{N_{n+1}}^U\right)^{w_{n+1}}+\left(T_{N_{n+1}}^U\right)^{w_{n+1}}}\right], \\ \left[\frac{2\left(I_{N_{n+1}}^L\right)^{w_{n+1}}}{\left(2-I_{N_{n+1}}^L\right)^{w_{n+1}}+\left(I_{N_{n+1}}^L\right)^{w_{n+1}}}, \frac{2\left(I_{N_{n+1}}^U\right)^{w_{n+1}}}{\left(2-I_{N_{n+1}}^U\right)^{w_{n+1}}+\left(I_{N_{n+1}}^U\right)^{w_{n+1}}}\right], \\ \left[\frac{(1+F_{N_{n+1}}^L)^{w_{n+1}}-(1-F_{N_{n+1}}^L)^{w_{n+1}}}{(1+F_{N_{n+1}}^L)^{w_{n+1}}+(1-F_{N_{n+1}}^L)^{w_{n+1}}}, \frac{(1+F_{N_{n+1}}^U)^{w_{n+1}}-(1-F_{N_{n+1}}^U)^{w_{n+1}}}{(1+F_{N_{n+1}}^U)^{w_{n+1}}+(1-F_{N_{n+1}}^U)^{w_{n+1}}}\right], \\ \frac{(1+T_{N_{n+1}})^{w_{n+1}}-(1-T_{N_{n+1}})^{w_{n+1}}}{(1+T_{N_{n+1}})^{w_{n+1}}+(1-T_{N_{n+1}})^{w_{n+1}}}, \\ \frac{(1+I_{N_{n+1}})^{w_{n+1}}-(1-I_{N_{n+1}})^{w_{n+1}}}{(1+I_{N_{n+1}})^{w_{n+1}}+(1-I_{N_{n+1}})^{w_{n+1}}}, \\ \frac{2\left(F_{N_{n+1}}\right)^{w_{n+1}}}{\left(2-F_{N_{n+1}}\right)^{w_{n+1}}+\left(F_{N_{n+1}}\right)^{w_{n+1}}} \end{array}\right)$$

$$\overset{n+1}{\underset{j=1}{\otimes}}\left(N_j^E\right)^{w_j}=\left(\begin{array}{c}\left[\frac{2\prod_{j=1}^{n+1}\left(T_{N_j}^L\right)^{w_j}}{\prod_{j=1}^{n+1}\left(2-T_{N_j}^L\right)^{w_j}+\prod_{j=1}^{n+1}\left(T_{N_j}^L\right)^{w_j}},\frac{2\prod_{j=1}^{n+1}\left(T_{N_j}^U\right)^{w_j}}{\prod_{j=1}^{n+1}\left(2-T_{N_j}^U\right)^{w_j}+\prod_{j=1}^{n+1}\left(T_{N_j}^U\right)^{w_j}}\right],\\ \left[\frac{2\prod_{j=1}^{n+1}\left(I_{N_j}^L\right)^{w_j}}{\prod_{j=1}^{n+1}\left(2-I_{N_j}^L\right)^{w_j}+\prod_{j=1}^{n+1}\left(I_{N_j}^L\right)^{w_j}},\frac{2\prod_{j=1}^{n+1}\left(I_{N_j}^U\right)^{w_j}}{\prod_{j=1}^{n+1}\left(2-I_{N_j}^U\right)^{w_j}+\prod_{j=1}^{n+1}\left(I_{N_j}^U\right)^{w_j}}\right],\\ \left[\frac{\prod_{j=1}^{n+1}\left(1+F_{N_j}^L\right)^{w_j}-\prod_{j=1}^{n+1}\left(1-F_{N_j}^L\right)^{w_j}}{\prod_{j=1}^{n+1}\left(1+F_{N_j}^L\right)^{w_j}+\prod_{j=1}^{n+1}\left(1-F_{N_j}^L\right)^{w_j}},\frac{\prod_{j=1}^{n+1}\left(1+F_{N_j}^U\right)^{w_j}-\prod_{j=1}^{n+1}\left(1-F_{N_j}^U\right)^{w_j}}{\prod_{j=1}^{n+1}\left(1+F_{N_j}^U\right)^{w_j}+\prod_{j=1}^{n+1}\left(1-F_{N_j}^U\right)^{w_j}}\right],\\ \frac{\prod_{j=1}^{n+1}\left(1+T_{N_j}\right)^{w_j}-\prod_{j=1}^{n+1}\left(1-T_{N_j}\right)^{w_j}}{\prod_{j=1}^{n+1}\left(1+T_{N_j}\right)^{w_j}+\prod_{j=1}^{n+1}\left(1-T_{N_j}\right)^{w_j}},\frac{\prod_{j=1}^{n+1}\left(1+I_{N_j}\right)^{w_j}-\prod_{j=1}^{n+1}\left(1-I_{N_j}\right)^{w_j}}{\prod_{j=1}^{n+1}\left(1+I_{N_j}\right)^{w_j}+\prod_{j=1}^{n+1}\left(1-I_{N_j}\right)^{w_j}},\\ \frac{2\prod_{j=1}^{n+1}\left(F_{N_j}\right)^{w_j}}{\prod_{j=1}^{n+1}\left(2-F_{N_j}\right)^{w_j}+\prod_{j=1}^{n+1}\left(F_{N_j}\right)^{w_j}}\end{array}\right)$$

so result holds for all values of m. □

Theorem 7. *Let* $N_j = \left(\widetilde{T}_{N_j}, \widetilde{I}_{N_j}, \widetilde{F}_{N_j}, T_{N_j}, I_{N_j}, F_{N_j}\right)$, *where* $\widetilde{T}_{N_j} = \left[T_{N_j}^L, T_{N_j}^U\right]$, $\widetilde{I}_{N_j} = \left[I_{N_j}^L, I_{N_j}^U\right], \widetilde{F}_{N_j} = \left[F_{N_j}^L, F_{N_j}^U\right]$, $(j = 1,2,\ldots,m)$ *is a collection of neutrosophic cubic values and* $W = (w_1, w_2, \ldots, w_m)^T$ *is a weight vector of* $N_j (j = 1,2,3,\ldots,m)$, *with* $w_j \in [0,1]$ *and* $\sum_{j=1}^{m} w_j = 1$.

1. ***Idempotency:*** *If for all* $N_j = \left(\widetilde{T}_{N_j}, \widetilde{I}_{N_j}, \widetilde{F}_{N_j}, T_{N_j}, I_{N_j}, F_{N_j}\right)$, *where* $\widetilde{T}_{N_j} = \left[T_{N_j}^L, T_{N_j}^U\right]$, $\widetilde{I}_{N_j} = \left[I_{N_j}^L, I_{N_j}^U\right], \widetilde{F}_{N_j} = \left[F_{N_j}^L, F_{N_j}^U\right]$, $(j = 1,2,\ldots,m)$ *are equal, that is,* $N_j = N$ *for all k, then* $NCEWG_w(N_1, N_2, \ldots, N_m) = N$
2. ***Monotonicity:*** *Let* $B_j = \left(\widetilde{T}_{B_j}, \widetilde{I}_{B_j}, \widetilde{F}_{B_j}, T_{B_j}, I_{B_j}, F_{B_j}\right)$, *where* $\widetilde{T}_{B_j} = \left[T_{B_j}^L, T_{B_j}^U\right]$, $\widetilde{I}_{B_j} = \left[I_{B_j}^L, I_{B_j}^U\right], \widetilde{F}_{B_j} = \left[F_{B_j}^L, F_{B_j}^U\right]$ $(j = 1,2,\ldots,m)$ *be the collection of cubic values. If* $S_B(u) \geq S_N(u)$ *and* $B_j(u) \geq N_j(u)$ *then* $NCW\ G_w(N_1, N_2, \ldots, N_m) \leq NCWG_w(B_1, B_2, \ldots, B_m)$
3. ***Boundary:*** $N^- \leq NCWG_w\{(N_1)_T, (N_2)_T, \ldots, (N_m)_T\} \leq N^+$, *where*

$$N^- = \left\{\min_j T_{N_j}^L, \min_j I_{N_j}^L, 1 - \max_j F_{N_j}^L, \min_j T_{N_j}, \min_j I_{N_j}, 1 - \max_j F_{N_j}^L\right\},$$
$$N^+ = \left\{\max_j T_{N_j}^U, \max_j I_{N_j}^U, 1 - \min_j F_{N_j}^U, \max_j T_{N_j}, \max_j I_{N_j}, 1 - \min_j F_{N_j}\right\}$$

Proof. Followed by Theorem 2. □

Theorem 8. *Let* $N_j = \left(\widetilde{T}_{N_j}, \widetilde{I}_{N_j}, \widetilde{F}_{N_j}, T_{N_j}, I_{N_j}, F_{N_j}\right)$, *where* $\widetilde{T}_{N_j} = \left[T_{N_j}^L, T_{N_j}^U\right]$, $\widetilde{I}_{N_j} = \left[I_{N_j}^L, I_{N_j}^U\right], \widetilde{F}_{N_j} = \left[F_{N_j}^L, F_{N_j}^U\right]$, $(j = 1,2,\ldots,m)$ *be a collection of neutrosophic cubic values and* $W = (w_1, w_2, \ldots, w_m)^T$ *is a weight vector of the NCOWA, with* $w_j \in [0,1]$ *and* $\sum_{j=1}^{m} w_j = 1$.

1. *If* $w = (1,0,\ldots,0)^T$, *then NCEOWG* $(N_1, N_2, \ldots, N_m) = \max N_j$
2. *If* $w(0,0,\ldots,1)^T$, *then NCEOWG* $(N_1, N_2, \ldots, N_m) = \min N_j$
3. *If* $w_j = 1$, $w_j = 0$, *and* $j \neq j$, *then NCEOWG* $(N_1, N_2, \ldots, N_m) = N_j$

where N_j *is the jth largest of* $(N_1, N_2, \ldots, N_m)$.

Proof. Followed by Theorem 3. □

5. An Application of Neutrosophic cubic Geometric and Einstein Geometric Aggregation Operator to Group Decision Making Problems

Group decision making is an important factor of decision making theory. We are often in a situation with more then one expert, attribute and alternative to deal with. Motivated by such situations, a multi-attribute decision making method for more then one expert is proposed in this section.

In this section, we develop an algorithm for group decision making problems using the geometric and Einstein geometric aggregations (NCWG and NCEWG) under the neutrosophic cubic environment.

Algorithm. *Let* $F = \{F_1, F_2, \ldots, F_n\}$ *be the set of* n *alternatives,* $H = \{H_1, H_2, \ldots, H_m\}$ *be the* m *attributes subject to their corresponding weight* $W = \{w_1, w_2, \ldots, w_m\}$ *such that* $w_j \in [0,1]$ *and* $\sum_{j=1}^{m} w_j = 1$*, and* $D = \{D_1, D_2, ...D_r\}$ *be the* r *decision makers with their corresponding weight* $V = \{v_1, v_2, \ldots, v_r\}$*. such that* $v_j \in [0,1]$ *and* $\sum_{j=1}^{r} v_j = 1$ The method has the following steps:

Step1. *First, we construct neutrosophic cubic decision matrices for each decision maker* $D^{(s)} = \left[N_{ij}^{(s)}\right]_{n\times m} (s = 1, 2, \ldots, r)$.

Step2. *All decision matrices are aggregated to a single matrix consisting of m attributes, by NCWG and NCEWG corresponding to the weight assigned to the decision maker.*

Step3. *By using aggregation operators like NCWG and NCEWG, the decision matrix is aggregated by the weight assigned to the m attributes.*

Step4. *The n alternatives are ranked according to their scores and arranged in descending order to select the alternative with highest score.*

6. Application

Mobile companies play a vital role in Pakistan's stock market. The performance of these companies affects resources of capital market and have become a common concern of shareholders, government authorities, creditors and other stakeholders. In this example, an investor company wants to invest his capital levy in listed companies. They acquire two types of experts: Attorney and market maker. The attorney is acquired to look at the legal matters and the market maker is aquired to provide his expertise in capital market matters. Data are collected on the basis of stock market analysis and growth in different areas. Let the listed mobile companies be (x_1) Zong, (x_2) Jazz, (x_3) Telenor and (x_4) Ufone, which have higher ratios of earnings than the others available in the market, from the three alternatives of (A_1) stock market trends, (A_2) policy directions and (A_3) the annual performance. The two experts evaluated the mobile companies $(x_j, j = 1, 2, 3, 4)$ with respect to the corresponding attributes $(A_i, i = 1, 2, 3)$, and proposed their decision making matrices consisting of neutrosophic cubic values in Equation (1) and Equation (2). The Equation (3) represents the single matrix as the aggregation of Equtional1 and Equation (2) by NCWG or NCEWG. The Equation (4) is obtained by applying NCWG or NCEWG on attributes. The decision matrices are aggregated to a single decision matrix. At the end we rank the alternatives according to their score to get the desirable alternative(s).

Step 1. We construct the decision maker matrices in Equations (1) and (2).

Equation (1): Decision making matrix for the first expert(attorney) D_a is

$$\begin{pmatrix} & A_1 & A_2 & A_3 \\ X_1 & \begin{pmatrix} [0.2,0.6],[0.4,0.6], \\ [0.5,0.8],0.7,0.4,0.3 \end{pmatrix} & \begin{pmatrix} [0.1,0.4],[0.5,0.8], \\ [0.4,0.8],0.6,0.7,0.5 \end{pmatrix} & \begin{pmatrix} [0.4,0.6],[0.2,0.7], \\ [0.5,0.9],0.4,0.5,0.3 \end{pmatrix} \\ X_2 & \begin{pmatrix} [0.3,0.5],[0.6,0.9], \\ [0.3,0.6],0.3,0.6,0.7 \end{pmatrix} & \begin{pmatrix} [0.5,0.9],[0.1,0.3], \\ [0.4,0.8],0.8,0.3,0.6 \end{pmatrix} & \begin{pmatrix} [0.2,0.7],[0.1,0.6], \\ [0.4,0.7],0.5,0.4,0.7 \end{pmatrix} \\ X_3 & \begin{pmatrix} [0.6,0.9],[0.2,0.7], \\ [0.4,0.9],0.5,0.5,0.6 \end{pmatrix} & \begin{pmatrix} [0.2,0.6],[0.7,0.3], \\ [0.3,0.8],0.4,0.6,0.5 \end{pmatrix} & \begin{pmatrix} [0.5,0.9],[0.7,0.9], \\ [0.1,0.5],0.5,0.6,0.4 \end{pmatrix} \\ X_4 & \begin{pmatrix} [0.4,0.8],[0.5,0.9], \\ [0.3,0.8],0.5,0.8,0.5 \end{pmatrix} & \begin{pmatrix} [0.2,0.7],[0.4,0.9], \\ [0.5,0.7],0.6,0.4,0.5 \end{pmatrix} & \begin{pmatrix} [0.3,0.5],[0.5,0.9], \\ [0.7,0.3],0.3,0.3,0.8 \end{pmatrix} \end{pmatrix} \tag{1}$$

Equation (2): Decision making matrix for the second expert(market maker) D_m is

$$\begin{pmatrix} & A_1 & A_2 & A_3 \\ X_1 & \begin{pmatrix} [0.3,0.6],[0.2,0.6], \\ [0.2,0.6],0.8,0.7,0.2 \end{pmatrix} & \begin{pmatrix} [0.3,0.8],[0.4,0.8], \\ [0.3,0.8],0.6,0.7,0.4 \end{pmatrix} & \begin{pmatrix} [0.2,0.7],[0.2,0.6], \\ [0.3,0.8],0.5,0.3,0.5 \end{pmatrix} \\ X_2 & \begin{pmatrix} [0.2,0.5],[0.6,0.9], \\ [0.7,0.3],0.4,0.8,0.7 \end{pmatrix} & \begin{pmatrix} [0.4,0.9],[0.1,0.4], \\ [0.5,0.8],0.6,0.5,0.7 \end{pmatrix} & \begin{pmatrix} [0.4,0.9],[0.1,0.4], \\ [0.5,0.8],0.6,0.5,0.7 \end{pmatrix} \\ X_3 & \begin{pmatrix} [0.5,0.9],[0.2,0.6], \\ [0.3,0.8],0.7,0.7,0.8 \end{pmatrix} & \begin{pmatrix} [0.2,0.5],[0.2,0.7], \\ [0.5,0.8],0.6,0.7,0.2 \end{pmatrix} & \begin{pmatrix} [0.3,0.5],[0.3,0.9], \\ [0.2,0.5],0.6,0.5,0.4 \end{pmatrix} \\ X_4 & \begin{pmatrix} [0.3,0.5],[0.3,0.9], \\ [0.2,0.5],0.6,0.5,0.4 \end{pmatrix} & \begin{pmatrix} [0.4,0.7],[0.2,0.8], \\ [0.7,0.3],0.6,0.7,0.7 \end{pmatrix} & \begin{pmatrix} [0.2,0.6],[0.5,0.9], \\ [0.2,0.8],0.4,0.4,0.8 \end{pmatrix} \end{pmatrix} \tag{2}$$

Step2. Let $W = (0.4, 0.6)^T$, then the single matrix corresponding to weight W by use of NCWG operator is

Equation (3): The single decision matrix.

$$\begin{pmatrix} & A_1 & A_2 & A_3 \\ X_1 & \begin{pmatrix} [0.2551,0.6000], \\ [0.2885,0.6732], \\ [0.3371,0.6968], \\ 0.7647,0.6041,0.2352 \end{pmatrix} & \begin{pmatrix} [0.1933,0.6062], \\ [0.4430,0.8001], \\ [0.3418,0.8680], \\ 0.6000,0.7000,0.4772 \end{pmatrix} & \begin{pmatrix} [0.2638,0.6581], \\ [0.1999,0.6381], \\ [0.3881,0.8484], \\ 0.4621,0.3881,0.2223 \end{pmatrix} \\ X_2 & \begin{pmatrix} [0.2352,0.5577], \\ [0.6000,0.9000], \\ [0.3000,0.6634], \\ 0.3618,0.7360,0.7000 \end{pmatrix} & \begin{pmatrix} [0.2352,0.5577], \\ [0.6000,0.9000], \\ [0.3000,0.6634], \\ 0.3618,0.7360,0.7000 \end{pmatrix} & \begin{pmatrix} [0.5253,0.8670], \\ [0.1515,0.6000], \\ [0.4621,0.8448], \\ 0.3371,0.4621,0.3301 \end{pmatrix} \\ X_3 & \begin{pmatrix} [0.5378,0.9000], \\ [0.3565,0.8385], \\ [0.3418,0.8484], \\ 0.6319,0.6319,0.7130 \end{pmatrix} & \begin{pmatrix} [0.2000,0.5378], \\ [0.2352,0.7000], \\ [0.4279,0.8000], \\ 0.5295,0.6634,0.2885 \end{pmatrix} & \begin{pmatrix} [0.3680,0.6325], \\ [0.4210,0.9000], \\ [0.1614,0.5000], \\ 0.5626,0.5426,0.4000 \end{pmatrix} \\ X_4 & \begin{pmatrix} [0.5101,0.8000], \\ [0.3465,0.6325], \\ [0.2416,0.7449], \\ 0.5000,0.6133,0.3807 \end{pmatrix} & \begin{pmatrix} [0.3031,0.7000], \\ [0.2639,0.8385], \\ [0.3881,0.7000], \\ 0.6000,0.6041,0.6118 \end{pmatrix} & \begin{pmatrix} [0.2352,0.5578], \\ [0.5000,0.9000], \\ [0.2416,0.7647], \\ 0.3618,0.3618,0.8000 \end{pmatrix} \end{pmatrix} \tag{3}$$

Step3. Let the weight of attributes are $W = \{0.35, 0.30, 0.35\}$, using NCWG operators on attributes A's we get Equation (4),

$$NCWG = \begin{pmatrix} X_1 & \begin{pmatrix} [0.2375, 0.6195], \\ [0.2885, 0.7916], \\ [0.3567, 0.8146], \\ 0.6315, 0.5757, 0.2851 \end{pmatrix} \\ X_2 & \begin{pmatrix} [0.4426, 0.7657], \\ [0.2165, 0.5915], \\ [0.5382, 0.7804], \\ 0.4827, 0.5729, 0.5282 \end{pmatrix} \\ X_3 & \begin{pmatrix} [0.3500, 0.6616], \\ [0.3335, 0.8142], \\ [0.3131, 0.7498], \\ 0.5791, 0.6133, 0.4439 \end{pmatrix} \\ X_4 & \begin{pmatrix} [0.3327, 0.6774], \\ [0.3630, 0.7787], \\ [0.2888, 0.7396], \\ 0.4906, 0.5359, 0.5692 \end{pmatrix} \end{pmatrix} \quad (4)$$

Step4. Using the score function we rank the alternatives as:
$S(X_1) = 0.0321$, $S(X_2) = 0.0548$, $S(X_3) = 0.0839$ and $S(X_4) = -0.0969$, $X_3 > X_2 > X_1 > X_4$
The most desirable alternative is X_3.

7. Conclusions

Dealing with real life problems, decision makers encounter incomplete and vague data. The characteristics of neutrosophic cubic sets enablesdecision makers to deal with such a situation. Consequently, for each situation we defined the algebraic and Einstein sum, product and scalar multiplication. It is often difficult to compare two or more neutrosophic cubic values. The score and accuracy functions are defined to compare the neutrosophic cubic values values. Using these operations we defined neutrosophic cubic geometric, neutrosophic cubic weighted geometric, neutrosophic cubic Einstein geometric, and neutrosophic cubic Einstein weighted geometric aggregation operators with some useful properties. In the next section, a multi-criteria decision making algorithm was constructed. In the last section, a daily life problem was solved usingmulti-criteria decision making method (MCDM). This paper is based on some basic definitions and aggregation operators, which can be further extended to new horizons, like neutrosophic cubic hybrid geometric and neutrosophic cubic Einstein hybrid geometric aggregation operators.

Author Contributions: All authors contributed equally.

Acknowledgments: The 3rd author would like to thank the Deanship of Scientific Research at Majmaah University for supporting this work under Project Number 1440-52.

Conflicts of Interest: The authors declare no conflict of interest.

References

1. Zadeh, L.A. Fuzzy Sets. *Inf. Control* **1965**, *8*, 338–353. [CrossRef]
2. Turksen, I.B. Interval valued strict preferences with Zadeh triplet. *Fuzzy Sets Syst.* **1996**, *78*, 183–195. [CrossRef]
3. Zadeh, L.A. Outlines of new approach to the analysis of complex system and dicision procosses interval valued fuzzy sets. *IEEE Trans. Syst. Man Cybernet.* **1968**, *1*, 28–44.
4. Atanassov, K.T. Intuitionistic fuzzy sets. *Fuzzy Sets Syst.* **1986**, *20*, 87–96. [CrossRef]
5. Atanassov, K.T.; Gargov, G. Interval intuitionistic fuzzy sets. *Fuzzy Sets Syst.* **1989**, *31*, 343–349. [CrossRef]
6. Jun, Y.B.; Kim, C.S.; Yang, K.O. Cubic sets. *Ann. Fuzzy Math. Inform.* **2012**, *1*, 83–98.
7. Smarandache, F. *A Unifying Field in Logics, Neutrosophic Logic, Neutrosophy, Neutrosophic Set and Neutrosophic Probabilty*, 4th ed.; American Research Press: Rehoboth, DE, USA, 1999.

8. Wang, H.; Smarandache, F.; Zhang, Y.Q.; Sunderraman, R. Interval neutrosophic sets and loics. In *Theory and Application in Computing*; Hexis: Phoenox, AZ, USA, 2005.
9. Jun, Y.B.; Smarandache, F.; Kim, C.S. Neutrosophic cubic sets. *New. Math. Nat. Comput.* **2015**, *13*, 41–54. [CrossRef]
10. Jun, Y.B.; Smarandache, F.; Kim, C.S. P-union and P-intersection of neutrosophic cubic sets. *An. St. Univ. Ovidius Constanta* **2017**, *25*, 99–115. [CrossRef]
11. Zhan, J.; Khan, M.; Gulistan, M.; Ali, A. Applications of neutrosophic cubic sets in multi-criteria decision making. *Int. J. Uncertain. Quabtif.* **2017**, *7*, 377–394. [CrossRef]
12. Banerjee, D.; Giri, B.C.; Pramanik, S.; Smarandache, F. GRA for multi attribute decision making in neutrosophic cubic set environment. *Neutrosophic Sets Syst.* **2017**, *15*, 64–73.
13. Lu, Z.; Ye, J. Cosine measure for neutrosophic cubic sets for multiple attribte decision making. *Symmetry* **2017**, *9*, 121.
14. Pramanik, S.; Dalapati, S.; Alam, S.; Roy, S.; Smarandache, F. Neutrosophic cubic MCGDM method based on similarity measure. *Neutrosophic Sets Syst.* **2017**, *16*, 44–56.
15. Shi, L.; Ye, J. Dombi Aggregation Operators of Neutrosophic Cubic Set for Multiple Attribute Deicision Making. *Algorithms* **2018**, *11*, 29. [CrossRef]
16. Li, B.; Wang, J.; Yang, L.; Li, X. A Novel Generalized Simplified Neutrosophic Number Einstein Aggregation Operator. *Int. J. Appl. Math.* **2018**, *48*, 67–72.

symmetry

MDPI

Article

Inspection Strategy under Indeterminacy Based on Neutrosophic Coefficient of Variation

Muhammad Aslam * and Mansour Sattam Aldosari

Department of Statistics, Faculty of Science, King Abdulaziz University, Jeddah 21589, Saudi Arabia; msattam@hotmail.com

* Correspondence: magmuhammad@kau.edu.sa or aslam_ravian@hotmail.com; Tel.: +966-593-329-841

Received: 4 January 2019; Accepted: 30 January 2019; Published: 9 February 2019

Abstract: The existing sampling plans which use the coefficient of variation (CV) are designed under classical statistics. These available sampling plans cannot be used for sentencing if the sample or the population has indeterminate, imprecise, unknown, incomplete or uncertain data. In this paper, we introduce the neutrosophic coefficient of variation (NCV) first. We design a sampling plan based on the NCV. The neutrosophic operating characteristic (NOC) function is then given and used to determine the neutrosophic plan parameters under some constraints. The neutrosophic plan parameters such as neutrosophic sample size and neutrosophic acceptance number are determined through the neutrosophic optimization solution. We compare the efficiency of the proposed plan under the neutrosophic statistical interval method with the sampling plan under classical statistics. A real example which has indeterminate data is given to illustrate the proposed plan.

Keywords: neutrosophic statistical interval method; classical statistics; fuzzy logic; producer's risk; consumer's risk

1. Introduction

The sampling plan is an important tool for statistical quality control (SQC) which is used for the inspection of finished product batches before they are sent to the market [1]. Cheap inspections, which require less effort and minimized risks, are the main targets during the inspection process. These targets can be only achieved by applying a well-designed sampling plan. The decision regarding the product lot is made based on sample information. Ineffective sampling plans may mislead the experimenter, causing them to reject a good lot (producer's risk) and accepting a bad lot (consumer's risk). The supplier wishes to protect lots produced at permissible standards and the customer wants to reduce the chance of accepting a bad lot of the product. A well-designed sampling plan is based on the optimal plan parameters where the sample size required for the inspection and risks involved are at a minimum level.

Two major types of sampling plans are the attribute sampling plan and variable sampling plan. Variable sampling plans are designed for when the quality characteristic is measurable and attribute sampling plans are designed for a quality characteristic expressed as 'go, no-go'. The attribute sampling plans are easy to apply in practice but yield less information than variable sampling. The variable sampling plans can be applied to inspection lots using smaller sample sizes, while attribute plans require larger sample sizes to attain the same protection for the supplier and customer. Based on the advantages of the variable sampling over the attribute sampling plans, several authors from a variety of fields worked on designing variable sampling plans. For example, the authors of Reference [2] introduced the plan for the coefficient of variation (CV), and they proposed a CV-based plan for two-stage sampling. In addition, the authors of Reference [3] presented the design of a CV-based plan using repetitive sampling.

According to Yan et al. [4], the relative measure is called the coefficient of variation (CV). This is the ratio of standard deviation (SD), meaning it has been widely used to measure the relative variation to its mean in a variety of fields. For example, References [5–8] applied the CV in the testing of material reliability and the average quality characteristic used to measure the product quality. According to Reference [9], "in certain scenarios, the practitioner is not interested in the changes in the mean or the standard deviation but is instead interested in the relative variability compared with the mean". Due to the wide applications of the CV, several authors have designed sampling plans, this includes a proposed multiple dependent state sampling plan for the CV [4], a proposed CV-based plan for two stage sampling [2], and presenting the designing of CV based planning using the repetitive sampling [3].

The sampling plans in the literature are applied for the inspection of the submitted lot of products under the assumption that all the observations are determined. The fuzzy approach is applied when there is uncertainty in the proportion parameters. The sampling plans which use the fuzzy approach can be applied for the inspection of a product lots with ambiguous proportion parameters. In recent years, several authors contributed excellent work on the design of sampling plan using the fuzzy approach, the authors of Reference [10] proposed fuzzy attribute plans for multiple dependent state sampling, the authors of Reference [11] designed the fuzzy plan using double sampling, the authors of Reference [12] proposed the fuzzy sampling plan using the Poisson distribution, the authors of Reference [13] studied the fuzzy operating characteristics curve, the authors of References [14,15] proposed the double and sequential sampling plan using the fuzzy logic, and the authors of Reference [16] proposed the fuzzy plan using gamma distribution. More details can be seen in Reference [17].

Neutrosophic logic is the generalization of fuzzy logic as introduced by Reference [18]. Neutrosophic logic consists of measures of the truth, falsehood, and indeterminacy. Smarandache [19] introduced neutrosophic statistics using neutrosophic numbers. Neutrosophic statistics are the generalization of classical statistics, which can be applied when the data is indeterminate. This means that neutrosophic statistics can be applied to analyze data which may have some uncertain or unclear observations. The neutrosophic analysis provides the output in an indeterminacy interval rather than the determined values. Therefore, it is effective when applied under uncertain conditions. Recently, the authors of References [20,21] introduced the neutrosophic statistical interval method for the rock measuring problem. According to Reference [22], "observations include human judgments, and evaluations and decisions, a continuous random variable of a production process should include the variability caused by human subjectivity or measurement devices, or environmental conditions. These variabilities can create vagueness in the measurement system." In this situation, when there is indeterminacy in the data, the lot sentencing cannot be done using the traditional sampling plans. To handle this situation, the authors of Reference [23] introduced the neutrosophic statistics in the area of acceptance sampling plans, as well as proposed sampling plans based on the process loss function under the neutrosophic statistical interval method, and the authors of Reference [24] designed the neutrosophic sampling plan using sudden death testing. Aslam and Raza's [25] design proposed a neutrosophic sampling plan for multiple manufacturing lines, the authors of Reference [26] proposed a plan for exponential distribution under neutrosophic statistics, and the authors of Reference [27] proposed a neutrosophic plan for using the regression estimator.

By exploring the literature and to the best of our knowledge, there is no work on the sampling plan for the CV using neutrosophic statistics. In this paper, we introduce the neutrosophic coefficient of variation (NCV). We first design a sampling plan based on the CV according to the neutrosophic statistical interval method. The neutrosophic operating characteristic (NOC) is given. We expect that the proposed sampling under the neutrosophic statistical interval method will be more effective and adequate than the sampling plan based on the classical statistics under the indeterminacy environment.

2. Neutrosophic Coefficient of Variation

In this section, we first define the NCV and then design the sampling plan based on it.

Suppose that $X_{Ni}\epsilon\{X_L, X_U\} = i = 1, 2, 3, \ldots, n_N$ is a neutrosophic random variable of size n_N distributed as the neutrosophic normal distribution with neutrosophic mean $\mu_N\epsilon\{\mu_L, \mu_U\}$ and neutrosophic standard deviation $\sigma_N\epsilon\{\sigma_L, \sigma_U\}$; for details, see [19]. The NCV is defined by

$$CV_N = \sigma_N/\mu_N;\ \mu_N\epsilon\{\mu_L, \mu_U\},\ \sigma_N\epsilon\{\sigma_L, \sigma_U\},\ CV_N\epsilon\{CV_L, CV_U\}. \tag{1}$$

It is important to note that $\mu_N\epsilon\{\mu_L, \mu_U\}$ and $\sigma_N\epsilon\{\sigma_L, \sigma_U\}$ are usually unknown in practice and are estimated using sample information. The best linear unbiased estimates (BLUE) of $\mu_N\epsilon\{\mu_L, \mu_U\}$ and $\sigma_N\epsilon\{\sigma_L, \sigma_U\}$ are $\overline{X}_N\epsilon\{\overline{X}_L, \overline{X}_U\} = \left\{\sum_{i=1}^{n} x_i^L/n_L, \sum_{i=1}^{n} x_i^U/n_U\right\}$ and $s_N = \{s_L, s_U\} = \left\{\sqrt{\sum_{i=1}^{n}\left(x_i^L - \overline{X}_L\right)^2/n_L}, \sqrt{\sum_{i=1}^{n}\left(x_i^U - \overline{X}_U\right)^2/n_U}\right\}$, respectively. The natural estimate of NCV is defined by

$$\widehat{CV}_N = \frac{s_N}{\overline{X}_N};\overline{X}_N\epsilon\{\overline{X}_L, \overline{X}_U\},\ s_N = \{s_L, s_U\},\ CV_N\epsilon\{CV_L, CV_U\}. \tag{2}$$

By following Reference [28], the neutrosophic statistic $\sqrt{\overline{X}_N}/\widehat{CV}_N;\ X_{Ni}\epsilon\{X_L, X_U\}, \widehat{CV}_N\epsilon\left\{\widehat{CV}_L, \widehat{CV}_U\right\}$ follows the neutrosophic non-central chi-square distribution with a neutrosophic degree of freedom $n_N - 1$ and neutrosophic non-centrality parameters $\delta_N = \sqrt{n_N}/\widehat{CV}_N$. The neutrosophic chi-square distribution is used to test the association between the qualitative data and having some uncertain observations. The neutrosophic chi-square distribution is reduced to chi-square distribution under classical statistics if all observations in the data are precise. Some more details on the neutrosophic distributions can be read in [19,24].

The sampling plans based on the NCV are stated as follows:

Step-1: Select a random sample of size $n_N\epsilon\{n_L, n_U\}$ and compute $\widehat{CV}_N\epsilon\left\{\widehat{CV}_L, \widehat{CV}_U\right\}$.

Step-2: Accept a lot of product if $\widehat{CV}_N \le k_N$, otherwise reject a lot of the product.

Note that $k_N\epsilon\{k_{aL}, k_{aU}\}$ is a neutrosophic acceptance number.

The sampling plan based on the NCV has two neutrosophic plan parameters, $n_N\epsilon\{n_L, n_U\}$ and $k_N\epsilon\{k_{aL}, k_{aU}\}$, that will be determined through the neutrosophic optimization solution. The proposed plan is a progression of the sampling plan based on CV under classical statistics reported in [29]. The proposed sampling plan is reduced to the aforementioned sampling plan [29] when there is no indeterminacy in the data.

The neutrosophic operating characteristic (NOC) function is derived as follows:

$$P_{Na} = P\left\{\widehat{CV}_N \le k_N\right\};\ \widehat{CV}_N\epsilon\left\{\widehat{CV}_L, \widehat{CV}_U\right\},\ k_N\epsilon\{k_{aL}, k_{aU}\} \tag{3}$$

$$P_{Na} = P\left\{\widehat{CV}_N \le k_N\right\} = \mathrm{P}\left(\mathrm{t}_{n_N-1,\sqrt{n_N}/CV_N} > \sqrt{\mathrm{n}_\mathrm{N}}/k_N\right);\ \widehat{CV}_N\epsilon\left\{\widehat{CV}_L, \widehat{CV}_U\right\}, k_N\epsilon\{k_{aL}, k_{aU}\},\ n_N\epsilon\{n_L, n_U\}. \tag{4}$$

Suppose that α and β are the producer' risk and consumer's risk, respectively. To meet the producer and consumer requirements, the NOC should pass through $(p_1, 1-\alpha)$ and (p_2, β), where p_1 and p_2 are an acceptable quality level (AQL) and limiting quality level (LQL), respectively. The neutrosophic plan parameters of the proposed plan are determined through following neutrosophic optimization solution (NOS):

$$\text{minimize } n_N\epsilon\{n_L, n_U\} \tag{5}$$

subject to

$$P_{Na}\left(p_1 = \widehat{CV}_N\right) = \mathrm{P}\left(\mathrm{t}_{n_N-1,\sqrt{n_N}/\mathrm{CV_{AQL}}} > \sqrt{\mathrm{n}_\mathrm{N}}/k_N\right) \ge 1-\alpha;\ k_N\epsilon\{k_{aL}, k_{aU}\};\ n_N\epsilon\{n_L, n_U\} \tag{6}$$

and

$$P_{Na}\left(p_2 = \widehat{CV}_N\right) = \mathrm{P}\left(\mathrm{t}_{n_N-1,\sqrt{n_N}/\mathrm{CV}_{\mathrm{LQL}}} > \sqrt{\mathrm{n}_\mathrm{N}}/k_N\right) \leq \beta;\ k_N\epsilon\{k_{aL},k_{aU}\};\ n_N\epsilon\{n_L,n_U\}. \quad (7)$$

The plan parameters $n_N\epsilon\{n_L,n_U\}$ and $k_N\epsilon\{k_{aL},k_{aU}\}$ are determined through the above NOS using the search grid method. It is noted that several combinations of $n_N\epsilon\{n_L,n_U\}$ and $k_N\epsilon\{k_{aL},k_{aU}\}$ met the given conditions. The plan parameters $n_N\epsilon\{n_L,n_U\}$ and $k_N\epsilon\{k_{aL},k_{aU}\}$ are selected at smaller values of $n_N\epsilon\{n_L,n_U\}$. We used the following algorithm to find the neutrosophic plan parameters. The codes to make the tables are available from the authors upon request.

Step-1: Pre-fix the values of α,β, AQL, and LQL.
Step-2: Define a range for n_N, say $2 < n_N < 1000$. Determine the probabilities using Equations (6) and (7).
Step-3: Using search grid method, record those combinations from 10,000 combinations of plan parameters where $n_N\epsilon\{n_L,n_U\}$ is minimum.

Note here that the proposed NOS approaches the optimization solution under classical statistics if $k_{aL} = k_{aU}$ and $n_L = n_U$. Table 1 shows that plan parameters $n_N\epsilon\{n_L,n_U\}$ and $k_N\epsilon\{k_{aL},k_{aU}\}$ when $\alpha = 5\%$ and $\beta = 10\%$. Table 2 shows that plan parameters $n_N\epsilon\{n_L,n_U\}$ and $k_N\epsilon\{k_{aL},k_{aU}\}$ when $\alpha = 5\%$ and $\beta = 5\%$. We note from Tables 1 and 2 that when the AQL is fixed, $n_N\epsilon\{n_L,n_U\}$ decreases while $k_N\epsilon\{k_{aL},k_{aU}\}$ increases. Further, for the fixed values of α, $n_N\epsilon\{n_L,n_U\}$ decreases as β increses.

Table 1. The neutrosophic fuzzy plan parameters when $\alpha = 0.05$, $\beta = 0.10$.

CV_{AQL}	CV_{LQL}	$n_N\epsilon\{n_L,n_U\}$	$k_N\epsilon\{k_{aL},k_{aU}\}$	$P_{Na}(p_1)$	$P_{Na}(p_2)$
0.05	0.06	{161,265}	{0.05546,0.05637}	{0.9759,0.9983}	{0.0933,0.0861}
	0.07	{61,167}	{0.06092,0.06456}	{0.9924,0.9999}	{0.0854,0.08353}
	0.08	{37,92}	{0.06506,0.06616}	{0.9954,0.9999}	{0.0642,0.0111}
	0.09	{21,40}	{0.06858,0.06953}	{0.9922,0.9997}	{0.0778,0.0260}
	0.10	{43,58}	{0.06923,0.07187}	{0.9998,0.9999}	{0.00296,0.00163}
0.06	0.07	{332,407}	{0.06630,0.06641}	{0.9966,0.9988}	{0.0910,0.0752}
	0.08	{100,116}	{0.06766,0.07080}	{0.9659,0.9970}	{0.0168,0.0443}
	0.09	{72,146}	{0.07393,0.07485}	{0.9974,0.9999}	{0.0190,0.0024}
	0.10	{23,72}	{0.07674,0.07949}	{0.9723,0.9999}	{0.072,0.0085}
	0.11	{23,37}	{0.07711,0.08012}	{0.9747,0.9980}	{0.0289,0.0128}
0.07	0.08	{323,412}	{0.07564,0.07604}	{0.9799,0.9934}	{0.08757,0.0818}
	0.09	{125,145}	{0.07925,0.08196}	{0.9821,0.9982}	{0.0330,0.0698}
	0.10	{120,150}	{0.08340,0.08485}	{0.9985,0.9998}	{0.0060,0.0051}
	0.11	{61,97}	{0.09125,0.09208}	{0.9996,0.9999}	{0.0354,0.0138}
	0.12	{21,90}	{0.09236,0.10763}	{0.9814,1.000}	{0.0859,0.0930}
0.08	0.09	{748,928}	{0.08431,0.08450}	{0.9814,0.9961}	{0.0079,0.0086}
	0.10	{115,152}	{0.08918,0.09061}	{0.9601,0.9897}	{0.0562,0.0559}
	0.11	{68,80}	{0.09145,0.09623}	{0.9541,0.9949}	{0.0291,0.0643}
	0.12	{47,70}	{0.1001,0.1042}	{0.9926,0.9998}	{0.0639,0.0677}
	0.13	{33,60}	{0.10309,0.11169}	{0.9905,0.9999}	{0.0573,0.0713}
0.09	0.10	{655,820}	{0.09459,0.09559}	{0.9672,0.9963}	{0.0270,0.0553}
	0.11	{126,181}	{0.10066,0.10138}	{0.9703,0.9918}	{0.0971,0.0740}
	0.12	{77,107}	{0.10393,0.10514}	{0.9731,0.9929}	{0.0555,0.0401}
	0.13	{62,85}	{0.10558,0.11026}	{0.9736,0.9983}	{0.0223,0.0283}
	0.14	{41,74}	{0.10848,0.11624}	{0.9694,0.9998}	{0.0265,0.0237}

Table 2. The neutrosophic fuzzy plan parameters when $\alpha = 0.05$, $\beta = 0.05$.

CV_{AQL}	CV_{LQL}	$n_N\epsilon\{n_L,n_U\}$	$k_N\epsilon\{k_{aL},k_{aU}\}$	$P_{Na}(p_1)$	$P_{Na}(p_2)$
0.05	0.06	{320,545}	{0.05432,0.05689}	{0.9859,0.9999}	{0.0090,0.0456}
	0.07	{81,143}	{0.05806,0.06016}	{0.9809,0.9997}	{0.0403,0.0099}
	0.08	{55,102}	{0.06271,0.06716}	{0.9963,0.9999}	{0.0143,0.0126}
	0.09	{24,85}	{0.06492,0.07410}	{0.9817,1.000}	{0.0352,0.0126}
	0.10	{18,65}	{0.06857,0.07824}	{0.9875,1.000}	{0.0410,0.0081}
0.06	0.07	{289,490}	{0.06473,0.06561}	{0.9519,0.9986}	{0.0003,0.0113}
	0.08	{109,163}	{0.07068,0.07222}	{0.9958,0.9998}	{0.0474,0.0434}
	0.09	{51,137}	{0.07101,0.07700}	{0.9696,0.9999}	{0.0202,0.0970}
	0.10	{39,79}	{0.07241,0.07259}	{0.9679,0.9959}	{0.0097,0.0004}
	0.11	{23,76}	{0.07624,0.07805}	{0.9686,0.9999}	{0.0256,0.0002}
0.07	0.08	{412,689}	{0.07509,0.07618}	{0.9818,0.9994}	{0.0418,0.0401}
	0.09	{110,277}	{0.07967,0.08018}	{0.9803,0.9996}	{0.0494,0.0058}
	0.10	{50,104}	{0.08274,0.08277}	{0.9671,0.9958}	{0.0498,0.008}
	0.11	{41,102}	{0.08300,0.09190}	{0.9560,0.9999}	{0.0168,0.0112}
	0.12	{28,118}	{0.09056,0.10651}	{0.9864,1.000}	{0.0426,0.0476}
0.08	0.09	{431,789}	{0.08458,0.08571}	{0.9538,0.9976}	{0.0411,0.0311}
	0.10	{179,204}	{0.08845,0.08982}	{0.9775,0.9934}	{0.0163,0.0222}
	0.11	{80,177}	{0.09264,0.09787}	{0.9778,0.9999}	{0.0269,0.0215}
	0.12	{54,91}	{0.09739,0.09822}	{0.9882,0.9989}	{0.0305,0.0088}
	0.13	{35,106}	{0.09947,0.10929}	{0.9797,0.9999}	{0.0315,0.0123}
0.09	0.10	{673,906}	{0.09487,0.09513}	{0.9762,0.9921}	{0.0322,0.0206}
	0.11	{154,229}	{0.09869,0.09937}	{0.9555,0.9869}	{0.0398,0.0217}
	0.12	{93,167}	{0.10448,0.10806}	{0.9859,0.9998}	{0.0446,0.0388}
	0.13	{69,125}	{0.10876,0.11385}	{0.9928,0.9999}	{0.0328,0.0287}
	0.14	{41,111}	{0.10638,0.12125}	{0.9521,0.9999}	{0.0193,0.0271}

3. Advantage of the Proposed Plan

In this section, we discuss the advantage of the proposed plan under the neutrosophic interval method over the plan designed under classical statistics. The neutrosophic sample size has the form $n_N = n + uI$, where n is the determined part and also the parameter of the plan under classical statistics, and uI is an indeterminate part for n and $I \in \{infI, supI\}$. The proposed sampling plan reduces the plan under classical statistics when $n_L = n_U = n$. According to References [20,21], a method which provides the parameters in range/interval is considered more effective than the method which provides the determined value under the uncertainty environment. The values of $n_N\epsilon\{n_L, n_U\}$ and n are reported in Table 3 when $\alpha = 0.05$ and $\beta = 0.05$.

From Table 4, the proposed plan has the parameter in the indeterminacy interval while the existing sampling plan under classical statistics has determined values. For example, when AQL = 0.05 and LQL = 0.06, the proposed plan has the neutrosophic interval in sample size $n_N = 320 + 320I$, $I \in \{0, 1.7031\}$. Then, the neutrosophic sample size is $n_N\epsilon\{320, 545\}$ for $I \in \{0, 1.7031\}$. So, under conditions of uncertainty, the experimenter should select a sample size between 320 and 545. The existing sampling plan provides the determined value of 320. This means that the proposed plan is more effective and adequate under uncertainty and this theory coincides with the findings of [20,21].

Table 3. The neutrosophic fuzzy plan parameters when $\alpha = 0.05$, $\beta = 0.05$.

CV_{AQL}	CV_{LQL}	$n_N \epsilon \{n_L, n_U\}$	n
0.05	0.06	{320,545}	320
	0.07	{81,143}	81
	0.08	{55,102}	55
	0.09	{24,85}	24
	0.10	{18,65}	18
0.06	0.07	{289,490}	289
	0.08	{109,163}	109
	0.09	{51,137}	51
	0.10	{39,79}	39
	0.11	{23,76}	23
0.07	0.08	{412,689}	412
	0.09	{110,277}	110
	0.10	{50,104}	50
	0.11	{41,102}	41
	0.12	{28,118}	28
0.08	0.09	{431,789}	431
	0.10	{179,204}	179
	0.11	{80,177}	80
	0.12	{54,91}	54
	0.13	{35,106}	35
0.09	0.10	{673,906}	673
	0.11	{154,229}	154
	0.12	{93,167}	93
	0.13	{69,125}	69
	0.14	{41,111}	41

4. Case Study

In this section, we discuss the application of the proposed sampling plan using real data from the concrete industry. Concrete is an important material in building and road construction. The comprehensive strength is the maximum load which the material can sustain without any fracture. This quality characteristic is measured for engineering buildings and other structures. Therefore, engineers are more concerned about the stability of concrete strength. The comprehensive strength of concrete is not always a crisp value in practice. According to Reference [30] "Input fields of the fuzzy expert system are the weight percent of cement, water, blast furnace slag, fly ash, super plasticizer, fine aggregate, coarse aggregate, and age of the concrete". Measuring the strength of concrete is a complex system and there is a chance of having some indeterminate, incomplete, imprecise, or vague (unclear) observations. In this situation, the presentation of data in a set is called neutrosophication, making the proposed plan more effective in application than the existing sampling plan. The existing plan can be applied if data are deneutrosophied by replacing the neutrosophic data with the mid values of data, see [19]. Imprecise observations lead to uncertainty in the plan parameters needed for the inspection of a concrete product. Suppose that for this lot sentencing, AQL = 0.05, LQL = 0.09, $\alpha = 5\%$, and $\beta = 10\%$. The neutrosophic plan parameters from Table 1 are $n_N \epsilon \{21, 40\}$ and $k_N \epsilon \{0.0685, 0.0695\}$. The comprehensive strength of 25 concrete mixture specimens having some indeterminate, incomplete, imprecise, and vague observations is shown in Table 4.

Table 4. The data of concrete mixture specimens.

Column 1	Column 2	Column 3	Column 4	Column 5
(36.3, 36.9)	(40.1, 40.1)	(31.8, 32.1)	(33.6, 33.6)	(34.9, 35.2)
(31.2, 31.2)	(32.8, 32.8)	(25.8, 25.8)	(30.8, 32.2)	(32.9, 32.9)
(30.9, 30.9)	(31.9, 32.4)	(35.6, 35.6)	(30.9, 30.9)	(27.8,29.1)
(24.9, 24.9)	(31.6, 31.6)	(27.9, 28.2)	(33.7, 33.7)	(38.4, 38.4)
(28.5, 28.9)	(31.4, 31.8)	(26.9, 26.9)	(32.7, 32.7)	(34.1,34.6)

The NCV for the real data is computed as follows:
$\overline{X}_N \epsilon \{31.89, 32.13\}$ and $s_N \epsilon \{3.64, 3.67\}$.
So,
$\widehat{CV}_N \epsilon \{0.1133, 0.1152\}$, $\overline{X}_N \epsilon \{31.89, 32.13\}$, and $s_N \epsilon \{3.64, 3.67\}$.
The proposed plan is implemented as follows:

Step-1: Select a random sample of size $n_N \epsilon \{21, 40\}$ and compute $\widehat{CV}_N \epsilon \{0.1133, 0.1152\}$.
Step-2: Reject a lot of concrete product since $\widehat{CV}_N \epsilon \{0.1133, 0.1152\} > k_N \epsilon \{0.0685, 0.0695\}$.

5. Concluding Remarks

In this manuscript, we originally designed a sampling plan based on the neutrosophic coefficient of variation (NCV) which can be applied when observations/parameters are indeterminate in practice. The structure of the proposed plan is given to find the neutrosophic plan parameters. The application of the proposed neutrosophic plan is given with the aid of company data. The proposed plan under the neutrosophic interval method is more effective and adequate under the uncertainty environment. The proposed plan using big data from the industry can be studied as future research. The estimation and properties of neutrosophic normal distribution and neutrosophic chi-square distribution can be considered as future research.

Author Contributions: Conceived and designed the experiments, M.A.; Performed the experiments, M.A.; Analyzed the data, M.A. and M.S.A.; Contributed reagents/materials/analysis tools, M.A.; Wrote the paper, M.A. and M.S.A.

Funding: This article was funded by the Deanship of Scientific Research (DSR) at King Abdulaziz University, Jeddah. The authors, therefore, acknowledge with thanks DSR technical and financial support.

Acknowledgments: The authors are deeply thankful to editors and reviewers for their valuable suggestions to improve the quality of this manuscript.

Conflicts of Interest: The authors declare no conflict of interest regarding this paper.

References

1. Montgomery, D.C. *Introduction to Statistical Quality Control*; John Wiley & Sons: Hoboken, NJ, USA, 2007.
2. Yan, A.; Liu, S.; Dong, X. Variables two stage sampling plans based on the coefficient of variation. *J. Adv. Mech. Des. Syst. Manuf.* **2016**, *10*, JAMDSM0002. [CrossRef]
3. Yan, A.-J.; Aslam, M.; Azam, M.; Jun, C.-H. Developing a variable repetitive group sampling plan based on the coefficient of variation. *J. Ind. Prod. Eng.* **2017**, *34*, 398–405. [CrossRef]
4. Yan, A.; Liu, S.; Dong, X. Designing a multiple dependent state sampling plan based on the coefficient of variation. *SpringerPlus* **2016**, *5*, 1447. [CrossRef] [PubMed]
5. Reed, G.F.; Lynn, F.; Meade, B.D. Use of coefficient of variation in assessing variability of quantitative assays. *Clin. Diagn. Lab. Immunol.* **2002**, *9*, 1235–1239. [CrossRef] [PubMed]
6. Pereira, M.A.; Weggemans, R.M.; Jacobs, D.R., Jr.; Hannan, P.J.; Zock, P.L.; Ordovas, J.M.; Katan, M.B. Within-person variation in serum lipids: Implications for clinical trials. *Int. J. Epidemiol.* **2004**, *33*, 534–541. [CrossRef] [PubMed]
7. Parsons, H.M.; Ekman, D.R.; Collette, T.W.; Viant, M.R. Spectral relative standard deviation: A practical benchmark in metabolomics. *Analyst* **2009**, *134*, 478–485. [CrossRef] [PubMed]
8. He, X.; Oyadiji, S. Application of coefficient of variation in reliability-based mechanical design and manufacture. *J. Mater. Process. Technol.* **2001**, *119*, 374–378. [CrossRef]
9. Yeong, W.C.; Khoo, M.B.C.; Teoh, W.L.; Castagliola, P. A control chart for the multivariate coefficient of variation. *Qual. Reliab. Eng. Int.* **2016**, *32*, 1213–1225. [CrossRef]
10. Afshari, R.; Sadeghpour Gildeh, B. Designing a multiple deferred state attribute sampling plan in a fuzzy environment. *Am. J. Math. Manag. Sci.* **2017**, *36*, 328–345. [CrossRef]
11. Sadeghpour Gildeh, B.; Baloui Jamkhaneh, E.; Yari, G. Acceptance single sampling plan with fuzzy parameter. *Iran. J. Fuzzy Syst.* **2011**, *8*, 47–55.

12. Divya, P. Quality interval acceptance single sampling plan with fuzzy parameter using poisson distribution. *Int. J. Adv. Res Technol.* **2012**, *1*, 115–125.
13. Turanoğlu, E.; Kaya, İ.; Kahraman, C. Fuzzy acceptance sampling and characteristic curves. *Int. J. Comput. Intell. Syst.* **2012**, *5*, 13–29. [CrossRef]
14. Jamkhaneh, E.B.; Gildeh, B.S. Acceptance double sampling plan using fuzzy poisson distribution. *World Appl. Sci. J.* **2011**, *15*, 1692–1702.
15. Jamkhaneh, E.B.; Gildeh, B.S. Sequential sampling plan using fuzzy sprt. *J. Intell. Fuzzy Syst.* **2013**, *25*, 785–791.
16. Venkateh, A.; Elango, S. Acceptance sampling for the influence of trh using crisp and fuzzy gamma distribution. *Aryabhatta J. Math. Inform.* **2014**, *6*, 119–124.
17. Elango, S.; Venkatesh, A.; Sivakumar, G. A fuzzy mathematical analysis for the effect of trh using acceptance sampling plans. *Int. J. Pure Appl. Math.* **2017**, *117*, 1–11.
18. Smarandache, F. Neutrosophic logic—A generalization of the intuitionistic fuzzy logic. In *Multispace & Multistructure. Neutrosophic Transdisciplinarity (100 Collected Papers of Science)*; Infinite Study: El Segundo, CA, USA, 2010; Volume 4, p. 396.
19. Smarandache, F. *Introduction to Neutrosophic Statistics*; Infinite Study: El Segundo, CA, USA, 2014.
20. Chen, J.; Ye, J.; Du, S. Scale effect and anisotropy analyzed for neutrosophic numbers of rock joint roughness coefficient based on neutrosophic statistics. *Symmetry* **2017**, *9*, 208. [CrossRef]
21. Chen, J.; Ye, J.; Du, S.; Yong, R. Expressions of rock joint roughness coefficient using neutrosophic interval statistical numbers. *Symmetry* **2017**, *9*, 123. [CrossRef]
22. Senturk, S.; Erginel, N. Development of fuzzy x¯∼-r∼ and x¯∼-s∼ control charts using α-cuts. *Inf. Sci.* **2009**, *179*, 1542–1551. [CrossRef]
23. Aslam, M. A new sampling plan using neutrosophic process loss consideration. *Symmetry* **2018**, *10*, 132. [CrossRef]
24. Aslam, M.; Arif, O. Testing of grouped product for the weibull distribution using neutrosophic statistics. *Symmetry* **2018**, *10*, 403. [CrossRef]
25. Aslam, M.; Raza, M.A. Design of new sampling plans for multiple manufacturing lines under uncertainty. *Int. J. Fuzzy Syst.* **2018**, 1–15. [CrossRef]
26. Aslam, M. Design of sampling plan for exponential distribution under neutrosophic statistical interval method. *IEEE Access* **2018**, *6*, 64153–64158. [CrossRef]
27. Aslam, M.; AL-Marshadi, A. Design of sampling plan using regression estimator under indeterminacy. *Symmetry* **2018**, *10*, 754. [CrossRef]
28. Iglewicz, B.; Myers, R.H.; Howe, R.B. On the percentage points of the sample coefficient of variation. *Biometrika* **1968**, *55*, 580–581. [CrossRef]
29. Tong, Y.; Chen, Q. Sampling inspection by variables for coefficient of variation. *Theor. Appl. Probab.* **1991**, *3*, 315–327.
30. Abolpour, B.; Abolpour, B.; Abolpour, R.; Bakhshi, H. Estimation of concrete compressive strength by a fuzzy logic model. *Res. Chem. Intermed.* **2013**, *39*, 707–719. [CrossRef]

symmetry

MDPI

Article

Group Decision Making Based on Triangular Neutrosophic Cubic Fuzzy Einstein Hybrid Weighted Averaging Operators

Aliya Fahmi [1,*], Fazli Amin [1], Madad Khan [2] and Florentin Smarandache [3]

1. Department of mathematics, Hazara University Mansehra, Dhodial 21130, Pakistan; fazliamin@hu.edu.pk
2. Department of Mathematics, COMSATS University Islamabad, Abbottabad Campus 22060, Pakistan; madadmath@yahoo.com
3. Mathematics Department, University of New Mexico, 705 Gurley Ave., Gallup, NM 87301, USA; fsmarandache@gmail.com

* Correspondence: aliyafahmi@gmail.com

Received: 17 December 2018; Accepted: 28 January 2019; Published: 2 February 2019

Abstract: In this paper, a new concept of the triangular neutrosophic cubic fuzzy numbers (TNCFNs), their score and accuracy functions are introduced. Based on TNCFNs, some new Einstein aggregation operators, such as the triangular neutrosophic cubic fuzzy Einstein weighted averaging (TNCFEWA), triangular neutrosophic cubic fuzzy Einstein ordered weighted averaging (TNCFEOWA) and triangular neutrosophic cubic fuzzy Einstein hybrid weighted averaging (TNCFEHWA) operators are developed. Furthermore, their application to multiple-attribute decision-making with triangular neutrosophic cubic fuzzy (TNCF) information is discussed. Finally, a practical example is given to verify the developed approach and to demonstrate its practicality and effectiveness.

Keywords: triangular neutrosophic cubic fuzzy number; Einstein t-norm; arithmetic averaging operator; Multi-attribute decision making; numerical application

1. Introduction

Atanassov [1] introduced the IFS, which is a generalization of FS. Atanassov [2] introduced operations and relations over IFSs taking as a point of departure respective definitions of relations and operations over fuzzy sets. Bustince et al. [3] introduced the characterization of certain structures of intuitionistic relations according to the structures of two concrete fuzzy relations. Deschrijver et al. [4] established the relationships between intuitionistic fuzzy sets (Atanassov, VII ITKR's Session, Sofia, June 1983 (Deposed in Central Sci.-Techn. Library of Bulg. Acad. of Sci., 1697/84) (in Bulgarian)), L-fuzzy sets. Deschrijver et al. [5] defined the mathematical relationship between intuitionistic fuzzy sets and other models of imprecision. Jun et al. [6] introduced the cubic set. Mohiuddin et al. [7] showed that the union of two internal cubic soft sets might not be internal. Turksen [8] showed that the proposed representation (1) exists for certain families of the conjugate pairs of t-norms and t-norms, and (2) resolves some of the difficulties associated with particular interpretations of conjunction, disjunction, and implication in fuzzy set theories.

Xu [9] developed some aggregation operators, such as the intuitionistic fuzzy weighted averaging operator, intuitionistic fuzzy ordered weighted averaging operator, and intuitionistic fuzzy hybrid aggregation operator, to aggregate intuitionistic fuzzy values. Xu et al. [10] developed some new geometric aggregation operators, such as the intuitionistic fuzzy weighted geometric (IFWG) operator and the intuitionistic fuzzy ordered weighted geometric (IFOWG) operator. Xu et al. [11] provided a survey of the aggregation techniques of intuitionistic fuzzy information and their applications in various fields, such as decision making, cluster analysis, medical diagnosis, forecasting, and manufacturing grid. Liu et al. [12] introduced and discussed the concept of intuitionistic fuzzy point operators. Zeng et al. [13] defined the situation with intuitionistic fuzzy information and developed an intuitionistic fuzzy ordered weighted distance (IFOWD) operator. The fuzzy set was introduced by Zadeh [14]. Zadeh [15] introduced the interval-valued fuzzy set Li et al. [16] proposed group decision-making methods of the interval-valued intuitionistic uncertain linguistic variable based on Archimedean t-norm and Choquet integral. Zhao et al. [17] developed some hesitant triangular fuzzy aggregation operators based on the Einstein operation: the hesitant triangular fuzzy Einstein weighted averaging (HTFEWA) operator. Xu et al. [18] introduced two new aggregation operators: dynamic intuitionistic fuzzy weighted averaging (DIFWA) operator and uncertain dynamic intuitionistic fuzzy weighted averaging (UDIFWA) operator.

The Neutrosophic Set (NS) was projected by Smarandache [19,20]. Neutrosophic sets are characterized by fact participation, an indeterminacy-enrollment work and misrepresentation participation, which are inside the ordinary or nonstandard unit interim $]^{-}0, 1^{+}[$ in order to apply NS to genuine applications. In order to apply NS to real-world applications, Aliya et al. [21] introduced the concept of the triangular cubic fuzzy number. Aliya et al. [22] introduced the triangular cubic hesitant fuzzy Einstein weighted averaging (TCHFEWA) operator, triangular cubic hesitant fuzzy Einstein ordered weighted averaging (TCHFEOWA) operator and triangular cubic hesitant fuzzy Einstein hybrid weighted averaging (TCHFEHWA) operator.

Beg et al. [23] introduced a computational means to manage situations in which experts assess alternatives in possible membership and non-membership values. Przemyslaw et al. [24] introduced a simple test that sometimes might be helpful in detecting non-separability at a glance.

The differences between Reference 21, 22 and the current paper are as Table 1:

Table 1. Difference between references 21, 22 and current paper.

Reference 21	Reference 22	Current Paper
Defines a new extension of the triangular cubic fuzzy number by using a cubic set.	Defines a new extension of the triangular cubic hesitant fuzzy number by using a cubic set.	Defines a new extension of the triangular neutrosophic cubic fuzzy number by using a neutrosophic set.
Introduced the triangular cubic fuzzy number, operational laws, and their score and accuracy functions.	Introduced the triangular cubic hesitant fuzzy number, operational laws, and their score, accuracy functions, membership uncertainty index and hesitation index.	Introduced the triangular cubic fuzzy number, operational laws, and their score and accuracy functions, membership uncertainty index and hesitation index.
Introduced the triangular cubic fuzzy hybrid aggregation operator.	Introduced three Einstein aggregation operators, such as the triangular cubic fuzzy hybrid aggregation operator, and the TCHFEWA, TCHFEOWA and TCHFEHWA operators	Introduced three Einstein aggregation operators, such as the triangular neutrosophic cubic fuzzy hybrid aggregation operator, and the TNCFEWA, TNCFEOWA and TNCFEHWA operators

Based on the above analysis, in this paper we develop TNCFNs, which is the generalization of the triangular neutrosophic intuitionistic fuzzy number and triangular neutrosophic interval fuzzy number. We perform some operations based on Einstein T-norm and Einstein T-conorm for TNCFNs. We also develop score and accuracy functions to compare two TNCFNs. Due to the developed operation, we propose the TNCFEWA operator, TNCFEOWA operator, and TNCFEHWA operator, to aggregate a collection of TNCFNs.

This paper is organized as follows. In Section 2, we define some concepts of FS, CS, and TNCFNs. In Section 3, we discuss some Einstein operations on TNCFNs and their properties. In Section 4, we first develop some novel arithmetic averaging operators, such as the TNCFEWA operator, TNCFEOWA operator, and TNCFEHWA operator, for aggregating a group of TNCFNs. In Section 5, we apply the

TNCFEHWA operator to MADM with TNCFNs material. In Section 6, we offer a numerical example consistent with our approach. In Section 7, we discuss comparison analysis. In Section 8, we present a conclusion.

2. Preliminaries

Definition 1. [15]. *Let H be a fixed set, a FS F in H is defined as:*$F = \{(h, \Gamma_F(h)|h \in H\}$ *where* $\Gamma_F(h)$ *is a mapping from h to the closed interval* $[0, 1]$ *and for each* $h \in H, \Gamma_F(h)$ *is called the degree of membership of h in H.*

Definition 2. *Let H is a fixed set and an interval-valued fuzzy set I in H is defined as* $I = \{h, R_I^-(h), R_I^+(h)|h \in H\}$, *where* $R_I^- : H \to [0, 1]$ *and* $R_I^+ : H \to [0, 1]$. *The* $R_I^-(h)$ *is lower membership and* $R_I^+(h)$ *is upper membership such that* $0 \le R_I^-(h) \le R_I^+(h) \le 1$.

Definition 3. [1]. *An IFS Đ in H* is given by $Đ = \{(h, R_Đ(h), \Omega_Đ(h)|h \in H\}$, *where* $R_Đ : H \to [0, 1]$ *and* $\Omega_Đ : H \to [0, 1]$, *with the condition* $0 \le R_Đ(h) + \Omega_Đ(h) \le 1$.

The numbers $R_Đ(h)$ *and* $\Omega_Đ(h)$ *represent, respectively, the membership degree and non-membership degree of the element h to the set Đ.*

Triangular Neutrosophic Cubic Fuzzy Number

Definition 4. *Let* $A_1 = \left\{\begin{array}{c} [p_1(h), q_1(h), \\ r_1(h)], \\ \langle[Y_1^-(h), \\ R_1^-(h), \delta_1^-(h)], \\ [Y_1^+(h), \\ R_1^+(h), \delta_1^+(h)], \\ [Y_1(h), \\ R_1(h), \delta_1(h)]\rangle \\ |h \in H \end{array}\right\}$ and $A_2 = \left\{\begin{array}{c} [p_2(h), q_2(h), \\ r_2(h)], \\ \langle[Y_2^-(h), \\ R_2^-(h), \delta_2^-(h)], \\ [Y_2^+(h), \\ R_2^+(h), \delta_2^+(h)], \\ [Y_2(h), \\ R_2(h), \delta_2(h)]\rangle \\ |h \in H \end{array}\right\}$ *are two TNCFNs, some operations on TNCFNs are defined as follows:*

(a) $A_1 \subseteq A_2$ *iff* $\forall h \in H$, $p_1(h)) \ge p_2(h)$ $q_1(h) \ge q_2(h)$, $r_1(h) \ge r_2(h)$, $Y_1^-(h) \ge Y_2^-(h)$, $R_1^-(h) \ge R_2^-(h)$, $\delta_1^-(h) \ge \delta_1^-(h)$, $Y_1^+(h) \ge Y_2^+(h)$ $R_1^+(h) \ge R_2^+(h)$,, $\delta_1^+(h) \ge \delta_2^+(h)$ and $Y_1(h) \ge Y_2(h)$ $\delta_1(h) \le \delta_2(h)$.

(b) $A_1 \cap_{T,S} A_2 = T[p_1(h), p_2(h)]$, $T[q_1(h), q_2(h)]$, $T[r_1(h), r_2(h)]$, $\langle T[Y_1^-(h), Y_2^-(h)]$, $T[R_1^-(h), R_2^-(h)]$, $T[\delta_1^-(h), \delta_2^-(h)]$, $T[Y_1^+(h), Y_2^+(h)]$, $T[R_1^+(h), R_2^+(h)]$, $T[\delta_1^+(h), \delta_2^+(h)]$, $S[Y_1(h), Y_2(h)]$, $S[R_1(h), R_2(h)]$, $S[\delta_1(h), \delta_2(h)]\rangle$

Example 1. *Let* $\ddot{A}_1 = \left\{\begin{array}{c} \langle[0.1, 0.2, 0.3], \\ [0.2, 0.4, 0.6], \\ [0.4, 0.6, 0.8], \\ [0.3, 0.5, 0.7]\rangle \end{array}\right\}$ *and* $\ddot{A}_2 = \left\{\begin{array}{c} \langle[0.103, 0.104, 0.105], \\ [0.100, 0.102, 0.104], \\ [0.102, 0.104, 0.106], \\ [0.101, 0.103, 0.105]\rangle \end{array}\right\}$ *be two TNCFSs*

(a) $\ddot{A}_1 \subseteq \ddot{A}_2$, *if* $\forall \ddot{z} \in Z$, $0.1 \ge 0.103, 0.2 \ge 0.104, 0.3 \ge 0.105, 0.2 \ge 0.100, 0.4 \ge 0.102, 0.6 \ge 0.104, 0.4 \ge 0.102, 0.6 \ge 0.104, 0.8 \ge 0.106$ *and* $0.3 \le 0.101, 0.5 \le 0.103, 0.7 \le 0.105$.

(b) $\ddot{A}_1 \cap_{T,S} \ddot{A}_2 = T[0.1, 0.103], T[0.2, 0.104], T[0.3, 0.105], [T[0.2, 0.10], T[0.4, 0.12], T[0.6, 0.14], T[0.4, 0.12], T[0.6, 0.14], T[0.8, 0.16]$ *and* $S[0.3, 0.11], S[0.7, 0.15]$.

Definition 5. *Let* $C = \left\{ \begin{array}{c} [p_C(h), q_C(h), r_C(h)] \\ \Big\langle [A_C^-(h), R_C^-(h), \widetilde{U}_C^-(h)], \\ [A_C^+(h), R_C^+(h), \widetilde{U}_C^+(h)], \\ [A_C(h), R_C(h), \widetilde{U}_C(h)] \Big\rangle \Big| h \in H \end{array} \right\}$ *be a TNCFN and then the score function* $S(C)$*, accuracy function* $N(C)$*, membership uncertainty index* $T(C)$ *and hesitation uncertainty index* $G(C)$ *of a TNCFN C are defined by*

$$S(C) = \frac{\begin{array}{c} \Big\langle [p_C(h) + q_C(h) + r_C(h)][[A_C^-(h) + R_C^-(h) + \widetilde{U}_C^-(h)] \\ +[A_C^+(h) + R_C^+(h) + \widetilde{U}_C^+(h)]] \\ -[A_C(h) + R_C(h) + \widetilde{U}_C(h)] \Big\rangle \end{array}}{27},$$

$$N(C) = \frac{\begin{array}{c} \Big\langle [p_C(h) + q_C(h) + r_C(h)][[A_C^-(h) + R_C^-(h) + \widetilde{U}_C^-(h)] \\ +[A_C^+(h) + R_C^+(h) + \widetilde{U}_C^+(h)]] + [A_C(h) + R_C(h) + \widetilde{U}_C(h)] \Big\rangle \end{array}}{27}$$

$$T(C) = \Big\langle [p_C(h) + q_C(h) + r_C(h)][[A_C^+(h) + R_C^+(h) + \widetilde{U}_C^+(h)] + [A_C(h) + R_C(h) + \widetilde{U}_C(h)] - [A_C^-(h) + R_C^-(h) + \widetilde{U}_C^-(h)] \Big\rangle,$$

$$G(C) = \Big\langle [p_C(h) + q_C(h) + r_C(h)][[A_C^+(h) + R_C^+(h) + \widetilde{U}_C^+(h)] + [A_C^-(h) + R_C^-(h) + \widetilde{U}_C^-(h)] - [A_C(h) + R_C(h) + \widetilde{U}_C(h)] \Big\rangle.$$

Example 2. *Let* $C = \left\{ \begin{array}{c} \langle [0.101, 0.102, \\ 0.103], [0.5, 0.7, 0.9], \\ [0.7, 0.9, 0.11], [0.6, \\ 0.8, 0.10] \rangle \end{array} \right\}$ *be a TNCFN. Then the score function* $S(C)$*, accuracy function* $H(C)$*, membership uncertainty index* $T(C)$ *and hesitation uncertainty index* $G(C)$ *of a TNCFN C are defined by*

$$S(C) = \frac{\begin{array}{c} \langle [0.101 + 0.102 + 0.103][0.5 + 0.7 + 0.9]] \\ +[0.7 + 0.9 + 0.11] - [0.6 + 0.8 + 0.10] \rangle \end{array}}{27} = \frac{0.306(3.81 - 1.5)}{27} = \frac{0.7068}{27} = 0.0261,$$

$$H(C) = \frac{\begin{array}{c} \langle [0.101 + 0.102 + 0.103][0.5 + 0.7 + 0.9]] \\ +[0.7 + 0.9 + 0.11] + [0.6 + 0.8 + 0.10] \rangle \end{array}}{27} = \frac{0.306(3.81 + 1.5)}{27} = \frac{1.6248}{27} = 0.0601,$$

$$T(C) = \left\{ \begin{array}{c} [0.101 + 0.102 + 0.103] \langle [0.7 + 0.9 + 0.11] \\ +[0.6 + 0.8 + 0.10] - [0.5 + 0.7 + 0.9] \rangle \end{array} \right\} = 0.6(1.71 + 1.5 - 2.1) = 0.306(3.21 - 2.1) = 0.3396,$$

$$G(C) = \left\{ \begin{array}{c} \langle [0.101 + 0.102 + 0.103][0.7 + 0.9 + 0.11] \\ +[0.5 + 0.7 + 0.9] - [0.6 + 0.8 + 0.10] \rangle \end{array} \right\} = 0.306(1.71 + 2.1 - 1.5) = 0.7068.$$

See Figure 1.

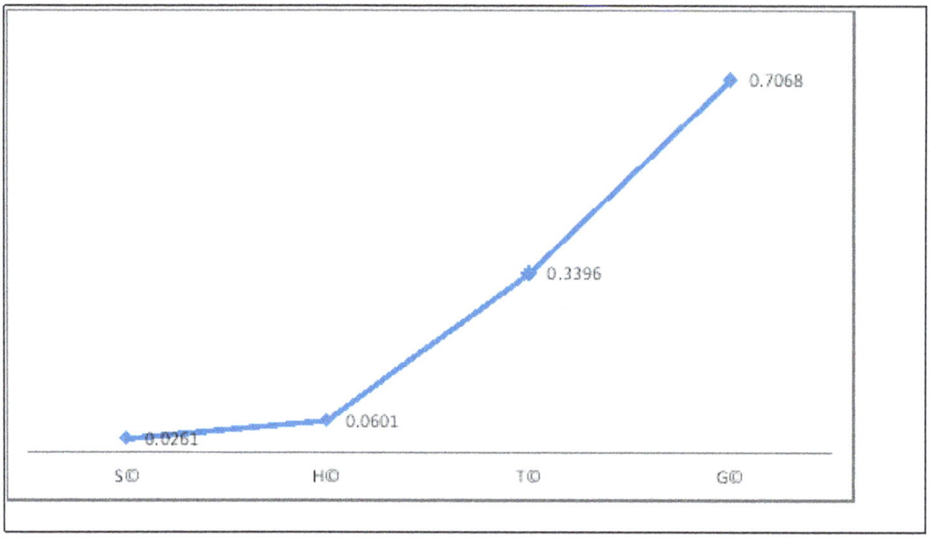

Figure 1. The score function, accuracy function, membership uncertainty index and hesitation uncertainty index are ranking of TNCFN.

3. Some Einstein Operations on TNCFNs

Definition 6. *Let* $C = \left\{ \begin{array}{c} [a(h), e(h), \\ G(h)], \langle [Y=^{-}(h), \\ k^{-}(h), \Gamma^{-}(h)], \\ [Y=^{+}(h), k^{+}(h), \\ \Gamma^{+}(h)], [Y=(h), \\ k(h), \Gamma(h)] \rangle \\ |h \in H \end{array} \right\}$, $C_1 = \left\{ \begin{array}{c} [a_1(h), e_1(h), \\ G_1(h)] \langle [Y=_1^{-}(h), \\ k_1^{-}(h), \Gamma_1^{-}(h)], \\ [Y=_1^{+}(h), k_1^{+}(h), \\ \Gamma_1^{+}(h)], [Y=_1(h), \\ k_1(h), \Gamma_1(h)] \rangle \\ |h \in H \end{array} \right\}$ *and* $C_2 = \left\{ \begin{array}{c} [a_2(h), e_2(h), \\ G_2(h)] \langle [Y=_2^{-}(h), \\ k_2^{-}(h), \Gamma_2^{-}(h)], \\ [Y=_2^{+}(h), k_2^{+}(h), \\ \Gamma_2^{+}(h)], [Y=_2(h), \\ k_2(h), \Gamma_2(h)] \rangle \\ |h \in H \end{array} \right\}$ *be any three TNCFNs. Then some Einstein operations of* C_1 *and* C_2 *can be defined as:* $C_1 + C_2 = \left\langle \left[\begin{array}{c} \frac{a_1(h)+a_2(h)}{1+a_1(h)(1-a_2(h))}, \\ \frac{e_1(h)+e_2(h)}{1+e_1(h)(1-e_2(h))}, \\ \frac{G_1(h)+G_2(h)}{1+G_1(h)(1-G_2(h))} \end{array} \right], \left[\begin{array}{c} \frac{Y=_1^{-}(h)+Y=_2^{-}(h)}{1+Y=_1^{-}(h)(1-Y=_2^{-}(h))}, \\ \frac{Y=_1^{+}(h)+Y=_2^{+}(h)}{1+Y=_1^{+}(h)(1-Y=_2^{+}(h))} \end{array} \right], \left[\begin{array}{c} \frac{k_1^{-}(h)+k_2^{-}(h)}{1+k_1^{-}(h)(1-k_2^{-}(h))}, \\ \frac{k_1^{+}(h)+k_2^{+}(h)}{1+k_1^{+}(h)(1-k_2^{+}(h))} \end{array} \right], \left[\begin{array}{c} \frac{\Gamma_1^{-}(h)+\Gamma_2^{-}(h)}{1+\Gamma_1^{-}(h)(1-\Gamma_2^{-}(h))}, \\ \frac{\Gamma_1^{+}(h)+\Gamma_2^{+}(h)}{1+\Gamma_1^{+}(h)(1-\Gamma_2^{+}(h))} \end{array} \right] \right.$,

$$\left[\begin{array}{c} \frac{Y{=}_1(h)Y{=}_2(h)}{(1+(1-Y{=}_1(h))(1-Y{=}_2(h)))}, \\ \frac{\text{ķ}_1(h)\text{ķ}_2(h)}{(1+(1-\text{ķ}_1(h))(1-\text{ķ}_2(h)))}, \\ \frac{\Gamma_1(h)\Gamma_2(h)}{(1+((1-\Gamma_1(h))(1-\Gamma_2(h)))} \end{array}\right]\Bigg\rangle, \lambda C = \Bigg\langle \left[\begin{array}{c} \frac{[1+\text{ą}_C(h)]^\lambda-[1-\text{ą}_C(h)]^\lambda}{[1+\text{ą}_C(h)]^\lambda+[1-\text{ą}(h)]^\lambda}, \\ \frac{[1+\text{ę}_C(h)]^\lambda-[1-\text{ę}_C(h)]^\lambda}{[1+\text{ę}_C(h)]^\lambda+[1-\text{ę}(h)]^\lambda}, \\ \frac{[1+\text{Ģ}_C(h)]^\lambda-[1-\text{Ģ}_C(h)]^\lambda}{[1+\text{Ģ}_C(h)]^\lambda+[1-\text{Ģ}(h)]^\lambda} \end{array}\right], \left[\begin{array}{c} \frac{[1+Y{=}_C^-(h)]^\lambda-[1-Y{=}_C^-(h)]^\lambda}{[1+Y{=}_C^-(h)]^\lambda+[1-Y{=}_C^-(h)]^\lambda}, \\ \frac{[1+Y{=}_C^+(h)]^\lambda-[1-Y{=}_C^+(h)]^\lambda}{[1+Y{=}_C^+(h)]^\lambda+[1-Y{=}_C^+(h)]^\lambda}, \\ \frac{[1+\text{ķ}_C^-(h)]^\lambda-[1-\text{ķ}_C^-(h)]^\lambda}{[1+\text{ķ}_C^-(h)]^\lambda+[1-\text{ķ}_C^-(h)]^\lambda}, \\ \frac{[1+\text{ķ}_C^+(h)]^\lambda-[1-\text{ķ}_C^+(h)]^\lambda}{[1+\text{ķ}_C^+(h)]^\lambda+[1-\text{ķ}_C^+(h)]^\lambda}, \\ \frac{[1+\Gamma_C^-(h)]^\lambda-[1-\Gamma_C^-(h)]^\lambda}{[1+\Gamma_C^-(h)]^\lambda+[1-\Gamma_C^-(h)]^\lambda}, \\ \frac{[1+\Gamma_C^+(h)]^\lambda-[1-\Gamma_C^+(h)]^\lambda}{[1+\Gamma_C^+(h)]^\lambda+[1-\Gamma_C^+(h)]^\lambda} \end{array}\right],$$

$$\left[\begin{array}{c} \frac{2[Y{=}_C(h)]^\lambda}{[(2-Y{=}_C(h)]^\lambda+[Y{=}_C(h)]^\lambda}, \\ \frac{2[\text{ķ}_C(h)]^\lambda}{[(2-\text{ķ}_C(h)]^\lambda+[\text{ķ}_C(h)]^\lambda}, \\ \frac{2[\Gamma_C(h)]^\lambda}{[(2-\Gamma_C(h)]^\lambda+[\Gamma_C(h)]^\lambda} \end{array}\right]\Bigg\rangle.$$

Proposition 1. *Let $\ddot{A}$, $\ddot{A}_1$ and $\ddot{A}_2$ be three TNCFNs, λ, λ_1, $\lambda_2 > 0$, then we have:*

(1) $\ddot{A}_1 + \ddot{A}_2 = \ddot{A}_2 + \ddot{A}_1$,

(2) $\lambda(\ddot{A}_1 + \ddot{A}_2) = \lambda\ddot{A}_2 + \lambda\ddot{A}_1$,

(3) $\lambda_1\ddot{A} + \lambda_2\ddot{A} = (\lambda_1 + \lambda_2)\ddot{A}$.

Proof. The proof of these propositions is provided in Appendix A. □

Remark 1. *If $\alpha_1 \leq_{L_{\mathrm{TNCFN}}} \alpha_2$, then $\alpha_1 \leq \alpha_2$, the total order is the partial order on L_{TNCFN}, see Figure 2.*

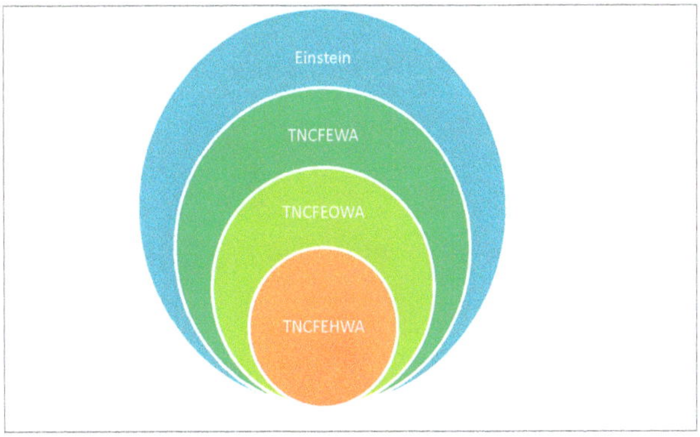

Figure 2. New extend aggregation operators, such as TNCFEWA, TNCFEOWA and TNCFEHWA operators.

4. Triangular Neutrosophic Cubic Fuzzy Averaging Operators Based on Einstein Operations

In this section, we define the aggregation operators.

4.1. Triangular Neutrosophic Cubic Fuzzy Einstein Weighted Averaging Operator

Definition 7. Let $\ddot{A} = \left\{ \begin{array}{c} [\alpha(h), \beta(h), \\ \Delta(h)]\langle[\xi_1^-(h), \\ \xi_2^-(h), \xi_3^-(h)], \\ [\xi_1^+(h), \xi_2^+(h), \\ \xi_3^+(h)], [\xi_1(h), \\ \xi_2(h), \xi_3(h)]\rangle \\ |h \in H \end{array} \right\}$ be a collection of TNCFNs in L_{TNCFN} and $\ddot{\omega} = (\ddot{\omega}_1, \ddot{\omega}_2, \ldots, \ddot{\omega}_n)^T$ be the weight vector, with $\ddot{\omega}_j \in [0, 1]$, $\sum_{j=1}^{n} \ddot{\omega}_j = 1$. Hence TNCFEWA operator of dimension n is a mapping TNCFEWA : $L^n_{\text{TNCFN}} \rightarrow L_{\text{TNCFN}}$ and defined by $\text{TNCFEWA}(\ddot{A}_1, \ddot{A}_2, \ldots, \ddot{A}_n) = \ddot{\omega}_1 \ddot{A}_1 + \ddot{\omega}_2 \ddot{A}_2, \ldots, \ddot{\omega}_n \ddot{A}_n$.

If $\ddot{\omega} = (\frac{1}{n}, \frac{1}{n}, \ldots, \frac{1}{n})^T$. *Hence the TNCFEWA operator is reduced to TNCFEA operator of dimension n. It can be defined as follows:* $\text{TNCFEA}(\ddot{A}_1, \ddot{A}_2, \ldots, \ddot{A}_n) = \frac{1}{n}(\ddot{A}_1, \ddot{A}_2, \ldots, \ddot{A}_n)$.

Theorem 1. *Let* $\ddot{A} = \left\{ \begin{array}{c} [\alpha_1(h), \beta_1(h), \\ \Delta_1(h)], \langle[p_1^-(h), \\ q_1^-(h), r_1^-(h)], \\ [p_1^+(h), q_1^+(h), \\ r_1^+(h)], [p_1(h), \\ q_1(h), r_1(h)]\rangle \\ |h \in H \end{array} \right\}$ *be a collection of TNCFNs in* L_{TNCFN}. *The amassed an incentive by utilizing the TNCFEWA operator is additionally a TNCFN and TNCFEWA.*

$$(\ddot{A}_1, \ddot{A}_2, \ldots, \ddot{A}_n) = \left\langle \begin{bmatrix} \frac{\prod_{j=1}^{n}[1+\alpha_1(h)]^{\ddot{\omega}} - \prod_{j=1}^{n}[1-\alpha_1(h)]^{\ddot{\omega}}}{\prod_{j=1}^{n}[1+\alpha_1(h)]^{\ddot{\omega}} + \prod_{j=1}^{n}[1-\alpha_1(h)]^{\ddot{\omega}}}, \\ \frac{\prod_{j=1}^{n}[1+\beta_1(h)]^{\ddot{\omega}} - \prod_{j=1}^{n}[1-\beta_1(h)]^{\ddot{\omega}}}{\prod_{j=1}^{n}[1+\beta_1(h)]^{\ddot{\omega}} + \prod_{j=1}^{n}[1-\beta_1(h)]^{\ddot{\omega}}}, \\ \frac{\prod_{j=1}^{n}[1+\Delta_1(h)]^{\ddot{\omega}} - \prod_{j=1}^{n}[1-\Delta_1(h)]^{\ddot{\omega}}}{\prod_{j=1}^{n}[1+\Delta_1(h)]^{\ddot{\omega}} + \prod_{j=1}^{n}[1-\Delta_1(h)]^{\ddot{\omega}}}, \end{bmatrix}, \begin{bmatrix} \frac{\prod_{j=1}^{n}[1+p_1^-(h)]^{\ddot{\omega}} - \prod_{j=1}^{n}[1-p_1^-(h)]^{\ddot{\omega}}}{\prod_{j=1}^{n}[1+p_1^-(h)]^{\ddot{\omega}} + \prod_{j=1}^{n}[1-p_1^-(h)]^{\ddot{\omega}}}, \\ \frac{\prod_{j=1}^{n}[1+q_2^-(h)]^{\ddot{\omega}} - \prod_{j=1}^{n}[1-q_2^-(h)]^{\ddot{\omega}}}{\prod_{j=1}^{n}[1+q_2^-(h)]^{\ddot{\omega}} + \prod_{j=1}^{n}[1-q_2^-(h)]^{\ddot{\omega}}}, \\ \frac{\prod_{j=1}^{n}[1+r_3^-(h)]^{\ddot{\omega}} - \prod_{j=1}^{n}[1-r_3^-(h)]^{\ddot{\omega}}}{\prod_{j=1}^{n}[1+r_3^-(h)]^{\ddot{\omega}} + \prod_{j=1}^{n}[1-r_3^-(h)]^{\ddot{\omega}}} \end{bmatrix}, \right.$$
$$\left. \begin{bmatrix} \frac{\prod_{j=1}^{n}[1+p_1^+(h)]^{\ddot{\omega}} - \prod_{j=1}^{n}[1-p_1^+(h)]^{\ddot{\omega}}}{\prod_{j=1}^{n}[1+p_1^+(h)]^{\ddot{\omega}} + \prod_{j=1}^{n}[1-p_1^+(h)]^{\ddot{\omega}}}, \frac{\prod_{j=1}^{n}[1+q_2^+(h)]^{\ddot{\omega}} - \prod_{j=1}^{n}[1-q_2^+(h)]^{\ddot{\omega}}}{\prod_{j=1}^{n}[1+q_2^+(h)]^{\ddot{\omega}} + \prod_{j=1}^{n}[1-q_2^+(h)]^{\ddot{\omega}}}, \\ \frac{\prod_{j=1}^{n}[1+r_3^+(h)]^{\ddot{\omega}} - \prod_{j=1}^{n}[1-r_3^+(h)]^{\ddot{\omega}}}{\prod_{j=1}^{n}[1+r_3^+(h)]^{\ddot{\omega}} + \prod_{j=1}^{n}[1-r_3^+(h)]^{\ddot{\omega}}} \end{bmatrix}, \begin{bmatrix} \frac{2\prod_{j=1}^{n}[p_1(h)]^{\ddot{\omega}}}{\prod_{j=1}^{n}[(2-p_1(h)]^{\ddot{\omega}} + \prod_{j=1}^{n}[p_1(h)]^{\ddot{\omega}}}, \frac{2\prod_{j=1}^{n}[q_2(h)]^{\ddot{\omega}}}{\prod_{j=1}^{n}[(2-q_2(h)]^{\ddot{\omega}} + \prod_{j=1}^{n}[q_2(h)]^{\ddot{\omega}}}, \\ , \frac{2\prod_{j=1}^{n}[r_3(h)]^{\ddot{\omega}}}{\prod_{j=1}^{n}[(2-r_3(h)]^{\ddot{\omega}} + \prod_{j=1}^{n}[r_3(h)]^{\ddot{\omega}}} \end{bmatrix} \right\rangle$$

where $\ddot{\omega} = (\ddot{\omega}_1, \ddot{\omega}_2, \ldots, \ddot{\omega}_n)^T$ *be the weight vector of* $\ddot{A}_j (j = 1, 2, \ldots, n)$ *such that* $\ddot{\omega}_j \in [0, 1]$ *and* $\sum_{j=1}^{n} \ddot{\omega}_j = 1$. *If* $\alpha_1(h) = \alpha_1(h)$, $\beta_1(h) = \beta_1(h)$, $\Delta_1(h) = \Delta_1(h)$, $p_1^-(h) = p_1^-(h)$, $q_2^-(h) = q_2^-(h)$, $r_3^-(h) = r_3^-(h)$, $p_1^+(h) = p_1^+(h)$, $q_2^+(h) = q_2^+(h)$, $r_3^+(h) = r_3^+(h)$ *and* $p_1(h) = p_1(h)$, $q_2(h) = q_2(h)$,

$r_3(h) = r_3(h)$. Then the TNCFN $\ddot{A} = \left\{ \begin{array}{c} [\alpha_1(h),\, \beta_1(h), \\ \Delta_1(h)],\, \langle [p_1^-(h), \\ q_1^-(h),\, r_1^-(h)], \\ [p_1^+(h),\, q_1^+(h), \\ r_1^+(h)], [p_1(h), \\ q_1(h),\, r_1(h)] \rangle \\ |h \in H \end{array} \right\}$ *are reduced to the triangular neutrosophic cubic fuzzy numbers* $\ddot{A} = \left\{ \begin{array}{c} [\alpha_1(h),\, \beta_1(h), \\ \Delta_1(h)],\, \langle [p_1^-(h), \\ q_1^-(h),\, r_1^-(h)], \\ [p_1^+(h),\, q_1^+(h), \\ r_1^+(h)], [p_1(h), \\ q_1(h),\, r_1(h)] \rangle \\ |h \in H \end{array} \right\}$ *and the TNCFEWA operator is reduced to the TNCFEWA operator.*

Proof. The proof of this theorem is provided in Appendix B. □

Example 3. *Let* $C_1 = \left\{ \begin{array}{c} \langle [0.02,\, 0.03,\, 0.04], \\ [0.02,\, 0.04,\, 0.06], \\ [0.04,\, 0.06,\, 0.08], \\ [0.03,\, 0.05,\, 0.07] \rangle \end{array} \right\}$, $C_2 = \left\{ \begin{array}{c} \langle [0.205,\, 0.207,\, 0.209], \\ [0.211,\, 0.213,\, 0.215], \\ [0.213,\, 0.215,\, 0.217], \\ [0.212,\, 0.214,\, 0.216] \rangle \end{array} \right\}$ *and* $C_3 = \left\{ \begin{array}{c} \langle [0.004,\, 0.005,\, 0.006], \\ [0.102,\, 0.104,\, 0.106], \\ [0.104,\, 0.106,\, 0.108], \\ [0.103,\, 0.105,\, 0.107] \rangle \end{array} \right\}$ *be a TNCFN. Then the score function is defined by* $S(C_1) = \frac{\left\{ \begin{array}{c} \langle [0.02 + 0.03 + 0.04], \\ [0.02 + 0.04 + 0.06] + \\ [0.04 + 0.06 + 0.08] - \\ [0.03 + 0.05 + 0.07] \rangle \end{array} \right\}}{27} = 0.0005$, $S(C_2) = \frac{\left\{ \begin{array}{c} \langle [0.205 + 0.207 + 0.209], \\ [0.211 + 0.213 + 0.215] + \\ [0.213 + 0.215 + 0.217] - \\ [0.212 + 0.214 + 0.216] \rangle \end{array} \right\}}{27} = 0.0147$, $S(C_3) = \frac{\left\{ \begin{array}{c} \langle [0.004,\, 0.005,\, 0.006], \\ [0.102,\, 0.104,\, 0.106], \\ [0.104,\, 0.106,\, 0.108], \\ [0.103,\, 0.105,\, 0.107] \rangle \end{array} \right\}}{27} = \frac{0.015(0.63 - 0.315)}{27} = 0.0001$.

See Figure 3.

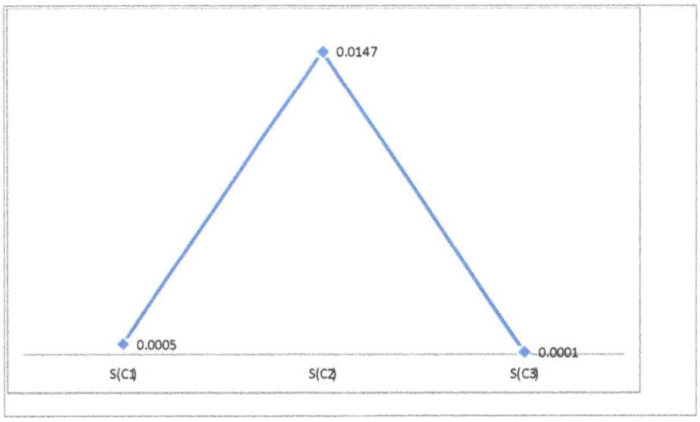

Figure 3. S(C2) is the first score value, S(C1) is the second score value and S(C3) is the third score value.

Example 4. *Let* $C_1 = \left\{\begin{array}{c}\langle[0.04, 0.05, 0.06], \\ [0.01, 0.03, 0.05], \\ [0.03, 0.05, 0.07], \\ [0.02, 0.04, 0.06]\rangle\end{array}\right\}$, $C_2 = \left\{\begin{array}{c}\langle[0.05, 0.07, 0.09], \\ [0.11, 0.13, 0.15], \\ [0.13, 0.15, 0.17], \\ [0.12, 0.14, 0.16]\rangle\end{array}\right\}$ *and* $C_3 = \left\{\begin{array}{c}\langle[0.2004, 0.2005, 0.2006], \\ [0.2102, 0.2104, 0.2106], \\ [0.2104, 0.2106, 0.2108], \\ [0.2103, 0.2105, 0.2107]\rangle\end{array}\right\}$ *be a TNCFN. Then the accuracy function is defined by* $H(C_1) = \frac{\left\{\begin{array}{c}\langle[0.04+0.05+0.06], \\ [0.01+0.03+0.05]+ \\ [0.03+0.05+0.07]+ \\ [0.02+0.04+0.06]\rangle\end{array}\right\}}{27} = 0.002$, $H(C_2) = \frac{\left\{\begin{array}{c}\langle[0+05+0.07+0.09], \\ [0.11+0.13+0.15]+ \\ [0.13+0.15+0.17]+ \\ [0.12+0.14+0.16]\rangle\end{array}\right\}}{27} = 0.0098$, $H(C_3) = \frac{\left\{\begin{array}{c}\langle[0.2004+0.2005+0.2006], \\ [0.2102+0.2104+0.2106]+ \\ [0.2104+0.2106+0.2108]+ \\ [0.2103+0.2105+0.2107]\rangle\end{array}\right\}}{27} = 0.0422.$

See Figure 4.

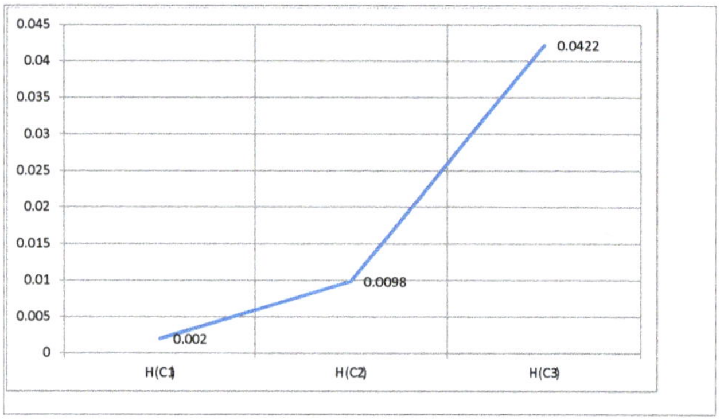

Figure 4. H(C2) is the first score value, H(C1) is the second score value and H(C3) is the third score value.

Proposition 2. *Let* $\ddot{A} = \left\{\begin{array}{c}[\alpha_1(h), \beta_1(h), \Delta_1(h)] \\ \langle[p_1^-(h), q_1^-(h), r_1^-(h)], \\ [p_1^+(h), q_1^+(h), r_1^+(h)], \\ [p_1(h), q_1(h), r_1(h)]\rangle \\ |h \in H\end{array}\right\}$ *be a collection of TNCFNs in* L_{TNCFN} *and where* $\ddot{\omega} = (\ddot{\omega}_1, \ddot{\omega}_2, \ldots, \ddot{\omega}_n)^T$ *is the weight vector of* $\ddot{A}_j (j = 1, 2, \ldots, n)$ *with* $\ddot{\omega}_j \in [0, 1]$ *and* $\sum_{j=1}^{n} \ddot{\omega}_j = 1$.

Then (1) (Idempotency): If all A_j, $j = 1, 2, \ldots, n$ *are equal, i.e.,* $A_j = A$, *for all* $j = 1, 2, \ldots, n$, *then* $TNCFEWA\ (A_1, A_2, \ldots, A_n) = A$.

(2) (Boundary): If $\alpha_{\min} = \min_{1\le j\le n}\alpha_j$, $\beta_{\min} = \min_{1\le j\le n}\beta_j$, $\Delta_{\min} = \min_{1\le j\le n}\Delta_j$, $p_{\min}^- = \min_{1\le j\le n}p_j^-$, $q_{\min}^- = \min_{1\le j\le n}q_j^-$, $r_{\min}^- = \min_{1\le j\le n}r_j^-$, $p_{\min}^+ = \min_{1\le j\le n}p_j^+$, $q_{\min}^+ = \min_{1\le j\le n}q_j^+$, $r_{\min}^+ = \min_{1\le j\le n}r_j^+$, $p_{\max} = \max_{1\le j\le n}p_j$, $q_{\max} = \max_{1\le j\le n}q_j$, $r_{\max} = \max_{1\le j\le n}r_j$, $\mu_{\max} = \max_{1\le j\le n}\mu_j$, $p_{\max}^- = \max_{1\le j\le n}p_j^-$, $q_{\max}^- = \max_{1\le j\le n}q_j^-$, $r_{\max}^- = \max_{1\le j\le n}r_j^-$, $\mu_{\max}^- = \max_{1\le j\le n}\mu_j^-$, $p_{\max}^+ = \max_{1\le j\le n}p_j^+$, $q_{\max}^+ = \max_{1\le j\le n}q_j^+$, $r_{\max}^+ = \max_{1\le j\le n}r_j^+$, $p_{\min} = \min_{1\le j\le n}p_j$,

$q_{\min} = \min_{1\leq j\leq n} q_j$, $r_{\min} = \min_{1\leq j\leq n} r_j$, $\mu_{\min} = \min_{1\leq j\leq n} \mu_j$ *for all* $j = 1, 2, .., n$, *we can determine that*

$$\left\{\begin{array}{c} [\alpha_{\min}(h), \\ \beta_{\min}(h), \Delta_{\min}(h)] \\ \langle [p^-_{\min}(h), \\ q^-_{\min}(h), r^-_{\min}(h)] \\ [p^+_{\min}(h), \\ q^+_{\min}(h), r^+_{\min}(h)] \\ [p_{\max}(h), q_{\max}(h), \\ r_{\max}(h), s_{\max}(h)]\rangle | \\ h \in H \end{array}\right\} \leq TNCFEWA(A_1, A_2, \ldots, A_n) \leq \left\{\begin{array}{c} [\alpha_{\max}(h), \beta_{\max}(h), \\ \Delta_{\max}(h)], \langle [p^-_{\max}(h), \\ q^-_{\max}(h), r^-_{\max}(h)], \\ [p^+_{\max}(h), q^+_{\max}(h), \\ r^+_{\max}(h)], [p_{\min}(h), \\ q_{\min}(h), r_{\min}(h)]\rangle | \\ h \in H \end{array}\right\}$$

(3) *(Monotonicity)*: $A = \left\{\begin{array}{c} [\alpha_A(h), \\ \beta_A(h), \Delta_A(h)], \\ \langle [\{p^-_A(h), \\ q^-_A(h), r^-_A(h)] \\ , [p^+_A(h), \\ q^+_A(h), r^+_A(h)] \\ , [p_A(h), \\ q_A(h), r_A(h)]\rangle | \\ h \in H \end{array}\right\}$ *and* $B = \left\{\begin{array}{c} [\alpha_B(h), \\ \beta_B(h), \Delta_B(h)], \\ \langle [p^-_B(h), \\ q^-_B(h), r^-_B(h)] \\ [p^+_B(h), \\ q^+_B(h), r^+_B(h)] \\ [p_B(h), \\ q_B(h), r_B(h)\rangle \\ |h \in H \end{array}\right\}$ *be two collection of TNCFNs in* L_{TNCFN} *and* $A_j \leq L_{\text{TNCFN}} B_j$ *i.e.*, $\alpha_A(h) \leq \alpha_B(h)$, $\beta_A(h) \leq \beta_B(h)$, $\Delta_A(h) \leq \Delta_B(h)$, $p^-_A(h) \leq p^-_B(h)$, $q^-_A(h) \leq q^-_B(h)$, $r^-_A(h) \leq r^-_B(h)$, $p^+_A(h) \leq p^+_B(h)$, $q^+_A(h) \leq q^+_B(h)$, $r^+_A(h) \leq r^+_B(h)$ *and* $p_A(h) \leq p_B(h)$, $q_A(h) \leq q_B(h)$, $r_A(h) \leq r_B(h)$ *then TNCFEWA* $(A_1, A_2, \ldots, A_n) \leq$ *TNCFEWA* $(B_1, B_2, \ldots, B_n)$.

4.2. Triangular Neutrosophic Cubic Fuzzy Einstein Ordered Weighted Averaging Operator

Definition 8. *Let* $\ddot{A} = \left\{\begin{array}{c} [\alpha_{\ddot{A}}(h), \beta_{\ddot{A}}(h), \\ \Delta_{\ddot{A}}(h)], \langle [p^-_{\ddot{A}}(h), \\ q^-_{\ddot{A}}(h), r^-_{\ddot{A}}(h)], \\ [p^+_{\ddot{A}}(h), q^-_{\ddot{A}}(h), \\ r^-_{\ddot{A}}(h)], p_{\ddot{A}}(h), \\ q_{\ddot{A}}(h), r_{\ddot{A}}(h)\rangle \\ |h \in H \end{array}\right\}$ *be a collection of TNCFNs in* L_{TNCFN}, *a TNCFEOWA operator of dimension n is a mapping TNCFEOWA:* $L^n_{\text{TNCFN}} \to L_{\text{TNCFN}}$, *that has an associated vector* $w = (w_1, w_2, \ldots, w_n)^T$ *such that* $w_j \in [0, 1]$ *and* $\sum_{j=1}^{n} w_j = 1$. *TNCFEOWA* $(\ddot{A}_1, \ddot{A}_2, \ldots, \ddot{A}_n) = \eth_1 \ddot{A}_{(\sigma)1} + \eth_2 \ddot{A}_{(\sigma)2}, \ldots, + \eth_n \ddot{A}_{(\sigma)n}$, *where* $(\sigma(1), \sigma(2), \ldots, \sigma(n))$ *is a permutation of* $(1, 2, \ldots, n)$ *such that* $\ddot{A}_{\sigma(1)} \leq \ddot{A}_{\sigma(j-1)}$ *for all* $j = 2, 3, \ldots, n$ *(i.e.,* $\ddot{A}_{\sigma(j)}$ *is the j the largest value in the collection* $(\ddot{A}_1, \ddot{A}_2, \ldots, \ddot{A}_n)$. *If* $w = (w_1, w_2, \ldots, w_n)^T = (\frac{1}{n}, \frac{1}{n}, \ldots, \frac{1}{n})^T$. *Then the TNCFEOWA operator is reduced to the TCFA operator (2) of dimension n.*

Theorem 2. *Let* $\ddot{A} = \left\{ \begin{array}{c} [\alpha_{\ddot{A}}(h), \beta_{\ddot{A}}(h), \\ \Delta_{\ddot{A}}(h)], \langle [p^{-}_{\ddot{A}}(h), \\ q^{-}_{\ddot{A}}(h), r^{-}_{\ddot{A}}(h)], \\ [p^{+}_{\ddot{A}}(h), q^{-}_{\ddot{A}}(h), \\ r^{-}_{\ddot{A}}(h)], p_{\ddot{A}}(h), \\ q_{\ddot{A}}(h), r_{\ddot{A}}(h) \rangle \\ |h \in H \end{array} \right\}$ *be a collection of TNCFNs in* L_{TNCFN}. *Then their aggregated value by using the TNCFEOWA operator is also a TNCFN and TNCFEOWA*

$$(\ddot{A}_1, \ddot{A}_2, \ldots, \ddot{A}_n) = \left\langle \begin{array}{l} \left[\begin{array}{c} \frac{\prod_{j=1}^{n}[1+\alpha_1(h)]^{\ddot{\omega}} - \prod_{j=1}^{n}[1-\alpha_1(h)]^{\ddot{\omega}}}{\prod_{j=1}^{n}[1+\alpha_1(h)]^{\ddot{\omega}} + \prod_{j=1}^{n}[1-\alpha_1(h)]^{\ddot{\omega}}}, \\ \frac{\prod_{j=1}^{n}[1+\beta_1(h)]^{\ddot{\omega}} - \prod_{j=1}^{n}[1-\beta_1(h)]^{\ddot{\omega}}}{\prod_{j=1}^{n}[1+\beta_1(h)]^{\ddot{\omega}} + \prod_{j=1}^{n}[1-\beta_1(h)]^{\ddot{\omega}}}, \\ \frac{\prod_{j=1}^{n}[1+\Delta_1(h)]^{\ddot{\omega}} - \prod_{j=1}^{n}[1-\Delta_1(h)]^{\ddot{\omega}}}{\prod_{j=1}^{n}[1+\Delta_1(h)]^{\ddot{\omega}} + \prod_{j=1}^{n}[1-\Delta_1(h)]^{\ddot{\omega}}} \end{array} \right], \left[\begin{array}{c} \frac{\prod_{j=1}^{n}[1+p^{-}_{\sigma(j)}(h)]^{\ddot{\omega}} - \prod_{j=1}^{n}[1-p^{-}_{\sigma(j)}(h)]^{\ddot{\omega}}}{\prod_{j=1}^{n}[1+p^{-}_{\sigma(j)}(h)]^{\ddot{\omega}} + \prod_{j=1}^{n}[1-p^{-}_{\sigma(j)}(h)]^{\ddot{\omega}}}, \\ \frac{\prod_{j=1}^{n}[1+q^{-}_{\sigma(j)}(h)]^{\ddot{\omega}} - \prod_{j=1}^{n}[1-q^{-}_{\sigma(j)}(h)]^{\ddot{\omega}}}{\prod_{j=1}^{n}[1+q^{-}_{\sigma(j)}(h)]^{\ddot{\omega}} + \prod_{j=1}^{n}[1-q^{-}_{\sigma(j)}(h)]^{\ddot{\omega}}}, \\ \frac{\prod_{j=1}^{n}[1+r^{-}_{\sigma(j)}(h)]^{\ddot{\omega}} - \prod_{j=1}^{n}[1-r^{-}_{\sigma(j)}(h)]^{\ddot{\omega}}}{\prod_{j=1}^{n}[1+r^{-}_{\sigma(j)}(h)]^{\ddot{\omega}} + \prod_{j=1}^{n}[1-r^{-}_{\sigma(j)}(h)]^{\ddot{\omega}}} \end{array} \right], \\ \left[\begin{array}{c} \frac{\prod_{j=1}^{n}[1+p^{+}_{\sigma(j)}(h)]^{\ddot{\omega}} - \prod_{j=1}^{n}[1-p^{+}_{\sigma(j)}(h)]^{\ddot{\omega}}}{\prod_{j=1}^{n}[1+p^{+}_{\sigma(j)}(h)]^{\ddot{\omega}} + \prod_{j=1}^{n}[1-p^{+}_{\sigma(j)}(h)]^{\ddot{\omega}}}, \frac{\prod_{j=1}^{n}[1+q^{+}_{\sigma(j)}(h)]^{\ddot{\omega}} - \prod_{j=1}^{n}[1-q^{+}_{\sigma(j)}(h)]^{\ddot{\omega}}}{\prod_{j=1}^{n}[1+q^{+}_{\sigma(j)}(h)]^{\ddot{\omega}} + \prod_{j=1}^{n}[1-q^{+}_{\sigma(j)}(h)]^{\ddot{\omega}}}, \\ \frac{\prod_{j=1}^{n}[1+r^{+}_{\sigma(j)}(h)]^{\ddot{\omega}} - \prod_{j=1}^{n}[1-r^{+}_{\sigma(j)}(h)]^{\ddot{\omega}}}{\prod_{j=1}^{n}[1+r^{+}_{\sigma(j)}(h)]^{\ddot{\omega}} + \prod_{j=1}^{n}[1-r^{+}_{\sigma(j)}(h)]^{\ddot{\omega}}} \end{array} \right], \\ \left[\begin{array}{c} \frac{2\prod_{j=1}^{n}[p_{\sigma(j)}(h)]^{\ddot{\omega}}}{\prod_{j=1}^{n}[(2-p_{\sigma(j)}(h)]^{\ddot{\omega}} + \prod_{j=1}^{n}[p_{\sigma(j)}(h)]^{\ddot{\omega}}}, \frac{2\prod_{j=1}^{n}[q_{\sigma(j)}(h)]^{\ddot{\omega}}}{\prod_{j=1}^{n}[(2-q_{\sigma(j)}(h)]^{\ddot{\omega}} + \prod_{j=1}^{n}[q_{\sigma(j)}(h)]^{\ddot{\omega}}}, \\ \frac{2\prod_{j=1}^{n}[r_{\sigma(j)}(h)]^{\ddot{\omega}}}{\prod_{j=1}^{n}[(2-r_{\sigma(j)}(h)]^{\ddot{\omega}} + \prod_{j=1}^{n}[r_{\sigma(j)}(h)]^{\ddot{\omega}}} \end{array} \right] \end{array} \right\rangle$$

where $(\sigma(1), \sigma(2), \ldots, \sigma(n))$ *is a permutation of* $(1, 2, \ldots, n)$ *with* $\ddot{A}_{\sigma(1)} \leq \ddot{A}_{\sigma(j-1)}$ *for all* $j = 2, 3, \ldots, n$, $\ddot{\omega} = (\ddot{\omega}_1, \ddot{\omega}_2, \ldots, \ddot{\omega}_n)^T$ *is the weight vector of* $\ddot{A}_j (j = 1, 2, \ldots, n)$ *such that* $\ddot{\omega}_j \in [0, 1]$, *and* $\sum_{j=1}^{n} \ddot{\omega}_j = 1$. *If* $\ddot{\omega} = (\ddot{\omega}_1, \ddot{\omega}_2, \ldots, \ddot{\omega}_n)^T = (\frac{1}{n}, \frac{1}{n}, \ldots, \frac{1}{n})^T$. *Then the TNCFEOWA operator is reduced to the TNCFA operator of dimension n. Where* $\ddot{\omega} = (\ddot{\omega}_1, \ddot{\omega}_2, \ldots, \ddot{\omega}_n)^T$ *is the weight vector of* $\ddot{A}_j (j = 1, 2, \ldots, n)$ *such that* $\ddot{\omega}_j \in [0, 1]$ *and* $\sum_{j=1}^{n} \ddot{\omega}_j = 1$. *If* $\alpha_1(h) = \alpha_1(h)$, $\beta_1(h) = \beta_1(h)$, $\Delta_1(h) = \Delta_1(h)$, $p^{-}_{\ddot{A}}(h) = p^{-}_{\ddot{A}}(h)$, $q^{-}_{\ddot{A}}(h) = q^{-}_{\ddot{A}}(h)$, $r^{-}_{\ddot{A}}(h) = r^{-}_{\ddot{A}}(h)$, $p^{+}_{\ddot{A}}(h) = p^{+}_{\ddot{A}}(h)$, $q^{+}_{\ddot{A}}(h) = q^{+}_{\ddot{A}}(h)$, $r^{+}_{\ddot{A}}(h) = r^{+}_{\ddot{A}}(h)$ *and* $p_{\ddot{A}}(h) = p_{\ddot{A}}(h)$,

$q_{\ddot{A}}(h) = q_{\ddot{A}}(h), r_{\ddot{A}}(h) = r_{\ddot{A}}(h)$. *The TNCFN* $\ddot{A} = \left\{ \begin{array}{c} [\alpha_{\ddot{A}}(h), \beta_{\ddot{A}}(h), \\ \Delta_{\ddot{A}}(h)]\langle [p^{-}_{\ddot{A}}(h), \\ q^{-}_{\ddot{A}}(h), r^{-}_{\ddot{A}}(h)], \\ [p^{+}_{\ddot{A}}(h), q^{-}_{\ddot{A}}(h), \\ r^{-}_{\ddot{A}}(h)], p_{\ddot{A}}(h), \\ q_{\ddot{A}}(h), r_{\ddot{A}}(h)\rangle \\ |h \in H \end{array} \right\}$ *are reduced to the triangular neutrosophic cubic fuzzy numbers* $\ddot{A} = \left\{ \begin{array}{c} [\alpha_{\ddot{A}}(h), \beta_{\ddot{A}}(h), \\ \Delta_{\ddot{A}}(h)], \langle [p^{-}_{\ddot{A}}(h), \\ q^{-}_{\ddot{A}}(h), r^{-}_{\ddot{A}}(h)], \\ [p^{+}_{\ddot{A}}(h), q^{-}_{\ddot{A}}(h), \\ r^{-}_{\ddot{A}}(h)], p_{\ddot{A}}(h), \\ q_{\ddot{A}}(h), r_{\ddot{A}}(h)\rangle \\ |h \in H \end{array} \right\}$. *Then the TNCFEWA operator is reduced to the triangular neutrosophic cubic fuzzy Einstein ordered weighted averaging operator.*

Proof. The process of this proof is the same as Theorem 1. □

Example 5. *Let* $C_1 = \left\{ \begin{array}{c} \langle [0.01, 0.02, 0.03], \\ [0.103, 0.105, 0.107], \\ [0.105, 0.107, 0.109], \\ [0.104, 0.106, 0.108]\rangle \end{array} \right\}$, $C_2 = \left\{ \begin{array}{c} \langle [0.306, 0.308, 0.310], \\ [0.310, 0.313, 0.315], \\ [0.313, 0.315, 0.317], \\ [0.312, 0.314, 0.316]\rangle \end{array} \right\}$ *and* $C_3 = \left\{ \begin{array}{c} \langle [0.44, 0.55, 0.66], \\ [0.122, 0.124, 0.126], \\ [0.124, 0.126, 0.128], \\ [0.123, 0.125, 0.127]\rangle \end{array} \right\}$ *be a TNCFN. Then the score function is defined by* $S(C_1) = \frac{\left\{ \begin{array}{c} \langle [0.01 + 0.02 + 0.03], \\ [0.103 + 0.105 + 0.107] + \\ [0.105 + 0.107 + 0.109] - \\ [0.104 + 0.106 + 0.108]\rangle \end{array} \right\}}{27} = 0.00004$, $S(C_2) = \frac{\left\{ \begin{array}{c} \langle [0.306 + 0.308 + 0.310], \\ [0.310 + 0.313 + 0.315] + \\ [0.313 + 0.315 + 0.317] - \\ [0.312 + 0.314 + 0.316]\rangle \end{array} \right\}}{27} = 0.0322$, $S(C_3) = \frac{\left\{ \begin{array}{c} \langle [0.44 + 0.55 + 0.66], \\ [0.122 + 0.124 + 0.126] + \\ [0.124 + 0.126 + 0.128] - \\ [0.123 + 0.125 + 0.127]\rangle \end{array} \right\}}{27} = 0.0229$.

See Figure 5.

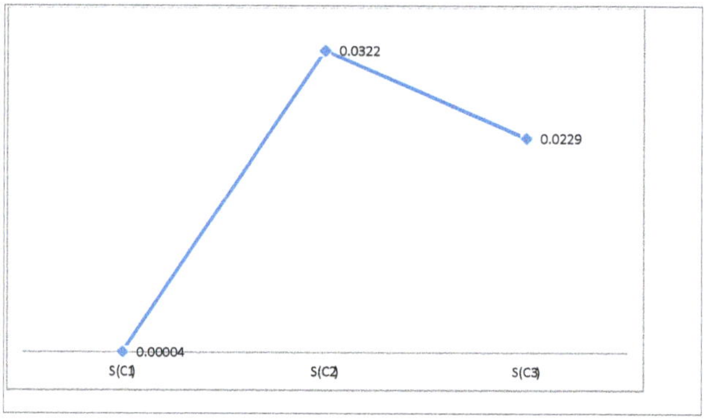

Figure 5. Different score ranking of TNCFEOWA operator.

Example 6. *Let* $C_1 = \left\{ \begin{array}{c} \langle [0.02, 0.03, 0.04], \\ [0.06, 0.08, 0.010], \\ [0.08, 0.010, 0.012], \\ [0.07, 0.09, 0.011] \rangle \end{array} \right\}$, $C_2 = \left\{ \begin{array}{c} \langle [0.1105, 0.1107, 0.1109], \\ [0.1111, 0.1113, 0.1115], \\ [0.1113, 0.1115, 0.1117], \\ [0.1112, 0.1114, 0.1116] \rangle \end{array} \right\}$ *and* $C_3 = \left\{ \begin{array}{c} \langle [0.214, 0.215, 0.216], \\ [0.2202, 0.2204, 0.2206], \\ [0.2204, 0.2206, 0.2208], \\ [0.2203, 0.2205, 0.2207] \rangle \end{array} \right\}$ *be a TNCFN. Then the accuracy function is defined by* $H(C_1) = \frac{\left\{ \begin{array}{c} \langle [0.02, 0.03, 0.04], \\ [0.06, 0.08, 0.010], \\ [0.08, 0.010, 0.012], \\ [0.07, 0.09, 0.011] \rangle \end{array} \right\}}{27} = -0.0021$, $H(C_2) = \frac{\left\{ \begin{array}{c} \langle [0.1105, 0.1107, 0.1109], \\ [0.1111, 0.1113, 0.1115], \\ [0.1113, 0.1115, 0.1117], \\ [0.1112, 0.1114, 0.1116] \rangle \end{array} \right\}}{27} = 0.0041$, $H(C_3) = \frac{\left\{ \begin{array}{c} \langle [0.214, 0.215, 0.216], \\ [0.2202, 0.2204, 0.2206], \\ [0.2204, 0.2206, 0.2208], \\ [0.2203, 0.2205, 0.2207] \rangle \end{array} \right\}}{27} = 0.0158.$

See Figure 6.

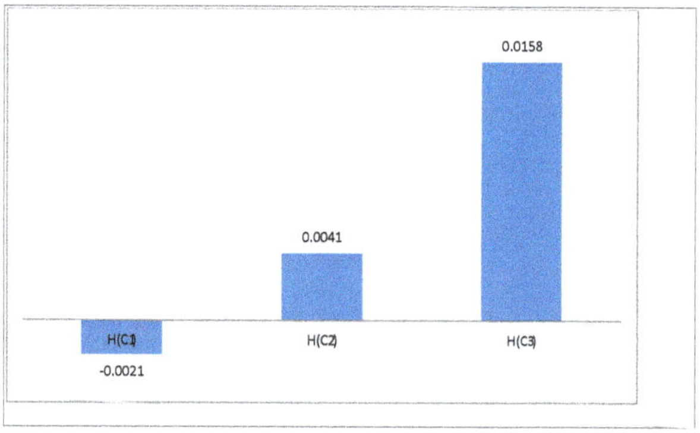

Figure 6. Different accuracy ranking of TNCFEOWA operator.

4.3. Triangular Neutrosophic Cubic Fuzzy Einstein Hybrid Weighted Averaging Operator

Definition 9. *Let* $\ddot{A} = \left\{ \begin{array}{c} [\Gamma_{\ddot{A}}(h), \Omega_{\ddot{A}}(h), \\ _{\ddot{A}}(h)], \langle [p^{-}_{\ddot{A}}(h), \\ q^{-}_{\ddot{A}}(h), r^{-}_{\ddot{A}}(h)], \\ [p^{+}_{\ddot{A}}(h), q^{-}_{\ddot{A}}(h), \\ r^{-}_{\ddot{A}}(h)], p_{\ddot{A}}(h), \\ q_{\ddot{A}}(h), r_{\ddot{A}}(h) \rangle \\ |h \in H \end{array} \right\}$ *be a collection of TNCFNs in* L_{TNCFN} *and* $\ddot{\omega} = (\ddot{\omega}_1, \ddot{\omega}_2, \ldots, \ddot{\omega}_n)^T$ *is the weight vector of* $\ddot{A}_j (j = 1, 2, \ldots, n)$ *such that* $\ddot{\omega}_j \in [0, 1]$ *and* $\sum_{j=1}^{n} \ddot{\omega}_j = 1$. *Then TNCFEHWA operator of dimension n is a mapping TNCFEHWA* : $L^n_{\text{TNCFN}} \to L_{\text{TNCFN}}$, *that is an associated vector* $w = (w_1, w_2, \ldots, w_n)^T$ *such that* $w_j \in [0, 1]$ *and* $\sum_{j=1}^{n} w_j = 1$. *TNCFEHWA* $(\ddot{A}_1, \ddot{A}_2, \ldots, \ddot{A}_n) = p_1 \ddot{A}_{\sigma(1)} + p_2 \ddot{A}_{\sigma(1)}, \ldots, p_n \ddot{A}_{\sigma(1)}$. *If* $p = \theta \ddot{\omega}_{\sigma(j)} + (1 - \theta) w_{\sigma(j)}$ *with a balancing coefficient* $\theta \in [0, 1]$, $(\sigma(1), \sigma(2), \ldots, \sigma(n))$ *is a permutation of* $(1, 2, \ldots, n)$ *such that* $\ddot{A}_{\sigma(j)} \leq \ddot{A}_{\sigma(j-1)}$ *for all* $j = 2, 3, \ldots, n$ *(i.e.,* $\ddot{A}_{\sigma(j)}$ *is the j th largest value in the collection* $(\ddot{A}_1, \ddot{A}_2, \ldots, \ddot{A}_n)$.

Example 7. *Let* $C_1 = \left\{ \begin{array}{c} \langle [0.11, 0.13, 0.14], \\ [0.62, 0.64, 0.66], \\ [0.64, 0.66, 0.68], \\ [0.63, 0.65, 0.67] \rangle \end{array} \right\}$, $C_2 = \left\{ \begin{array}{c} \langle [0.51, 0.52, 0.53], \\ [0.311, 0.313, 0.315], \\ [0.313, 0.315, 0.317], \\ [0.312, 0.314, 0.316] \rangle \end{array} \right\}$ *and* $C_3 = \left\{ \begin{array}{c} \langle [0.1004, 0.1005, 0.1006], \\ [0.3102, 0.3104, 0.3106], \\ [0.3104, 0.3106, 0.3108], \\ [0.3103, 0.3105, 0.3107] \rangle \end{array} \right\}$ *be a TNCFN. Then the score function is defined by* $S(C_1) = \frac{\left\{ \begin{array}{c} \langle [0.11 + 0.13 + 0.14], \\ [0.62 + 0.64 + 0.66] + \\ [0.64 + 0.66 + 0.68] - \\ [0.63 + 0.65 + 0.67] \rangle \end{array} \right\}}{27} = 0.0281$, $S(C_2) = \frac{\left\{ \begin{array}{c} \langle [0.51 + 0.52 + 0.53], \\ [0.311 + 0.313 + 0.315] + \\ [0.313 + 0.315 + 0.317] - \\ [0.312 + 0.314 + 0.316] \rangle \end{array} \right\}}{27} = 0.0543$, $S(C_3) = \frac{\left\{ \begin{array}{c} \langle [0.1004, 0.1005, 0.1006], \\ [0.3102, 0.3104, 0.3106], \\ [0.3104, 0.3106, 0.3108], \\ [0.3103, 0.3105, 0.3107] \rangle \end{array} \right\}}{27} = 0.0104.$
See Figure 7.

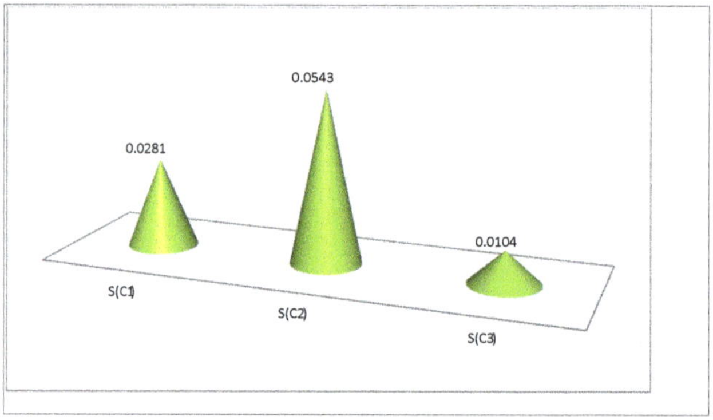

Figure 7. Different score ranking of TNCFEHWA operator.

Example 8. *Let* $C_1 = \left\{ \begin{array}{c} \langle [0.112, 0.113, 0.114], \\ [0.21, 0.24, 0.28], \\ [0.24, 0.28, 0.32], \\ [0.23, 0.26, 0.30] \rangle \end{array} \right\}$, $C_2 = \left\{ \begin{array}{c} \langle [0.0019, 0.0021, 0.0034], \\ [0.1231, 0.1233, 0.1235], \\ [0.1233, 0.1235, 0.1237], \\ [0.1232, 0.1234, 0.1236] \rangle \end{array} \right\}$ *and* $C_3 = \left\{ \begin{array}{c} \langle [0.2554, 0.2555, 0.2556], \\ [0.2662, 0.2664, 0.2666], \\ [0.2664, 0.2666, 0.2668], \\ [0.2663, 0.2665, 0.2667] \rangle \end{array} \right\}$ *be a TNCFN. Then the accuracy function is defined by* $H(C_1) = \frac{\left\{ \begin{array}{c} \langle [0.112 + 0.113 + 0.114], \\ [0.21 + 0.24 + 0.28] + \\ [0.24 + 0.28 + 0.32] - \\ [0.23 + 0.26 + 0.30] \rangle \end{array} \right\}}{27} = 0.0111$, $H(C_2) = \frac{\left\{ \begin{array}{c} \langle [0.0019 + 0.0021 + 0.0034] \\ [0.1231 + 0.1233 + 0.1235] + \\ [0.1233 + 0.1235 + 0.1237] - \\ [0.1232 + 0.1234 + 0.1236] \rangle \end{array} \right\}}{27} = 0.1124$, $H(C_3) = \frac{\left\{ \begin{array}{c} \langle [0.2554, 0.2555, 0.2556], \\ [0.2662, 0.2664, 0.2666], \\ [0.2664, 0.2666, 0.2668], \\ [0.2663, 0.2665, 0.2667] \rangle \end{array} \right\}}{27} = 0.0226.$

See Figure 8.

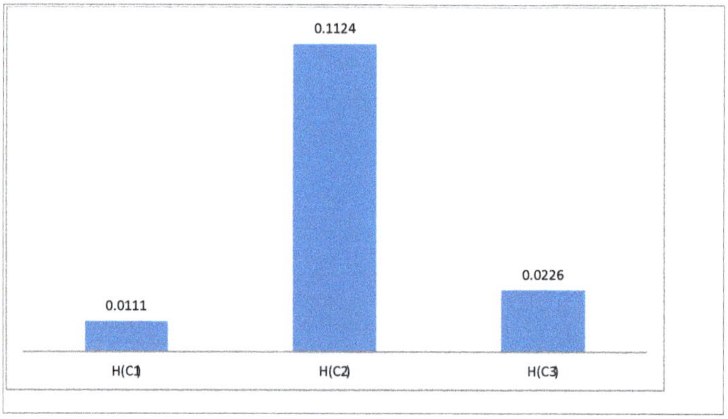

Figure 8. Different accuracy ranking of TNCFEHWA operator.

5. An Approach to MADM with TNCF Data

Let us suppose the discrete set is $h = \{h_1, h_2, \ldots, h_n\}$ and $G = \{g_1, g_2, \ldots, g_n\}$ are the attributes. Consider that the value of alternatives h_i $(i = 1, 2, \ldots, n)$ on attributes g_j $(j = 1, 2, \ldots, m)$ given by decision maker are TNCFNs in L_{TNCFN}: $\ddot{A} = \left\{ \begin{array}{c} [\Gamma_{\ddot{A}}(h), \Omega_{\ddot{A}}(h), \\ {}_{\ddot{A}}(h)], \left\langle [p^-_{\ddot{A}}(h), \right. \\ q^-_{\ddot{A}}(h), r^-_{\ddot{A}}(h)], \\ [p^+_{\ddot{A}}(h), q^+_{\ddot{A}}(h), \\ r^+_{\ddot{A}}(h)], p_{\ddot{A}}(h), \\ \left. q_{\ddot{A}}(h), r_{\ddot{A}}(h) \right\rangle \\ |h \in H \end{array} \right\}$, a MADM problem is expressed in the TNCF-decision matrix $\ddot{D} = (\ddot{A}_{ij})_{m \times n} = \left\{ \begin{array}{c} [\Gamma_{\ddot{A}}(h), \Omega_{\ddot{A}}(h), {}_{\ddot{A}}(h)], \\ \left\langle [p^-_{\ddot{A}}(h), q^-_{\ddot{A}}(h), r^-_{\ddot{A}}(h)], \right. \\ [p^+_{\ddot{A}}(h), q^+_{\ddot{A}}(h), r^+_{\ddot{A}}(h)], \\ \left. p_{\ddot{A}}(h), q_{\ddot{A}}(h), r_{\ddot{A}}(h) \right\rangle \\ |h \in H \end{array} \right\}$.

Step 1: Calculate the TNCF decision matrix.

Step 2: Utilize the TNCFEWA operator to mix all values $\ddot{\beta}_{ij}$ $(j = 1, 2, \ldots, m)$ and $\ddot{\omega} = (\ddot{\omega}_1, \ddot{\omega}_2, \ldots, \ddot{\omega}_n)^T$ is the weight vector.

Step 3: Calculate the score function.

Step 4: Find the ranking.

See Figure 9.

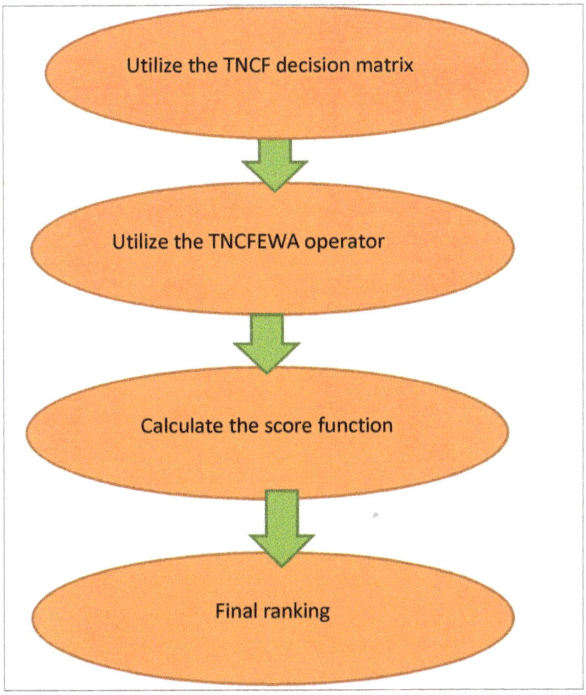

Figure 9. Proposed method.

6. Numerical Application

The inspiration structure is designed to be dependent upon an assessment that has been devised for the purpose of a stimulus/influencing technique of a twofold entire traveler dispersion to work over the Lahore in Faisalabad by lessening the adventure stage in extraordinarily brimful waterway movement. Inspiration structure choices are sure the settled of options $A = \{A_1, A_2, A_3, A_4\}$

A_1 : Old-style propeller and high trundle

A_2 : Get-up-and-go,

A_3 : Cyclonical propeller,

A_4 : Outmoded

See Figure 10.

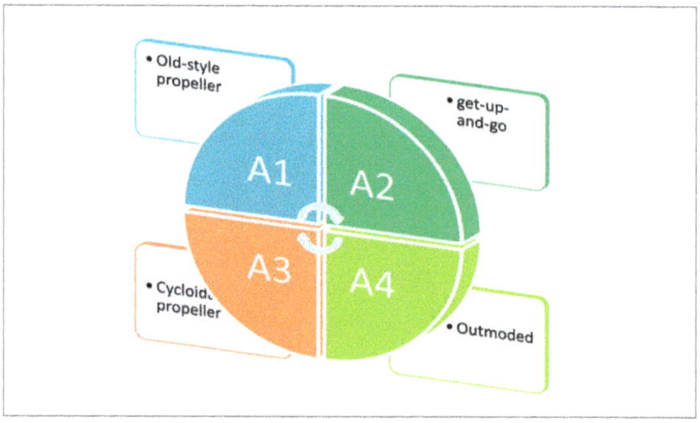

Figure 10. Four alternatives.

The ideal is prepared on the possibility of lone zone and four issue characteristics, which are as follows:

c_1 : Theory rate

c_2 : Reparation and support uses

c_3 : Agility

c_4 : Tremor and unrest.

See Figure 11.

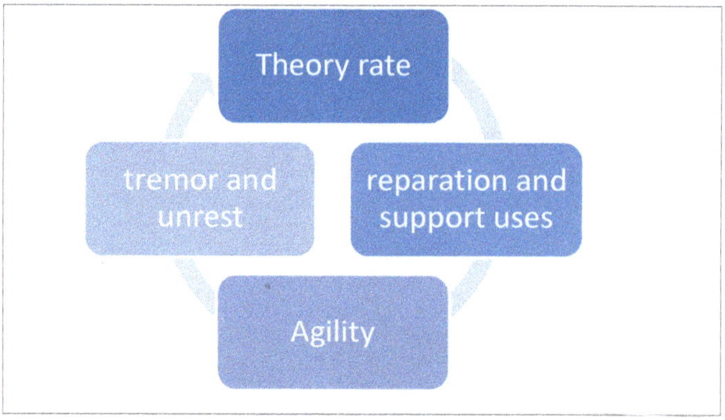

Figure 11. Different criteria.

The weight vector is $\ddot{\omega} = (0.25, 0.50, 0.25)$. So, the triangular neutrosophic cubic fuzzy MADM issue is intended to choose the appropriate energy structure from between 3 choices.

Step 1: Calculate the TNCF decision matrix.
The TNCF decision matrix is as Table 2

Table 2. Triangular Neutrosophic Cubic Fuzzy Decision Matrix.

	c_1	c_2
A_1	$\left\langle \begin{array}{c} [0.1, 0.2, 0.3], \\ [0.4, 0.6, 0.8], \\ [0.6, 0.8, 0.10], \\ [0.5, 0.7, 0.9] \end{array} \right\rangle$	$\left\langle \begin{array}{c} [0.21, 0.22, 0.23], \\ [0.5, 0.7, 0.9], \\ [0.7, 0.9, 0.11], \\ [0.6, 0.8, 0.10] \end{array} \right\rangle$
A_2	$\left\langle \begin{array}{c} [0.21, 0.22, 0.23], \\ [0.5, 0.7, 0.9], \\ [0.7, 0.9, 0.11], \\ [0.6, 0.8, 0.10] \end{array} \right\rangle$	$\left\langle \begin{array}{c} [0.1, 0.2, 0.3], \\ [0.4, 0.6, 0.8], \\ [0.6, 0.8, 0.10], \\ [0.5, 0.7, 0.9] \end{array} \right\rangle$
A_3	$\left\langle \begin{array}{c} [0.15, 0.16, 0.17], \\ [0.12, 0.14, 0.16], \\ [0.14, 0.16, 0.18], \\ [0.13, 0.15, 0.17] \end{array} \right\rangle$	$\left\langle \begin{array}{c} [0.3, 0.4, 0.5], \\ [0.2, 0.4, 0.6], \\ [0.4, 0.6, 0.8], \\ [0.3, 0.5, 0.7] \end{array} \right\rangle$
A_4	$\left\langle \begin{array}{c} [0.3, 0.4, 0.5], \\ [0.2, 0.4, 0.6], \\ [0.4, 0.6, 0.8], \\ [0.3, 0.5, 0.7] \end{array} \right\rangle$	$\left\langle \begin{array}{c} [0.15, 0.16, 0.17], \\ [0.12, 0.14, 0.16], \\ [0.14, 0.16, 0.18], \\ [0.13, 0.15, 0.17] \end{array} \right\rangle$
	c_3	c_4
A_1	$\left\langle \begin{array}{c} [0.15, 0.16, 0.17], \\ [0.12, 0.14, 0.16], \\ [0.14, 0.16, 0.18], \\ [0.13, 0.15, 0.17] \end{array} \right\rangle$	$\left\langle \begin{array}{c} [0.21, 0.22, 0.23], \\ [0.5, 0.7, 0.9], \\ [0.7, 0.9, 0.11], \\ [0.6, 0.8, 0.10] \end{array} \right\rangle$
A_2	$\left\langle \begin{array}{c} [0.3, 0.4, 0.5], \\ [0.4, 0.6, 0.8], \\ [0.6, 0.8, 0.10], \\ [0.5, 0.7, 0.9] \end{array} \right\rangle$	$\left\langle \begin{array}{c} [0.15, 0.16, 0.17], \\ [0.12, 0.14, 0.16], \\ [0.14, 0.16, 0.18], \\ [0.13, 0.15, 0.17] \end{array} \right\rangle$
A_3	$\left\langle \begin{array}{c} [0.1, 0.2, 0.3], \\ [0.4, 0.6, 0.8], \\ [0.6, 0.8, 0.10], \\ [0.5, 0.7, 0.9] \end{array} \right\rangle$	$\left\langle \begin{array}{c} [0.3, 0.4, 0.5], \\ [0.2, 0.4, 0.6], \\ [0.4, 0.6, 0.8], \\ [0.3, 0.5, 0.7] \end{array} \right\rangle$
A_4	$\left\langle \begin{array}{c} [0.21, 0.22, 0.23], \\ [0.5, 0.7, 0.9], \\ [0.7, 0.9, 0.11], \\ [0.6, 0.8, 0.10] \end{array} \right\rangle$	$\left\langle \begin{array}{c} [0.1, 0.2, 0.3], \\ [0.4, 0.6, 0.8], \\ [0.6, 0.8, 0.10], \\ [0.5, 0.7, 0.9] \end{array} \right\rangle$

Step 2: Calculate the TNCFEWA operator to total all the rating values and $w = (0.1,\ 0.2,\ 0.4,\ 0.3)^T$. The TNCFEWA operator are defined in Table 3.

Table 3. TNCFEWA Operator.

A_1	$\left\{ \left\langle \begin{array}{c} [0.2539, 0.2751, 0.2965], [0.1628, 0.2513, 0.3973], \\ [0.2513, 0.3973, 0.0503], [0.7335, 0.8054, 0.6003] \end{array} \right\rangle \right\}$
A_2	$\left\{ \left\langle \begin{array}{c} [0.1536, 0.1995, 0.2477], [0.2944, 0.3597, 0.6447], \\ [0.3597, 0.6447, 0.0988], [0.4838, 0.6067, 0.4831] \end{array} \right\rangle \right\}$
A_3	$\left\{ \left\langle \begin{array}{c} [0.3481, 0.4499, 0.5594], [0.3626, 0.5867, 0.7852], \\ [0.5867, 0.7852, 0.7582], [0.1049, 0.2122, 0.3571] \end{array} \right\rangle \right\}$
A_4	$\left\{ \left\langle \begin{array}{c} [0.2282, 0.2945, 0.3622], [0.3704, 0.5631, 0.7729], \\ [0.5631, 0.7729, 0.4197], [0.2593, 0.3985, 0.2735] \end{array} \right\rangle \right\}$

Step 3: The score value are calculated as $s_1 = -0.0192$, $s_2 = 0.0184$, $s_3 = 0.1603$, $s_4 = 0.0829$.

Step 4: Ranking $\ddot{s}_3 > \ddot{s}_4 > \ddot{s}_2 > \ddot{s}_1$.
See Figure 12.

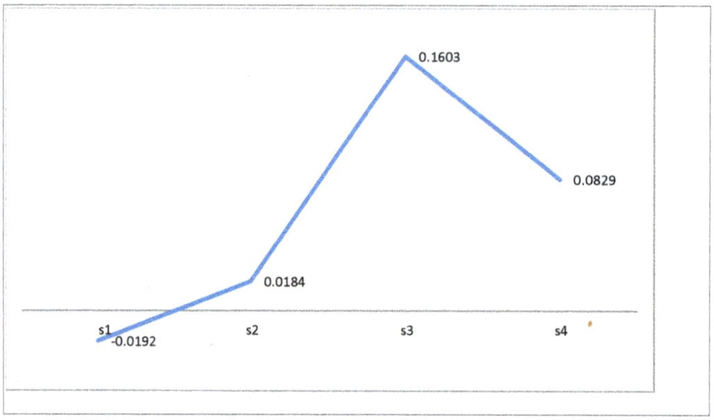

Figure 12. Rating value different range of values.

7. Comparsion Analysis

So as to check the legitimacy and viability of the proposed methodology, a near report is led utilizing the techniques triangular cubic fuzzy number [21], which are unique instances of TNCFNs, to the equivalent illustrative model.

A Comparison Analysis with the Existing MCDM Method Triangular Cubic Fuzzy Number

Aliya et al [21] after transformation, the triangular cubic fuzzy information is given in Table 4.

Table 4. Triangular cubic fuzzy decision matrix.

	c_1	c_2	c_3	c_4
A_1	$\left\{\begin{array}{c}[0.1,0.2,0.3],\\ \langle[0.4,0.8],0.7\rangle\end{array}\right\}$	$\left\{\begin{array}{c}[0.21,0.22,0.23],\\ \langle[0.5,0.9],0.8\rangle\end{array}\right\}$	$\left\{\begin{array}{c}[0.15,0.16,0.17],\\ \langle[0.12,0.16],0.15\rangle\end{array}\right\}$	$\left\{\begin{array}{c}[0.21,0.22,0.23],\\ \langle[0.5,0.9],0.8\rangle\end{array}\right\}$
A_2	$\left\{\begin{array}{c}[0.21,0.22,0.23],\\ \langle[0.5,0.9],0.8\rangle\end{array}\right\}$	$\left\{\begin{array}{c}[0.1,0.2,0.3],\\ \langle[0.4,0.8],0.7\rangle\end{array}\right\}$	$\left\{\begin{array}{c}[0.3,0.4,0.5],\\ \langle[0.2,0.6],0.4\rangle\end{array}\right\}$	$\left\{\begin{array}{c}[0.15,0.16,0.17],\\ \langle[0.12,0.16],0.15\rangle\end{array}\right\}$
A_3	$\left\{\begin{array}{c}[0.15,0.16,0.17],\\ \langle[0.12,0.16],0.15\rangle\end{array}\right\}$	$\left\{\begin{array}{c}[0.3,0.4,0.5],\\ \langle[0.2,0.6],0.4\rangle\end{array}\right\}$	$\left\{\begin{array}{c}[0.1,0.2,0.3],\\ \langle[0.4,0.8],0.7\rangle\end{array}\right\}$	$\left\{\begin{array}{c}[0.3,0.4,0.5],\\ \langle[0.2,0.6],0.5\rangle\end{array}\right\}$
A_4	$\left\{\begin{array}{c}[0.3,0.4,0.5],\\ \langle[0.2,0.6],0.4\rangle\end{array}\right\}$	$\left\{\begin{array}{c}[0.15,0.16,0.17],\\ \langle[0.12,0.16],0.15\rangle\end{array}\right\}$	$\left\{\begin{array}{c}[0.21,0.22,0.23],\\ \langle[0.5,0.9],0.8\rangle\end{array}\right\}$	$\left\{\begin{array}{c}[0.1,0.2,0.3],\\ \langle[0.4,0.8],0.7\rangle\end{array}\right\}$

Calculate the TCFA operator and $w = (0.1,\ 0.2,\ 0.4,\ 0.3)^T$.
The TCFA operator is presented in Table 5.

Table 5. TCFA operator.

A_1	$\langle[0.067, 0.081, 0.093], [0.1833, 0.4721], 0.1384\rangle$
A_2	$\langle[0.152, 0.196, 0.24], [0.2672, 0.6329], 0.5073\rangle$
A_3	$\langle[0.42, 0.464, 0.588], [0.4631, 0.7646], 0.2132\rangle$
A_4	$\langle[0.228, 0.294, 0.36], [0.3727, 0.7771], 0.3613\rangle$

Calculate the score function $s_1 = 0.0138$, $s_2 = 0.0256$, $s_3 = 0.1659$, $s_4 = 0.0772$.
See Figure 13.

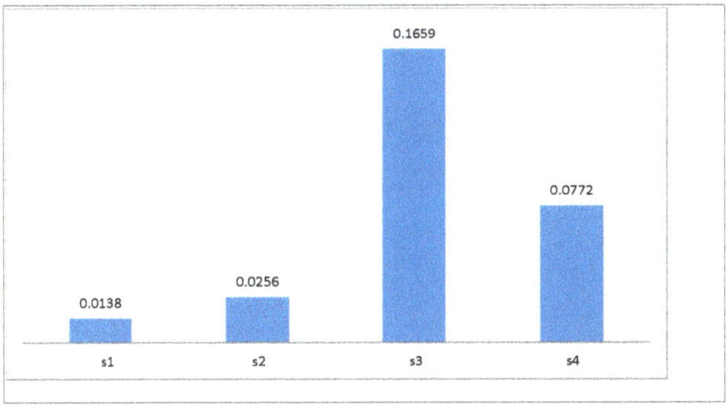

Figure 13. s_3 is the first ranking, s_4 is the 2nd ranking, s_2 is the third ranking and s_1 is the 4th ranking in the TCFN.

The existing Table 6 is as

Table 6. Comparison method with existing methods.

Method	Ranking
TNCFNs	$\ddot{s}_3 > \ddot{s}_4 > \ddot{s}_2 > \ddot{s}_1$
TCFN [21]	$\ddot{s}_3 > \ddot{s}_4 > \ddot{s}_2 > \ddot{s}_1$

See Figure 14.

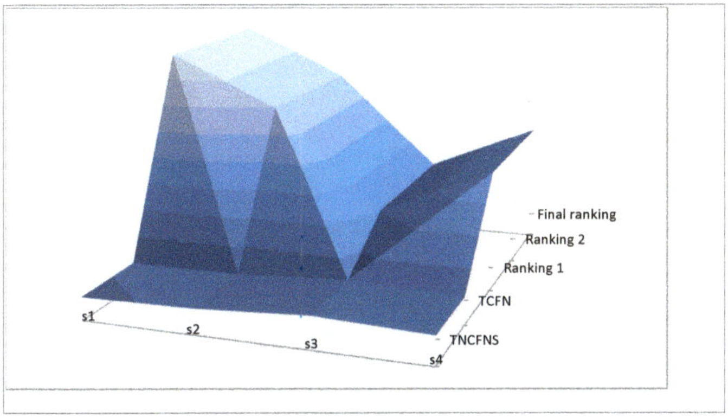

Figure 14. Comparison analysis with existing method.

The comparison method of score function is presented in Table 7.

Table 7. Comparison method with score function.

Score function	Ranking
TNCFEWA operator	$S(C_3) > S(C_2) > S(C_1)$
TNCFEOWA operator	$S(C_1) > S(C_2) > S(C_3)$
TNCFEHWA operator	$S(C_2) > S(C_1) > S(C_3)$

See Figure 15.

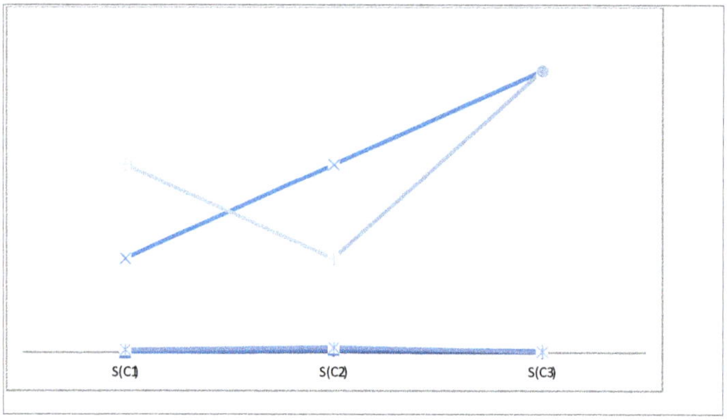

Figure 15. Different score value.

8. Conclusions

In this paper, we introduce a new concept of TNCFNs and operational laws. We introduce three aggregation operators, namely, the TNCFEWA operator, TNCFEOWA operator and TNCFEWA operator. We introduce group decision making under TNCFNs. Finally, a numerical example is provided to demonstrate the utility of the established approach. In cluster decision-making issues, consultants sometimes return from completely different specialty fields and have different backgrounds and levels of data; as such, they sometimes have branching opinions. These operators may be applied to several different fields, like data fusion, data processing, and pattern recognition, triangular neutrosophic cube like linguistic fuzzy Vikor methodology and quadrangle neutrosophic cube linguistic fuzzy Vikor methodology, which may be a suitable topic for longer term analysis, see Figure 16.

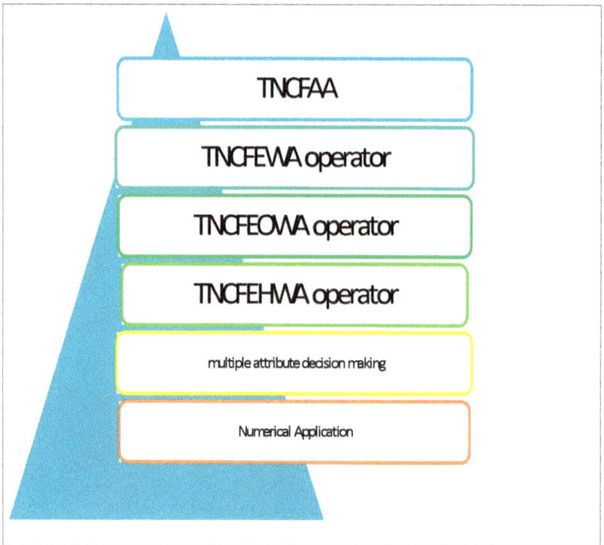

Figure 16. Flowcharts of whole papers.

Author Contributions: All authors contributed equally to this paper.

Funding: This research received no external funding.

Conflicts of Interest: The authors declare no conflict of interest.

Appendix A. Proof of Proposition 1

(1) $A_1 + A_2 = A_2 + A_1$;

$$A_1 + A_2 = \left\{ \begin{array}{c} \left\langle [\frac{\alpha_1(h)+\alpha_2(h)}{(1+\alpha_1(h))(1-\alpha_2(h))}, \frac{\beta_1(h)+\beta_2(h)}{(1+\beta_1(h))(1-\beta_2(h))}, \frac{\Delta_1(h)+\Delta_2(h)}{(1+\Delta_1(h))(1-\Delta_2(h))}], \right. \\ \left[\begin{array}{c} [\frac{p_1^-(h)+p_2^-(h)}{(1+p_1^-(h))(1-p_2^-(h))}, \frac{q_1^-(h)+q_2^-(h)}{(1+q_1^-(h))(1-q_2^-(h))}, \frac{r_1^-(h)+r_2^-(h)}{(1+r_1^-(h))(1-r_2^-(h))}], \\ [\frac{p_1^+(h)+p_2^+(h)}{(1+p_1^+(h))(1-p_2^+(h))}, \frac{q_1^+(h)+q_2^+(h)}{(1+q_1^+(h))(1-q_2^+(h))}, \frac{r_1^+(h)+r_2^+(h)}{(1+r_1^+(h))(1-r_2^+(h))}] \end{array} \right], \\ \left. \left[[\frac{p_1(h)p_2(h)}{1+((1-p_1(h))(1-p_2(h)))}, \frac{q_1(h)q_2(h)}{1+((1-q_1(h))(1-q_2(h)))}, \frac{r_1(h)r_2(h)}{1+((1-r_1(h))(1-r_2(h)))}]\right] \right\rangle \end{array} \right\} =$$

$$\left\{ \begin{array}{c} \left\langle [\frac{\alpha_2(h)+\alpha_1(h)}{(1+\alpha_2(h))(1-\alpha_1(h))}, \frac{\beta_2(h)+\beta_1(h)}{(1+\beta_2(h))(1-\beta_1(h))}, \frac{\Delta_2(h)+\Delta_1(h)}{(1+\Delta_2(h))(1-\Delta_1(h))}], \right. \\ \left[\begin{array}{c} [\frac{p_2^-(h)+p_1^-(h)}{(1+p_2^-(h))(1-p_1^-(h))}, \frac{q_2^-(h)+q_1^-(h)}{(1+q_2^-(h))(1-q_1^-(h))}, \frac{r_2^-(h)+r_1^-(h)}{(1+r_2^-(h))(1-r_1^-(h))}], \\ [\frac{p_2^+(h)+p_1^+(h)}{(1+p_2^+(h))(1-p_1^+(h))}, \frac{q_2^+(h)+q_1^+(h)}{(1+q_2^+(h))(1-q_1^+(h))}, \frac{r_2^+(h)+r_1^+(h)}{(1+r_2^+(h))(1-r_1^+(h))}], \end{array} \right], \\ \left. \left[[\frac{p_2(h)p_1(h)}{1+((1-p_2(h))(1-p_1(h)))}, \frac{q_2(h)q_1(h)}{1+((1-q_2(h))(1-q_1(h)))}, \frac{r_2(h).r_1(h)}{1+((1-r_2(h))(1-r_1(h)))}]\right] \right\rangle \end{array} \right\} = A_2 + A_1$$

Hence $A_1 + A_2 = A_2 + A_1$.
(2) $\lambda(A_1 + A_2) = \lambda A_2 + \lambda A_1$

$$\lambda(A_1 + A_2) = \left\langle \left[\begin{array}{c} \frac{[(1+\alpha_1(h))(1-\alpha_1(h))]^\lambda [((1+\alpha_2(h))(1-\alpha_2(h))]^\lambda}{[(1+\alpha_1(h))(1-\alpha_1(h))]^\lambda [(1+\alpha_2(h))(1-\alpha_2(h))]^\lambda}, \\ \frac{[(1+\beta_1(h))(1-\beta_1(h))]^\lambda [((1+\beta_2(h))(1-\beta_2(h))]^\lambda}{[(1+\beta_1(h))(1-\beta_1(h))]^\lambda [(1+\beta_2(h))(1-\beta_2(h))]^\lambda}, \\ \frac{[(1+\Delta_1(h))(1-\Delta_1(h))]^\lambda [((1+\Delta_2(h))(1-\Delta_2(h))]^\lambda}{[(1+\Delta_1(h))(1-\Delta_1(h))]^\lambda [(1+\Delta_2(h))(1-\Delta_2(h))]^\lambda} \end{array} \right], \left[\begin{array}{c} [\frac{[(1+p_1^-(h))(1-p_1^-(h))]^\lambda [(1+p_2^-(h))(1-p_2^-(h))]^\lambda}{[(1+p_1^-(h))(1-p_1^-(h))]^\lambda [(1+p_2^-(h))(1-p_2^-(h))]^\lambda}, \\ \frac{[(1+q_1^-(h))(1-q_1^-(h))]^\lambda [(1+q_2^-(h))(1-q_2^-(h))]^\lambda}{[(1+q_1^-(h))(1-q_1^-(h))]^\lambda [(1+q_2^-(h))(1-q_2^-(h))]^\lambda}, \\ \frac{[(1+r_1^-(h))(1-r_1^-(h))]^\lambda [(1+r_2^-(h))(1-r_2^-(h))]^\lambda}{[(1+r_1^-(h))(1-r_1^-(h))]^\lambda [(1+r_2^-(h))(1-r_2^-(h))]^\lambda}] \\ [\frac{[(1+p_1^+(h))(1-p_1^+(h))]^\lambda [(1+p_2^+(h))(1-p_2^+(h))]^\lambda}{[(1+p_1^+(h))(1-p_1^+(h))]^\lambda [(1+p_2^+(h))(1-p_2^+(h))]^\lambda}, \\ \frac{[(1+q_1^+(h))(1-q_1^+(h))]^\lambda [(1+q_2^+(h))(1-q_2^+(h))]^\lambda}{[(1+q_1^+(h))(1-q_1^+(h))]^\lambda [(1+q_2^+(h))(1-q_2^+(h))]^\lambda}, \\ \frac{[(1+r_1^+(h))(1-r_1^+(h))]^\lambda [(1+r_2^+(h))(1-r_2^+(h))]^\lambda}{[(1+r_1^+(h))(1-r_1^+(h))]^\lambda [(1+r_2^+(h))(1-r_2^+(h))]^\lambda}] \end{array} \right], \right.$$

$$\left. \left[\begin{array}{c} [\frac{2[p_1(h)p_2(h)]^\lambda}{[(4-2p_1(h)-2p_2(h)-p_1(h)p_2(h)]^\lambda + [p_1(h)p_2(h)]^\lambda}, \\ \frac{2[q_1(h)q_2(h)]^\lambda}{[(4-2q_1(h)-2q_2(h)-q_1(h)q_2(h)]^\lambda + [q_1(h)q_2(h)]^\lambda}, \\ \frac{2[r_1(h)r_2(h)]^\lambda}{[(4-2r_1(h)-2r_2(h)-r_1(h)r_2(h)]^\lambda + [r_1(h)r_2(h)]^\lambda}] \end{array} \right] \right\rangle$$

and we have

$$\lambda A_1 = \left\langle \left[\begin{array}{c} \frac{[(1+\alpha_1(h))]^\lambda - [(1-\alpha_1(h))]^\lambda}{[(1+\alpha_1(h))]^\lambda + [(1-\alpha_1(h))]^\lambda}, \\ \frac{[(1+\beta_1(h))]^\lambda - [(1-\beta_1(h))]^\lambda}{[(1+\beta_1(h))]^\lambda + [(1-\beta_1(h))]^\lambda}, \\ \frac{[(1+\Delta_1(h))]^\lambda - [(1-\Delta_1(h))]^\lambda}{[(1+\Delta_1(h))]^\lambda + [(1-\Delta_1(h))]^\lambda} \end{array} \right], \left[\begin{array}{c} [\frac{[(1+p_1^-(h))^\lambda - (1-p_1^-(h))^\lambda]}{[(1+p_1^-(h))^\lambda + (1-p_1^-(h))^\lambda]}, \\ \frac{[(1+q_1^-(h))^\lambda - (1-q_1^-(h))^\lambda]}{[(1+q_1^-(h))^\lambda + (1-q_1^-(h))^\lambda]}, \frac{[(1+r_1^-(h))^\lambda - (1-r_1^-(h))^\lambda]}{[(1+r_1^-(h))^\lambda + (1-r_1^-(h))^\lambda]}] \\ [\frac{[(1+p_1^+(h))^\lambda - (1-p_1^+(h))^\lambda]}{[(1+p_1^+(h))^\lambda + (1-p_1^+(h))^\lambda]}, \\ \frac{[(1+q_1^+(h))^\lambda - (1-q_1^+(h))^\lambda]}{[(1+q_1^+(h))^\lambda + (1-q_1^+(h))^\lambda]}, \frac{[(1+r_1^+(h))^\lambda - (1-r_1^+(h))^\lambda]}{[(1+r_1^+(h))^\lambda + (1-r_1^+(h))^\lambda]}] \end{array} \right], \right.$$

$$\left. \left[\begin{array}{c} [\frac{2p_1^\lambda(h)}{[(2-p_1(h)]^\lambda + [p_1(h)]^\lambda}, \\ \frac{2q_1^\lambda(h)}{[(2-q_1(h)]^\lambda + [q_1(h)]^\lambda}, \frac{2r_1^\lambda(h)}{[(2-r_1(h)]^\lambda + [r_1(h)]^\lambda}] \end{array} \right] \right\rangle$$

$$\lambda A_2 = \left\langle \begin{bmatrix} \frac{[(1+\alpha_2(h))]^{\lambda}-[(1-\alpha_2(h))]^{\lambda}}{[(1+\alpha_2(h))]^{\lambda}+[(1-\alpha_2(h))]^{\lambda}}, \\ \frac{[(1+\beta_2(h))]^{\lambda}-[(1-\beta_2(h))]^{\lambda}}{[(1+\beta_2(h))]^{\lambda}+[(1-\beta_2(h))]^{\lambda}}, \\ \frac{[(1+\Delta_2(h))]^{\lambda}-[(1-\Delta_2(h))]^{\lambda}}{[(1+\Delta_2(h))]^{\lambda}+[(1-\Delta_2(h))]^{\lambda}} \end{bmatrix}, \begin{bmatrix} [\frac{[(1+p_2^-(h))^{\lambda}-(1-p_2^-(h))^{\lambda}]}{[(1+p_2^-(h))^{\lambda}+(1-p_2^-(h))^{\lambda}]}, \\ \frac{[(1+q_2^-(h))^{\lambda}-(1-q_2^-(h))^{\lambda}]}{[(1+q_2^-(h))^{\lambda}+(1-q_2^-(h))^{\lambda}]}, \frac{[(1+r_2^-(h))^{\lambda}-(1-r_2^-(h))^{\lambda}]}{[(1+r_2^-(h))^{\lambda}+(1-r_2^-(h))^{\lambda}]}], \\ [\frac{[(1+p_2^+(h))^{\lambda}-(1-p_2^+(h))^{\lambda}]}{[(1+p_2^+(h))^{\lambda}+(1-p_2^+(h))^{\lambda}]}, \\ \frac{[(1+q_2^+(h))^{\lambda}-(1-q_2^+(h))^{\lambda}]}{[(1+q_2^+(h))^{\lambda}+(1-q_2^+(h))^{\lambda}]}, \frac{[(1+r_2^+(h))^{\lambda}-(1-r_2^+(h))^{\lambda}]}{[(1+r_2^+(h))^{\lambda}+(1-r_2^+(h))^{\lambda}]}] \end{bmatrix}, \begin{bmatrix} [\frac{2p_2^{\lambda}(h)}{[(2-p_2(h)]^{\lambda}+[p_2(h)]^{\lambda}}, \\ \frac{2q_2^{\lambda}(h)}{[(2-q_2(h)]^{\lambda}+[q_2(h)]^{\lambda}}, \frac{2r_2^{\lambda}(h)}{[(2-r_2(h)]^{\lambda}+[r_2(h)]^{\lambda}}] \end{bmatrix} \right\rangle$$

$$\lambda A_2 + \lambda A_1 = \left\langle \begin{bmatrix} \frac{[(1+\alpha_2(h))(1-\alpha_2(h))]^{\lambda}[((1+\alpha_1(h))(1-\alpha_1(h))]^{\lambda}}{[(1+\alpha_2(h))(1-\alpha_2(h))]^{\lambda}[(1+\alpha_1(h))(1-\alpha_1(h))]^{\lambda}}, \\ \frac{[(1+\beta_2(h))(1-\beta_2(h))]^{\lambda}[((1+\beta_1(h))(1-\beta_1(h))]^{\lambda}}{[(1+\beta_2(h))(1-\beta_2(h))]^{\lambda}[(1+\beta_1(h))(1-\beta_1(h))]^{\lambda}}, \\ \frac{[(1+\Delta_2(h))(1-\Delta_2(h))]^{\lambda}[((1+\Delta_1(h))(1-\Delta_1(h))]^{\lambda}}{[(1+\Delta_2(h))(1-\Delta_2(h))]^{\lambda}[(1+\Delta_1(h))(1-\Delta_1(h))]^{\lambda}} \end{bmatrix}, \begin{bmatrix} [\frac{[(1+p_2^-(h))(1-p_2^-(h))]^{\lambda}[(1+p_1^-(h))(1-p_1^-(h))]^{\lambda}}{[(1+p_2^-(h))(1-p_2^-(h))]^{\lambda}[(1+p_1^-(h))(1-p_1^-(h))]^{\lambda}}, \\ \frac{[(1+q_2^-(h))(1-q_2^-(h))]^{\lambda}[(1+q_1^-(h))(1-q_1^-(h))]^{\lambda}}{[(1+q_2^-(h))(1-q_2^-(h))]^{\lambda}[(1+q_1^-(h))(1-q_1^-(h))]^{\lambda}}, \\ \frac{[(1+r_2^-(h))(1-r_2^-(h))]^{\lambda}[(1+r_1^-(h))(1-r_1^-(h))]^{\lambda}}{[(1+r_2^-(h))(1-r_2^-(h))]^{\lambda}[(1+r_1^-(h))(1-r_1^-(h))]^{\lambda}}] \\ [\frac{[(1+p_2^+(h))(1-p_2^+(h))]^{\lambda}[(1+p_1^+(h))(1-p_1^+(h))]^{\lambda}}{[(1+p_2^+(h))(1-p_2^+(h))]^{\lambda}[(1+p_1^+(h))(1-p_1^+(h))]^{\lambda}}, \\ \frac{[(1+q_2^+(h))(1-q_2^+(h))]^{\lambda}[(1+q_1^+(h))(1-q_1^+(h))]^{\lambda}}{[(1+q_2^+(h))(1-q_2^+(h))]^{\lambda}[(1+q_1^+(h))(1-q_1^+(h))]^{\lambda}}, \\ \frac{[(1+r_2^+(h))(1-r_2^+(h))]^{\lambda}[(1+r_1^+(h))(1-r_1^+(h))]^{\lambda}}{[(1+r_2^+(h))(1-r_2^+(h))]^{\lambda}[(1+r_1^+(h))(1-r_1^+(h))]^{\lambda}}] \end{bmatrix}, \begin{bmatrix} [\frac{2[p_2(h)p_1(h)]^{\lambda}}{[(4-2p_2(h)-2p_1(h)-p_2(h)p_1(h)]^{\lambda}+[p_2(h)p_1(h)]^{\lambda}}, \\ \frac{2[q_2(h)q_1(h)]^{\lambda}}{[(4-2q_2(h)-2q_1(h)-q_2(h)q_1(h)]^{\lambda}+[q_2(h)q_1(h)]^{\lambda}}, \\ \frac{2[r_2(h)r_1(h)]^{\lambda}}{[(4-2r_2(h)-2r_1(h)-r_2(h)r_1(h)]^{\lambda}+[r_2(h)r_1(h)]^{\lambda}}] \end{bmatrix} \right\rangle$$

so, we have $\lambda(A_1 + A_2) = \lambda A_2 + \lambda A_1$.

(3) $\lambda_1 A + \lambda_2 A = (\lambda_1 + \lambda_2)A$

$$\lambda_1 A = \left\langle \begin{bmatrix} \frac{[(1+\alpha_A(h))]^{\lambda_1}-[(1-\alpha_A(h))]^{\lambda_1}}{[(1+\alpha_A(h))]^{\lambda_1}+[(1-\alpha_A(h))]^{\lambda_1}}, \\ \frac{[(1+\beta_A(h))]^{\lambda_1}-[(1-\beta_A(h))]^{\lambda_1}}{[(1+\beta_A(h))]^{\lambda_1}+[(1-\beta_A(h))]^{\lambda_1}}, \\ \frac{[(1+\Delta_A(h))]^{\lambda_1}-[(1-\Delta_A(h))]^{\lambda_1}}{[(1+\Delta_A(h))]^{\lambda_1}+[(1-\Delta_A(h))]^{\lambda_1}} \end{bmatrix}, \begin{bmatrix} [\frac{[1+p_A^-(h)]^{\lambda_1}-[1-p_A^-(h)]^{\lambda_1}}{[1+p_A^-(h)]^{\lambda_1}+[1-p_A^-(h)]^{\lambda_1}}, \\ \frac{[1+q_A^-(h)]^{\lambda_1}-[1-q_A^-(h)]^{\lambda_1}}{[1+q_A^-(h)]^{\lambda_1}+[1-q_A^-(h)]^{\lambda_1}}, \frac{[1+r_A^-(h)]^{\lambda_1}-[1-r_A^-(h)]^{\lambda_1}}{[1+r_A^-(h)]^{\lambda_1}+[1-r_A^-(h)]^{\lambda_1}}] \\ [\frac{[1+p_A^+(h)]^{\lambda_1}-[1-p_A^+(h)]^{\lambda_1}}{[1+p_A^+(h)]^{\lambda_1}+[1-p_A^+(h)]^{\lambda_1}}, \\ \frac{[1+q_A^+(h)]^{\lambda_1}-[1-q_A^+(h)]^{\lambda_1}}{[1+q_A^+(h)]^{\lambda_1}+[1-q_A^+(h)]^{\lambda_1}}, \frac{[1+r_A^+(h)]^{\lambda_1}-[1-r_A^+(h)]^{\lambda_1}}{[1+r_A^+(h)]^{\lambda_1}+[1-r_A^+(h)]^{\lambda_1}}] \end{bmatrix}, \begin{bmatrix} [\frac{2[p_A(h)]^{\lambda_1}}{[(2-p_A(h)]^{\lambda_1}+[p_A(h)]^{\lambda_1}}, \\ \frac{2[q_A(h)]^{\lambda_1}}{[(2-q_A(h)]^{\lambda_1}+[q_A(h)]^{\lambda_1}}, \frac{2[r_A(h)]^{\lambda_1}}{[(2-r_A(h)]^{\lambda_1}+[r_A(h)]^{\lambda_1}}] \end{bmatrix} \right\rangle$$

$$\text{and } \lambda_2 A = \left\langle \begin{bmatrix} \frac{[(1+\alpha_A(h))]^{\lambda_2}-[(1-\alpha_A(h))]^{\lambda_2}}{[(1+\alpha_A(h))]^{\lambda_2}+[(1-\alpha_A(h))]^{\lambda_2}}, \\ \frac{[(1+\beta_A(h))]^{\lambda_2}-[(1-\beta_A(h))]^{\lambda_2}}{[(1+\beta_A(h))]^{\lambda_2}+[(1-\beta_A(h))]^{\lambda_2}}, \\ \frac{[(1+\Delta_A(h))]^{\lambda_2}-[(1-\Delta_A(h))]^{\lambda_2}}{[(1+\Delta_A(h))]^{\lambda_2}+[(1-\Delta_A(h))]^{\lambda_2}} \end{bmatrix}, \begin{bmatrix} [\frac{[1+p_A^-(h)]^{\lambda_2}-[1-p_A^-(h)]^{\lambda_2}}{[1+p_A^-(h)]^{\lambda_2}+[1-p_A^-(h)]^{\lambda_2}}, \frac{[1+q_A^-(h)]^{\lambda_2}-[1-q_A^-(h)]^{\lambda_2}}{[1+q_A^-(h)]^{\lambda_2}+[1-q_A^-(h)]^{\lambda_2}}, \\ \frac{[1+r_A^-(h)]^{\lambda_2}-[1-r_A^-(h)]^{\lambda_2}}{[1+r_A^-(h)]^{\lambda_2}+[1-r_A^-(h)]^{\lambda_2}}], [\frac{[1+p_A^+(h)]^{\lambda_2}-[1-p_A^+(h)]^{\lambda_2}}{[1+p_A^+(h)]^{\lambda_2}+[1-p_A^+(h)]^{\lambda_2}}, \\ \frac{[1+q_A^+(h)]^{\lambda_2}-[1-q_A^+(h)]^{\lambda_2}}{[1+q_A^+(h)]^{\lambda_2}+[1-q_A^+(h)]^{\lambda_2}}, \frac{[1+r_A^+(h)]^{\lambda_2}-[1-r_A^+(h)]^{\lambda_2}}{[1+r_A^+(h)]^{\lambda_2}+[1-r_A^+(h)]^{\lambda_2}}] \end{bmatrix}, \\ \begin{bmatrix} [\frac{2[p_A(h)]^{\lambda_2}}{[(2-p_A(h)]^{\lambda_2}+[p_A(h)]^{\lambda_2}}, \frac{2[q_A(h)]^{\lambda_2}}{[(2-q_A(h)]^{\lambda_2}+[q_A(h)]^{\lambda_2}}, \frac{2[r_A(h)]^{\lambda_2}}{[(2-r_A(h)]^{\lambda_2}+[r_A(h)]^{\lambda_2}}] \end{bmatrix} \right\rangle =$$

$$\left\langle \begin{bmatrix} \frac{[1+\alpha_A(h)]^{\lambda_1+\lambda_2}-[1-\alpha_A(h)]^{\lambda_1+\lambda_2}}{[1+\alpha_A(h)]^{\lambda_1+\lambda_2}+[1-\alpha_A(h)]^{\lambda_1+\lambda_2}}, \frac{[1+\beta_A(h)]^{\lambda_1+\lambda_2}-[1-\beta_A(h)]^{\lambda_1+\lambda_2}}{[1+\beta_A(h)]^{\lambda_1+\lambda_2}+[1-\beta_A(h)]^{\lambda_1+\lambda_2}}, \\ \frac{[1+\Delta_A(h)]^{\lambda_1+\lambda_2}-[1-\Delta_A(h)]^{\lambda_1+\lambda_2}}{[1+\Delta_A(h)]^{\lambda_1+\lambda_2}+[1-\Delta_A(h)]^{\lambda_1+\lambda_2}} \end{bmatrix}, \begin{bmatrix} [\frac{[1+p_A^-(h)]^{\lambda_1+\lambda_2}-[1-p_A^-(h)]^{\lambda_1+\lambda_2}}{[1+p_A^-(h)]^{\lambda_1+\lambda_2}+[1-p_A^-(h)]^{\lambda_1+\lambda_2}}, \frac{[1+q_A^-(h)]^{\lambda_1+\lambda_2}-[1-q_A^-(h)]^{\lambda_1+\lambda_2}}{[1+q_A^-(h)]^{\lambda_1+\lambda_2}+[1-q_A^-(h)]^{\lambda_1+\lambda_2}}, \\ \frac{[1+r_A^-(h)]^{\lambda_1+\lambda_2}-[1-r_A^-(h)]^{\lambda_1+\lambda_2}}{[1+r_A^-(h)]^{\lambda_1+\lambda_2}+[1-r_A^-(h)]^{\lambda_1+\lambda_2}}], [\frac{[1+p_A^+(h)]^{\lambda_1+\lambda_2}-[1-p_A^+(h)]^{\lambda_1+\lambda_2}}{[1+p_A^+(h)]^{\lambda_1+\lambda_2}+[1-p_A^+(h)]^{\lambda_1+\lambda_2}}, \\ \frac{[1+q_A^+(h)]^{\lambda_1+\lambda_2}-[1-q_A^+(h)]^{\lambda_1+\lambda_2}}{[1+q_A^+(h)]^{\lambda_1+\lambda_2}+[1-q_A^+(h)]^{\lambda_1+\lambda_2}}, \frac{[1+r_A^+(h)]^{\lambda_1+\lambda_2}-[1-r_A^+(h)]^{\lambda_1+\lambda_2}}{[1+r_A^+(h)]^{\lambda_1+\lambda_2}+[1-r_A^+(h)]^{\lambda_1+\lambda_2}}] \end{bmatrix}, \begin{bmatrix} [\frac{2[p_A(h)]^{\lambda_1+\lambda_2}}{[(2-p_A(h)]^{\lambda_1+\lambda_2}+[p_A(h)]^{\lambda_1+\lambda_2}}, \\ \frac{2[q_A(h)]^{\lambda_1+\lambda_2}}{[(2-q_A(h)]^{\lambda_1+\lambda_2}+[q_A(h)]^{\lambda_1+\lambda_2}}, \frac{2[r_A(h)]^{\lambda_1+\lambda_2}}{[(2-r_A(h)]^{\lambda_1+\lambda_2}+[r_A(h)]^{\lambda_1+\lambda_2}}] \end{bmatrix} \right\rangle = (\lambda_1+\lambda_2)A.$$

Appendix B. Proof of Theorem 1

Assume that $n = 1$, TCFEWA $(A_1, A_2, \ldots, A_n) = \overset{k}{\underset{j=1}{\oplus}} w_1 A_1$ $\langle(\lambda(A_1 + A_2) = \lambda A_2 + \lambda A_1$

$$\lambda(A_1+A_2) = \begin{bmatrix} \frac{[(1+\alpha_1(h))(1-\alpha_1(h))]^{\lambda}[((1+\alpha_2(h))(1-\alpha_2(h))]^{\lambda}}{[(1+\alpha_1(h))(1-\alpha_1(h))]^{\lambda}[(1+\alpha_2(h))(1-\alpha_2(h))]^{\lambda}}, \\ \frac{[(1+\beta_1(h))(1-\beta_1(h))]^{\lambda}[((1+\beta_2(h))(1-\beta_2(h))]^{\lambda}}{[(1+\beta_1(h))(1-\beta_1(h))]^{\lambda}[(1+\beta_2(h))(1-\beta_2(h))]^{\lambda}}, \\ \frac{[(1+\Delta_1(h))(1-\Delta_1(h))]^{\lambda}[((1+\Delta_2(h))(1-\Delta_2(h))]^{\lambda}}{[(1+\Delta_1(h))(1-\Delta_1(h))]^{\lambda}[(1+\Delta_2(h))(1-\Delta_2(h))]^{\lambda}} \end{bmatrix}, \left\langle \begin{bmatrix} [\frac{[(1+p_1^-(h))(1-p_1^-(h))]^{\lambda}[(1+p_2^-(h))(1-p_2^-(h))]^{\lambda}}{[(1+p_1^-(h))(1-p_1^-(h))]^{\lambda}[(1+p_2^-(h))(1-p_2^-(h))]^{\lambda}}, \\ \frac{[(1+q_1^-(h))(1-q_1^-(h))]^{\lambda}[(1+q_2^-(h))(1-q_2^-(h))]^{\lambda}}{[(1+q_1^-(h))(1-q_1^-(h))]^{\lambda}[(1+q_2^-(h))(1-q_2^-(h))]^{\lambda}}, \\ \frac{[(1+r_1^-(h))(1-r_1^-(h))]^{\lambda}[(1+r_2^-(h))(1-r_2^-(h))]^{\lambda}}{[(1+r_1^-(h))(1-r_1^-(h))]^{\lambda}[(1+r_2^-(h))(1-r_2^-(h))]^{\lambda}}] \\ ,[\frac{[(1+p_1^+(h))(1-p_1^+(h))]^{\lambda}[(1+p_2^+(h))(1-p_2^+(h))]^{\lambda}}{[(1+p_1^+(h))(1-p_1^+(h))]^{\lambda}[(1+p_2^+(h))(1-p_2^+(h))]^{\lambda}}, \\ \frac{[(1+q_1^+(h))(1-q_1^+(h))]^{\lambda}[(1+q_2^+(h))(1-q_2^+(h))]^{\lambda}}{[(1+q_1^+(h))(1-q_1^+(h))]^{\lambda}[(1+q_2^+(h))(1-q_2^+(h))]^{\lambda}}, \\ \frac{[(1+r_1^+(h))(1-r_1^+(h))]^{\lambda}[(1+r_2^+(h))(1-r_2^+(h))]^{\lambda}}{[(1+r_1^+(h))(1-r_1^+(h))]^{\lambda}[(1+r_2^+(h))(1-r_2^+(h))]^{\lambda}}] \end{bmatrix}, \right.$$

$$\left. \begin{bmatrix} [\frac{2[p_1(h)p_2(h)]^{\lambda}}{[(4-2p_1(h)-2p_2(h)-p_1(h)p_2(h)]^{\lambda}+[p_1(h)p_2(h)]^{\lambda}}, \\ \frac{2[q_1(h)q_2(h)]^{\lambda}}{[(4-2q_1(h)-2q_2(h)-q_1(h)q_2(h)]^{\lambda}+[q_1(h)q_2(h)]^{\lambda}}, \\ \frac{2[r_1(h)r_2(h)]^{\lambda}}{[(4-2r_1(h)-2r_2(h)-r_1(h)r_2(h)]^{\lambda}+[r_1(h)r_2(h)]^{\lambda}}] \end{bmatrix} \right\rangle$$

and we have

$$\lambda A_1=\left\langle\begin{bmatrix}\frac{[(1+\alpha_1(h))]^{\lambda}-[(1-\alpha_1(h))]^{\lambda}}{[(1+\alpha_1(h))]^{\lambda}+[(1-\alpha_1(h))]^{\lambda}},\\ \frac{[(1+\beta_1(h))]^{\lambda}-[(1-\beta_1(h))]^{\lambda}}{[(1+\beta_1(h))]^{\lambda}+[(1-\beta_1(h))]^{\lambda}},\\ \frac{[(1+\Delta_1(h))]^{\lambda}-[(1-\Delta_1(h))]^{\lambda}}{[(1+\Delta_1(h))]^{\lambda}+[(1-\Delta_1(h))]^{\lambda}}\end{bmatrix},\begin{bmatrix}[\frac{[(1+p_1^-(h))^{\lambda}-(1-p_1^-(h))^{\lambda}]}{[(1+p_1^-(h))^{\lambda}+(1-p_1^-(h))^{\lambda}]},\\ \frac{[(1+q_1^-(h))^{\lambda}-(1-q_1^-(h))^{\lambda}]}{[(1+q_1^-(h))^{\lambda}+(1-q_1^-(h))^{\lambda}]},\frac{[(1+r_1^-(h))^{\lambda}-(1-r_1^-(h))^{\lambda}]}{[(1+r_1^-(h))^{\lambda}+(1-r_1^-(h))^{\lambda}]}]\\ [\frac{[(1+p_1^+(h))^{\lambda}-(1-p_1^+(h))^{\lambda}]}{[(1+p_1^+(h))^{\lambda}+(1-p_1^+(h))^{\lambda}]},\\ \frac{[(1+q_1^+(h))^{\lambda}-(1-q_1^+(h))^{\lambda}]}{[(1+q_1^+(h))^{\lambda}+(1-q_1^+(h))^{\lambda}]},\frac{[(1+r_1^+(h))^{\lambda}-(1-r_1^+(h))^{\lambda}]}{[(1+r_1^+(h))^{\lambda}+(1-r_1^+(h))^{\lambda}]}]\end{bmatrix},\begin{bmatrix}[\frac{2p_1^{\lambda}(h)}{[(2-p_1(h)]^{\lambda}+[p_1(h)]^{\lambda}},\\ \frac{2q_1^{\lambda}(h)}{[(2-q_1(h)]^{\lambda}+[q_1(h)]^{\lambda}},\\ \frac{2r_1^{\lambda}(h)}{[(2-r_1(h)]^{\lambda}+[r_1(h)]^{\lambda}}]\end{bmatrix}\right\rangle$$

$$\lambda A_2=\left\langle\begin{bmatrix}\frac{[(1+\alpha_2(h))]^{\lambda}-[(1-\alpha_2(h))]^{\lambda}}{[(1+\alpha_2(h))]^{\lambda}+[(1-\alpha_2(h))]^{\lambda}},\\ \frac{[(1+\beta_2(h))]^{\lambda}-[(1-\beta_2(h))]^{\lambda}}{[(1+\beta_2(h))]^{\lambda}+[(1-\beta_2(h))]^{\lambda}},\\ \frac{[(1+\Delta_2(h))]^{\lambda}-[(1-\Delta_2(h))]^{\lambda}}{[(1+\Delta_2(h))]^{\lambda}+[(1-\Delta_2(h))]^{\lambda}}\end{bmatrix},\begin{bmatrix}[\frac{[(1+p_2^-(h))^{\lambda}-(1-p_2^-(h))^{\lambda}]}{[(1+p_2^-(h))^{\lambda}+(1-p_2^-(h))^{\lambda}]},\\ \frac{[(1+q_2^-(h))^{\lambda}-(1-q_2^-(h))^{\lambda}]}{[(1+q_2^-(h))^{\lambda}+(1-q_2^-(h))^{\lambda}]},\frac{[(1+r_2^-(h))^{\lambda}-(1-r_2^-(h))^{\lambda}]}{[(1+r_2^-(h))^{\lambda}+(1-r_2^-(h))^{\lambda}]}]\\ [\frac{[(1+p_2^+(h))^{\lambda}-(1-p_2^+(h))^{\lambda}]}{[(1+p_2^+(h))^{\lambda}+(1-p_2^+(h))^{\lambda}]},\\ \frac{[(1+q_2^+(h))^{\lambda}-(1-q_2^+(h))^{\lambda}]}{[(1+q_2^+(h))^{\lambda}+(1-q_2^+(h))^{\lambda}]},\frac{[(1+r_2^+(h))^{\lambda}-(1-r_2^+(h))^{\lambda}]}{[(1+r_2^+(h))^{\lambda}+(1-r_2^+(h))^{\lambda}]}]\end{bmatrix},$$

$$\begin{bmatrix}[\frac{2p_2^{\lambda}(h)}{[(2-p_2(h)]^{\lambda}+[p_2(h)]^{\lambda}},\frac{2q_2^{\lambda}(h)}{[(2-q_2(h)]^{\lambda}+[q_2(h)]^{\lambda}},\\ \frac{2r_2^{\lambda}(h)}{[(2-r_2(h)]^{\lambda}+[r_2(h)]^{\lambda}}]\end{bmatrix}\right\rangle$$

$$\lambda A_2+\lambda A_1=\left\langle\begin{bmatrix}\frac{[(1+\alpha_2(h))(1-\alpha_2(h))]^{\lambda}[((1+\alpha_1(h))(1-\alpha_1(h))]^{\lambda}}{[(1+\alpha_2(h))(1-\alpha_2(h))]^{\lambda}[(1+\alpha_1(h))(1-\alpha_1(h))]^{\lambda}},\\ \frac{[(1+\beta_2(h))(1-\beta_2(h))]^{\lambda}[((1+\beta_1(h))(1-\beta_1(h))]^{\lambda}}{[(1+\beta_2(h))(1-\beta_2(h))]^{\lambda}[(1+\beta_1(h))(1-\beta_1(h))]^{\lambda}},\\ \frac{[(1+\Delta_2(h))(1-\Delta_2(h))]^{\lambda}[((1+\Delta_1(h))(1-\Delta_1(h))]^{\lambda}}{[(1+\Delta_2(h))(1-\Delta_2(h))]^{\lambda}[(1+\Delta_1(h))(1-\Delta_1(h))]^{\lambda}}\end{bmatrix},\begin{bmatrix}[\frac{[(1+p_2^-(h))(1-p_2^-(h))]^{\lambda}[(1+p_1^-(h))(1-p_1^-(h))]^{\lambda}}{[(1+p_2^-(h))(1-p_2^-(h))]^{\lambda}[(1+p_1^-(h))(1-p_1^-(h))]^{\lambda}},\\ \frac{[(1+q_2^-(h))(1-q_2^-(h))]^{\lambda}[(1+q_1^-(h))(1-q_1^-(h))]^{\lambda}}{[(1+q_2^-(h))(1-q_2^-(h))]^{\lambda}[(1+q_1^-(h))(1-q_1^-(h))]^{\lambda}},\\ \frac{[(1+r_2^-(h))(1-r_2^-(h))]^{\lambda}[(1+r_1^-(h))(1-r_1^-(h))]^{\lambda}}{[(1+r_2^-(h))(1-r_2^-(h))]^{\lambda}[(1+r_1^-(h))(1-r_1^-(h))]^{\lambda}}]\\ [\frac{[(1+p_2^+(h))(1-p_2^+(h))]^{\lambda}[(1+p_1^+(h))(1-p_1^+(h))]^{\lambda}}{[(1+p_2^+(h))(1-p_2^+(h))]^{\lambda}[(1+p_1^+(h))(1-p_1^+(h))]^{\lambda}},\\ \frac{[(1+q_2^+(h))(1-q_2^+(h))]^{\lambda}[(1+q_1^+(h))(1-q_1^+(h))]^{\lambda}}{[(1+q_2^+(h))(1-q_2^+(h))]^{\lambda}[(1+q_1^+(h))(1-q_1^+(h))]^{\lambda}},\\ \frac{[(1+r_2^+(h))(1-r_2^+(h))]^{\lambda}[(1+r_1^+(h))(1-r_1^+(h))]^{\lambda}}{[(1+r_2^+(h))(1-r_2^+(h))]^{\lambda}[(1+r_1^+(h))(1-r_1^+(h))]^{\lambda}}]\end{bmatrix},$$

$$\begin{bmatrix}[\frac{2[p_2(h)p_1(h)]^{\lambda}}{[(4-2p_2(h)-2p_1(h)-p_2(h)p_1(h)]^{\lambda}+[p_2(h)p_1(h)]^{\lambda}},\\ \frac{2[q_2(h)q_1(h)]^{\lambda}}{[(4-2q_2(h)-2q_1(h)-q_2(h)q_1(h)]^{\lambda}+[q_2(h)q_1(h)]^{\lambda}},\\ \frac{2[r_2(h)r_1(h)]^{\lambda}}{[(4-2r_2(h)-2r_1(h)-r_2(h)r_1(h)]^{\lambda}+[s_2(h)s_1(h)]^{\lambda}}]\end{bmatrix}\right\rangle$$

so, we have $\lambda(A_1+A_2)=\lambda A_2+\lambda A_1$.

$\lambda_1 A+\lambda_2 A=(\lambda_1+\lambda_2)A$

$$\lambda_1 A=\left\langle\begin{bmatrix}\frac{[1+\alpha_A(h)]^{\lambda_1}-[1-\alpha_A(h)]^{\lambda_1}}{[1+\alpha_A(h)]^{\lambda_1}+[1-\alpha_A(h)]^{\lambda_1}},\frac{[1+\beta_A(h)]^{\lambda_1}-[1-\beta_A(h)]^{\lambda_1}}{[1+\beta_A(h)]^{\lambda_1}+[1-\beta_A(h)]^{\lambda_1}},\\ \frac{[1+\Delta_A(h)]^{\lambda_1}-[1-\Delta_A(h)]^{\lambda_1}}{[1+\Delta_A(h)]^{\lambda_1}+[1-\Delta_A(h)]^{\lambda_1}}\end{bmatrix},\right.$$

$$\begin{bmatrix}[\frac{[1+p_A^-(h)]^{\lambda_1}-[1-p_A^-(h)]^{\lambda_1}}{[1+p_A^-(h)]^{\lambda_1}+[1-p_A^-(h)]^{\lambda_1}},\frac{[1+q_A^-(h)]^{\lambda_1}-[1-q_A^-(h)]^{\lambda_1}}{[1+q_A^-(h)]^{\lambda_1}+[1-q_A^-(h)]^{\lambda_1}},\frac{[1+r_A^-(h)]^{\lambda_1}-[1-r_A^-(h)]^{\lambda_1}}{[1+r_A^-(h)]^{\lambda_1}+[1-r_A^-(h)]^{\lambda_1}}]\\ [\frac{[1+p_A^+(h)]^{\lambda_1}-[1-p_A^+(h)]^{\lambda_1}}{[1+p_A^+(h)]^{\lambda_1}+[1-p_A^+(h)]^{\lambda_1}},\frac{[1+q_A^+(h)]^{\lambda_1}-[1-q_A^+(h)]^{\lambda_1}}{[1+q_A^+(h)]^{\lambda_1}+[1-q_A^+(h)]^{\lambda_1}},\frac{[1+r_A^+(h)]^{\lambda_1}-[1-r_A^+(h)]^{\lambda_1}}{[1+r_A^+(h)]^{\lambda_1}+[1-r_A^+(h)]^{\lambda_1}}]\end{bmatrix},$$

$$\left.\begin{bmatrix}\frac{2[p_A(h)]^{\lambda_1}}{[(2-p_A(h)]^{\lambda_1}+[p_A(h)]^{\lambda_1}},\frac{2[q_A(h)]^{\lambda_1}}{[(2-q_A(h)]^{\lambda_1}+[q_A(h)]^{\lambda_1}},\frac{2[r_A(h)]^{\lambda_1}}{[(2-r_A(h)]^{\lambda_1}+[r_A(h)]^{\lambda_1}}\end{bmatrix}\right\rangle$$

and $$\lambda_2 A=\left\langle\begin{bmatrix}\frac{[1+\alpha_A(h)]^{\lambda_2}-[1-\alpha_A(h)]^{\lambda_2}}{[1+\alpha_A(h)]^{\lambda_2}+[1-\alpha_A(h)]^{\lambda_2}},\\ \frac{[1+\beta_A(h)]^{\lambda_2}-[1-\beta_A(h)]^{\lambda_2}}{[1+\beta_A(h)]^{\lambda_2}+[1-\beta_A(h)]^{\lambda_2}},\\ \frac{[1+\Delta_A(h)]^{\lambda_2}-[1-\Delta_A(h)]^{\lambda_2}}{[1+\Delta_A(h)]^{\lambda_2}+[1-\Delta_A(h)]^{\lambda_2}}\end{bmatrix},\begin{bmatrix}[\frac{[1+p_A^-(h)]^{\lambda_2}-[1-p_A^-(h)]^{\lambda_2}}{[1+p_A^-(h)]^{\lambda_2}+[1-p_A^-(h)]^{\lambda_2}},\frac{[1+q_A^-(h)]^{\lambda_2}-[1-q_A^-(h)]^{\lambda_2}}{[1+q_A^-(h)]^{\lambda_2}+[1-q_A^-(h)]^{\lambda_2}},\\ \frac{[1+r_A^-(h)]^{\lambda_2}-[1-r_A^-(h)]^{\lambda_2}}{[1+r_A^-(h)]^{\lambda_2}+[1-r_A^-(h)]^{\lambda_2}}],[\frac{[1+p_A^+(h)]^{\lambda_2}-[1-p_A^+(h)]^{\lambda_2}}{[1+p_A^+(h)]^{\lambda_2}+[1-p_A^+(h)]^{\lambda_2}},\\ \frac{[1+q_A^+(h)]^{\lambda_2}-[1-q_A^+(h)]^{\lambda_2}}{[1+q_A^+(h)]^{\lambda_2}+[1-q_A^+(h)]^{\lambda_2}},\frac{[1+r_A^+(h)]^{\lambda_2}-[1-r_A^+(h)]^{\lambda_2}}{[1+r_A^+(h)]^{\lambda_2}+[1-r_A^+(h)]^{\lambda_2}}]\end{bmatrix},$$

$$\left[\begin{array}{c}[\frac{2[p_A(h)]^{\lambda_2}}{[(2-p_A(h)]^{\lambda_2}+[p_A(h)]^{\lambda_2}},\\ \frac{2[q_A(h)]^{\lambda_2}}{[(2-q_A(h)]^{\lambda_2}+[q_A(h)]^{\lambda_2}},\frac{2[r_A(h)]^{\lambda_2}}{[(2-r_A(h)]^{\lambda_2}+[r_A(h)]^{\lambda_2}}]\end{array}\right]\Bigg\rangle=$$

$$\Bigg\langle\left[\begin{array}{c}\frac{[1+\alpha_A(h)]^{\lambda_1+\lambda_2}-[1-\alpha_A(h)]^{\lambda_1+\lambda_2}}{[1+\alpha_A(h)]^{\lambda_1+\lambda_2}+[1-\alpha_A(h)]^{\lambda_1+\lambda_2}},\frac{[\beta_A(h)]^{\lambda_1+\lambda_2}-[1-\beta_A(h)]^{\lambda_1+\lambda_2}}{[1+\beta_A(h)]^{\lambda_1+\lambda_2}+[1-\beta_A(h)]^{\lambda_1+\lambda_2}},\\ \frac{[1+\Delta_A(h)]^{\lambda_1+\lambda_2}-[1-\Delta_A(h)]^{\lambda_1+\lambda_2}}{[1+\Delta_A(h)]^{\lambda_1+\lambda_2}+[1-\Delta_A(h)]^{\lambda_1+\lambda_2}}\end{array}\right],$$

$$\left[\begin{array}{c}[\frac{[1+p_A^-(h)]^{\lambda_1+\lambda_2}-[1-p_A^-(h)]^{\lambda_1+\lambda_2}}{[1+p_A^-(h)]^{\lambda_1+\lambda_2}+[1-p_A^-(h)]^{\lambda_1+\lambda_2}},\frac{[1+q_A^-(h)]^{\lambda_1+\lambda_2}-[1-q_A^-(h)]^{\lambda_1+\lambda_2}}{[1+q_A^-(h)]^{\lambda_1+\lambda_2}+[1-q_A^-(h)]^{\lambda_1+\lambda_2}},\\ \frac{[1+r_A^-(h)]^{\lambda_1+\lambda_2}-[1-r_A^-(h)]^{\lambda_1+\lambda_2}}{[1+r_A^-(h)]^{\lambda_1+\lambda_2}+[1-r_A^-(h)]^{\lambda_1+\lambda_2}}],[\frac{[1+p_A^+(h)]^{\lambda_1+\lambda_2}-[1-p_A^+(h)]^{\lambda_1+\lambda_2}}{[1+p_A^+(h)]^{\lambda_1+\lambda_2}+[1-p_A^+(h)]^{\lambda_1+\lambda_2}},\\ \frac{[1+q_A^+(h)]^{\lambda_1+\lambda_2}-[1-q_A^+(h)]^{\lambda_1+\lambda_2}}{[1+q_A^+(h)]^{\lambda_1+\lambda_2}+[1-q_A^+(h)]^{\lambda_1+\lambda_2}},\frac{[1+r_A^+(h)]^{\lambda_1+\lambda_2}-[1-r_A^+(h)]^{\lambda_1+\lambda_2}}{[1+r_A^+(h)]^{\lambda_1+\lambda_2}+[1-r_A^+(h)]^{\lambda_1+\lambda_2}}]\end{array}\right],$$

$$\left[\begin{array}{c}[\frac{2[p_A(h)]^{\lambda_1+\lambda_2}}{[(2-p_A(h)]^{\lambda_1+\lambda_2}+[p_A(h)]^{\lambda_1+\lambda_2}},\\ \frac{2[q_A(h)]^{\lambda_1+\lambda_2}}{[(2-q_A(h)]^{\lambda_1+\lambda_2}+[q_A(h)]^{\lambda_1+\lambda_2}},\frac{2[r_A(h)]^{\lambda_1+\lambda_2}}{[(2-r_A(h)]^{\lambda_1+\lambda_2}+[r_A(h)]^{\lambda_1+\lambda_2}}]\end{array}\right]\Bigg\rangle=$$

$$\left[\frac{[[1+\alpha_1(h)]^{\lambda_1}-[1-\alpha_1(h)]^{\lambda_1}}{[1+\alpha_1(h)]^{\lambda_1}+[1-\alpha_1(h)]^{\lambda_1}},\frac{[[1+\beta_1(h)]^{\lambda_1}-[1-\beta_1(h)]^{\lambda_1}}{[1+\beta_1(h)]^{\lambda_1}+[1-\beta_1(h)]^{\lambda_1}},\frac{[[1+\Delta_1(h)]^{\lambda_1}-[1-\Delta_1(h)]^{\lambda_1}}{[1+\Delta_1(h)]^{\lambda_1}+[1-\Delta_1(h)]^{\lambda_1}}\right]$$

$$\left[\begin{array}{c}\frac{[[1+p_1^-(h)]^{\lambda_1}-[1-p_1^-(h)]^{\lambda_1}}{[1+p_1^-(h)]^{\lambda_1}+[1-p_1^-(h)]^{\lambda_1}},\frac{[1+q_1^-(h)]^{\lambda_1}-[1-q_1^-(h)]^{\lambda_1}}{[1+q_1^-(h)]^{\lambda_1}+[1-q_1^-(h)]^{\lambda_1}},\\ \frac{[1+r_1^-(h)]^{\lambda_1}-[1-r_1^-(h)]^{\lambda_1}}{[1+r_1^-(h)]^{\lambda_1}+[1-r_1^-(h)]^{\lambda_1}}\end{array}\right];\left[\begin{array}{c}\frac{[1+p_1^+(h)]^{\lambda_1}-[1-p_1^+(h)]^{\lambda_1}}{[1+p_1^+(h)]^{\lambda_1}+[1-p_1^+(h)]^{\lambda_1}},\frac{[1+q_1^+(h)]^{\lambda_1}-[1-q_1^+(h)]^{\lambda_1}}{[1+q_1^+(h)]^{\lambda_1}+[1-q_1^+(h)]^{\lambda_1}},\\ \frac{[1+r_1^+(h)]^{\lambda_1}-[1-r_1^+(h)]^{\lambda_1}}{[1+r_1^+(h)]^{\lambda_1}+[1-r_1^+(h)]^{\lambda_1}}\end{array}\right];$$

$$\left[\begin{array}{c}\frac{2[p_1(h)]^{\lambda_1}}{[(2-p_1(h)]^{\lambda_1}+[p_1(h)]^{\lambda_1}},\frac{2[q_1(h)]^{\lambda_1}}{[(2-q_1(h)]^{\lambda_1}+[q_1(h)]^{\lambda_1}},\\ \frac{2[r_1(h)]^{\lambda_1}}{[(2-r_1(h)]^{\lambda_1}+[r_1(h)]^{\lambda_1}}\end{array}\right].$$

Assume that $n=k$, TCFEWA $(A_1, A_2, \ldots, A_n)=\overset{k}{\underset{j=1}{\oplus}} w_j A_j$

$$\Bigg\langle\left[\frac{[\prod_{j=1}^{k}[1+\alpha_1(h)]^{\varpi}-\prod_{j=1}^{k}[1-\alpha_1(h)]^{\varpi}}{\prod_{j=1}^{k}[1+\alpha_1(h)]^{\varpi}+\prod_{j=1}^{k}[1-\alpha_1(h)]^{\varpi}},\frac{[\prod_{j=1}^{k}[1+\beta_1(h)]^{\varpi}-\prod_{j=1}^{k}[1-\beta_1(h)]^{\varpi}}{\prod_{j=1}^{k}[1+\beta_1(h)]^{\varpi}+\prod_{j=1}^{k}[1-\beta_1(h)]^{\varpi}},\frac{[\prod_{j=1}^{k}[1+\Delta_1(h)]^{\varpi}-\prod_{j=1}^{k}[1-\Delta_1(h)]^{\varpi}}{\prod_{j=1}^{k}[1+\Delta_1(h)]^{\varpi}+\prod_{j=1}^{k}[1-\Delta_1(h)]^{\varpi}}\right]$$

$$\left[\frac{[\prod_{j=1}^{k}[1+p_1^-(h)]^{\varpi}-\prod_{j=1}^{k}[1-p_1^-(h)]^{\varpi}}{\prod_{j=1}^{k}[1+p_1^-(h)]^{\varpi}+\prod_{j=1}^{k}[1-p_1^-(h)]^{\varpi}},\frac{\prod_{j=1}^{k}[1+q_1^-(h)]^{\varpi}-\prod_{j=1}^{k}[1-q_1^-(h)]^{\varpi}}{\prod_{j=1}^{k}[1+q_1^-(h)]^{\varpi}+\prod_{j=1}^{k}[1-q_1^-(h)]^{\varpi}},\frac{\prod_{j=1}^{k}[1+r_1^-(h)]^{\varpi}-\prod_{j=1}^{k}[1-r_1^-(h)]^{\varpi}}{\prod_{j=1}^{k}[1+r_1^-(h)]^{\varpi}+\prod_{j=1}^{k}[1-r_1^-(h)]^{\varpi}}\right],$$

$$\left[\frac{\prod_{j=1}^{k}[1+p_1^+(h)]^{\varpi}-\prod_{j=1}^{k}[1-p_1^+(h)]^{\varpi}}{\prod_{j=1}^{k}[1+p_1^+(h)]^{\varpi}+\prod_{j=1}^{k}[1-p_1^+(h)]^{\varpi}},\frac{\prod_{j=1}^{k}[1+q_1^+(h)]^{\varpi}-\prod_{j=1}^{k}[1-q_1^+(h)]^{\varpi}}{\prod_{j=1}^{k}[1+q_1^+(h)]^{\varpi}+\prod_{j=1}^{k}[1-q_1^+(h)]^{\varpi}},\frac{\prod_{j=1}^{k}[1+r_1^+(h)]^{\varpi}-\prod_{j=1}^{k}[1-r_1^+(h)]^{\varpi}}{\prod_{j=1}^{k}[1+r_1^+(h)]^{\varpi}+\prod_{j=1}^{k}[1-r_1^+(h)]^{\varpi}}\right];$$

$$\left[\frac{2\prod_{j=1}^{k}[p_1(h)]^{\varpi}}{\prod_{j=1}^{k}[(2-p_1(h)]^{\varpi}+\prod_{j=1}^{k}[p_1(h)]^{\varpi}},\frac{2\prod_{j=1}^{k}[q_1(h)]^{\varpi}}{\prod_{j=1}^{k}[(2-q_1(h)]^{\varpi}+\prod_{j=1}^{k}[q_1(h)]^{\varpi}},\frac{2\prod_{j=1}^{k}[r_1(h)]^{\varpi}}{\prod_{j=1}^{k}[(2-r_1(h)]^{\varpi}+\prod_{j=1}^{k}[r_1(h)]^{\varpi}}\right].$$

Then when $n=k+1$, we have TCFEWA $(A_1, A_2, \ldots, A_{k+1})=$ TCFEWA $(A_1, A_2, \ldots, A_k)\oplus A_{k+1})$

$$\left\langle \left[\frac{[\prod_{j=1}^{k}[1+\alpha_1(h)]^{\varpi}-\prod_{j=1}^{k}[1-\alpha_1(h)]^{\varpi}}{\prod_{j=1}^{k}[1+\alpha_1(h)]^{\varpi}+\prod_{j=1}^{k}[1-\alpha_1(h)]^{\varpi}}, \frac{[\prod_{j=1}^{k}[1+\beta_1(h)]^{\varpi}-\prod_{j=1}^{k}[1-\beta_1(h)]^{\varpi}}{\prod_{j=1}^{k}[1+\beta_1(h)]^{\varpi}+\prod_{j=1}^{k}[1-\beta_1(h)]^{\varpi}}, \frac{[\prod_{j=1}^{k}[1+\Delta_1(h)]^{\varpi}-\prod_{j=1}^{k}[1-\Delta_1(h)]^{\varpi}}{\prod_{j=1}^{k}[1+\Delta_1(h)]^{\varpi}+\prod_{j=1}^{k}[1-\Delta_1(h)]^{\varpi}} \right] \right.$$

$$\left[\frac{[\prod_{j=1}^{k}[1+p_1^-(h)]^{\varpi}-\prod_{j=1}^{k}[1-p_1^-(h)]^{\varpi}}{\prod_{j=1}^{k}[1+p_1^-(h)]^{\varpi}+\prod_{j=1}^{k}[1-p_1^-(h)]^{\varpi}}, \frac{\prod_{j=1}^{k}[1+q_1^-(h)]^{\varpi}-\prod_{j=1}^{k}[1-q_1^-(h)]^{\varpi}}{\prod_{j=1}^{k}[1+q_1^-(h)]^{\varpi}+\prod_{j=1}^{k}[1-q_1^-(h)]^{\varpi}}, \frac{\prod_{j=1}^{k}[1+r_1^-(h)]^{\varpi}-\prod_{j=1}^{k}[1-r_1^-(h)]^{\varpi}}{\prod_{j=1}^{k}[1+r_1^-(h)]^{\varpi}+\prod_{j=1}^{k}[1-r_1^-(h)]^{\varpi}} \right];$$

$$\left[\frac{\prod_{j=1}^{k}[1+p_1^+(h)]^{\varpi}-\prod_{j=1}^{k}[1-p_1^+(h)]^{\varpi}}{\prod_{j=1}^{k}[1+p_1^+(h)]^{\varpi}+\prod_{j=1}^{k}[1-p_1^+(h)]^{\varpi}}, \frac{\prod_{j=1}^{k}[1+q_1^+(h)]^{\varpi}-\prod_{j=1}^{k}[1-q_1^+(h)]^{\varpi}}{\prod_{j=1}^{k}[1+q_1^+(h)]^{\varpi}+\prod_{j=1}^{k}[1-q_1^+(h)]^{\varpi}}, \frac{\prod_{j=1}^{k}[1+r_1^+(h)]^{\varpi}-\prod_{j=1}^{k}[1-r_1^+(h)]^{\varpi}}{\prod_{j=1}^{k}[1+r_1^+(h)]^{\varpi}+\prod_{j=1}^{k}[1-r_1^+(h)]^{\varpi}} \right];$$

$$\left[\frac{2\prod_{j=1}^{k}[p_1(h)]^{\varpi}}{\prod_{j=1}^{k}[(2-p_1(h)]^{\varpi}+\prod_{j=1}^{k}[p_1(h)]^{\varpi}}, \frac{2\prod_{j=1}^{k}[q_1(h)]^{\varpi}}{\prod_{j=1}^{k}[(2-q_1(h)]^{\varpi}+\prod_{j=1}^{k}[q_1(h)]^{\varpi}}, \frac{2\prod_{j=1}^{k}[r_1(h)]^{\varpi}}{\prod_{j=1}^{k}[(2-r_1(h)]^{\varpi}+\prod_{j=1}^{k}[r_1(h)]^{\varpi}} \right] \oplus_{k+1}$$

$$\left\langle \left[\frac{[\prod_{j=1}^{k+1}[1+\alpha_1(h)]^{\varpi}-\prod_{j=1}^{k+1}[1-\alpha_1(h)]^{\varpi}}{\prod_{j=1}^{k+1}[1+\alpha_1(h)]^{\varpi}+\prod_{j=1}^{k+1}[1-\alpha_1(h)]^{\varpi}}, \frac{[\prod_{j=1}^{k+1}[1+\beta_1(h)]^{\varpi}-\prod_{j=1}^{k+1}[1-\beta_1(h)]^{\varpi}}{\prod_{j=1}^{k+1}[1+\beta_1(h)]^{\varpi}+\prod_{j=1}^{k+1}[1-\beta_1(h)]^{\varpi}}, \frac{[\prod_{j=1}^{k+1}[1+\Delta_1(h)]^{\varpi}-\prod_{j=1}^{k+1}[1-\Delta_1(h)]^{\varpi}}{\prod_{j=1}^{k+1}[1+\Delta_1(h)]^{\varpi}+\prod_{j=1}^{k+1}[1-\Delta_1(h)]^{\varpi}} \right] \right.$$

$$\left[\frac{[\prod_{j=1}^{k+1}[1+p_1^-(h)]^{\varpi}-\prod_{j=1}^{k+1}[1-p_1^-(h)]^{\varpi}}{\prod_{j=1}^{k+1}[1+p_1^-(h)]^{\varpi}+\prod_{j=1}^{k+1}[1-p_1^-(h)]^{\varpi}}, \frac{\prod_{j=1}^{k+1}[1+q_1^-(h)]^{\varpi}-\prod_{j=1}^{k+1}[1-q_1^-(h)]^{\varpi}}{\prod_{j=1}^{k+1}[1+q_1^-(h)]^{\varpi}+\prod_{j=1}^{k+1}[1-q_1^-(h)]^{\varpi}}, \frac{\prod_{j=1}^{k+1}[1+r_1^-(h)]^{\varpi}-\prod_{j=1}^{k+1}[1-r_1^-(h)]^{\varpi}}{\prod_{j=1}^{k+1}[1+r_1^-(h)]^{\varpi}+\prod_{j=1}^{k+1}[1-r_1^-(h)]^{\varpi}} \right];$$

$$\left[\frac{\prod_{j=1}^{k+1}[1+p_1^+(h)]^{\varpi}-\prod_{j=1}^{k+1}[1-p_1^+(h)]^{\varpi}}{\prod_{j=1}^{k+1}[1+p_1^+(h)]^{\varpi}+\prod_{j=1}^{k+1}[1-p_1^+(h)]^{\varpi}}, \frac{\prod_{j=1}^{k+1}[1+q_1^+(h)]^{\varpi}-\prod_{j=1}^{k+1}[1-q_1^+(h)]^{\varpi}}{\prod_{j=1}^{k+1}[1+q_1^+(h)]^{\varpi}+\prod_{j=1}^{k+1}[1-q_1^+(h)]^{\varpi}}, \frac{\prod_{j=1}^{k+1}[1+r_1^+(h)]^{\varpi}-\prod_{j=1}^{k+1}[1-r_1^+(h)]^{\varpi}}{\prod_{j=1}^{k+1}[1+r_1^+(h)]^{\varpi}+\prod_{j=1}^{k+1}[1-r_1^+(h)]^{\varpi}} \right];$$

$$\left[\frac{2\prod_{j=1}^{k+1}[p_1(h)]^{\varpi}}{\prod_{j=1}^{k+1}[(2-p_1(h)]^{\varpi}+\prod_{j=1}^{k+1}[p_1(h)]^{\varpi}}, \frac{2\prod_{j=1}^{k+1}[q_1(h)]^{\varpi}}{\prod_{j=1}^{k+1}[(2-q_1(h)]^{\varpi}+\prod_{j=1}^{k+1}[q_1(h)]^{\varpi}}, \frac{2\prod_{j=1}^{k+1}[r_1(h)]^{\varpi}}{\prod_{j=1}^{k+1}[(2-r_1(h)]^{\varpi}+\prod_{j=1}^{k+1}[r_1(h)]^{\varpi}} \right]$$

$$= \left[\frac{[\prod_{j=1}^{k+1}[1+p_1^-(h)]^{\varpi}-\prod_{j=1}^{k}[1-p_1^-(h)]^{\varpi}}{\prod_{j=1}^{k+1}[1+p_1^-(h)]^{\varpi}+\prod_{j=1}^{k}[1-p_1^-(h)]^{\varpi}}, \frac{\prod_{j=1}^{k+1}[1+q_1^-(h)]^{\varpi}-\prod_{j=1}^{k}[1-q_1^-(h)]^{\varpi}}{\prod_{j=1}^{k+1}[1+q_1^-(h)]^{\varpi}+\prod_{j=1}^{k}[1-q_1^-(h)]^{\varpi}}, \frac{\prod_{j=1}^{k+1}[1+r_1^-(h)]^{\varpi}-\prod_{j=1}^{k}[1-r_1^-(h)]^{\varpi}}{\prod_{j=1}^{k+1}[1+r_1^-(h)]^{\varpi}+\prod_{j=1}^{k}[1-r_1^-(h)]^{\varpi}} \right],$$

$$\left[\frac{\prod_{j=1}^{k+1}[1+p_1^+(h)]^{\varpi}-\prod_{j=1}^{k}[1-p_1^+(h)]^{\varpi}}{\prod_{j=1}^{k+1}[1+p_1^+(h)]^{\varpi}+\prod_{j=1}^{k}[1-p_1^+(h)]^{\varpi}}, \frac{\prod_{j=1}^{k+1}[1+q_1^+(h)]^{\varpi}-\prod_{j=1}^{k}[1-q_1^+(h)]^{\varpi}}{\prod_{j=1}^{k+1}[1+q_1^+(h)]^{\varpi}+\prod_{j=1}^{k}[1-q_1^+(h)]^{\varpi}}, \frac{\prod_{j=1}^{k+1}[1+r_1^+(h)]^{\varpi}-\prod_{j=1}^{k}[1-r_1^+(h)]^{\varpi}}{\prod_{j=1}^{k+1}[1+r_1^+(h)]^{\varpi}+\prod_{j=1}^{k}[1-r_1^+(h)]^{\varpi}} \right],$$

$$\left[\frac{2\prod_{j=1}^{k+1}[p_1(h)]^{\varpi}}{\prod_{j=1}^{k+1}[(2-p_1(h)]^{\varpi}+\prod_{j=1}^{k+1}[p_1(h)]^{\varpi}}, \frac{2\prod_{j=1}^{k+1}[q_1(h)]^{\varpi}}{\prod_{j=1}^{k+1}[(2-q_1(h)]^{\varpi}+\prod_{j=1}^{k+1}[q_1(h)]^{\varpi}}, \frac{2\prod_{j=1}^{k+1}[r_1(h)]^{\varpi}}{\prod_{j=1}^{k+1}[(2-r_1(h)]^{\varpi}+\prod_{j=1}^{k+1}[r_1(h)]^{\varpi}} \right].$$

Especially, if $w = \left(\frac{1}{n}, \frac{1}{n}, \ldots, \frac{1}{n}\right)^T$, then the the TNCFEWA operator is reduced to the triangular neutrosophic cubic fuzzy einstein averaging operator, which is shown as follows:

$$\left\langle \left[\frac{[\prod_{j=1}^{n}[1+\alpha_1(h)]^{\frac{1}{n}} - \prod_{j=1}^{n}[1-\alpha_1(h)]^{\frac{1}{n}}}{\prod_{j=1}^{n}[1+\alpha_1(h)]^{\frac{1}{n}} + \prod_{j=1}^{n}[1-\alpha_1(h)]^{\frac{1}{n}}}, \frac{[\prod_{j=1}^{n}[1+\beta_1(h)]^{\frac{1}{n}} - \prod_{j=1}^{n}[1-\beta_1(h)]^{\frac{1}{n}}}{\prod_{j=1}^{n}[1+\beta_1(h)]^{\frac{1}{n}} + \prod_{j=1}^{n}[1-\beta_1(h)]^{\frac{1}{n}}}, \frac{[\prod_{j=1}^{n}[1+\Delta_1(h)]^{\frac{1}{n}} - \prod_{j=1}^{n}[1-\Delta_1(h)]^{\frac{1}{n}}}{\prod_{j=1}^{n}[1+\Delta_1(h)]^{\frac{1}{n}} + \prod_{j=1}^{n}[1-\Delta_1(h)]^{\frac{1}{n}}} \right] \right.$$

$$\left[\frac{[\prod_{j=1}^{n}[1+p_1^-(h)]^{\frac{1}{n}} - \prod_{j=1}^{n}[1-p_1^-(h)]^{\frac{1}{n}}}{\prod_{j=1}^{n}[1+p_1^-(h)]^{\frac{1}{n}} + \prod_{j=1}^{n}[1-p_1^-(h)]^{\frac{1}{n}}}, \frac{\prod_{j=1}^{n}[1+q_1^-(h)]^{\frac{1}{n}} - \prod_{j=1}^{n}[1-q_1^-(h)]^{\frac{1}{n}}}{\prod_{j=1}^{n}[1+q_1^-(h)]^{\frac{1}{n}} + \prod_{j=1}^{n}[1-q_1^-(h)]^{\frac{1}{n}}}, \frac{\prod_{j=1}^{n}[1+r_1^-(h)]^{\frac{1}{n}} - \prod_{j=1}^{n}[1-r_1^-(h)]^{\frac{1}{n}}}{\prod_{j=1}^{n}[1+r_1^-(h)]^{\frac{1}{n}} + \prod_{j=1}^{n}[1-r_1^-(h)]^{\frac{1}{n}}} \right];$$

$$\left[\frac{\prod_{j=1}^{n}[1+p_1^+(h)]^{\frac{1}{n}} - \prod_{j=1}^{n}[1-p_1^+(h)]^{\frac{1}{n}}}{\prod_{j=1}^{n}[1+p_1^+(h)]^{\frac{1}{n}} + \prod_{j=1}^{n}[1-p_1^+(h)]^{\frac{1}{n}}}, \frac{\prod_{j=1}^{n}[1+q_1^+(h)]^{\frac{1}{n}} - \prod_{j=1}^{n}[1-q_1^+(h)]^{\frac{1}{n}}}{\prod_{j=1}^{n}[1+q_1^+(h)]^{\frac{1}{n}} + \prod_{j=1}^{n}[1-q_1^+(h)]^{\frac{1}{n}}}, \frac{\prod_{j=1}^{n}[1+r_1^+(h)]^{\frac{1}{n}} - \prod_{j=1}^{n}[1-r_1^+(h)]^{\frac{1}{n}}}{\prod_{j=1}^{n}[1+r_1^+(h)]^{\frac{1}{n}} + \prod_{j=1}^{n}[1-r_1^+(h)]^{\frac{1}{n}}} \right];$$

$$\left[\frac{2\prod_{j=1}^{n}[p_1(h)]^{\frac{1}{n}}}{\prod_{j=1}^{n}[(2-p_1(h)]^{\frac{1}{n}} + \prod_{j=1}^{n}[p_1(h)]^{\frac{1}{n}}}, \frac{2\prod_{j=1}^{n}[q_1(h)]^{\frac{1}{n}}}{\prod_{j=1}^{n}[(2-q_1(h)]^{\frac{1}{n}} + \prod_{j=1}^{n}[q_1(h)]^{\frac{1}{n}}}, \frac{2\prod_{j=1}^{n}[r_1(h)]^{\frac{1}{n}}}{\prod_{j=1}^{n}[(2-r_1(h)]^{\frac{1}{n}} + \prod_{j=1}^{n}[r_1(h)]^{\frac{1}{n}}} \right].$$

References

1. Atanassov, K.T. Intuitionistic Fuzzy Sets. *Fuzzy Sets Syst.* **1986**, *20*, 87–96. [CrossRef]
2. Atanassov, K.T. New Operations Defined Over the Intuitionistic Fuzzy Sets. *Fuzzy Sets Syst.* **1994**, *61*, 137–142. [CrossRef]
3. Bustince, H.; Burillo, P. Structures on intuitionistic fuzzy relations. *Fuzzy Sets Syst.* **1996**, *78*, 293–303. [CrossRef]
4. Deschrijver, G.; Kerre, E.E. On the relationship between some extensions of fuzzy set theory. *Fuzzy Sets Syst.* **2003**, *133*, 227–235. [CrossRef]
5. Deschrijver, G.; Kerre, E.E. On the position of intuitionistic fuzzy set theory in the framework of theories modelling imprecision. *Inf. Sci.* **2007**, *177*, 1860–1866. [CrossRef]
6. Jun, Y.B.; Kim, C.S.; Yang, K.O. Annals of Fuzzy Mathematics and Informatics. *Cubic Sets* **2011**, *4*, 83–98.
7. Muhiuddin, G.; Feng, F.; Jun, Y.B. Subalgebras of BCK/BCI-Algebras Based on Cubic Soft Sets. *Sci. World J.* **2014**, *2014*, 458638. [CrossRef]
8. Turksen, I.B. Interval valued fuzzy sets based on normal forms. *Fuzzy Sets Syst.* **1986**, *20*, 191–210. [CrossRef]
9. Xu, Z.S. Intuitionistic fuzzy aggregation operators. *IEEE Trans. Fuzzy Syst.* **2007**, *15*, 1179–1187.
10. Xu, Z.S.; Yager, R.R. Some geometric aggregation operators based on intuitionistic fuzzy sets. *Int. J. Gener. Syst.* **2006**, *35*, 417–433. [CrossRef]
11. Xu, Z.S.; Cai, X. Recent advances in intuitionistic fuzzy information aggregation. *Fuzzy Optim. Decis. Mak.* **2010**, *9*, 359–381. [CrossRef]
12. Liu, H.W.; Wang, G.-J. Multi-criteria decision-making methods based on intuitionistic fuzzy sets. *Eur. J. Oper. Res.* **2007**, *179*, 220–233. [CrossRef]
13. Zeng, S.Z.; Su, W.H. Intuitionistic fuzzy ordered weighted distance operator. *Knowl.-Based Syst.* **2011**, *24*, 1224–1232. [CrossRef]
14. Zadeh, L.A. Fuzzy sets. *Inform. Contr.* **1965**, *8*, 338–353. [CrossRef]
15. Zadeh, L.A. Outline of a new approach to analysis of complex systems and decision processes interval-valued fuzzy sets. *IEEE Trans. Syst. Man Cybern.* **1973**, *3*, 28–44. [CrossRef]
16. Li, J.; Zhang, X.L.; Gong, Z.-T. Aggregating of Interval-valued Intuitionistic Uncertain Linguistic Variables based on Archimedean t-norm and It Applications in Group Decision Makings. *J. Comput. Anal. Appl.* **2018**, *24*, 874–885.
17. Zhao, X.; Lin, R.; Wei, G. Hesitant triangular fuzzy information aggregation based on Einstein operations and their application to multiple attribute decision making. *Expert Syst. Appl.* **2014**, *41*, 1086–1094. [CrossRef]
18. Xu, Z.S.; Yager, R.R. Dynamic intuitionistic fuzzy multi-attribute decision making. *Int. J. Approx. Reason.* **2008**, *48*, 246–262. [CrossRef]

19. Smarandache, F. Neutrosophic set—A generalization of the intuitionistic fuzzy set. In Proceedings of the 2006 IEEE International Conference on Granular Computing, Atlanta, GA, USA, 10–12 May 2006; pp. 38–42.
20. Smarandache, F. A geometric interpretation of the neutrosophic set—A generalization of the intuitionistic fuzzy set. In Proceedings of the 20011 IEEE International Conference on Granular Computing (GrC), Kaohsiung, Taiwan, 8–10 November 2011; pp. 602–606.
21. Fahmi, A.; Abdullah, S.; Amin, F.; Ali, A. Weighted average rating (WAR) method for solving group decision making problem using triangular cubic fuzzy hybrid aggregation (TCFHA). *Punjab Univ. J. Math.* **2018**, *50*, 23–34.
22. Fahmi, A.; Amin, F.; Smarandache, F.; Khan, M.; Hassan, N. Triangular Cubic Hesitant Fuzzy Einstein Hybrid Weighted Averaging Operator and Its Application to Decision Making. *Symmetry* **2018**, *10*, 658. [CrossRef]
23. Beg, I.; Rashid, T. Group decision making using intuitionistic hesitant fuzzy sets. *Int. J. Fuzzy Log. Intell. Syst.* **2014**, *14*, 181–187. [CrossRef]
24. Grzegorzewski, P. On Separability of Fuzzy Relations. *Int. J. Fuzzy Log. Intell. Syst.* **2017**, *17*, 137–144. [CrossRef]

symmetry

MDPI

Article

Neutrosophic Compound Orthogonal Neural Network and Its Applications in Neutrosophic Function Approximation

Jun Ye * and Wenhua Cui

Department of Electrical and Information Engineering, Shaoxing University, 508 Huancheng West Road, Shaoxing 312000, China; wenhuacui@usx.edu.cn

* Correspondence: yehjun@aliyun.com; Tel.: +86-575-88327323

Received: 16 January 2019; Accepted: 28 January 2019; Published: 29 January 2019

Abstract: Neural networks are powerful universal approximation tools. They have been utilized for functions/data approximation, classification, pattern recognition, as well as their various applications. Uncertain or interval values result from the incompleteness of measurements, human observation and estimations in the real world. Thus, a neutrosophic number (NsN) can represent both certain and uncertain information in an indeterminate setting and imply a changeable interval depending on its indeterminate ranges. In NsN settings, however, existing interval neural networks cannot deal with uncertain problems with NsNs. Therefore, this original study proposes a neutrosophic compound orthogonal neural network (NCONN) for the first time, containing the NsN weight values, NsN input and output, and hidden layer neutrosophic neuron functions, to approximate neutrosophic functions/NsN data. In the proposed NCONN model, single input and single output neurons are the transmission notes of NsN data and hidden layer neutrosophic neurons are constructed by the compound functions of both the Chebyshev neutrosophic orthogonal polynomial and the neutrosophic sigmoid function. In addition, illustrative and actual examples are provided to verify the effectiveness and learning performance of the proposed NCONN model for approximating neutrosophic nonlinear functions and NsN data. The contribution of this study is that the proposed NCONN can handle the approximation problems of neutrosophic nonlinear functions and NsN data. However, the main advantage is that the proposed NCONN implies a simple learning algorithm, higher speed learning convergence, and higher learning accuracy in indeterminate/NsN environments.

Keywords: Neutrosophic compound orthogonal neural network; Neutrosophic number; Neutrosophic function; Function approximation

1. Introduction

Neural networks are powerful universal approximation tools. They have been utilized for data modeling, function approximation, classification analysis, pattern recognition, as well as their various applications. Uncertain or interval values result from the incompleteness of measurements, human observation and estimations in the real world. Hence, Baker and Patil [1] proposed an interval neural network (INN), which used interval weights rather than interval data input to approximate an interval function. Then, Hu et al. [2] presented an INN with interval weights, where the network is modeled like the problem of the solution equations, implying its complexity in the solution process. Rossi and Conan-Guez [3] introduced a multilayer perceptron neural network on interval data for the classification analysis of interval data. For processing the interval neural network, Patiño-Escarcina [4] presented an INN, where one of its input, output, and weight sets is interval values and the output set is a binary one, therefore its outputs are binaries too for classifiers. Recently, Lu et al. [5] introduced a neural network-based interval matcher corresponding to linguistic IF-THEN constructions, which is an interval pattern matcher to identify patterns with interval elements using the neural network, which can handle interval inputs values and interval output values based on a traditional neural network and is only suitable for interval pattern matching. Kowalski and Kulczycki [6] presented the interval probabilistic neural network (IPNN) for the classification of interval data, where the IPNN structure is based on Specht's probabilistic network [7].

In indeterminate environments, neutrosophic theory [8–10] has been used for various applications [11–14]. Since a neutrosophic number (NsN) [8–10] can represent both certain and uncertain information in indeterminate settings and contain a changeable interval depending on its indeterminate ranges, NsNs have been wildly applied to decision making [15–17], fault diagnoses [18,19], linear and nonlinear optimization problems [20–23], expression and analysis of the rock joint roughness coefficient (JRC) [24–27]. However, there is no study on neutrosophic neural networks with NsNs in existing literature, while existing INNs also cannot deal with uncertain problems with NsNs. Therefore, this original study proposes a neutrosophic compound orthogonal neural network (NCONN) for the first time, which contains the NsN weight values, NsN input and output neurons and hidden layer neutrosophic neurons, to approximate neutrosophic functions and NsN data. In the proposed NCONN model, single input and single output data are NsNs (changeable interval numbers) and hidden layer neutrosophic neuron functions are composed of the Chebyshev neutrosophic orthogonal polynomial and neutrosophic sigmoid function. In addition, illustrative and actual examples are provided to verify the effectiveness and performance of the proposed NCONN model in approximating neutrosophic nonlinear functions and NsN data. The contribution of this study is that the proposed NCONN can handle the approximating and modelling problems of neutrosophic functions and NsN data for the first time. The main advantage is that the proposed NCONN implies a simple learning algorithm, higher speed learning convergence, and higher learning accuracy in indeterminate/NsN environments.

This study was formed as the following framework. The second section introduces the basic concepts and operations of NsNs. The third section proposes a NCONN structure and its learning algorithm. Then, two illustrative examples about neutrosophic nonlinear function approximations and an actual example (a real case) about the approximation problem of rock JRC NsNs are presented in the fourth section and the fifth section, respectively, to verify the effectiveness and performance of the proposed NCONN in approximating neutrosophic nonlinear functions and NsN data under indeterminate/NsN environments. The last section contains conclusions and future work.

2. Basic Concepts and Operations of NsNs

In an uncertain setting, Smarandache [8–10] introduced the NsN concept represented by the mathematical form $N = c + uI$ for $a, b \in R$ (all real numbers) and I (indeterminacy), in which the certain part c with its uncertain part uI for $I \in [I^-, I^+]$ are combined. Hence, it can depict and express the certain and/or uncertain information in indeterminate problems.

Provided there is the NsN $N = 5 + 3I$, it depicts that the certain value is five and its uncertain value is $3I$. Then, some interval range of the indeterminacy $I \in [I^-, I^+]$ is possibly specified in actual applications to satisfy some applied requirement. For instance, the indeterminacy I is specified as such a possible interval $I \in [0, 2]$. Thus, it is equivalent to $N = [5, 11]$. If $I \in [1, 3]$, then there is $N = [8, 14]$. It is obvious that it is a changeable interval depending on the specified indeterminate range of $I \in [I^-, I^+]$, which is also denoted by $N = [c + uI^-, c + uI^+]$.

In some special cases, a NsN $N = c + uI$ for $N \in U$ (U is all NsNs) may be represented as either a certain number $N = c$ for $uI = 0$ (the best case) or an uncertain number $N = uI$ for $c = 0$ (the worst case).

Provided that there are two NsNs $N_1 = c_1 + u_1I$ and $N_2 = c_2 + u_2I$ for $N_1, N_2 \in U$ and $I \in [I^-, I^+]$, then their operational laws are introduced as follows [21]:

$$N_1 + N_2 = c_1 + c_2 + (u_1 + u_2)I = [c_1 + c_2 + u_1I^- + u_2I^-, c_1 + c_2 + u_1I^+ + u_2I^+] \quad (1)$$

$$N_1 - N_2 = c_1 - c_2 + (u_1 - u_2)I = [c_1 - c_2 + u_1I^- - u_2I^-, c_1 - c_2 + u_1I^+ - u_2I^+] \quad (2)$$

$$N_1 \times N_2 = c_1c_2 + (c_1u_2 + c_2u_1)I + u_1u_2I^2 = \begin{bmatrix} \min\begin{pmatrix} (c_1+u_1I^-)(c_2+u_2I^-), (c_1+u_1I^-)(c_2+u_2I^+), \\ (c_1+u_1I^+)(c_2+u_2I^-), (c_1+u_1I^+)(c_2+u_2I^+) \end{pmatrix}, \\ \max\begin{pmatrix} (c_1+u_1I^-)(c_2+u_2I^-), (c_1+u_1I^-)(c_2+u_2I^+), \\ (c_1+u_1I^+)(c_2+u_2I^-), (c_1+u_1I^+)(c_2+u_2I^+) \end{pmatrix} \end{bmatrix} \quad (3)$$

$$\frac{N_1}{N_2} = \frac{c_1 + u_1I}{c_2 + u_2I} = \frac{[c_1 + u_1I^-, c_1 + u_1I^+]}{[c_2 + u_2I^-, c_2 + u_2I^+]} = \begin{bmatrix} \min\left(\frac{c_1+u_1I^-}{c_2+u_2I^+}, \frac{c_1+u_1I^-}{c_2+u_2I^-}, \frac{c_1+u_1I^+}{c_2+u_2I^+}, \frac{c_1+u_1I^+}{c_2+u_2I^-}\right), \\ \max\left(\frac{c_1+u_1I^-}{c_2+u_2I^+}, \frac{c_1+u_1I^-}{c_2+u_2I^-}, \frac{c_1+u_1I^+}{c_2+u_2I^+}, \frac{c_1+u_1I^+}{c_2+u_2I^-}\right) \end{bmatrix} \quad (4)$$

Regarding an uncertain function containing NsNs, Ye [21,22] defined a neutrosophic function in n variables (unknowns) as $y(\boldsymbol{x}, \boldsymbol{I})$: $U^n \rightarrow U$ for $\boldsymbol{x} = [x_1, x_2, \ldots, x_n]^{\mathrm{T}} \in U^n$ and $I \in [I^-, I^+]$, which is then a neutrosophic nonlinear or linear function.

For example, $y_1(x, I) = N_1x\cos(x) = (c_1 + u_1I)x\cos(x)$ for $x \in U$ and $I \in [I^-, I^+]$ is a neutrosophic nonlinear function, while $y_2(\boldsymbol{x}, I) = N_1x_1 + N_2x_2 + N_3 = (c_1 + u_1I)x_1 + (c_2 + u_2I)x_2 + (c_3 + u_3I)$ for $\boldsymbol{x} = [x_1, x_2]^{\mathrm{T}} \in U^2$ and $I \in [I^-, I^+]$ is a neutrosophic linear function.

Generally, the values of $\boldsymbol{x}$ and $y(\boldsymbol{x})$ are NsNs (usually, but not always).

3. NCONN with NsNs

This section proposes a NCONN structure and its learning algorithm based on the NsN concept for the first time.

A three-layer feedforward NCONN structure with a single input, single output, and hidden layer neutrosophic neurons are indicated in Figure 1. In Figure 1, the weight values between the input layer neuron and the hidden layer neutrosophic neurons are equal to the constant value 1 and the NsN weight values between the hidden layer neutrosophic neurons and the output layer neuron are w_j ($j = 1, 2, \ldots, p$); x_k ($k = 1, 2, \ldots, n$) is the kth NsN input signal; y_k is the kth NsN output signal; and p is the number of the hidden layer neutrosophic neurons.

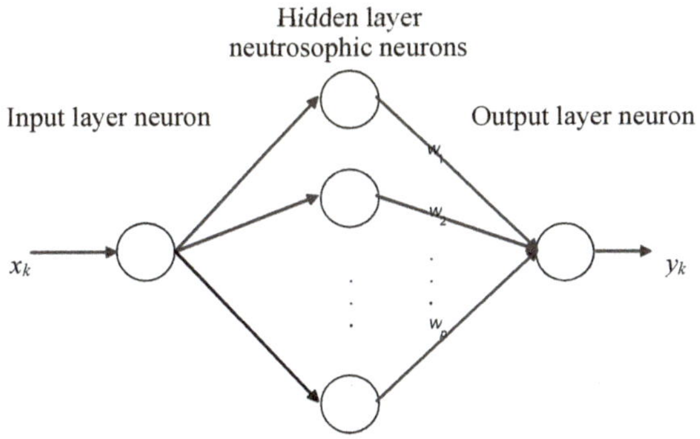

Figure 1. A three-layer feedforward neutrosophic compound orthogonal neural network (NCONN structure).

In the learning process, when each NsN input signal is given by $x_k = c_k + u_k I = [c_k + u_k I^-, c_k + u_k I^+]$ ($k = 1, 2, \ldots, n$) for $I \in [I^-, I^+]$, the actual output value is given as:

$$y_k = \sum_{j=1}^{p} w_j \widetilde{q}_j, k = 1, 2, \ldots, n \tag{5}$$

where the neutrosophic neuron functions of the hidden layer $\widetilde{q}_j$ for $j = 1, 2, \ldots, p$ are the Chebyshev compound neutrosophic orthogonal polynomial: $\widetilde{q}_1 = [1, 1]$, $\widetilde{q}_2 = \widetilde{X}$, $\widetilde{q}_j = 2\widetilde{X} \cdot \widetilde{q}_{j-1} - \widetilde{q}_{j-2}$, and $\widetilde{X}$ is specified as the following unipolar neutrosophic sigmoid function (the neutrosophic S-function):

$$\widetilde{X} = \frac{1}{1 + e^{-\alpha x_k}} \tag{6}$$

The neutrosophic S-function can transform NsN into the interval (0, 1) and the different scalar parameters of α can change the slant degree of the neutrosophic S-function curve.

Then, the square interval of output errors between the desired output $y_k^d = c_k^d + u_k^d I$ and the actual output $y_k = c_k + u_k I$ for $I \in [I^-, I^+]$ is given as follows:

$$\widetilde{E}_k^2 = [(c_k^d + u_k^d I^- - c_k - u_k I^-)^2, (c_k^d + u_k^d I^+ - c_k - u_k I^+)^2] \tag{7}$$

Whereas, the learning performance index of the proposed NCONN is specified as the following requirement:

$$\widetilde{E} = \frac{1}{2} \sum_{k=1}^{n} \widetilde{E}_k^2 \tag{8}$$

The NCONN weight values can be adjusted by the following formula:

$$\widetilde{W}_k(l+1) = \widetilde{W}_k(l) + \lambda \widetilde{E}_k \widetilde{Q}, k = 1, 2, \ldots, n \tag{9}$$

where $\widetilde{W}_k(l) = [\widetilde{w}_1(l), \widetilde{w}_2(l), \ldots, \widetilde{w}_q(l)]^T$ and $\widetilde{Q}_k(l) = [\widetilde{q}_1(l), \widetilde{q}_2(l), \ldots, \widetilde{q}_p(l)]^T$ is the NsN weight vector and the function vector of the hidden layer neutrosophic neurons, λ is the learning rate of the NCONN to determine the convergence velocity for $\lambda \in (0, 1)$, and l is the lth iteration learning of the NCONN.

Thus, this NCONN learning algorithm can be described below:

Step 1: Give $\widetilde{W}_k(0)$ by small random values,

Step 2: Input a NsN and calculate the actual output of a NCONN based on Equations (5) and (6),

Step 3: Calculate the output error by using Equations (7) and (8),

Step 4: Adjust weight values by using Equation (9),

Step 5: Input the next NsN and return to Step 2.

In the NCONN learning process, the learning termination condition depends on the requirement of the specified learning error or iteration number.

Since NsN can be considered as a changeable interval depending to its indeterminacy $I \in [I^-, I^+]$, the learning algorithm of NCONN permits changeable interval operations, which are different from existing neural network algorithms and show its advantage of approximating neutrosophic nonlinear functions/NsN data in an uncertain/NsN setting.

Generally, the more the hidden layer neutrosophic neurons are, the higher the approximation accuracy of the proposed NCONN is. Then, the number of the hidden layer neutrosophic neurons determinated in actual applications will depend on the accuracy requirements of actual approximation models.

4. NsN Nonlinear Function Approximation Applied by the Proposed NCONN

To prove the effectiveness of approximating any neutrosophic nonlinear function based on the proposed NCONN model, we present two illustrative examples in this section.

Example 1. Supposing there is a neutrosophic nonlinear function:

$$y_1(x, I) = 1 + 0.3I + (0.5 + 0.2I)x\cos(\pi x) \text{ for } I \in [0,\ 1].$$

For $x \in [-1{+}0.02I, 1{+}0.023I]$ and $I \in [0, 1]$, the proposed NCONN needs to approximate the above neutrosophic nonlinear function.

To prove the approximation ability of the proposed NCONN, we give the proposed NCONN structure with eight hidden layer neutrosophic neurons (p = 8) and learning parameters, which are indicated in Table 1.

Table 1. The NCONN structure and learning parameters.

NCONN Structure	α	λ	The Number of the Specified Learning Iteration	$\widetilde{E}$
$1 \times 8 \times 1$	2.5	0.25	20	[3.2941, 8.5088]

Then, the desired output $y_{1d} = [y_{1d}^-, y_{1d}^+]$ and actual output $y_1 = [y_1^-, y_1^+]$ of the proposed NCONNs are shown in Figure 2. Obviously, the desired output curves and the actual output curves were very close to each other, to demonstrate the better approximation accuracy in the neutrosophic nonlinear function approximation of the proposed NCONN. Hence, the proposed NCONN indicated the better approximation performance regarding the neutrosophic nonlinear function.

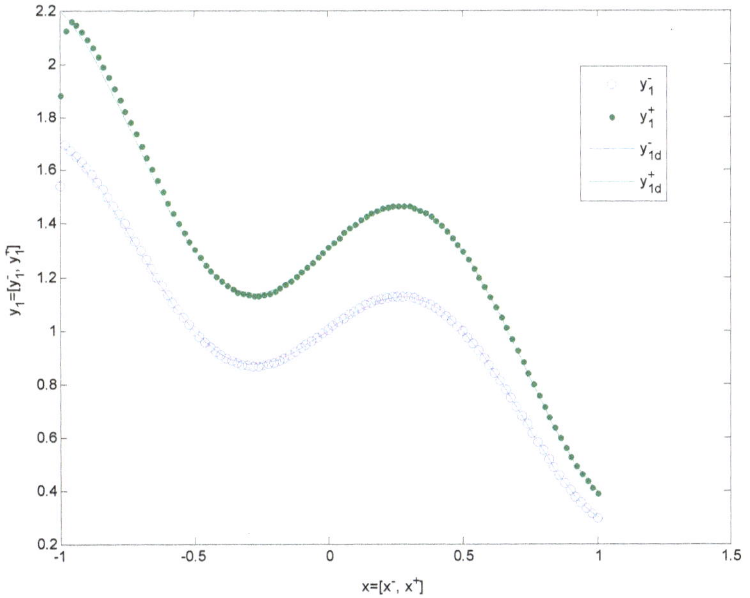

Figure 2. The desired output $y_{1d} = [y_{1d}^-, y_{1d}^+]$ and actual output $y_1 = [y_1^-, y_1^+]$ of the proposed NCONN.

Example 2. Considering a neutrosophic nonlinear function:
$y_2(x, I) = (0.6 + 0.3I)\sin(\pi x) + (0.3 + 0.15I)\sin(3\pi x) + (0.1 + 0.05I)\sin(5\pi x)$ for $I \in [0, 1]$.

For $x \in [0 + 0.002I, 1 + 0.002I]$ and $I \in [0, 1]$, the proposed NCONN needs to approximate the above neutrosophic nonlinear function.

To prove the approximation ability of the proposed NCONN model, we also give the NCONN structure with eight hidden layer neutrosophic neurons (p = 8) and learning parameters, which are indicated in Table 2.

Table 2. The NCONN structure and learning parameters.

NCONN Structure	α	λ	The Number of the Specified Learning Iteration	$\tilde{E}$
$1 \times 8 \times 1$	8	0.3	20	[0.5525, 1.1261]

Thus, the desired output $y_{2d} = [y_{2d}^-, y_{2d}^+]$ and actual output $y_2 = [y_2^-, y_2^+]$ of the proposed NCONN are indicated in Figure 3. It was obvious that the desired output curves and the actual output curves were also very close, so as to demonstrate the better approximating accuracy and performance in the neutrosophic nonlinear function approximation of the proposed NCONN.

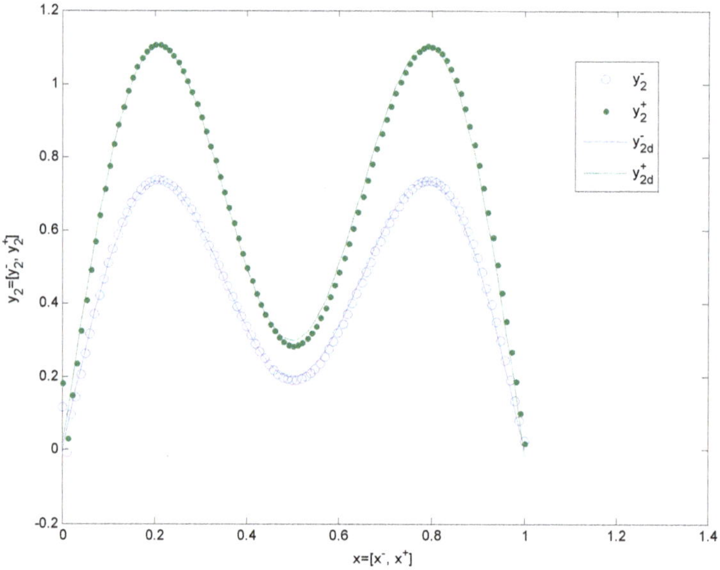

Figure 3. The desired output $y_{2d} = [y_{2d}^-, y_{2d}^+]$ and actual output $y_2 = [y_2^-, y_2^+]$ of the proposed NCONN.

Corresponding to the learning results obtained from the above two illustrative examples, we could see that the proposed NCONN showed faster learning velocity and a higher learning accuracy, which indicated a better approximation performance regarding the neutrosophic nonlinear functions.

5. Actual Example on the Approximation of the JRC NsNs Based on the Proposed NCONN

In rock machanics, the JRC of rock joints implies uncertainty in different sampling lengths and directions of rock joints. Therefore, JRC uncertainty may make the shear strength of joints uncertain because of the corresponding relationship between JRC and the shear strength, which results in the difficulty of making assessments of side stability [25–27]. However, the lengths of the testing samples can affect JRC values, which indicates their scale effect. To establish a relationship between the sampling lengths L and the JRC values in an uncertain/NsN setting, existing literature [25–27] used the uncertain/neutrosophic statistic method and fitting functions to establish some related model of L and the JRC. Since the proposed NCONN is able to approximate NsN data, the proposed NCONN could be applied to the relative approximation model between the sampling length L and the NsN data of the JRC by an actual example (a real case) in this section, to show its effectiveness.

According to the testing samples of the specified area in Shaoxing city, China and data analysis, we found a relationship between the sampling length L and the NsN data of JRC, which are shown in Table 3.

Table 3. NsN data of rock joint roughness coefficient (JRC) regarding different sampling lengths for $I \in [0, 1]$.

Sample Length L (cm)	x_k	JRC	y_k
9.8 + 0.4I	[9.8, 10.2]	8.321 + 6.231I	[8.321, 14.552]
19.8 + 0.4I	[19.8, 20.2]	7.970 + 6.419I	[7.970, 14.389]
29.8 + 0.4I	[29.8, 30.2]	7.765 + 6.529I	[7.765, 14.294]
39.8 + 0.4I	[39.8, 40.2]	7.762 + 6.464I	[7.762, 14.226]
49.8 + 0.4I	[49.8, 50.2]	7.507 + 6.64I	[7.507, 14.147]
59.8 + 0.4I	[59.8, 60.2]	7.417 + 6.714I	[7.417, 14.131]
69.8 + 0.4I	[69.8, 70.2]	7.337 + 6.758I	[7.337, 14.095]
79.8 + 0.4I	[79.8, 80.2]	7.269 + 6.794I	[7.269, 14.063]
89.8 + 0.4I	[89.8, 90.2]	7.210 + 6.826I	[7.210, 14.036]
99.8 + 0.4I	[99.8, 100.2]	7.156 + 6.855I	[7.156, 14.011]

To establish the approximation model of the proposed NCONN regarding the actual example, we took the NCONN structure with eight hidden layer neutrosophic neurons ($p = 8$) and indicated the learning parameters in Table 4.

Table 4. The NCONN structure and learning parameters regarding the actual example.

NCONN Structure	α	λ	The Number of the Specified Learning Iteration	$\widetilde{E}$
1 × 8 × 1	8	0.11	5	[3.2715, 22.3275]

From Figure 4, we can see that the proposed NCONN could approximate the JRC NsN data regarding different sampling lengths L and showed a higher speed convergence and higher approximating accuracy in its learning process for the actual example. Obviously, the proposed NCONN could find the approximating model between different sampling lengths L and JRC NsN data, while existing neural networks cannot do them in the uncertain/NsN setting.

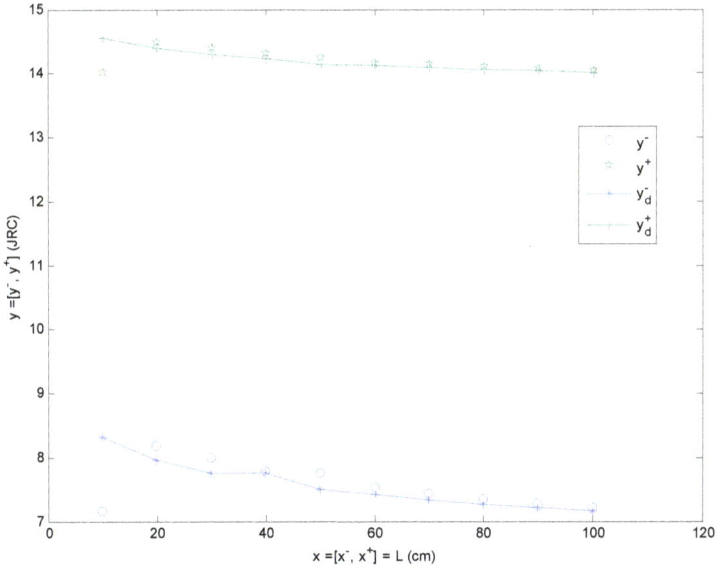

Figure 4. The proposed NCONN approximation results of the JRC NsN data regarding different sampling lengths L.

6. Conclusions

In a NsN setting, this original study presented a NCONN to approximate neutrosophic functions/NsN data for the first time. It is a three-layer feedforward neutrosophic network structure composed of a single input, a single output, and hidden layer neutrosophic neurons, where the single input and single output information are NsNs and hidden layer neutrosophic neuron functions are composed of both the Chebyshev neutrosophic orthogonal polynomial and the neutrosophic sigmoid function. Illustrative and actual examples were provided to verify the effectiveness and rationality of the proposed NCONN model for approximating neutrosophic nonlinear functions and establishing the approximation model of NsN data. Therefore, the contribution of this study is that the proposed NCONN could handle the approximating and modeling problems of uncertain/interval/neutrosophic functions and NsN data. Here, the main advantage is that the proposed NCONN implies a simpler learning algorithm, higher speed learning convergence, and higher learning accuracy in indeterminate/NsN environments.

In the future work, we shall propose further NCONNs with multi-inputs and multi-outputs and apply them to the modeling and approximating problems of neutrosophic functions and NsN data, the clustering analysis of NsNs, medical diagnosis problems, and possible applications for decision-making and control in robotics [14,28] in an indeterminate/NsN setting.

Author Contributions: J.Y. proposed the NCONN model and learning algorithm; W.H.C. provided the simulation analysis of illustrative and actual examples; J.Y. and W.H.C. wrote the paper together.

Funding: This paper was supported by the National Natural Science Foundation of China (Nos. 71471172, 61703280).

Conflicts of Interest: The authors declare that we have no conflict of interest regarding the publication of this paper.

References

1. Baker, M.R.; Patil, R.B. Universal approximation theorem for interval neural networks. *Reliab. Comput.* **1998**, *4*, 235–239. [CrossRef]
2. Beheshti, M.; Berrached, A.; Korvin, A.D.; Hu, C.; Sirisaengtaksin, O. On interval weighted three-layer neural networks. In Proceedings of the 31st Annual Simulation Symposium, Boston, MA, USA, 5–9 April 1998; pp. 188–195.
3. Rossi, F.; Conan-Guez, B. Multi-layer perceptron on interval data. Classification, clustering, and data analysis. In *Studies in Classification, Data Analysis, and Knowledge Organization*; Springer: Berlin, Germany, 2002; pp. 427–434.
4. Patiño-Escarcina, R.E.; Callejas Bedregal, B.R.; Lyra, A. Interval computing in neural networks: One layer interval neural networks. In *Intelligent Information Technology. Lecture Notes in Computer Science*; Das, G., Gulati, V.P., Eds.; Springer: Berlin/Heidelberg, Germany, 2004; Volume 3356, pp. 68–75.
5. Lu, J.; Xue, S.; Zhang, X.; Han, Y. A neural network-based interval pattern matcher. *Information* **2015**, *6*, 388–398. [CrossRef]
6. Piotr, A.; Kowalski, P.K. Interval probabilistic neural network. *Neural Comput. Appl.* **2017**, *28*, 817–834.
7. Specht, D.F. Probabilistic neural networks. *Neural Netw.* **1990**, *3*, 109–118. [CrossRef]
8. Smarandache, F. *Neutrosophy: Neutrosophic Probability, Set, and Logic*; American Research Press: Rehoboth, DE, USA, 1998.
9. Smarandache, F. *Introduction to Neutrosophic Measure, Neutrosophic Integral, and Neutrosophic Probability*; Sitech & Education Publisher: Craiova, Columbus, 2013.
10. Smarandache, F. *Introduction to Neutrosophic Statistics*; Sitech & Education Publishing: Columbus, OH, USA, 2014.
11. Ye, J. A multicriteria decision-making method using aggregation operators for simplified neutrosophic sets. *J. Intell. Fuzzy Syst.* **2014**, *26*, 2459–2466.

12. Broumi, S.; Bakali, A.; Talea, M.; Smarandache, F.; Uluçay, V.; Sahin, M.; Dey, A.; Dhar, M.; Tan, R.P.; Bahnasse, A.; et al. Neutrosophic sets: An overview. In *New Trends in Neutrosophic Theory and Applications*; Smarandache, F., Pramanik, S., Eds.; Pons Publishing House: Brussels, Belgium, 2018; Volume II, pp. 403–434.
13. Broumi, S.; Bakali, A.; Talea, M.; Smarandache, F.; Vladareanu, L. Computation of shortest path problem in a network with SV-trapezoidal neutrosophic numbers. In Proceedings of the 2016 International Conference on Advanced Mechatronic Systems, Melbourne, Australia, 30 November–3 December 2016; pp. 417–422.
14. Gal, I.A.; Bucur, D.; Vladareanu, L. DSmT decision-making algorithms for finding grasping configurations of robot dexterous hands. *Symmetry* **2018**, *10*, 198. [CrossRef]
15. Ye, J. Multiple-attribute group decision-making method under a neutrosophic number environment. *J. Intell. Syst.* **2016**, *25*, 377–386. [CrossRef]
16. Chen, J.Q.; Ye, J. A projection model of neutrosophic numbers for multiple attribute decision making of clay-brick selection. *Neutrosophic Sets Syst.* **2016**, *12*, 139–142.
17. Ye, J. Bidirectional projection method for multiple attribute group decision making with neutrosophic numbers. *Neural Comput. Appl.* **2017**, *28*, 1021–1029. [CrossRef]
18. Kong, L.W.; Wu, Y.F.; Ye, J. Misfire fault diagnosis method of gasoline engines using the cosine similarity measure of neutrosophic numbers. *Neutrosophic Sets Syst.* **2015**, *8*, 43–46.
19. Ye, J. Fault diagnoses of steam turbine using the exponential similarity measure of neutrosophic numbers. *J. Intell. Fuzzy Syst.* **2016**, *30*, 1927–1934. [CrossRef]
20. Jiang, W.Z.; Ye, J. Optimal design of truss structures using a neutrosophic number optimization model under an indeterminate environment. *Neutrosophic Sets Syst.* **2016**, *14*, 93–975.
21. Ye, J. Neutrosophic number linear programming method and its application under neutrosophic number environments. *Soft Comput.* **2018**, *22*, 4639–4646. [CrossRef]
22. Ye, J. An improved neutrosophic number optimization method for optimal design of truss structures. *New Math. Nat. Comput.* **2018**, *14*, 295–305. [CrossRef]
23. Ye, J.; Cui, W.H.; Lu, Z.K. Neutrosophic number nonlinear programming problems and their general solution methods under neutrosophic number environments. *Axioms* **2018**, *7*, 13. [CrossRef]
24. Ye, J.; Chen, J.Q.; Yong, R.; Du, S.G. Expression and analysis of joint roughness coefficient using neutrosophic number functions. *Information* **2017**, *8*, 69. [CrossRef]
25. Chen, J.Q.; Ye, J.; Du, S.G. Scale effect and anisotropy analyzed for neutrosophic numbers of rock joint roughness coefficient based on neutrosophic statistics. *Symmetry* **2017**, *9*, 208. [CrossRef]
26. Chen, J.Q.; Ye, J.; Du, S.G.; Yong, R. Expressions of rock joint roughness coefficient using neutrosophic interval statistical numbers. *Symmetry* **2017**, *9*, 123. [CrossRef]
27. Ye, J.; Yong, R.; Liang, Q.F.; Huang, M.; Du, S.G. Neutrosophic functions of the joint roughness coefficient (JRC) and the shear strength: A case study from the pyroclastic rock mass in Shaoxing City, China. *Math. Probl. Eng.* **2016**, 4825709. [CrossRef]
28. Vlădăreanu, V.; Dumitrache, I.; Vlădăreanu, L.; Sacală, I.S.; Tonţ, G.; Moisescu, M.A. Versatile intelligent portable robot control platform based on cyber physical systems principles. *Stud. Inform. Control* **2015**, *24*, 409–418. [CrossRef]

symmetry

MDPI

Article

An Approach toward a Q-Neutrosophic Soft Set and Its Application in Decision Making

Majdoleen Abu Qamar and Nasruddin Hassan *

School of Mathematical Sciences, Faculty of Science and Technology, Universiti Kebangsaan Malaysia, Bangi 43600, Selangor, Malaysia; mjabuqamar@gmail.com

* Correspondence: nas@ukm.edu.my; Tel.: +60-1921-457-50

Received: 20 December 2018; Accepted: 23 January 2019; Published: 27 January 2019

Abstract: A neutrosophic set was proposed as an approach to study neutral uncertain information. It is characterized through three memberships, T, I and F, such that these independent functions stand for the truth, indeterminate, and false-membership degrees of an object. The neutrosophic set presents a symmetric form since truth enrolment T is symmetric to its opposite false enrolment F with respect to indeterminacy enrolment I that acts as an axis of symmetry. The neutrosophic set was further extended to a Q-neutrosophic soft set, which is a hybrid model that keeps the features of the neutrosophic soft set in dealing with uncertainty, and the features of a Q-fuzzy soft set that handles two-dimensional information. In this study, we discuss some operations of Q-neutrosophic soft sets, such as subset, equality, complement, intersection, union, AND operation, and OR operation. We also define the necessity and possibility operations of a Q-neutrosophic soft set. Several properties and illustrative examples are discussed. Then, we define the Q-neutrosophic-set aggregation operator and use it to develop an algorithm for using a Q-neutrosophic soft set in decision-making issues that have indeterminate and uncertain data, followed by an illustrative real-life example.

Keywords: decision making; neutrosophic set; Q-neutrosophic set; Q-neutrosophic soft set; soft set

1. Introduction

Fuzzy-set theory was established by Zadeh in 1965 [1]. Since then, fuzzy logic has been utilized in several real-world problems in uncertain environments. Consequently, numerous analysts discussed many results using distinct directions of fuzzy-set theory, for instance, interval valued fuzzy set [2] and intuitionistic fuzzy set [3]. These extensions can deal with uncertain real-world problems. An intuitionistic fuzzy set can only cope with incomplete data through its truth and falsity membership values, but it does not cope with indeterminate data. Thus, Smarandache [4] initiated the neutrosophic idea to overcome this problem. A neutrosophic set (NS) [5] is a mathematical notion serving issues containing inconsistent, indeterminate, and imprecise data. Recent studies on NS include a single-valued neutrosophic set [6] and complex neutrosophic set [7].

Molodtsov [8] proposed the notion of the soft set as an important mathematical notion for handling uncertainties. The main advantage of this notion in data analysis is that it does not need any grade of membership as in fuzzy-set theory. Maji et.al. [9] presented the fuzzy soft set, which is a combination between a soft set and a fuzzy set. Later, many researchers developed several extensions of the soft-set model, such as vague soft set [10], interval-valued vague soft set [11–13], soft expert set [14], and soft multiset theory [15]. Maji [16] extended the notion of the fuzzy soft set to the neutrosophic soft set (NSS) and defined some of its properties.

NS is very appropriate for handling inconsistent, indeterminate, and incomplete information in real applications. Recently, many studies have been done on NSS [17–24], the most recent being on vague soft sets [25], neutrosophic vague soft expert sets [26], n-valued refined neutrosophic soft sets [27], complex neutrosophic soft expert sets [28,29] and time-neutrosophic soft sets [30]. Many researchers [23,31–36] have constructed several aggregation operators, such as simplified neutrosophic prioritized aggregation operators, single-valued neutrosophic Dombi weighted aggregation operators, simplified neutrosophic weighted aggregation operators, interval neutrosophic exponential weighted aggregation operators, and used them in decision-making issues. Aggregation operators perform a vital role in multicriteria decision making (MCDM) issues whose principle target is to aggregate a collection of inputs to a single number. Thus, aggregation operators give us effective tools to handle neutrosophic data in the decision process.

For a two-dimensional universal set, Adam and Hassan [37,38] introduced the Q-fuzzy soft set (Q-FSS) and multi-Q-FSS, which includes a Q-fuzzy soft aggregation operator that allows constructing more efficient decision-making methods. Broumi [39] presented the notion of the Q-intuitionistic fuzzy soft set (Q-IFSS), and defined some basic properties and basic operations. Actually, these notions cannot handle indeterminate data that appear in two universal sets. Inspired by this, Abu Qamar and Hassan [40] initiated the concept of the Q-neutrosophic soft set (Q-NSS) by upgrading the membership functions of the NSS to a two-dimensional entity. As a result, Q-NSS is premium to these models with three two-dimensional independent membership functions. Hence, this concept serves indeterminacy and two-dimensionality at the same time. Moreover, the Q-neutrosophic set (Q-NS) is basically an NS defined over a two-dimensional set. Thus, it has added advantages to NS by treating a two-dimensional universal set, which makes it more valid in modeling real-life problems where two-dimensional sets and indeterminacy majorly appear. Q-NSS is created to keep the advantages of Q-FSS while holding NSS features. Therefore, it incorporates the benefits of these models. Abu Qamar and Hassan studied different aspects of Q-NSS, such as their relations [40], measures of Q-NSS information [41], generalized Q-neutrosophic soft expert set [42], and different decision problems of these concepts.

Motivated by these studies, in this study we discuss the different operations and properties of Q-NSS. To facilitate the discussion, we arranged this article as follows. Section 2 contains a review of basic definitions pertaining to this work. In Section 3, we discuss some operations of Q-NSS, such as subset, equality, complement, intersection, union, AND operation, and OR operation. We discuss necessity and possibility operations in Section 4 along with properties and illustrative examples. In Section 5, we define the aggregation Q-NS in order to utilize it in the algorithm that we constructed to solve decision-making problems using Q-NSS. In Section 6, comparison analysis is presented to validate the proposed approach. Finally, conclusions and future work are in Section 7. Consequently, the concept of Q-NSS will enrich current NSS studies.

2. Preliminaries

In this section, we recall the concepts of soft set, NS, and Q-NS, which are relevant to this paper. Soft-set theory was proposed by Molodtsov [8].

Definition 1 ([8]). *A pair (F,E) is a soft set over U if and only if $F: E \to \mathcal{P}(U)$ is mapping. That is, the soft set is a parameterized family of subsets of U.*

In order to handle inconsistent and indeterminate information that exists in some real-world issues, Smarandache [5] initiated the NS concept as follows:

Definition 2 ([5]). *An NS Γ on universe U is defined as $\Gamma = \{\langle u, (T_\Gamma(u), I_\Gamma(u), F_\Gamma(u))\rangle : u \in U\}$, where $T, I, F : U \to]^-0, 1^+[$ and $^-0 \leq T_\Gamma(u) + I_\Gamma(u) + F_\Gamma(u) \leq 3^+$.*

Now, we recall some basic NS operations proposed by Smarandache [43].

Definition 3 ([43]). *Let Γ and Ψ be two NSs. Then, Γ is a subset of Ψ, written as $\Gamma \subseteq \Psi$, if and only if $T_\Gamma(u) \leq T_\Psi(u)$, $I_\Gamma(u) \geq I_\Psi(u)$ and $F_\Gamma(u) \geq F_\Psi(u) \forall u \in U$.*

Definition 4 ([43]). *The union of two NSs* Γ *and* Ψ *in* U*, written as* $\Gamma \cup \Psi = \Lambda$*, where* $\Lambda = \{\langle u, (\max\{T_\Gamma(u), T_\Psi(u)\}, \min\{I_\Gamma(u), I_\Psi(u)\}, \min\{F_\Gamma(u), F_\Psi(u)\})\rangle : u \in U\}$.

Definition 5 ([43]). *The intersection of two NSs* Γ *and* Ψ *in* U*, written as* $\Gamma \cap \Psi = \Lambda$*, where* $\Lambda = \{\langle u, (\min\{T_\Gamma(u), T_\Psi(u)\}, \max\{I_\Gamma(u), I_\Psi(u)\}, \max\{F_\Gamma(u), F_\Psi(u)\})\rangle : u \in U\}$.

Definition 6 ([43]). *The complement of an NS* Γ *in* U*, denoted by* Γ^c*, where*

$$\Gamma^c = \{\langle u, (1 - T_\Gamma(u), 1 - I_\Gamma(u), 1 - F_\Gamma(u))\rangle : u \in U\}.$$

Abu Qamar and Hassan [40] introduced the idea of Q-NS as follows:

Definition 7 ([40]). *A Q-NS* Γ_Q *in* U *is an object of form*

$$\Gamma_Q = \left\{ \left\langle (u,t), T_{\Gamma_Q}(u,t), I_{\Gamma_Q}(u,t), F_{\Gamma_Q}(u,t) \right\rangle : u \in U, t \in Q \right\},$$

where $Q \neq \phi$ *and* $T_{\Gamma_Q}, I_{\Gamma_Q}, F_{\Gamma_Q} : U \times Q \rightarrow]^-0, 1^+[$ *are the true, indeterminacy, and false membership functions, respectively, with* $^-0 \leq T_{\Gamma_Q} + I_{\Gamma_Q} + F_{\Gamma_Q} \leq 3^+$.

3. Q-Neutrosophic Soft Sets

In this section, we discuss numerous properties and operations concerning Q-NSSs, namely, union, intersection, and AND and OR operations. The concept of Q-NSS was briefly mentioned in the following definition without any algebraic operations in the discussion on Q-neutrosophic soft relations [40].

Definition 8 ([40]). *Let* U *be a universal set,* Q *be a nonempty set and* $A \subseteq E$ *be a set of parameters. Let* $\mu^l QNS(U)$ *be the set of all multi-Q-NSs on* U *with dimension* $l = 1$*. A pair* (Γ_Q, A) *is called a Q-NSS over* U*, where* $\Gamma_Q : A \rightarrow \mu^l QNS(U)$ *is a mapping, such that* $\Gamma_Q(e) = \phi$ *if* $e \notin A$.

A Q-NSS can be presented as

$$(\Gamma_Q, A) = \{(e, \Gamma_Q(e)) : e \in A, \Gamma_Q \in \mu^l QNS(U)\}.$$

The set of all Q-NSSs in U *is denoted by Q-NSS(*U*).*

The following example shows how Q-NSS can represent real-world problems.

Example 1. *Suppose we want to examine the attractiveness of a cell phone that a person is considering buying. Suppose there are two choices in the universe* $U = \{u_1, u_2\}$, $Q = \{s = black, t = white\}$ *is the set of colors under consideration and* $E = \{e_1 = price, e_2 = version, e_3 = device\ specification\}$ *is a set of decision parameters. Then, the Q-NSS* (Γ_Q, A) *is given by:*

$$\begin{aligned}(\Gamma_Q, A) = \Big\{ &\Big\langle e_1, [(u_1,s), 0.5, 0.1, 0.2], [(u_1,t), 0.7, 0.5, 0.3], [(u_2,s), 0.7, 0.5, 0.4], [(u_2,t), 0.4, 0.3, 0.1] \Big\rangle, \\ &\Big\langle e_2, [(u_1,s), 0.1, 0.2, 0.6], [(u_1,t), 0.8, 0.1, 0.5], [(u_2,s), 0.4, 0.6, 0.9], [(u_2,t), 0.4, 0.6, 0.7] \Big\rangle, \\ &\Big\langle e_3, [(u_1,s), 0.6, 0.2, 0.2], [(u_1,t), 0.1, 0.8, 0.5], [(u_2,s), 0.6, 0.4, 0.9], [(u_2,t), 0.3, 0.1, 0.5] \Big\rangle \Big\}.\end{aligned}$$

Each element in (Γ_Q, A) *represents the degree of attractiveness of each cell phone with a specific color based on each parameter. For example, element* $[(u_1, s), 0.6, 0.2, 0.2]$ *under parameter* e_3 *represents the degree of true, indeterminacy, and falsity attractiveness of device specification of cell phone* u_1 *with a black color, and they are 0.6, 0.2, and 0.2, respectively.*

Now, we introduce some basic definitions of Q-NSSs.

Definition 9. *Let* $(\Gamma_Q, A) \in$ *Q-NSS(U). If* $\Gamma_Q(e) = \phi$ *for all* $e \in A$*, then* (Γ_Q, A) *is called a null Q-NSS(U) denoted by* (ϕ, A).

Definition 10. *Let* $(\Gamma_Q, A), (\Psi_Q, B) \in$ *Q-NSS(U), then*

1. (Γ_Q, A) *is a Q-neutrosophic soft subset of* (Ψ_Q, B)*, denoted by* $(\Gamma_Q, A) \subseteq (\Psi_Q, B)$*, if* $A \subseteq B$ *and* $\Gamma_Q(e) \subseteq \Psi_Q(e)$ *for all* $e \in A$*, that is* $T_{\Gamma_Q(e)}(u,t) \leq T_{\Psi_Q(e)}(u,t)$, $I_{\Gamma_Q(e)}(u,t) \geq I_{\Psi_Q(e)}(u,t), F_{\Gamma_Q(e)}(u,t) \geq F_{\Psi_Q(e)}(u,t)$*, for all* $(u,t) \in U \times Q$.
2. (Γ_Q, A) *and* (Ψ_Q, B) *are equal, denoted by* $(\Gamma_Q, A) = (\Psi_Q, B)$*, if and only if* $(\Gamma_Q, A) \subseteq (\Psi_Q, B)$ *and* $(\Psi_Q, B) \subseteq (\Gamma_Q, A)$ *for all* $u \in U$.

Now, we define the complement of $Q - NSS(U)$:

Definition 11. *Let* $(\Gamma_Q, A) \in$ *Q-NSS(U). Then, its complement written as* $(\Gamma_Q, A)^c = (\Gamma^c_Q, A)$*, and is defined as*

$$(\Gamma_Q, A)^c = \left\{\left\langle e, T^c_{\Gamma_Q(e)}(u,t), I^c_{\Gamma_Q(e)}(u,t), F^c_{\Gamma_Q(e)}(u,t)\right\rangle : e \in A, (u,t) \in U \times Q\right\},$$

such that $\forall e \in A, (u,t) \in U \times Q$

$$\begin{aligned} T^c_{\Gamma_Q(e)}(u,t) &= 1 - T_{\Gamma_Q(e)}(u,t), \\ I^c_{\Gamma_Q(e)}(u,t) &= 1 - I_{\Gamma_Q(e)}(u,t), \\ F^c_{\Gamma_Q(e)}(u,t) &= 1 - F_{\Gamma_Q(e)}(u,t). \end{aligned}$$

Example 2. *Let* $U = \{u_1, u_2\}$ *be a universal set,* $A = \{e_1, e_2\}$ *and* $Q = \{s, t\}$.

$$\begin{aligned} (\Gamma_Q, A) = \Big\{ &\Big\langle e_1, [(u_1,s),0.2,0.3,0.5],[(u_1,t),0.6,0.4,0.1],[(u_2,s),0.1,0.4,0.7],[(u_2,t),0.3,0.2,0.8]\Big\rangle, \\ &\Big\langle e_2, [(u_1,s),0.4,0.1,0.9],[(u_1,t),0.6,0.2,0.1],[(u_2,s),0.5,0.3,0.7],[(u_2,t),0.1,0.3,0.2]\Big\rangle\Big\}. \end{aligned}$$

is a Q-NSS and the complement of (Γ_Q, A) *is*

$$\begin{aligned} (\Gamma^c_Q, A) = \Big\{ &\Big\langle e_1, [(u_1,s),0.8,0.7,0.5],[(u_1,t),0.4,0.6,0.9],[(u_2,s),0.9,0.6,0.3],[(u_2,t),0.7,0.8,0.2]\Big\rangle, \\ &\Big\langle e_2, [(u_1,s),0.6,0.9,0.1],[(u_1,t),0.4,0.8,0.9],[(u_2,s),0.5,0.7,0.3],[(u_2,t),0.9,0.7,0.8]\Big\rangle\Big\}. \end{aligned}$$

Proposition 1. *If* $(\Gamma_Q, A) \in$ *Q-NSS(U), then* $((\Gamma_Q, A)^c)^c = (\Gamma_Q, A)$.

Proof. From Definition 11, we have

$$\begin{aligned} (\Gamma_Q, A)^c &= \left\{\left\langle e, T^c_{\Gamma_Q(e)}(u,t), I^c_{\Gamma_Q(e)}(u,t), F^c_{\Gamma_Q(e)}(u,t)\right\rangle : e \in A, (u,t) \in U \times Q\right\} \\ &= \left\{\left\langle e, 1 - T_{\Gamma_Q(e)}(u,t), 1 - I_{\Gamma_Q(e)}(u,t), 1 - F_{\Gamma_Q(e)}(u,t)\right\rangle : e \in A, (u,t) \in U \times Q\right\}. \end{aligned}$$

Thus,

$$
\begin{aligned}
((\Gamma_Q, A)^c)^c = & \Big\{\Big\langle e, 1-(1-T_{\Gamma_Q(e)}(u,t)), 1-(1-I_{\Gamma_Q(e)}(u,t)), 1-(1-F_{\Gamma_Q(e)}(u,t))\Big\rangle : \\
& e \in A, (u,t) \in U \times Q\Big\} \\
= & \Big\{\Big\langle e, T_{\Gamma_Q(e)}(u,t), I_{\Gamma_Q(e)}(u,t), F_{\Gamma_Q(e)}(u,t)\Big\rangle : e \in A, (u,t) \in U \times Q\Big\}
\end{aligned}
$$

This completes the proof. □

Next, we discuss the operations of union, intersection, and AND and OR operations for Q-NSSs, along with some results and examples.

Definition 12. *The union of two Q-NSSs (Γ_Q, A) and (Ψ_Q, B) is the Q-NSS (Λ_Q, C) written as $(\Gamma_Q, A) \cup (\Psi_Q, B) = (\Lambda_Q, C)$, where $C = A \cup B$ and for all $c \in C$, $(u,t) \in U \times Q$, the truth membership, indeterminacy membership, and falsity membership of (Λ_Q, C) are as follows:*

$$
T_{\Lambda_Q(c)}(u,t) = \begin{cases} T_{\Gamma_Q(c)}(u,t) & \text{if } c \in A - B, \\ T_{\Psi_Q(c)}(u,t) & \text{if } c \in B - A, \\ \max\{T_{\Lambda_Q(c)}(u,t), T_{\Psi_Q(c)}(u,t)\} & \text{if } c \in A \cap B, \end{cases}
$$

$$
I_{\Lambda_Q(c)}(u,t) = \begin{cases} I_{\Gamma_Q(c)}(u,t) & \text{if } c \in A - B, \\ I_{\Psi_Q(c)}(u,t) & \text{if } c \in B - A, \\ \min\{I_{\Gamma_Q(c)}(u,t), I_{\Psi_Q(c)}(u,t)\} & \text{if } c \in A \cap B, \end{cases}
$$

$$
F_{\Lambda_Q(c)}(u,t) = \begin{cases} F_{\Gamma_Q(c)}(u,t) & \text{if } c \in A - B, \\ F_{\Psi_Q(c)}(u,t) & \text{if } c \in B - A, \\ \min\{F_{\Gamma_Q(c)}(u,t), F_{\Psi_Q(c)}(u,t)\} & \text{if } c \in A \cap B. \end{cases}
$$

Definition 13. *The intersection of two Q-NSSs (Γ_Q, A) and (Ψ_Q, B) is the Q-NSS (Λ_Q, C) written as $(\Gamma_Q, A) \cap (\Psi_Q, B) = (\Lambda_Q, C)$, where $C = A \cap B$ and for all $c \in C$ and $(u,t) \in U \times Q$ the truth membership, indeterminacy membership, and falsity membership of (Λ_Q, C) are as follows:*

$$
\begin{aligned}
T_{\Lambda_Q(c)}(u,t) &= \min\{T_{\Gamma_Q(c)}(u,t), T_{\Psi_Q(c)}(u,t)\}, \\
I_{\Lambda_Q(c)}(u,t) &= \max\{I_{\Gamma_Q(c)}(u,t), I_{\Psi_Q(c)}(u,t)\}, \\
F_{\Lambda_Q(c)}(u,t) &= \max\{F_{\Gamma_Q(c)}(u,t), F_{\Psi_Q(c)}(u,t)\}.
\end{aligned}
$$

Example 3. *Let $U = \{u_1, u_2\}$ be a universal set, $E = \{e_1, e_2\}$ and $Q = \{s, t\}$. If $A = B = \{e_1, e_2\} \subseteq E$,*

$$
\begin{aligned}
(\Gamma_Q, A) = \Big\{ & \Big\langle e_1, [(u_1,s), 0.2, 0.3, 0.5], [(u_1,t), 0.6, 0.4, 0.1], [(u_2,s), 0.1, 0.4, 0.7], [(u_2,t), 0.3, 0.2, 0.8]\Big\rangle, \\
& \Big\langle e_2, [(u_1,s), 0.4, 0.1, 0.9], [(u_1,t), 0.6, 0.2, 0.1], [(u_2,s), 0.5, 0.3, 0.7], [(u_2,t), 0.1, 0.3, 0.2]\Big\rangle\Big\}
\end{aligned}
$$

and

$$
\begin{aligned}
(\Psi_Q, B) = \Big\{ & \Big\langle e_1, [(u_1,s), 0.4, 0.5, 0.2], [(u_1,t), 0.2, 0.3, 0.5], [(u_2,s), 0.3, 0.1, 0.7], [(u_2,t), 0.2, 0.4, 0.6]\Big\rangle, \\
& \Big\langle e_2, [(u_1,s), 0.9, 0.3, 0.2], [(u_1,t), 0.4, 0.3, 0.1], [(u_2,s), 0.8, 0.6, 0.3], [(u_2,t), 0.2, 0.4, 0.6]\Big\rangle\Big\},
\end{aligned}
$$

then

$$
\begin{aligned}
&(\Gamma_Q, A) \cup (\Psi_Q, B) = \\
&\qquad \Big\{ \Big\langle e_1, [(u_1, s), 0.4, 0.3, 0.2], [(u_1, t), 0.6, 0.3, 0.1], [(u_2, s), 0.3, 0.1, 0.7], [(u_2, t), 0.3, 0.2, 0.6] \Big\rangle, \\
&\qquad \Big\langle e_2, [(u_1, s), 0.9, 0.1, 0.2], [(u_1, t), 0.6, 0.2, 0.1], [(u_2, s), 0.8, 0.3, 0.3], [(u_2, t), 0.2, 0.3, 0.2] \Big\rangle \Big\},
\end{aligned}
$$

and

$$
\begin{aligned}
&(\Gamma_Q, A) \cap (\Psi_Q, B) = \\
&\qquad \Big\{ \Big\langle e_1, [(u_1, s), 0.2, 0.5, 0.5], [(u_1, t), 0.2, 0.4, 0.5], [(u_2, s), 0.1, 0.4, 0.7], [(u_2, t), 0.2, 0.4, 0.8] \Big\rangle, \\
&\qquad \Big\langle e_2, [(u_1, s), 0.4, 0.3, 0.9], [(u_1, t), 0.4, 0.3, 0.1], [(u_2, s), 0.5, 0.6, 0.7], [(u_2, t), 0.1, 0.4, 0.6] \Big\rangle \Big\}.
\end{aligned}
$$

Here, we give a proposition concerning the union and intersection of Q-NSSs.

Proposition 2. *Let* $(\Gamma_Q, A), (\Psi_Q, B)$ *and* $(\Lambda_Q, C) \in$ *Q-NSS(U). Then,*

1. $(\Gamma_Q, A) \cup (\Psi_Q, B) = (\Psi_Q, B) \cup (\Gamma_Q, A)$.
2. $(\Gamma_Q, A) \cap (\Psi_Q, B) = (\Psi_Q, B) \cap (\Gamma_Q, A)$.
3. $(\Gamma_Q, A) \cup ((\Psi_Q, B) \cup (\Lambda_Q, C)) = ((\Gamma_Q, A) \cup (\Psi_Q, B)) \cup (\Lambda_Q, C)$.
4. $(\Gamma_Q, A) \cap ((\Psi_Q, B) \cap (\Lambda_Q, C)) = ((\Gamma_Q, A) \cap (\Psi_Q, B)) \cap (\Lambda_Q, C)$.

Proof. 1. We show that $(\Gamma_Q, A) \cup (\Psi_Q, B) = (\Psi_Q, B) \cup (\Gamma_Q, A)$ by using Definition 12. Consider case $c \in A \cap B$, as other cases are trivial.

$$
\begin{aligned}
(\Gamma_Q, A) \cup (\Psi_Q, B) &= \Big\{ \Big\langle c, \big(\max\{T_{\Gamma_{Q}(c)}(u,t), T_{\Psi_{Q}(c)}(u,t)\}, \min\{I_{\Gamma_{Q}(c)}(u,t), I_{\Psi_{Q}(c)}(u,t)\}, \\
&\qquad \min\{F_{\Gamma_{Q}(c)}(u,t), F_{\Psi_{Q}(c)}(u,t)\}\}\} \Big\rangle : (u,t) \in U \times Q \Big\} \\
&= \Big\{ \Big\langle c, \max\{T_{\Psi_{Q}(c)}(u,t), T_{\Gamma_{Q}(c)}(u,t)\}, \min\{I_{\Psi_{Q}(c)}(u,t), I_{\Gamma_{Q}(c)}(u,t)\}, \\
&\qquad \min\{F_{\Psi_{Q}(c)}(u,t), F_{\Gamma_{Q}(c)}(u,t)\} \Big\rangle : (u,t) \in U \times Q \Big\} \\
&= (\Psi_Q, B) \cup (\Gamma_Q, A).
\end{aligned}
$$

2. The proof is similar to that of Part (1).

3. We show that $((\Gamma_Q, A) \cup (\Psi_Q, B)) \cup (\Upsilon_Q, C) = (\Gamma_Q, A) \cup ((\Psi_Q, B) \cup (\Upsilon_Q, C))$ by using Definition 12. Consider case $c \in A \cap B$, as other cases are trivial.

$$
\begin{aligned}
(\Gamma_Q, A) \cup (\Psi_Q, B) &= \Big\{ \Big\langle c, \big(\max\{T_{\Gamma_{Q}(c)}(u,t), T_{\Psi_{Q}(c)}(u,t)\}, \min\{I_{\Gamma_{Q}(c)}(u,t), I_{\Psi_{Q}(c)}(u,t)\}, \\
&\qquad \min\{F_{\Gamma_{Q}(c)}(u,t), F_{\Psi_{Q}(c)}(u,t)\}\} \big) \Big\rangle : (u,t) \in U \times Q \Big\}.
\end{aligned}
$$

$$
\begin{aligned}
&((\Gamma_Q, A) \cup (\Psi_Q, B)) \cup (\Upsilon_Q, C) \\
&\qquad = \Big\{ \Big\langle c, \Big(\big(\max\big\{ \max\{T_{\Gamma_{Q}(c)}(u,t), T_{\Psi_{Q}(c)}(u,t)\}, T_{\Upsilon_{Q}(c)}(u,t)\big\}, \\
&\qquad\qquad \min\big\{ \min\{I_{\Gamma_{Q}(c)}(u,t), I_{\Psi_{Q}(c)}(u,t)\}, I_{\Upsilon_{Q}(c)}(u,t)\big\}, \\
&\qquad\qquad \min\big\{ \min\{F_{\Gamma_{Q}(c)}(u,t), F_{\Psi_{Q}(c)}(u,t)\}, F_{\Upsilon_{Q}(c)}(u,t)\big\}\big) \Big) \Big\rangle : (u,t) \in U \times Q \Big\}
\end{aligned}
$$

$$= \Big\{\Big\langle c, \Big(\big(\max\{T_{\Gamma_Q(c)}(u,t), T_{\Psi_Q(c)}(u,t), T_{\Upsilon_Q(c)}(u,t)\},$$
$$\min\{I_{\Gamma_Q(c)}(u,t), I_{\Psi_Q(c)}(u,t), I_{\Upsilon_Q(c)}(u,t)\},$$
$$\min\{F_{\Gamma_Q(c)}(u,t), F_{\Psi_Q(c)}(u,t), F_{\Upsilon_Q(c)}(u,t)\}\big)\Big)\Big\rangle : (u,t) \in U \times Q\Big\}$$
$$= \Big\{\Big\langle c, \Big(\big(\max\{T_{\Gamma_Q(c)}(u,t), \max\{T_{\Psi_Q(c)}(u,t), T_{\Upsilon_Q(c)}(u,t)\}\},$$
$$\min\{I_{\Gamma_Q(c)}(u,t), \min\{I_{\Psi_Q(c)}(u,t), I_{\Upsilon_Q(c)}(u,t)\}\},$$
$$\min\{F_{\Gamma_Q(c)}(u,t), \min\{F_{\Psi_Q(c)}(u,t), F_{\Upsilon_Q(c)}(u,t)\}\}\big)\Big)\Big\rangle : (u,t) \in U \times Q\Big\}$$
$$= (\Gamma_Q, A) \cup ((\Psi_Q, B) \cup (\Upsilon_Q, C)).$$

4. The proof is similar to that of Part (3). □

Next, we introduce the AND and OR operations of Q-NSSs.

Definition 14. *If (Γ_Q, A) and (Ψ_Q, B) are two Q-NSSs on U, then (Γ_Q, A) AND (Ψ_Q, B) is the Q-NSS denoted by $(\Gamma_Q, A) \wedge (\Psi_Q, B)$ and defined by $(\Gamma_Q, A) \wedge (\Psi_Q, B) = (\Lambda_Q, A \times B)$, where $\Lambda_Q(a,b) = \Gamma_Q(a) \cap \Psi_Q(b)$ for all $(a,b) \in A \times B$ is the operation of intersection of two Q-NSs on U. That is, the truth, indeterminacy, and falsity memberships of (Γ_Q, A) and (Ψ_Q, B) are as follows:*

$$T_{\Lambda_Q(a,b)}(u,t) = \min\{T_{\Gamma_Q(a)}(u,t), T_{\Psi_Q(b)}(u,t)\},$$
$$I_{\Lambda_Q(a,b)}(u,t) = \max\{I_{\Gamma_Q(a)}(u,t), I_{\Psi_Q(b)}(u,t)\},$$
$$F_{\Lambda_Q(a,b)}(u,t) = \max\{F_{\Gamma_Q(a)}(u,t), F_{\Psi_Q(b)}(u,t)\},$$

Definition 15. *If (Γ_Q, A) and (Ψ_Q, B) are two Q-NSSs on U, then (Γ_Q, A) OR (Ψ_Q, B) is the Q-NSS denoted by $(\Gamma_Q, A) \vee (\Psi_Q, B)$ and defined by $(\Gamma_Q, A) \vee (\Psi_Q, B) = (\Lambda_Q, A \times B)$, where $\Lambda_Q(a,b) = \Gamma_Q(a) \cup \Psi_Q(b)$ for all $(a,b) \in A \times B$ is the operation of union of two Q-NSs on U. The truth, indeterminacy, and falsity memberships of $(\widehat{\Lambda}_Q^h, A \times B)$ are as follows:*

$$T_{\Upsilon_Q(a,b)}(u,t) = \max\{T_{\Gamma_Q(a)}(u,t), T_{\Psi_Q(b)}(u,t)\},$$
$$I_{\Upsilon_Q(a,b)}(u,t) = \min\{I_{\Gamma_Q(a)}(u,t), I_{\Psi_Q(b)}(u,t)\},$$
$$F_{\Upsilon_Q(a,b)}(u,t) = \min\{F_{\Gamma_Q(a)}(u,t), F_{\Psi_Q(b)}(u,t)\},$$

Here, we present an example of AND and OR operations followed by the corresponding propositions.

Example 4. *Reconsider Example 3, then*

$$(\Gamma_Q, A) \wedge (\Psi_Q, B)$$
$$= \Big\{\Big\langle (e_1,e_1), [(u_1,s), 0.2, 0.5, 0.5], [(u_1,t), 0.2, 0.4, 0.5], [(u_2,s), 0.1, 0.4, 0.7], [(u_2,t), 0.2, 0.4, 0.8]\Big\rangle,$$
$$\Big\langle (e_1,e_2), [(u_1,s), 0.2, 0.3, 0.5], [(u_1,t), 0.4, 0.4, 0.1], [(u_2,s), 0.1, 0.6, 0.7], [(u_2,t), 0.2, 0.4, 0.8]\Big\rangle,$$
$$\Big\langle (e_2,e_1), [(u_1,s), 0.4, 0.5, 0.9], [(u_1,t), 0.2, 0.3, 0.5], [(u_2,s), 0.3, 0.3, 0.7], [(u_2,t), 0.1, 0.4, 0.6]\Big\rangle,$$
$$\Big\langle (e_2,e_2), [(u_1,s), 0.4, 0.3, 0.9], [(u_1,t), 0.4, 0.3, 0.1], [(u_2,s), 0.5, 0.6, 0.7], [(u_2,t), 0.1, 0.4, 0.6]\Big\rangle\Big\}.$$

$$
\begin{aligned}
&(\Gamma_Q, A) \vee (\Psi_Q, B) \\
&= \Big\{ \Big\langle (e_1, e_1), [(u_1, s), 0.4, 0.3, 0.2], [(u_1, t), 0.6, 0.3, 0.1], [(u_2, s), 0.3, 0.1, 0.7], [(u_2, t), 0.3, 0.2, 0.6] \Big\rangle, \\
&\quad \Big\langle (e_1, e_2), [(u_1, s), 0.9, 0.3, 0.2], [(u_1, t), 0.6, 0.4, 0.1], [(u_2, s), 0.8, 0.4, 0.3], [(u_2, t), 0.3, 0.2, 0.6] \Big\rangle, \\
&\quad \Big\langle (e_2, e_1), [(u_1, s), 0.4, 0.1, 0.2], [(u_1, t), 0.6, 0.2, 0.1], [(u_2, s), 0.5, 0.1, 0.7], [(u_2, t), 0.2, 0.3, 0.2] \Big\rangle, \\
&\quad \Big\langle (e_2, e_2), [(u_1, s), 0.9, 0.1, 0.2], [(u_1, t), 0.6, 0.2, 0.1], [(u_2, s), 0.8, 0.3, 0.3], [(u_2, t), 0.2, 0.3, 0.2] \Big\rangle \Big\}.
\end{aligned}
$$

Proposition 3. *Let $(\Gamma_Q, A), (\Psi_Q, B)$, and (Λ_Q, C) be three Q-NSSs on U. Then, we have the following associative properties:*

1. $(\Gamma_Q, A) \wedge ((\Psi_Q, B) \wedge (\Lambda_Q, C)) = ((\Gamma_Q, A) \wedge (\Psi_Q, B)) \wedge (\Lambda_Q, C)$.
2. $(\Gamma_Q, A) \vee ((\Psi_Q, B) \vee (\Lambda_Q, C)) = ((\Gamma_Q, A) \vee (\Psi_Q, B)) \vee (\Lambda_Q, C)$.

Proof. 1. Let $(\Psi_Q, B) \wedge (\Lambda_Q, C) = (Y_Q, B \times C)$, where $Y_Q(b, c) = \Psi_Q(b) \cap \Lambda_Q(c)$.

Now, $(\Gamma_Q, A) \wedge ((\Psi_Q, B) \wedge (\Lambda_Q, C)) = (\Gamma_Q, A) \wedge (Y_Q, B \times C) = (\Omega_Q, A \times B \times C)$, where $\Omega_Q(a, b, c) = \Gamma_Q(a) \cap Y_Q(b, c) = \Gamma_Q(a) \cap \Psi_Q(b) \cap \Lambda_Q(c)$.

Also, $(\Gamma_Q, A) \wedge (\Psi_Q, B) = (\Theta_Q, A \times B)$, where $\Theta_Q(a, b) = \Gamma_Q(A) \cap \Psi_Q(B)$. Therefore, $((\Gamma_Q, A) \wedge (\Psi_Q, B)) \wedge (\Lambda_Q, C) = (\Theta_q, A \times B) \wedge (\Lambda_Q, C) = (\Delta_Q, A \times B \times C)$, where $\Delta_Q(a, b, c) = \Theta_Q(a, b) \cap \Lambda_Q(c) = \Gamma_Q(a) \cap \Psi_Q(b) \cap \Lambda_Q(c)$. Hence, $(\Gamma_Q, A) \wedge ((\Psi_Q, B) \wedge (\Lambda_Q, C)) = ((\Gamma_Q, A) \wedge (\Psi_Q, B)) \wedge (\Lambda_Q, C)$.

2. The result can be proved in a similar fashion as in Assertion 1. □

4. Necessity and Possibility Operations on Q-Neutrosophic Soft Sets with Some Properties

In this section, we introduce necessity and possibility operations on Q-NSSs.

Definition 16. *The necessity operation on a Q-NSS on U (Γ_Q, A) is denoted by $\oplus(\Gamma_Q, A)$ and is defined as, for all $e \in A$,*

$$\oplus(\Gamma_Q, A) = \Big\{ \Big\langle e, [(u, t), T_{\Gamma_Q}(u, t), I_{\Gamma_Q}(u, t), 1 - T_{\Gamma_Q}(u, t)] \Big\rangle : (u, t) \in U \times Q \Big\}.$$

Example 5. *Reconsider Example 2, then*

$$
\begin{aligned}
\oplus(\Gamma_Q, A) = \Big\{ &\Big\langle e_1, [(u_1, s), 0.2, 0.3, 0.8], [(u_1, t), 0.6, 0.4, 0.4], [(u_2, s), 0.1, 0.4, 0.9], [(u_2, t), 0.3, 0.2, 0.7] \Big\rangle, \\
&\Big\langle e_2, [(u_1, s), 0.4, 0.1, 0.6], [(u_1, t), 0.6, 0.2, 0.4], [(u_2, s), 0.5, 0.3, 0.5], [(u_2, t), 0.1, 0.3, 0.9] \Big\rangle \Big\}
\end{aligned}
$$

Proposition 4. *Let (Γ_Q, A) and (Ψ_Q, B) be two Q-NSSs on U. Then,*

1. $\oplus((\Gamma_Q, A) \cup (\Psi_Q, B)) = \oplus(\Gamma_Q, A) \cup \oplus(\Psi_Q, B)$.
2. $\oplus((\Gamma_Q, A) \cap (\Psi_Q, B)) = \oplus(\Gamma_Q, A) \cap \oplus(\Psi_Q, B)$.
3. $\oplus(\oplus(\Gamma_Q, A)) = \oplus(\Gamma_Q, A)$.

Proof. 1. $(\Gamma_Q, A) \cup (\Psi_Q, B) = (\Lambda_Q, C)$, where $C = A \cup B$
$(\Lambda_Q, C) = \left\{ \left\langle e, [(u,t), T_{\Lambda_Q}(u,t), I_{\Lambda_Q}(u,t), F_{\Lambda_Q}(u,t)] \right\rangle : (u,t) \in U \times Q \right\}$, such that

$$T_{\Lambda_Q}(u,t) = \begin{cases} T_{\Gamma_Q}(u,t) & \text{if } e \in A - B, \\ T_{\Psi_Q}(u,t) & \text{if } e \in B - A, \\ \max\{T_{\Gamma_Q}(u,t), T_{\Psi_Q}(u,t)\} & \text{if } e \in A \cap B, \end{cases}$$

$$I_{\Lambda_Q}(u,t) = \begin{cases} I_{\Gamma_Q}(u,t) & \text{if } e \in A - B, \\ I_{\Psi_Q}(u,t) & \text{if } e \in B - A, \\ \min\{I_{\Gamma_Q}(u,t), I_{\Psi_Q}(u,t)\} & \text{if } e \in A \cap B, \end{cases}$$

and

$$F_{\Lambda_Q}(u,t) = \begin{cases} F_{\Gamma_Q}(u,t) & \text{if } e \in A - B, \\ F_{\Psi_Q}(u,t) & \text{if } e \in B - A, \\ \min\{F_{\Gamma_Q}(u,t), F_{\Psi_Q}(u,t)\} & \text{if } e \in A \cap B. \end{cases}$$

Now, by Definition 16, for all $e \in C$
$\oplus(\Lambda_Q, C) = \left\{ \left\langle e, [(u,t), \oplus T_{\Lambda_Q}(u,t), \oplus I_{\Lambda_Q}(u,t), \oplus F_{\Lambda_Q}(u,t)] \right\rangle : (u,t) \in U \times Q \right\}$, where

$$\oplus T_{\Lambda_Q}(u,t) = \begin{cases} T_{\Gamma_Q}(u,t) & \text{if } e \in A - B, \\ T_{\Psi_Q}(u,t) & \text{if } e \in B - A, \\ \max\{T_{\Gamma_Q}(u,t), T_{\Psi_{Qi}}(u,t)\} & \text{if } e \in A \cap B, \end{cases}$$

$$\oplus I_{\Lambda_Q}(u,t) = \begin{cases} I_{\Gamma_Q}(u,t) & \text{if } e \in A - B, \\ I_{\Psi_Q}(u,t) & \text{if } e \in B - A, \\ \min\{I_{\Gamma_Q}(u,t), I_{\Psi_Q}(u,t)\} & \text{if } e \in A \cap B, \end{cases}$$

and

$$\oplus F_{\Lambda_Q}(u,t) = \begin{cases} 1 - T_{\Gamma_Q}(u,t) & \text{if } e \in A - B, \\ 1 - T_{\Psi_Q}(u,t) & \text{if } e \in B - A, \\ 1 - \max\{T_{\Gamma_Q}(u,t), T_{\Psi_Q}(u,t)\} & \text{if } e \in A \cap B. \end{cases}$$

Assume for all $e \in A$
$\oplus(\Gamma_Q, A) = \left\{ \left\langle e, [(u,t), T_{\Gamma_Q}(u,t), I_{\Gamma_Q}(u,t), 1 - T_{\Gamma_Q}(u,t)] \right\rangle : (u,t) \in U \times Q \right\}$, and
$\oplus(\Psi_Q, B) = \left\{ \left\langle e, [(u,t), T_{\Psi_Q}(u,t), I_{\Psi_Q}(u,t), 1 - T_{\Psi_Q}(u,t)] \right\rangle : (u,t) \in U \times Q \right\}$. Now, $\oplus(\Gamma_Q, A) \cup \oplus(\Psi_Q, B) = (\Upsilon_Q, C)$, where
$(\Upsilon_Q, C) = \left\{ \left\langle e, [(u,t), T_{\Upsilon_Q}(u,t), I_{\Upsilon_Q}(u,t), F_{\Upsilon_Q}(u,t)] \right\rangle : (u,t) \in U \times Q \right\}$, such that

$$T_{\Upsilon_Q}(u,t) = \begin{cases} T_{\Gamma_Q}(u,t) & \text{if } e \in A - B, \\ T_{\Psi_Q}(u,t) & \text{if } e \in B - A, \\ \max\{T_{\Gamma_Q}(u,t), T_{\Psi_Q}(u,t)\} & \text{if } e \in A \cap B, \end{cases}$$

$$I_{Y_Q}(u,t) = \begin{cases} I_{\Gamma_Q}(u,t) & \text{if } e \in A-B, \\ I_{\Psi_Q}(u,t) & \text{if } e \in B-A, \\ \min\{I_{\Gamma_Q}(u,t), I_{\Psi_Q}(u,t)\} & \text{if } e \in A \cap B, \end{cases}$$

and

$$\begin{aligned} F_{Y_Q}(u,t) &= \begin{cases} 1 - T_{\Gamma_Q}(u,t) & \text{if } e \in A-B, \\ 1 - T_{\Psi_Q}(u,t) & \text{if } e \in B-A, \\ \min\{1 - T_{\Gamma_Q}(u,t), 1 - T_{\Psi_Q}(u,t)\} & \text{if } e \in A \cap B, \end{cases} \\ &= \begin{cases} 1 - T_{\Gamma_Q}(u,t) & \text{if } e \in A-B, \\ 1 - T_{\Psi_Q}(u,t) & \text{if } e \in B-A, \\ 1 - \max\{T_{\Gamma_Q}(u,t), T_{\Psi_Q}(u,t)\} & \text{if } e \in A \cap B. \end{cases} \end{aligned}$$

Consequently, $\oplus(\Lambda_Q, C)$ and (Y_Q, C) are the same. Thus, $\oplus((\Gamma_Q, A) \cup (\Psi_Q, B)) = \oplus(\Gamma_Q, A) \cup \oplus(\Psi_Q, B)$.
2. Can be analogously proven.
3. Assume for all $e \in A$

$$\oplus(\Gamma_Q, A) = \left\{\left\langle e, [(u,t), T_{\Gamma_Q}(u,t), I_{\Gamma_Q}(u,t), 1 - T_{\Gamma_Q}(u,t)]\right\rangle : (u,t) \in U \times Q\right\}.$$

Now,

$$\begin{aligned} \oplus(\oplus(\Gamma_Q, A)) &= \left\{\left\langle e, [(u,t), T_{\Gamma_Q}(u,t), I_{\Gamma_Q}(u,t), 1 - T_{\Gamma_Q}(u,t)]\right\rangle : (u,t) \in U \times Q\right\} \\ &= \oplus(\Gamma_Q, A). \end{aligned}$$

□

Definition 17. *The possibility operation on a Q-NSS on U* (Γ_Q, A) *is denoted by* $\otimes(\Gamma_Q, A)$ *and is defined as, for all* $e \in A$,

$$\otimes(\Gamma_Q, A) = \left\{\left\langle e, [(u,t), 1 - F_{\Gamma_Q}(u,t), I_{\Gamma_Q}(u,t), F_{\Gamma_Q}(u,t)]\right\rangle : (u,t) \in U \times Q\right\}.$$

Example 6. *Reconsider Example 2, then*

$$\begin{aligned} (\Gamma_Q, A) = \Big\{ &\Big\langle e_1, [(u_1,s), 0.5, 0.3, 0.5], [(u_1,t), 0.9, 0.4, 0.1], [(u_2,s), 0.3, 0.4, 0.7], [(u_2,t), 0.2, 0.2, 0.8] \Big\rangle, \\ &\Big\langle e_2, [(u_1,s), 0.1, 0.1, 0.9], [(u_1,t), 0.9, 0.2, 0.1], [(u_2,s), 0.3, 0.3, 0.7], [(u_2,t), 0.8, 0.3, 0.2] \Big\rangle \Big\}. \end{aligned}$$

Proposition 5. *Let* $(\Gamma_Q, A), (\Psi_Q, B)$ *be two Q-NSSs on U. Then,*

1. $\otimes((\Gamma_Q, A) \cup (\Psi_Q, B)) = \otimes(\Gamma_Q, A) \cup \otimes(\Psi_Q, B)$.
2. $\otimes((\Gamma_Q, A) \cap (\Psi_Q, B)) = \otimes(\Gamma_Q, A) \cap \otimes(\Psi_Q, B)$.
3. $\otimes(\otimes(\Gamma_Q, A)) = \otimes(\Gamma_Q, A)$.

Proof. 1. $(\Gamma_Q, A) \cup (\Psi_Q, B) = (\Lambda_Q, C)$, where $C = A \cup B$ and

$$(\Lambda_Q, C) = \left\{\left\langle e, [(u,t), T_{\Lambda_Q}(u,t), I_{\Lambda_Q}(u,t), F_{\Lambda_Q}(u,t)]\right\rangle : (u,t) \in U \times Q\right\},$$

such that

$$T_{\Lambda_Q}(u,t) = \begin{cases} T_{\Gamma_Q}(u,t) & \text{if } e \in A - B, \\ T_{\Psi_Q}(u,t) & \text{if } e \in B - A, \\ \max\{T_{\Gamma_Q}(u,t), T_{\Psi_Q}(u,t)\} & \text{if } e \in A \cap B, \end{cases}$$

$$I_{\Lambda_Q}(u,t) = \begin{cases} I_{\Gamma_Q}(u,t) & \text{if } e \in A - B, \\ I_{\Psi_Q}(u,t) & \text{if } e \in B - A, \\ \min\{I_{\Gamma_{Qi}}(u,t), I_{\Psi_Q}(u,t)\} & \text{if } e \in A \cap B, \end{cases}$$

and

$$F_{\Lambda_Q}(u,t) = \begin{cases} F_{\Gamma_Q}(u,t) & \text{if } e \in A - B, \\ F_{\Psi_Q}(u,t) & \text{if } e \in B - A, \\ \min\{F_{\Gamma_Q}(u,t), F_{\Psi_Q}(u,t)\} & \text{if } e \in A \cap B. \end{cases}$$

Now, by Definition 17, for all $e \in A$

$$\otimes(\Lambda_Q, C) = \left\{\left\langle e, [(u,t), \otimes T_{\Lambda_Q}(u,t), \otimes I_{\Lambda_Q}(u,t), \otimes F_{\Lambda_Q}(u,t)]\right\rangle : (u,t) \in U \times Q\right\}, \text{ where}$$

$$\otimes T_{\Lambda_Q}(u,t) = \begin{cases} 1 - F_{\Gamma_Q}(u,t) & \text{if } e \in A - B, \\ 1 - F_{\Psi_Q}(u,t) & \text{if } e \in B - A, \\ 1 - \min\{F_{\Gamma_Q}(u,t), F_{\Psi_Q}(u,t)\} & \text{if } e \in A \cap B, \end{cases}$$

$$\otimes I_{\Lambda_Q}(u,t) = \begin{cases} I_{\Gamma_Q}(u,t) & \text{if } e \in A - B, \\ I_{\Psi_Q}(u,t) & \text{if } e \in B - A, \\ \min\{I_{\Gamma_Q}(u,t), I_{\Psi_Q}(u,t)\} & \text{if } e \in A \cap B, \end{cases}$$

and

$$\otimes F_{\Lambda_Q}(u,t) = \begin{cases} F_{\Gamma_Q}(u,t) & \text{if } e \in A - B, \\ F_{\Psi_Q}(u,t) & \text{if } e \in B - A, \\ \min\{F_{\Gamma_Q}(u,t), F_{\Psi_Q}(u,t)\} & \text{if } e \in A \cap B. \end{cases}$$

Assume for all $e \in A$

$$\otimes(\Gamma_Q, A) = \left\{\left\langle e, [(u,t), 1 - F_{\Gamma_Q}(u,t), I_{\Gamma_Q}(u,t), F_{\Gamma_Q}(u,t)]\right\rangle : (u,t) \in U \times Q\right\},$$

and

$$\otimes(\Psi_Q, B) = \left\{\left\langle e, [(u,t), 1 - F_{\Psi_Q}(u,t), I_{\Psi_Q}(u,t), F_{\Psi_Q}(u,t)]\right\rangle : (u,t) \in U \times Q\right\}.$$

Now,
$\otimes(\Gamma_Q, A) \cup \otimes(\Psi_Q, B) = (\Upsilon_Q, C)$, where

$$(\Upsilon_Q, C) = \left\{\left\langle e, [(u,t), T_{\Upsilon_Q}(u,t), I_{\Upsilon_Q}(u,t), F_{\Upsilon_Q}(u,t)]\right\rangle : (u,t) \in U \times Q\right\},$$

such that

$$T_{\Upsilon_Q}(u,t) = \begin{cases} 1 - F_{\Gamma_Q}(u,t) & \text{if } e \in A - B, \\ 1 - F_{\Psi_Q}(u,t) & \text{if } e \in B - A, \\ \max\{1 - F_{\Gamma_Q}(u,t), 1 - F_{\Psi_Q}(u,t)\} & \text{if } e \in A \cap B, \end{cases}$$
$$= \begin{cases} 1 - F_{\Gamma_Q}(u,t) & \text{if } e \in A - B, \\ 1 - F_{\Psi_Q}(u,t) & \text{if } e \in B - A, \\ 1 - \min\{F_{\Gamma_Q}(u,t), F_{\Psi_Q}(u,t)\} & \text{if } e \in A \cap B, \end{cases}$$

$$I_{\Upsilon_Q}(u,t) = \begin{cases} I_{\Gamma_Q}(u,t) & \text{if } e \in A - B, \\ I_{\Psi_Q}(u,t) & \text{if } e \in B - A, \\ \min\{I_{\Gamma_Q}(u,t), I_{\Psi_Q}(u,t)\} & \text{if } e \in A \cap B, \end{cases}$$

and

$$F_{\Upsilon_Q}(u,t) = \begin{cases} F_{\Gamma_Q}(u,t) & \text{if } e \in A - B, \\ F_{\Psi_Q}(u,t) & \text{if } e \in B - A, \\ \min\{F_{\Gamma_Q}(u,t), F_{\Psi_Q}(u,t)\} & \text{if } e \in A \cap B. \end{cases}$$

Consequently, $\otimes(\Lambda_Q, C)$ and (Υ_Q, C) are the same. Thus, $\otimes((\Gamma_Q, A) \cup (\Psi_Q, B)) = \otimes(\Gamma_Q, A) \cup \otimes(\Psi_Q, B)$.
2. Can be analogously proven.
3. Assume for all $e \in A$ $\otimes(\Gamma_Q, A) = \left\{\left\langle e, [(u,t), T_{\Gamma_Q}(u,t), I_{\Gamma_Q}(u,t), 1 - T_{\Gamma_Q}(u,t)]\right\rangle : (u,t) \in U \times Q\right\}$.
Now,

$$\begin{aligned} \otimes(\otimes(\Gamma_Q, A)) &= \left\{\left\langle e, [(u,t), T_{\Gamma_Q}(u,t), I_{\Gamma_Q}(u,t), 1 - T_{\Gamma_Q}(u,t)]\right\rangle : (u,t) \in U \times Q\right\} \\ &= \otimes(\Gamma_Q, A). \end{aligned}$$

□

Proposition 6. *Let (Γ_Q, A) be a Q-NSS over U and Q; we have the following properties:*

1. $\otimes \oplus (\Gamma_Q, A) = \oplus(\Gamma_Q, A)$.
2. $\oplus \otimes (\Gamma_Q, A) = \otimes(\Gamma_Q, A)$.

Proof. 1. Suppose that, for any $e \in A$,

$$(\Gamma_Q, A) = \left\{\left\langle e, [(u,t), T_{\Gamma_Q}(u,t), I_{\Gamma_Q}(u,t), F_{\Gamma_Q}(u,t)]\right\rangle : (u,t) \in U \times Q\right\}. \text{ Then,}$$

$$\oplus(\Gamma_Q, A) = \left\{\left\langle e, [(u,t), T_{\Gamma_Q}(u,t), I_{\Gamma_Q}(u,t), 1 - T_{\Gamma_Q}(u,t)]\right\rangle : (u,t) \in U \times Q\right\} \text{ and}$$

$$\otimes(\Gamma_Q, A) = \left\{\left\langle e, [(u,t), 1 - F_{\Gamma_Q}(u,t), I_{\Gamma_Q}(u,t), F_{\Gamma_Q}(u,t)]\right\rangle : (u,t) \in U \times Q\right\}.$$

Thus, $\otimes \oplus (\Gamma_Q, A) = \left\{\left\langle e, [(u,t), T_{\Gamma_Q}(u,t), I_{\Gamma_Q}(u,t), 1 - T_{\Gamma_Q}(u,t)]\right\rangle : (u,t) \in U \times Q\right\} = \oplus(\Gamma_Q, A)$.

2. The proof is similar to that of Assertion 1. □

Proposition 7. *Let* (Γ_Q, A) *and* (Ψ_Q, B) *be two Q-NSSs over* U *and* Q*, we have the following:*

1. $\oplus((\Gamma_Q, A) \wedge (\Psi_Q, B)) = \oplus(\Gamma_Q, A) \wedge \oplus(\Psi_Q, B)$.
2. $\oplus((\Gamma_Q, A) \vee (\Psi_Q, B)) = \oplus(\Gamma_Q, A) \vee \oplus(\Psi_Q, B)$.
3. $\otimes((\Gamma_Q, A) \wedge (\Psi_Q, B)) = \otimes(\Gamma_Q, A) \wedge \otimes(\Psi_Q, B)$.
4. $\otimes((\Gamma_Q, A) \vee (\Psi_Q, B)) = \otimes(\Gamma_Q, A) \vee \otimes(\Psi_Q, B)$.

Proof. 1. Assume $(\Gamma_Q, A) \wedge (\Psi_Q, B) = (\Lambda_Q, A \times B)$, where for all $e_a \in A, e_b \in B$,

$$\Lambda_Q(e_a, e_b) = \Big\{\langle(e_a, e_b),[(u,t), \min\{T_{\Gamma_Q}(u,t), T_{\Psi_Q}(u,t)\}, \max\{I_{\Gamma_Q}(u,t), I_{\Psi_Q}(u,t)\}, \max\{F_{\Gamma_Q}(u,t), F_{\Psi_Q}(u,t)\}]\Big\rangle : (u,t) \in U \times Q\Big\}.$$

By Definition 16, and for all $e_a \in A, e_b \in B$, we have

$$\oplus((\Gamma_Q, A) \wedge (\Psi_Q, B)) = \Big\{\Big\langle(e_a, e_b),[(u,t), \min\{T_{\Gamma_Q}(u,t), T_{\Psi_Q}(u,t)\}, \max\{I_{\Gamma_Q}(u,t), I_{\Psi_Q}(u,t)\}, 1 - \min\{F_{\Gamma_Q}(u,t), F_{\Psi_Q}(u,t)\}]\Big\rangle : (u,t) \in U \times Q\Big\}.$$

Since $\oplus(\Gamma_Q, A) = \Big\{\Big\langle e_a, [(u,t), T_{\Gamma_Q}(u,t), I_{\Gamma_Q}(u,t), 1 - T_{\Gamma_Q}(u,t)]\Big\rangle : e_a \in A, (u,t) \in U \times Q\Big\}$, and $\oplus(\Psi_Q, B) = \Big\{\Big\langle e_b, [(u,t), T_{\Psi_Q}(u,t), I_{\Psi_Q}(u,t), 1 - T_{\Psi_Q}(u,t)]\Big\rangle : e_b \in B, (u,t) \in U \times Q\Big\}$. Then, for all $e_a \in A, e_b \in B$ we have

$$\begin{aligned}
&\oplus(\Gamma_Q, A) \wedge \oplus(\Psi_Q, B) \\
&= \Big\{\Big\langle(e_a, e_b), [(u,t), \min\{T_{\Gamma_Q}(u,t), T_{\Psi_Q}(u,t)\}, \max\{I_{\Gamma_Q}(u,t), I_{\Psi_Q}(u,t)\}, \\
&\qquad \max\{1 - T_{\Gamma_Q}(u,t), 1 - T_{\Psi_Q}(u,t)\}]\Big\rangle : (u,t) \in U \times Q\Big\}. \\
&= \Big\{\Big\langle(e_a, e_b), [(u,t), \min\{T_{\Gamma_Q}(u,t), T_{\Psi_Q}(u,t)\}, \max\{I_{\Gamma_Q}(u,t), I_{\Psi_Q}(u,t)\}, \\
&\qquad 1 - \min\{T_{\Gamma_Q}(u,t), T_{\Psi_Q}(u,t)\}]\Big\rangle : (u,t) \in U \times Q\Big\}. \\
&= \oplus((\Gamma_Q, A) \wedge (\Psi_Q, B))
\end{aligned}$$

2. The proof is similar to that of Assertion 1.

3. Since for all $e_a \in A, e_b \in B$

$$(\Gamma_Q, A) \wedge (\Psi_Q, B) = \Big\{\Big\langle(e_a, e_b), [(u,t), \min\{T_{\Gamma_Q}(u,t), T_{\Psi_Q}(u,t)\}, \max\{I_{\Gamma_Q}(u,t), I_{\Psi_Q}(u,t)\}, \max\{F_{\Gamma_Q}(u,t), F_{\Psi_Q}(u,t)\}]\Big\rangle : (u,t) \in U \times Q\Big\}.$$

By Definition 17,

$$(\Gamma_Q, A) \wedge (\Psi_Q, B) = \Big\{\Big\langle(e_a, e_b), [(u,t), 1 - \max\{F_{\Gamma_Q}(u,t), F_{\Psi_Q}(u,t)\}, \max\{I_{\Gamma_Q}(u,t), I_{\Psi_Q}(u,t)\}, \max\{F_{\Gamma_Q}(u,t), F_{\Psi_Q}(u,t)\}]\Big\rangle : (u,t) \in U \times Q\Big\}.$$

Since, $\otimes(\Gamma_Q, A) = \left\{\left\langle e_a, [(u,t), 1 - F_{\Gamma_Q}(u,t), I_{\Gamma_Q}(u,t), F_{\Gamma_Q}(u,t)]\right\rangle : e_a \in A, (u,t) \in U \times Q\right\}$, and $\otimes(\Psi_Q, B) = \left\{\left\langle e_b, [(u,t), 1 - F_{\Psi_Q}(u,t), I_{\Psi_Q}(u,t), F_{\Psi_Q}(u,t)]\right\rangle : e_b \in B, (u,t) \in U \times Q\right\}$. Then, for all $e_a \in A, e_b \in B$ we have

$$\begin{aligned}
\otimes(\Gamma_Q, A) \wedge \otimes(\Psi_Q, B) &= \Big\{\Big\langle (e_a, e_b), [(u,t), \min\{1 - F_{\Gamma_Q}(u,t), 1 - F_{\Psi_Q}(u,t)\}, \max\{I_{\Gamma_Q}(u,t), I_{\Psi_Q}(u,t)\}, \\
&\qquad \max\{F_{\Gamma_Q}(u,t), F_{\Psi_Q}(u,t)\}]\Big\rangle : (u,t) \in U \times Q\Big\} \\
&= \Big\{\Big\langle (e_a, e_b), [(u,t), 1 - \max\{F_{\Gamma_Q}(u,t), F_{B_Q}(u,t)\}, \max\{I_{\Gamma_Q}(u,t), I_{\Psi_Q}(u,t)\}, \\
&\qquad \max\{F_{\Gamma_Q}(u,t), F_{\Psi_Q}(u,t)\}]\Big\rangle : (u,t) \in U \times Q\Big\} \\
&= \otimes((\Gamma_Q, A) \wedge (\Psi_Q, B)).
\end{aligned}$$

4. The proof is similar to that of Assertion 3. □

5. An Application of Q-Neutrosophic Soft Sets

In this section, we present a Q-NS aggregation operator of Q-NSS that produces a Q-NS from a Q-NSS and then reduces it to a Q-fuzzy set in order to use it in a decision-making problem.

Definition 18. *Let (Γ_Q, A) be Q-NSS over U. Then, a Q-NS aggregation operator of (Γ_Q, A), denoted by Γ_Q^{agg}, is defined by $\Gamma_Q^{agg} = \{\langle (u,t), T_Q^{agg}(u,t), I_Q^{agg}(u,t), F_Q^{agg}(u,t)\rangle : (u,t) \in U \times Q\}$, which is a Q-NS over U, where $T_Q^{agg}, I_Q^{agg}, F_Q^{agg} : U \times Q \to [0,1]$*

$$\begin{aligned}
T_Q^{agg} &= \frac{1}{|A|} \sum_{(u,t)\in U\times Q} T_{\Gamma_Q}(u,t), \\
I_Q^{agg} &= \frac{1}{|A|} \sum_{(u,t)\in U\times Q} I_{\Gamma_Q}(u,t) \text{ and} \\
F_Q^{agg} &= \frac{1}{|A|} \sum_{(u,t)\in U\times Q} F_{\Gamma_Q}(u,t).
\end{aligned}$$

Definition 19. *The reduced Q-fuzzy set of a Q-NS Γ_Q is*

$$\widehat{\Gamma}_Q = \{(u,t), \mu_{\Gamma_Q}(u,t) : (u,t) \in U \times Q\},$$

where $\mu_{\Gamma_Q} : U \times Q \to [0,1]$ and given by $\mu_{\Gamma_Q} = \frac{1}{3}[T_Q(u,t) + 2 - I_Q(u,t) - F_Q(u,t)]$.

Now, using the definitions of a Q-NS aggregation operator and a reduced Q- fuzzy set, we construct the following algorithm for a decision method:

Step 1 Construct a Q-NSS over U.
Step 2 Compute the Q-NS aggregation operator.
Step 3 Compute the reduced Q-fuzzy set of the Q-NS aggregation operator.
Step 4 The decision is any element in M, where $M = \max_{(u,t)\in U\times Q}\{\mu_{\Gamma_Q}^{agg}\}$.

Now, we provide an example for Q-NSS decision-making method.

Example 7. *Suppose a university needs to fill a position in the mathematics department to be selected by expert committee. There are three candidates, $U = \{u_1, u_2, u_3\}$, with two types of scientific degree, $Q = \{s = \text{assistant professor}, t = \text{associate professor}\}$, and the hiring committee considers a set of parameters $E = \{e_1, e_2, e_3\}$ representing experience, language fluency, and computer knowledge, respectively.*

Now, we can apply the method to help the committee fill the position with the suitable candidate as follows:

Step 1 *The committee construct the following Q-NSS:*

$$(\Gamma_Q, A) = \Big\{\Big\langle e_1,[(u_1,s),0.2,0.3,0.5],[(u_1,t),0.6,0.4,0.1],[(u_2,s),0.1,0.4,0.7],[(u_3,t),0.3,0.2,0.8]\Big\rangle, \Big\langle e_2,[(u_1,s),0.4,0.1,0.9],[(u_2,s),0.6,0.2,0.1],[(u_3,s),0.5,0.3,0.7],[(u_3,t),0.1,0.3,0.2]\Big\rangle, \Big\langle e_3,((u_1,t),0.2,0.3,0.6),((u_2,t),0.9,0.1,0.2)\Big\rangle\Big\}.$$

Step 2 *The Q-NS aggregation operator is*

$$\Gamma_Q^{agg} = \Big\{[(u_1,s),0.2,0.133,0.467],[(u_1,t),0.267,0.233,0.233],[(u_2,s),0.233,0.2,0.267], [(u_2,t),0.3,0.033,0.067],[(u_3,s),0.166,0.1,0.233],[(u_3,t),0.133,0.167,0.333]\Big\}.$$

Step 3 *The reduced Q-fuzzy set of the* Γ_Q^{agg} *is*

$$\widehat{\Gamma}_Q = \Big\{[(u_1,s),0.533],[(u_1,t),0.6],[(u_2,s),0.589],[(u_2,t),0.733],[(u_3,s),0.611],[(u_3,t),0.544]\Big\}.$$

Step 4 *The largest membership grade is* $\mu_{\Gamma_Q}^{agg}(u_2,t) = 0.733$ *which implies that the committee is inclined to choose associate-professor candidate* u_2 *with a scientific degree for the job.*

6. Comparative Analysis

In this section, we compare the concept of the Q-neutrosophic soft method to the neutrosophic soft method [16], Q-FSS [38], and Q-IFSS [39].

In contrast to the neutrosophic soft method that uses the NSS to characterize decision-making data, the novel Q-neutrosophic soft method introduces a new descriptor, that is, Q-NSS, to provide actual decision-making information. From Example 7, it can be seen that the NSS was unable to represent variables in two dimensions. However, the framework of the Q-NSS offers the capacity to simultaneously handle these two dimensions.

On the other hand, Q-NSS is identified through three independent degrees of membership, namely, truth, indeterminacy, and falsity. Hence, it is more accurate than Q-FSS, which is identified by one truth value, and Q-IFSS, which is identified by two dependent memberships for truth and falsity. Thus, the proposed method has certain advantages, that is, this method uses the Q-NSS to represent decision information as an extension of Q-FSS and Q-IFSS. An effective aggregation formula is employed to convert the Q-NSS to Q-NS, which preserves the entirety of the original data without reducing or distorting them. Our method also provides decision making with a simple computational process.

As a result, the Q-NSS has the ability to deal with indeterminate and inconsistent data in two-dimensional sets. Consequently, it is capable to interact with deeper imprecise data. The basic characteristics of Q-NSS were compared to those of NSS, Q-FSS, and Q-IFSS, as shown in Table 1.

Table 1. Characteristic comparison of the Q-neutrosophic soft set (Q-NSS) with other variants.

Method	NSS	Q-FSS	Q-IFSS	Q-NSS
Authors	Maji [16]	Adam and Hassan [38]	Broumi [39]	Proposed Method
Domain	Universe of discourse	Universe of discourse	Universe of discourse	Universe of discourse
Codomain	$[0,1]^3$	$[0,1]$	$[0,1]^2$	$[0,1]^3$
Q	No	Yes	Yes	Yes
True	Yes	Yes	Yes	Yes
Indeterminacy	Yes	No	No	Yes
Falsity	Yes	No	Yes	Yes

7. Conclusions

Q-NSS is an NSS over a two-dimensional universal set. Thus, a Q-NSS is a tricomponent set that can simultaneously handle two-dimensional and indeterminate data. This study discussed some operations of Q-NSSs, namely, subset, equality, complement, intersection, union, AND operation, and OR operation. It discussed the necessity and possibility operations along with some properties and illustrative examples. Finally, the Q-NS aggregation operator was defined and applied to develop an algorithm for using Q-NSS in decision-making problems that involve uncertainty. This new model provides an important extension to existing studies that can handle indeterminacy, where two-dimensionality appears in the decision process, thus offering the opportunity for further relevant research. Q-NSS encourages the path to various scopes for future research since it deals with indeterminacy and two-dimensionality at the same time. Hence, it can be expanded by utilizing the n-valued refined neutrosophic set [44], possibility neutrosophic set [45], and numerous different structures. Moreover, different algebraic structures, for instance, the field, ring, and group of the Q-NSS and its extensions may be investigated.

Author Contributions: M.A.Q. and N.H. contributed equally to achieve this work.

Funding: Universiti Kebangsaan Malaysia GUP-2017-105.

Acknowledgments: We are indebted to Universiti Kebangsaan Malaysia for providing financial support and facilities for this research under grant GUP-2017-105.

Conflicts of Interest: The authors declare no conflict of interest.

References

1. Zadeh, L.A. Fuzzy sets. *Inf. Control* **1965**, *8*, 338–353. [CrossRef]
2. Turksen, I.B. Interval valued fuzzy sets based on normal forms. *Fuzzy Sets Syst.* **1986**, *20*, 191–210. [CrossRef]
3. Atanassov, K.T. Intuitionistic fuzzy sets. *Fuzzy Sets Syst.* **1986**, *20*, 87–96. [CrossRef]
4. Smarandache, F. *Neutrosophy. Neutrosophic Probability, Set and Logic*; American Research Press: Rehoboth, IL, USA, 1998.
5. Smarandache, F. Neutrosophic set, a generalisation of the intuitionistic fuzzy sets. *Int. J. Pure Appl. Math.* **2005**, *24*, 287–297.
6. Wang, H.; Smarandache, F.; Zhang, Y.; Sunderraman, R. Single valued neutrosophic sets. *Mult. Multistruct.* **2010**, *4*, 410–413.
7. Ali, M.; Smarandache, F. Complex neutrosophic set. *Neural Comput. Appl.* **2017**, *28*, 1817–1834. [CrossRef]
8. Molodtsov, D. Soft set theory-first results. *Comput. Math. Appl.* **1999**, *37*, 19–31. [CrossRef]
9. Maji, P.K.; Biswas, R.; Roy, A.R. Fuzzy soft set theory. *J. Fuzzy Math.* **2001**, *9*, 589–602.
10. Xu, W.; Ma, J.; Wang, S.; Hao, G. Vague soft sets and their properties. *Comput. Math. Appl.* **2010**, *59*, 787–794. [CrossRef]
11. Alhazaymeh, K.; Hassan, N. Generalized interval-valued vague soft set. *Appl. Math. Sci.* **2013**, *7*, 6983–6988. [CrossRef]
12. Alhazaymeh, K.; Hassan, N. Interval-valued vague soft sets and its application. *Adv. Fuzzy Syst.* **2012**, *2012*, 208489. [CrossRef]
13. Alhazaymeh, K.; Hassan, N. Possibility interval-valued vague soft set. *Appl. Math. Sci.* **2013**, *7*, 6989–6994. [CrossRef]
14. Alkhazaleh, S.; Salleh, A.R. Soft expert sets. *Adv. Decis. Sci.* **2011**, *2011*, 757868. [CrossRef]
15. Alkhazaleh, S.; Salleh, A.R.; Hassan, N. Soft multi sets theory. *Appl. Math. Sci.* **2011**, *5*, 3561–3573.
16. Maji, P.K. Neutrosophic soft set. *Ann. Fuzzy Math. Inform.* **2013**, *5*, 157–168.
17. Broumi, S. Generalized neutrosophic soft set. *Int. J. Comput. Sci. Eng. Inf. Technol.* **2013**, *3*, 17–30. [CrossRef]
18. Broumi, S.; Smarandache, F. Intuitionistic neutrosophic soft set. *J. Inf. Comput. Sci.* **2013**, *8*, 130–140.
19. Deli, I. Interval-valued neutrosophic soft sets and its decision making. *Int. J. Mach. Learn. Cybern.* **2017**, *8*, 665–676. [CrossRef]
20. Deli, I; Broumi, S. Neutrosophic soft matrices and NSM decision making. *J. Intell. Fuzzy Syst.* **2015**, *28*, 2233–2241. [CrossRef]

21. Karaaslan, F. Neutrosophic soft sets with applications in decision making problem. *Int. J. Inf. Sci. Intell. Syst.* **2015**, *4*, 1–20.
22. Ye, J.; Cui, W. Exponential entropy for simplified neutrosophic sets and its application in decision making. *Entropy* **2018**, *20*, 357. [CrossRef]
23. Ye, J. Exponential operations and aggregation operators of interval neutrosophic sets and their decision making methods. *SpringerPlus* **2016**, *5*, 1488. [CrossRef]
24. Hassan, N.; Uluçay, V.; Şahin, M. Q-neutrosophic soft expert set and its application in decision making. *Int. J. Fuzzy Syst. Appl.* **2018**, *7*, 37–61. [CrossRef]
25. Al-Quran, A.; Hassan, N. Neutrosophic vague soft set and its applications. *Malays. J. Math. Sci.* **2017** *11*, 141–163.
26. Al-Quran, A.; Hassan, N. Neutrosophic vague soft expert set theory. *J. Intell. Fuzzy Syst.* **2016**, *30*, 3691–3702. [CrossRef]
27. Alkhazaleh, S. n-Valued refined neutrosophic soft set theory. *J. Intell. Fuzzy Syst.* **2017** *32*, 4311–4318. [CrossRef]
28. Al-Quran, A.; Hassan, N. The complex neutrosophic soft expert set and its application in decision making. *J. Intell. Fuzzy Syst.* **2018**, *34*, 569–582. [CrossRef]
29. Al-Quran, A.; Hassan, N. The complex neutrosophic soft expert relation and its multiple attribute decision-making method. *Entropy* **2018**, *20*, 101. [CrossRef]
30. Alkhazaleh, S. Time-neutrosophic soft set and its application. *J. Intell. Fuzzy Syst.* **2016**, *30*, 1087–1098. [CrossRef]
31. Wu, X.H.; Wang, J.Q.; Peng, J.J.; Chen, X.H. Cross-entropy and prioritized aggregation operator with simplified neutrosophic sets and their application in multi-criteria decision-making problems. *J. Intell. Fuzzy Syst.* **2016**, *18*, 1104–1116. [CrossRef]
32. Chen, J.Q.; Ye, J. Some single-valued neutrosophic Dombi weighted aggregation operators for multiple attribute decision-making. *Symmetry* **2017**, *9*, 82. [CrossRef]
33. Peng, J.J.; Wang, J.Q.; Wu, X.H.; Wang, J.; Chen, X.H. Multi-valued neutrosophic sets and power aggregation operators with their applications in multi-criteria group decision-making problems. *Int. J. Comput. Intell. Syst.* **2015**, *8*, 345–363. [CrossRef]
34. Ye, J. A multicriteria decision-making method using aggregation operators for simplified neutrosophic sets. *Int. J. Fuzzy Syst.* **2014**, *26*, 2459–2466.
35. Li, X.; Zhang, X. Single-valued neutrosophic hesitant fuzzy Choquet aggregation operators for multi-attribute decision making. *Symmetry* **2018**, *10*, 50. [CrossRef]
36. Lu, Z.; Ye, J. Single-valued neutrosophic hybrid arithmetic and geometric aggregation operators and their decision-making method. *Information* **2017**, *8*, 84. [CrossRef]
37. Adam, F.; Hassan, N. Multi Q-fuzzy soft set and its application. *Far East J. Math. Sci.* **2015**, *97*, 871–881. [CrossRef]
38. Adam, F.; Hassan, N. Operations on Q-fuzzy soft set. *Appl. Math. Sci.* **2014**, *8*, 8697–8701. [CrossRef]
39. Broumi, S. Q-intuitionistic fuzzy soft sets. *J. New Theory* **2015**, *5*, 80–91.
40. Abu Qamar, M.; Hassan, N. Q-Neutrosophic soft relation and its application in decision making. *Entropy* **2018**, *20*, 172. [CrossRef]
41. Abu Qamar, M.; Hassan, N. Entropy, measures of distance and similarity of Q-neutrosophic soft sets and some applications. *Entropy* **2018**, *20*, 672. [CrossRef]
42. Abu Qamar, M.; Hassan, N. Generalized Q-neutrosophic soft expert set for decision under uncertainty. *Symmetry* **2018**, *10*, 621. [CrossRef]
43. Smarandache, F. *A Unifying Field in Logics. Neutrosophy: Neutrosophic Probability, Set and Logic*; American Research Press: Rehoboth, MA, USA, 1999.

44. Smarandache, F. n-Valued Refined Neutrosophic Logic and Its Applications in Physics. *Prog. Phys.* **2013**, *4*, 143–146.
45. Karaaslan, F. Possibility neutrosophic soft sets and PNS-decision making method. *Appl. Soft Comput.* **2017**, *54*, 403–414. [CrossRef]

symmetry

MDPI

Article

A Variable Acceptance Sampling Plan under Neutrosophic Statistical Interval Method

Muhammad Aslam

Department of Statistics, Faculty of Science, King Abdulaziz University, Jeddah 21551, Saudi Arabia; magmuhammad@kau.edu.sa or aslam_ravian@hotmail.com

Received: 3 January 2019; Accepted: 15 January 2019; Published: 19 January 2019

Abstract: The acceptance sampling plan plays an important role in maintaining the high quality of a product. The variable control chart, using classical statistics, helps in making acceptance or rejection decisions about the submitted lot of the product. Furthermore, the sampling plan, using classical statistics, assumes the complete or determinate information available about a lot of product. However, in some situations, data may be ambiguous, vague, imprecise, and incomplete or indeterminate. In this case, the use of neutrosophic statistics can be applied to guide the experimenters. In this paper, we originally proposed a new variable sampling plan using the neutrosophic interval statistical method. The neutrosophic operating characteristic (NOC) is derived using the neutrosophic normal distribution. The optimization solution is also presented for the proposed plan under the neutrosophic interval method. The effectiveness of the proposed plan is compared with the plan under classical statistics. The tables are presented for practical use and a real example is given to explain the neutrosophic fuzzy variable sampling plan in the industry.

Keywords: optimization solution; sampling plan; producer's risk'; consumer's risk; sample size

1. Introduction

In this modern era, there is a strict competition between the companies to earn good reputation in the market. So, quality is considered as a benchmark for the well-reputed company. The good quality of the product means a good reputation of the company in the market. To maintain the high quality of the product, the inspection of the product from the raw material to the finished product should be done. Inspection of a finished lot of the products should be done before sending them to the market. Therefore, the inspection of the finished product is aimed at the high quality of the product. At the time of inspection, it may not possible to inspect 100% of the items and the entire submitted lot of product. Therefore, inspection of a lot of product is done using the acceptance sampling plans. A well-designed sampling plan reduces the cost and time of the inspection. A sampling plan also pressures the producer to increase the quality of the product. As the decision about the submitted lot of product is taken based on the sample information, there is a chance of committing errors. The chance of rejecting a lot meets the given specification is called the producer's risk, and accepting a bad lot is known as the consumer's risk. Therefore, sampling plans are designed to give those parameters for the inspection of a lot of product, where these two risks are satisfied. More details about the sampling plans can be seen in [1,2].

The variable sampling plan is applied when the data is continuous, such as the diameter of the ball bearing. Several variable acceptance sampling plans are available in the literature using classical statistics [1,3–5]. The variable sampling plans designed under classical statistics can only be applied when there is no certainty in the observations. According to [6] "observations include human judgments, and evaluations and decisions, a continuous random variable of a production process should include the variability caused by human subjectivity or measurement devices, or environmental conditions. These variability causes create vagueness in the measurement system". In this situation, the sampling plans designed use the fuzzy logic. Several authors presented excellent work to design the sampling plan under the fuzzy environment. Kanagawa and Ohta [7] designed the fuzzy attribute sampling plan. Jamkhaneh et al. [8] studied the rectifying fuzzy single sampling plan. Sadeghpour et al. [9] presented the plan with fuzzy parameters. Jamkhaneh et al. [10] discussed the effect of inspection errors on the single fuzzy plan. Tong and Wang [11] proposed the fuzzy sampling plan for the geospatial data. Turanoğlu et al. [12] presented the characteristic curve for the fuzzy plan. Uma and Ramya [13] presented a review of the fuzzy sampling plans. Kahraman et al. [14] worked on the single and double sampling using the fuzzy approach, and Afshari and Gildeh [15] designed a fuzzy multiple dependent state sampling plan.

According to [16], the neutrosophic logic is the generalization of classical fuzzy logic. The neutrosophic statistics developed by [17] is the generalization of the classical statistics. The neutrosophic statistics can be applied under the uncertainty environment. References [18,19] introduced the neutrosophic interval method in rock measurement. Recently, references [20–25] introduced the neutrosophic statistics in the area of the acceptance sampling plan. Aslam [26] proposed a sampling plan for the exponential distribution using the neutrosophic statistics.

Although a rich variety of variable sampling plans under the fuzzy approach and classical statistics is available in the literature, according to the best of the author's knowledge, there is no work on the design of a variable sampling plan using the neutrosophic interval method. In this paper, we originally proposed a new variable sampling plan using the neutrosophic interval statistical method. The neutrosophic operating characteristic (NOC) is derived using the neutrosophic normal distribution. The optimization solution is also presented for the proposed plan under the neutrosophic statistical interval method. The effectiveness of the proposed plan is compared with the plan using classical statistics. The tables presented for practical use and a real example is given to explain the neutrosophic fuzzy variable sampling plan in the industry. A brief introduction to the neutrosophic approach is given in Section 2. The design of the proposed plan is given in Section 3. The advantages of the proposed plan are discussed in Section 4. An example is given in Section 5, and some concluding remarks are given in the last section.

2. Neutrosophic Approach

According to [16], the neutrosophic logic is an extension of fuzzy logic. The neutrosophic logic considers the measures of truth, false, and indeterminacy. The neutrosophic statistics using the neutrosophic logic is introduced by [17]. Classical statistics is the special case of the neutrosophic statistics. The latter one is applied when the sample is selected from the population having uncertain observations. According to [17] "neutrosophic statistics is an extension of the classical statistics. In the neutrosophic statistics, the data may be ambiguous, vague, imprecise, incomplete, even unknown. Instead of crisp numbers used in classical statistics, one uses sets in neutrosophic statistics". Suppose that $X_N \in \{X_L, X_U\}$ denotes the neutrosophic random variable, where X_L and X_U are lower and upper values of indeterminacy interval, respectively. Let $n_N \in \{n_L, n_U\}$ represent the neutrosophic sample size selected from the population having indeterminate observations. Let $\mu_N \in \{\mu_L, \mu_U\}$ and $\sigma_N \in \{\sigma_L, \sigma_U\}$ be the corresponding neutrosophic population mean and variance, respectively. Suppose that $s_N \in \{s_L, s_U\}$ $\overline{X}_N \in \{\overline{X}_L, \overline{X}_U\}$ represent the neutrosophic sample mean and variance, respectively.

3. Design of the Proposed Plan Neutrosophic Interval Method

Based on the above information, in this section, we present the design of the proposed sampling under the neutrosophic environment. The operational procedure of the proposed plan is given as:

Step-1: Select a random sample of size $n_L \le n_N \le n_U$; $n_N \in \{n_L, n_U\}$ from the lot of product. Compute the statistic $v_N = \frac{U-\overline{X}_N}{s_N}$, where $\overline{X}_N \in \{\overline{X}_L, \overline{X}_U\}$; $\overline{X}_L = \sum_{i=1}^{n} x_i^L / n_L$, $\overline{X}_U = \sum_{i=1}^{n} x_i^U / n_U$, and $s_N \in \{s_L, s_U\}$, where $s_L = \sqrt{\sum_{i=1}^{n}\left(x_i^L - \overline{X}_L\right)^2/n_L}$ and $s_U = \sqrt{\sum_{i=1}^{n}\left(x_i^U - \overline{X}_U\right)^2/n_U}$; i =1,2,3, …, n.

Step-2: Accept the lot of product of $v \geq k_{Na}$; $k_N \in \{k_{aL}, k_{aU}\}$ where k_{Na} is the neutrosophic acceptance number.

The proposed plan is applied to test the hypothesis that the product is good versus the alternative hypothesis that the product is bad, on the basis of sample information. The null hypothesis is accepted if $v \geq k_N$, otherwise, the alternative hypothesis is accepted. The proposed plan has two parameters, $n_N \in \{n_L, n_U\}$ and $k_{Na} \in \{k_{aL}, k_{aU}\}$. The neutrosophic normal distribution, with mean $\mu_N \in \{\mu_L, \mu_U\}$ and standard deviation $\sigma_N \in \{\sigma_L, \sigma_U\}$, is defined by

$$X_N \sim N_N(\mu_N, \sigma_N) = \frac{1}{\sigma_N\sqrt{2\pi}}\exp\left(-\frac{(x-\mu_N)^2}{2\sigma_N^2}\right) \tag{1}$$

where $N_N(\mu_N, \sigma_N)$ denotes the neutrosophic normal distribution. The neutrosophic operating characteristic (NOC) of the proposed sampling plan is derived as follows:

Following [27], $\overline{X}_N \pm k_{Na}s_N$; $\overline{X}_N \in \{\overline{X}_L, \overline{X}_U\}$ and $s_N \in \{s_L, s_U\}$ the approximate neutrosophic normal distribution with mean $\mu_N \pm c\sigma_N$; $\mu_N \in \{\mu_L, \mu_U\}$; $\sigma_N \in \{\sigma_L, \sigma_U\}$ and $\frac{\sigma_N^2}{n_N} + \frac{c^2\sigma_N^2}{2n_N}$; $\mu_N \in \{\mu_L, \mu_U\}$; $\sigma_N \in \{\sigma_L, \sigma_U\}$ and $n_N \in \{n_L, n_U\}$. The lot acceptance probability is given by

$$L(p) = P(v_N \geq k_{Na}) = P\{\overline{X}_N + k_{Na}s_N \leq U\};\ \overline{X}_N \in \{\overline{X}_L, \overline{X}_U\} \tag{2}$$

Therefore, the lot acceptance probability is given by

$$L(p) = \Phi\left(\frac{U - \mu_N - k_{Na}s_N}{\left(\frac{\sigma_N}{\sqrt{n_N}}\right)\sqrt{1+\frac{k_{Na}^2}{2}}}\right); \overline{X}_N \in \{\overline{X}_L, \overline{X}_U\}, n_N \in \{n_L, n_U\},\ k_{Na} \in \{k_{aL}, k_{aU}\} \text{ and } s_N \in \{s_L, s_U\} \tag{3}$$

Suppose p_U is the probability that defective items beyond U, so $p_U = P(X_N > U|\mu_N)$; $X_N \in \{x_i^L, x_i^U\}$ and $\mu_N \in \{\mu_L, \mu_U\}$, where $Z_{Np_U} = \frac{U-\mu_N}{\sigma_N}$; $\mu_N \in \{\mu_L, \mu_U\}$ and Z_{Np_U} is the neutrosophic standard normal distribution. After some simplification, the NFOC is given by

$$L_N(p) = \Phi\left((Z_{Np_U} - k_{Na})\sqrt{\frac{n_N}{1+(k_{Na}^2/2)}}\right);\ k_{Na} \in \{k_{aL}, k_{aU}\};\ n_N \in \{n_L, n_U\} \tag{4}$$

where α and β are the producer's risk and consumer's risk, respectably. The plan parameters $k_{Na} \in \{k_{aL}, k_{aU}\}$; $n_N \in \{n_L, n_U\}$ of the neutrosophic plan will be determined, such that the lot acceptance probability should be larger than $1-\alpha$ at acceptable quality level (AQL), and p_1 and bad lot acceptance probability should be smaller than β at limiting quality level (LQL), say p_2. The neutrosophic plan parameters of the proposed sampling plans will be determined by the following non-linear optimization problem.

$$\text{minimize } n_N \in \{n_L, n_U\} \tag{5a}$$

subject to

$$L_N(p_1) = \Phi\left((Z_{Np_{U1}} - k_{Na})\sqrt{\frac{n_N}{1+(k_N^2/2)}}\right) \geq 1-\alpha;\ k_{Na} \in \{k_{aL}, k_{aU}\};\ n_N \in \{n_L, n_U\} \tag{5b}$$

and

$$L_N(p_2) = \Phi\left((Z_{Np_{U2}} - k_{Na})\sqrt{\frac{n_N}{1+(k_N^2a/2)}}\right) \leq \beta;\ k_N \in \{k_{aL}, k_{aU}\};\ n_N \in \{n_L, n_U\} \tag{5c}$$

The neutrosophic plan parameters such as $n_N \in \{n_L, n_U\}$, $k_{Na} \in \{k_{aL}, k_{aU}\}$, $L_N(p_1)$, and $L_N(p_2)$ for various values of AQL and LQL are placed in Table 1.

Table 1. The neutrosophic plan parameter when $\alpha = 0.05$, $\beta = 0.05$.

p_1	p_2	n_N	k_{Na}	$L_N(p_1)$	$L_N(p_2)$
0.001	0.002	{388,569}	{1.062,1.068}	{0.0151,0.065}	{0.9501,0.9502}
	0.003	{213,268}	{1.05,1.055}	{0.000,0.000}	{0.9501,0.9502}
	0.004	{180,213}	{1.046,1.05}	{0.000,0.000}	{0.9500,0.9502}
	0.006	{139,150}	{1.039,1.041}	{0.000,0.000}	{0.9504,0.9508}
	0.008	{103,130}	{1.03,1.037}	{0.000,0.000}	{0.9500,0.9506}
	0.010	{78,117}	{1.02,1.034}	{0.000,0.026}	{0.9501,0.9506}
	0.015	{61,100}	{1.01,1.029}	{0.000,0.007}	{0.9500,0.9512}
	0.020	{50,80}	{0.99,1.02}	{0.000,0.010}	{0.9527,0.6983}
0.0025	0.030	{143,233}	{0.696,0.706}	{0.0173,0.0847}	{0.9501,0.9504}
	0.050	{138,213}	{0.695,0.705}	{0.0214,0.0937}	{0.9504,0.9506}
0.005	0.050	{77,83}	{0.708,0.711}	{0.9999,1.000}	{0.0791,0.0990}
	0.100	{15,22}	{0.77,0.795}	{0.0012,0.0118}	{0.9545,0.9577}
0.01	0.020	{184,210}	{0.812,0.814}	{0.0731,0.0985}	{0.9500,0.9536}
	0.030	{95,102}	{0.794,0.796}	{0.0275,0.0358}	{0.9502,0.9508}
0.03	0.060	{127,149}	{0.669,0.673}	{0.0638,0.0992}	{0.9516,0.9520}
	0.090	{112,121}	{0.666,0.668}	{0.0010,0.0018}	{0.9502,0.9507}
0.05	0.100	{125,132}	{0.60,0.614}	{0.0029,0.0551}	{0.9501,0.9503}
	0.150	{38,40}	{0.557,0.559}	{0.0812,0.0921}	{0.9512,0.9519}

From Table 1, we note following trends in neutrosophic plan parameters

1. For the fixed values AQL, $n_N \in \{n_L, n_U\}$ decreases as LQL increases.
2. For the fixed values AQL, $k_{Na} \in \{k_{aL}, k_{aU}\}$ decreases as LQL increases.

Comparative Study

Now we compare the proposed plan with the sampling plan under classical statistics in [19], a method which provides the range of the parameters under the uncertainty is called the most effective and adequate method. We preened the values of both sampling plans for some combinations of AQL and LQL in Table 2. From Table 2, it can be noted that the plan under classical statistics provides the determined value, while the proposed plan provides the plan parameter in the indeterminacy interval. For example, when AQL = 0.001 and LQL = 0.002, the proposed plan has indeterminacy interval $n_N \in \{388{,}569\}$, while the plan under classical statistics has a determined value $n = 388$. Under the uncertainty, when AQL = 0.001 and LQL = 0.002, the suitable sample size should be selected between 388 and 569. Therefore, the sampling plan under the neutrosophic interval method has the advantage over the plan under classical statistics under the uncertainty environment. The proposed plan is more effective, informative, flexible, and adequate to be applied in uncertainty than the plan based on classical statistics.

Table 2. The comparison of neutrosophic plan with plan under classical Statistics, when $\alpha = 0.05$, $\beta = 0.05$.

p_1	p_2	Proposed Plan	Existing Plan
		n_N	n
0.001	0.002	{388,569}	388
0.001	0.010	{78,117}	78
0.001	0.020	{50,80}	50
0.005	0.050	{77,83}	77
0.005	0.100	{15,22}	15
0.01	0.020	{184,210}	184
0.01	0.030	{95,102}	95
0.05	0.100	{125,132}	125

4. Application of the Proposed Plan

In this section, we present the application of the proposed neutrosophic plan using color STN display data collected from the industry of LCD. According to [28], "Color STN displays are created by adding color filters to traditional monochrome. In color STN displays, each pixel is divided into R, G, and B sub-pixels. In this study, the membrane thickness of each pixel is the quality characteristic". The data for this variable of study is obtained from the measurement process. Senturk and Erginel [6] pointed out that the observations obtained from the measurement devices have variability. The present state of this variation makes some observations imprecise. During the sample study, we found that some observations about each pixel are determinate or clear, and some are indeterminate or unclear. For this experiment, the experimenter did not determinate the sample size that should be selected for the inspection of this LCD product. Suppose we fixed $\alpha = 0.05$, $\beta = 0.05$, AQL = 0.001, and LQL = 0.020. From Table 1, it can be noted that the optimal sample size n_N for this case should be $n_N \in \{50, 80\}$. Therefore, we need to collect the data for a sample size having between 50 and 80. Let $n_N = 55$ once $n_N \in \{50, 80\}$. The data of 55 observations, including determinate and indeterminate observations about each pixel, is reported in Table 3.

Table 3. Data on color Super Twisted Nematic (STN) displays.

[11,816.7,11,816.7]	[11,710.1,11,710.1]	[11,722.6,11,823.5]	[11,744.1,11,744.1]	[11,681.1,11,681.1]	[11,728.4,11,728.4]
[11,712.6,11,712.6]	[11,775.2,11,775.2]	[11,743.3,11,743.3]	[11,786.1,11,786.1]	[11,760.6,11,760.6]	[11,723.6,11,723.6]
[11,721.7,11,721.7]	[11,698,11,698]	[11,695.9,11,695.9]	[11,726.4,11,726.4]	[11,797.2,11,797.2]	[11,773.1,11,773.1]
[11,769.1,11,769.1]	[11,800.8,11,800.8]	[11,780.7,11,780.7]	[11,670.9,11,675.9]	[11,692.3,11,692.3]	[11,666.2,11,666.2]
[11,755.2,11,762.5]	[11,712.7,11,712.7]	[11,775.5,11,775.5]	[11,731.2,11,731.2]	[11,625.6,11,625.6]	[11,757.5,11,757.5]
[11,674.7,11,674.7]	[11,729.2,11,729.2]	[11,681.3,11,681.3]	[11,636.4,11,636.4]	[11,682.1,11,690.7]	[11,667.9,11,667.9]
[11,722.9,11,722.9]	[11,655.3,11,655.3]	[11,700.2,11,700.2]	[11,754.2,11,754.2]	[11,769.9,11,769.9]	[11,705.9,11,705.9]
[11,589.8,11,589.8]	[11,738.4,11,745.6]	[11,745.4,11,745.4]	[11,727.7,11,727.7]	[11,664.3,11,664.3]	[11,647.2,11,647.2]
[11,755,11,755]	[11,671.8,11,671.8]	[11,705.8,11,705.8]	[11,664.2,11,664.2]	[11,677.0,11,695.2]	[11,680.5,11,687.4]
[11,633.6,11,633.6]					

As the data given in Table 3 is neutrosophic, therefore, the sampling plan under classical statistics cannot be applied for the inspection of this product. The proposed sampling plan for the neutrosophic data is explained as follows.

By following [17], the sample mean $\overline{X}_N \in \{\overline{X}_L, \overline{X}_U\}$ and $s_N \in \{s_L, s_U\}$ for this data are calculated as follows

$$\overline{X}_N = \frac{\begin{array}{c}[11,816.7, 11,816.7] + [11,710.1, 11,710.1] + [11,722.6, 11,722.6] + \ldots + [11,633.6, 11,633.6], [11,816.7, 11,816.7] + \ldots \\ + [11,687.4, 11687.4] + [11,633.6, 11,633.6]\end{array}}{55}$$

Or

$$\overline{X}_N \in \{11,715.2, 11,719.6\} \text{ and } s_N = \{49.21, 49.70\}$$

Suppose that U = 12,500 for the submitted LCD product. The proposed sampling plan implemented as follows.

Step-1: select a random sample of size $n_N = \{n_L, n_U\}$ from a lot of product. Compute the statistic $v_N = \frac{U - \overline{X}_N}{s_N} = \frac{[12{,}500, 12{,}500] - [11{,}715.2, 11{,}719.6]}{[49.21, 49.70]}$, so $v_N \in [15.70, 15.94]$.

Step-2: Accept a lot of the product of $v_N \geq k_N$; $k_{Na} \in \{k_{aL}, k_{aU}\}$. From Table 1, we have $k_{Na} \in \{0.99, 1.02\}$. So, $v_N \geq k_{Na}$, the lot of product, should be accepted to send to the market.

From this real example, it is concluded that the proposed sampling is quite reasonable and adequate to apply when observations are imprecise.

5. Concluding Remarks

A new variable neutrosophic variable sampling plan is proposed in this paper. The proposed plan is the extension of the variable sampling plan based on classical statistics. The proposed plan can be applied in those situations where data is incomplete or indeterminate came from the complex process. The non-linear optimization problem was developed under the neutrosophic approach, and some results are presented for the practical use of the proposed plan. A real example shows the application of the proposed plan in the industry. From the comparison study, it was concluded that the proposed plan under the neutrosophic interval method is an adequate, flexible, effective, and reasonable method in the uncertainty environment. The proposed plan has the limitation that it can be applied only when the data follows the neutrosophic normal distribution. The proposed plan, using some other sampling scheme, will be considered for future research. The proposed plan for some non-normal distribution can be extended as future research.

Funding: This work was supported by the Deanship of Scientific Research (DSR), King Abdulaziz University, Jeddah. The author, Muhammad Aslam, therefore, acknowledges with thanks DSR technical support.

Acknowledgments: The author is deeply thankful to the editor and reviewers for their valuable suggestions to improve the quality of this manuscript.

Conflicts of Interest: The author declares no conflict of interest regarding this paper.

References

1. Montgomery, D.C. *Introduction to Statistical Quality Control*; John Wiley & Sons: Hoboken, NJ, USA, 2007.
2. Aslam, M.; Wu, C.-W.; Azam, M.; Jun, C.-H. Variable sampling inspection for resubmitted lots based on process capability index Cpk for normally distributed items. *Appl. Math. Model.* **2013**, *37*, 667–675. [CrossRef]
3. Balamurali, S.; Jun, C.-H. Repetitive group sampling procedure for variables inspection. *J. Appl. Stat.* **2006**, *33*, 327–338. [CrossRef]
4. Jun, C.-H.; Balamurali, S.; Lee, S.-H. Variables sampling plans for Weibull distributed lifetimes under sudden death testing. *IEEE Trans. Reliab.* **2006**, *55*, 53–58. [CrossRef]
5. Wu, C.-W.; Aslam, M.; Jun, C.-H. Developing a variables two-plan sampling system for product acceptance determination. *Commun. Stat. Theory Methods* **2017**, *46*, 706–720. [CrossRef]
6. Senturk, S.; Erginel, N. Development of fuzzy X¯~-R~ and X¯~-S~ control charts using α-cuts. *Inf. Sci.* **2009**, *179*, 1542–1551. [CrossRef]
7. Kanagawa, A.; Ohta, H. A design for single sampling attribute plan based on fuzzy sets theory. *Fuzzy Sets Syst.* **1990**, *37*, 173–181. [CrossRef]
8. Jamkhaneh, E.B.; Sadeghpour-Gildeh, B.; Yari, G. Important criteria of rectifying inspection for single sampling plan with fuzzy parameter. *Int. J. Contemp. Math. Sci.* **2009**, *4*, 1791–1801.
9. Sadeghpour Gildeh, B.; Baloui Jamkhaneh, E.; Yari, G. Acceptance single sampling plan with fuzzy parameter. *Iran. J. Fuzzy Syst.* **2011**, *8*, 47–55.
10. Jamkhaneh, E.B.; Sadeghpour-Gildeh, B.; Yari, G. Inspection error and its effects on single sampling plans with fuzzy parameters. *Struct. Multidiscip. Optim.* **2011**, *43*, 555–560. [CrossRef]
11. Tong, X.; Wang, Z. Fuzzy acceptance sampling plans for inspection of geospatial data with ambiguity in quality characteristics. *Comput. Geosci.* **2012**, *48*, 256–266. [CrossRef]
12. Turanoğlu, E.; Kaya, İ.; Kahraman, C. Fuzzy acceptance sampling and characteristic curves. *Int. J. Comput. Intell. Syst.* **2012**, *5*, 13–29. [CrossRef]
13. Uma, G.; Ramya, K. Impact of Fuzzy Logic on Acceptance Sampling Plans—A Review. *Autom. Auton. Syst.* **2015**, *7*, 181–185.
14. Kahraman, C.; Bekar, E.T.; Senvar, O. A Fuzzy Design of Single and Double Acceptance Sampling Plans. In *Intelligent Decision Making in Quality Management*; Springer: Berlin/Heidelberg, Germany, 2016; pp. 179–211.
15. Afshari, R.; Gildeh, B.S. Construction of fuzzy multiple deferred state sampling plan. In Proceedings of the Fuzzy Systems Association and 9th International Conference on Soft Computing and Intelligent Systems (IFSA-SCIS), 2017 Joint 17th World Congress of International, Otsu, Japan, 27–30 June 2017; pp. 1–7.

16. Smarandache, F. Neutrosophic Logic-A Generalization of the Intuitionistic Fuzzy Logic. In *Multispace & Multistructure Neutrosophic Transdisciplinarity (100 Collected Papers of Science)*; Infinite Study: Conshohocken, PA, USA, 2010; Volume 4, p. 396.
17. Smarandache, F. *Introduction to Neutrosophic Statistics*; Infinite Study: Conshohocken, PA, USA, 2014.
18. Chen, J.; Ye, J.; Du, S. Scale effect and anisotropy analyzed for neutrosophic numbers of rock joint roughness coefficient based on neutrosophic statistics. *Symmetry* **2017**, *9*, 208. [CrossRef]
19. Chen, J.; Ye, J.; Du, S.; Yong, R. Expressions of rock joint roughness coefficient using neutrosophic interval statistical numbers. *Symmetry* **2017**, *9*, 123. [CrossRef]
20. Aslam, M. A New Sampling Plan Using Neutrosophic Process Loss Consideration. *Symmetry* **2018**, *10*, 132. [CrossRef]
21. Aslam, M.; Arif, O. Testing of Grouped Product for the Weibull Distribution Using Neutrosophic Statistics. *Symmetry* **2018**, *10*, 403. [CrossRef]
22. Aslam, M.; Raza, M.A. Design of New Sampling Plans for Multiple Manufacturing Lines Under Uncertainty. *Int. J. Fuzzy Syst.* **2018**, 1–15. [CrossRef]
23. Aslam, M. A new attribute sampling plan using neutrosophic statistical interval method. *Complex. Intell. Syst.* **2019**, 1–6. [CrossRef]
24. Aslam, M. A New Failure-Censored Reliability Test Using Neutrosophic Statistical Interval Method. *Int. J. Fuzzy Syst.* **2019**, 1–7. [CrossRef]
25. Aslam, M. Product Acceptance Determination with Measurement Error Using the Neutrosophic Statistics. *Adv. Fuzzy Syst.* **2019**, *2019*, 8953051. [CrossRef]
26. Aslam, M. Design of Sampling Plan for Exponential Distribution under Neutrosophic Statistical Interval Method. *IEEE Access* **2018**, *6*, 64153–64158. [CrossRef]
27. Duncan, A.J. *Quality Control and Industrial Statistics*; R. D. Irwin: Homewood, IL, USA, 1974.
28. Aslam, M.; Yen, C.-H.; Chang, C.-H.; Jun, C.-H.; Ahmad, M.; Rasool, M. Two-stage variables acceptance sampling plans using process loss functions. *Commun. Stat. Theory Methods* **2012**, *41*, 3633–3647. [CrossRef]

symmetry

MDPI

Article

A Robust Single-Valued Neutrosophic Soft Aggregation Operators in Multi-Criteria Decision Making

Chiranjibe Jana * and **Madhumangal Pal**

Department of Applied Mathematics with Oceanology and Computer Programming, Vidyasagar University, Midnapore 721102, India; mmpalvu@gmail.com

* Correspondence: jana.chiranjibe7@gmail.com; Tel.: +91-9647135650

Received: 14 December 2018; Accepted: 17 January 2019; Published: 18 January 2019

Abstract: Molodtsov originated soft set theory that was provided a general mathematical framework for handling with uncertainties in which we meet the data by affix parameterized factor during the information analysis as differentiated to fuzzy as well as neutrosophic set theory. The main object of this paper is to lay a foundation for providing a new approach of single-valued neutrosophic soft tool which is considering many problems that contain uncertainties. In present study, a new aggregation operators of single-valued neutrosophic soft numbers have so far not yet been applied for ranking of the alternatives in decision-making problems. To this propose work, single-valued neutrosophic soft weighted arithmetic averaging (SVNSWA) operator, single-valued neutrosophic soft weighted geometric averaging (SVNSWGA) operator have been used to compare two single-valued neutrosophic soft numbers (SVNSNs) for aggregating different single-valued neutrosophic soft input arguments in neutrosophic soft environment. Then, its related properties have been investigated. Finally, a practical example for Medical diagnosis problems provided to test the feasibility and applicability of the proposed work.

Keywords: single-valued neutrosophic soft number and its operations; SVN soft weighted arithmetic averaging operator; SVN soft weighted geometric averaging operator; decision-making

1. Introduction

Multi-criteria decision-making (MCDM) problems seek great attention in modern decision science. The method is addressed to select the best alternative among the finite set of alternatives as claimed by decision makers under the preference values of the alternatives. MCDM problems extensively applied with quantitative or qualitative attribute values and have a board application in medical diagnosis [1,2], ecology [3], sensor network [4] management science and engineering [5,6], economic [7], market prediction and engineering technology [8], transport service problem [9] etc. As our modern society move forward with the decision-making process, so it always faces imprecise, vague and uncertain facts to take a decision in solving decision-making problems. In order to solve imprecise and uncertain data, [10] initiated the idea of intuitionistic fuzzy set (IFS), a powerful extension of fuzzy set (FS) [11]. Even though (FS) and (IFS) are very powerful set to model decision problems containing uncertainties, in some cases these sets are not sufficient to overcome indeterminate and inconsistent statistics experience in real world problems. As SVNS [12] have strong acceptance for modeling of problems including the incomplete, indeterminate and inconsistent data. The aggregated information for the execution of the criteria for alternatives, weighted and order weighted aggregation operators [13–22] takes a significant role during the combination of the information process. The above aggregation operators based decision-making problems are not enough for the solution of real-world problems because they have insufficient of parameterizations. In real life problems involve different parameters.

Most of MCDM problems, researchers can not consider parameterizations factor when they aggregate the information of the alternatives. Therefore, there are lack of information of the alternatives about the involves parameter. Motivated by the aforementioned limitations, this research developed a novel fuzzy-based MCDM approach for the evaluation of medical diagnosis problems. Initially, we introduced the SVSNs to quantify evaluation information on criteria and alternatives. We combined the SVNNs with the concept of soft set. Subsequently, this study proposed single-valued neutrosophic soft weighted averaging (SVNSWA) operator and single-valued neutrosophic soft weighted geometric (SVNSWG) operator to aggregate criteria by considering their parameterizations factors. We haved studied idempotency, boundedness, shift-invariance and Homogeneity property of these two kinds of soft weighted aggregation operators. The main advantage of these operators is that they are able to make smooth description of the real-world problems by the use of parameterizations factor. In order to rank the alternatives, aggregation operators lead to aggregate the over all information of the objects for the preferences of the decision maker into a collective one and hence find to a desirable according to its score values. To the best of our knowledge, the research developed on FSS and SVNSS is only about their basic theory and its applications, but, there have been no research done on single-valued neutrosophic soft aggregation numbers. So, it is a new issue and have a scope for future development in decision science. Therefore, decision-making problems in single-valued neutrosophic soft environment under proposed aggregation operators which makes us enough motivation to develop the propose problems. The main object of this article is to exhibit some aggregation operators under SVN data called as single-valued neutrosophic soft aggregation for collect the distinct priorities of the choices of this technique.

The remainder of this paper is organized as: In next Section, briefly survey some essential ideas of the FSS and SVNSS. In Section 3, we define some operational principles of single-valued neutrosophic soft numbers and then define single-valued neutrosophic soft weighted averaging (SVNSWA) operator, single-valued neutrosophic soft weighted geometric (SVNSWGA) operator and established its related properties. In next Section, we utilize those operators to create single-valued neutrosophic soft multi-criteria group decision-making problems. An interpretative case is specified for the selection of most illness patient in Section 5. In Section 6, a comparative analysis has been made between the existing works and the proposed study. Finally, in Section 7, follows a remark.

2. Literature Review

Neutrosophic set (NS) a tremendous branch of philosophy was proposed by Smarandache [23,24]. This proposed approach is characterized by three functions called (truth-, indeterminacy-,falsity)-membership functions. Therefore, (NS) has strong acceptance to develop models carrying indeterminate and inconsistent data. However, since codomain of membership functions of a (NS) is real standard or nonstandard subsets of $]^{-}0, 1^{+}[$, in some applications areas engineering and real scientific fields they have some difficulties in modeling of problems. To overcome difficulties in these areas, Wang et al. [12] defined the view of single-valued neutrosophic set (SVNs). As (SVNs) have strong acceptance for modeling of problems including the incomplete, indeterminate and inconsistent data. So, scholars have been investigating on how to find a proper one alternative and have obtained some achievements. Ye utilized [25] arithmetic and geometric aggregation functions under simplified neutrosophic numbers to develop MCDM problems. Garg and Nancy [26] have followed to the study of linguistic SVN prioritized aggregation function to propose a MADM problem. Wan et al. [27] introduced Frank Choquet Bonferroni mean operators and utilized this operator develop MCDM problems in single-valued bipolar neutrosophic environment. Shi and Ye [28] introduced Dombi aggregation operator to originate neutrosophic cubic Dombi (NCD) aggregation functions to study a decision-making problems. Wei and Zhang [29] utilized combination of power averaging and Bonferroni mean operator to developed SVN Bonferroni power aggregation operators to develop a MADM problem. Ulucay et al. [30] developed a decision-making problem using similarity measure method under bipolar neutrosophic environment. Abdel-Basset et al. [31] studied

MCGDM based on neutrosophic hierarchy method. Abdel-Basset et al. [32] proposed strategic planning and decision-making based on neutrosophic AHP-SWOT analysis. Dalapati et al. [33] proposed cross entropy based MAGDM based on interval neutrosophic information. In [34], Bausys and Zavadskas provided VIKOR method based MCDM problems using interval neutrosophic numbers. Biswas et al. [35] utilized TOPSIS method for MCDM problems under SVN environment. Broumi et al. [36] introduced an algorithm to solve a neutrosophic shortest problems from source node to destination node. Sahin and Liu [37] derived correlation coefficient between two SVN hesitant fuzzy numbers. Jana et al. [38] studied trapezoidal neutrosophic aggregation functions and utilized these operators develop MADM problems. Recently, researchers have drawn attention to model interval rough sets with their application problems [39].

But the technique of the above papers are not enough for the solution of real-world problems because they have insufficient of parameterizations. In that context, soft set theory plays an important role to overcome such barrier and effectively applied to solve the conditions. Maji et al. [40,41] provided with the bridge connection between FS and IFS with soft sets theory [42]. Some hybrid models together with soft set theory have been develop in various uncertain environments such as on fuzzy soft set theory with parameterizations [43,44], fuzzy soft expert sets [45], generalized intuitionsitic fuzzy soft sets [46], IVIF soft sets [47,48] and its applications, bipolar intuitionistic fuzzy soft sets and decision-making [49], Hesitant fuzzy soft sets [50]. Jana and pal [51,52] have studied soft intersection BCK/BCI-algebras, and soft intersection group structure based on (α, β)-soft intersectional sets. Selvachandran and Peng [53] has found a modified TOPSIS method using vauge parameterized vegue soft set and gave its application in decision making. Recently, Arora and Garg [54] provided a new approach of aggregation operator using parameterized factor in intuitionistic fuzzy soft environment. In the same time, a tool combination of neutrosophic set and soft set have gave a momentum to the solution of real life problems in many directions. Karaaslan [55] used possibility theory to develop PNS-decision-making method using neutrosophic soft OR-product and AND-product. Broumi and Samarandache [56] proposed single-valued neutrosophic soft expert set and its application in decision-making. Ali et al. [57] gave an application of bipolar neutrosophic soft sets in decision making in the environment of bipolar neutrosophic set. Deli et al. [58] motivated to develop a decision-making method called ivnpivn-soft sets using neutrosophic information. In [59], Khalid and Abbas used soft set theory in distance measure.

In this study, multi-criteria decision-making approach is characterized by single-valued neutrosophic soft numbers (SVNSNs). The SVNSWA and SVNSWG aggregation operators are presented. Then, a medical diagnosis problems is solved by using these proposed operators.

3. Basic Concepts of FSS and SVNSS

In what follows, U, E and $\mathcal{P}(U)$ respectively denote universal set, parameter set and power set of U. Also, $A \subseteq E$.

Definition 1 ([11])**.** *Let X be a non-empty set. A fuzzy set μ of X is defined as a mapping $\mu : X \to [0,1]$, where $[0,1]$ is the usual interval of real numbers. We take $\mathbb{F}(X)$ as the set of all fuzzy subsets of X.*

Definition 2 ([42])**.** *A pair $(\mathcal{F}, E)$ is called a soft set over U if $\mathcal{F}$ is a mapping given by $\mathcal{F} : E \to \mathcal{P}(U)$. In other words, a soft over the universe U is a parameterized family of subsets of the universal set U. For $\varepsilon \in A$, $\mathcal{F}(\varepsilon)$ may be considered as the set of ε-elements of the soft $(\mathcal{F}, A)$, or as the set of ε-approximate elements of the soft set.*

The following example illustrate the above idea.

Example 1. *Let (X, τ) be a topological space, i.e. τ is a family of subsets of the set X called the open sets of X. Then, the family of open neighborhood $N(x)$ of point x, where $N(x) = \{V \in \tau | x \in V\}$, may be consider as the soft set $(N(x), \tau)$.*

Definition 3 ([40]). *Let U be the universe set and E be the set of parameters. Let $\mathcal{P}(U)$ be the power set of U and $A \subseteq E$, and $\mathcal{P}(U)$ is the collection of all fuzzy subsets of U, then $(\mathcal{F}, A)$ is called fuzzy soft set, where $\mathcal{F} : A \rightarrow \mathcal{P}(U)$.*

Example 2. *Let $U = \{M_1, M_2, M_3, M_4\}$ be the set of four mobiles under consideration and $E = \{beautiful(e_1),\ costly(e_2),\ batterybackup(e_3)$ and $apps(e_4)\}$ be a set of parameters then FSS for describing "attractiveness of the mobiles"*
is $(\mathcal{F}, A) = \{\mathcal{F}_{e_1}, \mathcal{F}_{e_2}, \mathcal{F}_{e_3}\}$, where $A = \{e_1, e_2, e_3\} \subseteq E$ and (F, A) can be defined as:
$\mathcal{F}_{e_1} = \{(M_1, 0.6), (M_2, 0.4), (M_3, 0.5), (M_4, 0.3)\}$,
$\mathcal{F}_{e_2} = \{(M_1; 0.7), (M_2, 0.6), (M_3, 0.5), (M_4, 0.4)\}$ *and*
$\mathcal{F}_{e_3} = \{(M_1, 0.9), (M_2, 0.5), (M_3, 0.3), (M_4, 0.6)\}$.

Definition 4 ([23]). *Let X be finite, with a generic element in X denoted by x. A NS $\tilde{c}$ in X is defined by*

$$\tilde{C} = \left\{ \langle T_C(x), I_C(x), F_C(x) \rangle | x \in X \right\},$$

where its truth-function T_C is presented by $T_C : X \rightarrow]0^-, 1^+[$, indeterminacy-function I_C presented $I_C : X \rightarrow]0^-, 1^+[$, and falsity- function $\hat{F}_C$ interpreted as $F_C : X \rightarrow]0^-, 1^+[$. Also, T_C, I_C and F_C are real standard or non-standard subsets of $]0^-, 1^+[$. There is no restriction on the sum of T_C, I_C and F_C, and so $0^- \leq T_C + I_C + F_C \leq 3^+$.

For real applications of NS, Wang et al. [12] introduced SVNs in the following definition.

Definition 5 ([12]). *Let X be a finite set, with a generic element in X denoted by x. A SVNS is defined as:*

$$\tilde{C} = \left\{ \langle T_C(x), I_C(x), F_C(x) \rangle | x \in X \right\},$$

where $T_C : X \rightarrow [0, 1]$ indicated the truth, $I_C : X \rightarrow [0, 1]$ is the indeterminacy and $F_C : X \rightarrow [0, 1]$ is the falsity function of x to C with the condition $0 \leq T_C + I_C + F_C \leq 3$.

Definition 6. *Let U be universal set and E be the parameter set. For $N \subset E$. Let $\mathcal{P}(U)$ called the subsets of single-valued neutrosophic sets of U. The term $(\mathcal{F}_C, C)$ is called single-valued neutrosophic soft sets of U, where $\mathcal{F}_C$ is a function follows as, $\mathcal{F}_C : N \rightarrow \mathcal{P}(U)$.*

Example 3. *Let $U = \{O_1, O_2, O_3, O_4\}$ be the set of four mobiles under consideration and*
$E = \{beautiful(e_1),\ costly(e_2),\ batterybackup(e_3)$ and $apps(e_4)\}$ be a set of parameters under SVNSS for describing "attractiveness of the mobiles" is $(C, N) = \{\mathcal{F}_{e_1}, \mathcal{F}_{e_2}, \mathcal{F}_{e_3}\}$, where $A = \{e_1, e_2, e_3\} \subseteq E$ and $(\tilde{C}, N)$ can be defined as:
$\tilde{C}_{e_1} = \{(O_1, 0.6, 0.4, 0.2), (O_2, 0.4, 0.5, 0.1), (O_3, 0.5, 0.2, 0.3), (O_4, 0.3, 0.6, 0.1)\}$,
$\tilde{C}_{e_2} = \{(O_1, 0.7, 0.1, 0.2), (O_2, 0.6, 0.3, 0.1), (O_3, 0.5, 0.3, 0.3), (O_4, 0.4, 0.4, 0.1)\}$ *and*
$\tilde{C}_{e_3} = \{(O_1, 0.9, 0.1, 0.3), (O_2, 0.5, 0.2, 0.2), (O_3, 0.3, 0.5, 0.1), (O_4, 0.6, 0.4, 0.4)\}$.

For the sake of simplicity, we denote the pair of $\tilde{C}_{e_t}(x_c) = \{\langle T_c(x), I_c(x), F_c(x) \rangle | x_s \in U\}$, i.e., $\tilde{C}_{e_{st}} = \langle T_{st}, I_{st}, F_{st} \rangle$ is called as single-valued neutrosophic soft (SVNSN) numbers. For the application purpose, it is necessary to define score function for ranking it. For this, a score function of $\tilde{C}_{e_{st}}$ is defined as

$$\Psi(\tilde{C}_{e_{st}}) = T_{st} - F_{st} \tag{1}$$

where, $\Psi(\tilde{C}_{e_{st}}) \in [0, 1]$. By this definition, it is clear that the larger the $\Psi(\tilde{C}_{e_{st}})$, the larger is SVNSN $\tilde{C}_{e_{st}}$.

Example 4. *Let $\tilde{C}_{e_{11}} = \langle 0.6, 0.2, 0.2\rangle$ and $\tilde{C}_{e_{12}} = \langle 0.3, 0.5, 0.5\rangle$ be two SVNSNs, then by Equation (2), we get $\Psi(\tilde{C}_{e_{11}}) = 0.4$ and $\Psi(\tilde{C}_{e_{12}}) = -0.2$. Since $\Psi(\tilde{C}_{e_{11}}) > \Psi(\tilde{C}_{e_{12}})$ which imply $\tilde{C}_{e_{11}} > \tilde{C}_{e_{12}}$.*

However, there are some situation, where above function can not be used to compare SVNSNs. For example, let $\tilde{C}_{e_{11}} = \langle 0.6, 0.2, 0.2\rangle$ and $\tilde{C}_{e_{12}} = \langle 0.5, 0.1, 0.1\rangle$, then it is not possible to compare SVNSNs, which one of them is bigger as $\Psi(\tilde{C}_{e_{11}}) = \Psi(\tilde{C}_{e_{12}})$. To overcome this situation, we define accuracy function of $\tilde{C}_{e_{st}}$ as follows:

$$\mathcal{H}(\tilde{C}_{e_{st}}) = T_{st} + I_{st} + F_{st} \tag{2}$$

where, $\mathcal{H}(\tilde{C}_{e_{st}}) \in [0,1]$. Based on score function Ψ and accuracy function $\mathcal{H}$, defined order relation on two SVNSNs $\tilde{P}_{e_{st}}$ and $\tilde{Q}_{e_{st}}$ as follows:

(i) If $\Psi(\tilde{P}_{e_{st}}) < \Psi(\tilde{Q}_{e_{st}})$, then $\tilde{P}_{e_{st}} \prec \tilde{Q}_{e_{st}}$
(ii) If $\Psi(\tilde{P}_{e_{st}}) > \Psi(\tilde{Q}_{e_{st}})$, then $\tilde{P}_{e_{st}} \succ \tilde{Q}_{e_{st}}$
(iii) If $\Psi(\tilde{P}_{e_{st}}) = \Psi(\tilde{Q}_{e_{st}})$, then

  (1) If $\mathcal{H}(\tilde{P}_{e_{st}}) < \mathcal{H}(\tilde{Q}_{e_{st}})$, then $\tilde{P}_{e_{st}} \prec \tilde{Q}_{e_{st}}$.
  (2) If $\mathcal{H}(\tilde{P}_{e_{st}}) > \mathcal{H}(\tilde{Q}_{e_{st}})$, then $\tilde{P}_{e_{st}} \succ \tilde{Q}_{e_{st}}$.
  (3) If $\mathcal{H}(\tilde{P}_{e_{st}}) = \mathcal{H}(\tilde{Q}_{e_{st}})$, then $\tilde{P}_{e_{st}} \sim \tilde{Q}_{e_{st}}$.

4. Single-Valued Neutrosophic Soft Weighted Arithmetic Averaging (SVNSWAA) Operator

In this Section, an aggregation operators namely single-valued neutrosophic soft weighted averaging (SVNSWA) operator and single-valued neutrosophic soft weighted geometric averaging (BFSWGA) operator for neutrosophic soft numbers (SVNSNs) are proposed.

4.1. Operational Law for SVNSNs

Definition 7. *Let $\tilde{C}_e = \langle T, I, F\rangle$ and $\tilde{C}_{e_{11}} = \langle T_{11}, I_{11}, F_{11}\rangle)$ and $\tilde{C}_{e_{12}} = \langle T_{12}, I_{12}, F_{12}\rangle)$ be the three SVNSNs over the universe X, then following operations are defined as follows:*

(i) $\tilde{C}_{e_{11}} \oplus \tilde{C}_{e_{12}} = (\langle T_{11} + T_{12} - T_{11}T_{12}, I_{11}I_{12}, F_{11}F_{12}\rangle)$
(ii) $\tilde{C}_{e_{11}} \otimes \tilde{C}_{e_{12}} = (\langle T_{11}T_{12}, I_{11} + I_{12} - I_{11}I_{12}, F_{11} + F_{12} - F_{11}F_{12}\rangle)$
(iii) $\lambda\tilde{C}_e = (1 - (1-T)^{\lambda}, I^{\lambda}, F^{\lambda})$
(iv) $\tilde{C}_e^{\lambda} = (T^{\lambda}, 1 - (1-I)^{\lambda}, 1 - (1-F)^{\lambda})$.

Definition 8. *Let $\tilde{C}_{e_{st}} = (T_{st}, I_{st}, F_{st})$ $(s = 1, 2, \ldots, m; t = 1, 2, \ldots, n)$ be a number of SVNSNs and ϕ_t, θ_s are the are weight vectors for the parameter e_t's and expert y_s's respectively, satisfying $\phi_t \geq 0$, $\theta_s \geq 0$ such that $\sum_{t=1}^{n} \phi_t = 1$ and $\sum_{s=1}^{m} \theta_s = 1$. Then single-valued neutrosophic soft weighted averaging (SVNSWA) operator is function $SVNSWA : \tilde{C}^n \to \tilde{C}$ such that*

$$SVNSWA(\tilde{C}_{e_{11}}, \tilde{C}_{e_{12}}, \ldots, \tilde{C}_{e_{mn}}) = \bigoplus_{t=1}^{n} \phi_t \Big(\bigoplus_{s=1}^{m} \theta_s \tilde{C}_{e_{st}} \Big). \tag{3}$$

We get the following theorem that follows on SVNSWA operator.

Theorem 1. *$\tilde{C}_{e_{st}} = (T_{st}, I_{st}, F_{st})$ $(s = 1, 2, \ldots, m; t = 1, 2, \ldots, n)$ be a number of (SVNSNs), then aggregated value of them using the SVNSWA operator is also a SVNSNs, and $SVNSWA(\tilde{C}_{e_{11}}, \tilde{C}_{e_{12}}, \ldots, \tilde{C}_{e_{mn}})$*

$$= \Big\langle 1 - \prod_{t=1}^{n} \Big(\prod_{s=1}^{m} (1 - T_{st})^{\theta_s} \Big)^{\phi_t}, \prod_{t=1}^{n} \Big(\prod_{s=1}^{m} \big(I_{st}\big)^{\theta_s} \Big)^{\phi_t}, \prod_{t=1}^{n} \Big(\prod_{s=1}^{m} \big(F_{st}\big)^{\theta_s} \Big)^{\phi_t} \Big\rangle. \tag{4}$$

Theorem 1 can be proved by the method of mathematical induction as follows:

Proof. For $m = 1$, we get $\theta_1 = 1$. Then by Definition 7 of operational law,

$$\begin{aligned} SVNSWA(\tilde{C}_{e_{11}}, \tilde{C}_{e_{12}}, \ldots, \tilde{C}_{e_{mn}}) &= \bigoplus_{t=1}^{n} \phi_t \left(\tilde{C}_{e_{1t}}\right) \\ &= \left\langle 1 - \prod_{t=1}^{n} \left(1 - T_{1t}\right)^{\phi_t}, \prod_{t=1}^{n} (I_{1t})^{\phi_t}, \prod_{t=1}^{n} (F_{1t})^{\phi_t} \right\rangle \\ &= \left\langle 1 - \prod_{t=1}^{n} \left(\prod_{s=1}^{1} (1 - F_{st})^{\theta_s}\right)^{\phi_t}, \prod_{t=1}^{n} \left(\prod_{s=1}^{1} (I_{st})^{\theta_s}\right)^{\phi_t}, \prod_{t=1}^{n} \left(\prod_{s=1}^{1} (F_{st})^{\theta_s}\right)^{\phi_t} \right\rangle. \end{aligned}$$

Again, for $n = 1$ and $\phi_1 = 1$ and hence,

$$\begin{aligned} SVNSWA(\tilde{C}_{e_{11}}, \tilde{C}_{e_{12}}, \ldots, \tilde{C}_{e_{mn}}) &= \left(\bigoplus_{s=1}^{m} \theta_{s1} \tilde{C}_{e_{1t}}\right) \\ &= \left\langle 1 - \prod_{t=1}^{m} \left(1 - T_{s1}\right)^{\theta_s}, \prod_{s=1}^{m} (I_{s1})^{\theta_s}, \prod_{s=1}^{m} (F_{s1})^{\theta_s} \right\rangle \\ &= \left\langle 1 - \prod_{t=1}^{1} \left(\prod_{s=1}^{m} (1 - T_{st})^{\theta_s}\right)^{\phi_t}, \prod_{t=1}^{1} \left(\prod_{s=1}^{m} (I_{st})^{\theta_s}\right)^{\phi_t}, \prod_{t=1}^{1} \left(\prod_{s=1}^{m} (F_{st})^{\theta_s}\right)^{\phi_t} \right\rangle. \end{aligned}$$

Thus, (5) is true for $m = 1$ and $n = 1$. Assume that (5) is true for $n = p_1 + 1$, $m = p_2$ and $n = p_1$, $m = p_2 + 1$, then it follows that

$$\bigoplus_{t=1}^{p_1+1} \phi_t \left(\bigoplus_{s=1}^{p_2} \theta_s \tilde{C}_{e_{st}}\right) = \left\langle 1 - \prod_{t=1}^{p_1+1} \left(\prod_{s=1}^{p_2} (1 - T_{st})^{\theta_s}\right)^{\phi_t}, \prod_{t=1}^{p_1+1} \left(\prod_{s=1}^{p_2} (I_{st})^{\theta_s}\right)^{\phi_t}, \prod_{t=1}^{p_1+1} \left(\prod_{s=1}^{p_2} (F_{st})^{\theta_s}\right)^{\phi_t} \right\rangle.$$

Also,

$$\bigoplus_{t=1}^{p_1} \phi_t \left(\bigoplus_{s=1}^{p_2+1} \theta_s \tilde{C}_{e_{st}}\right) = \left\langle 1 - \prod_{t=1}^{p_1} \left(\prod_{s=1}^{p_2+1} (1 - T_{st})^{\theta_s}\right)^{\phi_t}, \prod_{t=1}^{p_1} \left(\prod_{s=1}^{p_2+1} (I_{st})^{\theta_s}\right)^{\phi_t}, \prod_{t=1}^{p_1} \left(\prod_{s=1}^{p_2+1} (F_{st})^{\theta_s}\right)^{\phi_t} \right\rangle.$$

Now for $n = p_1 + 1$ and $m = p_2 + 1$, we obtained

$$\begin{aligned} \bigoplus_{t=1}^{p_1+1} \phi_t \left(\bigoplus_{s=1}^{p_2+1} \theta_s \tilde{C}_{e_{st}}\right) &= \bigoplus_{t=1}^{p_1+1} \phi_t \left(\bigoplus_{s=1}^{p_2} \theta_s \tilde{C}_{e_{st}} \oplus \theta_{p_2+1} \tilde{C}_{e_{(p_2+1)t}}\right) \\ &= \bigoplus_{t=1}^{p_1+1} \bigoplus_{s=1}^{p_2} \phi_t \theta_s \tilde{C}_{e_{st}} \bigoplus_{t=1}^{p_1+1} \phi_t \theta_{p_2+1} \tilde{C}_{e_{(p_2+1)t}}\Big) \end{aligned}$$

$$= \left\langle 1 - \prod_{t=1}^{p_1+1} \left(\prod_{s=1}^{p_2} (1 - T_{st})^{\theta_s}\right)^{\phi_t} \oplus 1 - \prod_{t=1}^{p_1+1} \left((1 - T_{(p_2+1)t})^{\theta_{p_2+1}}\right)^{\phi_t}, \right.$$

$$\left. \prod_{t=1}^{p_1+1} \left(\prod_{s=1}^{p_2} (I_{st})^{\theta_s}\right)^{\phi_t} \oplus \prod_{t=1}^{p_1+1} \left((I_{(p_2+1)t})^{\theta_{p_2+1}}\right)^{\phi_t}, \prod_{t=1}^{p_1+1} \left(\prod_{s=1}^{p_2} (F_{st})^{\theta_s}\right)^{\phi_t} \oplus \prod_{t=1}^{p_1+1} \left((F_{(p_2+1)t})^{\theta_{p_2+1}}\right)^{\phi_t} \right\rangle$$

$$= \left\langle 1 - \prod_{t=1}^{p_1+1} \left(\prod_{s=1}^{p_2+1} (1 - T_{st})^{\theta_s}\right)^{\phi_t}, \prod_{t=1}^{p_1+1} \left(\prod_{s=1}^{p_2+1} (I_{st})^{\theta_s}\right)^{\phi_t}, \prod_{t=1}^{p_1+1} \left(\prod_{s=1}^{p_2+1} (F_{st})^{\theta_s}\right)^{\phi_t} \right\rangle.$$

Thus, (5) is true for $n = p_1 + 1$, $m = p_2 + 1$, therefore by induction the results is hold for all $m, n \geq 1$.

Since, $0 \leq T_{st} \leq 1 \Leftrightarrow 0 \leq \prod_{s=1}^{m} (1 - T_{st})^{\theta_s} \leq 1$ and hence, $0 \leq 1 - \prod_{t=1}^{n} (\prod_{s=1}^{m} (1 - T_{st})^{\theta_s})^{\phi_t} \leq 1$. Also, $0 \leq I_{st} \leq 1 \Leftrightarrow 0 \leq \prod_{s=1}^{m} (I_{st})^{\theta_s} \leq 1 \Leftrightarrow 0 \leq \prod_{t=1}^{n} (\prod_{s=1}^{m} (I_{st})^{\theta_s})^{\phi_t} \leq 1$, and $0 \leq F_{st} \leq 1 \Leftrightarrow 0 \leq \prod_{s=1}^{m} (F_{st})^{\theta_s} \leq 1 \Leftrightarrow$

$0 \leq \prod\limits_{t=1}^{n}(\prod\limits_{s=1}^{m}(F_{st})^{\theta_s})^{\phi_t} \leq 1$. Thus, $0 \leq 1 - \prod\limits_{t=1}^{n}(\prod\limits_{s=1}^{m}(1-T_{st})^{\theta_s})^{\phi_t} + \prod\limits_{t=1}^{n}(\prod\limits_{s=1}^{m}(I_{st})^{\theta_s})^{\phi_t} + \prod\limits_{t=1}^{n}(\prod\limits_{s=1}^{m}(F_{st})^{\theta_s})^{\phi_t} \leq 3$.
Hence, aggregated value obtained by SVNSWA is again a SVNSN. □

Corollary 1 ([60]). *For only one parameter e_1, i.e., $n = 1$, then SVNSWA operator reduces to SVNWA.*

$$SVNSWA(\tilde{C}_{e_{11}}, \tilde{C}_{e_{21}}, \ldots, \tilde{C}_{e_{m1}}) = \left\langle 1 - \prod_{s=1}^{m}(1-T_s)^{\theta_s}, \prod_{s=1}^{m}(I_s)^{\theta_s}, \prod_{s=1}^{m}(F_s)^{\theta_s} \right\rangle. \quad (5)$$

Therefore, it is justified that aggregation operator defined under SVNS environment is taken as a special case of the proposed operator.

Example 5. *Let $Y = \{y_1, y_2, y_3, y_4\}$ be the set of experts which are going to narrate the "attractiveness of two-wheeler bikes" under the set of parameters $E = \{e_1 = stylish,\ e_2 = weight,\ e_3 = milage,\ e_4 = price\}$. The rating value of the experts is assumed to be given in the form of SVNSNs$(C, E) = (T_{st}, I_{st}, F_{st})_{4\times 3}$ for each parameters which are given in the following table (Table 1).*

Table 1. Neutrosophic soft numbers.

Experts	e_1	e_2	e_3
y_1	$\langle 0.6, 0.2, 0.3\rangle$	$\langle 0.5, 0.5, 0.2\rangle$	$\langle 0.6, 0.3, 0.5\rangle$
y_2	$\langle 0.5, 0.4, 0.4\rangle$	$\langle 0.3, 0.2, 0.1\rangle$	$\langle 0.5, 0.4, 0.5\rangle$
y_3	$\langle 0.7, 0.1, 0.4\rangle$	$\langle 0.4, 0.3, 0.6\rangle$	$\langle 0.3, 0.1, 0.6\rangle$
y_4	$\langle 0.4, 0.5, 0.2\rangle$	$\langle 0.7, 0.2, 0.1\rangle$	$\langle 0.2, 0.6, 0.3\rangle$

Let $\phi = (0.3, 0.2, 0.5)^T$ and $\theta = (0.2, 0.1, 0.3, 0.4)^T$ be the weight vectors for the parameters and experts respectively. Then, we get by using Theorem 1 as:
$SVNSWA(\tilde{B}_{e_{11}}, \tilde{B}_{e_{12}}, \ldots, \tilde{B}_{e_{43}})$

$$= \left\langle 1 - \prod_{t=1}^{3}\left(\prod_{s=1}^{4}(1-T_{st})^{\theta_s}\right)^{\phi_t}, \prod_{t=1}^{3}\left(\prod_{s=1}^{4}(I_{st})^{\theta_s}\right)^{\phi_t}, \prod_{t=1}^{3}\left(\prod_{s=1}^{4}(I_{st})^{\theta_s}\right)^{\phi_t} \right\rangle$$

$$\Big\langle 1 - \Big(\Big\{(1-0.6)^{0.2}(1-0.5)^{0.1}(1-0.7)^{0.3}(1-0.4)^{0.4}\Big\}^{0.3} \ \Big\{(1-0.5)^{0.2}(1-0.3)^{0.1}(1-0.4)^{0.3}(1-0.7)^{0.4}\Big\}^{0.2} \ \Big\{(1-0.6)^{0.2}(1-0.5)^{0.1}(1-0.3)^{0.3}(1-0.2)^{0.4}\Big\}^{0.5}\Big), \Big\{(0.2)^{0.2}\ (0.4)^{0.1}\ (0.1)^{0.3}\ (0.5)^{0.4}\Big\}^{0.3}$$
$$\Big\{(0.5)^{0.2}\ (0.2)^{0.1}\ (0.3)^{0.3}\ (0.2)^{0.4}\Big\}^{0.2}$$
$$\Big\{(0.3)^{0.2}\ (0.4)^{0.1}\ (0.1)^{0.3}\ (0.6)^{0.4}\Big\}^{0.5}, \Big\{(0.3)^{0.2}\ (0.4)^{0.1}\ (0.4)^{0.3}\ (0.2)^{0.4}\Big\}^{0.3}$$
$$\Big\{(0.2)^{0.2}\ (0.1)^{0.1}\ (0.6)^{0.3}\ (0.1)^{0.4}\Big\}^{0.2}$$
$$\Big\{(0.5)^{0.2}\ (0.5)^{0.1}\ (0.6)^{0.3}\ (0.3)^{0.4}\Big\}^{0.5}\Big\rangle$$
$$= \Big\langle (0.5317, 0.2755, 0.3256) \Big\rangle.$$

We prove easily the following properties by using the operator SVNSWA.

Theorem 2 (Idempotency Property). *Let $\tilde{C}_{e_{st}} = (T_{st}, I_{st}, F_{st})$ $(s = 1, 2, \ldots, m; t = 1, 2, \ldots, n)$ be a number of SVNSNs are all equal, i.e., $\tilde{C}_{e_s}t = \tilde{C}_e$ for all s, t, then*

$$SVNSWA(\tilde{C}_{e_{11}}, \tilde{C}_{e_{12}}, \ldots, \tilde{C}_{e_{mn}}) = \tilde{C}_e. \quad (6)$$

Proof. Since $\tilde{C}_{e_{st}} = \tilde{C}_e = \langle T, I, F\rangle$ for all s, t. Then,

$$\begin{aligned} SVNSWA(\tilde{C}_{e_{11}}, \tilde{C}_{e_{12}}, \ldots, \tilde{C}_{e_{mn}}) &= \Big\langle 1 - \prod_{t=1}^{n}\Big(\prod_{s=1}^{m}(1-T)^{\theta_s}\Big)^{\phi_t}, \prod_{t=1}^{n}\Big(\prod_{s=1}^{m}(I)^{\theta_s}\Big)^{\phi_t} \prod_{t=1}^{n}\Big(\prod_{s=1}^{m}(F)^{\theta_s}\Big)^{\phi_t}\Big\rangle \\ &= \Big\langle 1 - \Big((1-T)^{\sum_{s=1}^{m}\theta_s}\Big)^{\sum_{t=1}^{n}\phi_t}, \Big((I)^{\sum_{s=1}^{m}\theta_s}\Big)^{\sum_{t=1}^{n}\phi_t} \Big((F)^{\sum_{s=1}^{m}\theta_s}\Big)^{\sum_{t=1}^{n}\phi_t}\Big\rangle \\ &= \Big\langle 1-(1-T), I, F\Big\rangle \\ &= \Big\langle T, I, F\Big\rangle. \end{aligned}$$

The proof is completed. □

Theorem 3 (Boundedness Property). *Let $\tilde{C}_{e_{st}} = (T_{st}, I_{st}, F_{st})$ $(s = 1, 2, \ldots, m; t = 1, 2, \ldots, n)$ be a collection of SVNSNs. Let $\tilde{C}^{-}_{e_{st}} = \langle \min_t \min_s\{T_{st}\}, \max_t \max_s\{I_{st}\}, \max_t \max_s\{F_{st}\}\rangle$ and $\tilde{C}^{+}_{e_{st}} = \langle \max_t \max_s\{T_{st}\}, \min_t \min_s\{I_{st}\}, \min_t \min_s\{F_{st}\}\rangle$. Then,*

$$\tilde{C}^{-}_{e_{st}} \leq SVNSWA(\tilde{C}_{e_{11}}, \tilde{C}_{e_{12}}, \ldots, \tilde{C}_{e_{mn}}) \leq \tilde{C}^{+}_{e_{st}}.$$

Proof. Since, $\tilde{C}_{e_{st}} = (T_{st}, I_{st}, F_{st})$ be a SVNSNs then $\min_t \min_s\{T_{st}\} \leq T_{st} \leq \max_t \max_s\{T_{st}\}$ which implies that $1 - \max_t \max_s\{T_{st}\} \leq 1 - T_{st} \leq 1 - \min_t \min_s\{T_{st}\} \Leftrightarrow (1 - \max_t \max_s\{T_{st}\})^{\theta_s} \leq (1 - T_{st} \leq (1 - \min_t \min_s\{T_{st})^{\theta_s} \Leftrightarrow 1 - \max_t \max_s\{T_{st}\} \leq \prod_{s=1}^{m}(1 - T_{st})^{\theta_s} \leq 1 - \min_t \min_s\{T_{st}\} \Leftrightarrow (1 - \max_t \max_s\{T_{st}\})^{\sum_{t=1}^{n}\phi_t} \leq \prod_{t=1}^{n}(\prod_{s=1}^{m}(1 - T_{st})^{\theta_s})^{\sum_{t=1}^{n}\phi_t} \leq (1 - \min_t \min_s\{T_{st}\})^{\sum_{t=1}^{n}\phi_t} \Leftrightarrow 1 - \max_t \max_s\{T_{st}\} \leq \prod_{t=1}^{n}(\prod_{s=1}^{m}(1 - T_{st})^{\theta_s})^{\sum_{t=1}^{n}\phi_t} \leq 1 - \min_t \min_s\{T_{st}\}$. Therefore,

$$\max_t \max_s\{T_{st}\} \leq 1 - \prod_{t=1}^{n}\Big(\prod_{s=1}^{m}(1 - T_{st})^{\theta_s}\Big)^{\sum_{t=1}^{n}\phi_t} \leq \min_t \min_s\{T_{st}\}. \tag{7}$$

Again,

$$\min_t \min_s\{I_{st}\} \leq I_{st} \leq \max_t \max_s\{I_{st}\}$$

which finds $(\min_t \min_s\{I_{st}\})^{\sum_{s=1}^{m}\theta_s} \leq \prod_{s=1}^{m}(I_{st})^{\theta_s} \leq (\max_t \max_s\{I_{st}\})^{\sum_{s=1}^{m}\theta_s} \Leftrightarrow \min_t \min_s\{I_{st}\} \leq \prod_{s=1}^{m}(I_{st})^{\theta_s} \leq \max_t \max_s\{I_{st}\} \Leftrightarrow (\min_t \min_s\{I_{st}\})^{\phi_t} \leq (\prod_{s=1}^{m}(I_{st})^{\theta_s})^{\phi_t} \leq (\max_t \max_s\{I_{st}\})^{\phi_t} \Leftrightarrow (\min_t \min_s\{I_{st}\})^{\sum_{t=1}^{n}\phi_t} \leq \prod_{t=1}^{n}(\prod_{s=1}^{m}(I_{st})^{\theta_s})^{\phi_t} \leq (\max_t \max_s\{I_{st}\})^{\sum_{t=1}^{n}\phi_t}$, hence we get,

$$\min_t \min_s\{I_{st}\} \leq \prod_{t=1}^{n}(\prod_{s=1}^{m}(I_{st})^{\theta_s})^{\phi_t} \leq \max_t \max_s\{I_{st}\}. \tag{8}$$

and,

$$\min_t \min_s\{F_{st}\} \leq F_{st} \leq \max_t \max_s\{F_{st}\}$$

which follows $(\min_t \min_s\{F_{st}\})^{\sum_{s=1}^{m}\theta_s} \leq \prod_{s=1}^{m}(F_{st})^{\theta_s} \leq (\max_t \max_s\{F_{st}\})^{\sum_{s=1}^{m}\theta_s} \Leftrightarrow \min_t \min_s\{F_{st}\} \leq \prod_{s=1}^{m}(F_{st})^{\theta_s} \leq \max_t \max_s\{F_{st}\} \Leftrightarrow (\min_t \min_s\{F_{st}\})^{\phi_t} \leq (\prod_{s=1}^{m}(F_{st})^{\theta_s})^{\phi_t} \leq (\max_t \max_s\{F_{st}\})^{\phi_t} \Leftrightarrow (\min_t \min_s\{F_{st}\})^{\sum_{t=1}^{n}\phi_t} \leq \prod_{t=1}^{n}(\prod_{s=1}^{m}(F_{st})^{\theta_s})^{\phi_t} \leq (\max_t \max_s\{F_{st}\})^{\sum_{t=1}^{n}\phi_t}$, hence we get,

$$\min_t \min_s\{F_{st}\} \leq \prod_{t=1}^{n}(\prod_{s=1}^{m}(F_{st})^{\theta_s})^{\phi_t} \leq \max_t \max_s\{F_{st}\}. \tag{9}$$

Let $\beta \equiv SVNSWA(\tilde{C}_{e_{11}}, \tilde{C}_{e_{12}}, \ldots, \tilde{C}_{e_{mn}}) = \langle T_\beta, I_\beta, F_\beta \rangle$, then from Equations (7)–(9), $\min_t \min_s\{T_{st}\} \leq T_\beta \leq \max_t \max_s\{T_{st}\}$ and $\min_t \min_s\{I_{st}\} \leq I_\beta \leq \max_t \max_s\{I_{st}\}$, and $\min_t \min_s\{F_{st}\} \leq F_\beta \leq \max_t \max_s\{F_{st}\}$. Then by definition of score function

$$\Psi(\beta) = T_\beta - F_\beta \leq \max_t \max_s\{T_{st}\} - \min_t \min_s\{F_{st}\} = \Psi(\tilde{C}^+_{e_{st}})$$

$$\Psi(\beta) = T_\beta - F_\beta \geq \min_t \min_s\{T_{st}\} - \max_t \max_s\{F_{st}\} = \Psi(\tilde{C}^-_{e_{st}}).$$

Now, there are three cases arises:

Case 1. If $\Psi(\tilde{C}_{e_{st}}) < \Psi(\tilde{C}^+_{e_{st}})$ and $\Psi(\tilde{C}_{e_{st}}) > \Psi(\tilde{C}^-_{e_{st}})$, then by comparison of two SVNSNs, we have

$$\tilde{C}^-_{e_{st}} \leq SVNSWA(\tilde{C}_{e_{11}}, \tilde{C}_{e_{12}}, \ldots, \tilde{C}_{e_{mn}}) \leq \tilde{C}^+_{e_{st}}.$$

Case 2. If $\Psi(\tilde{C}_{e_{st}}) = \Psi(\tilde{C}^+_{e_{st}})$, i.e., $T_\beta + F_\beta + F_{st} = \max_t \max_s\{T_{st}\} + \min_t \min_s\{I_{st}\} + \min_t \min_s\{F_{st}\}$, then by above inequalities $T_\beta = \max_t \max_s\{T_{st}\}$ and $I_\beta = \min_t \min_s\{I_{st}\}$, and $F_\beta = \min_t \min_s\{F_{st}\}$

Therefore,

$$\mathcal{H} = T_\beta + I_\beta + F_\beta = \max_t \max_s\{T_{st}\} + \min_t \min_s\{I_{st}\} + \min_t \min_s\{F_{st}\} = \mathcal{H}(\tilde{C}^+_{e_{st}}),$$

then by comparison of two SVNSNs, we have

$$SVNSWA(\tilde{C}_{e_{11}}, \tilde{C}_{e_{12}}, \ldots, \tilde{C}_{e_{mn}}) = C^+_{e_{st}}.$$

Case 3. If $\Psi(\tilde{C}_{e_{st}}) = \Psi(\tilde{C}^-_{e_{st}})$, i.e., $T_\beta + I_\beta + F_\beta = \min_t \min_s\{T_{st}\} + \max_t \max_s\{I_{st}\} + \max_t \max_s\{F_{st}\}$, then by above inequalities $T_\beta = \min_t \min_s\{T_{st}\}$, $I_\beta = \max_t \max_s\{I_{st}\}$, and $F_\beta = \max_t \max_s\{F_{st}\}$.

Hence,

$$\mathcal{H} = T_\beta + I_\beta + F_{st} = \min_t \min_s\{T_{st}\} + \max_t \max_s\{I_{st}\} + \max_t \max_s\{F_{st}\} = \mathcal{H}(\tilde{C}^-_{e_{st}}),$$

then by comparison of two SVNSNs, we have

$$SVNSWA(\tilde{C}_{e_{11}}, \tilde{C}_{e_{12}}, \ldots, \tilde{C}_{e_{mn}}) = C^-_{e_{st}}.$$

Thus, proof is completed. □

Theorem 4 (Shift-invariance property). *If $\tilde{C}_e = \langle T, I, F \rangle$ be another SVNSN, then*

$$SVNSWA(\tilde{C}_{e_{11}} \oplus \tilde{C}_e, \tilde{C}_{e_{12}} \oplus \tilde{C}_e, \ldots, \tilde{C}_{e_{mn}} \oplus \tilde{C}_e) = SVNSWA(\tilde{C}_{e_{11}}, \tilde{C}_{e_{12}}, \ldots, \tilde{C}_{e_{mn}}) \oplus \tilde{C}_e$$

Proof. Since $\tilde{C}_e$ and $\tilde{C}_{e_{st}}$ are SVNSNs. Then, we have $\tilde{C}_e \oplus \tilde{C}_{e_{st}} = \left\langle 1-(1-T)(1-T_{st}), II_{st}, FF_{st} \right\rangle$. Hence, $SVNSWA(\tilde{C}_{e_{11}} \oplus \tilde{C}_e, \tilde{C}_{e_{12}} \oplus \tilde{C}_e, \ldots, \tilde{C}_{e_{mn}} \oplus \tilde{C}_e)$

$$\begin{aligned}
&= \bigoplus_{t=1}^{n} \phi_t \Big(\bigoplus_{s=1}^{m} \theta_t (\tilde{C}_{e_{st}} \oplus \tilde{C}_e) \Big) \\
&= \Big\langle 1-\prod_{t=1}^{n}\Big(\prod_{s=1}^{m}(1-T_{st})^{\theta_s}(1-T)^{\theta_s}\Big)^{\phi_t}, \prod_{t=1}^{n}\Big(\prod_{s=1}^{m}(I_{st})^{\theta_s}(I)^{\theta_s}\Big)^{\phi_t}, \prod_{t=1}^{n}\Big(\prod_{s=1}^{m}(F_{st})^{\theta_s}(F)^{\theta_s}\Big)^{\phi_t} \Big\rangle \\
&= \Big\langle 1-(1-I)\prod_{t=1}^{n}\Big(\prod_{s=1}^{m}(1-T_{st})^{\theta_s}\Big)^{\phi_t}, I\prod_{t=1}^{n}\Big(\prod_{s=1}^{m}(I_{st})^{\theta_s}\Big)^{\phi_t}, F\prod_{t=1}^{n}\Big(\prod_{s=1}^{m}(F_{st})^{\theta_s}\Big)^{\phi_t} \Big\rangle \\
&= \Big\langle 1-\prod_{t=1}^{n}\Big(\prod_{s=1}^{m}(1-T_{st})^{\theta_s}\Big)^{\phi_t}, \prod_{t=1}^{n}\Big(\prod_{s=1}^{m}(I_{st})^{\theta_s}\Big)^{\phi_t}, \prod_{t=1}^{n}\Big(\prod_{s=1}^{m}(F_{st})^{\theta_s}\Big)^{\phi_t} \Big\rangle \oplus \Big\langle T, I, F \Big\rangle \\
&= SVNSWA(\tilde{C}_{e_{11}}, \tilde{C}_{e_{12}}, \ldots, \tilde{C}_{e_{mn}}) \oplus \tilde{C}_e.
\end{aligned}$$

Hence the result. □

Theorem 5 (Homogeneity property). *For any real number $\lambda > 0$, we have*

$$SVNSWA(\lambda\tilde{C}_{e_{11}}, \lambda\tilde{C}_{e_{12}}, \ldots, \lambda\tilde{C}_{e_{mn}}) = \lambda SVNSWA(\tilde{C}_{e_{11}}, \tilde{C}_{e_{12}}, \ldots, \tilde{C}_{e_{mn}}).$$

Proof. Let $\tilde{C}_{e_{st}} = (T_{st}, I_{st}, F_{st})$ $(s = 1, 2, \ldots, m; t = 1, 2, \ldots, n)$ be a number of SVNSNs and $\lambda > 0$ be any real number. Then, $\lambda\tilde{C}_{e_{st}} = \left\langle 1-(1-T_{st})^{\lambda}, (I_{st})^{\lambda}, (F_{st})^{\lambda} \right\rangle$. Thus,

$$\begin{aligned}
SVNSWA(\lambda\tilde{C}_{e_{11}}, \lambda\tilde{C}_{e_{12}}, \ldots, \lambda\tilde{C}_{e_{mn}}) &= \Big\langle 1-\prod_{t=1}^{n}\Big(\prod_{s=1}^{m}(1-T_{st})^{\lambda\theta_s}\Big)^{\phi_t}, \prod_{t=1}^{n}\Big(\prod_{s=1}^{m}(I_{st})^{\lambda\theta_s}\Big)^{\phi_t}, \prod_{t=1}^{n}\Big(\prod_{s=1}^{m}(F_{st})^{\lambda\theta_s}\Big)^{\phi_t} \Big\rangle \\
&= \Big\langle 1-\Big(\prod_{t=1}^{n}\Big(\prod_{s=1}^{m}(1-T_{st})^{\theta_s}\Big)^{\phi_t}\Big)^{\lambda}, \Big(\prod_{t=1}^{n}\Big(\prod_{s=1}^{m}(I_{st})^{\theta_s}\Big)^{\phi_t}\Big)^{\lambda}, \\
&\quad \Big(\prod_{t=1}^{n}\Big(\prod_{s=1}^{m}(F_{st})^{\theta_s}\Big)^{\phi_t}\Big)^{\lambda} \Big\rangle \\
&= \lambda\, SVNSWA(\tilde{C}_{e_{11}}, \tilde{C}_{e_{12}}, \ldots, \tilde{C}_{e_{mn}}).
\end{aligned}$$

Hence the proof is completed. □

4.2. Single-Valued Neutrosophic Soft Weighted Geometric Averaging (SVNSWGA) Operator

In this Section, we defined single-valued neutrosophic soft weighted geometric averaging (SVNSWGA) operator and studied

Definition 9. *Let $\tilde{C}_{e_{st}} = (T_{st}, I_{st}, F_{st})$ $(s = 1, 2, \ldots, m; t = 1, 2, \ldots, n)$ be a number of SVNSNs and ϕ_t, θ_s are the are weight vectors for the parameter e_t's and expert y_s's respectively, satisfying $\phi_t \geq 0$, $\theta_s \geq 0$ such that $\sum_{t=1}^{m} \phi_t = 1$ and $\sum_{s=1}^{n} \theta_s = 1$. Then single-valued neutrosophic soft weighted geometric (SVNSWGA) operator is a function $SVNSWGA : \tilde{C}^n \to \tilde{C}$ such that*

$$SVNSWGA(\tilde{C}_{e_{11}}, \tilde{C}_{e_{12}}, \ldots, \tilde{C}_{e_{mn}}) = \bigotimes_{t=1}^{n} \Big(\bigotimes_{s=1}^{m} \tilde{C}_{e_{st}}^{\theta_s} \Big)^{\phi_t}.$$

Theorem 6. *Then single-valued neutrosophic soft weighted geometric (SVNSWGA) operator is a function $SVNSWGA : \tilde{C}^n \to \tilde{C}$ such that*

$$SVNSWA(\tilde{C}_{e_{11}}, \tilde{C}_{e_{12}}, \ldots, \tilde{C}_{e_{1n}}) = \Big\langle \prod_{t=1}^{n}\Big(\prod_{s=1}^{m}(T_{st})^{\theta_s}\Big)^{\phi_t}, 1-\prod_{t=1}^{n}\Big(\prod_{s=1}^{m}(1-I_{st})^{\theta_s}\Big)^{\phi_t}, 1-\prod_{t=1}^{n}\Big(\prod_{s=1}^{m}(1-F_{st})^{\theta_s}\Big)^{\phi_t} \Big\rangle. \quad (10)$$

Proof. For $m = 1$ and $\theta_1 = 1$ then by Definition 7, we have

$$\begin{aligned} SVNSWGA(\tilde{C}_{e_{11}}, \tilde{C}_{e_{12}}, \ldots, \tilde{C}_{e_{1n}}) &= \bigotimes_{t=1}^{n} \tilde{C}_{e_{st}}^{\phi_t} \\ &= \Big\langle \prod_{t=1}^{n} (T_{1t})^{\phi_t}, 1 - \prod_{t=1}^{n} (1 - I_{1t})^{\phi_t}, 1 - \prod_{t=1}^{n} (1 - F_{1t})^{\phi_t} \Big\rangle \\ &= \Big\langle \prod_{t=1}^{n} \Big(\prod_{s=1}^{1} (T_{st})^{\theta_s} \Big)^{\phi_t}, 1 - \prod_{t=1}^{n} \Big(\prod_{s=1}^{1} (1 - I_{st})^{\theta_s} \Big)^{\phi_t}, 1 - \prod_{t=1}^{n} \Big(\prod_{s=1}^{1} (1 - F_{st})^{\theta_s} \Big)^{\phi_t} \Big\rangle. \end{aligned}$$

For $n = 1$ and $\phi_1 = 1$, then by Definition 9, we get

$$\begin{aligned} SVNSWGA(\tilde{C}_{e_{11}}, \tilde{C}_{e_{21}}, \ldots, \tilde{C}_{e_{m1}}) &= \bigotimes_{t=1}^{n} \tilde{C}_{e_{st}}^{\theta_s} \\ &= \Big\langle \prod_{s=1}^{m} (T_{s1})^{\theta_s}, 1 - \prod_{s=1}^{m} (1 - I_{s1})^{\theta_s}, 1 - \prod_{s=1}^{m} (1 - F_{s1})^{\theta_s} \Big\rangle \\ &= \Big\langle \prod_{t=1}^{1} \Big(\prod_{s=1}^{m} (T_{st})^{\theta_s} \Big)^{\phi_t}, 1 - \prod_{t=1}^{1} \Big(\prod_{s=1}^{m} (1 - I_{st})^{\theta_s} \Big)^{\phi_t}, 1 - \prod_{t=1}^{1} \Big(\prod_{s=1}^{m} (1 - F_{st})^{\theta_s} \Big)^{\phi_t} \Big\rangle. \end{aligned}$$

Assume that (10) is true for $n = p_1 + 1, m = p_2$ and $n = p_1, m = p_2 + 1$, then it follows that

$$\bigotimes_{t=1}^{p_1+1} \Big(\bigotimes_{s=1}^{p_2} \tilde{C}_{e_{st}}^{\theta_s} \Big)^{\phi_t} = \Big\langle \prod_{t=1}^{p_1+1} \Big(\prod_{s=1}^{p_2} (T_{st})^{\theta_s} \Big)^{\phi_t}, 1 - \prod_{t=1}^{p_1+1} \Big(\prod_{s=1}^{p_2} (1 - I_{st})^{\theta_s} \Big)^{\phi_t}, 1 - \prod_{t=1}^{p_1+1} \Big(\prod_{s=1}^{p_2} (1 - F_{st})^{\theta_s} \Big)^{\phi_t} \Big\rangle.$$

Also,

$$\bigotimes_{t=1}^{p_1} \Big(\bigotimes_{s=1}^{p_2+1} \tilde{C}_{e_{st}}^{\theta_s} \Big)^{\phi_t} = \Big\langle \prod_{t=1}^{p_1} \Big(\prod_{s=1}^{p_2+1} (T_{st})^{\theta_s} \Big)^{\phi_t}, 1 - \prod_{t=1}^{p_1} \Big(\prod_{s=1}^{p_2+1} (1 - I_{st})^{\theta_s} \Big)^{\phi_t}, 1 - \prod_{t=1}^{p_1} \Big(\prod_{s=1}^{p_2+1} (1 - F_{st})^{\theta_s} \Big)^{\phi_t} \Big\rangle.$$

Now for $n = p_1 + 1$ and $m = p_2 + 1$, we obtained

$$\begin{aligned} \bigotimes_{t=1}^{p_1+1} \Big(\bigotimes_{s=1}^{p_2+1} \tilde{C}_{e_{st}}^{\theta_s} \Big)^{\phi_t} &= \bigotimes_{t=1}^{p_1+1} \Big(\bigotimes_{s=1}^{p_2} \tilde{C}_{e_{st}}^{\theta_s} \otimes \tilde{C}_{e_{(p_2+1)t}}^{\theta_{p_2+1}} \Big)^{\phi_t} \\ &= \bigotimes_{t=1}^{p_1+1} \Big(\bigotimes_{s=1}^{p_2} \tilde{C}_{e_{st}}^{\theta_s} \Big)^{\phi_t} \bigotimes_{t=1}^{p_1+1} \Big(\big(\tilde{C}_{e_{(p_2+1)t}}^{\theta_{p_2+1}} \big)^{\phi_t} \Big) \\ &= \Big\langle \prod_{t=1}^{p_1+1} \Big(\prod_{s=1}^{p_2} (T_{st})^{\theta_s} \Big)^{\phi_t} \otimes \prod_{t=1}^{p_1+1} \Big((T_{(p_2+1)t})^{\theta_{p_2+1}} \Big)^{\phi_t}, \\ &\quad 1 - \prod_{t=1}^{p_1+1} \Big(\prod_{s=1}^{p_2} (1 - I_{st})^{\theta_s} \Big)^{\phi_t} \otimes 1 - \prod_{t=1}^{p_1+1} \Big((1 - I_{(p_2+1)t})^{\theta_{p_2+1}} \Big)^{\phi_t}, \\ &\quad 1 - \prod_{t=1}^{p_1+1} \Big(\prod_{s=1}^{p_2} (1 - F_{st})^{\theta_s} \Big)^{\phi_t} \otimes 1 - \prod_{t=1}^{p_1+1} \Big((1 - F_{(p_2+1)t})^{\theta_{p_2+1}} \Big)^{\phi_t} \Big\rangle \\ &= \Big\langle \prod_{t=1}^{p_1+1} \Big(\prod_{s=1}^{p_2+1} (T_{st})^{\theta_s} \Big)^{\phi_t}, 1 - \prod_{t=1}^{p_1+1} \Big(\prod_{s=1}^{p_2+1} (1 - I_{st})^{\theta_s} \Big)^{\phi_t}, \\ &\quad 1 - \prod_{t=1}^{p_1+1} \Big(\prod_{s=1}^{p_2+1} (1 - F_{st})^{\theta_s} \Big)^{\phi_t} \Big\rangle. \end{aligned}$$

Thus, (10) is true for $n = p_1 + 1, m = p_2 + 1$, therefore by induction the results is hold for all $m, n \geq 1$.

Since, $0 \leq I_{st} \leq 1 \Leftrightarrow 0 \leq \prod_{s=1}^{m} (1 - I_{st})^{\theta_s} \leq 1 \Leftrightarrow 0 \leq \prod_{t=1}^{n} \Big(\prod_{s=1}^{m} (1 - I_{st})^{\theta_s} \Big)^{\phi_t} \leq 1 \Leftrightarrow 0 \leq 1 - \prod_{t=1}^{n} \Big(\prod_{s=1}^{m} (1 - I_{st})^{\theta_s} \Big)^{\phi_t} \leq 1$, and $0 \leq F_{st} \leq 1 \Leftrightarrow 0 \leq \prod_{s=1}^{m} (1 - F_{st})^{\theta_s} \leq$

$1 \Leftrightarrow 0 \leq \prod_{t=1}^{n}\left(\prod_{s=1}^{m}(1-F_{st})^{\theta_s}\right)^{\phi_t} \leq 1 \Leftrightarrow 0 \leq 1-\prod_{t=1}^{n}\left(\prod_{s=1}^{m}(1-F_{st})^{\theta_s}\right)^{\phi_t} \leq 1$ On the other hand, $0 \leq T_{st} \leq 1 \Leftrightarrow 0 \leq \prod_{s=1}^{m}(T_{st})^{\theta_s} \leq 1 \Leftrightarrow 0 \leq \prod_{t=1}^{n}\left(\prod_{s=1}^{m}(T_{st})^{\theta_s}\right)^{\phi_t} \leq 1$. Therefore, $0 \leq 1-\prod_{t=1}^{n}\left(\prod_{s=1}^{m}(1-I_{st})^{\theta_s}\right)^{\phi_t} + 1-\prod_{t=1}^{n}\left(\prod_{s=1}^{m}(1-F_{st})^{\theta_s}\right)^{\phi_t} + \prod_{t=1}^{n}\left(\prod_{s=1}^{m}(T_{st})^{\theta_s}\right)^{\phi_t} \leq 3$.

Thus, aggregated value obtained by SVNSWG operator is again a SVNSN. □

Example 6. *Let* $Y = \{y_1, y_2, y_3, y_4\}$ *be the set of experts which are going to narrate the "attractiveness of two-wheeler bikes" under the set of parameters* $E = \{e_1 = stylish,\ e_2 = weight,\ e_3 = milage,\ e_4 = price\}$*. The rating value of the experts is assumed to be given in the form of* $SVNSNs(B, E) = (T_{st}, I_{st}, F_{st})_{4\times 3}$ *for each parameters which are given in the following table (Table 2).*

Table 2. Neutrosophic soft numbers.

Experts	e_1	e_2	e_3
y_1	$\langle 0.6, 0.3, 0.3\rangle$	$\langle 0.7, 0.2, 0.5\rangle$	$\langle 0.4, 0.2, 0.2\rangle$
y_2	$\langle 0.5, 0.4, 0.2\rangle$	$\langle 0.5, 0.3, 0.2\rangle$	$\langle 0.8, 0.1, 0.1\rangle$
y_3	$\langle 0.4, 0.1, 0.2\rangle$	$\langle 0.7, 0.1, 0.5\rangle$	$\langle 0.5, 0.2, 0.3\rangle$
y_4	$\langle 0.6, 0.2, 0.4\rangle$	$\langle 0.6, 0.3, 0.4\rangle$	$\langle 0.6, 0.3, 0.3\rangle$

Let $\phi = (0.3, 0.2, 0.5)^T$ *and* $\theta = (0.2, 0.1, 0.3, 0.4)^T$ *be the weight vectors for the parameters and experts respectively. Then, we get by using Theorem 6 as:*

$SVNSWGA(\tilde{C}_{e_{11}}, \tilde{C}_{e_{12}}, \ldots, \tilde{C}_{e_{43}})$

$$= \left\langle \prod_{t=1}^{3}\left(\prod_{s=1}^{4}(T_{st})^{\theta_s}\right)^{\phi_t}, 1-\prod_{t=1}^{3}\left(\prod_{s=1}^{4}(1-I_{st})^{\theta_s}\right)^{\phi_t}, 1-\prod_{t=1}^{3}\left(\prod_{s=1}^{4}(1-F_{st})^{\theta_s}\right)^{\phi_t}\right\rangle$$

$$\Big\langle \Big(\Big\{(0.6)^{0.2}(0.5)^{0.1}(0.4)^{0.3}(0.6)^{0.4}\Big\}^{0.3} \Big\{(0.7)^{0.2}(0.5)^{0.1}(0.7)^{0.3}(0.6)^{0.4}\Big\}^{0.2}$$
$$\Big\{(0.4)^{0.2}(0.8)^{0.1}(0.5)^{0.3}(0.6)^{0.4}\Big\}^{0.5}\Big),$$
$$1-\Big\{(1-0.3)^{0.2}\,(1-0.4)^{0.1}\,(1-0.1)^{0.3}\,(1-0.2)^{0.4}\Big\}^{0.3}$$
$$\Big\{(1-0.2)^{0.2}\,(1-0.3)^{0.1}\,(1-0.1)^{0.3}\,(1-0.3)^{0.4}\Big\}^{0.2}$$
$$\Big\{(1-0.2)^{0.2}\,(1-0.1)^{0.1}\,(1-0.2)^{0.3}\,(1-0.3)^{0.4}\Big)^{0.5}, 1-\Big\{(1-0.3)^{0.2}\,(1-0.2)^{0.1}\,(1-0.2)^{0.3}\,(1-0.4)^{0.4}\Big\}^{0.3}$$
$$\Big\{(1-0.5)^{0.2}\,(1-0.2)^{0.1}\,(1-0.5)^{0.3}\,(1-0.4)^{0.4}\Big\}^{0.2}$$
$$\Big\{(1-0.2)^{0.2}\,(1-0.1)^{0.1}\,(1-0.3)^{0.3}\,(1-0.3)^{0.4}\Big)^{0.5}\Big\rangle$$
$$= \Big\langle (0.5518, 0.2261, 0.3138) \Big\rangle.$$

SVNSWGA operator satisfies the following properties as similar as SVNSWA operator

- **(Idempotency Property)** If $\tilde{C}_{e_{st}} = \tilde{C}_e = \langle T, I, F\rangle$ for all s, t, then

$$SVNSWGA(\tilde{C}_{e_{11}}, \tilde{C}_{e_{12}}, \ldots, \tilde{C}_{e_{mn}}) = \tilde{C}_e.$$

- **(Boundedness Property)** If $\tilde{C}^{-}_{e_{st}} = \langle \min_t \min_s\{T_{st}\}, \max_t \max_s\{I_{st}\}, \max_t \max_s\{F_{st}\}\rangle$ and if

$$\tilde{C}^{+}_{e_{st}} = \langle \max_t \max_s\{T_{st}\}, \min_t \min_s\{I_{st}\}, \min_t \min_s\{F_{st}\}\rangle,$$

then

$$\tilde{C}_{e_{st}}^{-} \leq SVNSWGA(\tilde{C}_{e_{11}}, \tilde{C}_{e_{12}}, \ldots, \tilde{C}_{e_{mn}}) \leq \tilde{C}_{e_{st}}^{+}.$$

- **(Shift-invariance Property)** Let $\tilde{C}_e = \langle T, I, F \rangle$ be another SVNSN then

$$SVNSWGA(\tilde{C}_{e_{11}} \otimes \tilde{C}_e, \tilde{C}_{e_{12}} \otimes \tilde{C}_e, \ldots, \tilde{C}_{e_{mn}} \otimes \tilde{C}_e) = SVNSWGA(\tilde{C}_{e_{11}}, \tilde{C}_{e_{12}}, \ldots, \tilde{C}_{e_{mn}}) \otimes \tilde{C}_e.$$

- **(Homogeneity Property)** For any real number $\lambda > 0$, we have

$$SVNSWGA(\tilde{C}_{e_{11}}^{\lambda}, \tilde{C}_{e_{12}}^{\lambda}, \ldots, \tilde{C}_{e_{mn}}^{\lambda}) = \left(SVNSWGA(\tilde{C}_{e_{11}}, \tilde{C}_{e_{12}}, \ldots, \tilde{C}_{e_{mn}})\right)^{\lambda}.$$

5. Model for MCDM Method Using Single-Valued Soft Information

In this Section, we shall present multi-criteria decision making (MCDM) method using single-valued neutrosophic soft weighted averaging operator (SVNSWA) and single-valued neutrosophic soft weighted geometric (SVNSWGA) operator in the environment of single-valued neutrosophic soft numbers.

An Approach Based on Proposed Operators

Let $\tilde{Y} = \{\tilde{Y}_1, \tilde{Y}_2, \ldots, \tilde{Y}_l\}$ be the discrete set of alternatives is evaluated by the set of m experts $\{y_1, y_2, \ldots, y_m\}$ under the constraints of n parameters $E = \{e_1, e_2, \ldots, e_n\}$. Let $\theta = (\theta_1, \theta_2, \ldots, \theta_m)^T$ and $\phi = (\phi_1, \phi_2, \ldots, \phi_n)^T$ are respectively denote the weight vectors of the m experts $x_s's$ and n parameters $e_t's$ that $\theta_s > 0, \theta \in [0,1]$ such that $\sum_{s=1}^{m} \theta_s = 1$ and $\phi_t > 0, \phi \in [0,1]$ such that $\sum_{t=1}^{n} \phi_s = 1$. In order to choice the best l alternates by the preference of n experts in the form of SVNSNs $\tilde{C}_{e_{st}} = \langle T_{st}, I_{st}, F_{st} \rangle$ where $0 \leq T_{st} + I_{st} + F_{st} \leq 3$ and collective over all decision matrix is expressed as $\tilde{M} = (\tilde{C}_{e_{st}})_{m \times n}$. By these preference values of the experts, the aggregated SVNSN $\tilde{C}_{e_k}$ for the alternatives $\tilde{p}_k$ $(k = 1, 2, \ldots, l)$ is $\tilde{C}_{e_k} = \langle T_k, I_k, F_k \rangle$ by applying weighted averaging or geometric averaging operators which is given in Equations (5) and (8). Ranking order of the alternatives is determine based on the score function of the aggregated values of SVNSNs $\tilde{C}_{e_k}$ $(k = 1, 2, \ldots, l)$.

In the following algorithm we propose to solve MCDM problem with single-valued neutrosophic soft information using SVNSWA and SVNSWGA operators.

Step 1. Collect all the information in the form of single-valued neutrosophic soft matrix $C = \langle T_{st}, I_{st}, F_{st} \rangle$ $(s = 1, 2, \ldots, m; t = 1, 2, \ldots, n)$ related to each alternatives under different parameters e_k $(k = 1, 2, \ldots, l)$ as

$$\tilde{C}_{m \times n} = M = \begin{bmatrix} (T_{11}, I_{11}, F_{st}) & (T_{12}, I_{12}, F_{12}) & \ldots & (T_{1n}, I_{1n}, F_{1n}) \\ (T_{21}, I_{21}, F_{21}) & (T_{22}, I_{22}, F_{22}) & \ldots & (T_{2n}, I_{2n}, F_{2n}) \\ \vdots & \vdots & \ddots & \vdots \\ (T_{m1}, I_{m1}, F_{m1}) & (T_{m2}, I_{m2}, F_{m2}) & \ldots & (T_{mn}, I_{mn}, F_{mn}) \end{bmatrix}$$

Step 2. To normalize the aggregated decision matrix by transforming values of benefit type (B) into cost (C) type by using the formula depicted in [61].

$$g_{ij} = \begin{cases} \tilde{C}_{e_{st}}^{c}, & \text{if} e_t \in \tilde{B} \\ \tilde{C}_{e_{st}}, & \text{if} e_t \in \tilde{C} \end{cases}$$

where $\tilde{C}_{e_{st}}^{c} = \langle 1 - F_{st}, I_{st}, T_{st} \rangle$ is the complement of $\tilde{C}_{e_{st}} = \langle T_{st}, I_{st}, F_{st} \rangle$.

Step 3. Aggregate the SVNSNs $\tilde{C}_{e_{st}}$ $(s = 1, 2, \ldots, m; t = 1, 2, \ldots, n)$ for each alternatives Y_k $(k = 1, 2, \ldots, l)$ into collective decision matrix Ψ_k using SVNSWA or (SVNSWGA) operators.

Step 4. Using Equation (1) we get the score value of Ψ_k $(k = 1, 2, \ldots, l)$ for each alternatives A_k $(k = 1, 2, \ldots, l)$.

Step 5. Rank all the alternative A_k $(k = 1, 2, \ldots, l)$ in order to choice the best one(s) in accordance with Ψ_k $(k = 1, 2, \ldots, l)$.

Step 6. End.

6. Numerical Example

In the above described decision-making method has been demonstrated with a practical example about the Medical diagnosis. The experts of five doctors m_1, m_2, m_3, m_4, m_5 whose weight vector is $\theta = (0.2, 0.15, 0.2, 0.3, 0.15)^T$, will give their judgement based on the diagnosis of four patients Y_1, Y_2, Y_3, Y_4 under the parameters

$$E = \{Temparature(e_1), Headache(e_2), Stomachpain(e_3), Cough(e_4), Chestpain(e_5)\}$$

with weight vector $\phi = (0.2, 0.1, 0.3, 0.15, 0.25)^T$. Then, we utilize the developed method to get most desirable candidate(s).

6.1. By SVNSWA Operator

The steps of the proposed approach performed and their corresponding details are reviewed as follows:

Step 1. The given patients are being evaluated by five experts doctors to give their grades in terms of SVNSNs and are found in Tables 3–6 respectively for each candidate.

Step 2. All the parameters are of same type, so, there is no required for normalization.

Table 3. Single-valued neutrosophic soft matrix for the patient Y_1.

Experts	e_1	e_2	e_3	e_4	e_5
m_1	$\langle 0.6, 0.2, 0.3 \rangle$	$\langle 0.5, 0.5, 0.5 \rangle$	$\langle 0.6, 0.3, 0.2 \rangle$	$\langle 0.3, 0.5, 0.7 \rangle$	$\langle 0.3, 0.4, 0.4 \rangle$
m_2	$\langle 0.6, 0.4, 0.5 \rangle$	$\langle 0.3, 0.2, 0.4 \rangle$	$\langle 0.5, 0.4, 0.5 \rangle$	$\langle 0.4, 0.6, 0.4 \rangle$	$\langle 0.4, 0.1, 0.2 \rangle$
m_3	$\langle 0.7, 0.1, 0.4 \rangle$	$\langle 0.4, 0.3, 0.4 \rangle$	$\langle 0.3, 0.1, 0.4 \rangle$	$\langle 0.7, 0.3, 0.2 \rangle$	$\langle 0.6, 0.2, 0.6 \rangle$
m_4	$\langle 0.4, 0.5, 0.3 \rangle$	$\langle 0.7, 0.2, 0.1 \rangle$	$\langle 0.2, 0.6, 0.4 \rangle$	$\langle 0.6, 0.1, 0.5 \rangle$	$\langle 0.5, 0.1, 0.5 \rangle$
m_5	$\langle 0.5, 0.2, 0.2 \rangle$	$\langle 0.6, 0.1, 0.6 \rangle$	$\langle 0.6, 0.2, 0.2 \rangle$	$\langle 0.4, 0.1, 0.1 \rangle$	$\langle 0.4, 0.2, 0.3 \rangle$

Table 4. Single-valued neutrosophic soft matrix for the patient Y_2.

Experts	e_1	e_2	e_3	e_4	e_5
m_1	$\langle 0.3, 0.4, 0.4 \rangle$	$\langle 0.8, 0.1, 0.1 \rangle$	$\langle 0.7, 0.1, 0.1 \rangle$	$\langle 0.4, 0.1, 0.3 \rangle$	$\langle 0.2, 0.3, 0.4 \rangle$
m_2	$\langle 0.5, 0.1, 0.3 \rangle$	$\langle 0.5, 0.2, 0.2 \rangle$	$\langle 0.4, 0.2, 0.3 \rangle$	$\langle 0.6, 0.1, 0.2 \rangle$	$\langle 0.2, 0.1, 0.3 \rangle$
m_3	$\langle 0.2, 0.1, 0.2 \rangle$	$\langle 0.4, 0.1, 0.4 \rangle$	$\langle 0.5, 0.4, 0.5 \rangle$	$\langle 0.4, 0.2, 0.6 \rangle$	$\langle 0.5, 0.2, 0.2 \rangle$
m_4	$\langle 0.7, 0.2, 0.3 \rangle$	$\langle 0.5, 0.1, 0.4 \rangle$	$\langle 0.6, 0.2, 0.4 \rangle$	$\langle 0.4, 0.2, 0.1 \rangle$	$\langle 0.7, 0.1, 0.1 \rangle$
m_5	$\langle 0.5, 0.2, 0.2 \rangle$	$\langle 0.5, 0.3, 0.2 \rangle$	$\langle 0.4, 0.1, 0.5 \rangle$	$\langle 0.3, 0.2, 0.2 \rangle$	$\langle 0.6, 0.4, 0.2 \rangle$

Table 5. Single-valued neutrosophic soft matrix for the patient Y_3.

Experts	e_1	e_2	e_3	e_4	e_5
m_1	$\langle 0.4, 0.3, 0.2 \rangle$	$\langle 0.8, 0.1, 0.4 \rangle$	$\langle 0.5, 0.2, 0.3 \rangle$	$\langle 0.6, 0.1, 0.2 \rangle$	$\langle 0.2, 0.3, 0.3 \rangle$
m_2	$\langle 0.5, 0.1, 0.2 \rangle$	$\langle 0.4, 0.2, 0.3 \rangle$	$\langle 0.3, 0.2, 0.4 \rangle$	$\langle 0.4, 0.2, 0.3 \rangle$	$\langle 0.5, 0.2, 0.2 \rangle$
m_3	$\langle 0.2, 0.1, 0.1 \rangle$	$\langle 0.4, 0.2, 0.3 \rangle$	$\langle 0.4, 0.2, 0.3 \rangle$	$\langle 0.5, 0.1, 0.4 \rangle$	$\langle 0.5, 0.1, 0.1 \rangle$
m_4	$\langle 0.7, 0.2, 0.4 \rangle$	$\langle 0.5, 0.2, 0.2 \rangle$	$\langle 0.2, 0.1, 0.5 \rangle$	$\langle 0.6, 0.2, 0.5 \rangle$	$\langle 0.6, 0.2, 0.2 \rangle$
m_5	$\langle 0.5, 0.3, 0.3 \rangle$	$\langle 0.5, 0.4, 0.4 \rangle$	$\langle 0.4, 0.1, 0.2 \rangle$	$\langle 0.4, 0.1, 0.5 \rangle$	$\langle 0.7, 0.2, 0.3 \rangle$

Table 6. Single-valued neutrosophic soft matrix for the patient Y_4.

Experts	e_1	e_2	e_3	e_4	e_5
m_1	$\langle 0.4, 0.1, 0.2\rangle$	$\langle 0.5, 0.4, 0.4\rangle$	$\langle 0.5, 0.2, 0.1\rangle$	$\langle 0.5, 0.1, 0.3\rangle$	$\langle 0.2, 0.3, 0.7\rangle$
m_2	$\langle 0.5, 0.1, 0.3\rangle$	$\langle 0.3, 0.2, 0.2\rangle$	$\langle 0.3, 0.2, 0.3\rangle$	$\langle 0.4, 0.2, 0.2\rangle$	$\langle 0.3, 0.2, 0.2\rangle$
m_3	$\langle 0.5, 0.3, 0.4\rangle$	$\langle 0.5, 0.1, 0.5\rangle$	$\langle 0.4, 0.2, 0.2\rangle$	$\langle 0.2, 0.2, 0.4\rangle$	$\langle 0.5, 0.2, 0.4\rangle$
m_4	$\langle 0.6, 0.2, 0.3\rangle$	$\langle 0.4, 0.5, 0.2\rangle$	$\langle 0.3, 0.2, 0.4\rangle$	$\langle 0.7, 0.2, 0.1\rangle$	$\langle 0.3, 0.1, 0.1\rangle$
m_5	$\langle 0.5, 0.3, 0.1\rangle$	$\langle 0.4, 0.6, 0.1\rangle$	$\langle 0.4, 0.2, 0.3\rangle$	$\langle 0.3, 0.1, 0.4\rangle$	$\langle 0.6, 0.3, 0.3\rangle$

Step 3. The opinion of doctors for each patient Y_k $(k = 1, 2, 3, 4)$ are aggregated by using Equation (5) given as follows: $\Omega_1 = \langle 0.4918, 0.2326, 0.3404\rangle$, $\Omega_2 = \langle 0.5154, 0.1700, 0.2522\rangle$, $\Omega_3 = \langle 0.4800, 0.1656, 0.2753\rangle$ and $\Omega_4 = \langle 0.4319, 0.1942, 0.2444\rangle$.

Step 4. The values of score functions are: $\Psi(\Omega_1) = 0.1514$, $\Psi(\Omega_2) = 0.2632$, $\Psi(\Omega_3) = 0.2047$ and $\Psi(\Omega_4) = 0.1875$.

Step 5. Ranking all the patients Y_k $(k = 1, 2, 3, 4)$ in accordance with the value of the score $\Psi(\Omega_k)$ $(k = 1, 2, 3, 4)$ of the overall single-valued neutrosophic soft numbers as $Y_2 \succ Y_3 \succ Y_4 \succ Y_1$.

Step 6. Therefore, Y_2 is the more illness patient than other patients.

6.2. By Using SVNSWGA Operator

If we apply SVNSWGA operator on the proposed problem for the selection of appropriate candidate(s) that follows the following steps:

Step 3. The aggregated values for each patients Y_k $(k = 1, 2, 3, 4)$ using SVNSWGA operator are as follows from Equation (10): $\Omega_1 = \langle 0.4432, 0.3084, 0.3913\rangle$, $\Omega_2 = \langle 0.4515, 0.1999, 0.3044\rangle$, $\Omega_3 = \langle 0.4224, 0.1825, 0.3079\rangle$ and $\Omega_4 = \langle 0.3960, 0.2204, 0.3030\rangle$.

Step 4. The values of score functions are: $\Psi(\Omega_1) = 0.0519$, $\Psi(\Omega_2) = 0.1471$, $\Psi(\Omega_3) = 0.1145$ and $\Psi(\Omega_4) = 0.0930$.

Step 5. Ranking all the candidates Y_k $(k = 1, 2, 3, 4)$ in accordance with the value of the score $\Psi(\Omega_k)$ $(k = 1, 2, 3, 4)$ of the overall single-valued neutrosophic soft numbers as $Y_2 \succ Y_3 \succ Y_4 \succ Y_1$.

Step 6. Hence, Y_2 is the most illness patient diagnosed by the expert doctors .

From the above analysis, it is clear that although overall rating values of the alternatives are different by using two operators, but ranking order of the alternatives are similar, the most illness patient is Y_2 among four patients.

7. Comparative Analysis

To compare the proposed work with the existing approach, an analysis has been made based on aggregation operator (see [25,60,62]). In that reason, the different parameters of single-valued neutrosophic soft numbers are aggregated by using weighted averaging operator corresponding to the weighted vector $(0.2, 0.1, 0.3, 0.15, 0.25)^T$ and then obtained aggregated single-valued neutrosophic soft matrix for the different candidates Y_k $(k = 1, 2, 3, 4)$ given in Table 7. From this evaluated matrix, a comparative study has been established with the existing work based on weighted aggregation operator on simplified neutrosophic numbers, single-valued neutrosophic Domi weighted aggregation operators and single-valued neutrosophic weighted averaging operators developed by researchers [25,60,62] for each candidate are shown in Table 8. It also shows that proposed method is stable with compare the existing methods. From the Table 7, we can see that the candidate Y_2 is most illness person diagnosed by the experts doctors. The characteristic comparison between propose study with existing methods are given in Table 9. The propose method utilize advance technique to compare the existing works [56,57] where a decision making method has been develop based on some soft algebraic operations in neutrosophic soft environment but present paper leads a decision making method based on aggregating single-valued neutrosophic soft arguments in the environment of SVN soft numbers.The advantages of this paper is that they are capable to facilitate the descriptions of the real-world problems situation with the help of their parameterizations property. Therefore, proposed method can be utilize to solve decision-making problems instead of other existing operator in the environment of SVN soft numbers.

Table 7. Aggregated value of single-valued neutrosophic soft matrix for the patients.

Experts	C_1	C_2	C_3	C_4
m_1	$\langle 0.4884, 0.3378, 0.3411 \rangle$	$\langle 0.5161, 0.1737, 0.2200 \rangle$	$\langle 0.4854, 0.2018, 0.2679 \rangle$	$\langle 0.4168, 0.1861, 0.2531 \rangle$
m_2	$\langle 0.4680, 0.2805, 0.3761 \rangle$	$\langle 0.4256, 0.1320, 0.2711 \rangle$	$\langle 0.4211, 0.1741, 0.2725 \rangle$	$\langle 0.3605, 0.1741, 0.2449 \rangle$
m_3	$\langle 0.5545, 0.1565, 0.3990 \rangle$	$\langle 0.4251, 0.2000, 0.3327 \rangle$	$\langle 0.4092, 0.1320, 0.1911 \rangle$	$\langle 0.4333, 0.2024, 0.3322 \rangle$
m_4	$\langle 0.4513, 0.2531, 0.3594 \rangle$	$\langle 0.6182, 0.1569, 0.2169 \rangle$	$\langle 0.5246, 0.1625, 0.3470 \rangle$	$\langle 0.4573, 0.1843, 0.2024 \rangle$
m_5	$\langle 0.5081, 0.1682, 0.2226 \rangle$	$\langle 0.4747, 0.2012, 0.2633 \rangle$	$\langle 0.5223, 0.1702, 0.2952 \rangle$	$\langle 0.4650, 0.2415, 0.2253 \rangle$

Table 8. Comparison analysis with the existing method.

Methods	$\Upsilon(\Omega_1)$	$\Upsilon(\Omega_2)$	$\Upsilon(\Omega_3)$	$\Upsilon(\Omega_4)$	**Ranking Order**
Proposed SVNSWA operator	0.1514	0.2632	0.2047	0.1875	$Y_2 \succ Y_3 \succ Y_4 \succ Y_1$
Proposed SVNSWGA operator	0.0519	0.1417	0.1145	0.0930	$Y_2 \succ Y_3 \succ Y_4 \succ Y_1$
Ye [25] by SNWAA operator	0.1440	0.2583	0.1969	0.1822	$Y_2 \succ Y_3 \succ Y_4 \succ Y_1$
Ye [25] by SNWGA operator	0.1487	0.2506	0.1999	0.1852	$Y_2 \succ Y_3 \succ Y_4 \succ Y_1$
Chen and Ye [62] SVNDWA operator	0.1594	0.2732	0.2131	0.1915	$Y_2 \succ Y_3 \succ Y_4 \succ Y_1$
Chen and Ye [62] SVNDWG operator	0.1378	0.2383	0.1875	0.1760	$Y_2 \succ Y_3 \succ Y_4 \succ Y_1$
Sahin [60] SVNWAA operator	0.1515	0.2632	0.2047	0.1869	$Y_2 \succ Y_3 \succ Y_4 \succ Y_1$
Sahin [60] SVNWGA operator	0.1412	0.2445	0.1921	0.1791	$Y_2 \succ Y_3 \succ Y_4 \succ Y_1$

Table 9. Characteristic comparisons of different methods.

Methods	*Fuzzy Information Easier*	*Weather Aggregate Parameter Information*
Ye [25]	Yes	No
Chen and Ye [62]	Yes	No
Sahin [60]	Yes	No
Proposed operators	Yes	Yes

8. Conclusions

In this article, we have studied multi-criteria group decision-making problem using in the environment of single-valued neutrosophic soft information. We have introduced two new operators namely (SVNSWA) operator, (SVNSWGA) operators in SVN soft environment. The different features of those recommended operators is deliberated. For this purpose, firstly some algebraic structures of two SVNSNs are given and their operational rules are defined. The two aggregation operators have been proposed in the environment of SVNS numbers. Some properties of these two kinds of operators have been established. We justify the propose method with the existing methods and a characteristic comparison also shown to demonstrate advantage and applicability of the proposed method. A Medical decision-making problems has been studied based on SVNSWA and SVNSWGA operators under the environment of SVN soft information. The main advantages of these operators is that they are able to make smooth description of the real-world problems by the use of parameterizations factor. Ultimately, a realistic example for the selection of most illness patient is provided to develop a strategy and in accordance with expounding the utility and effectiveness of the proposed method. In the future work, the propose model further develop new soft aggregation operators for simplified neutrosophic sets and apply them to solve practical applications like engineering [63], group decision-making [64], expert system, information fusion system, fault diagnosis, robotics design [65] and other domains under different fuzzy soft environments.

Author Contributions: Conceptualization, C.J.; formal analysis, M.P.; writing—original draft preparation, C.J.; writing—review and editing, C.J. and M.P.; supervision, M.P.

Funding: This research received no external funding.

Conflicts of Interest: The authors declare no conflict of interest.

References

1. Kurvers, R.H.J.M.; Krause, J.; Argenziano, G.; Zalaudek, I.; Wolf, M. Detection accuracy of collective intelligence assessments for skin cancer diagnosis. *JAMA Dermatol.* **2015**, *151*, 1346–1353. [CrossRef] [PubMed]
2. Wolf, M.; Krause, J.; Carney, P.A.; Bogart, A.; Kurvers, R.H.J.M. Collective intelligence meets medical decision-making: The collective outperforms the best radiologist. *PLoS ONE* **2015**, *10*, e0134269. [CrossRef] [PubMed]
3. Marshall, J.A.R.; Brown, G.; Radford, A.N. Individual confidence-weighting and group decision-making. *Trends Ecol. Evol.* **2017**, *32*, 636–645. [CrossRef] [PubMed]
4. Olfati-Saber, R.; Franco, E.; Frazzoli, E.; Shamma, J.S. Belief consensus and distributed hypothesis testing in sensor networks. In *Networked Embedded Sensing and Control*; Springer, Berlin/Heidelberg, Germany, 2006; pp. 169–182.
5. Mukhametzyanov, I.; Pamuear, D. A sensistivity analysis in MCDM problems: A statistical approach. *Decis. Mak. Appl. Manag. Eng.* **2018**, *1*, 51–80. [CrossRef]
6. Teixeira, C.; Lopes, I.; Figueiredo, M. Classification methodology for spare parts management combining maintenance and logistics perspectives. *J. Manag. Anal.* **2018**, *5*, 116–135. [CrossRef]
7. Ronaynea, D.; Brown Gordon, D.A. Multi-attribute decision by sampling: An account of the attraction, compromise and similarity effects. *J. Math. Psychol.* **2017**, *81*, 11–27. [CrossRef]
8. Abbasian, N.S.; Salajegheh Gaspar, A.H.; Brett, P.O. Improving early OSV design robustness by applying Multivariate big data analytics on a ship's life cycle. *J. Ind. Inf. Integr.* **2018**, *10*, 29–38. [CrossRef]
9. Liu, F.; Aiwu, G.; Lukovac, V.; Vukic, M. A multicriteria model for the selection of the transport service provider: A single valued neutrosophic DEMATEL multicriteria model. *Decis. Mak. Appl. Manag. Eng.* **2018**, *1*, 121–130. [CrossRef]
10. Atanassov, K.T. *Intuitionistic Fuzzy Sets: Theory and Applications, Studies in Fuzziness and Soft Computing*; Physica-Verlag: Heidelberg, Germany; New York, NY, USA, 1999; Volume 35.
11. Zadeh, L.A. Fuzzy sets. *Inf. Control* **1965**, *8*, 338–353. [CrossRef]
12. Wang, H.; Smarandache, F.; Zhang, Y.Q.; Sunderraman, R. Single valued neutrosophic sets. *Multispace Multistruct.* **2010**, *4*, 410–413.
13. Beliakov, G.; Pradera, A.; Calvo, T. *Aggregation Functions: A Guide for Practitioners*; Springer: Heidelberg/Berlin, Germany; New York, NY, USA, 2007.
14. Beliakov, G.; Bustince, H.; Goswami, D.P.; Mukherjee, U.K.; Pal, N.R. On averaging operators for Atanassov's intuitionistic fuzzy sets. *Inf. Sci.* **2011**, *181*, 1116–1124. [CrossRef]
15. Chen, S.M.; Chiou, C.H. Multiattribute decision making based on interval-valued intuitionistic fuzzy sets, PSO techniques, and evidential reasoning methodology. *IEEE Trans. Fuzzy Syst.* **2015**, *23*, 1905–1916. [CrossRef]
16. Jana, C.; Pal, M.; Karaaslan, F.; Wang, J.Q. Trapezoidal neutrosophic aggregation operators and its application in multiple attribute decision-making process. *Sci. Iran. E* **2018**, in accepted.
17. Jana, C.; Senapati, T.; Pal, M.; Yager, R.R. Picture fuzzy Dombi aggregation operators: Application to MADM process. *Appl. Soft Comput.* **2018**, *74*, 99–109. [CrossRef]
18. Peng, J.J.; Wang, J.Q.; Wu, X.H.; Zhang, H.Y.; Chen, X.H. The fuzzy cross-entropy for intuitionistic hesitant fuzzy sets and its application in multi-criteria decision-making. *Int. J. Syst. Sci.* **2015**, *46*, 2335–2350. [CrossRef]
19. Wang, P.; Xu, X.H.; Wang, J.Q.; Cai, C.G. Some new operation rules and a new ranking method for interval-valued intuitionistic linguistic numbers. *J. Int. Fuzzy syst.* **2017**, *32*, 1069–1078. [CrossRef]
20. Xu, Z.S.; Yager, R.R. Some geometric aggregation operators based on intuitionistic fuzzy sets. *Int. J. Gen. Syst.* **2006**, *35*, 417–433. [CrossRef]
21. Xu, Z.S. Intuitionistic fuzzy aggregation operators. *IEEE Trans. Fuzzy Syst.* **2007**, *15*, 1179–1187.
22. Ye, J. Multicriteria fuzzy decision-making method based ona novel accuracy function under interval-valued intuitionistic fuzzy environment. *Expert Syst. Appl.* **2009**, *36*, 899–6902. [CrossRef]

23. Smarandache, F. *A Unifying Field in Logics. Neutrosophy: Neutrosophic Probability, Set and Logic*; American Research Press: Rehoboth, DE, USA, 1999.
24. Smarandache, F. Neutrosophic set-a generalization of the intuitionistic fuzzy set. *Int. J. Pure Appl. Math.* **2005**, *24/3*, 287–297.
25. Ye, J. A multicriteria decision-making method using aggregation operators for simplified neutrosophic sets. *J. Int. Fuzzy Syst.* **2014**, *27*, 2459–2466.
26. Garg, H.; Nancy. Linguistic single-valued neutrosophic prioritized aggregation operators and their applications to multiple-attribute group decision-making. *J. Ambient Intell. Hum. Comput.* **2018**, *9*, 1975–1997. [CrossRef]
27. Wang, L.; Zhang, H.Y.; Wang, J.Q. Frank Choquet Bonferroni mean operators of bipolar neutrosophic sets and their application to multi-criteria decision-making problems. *Int. J. Fuzzy Syst.* **2018**, *20*, 13–28. [CrossRef]
28. Shi, L.; Ye, J. Dombi aggregation operators of neutrosophic cubic sets for multiple attribute decision-making. *Algorithms* **2018**, *11*, 29. [CrossRef]
29. Wei, G.W.; Zhang, Z. Some single-valued neutrosophic Bonferroni power aggregation operators in multiple attribute decision making. *J. Ambient Intell. Hum. Comput.* **2018**. [CrossRef]
30. Ulucay, V.; Deli, I.; Sahin, M. Similarity measures of bipolar neutrosophic sets and their application to multiple criteria decision-making. *Neural Comput. Appl.* **2018**, *29*, 739–748. [CrossRef]
31. Abdel-Basset, M.; Mohamed, M.; Zhou, Y.; Hezam, I. Multi-criteria group decision making based on neutrosophic analytic hierarchy process. *J. Int. Fuzzy Syst.* **2017**, *33*, 4055–4066. [CrossRef]
32. Abdel-Basset, M.; Mohamed, M.; Smarandache, F. An extension of neutrosophic AHP-SWOT analysis for strategic planning and decision-making. *Symmetry* **2018**, *10*, 116. [CrossRef]
33. Dalapati, S.; Pramanik, S.; Alam, S.; Smarandache, F.; Roy, T.K. IN-cross entropy based magdm strategy under interval neutrosophic set environment. *Neutrosophic Sets Syst.* **2017**, *18*, 43–57.
34. Bausys, R.; Zavadskas, E.K. Multicriteria decision making approach by VIKOR under interval neutrosophic set environment. *Econ. Comput. Econ. Cybern. Stud. Res.* **2015**, *4*, 33–48.
35. Biswas, P.; Pramanik, S.; Giri, B.C. TOPSIS method for multi-attribute group decision-making under single valued neutrosophic environment. *Neural Comput. Appl.* **2016**, *27*, 727–737. [CrossRef]
36. Broumi, S.; Bakali, A.; Talea, M.; Smarandache, F.; Krishnan Kishore, K.P.; Sahin, R. Shortest path problem under interval valued neutrosophic setting. *J. Fundam. Appl. Sci.* **2018**, *10*, 168–174.
37. Sahin, R.; Liu, P. Correlation coefficient of single-valued neutrosophic hesitant fuzzy sets and its applications in decision making. *Neural Comput. Appl.* **2017**, *28*, 1387–1395. [CrossRef]
38. Jana, C.; Pal, M.; Wang, J.Q. Bipolar fuzzy Dombi aggregation operators and its application in multiple attribute decision making process. *J. Ambient Intell. Hum. Comput.* **2018**. [CrossRef]
39. Pamuear, D.; Bozaniae, D.; Lukovac, V.; Komazec, N. Normalized weighted geometric bonferroni mean operator of interval rough numbers- application in interval rough DEMATEL-COPRAS. *Facta Univ. Ser. Mech. Eng.* **2018**, 1–22. [CrossRef]
40. Maji, P.K.; Biswas, R.; Roy, A.R. Fuzzy soft sets. *J. Fuzzy Math.* **2001**, *9*, 589–602.
41. Maji, P.K.; Biswas, R.; Roy, A.R. Intuitionistic fuzzy soft sets. *J. Fuzzy Math.* **2001**, *9*, 677–692.
42. Molodtsov, D. Soft set theory-first results. *Comput. Math. Appl.* **1999**, *27*, 19–31. [CrossRef]
43. Cagman, N.; Citak, F.; Enginoglu, S. Fuzzy parameterized fuzzy soft set theory and its applications. *Turk. J. Fuzzy Syst.* **2001**, *1*, 21–35.
44. Cagman, N.; Deli, I. Intuitionistic fuzzy parameterized soft set theory and its decision making. *Appl. Soft. Comput.* **2015**, *28*, 109–113.
45. Alkhazaleh, S.; Salleh, A.R. Fuzzy soft expert set and its application. *Appl. Math.* **2014**, *5*, 1349–1368. [CrossRef]
46. Garg, H.; Arora, R. Generalized and group-based generalized intuitionistic fuzzy soft sets with applications in decision-making. *Appl. Intell.* **2018**, *48*, 343–356. [CrossRef]
47. Jiang, Y.; Tang, Y.; Chen, Q.; Liu, H.; Tang, J. Interval-valued intuitionistic fuzzy soft sets and their properties. *Comput. Math. Appl.* **2010**, *60*, 906–918. [CrossRef]
48. Yang, X.; Lin, T.Y.; Yang, J.; Li, Y.; Yu, D. Combination of interval-valued fuzzy set and soft set. *Comput. Math. Appl.* **2009**, *58*, 521–527. [CrossRef]
49. Jana, C.; Pal, M. Application of bipolar intuitionistic fuzzy soft sets in decision making problem. *Int. J. Fuzzy Syst. Appl.* **2018**, *7*, 32–55. [CrossRef]

50. Babitha, K.V.; John, S.J. Hesistant fuzzy soft sets. *J. New Results Sci.* **2013**, *3*, 98–107.
51. Jana, C.; Pal, M. Applications of new soft intersection set on groups. *Ann. Fuzzy Math. Inf.* **2013**, *6*, 17–31.
52. Jana, C.; Pal, M. Application of (α, β)-soft intersectional sets on BCK/BCI-algebras. *Int. J. Intell. Syst. Technol. Appl.* **2017**, *16*, 269–288. [CrossRef]
53. Selvachandran, G.; Peng, X. A modified TOPSIS method based on vague parameterized vague soft sets and its application to supplier selection problems. *Neural Comput. Appl.* **2018**. [CrossRef]
54. Arora, R.; Garg, H. Prioritized averaging/geometric aggregation operators under the intuitionistic fuzzy soft set environment. *Sci. Iran. E* **2018**, *25*, 466–482. [CrossRef]
55. Karaaslan, F. Possibility neutrosophic soft sets and PNS-decision making method. *Appl. Soft Comput.* **2017**, *54*, 403–414. [CrossRef]
56. Broumi, S.; Samarandache, F. Single-valued neutrosophic soft expert sets and their application in decision-making. *J. New Theory* **2015**, *3*, 67–88.
57. Ali, M.; Son, L.; Deli, I.; Tien, N.D. Bipolar Neutrosophic Soft Sets and Applications in Decision Making. *J. Int. Fuzzy Syst.* **2017**, *33*, 4077–4087. [CrossRef]
58. Deli, I.; Eraslan, S.; Cagman, N. ivnpiv-Neutrosophic soft sets and their decision making based on similarity measure. *Neural Comput. Appl.* **2018**, *29*, 187–203. [CrossRef]
59. Khalid, A.; Abbas, M. Distance measures and operations in intuitionistic and interval- valued intuitionistic fuzzy soft set theory. *Int. J. Fuzzy Syst.* **2015**, *17*, 490–497. [CrossRef]
60. Sahin, R. Multi-criteria neutrosophic decision making method based on score and accuracy functions under neutrosophic environment. *arXiv* **2014**, arXiv:412.5202.
61. Xu, Z.S.; Hu, H. Projection models for intuitionistic fuzzy multiple attribute decision- making. *Int. J. Inf. Technol. Dec. Mak.* **2010**, *9*, 267–280. [CrossRef]
62. Chen, J.; Ye, J. Some Single-Valued Neutrosophic Dombi Weighted Aggregation Operators for Multiple Attribute Decision-Making. *Symmetry* **2017**, *9*, 82. [CrossRef]
63. Pamuear, D.; Badi, I.; Sanja, S.; Obradovic, R. A novel approach for the selection of power-generation technology using a linguistic neutrosophic CODAS method: A case study in Libya. *Energies* **2018**, *11*, 2489. [CrossRef]
64. Maio, C.D.; Fenza, G.; Loia, V.; Orciuoli, F.; Herrera-Viedma, E. A framework for context-aware heterogeneous group decision making in business processes. *Knowl.-Based Syst.* **2016**, *102*, 39–50. [CrossRef]
65. Smarandache, F.; Vladareanu, L. Applications of Neutrosophic Logic to Robotics. Available online: https://www.researchgate.net/publication/268443363_Applications_of_Neutrosophic_Logic_to_Robotics_An_Introduction (accessed on 21 March 2015).

MDPI
St. Alban-Anlage 66
4052 Basel
Switzerland
Tel. +41 61 683 77 34
Fax +41 61 302 89 18
www.mdpi.com

Symmetry Editorial Office
E-mail: symmetry@mdpi.com
www.mdpi.com/journal/symmetry

www.ingramcontent.com/pod-product-compliance
Lightning Source LLC
Chambersburg PA
CBHW041912220326
R18017400002B/R180174PG41597CBX00010B/3
* 9 7 8 3 0 3 9 2 1 9 3 8 4 *